ADVANCED INORGANIC CHEMISTRY

Volume II

For B.Sc. (Pass & Honours) IIIrd Year Students of Indian Universities

SATYA PRAKASH
D.Sc., F.A.Sc.
Formerly Professor of Chemistry
Allahabad University, Allahabad

G.D. TULI
M.Sc., Ph.D.
Formerly Principal
Shyam Lal College
University of Delhi, Delhi

S.K. BASU
M.Sc., Ph.D.
Formerly Professor of Chemistry
Banaras Hindu University, Varanasi

R.D. MADAN
M.Sc., Ph.D.
Reader
Department of Chemistry
N.R.E.C. College, Khurja

S Chand And Company Limited

(ISO 9001 Certified Company)

S Chand And Company Limited

(ISO 9001 Certified Company)

Head Office: Block B-1, House No. D-1, Ground Floor, Mohan Co-operative Industrial Estate, New Delhi – 110 044 | Phone: 011-66672000

Registered Office: A-27, 2nd Floor, Mohan Co-operative Industrial Estate, New Delhi – 110 044 Phone: 011-49731800

www.**schandpublishing.com**; e-mail: **info@schandpublishing.com**

Branches

Chennai	:	Ph: 23632120; chennai@schandpublishing.com
Guwahati	:	Ph: 2738811, 2735640; guwahati@schandpublishing.com
Hyderabad	:	Ph: 40186018; hyderabad@schandpublishing.com
Jalandhar	:	Ph: 4645630; jalandhar@schandpublishing.com
Kolkata	:	Ph: 23357458, 23353914; kolkata@schandpublishing.com
Lucknow	:	Ph: 4003633; lucknow@schandpublishing.com
Mumbai	:	Ph: 25000297; mumbai@schandpublishing.com
Patna	:	Ph: 2260011; patna@schandpublishing.com

First Edition 1944
Subsequent Editions and Reprints 1945, 50, 52, 54, 56, 58, 60, 61, 63, 64, 65, 66, 67, 68, 69, 70, 71, 73, 74, 76, 79, 80, 81, 82, 83, 84, 85 (Twice), 87 (Twice), 88, 89, 90, 92, 93, 94, 96, 98, 99, 2000, 2003, 2004, 2005, 2006 (Twice), 2007, 2008, 2012, 2013, 2014 (Twice), 2016 (Twice), 2018, 2020 (Twice)
Reprint 2021

ISBN : 978-81-219-1787-2 **Product Code :** H6AIC68CHEM10ENAR0XR

PRINTED IN INDIA

By Vikas Publishing House Private Limited, Plot 20/4, Site-IV, Industrial Area Sahibabad, Ghaziabad – 201 010 and Published by S Chand And Company Limited, A-27, 2nd Floor, Mohan Co-operative Industrial Estate, New Delhi – 110 044.

PREFACE TO THE EIGHTEENTH EDITIONS

In this edition the style of the arrangement and presentation of different chapters of the previous edition has been kept unchanged. The printing mistakes and errors that have been reported to us, from time to time, by the students and teachers have been corrected.

Main Features :

In this edition we have given some selected problems with their solution at the end of some chapters. In some chapters latest questions asked in B.Sc. examination of different Indian Universities have also been added at the end of the chapter.

Acknowledgment:

We are extremely grateful to Shri Ravindra Kumar Gupta (Managing Director), Shri Navin Joshi, General Manager (Sales & Marketing), Shri R.S. Saxena (Advisor) for their whole-hearted cooperation in bringing out this new edition.

To our fellow-teachers and students

We humbly request our students and teachers to send us their constructive criticism and suggestions to me in order to enable us to incorporate them in the next edition to make this book error-free and more useful from users' point of view.

Dr. R.D. Madan
37, Bhagirithi Apartments
Sector – 9, Rohini
Delhi – 110085
Phone: 7860313

AUTHORS

Preface to the Seventeenth Edition

We are immensely happy to state that the Volume I of this book has been well received and appreciated by both students and teachers alike. In this (Volume I), we have discussed *Basic Principles of Chemistry* and *Chemistry of Main Group Elements (s and p-Block Elements) and their Compounds.*

The subject matter of the present Volume II has been divided into four parts. **Part I** deals with *Coordination Chemistry*, **Part II** discusses *Transition Elements* (*d-Block Elements and their Compounds,* **Part III** is devoted to the study of *Inner-transition Elements*: *f-Block Elements* (*Lanthanides and Actinides*) *and their Compounds* and **Part IV** explains some special topics like: (*i*) *Transition Metal Complexes with Carbon Monoxide,* (*ii*) *Metallic Carbonyls,* (*iii*) *Transition Metal Complexes with Nitric Oxide,* (*iv*) *Metallic Nitrosyls,* (*v*) *Pollution and its Adverse Effects,* (*vi*) *Organo-metallic Compounds,* (*vii*) *Chromatography* (*viii*) *Uses of Organic and Inorganic Reagents in Inorganic Analysis,* (*ix*) *Role of Metal Ions in Biological Systems* and (*x*) *Non-aqueous Solvents.*

We invite constructive suggestions from our fellow-teachers and students for further improvement of the book.

We express our gratitude to Shri Ravindra Kumar Gupta, Director, S. Chand & Company Ltd. for his full cooperation in bringing out this voluminous work in a short time and in such a fine format.

One of the authors, Dr. R.D. Madan is extremely thankful to his son, Vikas Madan and his wife, Rukmani Madan without whose help and inspiration this work would never have been completed.

Authors

CONTENTS

Part I: Coordination Chemistry

Part II: Transition Metals (*d*-Block Elements) and Their Compounds

mate ($K_6Cr_2O_6$) – 383; Red tetraperoxy chromates, $M_3^+ [Cr (O_2)_4]^{3-}$ ($M^+ = K^+$, NH_4^+) – 383; Red-brown diperoxy chromate, $(NH_3)_3Cr (O_2)_2$ – 384.

Compounds of molybdenum: Molybdenum trioxide (MoO_3) – 385; Molybdic acid ($H_2MoO_4.H_2O$) – 385; Ammonium molybdate [$(NH_4)_2MoO_4$] – 385; Molybdenum blue – 386.

Compounds of tungsten: Tunsgen trioxide or tungstic oxide (WO_3) – 387; Tungstic acid (H_2WO_4) –387; tungsten bronzes (M_xWO_3) – 388; Tunsten halides – 388;

Compounds of manganese: Manganeous oxide (MnO) – 389; Manganic oxide or manganese sesquoxide (Mn_2O_3) – 390; Manganese tetroxide or mangano-manganic oxide or red oxide of manganese ($Mn_3O_4 \equiv MnO.Mn_2O_3$) – 391; Manganese dioxide (MnO_2) – 391; Manganates (MnO_4^{2-}) – 392; Mangenese heptoxide (Mn_2O_7) – 393; Permanganic acid ($HMnO_4$) – 394; Potassium permanganate ($KMnO_4$) – 394.

Compounds of iron: Hexamine ferrate (II) chloride, $[Fe (NH_3)_6] Cl_2$ – 405; Sodium pentacyanonitrosyl ferrate (II) (also called nitroprusside), $Na_2 [Fe^{2+} (CN)_5 (NO^+)]. 2H_2O$ – 405; Potassium ferrocyanide or potassium hexacyanoferrate (II), $K_4 [Fe (CN)_6]$ – 405; Potassium ferrous ferro hexacyanide, $K_2Fe^{2+} [Fe^{2+} (CN)_6]$ – 408; Hydroferricyanic acid, $H_3 [Fe (CN)_6]$ – 408; Potassium ferricyanide or potassium hexacynoferrate (III), $K_3 [Fe (CN)_6]$ – 408; Potassium ferric hexacyanoferrate (II) or potassium ferric ferro hexacyanide, $K Fe^{3+} [Fe^{2+} (CN)_6]$ – 411; Ferric hexacyano ferrate (II) or ferric ferro hexácyanide, $Fe_4^{3+} [Fe^{2+} (CN)_6]_3$ – 411; Prussion blue or Turnbull's blue, $KFe^{3+} [Fe^{2+} (CN)_6]$ or $Fe_4^{3+} [Fe^{2+} (CN)_6]_3$ – 411

Compounds of osmium: Osmium dioxide (OsO_2) – 414; Osmium tetroxide (OsO_4) – 414; Osmium hexafluoride (OsF_6) – 415; Osmimium tetrachloride ($OsCl_4$) – 415.

Compounds of cobalt: Cobatious oxide (CoO) – 415; Cobalto-cobaltic oxide ($Co_3O_4 \equiv CoO.Co_2O_3$) – 416; Sodium hexanitrocobaltate (III) or sodium cobaltinitrite, $Na_3 [Co (NO_2)_6]$ – 416; Potassium hexanitro cobaltate (III) or potassium cobaltinite (also called Fischer's salt), $K_3 [Co (NO_2)_6]$ – 417; Hexammine cobalt (III) chloride or hexammine cobaltic chloride, $[Co (NH_3)_6] Cl_3$ – 417.

Compounds of nickel: Nickel sulphate ($NiSO_4.7H_2O$) – 418; Nickel ammonium sulphate [$NiSO_4.(NH_4)_2SO_4.6H_2O$] – 419; Bis (dimethylglyoximato) nickel (II), $[Ni (C_4H_7N_2O_2)_2]^0$ or $[Ni (dmg)_2]^0$ – 419.

Compounds of palladium: Flourides – 420; Chlorides – 420; Palladium bromide ($PdBr_2$) – 421, Palladium iodide (PdI_2) – 421; Oxides – 421.

Compounds of platinum: Hydrogen tetrachloroplatinate (II) or tetrachloroplatinuous acid, $H_2 [Pt Cl_4]$ – 421; Potassium tetra cyano platinate (II) or potassium platino cyanide, $K_2[Pt(CN)_4].3H_2O$ – 422; Barium tetracyanoplatinate (II) or barium platinocyanide, $Ba [Pt (CN)_4].4H_2O$ – 422; Platinum (IV) chloride or platinic chloride, $PtCl_4$ – 422; Hydrogen hexachloro platinate (IV) or hexachloro platinic acid, $H_2 [Pt Cl_6]$ – 423; Potassium hexachloroplatinate (IV), $K_2[Pt Cl_6]$ – 425; Ammonium hexachloroplatinate (IV) or ammonium chloroplatinate, $(NH_4)_2 [Pt Cl_6]$ – 426.

Part III: Inner-transition Elements and Their Compounds

Part I : Coordination Chemistry

1

Basic Concepts of Coordination Chemistry

Addition or molecular compounds

When solutions of two or more simple stable salts are mixed together in simple molecular proportion and the solution thus obtained is allowed to evaporate, crystals of a new compound are obtained. This new compound is called addition or molecular compound. The formation of some addition compounds has been shown below :

Simple compounds		*Addition compounds*
$KCl + MgCl_2 + 6H_2O$	$\longrightarrow$	$KCl.MgCl_2.6H_2O$ Carnollite
$K_2SO_4 + Al_2(SO_4)_3 + 24H_2O$	$\longrightarrow$	$K_2SO_4.Al_2(SO_4)_3.24H_2O$ Potash alum
$(NH_4)_2SO_4 + FeSO_4 + 6H_2O$	$\longrightarrow$	$FeSO_4.(NH_4)_2SO_4.6H_2O$ Mohr's salt
$NiCl_2 + 6NH_3$	$\longrightarrow$	$NiCl_2.6NH_3$ Hexammine nickel chloride
$CuSO_4 + 4NH_3$	$\longrightarrow$	$CuSO_4.4NH_3$ Tetrammine copper sulphate
$Fe(CN)_2 + 4KCN$	$\longrightarrow$	$Fe(CN)_2.4KCN$ Potassium ferrocyanide

Hydrates, ammoniates and peroxohydrates

Many salts crystallise with solvent molecules like H_2O, NH_3 and H_2O_2. The salts obtained in this way are also *addition compounds.* When the added solvent molecule is H_2O, NH_3 or H_2O_2, the addition compounds formed are called *hydrates, ammoniates* and *peroxohydrates.* In naming these compounds the number of solvent molecules is indicated by Arabic numbers (1,2,3 etc.). For example $CuSO_4.5H_2O$ [Copper (II) sulphate 5-hydraté], $CaCl_2.6NH_3$ (Calcium chloride 6-ammoniate), $NaO_2H.H_2O_2$ (Sodium hydrogen peroxide 1-peroxohydrate).

Types of addition compounds : Double salts and coordination (or complex) compounds

Addition compounds are of two types:

1. Double salts. These are the addition compounds which are stable in the solid state but give their constituent ions when dissolved in water or in any other ionic solvent. In these compounds the individual properties of the constituent ions are *not* lost. For example, when carnollite, $KCl.MgCl_2.6H_2O$ which is a double salt gives its constuent ions namely K^+, Mg^{2+} and Cl^- ions when dissolved in water and hence the aqueous solution of $KCl.MgCl_2.6H_2O$ gives the tests of K^+, Mg^{2+} and Cl^- ions.

2. Coordination (or complex) compounds. These are the addition compounds which donot give all their constituent ions when dissolved in water. In these compounds the individual properties of some constituent ions are lost. For example $CuSO_4.4\ NH_3$ which is a complex compound does not give all its constituent ions *viz.* Cu^{2+} and SO_4^{2-} in solution. It gives only SO_4^{2-} ion and Cu^{2+} ion is obtained as $[Cu(NH_3)_4]^{2+}$ ion which is called *complex ion.*

$$CuSO_4.4NH_3 \rightleftharpoons [Cu(NH_3)_4]^{2+} + SO_4^{2-}$$

Thus $CuSO_4\ .\ 4NH_3$ is formulated as $[Cu(NH_3)_4]SO_4$. It follows from this discussion that $CuSO_4\ .\ 4NH_3$ gives the tests of SO_4^{2-} only and does not show the tests of Cu^{2+} ions. $[Cu(NH_3)_4]SO_4$ is called *complex* (or *coordination*) *compound.* A complex compound may be composed of simple cation and complex anion. complex cation and simple anion or complex cation and complex anion as shown below :

$$K_3[Fe(CN)_6] \rightleftharpoons \underset{\text{Simple cation}}{3K^+} + \underset{\text{Complex anion}}{[Fe(CN)_6]^{3-}}$$

$$[Co(NH_3)_6]Cl_3 \rightleftharpoons \underset{\text{Complex cation}}{[Co(NH_3)_6]^{3+}} + \underset{\text{Simple anion}}{3Cl^-}$$

$$[Pt(NH_3)_4]\ [PtCl_4] \rightleftharpoons \underset{\text{Complex cation}}{[Pt(NH_3)_4]^{2+}} + \underset{\text{Complex anion}}{[PtCl_4]^{2-}}$$

A neutral complex compound like $[Ni^0(CO)_4]^0$, $[Ni^{2+}(dmg)_2]^0$, $[Pt^{2+}(NH_3)_2Cl_2]^0$ etc. does not undergo ionisation and hence does not give any ions in aqueous solution.

Nomenclature of double salts

(*a*) Double salts are named by connecting the names of the individual compounds by hyphens (–). The number of the molecules of individual compounds is indicated by using *Arabic numerals* (*e.g.* 1,2,3 etc.) separated by solid lines and enclosed in brackets.

For example:

$K_2SO_4.Al_2(SO_4)_3.24H_2O$ ... Potassium sulphate-aluminium sulphate-water (1/1/24)
$FeSO_4.(NH_4)_2SO_4.6H_2O$ ... Ferrous sulphate ammonium sulphate-water (1/1/6)
$Na_2CO_3.10H_2O$ ... Sodium carbonate water (1/10)
$3CdSO_4.8H_2O$ ... Cadmium sulphate water (3/8)
$AlCl_3.4C_2H_5OH$ ... Aluminium chloride-ethanol (1/4)
$NH_3.BF_3$ ... Ammonia-boron trifluoride (1/1)
$8H_2S.46H_2O$ ... Hydrogen sulphide water (8/46)

(*b*) If a double salt contains the cations in the same oxidation state, the cations are written alphabetically. Hydrogen is an exception which is written in the end. For example, $KHCO_3$ (Potassium hydrogen carbonate), $KNaCO_3$ (Potassium sodium carbonate), $NaNH_4HPO_4.4H_2O$ (Ammonium sodium hydrogen phosphate 4-hydrate) etc.

(*c*) If a double salt contains the cations in different oxidation states, the cations are written in the increasing order of their oxidation states, hydrogen being an exception which is written at the end. Examples are $KMgCl_3.6H_2O$ (potassium magnesium chloride 6-hydrate), $KAl(SO_4)_2.12H_2O$ (potassium aluminium sulphate 12-hydrate), $NH_4MgPO_4.6H_2O$ (ammonium magnesium phosphate 6-hydrate), $(NH_4)_2Co(SO_4)_2.6H_2O$ [ammonium cobalt (II) sulphate 6-hydrate], $NaZn(UO_2)_3(CH_3COO)_9.6H_2O$ (sodium zinc triuranyl acetate 6-hydrate).

(*d*) If a double salt contains the anions in the same oxidation state, the anions are written alphabetically. For example PbClF is called lead (II) chloride fluoride.

(*e*) If a double salt contains the anions in different oxidation states, the anions are arranged in the order : O^{2-}, OH^-, other simple inorganic anions with two elements, organic anions, H^-.

The number of ions is indicated by writing the prefix *bis, tris* etc. in parentheses. For example $NaCl.NaF.2Na_2SO_4$ is called (*hexa*) sodium chloride fluoride (*bis*) sulphate.

(*f*) The hydrolysed or basic salts are also treated as double salts containing more than one anion and hence are named by the same rules as used for naming double salts. For example BiOCl [bismuth (III) oxide chloride], Sn(OH)Cl [tin (IV) hydroxide chloride], $ZrOCl_2.8H_2O$ [zirconium (IV) oxide (di)chloride 8-hydrate], $CuCl_2.3Cu(OH)_2$ or $Cu_2(OH)_2Cl$ [(di) copper (II) trihydroxide chloride].

Note: It may be noted that for H_2O molecule we have used *hydrate* and the number of H_2O molecules has been indicated by Arabic numbers.

Coordination chemistry

The branch of inorganic chemistry that deals with the study of coordination compounds is called *coordination chemistry.*

Complex ion

Complex ion is a positively or negatively charged species which contains a central metalic atom and a suitable number of ligands surrounding the central metallic atom. The central metallic atom may be in zero, positive or negative oxidation state. The ligands may be either neutral molecules or anions (generally) or a combination of the neutral molecules and anions. Examples of complex ions are $[Co^{3+}(NH_3)_6]^{3+}$, $[Ni^0(CN)_4]^{-4}$, $[Fe^{2+}(CN)_6]^{4-}$, $[Co^-(CO)_4]^-$, $[Co^{3+}(NH_3)_5Cl]^{2+}$ *etc.*

It may be seen that, depending on the nature of the charge on the complex ion, complex ion may be *cationic complex ion* or *anionic complex ion.* Cationic complex ion has positive charge on it while anionic complex ion has negative charge. Thus $[Co^{3+}(NH_3)_6]^{3+}$ and $[Co^{3+}(NH_3)_5Cl]^{2+}$ are cationic complex ions and $[Ni^0(CN)_4]^{4-}$, $[Fe^{2+}(CN)_6]^{4-}$ and $[Co^-(CO)_4]^-$ are anionic complex ions.

Neutral complex

A complex which has no charge on it is called neutral complex. It is a non electrolyte, does not undergo ionisation and hence does not give any ions in aqueous solution. Examples of neutral complexes are $[Co^{3+}(NH_3)_3Cl_3]^0$, $[Ni^0(CO)_4]^0$, $[Pt^{2+}(NH_3)_2Cl_2]^0$, $[Ni^{2+}(dmg)_2]^0$ *etc.*

Ligands or coordinating groups and central metallic atom

The neutral molecules or ions (usually anions) which are attached with the central ion in complex compounds are called ligands or co-ordinating groups. For example in the complex ion, $[Fe(CN_6)]^{3-}$ the six CN^- ions which are attached with the central Fe^{3+} ion as shown in the margin act as ligands. In Lewis sense, in most of the complex compounds the ligands act as Lewis bases (electron pair donors) and the central metal ion acts as a Lewis acid (electron pair acceptor), *i.e.* in most of the complex compounds the ligands donate one or more electron pairs to the central metal ion.

$$\left[Fe^{3+} \leftarrow 6\,CN^- \right]^{3-}$$

(Margin structure: Fe^{3+} surrounded by six CN^- / NC^- ligands, each with an arrow pointing to Fe^{3+}; overall charge 3−.)

$$\underset{\text{Central metal ion(Lewis acid or electron pair acceptor)}}{M^{n+}} + \underset{\text{Ligand (Lewis base or electron pair donor)}}{x\,L} \longrightarrow \underset{\text{Complex compound}}{[ML_x]^{n+}}$$

In some complex compounds the ligand acts both as donor and acceptor. For example, in metallic carbonyls, CO molecules which act as ligands act both as donor and acceptor ($M \rightleftharpoons CO$). In a ligand the atom which actually donates the electron pair to the central metal ion is called *donor* or *coordinating atom.* The ligands are attached with the central metal ion through their donor atom (or atoms).

The metallic atom with which the ligands are attached through coordinating bonds is called *central metallic atom.* This metallic atom may be in *zero, positive* or *negative oxidation state.*

Coordination number of the central metal atom/ion

Coordination number of the central metal atom/ion in a given complex compound is equal to the total number of donor atoms which are actually attached with the central metallic atom. In other words we can say that the coordination number of the central metallic atom is equal to the number of sites at which the ligand(s) is attached with the central metallic atom.

In case of complex compounds which contain only monodentate ligands, the coordination number of the central metallic atom is equal to the number of monodentate ligands coordinated to the metallic atom. This rule does not hold good for the complexes containing polydentate (*i.e.* bidentate, tridentate, tetradentate *etc.*) ligands.

Coordination number of the metallic atom predicts the geometry of the complex compound Thus for coordination number equal to 2,3,4,5 and 6, the geometry of the complex compound is *linear, trigonal planar, tetrahedral* (or *square planar), trigonal bipyramidal* and *octahedral* respectively. This discussion shows that the coordination number gives us an idea about the way in which the ligands are arranged round the central metallic atom.

Coordination sphere and ionisation sphere

While writting the structural formula of a given complex compound the central metal atom and the ligands attached with it are always written in a square bracket, [ ]. This square bracket is called *coordination* (or *inner*) *sphere.* The portion outside the coordination sphere is called *ionisation* (or *outer*) *sphere.* Thus in $[Co(NH_3)_5Cl]Cl_2$, the square bracket which contains the central metal ion (Co^{3+} ion) and the ligands (five NH_3 molecules and one Cl^- ion) is coordination sphere and the portion that contains two Cl^- ions is ionisation sphere.

The species written in ionisation sphere are ionisable and hence can be precipitated by means of a suitable precipitating agent while those given in the coordination sphere (*i.e.* metallic atom and ligands) are non-ionisable and hence cannot be precipitated. This is shown below for $[Co(NH_3)_5Cl]Cl_2$

$$[Co(NH_3)_5Cl]Cl_2 \rightleftharpoons [Co(NH_3)_5Cl]^{2+} + 2Cl^-$$

Two Cl^- ions present in ionisation sphere can be precipitated as AgCl (white ppt) by adding $AgNO_3$ solution.

$$2Ag^+ + 2Cl^- \longrightarrow 2AgCl\downarrow$$

Cl^- ion written in coordination sphere is not ionisable and can, therefore, not be precipated.

Classification of ligands

Depending on the number of sites at which one molecule of a ligand is coordinated to the central metallic atom, the ligands have been classified as monodentate (or unidentate) and polydentate (or multidentate) ligands.

1. Monodentate or unidentate ligands. The ligands which have only one donor atom or are co-ordinated through one electron pair are called monodentate or unidentate ligands. Such ligands are coordinated to the central metal ion at one site or by one metal–ligand bond only. These ligands may be neutral molecules or in anionic form.

2. Polydentate or multidentate ligands. These may be *bidentate, tridentate, tetradentate, pentadentate* and *hexadentate,* if the number of donor atoms present in one molecule of the ligand attached with the central metallic atom is 2, 3, 4, 5 and 6 respectively. Thus one molecule of these ligands is coordinated to the central metallic atom at 2, 3, 4, 5 and 6 sites respectively. In other words, we can say that one molecule of these ligands makes 2, 3, 4, 5 and 6 metal—ligand coordinate bonds respectively.

Examples of monodentate ligands

Examples of monodentate ligands are given in Table 1.1.

Table 1.1. *Examples of monodentate ligands.*

(*a*) **Neutral ligands** which are named as such. Their abbreviations, if any, are also given.

$(C_2H_5)_3N$	... Triethyl amine	$(C_6H_5)_3P$	... Triphenyl phosphine
$NH_2{-}NH_2$ or N_2H_4	... Hydrazine	CH_3CN	... Acetonitrile
CH_3NH_2	... Methyl amine	O_2 or $:\ddot{O}=\ddot{O}:$	... Dioxygen
NH_2OH	... Hydroxylamine	N_2 or $:N\equiv N:$	... Dinitrogen
$(CH_3)_2NH$	... Dimethylamine	PF_3	... Phosphorus trifluoride
C_5H_5N or py	... Pyridine	$(C_2H_5)_3P$	... Triethyl phosphine

(*b*) **Neutral ligands** which are given special names, *e.g.*

CO	... Carbonyl	NO	... Nitrosyl
CS	... Thiocarbonyl	NS	... Thionitrosyl
H_2O	... Aquo or aqua	NH_3	... Ammine

According to latest system of nomenclature, the word "aqua" is used for H_2O molecule.

(*c*) **Anionic (negative) ligands.** The names of negative ligands end in *o*. Examples are given below:

F^-	... Fluoro	Cl^-	... Chloro
Br^-	... Bromo	I^-	... Iodo
H^-	... Hydro or hydrido	CH_3COO^- or $CH_3-\underset{\underset{O}{\parallel}}{C}-\ddot{O}:^-$	... Acetato
NH_2^-	... Amido	NH^{2-}	... Imido
N^{3-}	... Nitrido	$[NH(OH)]^-$	... Hydroxo amido
$[(CH_3)_2N]^-$	... Dimethyl amido	OH^-	... Hydroxo or hydroxyl
O^{2-}	... Oxo	HO_2^-	... Perhydroxo
O_2^{2-} or $:\ddot{O}^- - \ddot{O}:^-$	... Peroxo	O_2^- or $\dot{O}\overset{\cdot\cdot}{-}\ddot{O}$	... Superoxo
HS^-	... Mercapto	S^{2-}	... Sulphido or thio
SO_3^{2-}	... Sulphito	SO_4^{2-}	... Sulphato
CO_3^{2-}	... Carbonato	CN^-	... Cyno(coordination through C–atom)
NC^-	... Iso-cyano (coordination through N-atom)	$S_2O_3^{2-}$	... Thiosulphato
CH_3O^-	.. Methoxo	$C_2H_5O^-$	... Ethoxo
N_3^-	... Azido	NO_2^-	... Nitro (coordination thorough N-atom)
ONO^-	... Nitrito (coordination through O-atom)	SCN^-	... Thiocyanato
NCS^-	... Iso-thiocyanato	NO_3^-	... Nitrato

Examples of bidentate ligands

Bidentate ligends may be neutral molecules or anions.

1. Alkenes like ethylene, propylene, butylene and isobutylene give a number of diamine derivatives which have been found to act as bidentate ligands as shown in Table 1.2.

Table 1.2. *Bidentate ligands derived from alkenes.*

Alkenes	*Bidentate ligands*
Ethylene $H_2C{=}CH_2$ (H H / C=C / H H)	Ethylene diamine (*en*) $CH_2—CH_2$ with $\ddot{N}H_2$ on each C Ethylene diphosphine (*diphos*) $CH_2—CH_2$ with $\ddot{P}H_2$ on each C
Propylene (CH₃ H / C=C / H H)	Propylene diamine (*pn*) CH_3 on HC; $HC—CH_2$ with $\ddot{N}H_2$ $\ddot{N}H_2$ This compound is also called 1, 2-diamino propane (*pn*) $\overset{1}{C}H_2—\overset{2}{C}H—\overset{3}{C}H_3$ (with NH_2 on C-1 and C-2), since the structure of propane is $CH_3—CH_2—CH_3$ 1, 3-diamino propane or trimethylene diamine (*tn*) $\overset{1}{C}H_2—\overset{2}{C}H—\overset{3}{C}H_2$ with $\ddot{N}H_2$ on C-1 and C-3
Butylene (CH₃ CH₃ / C=C / H H)	Butylene diamine (*bn*) CH_3 CH_3 / $HC—CH$ / $\ddot{N}H_2$ $\ddot{N}H_2$
Isobutylene (CH₃ H / C=C / CH₃ H)	Isobutylene diamine (*i-bn*) CH_3 H / $H_3C—C—C—H$ / $\ddot{N}H_2$ $\ddot{N}H_2$
Tetramethyl ethylene (CH₃ CH₃ / C=C / CH₃ CH₃)	Tetramethyl ethylene diamine (*tetrameen*) CH_3 H / $H_3C—C—C—CH_3$ / $\ddot{N}H_2$ $\ddot{N}H_2$

Ethylene diphosphine (*diphos*) has two P-atoms which act as donor atoms, since they have one lone pair of electrons each. The coordination of this ligand to the central metal atom takes place through these two P-atoms, leading to the formation of one 5-membered ring.

Each of the diamines gets attached with the central metal atom through its two N-atoms (donor atoms), forming one 5-membered ring including the central metal atom. This is evident from the structures of $[Co^{3+}(en)_2Cl_2]^+$ and $[Pt^{2+}(bn)_2]^{2+}$ ions given in Fig. 1.20 (see symmetrical and unsymmetrical bidentate ligands).

2. 2, 2-bipyridine (bipy) (also called **bipyridyl**). It is a neutral molecule and has two N-atoms which act as donor atoms as shown at (*a*) of Fig. 1.1. When this ligand gets linked with a metal atom M, 5-membered ring is obtained [See (*b*) of Fig. 1.1].

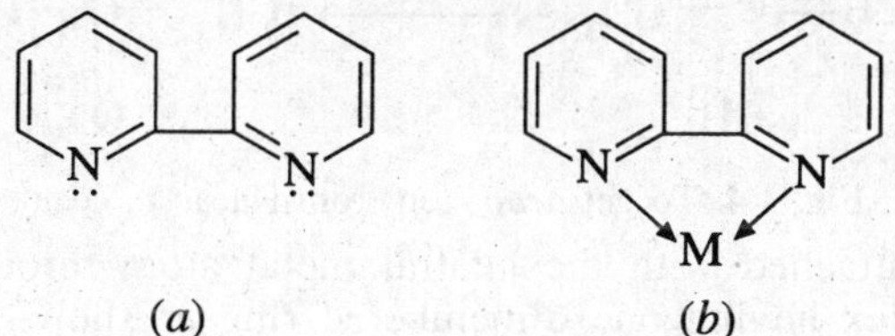

Fig. 1.1. (*a*) Structure of 2,2′-bipyridine (bipy) or bipyridyl (*b*) Attachment of bipy with the central metal atom, M.

3. Hydrazine, $\ddot{N}H_2—\ddot{N}H_2$. Its attachment to a metal atom occurs through its two N-atoms, forming one 3-membered ring which can be represented as $\begin{matrix} H_2N \searrow \\ | \quad M \\ H_2N \nearrow \end{matrix}$. This ring is very strained and hence is unstable.

4. o-phenanthroline or 1, 10-phenanthroline (o-phen or phen or phenan). It is a neutral symmetrical bidentate ligand, since its two donor atoms *viz.* N-atoms are similar atoms. Its structure is given at (*a*) of Fig. 1.2. It is abbreviated as *phen* or *phenan*. When it gets attached with a metal atom, it is denoted as $\begin{matrix} N \\ N \end{matrix}\Big)$ *phen* and its attachment to a metal atom, *M* is indicated as $M \begin{matrix} \nwarrow N \\ \swarrow N \end{matrix}\Big)$*phen.* For example this ligand gets attached with Fe^{2+} ion and gives $[Fe(phen)_3]^{2+}$ complex ion whose structure is given at (*b*) of Fig. 1.2.

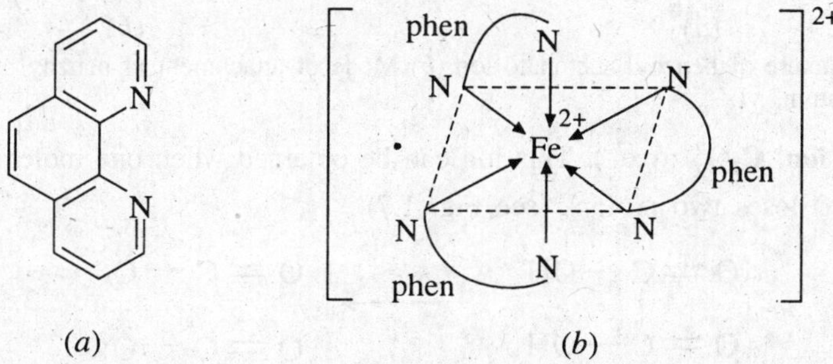

Fig. 1.2. (*a*) Structure of o-phenonthroline (phen or phenon) (*b*) structure of $[Fe(phen)_3]^{2+}$ ion.

5. o-phenylene bis dimethyl arsine (diars or D). The structure and the mode of its attachment with the central metal atom in complexes are shown in Fig. 1.3. Here two As atoms are donor atoms.

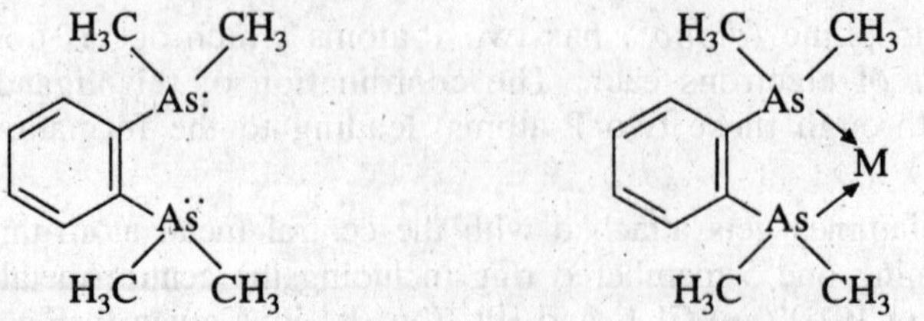

Fig. 1.3. (*a*) Structure of *o*-phenylene *bis* dimethyl arsine (diars or D) (*b*) Mode of attachment of diars with a metal atom, M in complex compounds.

6. Acetylacetonato ion (acac⁻). This ion is obtained when one molecule of the enol form of acetyl acetone (Hacac) loses a proton, H^+ (see Fig. 1.4).

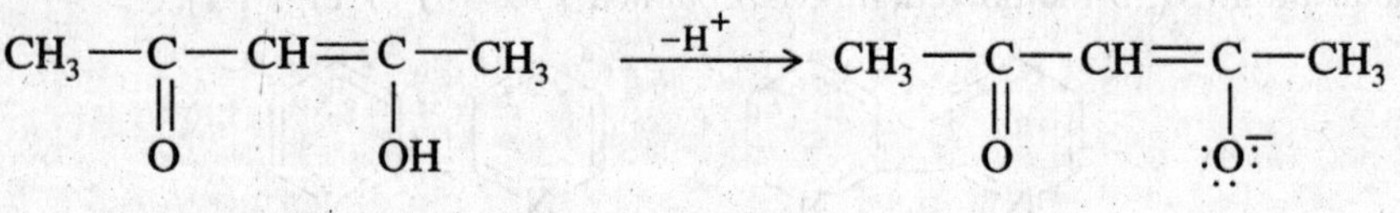

$$CH_3-\underset{\underset{O}{\|}}{C}-CH=\underset{\underset{OH}{|}}{C}-CH_3 \xrightarrow{-H^+} CH_3-\underset{\underset{O}{\|}}{C}-CH=\underset{\underset{:\ddot{O}:^-}{|}}{C}-CH_3$$

Fig. 1.4. To get *acac*⁻ ion from Hacac molecule.

This ligand gets attached with the central metal atom through its two O-atoms and forms a chelated complex having one 6-membered ring as shown in Fig. 1.5.

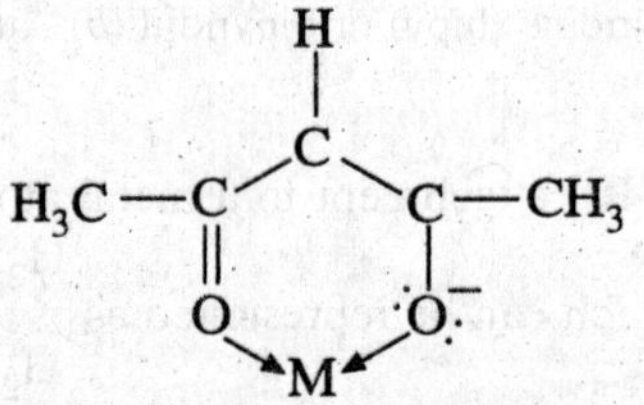

Fig. 1.5. Mode of attachment of acetylacetonato ion to the metal atom, M.

7. Benzoyl acetonato ion. The structure of this ligand and the mode of its attachment with the cental metal atom can be shown as depicted in Fig. 1.6.

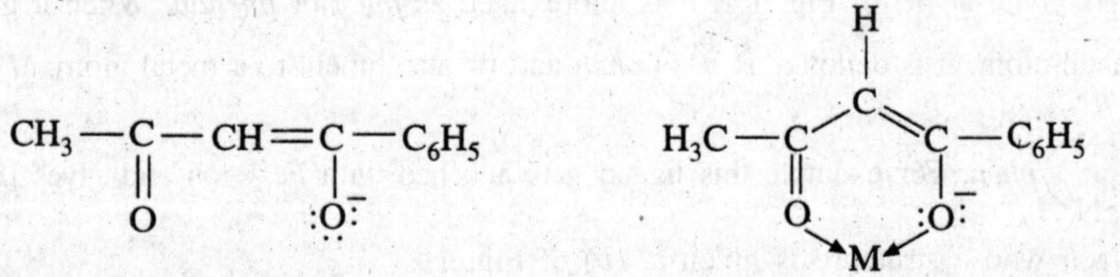

$$CH_3-\underset{\underset{O}{\|}}{C}-CH=\underset{\underset{:\ddot{O}:^-}{|}}{C}-C_6H_5$$

(*a*) (*b*)

Fig. 1.6. (*a*) Structure of benzoyl acetonato ion (*b*) Mode of attachment of benzoyl acetonato ion to the metal atom, M.

8. Oxalato ion, $C_2O_4^{2-}$ (ox^{2-}). This ion can be obtained when one molecule of oxalic acid ($H_2C_2O_4$ or H_2ox) loses two protons (see Fig. 1.7).

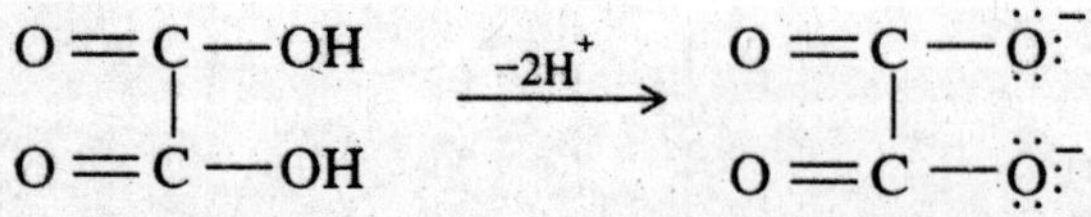

$$\begin{array}{l} O=C-OH \\ \quad\;\, | \\ O=C-OH \end{array} \xrightarrow{-2H^+} \begin{array}{l} O=C-\ddot{O}:^- \\ \quad\;\, | \\ O=C-\ddot{O}:^- \end{array}$$

H_2ox molecule ox^{2-} molecule

Fig. 1.7. To get ox^{2-} ion from H_2ox molecule.

$C_2O_4^{2-}$ (ox^{2-}) is a bidentate dibasic ligand and its coordination to the central metal atom occurs through its both negatively charged O-atoms (acidic donor atoms), forming 5-membered chelated complex as is evident from the structure of dioxalato platinate (II) ion, $[Pt^{2+}(C_2O_4)_2]^{2-}$ or $[Pt^{2+}(ox)_2]^{2-}$ as shown in Fig. 1.8.

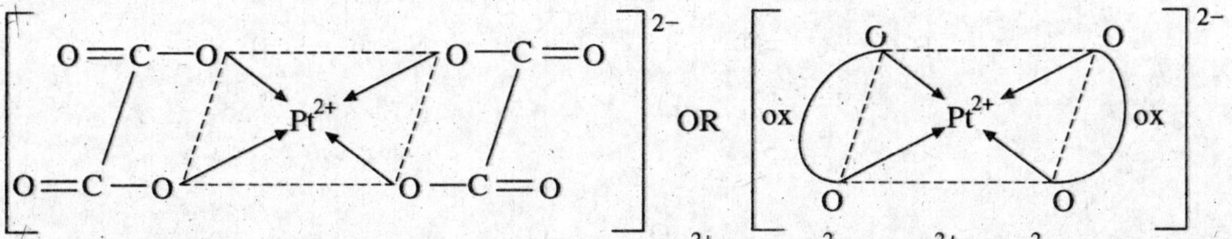

Fig. 1.8. Structure of dioxalato platinate (II) ion, $[Pt^{2+}(C_2O_4)_2]^{2-}$ or $[Pt^{2+}(ox)_2]^{2-}$ (square planar complex ion).

9. 8-hydroxyl quinolinato ion (oxine⁻ or oxinato⁻). This ion is obtained when OH group attached with the ring of 8-hydroxyl quinoline molecule, Hoxine loses a proton (H^+) as shown in Fig. 1.9.

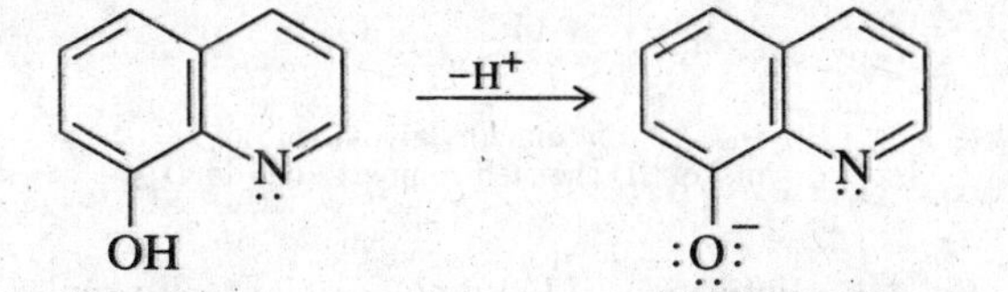

8-hydroxy quinoline molecule, Hoxine

8-hydroxy quino linato ion, oxine⁻

Fig. 1.9. To get oxine⁻ ion from H oxine molecule.

The attachment of oxine⁻ ion with a metal atom (M) in complex compounds takes place as shown in Fig. 1.10. By this type of attachment a 5-membered ring is formed.

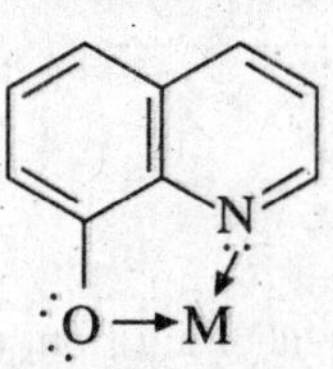

Fig. 1.10. Attachment of oxine⁻ ion to a metal atom (M) in complex compounds.

10. Dimethyl glyoximato ion (dmg⁻ or DMG⁻). This ion is obtained when one OH group attached with N-atom in Hdmg molecule loses a proton (Fig. 1.11).

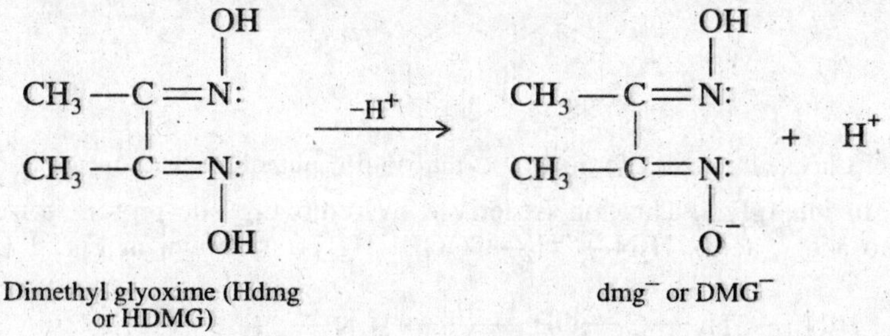

Dimethyl glyoxime (Hdmg or HDMG)

dmg⁻ or DMG⁻

Fig. 1.11. To get dmg⁻ ion from Hdmg molecule.

Two dmg⁻ ions react with Ni^{2+} or Pd^{2+} ion in ammoniacal solution and form an insoluble neutral red chelated complex, $[M(dmg)_2]^0$ ($M = Ni^{2+}$, Pd^{2+}) which has two 5-membered and two 6-membered rings. The formation of $[Ni(dmg)_2]^0$ has been shown in Fig. 1.12.

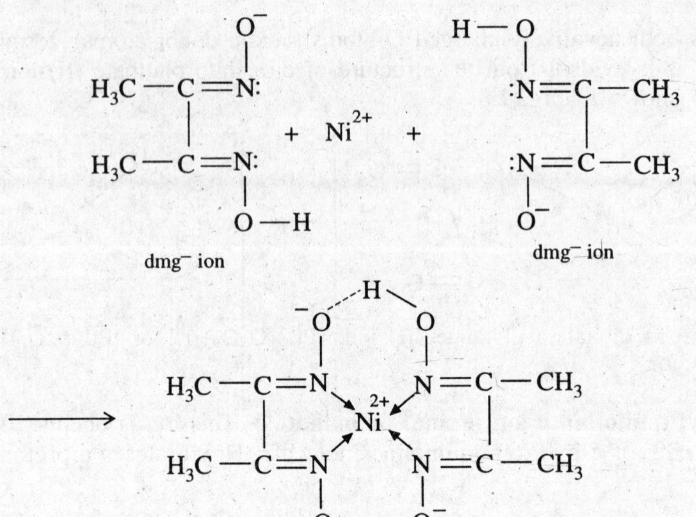

dmg$^-$ ion + Ni^{2+} + dmg$^-$ ion

Bis-(dimethylglyoximato) nickel (II) chelated complex, $[Ni(dmg)_2]^0$ (red ppt.)

Fig. 1.12. Formation of $[Ni(dmg)_2]^0$ (chelated complex).

11. α-nitroso β-naphthol anion. This anion is obtained when α-nitroso β-naphthol molecule loses a proton (H^+) from OH group (see Fig. 1.13).

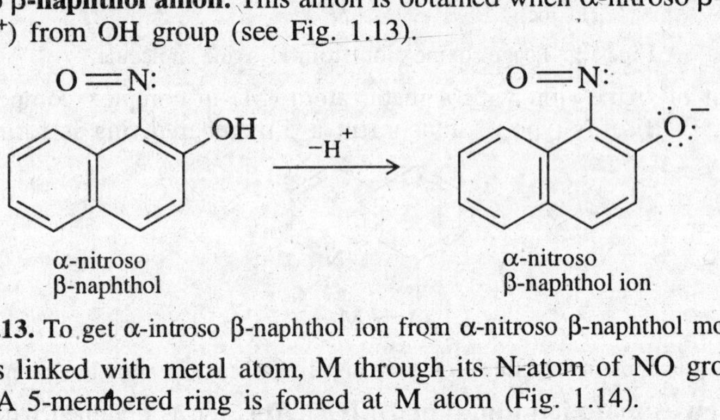

α-nitroso β-naphthol $\xrightarrow{-H^+}$ α-nitroso β-naphthol ion

Fig. 1.13. To get α-introso β-naphthol ion from α-nitroso β-naphthol molecule.

This ion gets linked with metal atom, M through its N-atom of NO group and negatively charged O-atom. A 5-membered ring is fomed at M atom (Fig. 1.14).

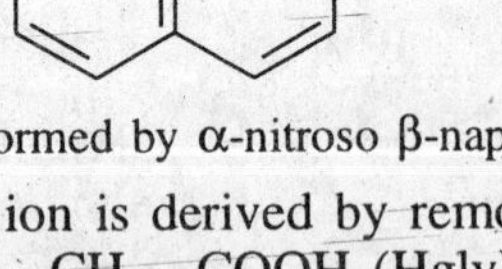

Fig. 1.14. Chelated complex formed by α-nitroso β-naphthol ion with metal atom, M.

12. Glycinato ion (gly$^-$). This ion is derived by removing one proton from one molecule of glycine (amino acetic acid), $H_2N—CH_2—COOH$ (Hgly) as shown in Fig. 1.15.

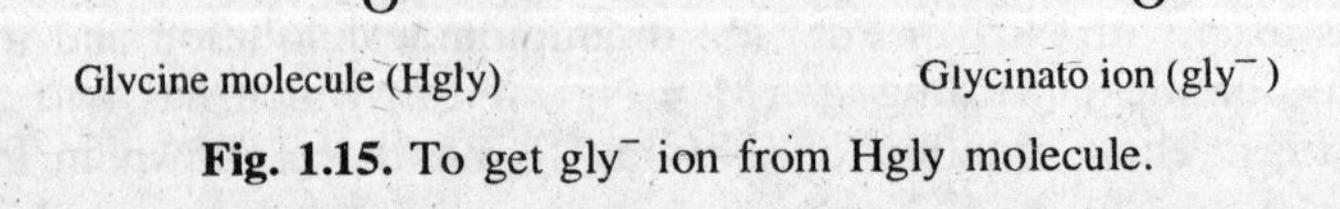

$$H_2\ddot{N}—CH_2—\underset{\underset{O}{\|}}{C}—OH \xrightarrow{-H^+} H_2\ddot{N}—CH_2—\underset{\underset{O}{\|}}{C}—\ddot{O}:^-$$

Glycine molecule (Hgly) Glycinato ion (gly$^-$)

Fig. 1.15. To get gly$^-$ ion from Hgly molecule.

Glycinato ion (gly$^-$) is a bidentate monobasic ligand and its co-ordination to the metal ion occurs through neutral N-atom (neutral donor atom) and negatively-charged O-atom (acidic donor atom), forming a 5-membered ring as is evident from the strucuture of *cis*-and *trans*-isomers of octahedral triglycinato chromium (III), $[Cr(NH_2{-}CH_2{-}\underset{\underset{O}{\|}}{C}{-}O)_3]^0$ or $[Cr(gly)_3]^0$ shown in Fig. 1.21 (see symmetrical and unsymmetrical bidentate ligands).

13. Biguanido ion (big$^-$). This ion is obtained when biguanide molecule loses H-atom of NH group in C=NH unit (see Fig. 1.16).

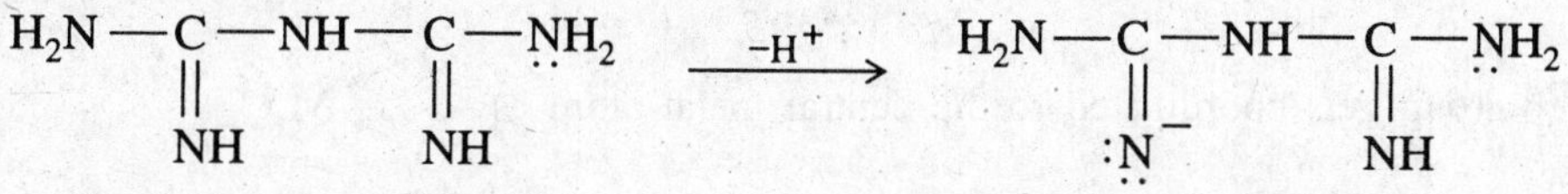

$$H_2N{-}\underset{\underset{NH}{\|}}{C}{-}NH{-}\underset{\underset{NH}{\|}}{C}{-}\ddot{N}H_2 \xrightarrow{-H^+} H_2N{-}\underset{\underset{:\ddot{N}^-}{\|}}{C}{-}NH{-}\underset{\underset{NH}{\|}}{C}{-}\ddot{N}H_2$$

Biguanide molecule (big H) Biguanido ion (big$^-$).

Fig. 1.16. To get big$^-$ ion from Hbig molecule.

This ion gets attached with the central metal atom through its negatively-charged N-atom and N-atom of NH_2 group and forms a chelated complex containing 6-membered ring as shown in Fig. 1.17.

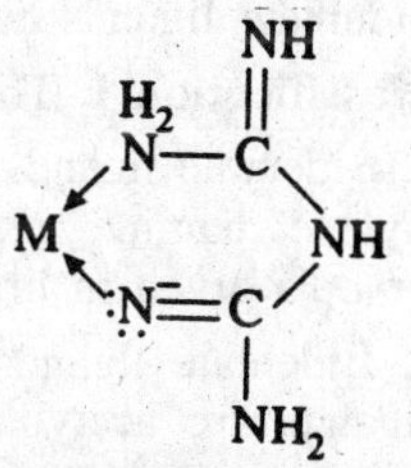

Fig. 1.17. Attachment of biguanido ion with metallic atom

14. Carbonato (CO_3^{2-}), nitrato (NO_3^-) and sulphato (SO_4^{2-}). The structure of these ions and the way in which they are coordinated to the metal atom are shown below in Fig. 1.18.

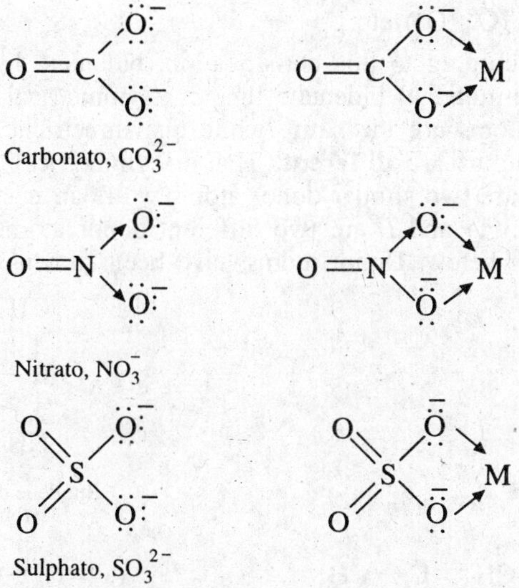

Carbonato, CO_3^{2-}

Nitrato, NO_3^-

Sulphato, SO_3^{2-}

Fig. 1.18. Structure of CO_3^{2-}, NO_3^- and SO_4^{2-} ions and their attachment with metal atom, M.

Note. Carbonato, nitrato and sulphato ligands also act as monodentate ligands as is evident from the following complex ions.

$[Co^{3+}(NH_3)_5(CO_3)]^+$ ... Pentammine carbonato cobalt (III) ion (octahedral)

$[Co^{3+}(NH_3)_5(SO_4)]^+$ ... Pentammine sulphato cobalt (III) ion (octahedral)

$[Ti^{4+}(NO_3)_4]^0$ ... Tetranitrato titanium (IV) (square planar)

$[Co^{3+}(en)_2(NO_3)_2]^+$ ... Bis-(ethylene diamine) dinitrato cobalt (III) ion (octahedral)

15. Peroxo, O_2^{2-}. This ligand can be supposed to have been obtained by eliminating two protons from H_2O_2 molecule (Fig. 1.19).

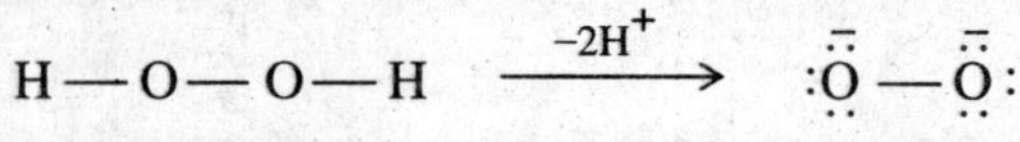

$$H—O—O—H \xrightarrow{-2H^+} :\ddot{O}—\ddot{O}:$$

Hydrogen peroxide molecule, H_2O_2 — Peroxo group (O_2^{2-})

Fig. 1.19. To get peroxo ion (O_2^{2-}) from hydrogen peroxide molecule (H_2O_2)

O_2^{2-} group gets coordinated to the central metal atom as $:\ddot{O}$—$:\ddot{O}$ → M (both O atoms donating to M)

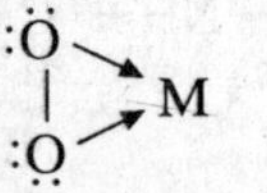

Classification of bidentate ligands

Bidentate ligands have been clasified into the following two ways:

Classification I. This classification gives the following three types of bidentate ligands.

1. Bidentate ligands whose both donor atoms are neutral atoms. Examples of such ligands are ethylene diamine (N, N), dipyridyl (N, N), o-phenylene bis dimethyl arsine (As, As) etc. The symbols given in bracket are the donor atoms.

2. Bidentate ligands which have one neutral and one anionic donor atom. Examples of such ligands are acetyl acetonate ion (O, O^-), 8-hydroxy quinolinate ion (N, O^-), α-nitroso β-naphthol ion (N, O^-) etc.

3. Bidentate ligand which have both anionic donor atoms, *e.g.* oxalate ion (O^-, O^-), carbonate ion (O^-, O^-), sulphate ion (O^-, O^-) etc.

Classification II. According to this classification, bidentate ligands may be **symmetrical** bidentate ligands or **unsymmetrical** bidentate ligands. Symmetrical bidentate ligands are those in which the two donor atoms are the **same** while unsymmetrical bidentate ligands are those in which the two donor atoms are **different.** Thus a symmetrical bidentate ligand is denoted by (*AA*) in which two *A*'s are two similar donor atoms while an unsymmetrical bidentate ligand is denoted as (*AB*) in which *A* and *B* are two different donor atoms. Examples of symmetrical bidentate ligands are given below. Donor atoms have been shown in bracket.

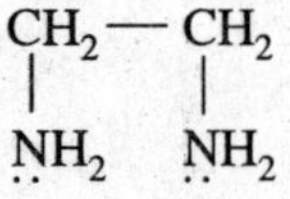

$CH_2—CH_2$ with $\ddot{N}H_2$ on each carbon

Ethylene diamine, *en*
(N, N)

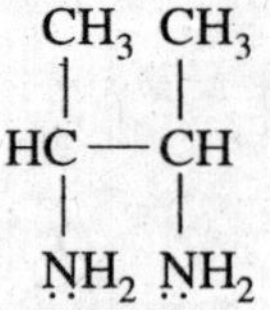

CH_3–CH(–$\ddot{N}H_2$)–CH(–$\ddot{N}H_2$)–CH_3 (HC—CH bearing CH_3 and NH_2 on each carbon)

Butylene diamine, *bn*
(N, N)

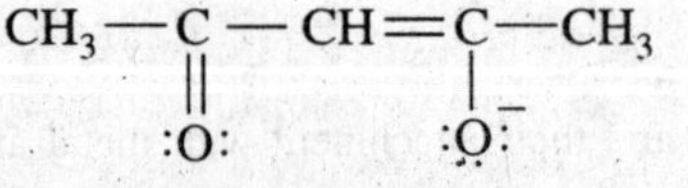

$CH_3—C(=\ddot{O}:)—CH=C(—:\ddot{O}:^-)—CH_3$

Acetylacetonato, *acac*$^-$
(O, O^-)

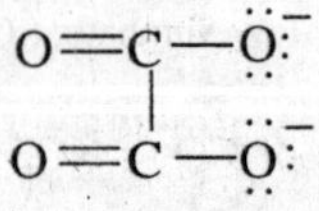

$O=C(—\ddot{O}:^-)—C(—\ddot{O}:^-)=O$ *or* $C_2O_4^{2-}$

Oxatato, *ox*$^{2-}$
(O, O^-)

The attachment of a symmetrical bidentate ligand to the central metal atom is shown by a curve $\left.\begin{matrix}A\\A\end{matrix}\right)L$ where where *L* is the abbrevation of the symmetrical bidentate ligand and two *A*'s are two (similar) donor atoms.

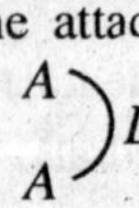

For example the attachment of two molecules of ethylene diamine, $NH_2—CH_2—CH_2—NH_2$ (en) to Co^{3+} ion to form *trans*-isomer of $[Co(en)_2Cl_2]^+$ ion (octahedral ion) can be shown as depicted at (*a*) of Fig. 1.20. Similarly, the attachment of two molecules of butylene diamine, $NH_2—CH(CH_3)—CH(CH_3)—NH_2$ (bn) to Pt^{2+} ion to form square planar $[Pt(en)_2]^{2+}$ ion can be shown at (*b*) of Fig. 1.20.

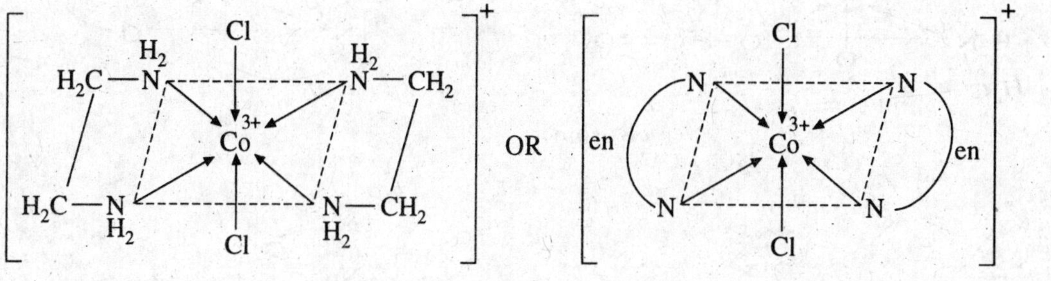

(*a*)

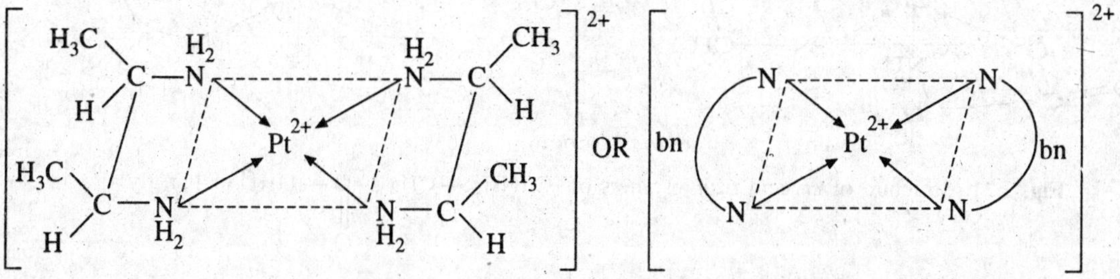

(*b*)

Fig. 1.20. Structure of trans-isomer of bis (ethylene diamine) dichloro cobalt (III) ion, $[Co(NH_2—CH_2—CH_2—NH_2)_2Cl_2]^+$ or $[Co(en)_2\ Cl_2]^+$ (octahedral ion) [shown at (*a*)] and bis (butylene diamine) platinum (II) ion, $[Pt\{NH_2—CH(CH_3)—CH(CH_3)—NH_2\}_2]^{2+}$ or $[Pt(bn)_2]^{2+}$ (square planar ion) [shown at (*b*)].

Examples of unsymmetrical bidentate ligands are given below. Donor atoms are given in bracket.

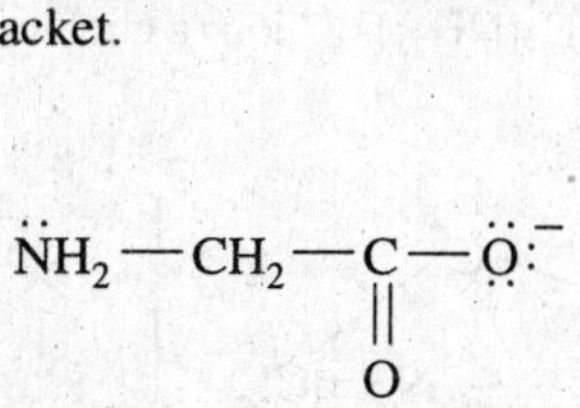

Glycinato ion, *gly*$^-$ (N, O^-)

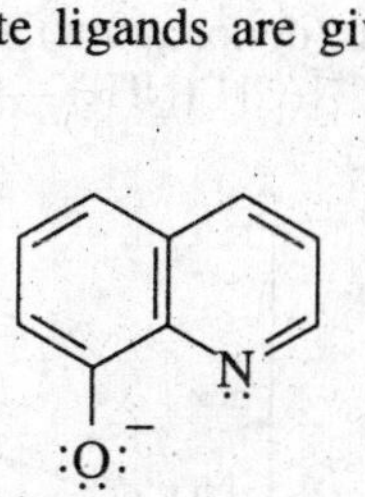

8-hydroxy quinolinato ion (N, O^-)

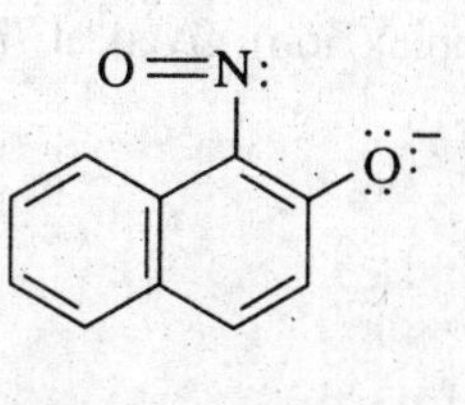

α-nitroso β-naphthol ion (N, O^-)

The attachment of an unsymmetrical bidentate ligand to the central metal atom is shown by a curve $\left.\begin{matrix}A\\B\end{matrix}\right) L$ where L is the abbreviation of the ligand and A and B are two (different) donor atoms. For example, the attachment of three glycinate ions, $NH_2—CH_2—C(=O)—O^-$ (gly^-) to Cr^{3+} ion to form *cis*-and *trans*-isomers of octahedral $[Cr(gly)_3]^0$ can be shown as depicted in Fig. 1.21.

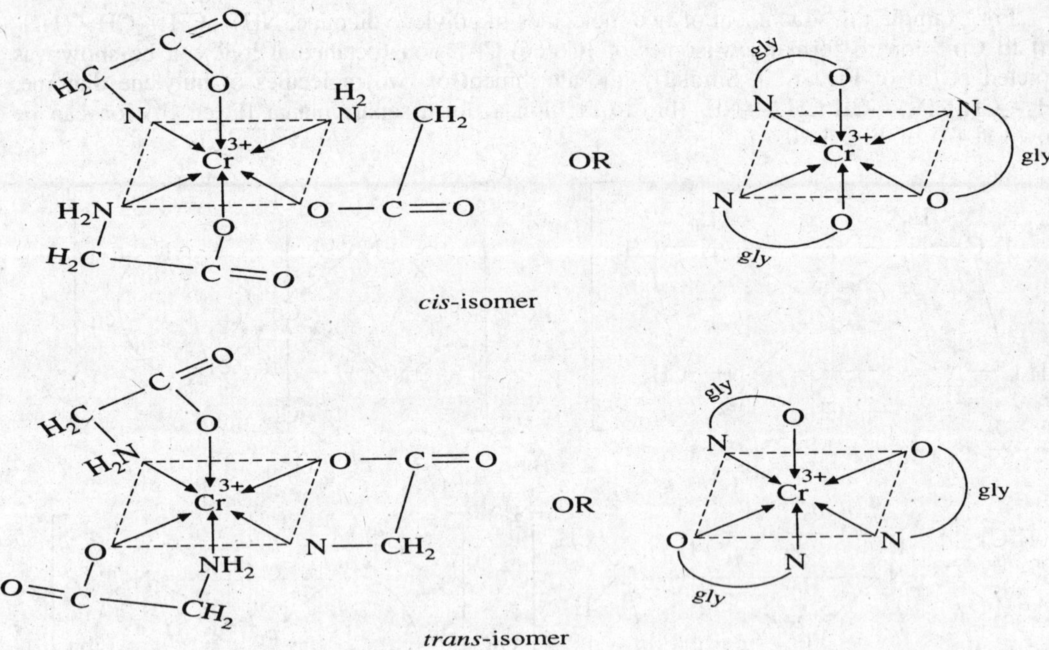

Fig. 1.21. Structure of *cis*-and *trans*-isomers of $[Cr(NH_2-CH_2-C(=O)-O)_3]^0$ or $[Cr(gly)_3]^0$

Examples of tridentate ligands

These ligands may be neutral molecules or anionic species. Important examples are given below:

1. Triaminopropane. The structure of this ligand is given at (*a*) of Fig. 1.22 which shows that it has three N-atoms as its donor atoms. The coordination of this ligand to the central metal atom takes place through its three N-atoms (donor atoms), leading to the formation of two 5-membered rings as is evident from the structure of $[Co^{3+}\{(CH_2(NH_2)-CH(NH_2)-CH_2(NH_2)\}_2]^{3+}$ ion (a chelated complex ion) given at (*b*) of Fig. 1.22.

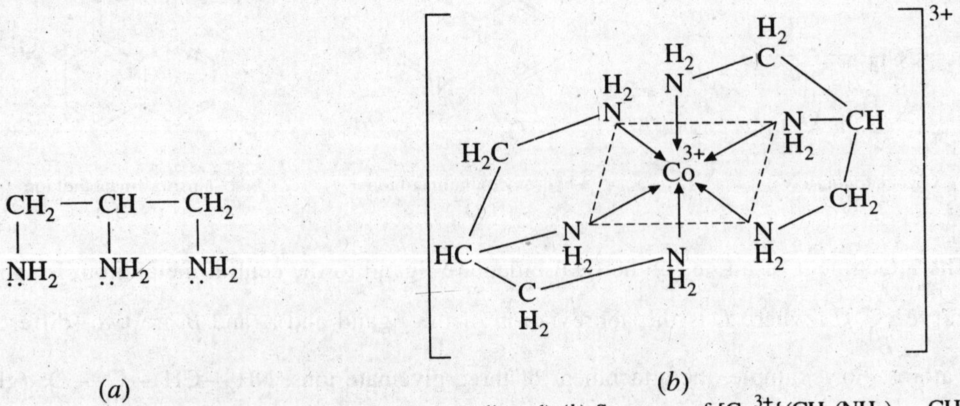

Fig. 1.22. (*a*) Structure of triaminopropane (tridentate ligand) (*b*) Structure of $[Co^{3+}\{(CH_2(NH_2)-CH(NH_2)-CH_2(NH_2)\}_2]^{3+}$ ion (octahedral chelated complexion).

2. Diethylene triamine (dien). Its structure is given in Fig. 1.23 (*a*). This ligand gets attached with the central metal atom through its three N-atoms (donor atoms) and forms a chelate which contains two 5-membered rings. [See Fig. 1.23 (*b*)].

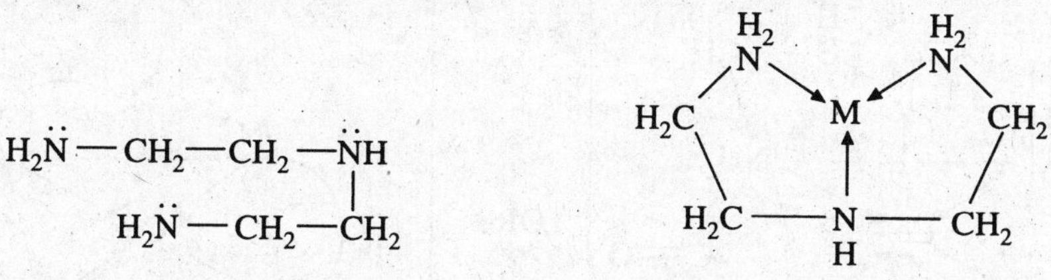

$$H_2\ddot{N}—CH_2—CH_2—\ddot{N}H$$
$$H_2\ddot{N}—CH_2—CH_2$$

(*a*)

H_2N, H_2N, H_2C, M, CH_2, H_2C—N—CH_2, H

(*b*)

Fig. 1.23. (*a*) Structure of diethylene triamine (dien) (*b*) chelated complex formed by dien with a metallic atom, M.

3. 2, 2′, 2″-terpyridine or terpyridyl (terpy). Its structure and its mode of linking with the central metal atom are shown below in Fig. 1.24.

N̈, N:, :N

(*a*)

N, N, N, M

(*b*)

Fig. 1.24. (*a*) Structure of 2, 2′, 2″-terpyridine (terpy) (*b*) Mode of attachment of terpy with the central metal atom, M.

4. Imido di-acetato ion (imda^{-2}). This ion is obtained when imido di-acetic acid loses two protons (See Fig. 1.25).

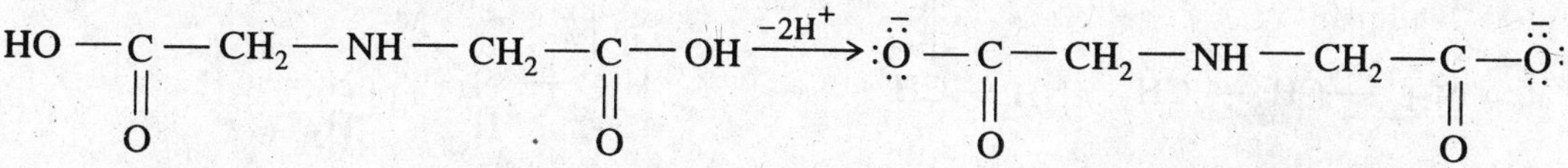

$$HO—\underset{\underset{O}{\|}}{C}—CH_2—NH—CH_2—\underset{\underset{O}{\|}}{C}—OH \xrightarrow{-2H^+} :\ddot{\bar{O}}—\underset{\underset{O}{\|}}{C}—CH_2—NH—CH_2—\underset{\underset{O}{\|}}{C}—\ddot{\bar{O}}:$$

Imido di-acetic acid (H2imda) Imido di-acetato ion (imda^{2-})

Fig. 1.25. To get imda^{2-} ion from H2imda moelcule.

This ligand links to the central metal atom, M through its N-atom and two negatively-charged O-atoms, forming a chelated complex which contains two 5-membered rings as shown in Fig. 1.26.

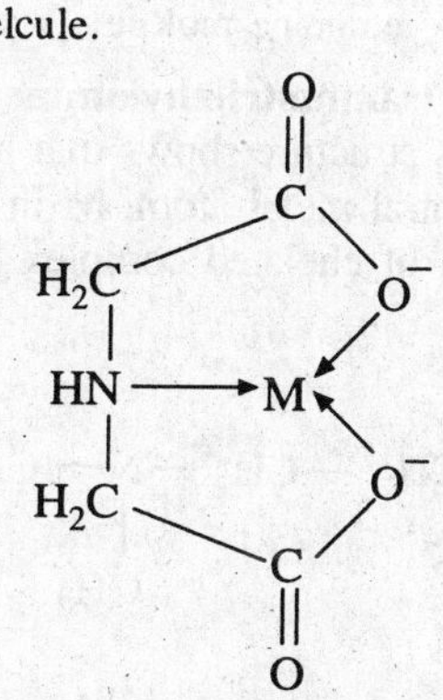

5. Anion of diamino propionic acid. This ligand is obtained by removing one proton from diaminopropionic acid as shown in Fig. 1.27.

Obviously it is a tridentate monobasic ligand and its coordination to the metal ion takes place through its two N-atoms (natural donor atoms) and one negatively-charged

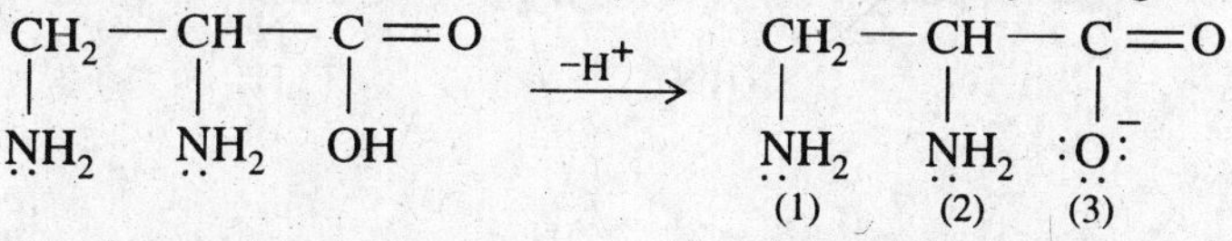

O, C, H_2C, O^-, HN, M, H_2C, O^-, C, O

Fig. 1.26. Structure of the chelated complex formed by imido diacetato with metallic atom, *M*.

$$\underset{\ddot{N}H_2}{CH_2}—\underset{\ddot{N}H_2}{CH}—\underset{OH}{C}=O \xrightarrow{-H^+} \underset{\underset{(1)}{\ddot{N}H_2}}{CH_2}—\underset{\underset{(2)}{\ddot{N}H_2}}{CH}—\underset{\underset{(3)}{:\ddot{O}:^-}}{C}=O$$

Diamino propionic acid Anion of diamino propionic acid

Fig. 1.27. To get an anion from diamino propionic acid molecule by removing one proton from it.

O-atom (acidic donor atom) as is evident from the structure of the chelate formed by this ligand with Co^{3+} ion as shown in Fig. 1.28.

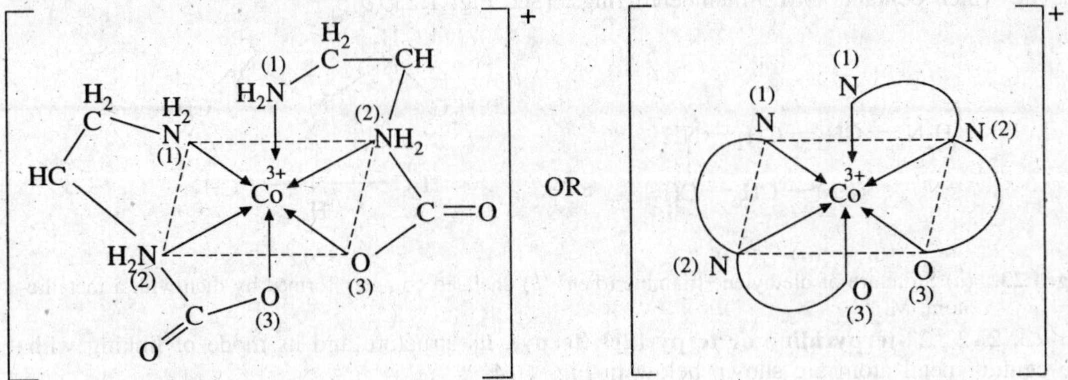

Fig. 1.28. Structure of $[Co^{3+}\{(CH_2(NH_2)—CH(NH_2)—COO\}_2]^+$ ion (chelated complex ion).

Examples of tetradentate ligands

Important examples of these ligands are given below:

1. Triethylene tetramine (trien). The structure of this ligand is given at (*a*) of Fig. 1.29. This structure shows that this ligand has four N-atoms as its donor atoms (neutral donor atoms). Coordination of this ligand to the metal ion takes place through these N-atoms, forming a chelated complex containing three 5-membered rings as shown at (*b*) of Fig. 1.29.

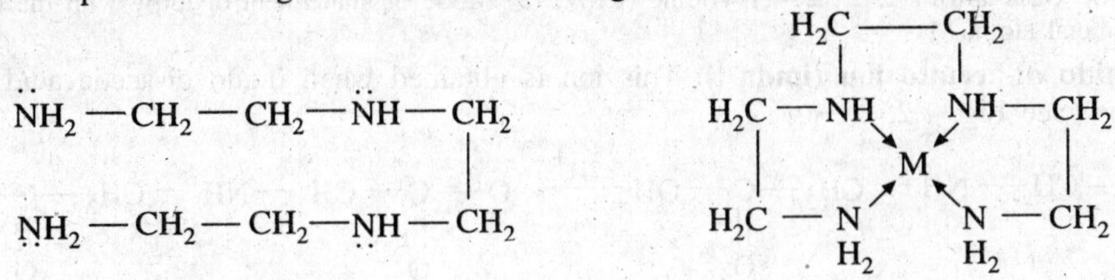

$\ddot{N}H_2—CH_2—CH_2—\ddot{N}H—CH_2$

$\ddot{N}H_2—CH_2—CH_2—\ddot{N}H—CH_2$

(*a*)

$H_2C—CH_2$

$H_2C—NH \quad NH—CH_2$

M

$H_2C—N \quad N—CH_2$

$H_2 \quad H_2$

(*b*)

Fig. 1.29. (*a*) Structure of triethylene tetramine (trien) molecule (*b*) Chelated complex formed by triethylene tetramine molecule with the metallic atom, M.

2. Triaminotriethylamine (tren), $N(CH_2—CH_2—NH_2)_3$. Its structure is given at (*a*) of Fig. 1.30. This structure shows that it has four N-atoms as its donor atoms. The attachment of this ligand to the central metal atom, M in its complexes takes place through its four N-atoms, resulting in the formation of chelated complex having three 5-membered rings as shown at (*b*) of Fig. 1.30.

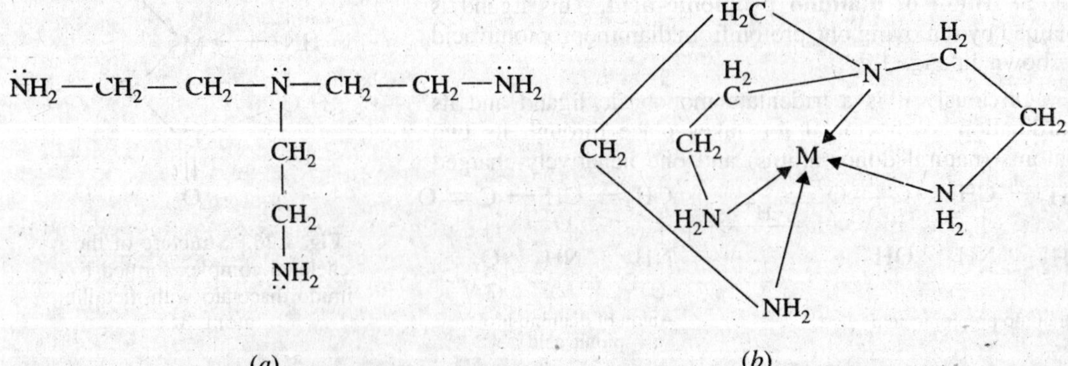

Fig. 1.30. (*a*) Structure of triaminotriethylamine (tren), $N(CH_2—CH_2—NH_2)_3$ (*b*) Structure of the chelated complex formed by the attachment of triaminotriethylamine molecule with metallic atom, M.

3. Nitrilotriacetato ion (nta^{-3} or NTA^{-3}). The structure of this ion is given at (*a*) of Fig. 1.31. This structure shows that this ligand is a tribasic tetradentate ligand. The coordination of this ligand to the metal ion takes place through N-atom (neutral donor atom) and three negatively-charged O-atoms (acidic donor atoms) of COO^- groups. This type of coordination gives rise to the formation of a chelate having three 5-membered rings as shown at (*b*) of Fig. 1.31.

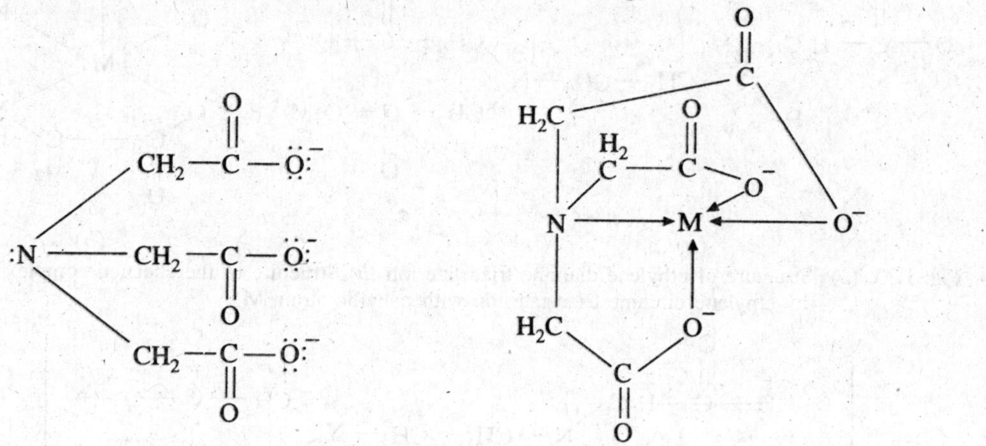

Fig. 1.31. (*a*) Structure of nitrilotriacetato ion (nta^{-3} or NTA^{-3}) (*b*) Structure of the chelated complex formed by nta^{-3} ion.

Examples of pentadentate ligands

Important examples are given below:

1. Tetraethylene pentamine (tetraen). The structure of this molecule is shown at (*a*) of Fig. 1.32. This structure shows that it has five N-atoms as its donor atoms (neutral donor atoms). Thus this ligand coordinates to the metal atom through these five N-atoms and forms four 5-membered rings as shown at (*b*) of Fig. 1.32.

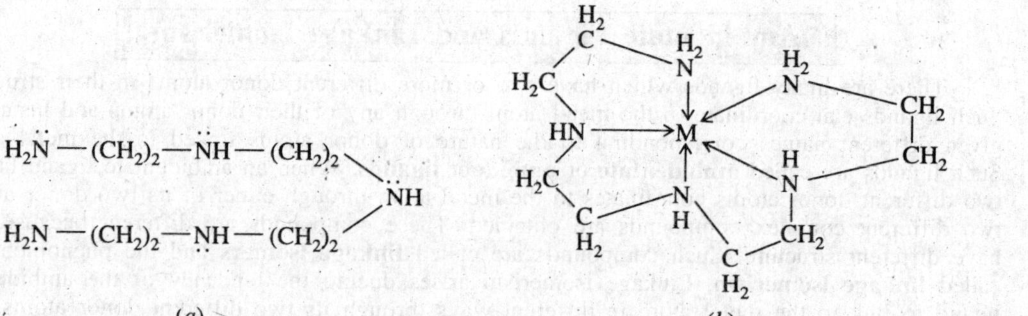

Fig. 1.32. (*a*) Structure of tetraethylene pentamine (tetraen) (*b*) Structure of the chelated complex formed by tetraethylene pentamine with metallic atom, M.

2. Ethylene diamine triacetate ion. It is an anionic ligand and its structure is shown at (*a*) of Fig. 1.33. This structure shows that it has three negative charges and five donor atoms *viz* two N-atoms (numbered as 1 and 2) and three negatively-charged O-atoms (numbered as 3, 4 and 5). Its coordination to the metal atom occurs through two N-atoms (neutral donor atoms) and three negatively-charged O-atoms (acidic donor atoms) of COO^- groups, forming four 5-membered rings as shown at (*b*) of Fig. 1.33.

Example of hexadentate ligand

Ethylene diamine tetracetate ion (edta^{-4} or EDTA^{-4} or Y^{-4}) is an important example of hexadentate ligand. It structure is given in Fig. 1.34.

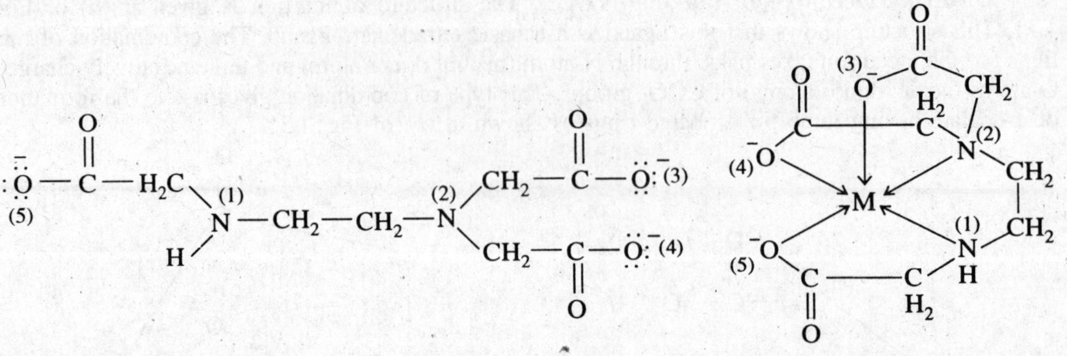

$:\ddot{O}^{-}—C(=O)—H_2C—N(H)—CH_2—CH_2—N(CH_2—C(=O)—\ddot{O}^{-})(CH_2—C(=O)—\ddot{O}^{-})$

(*a*) (*b*)

Fig. 1.33. (*a*) Structure of ethylene diamine triacetate ion (*b*) Structure of the chelated complex formed by ethylene diamine triacetate ion with metallic atom, M.

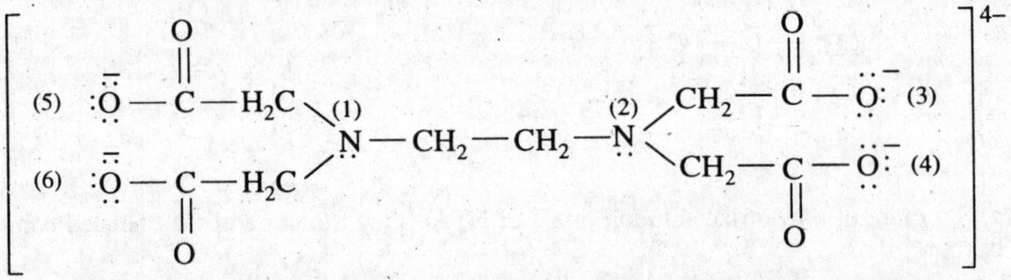

$[(\ddot{O}^{-}—C(=O)—H_2C)_2\ddot{N}—CH_2—CH_2—\ddot{N}(CH_2—C(=O)—\ddot{O}^{-})_2]^{4-}$

Fig. 1.34. Structure of ethylene diamine tetracetate ion (edta^{4-} or EDTA^{4-})

The structure shows that it has two neutral N-atoms (numbered as 1 and 2) and four negatively-charged O-atoms (numbered as 3, 4, 5 and 6) as its donor atoms. Thus it is tetrabasic hexadentate ligand. The co-ordination of this ligand to the metal atom takes place thorugh two N-atoms and four O-atoms, forming five 6-membered rings as is evident from the structure of *d* and *l*-forms of [Co^{3+}(EDTA)]$^-$ ion given in Figures 4.33 and 4.34 on page 91of chapter 4 (See *optical isomerism exhibited by octahedral complexes containing polydentate ligands*).

Ambidentate Ligands and Linkage Isomerism

There are many ligands which have two or more different donor atoms in their structure. Such ligands can coordinate to the metal atom through any of their donor atoms and hence are given different names corresponding to the nature of donor atoms linked to the metal atom. Such ligands are called **ambidentate** or **ambident ligands.** When an ambidentate ligand having two different donor atoms coordinates to the metal atom through either of its two donor atoms, two different complex compounds are obtained. These compounds are different because they have different structure. Such compounds are called **linkage isomers** and the phenomenon is called **linkage isomerism**. Linkage isomerism arises due to the tendency of the ambidentate ligand to link to the metal atom in different ways through its two different donor atoms. The less stable form of a pair of linkage isomers often reverts to the more stable form. The less stable form is likely to exist at low temperature and with those metal ions which are known to form kinetically inert complexes (*e.g.*, Co^{2+}, Pt^{2+}, Pd^{2+} etc.).

Examples of some ambidentate ligands and the pair of linkage isomers given by them

1. NO_2^- ion. We know that this ion can have any of the following two structures.

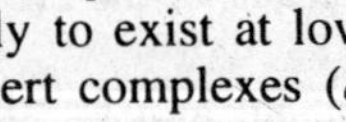

$:\ddot{O}\leftarrow \bar{N}=\ddot{O}:$ $\qquad$ $:\bar{\ddot{O}}—N=\ddot{O}:$

Structure (I) (Nitro group) 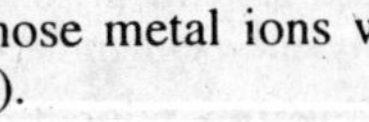 $\qquad$ Structure (II) (Nitrito group)

In structure (I) N-atom is negatively-charged and contains one unshared pair of electrons while in structure (II), one O-atom is negatively-charged and this O-atom contains three unshared pairs of electrons. Thus when NO_2^- ion acts as a monodentate ligand, it may be linked to the central metal atom either through the lone pair of electrons on negatively-charged N atom of structure (I) or through the lone pair of electrons on negatively-charged on O-atom of structure (II). Thus:

(*a*) When NO_2^- ion gets coordinated to the metal atom through a lone pair of electrons on $\bar{N}$ $\left[\ddot{O} \nwarrow \bar{N} \rightarrow M \text{ (with } =\ddot{O}) \text{ or } OO\bar{N} \rightarrow M \text{ or } O_2\bar{N} \rightarrow M \right]$, it is called nitro ligand and is written as NO_2^-. The complexes containing nitro ligand are called **nitro complexes.** For example $[Co^{3+}(NH_3)_5(NO_2)]Cl_2$ (yellow brown) is called pentammine nitro cobalt (III) chloride, since the coordination of Co^{3+} ion to NO_2^- ion occurs through N-atom.

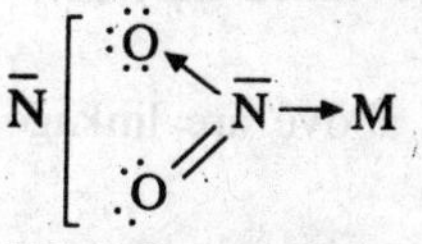

(*b*) When NO_2^- ion gets coordinated to the metal atom through a lone a pair of electrons on $\bar{O}$ $[M \leftarrow \bar{\ddot{O}} - \ddot{N} = \ddot{O}:$ *or* $M \leftarrow \bar{O}NO]$, it is called **nitrito ligand** which is written as $ON\bar{O}$. The complexes containing nitrito ligand are called **nitrito complexes.** For example $[Co^{3+}(NH_3)_5(ONO]Cl_2$ (red) is called pentammine nitrito cobalt (III) chloride, because the coordination of NO_2^- ion to Co^{3+} ion takes place through O-atom.

Sometimes ambidentate ligands are distinguished by putting the symbol of the atom through which the coordination of the ligand to the central metal atom takes place. For example $[Co^{3+}(NH_3)_5(NO_2)]Cl_2$ (yellow-brown) is called pentammine nitrito (N) cobalt (III) chloride and $[Co^{3+}(NH_3)_5(ONO)]Cl_2$ (red) is called *pentammine nitrito* (O) *cobalt* (III) *chloride*.

Both nitro and nitrito complex compounds mentioned above are called **linkage isomers** and the isomerism shown by them is called **linkage isomerism.**

Another example of a pair of linkage isomers is furnished by *diammine bis* (*pyridine*) *dinitro cobalt* (III) *nitrate*, $[Co^{3+}(NH_3)_2(py)_2(NO_2)_2]NO_3$ and *diammine bis* (*pyridine*) *dinitrito cobalt* (III) *nitrate*, $[Co^{3+}(NH_3)_2(py)_2(ONO)_2]NO_3$.

It has been found that nitro (NO_2^-) is a stronger ligand than the nitrite (ONO^-). Linkage isomers can be identified by infra-red spectroscopy.

2. SCN^- ion. We know that this ion can be represented by any of the following two structures:

$:\ddot{\bar{S}} - C \equiv N:$	$:S = C = \ddot{\bar{N}}:$
Structure (I) (Thio cyanato group)	Structure (II) (Iso-thiocyanato group)

Structures given above show that when SCN^- ion acts as a monodentate ligand , it may be coordinated to the central metal atom either through the lone pair of electrons on $\bar{S}$ atom [Structure (I)] or through the lone pair of electrons on N^- atom [Structure (II)]. Thus:

(*a*) When SCN^- ion gets coordinated to the metal atom through the lone pair on negatively-charged S-atom $[M \leftarrow \ddot{\bar{S}} - C \equiv N:$ *or* $M \leftarrow \bar{S}CN]$, it is called **thiocyanato ligand** and is written as $\bar{S}CN$. The complexes containing thiocyanato ligand are called **thiocyanato complexes.** For example $[Pd^{2+}(dipy)(SCN)_2]^0$ (orange yellow) is called *dipyridyl dithiocyanato palladium* (II), since the coordination of SCN ligand to Pd^{2+} ion takes place through S-atom. This compound is also called *dipyridyl dithiocyanato* (S) *palladium* (II).

(*b*) When the coordination of SCN ion to the metal atom takes place through N-atom ($M \leftarrow \ddot{\bar{N}}{=}C{=}S$ *or* $M \leftarrow \bar{N}CS$), this ion is called **isothiocyanato ligand** and is written as $\bar{N}CS$.

The complex compounds having $\bar{N}CS$ ion as a monodentate ligand are called **isothiocyanato complexes.** For example: $[Pd^{2+}(dipy)(NCS)_2]^0$ (light yellow) is called *dipyridyl di isothiocyanato palladium* (II), since Pd^{2+} is linked with SCN ligand through N-atom. It is also called *dipyridyl dithio cyanato* (N) *palladium* (II).

Both the complex compounds (orange yellow and light yellow) shown above are linkage isomers to each other.

Other examples of pairs of linkage isomers containing $SC\bar{N}/NC\bar{S}$ ions are:

(*a*) $[Pd^{2+}(Ph_3P)_2(SCN)_2]^0$ and $[Pd^{2+}(Ph_3P)_2(NCS)_2]^0$

(*b*) $[Mn(CO)_5(SCN)]$ and $[Mn(CO)_5(NCS)]$

It has been observed that first row transition metal's form N-bonded thiocyanate (isothiocynate) complexes while second and third row transition elements form S-bonded thiocyanate (thiocyanate) complexes.

Soft acid ions (*e.g.*, Pd^{2+}, Pt^{2+} etc.) prefer to form thiocyanato complexes with SCN^- ion while hard acid ions (*e.g.*, Co^{3+}, Cr^{3+} etc.) prefer to give isothiocyanato complexes with this ion.

In spectrochemical series of ligands $SC\bar{N}$ (thiocyanate ion) lies close to Cl^- ion while NCS^- (isothiocyanate ion) is close to H_2O.

3. Cyanide ion, CN^-. This ion can be represented by any of the following two structures.

$$:\bar{C}{\equiv}N: \qquad\qquad :C{\equiv}\bar{N}:$$

Structure (I) (Cyano group) — Structure (II) (Iso-cyano group)

Structures given above show that when CN^- ion acts as a monodentate ligand, it may be linked to the central metal atom either through the lone pair of electrons on $\bar{C}$-atom [Structure (I)] or through the lone pair of electrons on $\bar{N}$ atom [Structure (II)]. Thus:

(*a*) When this ion gets linked to the metal atom as a monodentate ligand through C-atom ($M \leftarrow \bar{C}{\equiv}N$: *or* $M \leftarrow CN$), this ion is called cyano ligand which is written as CN^-.

(*b*) When this ion is linked to the metal atom as a monodentate ligand through N-atom ($M \leftarrow \ddot{\bar{N}}{=}C$: *or* $M \leftarrow \bar{N}C$), this ion is called isocyano ligand.

Actual isomeric compounds containing cyano ($\bar{C}N$) and isocyano ($\bar{N}C$) ligands have not yet been fully established.

4. Thiosulphate ion, $S_2O_3^{2-}$. This ion can have any of the following two structures:

$$\begin{array}{c} :\ddot{O}: \\ \| \\ :\ddot{\bar{O}}{-}S{-}\ddot{\bar{O}}: \\ \| \\ :\ddot{S}: \end{array} \qquad\qquad \begin{array}{c} :\ddot{O}: \\ \| \\ :\ddot{\bar{O}}{-}S{-}\ddot{\bar{S}}: \\ \| \\ :\ddot{O}: \end{array}$$

[Thiosulphato – O ion, OSO_2S^{2-}] — [Thiosulphato – S ion, SSO_3^{2-}]

Structures given above show that when $S_2O_3^{2-}$ ion acts as a monodentate ligand, it may be linked to the central metal atom either through the lone pair of electrons on O^- [Structure (I)] or through the lone pair of electrons on S^- [Structure (II)]. The complex compounds in which $S_2O_3^{2-}$ ion is linked to the metal atom through O-atom are called *thiosulphate-O complex compounds* (M—O—S_2O_2) while those in which $S_2O_3^{2-}$ ion is attached to the metal through S-atom are called *thiosulphate-S complex compounds* (M—S—SO_3). Both these complexes are linkage isomers to each other.

Examples: *(a)* $H_4[(CN)_5Co^{3+}—O—S_2O_2]$ (yellow) (O-bonded form) and $H_4[(CN)_5Co^{3+}—S—SO_3]$ (red) (S-bonded form) are linkage isomers to each other. O-bonded form is decomposed by H_2O to give sulphur and H_2SO_3 while S-bonded form gives H_2S and H_2SO_4.

$$H_4[(CN)_5Co^{3+}—O—S_2O_2] + H_2O \longrightarrow H_2[Co^{3+}(CN)_5 (H_2O)] + S + H_2SO_3$$

$$H_4[(CN)_5Co^{3+}—S—SO_3] + 2H_2O \longrightarrow H_2[Co^{3+}(CN)_5 (H_2O)] + H_2S + H_2SO_4$$

(b) $[(NH_3)_5Co^{3+}—O—S_2O_2]^+$ (O-bonded form) and $[(NH_3)_5Co^{3+}—S—SO_3]^+$ (S-bonded form) are linkage isomers.

Bridging Ligands and Bridged Complexes

Although the ligands like N_2 or :N≡N: (dinitrogen), $NH_2—NH_2$ or N_2H_4 (hydrazine), OH^- (hydroxo), NH_2^- (amido or amino), NH^{2-} (imido), Cl^-, F^-, O^{2-}(oxo), O_2^{2-} or $:\ddot{O}—\ddot{O}:$ (peroxo), O_2^- or :O ⋯ O: (superoxo), SO_4^{2-}, NO_2^-, CO etc. are monodentate ligands, they also act as bidendate ligands when they are attached with two separate metals atoms, making a bridge between them. Such ligands are called *bridging ligands* and the complexes thus formed are called *bridged* (or *polynuclear* or *multinuclear*) *complexes.* Each bridging ligand makes two σ-bonds with two metal atoms. A bridging ligand must have at least two lone pairs of electrons which the ligand uses to get coordinated to two metal atoms. The polynuclear complex may be *dinuclear, trinuclear, tetranuclear* etc.

Nomenclature of bridging ligands

While naming the bridging ligands, the prefex μ is used before the name of each of the bridging ligands. For two similar bridging ligands *di*–μ or μ–*di* is used [Also see "*Nomenclature of Coordination Compounds*" discussed in chapter 2].

Bridged complexes of Co(III)

Co (III) forms a number of bridged complexes. These complexes are of the following types:

1. Bridged complexes containing hydroxo ions (OH^-) as bridging ligands. Examples of such complexes are *(i)*, *(ii)*, *(iii)* and *(iv)* given below in Fig. 1.35.

(i) $\left[(NH_3)_4 Co^{3+} \langle {}^{OH}_{OH} \rangle Co^{3+} (NH_3)_4\right](SO_4)_2$

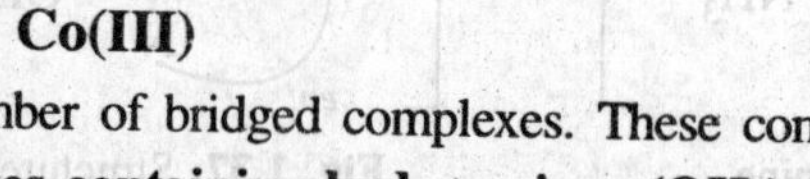

Octammine μ-dihydroxo dicobalt (III) sulphate

(ii) $\left[(en)_2 Co^{3+} \langle {}^{OH}_{OH} \rangle Co^{3+} (en)_2\right](S_2O_6)_2$

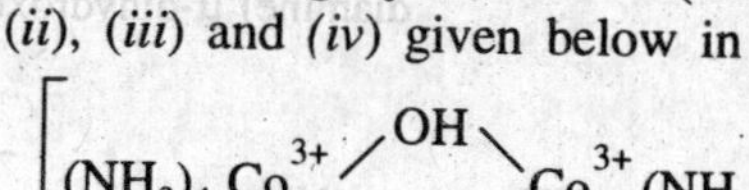

Tetrakis-(ethylenediamine) μ-dihydroxo dicobalt (III) dithionate

(iii) $\left[(NH_3)_4 Co^{3+} \langle {}^{OH}_{OH} \rangle Co^{3+}(NH_3)_2 \langle {}^{OH}_{OH} \rangle Co^{3+} (NH_3)_4\right]^{5+}$ (with NH_3 above and below the central Co)

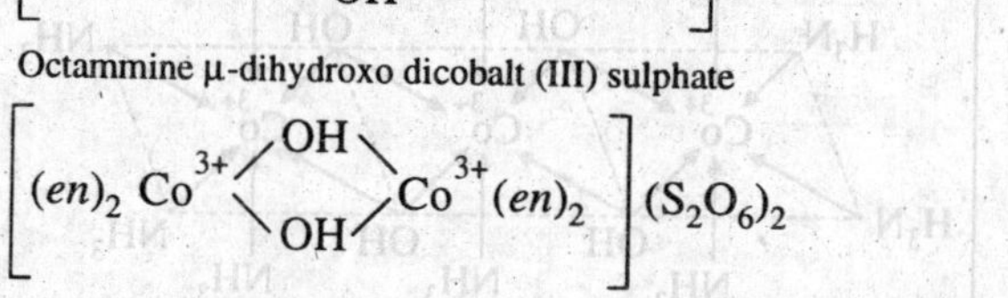

Tetra- μ-hydroxo decammine tricobalt (III) ion

$$(iv)\quad \left[(NH_3)_4\,\overset{3+}{Co} \begin{matrix} OH \\ OH \end{matrix} \overset{3+}{\underset{NH_3}{\overset{NH_3}{Co}}} \begin{matrix} OH \\ OH \end{matrix} \overset{3+}{\underset{NH_3}{\overset{NH_3}{Co}}} \begin{matrix} OH \\ OH \end{matrix} \overset{3+}{Co}\,(NH_3)_4 \right]^{6+}$$

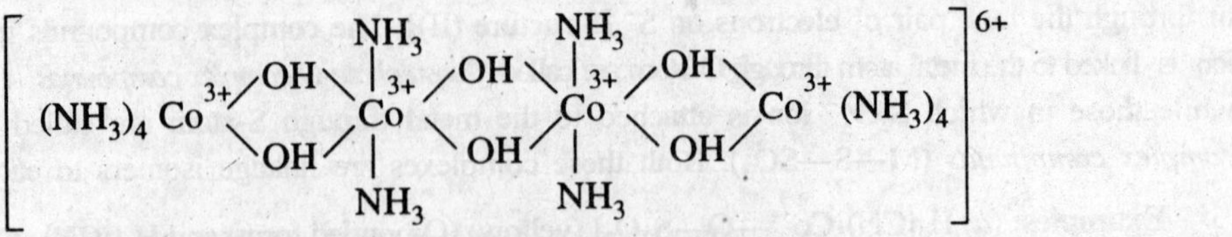

Dodeca ammine hexa-μ-hydroxo tetra cobalt (III) ion

Fig. 1.35. Examples of hydroxy-bridged complexes of Co (III) containing similar terminal ligands.

Complexes (*i*) and (*ii*) are binuclear complexes, complex (*iii*) is a trinuclear complex while complex (*iv*) is a tetranuclear complex. (*i*) and (*ii*) are obtained by the hydrolytic reactions of Co (III) compound in solution.

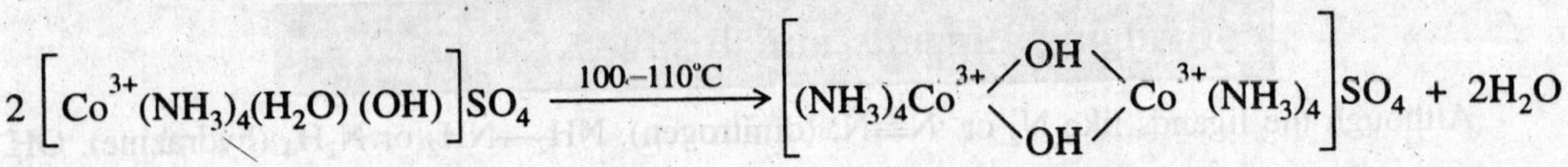

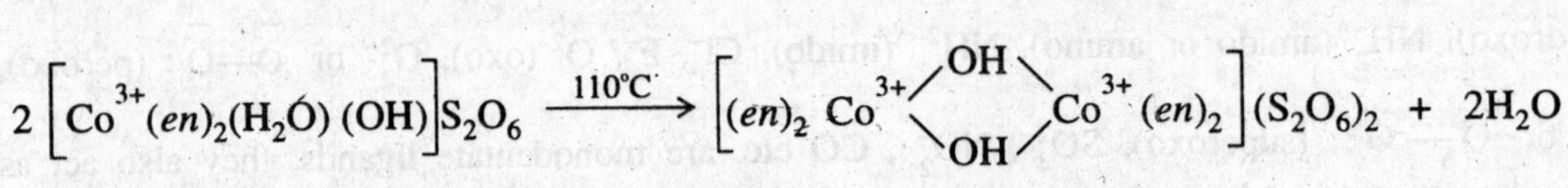

$$2\left[\overset{3+}{Co}(NH_3)_4(H_2O)(OH)\right]SO_4 \xrightarrow{100-110^\circ C} \left[(NH_3)_4\overset{3+}{Co} \begin{matrix} OH \\ OH \end{matrix} \overset{3+}{Co}(NH_3)_4\right]SO_4 + 2H_2O$$

$$2\left[\overset{3+}{Co}(en)_2(H_2O)(OH)\right]S_2O_6 \xrightarrow{110^\circ C} \left[(en)_2\,\overset{3+}{Co} \begin{matrix} OH \\ OH \end{matrix} \overset{3+}{Co}\,(en)_2\right](S_2O_6)_2 + 2H_2O$$

Complex (*iii*) is obtained by the treatment of $[Co^{3+}(NH_3)_4(H_2O)Cl]SO_4$ with cold alkali. Structures of complexes (*i*) (*ii*), (*iii*) and (*iv*) shown in Fig. 1.35 are given in Figures 1.36, 1.37, 1.38 and 1.39 respectively.

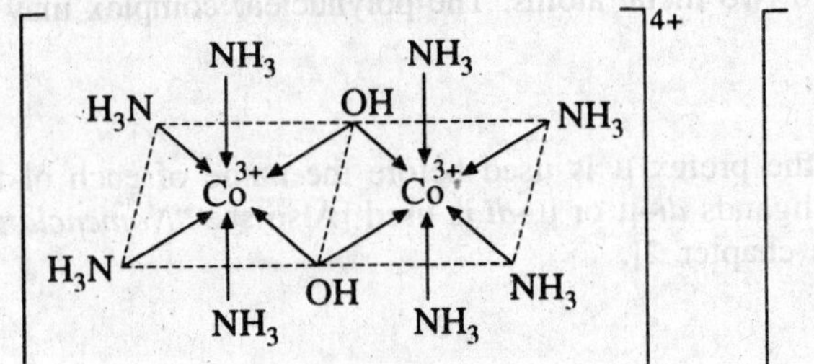

Fig. 1.36. Structure of octammine μ-dihydroxo dicobalt (III) ion.

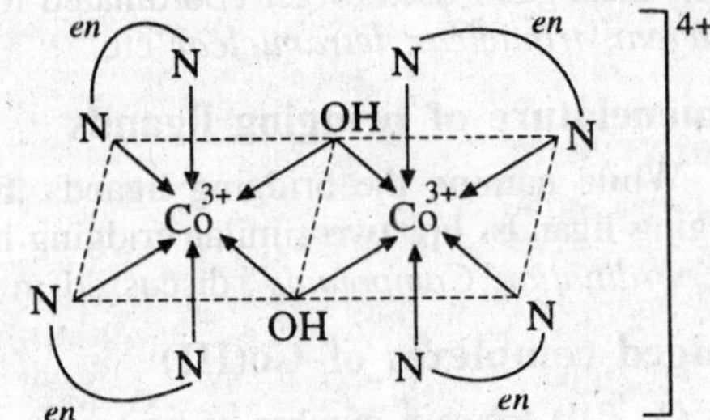

Fig. 1.37. Structure of tetrakis-(ethylene diamine) μ-dihydroxo dicobalt (III) ion.

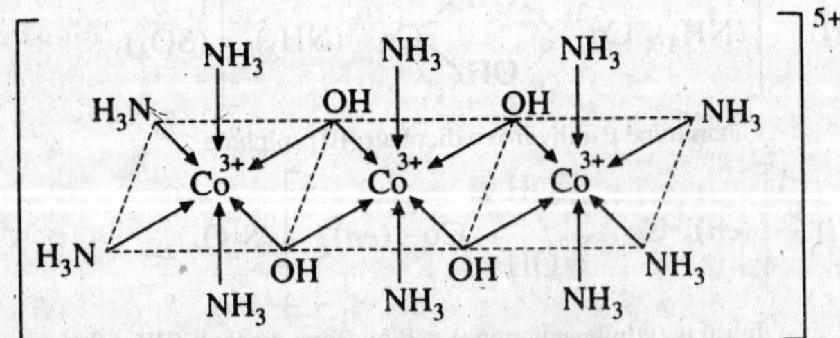

Fig. 1.38. Structure of μ-tetrahydroxo-decammine tricoblat (III) ion.

So far we have discussed hydroxy-bridged polynuclear complexes of Co(III) which contain similar terminal ligands. Co(III), however, also forms hydroxy-bridged binuclear complexes in

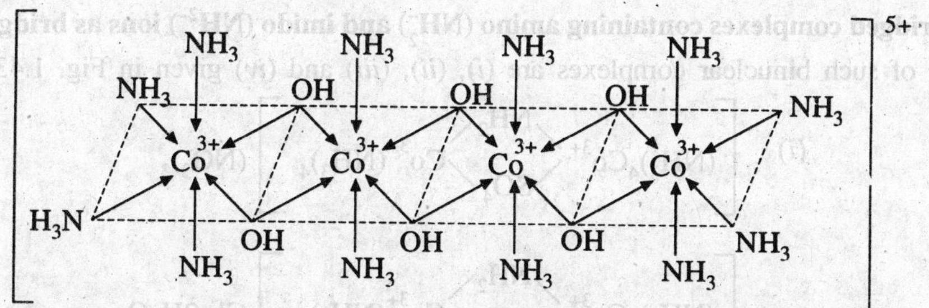

Fig. 1.39. Structure of dodeca ammine μ-hexa-hydroxo tetracobalt (III) ion.

which the terminal ligands are of different type. Examples of two such complexes are (*i*) and (*ii*) given in Fig. 1.40.

$$(i)\quad \left[(NH_3)_2Cl_2Co^{3+} \begin{matrix} OH \\ \\ OH \end{matrix} Co^{3+}Cl_2\,(NH_3)_2 \right]^0$$

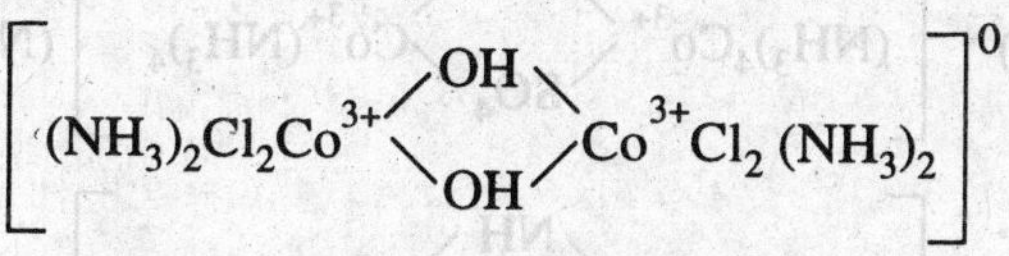

μ-dihydroxo tetrachlorotetrammine dicobalt (III).

$$(ii)\quad \left[(NH_3)_4Co^{3+} \begin{matrix} OH \\ \\ OH \end{matrix} Co^{3+}\,(NH_3)_2\,Cl_2 \right]^{2+}$$

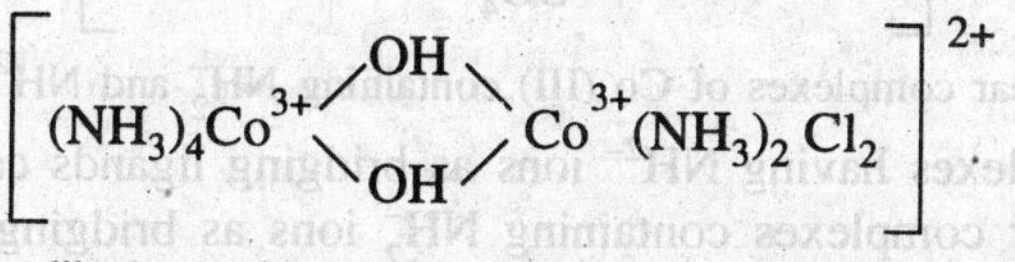

μ-dihydroxo dichlorohexammine dicobalt (III) ion.

Fig. 1.40. Examples of hydroxy-bridged binuclear complexes of Co (III) containing different terminal ligands.

The structures of complexes (*i*) and (*ii*) shown in Fig. 1.40 are given in Fig. 1.41 and 1.42 respectively.

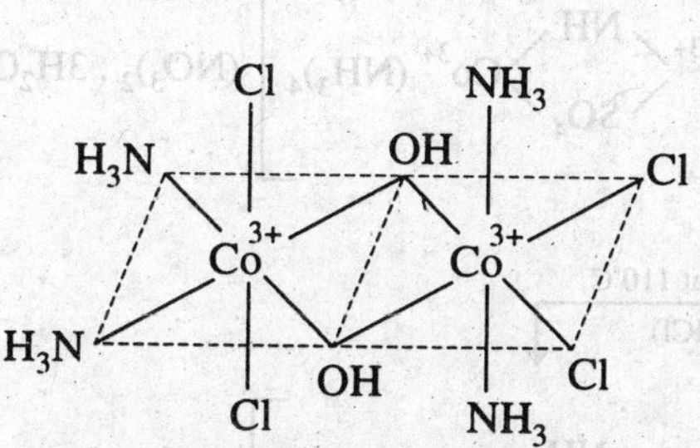

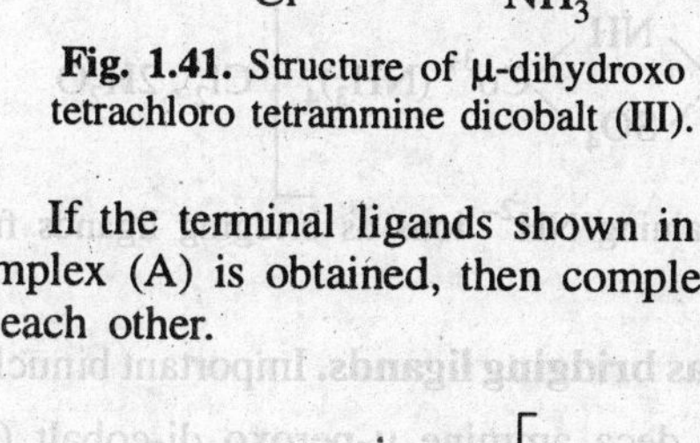

Fig. 1.41. Structure of μ-dihydroxo tetrachloro tetrammine dicobalt (III).

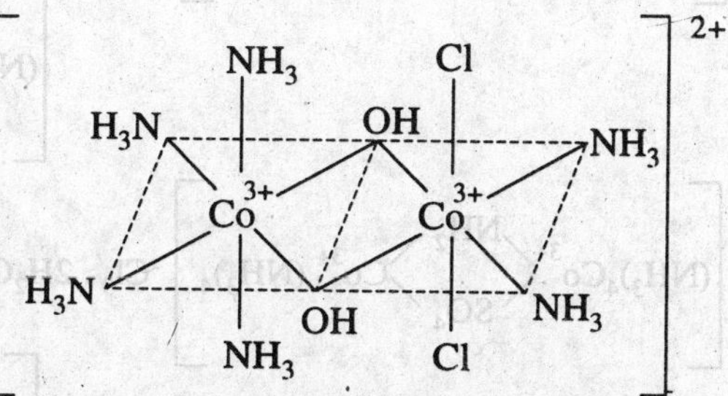

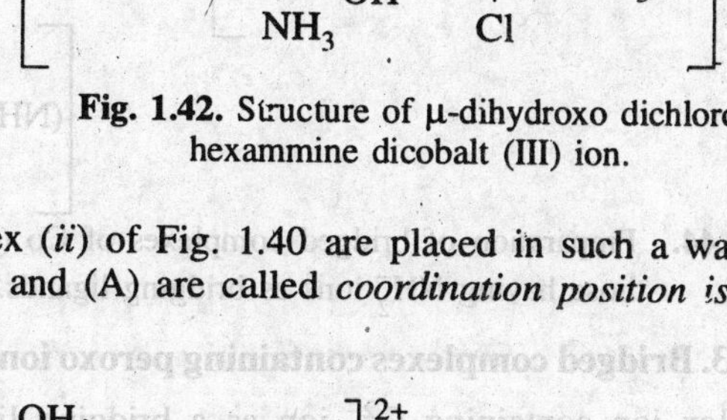

Fig. 1.42. Structure of μ-dihydroxo dichloro hexammine dicobalt (III) ion.

If the terminal ligands shown in complex (*ii*) of Fig. 1.40 are placed in such a way that complex (A) is obtained, then complexes (*ii*) and (A) are called *coordination position isomers* to each other.

$$(A)\quad \left[(NH_3)_3ClCo^{3+} \begin{matrix} OH \\ \\ OH \end{matrix} Co^{3+}(NH_3)_3Cl \right]^{2+}$$

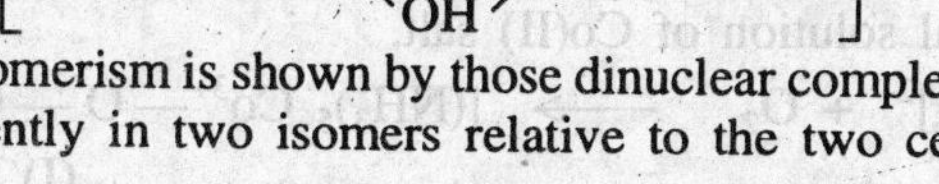

Coordination position isomerism is shown by those dinuclear complexes in which the terminal ligands can be placed differently in two isomers relative to the two central metal atoms.

2. Bridged complexes containing amino (NH_2^-) and imido (NH^{2-}) ions as bridging ligands. Examples of such binuclear complexes are (*i*), (*ii*), (*iii*) and (*iv*) given in Fig. 1.43.

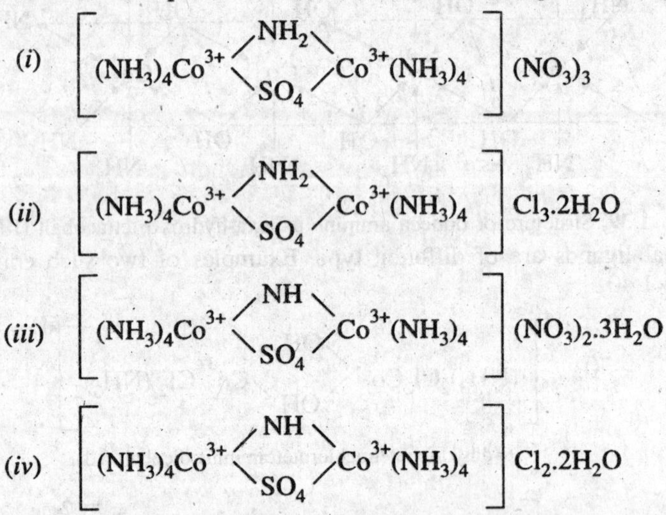

(*i*) $\left[(NH_3)_4Co^{3+}\begin{matrix} NH_2 \\ SO_4 \end{matrix}Co^{3+}(NH_3)_4\right](NO_3)_3$

(*ii*) $\left[(NH_3)_4Co^{3+}\begin{matrix} NH_2 \\ SO_4 \end{matrix}Co^{3+}(NH_3)_4\right]Cl_3.2H_2O$

(*iii*) $\left[(NH_3)_4Co^{3+}\begin{matrix} NH \\ SO_4 \end{matrix}Co^{3+}(NH_3)_4\right](NO_3)_2.3H_2O$

(*iv*) $\left[(NH_3)_4Co^{3+}\begin{matrix} NH \\ SO_4 \end{matrix}Co^{3+}(NH_3)_4\right]Cl_2.2H_2O$

Fig. 1.43. Binuclear complexes of Co (III) containing NH_2^- and NH^{2-} ions as bridging ligands.

Binuclear complexes having NH^{2-} ions as bridging ligands can sometimes be obtained by treating the binuclear complexes containing NH_2^- ions as bridging ligands with conc. aq. NH_3 or by heating as shown in Fig. 1.44.

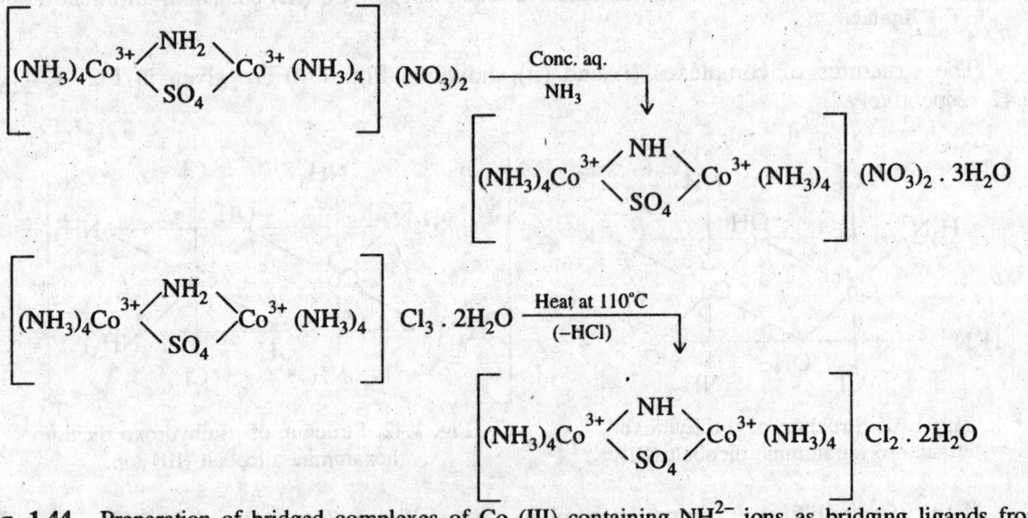

$$\left[(NH_3)_4Co^{3+}\begin{matrix} NH_2 \\ SO_4 \end{matrix}Co^{3+}(NH_3)_4\right](NO_3)_2 \xrightarrow[NH_3]{\text{Conc. aq.}} \left[(NH_3)_4Co^{3+}\begin{matrix} NH \\ SO_4 \end{matrix}Co^{3+}(NH_3)_4\right](NO_3)_2 \cdot 3H_2O$$

$$\left[(NH_3)_4Co^{3+}\begin{matrix} NH_2 \\ SO_4 \end{matrix}Co^{3+}(NH_3)_4\right]Cl_3 \cdot 2H_2O \xrightarrow[(-HCl)]{\text{Heat at } 110^\circ C} \left[(NH_3)_4Co^{3+}\begin{matrix} NH \\ SO_4 \end{matrix}Co^{3+}(NH_3)_4\right]Cl_2 \cdot 2H_2O$$

Fig. 1.44. Preparation of bridged complexes of Co (III) containing NH^{2-} ions as bridging ligands from those having NH_2^- ions as bridging ligands.

3. Bridged complexes containing peroxo ions (O_2^{2-}) as bridging ligands. Important binuclear complex ion containing O_2^{2-} ion as a bridging ligand is deca ammine μ-peroxo di-cobalt (III) ion, $[(NH_3)_5Co^{3+}—O—O—Co^{3+}(NH_3)_5]^{4+}$ (brown ion). This ion is formed by the careful air oxidation of ammoniacal solution of Co(II) salt.

$$2[Co(NH_3)_6]^{2+} + O_2 \longrightarrow \underset{\text{(I)}}{[(NH_3)_5\,Co^{3+}—O—O—Co^{3+}(NH_3)_5]^{4+}} + 2NH_3$$

When the solution containing (I) is further oxidised (still carefully) by warming with dil. (1:1) HNO_3, a green complex ion, $[(NH_3)_5Co^{3+}—O—O—Co^{4+}(NH_3)_5]^{5+}$ is obtained.

4. Bridged binuclear octahedral complex containing NH_2^- and NO_2^- ions as bridging ligands. Co(III) forms a binuclear octahedral complex which contains NH_2^- and NO_2^- ions as the bridging ligands. It is formulated as $[Co_2^{3+}(en)_4(NH_2)(NO_2)]^{4+}$. Its structure is given in Fig. 1.45.

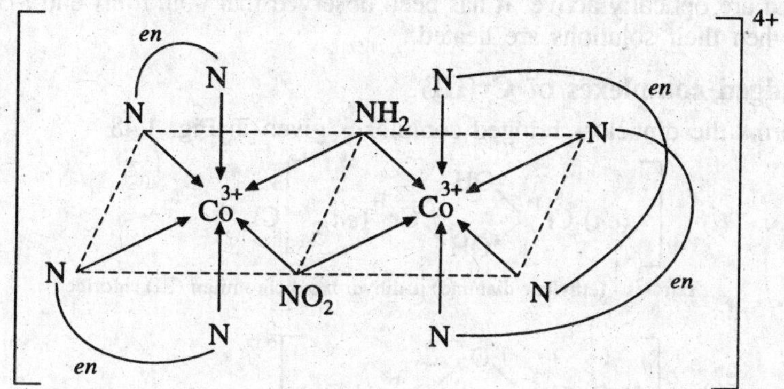

Fig. 1.45. Structure of $[Co_2^{3+}(en)_4(NH_2)(NO_2)]^{4+}$ ion.

It has been shown that this complex ion exists in *cis*- and *trans*- isomers (geometrical isomerism) whose structures are given in Fig. 1.46.

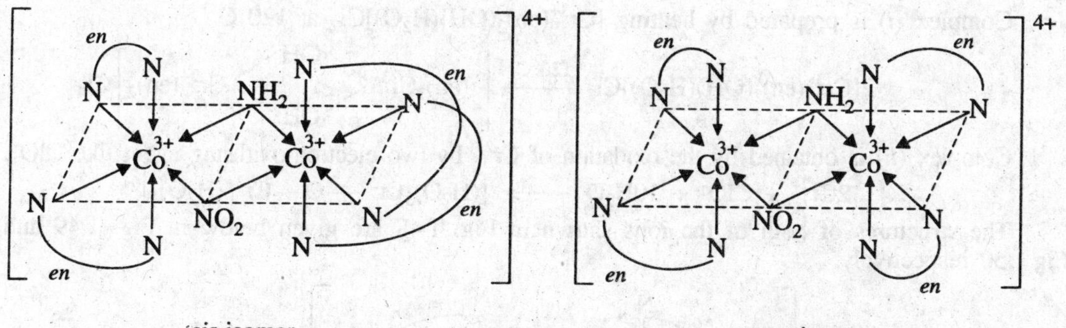

Fig. 1.46. *cis*-and *trans*- isomers of $[Co_2(en)_4(NH_2)(NO_2)]^{4+}$ ion.

Since *cis*-isomer is asymmetrical and hence is optically active, *i.e., cis*-form has been resolved into two optical isomers *viz. d-cis*-isomer and *l-cis*-isomer. The structures of these optical isomers are given in Fig. 1.47. *Trans*-isomer, being symmetrical, is optically inactive

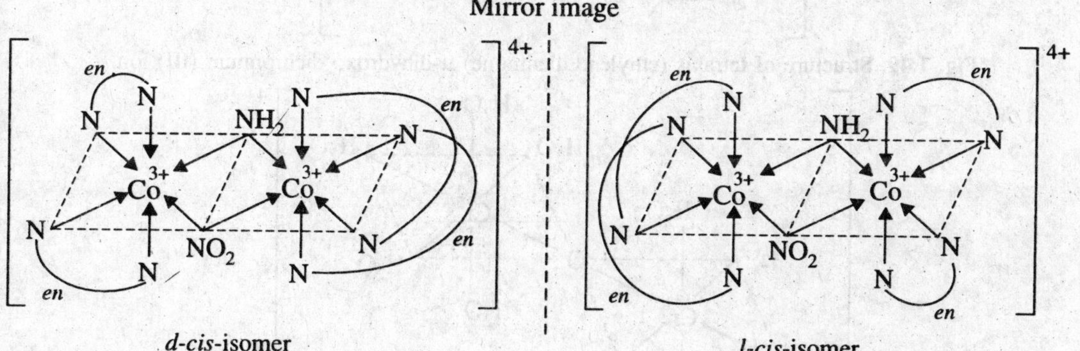

Fig. 1.47. *d-cis*-and *l-cis*-isomers (optically active isomers) of $[Co_2^{3+}(en)_4(NH_2)(NO_2)]^{4+}$ ion.

and hence does not give optical isomers. Thus we see that $[Co_2^{3+}(en)_4(NH_2)(NO_2)]^{4+}$ ion exists in three optical isomers namely *d-cis*-form, *l-cis*-form and *meso* (*trans*) form. Meso form is optically inactive and is internally compensated form. *d-cis* and *l-cis* forms are mirror images to each other and are optically active. It has been observed that both *d-cis* and *l-cis* forms revert to meso form when their solutions are heated.

Dinuclear bridged complexes of Cr(III)

Cr (III) forms the dinuclear bridged complexes given in Fig. 1.48.

(*i*) $$\left[(en)_2Cr^{3+}\begin{smallmatrix}\diagup OH \diagdown \\ \diagdown OH \diagup\end{smallmatrix}Cr^{3+}(en)_2\right]Cl_4$$

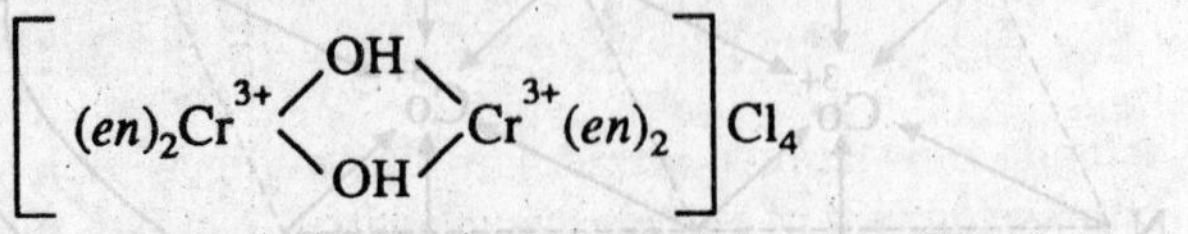

Tetrakis - (ethylene diamine) μ-dihydroxo dichromium (III) chloride

(*ii*) $$\left[(H_2O)_5Cr^{3+}\diagup\overset{O^{2-}}{}\diagdown Cr^{3+}(H_2O)_5\right]^{4+}$$

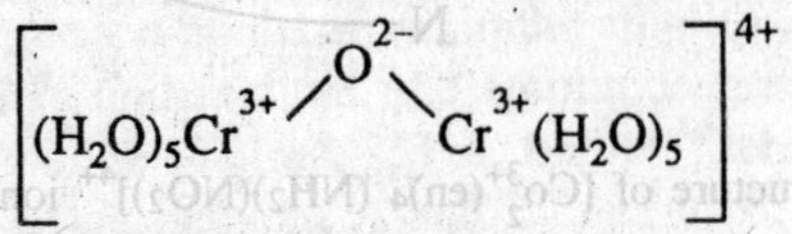

Deca aquo μ-oxodichromium (III) ion.

Fig. 1.48. Dinuclear bridged complexes formed by Cr(III).

Complex (*i*) is prepared by heating $[Cr^{3+}(en)_2(OH)(H_2O)]Cl_2$ at 120°C.

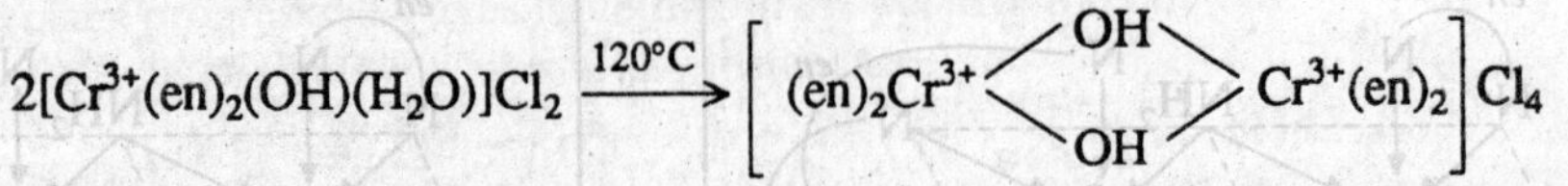

$$2[Cr^{3+}(en)_2(OH)(H_2O)]Cl_2 \xrightarrow{120°C} \left[(en)_2Cr^{3+}\begin{smallmatrix}\diagup OH \diagdown \\ \diagdown OH \diagup\end{smallmatrix}Cr^{3+}(en)_2\right]Cl_4$$

Complex (*ii*) is obtained by the oxidation of Cr^{2+} by two electron-oxidising agent like ClO^-.

$$2Cr^{2+} + ClO^- + 10H_2O \longrightarrow [(H_2O)_5Cr^{3+}—O—Cr^{3+}(H_2O)_5]^{4+}$$

The structures of both of the ions shown in Fig. 1.48 are given below in Fig. 1.49 and Fig.1.50 respectively.

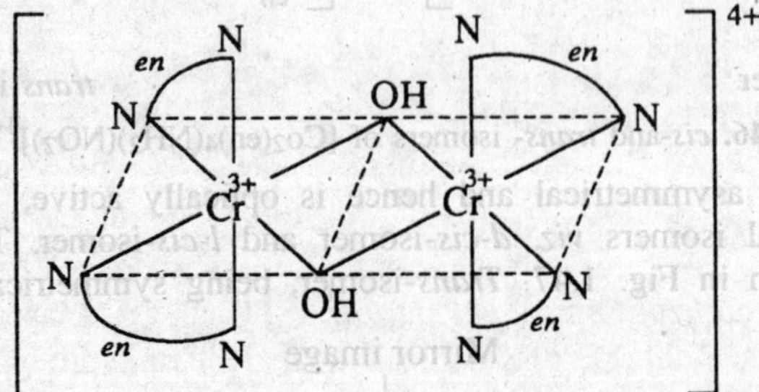

Fig. 1.49. Structure of tetrakis (ethylene diammine) μ-dihydroxo dichromium (III) ion.

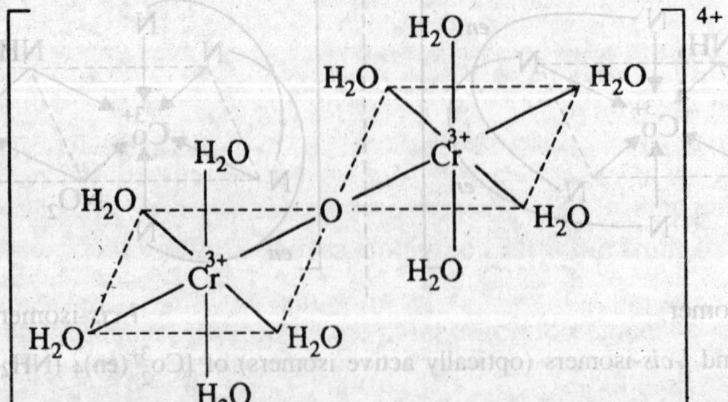

Fig. 1.50. Structure of decaquo μ-oxo dichromium (III) ion.

Dinuclear bridged complex of Ni(I)

$[Ni^{2+}(CN)_4]^{2-}$ ion can be reduced by sodium amalgam to form $[Ni^{+}(CN)_3]^{2-}$ ion (red). X-ray and magnetic studies have shown that $[Ni^{+}(CN)_3]^{2-}$ ion is dimeric in its crystalline potassium salt. Dimer of $[Ni(CN)_3]^{2-}$ ion has a bridged structure in which two CN^- ions make bridges between two Ni^+ ions. Thus the structure of dimeric $[Ni(CN)_3]^{4-}$ can be written as shown in Fig. 1.51.

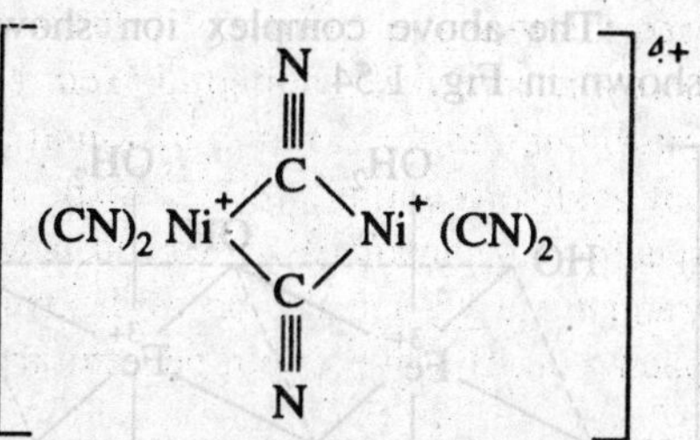

Fig. 1.51. Structure of tetracyano μ-dicyano dinickel (I) ion.

Tetranuclear bridged complex of Au(I)

R_2AuCN is tetrameric and hence is formulated as $(R_2AuCN)_4$. This tetrameric compound contains four CN^- ions as bridging ligands as shown in Fig. 1.52. It may be noted that the four groups *viz.* two C_2H_5 groups and two CN^- ions surround Au(I) in a square planar fashion.

```
          C2H5                  C2H5
           |                     |
 C2H5 —Au — C≡N → Au — C2H5
           ↑                     |
           N                     C
           |||                   |||
           C                     N
           |                     ↓
 C2H5 —Au ← N≡C — Au — C2H5
           |                     |
          C2H5                  C2H5
```

Fig. 1.52. Structure of tetramer, $[(C_2H_5)_2\ AuCN]_4$.

Dinuclear bridged complexes of Mn

Following are important dinuclear complexes formed by manganese.

(*i*) $\left[(dipy)_2\ Mn^{3+} \begin{matrix} O^{2-} \\ O^{2-} \end{matrix} Mn^{4+}\ (dipy)_2\right]^{3+}$

(*ii*) $\left[(big)_2\ Mn^{3+} \begin{matrix} OH \\ OH \end{matrix} Mn^{3+}\ (big)_2\right]^{0}$

Dinuclear bridged octahedral complex of Fe (III)

Fe(III) forms hydroxy-bridged dinuclear complex of $[Fe^{3+}(OH)_2(H_2O)_3]_2^{2+}$ or $[Fe_2^{3+}(OH)_4(H_2O)_6]^{2+}$ type. The structure of this ion is given below in Fig. 1.53. This structure shows that two octahedral units are linked together through two bridging OH^- groups

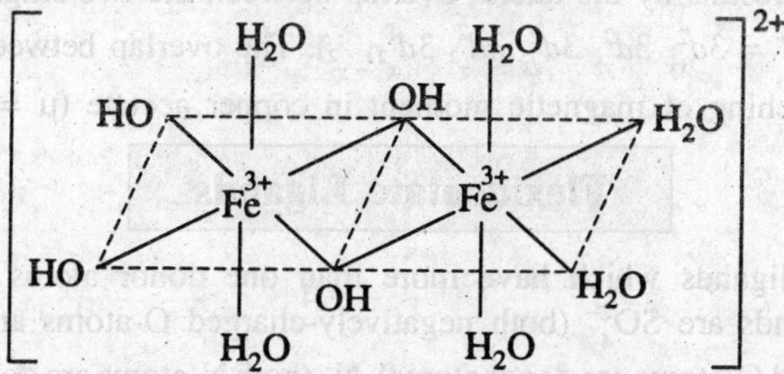

Fig. 1.53. Structure of $[Fe_2^{3+}(OH)_4(H_2O)_6]^{2+}$ ion.

The above complex ion shows geometrical isomerism. Its three geometric isomers are shown in Fig. 1.54.

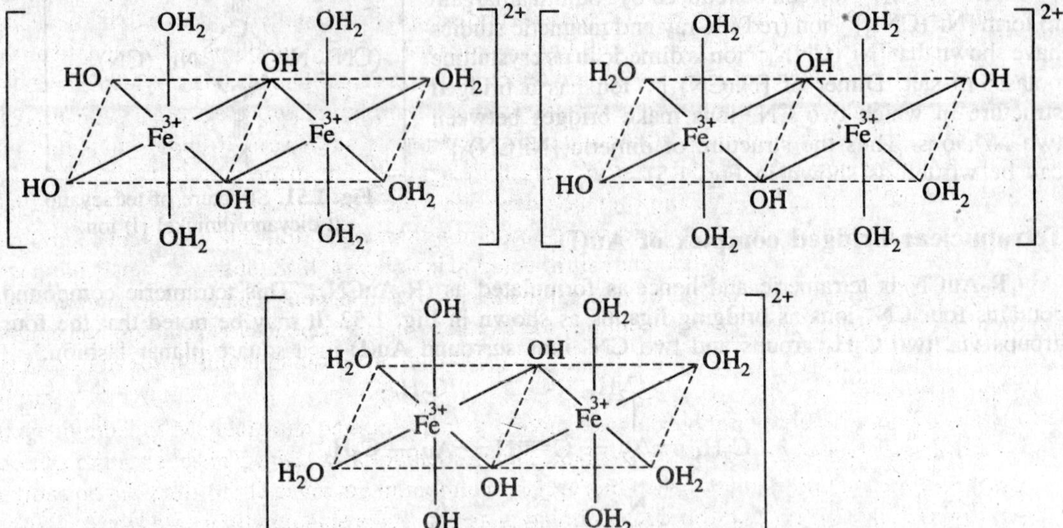

Fig. 1.54. Geometrical isomers of $[Fe(OH)_2(H_2O)_3]_2^{2+}$ ion which is binuclear complex of Fe (III).

Bridged binuclear complexes of transition metals in which carbonyl groups (CO) act as bridging ligands

Iron, osmium and cobalt in their zero oxidation state form bridged binuclear complexes in which CO groups act as bridging ligands. These complexes are called bridged binuclear carbonyls. Examples of such carbonyls are $Fe_2(CO)_9$, $Os_2(CO)_9$ and $Co_2(CO)_8$ (in the solid state). $Fe_2(CO)_9$ and $Os_2(CO)_9$ both have six terminal carbonyl groups :C $\rightleftarrows$ O:), three bridging (ketonic) carbonyls groups (:Ö = C<) and one M—M bond (M=Fe, Os). $Co_2(CO)_8$ (in the solid state) has six terminal carbonyl groups, one Co—Co bond and two bridging carbonyl groups. These facts are evident from the structure of these carbonyls as given in chapter 13.

Dinuclear complexes of Cu(II)

Hydrated copper (II) acetate, $Cu(CH_3COO)_2.H_2O$ is dimeric in both solid and solution state. The structure of dimeric molecule, $[Cu(CH_3COO)_2.H_2O]_2$ is shown in Fig. 1.55. This figure shows that each Cu^{2+} ion is placed at the centre of a square plane and is linked with four O-atoms of four bridging acetate ions (CH_3—C(=O)—O^-), one H_2O molecule and other Cu^{2+} ion. Thus each Cu^{2+} ion has octahedral environment. Two Cu-atoms are attached with each other by δ-bond which results by the lateral overlap between the two singly-filled $d_{x^2-y^2}$ orbitals on each Cu^{2+} ion. ($Cu^{2+} = 3d_{xy}^2\ 3d_{yz}^2\ 3d_{zx}^2\ 3d_{z^2}^2\ 3d_{x^2-y^2}^1$). The overlap between two $3d_{x^2-y^2}$ orbitals leads to a partial quenching of magnetic moment in copper acetate (μ = 1.43 B.M.).

Flexidentate Ligands

There are many ligands which have more than one donor atoms (polydentate ligands). Examples of such ligands are SO_4^{2-} (both negatively-charged O-atoms are donor atoms), CO_3^{2-} (both negatively-charged O-atoms are donor atoms), N_2 (both N-atoms are donor atoms), NH_2—NH_2

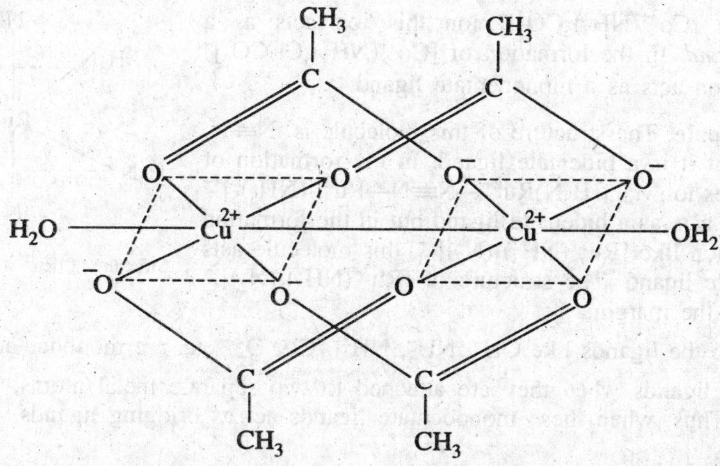

Fig. 1.55. Structure of dimeric copper (II) acetate, $[Cu(CH_3COO)]_2.H_2O]_2$

(both N-atoms are donor atoms) etc. All these ligands have two donor atoms. When these ligands get coordinated to the metal atom/ion, they give rise to the formation of two types of complex compounds. In one type they get linked to the metal atom through one donor atom only while in the formation of another type of the complex compound, they get coordinated to the metal atom through both of its donor atoms. Thus we see that it is not necessary for a polydentate ligand that it always uses all its donor atoms for coordination. Such ligands are called *flexidentate ligands.*

Examples of flexidentate ligands and complexes formed by them

Examples of flexidentate ligands and the complexes given by them are given below:

1. Sulphate (SO_4^{2-}) and carbonate (CO_3^{2-}) ions. SO_4^{2-} ion $\left[O{=}S(O^-)(O^-){=}O \right]$ has two negatively-charged O-atoms as its donor atoms and hence behaves as a bidentate ligand in complex ions like $[Co^{3+}(NH_3)_4\,(SO_4)]^+$ (I). This ion uses only one of its two negatively-charged O-atoms as donor atom in the formation of complex ions like $[Co^{3+}(NH_3)_5(SO_4)]^+$ (II). Thus in this complex ion SO_4^{2-} ion behaves as a monodentate ligand. Note that in both the complex ions given above coordination number of Co^{3+} ion is six. In (I) this coordination number is satisfied by four NH_3 and one SO_4^{2-} ion (bidentate ligand) while in (II) this number is satisfied by five NH_3 molecules and one SO_4^{2-} ion (monodentate ligand).

NH3
H3N
O
O
Co3+
S
H3N
O
O
NH3
+

Structure of $[Co^{3+}(NH_3)_4(SO_4)]^+$ ion (I)

NH3
H3N
NH3
Co3+
O−
H3N
O — S = O
O
NH3
+

Structure of $[Co^{3+}(NH_3)_5(SO_4)]^+$ ion. (II)

In the formation of $[Co^{3+}(NH_3)_4CO_3]^+$ ion CO_3^{2-} ion $\left[O{=}C(O^-)(O^-) \right]$ acts as a *bidentate ligand* while in $[Co^{3+}(NH_3)_5CO_3]^+$ ion this ion acts as a *monodentate ligand.* In the formation of $[Co^{3+}(NH_3)_3Cl_2CO_3]^-$ ion also, CO_3^{2-} ion acts as a monodentate ligand.

2. N_2 molecule. The structure of this molecule is :N≡N: which shows that it is a bidentate ligand. In the formation of a bridged complex ion *viz.* $[(H_3N)Ru^{2+}\leftarrow N\equiv N\rightarrow Ru^{2+}(NH_3)_5]^{4+}$ ion N_2 molecule acts as a bidentate ligand but in the formation of the complex ion like $[Ru^{2+}(NH_3)_5(N_2)]^{2+}$, this molecule acts as a monodentate ligand. The structure of $[Ru^{2+}(NH_3)_5(N_2)]^{2+}$ ion is shown in the margin.

NH₃, H₃N, NH₃, Ru^{2+}, H₃N, NH₃, N≡N:

Structure of $[Ru^{2+}(NH_3)_5(N_2)]^{2+}$ ion

3. Although the ligands like OH^-, NH_2^-, NH^{2-}, Cl^-, O_2^{2-} etc. are monodentate ligands, they act as bidentate ligands when they are attached to two separate metal atoms, forming bridge between them. Thus, when these monodentate ligands act as bridging ligands, they behave as bidentate ligands.

4. Although $EDTA^{4-}$ ion is a hexadentate ligand, it may act as a tetra or pentadentate ligand depending on the size and stereochemistry of the complex formed.

Chelating Ligands and Chelates

A chelating ligand is a bidendate or polydentate ligand which is attached to the same central metal atom by two or more of its donor atoms resulting in the formation of a complex having a strain-free ring structure. The complex having the ring structure is called *chelate* or *chelated complex.* The chelate is also called by various other names like *cyclic complex, ring-type complex* etc. The formation of a chelate is called *chelation* or *cyclisation.*

Examples of the formation of chelates

(*i*) When two molecules of ethylene diamine, $\ddot{N}H_2—CH_2—CH_2—\ddot{N}H_2$ (*en*) which is a bidentate ligand, get attached with one Cu^{2+} ion through its two N-donor atoms of each molecule, complex ion *viz.* $[Cu(NH_2—CH_2—CH_2—NH_2)_2]^{2+}$ or $[Cu(en)_2]^{2+}$ which contains two 5-membered rings is obtained. In the formation of this complex ion, ethylene diamine molecule (bidentate ligand) acts as a chelating ligand and the complex ion thus formed is called a chelate [see Fig. 1.56 (*a*)].

(*ii*) When three oxalate ions, $\begin{matrix} COO^- \\ | \\ COO^- \end{matrix}$ or $C_2O_4^{2-}$ or ox^{2-} get attached with one Fe^{3+} ion, $[Fe(C_2O_4)_3]^{3-}$ or $[Fe(ox)_3]^{3-}$ ion is obtained. This complex ion is a chelate. This chelate has three 5-membered rings. In the formation of this chelate, $C_2O_4^{2-}$ ion which is a bidentate ligand acts as a chelating ligand [see Fig. 1.56 (*b*)].

It may be noted that OH^-, Cl^- and NH_2^- ions cannot serve as chelating ligands, because the angle between the lone pairs donot allow them to span adjacent coordination positions of one and the same metal atom. Hydrazine ($H_2\ddot{N}—\ddot{N}H_2$) cannot function as a chelating ligand, since a 3-membered ring-type complex formed $\left[\begin{matrix} H_2N \searrow \\ | \quad\; M \\ H_2N \nearrow \end{matrix}\right]$ is very strained and hence is unstable. Thus every polydentate ligand is not necessarily a chelating ligand.

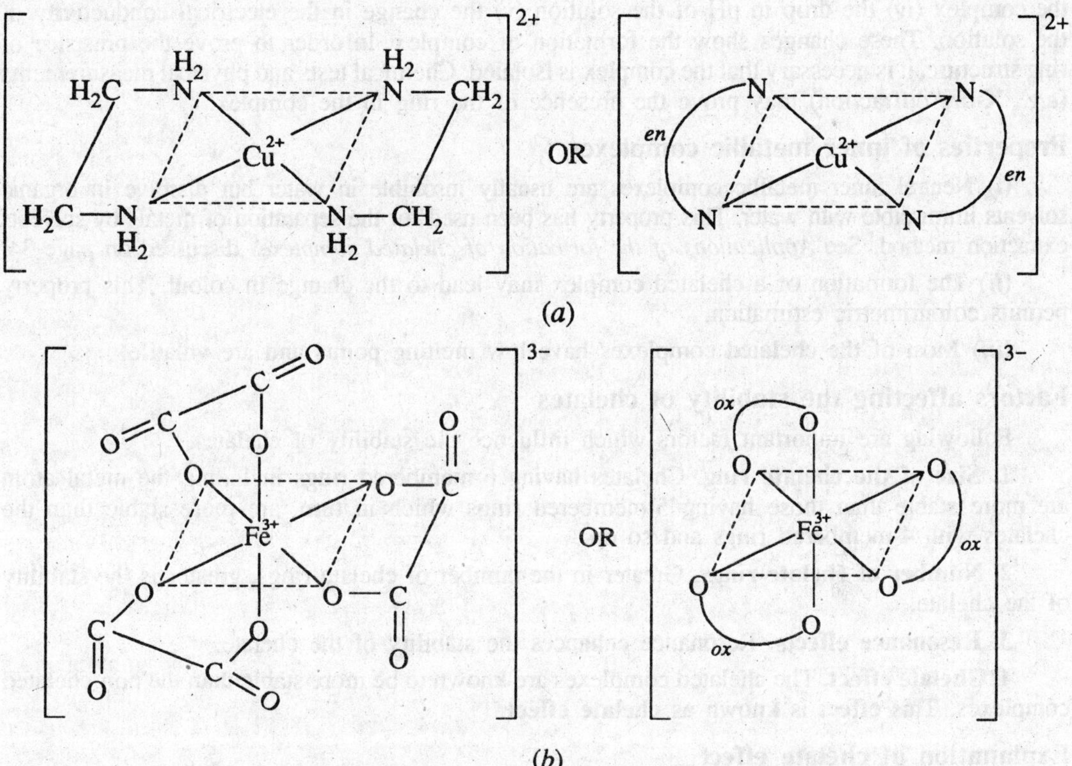

(a)

(b)

Fig. 1.56. Structure of (*a*) $[Cu(NH_2—CH_2—CH_2—NH_2)_2]^{2+}$ or $[Cu(en)_2]^{2+}$ and of (*b*) $[Fe(C_2O_4)_3]^{3-}$ or $[Fe(ox)_3]^{3-}$ chelated complex ions.

Classification of chelates

Chelates which donot carry any charge are called *inner metallic complexes of the first order.* These chelates are neutral (non-electrolyte) in nature. Examples of such chelates are $[Cr^{3+}(gly)_3]^0$, $[Co^{3+}(acac)_3]^0$, $[Cu^{2+}(big)_2]^0$ etc.

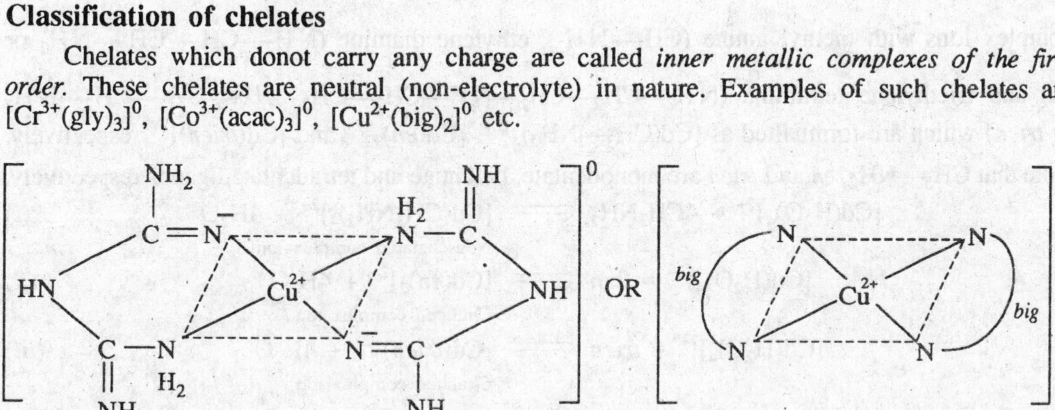

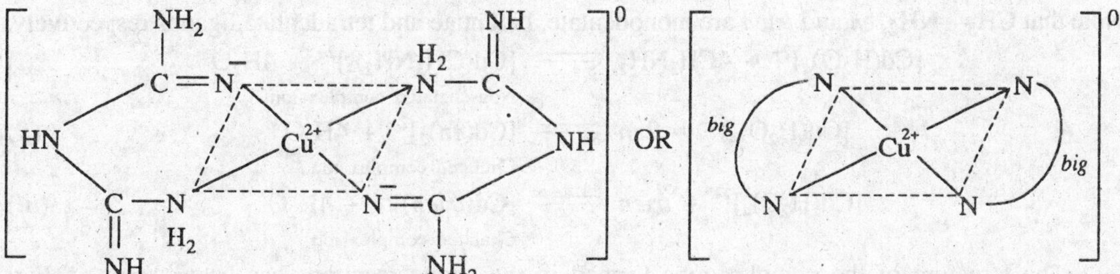

Fig. 1.57. Structure of $[Cu^{2+}(big)_2]^0$. Here big is the abbreviation for biguanide ion which is a bidentate ligand.

The chelates which carry some charge are called *inner-metallic complexes of the second order.* These chelates are ionic in nature. Examples of such chelates are $[Co^{3+}(en)_3]^{3+}$, $[Co^{2+}(acac)_3]^-$ etc. The structure of $[Cu^{2+}(big)_2]^0$ is shown in Fig. 1.57.

Detection of chelate formation

Since the formation of a chelate is also complex formation, the formation of a chelate may be detected by all those methods that are used for the detection of complex formation. Thus the formation of a chelate may lead to the changes namely (*i*) the change in the colour of the metal ion (*ii*) the dissolution of the precipitate (*iii*) the enhancement of the stability of the complex (*iv*) the drop in pH of the solution (*v*) the change in the electrical conductivity of the solution. These changes show the formation of complex. In order to prove the presence of ring structure, it is necessary that the complex is isolated. Chemical tests and physical measurements (*e.g.*, X-ray diffraction) may prove the presence of the ring in the complex.

Properties of inner metallic complexes

(*i*) Neutral inner metallic complexes are usually insoluble in water but dissolve in organic solvents immiscible with water. This property has been used for the separation of metals by solvents extraction method. See *Applications of the formation of chelated complexes* discussed on page 34.

(*ii*) The formation of a chelated complex may lead to the change in colour. This property permits colourimetric estimation.

(*iii*) Most of the chelated complexes have low melting points and are volatile.

Factors affecting the stability of chelates

Following are important factors which influence the stability of chelates.

1. Size of the chelate ring. Chelates having 6-membered rings including the metal atom are more stable than those having 5-membered rings which in turn, are more stable than the chelates with 4-membered rings and so on.

2. Number of chelate rings. Greater in the number of chelate rings, greater is the stability of the chelate.

3. Resonance effects. Resonance enhances the stability of the chelate.

4. Chelate effect. The chelated complexes are known to be more stable than the non-chelated complexes. This effect is known as **chelate effect.**

Explanation of chelate effect

Chelate effect can be explained by considering the formation of some non-chelated and chelated complex ions of the same metal. Hydrated cadmium ion, $[Cd(H_2O)_4]^{2+}$ forms 4-coordinated complex ions with methyl amine ($CH_3—\ddot{N}H_2$), ethylene diamine ($\ddot{N}H_2—CH_2—CH_2—\ddot{N}H_2$ or *en*) and triethylene tetramine ($\ddot{N}H_2—CH_2—CH_2—\ddot{N}H—CH_2—CH_2—\ddot{N}H—CH_2—CH_2—\ddot{N}H_2$ or *trien*) which are formualted as $[Cd(CH_3—NH_2)_4]^{2+}$, $[Cd(en)_2]^{2+}$ and $[Cd(trien)]^{2+}$ respectively. Note that $CH_3—NH_2$, *en* and *trien* are monodentate, bidentate and tetradentate ligands respectively.

$$[Cd(H_2O)_4]^{2+} + 4CH_3NH_2 \rightleftharpoons \underset{\text{Non-chelated complex ion}}{[Cd(CH_3NH_2)_4]^{2+}} + 4H_2O \quad ...(i)$$

$$[Cd(H_2O)_4]^{2+} + 2en \rightleftharpoons \underset{\text{Chelated complex ion}}{[Cd(en)_2]^{2+}} + 4H_2O \quad ...(ii)$$

$$[Cd(H_2O)_4]^{2+} + trien \rightleftharpoons \underset{\text{Chelated complex ion}}{[Cd(trien)]^{2+}} + 4H_2O \quad ...(iii)$$

The structure of the complex ions formed in the above reactions has been shown in Fig. 1.58. These structures show that (*i*) $[Cd(CH_3—NH_2)_4]^{2+}$ ion has no rings while $[Cd(en)_2]^{2+}$ and $[Cd(trien)]^{2+}$ ions have two and three 5-membered rings including Cd^{2+} ion. (*ii*) The number of coordination centres of one molecule of $CH_3.NH_2$, *en* and *trien* (ligands) is equal to 1, 2 and 4 respectively. In each complex ion the coordination number of Cd^{2+} ion is equal to 4.

Since $[Cd(en)_2]^{2+}$ and $[Cd(trien)]^{2+}$ ions are chelated ions, each of these is more stable than $[Cd(CH_3NH_2)_4]^{2+}$ ion which is a non-chelated ion. It has been found that the stability of these ions is in the order:

$$[Cd(CH_3NH_2)_4]^{2+} < [Cd(en)_2]^{2+} < [Cd(trien)]^{2+}$$

The above order can be explanied on the basis of $\Delta G°$ value of reactions (*i*), (*ii*) and (*iii*) which lead to the formation of the complex ions. The value $\Delta G°$ for a reaction is given by:

$$\Delta G^\circ = \Delta H^\circ - T\Delta S^\circ$$

The value of ΔH° for all the three reactions has been found to be negative. It has been found that larger is the negative value of ΔG°, greater is the stability of the complex ion.

In reaction (*i*), since the number of molecules on the RHS is the same as that on LHS, there is no change in the randomness or entropy in going from reactants to products, *i.e.*, for reaction (*i*), $\Delta S^\circ = 0$ and hence

$$\Delta G^\circ = \Delta H^\circ - T \times 0 = \Delta H^\circ$$

$$= -\text{ Negative quantity, since } \Delta H^\circ \text{ is negative.}$$

In reaction (*ii*), since four molecules of H_2O are replaced by two molecules of *en*, there is an increase of two molecules on going from reactants to products and hence there is an increase in the value of ΔS° in the same direction. Thus ΔS° would have a large positive value. This implies that since ΔH° is negative, ΔG° would be a large negative quantity.

In reaction (*iii*), since four molecules of H_2O are replaced by one molecule of *trien*, there is an increase of three molecules on proceeding from reactants to the products and hence there is a greater increase in the magnitude of ΔS° in the same direction. Thus ΔS° would have a larger positive value. This implies that, since ΔH° is negative, ΔG° would be a larger negative quantity. Thus the value of ΔG° would be larger than that (*i.e.*, ΔG°) of reaction (*ii*).

The above discussion shows that ΔG° has a negative value for all the three reactions and the magnitude of this negative value increases in the order $[Cd(CH_3NH_2)_4]^{2+} < [Cd(en)_2]^{2+} < [Cd(trien)]^{2+}$. Thus, $[Cd(trien)]^{2+}$ is more stable than $[Cd(en)_2]^{2+}$ which is more stable than $[Cd(NH_2CH_3)_4]^{2+}$. It also follows from this discussion that as the number of coordination centres of CH_3NH_2, *en*, and *trien* (ligands) increases as 1→2→4, the stability of the complexes given by these ligands with the same metal also increases.

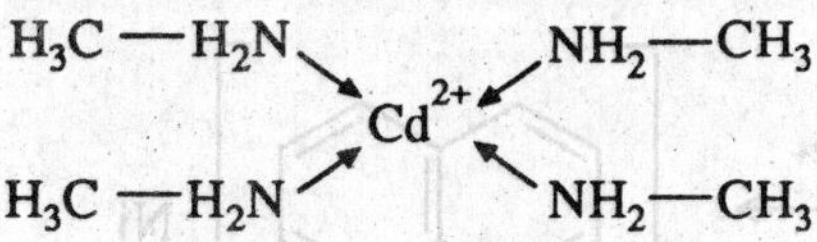

$[Cd(NH_2-CH_3)_4]^{2+}$ ion (Non-chelated complex ion)

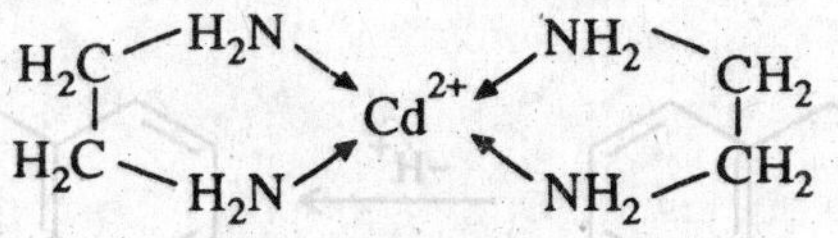

$[Cd(NH_2-CH_2-CH_2-NH_2)_2]^{2+}$ or $[Cd(en)_2]^{2+}$ ion (chelated complex ion)

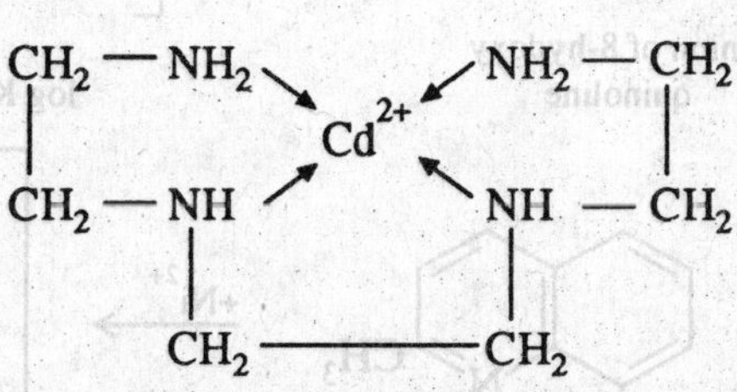

$[Cd(NH_2-CH_2-CH_2-NH-CH_2-CH_2-NH-CH_2-CH_2-NH_2)]^{2+}$ ion or $[Cd(trien)]^{2+}$ ion (chelated complex ion)

Fig. 1.58. Formation of $[Cd(NH_2.CH_3)_4]^{2+}$, $[Cd(en)_2]^{2+}$ and $[Cd(trien)]^{2+}$ ions from $[Cd(H_2O)_4]^{2+}$ ion.

5. Steric effect (Also called steric hindrance or steric strain). When a group is either attached with the donor atom of the ligand or is present near the donor atom, the metal–ligand bond gets weakened and hence the stability of the complex is lowered. This effect is called steric effect. This effect can be illustrated by the following examples.

(*a*) Ethylene diamine, $NH_2-CH_2-CH_2-NH_2$ (*en*) and its N-tetramethyl derivative *viz.* N-tetramethyl ethylene diamine, $(CH_3)_2N-CH_2-CH_2-N(CH_3)_2$ (*tetrameen*) both form chelated octahedral complex ions with Ni^{2+} ion. These complex ions are formulated as $[Ni(en)_3]^{2+}$ and $[Ni(tetrameen)_3]^{2+}$, since both the ligands are bidentate ligands. The structure of these complex ions is given below in Fig. 1.59 and 1.60 respectively.

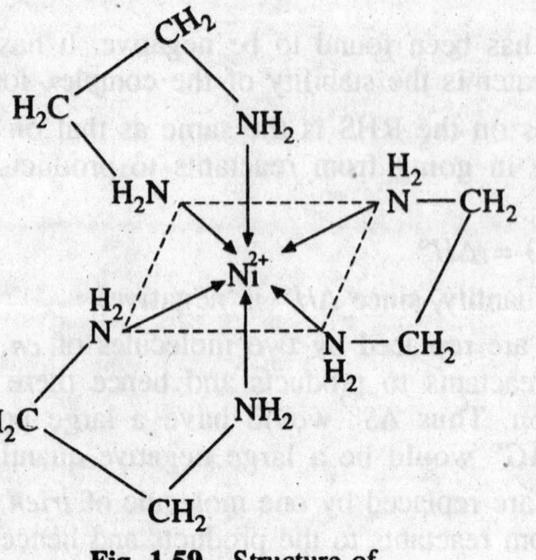

Fig. 1.59. Structure of $Ni(NH_2—CH_2—CH_2—NH_2)_3]^{2+}$ or $[Ni(en)_3]^{2+}$ ion (I).

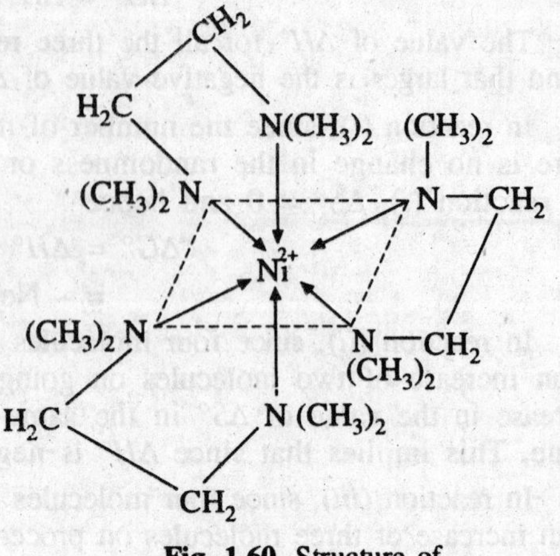

Fig. 1.60. Structure of $[Ni\{(CH_3)_2N—CH_2—CH_2—N(CH_3)_2\}]^{2+}$ or $[Ni(tetrameen)_3]^{2+}$ ion (II).

It has been found that complex ion (II) has lower stability than complex ion (I). Lower stability of (II) is due to the fact that the two bulky CH_3 groups attached with N-atom (donor atom) produces steric overcrowding round the central Ni^{2+} ion. Thus complex ion (II) has lower stability than the complex ion (I).

(*b*) 8-hydroxy quinoline and its 2-methyl derivative *viz.* 2-methyl 8-hydroxy quinoline both form chelated complexes with Ni^{2+} ion as shown in Fig. 1.61.

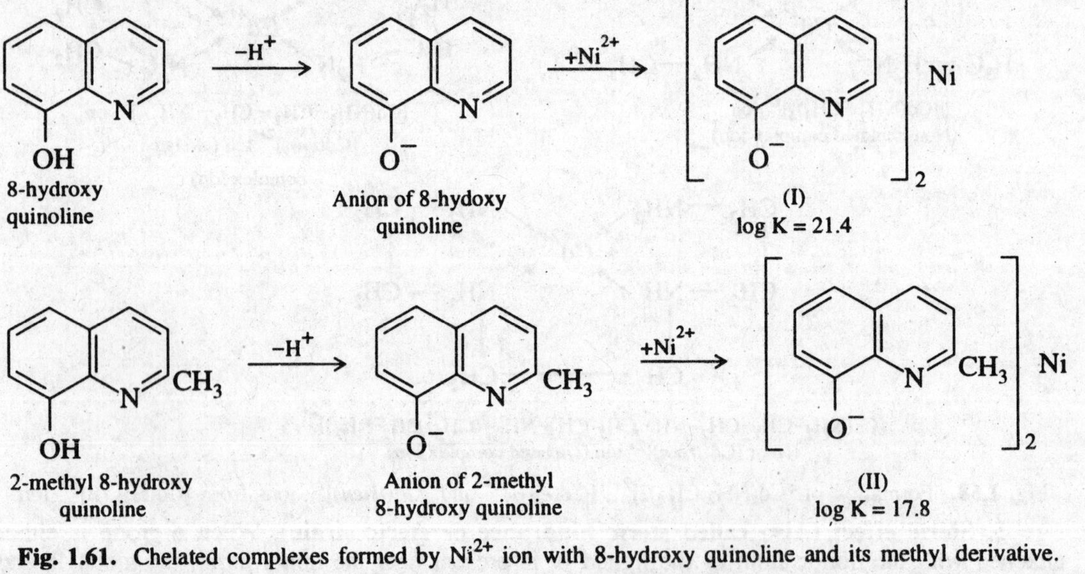

8-hydroxy quinoline $\xrightarrow{-H^+}$ Anion of 8-hydoxy quinoline $\xrightarrow{+Ni^{2+}}$ (I) log K = 21.4

2-methyl 8-hydroxy quinoline $\xrightarrow{-H^+}$ Anion of 2-methyl 8-hydroxy quinoline $\xrightarrow{+Ni^{2+}}$ (II) log K = 17.8

Fig. 1.61. Chelated complexes formed by Ni^{2+} ion with 8-hydroxy quinoline and its methyl derivative.

As is evident from the values of log*K* of both the complexes, complex (II) has lower stability than complex (I). Lower stability of (II) is due to the steric overcrowding produced by the presence of CH_3 group near N-atom (donor atom) in the ring in position number 2.

Applications of the formation of chelated complexes

Some applications of the formation of chelated complexes are given below :

1. Formation of chelates in analytical chemistry. Examples are :

(*i*) Gravimetric estimation and identification of Ni^{2+} ions by dimethyl glyoxime.

(*ii*) Separation of Co^{2+} ion from Ni^{2+} ion.

(*iii*) Estimation of Mg^{2+} and Ca^{2+} ion by EDTA.

2. Formation of chelates in softening water and estimation of hardness of water.
3. Formation of chelates in removing poisonous metals from the body.
4. Separation of ions by solvent extraction method.
5. Role of metal chelates in living systems.

Above applications are discussed one by one in detail.

Gravimetric estimation and identification of Ni^{2+} ions by dimethyl glyoxime

Dimethyl glyoxime solution reacts with Ni^{2+} ions in ammoniacal medium and forms an insoluble neutral red chelate of bis-(dimethyl glyoximato) nickel (II) (Fig. 1.62). Thus the formation of this chelate can be used for the identification, estimation of Ni^{2+} and also for its separation from Co^{2+} ions.

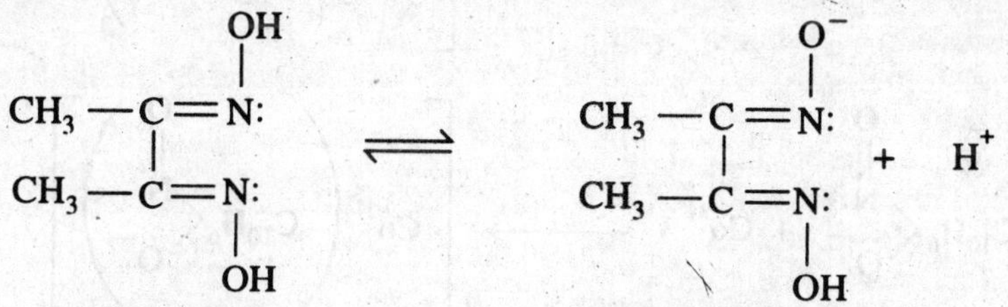

$$CH_3-C(=N\text{:}-OH)-C(=N\text{:}-OH)-CH_3 \rightleftharpoons CH_3-C(=N\text{:}-O^-)-C(=N\text{:}-OH)-CH_3 + H^+$$

Dimethyl glyoxime — Dimethyl glyoxime ion

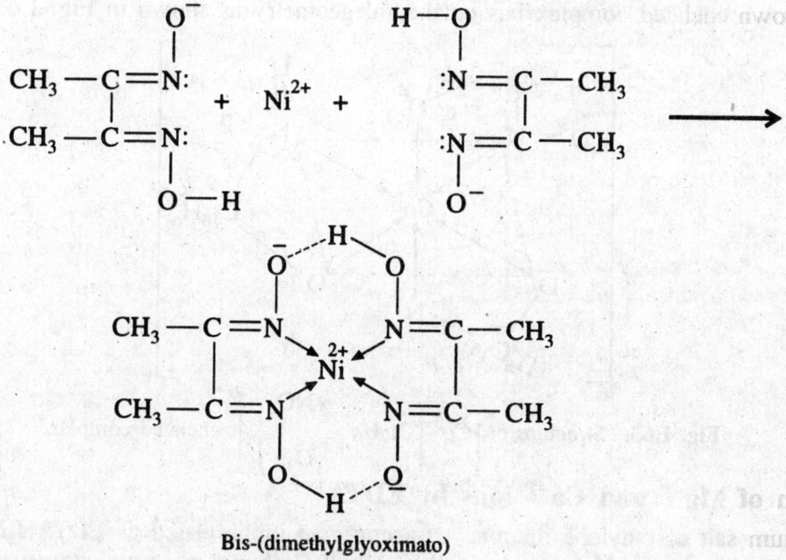

$$2\,[CH_3-C(=N\text{:}-O^-)-C(=N\text{:}-OH)-CH_3] + Ni^{2+} \longrightarrow [Ni(C_4H_7N_2O_2)_2]$$

Bis-(dimethylglyoximato) nickel (II) chelate (red ppt.)

Fig. 1.62. Formation of bis-(dimethyl oximato) nickel (II) chelate (red ppt.)

Separation of Co^{2+} ion from Ni^{2+} ion

When α-nitroso β-naphthol solution (chelating agent) is added to a slightly acidic solution of Co^{2+} ion and is warmed, red-brown chelate is obtained. This test is not shown by Ni^{2+} ion, since Ni^{2+} does not form any chelate.

$$2Co^{2+} + H_2O + \frac{1}{2}O_2 \ (\textit{From air}) \longrightarrow 2Co^{3+} + 2OH^-$$

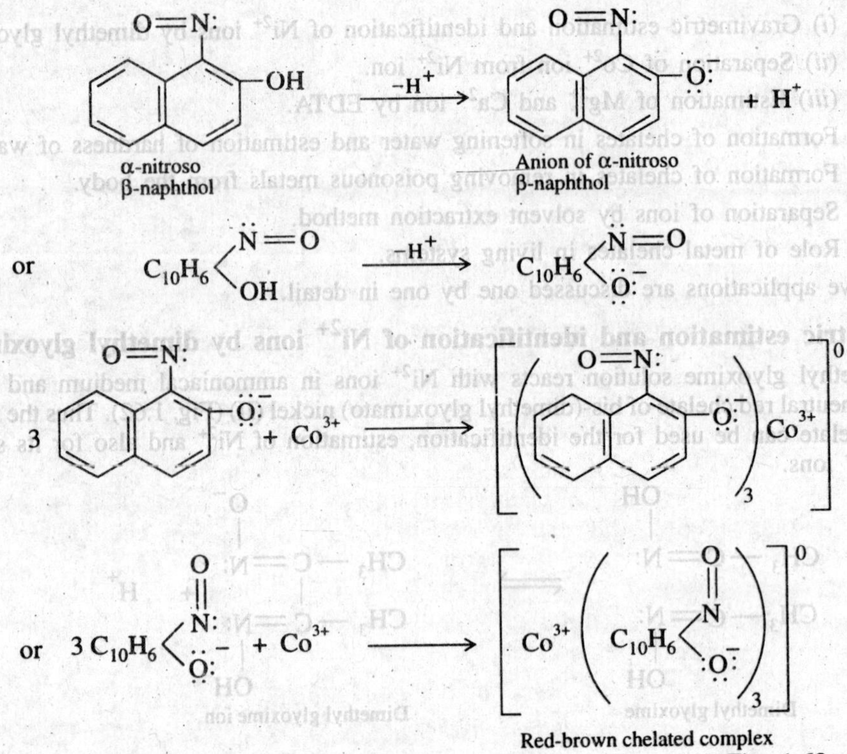

$$\text{α-nitroso β-naphthol} \xrightarrow{-H^+} \text{Anion of α-nitroso β-naphthol} + H^+$$

or

$$C_{10}H_6 \langle^{\ddot{N}=O}_{OH} \xrightarrow{-H^+} C_{10}H_6 \langle^{\ddot{N}=O}_{\ddot{O}:^-}$$

$$3\,(\text{anion}) + Co^{3+} \longrightarrow [(\text{anion})_3 Co^{3+}]^0$$

or

$$3\,C_{10}H_6 \langle^{\ddot{N}=O}_{\ddot{O}:^-} + Co^{3+} \longrightarrow \left[Co^{3+}\left(C_{10}H_6 \langle^{N=O}_{O^-}\right)_3\right]^0$$

Red-brown chelated complex

Red-brown chelated complex has octahedral geometry as shown in Fig. 1.63.

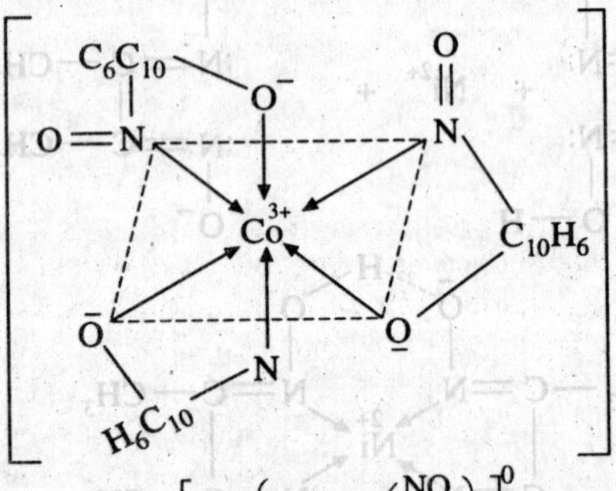

Fig. 1.63. Structure of $\left[Co^{3+}\left(C_{10}H_6 \langle^{NO}_{O}\right)_3\right]^0$ chelated complex.

Estimation of Mg^{2+} and Ca^{2+} ions by EDTA

Disodium salt of ethylene diamine tetracetic acid (abbreviated as $EDTAH_2Na_2$) has been used for the estimation of M^{2+} metal ions like Mg^{2+}, Ca^{2+} etc. by complexometric titrations. In these titrations an organic dye is used as an indicator. A good indicator should meet the following requirements: (*a*) The indicator must form a coloured complex with the metal ion at the same pH at which the $EDTA^{4-}$ ion forms complex with the same metal ion. (*b*) The colour of the metal-indicator complex should be different from that of the free indicator ion. (*c*) The stability of the metal-indicator complex should be low compared to that of metal-EDTA complex.

Suppose in the complexometric titration of Mg^{2+} or Ca^{2+} ions with $EDTAH_2Na_2$, we use Erio-chrome Black T (abbreviated as H_3D) as an indicator. In this trtration the solution containing Mg^{2+} or Ca^{2+} ions is buffered to pH = 10 (This pH is obtained by using NH_4Cl—NH_4OH

buffer), a few drops of the indicator are added and the solution is titrated with a standard solution of $EDTAH_2Na_2$. In this titration, indicator ions (D^{3-} ions) form red metal-indicator complex, $[MD]^-$ with M^{2+} ($M^{2+} = Mg^{2+}$ or Ca^{2+}) ions.

$$\underset{\substack{\text{Metal ion} \\ (M^{2+} = Mg^{2+} \text{ or } Ca^{2+})}}{M^{2+}} + \underset{\substack{\text{Indicator} \\ \text{ion (blue)}}}{D^{3-}} \xrightleftharpoons{pH=10} \underset{\substack{\text{Metal-indicator} \\ \text{complex (red)}}}{[M^{2+}D]^-}$$

Metal-indicator complex ion, $[M^{2+}D]^-$ reacts with $EDTA^{4-}$ ion and forms metal-EDTA complex.

$$\underset{\substack{\text{Metal-indicator} \\ \text{complex (red)}}}{[M^{2+}D]^-} + \underset{\text{EDTA ion}}{EDTA^{4-}} \rightleftharpoons \underset{\substack{\text{Metal-EDTA} \\ \text{complex}}}{[M^{2+}(EDTA)]^{2-}} + \underset{\substack{\text{Indicator} \\ \text{ion (blue)}}}{D^{3-}}$$

$EDTA^{4-}$ ion also forms metal-EDTA complex with the free Ca^{2+} or Mg^{2+} ions (present in the titration solution) that have not reacted with D^{3-} ions to form $[M^{2+}D]^-$ complex.

Thus, we see that free indicator ion, D^{3-} is obtained at the end point. The release of the free indicator ion is marked by a change from red (colour of $[M^{2+}D]^-$ complex) to blue (colour of D^{3-} ion) colour.

Formation of chelates in softening water and estimation of hardness of water

Ethylene diamine tetracetic acid ($EDTAH_4$) has been used for softening water which does a lot of harm in industrial boilers and other equipments by forming a scale in them. Ethylenediamine tetracetic acid is added to the boilers in which $EDTA^{-4}$ ion forms water-soluble chelated complex ion, $[M^{2+}(EDTA)]^{2-}$ with Ca^{2+} and Mg^{2+} ions present in hard water and thus, these ions are removed.

$$EDTAH_4 \rightleftharpoons EDTA^{4-} + 4H^+$$

$$\underset{(M^{2+} = Ca^{2+},\ Mg^{2+})}{M^{2+} + EDTA^{4-}} \longrightarrow \underset{\text{(Water soluble chelate)}}{[M(EDTA)]^{2-}}$$

Since the stability constants of $[Ca(EDTA)]^{2-}$ and $[Mg(EDTA)]^{2-}$ ions are different, the selective estimation of Ca^{2+} and Mg^{2+} ions can also be made, *i.e.*, hardness of water can be estimated.

Formation of chelates in removing poisonous metals from the body

This application can be illustrated by the following examples:

1. Removal of lead from the body. In this case, injection of Ca-EDTA chelate, $Ca[Na_2EDTA]$ is given to the patient. $Ca[Na_2EDTA]$ complex reacts with Pb in the body and forms $Pb[Na_2EDTA]$ chelate which is more stable than $Ca[Na_2EDTA]$ and goes out of the body in urine.

2. Removal of Cu^{2+} ions from the lever. The accumulation of excess of Cu^{2+} ions in the lever leads to the malfunctioning of the central nervous system. In order to remove the excess of Cu^{2+} ions, penicillamine (chelating agent) is given to the patient. This chelating ligand forms a soluble chelated complex with Cu^{2+} ions by coordinating with S, N and O donor atoms. This Cu^{2+}– chelate goes out of body through urine.

```
           H
           |
   HS — C — CH3
           |
  H2N — C — C = O
           |     |
           H     O
```

Penicilliamine

Separation of ions by solvent extraction method

Certain chelating ligands such as dimethyl glyoxime, acetylacetone, trifluoroacetyl acetone, 8-hydroxyquinoline (oxine), dithioxone, cupferron etc. form chelates with many metallic ions. These chelates carry no charge *i.e.*, these chelates are neutral complex compounds and are called **inner complexes**. These complexes are usually insoluble in water but dissolve in organic solvents at some suitable experimental conditions. Thus by extracting the inner complexes in organic solvents it is possible to separate a given ion from other ions. This method of separating a given ion from other ions is called **solvent extraction method.**

Examples (*a*) Cu^{2+} and Fe^{3+} ion can be separated from other ions by solvent extraction method as follows. When Cu^{2+} ion is treated with acetyl acetone, *acac*H(chelating ligand), Cu^{2+}–chelate (I) is formed. Cu^{2+}-chelate (I) formed as above is extracted with some suitable organic solvent and thus Cu^{2+} ion can be separated from other ions.

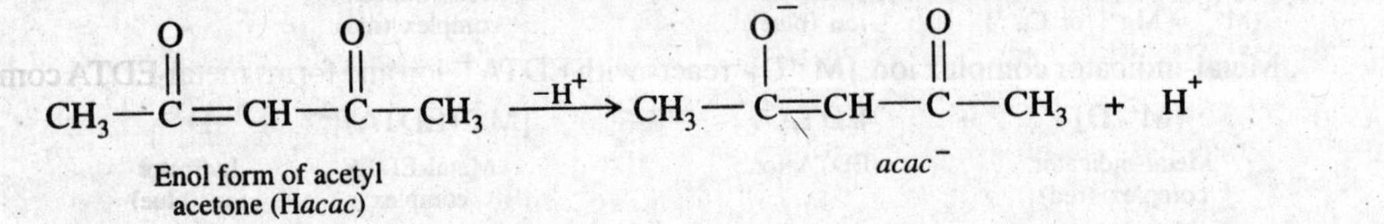

$$CH_3-\overset{O}{\overset{\|}{C}}=CH-\overset{O}{\overset{\|}{C}}-CH_3 \xrightarrow{-H^+} CH_3-\overset{O^-}{\overset{|}{C}}=CH-\overset{O}{\overset{\|}{C}}-CH_3 + H^+$$

Enol form of acetyl acetone (H*acac*) — *acac*$^-$

or H*acac* $\xrightarrow{-H^+}$ *acac*$^-$ + H^+

$$2\ CH_3-\overset{O^-}{\overset{|}{C}}=CH-\overset{O}{\overset{\|}{C}}-CH_3 + Cu^{2+} \longrightarrow$$

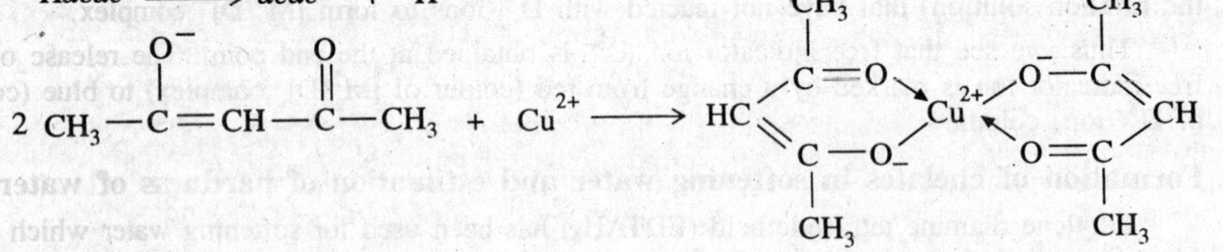

Cu^{2+} - chelate (neutral chelate) (I)

or $2\ acac^- + Cu^{2+} \longrightarrow [Cu(acac)_2]^0$

(*b*) Trifluoroacetylacetone (chelating ligand) has been used in the separation of Zr and Hf.

(*c*) At suitable pH, reasonably good separation of Ga^{3+} ion from other ions has been achieved by extracting the Ga^{3+}- oxine chelate into chloroform. Similarly, Fe^{3+} ion has also been separated from other ions by extracting Fe^{3+}- oxine chelate in chloroform at a suitable pH.

Role of metal chelates in living systems

The role of metal chelates in living systems can be illustrated by studying the follwing examples:

1. Role of hemoglobin. See chapter 19 (*Role of Metal Ions in Biological systems*)

2. Role of chlorophyll in plants : Photosynthesis. See chapter 19 (*Role of Metal Ions in Biological Systems*).

3. Function of trace metals in human body. At present it is believed that all the essential trace metals in human body probably function through processes involving chelate formation.

4. Function of Fe(II)-Fe(II) protein-based chelated complex. Within the animal cell, Fe(II)-Fe(III) protein-based coordination compound of hemoglobin family, called *cytochrome-C* acts as a catalyst for the utilisation of oxygen in the basic cell metabolism of the fundamental life process.

Reactions that are used for the preparation of complex compounds

Compelx compounds are generally prepared by the following reactions:

1. Ligand substitution reactions. In these reactions one or more ligands present in the coordination sphere are replaced by other ligands. For example the treatment of an aqueous solution of penta-hydrate copper sulphate, $CuSO_4.5H_2O$ or $[Cu(H_2O)_4]SO_4.H_2O$ with the excess of NH_3 gives tetrammine copper (III) sulphate, $[Cu(NH_3)_4]SO_4$ (*deep blue*).

$$CuSO_4.5H_2O \longrightarrow [Cu(H_2O)_4]SO_4 + H_2O$$

$$[Cu(H_2O)_4]SO_4 + 4NH_3 \longrightarrow [Cu(NH_3)_4]SO_4 + 4H_2O$$

or $[Cu(H_2O)_4]^{2+} + 4NH_3 \longrightarrow [Cu(NH_3)_4]^{2+} + 4H_2O$

Other examples are:

$$[Cd(H_2O)_4]Cl_2 + 4NH_3 \longrightarrow [Cd(NH_3)_4]Cl_2 + 4H_2O$$

$$[Ni(H_2O)_6]^{2+} + 3en \longrightarrow [Ni(en)_3]^{2+} + 6H_2O$$

$$[Cr(H_2O)_6]Cl_3 + 3CH_3COOH \longrightarrow [Cr(CH_3COO)_3]^0 + 6H_2O + 3HCl$$

$$K_3[Co^{3+}(NO_2)_6] + 2en \longrightarrow cis\text{-}[Co^{3+}(NO_2)_2(en)_2]NO_2 + 3KNO_2$$

or $[Co^{3+}(NO_2)_6]^{3-} + 2en \longrightarrow cis\text{-}[Co^{3+}(NO_2)_2(en)_2]^+ + 4NO_2^-$

$$[Co^{3+}en_2Cl_2]^+ + C_2O_4^{2-} \longrightarrow [Co^{3+}en_2(C_2O_4)]^+ + 2Cl^-$$

$$[Co^{3+}(dien)(NO_2)_3]^0 + en \longrightarrow [Co^{3+}(dien)(NO_2)(en)](NO_2)_2$$

2. Redox reactions. In these reactions an electron is transferred form one species to the other. Examples are:

(*a*) $[Co^{3+}(NH_3)_6]Cl_3$ can be prepared by the oxidation of an aqueous solution of $CoCl_2$ containing NH_4Cl and NH_3 with H_2O_2.

$$\underset{(Co\,=\,+2)}{2CoCl_2} + 2NH_4Cl + 10NH_3 + \underset{(O\,=\,-1)}{H_2O_2} \longrightarrow \underset{(Co\,=\,+3)}{2[Co(NH_3)_6]Cl_3} + \underset{(O\,=\,-2)}{2H_2O}$$

Similalry $[Co(NH_3)_5(NO_3)](NO_3)_2$ can be prepared by oxidising $[Co(H_2O)_6](NO_3)_2$ containing NH_4NO_3 and NH_3 with H_2O_2.

$$\underset{(Co\,=\,+2)}{2[Co(H_2O)_6](NO_3)_2} + 2NH_4NO_3 + 8NH_3 + \underset{(O\,=\,-2)}{H_2O_2} \longrightarrow \underset{(Co\,=\,+3)}{2[Co(NH_3)_5(NO_3)](NO_3)_2} + \underset{(O\,=\,-2)}{14H_2O}$$

(*b*) Potassium trioxalato chromate (III), $K_3[Cr(C_2O_4)_3]$ can be prepared by reducing $K_2Cr_2O_7$ with oxalic acid ($H_2C_2O_4$) and potassium oxalate ($K_2C_2O_4$)

$$\underset{(Cr\,=\,+6)}{K_2Cr_2O_7} + \underbrace{7H_2C_2O_4 + 2K_2C_2O_4}_{(C\,=\,+3)} \longrightarrow \underset{(Cr\,=\,+3)}{2K_3[Cr(C_2O_4)_3]} + \underset{(C\,=\,+4)}{6CO_2} + 7H_2O$$

Examples of other redox reactions are:

(*c*) $\underset{(Fe\,=\,+2)}{[Fe(CN)_6]^{4-}} + \underset{(Ir\,=\,+4)}{[IrCl_6]^{2-}} \longrightarrow \underset{(Fe\,=\,+3)}{[Fe(CN)_6]^{3-}} + \underset{(Ir\,=\,+3)}{[IrCl_6]^{3-}}$

(*d*) $\underset{(Fe\,=\,+2)}{[Fe(H_2O)_6]^{2+}} + \underset{(O\,=\,-1)}{H_2O_2} \longrightarrow \underset{(Fe\,=\,+3)}{[Fe(H_2O)(OH)]^{2+}} + \underset{(O\,=\,-2)}{H_2O} + OH^-$

3. Direct combination. Many metal ammines can be prepared by the direct addition of metal salt to liquid NH_3. For example:

$$NiCl_2 + 6NH_3 \longrightarrow [Ni(NH_3)_6]Cl_2$$

$$CoCl_3 + 6NH_3 \longrightarrow [Co(NH_3)_6]Cl_3$$

$$CdCl_2 + 4NH_3 \longrightarrow [Cd(NH_3)_4]Cl_2$$

$$AgCl + 2NH_3 \longrightarrow [Ag(NH_3)_2]Cl$$

Apart from ammines, other complex compounds have been prepared by direct combination method, *e.g.*,

$$PtCl_2 + 2en \longrightarrow [Pt(en)_2]Cl_2$$

$$PtCl_4 + 2KCl \longrightarrow K_2[PtCl_6]$$

$$Ni + 4CO \longrightarrow [Ni(CO)_4]$$

$$BF_3(g) + NH_3(g) \longrightarrow [BF_3(NH_3)]$$

$$SnCl_4(l) + 2NMe_3(l) \xrightarrow[\text{Petroleum ether}]{40\text{–}60°C} trans\text{-}[SnCl_4(NMe_3)_2]$$

$$NiCl_2 + 3en \longrightarrow [Ni(en)_3]Cl_2$$

4. Thermal dissociation reactions. Some complexes have been prepared by this method, *e.g.*,

$$[Cr(en)_3]Cl_3 \xrightarrow{210°C} cis\text{-}[Cr(en)_2Cl_2]Cl + en$$

$$[Cr(en)_3](SCN)_3 \xrightarrow{130°C} trans\text{-}[Cr(en)_2(SCN)_2]SCN + en$$

$$trans\text{-}[Ni(py)_4Cl_2] \xrightarrow[\text{in vacuum}]{180°C} [Ni(py)_2Cl_2] + 2py$$

Importance and uses of coordination compounds

Coordination (complex) compounds are of great importance. These compounds constitute the minerals, plants and are also present in animals. These play important functions. Besides these, the coordination compounds are also used in the field of metallurgy, biochemisty, ion exchange, texile dyeing, analytical chemistry etc.

Complex compounds of various metals have been used for various purposes. For example:

1. Uses of complex compounds as electrolytes in electroplating. Many complex compounds are used as electrolytes for electroplating. These complexes deliver the metal ions for reduction. For example for silver plating, $K[Ag(CN)_2]$ is used as electrolyte. Similarly $K_3[Cu(CN)_4]$ is used for cupper electroplating and $K[Au(CN)_2]$ is used for gold plating.

2. Uses of complex compounds as medicines. (*a*) Ca-EDTA chelate, $Ca[EDTANa_2]$ is used in the treatment of lead poisoning (*i.e.*, for removing lead from the body). See "*Applications of the formation of chelates*" discussed on page 34.

(*b*) *cis*-$[Pt(NH_3)_2Cl_2]$ (called cisplatin) is used as an antitumor agent in the treatment of cancer.

3. Uses of complex compounds as catalysts. Several complex compounds are used as catalysts for different reactions, *e.g.*, $Co_2(CO)_8$ acts as a catalyst in the conversion of olefins into alcohols.

4. Uses of complex compounds in softening water. Complex compounds like sodium metapolyphosphate are used to remove Ca^{2+} and Mg^{2+} ions from hard water by forming complexes with these ions. The formation of complex compounds prevents the scale formation in boilers.

Complex Formation in Various Fields

There are many processes in which complex compounds are formed. The complex formation has been utilised for various purposes. Examples are given below:

I. Complex formation in the formation of chelates

See page 34.

II. Complex formation in photography

The formation of sodium dithiosulphate argentate (I) takes place when the developed film is fixed by washing it with a solution of $Na_2S_2O_3$.

$$AgBr(s) + 2Na_2S_2O_3\ (aq) \longrightarrow Na_3[Ag(S_2O_3)_2]\ (aq) + NaBr\ (aq)$$

or

$$AgBr(s) + 2S_2O_3^{2-}\ (aq) \longrightarrow [Ag(S_2O_3)_2]^{3-}(aq) + Br^-(aq)$$

The process of fixing the developed film removes AgBr (which is not changed to Ag during the process of developing) as soluble $[Ag(S_2O_3)_2]^{3-}$ ion (complex ion).

III. Complex formation in analytical chemistry

The formation of complex compounds has been utilised for the identification (qualitative estimation), for the gravimetric estimation (quantitative estimation) and for the separation of metal ions. Quantitative (gravimetric) estimation of many metal ions has been made as their insoluble complex compounds by means of many complexing agents. Insoluble complex compounds formed are generally coloured and have usually high molecular weights so that a small amount of the metal ions gives a relatively large amount of the precipitate of complex compounds. Complexing agents (organic reagents) which are generally used for precipitating the metal ions in the form of their complex compounds are *cupron, dimethyl glyoxime, α-nitroso β-naphthol, cupferron*, 8-*hydroxy quinoline* (*oxine*) etc. The separation of metal ions has also been carried out by using organic reagents at suitable experimental conditions.

Examples of qualitative/quantitative estimations and separation of some metal ions are given below:

1. Preparation of Nessler's reagent, $K_2[HgI_4]$ and identification (qualitative estimation) of NH_3 and NH_4^+ salts. Complex formation has been used in the preparation of Nessler's reagent which is potassium mercuric tetraiodide (complex compound) and is used for the identification of NH_3 and ammonium salts (NH_4^+). Nessler's reagent is used as follows. When KI solution is added drop by drop to the solution of $HgCl_2$, red ppt. of AgI_2 is obtained. This ppt. gets dissolved in the excess of KI solution and forms colourless Nessler's reagent.

$$HgCl_2 + 2KI \longrightarrow \underset{\text{Red ppt}}{HgI_2} + 2KCl$$

$$HgI_2 + 2KI\ (excess) \longrightarrow \underset{\text{Nessler's reagent}}{K_2[HgI_4]}$$

For the identification of NH_3 and NH_4^+ salts, an alkaline solution of Nessler's reagent (The reagent should contain NaOH) is used. NH_3 and NH_4^+ salts both give a redish-brown ppt. of iodide of Millon's base.

$$NaOH\ (\textit{present in Nessler's reagent}) \longrightarrow Na^+ + OH^-$$

$$NH_4^+ + OH^- \longrightarrow NH_3\uparrow + H_2O$$

$$\underset{\text{Nessler's reagent}}{K_2[HgI_4]} \longrightarrow 2KI + HgI_2$$

$$HgI_2 + 2NH_3 \longrightarrow \underset{\text{Amino mercuric iodide}}{[Hg(NH_2)]I} + NH_4I$$

$$2[Hg\,(NH_2)]\,I + 2H_2O \longrightarrow \underset{\text{Iodide of Millon's base (Reddish-brown ppt.)}}{\left[O^{2-} \begin{matrix} Hg^{2+} \\ Hg^{2+} \end{matrix} NH_2^- \right] I \cdot H_2O} + NH_4I$$

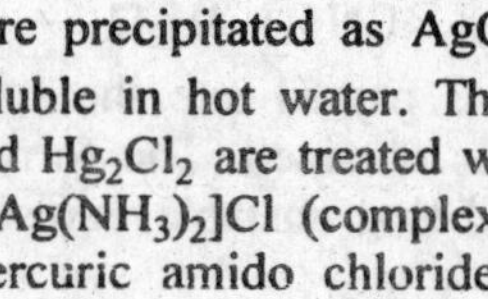

2. Separation of Ag^+ from Hg_2^{2+} in group I. When dil. HCl is added to the aqueous solution of Ag^+ and Hg_2^{2+} salts, these ions are precipitated as AgCl and Hg_2Cl_2 (both white precipitates). AgCl and Hg_2Cl_2 both are insoluble in hot water. These are separated by using the fact that when the precipitates of AgCl and Hg_2Cl_2 are treated with NH_4OH solution, AgCl forms soluble diammine silver (I) chloride, $[Ag(NH_3)_2]Cl$ (complex compound) while Hg_2Cl_2 gives a black mixture which consists of mercuric amido chloride, $[Hg^{2+}(NH_2)]Cl$ (complex compound) and Hg.

$$AgCl + 2NH_4OH\ (excess) \longrightarrow [Ag(NH_3)_2]Cl + 2H_2O$$

or $$Ag^+ + 2NH_4OH\ (excess) \longrightarrow [Ag(NH_3)_2]^+ + 2H_2O$$

$$Hg_2Cl_2 + 2NH_4OH \longrightarrow \underbrace{[Hg(NH_2)]Cl + Hg}_{\text{Black ppt.}} + NH_4Cl + 2H_2O$$

or $$Hg_2^{2+} + 2NH_4OH \longrightarrow \underbrace{[Hg(NH_2)]^+ + Hg}_{\text{Black ppt.}} + NH_4^+ + 2H_2O$$

3. Separation of IIA group sulphides (CuS, HgS, PbS, Bi_2S_3 and CdS) from II B group sulphides (As_2S_3, Sb_2S_3, SnS and SnS_2). This separation is based on the fact that the sulphides of group II A elements are insoluble in yellow ammonium sulphide, $[(NH_4)_2S + S]$ while those of group II B are converted into soluble thio complexes.

$$M_2S_3 + 3(NH_4)_2S + 2S \longrightarrow 2(NH_4)_3[MS_4]$$
(M = As, Sb) — Ammonium tetrathio arsenate (V) or ammonium tetrathio antimonate (V)

$$SnS + (NH_4)_2S + S \longrightarrow (NH_4)_2[SnS_3]$$
Ammonium trithio stannate (IV)

$$SnS_2 + (NH_4)_2S \longrightarrow (NH_4)_2[SnS_3]$$

4. Identification of Cu^{2+} ions. When NH_4OH is added to the solution containing Cu^{2+} ions, deep blue solution is obtained. This solution contains tetrammine cuprate (II) ion.

$$Cu^{2+} + 4NH_4OH \longrightarrow [Cu(NH_3)_4]^{2+} + 4H_2O$$
Tetrammine cupprate (II) ion (Deep blue solution)

Now acidify the solution containing $[Cu(NH_3)_4]^{2+}$ ions with CH_3COOH and add $K_4[Fe(CN)_6]$ so that chocolate coloured ppt. of cupric ferrohexacyanide, $Cu_2[Fe(CN)_6]$ is obtained

$$[Cu(NH_3)_4]^{2+} + 4CH_3COOH \longrightarrow Cu^{2+} + 4CH_3COONH_4$$
$$2Cu^{2+} + [Fe(CN)_6]^{4-} \longrightarrow Cu_2[Fe(CN)_6]\downarrow$$
Chocolate ppt. (Red brown)

5. Identification of Cd^{2+} ions in presence of Cu^{2+} ions. Following two methods are used:

Method 1. When NH_4OH is added to the solution containing Cu^{2+} and Cd^{2+} salts, soluble tetrammine complexes of Cu^{2+} and Cd^{2+} ions are obtained.

$$Cu^{2+} + 4NH_4OH \longrightarrow [Cu(NH_3)_4]^{2+} + 4H_2O$$
Deep blue solution

$$Cd^{2+} + 4NH_4OH \longrightarrow [Cd(NH_3)_4]^{2+} + 4H_2O$$
Colourless solution

Obviously, if the solution containing tetrammine complexes is deep blue in colour, Cu^{2+} ions are also present along with Cd^{2+} ions. Now the excess of KCN is added to the solution. By this operation, $[Cu^{2+}(NH_3)_4]^{2+}$ and $[Cd^{2+}(NH_3)_4]^{2+}$ ions are converted into $[Cu^+(CN)_3]^{2-}$ and $[Cd^{2+}(CN)_4]^{2-}$ ions respectively.

$$2[Cu(NH_3)_4]^{2+} + 7CN^- + H_2O \longrightarrow 2[Cu^+(CN)_3]^{2-} + CNO^- + 6NH_3 + 2NH_4^+$$
$$[Cd(NH_3)_4]^{2+} + 4CN^- + 4H_2O \longrightarrow [Cd^{2+}(CN)_4]^{2-} + 4NH_4OH$$

Now when H_2S gas is passed into the solution containing $[Cu(CN)_3]^{2-}$ and $[Cd(CN)_4]^{2-}$ ions, yellow ppt. of CdS is obtained. The formation of yellow ppt. of CdS is because of the fact that $[Cd(CN)_4]^{2-}$ ion, being unstable, ionises to give Cd^{2+} ions which react with H_2S and gives yellow ppt. of CdS

$$[Cd(CN)_4]^{2-} \longrightarrow Cd^{2+} + 4CN^-$$
$$Cd^{2+} + H_2S \longrightarrow CdS\downarrow \text{ (yellow ppt.)} + 2H^+$$

$[Cu(CN)_3]^{2-}$ ion, being stable, does not ionise to give Cu^{2+} and hence the ppt. of CuS is not formed.

Method 2. When KCN is added to the solution containing Cu^{2+} and Cd^{2+} ions, $Cu(CN)_2$ (yellow ppt.) and $Cd(CN)_2$ (white ppt.) are precipitated. $Cu(CN)_2$, being unstable, gets decomposed to give $Cu_2(CN)_2$ and $(CN)_2$. $Cu_2(CN)_2$ and $Cd(CN)_2$ thus formed react with the excess of KCN and form soluble complex compounds namely $K_2[Cu^+(CN)_3]$ and $K_2[Cd^{2+}(CN)_4]$ respectively.

$$Cu^{2+} + 2KCN \longrightarrow \underset{\text{yellow ppt.}}{Cu(CN)_2\downarrow} + 2K^+$$

$$2Cu^{2+}(CN)_2 \xrightarrow{\text{Decomposition}} Cu_2^+(CN)_2 + (CN)_2$$

$$Cu_2(CN)_2 + 4KCN \longrightarrow \underset{\text{soluble complex}}{2K_2[Cu^+(CN)_3]}$$

$$Cd^{2+} + 2KCN \longrightarrow \underset{\text{white ppt.}}{Cd(CN)_2\downarrow} + 2K^+$$

$$Cd(CN)_2 + 2KCN \longrightarrow \underset{\text{soluble complex}}{K_2[Cd^{2+}(CN)_4]}$$

Now H_2S gas is passed into the solution containing $[Cu(CN)_3]^{2-}$ and $[Cd(CN)_4]^{2-}$ ions, yellow ppt. of CdS is obtained. (See above).

6. Identification of Fe^{3+} ions. Following two methods are used:

Method 1. When $K_4[Fe(CN)_6]$ is added to the solution containing Fe^{3+} ions, precipitate or solution of prussian blue is obtained.

$$Fe^{3+}\ (excess) + K_4[Fe(CN)_6] \longrightarrow \underset{\text{solution of prussian blue}}{K^+Fe^{3+}[Fe^{2+}(CN)_6]} + 3K^+$$

or $$Fe^{3+}(excess) + [Fe(CN)_6]^{4-} \longrightarrow [Fe^{3+}\{Fe^{2+}(CN)_6\}]^-$$

$$4Fe^{3+} + 3K_4[Fe(CN)_6] \longrightarrow \underset{\text{ppt. of prussian blue}}{Fe_4^{3+}[Fe^{2+}(CN)_6]_3\downarrow} + 12K^+$$

or $$4Fe^{3+} + 3[Fe(CN)_6]^{4-} \longrightarrow Fe_4^{3+}[Fe^{2+}(CN)_6]_3\downarrow$$

Method 2. When potassium (or ammonium) thio-cyanate [KSCN or NH_4SCN] is added to the solution containing Fe^{3+} ions, ferrithiocyanate (deep red colour) is obtained.

$$Fe^{3+} + SCN^- \longrightarrow [Fe^{3+}(SCN)]^{2+} \quad (blood\text{-}red\ colour)$$

$$Fe^{3+} + 3SCN^- \longrightarrow Fe(SCN)_3 \quad (blood\text{-}red\ colour)$$

$$Fe^{3+} + 6SCN^-\ (excess) \longrightarrow [Fe^{3+}(SCN)_6]^{3-} \quad (blood\text{-}red\ colour)$$

$$Fe^{3+} + 6SCN^-\ (excess) \longrightarrow Fe^{3+}[Fe^{3+}(SCN)_6] \quad (blood\text{-}red\ colour)$$

7. Identification of Ni^{2+} ions. Ni^{2+} ions are identified as their insoluble red chelated complex with dimethyl glyoxime in ammonical medium. For details see page 35.

8. Identification of Co^{2+} ions. Black ppt. of CoS obtained by passing H_2S gas through the solution containing Co^{2+} ions is insoluble even in conc. HCl. It is dissolved in aqua regia to get $CoCl_2$ in solution.

$$[3HCl + HNO_3 \longrightarrow NO + 2H_2O + 3Cl] \times 2$$

$$[CoS + 2Cl \longrightarrow CoCl_2 + S\downarrow] \times 2$$

$$\underset{\text{Aqua regia}}{3COS + 6HCl + 2HNO_3} \longrightarrow \underset{\text{(In solution)}}{CoCl_2} + 2NO + 4H_2O + 3S\downarrow$$

Now the solution containing $CoCl_2$ is divided into two parts.

Part I. When crystals of ammonium thiocyanate, NH_4CNS are added to the solution, containing $CoCl_2$, in acetone, blue colour is produced in acetone layer. This blue colour is due to the formation of ammonium tetrathiocyanato cobaltate (II), $(NH_4)_2[Co^{2+}(CNS)_4]$

$$CoCl_2 + 4NH_4CNS \longrightarrow (NH_4)_2[Co(CNS)_4] + 2NH_4Cl$$

or $$Co^{2+} + 4NH_4CNS \longrightarrow \underset{\text{Blue colour}}{(NH_4)_2[Co(CNS)_4]} + 2NH_4^+$$

Part II. When a solution containing $CoCl_2$ is acidified by means of CH_3COOH and then crystals of KNO_2 are added, yellow ppt. of *potassium hexanitro cobaltate* (III), $K_3[Co^{3+}(NO_2)_6]$ is obtained. The reactions taking place are:

$$CoCl_2 + 2KNO_2 \longrightarrow Co(NO_2)_2 + 2KCl \qquad ...(i)$$

$$CH_3COOH + KNO_2 \longrightarrow CH_3COOK + HNO_2 \qquad ...(ii)$$

HNO_2 produced in reaction (*ii*) oxidises $Co(NO_2)_2$ formed in reaction (*i*) to $Co(NO_3)_3$ and is itself reduced to NO.

$$\underset{(N = +3)}{2HNO_2} + \underset{(Co = +2)}{Co(NO_2)_2} \longrightarrow \underset{(Co = +3)}{Co(NO_2)_3} + \underset{(N = +2)}{NO} + H_2O \qquad ...(iii)$$

$Co(NO_2)_3$ formed as above combines with KNO_2 to form $K_3[Co(NO_2)_6]$ *(yellow ppt)*

$$Co(NO_2)_3 + 3KNO_2 \longrightarrow K_3[Co^{3+}(NO_2)_6] \qquad ...(iv)$$

Now multiply equation (*i*) by two and then add all the four equations to get :

$$7KNO_2 + 2CH_3COOH + CoCl_2 \longrightarrow 2CH_3COOK + 2KCl + NO + H_2O + K_3[Co(NO_2)_6]$$

or $$7K^+ + 7NO_2^- + 2CH_3COO^- + 2H^+ + Co^{2+} + 2Cl^- \longrightarrow 2CH_3COO^- + 2K^+ + 2K^+ + 2Cl^- + NO + H_2O + 3K^+ + [Co(NO_2)_6]^{3-}$$

or $$Co^{2+} + 7NO_2^- + 2H^+ \longrightarrow \underset{\text{Hexanitro cobaltate (III) ion}}{[Co^{3+}(NO_2)_6]^{3-}} + H_2O + NO$$

9. Identification of Co^{2+} and Ni^{2+} ions given together. Black precipitate of CoS and NiS obtained by passing H_2S gas through the solution containing Co^{2+} and NI^{2+} ions is insoluble even in conc. HCl. It is dissolved in aqua regia.

$$\underset{(M = Co,\ NI)}{3MS} + \underbrace{6HCl + 2HNO_3}_{\text{Aqua regia}} \longrightarrow \underset{\text{(In solution)}}{3MCl_2} + 2NO + 3S + 4H_2O$$

The solution containing $CoCl_2$ and $NiCl_2$ is evaporated to dryness and the residue thus obtained is treated with dil. HCl. The solution which contains $CoCl_2$ and $NiCl_2$ is treated with $NaHCO_3$ and bromine water. Appearance of green colour confirms the presence of Co^{2+} ions. If no green colour is observed, the solution is heated when *black residue* of Ni_2O_3 is obtained which shows the presence of Ni^{2+} ions. The formation of green colour which confirms the presence of Co^{2+} ions is due to the formation of *sodium tricarbonato cobaltate* (III), $Na_3[Co^{3+}(CO_3)_3]$.

$$CoCl_2 + 6NaHCO_3 \longrightarrow Na_4[Co^{2+}(CO_3)_3] + 3CO_2 + 3H_2O + 2NaCl$$

$$Br_2 + H_2O \longrightarrow 2HBr + O$$

This nascent oxygen oxidises $Na_4[Co^{2+}(CO_3)_3]$ to $Na_3[Co^{3+}(CO_3)_3]$ (green colouration)

$$2Na_4[Co^{2+}(CO_3)_3] + H_2O + O \longrightarrow \underset{\text{(Green colouration)}}{2Na_3[Co^{3+}(CO_3)_3]} + 2NaOH$$

$$NiCl_2 + 2NaHCO_3 \longrightarrow NiCO_3 + 2NaCl + H_2O + CO_2$$

$$Br_2 + H_2O \longrightarrow 2HBr + O$$

$$\underset{(Ni = +2)}{2NiCO_3} + O \longrightarrow \underset{\substack{(Ni = +3) \\ \text{(Black residue)}}}{Ni_2O_3} + 2CO_2$$

10. Gravimetric estimation of Ni^{2+} ions. See page 34.

11. Gravimetric estimation of Mg^{2+} and Ca^{2+} ions by EDTA. See page 36.

12. Separation of Co^{2+} and Ni^{2+} ions. See page 35.

IV. Complex formation in volumetric analysis

As its name implies, volumetric analysis is a branch of qualitative analysis, which relies on methods involving accurate measurements of volumes of liquids, though one or more weighings may also be needed.

The formation of complex compounds has been utilised in various volumetric analyses. Following examples illustrate this point.

1. Titration of a cyanide with $AgNO_3$ (Leibig's method, 1851). In this titration, $AgNO_3$ solution is added to a solution of soluble cyanide (*e.g.* KCN) when a stable soluble complex ion, $[Ag(CN)_2]^-$ is formed.

$$AgNO_3 + 2CN^- \longrightarrow [Ag(CN)_2]^-$$

or

$$Ag^+ + 2CN^- \longrightarrow \underset{\text{(In solution)}}{[Ag(CN)_2]^-}$$

When all CN^- ions have reacted with Ag^+ ions, further addition of $AgNO_3$ results in the formation of an insoluble complex compound, $Ag[Ag(CN)_2]$.

$$Ag^+ + [Ag(CN)_2]^- \longrightarrow \underset{\text{Insoluble}}{Ag[Ag(CN)_2]}$$

The end point is, therefore, indicated by the appearance of slightly turbidity or by the production of a ppt. of $Ag[Ag(CN)_2]$.

2. Titration of an iodide with Hg^{2+} ions. When a solution containing Hg^{2+} ions (*e.g.*, $HgCl_2$) is added to a solution of an iodide (*e.g.* KI), a red ppt. of HgI_2 is first formed which dissolves in the excess of KI, on shaking, due to the formation of a soluble potassium tetraiodo mercurate (II), $K_2[HgI_4]$ (complex compound).

$$HgCl_2 + 2KI \longrightarrow HgI_2\downarrow + 2KCl$$

or

$$Hg^{2+} + 2I^- \longrightarrow \underset{\text{Red ppt}}{HgI_2\downarrow}$$

$$HgI_2 + 2KI\ (excess) \longrightarrow K_2[HgI_4]\ (soluble\ complex)$$

Further addition of $HgCl_2$ solution gives red ppt. of HgI_2.

$$K_2[HgI_4] + HgCl_2 \longrightarrow 2HgI_2\ (red\ ppt.) + 2KCl$$

Thus, the end point is indicated by the appearance of red ppt. of HgI_2.

V. Complex formation in metallurgy

The extraction/concentration of many metals involves complex formation. The following examples can be studied.

1. Extraction of Ag and Au by Mac-Arthur and Forest's cyanide process. In the extraction of Ag and Au by this process, the ore of the metal is treated with a dil. solution of NaOH in the presence of excess of air so that these metals present in the ore are converted into soluble cyanide complex, $Na[M(CN)_2]$ (M = Ag or Au)

$$\underset{\text{(M = Ag or Au)}}{M^+} + 2NaCN\ (aq) \longrightarrow \underset{\text{Soluble complex}}{Na[M(CN)_2]} + Na^+$$

The solution containing the complex compound is treated with Zn dust when Ag/Au gets precipitated.

$$2Na[M(CN)_2] + Zn \longrightarrow Na_2[Zn(CN)_4] + 2M\downarrow\ (M = Ag\ or\ Au)$$

2. Purification of crude Ni by Mond's process. In this process, crude nickel is treated with CO so that Ni is converted into the vapours of nickel tetracarbonyl, $Ni(CO)_4$ (complex compound) which, on being heated to 180°C, gives 99.9% pure nickel.

$$Ni(crude) + 4CO \longrightarrow Ni(CO)_4 \xrightarrow{180^\circ C} Ni\ (99.9\%\ pure) + 4CO\uparrow$$

3. Concentration of bauxite ore ($Al_2O_3.2H_2O$) by Baeyer's process. Aluminium is generally extracted from bauxite ore which is associated with the impurities of silica, SiO_2 and ferric oxide, Fe_2O_3 (also called red bauxite). These impurities are removed by Baeyer's process in which the bauxite ore is digested at 150°C with a concentrated solution of NaOH so that bauxite ore dissolves in NaOH due to the formation of souble sodium meta-aluminate, $Na[AlO_2]$ (complex compound) while the impurities of SiO_2 and Fe_2O_3 are left behind.

$$\underset{\text{Bauxite ore}}{Al_2O_3.2H_2O} + 2NaOH \longrightarrow \underset{\text{(soluble)}}{2Na[AlO_2]} + 3H_2O$$

VI. Complex formation in colourimetric and spectrophotometric determination of metal ions

The tendency of transition metals to form coloured complex compounds has been utilised in the colourimetric and spectrometric determination of transition metal ions. 1, 10-phenanthroline, 2, 2′-dipyridyl, dithiozone etc., which form coloured complexes with transition metal ions (*e.g.*, Fe^{2+}, Co^{2+}, Cu^{2+}, Ni^{2+} etc.) are used in the colourimetric estimation of these ions. The trace amount of a metal ion can be determined by spectrophotometric method simply by measuring the optical density of its coloured complex compound at an appropriate wavelength. For example trace amount of Fe(II) can be estimated by measuring the optical density of its complex compound with 1, 10-phenanthroline (pH = 2 – 9) at 515mμ.

Stability of Complex Compounds

Two types of stability

In studying the formation of complex compounds in solution, two kinds of stability have to be distinguished. These are:

1. Thermodynamic stability. This kind of stability deals with the properties like bond energies, stability constants and redox potentials that affect the equilibrium conditions.

On the basis of thermodynamic stability of complexes in solution, Biltz (1927) has classified the complex compounds into stable and unstable complexes. Stable complexes are those which possess sufficient stability to retain their identity in solution while unstable complexes are those which are reversibly dissociated in solution into their components. Stable and unstable complexes have also been called ***penetration*** and ***normal complexes*** respectively.

2. Kinetic stability. This kind of stability deals with the rates of reactions (*i.e.*, reactivity) of complexes in solution, the mechanisms of chemical reactions, formation of intermediate complexes, activation energies for the process etc. On the basis of the rate of reactions (*i.e.* kinetic stability) of the complex in solution, Taube (1950) has classified the complexes into *labile* and *inert* complexes. Labile complexes are those whose one or more ligands in the co-ordination sphere can be rapidly replaced by other ligands and the ability of a complex to replace its one or more ligands by other ligands is called its lability. Inert complexes are those whose one or more ligands can either not be replaced or can be replaced with difficulty by other ligands.

Stepwise formation of complexes : Stepwise formation constants and overall formation constants

According to J Bjerrum (1941) the formation of a complex in solution proceeds by the stepwise addition of the ligands to the metal ion. Thus the formation of the complex ML_n (M = central metal cation, L = monodentate ligand and n = maximum coordination number of the metal ion M for the ligand L. n varies from one ligand to another for the same metal ion) may be supposed to take place by the following n consecutive steps and equilibrium constants:

$$M + L \overset{K_1}{\rightleftharpoons} ML, \quad K_1 = \frac{[ML]}{[M]\,[L]}$$

$$ML + L \overset{K_2}{\rightleftharpoons} ML_2, \; K_2 = \frac{[ML_2]}{[ML]\,[L]}$$

$$\vdots \quad \vdots \quad \vdots \quad \vdots$$

$$ML_{n-1} + L \overset{K_n}{\rightleftharpoons} ML_n, \; K_n = \frac{[ML_n]}{[ML_{n-1}]\,[L]}$$

The equilibrium constants K_1, K_2,...... K_n are called **stepwise formation constants** or **stepwise stability constants.**

The formation of the complex ML_n may also be expressed by the following steps and equilibrium constants:

$$M + L \overset{\beta_1}{\rightleftharpoons} ML, \quad \beta_1 = \frac{[ML]}{[M]\,[L]}$$

$$ML + 2L \overset{\beta_2}{\rightleftharpoons} ML_2, \; \beta_2 = \frac{[ML_2]}{[M]\,[L]^2}$$

$$\vdots \quad \vdots \quad \vdots \quad \vdots$$

$$M + nL \overset{\beta_n}{\rightleftharpoons} ML_n, \; \beta_n = \frac{[ML_n]}{[ML]\,[L]^n}$$

The equilibrium constants β_1, β_2 ......β_n are called *overall (or cumulative) formation constants or overall (or cumulative) stability constants.* β_n is termed as nth overall (or cumulative) formation constant or overall (or cumulative) stability constant.

With a few exceptions the values of successive stability constants decrease regularly from K_1 to K_n. *i.e.*,

$$K_1 > K_2 > K_3 ..... > K_{n-1} > K_n.$$

The above order is illustrated by the data for the Cd^{II} —NH_3 system where the ligands are neutral molecules and by Cd^{II}—CN^- system where the ligands are charged.

$$Cd^{2+} + NH_3 \longrightarrow [Cd(NH_3)]^{2+} \quad K_1 = 10^{2.65}$$
$$[Cd(NH_3)]^{2+} + NH_3 \longrightarrow [Cd(NH_3)_2]^{2+} \quad K_2 = 10^{2.10}$$
$$[Cd(NH_3)_2]^{2+} + NH_3 \longrightarrow [Cd(NH_3)_3]^{2+} \quad K_3 = 10^{1.44}$$
$$[Cd(NH_3)_3]^{2+} + NH_3 \longrightarrow [Cd(NH_3)_4]^{2+} \quad K_4 = 10^{0.93}$$
$$\beta_4 = 10^{7.12}$$

$$Cd^{2+} + CN^- \longrightarrow [Cd(CN)]^+ \quad K_1 = 10^{5.48}$$
$$[Cd(CN)]^+ + CN^- \longrightarrow [Cd(CN)_2] \quad K_2 = 10^{5.12}$$
$$[Cd(CN)_2] + CN^- \longrightarrow [Cd(CN)_3]^- \quad K_3 = 10^{4.63}$$
$$[Cd(CN)]^{3-} + CN^- \longrightarrow [Cd(CN)_4]^{2-} \quad K_4 = 10^{3.65}$$
$$\beta_4 = 10^{18.8}$$

The steady decrease in the values of K_1, K_2, ... K_n with increasing number of ligands is due to the fact that as more and more ligands move into the coordination zone, less and less aqua-molecules are available to fresh ligands for replacement. With progressive intake of ligands the metal ion becomes less electron greedy. In case of the complexes of the charged ligands the more important factors responsible for the steady decrease are statistical, steric hindrance and coulombic factors.

The higher the value of stability constant for a complex ion, the greater will be its stability. Alternatively 1/K values sometimes called instability constant, give a measure of the extent to which the equilibrium representing the formation of a complex lies to the right.

Stepwise and cummulative stability constants are also expressed as $\log_{10} K_1$.... $\log_{10} K_n$ and $\log_{10} \beta_n$ respectively.

In all the above equilibria we have not specified the charge of metal ion and degree of solvation. The former omission is of no importance, since the equilibria may be expressed as above whatever the charges. Omission of water molecules is a matter of convention, since it is usually convenient and harmless. This omission should not be allowed whenever necessary. Square brackets indicate the concentration of the enclosed species.

Relationship between β_n and $K_1, K_2......K_n$

K's and β's are related to one another. Consider, for example, the expression for β_3 *viz.*

$$\beta_3 = \frac{[ML_3]}{[M]\,[L]^3}$$

On multiplying both numerator and denominator by $[ML][ML_2]$ and on rearranging, we get

$$\beta_3 = \frac{[ML_3]}{[M]\,[L]^3} \cdot \frac{[ML]\,[ML_2]}{[ML]\,[ML_2]}$$

$$= \frac{[ML]}{[M]\,[L]} \cdot \frac{[ML_2]}{[ML]\,[L]} \cdot \frac{[ML_3]}{[ML_2]\,[L]}$$

$$= K_1 \,.\, K_2 \,.\, K_3$$

Thus,
$$\beta_n = \frac{[ML]}{[M]\,[L]} \cdot \frac{[ML_2]}{[ML]\,[L]} \,.......\, \frac{[ML_n]}{[ML_{n-1}]\,[L]}$$

$$= K_1 \,.\, K_2 \,.......\, K_n \qquad ...(i)$$

or,
$$\beta_n = \sum_{n=1}^{n=n} \cdot K_n$$

From relation (*i*) it is evident that the overall stability constant, β_n (or simply β), is equal to the product of the successive (*i.e.*, stepwise) stability constants K_1, K_2,..... K_{n-1}, K_n. This is other words means that the value of stability constant for a given complex is actually made up of a number of step-wise stability constants as is evident from Table 1.3 in which the value of overall stability constant as β_n and $\log_{10}\beta_n$ for $[Ni(NH_3)_6]^{2+}$ ion and those of stepwise stability constants as K_1, K_2....., K_5, K_6 and $\log_{10} K_1$, $\log_{10} K_2$....., $\log_{10} K_6$ corresponding to the six equilibria given in the first column through which $[Ni(NH_3)_6]^{2+}$ ion can be regarded as being formed, are given. This table shows that

$$\beta_n = K_1 \times K_2 \times .... \times K_5 \times K_6$$

and
$$\log_{10} \beta_n = \log_{10} K_1 + \log_{10} K_2 + ... + \log_{10} K_6$$

In Table 1.3 we also see that log K_6 is negative (= – 0.10). This negative value shows that the pentammine complex, $[Ni(H_2O)(NH_3)_5]^{2+}$ is thermodynamically favoured with respect to the hexammine complex, $[Ni(NH_3)_6]^{2+}$ [See equilibrium No. (*vi*)].

Kinetic vs. thermodynamic stability

Since the terms *viz.* labile and inert show the speed or rates at which the substitution of one ligand by other occurs, these terms represent the kinetic stability of complexes. These terms should, therefore, not be confused with or used for thermodynamic stability terms viz. unstable and stable respectively. Although thermodynamically stable complexes may be labile or inert, and unstable complexes which are usually labile may also be inert, there is no correlation between thermodynamic and kinetic stability terms, *e.g.* $[Hg(CN)_4]^{2-}$ which is thermodynamically very stable (formation constant = 10^{42}) is labile, since in solution it exchanges CN^- ligands with labelled cyanide ions, $^{14}CN^-$ at a very fast rate.

Table 1.3. *Stability constant value (i.e., $\log_{10} K$ value) of $[Ni(NH_3)_6]^{2+}$ ion*

Six equilibria	Values of $K_1, K_2, \ldots K_6$ and β_n	Values of $\log_{10}K_1, \log_{10}K_2, \ldots \log_{10} K_6$ and $\log_{10} \beta_n$
(*i*) $[Ni(H_2O)_6]^{2+} + NH_3 \underset{}{\overset{K_1}{\rightleftharpoons}} [Ni(H_2O)_5(NH_3)]^{2+} + H_2O$	$K_1 = 5 \times 10^2$	$\log_{10} K_1 = 2.70$
(*ii*) $[Ni(H_2O)_5NH_3)]^{2+} + NH_3 \overset{K_2}{\rightleftharpoons} [Ni(H_2O)_4(NH_3)_2]^{2+} + H_2O$	$K_2 = 1.3 \times 10^2$	$\log_{10} K_2 = 2.11$
(*iii*) $[Ni(H_2O)_4(NH_3)_2]^{2+} + NH_3 \overset{K_3}{\rightleftharpoons} [Ni(H_2O)_3(NH_3)_3]^{2+} + H_2O$	$K_3 = 40$	$\log_{10} K_3 = 1.60$
(*iv*) $[Ni(H_2O)_3(NH_3)_3]^{2+} + NH_3 \overset{K_4}{\rightleftharpoons} [Ni(H_2O)_2(NH_3)_4]^{2+} + H_2O$	$K_4 = 12$	$\log_{10} K_4 = 1.08$
(*v*) $[Ni(H_2O)_2(NH_3)_4]^{2+} + NH_3 \overset{K_5}{\rightleftharpoons} [Ni(H_2O)(NH_3)_5]^{2+} + H_2O$	$K_5 = 4$	$\log_{10} K_5 = 0.60$
(*vi*) $[Ni(H_2O)(NH_3)_5]^{2+} + NH_3 \overset{K_6}{\rightleftharpoons} [Ni(NH_3)_6]^{2+} + H_2O$	$K_6 = 0.8$	$\log_{10} K_6 = -0.10$
On adding (*i*) to (*vi*) $[Ni(H_2O)_6]^{2+} + 6NH_3 \overset{\beta_n}{\rightleftharpoons} [Ni(NH_3)_6]^{2+} + 6H_2O$	$\beta_n = K_1.K_2 \ldots K_5.K_6$ $= 9.98 \times 10^7$	$\log_{10} \beta_n = \log_{10} K_1 + \log_{10} K_2 + \ldots + \log_{10} K_6$ $= 7.99$

$$[Hg(CN)_4]^{2-} + 4^{14}CN^- \rightleftharpoons [Hg(^{14}CN)_4]^{2-} + 4CN^-$$

Thus the stability of this complex does not ensure its inertness.

On the other hand the complex, $[Co(NH_3)_6]^{3+}$ which is thermodynamically unstable can remain unchanged in acid solution for weeks. Thus this complex is unstable but inert in acid solution.

It may be concluded from this description that the *inert complexes are not necessarily thermodynamically stable and that labile complexes are not necessarily thermodynamically unstable.*

The stability of a complex depends on the reaction energy while the lability of a compound depends on the activation energy. Larger the activation energy, lower will be the lability of the compound, *i.e.*, the compound with larger activation energy will react slowly.

Factors affecting the stability of metal complexes

The stability (or stability constants) of metal complexes depends on the following factors:

A. Properties of central metal ion

The following properties of central metal ion affect the stability of the metal complexes:

1. Size of central metal ion. For a given ligand the stability (or stability constant) of the complexes of the metallic ions having the same charge on them decreases with the increase of the size of the central metal ion. Thus the stability of complexes given by the cations belonging to the same group and having the same charge decreases as we proceed from top to bottom in the group, since the size of the metallic cations increases in the same order. For example:

(*i*) The stability of hydroxide complexes given by alkali metal ions (Li^+, Na^+ etc.), alkaline earth metal ions (Be^{2+}, Mg^{2+} etc.) and III B group ions (Sc^{3+}, Y^{3+} and La^{3+}) is in the order:

(*a*) Li^+ (r = 0.60 Å) > Na^+ (*r* = 0.95 Å) > K^+ (*r* = 1.33 Å) > Rb^+ (*r* = 1.48 Å) > Cs^+ (*r* = 1.69Å).

(*b*) Be^{2+} (*r* = 0.31 Å) > Mg^{2+} (*r* = 0.65 Å) > Ca^{2+} (*r* = 0.99 Å) > Sr^{2+} (*r* = 1.13 Å) > Ba^{2+} (*r* = 1.35 Å) > Ra^{2+} (*r* = 1.40 Å).

(*c*) Sc^{3+} (*r* = 0.81 Å) > Y^{3+} (*r* = 0.93 Å) > La^{3+} (*r* = 1.15 Å).

It may be noted that EDTA complex of Mg^{2+} (*r* = 0.65 Å) is less stable than that of Ca^{2+} (*r* = 0.99Å).

(*ii*) The inverse relation between the size of the central metal ion and the stability of the complexes formed is alo confirmed when we see that the stability of the complexes of Mn^{2+}, Fe^{2+}, Co^{2+}, Ni^{2+}, Cu^{2+} and Zn^{2+} ions with a given ligand increases from Mn^{2+} to Cu^{2+} and then decreases at Zn^{2+} (These ions are *divalent* ions of the elements of 1st transition series).

Ions:	Mn^{2+}	Fe^{2+}	Co^{2+}	Ni^{2+}	Cu^{2+}	Zn^{2+}
Ionic radii (Å):	0.91	0.83	0.82	0.78	0.69	0.74
Order of stability of complexes :	Mn^{2+} <	Fe^{2+} <	Co^{2+} <	Ni^{2+} <	Cu^{2+} >	Zn^{2+}

This sequence of stability is commonly known as **Irving-William order of stability** of complexes of M^{2+} ions.

2. Charge on the central metal ion. For a given ligand, the stability of the complexes of the metallic ions having almost the same size but different charges on the them decreases with the decrease of the charge on them. Thus the stability of complexes given by: (*a*) La^{3+}, Sr^{2+} and K^+ ions. (*b*) Co^{3+} and Co^{2+} ions. (*c*) Fe^{3+} and Fe^{2+} ions, and (*d*) Th^{4+}, Y^{3+}, Ca^{2+} and Na^+ ions with the same ligand is in the order:

(*a*) La^{3+} (*r* = 1.15 Å) > Sr^{2+} (= 1.13 Å) > K^+ (= 1.33 Å)

(*b*) Co^{3+} (*r* = 0.63 Å) > Co^{2+} (= 0.74 Å)

(*c*) Fe^{3+} (*r* = 0.64 Å) > Fe^{2+} (*r* = 0.76 Å)

(*d*) Th^{4+} (*r* = 0.95 Å) > Y^{3+} (= 0.93 Å) > Ca^{2+} (= 0.99 Å) > Na^+ (= 0.95 Å).

If the factors 1 and 2 mentioned above are combined, then we can say that with the increase of ionic potential of the central metal ion (Ionic potential of the metal ion = charge on the metal ion/size of the ion), the stability of the complexes with the same ligand also increases. For example, the stability of hydroxide complexes of Li^+, Ca^{2+}, Ni^{2+}, ... Be^{2+} ions whose ionic potential increase from Li^+ to Be^{2+} ions also increases in the same direction as shown below:

Central metal ion (charge on the ion is shown in parentheses)	*Ionic radius (Å)*	*Ionic potential*		*Order of stability of complexes*
Li^+ (+1)	0.60	1/0.60 = 1.6	Increasing ↓	Increasing ↓
Ca^{2+} (+2)	0.99	2/0.99 = 2.0		
Ni^{2+} (+2)	0.72	2/0.72 = 2.97		
Y^{3+} (+3)	0.93	3/0.93 = 3.22		
Th^{4+} (+4)	0.95	4/0.95 = 4.20		
Al^{3+} (+3)	0.50	3/0.50 = 6.0		
Be^{2+} (+2)	0.31	2/0.31 = 6.45		

3. Electronic configuration of the central metal ion. In case of metallic ions which have the same radius and same charge (*e.g.*, Na^+ and Cu^+; Ca^{2+} and Cd^{2+} or Sn^{2+}; Sc^{3+} and Ga^{3+} or Sb^{3+} ions), the outer electronic configuration of the ions plays an important role in determining the stability of the complexes. The complexes given by metal ions possessing inert gas (8-electron ions), pseudo inert gas (18-electron ions) and pseudo inert gas plus two (18 + 2 electron ions) electronic configuration are less stable than those given by transition metal ions which have $(n-1)d$ orbitals available for accepting the electron pairs donated by the ligands. The stability of transition metal complexes depends on, besides ionic potential, other factors like crystal field stabilisation energy, electron pairing energy, stereochemistry, availability of empty d_{z^2} and $d_{x^2-y^2}$ orbitals for π bonding or back donation. The use of inner *d*-orbitals gives stronger metal–ligand bonding than the use of *s*- and *p*-orbitals or outer *d*-orbitals. The various factors influencing the stability of complexes given by transiton metal ions ultimately depends on the number of electrons present in $(n-1)d$ orbitals. The stability constants of the complexes given by the transition metal d^n ions for the equilibrium:

$$[M(H_2O)_n] + nL \longrightarrow MLn + nH_2O$$

follow the order :

$$d^0 < d^1 < d^2 < d^3 \gtrless d^4 > d^5 < d^6 < d^7 < d^8 \gtrless d^9 > d^{10}.$$

The sequence $d^3 > d^4$ and $d^8 > d^9$, is expected when the John-Teller effect is low and the coordination number is 6. The sequenee $d^3 < d^4$ and $d^8 < d^9$, however, is expected when the John-Teller effect is large and the coordination number is 4.

4. Electronegativity and polorising power of the cental metal ion. The bonding between a central metal ion and a ligand is to some extent, due to the donation of electrons by the ligand to the central ion. Thus, it might be concluded that a central ion with higher electronegativity (*i.e.*, a central ion with a strong electron-attracting tendency) would give more stable complexes. Similarly, the central metal ion with high polarising power gives more stable complexes. For example, complexes given by Cu^+ and Cd^{2+} ions are more stable than those given by Na^+ and Ca^{2+} ions respectively.

5. Ionisation energy of the central metal ion. Generally the stability of complexes given by M^{2+} ions increases with the increase of first plus second ionisation potential for M^{2+} ions.

6. Class *a* and class *b* metals. Chatt and Ahrland have classified the metals into three categories: *a*, *b* and borderline, on the basis of their electron-acceptor properties. This classification is shown below (normal valence states are assumed).

(*a*) *Class a metals:* H, the alkali and alkaline earth metals, the elements Sc $\rightarrow$ Cr, Al $\rightarrow$ Cl, Zn $\rightarrow$ Br, In, Sn, Sb and I, the lanthanides and actinides

(*b*) *Class b metals:* Rh, Pd, Ag, Ir. Pt, Au, Hg

(*c*) *Borderline metals:* The elements Mn $\rightarrow$ Cu, Ti $\rightarrow$Po , Mo, Te, Ru, W, Re, Os, Cd.

Class *a* metals form more stable complexes with ligands having the coordinating atoms from the second period elements (*e.g.*, N, O, F) than those of an analogous ligand in which the donor atom is from third or later period (*e.g.*, P, S, Cl). Class *b* metals have the relative stabilities reversed. If the ligand contains the heavier donor atoms, class *a* and *b* metals are characterised by the stability order:

$$\left.\begin{array}{l} F^- > Cl^- > Br^- > I^- \\ O >> S > Se > Te \\ N >> P > As > Sb > Bi \end{array}\right\} \text{Class } a \text{ metals}$$

$$\left.\begin{array}{l} F^- < Cl^- < Br^- < I^- \\ O << S \simeq Se \simeq Te \\ N << P < As < Sb < Bi \end{array}\right\} \text{Class } b \text{ metals}$$

Class *b* metals are characterised by the presence of a number of *d*-electrons beyond an inert gas core. These *d*-electrons are used to form π-bond with ligand atoms. It is believed that the stability of the complexes of class *b* metals results from covalent contribution to metal–ligand bonds and from the transfer of electron density from the metal to the ligand via π-bonding. The most stable complexes of class *b* metals are formed with ligands like PMe_3, S^{2-} and I^- which have vacant *d*-orbitals or like CO, CN^- which have vacant molecular orbitals of low energy.

For *borderline metals* the stability constants do not display either class *a* or class *b* behaviour uniquely.

B. Properties of the ligand

The following properties of the ligand affect the stability of the metal complexes.

1. Size and charge of ligand. If a ligand is smaller, it can approach the metal ion more closely forming a stable bond. Similarly, a highly charged ligand would also form a strong bond with the metal. Thus the high charge and small size of a ligand leads to the formation of stable complexes. For example the stability of the complexes of a given metal ion with halide ions used as ligands is in the order: $F^- > Cl^- > Br^- > I^-$. This order is applicable for class *a* metals. When class *b* metals (*e.g.* Pd, Ag, Pt, Hg etc.) are used, the above order of stability is reversed, *i.e.*, for class *b* metals the order is : $F^- < Cl^- < Br^- < I^-$.

2. Dipole moment of ligands. For neutral ligands, the larger the magnitude of permanent dipole moment, the greater is the stability of the complexes. For example, the order of stability of complexes formed by some neutral ligands is as : ammonia > ethylamine > diethylamine > triethylamine.

3. Basic character of ligands. The more basic is the ligand, more easily it can denote electron pairs to the central ion and hence more easily it can form complexes of greater stability. The ligands that bind H^+ firmly form stable complexes with metal ions. Thus F^- should form more stable complexes than Cl^-, Br^- or I^-, and NH_3 should be better ligand than H_2O which in turn should be better than HF. ($NH_3 > H_2O > HF$). This behaviour is observed for alkali, alkaline earth and other electropositive metals like first row transition elements, lanthanides and actinides.

4. π -bonding capacity of ligands. The ligands like CN^-, CO, PR_3, AsR_3, SR_2, alkenes, alkynes which are capable of forming π -bonds with transition metal ions give more stable complexes.

5. Steric hindrance. When a bulky group is either attached to or is present near a donor atom of a ligand, repulsion between the donor atom of the ligand and the bulky group is produced and this mutual repulsion weakens the metal-ligand bonding and hence makes the complex less stable. For example the complex of Ni^{2+} ion with 2-methyl-8-hydroxy quinoline ($\log_{10} \beta = 17.8$) is less stable than that with 8-hydroxy quinoline ($\log_{10} \beta = 17.8$). The effect of the presence of bulky group on the stability of a complex is commonly called steric hindrance.

C. Stability of chelates

See "*Factors affecting the stability of chelates*" discussed on page 32.

SOME TYPICAL PROBLEMS WITH SOLUTIONS

Problem 1. *Out of the following pairs of complexes which one is more stable? Also given reason for your answer.*

(*i*) $K_4[Fe(CN)_6]$ *and* $K_3[Fe(CN)_6]$

(*ii*) $[Co(H_2O)_6]^{2+}$ *and* $[Co(NH_3)_6]^{2+}$

(*iii*) $[Cu(en)_2]Cl_2$ *and* $[Cu(NH_3)_4]Cl_2$

(*iv*) $[Co(NO_2)_6]^{4-}$ *and* $[Co(NO_2)_6]^{3-}$ (*Delhi 2000*)

Solution. (*i*) We know that for a given ligand, the stability of complexes of the metallic ions having different charges decreases with the decrease of the charge on them. Thus, since Fe has +2 charges in $K_4[Fe(CN)_6]$ and +3 charges in $K_3[Fe(CN)_6]$, $K_4[Fe(CN)_6]$ is less stable than $K_3[Fe(CN)_6]$.

(*ii*) Both the complex ions have Co in the same oxidation state (= +2). Here since NH_3 ligand is more basic than H_2O ligand, NH_3 will donate electron pair to the central Co^{2+} ion more easily than H_2O and hence will form more stable complex. Thus $[Co(NH_3)_6]^{2+}$ ion is more stable than $[Co(H_2O)_6]^{2+}$ ion.

(*iii*) We know that since *bis* (ethylene diamine) copper (II) chloride, $[Cu(en)_2]Cl_2$ has two 5-membered rings in its structure, it is a chelated (cyclic) complex. On the other hand, tetrammine copper (II) chloride, $[Cu(NH_3)_4]Cl_2$ is a non-chelated complex. Thus, being chelated complex, $[Cu(en)_2]Cl_2$ is more stable than $[Cu(NH_3)_4]Cl_2$ (Also see the figure geven below)

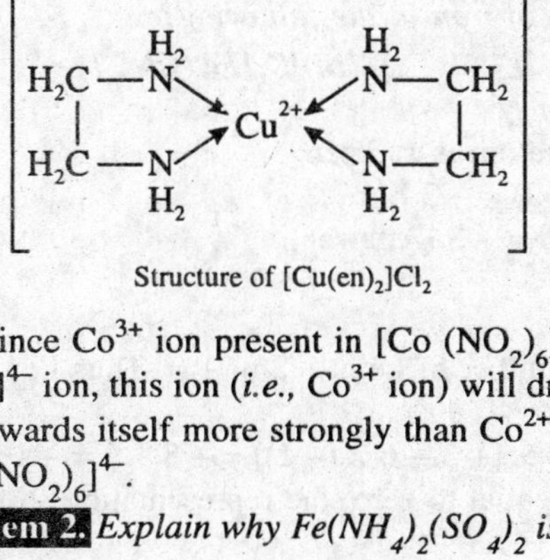

Structure of $[Cu(en)_2]Cl_2$

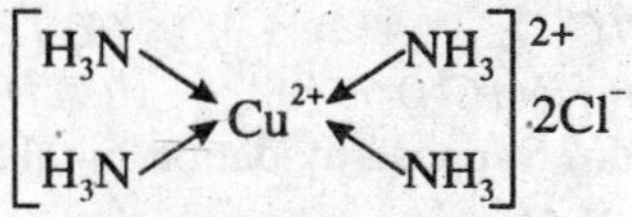

Structure of $[Cu(NH_3)_4]Cl_2$

(*iv*) Since Co^{3+} ion present in $[Co(NO_2)_6]^{3-}$ ion is smaller in size than Co^{2+} ion present in $[Co(NO_2)_6]^{4-}$ ion, this ion (*i.e.*, Co^{3+} ion) will draw the lone pair of electrons present on NO_2^- ion (ligands) towards itself more strongly than Co^{2+} ion and hence $[Co^{3+}(NO_2)_6]^{3-}$ ion is more stable than $[Co^{2+}(NO_2)_6]^{4-}$.

Problem 2. *Explain why* $Fe(NH_4)_2(SO_4)_2$ *is a salt but* $K_4[Fe(CN)_6]$ *is a complex compound.* (*M.D. Rohtak 97*)

Solution. $Fe(NH_4)_2(SO_4)_2$ is a salt, since its solution gives the reactions of all its constituent ions namely Fe^{2+}, NH_4^+ and SO_4^{2-} ions. On the other hand, $K_4[Fe(CN)_6]$ is not a salt, since its solution does not give the reactions of K^+, Fe^{2+} and CN^- ions. This compound gives the reactions of K^+ (simple ion) and $[Fe(CN)_6]^{2-}$ (complex ion) ions only.

Problem 3. *What are the differences between double salts and complex compounds?*

Solution. When a double salt is dissolved in water, it gets dissociated completely to give all its constituent ions. For example aqueous solution of Mohr's salt, $FeSO_4.(NH_4)_2SO_4.6H_2O$ which is a double salt gives all its constituent ions as shown below:

$$FeSO_4.(NH_4)_2SO_4.6H_2O \longrightarrow Fe^{2+} + 2SO_4^{2-} + 2NH_4^+ + 6H_2O$$

On the other hand, when a complex compound is dissolved in water, it does not give all its constituent ions. For example aqueous solution of $K_4[Fe(CN)_6]$ which is a complex compound does not give K^+, Fe^{2+} and CN^- ions. Rather it gives K^+ and $[Fe(CN)_6]^{4-}$ (complex ion) ions

$$K_4[Fe(CN)_6] \longrightarrow 4K^+ + [Fe(CN)_6]^{4-}$$

Problem 4. *$FeSO_4$ solution mixed with $(NH_4)_2SO_4$ solution (in the molar ratio of 1 : 1) gives the test of Fe^{2+} ion but $CuSO_4$ solution mixed with liquid NH_3 (in the molar ratio of 1 : 4) does not give the test of Cu^{2+}. Explain why?*

Solution. When $FeSO_4$ solution is mixed with that of $(NH_4)_2SO_4$, we get a double salt, $FeSO_4.(NH_4)_2SO_4.6H_2O$ (called Mohr's salt). This salt dissociates completely in aqueous solution to give all its constituent ions.

$$FeSO_4.(NH_4)_2SO_4.6H_2O \longrightarrow Fe^{2+} + 2NH_4^+ + SO_4^{2-} + 6H_2O$$

Thus the aqueous solution of $FeSO_4 . [NH_4)_2SO_4.6H_2O$ gives the test of Fe^{2+} ion.

When $CuSO_4$ solution is mixed with liquid NH_3, we get a complex compound, $[Cu(NH_3)_4]SO_4$ which dissociates to give $[Cu(NH_3)_4]^{2+}$ and SO_4^{2-} ions

$$[Cu(NH_3)_4]SO_4 \longrightarrow [Cu(NH_3)_4]^{2+} + SO_4^{2-}$$

Thus since Cu^{2+} ions are not obtained in the free state, the aqueous solution of $[Cu(NH_3)_4]SO_4$ does not give the test of Cu^{2+} ions.

Problem 5. *Explain why Be^{2+} ion has much stronger tendency to form complexes than Mg^{2+} ion.* *(Meerut 1980 S)*

Solution. Smaller ion has greater tendency to attract the lone pair of electrons on the ligands and hence forms the complex compound more readily than the larger ion. Now since Be^{2+} ion is smaller in size than Mg^{2+} ion, the former accepts the lone pair of electrons donated by F^- ion (ligand) and forms the complex ion, $[BeF_4]^{2-}$. On the other hand, since Mg^{2+} ion is larger in size than Be^{2+} ion, this ion does not form $[MgF_4]^{2-}$ ion.

Problem 6. *What is the oxidation number of iron in the following?*

(i) $Fe(CO)_5$ *(ii)* $[Fe(C_2O_4)_3]^{3-}$ *(iii)* $K_3[Fe(CN)_6]$

(iv) $Fe_4[Fe(CN)_6]_3$ *(v)* $[Fe(C_5H_5)_2]$

Solution. Suppose oxidation number of Fe atom is x. Then:

(i) $x + 5 \times 0 = 0$ *(ii)* $x + 3 \times (-2) = -3$

$\therefore x = 0$ **(Answer)** $\therefore x = -3 + 6 = +3$ **(Answer)**

(iii) $3 \times (+1) + x + 6 \times (-1) = 0$

$\therefore x = -3 + 6 = +3$ **(Answer)**

(iv) We know that the common oxidation number of Fe is + 2 or + 3. Thus $Fe_4[Fe(CN)_6]_3$ can be represented as shown at (*a*), (*b*), (*c*) or (*d*).

(*a*) $Fe_4^{2+}[Fe^{3+}(CN)_6]_3$. Here : $4 \times (+2) + 3 \times [+3 + 6 \times (-1)] = +8 - 9 = -1$

Since the sum of oxidation numbers is not equal to zero, the representation shown at (*a*) is wrong

(*b*) $Fe_4^{2+}[Fe^{2+}(CN)_6]_3$. Here : $4 \times (+2) + 3 \times [+2 + 6 \times (-1)] = +8 - 12 = -4$

Since the sum of oxidation numbers is not equal to zero, the representation shown at (*b*) is also wrong.

(*c*) $Fe_4^{3+} [Fe^{3+} (CN)_6]_3$. Here : $4 \times (+3) + 3 \times [+3 + 6 \times (-1)] = +12 - 9 = +3$

Since the sum of oxidation numbers is not equal to zero, the representation shown at (*c*) is also wrong.

(*d*) $Fe_4^{3+} [Fe^{2+} (CN)_6]_3$. Here : $4 \times (+3) + 3 \times [+2 + 6 \times (-1)] = +12 - 12 = 0$

Here since the sum of oxidation numbers is equal to zero, the representation shown at (*d*) is correct. Thus :

Fe atom written outside the coordination sphere has oxidation number = + 3 and Fe atom given inside the coordination sphere has oxidation number = + 2. (**Answer**)

(*v*) In $[Fe (C_6H_5)_2]$, the ligand is cyclopentadienyl ion, $C_5H_5^-$ whose O.N. = – 1.

Thus : $x + 2 \times (-1) = 0 \quad \therefore x = +2$ (**Answer**)

University Question

1. (*a*) What is meant by thermodynamic and kinetic stability of complexes ? Explain in short the different factors affecting the stability of complexes in solution.

   (*b*) Define EAN and calculate the value of EAN of : (*i*) $[PtCl_6]^{2-}$ (Z = 78)

   (*ii*) $[Pd(NH_3)_6]^{4+}$ (Z = 46)

   (*c*) Give any two industrial applications of chelates. *(Nagpur 2002)*

2. A complex compound with the formula, $M (CrO_4) Cl_2 (NH_3)_2$ forms two types of coloured crystals *viz.* red A and blue B. A (red) reacts with 1 mole of Ag_2CrO_4 to give white ppt. While B (blue) reacts with $AgNO_3$ to give red ppt. From these observations find (*i*) Coordination number of M (*ii*) Hybrid orbitals on M (*iii*) Stereo chemistry of red blue forms. *(Delhi 2002)*

Ans. Reactions can be showsn as :

$$[M^{4+}(NH_3)_2(CrO_4)]Cl_2 + Ag_2CrO_4 \longrightarrow 2AgCl + [M^{4+}(NH_3)_2(CrO_4)]CrO_4$$

A (red) White ppt

$$[M^{4+}(NH_3)_2Cl_2]CrO_4 + 2AgNO_3 \longrightarrow Ag_2CrO_4 + [M^{4+}(NH_3)_2Cl_2](NO_3)_2$$

B (blue) Red ppt

Obviously : (*i*) Coordination number (C. N.) of M = 4 (*ii*) Since C. N. of M is 4, the hybrid orbitals on M may be dsp^2 (square planar) or sp^3 (tetrahderal).

(*iii*) The stereochemistry may be square planer (dsp^2) or tetrahedral (sp^3).

3. What is oxidation state of Fe in $[Fe(H_2O)_5(NO)]SO_4$ *(VBS Purvanchal 2003)*

4. (*i*) How many primary and secondary valencies are there for Fe in $K_3[Fe(C_2O_4)_3]$

   (*ii*) Define : (*a*) Ligand (*b*) Coordination number. *(Rohilkhand 2003).*

5. Explain why no ligand is a three-electron donor. *(Rohilkhand 2003).*

6. Explain why stability constant value increases in the order :

   $[Cu(NH_3)_4]^{2+} < [Cu(en)_2]^{2+} < [Cu(trien)]^{2+}$. *(Nagpur 2003).*

7. (*a*) Explain the following :

   (*i*) Complexion (*ii*) Ligend

   (*iii*) Double salt (*iv*) Co-ordination number

   (*b*) Write down the name and structural formula of bidentate and tridentate ligand. *(Bhopal 2004)*

8. (*a*) Differentiate between a ligand and a chelate.

   (*b*) Differentiate between a double salt and a coordination compound. *(GND Amritsar, 2004)*

9. Write the structure of the complex compounds formed in the following reactions :

   (*i*) Ni^{2+} + Dimethyl glyoxime $\longrightarrow$ .....

   (*ii*) Ca^{2+} + Disodium salt of EDTA. $\longrightarrow$ ..... *(GND Amritsar, 2004)*

2

Nomenclature of Coordination Compounds

Coordination compounds are named according to the rules suggested by *International Union of Pure and Applied Chemistry, IUPAC* (1976). These rules are given below:

1. Order of naming cations and anions of an ionic complex compound

(*i*) If a complex compound is ionic, *i.e.*, if a complex compound is composed of cations and anions, the name of the cation is mentioned first and then the name of the anion is written. For example:

(*a*) In naming $K_2[PtCl_6]$, the name of the cation, *viz.* K^+ is written first and then the name of the anion namely $[PtCl_6]^{2-}$ is mentioned.

(*b*) In naming $[Co(NH_3)_5Cl]Cl_2$, the name of the cation, $[Co(NH_3)_5Cl]^{2+}$ is written first and then the name of the anion, Cl^- is written.

(*c*) In $[Pt^{4+}(NH_3)_4Cl_2]^{2+}$ $[Pt^{2+}Cl_4]^{2-}$, the name of the cation, $[Pt^{4+}(NH_3)_4Cl_2]^{2+}$ is written first and then the name of the anion, $[Pt^{2+}Cl_4]^{2-}$ is mentioned.

(*ii*) If the complex compound is non-ionic (neutral), the name of the complex compound is written as one word. Thus the name of $[Pt(NH_3)_4Cl_2]^0$ is written as one word.

2. How to name the species present in coordination sphere

In naming the species present in coordination sphere, the ligands are named first and then the name of the central metal atom is mentioned.

3. How to name the ligands?

(*i*) **How to name neutral ligands?** See Table 1.1, page 5 (chapter 1).

(*ii*) **How to name negative (anionic) ligands?** The names of negative ligands end in *o*. Examples of such ligands have already been given in Table 1.1. on page 5 (chapter 1).

(*iii*) **How to name positive ligands?** The names of positive ligands which are rare, end in *ium*. Examples of some positive ligands are given in Table 2.1.

(*iv*) **How to name free radicals?** Free radicals are given their own names. These are supposed to be neutral species, *i.e.*, they donot contain any charge on them. Examples of some free radicals are given in Table 2.1.

(*v*) **How to indicate the number of ligands.** (*a*) If a complex contains two or more simple ligands of the same type (*e.g.*, Cl^-, Br^-, NO_2^-, $C_2O_4^{2-}$, NH_3, H_2O etc.), their number is indicated by putting prefixes *di* (for 2), *tri* (for 3), *tetra* (for 4), *penta* (for 5), *hexa* (for 6), *septa* (for 7), *octa* (for 8), *nona* (for 9), *deca* (for 10), *undeca* (for 11), *dodeca* (for 12) etc. before their names.

(*b*) If a complex contains two or more complex ligands or organic molecules which already have *di*, *tri* etc. in their names, their number is indicated by putting the prefixes *bis* (for 2), *tris* (for 3), *tetrakis* (for 4), *pentakis* (for 5), *hexakis* (for 6), *hepta* (for 7), *octa* (for 8), *nona* (for 9), *deca* (for 10), *undeca* (for 11), *dodeca* (for 12) etc. before their names. The name of the complex ligand or organic molecule is written in bracket. Examples of complex ligands are ethylene diamine (*en*), diethylamine, phosphorus trifluoride, acetylacetonato ion, benzoyl acetonate ion, dipyridyl, trichloro stannyl ion ($SnCl_3^-$), pyridine, methyl amine, triphynyl phosphine etc.

Table 2.1. *Examples of positive ligands and free radicals.*

Positive (cationic) ligands. The names of positive ligands end in *ium*, *e.g.*			
NH_4^+	..... Ammonium	O_2^+	..... Oxygenyl
H_3O^+	..... Hydronium	NO_2^+	..... Nitronium
NO^+	..... Nitrosonium (also called nitrosyl, nitrolium or nitroso)	$[H_3N—NH_2]^+$	.... Hydrazinium
Free radicals. Free radicals are given their own names, *e.g.*,			
CH_3	..... Methyl	C_2H_5	..... Ethyl
C_6H_5	..... Phenyl	CH_3CN	..... Acetonitrile

(*vi*) **How to name bridging ligands.** (*a*) If a complex has bridging ligands, the prefix μ (mu) is used before the name of each of the ligands. For example

$$[(NH_3)_5Co—NH_2—Co(NH_3)_4(H_2O)]Cl_5$$

is called pentammine cobalt (III) μ-amido tetrammine aquo cobalt (III) chloride or nonammine aquo (μ-amido) dicobalt (III) chloride.

(*b*) If a complex has two similar bridging ligands, μ-*di* is used for its name. For example:

$$\left[(H_2O)_4\overset{3+}{Fe}\begin{matrix} OH^- \\ OH^- \end{matrix}\overset{3+}{Fe}(H_2O)_4\right](SO_4)_2$$

is called tetraquo iron (III) μ-dihydroxo tetraquo iron (III) sulphate *or* octaquo μ-di-hydroxo di-iron (III) sulphate.

(*c*) When the same ligand is present as a bridging ligand and as a non-bridging (terminal) ligand, the bridging ligand is named first. In case of the presence of three similar bridging ligands, μ-tri is used before their name. For example

$$\left[(CO)_3Fe\begin{matrix} CO \\ CO \\ CO \end{matrix}Fe(CO)_3\right]$$

is called μ-tricarbonyl hexacarbonyl di-iron (0).

(*d*) If the bridging ligand bridges more than two metal atoms, μ_3, μ_4 etc. is used before the name of bridging ligand. μ_3, μ_4 etc. indicates the number of metal atoms between which the bridging ligand makes bridges, *e.g.* since in basic berylium acetate, $Be_4^{2+}O^{2-}(CH_3COO)_6$, *oxo* ($O^{2-}$) group briges four Be-atoms, in the name of this compound μ_4 is used before the name of bridging ligand *viz.* O^{2-} group. Thus, the name of this compound is μ_4–*oxo*-μ-hexa acetato tetra beryllium (II).

(*vii*) **How to name ambidentate ligands.** See *Ambidentate ligands and linkage isomerism discussed* on page 18 (chapter 1).

(*viii*) **Order of naming ligands.** If the coordination sphere of a given complex compound contains various types of ligands, these ligands are named in alphabetical order. The prefixes di, tri etc. are not considered while determining this alphabetical order. For example $[Co(NH_3)_4Cl(NO_2)]^+$ ion is named as tetraaminechloronitrocobalt (III) ion.

(*ix*) **Use of abbreviations for complicated ligands.** Instead of writting the long names of complicated ligands. abbreviations are customarily used. Abbreviations of some ligands have already been given in chapter 1.

4. How to name the central metal atom

The following two cases arise:

(*i*) (*a*) If the coordination sphere of the complex compound has **negative charge**, the name of the central metal atom ends in **ate** and the oxidation state of the metal (whether positive, negative or zero) is written in Roman numerals (0, I, II, III... –I, –II, –III etc.) in bracket at the end of the name of the metal atom. The names of some metals are given below:

Cr ...... Chromate
Co ...... Cobaltate
Cu ...... Cuperate
Pd ...... Palladate
Re ...... Rhenate
Ir ...... Iridate
Os ...... Osmate
Hg ...... Mercurate

Cd ...... Cadmate
Zn ...... Zincate
Si ...... Silicate
Ni ...... Nicklate
Al ...... Aluminate
B ...... Borate
Pt ...... Platinate

The names of the following complex compounds illustrate the nomenclature of the central metallic atom.

$Na_2[CrOF_4]$ ...... Sodiumtetrafluorooxo**chromate (IV)**
$K_4[Ni(CN)_4]$ .... Potassiumtetracyano**nickelate (0)**
$[Co(CO)_4]^-$ .... Tetracarbonyl**cobaltate** (–I)

(*b*) If the complex compound having the negatively-charged coordination sphere is an acid, the name of the metal ends in **ic.** For example $H_2[PtCl_6]$ is called hexa chloro platin**ic** acid (IV).

(*c*) For some metals, their Latin names are used. For example:

Fe ...... Ferrate
Ag ...... Argentate
Sn ...... Stannate

Pb ...... Plumbate
Au ...... Aurate

The names of the following complex compounds illustrate the nomenclature of the central metal atom.

$K_4[Fe(CN)_6]$ ...... Potassiumhexacyno**ferrate (II)**
$Na_3[AgF_4]$ ...... Sodiumtetrafluoro**argentate (I)**

(*ii*) If the coordination sphere of the complex compound has **positive charge** or is **neutral** (*i.e.*, non-ionic), the name of the central metal atom remains as such and the oxidation state of the metal is written in Roman numerals in bracket at the end of the name of the metal.

The following examples illustrate this point:

$[Ag(NH_3)_2]^+$ ...... Diammine **silver (I)** ion
$[Ni(CO)_4]^0$ ...... Tetracarbonyl **nickel (0)**

5. How to name cis- and trans- isomers (geometrical isomers) of square planar and octahedral complexes

See "*Geometrical isomerism in square planar and octahedral complexes*" discused in chapter 4.

6. How to name optical isomers

We know that *dextro* and *levo* forms are optically active isomers while *meso* form in optically inactive. Dextro form is represented by putting *d* or (+) before its name or formula. Similarly levo form is designated by putting *l* or (–) before its name or formula. Meso form is called *meso* isomer. Thus dextro potassium trioxalatoiridate (III) is formulated as *d* or (+) $K_3[Ir^{3+}(C_2O_4)_3]$.

7. How to name complex compounds containing optically active ligands

See "*Optical isomerism in octahedral complexes containing active ligands*" discussed in chapter 4.

Solved Examples

Example 1. *Write the names of the complex compounds/ions having the following formula*

(i) $[Fe(H_2O)_5(NO)]^{2+}$ (ii) $Na_2[Fe(CN)_5(NO)]$

(iii) $[Co(N_3)(NH_3)_5]SO_4$ (iv) $K_2[OsNCl_5]$

(v) $K_2[Cr(CN)_2(O)_2(O_2)(NH_3)]$ (vi) $K[CrF_4O]$

(vii) $Na[BH(OCH_3)_3]$ (viii) $[Pt(NH_3)_4Cl_2][PtCl_4]$

(ix) $[Be(CH_3-CO-CH-CO-C_6H_5)_2]^0$ (x) $K_4[Ni(CN)_4]$

(xi) 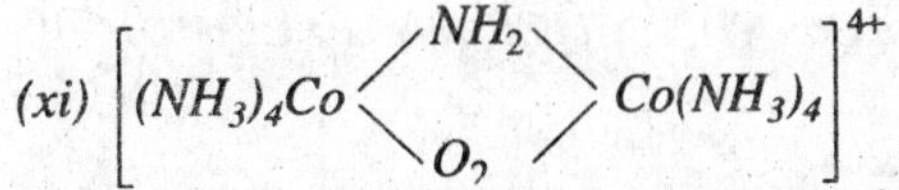

(xii) $H[AuCl_4]$ (xiii) $H[BF_4]$

Solution. (*i*) In $[Fe(H_2O)_5(NO)]^{2+}$ ion, Fe is in +1 oxidation state and NO is present as NO^+. +1 oxidation state of Fe ($Fe^+ = 3d^6 4s^1 = 3d^7$, $n = 3$) is confirmed by the fact that $[Fe(H_2O)_5(NO)]^{2+}$ ion is paramagnetic corresponding to the presence of three unpaired electrons ($\mu = 3.9$ B.M.). This ion results from sp^3d^2 hybridisation (high spin complex ion). Consequently the name of the given complex ion is *pentaqua nitrosyl iron* (I).

(*ii*) In $Na_2[Fe(CN)_5(NO)]$, Fe is in +2 oxidation state ($Fe^{2+} = 3d^6$, $n = 0$) and NO is present as NO^+ ion. +2 oxidation state of Fe is confirmed by the fact that $[Fe(CN)_5(NO)]^{2-}$ ion is diamagnetic ($n = 0$). Thus this ion results from d^2sp^3 hybridisation (low spin complex ion). Consequently the name of the given complex compound is *sodium pentacyano nitrosyl ferrate (II)*.

(*iii*) Pentammine azido cobalt (III) sulphate.

(*iv*) Potassium pentachloro nitrido osmate (VI).

(*v*) Potassium ammine dicyano dioxo peroxo chromate (VI).

(*vi*) Potassium tetrafluoro oxo chromate (V).

(*vii*) Sodium hydrotrimethoxo borate (III)

(*viii*) Tetrammine dichloro platinum (IV) tetra chloro platinate (II).

(*ix*) Bis-(benzoylacetonato) beryllium (II).

(*x*) Potassium tetracyano nicklate (0).

(*xi*) Tetrammine cobalt (III) μ-amido -μ-superoxo tetrammine cobalt (III) iron *or* octammine μ-amido -μ-superoxo dicobalt (III) ion.

(*xii*) Tetrachloro auric acid (III).

(*xiii*) Tetrafluoro boric acid (III).

Example 2. *Write the formulae of the following : (i) Tris (ethylene diamine) cobalt (III) nitrate (ii) Potassium hexacyano nickelate (0) (iii) Sodium tetrafluoro argentate (I).*

(Meerut 1994)

Solution. (*i*) $[Co(en)_3](NO_3)_3$ (*ii*) $K_6[Ni(CN)_6)$ (*iii*) $Na_3[AgF_4]$.

Example 3. *We are given three complex compounds namely $[Cr(H_2O)_6)Cl_3$, $[Cr(H_2O)_5Cl]Cl_2.H_2O$ and $[Cr(H_2O)_4Cl_2]Cl.2H_2O$. Answer the following questions (i) What is the oxidation number of Cr-atom in each complex compound (ii) Write down the coordination number of the central Cr-atom in each case (iii) Write the name of each complex compound.*

Solution. (*i*) Let the oxidation number of Cr-atom be equal to *x*. Then:

in $[Cr(H_2O)_6]Cl_3$ $x + 6 \times 0 = +3$ $\therefore$ $x = +3$

in $[Cr(H_2O)_5Cl].Cl_2.H_2O$ $x + 5 \times 0 - 1 = +2$ $\therefore$ $x = +3$

in $[Cr(H_2O)_4Cl_2]Cl_2.2H_2O$ $x + 4 \times 0 - 2 \times 1 = +1$ $\therefore$ $x = +3$

Thus oxidation number of Cr-atom in each complex compound is equal to + 3.

(*ii*) Since the number of ligands surrounding the central Cr^{3+} ion in each complex compound is six, coordination number of Cr^{3+} ion in complex compound is equal to 6.

(*iii*) The names of the given complex compounds are hexaquochromium (III) chloride, pentaquochlorochromium (III) chloride monohydrate and tetraquodichlorochromium (III) chloride dihydrate respectively.

SOME TYPICAL PROBLEMS WITH SOLUTIONS

Problem 1. *Write the names of the following complex compoundis according to IUPAC system of nomenclature.*

(A) (i) Ca $[Fe(CN)_6]$ (ii) $[Fe(H_2O)_6]SO_4$ (iii) $[Co(NH_3)_5Cl]Cl_2$ (iv) $[PtCl_2(NH_3)_2]$ *(Delhi 1996)*

(B) (i) $\left[(C_2O_4)_2Cr\begin{matrix}OH\\ \\OH\end{matrix}Cr(C_2O_4)_2\right]^{4-}$ *(ii) $[Pt(py)_4)][PtCl_4]$ (iii) $[Ir(en)_2Cl_2]^+$*

(iv) $[Ni(NH_3)_2(en)](CH_3COO)_2$. *(Delhi 1997)*

(C) (i) $[PdI_2(ONO)_2(H_2O)_2]^{2-}$ (ii) $[Co(en)_3]_2(SO_4)_3$ (iii) $(CH_3)_4N[Cr(SCN)_4(NO)_2]$

(iv) Na $[Au(CN)_2)$ *(Delhi 1998)*

(D) (i) $(CH_3)_4N[Cr(NH_3)_2(NCS)_4]$ (ii) $[Co(en)_2(ONO]Cl]Cl$

(iii) $\left[(NH_3)_4Co\begin{matrix}NH_2\\ \\OH\end{matrix}Co(NH_3)_4\right](NO_3)_4$ *(iv) $[Cr(NH_3)_6][CoF_6]$ (Delhi 1999)*

(E) (i) $[Co(NH_2)_2(NH_3)_4]OC_2H_5$ (ii) $[(NH_3)_5Cr\text{–}OH\text{–}Cr(NH_3)_5]Cl_5$

(iii) $[Co(NH_3)_5(ONO)]SO_4$ (iv) $[Cr(en)_3][Ni(CN)_5]$ (v) $[VO(acac)_2]$ *(Delhi 2000)*

(F) (i) $[Rh(NCS)_3(NH_2)_3]$ (ii) $[(CH_3)_4N]_3[Cr(C_2O_4)_3]$ (iii) $[Co(NH_3)_2(H_2O)_2(CN)_2)$ Cl.

(iv) $[Fe(NH_3)_6][Fe(CN)_6]$ (v) $[Br_2PdBr_2PdBr_2]^{2-}$ *(Delhi 2001)*

(G) (i) $[Co(NH_3)_5ONO]Cl_2$ (ii) $K_3[Cr(CN)_6]$ (iii) $[Cr(NH_3)_5CO_3]Cl$

(iv) $[Cr(NH_3)_5(NCS)][ZnCl_4]$ *(IIT 1995, 96, 97)*

(H) (i) Linkage isomer of $[Ni(NH_3)_2(H_2O)(NO_2)]Br$

(ii) Linkage isomer of $[Cr(H_2O)_5(SCN)]^{2+}$ (iii) $K[BF_4]$ (iv) $[Fe(C_5H_5)_2]$ (v) $Mn_3(CO)_{12}$.

Solutions. (A) (*i*) Calcium hexacyanoferrate (II) (ii) Hexaaqua iron (II) sulphate (*iii*) Pentammine chlorocobalt (II) chloride (iv) Diamminedichloroplatinum (II)

(B) (*i*) Dioxalato chromium (III)-μ-dihydroxo dioxalato chromium (III) ion or tetraoxalato μ-dihydroxo dichromium (III) ion. (*ii*) Tetrapyridine platinum (II) tetrachloro platinate (II) (*iii*) Bis (ethylene diamine) dichloro iridium (III) ion. (*iv*) Diammineethylene diammine nickel (II) acetate.

(C) (*i*) Diaquadiiododinitrito palladate (II) ion (*ii*) *Tris* (ethylene diamine) cobalt (III) sulphate (*iii*) Tetramethylammonium tetrathiocyanatodinitrosyl chromate (III) (*iv*) Sodium dicyanoaurate (I).

(D) (*i*) Tetramethylammomiundiammine tetrathiocyanato chromate (III) (*ii*) Dichloro *bis* (ethylene diamine) nitrito cobalt (III) chloride (*iii*) Tetrammine cobalt (III) μ-amido-μ-hydroxo tetraammine cobalt (III) nitrate (iv) Hexaammine chromium (III) hexafluorocobaltate (III).

(E) (*i*) Diamidotetraammine cobalt (III) ethoxide (*ii*) Pentaammine chromium (III) -μ-hydroxopentaamine chromium (III) chloride (iii) Pentaamine nitritocobalt (III) sulphate (*iv*) *Tris* (ethylenediamine) chromium (III) pentacyanonicklate (II). (*v*) *Bis* (acetylacetonato) vanadyl (II).

(F) (*i*) Triisothiocyanatotrioxalate rhodium (VI) (*ii*) Tetramethyl ammonium tri-oxalato chromate (III) (*iii*) Diammine diaqua dicyano cobalt (III) chloride (*iv*) Hexaammine iron (III) hexacyanoferrate (III) (*v*) Dibromo palladium (II) μ-dibromodibromo palladium (II) ion or tetrabromo μ-dibromodipalladium (II) ion.

(G) (*i*) Pentaammine nitrito cobalt (III) chloride (*ii*) Potasium hexacyano chromate (III) (*iii*) Pentaamine carbonato chromium (III) chloride (*iv*) Pentaammine isothiocyanato chromium (III) tetrachlorozincate (II)

(H) (*i*) Tetrammineaqua nitrito nickel (II) bromide, $[Ni (NH_3)_2 (H_2O) (ONO)] Br$ (*ii*) Pentaaqua isothiocyanato chromium (II) iron, $[Cr (H_2O)_5 (NCS)]^{2+}$ (*iii*) Potasiumtetrafluoroborate (III) (*iv*) Bis (cyclopentadienyl) ion (II) (*v*) Dodecacarbonyl trimanganese (0).

Problem 2. *Write the formula of the following compounds.*

(A) (i) Dichlorotetraaquachromium (III) nitrate (ii) Potassium tetracyanonickelate (II) (iii) Hexaamine cobalt (III) hexacyanochromate (III) (iv) Tetrachloro mono(ethylenediamine) platinum (IV). (Delhi 1996)

(B) (i) Octacarbonyl-μ-dicarbonyl diiron (0) (ii) Ammonium pentafluoroaquanickelate (IV) (iii) Tetrammineaqua bromo cobalt (III) chloride (iv) Sodium dithiosulphatoagentate (I) (v) Bromodichloro iodopalladate (III) ion. (Delhi 1997)

(C) (i) Tetraammine carbonato cobalt (III) chloride (ii) Calciumhexacyanoferrate (II) (iii) Potassium trioxalatoaluminate (III) (iv) Sodium oxotetrafluorochromate (IV). (Delhi 1998)

(D) (i) Potassium tetrafluoroargentate (I) (ii) Dichloro bis (ethylenediamine) cobalt (II) (iii) Trans-chloronitrodiammine cobalt (II) (iv) Octacarbonyl-μ-dicarbonyl diiron (0) (Delhi 1999)

(E) (i) Cis-dichloro tetraammine cobalt (III) chloride (ii) Diammine silver (I) tetracyanoferrate (III) (iii) Potassium tetraazidocobaltate (III) (Delhi 2000)

(F) (i) Octacarbonyl-μ-dicarbonyl diiron (0) (ii) Sodium dithiosulphatoargentate (I) (iii) Diammine silver (II) tetra chloroplatinate (II). (Delhi 2001)

(G) (i) Pentaammine chlorocobalt (III) ion (ii) Lithiumtetrahydroaluminate (III) (IIT 1997)

(H) (i) Bis(acetylacetonato) oxovanadium (IV) (ii) Dichloroplatinum (IV) tetrachloroplatinate (II)

Solution.

(A) (*i*) $[Cr (H_2O)_4 Cl_2] NO_3$ (*ii*) $K_2[Ni (CN)_4]$ (*iii*) $[Co (NH_3)_6] [Cr (CN)_6]$ (*iv*) $[Pt (en) Cl_4]$

(B) (*i*) $\left[(CO)_4Fe \begin{matrix} CO \\ \\ CO \end{matrix} Fe(CO)_4\right]$ (*ii*) $NH_4[Ni F_5 (H_2O)]$ (*iii*) $[Co (NH_3)_4 Br (H_2O)] Cl_2$

(*iv*) $Na_3[Ag(S_2O_3)_2]$ (*v*) $[Pd\,BrCl_2I]^-$

(C) (*i*) $[Co(NH_3)_4CO_3]Cl$ (*ii*) $Ca_2[Fe(CN)_6]$ (*iii*) $K_3[Al(C_2O_4)_3]$ (*iv*) $Na_2[Cr\,OF_4]$

(D) (*i*) $K_3[Ag\,F_4]$ (*ii*) $[Co(en)_2Cl_2]$ (*iii*) *Trans*-$[Pt(NH_3)_2Cl(NO_2)]$

(*iv*) $\left[(CO)_4Fe\begin{smallmatrix}CO\\ \\CO\end{smallmatrix}Fe(CO)_4\right]$

(E) (*i*) *Cis* $[Co(NH_3)_4Cl_2)]Cl$ (*ii*) $[Ag(NH_3)_2][Fe(CN)_4]$ (*iii*) $K[Co(N_3)_4]$

(F) (*i*) $\left[(CO)_4Fe\begin{smallmatrix}CO\\ \\CO\end{smallmatrix}Fe(CO)_4\right]$ (*ii*) $Na_3[Ag(S_2O_3)_2]$ (*iii*) $[Ag(NH_3)_2][Pt\,Cl_4)$

(G) (*i*) $[Co(NH_3)_5Cl]^{2+}$ (ii) $Li[AlH_4]$

(H) (*i*) $[V(acac)_2O]$ (*ii*) $[PtCl_2][PtCl_4]$

Problem 3. *The valency of Cr in* $[Cr(H_2O)_4Cl_2]^+$ *ion is:*

(*a*) 3 (*b*) 1 (*c*) 6 (*d*) 5 (*MLNR 1993*)

Solution and explanation. Answer is (*a*)

$$x + 4 \times 0 + 2 \times (-1) = +1$$
$$\therefore \quad x = +3$$

Problem 4. *The oxidation number of Pt in* $[Pt(C_2H_4)Cl_3]^-$ *ion is:*

(*a*) + 1 (*b*) + 2 (*c*) + 3 (*d*) + 4

Solution and explanation. Answer is (*b*)

$$x + 4 \times 0 + 3 \times (-1) = -1$$
$$\therefore \quad x = -1 + 3 = +2$$

Problems 5. *The formula of dichlorobis(urea) copper (II) is:*

(*a*) $[Cu\{O = C(NH_2)\}_2]Cl_2$ (*b*) $[CuCl_2\{O = C(NH_2)_2\}]$

(*c*) $[Cu\{O = C(NH_2)_2\}Cl]Cl$ (*d*) $[CuCl_2\{O = C(NH_2)_2\}_2]$

Solution. Answer is (*d*)

Problem 6. *The IUPAC name of* $[Pt(NH_3)_3Br(NO_2)Cl]Cl$ *is:*

(a) Triammine chloro bromonitro platinum (IV) chloride

(b) Triammine chlorobromonitro chloro platinum (IV) chloride

(c) Triammine bromochloronitroplatinum (IV) chloride

(d) Triammine nitro chlorobromoplatinum (IV) chloride

Solution and explanation Answer is (*c*). The names of ligands are written in the alphabetical order.

University Questions

1. Write the IUPAC name of the following compounds/ions.

(A) (*i*) $[Cr(NH_3)_5CO_3]Cl$ (*IIT 96*)

(*ii*) $[Cu(NH_3)_4]SO_4$ (*Meerut 95*)

(*iii*) $[Cr(H_2O)_5Cl]Cl_2$ (*Meerut 95*)

2. Write the formula of the following compounds .

(*a*) Octa aquo m-dihydroxo *di* iron (III) ion.

(*b*) Ammonium hexaisothiocyanato chromate (III).

(c) Ammonium hexathio-cyanato platinate (IV) (*Meerut 97*)

3. Write IUPAC names of the following coordination compounds .

(*i*) $NH_4\left[Cr(NCS)_4(NH_3)_2\right]$ (*ii*) $\left[Co(en)_2(ONO)Cl\right]Cl$

(iii) $[Cr(H_2O)_5 Cl]Cl_2$ (iv) $\left[(NH_3)_4Co \begin{matrix} NH_2 \\ OH \end{matrix} Co(NH_3)_4\right]^{+4}$

4. Write IUPAC name of $[Cr(H_2O)_4 Cl_2]NO_3$ (Nagpur 2002)

5. (a) Write the names of the following complexes as per the IUPAC system :

(i) $[CoCl(NO_2)(NH_3)_4]Cl$ (ii) $[CuCl_2\{O=C(NH_2)_2\}]$

(iii) $Na[PtBrCl(NO_2)(NH_3)]$ (iv) $\left[(NH_3)_4Co \begin{matrix} NH_2 \\ OH \end{matrix} Co(NH_3)_4\right]^{+4}$

(b) Write the formula of the following compounds :

(i) Pentaaquonitrosyl iron (I)

(ii) Sodium pentacyanonitrosyl ferrate (II)

(iii) Tetraammine dichloroplatinum (IV) tetrachloroplatinate (II)

(iv) Potassium hexacyanoferrate (II) *(Delhi 2002)*

6. Write down the IUPAC names of the following :

(i) $[Co(NH_3)_3 NO_2Cl\ CN]$ (ii) $[Cr(NH_3)_6][Co(CN)_6]$

(iii) $\left[(NH_3)_4Co \begin{matrix} NH_2 \\ OH \end{matrix} Co(NH_3)_4\right]^{4-}$ *(Rohilkhand 2003)*

5. (a) Write IUPAC names of the following complexes :

(i) $Na[Au(CN)_4]$ (ii) $[Pt(en)_2(SCN)_2(NO_3)_2]$

(iii) $(NH_4)_3[Co(NO_2)_2(C_2O_4)_2]$ (iv) $[Cr(H_2O)_4 ONOCl]Cl_2$

(v) $K_4[Zn(OH)_2(NCS)_4]$

(b) Write the formulae of the following :

(i) Aquopentammine cobalt (III) nitrate

(ii) Tetrammine platinum (II) tetrachloro platinate (II)

(iii) Potassium hexanitro cobaltate (III)

(iv) Chloroslulfato diammine platinum (I) bromide. *(Delhi 2003).*

7. Write the IUPAC name of the following complex compounds :–

(i) $[Co^{III}(NH_3)_4 Cl_2]Cl$ (ii) $K_2[Pt^{IV}Cl_6]$. *(Nagpur 2003).*

8. Write down the IUPAC name of the following complexes :

(i) $[Pt(py)_4][PtCl_4]$ (ii) $K_4[Ni(CN)_4]$ *(Bhopal 2004)*

9. (a) Write the IUPAC names of the following :

(i) $Fe[Fe(CN)_6]$ (ii) $K_3[Al(C_2O_4)_3]$ (iii) $[Co(en)_3]Cl_3$

(b) Give IUPAC name of $[PtCl_2(C_2H_4)(NH_3)]$. *(GND Amritsar, 2004)*

3

Werner's Theory of Coordination Compounds and Effective Atomic Number (EAN) Rule

Introduction

In 1893 Werner put forward a theory, called Werner's theory of coordination compounds to explain the structure and properties of Co(III) and Pt(IV) ammines. These ammines are given below:

Co(III) ammines. $CoCl_3.6NH_3$ (orange yellow), $CoCl_3.5NH_3.H_2O$ (pink), $CoCl_3.5NH_3$ (violet), *trans*-$CoCl_3.4NH_3$ (green), *cis*-$CoCl_3.4NH_3$ (violet), $CoCl_3.3NH_3$ (blue-green).

Pt (IV) ammines. $PtCl_4.6NH_3$, $PtCl_4.5NH_3$, $PtCl_4.4NH_3$, $PtCl_4.3NH_3$, $PtCl_4.2NH_3$.

Postulates of Werner's Coordination Theory

Different postulates of Werner's coordination theory are given below:

1. Two types of valency shown by the central metal atom in coordination compounds. In complex compounds, the central metal atom exhibits two types of valency *viz. primary valency* (also called *principal valency*) and *secondary valency* (also called *auxiliary valency*). The metal atom always tends to satisfy both of its valencies.

2. Characteristics of secondary valency. (*i*) In modern terminology, the secondary valency of the metallic atom in a complex compound is equal to the **coordination number** of that metal.

(*ii*) Secondary valency of a metal is satisfied either by anions or by neutral molecules alone or by anions and neutral molecules both. For example.

(*a*) In $4KCN.Fe(CN)_2$ or $K_4[Fe(CN)_6]$, the secondary valency of Fe-atom is six and is satisfied by six CN^- ions (anions).

(*b*) In $CoCl_3.6NH_3$ or $[Co(NH_3)_6]Cl_3$, secondary valency of Co-atom is six and is satisfied by six NH_3 molecules (neutral molecules).

(*c*) In $CoCl_3.5NH_3$ or $[Co(NH_3)_5Cl]Cl_2$, the secondary valency of Co-atom is six and is satisfied by five NH_3 molecules (neutral molecules) and one Cl^- ion (anion).

According to modern concept of complex compounds, the species satisfying the secondary valency are called **ligands.**

(*iii*) While writing the structure of a complex compound, the species satisfying the secondary valency and the metal are written inside the coordination sphere. For example in $CoCl_3.4NH_3$, since four NH_3 molecules and two Cl^- ions satisfy the secondary valency of Co-atom, its structure is written as $[CoCl_2.4NH_3]Cl$ or $[Co(NH_3)_4Cl_2]Cl$.

(*iv*) The secondary valencies have directional nature, since the species satisfying the secondary valency (*i.e.*, ligands) are directed towards the fixed positions in space.

(*v*) The number of species satisfying the secondary valency gives a definite geometry to the complex compound. For example, when the number of monodentate species (ligands) which satisfy the secondary valency of the metal in a complex compound is 2, 3, 4 or 6, the geometry of the complex compound would be linear, trigonal planar, tetrahedral or square planar and octahedral respectively. In each of the four Co (III) ammines *viz*., $CoCl_3.6NH_3$, $CoCl_3.5NH_3$, $CoCl_3.4NH_3$ and $CoCl_3.3NH_3$, since the secondary valency of Co-atom is 6, each ammine has octahedral geometry.

(*vi*) The species satisfying the secondary valency cannot be obtained in the free state, when the aqueous solution of the complex compound undergoes ionisation. For example, when the aqueous solution of $CoCl_3.4NH_3$ or $[Co(NH_3)_4Cl_2]Cl$ undergoes ionisation, four NH_3 molecules and two Cl^- ions which satisfy the secondary valency of Co-atom are not obtained in the free state as shown below:

$$[Co(NH_3)_4Cl_2]Cl \longrightarrow [Co(NH_3)_4Cl_2]^+ + Cl^-$$

The above discussion can be concluded by saying that the species satisfying the secondary valency are not ionisable (non-ionisable) and hence cannot be precipitated.

(*vii*) The attachment of the species satisfying the secondary valency with the metal atom is shown by a solid line (–).

3. Characteristics of primary valency. (*i*) In modern terminology, the primary valency of the metallic atom in a complex compound is equal to the oxidation state (or oxidation number) of that metal, *e.g.*, the primary valency of Co-atom in all the four Co(III) ammines is equal to +3.

(*ii*) Primary valency of a metal in a complex compound is always satisfied by *anions*, *e.g.*, the primary valency of Co-atom in each of the four ammines is equal to +3 and is satisfied by three Cl^- ions.

(*iii*) The anions satisfying the primary valency are written outside the coordination sphere while the anions which satisfy both the valencies (dual character) are written inside the coordination sphere. Thus the species satisfyinig primary valency may be present inside and/or outside the coordination sphere.

(*iv*) The species satisfying both the valencies (*i.e.*, the species placed inside the coordination sphere) are directed towards specific directions in space and hence they have directional characteristics. The species which satisfy primary valency and are placed outside the coordination sphere have no directional characteristics.

(*v*) The species satisfying the primary valency do not give any geometry to the complex compound.

(*vi*) The species satisfying the primary valency can be obtained either completely or partially in their free state, when the complex compound undergoes ionisation in aqueous solution.

Examples (*a*) When the aqueous solution of $CoCl_3.6NH_3$ or $[Co(NH_3)_6]Cl_3$ undergoes ionisation, all the three Cl^- ions which satisfy primary valency of Co-atom are obtained in the free state.

$$[Co(NH_3)_6]Cl_3 \longrightarrow [Co(NH_3)_6]^{3+} + 3Cl^-$$

(*b*) Although the primary valency of Co-atom in $CoCl_3.5NH_3$ or $[Co(NH_3)_5Cl]Cl_2$ is satisfied by three Cl^- ions, only two Cl^- ions are obtained in the free state, when the aqueous solution of this ammine undergoes ionisation.

$$[Co(NH_3)_5Cl]Cl_2 \longrightarrow [Co(NH_3)_5Cl]^{2+} + 2Cl^-$$

The above discussion shows that all or some of the species satisfying the primary valencies are ionisable and hence can be precipitated by means of a suitable precipitating agent. For example, in the above ammine two Cl^- ions can be precipitated as AgCl (white ppt.) by adding aqueous solution of $AgNO_3$.

$$[Co(NH_3)_5Cl]Cl_2 \longrightarrow [Co(NH_3)_5Cl]^{2+} + 2Cl^-$$

$$2Ag^+ + 2Cl^- \longrightarrow 2AgCl \text{ (\textit{white ppt.})}$$

From the above discussion we can conclude that all the anions placed outside the coordination sphere can be obtained in the free state and hence can be precipitated. Thus all the anions placed outside the coordination sphere are ionisable.

(*vii*) The attachment of the species satisfying the primary valency to the metal is shown by broken line (- - - -).

4. How to show the attachment between the central metal and the species satisfying both the valencies. The attachment between the metal and the species which satisfy both the valencies is shown by a combined solid-broken line (═══). For example in $CoCl_3.5NH_3$ or $[Co(NH_3)_5Cl]Cl_2$, since one Cl^- ion satisfies primary as well as secondary valency of Co-atom, the attachment of this Cl^- ion to Co-atom is shown as Co^{3+} ═══ Cl^-.

To explain the structure of Co(III) ammines on the basis of Werner's coordination theory

With the help of the postulates of Werner's coordination theory, the structure of Co (III) ammines *viz.* $CoCl_3.6NH_3$, $CoCl_3.5NH_3$, $CoCl_3.4NH_3$ and $CoCl_3.3NH_3$ can well be explained.

In all these ammines, the primary valency of cobalt is equal to 3 while the secondary valency is equal to 6. Primary valency of 3 is satisfied by three Cl^- ions and hence these ions are attached with Co^{3+} ion by dotted line (----). Secondary valency of six is satisfied either by six NH_3 molecules alone or by both NH_3 molecules and Cl^- ions. The species satisfying the secondary valency are attached with Co^{3+} ion by solid line (——).

1. $CoCl_3.6NH_3$. The primary valency of Co-atom which is equal to 3 is satisfied by three Cl^- ions and hence the attachment of these ions with Co^{3+} ion is shown by broken lines [Fig. 3.1 (*a*)].

When this ammine is acted upon by an aqueous solution of $AgNO_3$, all the three Cl^- ions present in this ammine are precipitated as AgCl (white ppt).

$$CoCl_3.6NH_3 \longrightarrow [Co.6NH_3]^{3+} + 3Cl^-$$
$$3Ag^+ + 3Cl^- \longrightarrow 3AgCl \text{ (white ppt)}$$

The precipitation of all the three Cl^- ions indicates that these ions are outside the coordination sphere and it is only six NH_3 molecules which, along with Co atom, are present inside coordination sphere. Thus the structure of $CoCl_3.6NH_3$ can be written as $[Co.6NH_3]Cl_3$ or $[Co(NH_3)_6]Cl_3$. This structure shows that the secondary valency of Co atom which is equal to six is satisfied by six NH_3 molecules placed inside the coordination sphere and hence the attachment of NH_3 molecules to Co-atom is shown by solid lines [See Fig. 3.1 (*a*)].

2. $CoCl_3.5NH_3$. The primary valency of Co-atom in this ammine is equal to 3 and is satisfied by three Cl^- ions. The attachment of these ions with Co^{3+} ion is, therefore, shown by broken lines [Fig. 3.1 (*b*)].

The treatment of this ammine with the aqueous solution of $AgNO_3$ precipitates only two Cl^- ions as AgCl (white ppt.)

$$CoCl_3.5NH_3 \longrightarrow [CoCl.5NH_3]^{2+} + 2Cl^-$$
$$2Ag^+ + 2Cl^- \longrightarrow 2AgCl \text{ (white ppt.)}$$

The precipitation of only two Cl^- ions indicates that two Cl^- ions are outside the coordination sphere and the remaining one Cl^- ion and five NH_3 molecules are present inside the coordination sphere. Thus on the basis of these observations, the structure of $CoCl_3.5NH_3$ can be written as $[CoCl.5NH_3]Cl_2$ or $[Co(NH_3)_5Cl]Cl_2$. This structure shows that the secondary valency of Co-atom which is equal to six is satisfied by one Cl^- ion and five NH_3 molecules placed inside the coordination sphere. The attachment of one Cl^- ion and five NH_3 molecules to Co-atom is, therefore, shown by solid lines. The above discussion shows that, on the basis of Werner's theory, the structure of $CoCl_3.5NH_3$ can be shown as depicted in Fig. 3.1 (*b*). This figure shows that out of the three Cl^- ions, one Cl^- ion is attached with Co-atom by a combined solid-broken line (═ ═ ═), since this ion satisfies both primary as well as secondary valency of Co-atom.

3. $CoCl_3.4NH_3$. The primary valency of Co-atom which is equal to 3 is satisfied by three Cl^- ions and hence the attachment of these ions to Co-atom is shown by broken lines [see Fig. 3.1 (*c*)].

The treatment of this ammine with the aqueous solution of $AgNO_3$ precipitates only one Cl^- ion as AgCl (white ppt.).

$$CoCl_3.4NH_3 \longrightarrow [CoCl_2.4NH_3]^+ + Cl^-$$

$$Ag^+ + Cl^- \longrightarrow AgCl \text{ (white ppt.)}$$

The precipitation of only one Cl^- ion indicates that one Cl^- ion is outside the coordination sphere and the remaining two Cl^- ions and four NH_3 molecules are present inside the coordination sphere. Thus the structure of $CoCl_3.4NH_3$ can be represented as $[CoCl_2.4NH_3]Cl$ or $[Co(NH_3)_4Cl_2]Cl$ which shows that the secondary valency of Co-atom which is equal to six is satisfied by two Cl^- ions and four NH_3 molecules placed inside the coordination sphere and hence the attachment of these six species with Co-atom has been shown by solid lines. Thus, on the basis of Werner's theory, the structure of $CoCl_3.4NH_3$ can be shown as given in Fig. 3.1 (*c*) in which the attachment of two Cl^- ions with Co-atom has been shown by solid-broken lines, since these ions satisfy both types of valency of Co-atom.

4. $CoCl_3.3NH_3$. The primary valency of Co-atom which is equal to 3 is satisfied by three Cl^- ions and hence the attachment of these ions with Co-atom is represented by broken lines [see Fig. 3.1 (*d*)]

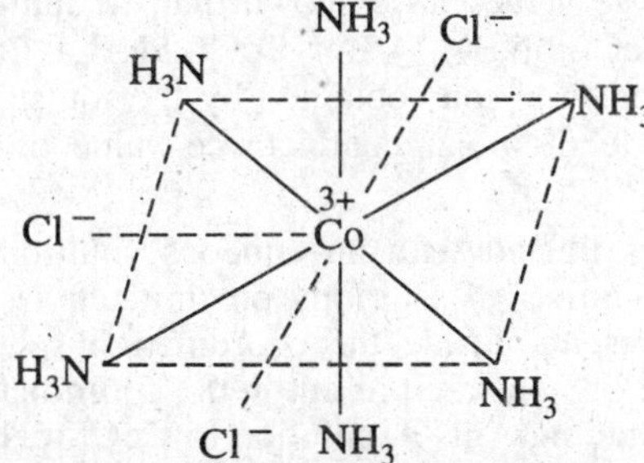

(*a*) Structure of $CoCl_3.6NH_3$ or $[Co(NH_3)_6]Cl_3$ Secondary valency of six is satisfied by six NH_3 molecules.

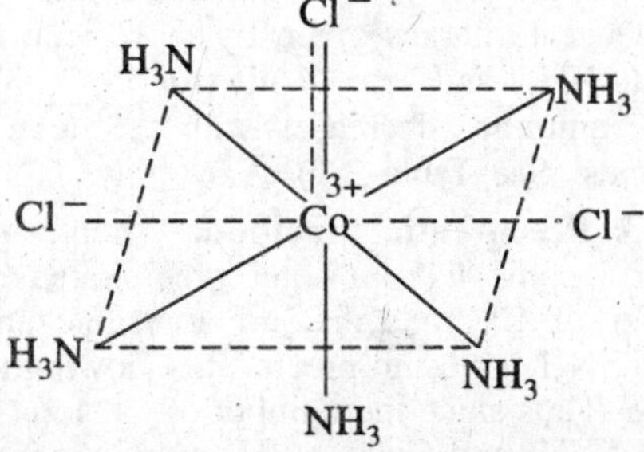

(*b*) Structure of $CoCl_3.5NH_3$. (or $[Co(NH_3)_5Cl]$ Cl_2. Secondary valency of six is satisfied by five NH_3 molecules and one Cl^- ion.

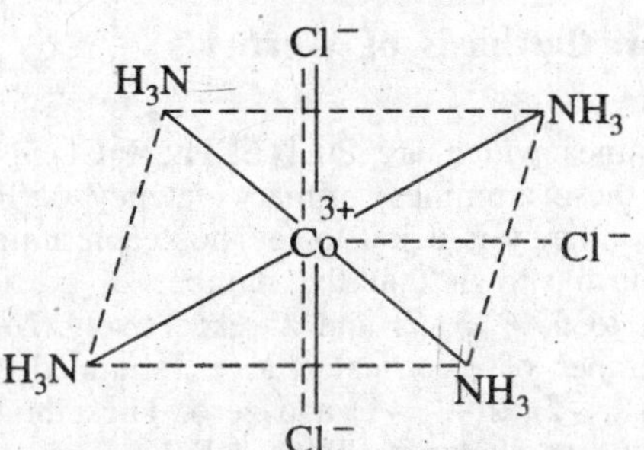

(*c*) Structure of $CoCl_3.4NH_3$ or $[Co(NH_3)_4Cl_2]Cl$. Secondary valency of six is satisfied by four NH_3 molecules and two Cl^- ions.

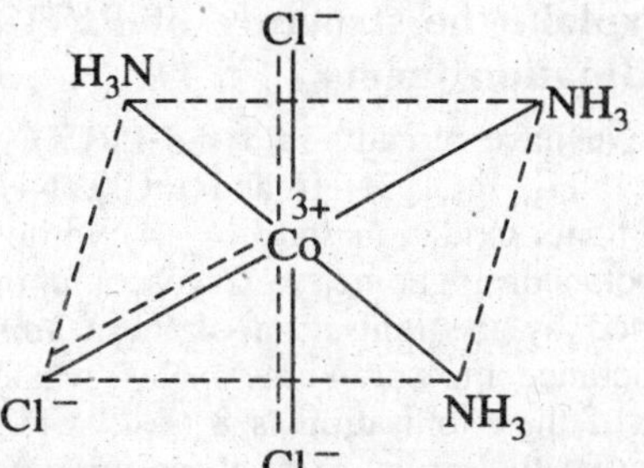

(*d*) Structure of $CoCl_3.3NH_3$ or $[Co(NH_3)_3Cl_3]^0$. Secondary valency of six is satisfied by three NH_3 molecules and three Cl^- ions.

Fig. 3.1. Structure of Co (III) ammines *viz.* $CoCl_3.6NH_3$, $CoCl_3.5NH_3$., $CoCl_3.4NH_3$ and $CoCl_3.3NH_3$ on the basis of Werner's coordination theory. In all these ammines the primary valency of Co-atom is equal to 3 and secondary valency is equal to six. Primary valency of three is satisfied by three Cl^- ions.

When this ammine is treated with the aqueous solution of $AgNO_3$, no precipitate is obtained. The non-formation of any precipitate shows that no Cl^- ion is outside the coordination sphere, *i.e.*, all the three Cl^- ions and three NH_3, along with Co-atom, are present inside the coordination sphere. Thus the structure of this ammine can be written as $[CoCl_3.3NH_3]^0$ or $[Co(NH_3)_3Cl_3]^0$. This structure shows that the secondary valency of Co-atom which is equal to six is satisfied by three Cl^- ions and three NH_3 molecules placed inside the co-ordination sphere. Thus the linking of these six species with Co-atom is represented by solid lines. This discussion indicates that, on the basis of Werner's theory, the structure of $CoCl_3.3NH_3$ can be shown as giver in Fig. 3.1 (*d*) in which the attachment of three Cl^- ions with Co-atom has been represented by solid-broken lines, since these ions satisfy both types of valency of Co-atom.

Experimental evidences in favour of Werner's theory

We have said that the Co(III) ammines given above can be represented as $[Co(NH_3)_6]Cl_3$, $[Co(NH_3)_5Cl]Cl_2$, $[Co(NH_3)_4Cl_2]Cl$ and $[Co(NH)_3Cl_3]$ respectively. These structures have been confirmed by the following studies :

1. Cryoscopic measurements. Cryoscopic measurements (*i.e.*, the measurement of the depression in freezing points of aqueous solutions) gives the number of ions furnished by the dissociation of one molecule of the compound. This measurement for Co (III) ammines has also indicated that these compounds dissociate to give 4 (for $CoCl_3.6NH_3$), 3 (for $CoCl_3.5NH_3$), 2(for $CoCl_3.4NH_3$) and 1 (for $CoCl_3.3NH_3$) particles (ionic or molecular) (See Table 3.1).

2. Molar electrolytic conductance (or conductivity) measurements. Measurement of molar conductance of a compound at its large dilution gives the sum of the electrical charges on the positive and negative ions furnished by each molecule of the compound. Molar conductance value for each Co (III) ammine has been determined. These values have shown that the number of electrical charges given by these ammines is 6 (for $CoCl_3.6NH_3$), 4 (for $CoCl_3.5NH_3$), 2(for $CoCl_3.4NH_3$) and zero or nil (for $CoCl_3.3NH_3$). Total number of ions obtained by the ionisation of the ammines decreases with the decrease of magnitude of molar conductance value of the ammines (See Table 3.1).

3. Precipitation method. When a given ammine is treated with an aqueous solution of $AgNO_3$, some Cl^- ions get precipitated as AgCl. The number of Cl^- ions precipitated is the number of Cl^- ions that are ionisable and hence are present outside the coordination sphere. $CoCl_3.3NH_3$ gives no precipitite, showing that all the three Cl^- ions are present in the coordination sphere. Thus since the number of Cl^- ions precipitated by adding $AgNO_3$ solution to $CoCl_3.6NH_3$, $CoCl_3.5NH_3$ and $CoCl_3.4NH_3$ ammines is 3, 2 and 1 respectively, the number of Cl^- ions present outside the coordination sphere is also equal to 3, 2 and 1 respectively. $CoCl_3.3NH_3$ gives no precipitate (*i.e.*, gives no Cl^- ions) with $AgNO_3$ solution and hence all the three Cl^- ions are present in the coordination sphere (See Table 3.1).

To explain the structure of Pt(IV) ammines on the basis of Werner's coordination theory

We have already said the Pt(IV) gives five ammines which are $PtCl_4.6NH_3$, $PtCl_4.5NH_3$, $PtCl_4.4NH_3$, $PtCl_4.3NH_3$ and $PtCl_4.2NH_3$. In each of these ammines, primary valency of Pt is equal to its oxidation state (= +4) while the secondary valency is equal to 6. Thus each ammine has octahedral geometry. Cryoscopic measurements have shown that the number of particles obtained by the ionisation of these ammines is equal to 5, 4, 3, 2, and 1 respectively. Molar conductance measurements have revealed that the number of electrical charges obtained as a result of their ionisation is 8 (+ 4, – 4), 6(+3, –3), 4(+2, –2), 2(+1, –1) and zero. These studies show that the structure of these ammines can be written as shown in Table 3.1.

Sidgwick's Electronic Concept of Coordinate Bond

In 1916 Lewis introduced a concept of two-electron covalent bond between two atoms in a molecule. According to this concept a covalent bond is formed between two atoms of a

Table 3.1. *Structure of Co (III) and Pt (IV) Ammines.*

Molecular and structural formula of ammines	*Ionisation of ammine*	*Number of particles (ionic or molecular) as determined by cryoscopic measurements*	*Molar conductance, Λ_m (ohm^{-1} cm^2)*	*Number of electrical charges determined by Λ_m*
Co (III) ammines				
$CoCl_3.6NH_3$ or $[Co(NH_3)_6]Cl_3$	$[Co(NH_3)_6]Cl_3 \rightarrow [Co(NH_3)_6]^{3+} + 3Cl^-$	4	390	6(+3, −3)
$CoCl_3.5NH_3$ or $[Co(NH_3)_5Cl]Cl_2$	$[Co(NH_3)_5Cl]Cl_2 \rightarrow [Co(NH_3)_5Cl]^{2+} + 2Cl^-$	3	262	4(+2, −2)
$CoCl_3.4NH_3$ or $[Co(NH_3)_4Cl_2]Cl$	$[Co(NH_3)_4Cl_2]Cl \rightarrow [Co(NH_3)_4Cl_2]^+ + Cl^-$	2	102	2 (+1, −1)
$CoCl_3.3NH_3$ or $[Co(NH_3)_3Cl_3]^0$ (Non-electrolyte)	$[Co(NH_3)_3Cl_3]^0 \rightarrow$ No ionisation	0	0	Nil
			Decreasing ↓	Decreasing ↓
Pt (IV) ammines				
$PtCl_4.6NH_3$ or $[Pt(NH_3)_6]Cl_4$	$[Pt(NH_3)_6]Cl_4 \rightarrow [Pt_5(NH_3)_6]^{4+} + 4Cl^-$	5	524	8(+4, −4)
$PtCl_4.5NH_3$ or $[Pt(NH_3)_5Cl]Cl_3$	$[Pt(NH_3)_5Cl]Cl_3 \rightarrow [Pt(NH_3)_5Cl]^{3+} + 3Cl^-$	4	404	6(+3, −3)
$PtCl_4.4NH_3$ or $[Pt(NH_3)_4Cl_2]Cl_2$	$[Pt(NH_3)_4Cl_2]Cl_2 \rightarrow [Pt(NH_3)_4Cl_2]^{2+} + 2Cl^-$	3	230	4(+2, −2)
$PtCl_4.3NH_3$ or $[Pt(NH_3)_3Cl_3]Cl$	$[Pt(NH_3)_3Cl_3]Cl \rightarrow [Pt(NH_3)_3Cl_3]^+ + Cl^-$	2	97	2(+1, −1)
$PtCl_4.2NH_3$ or $[Pt(NH_3)_2Cl_4]^0$ (Non-electrolyte)	$[Pt(NH_3)Cl_4]^0 \rightarrow$ No ionisation	0	0	Nil
			Decreasing ↓	Decreasing ↓

molecule by the sharing of an electron pair. Sidgwick (1927) extended this Lewis concept of covalent bond and introduced a new concept of *coordinate bond* which is also some times called *polar* or *dative bond.* According to this Sidgwick's concept the *ligands donate the electron pair to the central metal ion and thus form a coordinate bond,* M←L which indicates that the ligand, L (*i.e.*, donor) has donated an electron pair to the metal ion, M (*i.e.*, accept). This concept is based on the fact that *all the ligands contain at least one lone pair of electrons.* Thus on the basis of Sidgwick's concept, the structure of the complex ion, $[Co(NH_3)_6]^{3+}$, can be shown as given in the margin. Evidently, in the formation of the bonds between the six ligands (which are NH_3 molecules) and the central metal Co^{3+} ion, the N-atom of each NH_3 molecule donates an electron pair to Co^{3+} ion. Thus in all twelve electrons (*i.e.*, six electron pairs) are donated by six NH_3 ligands to Co^{3+} ion. The donation of an electron pair by NH_3 molecule has been represented by an arrow. Coordinate bond formed is not very different from a covalent bond. The only difference lies in the mode of its formation.

Limitations of Sidgwick's concept of coordinate bond

The donation of one electron pair by each of the ligands to the central metal ion to form L→M coordinate bond in complexes accumulates negative charge on the central metal ion which is most unlikely. For example in $[Co^{3+}(NH_3)_6]^{3+}$, six NH_3 molecules donate half share of 12 electrons (*i.e.*, 6 electrons) to Co^{3+} ion and thus +3 charge on cobalt atom is reduced to –3 (+3 – 6 = –3). Such an accumulation of negative charge on the central atom is unlikely.

Pauling suggested that the accumulation of undue negative charge on the central metal ion can be prevented, if it is assumed that *the metal–ligand bond has some ionic character (Electro-neutrality principle)* and *in addition to σ-bond, the central atom also forms a π-bond with the ligand (π-bonding in complexes).*

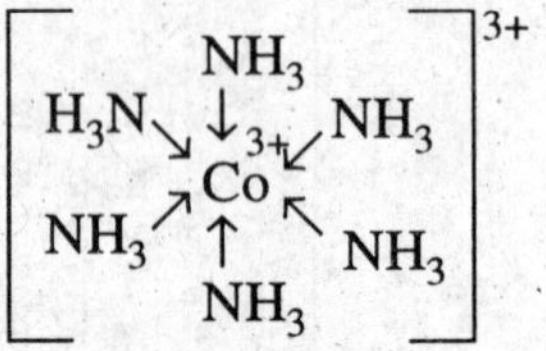

$$\left[\begin{array}{c} NH_3 \\ H_3N \searrow \downarrow \swarrow NH_3 \\ Co^{3+} \\ NH_3 \nearrow \uparrow \nwarrow NH_3 \\ NH_3 \end{array}\right]^{3+}$$

Formation of six $H_3N{\rightarrow}Co^{3+}$ bonds in $[Co(NH_3)_6]^{3+}$ ion.

Sidgwick's Effective Atomic Number (EAN) Rule

On the basis of his concept of coordinate bond, Sidgwick suggested that after the ligands have donated a certain number of electrons to the central metal ion through L → M bonding, the total number of electrons on the central atom, including those gained from ligands in the bonding, is called the effective atomic number (EAN) of the cental metal ion and in many cases this total number of electrons (*i.e.*, EAN) surrounding the coordinated metal ion is equal to the atomic number of the inert gas which follows the central metal atom in the periodic table. This is called *Effective Atomic Number Rule or Noble Gas Rule*. When the EAN is 36 (Kr), 54 (Xe) or 86 (Rn), the rule is said to be followed.

To calculate EAN of the central metal atom in complex ions

EAN of the central metal atom/ion in a given complex ion is given by:

$$EAN = (Z - x) + n \times y$$

Here Z = atomic number of the central metal atom, x = oxidation state of the central metal ion, n = number of ligands and y = number of electrons donated by one ligand. With the help of this formula, EAN of the central metal atom or ion of some 2-, 4- and 6-coordinated complex ions has been calculated as shown in Table 3.2.

This table shows that many complex ions donot obey EAN rule, *i.e.*, in case of many complex ions, the EAN of the central metal is some units more or less than the atomic number of the next inert gas.

To calculate EAN of the central metal atom in carbonyls

See chapter 13.

To calculate EAN of the central metal atom in metallic nitrosyls

See chapter 14.

Table 3.2. *To calculate the EAN of the central metal atom of some complex ions.*

Complex ion	Atomic number of the central metal atom (Z)	Oxidation state of the central metal atom (x)	Electrons donated by ligands = $n \times y$	EAN of the central metal ion = $(Z - x) + n \times y$
(A) Complex ions whose central metal ion obeys EAN rule				
$[Pd(NH_3)_6]^{4+}$	Pd = 46	+4 (Pd^{4+})	6 × 2 = 12	(46 – 4) + 12 = 54 (Xe_{54})
$[Fe(CN)_6]^{4-}$	Fe = 26	+2(Fe^{2+})	6 × 2 = 12	(26 – 2) + 12 = 36 (Kr_{36})
$[Co(NH_3)_6]^{3+}$	Co = 27	+3(Co^{3+})	6 × 2 = 12	(27 – 3) + 12 = 36 (Kr_{36})
$[Pt(NH_3)_6]^{4+}$	Pt = 78	+4(Pt^{4+})	6 × 2 = 12	(78 – 4) + 12 = 86 (Rn_{86})
$[Ag(NH_3)_4]^{+}$	Ag = 47	+1 (Ag^{+})	4 × 2 = 8	(47 – 1) + 8 = 54 (Xe_{54})
$[Cu(CN)_4]^{3-}$	Cu = 29	+1 (Cu^{+})	4 × 2 = 8	(29 – 1) + 8 = 36 (Kr_{36})
(B) Complex ions whose central metal ion does not obey EAN rule				
$[Fe(CN)_6]^{3-}$	Fe = 26	+3(Fe^{3+})	6 × 2 = 12	(26 – 3) + 12 = 35
$[Cr(NH_3)_6]^{3+}$	Cr = 24	+3(Cr^{3+})	6 × 2 = 12	(24 – 3) + 12 = 33
$[Ni(NH_3)_6]^{2+}$	Ni = 28	+2(Ni^{2+})	6 × 2 = 12	(28 – 2) + 12 = 38
$[Ni(en)_3]^{2+}$	Ni = 28	+2(Ni^{2+})	3 × 4 = 12	(28 – 2) + 12 = 38
$[Mn(H_2O)_6]^{2+}$	Mn = 25	+2(Mn^{2+})	6 × 2 = 12	(25 – 2) + 12 = 35
$[Co(CN)_6]^{4-}$	Co = 27	+2(Co^{2+})	6 × 2 = 12	(27 – 2) + 12 = 37
$[Pt(NH_3)_2Cl_2]^{0}$	Pt = 78	+2(Pt^{2+})	4 × 2 = 8	(78 – 2) + 8 = 84
$[Cu(NH_3)_4]^{2+}$	Cu = 29	+2(Cu^{2+})	4 × 2 = 8	(29 – 2) + 8 = 35
$[FeCl_4]^{-}$	Fe = 26	+3(Fe^{3+})	4 × 2 = 8	(26 – 3) + 8 = 31
$[Ni(CN)_4]^{2-}$	Ni = 28	+2(Ni^{2+})	4 × 2 = 8	(28 – 2) + 8 = 34
$[PdCl_4]^{2-}$	Pd = 46	+2(Pd^{2+})	4 × 2 = 8	(46 – 2) + 8 = 52
$[Pt(NH_3)_4]^{2+}$	Pt = 78	+2(Pt^{2+})	4 × 2 = 8	(78 – 2) + 8 = 84
$[AgX_2]^{-}$	Ag = 47	+1(Ag^{+})	2 × 2 = 4	(47 – 1) + 4 = 50
$[Ag(NH_3)_2]^{+}$	Ag = 47	+1(Ag^{+})	2 × 2 = 4	(47 – 1) + 4 = 50
$[CuCl_2]^{-}$	Cu = 29	+1(Cu^{+})	2 × 2 = 4	(29 – 1) + 4 = 32

Application of EAN rule

With the help of EAN rule, the magnetic property of the complex ions can be predicted. It has been observed that the complex ions whose central metal atom obeys EAN rule are diamagnetic. For example, since the EAN of Co^{3+} ion in $[Co(NH_3)_6]^{3+}$ ion is equal to [(27 – 3) + 6 × 2 = 36], this ion obeys EAN rule and hence $[Co(NH_3)_6]^{3+}$ ion is diamagnetic. Experimentally this ion has also been found diamagnetic.

Sidgwick and Bose have suggested that the complex ions whose central metal atom does not obey EAN rule are generally paramagnetic. The number of unpaired electrons present in the complex ion is equal to the difference between the EAN of the central metal atom and the atomic number of the inert gas which follows the central metal atom in the periodic table. With the help of these unpaired electrons the value of magnetic moment (μ) can be calculated. Calculated value of μ (μ_{cal}) has been found to be almost equal to the experimental value (μ_{exp}). The following examples illustrate this point.

1. $[Cu(NH_3)_4]^{2+}$ ion. EAN of $Cu^{2+} = (29 - 2) + 4 \times 2 = 35$

∴ No. of unpaired electrons = Atomic number of Kr – 35 = 36 – 35 = 1

∴ $\mu_{cal} = 1.73$ B.M., $\mu_{exp} = 1.82$ B.M.

2. **$[Cr(NH_3)_6]^{3+}$ ion.** EAN of $Cr^{3+} = (24 - 3) + 2 \times 6 = 21 + 12 = 33$

No. of unpaired electrons = Atomic number of Kr – 33 = 36 – 33 = 3

∴ $\mu_{cal} = 3.88$ B.M., $\mu_{exp} = 3.8$ B.M.

Although the central metal ion in $[Ni(dmg)_2]^{2+}$, $[Pt(NH_3)_2Cl_2]^0$, $[AuBr_4Cl]^-$ etc. does not obey EAN rule, these species are diamagnetic. Thus these complex ions donot obey the suggestion given by Sidgwick and Bose.

SOME TYPICAL PROBLEMS WITH SOLUTIONS

Problem 1. *The compound $CoCl_3.4NH_3$ contains only one Cl^- ion that is precipitated immediately on the addition of Ag^+ ions. Draw the structure of the compound on the basis of Werner's coordination theory.* *(Delhi1999)*

Solution. Since the C.N. of Co-atom is 6 and only one Cl^- ion is precipitated on the addition of Ag^+ ions, it contains only one Cl^- ion outside the coordination sphere and the structure of the given compound can be written as $[Co(NH_3)_4Cl_2]^+Cl^-$.

$$[Co(NH_3)_4Cl_2]^+Cl^- \longrightarrow [Co(NH_3)_4Cl_2]^+ + Cl^-$$

$$Ag^+ + Cl^- \longrightarrow AgCl \text{ (ppt)}$$

On the basis of Werner's coordination theory the structure of $CoCl_3.4NH_3$ can be drawn as shown in Fig. 3.2

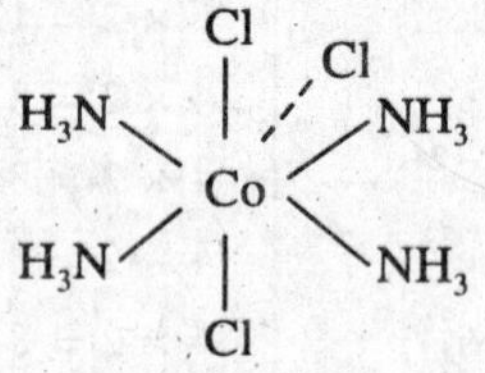

Cl, Cl, H_3N, NH_3, Co, H_3N, NH_3, Cl

Fig. 3.2. Structure of $CoCl_3.4NH_3$ or $[Co(NH_3)_4Cl_2]Cl$ on the basis of Werner's coordination theory.

Problem 2. *Molar conductance of the complex, $M(NH_3)_5X_4$ (X = monodentate ligand and M has C.N. = 6) is ~ 400 ohm^{-1}. Write the structure of the complex.* *[Delhi 2000]*

Solution. The given value of molar conductance indicates that the given complex, on ionisation in solution, gives four ions *viz* $[M(NH_3)_5X]^{3+}$ and $3X^-$.

$$M(NH_3)_5X_4 \longrightarrow \underbrace{[M(NH_3)_5X]^{3+} + 3X^-}_{\text{Four ions}}$$

Thus the structure of the complex can be written as $[M(NH_3)_5X]^{3+}X_3^-$.

Problem 3. *A coordination compound has the formula $CoCl_3.4NH_3$. It does not liberate NH_3 but precipitates Cl^- ions as AgCl. Give the IUPAC name of the compound and write its structural formula.*

Solution. Since the given compound does not give NH_3 but precipitates Cl^- ions as AgCl and the C.N. of Co-atom is 6, all the four NH_3 molecules and two Cl^- ions act as ligands and hence are present inside the coordination sphere. The remaining Cl^- ion is present outside the coordination sphere and is ionisable. Thus the structural formula of the compound is $[Co(NH_3)_4Cl_2]Cl$. Obviously, the IUPAC name of this compound is tetraminedichlorocobalt (III) chloride.

Problem 4. *A compound having the formula, M $(NH_3)_4$ (H_2O) Cl_3 has two isomers. The molar conductivity of these isomers is 110 and 250 ohm^{-1} mol^{-1} respectively. Write down the formula of both the isomers.*

Solution. Molar conductivity value equal to 110 ohm^{-1} mol^{-1} for one isomer shows that this isomer, on ionisation in solution, gives two ions. Thus this isomer can be represented by the formula $[M(NH_3)_4Cl_2]Cl.H_2O$. (C.N. of M = 6)

$$\underset{\text{(C.N. of M = 6)}}{[M^{3+}(NH_3)_4Cl_2]Cl.H_2O} \longrightarrow \underbrace{[M^{3+}(NH_3)_4Cl_2]^+ + Cl^- + H_2O}_{\text{Two ions}}$$

The molar conductivity value of 250 ohm^{-1} mol^{-1} for the other isomer shows that this isomer gives three ions on ionisation in solution. Thus this isomer has formula, $[M(NH_3)_4(H_2O)Cl]Cl_2$ (C.N. of M = 6)

$$\underset{\text{(C.N. of M = 6)}}{[M^{3+}(NH_3)_4(H_2O)Cl]Cl_2} \longrightarrow \underbrace{[(M^{3+}(NH_3)_4(H_2O)Cl]^{2+} + 2Cl^-}_{\text{Three ions.}}$$

Problem 5. *Two compounds have the empirical formula, Co $(NH_3)_3(NO_2)_3$. In aqueous solution one of these compounds does not conduct electricity while the other does. Write the possible structures of these two compounds.*

Solution. Compound 1. Since this compound does not conduct electricity in aqueous solution, it is neutral or uncharged (non-electrolyte) and hence can be represented by the structure, $[Co(NH_3)_3(NO_2)_3]$ (C.N. of Co = 6)

$$\underset{\text{Non-electrolyte}}{[Co(NH_3)_3(NO_2)_3]} \longrightarrow \text{No ions}$$

Compound 2. Since this compound conducts electricity,it gives ions in solution. Also since C.N. of Co-atom (central atom) is 6 and H_2O molecules can enter coordination sphere to make C.N. of Co-atom equal to 6, the possible structures of this compound are: (*i*) $[Co^{3+}(NH_3)_3(H_2O)_3]^{3+}$ $(NO_2)_3^-$ (*ii*) $[Co^{3+}(NH_3)_3(H_2O)_2(NO_2)]^{2+}$ $(NO_2)_2^-$ (*iii*) $[Co^{3+}(NH_3)_3(H_2O)(NO_2)_2]^+$ NO_2^-

Problem 6. *Explain the following, giving appropriate reasons:*

(*i*) *Out of K_4 [Fe $(CN)_6$] and K_3 [Fe $(CN)_6$] solutions, the former has higher value of molar conductivity.*

(*ii*) *[Pt $(NH_3)_2$ Cl_2] and [Pt $(NH_3)_6$] Cl_4 differ in their electrolytic conductance*

(*iii*) *The value of molar conductivity of the aqueous solution of [CoCl_3 $(NH_3)_3$] is zero.*

Solution. (*i*) On ionisation, since the solution of K_4 [Fe $(CN)_6$] gives greater number of ions (= five ions) than the solution of K_3 [Fe $(CN)_6$] (= four ions), the former has higher value of its molar conductivity.

$$K_4[Fe(CN)_6]\,(aq) \rightleftharpoons \underbrace{4K^+\,(aq) + [Fe(CN)_6]^{4-}\,(aq)}_{\text{Five ions}}$$

$$K_3[Fe(CN)_6]\,(aq) \rightleftharpoons \underbrace{3K^+\,(aq) + [Fe(CN)_6]^{3-}\,(aq)}_{\text{Four ions}}$$

(*ii*) When the aqueous solution of $[Pt(NH_3)_6]Cl_4$ gets ionised, one $[Pt(NH_3)_4]^{4+}$ ion (complex ion) and four Cl^- ions (simple ions) are obtained. On the other hand, since $[Pt(NH_3)_2Cl_2]$ is a non-electrolyte, it does not give any ion.

$$[Pt(NH_3)_6]Cl_4\,(aq) \longrightarrow \underbrace{[Pt(NH_3)_6]^{4+}\,(aq) + 4\,Cl^-\,(aq)}_{\text{Five ions}}$$

$$[Pt(NH_3)_2Cl_2] \longrightarrow \text{No ions are obtained.}$$

Thus $[Pt(NH_3)_2Cl_2]$ has no molar conductance while $[Pt(NH_3)_6]Cl_4$ has a certain value of molar conductance.

(*iii*) The given complex compound is a neutral compound and hence its aqueous solution does not undergo ionisation. Thus it does not give any ions on ionisation. Obviously, this complex compound has no value of molar conductivity.

Problem 7. *Arrange the following complexes in the increasing order of their electrical conductivity: $[Co(NH_3)_3Cl_3]$, $[Co(NH_3)_5Cl]Cl_2$, $[Co(NH_3)_6]Cl_3$ and $[Co(NH_3)_5Cl]Cl$.*

Solution. Let us consider the ionisation of the given complexes as given below:

$$[Co(NH_3)_3Cl_3] \longrightarrow \text{No ions}$$

$$[Co(NH_3)_5Cl]Cl \longrightarrow [Co(NH_3)_5Cl]^+ + Cl^-$$

(+ 1, – 1 electrolyte)

No. of electrical charges = 2

$$[Co(NH_3)_5Cl]Cl_2 \longrightarrow [Co(NH_3)_5Cl]^{2+} + 2Cl^-$$

(+ 2, – 2 electrolyte), 4

$$[Co(NH_3)_6]Cl_3 \longrightarrow [Co(NH_3)_6]^{3+} + 3Cl^-$$

(+ 3, – 3 electrolyte), 6

The electrical conductivity of complexes increases with the increase in the number of electrical charges given by the complexes on ionisation. Thus the electrical conductivity of the given complexes increases in the order:

$$[Co(NH_3)_3Cl_3] < [Co(NH_3)_5Cl]Cl < [Co(NH_3)_5Cl]Cl_2 < [Co(NH_3)_6]Cl_3.$$

Problem 8. *Use the EAN rule to prodict the molecular formulae for the simple carbonyls of Cr (Z = 24), Fe (Z = 26) and Ni (Z = 28). Assume that the oxidation state of the metals in these carbonyls is zero.*

Solution. Since Cr, Fe and Ni are present in 4th period of the periodic table, EAN of these metals in their respective carbonyls is equal to the atomic number of Kr (= 36). Let the number of CO groups present in one molecule of the carbonyls = x, then:

Carbonyl of Cr. Z of Cr = 24

No. of electrons denoted by x CO groups = $2 \times x$

$\therefore$ EAN of Cr-atom = $24 + 2x = 36$

$$\therefore \quad x = \frac{36-24}{2} = \frac{12}{2} = 6$$

$\therefore$ Formula = $\mathbf{Cr(CO)_6}$ **(Answer)**

Carbonyl of Fe. Z of Fe = 26

No. of electrons donated by x CO groups = $2 \times x$

$\therefore$ EAN of Fe atom = $26 + 2x = 36$

$$\therefore \quad x = \frac{36-26}{2} = \frac{10}{2} = 5$$

$\therefore$ Formula = $\mathbf{Fe(CO)_5}$ **(Answer)**

Carbonyl of Ni. Z of Ni = 28

No. of electrons denoted by x CO groups = $2 \times x$

$\therefore$ EAN of Ni atom = $28 + 2x = 36$

$$\therefore \quad x = \frac{36-28}{2} = \frac{8}{2} = 4$$

$\therefore$ Formula = **Ni $(CO)_4$ (Answer)**

Problem 9. *If EAN of each Mn-atom in $Mn_2\ (CO)_{10}$ is 36, draw the structure of the carbonyl. (Atomic number of Mn = 25).*

Solution. Since the given carbonyl contains one Mn – Mn bond and five CO groups are linked with each Mn-atom, its structure can be written as:

$$(CO)_5\ Mn - Mn\ (CO)_5$$

This structure is correct, since EAN of each Mn-atom comes out to be equal to 36 as shown below:

EAN of Mn-atom = Z of Mn + Electrons donated by five CO groups + Electrons donated by one Mn – Mn bond

= 25 + 5 × 2 + 1

= 36

Problem 10. *On the basis of EAN rule, predict the number of unpaired electrons and μ value of the following complexes : (i)* $[Cu\ (NH_3)_4]^{2+}$ *(ii)* $[Cr\ (NH_3)_6]Cl_3$.

Solution. It has been found that compounds in which EAN of the central metal ion is equal to the atomic number of the nearest noble gas are diamagnetic ($n = 0$) while those in which the EAN of the central metal ion is not equal to the atomic number of the nearest noble gas are paramagnetic. The number of unpaired electrons present in these paramagnetic compounds is given by:

No. of unpaired electrons (n) = At. No. of the nearest noble gas – EAN of the central metal ion.

(i) **$[Cu\ (NH_3)_4]^{2+}$ ion.** EAN of Cu^{2+} ion = No. of electrons in Cu^{2+} ion + No. of electrons donated by $4NH_3$ molecules.

= (29 – 2) + 4 × 2

= 27 + 8 = 35

$\therefore$ No. of unpaired electrons in $[Cu\ (NH_3)_4]^{2+}$ ion (n)

= At. No. of Kr – EAN of Cu^{2+} ion.

= 36 – 35 = 1 **(Answer)**

$$\therefore \quad \mu = \sqrt{n(n+2)}\ \text{BM} = \sqrt{1 \times (1+2)}\ \text{BM}$$

= **1.733 M (Answer)**

(ii) **$[Cu\ (NH_3)_6]^{3+}$ ion.** EAN of Cr^{3+} ion = No. of electrons in Cr^{3+} ion + No. of electrons donated by $6NH_3$ molecules

= (24 – 3) + 6 × 2

= 21 + 12 = 33

$\therefore$ No. of unpaired electrons in $[Cr\ (NH_3)_6]^{3+}$ ion (n)

= At. No. of Kr – EAN of Cr^{3+} ion

= 36 – 33 = 3 **(Answer)**

$$\therefore \quad \mu = \sqrt{n(n+2)}\ \text{BM} = \sqrt{3 \times (3+2)}\ \text{BM}$$

= **3.87 BM (Answer)**

Problem 11. *The number of ions present in $K_3\ [Fe\ (CN)_6]$ are:*

(a) 10 *(b)* 4 *(c)* 3 *(d)* 2 *[Karnataka 1996]*

Solution. The ionisation of $K_3\ [Fe\ (CN)_6]$ can be shown as:

$$K_3\ [Fe\ (CN)_6] \longrightarrow \underbrace{3K^+ + [Fe\ (CN)_6]^{3-}}_{\text{Four ions}}$$

Thus answer is (*b*)

Problem 12. *A coordination complex compound of Co has the molecular formula which contains five NH_3 molecules, one NO_2^- group, two Cl atoms and one Co-atom. One mole of this compound gives three mole ions in aqueous solution. On treating this solution with excess of $AgNO_3$ solution we get two moles of AgCl ppt. The formulae of this complex is :*

(a) $[Co(NH_3)_4(NO_2)Cl][NH_3Cl]$ *(b)* $[Co(NH_3)Cl][Cl\,NO_2]$

(c) $[Co(NH_3)_5(NO_2)]Cl_2$ *(d)* $[Co(NH_3)_5][(NO_2)_2Cl_2]$

Solution and explanation. Answer is (*c*). One mole of the given complex compound gives three mole ions in aqueous solution as shown below:

$$[Co(NH_3)_5(NO_2)]Cl_2 \rightleftharpoons \underbrace{[Co(NH_3)_5(NO_2)]^{2+} + 2Cl^-}_{\text{Three mole ions}}$$

On treating with $AgNO_3$ solution, $2Cl^-$ ions react with $2Ag^+$ ions and give two moles of AgCl ppt.

$$2Cl^- + 2Ag^+ \longrightarrow 2AgCl \downarrow \text{(ppt)}$$

University Questions

1. On the basis of Werner's theory explain why cobalt amine complex, $CoCl_3\,4NH_3$ when treated with $AgNO_3$ solution precipitates only one Cl^- ion although it has three Cl^- ions. *(Nagpur 2002)*

2. State in brief the experimental evidence in favour of Werner's theory. *(Delhi 2002)*

3. Write the structural formule of all the possible cationic complexes of Pt (IV) containing Cl^- ions and NH_3 molelcules as ligands. *(VBS Purvanchal 2003)*

Ans. $PtCl_4 \cdot 5NH_3$, $PtCl_4 \cdot 4NH_3$ and $PtCl_4 \cdot 3NH_3$ are the cationic complexes in which Pt (cenral atom) is in + 4 oxidation state and Cl^- ions and NH_3 molecules are present as ligands. The structural formulae of these compounds are $\left[Pt(NH_3)_5Cl\right]^{+3}\left[Cl^-\right]_3$, $\left[Pt(NH_3)_4Cl_2\right]^{2+}\left[Cl^-\right]_2$ and $\left[Pt(NH_3)_3Cl_3\right]^+Cl^-$ respectively.

4. Write the structure of $CoCl_3 \cdot 6NH_3$, $CoCl_3 \cdot 5NH_3$, $CoCl_3 \cdot 4NH_3$ and $CoCl_3 \cdot 3NH_3$ according to Werner's theory. *(Delhi 2003)*

5. Calculate the EAN of the central metal ion in the following complexes and write their IUPAC names : (*i*) $\left[Pt(NH_3)_4Cl_2\right]Cl_2$ (*ii*) $\left[Co(en)_2Cl_2\right]Cl$ (*ii*) $K_3\left[Fe(CN)_6\right]$. *(VBS Purvanchal 2003)*

6. Explain why $K_2[PtCl_6]$ does not give white precipitate of AgCl with $AgNO_3$. *(VBS Purvanchal 2003)*

7. Calculate the Effective Atomic Number (EAN) of the central metal atom in the following : (*i*) $\left[Cr(CO)_6\right]$ (*ii*) $K_3\left[Fe(CN)_6\right]$ *(Rohilkhand 2003)*

8. (*a*) What are the postulates of Werner's theory ?
(*b*) Calculate the Effective Atomic Number of the central metal ions in following complexes :
(*i*) $\left[Cu(NH_3)_4\right]^{++}$ $(Z = 29)$ (*ii*) $\left[Fe(CN)_6\right]^{4-}$ $(Z = 26)$. *(Nagpur 2003)*

9. List the basic postulates of Werner's coordination theory. How will you explain the bonding in $CoCl_3.3NH_3$ and $CoCl_3.5NH_3$? *(GND Amritsar, 2004)*

4

Isomerism in Coordination Compounds

Introduction

The coordination compounds which have the same molecular formula but have their ligands attached to the central metal atom in different ways are called **isomers.** These isomers have different properties. The phenomenon that gives rise to different isomers is called **isomerism.**

Isomerism may be of two types *viz., structural isomerism* and *stereo (space) isomerism.*

Structural Isomerism

This isomerism arises due to the difference in the structures of complex (coordination) compounds. This isomerism is of the following types:

1. Conformation isomerism

In this isomerism two isomers have different geometries but are otherwise identical. For example $[Ni^{2+}(P.Et.Ph_2)_2Br_2]$ complex (four coordinated complex) gives two conformation isomers. One of these is green and paramagnetic while the other is brown and diamagnetic. The green form is tetrahedral while the brown form is square planar. Thus these have different geometries but the coordination number of Ni^{2+} ion in both the isomers is the same (=4).

2. Ionisation isomerism

The complex compounds showing ionisation isomerism are called *ionisation isomers.* These isomers have the same molecular formulae and are produced when the ligands given in the coordination sphere and the anions present out side the coordination sphere are exchanged with each other. Thus both the complex compounds of each of the following pairs are ionisation isomers to each other.

(*i*) $[Co(NH_3)_5Br]SO_4$ and $[Co(NH_3)_5(SO_4)]Br$

(*ii*) $[Co(NH_3)_4Cl_2]NO_2$ and $[Co(NH_3)_4Cl(NO_2)]Cl$

(*iii*) $[Pt(NH_3)_4Cl_2]Br_2$ and $[Pt(NH_3)_4Br_2]Cl_2$

(*iv*) $[Co(NH_3)_5(NO_3)]SO_4$ and $[Co(NH_3)_5(SO_4)]NO_3$

(*v*) $[Cr(NH_3)_4I_2]Br$ and $[Cr(NH_3)_4IBr]I$

(*vi*) $[Fe(NH_3)_5(CN)]SO_4$ and $[Fe(NH_3)_5SO_4]CN$.

Ionisation isomers show different properties, *e.g.*, $[Co(NH_3)_5Br]SO_4$ (I) and $[Co(NH_3)_5(SO_4)]Br$ (II) which are ionisation isomers to each other show different properties as mentioned below:

(*i*) (I) is red-violet while (II) is red.

(*ii*) When these isomers are dissolved in water, they give different ions. For example :

$$\underset{\text{(I) (red-violet)}}{[Co(NH_3)_5Br]SO_4} \rightleftharpoons [Co(NH_3)_5Br]^{2+} + SO_4^{2-}$$

$$\underset{\text{(II) (red)}}{[Co(NH_3)_5SO_4]Br} \rightleftharpoons [Co(NH_3)_5SO_4]^{+} + Br^{-}$$

It is due to the formation of different ions in solution that these compounds show different behaviour when treated with an aqueous solution of $BaCl_2$ and $AgNO_3$.(I) gives a *white ppt.* of $BaSO_4$ with an aqueous solution of $BaCl_2$ indicating that SO_4^{2-} ions in (I) are out side the coordination sphere. This compound does not give any ppt. with the aqueous solution of $AgNO_3$. On the other hand (II) does not give any ppt. with a solution of $BaCl_2$ but forms a *yellow ppt.* of AgBr with $AgNO_3$ solution, indicating that Br^- ions in this compound are outside the coordination sphere.

$$\underset{\text{(I)}}{[Co(NH_3)_5Br]SO_4} + \underset{\text{(Solution)}}{BaCl_2} \longrightarrow \underset{\text{White ppt.}}{BaSO_4} + [Co(NH_3)_5Br]Cl_2$$

$$\underset{\text{(I)}}{[Co(NH_3)_5Br]SO_4} + \underset{\text{(Solution)}}{AgNO_3} \longrightarrow \text{No reaction}$$

$$\underset{\text{(II)}}{[Co(NH_3)_5SO_4]Br} + \underset{\text{(Solution)}}{BaCl_2} \longrightarrow \text{No reaction}$$

$$\underset{\text{(II)}}{[Co(NH_3)_5SO_4]} + \underset{\text{(Solution)}}{AgNO_3} \longrightarrow \underset{\text{Yellow ppt.}}{AgBr} + [Co(NH_3)_5SO_4]NO_3$$

It may be noted that the coordination number of the central atom in both the ionisation isomers is the same. Ionisation isomers can be detected by the determination of their conductance in solution.

3. Hydrate isomerism

This type of isomerism is similar to ionisation isomerism. Hydrate isomers are obtained when the H_2O molecules acting as ligands are replaced by the anions present outside the coordination sphere. $CrCl_3.6H_2O$ exists in three hydrate isomers which are $[Cr(H_2O)_6]Cl_3$ (I), $[Cr(H_2O)_5Cl]Cl_2.H_2O$ (II) and $[Cr(H_2O)_4Cl_2]Cl.2H_2O$ (III). These have different physical and chemical properties, *e.g.*,

(*i*) (I) is violet, does not lose water over conc.H_2SO_4 and all the three Cl^- ions as precipitated as AgCl with $AgNO_3$ solution.

$$\underset{\text{(I) (violet)}}{[Cr(H_2O)_6]Cl_3} + \underset{\text{(solution)}}{3AgNO_3} \longrightarrow \underset{\text{white ppt.}}{AgCl_2\downarrow} + [Cr(H_2O)_6](NO_3)_3$$

or $$3Cl^- + 3Ag^+ \longrightarrow 3AgCl\downarrow$$

(*ii*) (II) is blue-green, loses only one H_2O molecule over conc.H_2SO_4 and only two Cl^- ions are precipitated as AgCl with $AgNO_3$ solution.

$$\underset{\text{(II) (green)}}{[Cr(H_2O)_5Cl]Cl_2.H_2O} + \underset{\text{(solution)}}{2AgNO_3} \longrightarrow \underset{\text{white ppt.}}{2AgCl} + [Cr(H_2O)_5Cl](NO_3)_2.H_2O$$

or $$2Cl^- + 2Ag^+ \longrightarrow AgCl\downarrow$$

(*iii*) (III) is blue-green, loses two H_2O molecules over conc. H_2SO_4 and only one Cl^- ion is precipitated by $AgNO_3$ solution.

$$\underset{\text{(III) (green)}}{[Cr(H_2O)_4Cl_2]Cl.2H_2O} + \underset{\text{(solution)}}{AgNO_3} \longrightarrow \underset{\text{white ppt.}}{AgCl} + [Cr(H_2O)_4Cl_2]NO_3.2H_2O$$

or $$Cl^- + Ag^+ \longrightarrow AgCl\downarrow$$

The behaviour of these isomers over conc. H_2SO_4 as mentioned above indicates that the number of H_2O molecules outside the coordination sphere in (I), (II) and (III) is equal to zero, one and two respectively.

The action of $AgNO_3$ solution on these isomers indicates that (I), (II) and (III) have three, two and one Cl^- ions outside the coordination sphere.

Other examples of hydrate isomers are given below:

(*i*) $[Co(en)_2(H_2O)Cl]Cl_2$ and $[Co(en)_2Cl_2]Cl.H_2O$

(*ii*) $[Co(NH_3)_4(H_2O)Cl]Cl_2$ and $[Co(NH_3)_4Cl_2]Cl.H_2O$

(*iii*) $[Cr(py)_2(H_2O)_2Cl_2]Cl$ and $[Cr(py)_2(H_2O)Cl_3].H_2O$

(*iv*) $[Cr(en)_2(H_2O)_2]Br_3$ and $[Cr(en)_2(H_2O)Br]Br_2.H_2O$

4. Coordination isomerism

This type of isomerism is shown by those complex compounds which are composed of complex cation and complex anion. Coordination isomers are obtained when some or all ligands of both the coordination spheres are interchanged with each other. Thus both the complex compounds of each of the following pairs are coordination isomers to each other. In these pairs the central metallic atom in the two coordination spheres may be the same or different.

(*i*) $[Cr^{3+}(NH_3)_6]^{3+}$ $[Cr^{3+}(CN)_6]^{3-}$ and $[Cr^{3+}(NH_3)_4(CN)_2]^+$ $[Cr^{3+}(CN)_4(NH_3)_2]^-$
(*ii*) $[Pt^{2+}(NH_3)_4]^{2+}$ $[Pt^{2+}Cl_4]^{2-}$ and $[Pt^{2+}(NH_3)_3Cl]^+[Pt^{2+}Cl_3(NH_3)]^-$
(*iii*) $[Co^{3+}(NH_3)_6]^{3+}$ $[Cr^{3+}(CN)_6]^{3-}$ and $[Co^{3+}(CN)_6]^{3-}$ $[Cr(NH_3)_6]^{3+}$
(*iv*) $[Cu^{2+}(NH_3)_4]^{2+}$ $[Pt^{2+}Cl_4]^{2-}$ and $[Cu^{2+}Cl_4]^{2-}$ $[Pt^{2+}(NH_3)_4]^{2+}$
(*v*) $[Cr^{3+}(NH_3)_6]^{3+}$ $[Cr^{3+}(SCN)_6]^{3-}$ and $[Cr^{3+}(NH_3)_4(SCN)_2]^+$ $[Cr^{3+}(SCN)_4(NH_3)_2]^-$

5. Linkage isomerism

See "*Ambidentate ligands and linkage isomerism*" discussed in chapter 1.

6. Coordination position isomerism

This type of isomerism is shown by those complex compounds which contain bridging ligands and arises when the non-bridging ligands are differently placed round the central metal atom. Thus (I) and (II) are coordination position isomers to each other, since NH_3 molecules and Cl^- ions (non-bridging ligands) are differently placed round the two Co^{3+} ions.

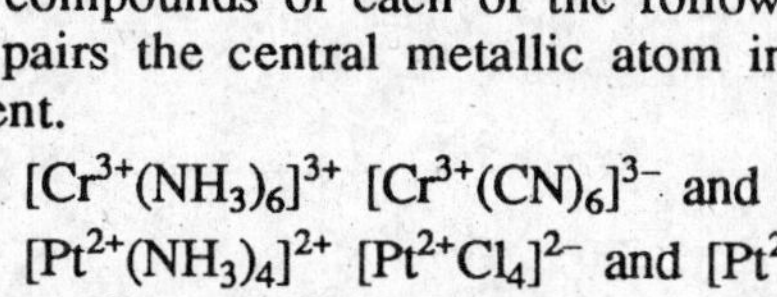

$$\left[(NH_3)_4\overset{3+}{Co}\underset{\underset{H}{O}}{\overset{\overset{H}{O}}{\diamond}}\overset{3+}{Co}(NH_3)_2Cl_2\right]^{2+} \qquad \left[Cl(NH_3)_3\overset{3+}{Co}\underset{\underset{H}{O}}{\overset{\overset{H}{O}}{\diamond}}\overset{3+}{Co}(NH_3)_3Cl\right]^{2+}$$

Tetrammine coblat (III) μ-dihydroxo diammine dichloro cobalt (III) ion. (I) (*unsymmetrical*)

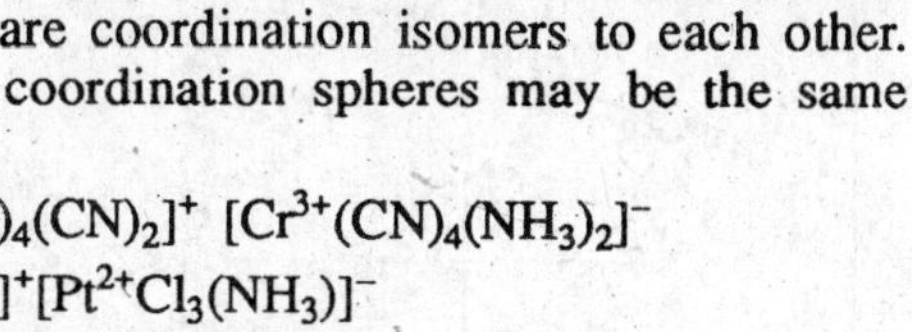

Chloro triammine cobalt (III) μ-dihydroxo chloro triammine cobalt (III) ion. (II) (*symmetrical*)

The following binuclear complexes of cobalt also show coordination position isomerism, since the terminal ligands *viz*. six NH_3 molecules and two Cl^- ions can be placed in an unsymmetrical and symmetrical manner round the central cobalt atom.

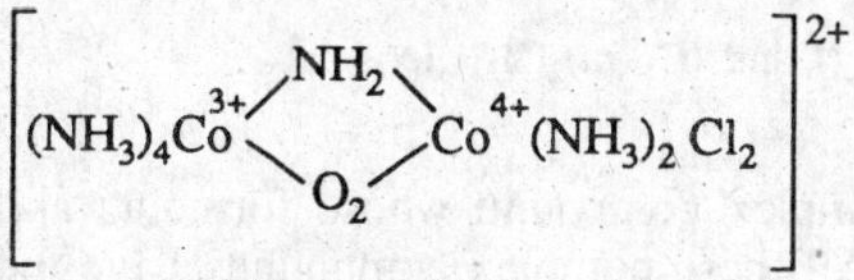

$$\left[(NH_3)_4\overset{3+}{Co}\underset{O_2}{\overset{NH_2}{\diamond}}\overset{4+}{Co}(NH_3)_2\,Cl_2\right]^{2+} \qquad \left[(NH_3)_3Cl\overset{3+}{Co}\underset{O_2}{\overset{NH_2}{\diamond}}\overset{4+}{Co}\,Cl(NH_3)_3\right]^{2+}$$

(*a*) unsymmetrical

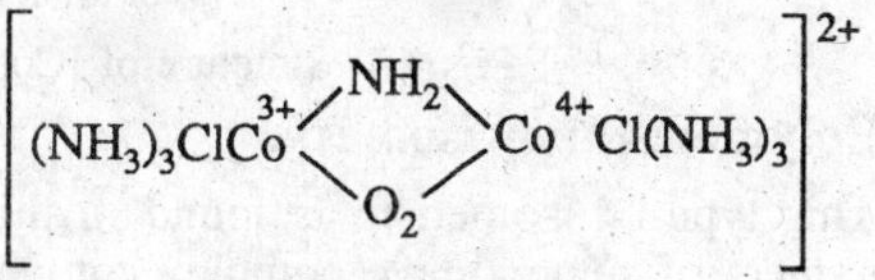

(*b*) symmetrical

7. Ligand isomerism

There are certain ligands which exist as isomers, *e.g.*, diamine derivatives of propane exist in two isomeric forms which are called 1, 2-diamino propane or propylene diamine (*pn*) and 1, 3-diamino propane or trimethylene diamine (*tn*). The structures of propane, *pn* and *tn* are given below :

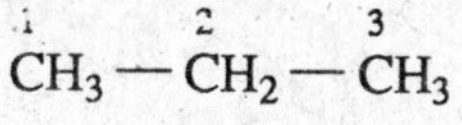

$$\overset{1}{C}H_3-\overset{2}{C}H_2-\overset{3}{C}H_3 \qquad \underset{NH_2}{\overset{1}{C}H_2}-\underset{NH_2}{\overset{2}{C}H_2}-\overset{3}{C}H_3 \qquad \underset{NH_2}{\overset{1}{C}H_2}-\overset{2}{C}H_2-\underset{NH_2}{\overset{3}{C}H_2}$$

Propane

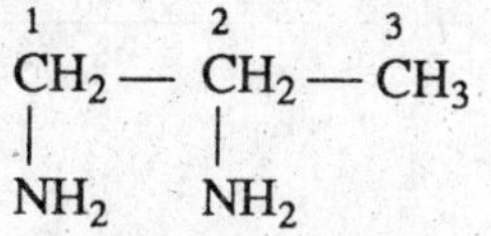

1, 2-diamino propane or propylene diamine (*pn*)

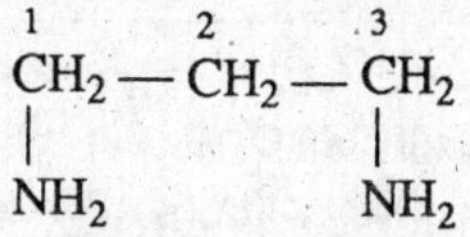

1, 3-diamino propane or trimethylene diamine (*tn*)

pn and *tn* both are bidentate (neutral) ligands. When these ligands get coordinated to the metal atom, two isomers are obtained. These are called ligand isomers and the isomerism is called ligand isomerism. Thus $[Co^{3+}(pn)_2Cl_2]^+$ and $[Co^{3+}(tn)_2Cl_2]^+$ are ligand isomers whose structures are given below in Fig. 4.1. Both these isomers have octahedral geometry.

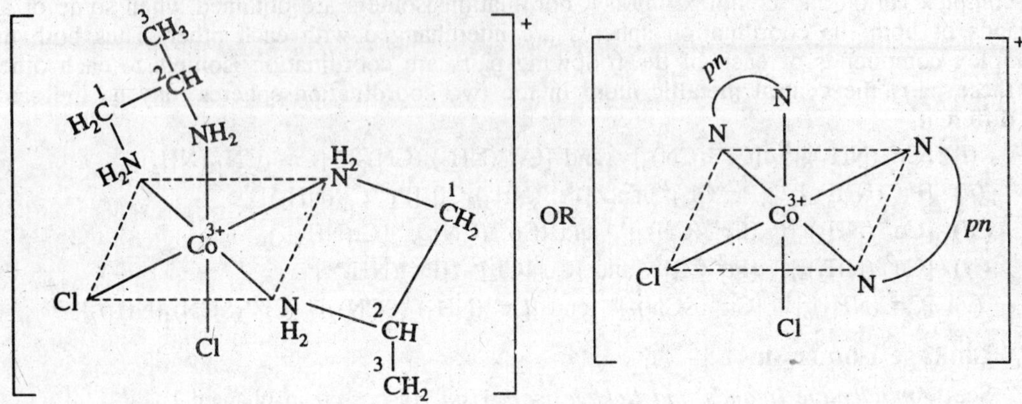

(*a*) Structure of $[Co^{3+}(pn)_2\,Cl_2]^+$ ion.

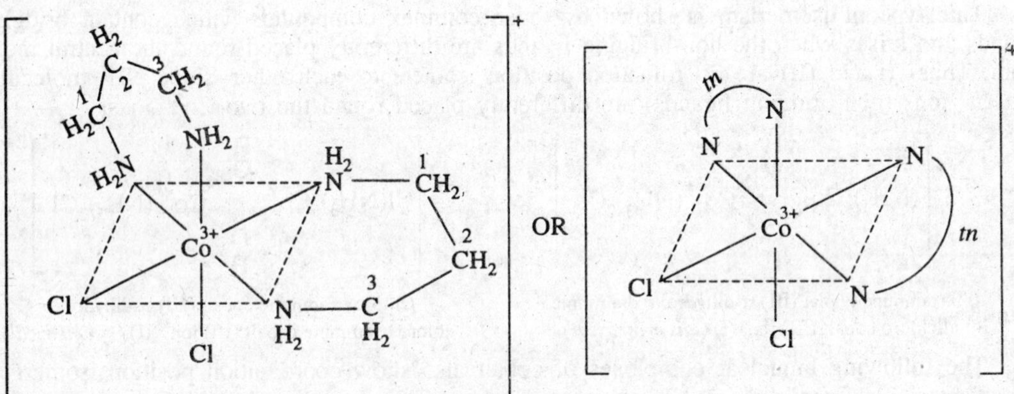

(*b*) Structure of $[Co^{3+}(tn)_2\,Cl_2]^+$ ion.

Fig. 4.1. Structure of $[Co(pn)_2Cl_2]^+$ and $[Co(tn)_2Cl_2]^+$ ions.

8. Polymerisation isomerism

This type of isomerism is found in those complex compound whose formulae appear to be polymers of some simple complex compound. All these complex compounds have the same ratio of different metal atoms and ligands in them. For example the following complex compounds are polymerisation isomers to each other, since (*ii*) and (*iii*) complexes appear to be dimer of (*i*) complex and (*iv*) complex appears to be the pentamer of (*i*) complex. The ratio Co^{3+} : NH_3 : NO_2^- in all the complexes is 1 : 3 : 3.

Complex compound	*Number of* Co^{3+}	*Number of* NH_3	*Number of* NO_2^-
(*i*) $[Co(NH_3)_3(NO_2)_3]$	1	3	3
(*ii*) $[Co(NH_3)_6][Co(NO_2)_6]$	2	6	6
(*iii*) $[Co(NH_3)_4(NO_2)_2][Co(NH_3)_2(NO_2)_4]$	2	6	6
(*iv*) $[Co(NH_3)_5(NO_2)]_3\ [Co(NO_2)_6]_2$	5	15	15

Stereo Isomerism

Stereo isomerism is exhibited by those compounds which have the same position of atoms or groups but these atoms or groups have different arrangement round the central atom. The compounds showing stereo isomerism are called **stereo isomers.** Stereo isomerism may be of two types *viz. geometrical* (or *cis-trans*) *isomerism and optical* (or *d–l* or *mirror-image*) *isomerism.*

Geometrical (or cis-trans) Isomerism

The complex compounds which have the same ligands in the coordination sphere but the relative position of the ligands round the central metal atom is different are called *geometrical isomers* and the phenomenon is called *geometrical isomerism.*

In a given complex compound the two ligands may occupy positions either adjacent to each other or opposite to each other. The complex compound having two ligands occupying the adjacent positions to each other is called *cis-isomer* while that in which the two ligands occupy opposite positions is called *trans-isomer.* Thus geometrical isomerism is also called *cis-trans isomerism.*

Geometrical isomerism is not found in complex compound with coordination number 2 and 3, since in these cases all the positions occupied by the ligands round the central metal atom are adjacent to one another. Geometrical isomerism is most common with the complex compounds having coordination number 4 and 6.

Geometrical isomerism in 4-coordinated complex compounds

Compelxes having central atom with 4-coordination number may have either tetrahedral or square planar geometry.

Geometrical isomerism in tetrahedral complexes

Geometrical isomerism cannot be shown by tetrahedral complexes, since all the four ligands in this geometry have adjacent position (*i.e.*, *cis* position) to one another and all the four bond angles are the same (= 109.5°).

Geometrical isomerism in square planar complexes

Before discussing the geometrical isomerism exhibited by various types of square planar complexes, we should know how *cis* and *trans* isomers of a square planar complex are named. A square planar complex having similar ligands at adjacent positions (90° apart) is called *cis*-isomer while a square planar complex having two similar ligands at opposite positions (180° apart) is called *trans*-isomer. Thus a square planar complex having two similar ligands at 1-2, 2-3,3-4 and 1-4 positions is called *cis*-isomer while that having two similar ligands at 1-3 and 2-4 positions is called *trans*-isomer. *cis*-and *trans*-isomers are also named by numbering system. Thus in $[Pd^{2+}Cl_2BrI]^{2+}$, if two Cl^- ions are placed cis to each other or at 1-2 positions, it is named as *cis-dichloro bromoiodo palladium* (II) *ion* or *1,2-dichloro bormoiodo palladium* (II) *ion*. On the other, if two Cl^- ions are placed *trans* to each other or at 1-3 positions, it is named as *trans-dichlorobromoiodo palladium (II) ion* or *1, 3-dichlorobromoiodopalladium* (II) ion (See Fig. 4.2.)

Here we shall consider geometrical isomerism in square planar complexes of $[Ma_4]$, $[Ma_3b]$, $[Mab_3]$, $[Ma_2b_2]$, $[Ma_2bc]$, $[Mabcd]$ and $[M(AB)_2]$ type in which M is the central metal atom; a, b, c and d are monodentate ligands and (AB) is an unsymmetrical bidentate ligand, since it has two different donor atoms namely A and B.

1. $[Ma_4]$, $[Ma_3b]$ and $[Mab_3]$ type complexes. Square planar complexes of this type donot show geometrical isomerism, since all the possible spatial arrangement of four ligands round the central metal atom is the same.

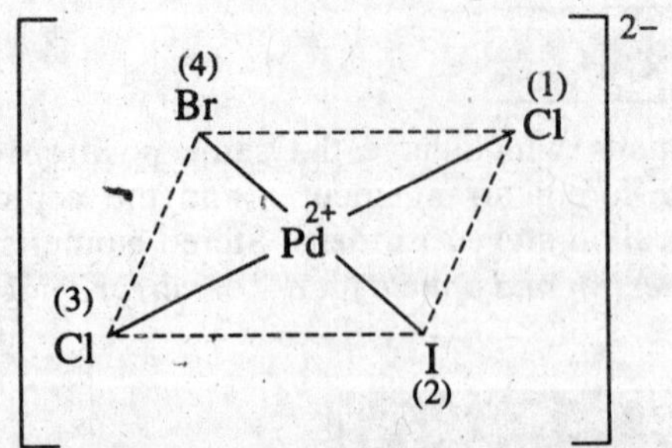

(*a*) *cis*-dichloro bromoiodo palladium (II) ion or 1, 2-dichloro bromoiodo palladium (II) ion, $[Pd^{2+}Cl_2BrI]^{2-}$

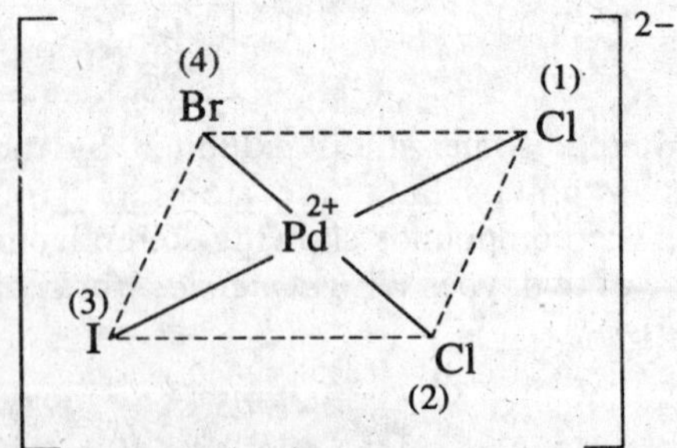

(*b*) *trans*-dichloro bromoiodo palladium (II) ion or 1, 2-dichloro bromoiodo palladium (II) ion, $[Pd^{2+}Cl_2BrI]^{2-}$

Fig. 4.2. Nomenclature of geometrical (*cis*-and *trans*) isomers of a square planar complex.

2. $[Ma_2b_2]$ type complexes. Important examples of square planar complexes of this type are $[Pt(NH_3)_2Cl_2]^0$, $[Pt(py)_2Cl_2]^0$, $[Pd(NH_3)_2(NO_2)_2]^0$ etc. These complexes exist in *cis*- and *trans*-isomers. These isomers of $[Pt^{2+}(NH_3)_2Cl_2]^0$ are shown in Fig. 4.3. In (*a*) since both NH_3 molecules and both Cl^- ions are *cis* to each other, it is called *cis*-isomer. On the other hand, in (*b*) since both NH_3 molecules and both Cl^- ions are *trans* with respect to each other, it is called *trans-isomer.*

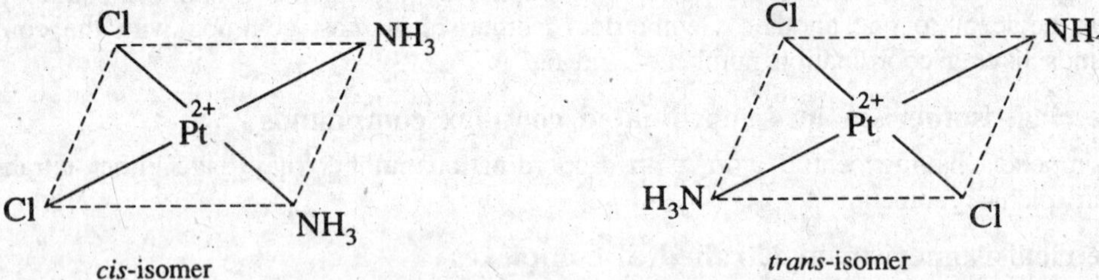

Fig. 4.3. *cis*-and *trans*-isomers of $[Pt^{2+}(NH_3)_2Cl_2]^0$.

3. [Mabcd] type compelxes. Square planar complexes of this type exist in three isomeric forms. For example $[Pt^{2+}(NH_3)(py)(Cl)(Br)]$ exists in three isomeric forms shown below in Fig. 4.4. These isomeric forms can be obtained by selecting one ligand, say NH_3, and then placing the remaining three ligands, one by one, *trans* to NH_3.

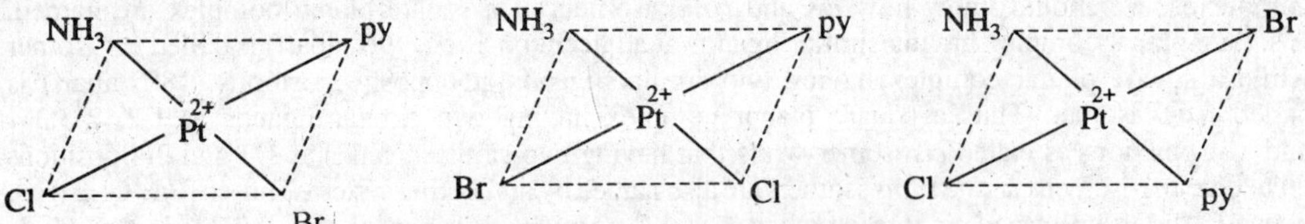

Fig. 4.4. Three isomeric forms of $[Pt^{2+}(NH_3)(py)(Cl)(Br)]$

$[Pt^{2+}(NO_2)(py)(NH_3)(NH_2OH)]^+$ and $[Pt^{2+}(C_2H_4)(NH_3)(Cl)Br)]$ are other examples of square planar complexes which exist in three isomeric forms.

The existence of three isomeric forms in case of the complexes mentioned above indicates that these complexes have square planar geometry.

4. $[Ma_2bc]$ type complexes. Square planar complexes of this type also show *cis-trans* isomerism. For example $[Pd^{2+}Cl_2BrI]^{2-}$ ion exists in *cis* and *trans*-isomers as shown below in Fig. 4.5.

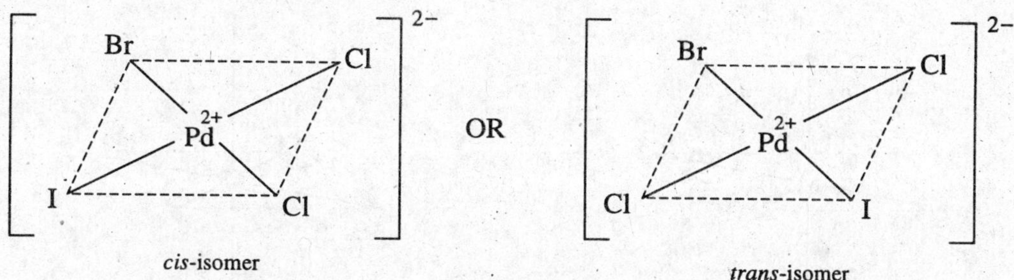

Fig. 4.5. *Cis*-and *trans*-isomers of $[Pd^{2+}Cl_2BrI]^{2-}$ ion.

$[Pt^{2+}(py)_2(NH_3)Cl]^+$ is another example of square planar complex of [Ma$_2$bc] type which exists in *cis*-and *trans*-isomers.

5. [M(AB)$_2$] type complexes. Here M is the central metal ion and [AB] represents an unsymmetrical bidentate ligand in which A and B are two different coordinating (donor) atoms. Square planar complexes of this type also exist in *cis*-and *trans*-isomers. For example $[Pt^{2+}(gly)_2]$ exists in *cis*-and *trans*- isomers as shown below in Fig. 4.6. Here gly represents the glycinato ligand, $NH_2.CH_2COO^-$ which has N and O atoms as its donor atoms (*i.e.*, A = N and B = O).

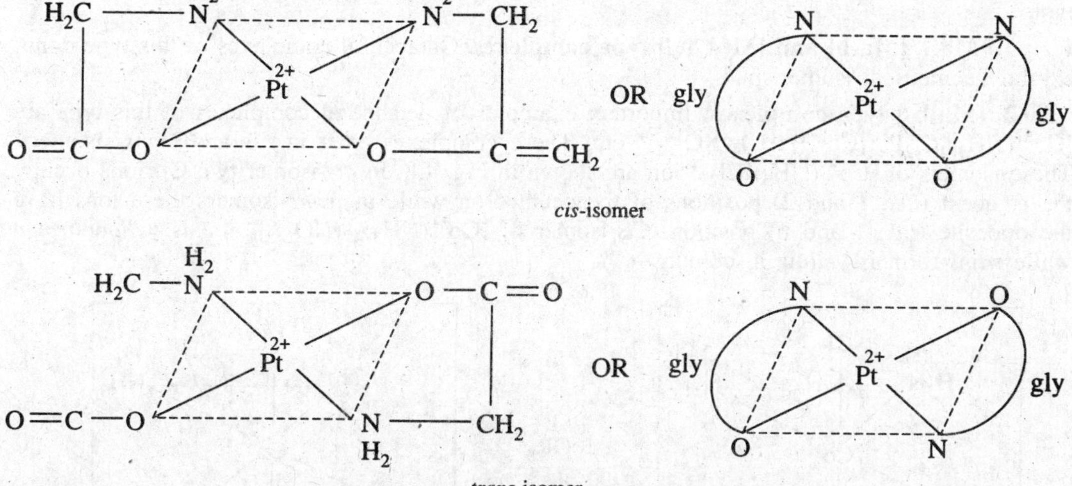

Fig. 4.6. *cis*- and *trans*-isomers of $[Pt^{2+}(gly)_2]$.

$[Cu^{2+}(gly)_2]^0$ is also an example of square planar complex of $[M(AA)_2]$ type. This complex also exhibits *cis-trans*-isomerism.

Geometrical isomerism in 6-coordinated complexes : Octahedral complexes

We know that a complex compound having central atom with coordination number equal to 6 is octahedral in shape. Before discussing the geometrical isomerism exhibited by various types of octahedral complexes, we should know how *cis*- and *trans*- isomers of an octahedral complex are named. In an octahedral complex, if two similar ligands are placed on any of the twelve edges of the octahedron, they are said to be in *cis* position. On the other hand, if two similar ligands are lying on a straight line which passes through the centre (where the metal ion is placed), they are said to be in *trans* position. Thus in an octahedral complex, the two ligands at positions 1-6, 2-4 and 3-5 are *trans* to each other and the two ligands occupying positions 1-2, 1-3, 2-3, 3-6, 6-4 etc. are *cis* to each other. (See Fig. 4.7)

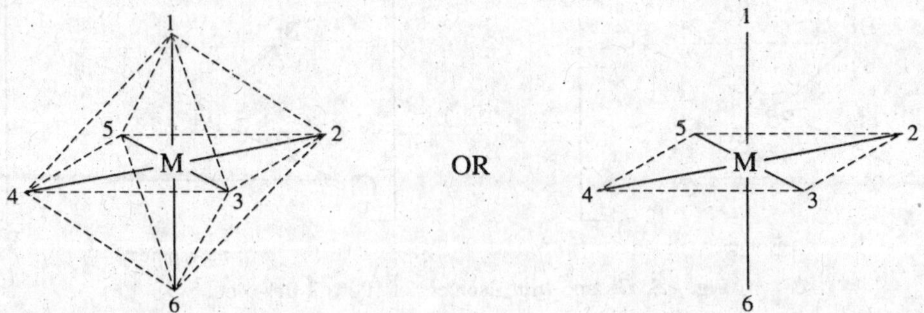

Fig. 4.7. Numbering of six ligands in a regular octahedral complex round the central metal ion, M.

The above numbering system of ligands can be exemplified by considering the nomenclature of *cis*- and *trans*-isomers of $[Co^{3+}(NH_3)_4Cl_2]^+$ ion which is an octahedral complex ion. *Cis*-isomer of this ion in which two Cl^- ions are *cis* to each other is also called *1, 2-dichlorotetrammine cobalt (III) ion* (see Fig. 4.8 (*a*)]. Similarly, *trans*-isomer of this ion in which two Cl^- ions are *trans* to each other is also called *1, 6-dichlorotetrammine cobalt (III) ion.* [See Fig. 4.8 (*b*)].

Here we shall discuss the geometrical isomerism in octahedral complexes of the following type:

1. $[Ma_6]$, $[Ma_5b]$ and $[M(AA)_3]$ type complexes. Octahedral complexes of this type donot exhibit geometrical isomerism.

2. $[Ma_4b_2]$ type complexes. Important examples of octahedral complexes of this type are $[Co^{3+}(NH_3)_4Cl_2]^+$, $[Co^{3+}(NH_3)_4(NO_2)_2]^+$ etc. These complexes exist in *cis* and *trans*- isomers. These isomers of $[Co^{3+}(NH_3)_4Cl_2]^+$ ion are shown in Fig. 4.8. In *cis*-isomer two Cl^- ions occupy the adjacent (*i.e.*, 1 and 2) positions of the octahedron while in *trans*-isomers these ions have the opposite (*i.e.*, 1 and 6) position. *Cis*-isomer of $[Co^{3+}(NH_3)_4(NO_2)_2]^+$ ion is *yellow-brown* while *trans*-form is *yellow* in colour.

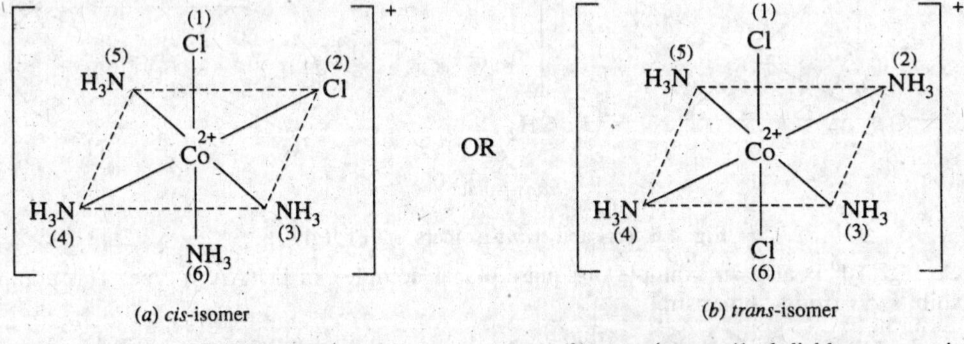

(*a*) *cis*-isomer (1,2-dichlorotetrammine cobalt (III) ion). (yellow-brown)

(*b*) *trans*-isomer (1, 6-dichlorotetrammine cobalt (III) ion). (yellow)

Fig. 4.8. *cis*- and *trans*-isomers of $[Co^{3+}(NH_3)_4Cl_2]^+$ ion.

3. $[Ma_4bc]$ type complexes. $[Co^{3+}(NH_3)_4(H_2O)Cl]^{2+}$ ion is an important example of octahedral complex of $[Ma_4bc]$ type. This ion has *cis*- and *trans*-isomers whose structures are given in Fig. 4.9. In *cis*-form two NH_3 molecules have *cis* positions to each other and in *trans*-form these ligands (*i.e.*, two NH_3 molecules) have trans positions with each other.

4. $[Ma_3b_3]$ type complexes. $[Co(NH_3)_3Cl_3]$, $[Co(NH_3)_3(NO_2)_3]$, $[Cr(H_2O)_3F_3]$, $[Cr(NH_3)_3Cl_3]$, $[Rh(py)_3Cl_3]$, $[Ru(H_2O)_3Cl_3]$, $[Ir(H_2O)_3Cl_3]$, $[Pt(NH_3)_3Br_3]^+$, $[Pt(NH_3)_3I_3]^+$ etc. are important

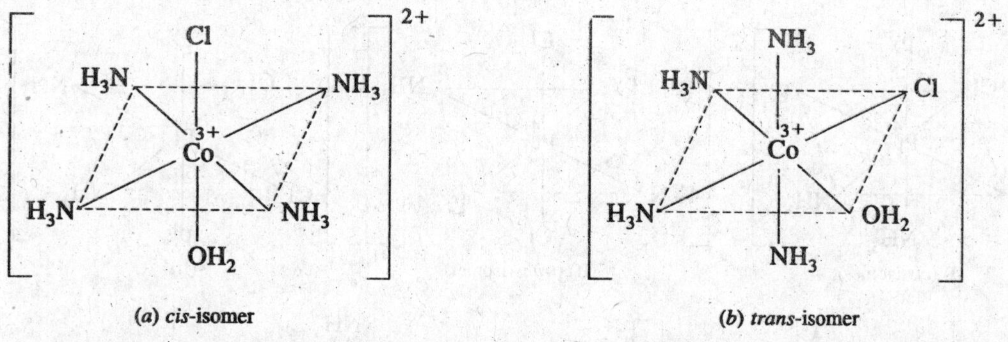

(*a*) *cis*-isomer (*b*) *trans*-isomer

Fig. 4.9. Structure of *cis*- and *trans*-isomers of $[Co^{3+}(NH_3)_4(H_2O)Cl]^{2+}$ octahedral complex ion.

examples of octahedral complexes of $[Ma_3b_3]$ type. Each of these complex exists in *cis*-and *trans*-isomers. As an example these isomers for $[Co^{3+}(NH_3)_3Cl_3]^0$ have been shown in Fig. 4.10.

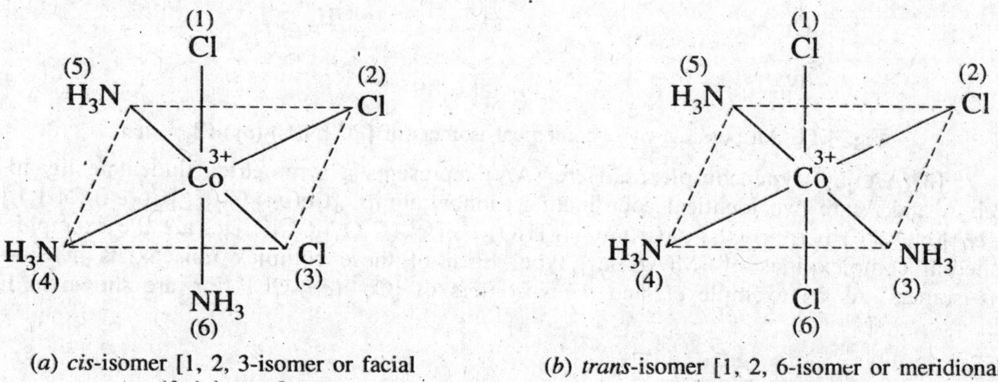

(*a*) *cis*-isomer [1, 2, 3-isomer or facial (*fac*) isomer] (*b*) *trans*-isomer [1, 2, 6-isomer or meridional (*mer*) isomer].

Fig. 4.10. *cis*- and *trans*-isomers of $[Co^{3+}(NH_3)_3Cl_3]$

It may be seen from the figure that in *cis*-isomer the three Cl^- ions lie on the corners of one of the triangular face of the octahedron and the three NH_3 molecules are lying on the corners of the opposite triangular face of the octahedron. Thus in cis-isomer the three Cl^- ions occupy 1, 2 and 3 positions and hence this isomer is also called *1, 2, 3 isomer* or *facial (fac) isomer*. In *trans*-isomer the three Cl^- ions are lying around one edge of the octahedron and the three NH_3 molecules are lying around the opposite edge of the octahedron. Thus in *trans*-isomer the three Cl^- ions occupy, 1, 2 and 6 positions and hence *trans*-isomer is also called *1, 2, 6 isomer or meridional (mer) isomer.*

5. $[Ma_2b_2c_2]$ type complexes. $[Pt^{4+}(NH_3)_2(py)_2Cl_2]^{2+}$ ion is an important example of octahedral complex of $[Ma_2b_2c_2]$ type. This ion can exist theoretically in five geometrical isomers that have been shown in Fig. 4.11, but it is only three isomers that have been actually isolated. It may be noted from the figure that (I) form is a *cis*-form since in this form two identical ligands are occupying adjacent positions. Form (II) is a *trans*-form because in it the two identical ligands are placed at opposite positions.

6. [Mabcdef] type complexes. Octahedral complexes which have different monodentate ligands have been prepared for Pt(IV). $[Pt^{4+}(py)(NH_3)(NO_2)(Cl)(Br)(I)]$ is the only complex of this type of complexes. Theoretically 15 geometrical isomers are possible for this complex compound. In fact only three geometrical isomers have been isolated.

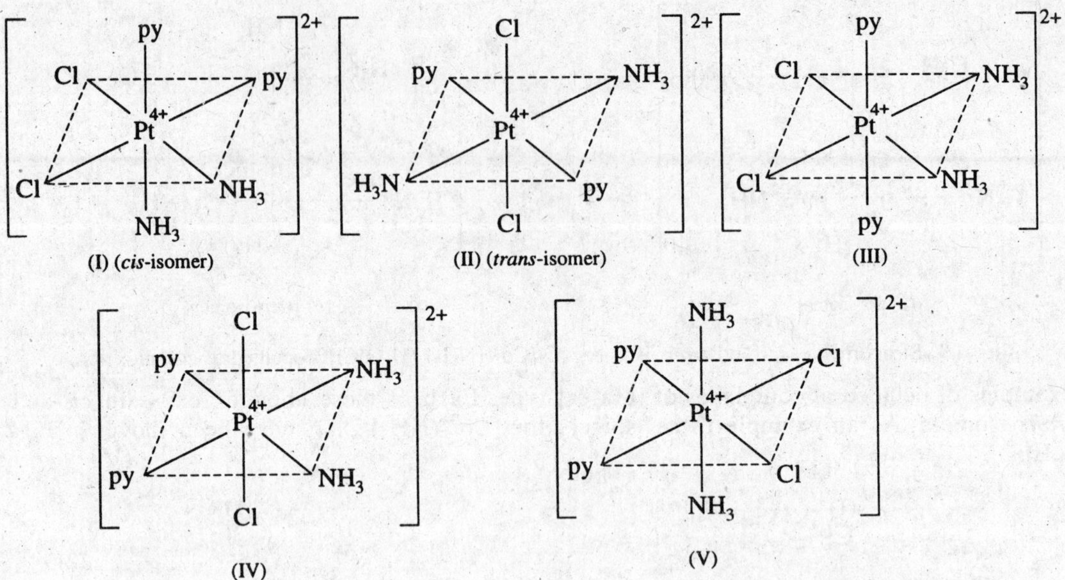

Fig. 4.11. All possible five geometrical isomers of $[Pt^{4+}(NH_3)_2(py)_2Cl_2]^{2+}$ ion.

7. $[M(AA)_2a_2]$ type complexes. Here (AA) represents a symmetrical bidentate ligand in which A and A are two identical coordinating (donor) atoms. $[Co(en)_2Cl_2]^+$, $[Co(en)_2(NH_3)_2]^{3+}$, $[Co(en)_2(NO_3)_2]^+$, $[Cr(en)_2Cl_2]^+$, $[Cr(C_2O_4)_2(H_2O)_2]^-$, $[Ir^{4+}(C_2O_4)_2Cl_2]^{2-}$ etc. are the examples of octahedral complex ions of $[M(AA)_2a_2]$ type. Each of these complex ions exists in *cis*-and *trans*-isomers. As an example *cis*-and *trans*-isomers of $[Co^{3+}(en)_2Cl_2]^+$ ion are shown in Fig. 4.12.

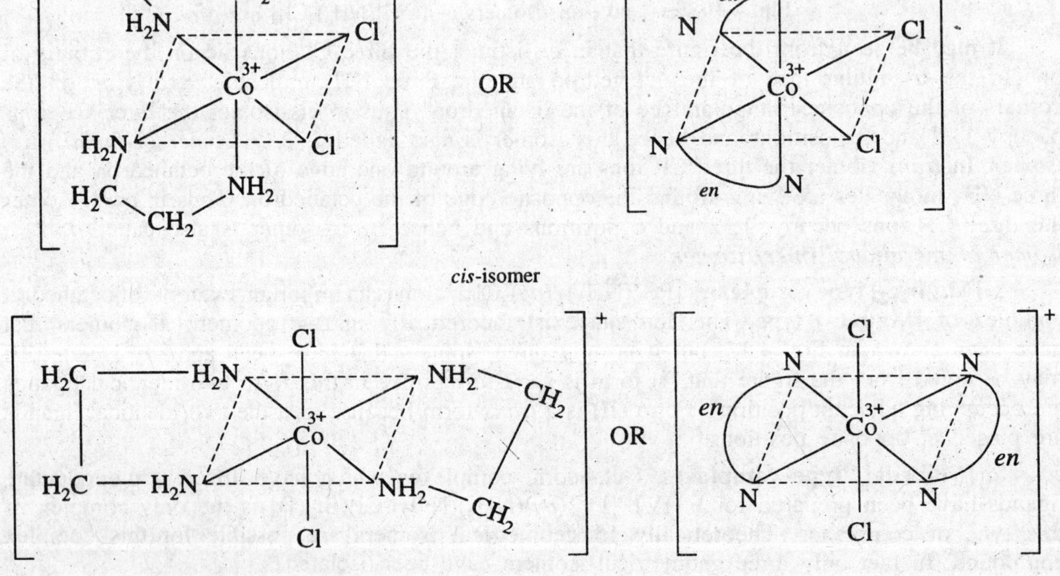

Fig. 4.12. *cis*- and *trans*-isomers of $[Co^{3+}(en)_2Cl_2]^+$ ion.

8. [M(AA)$_2$ab] type complexes. $[Co^{3+}(en)_2(NH_3)Cl]^{2+}$ is an important example of octahedral complex of [M(AA)$_2$ab] type. This complex ion exists in *cis*- and *trans*- isomers. In *cis*-isomer the two monodentate ligands *viz.* NH_3 and Cl^- occupy the adjacent (*i.e.*, *cis*) positions while in *trans*-isomer these ligands occupy opposite (*i.e.*, *trans*) positions (see Fig. 4.13). $[Ru^{3+}(C_2O_4)_2(py)(NO_2)]^{2-}$ also exists in *cis*- and *trans*- isomers.

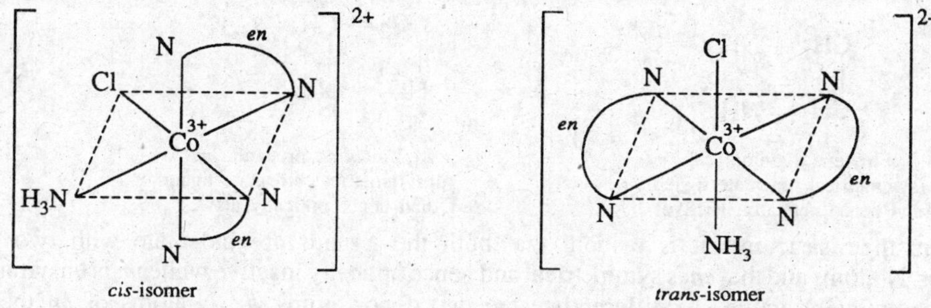

Fig. 4.13. *cis*- and *trans*-isomers of $[Co^{3+}(en)_2(NH_3)(Cl)]^{2+}$ ion.

9. Octahedral complexes of [M(AA)a$_2$b$_2$] type. $[Co^{3+}(en)(NH_3)_2Cl_2]^+$ ion is an important example of an octahedral complex ion of [M(AA)a$_2$b$_2$] type. In the *cis*-isomer of this ion, both NH_3 groups and both Cl^- ions occupy the *cis* positions while in *trans*-isomer both these pairs occupy the *trans* positions (see Fig. 4.14).

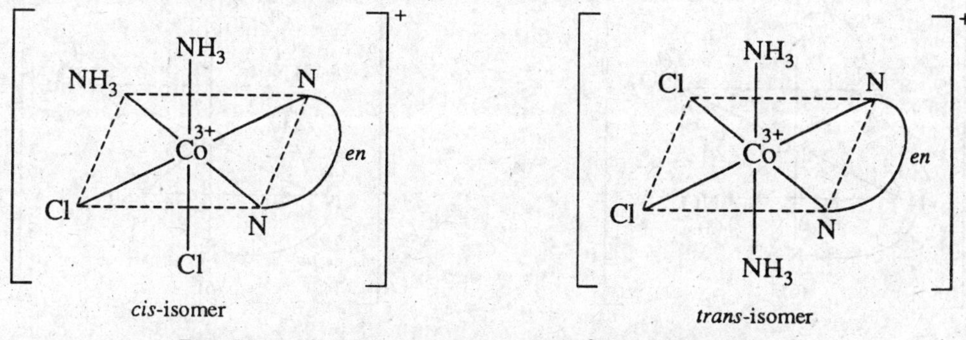

Fig. 4.14. *cis- and trans-isomers of* $[Co^{3+}(en)(NH_3)_2Cl_2]^+$ ion.

$[Co^{3+}(C_2O_4)(NH_3)_2(NO_2)_2]^-$ ion is another example of octahedral complex of [M(AA)a$_2$b$_2$] type.

10. M(AB)$_3$] type complexes. Here (AB) represents an unsymmetrical bidentate ligand in which A and B are two different coordinating (donor) atoms. Octahedral complexes of this type exist in *cis*- and *trans*-isomers. As an example the *cis*- and *trans*-isomers of $[Cr^{3+}(gly)_3]$ has been shown in Fig. 4.15. Each of these forms is optically active and hence each has a pair of optical isomers. In *cis*-isomer two N-atoms and two O-atoms of two *gly* ions are placed at *cis*-positions while in *trans*-isomer these atoms are occupying *trans*-positions.

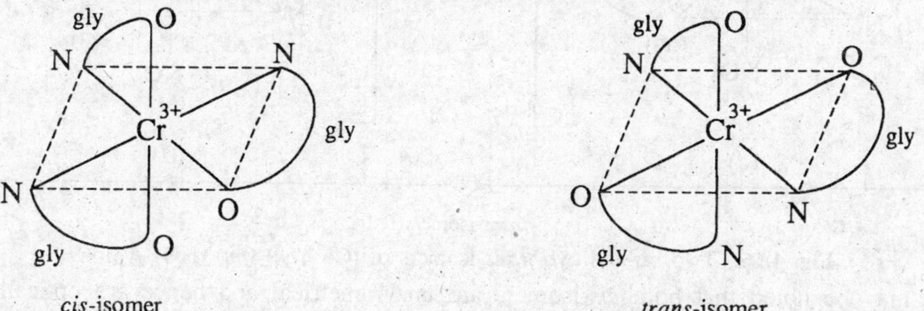

Fig. 4.15. *cis*- and *trans*-isomers of $[Cr^{3+}(gly)_3]$.

11. Octahedral complexes containing optically active bidentate ligands. $[Co^{3+}(en)(pn)(NO_2)_2]^+$ ion is an important example of such type of octahedral complexes. Here *en* and *pn* are ethylene diamine and 1, 2-diamino propane (also called propylene diamine) respectively. Both of these ligands are bidentate ligands and their structures are given below:

```
CH2—NH2
|
CH2—NH2
```

Ethylene diamine (*en*)
(symmetrical bidentate ligand
and hence optically inactive)

```
 1      2      3
CH2 —  CH  —  CH3
|       |
NH2     NH2
*       **
```

1,2-diamino propane (*pn*)
(unsymmetrical bidentate ligand
and hence optically active)

From their structures it is evident that both the ligands are bidentate with two neutral donors *viz.* N-atoms and that *en* is symmetrical and hence optically inactive while *pn* is unsymmetrical and hence optically active. To differentiate the two donor atoms *viz.* N-atoms of *pn* they have been shown with one and two asterisks. $[Co(en)(pn)(NO_2)_2]^+$ ion exists in four geometrical forms; two are *cis* and two are *trans*. (*a*) *cis*-isomer has been obtained by placing $\overset{*}{N}H_2$ and $\overset{**}{N}H_2$ at 4 and 6 positions of the octahedron and in (*b*) *cis*-isomer their positions have been reversed. (*c*) and (*d*) which are *trans*-isomers have been obtained similarly [see Fig. 4.16].

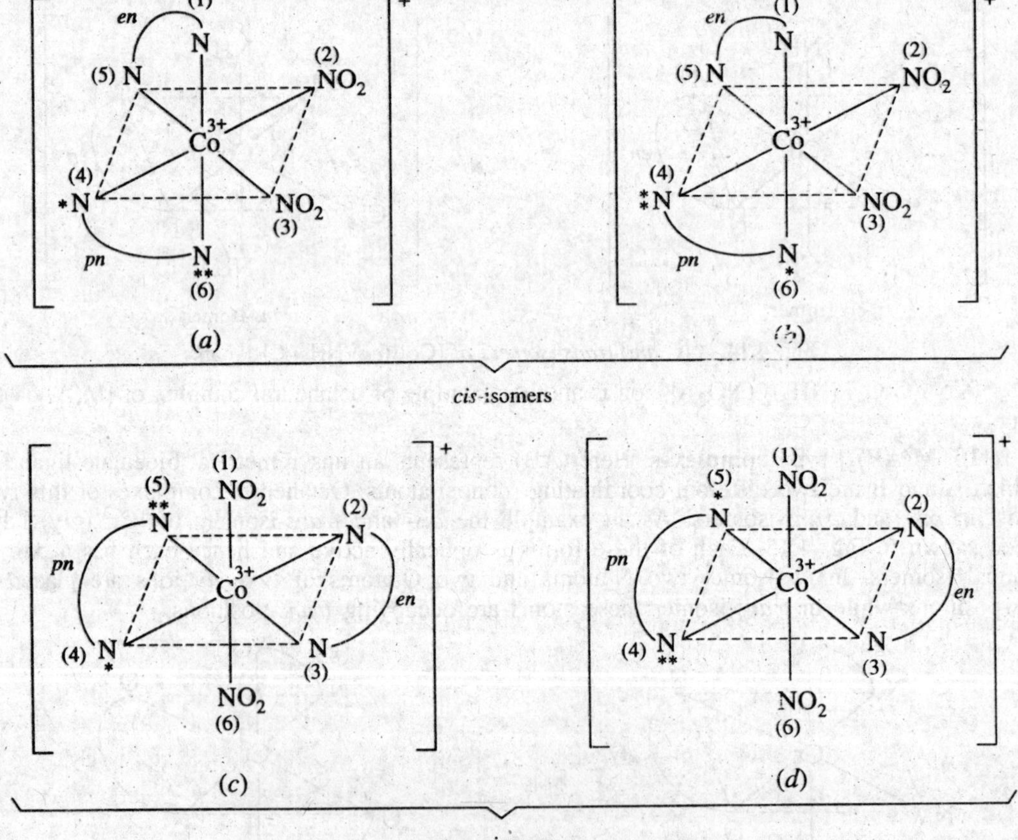

Fig. 4.16. Two *cis*-and two *trans*-isomers of $[Co^{3+}(en)(pn)(NO_2)_2]^+$ ion.

It may be noted that both *cis*-isomers are unsymmetrical and hence are optically active while both *trans*-isomers are symmetrical and hence are optically inactive.

12. Dinuclear bridged octahedral complexes. (*a*) Fe (III) forms hydroxy-bridged dinuclear octahedral complex of $[Fe_2^{3+}(OH)_4(H_2O)_6]^{2+}$ type. The structure of this ion has already been given at Fig. 1.53 (Chapter 1). This ion shows geometrical isomerism and gives three geometrical isomers whose structures are given in Fig. 1.54. in Chapter 1.

(*b*) Co (III) forms dinuclear bridged octahedral complex which contains NH_2^- and NO_2^- ions as bridging ligands. This complex is formulated as $[Co_2^{3+}(en)_4(NH_2)(NO_2)]^{4+}$ whose structure has already been given at Fig. 1.45 in chapter 1. This ion exists in *cis*-and *trans*-isomers. The structures of these isomers have been given at Fig. 1.46 in chapter 1.

To distinguish between *cis*- and *trans*-isomers

The following methods may be used to distinguish between *cis*- and *trans*-isomers.

1. Dipole moments measurements. Jensen has shown that the Pt (II) complexes of $[Pt^{2+}A_2X_2]$ type, where A = substituted phosphine, arsine or stilbine such as $(C_2H_5)_3P$, $(C_2H_5)_3As$ or $(C_2H_5)_3Sb$ and X = a halogen, have their dipole moments (μ) either equal to zero or between 8 and 12 Debye units. In the compounds with $\mu = 0$, the individual moments have cancelled one another and so these are *trans*-isomers. The compounds with $\mu = 8 - 12$ Debye units are cis-isomers.

2. X-ray crystal analysis. X-ray crystal analysis and dipole moment measurements of several Pt(II) complexes have confirmed the square planarity of the bonds around the central metal atom. This square planar arrangement has also been established for 4-coordinated Pt(II), Ag(II), Cu(II) and Au(III) complexes. Tetrahedral configurations have been asigned to Cu(I) Ag(I), Au(I), B(III), Al(III), Zn(II), Cd(II), Hg(II) and Co(II) in their coordinated compounds. Some metals such as Ni(II) appear to show either configuration for 4-coordination.

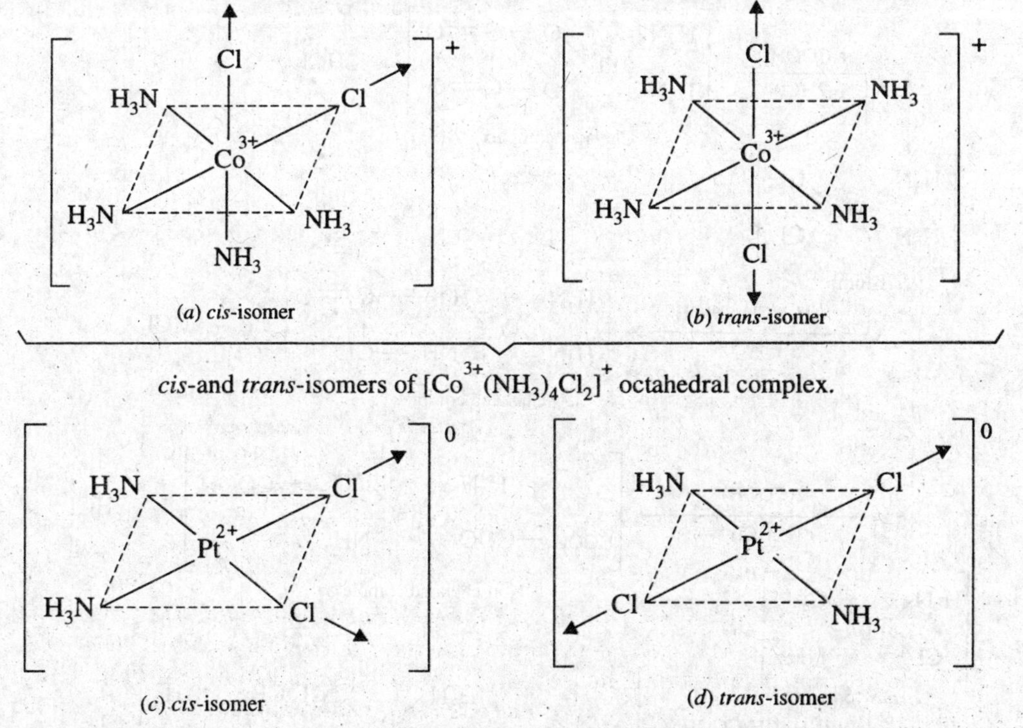

Fig. 4.17. Symmetrical Cl–metal–Cl stretching vibrations of *cis*-and *trans*- isomers of $[Co(NH_3)_4Cl_2]^+$ and $[Pt(NH_3)_2Cl_2]^0$

3. Infra-red spectroscopic technique. In a *trans* octahedral complex such as $[Co^{3+}(NH_2)_4Cl_2]^-$ or in *trans* square planar complex like $[Pt^{2+}(NH_3)_2Cl_2]^0$ the Cl–metal–Cl symmetrical stretching vibration produces no change in the dipole moment of the molecule [Fig. 4.17 (*b*) and (*d*)], and thus no band corresponding to this vibration is observed in the infra-red spectrum. However, in the *cis*-form of each compound, the symmetrical stretching vibration [Fig. 4.17 (*a*) and (*c*)] as well the unsymmetrical stretching vibration produce appreciable changes in the dipole moment; hence the infra-red spectrum of the *cis*-isomer will contain a large number of bands due to Cl–metal–Cl stretching.

4. A *cis*-isomer has a high molar extinction coefficient (absorption) than the corresponding *trans*-isomer and sometimes the two isomers may differ significantly in their absorption spectra.

5. The technique which consists of attempting to resolve the *cis*-isomer into the two possible optically active isomers is also commonly used to distinguish it from the *trans*-isomers.

6. Grinberg's Method. It is a chemical method and assumes that when a chelating ligand having two donor atoms separated by two to four other atoms reacts with *cis*-and *trans*-isomers separately, its two donor atoms coordinate to the central metal ion of the *cis*-form at two *cis*-positions and thus form five or six membered ring, while in the *trans*-form they coordinate to the central metal ion at *trans* positions and thus the ligand acts as a monodentate ligand, *i.e.*, the ligand cannot form a ring complex with *trans* form. Typical chelating ligands used for the purpose are oxalic acid, $(COOH)_2$, glycine ($H_2N—CH_2—COOH$) and ethylenediamine ($NH_2—CH_2—CH_2—NH_2$). The application of this method can by explained be considering the reaction of oxalic acid and glycine with *cis*- and *trans*-isomers of square planar complex of Pt(II), $[Pt^{2+}(NH_3)_2Cl_2]^0$ (Fig. 4.18). Quite obviously the *cis*- isomer gives a *chelated complex* while the *trans*-isomer yields a *non-chelated complex*.

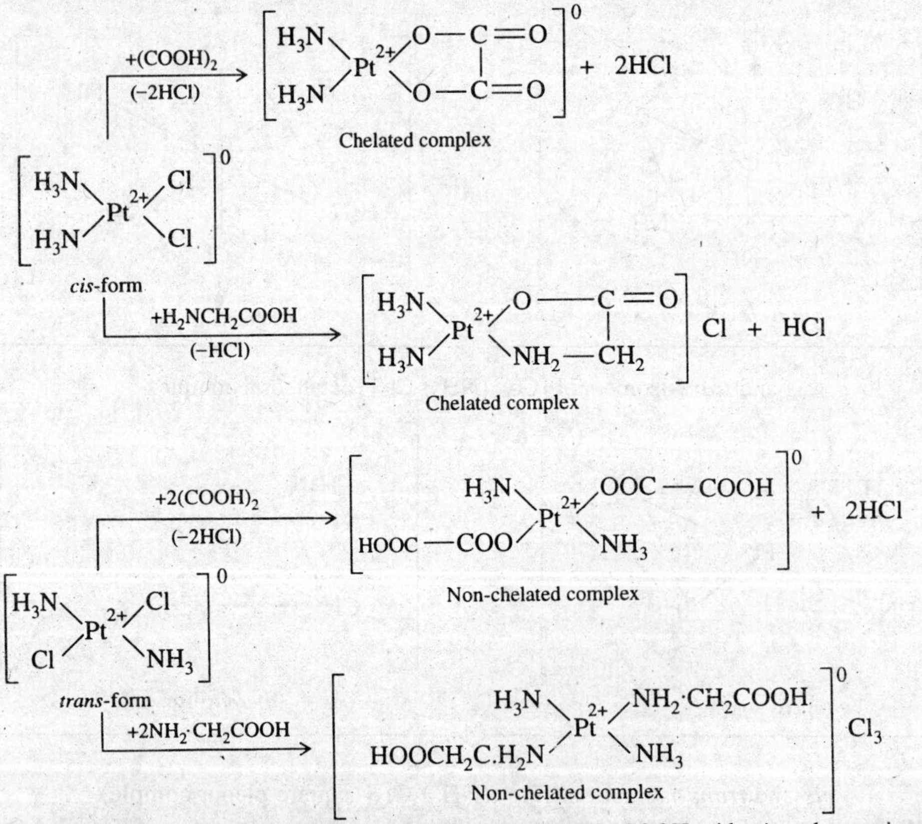

Fig. 4.18. Reaction of oxalic acid, $(COOH)_2$ and glycine, $NH_2.CH_2.COOH$ with *cis*-and *trans*-isomers of square planar complex of Pt(II), $[Pt(NH_2)_2Cl_2]^0$.

7. Kurnakov's reaction. Kurnakov utilised the phenomenon of trans-effect in distinguishing the pairs of *cis*- and *trans*-isomers of square planar complexes of $[PtA_2X_2]$ type by treating them with thiourea.

Optical (or *d-l* or mirror-image) Isomerism

Definition of some terms related with optical isomerism

Before discussing the optical isomerism shown by various types of complexes we will define some important terms related to this isomerism.

When the solutions of certain complex compounds are placed in the path of a *plane-polarised light* (the waves of the plane-polarised light vibrate only in one direction; vibrations in other directions are cut off), they rotate its plane through a certain angle which may be either to the left or to the right. *This property of a complex of rotating the plane of polarised light is called its optical activity and the complex possessing this property is said to be optically active.* Optically active complexes are said to exist in the following forms:

(*a*) One which rotates the plane of polarised light towards right (*i.e.*, in clockwise direction) is said to be *dextro-rotatory* or *d*-form. *d*-form is also represented by placing (+) sign before its name or formula.

(*b*) One which rotates the plane of polarised light towards left (*i.e.*, in anti-clockwise direction) is called *levo-rotatory* or *l*-form. *l*-form is also represented by putting (–) sign before its name or formula. (+), dextro, and (–), levo, refer to the sign of rotation of the optical isomer at the sodium D line wavelength. The *d*- and *l*-forms have the following characteristics: (*i*) Since *d*-and *l*-forms are capable of rotating the plane of polarised light, these are said to be *optically active forms* or *optical isomers* and this phenomenon is called *optical activity* or *optical isomerism.* These two forms have exactly identical physical and chemical properties, although they differ in their action on polarised light. (*ii*) *d*-and *l*-forms are mirror images to each other just as left hand is the mirror image of the right hand. Thus *d*- and *l*-forms can be superimposed on each other and hence *d*-and *l*-forms are also called *mirror-image isomers*. These optical isomers are also called *optical antipods* or *enantiomorphs* (Latin: *enantio* = opposite; *morphs* = forms) or *enantiomers.* From this description it follows that optical isomerism can also be called *mirror-image isomerism* or *enantiomorphism* or *enantiomerism.*

(*c*) The isomer which is not capable of rotating the plane of polarised light is called *optically inactive*. Such an isomer is called *meso, racemic, dl-or (±) form.* A racemic substance is composed of 50% *d*- and 50% *l*-form. The solution of a racemic (*dl* mixture) form in a solvent which contains equimolecular amounts of *d*- and *l*-form is symmetrical and hence optically inactive, because the rotation (of the plane of polarised light) produced by one isomer, say *d*-isomer, is balanced or compensated by equal but opposite rotation produced by the other (*i.e.*, *l*-isomer). Optical inactivity produced in *dl*-mixture is said to be due to *external compensation.*

Conditions for a molecule to show optical isomerism

A molecule in which the grouping of the atoms is asymmetric is called an *asymmetric* or *dissymmetric molecule*. An asymmetric molecule has the following features:

(*i*) An asymmetric molecule never has a plane of symmetry (also called *mirror-image plane*) which is defined as an imaginary plane dividing the molecule in such a way that the part of it on one side of the plane is the mirror image of that on the other side of the plane. The molecules possessing such a plane of symmetry are always *inactive while those having no plane of symmetry are optically active and hence show optical isomerism.*

(*ii*) An asymmetric molecule cannot be superimposed on its mirror image.

Thus the most necessary and sufficient condition for a molecule to show optical isomerism (*i.e.*, to exist in *d*-and *l*-forms) is that *the moelcule should be asymmetric, (i.e., it should have no plane of symmetry) and should not be superimposable on its mirror image.*

Optical isomerism in 4-coordinated complexes: Tetrahedral complexes

Here we shall discuss optical isomerism in the following types of tetrahedral complexes.

1. $[Ma_4]$ and $[Ma_3b]$ type complexes. Tetrahedral complexes of $[Ma_4]$ and $[Ma_3b]$ type are not able to show optical isomerism because all the possible arrangements of the ligands round the central metal ion, M are exactly equivalent.

2. [Mabcd] type complexes. Since the central atom (M) in tetrahedral complexes of [Mabcd] type is surrounded by four different ligands, the tetrahedral complex of this type is expected to produce a pair of enantiomarphs. For example $[As^{3+}(CH_3)(C_2H_5)(S)(C_6H_4COO)]^{2+}$ ion (tetrahedral) exists in two optical isomers as shown in Fig. 4.19.

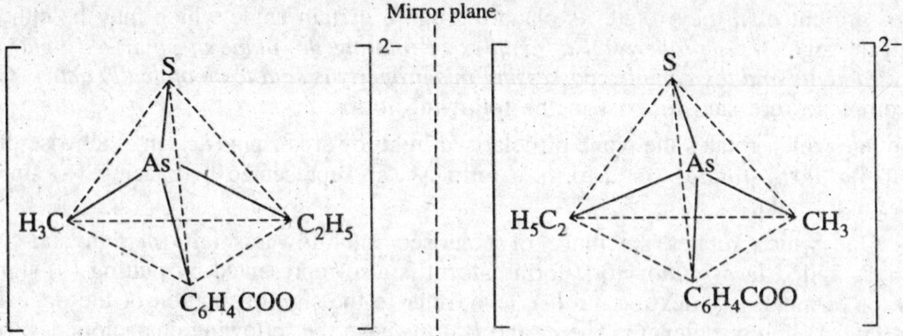

Fig. 4.19. Optical isomers of tetrahedral complex ion, $[As(CH_3)(C_2H_5)(S)(C_6H_4COO)]^{2-}$

Here it may be noted that 4 different groups round the central metal ion are not the only requirement to make the complex to show mirror-image isomerism. All that is required is that the molecule should be asymmetric (*i.e.*, unsymmetrical), *i.e.*, it should have no plane of symmetry so that it can exist in two mirror-image forms.

3. Tetrahedral complexes of Be(II), B(III) and Zn(II) with symmetrical bidentate ligands have been made and resolved into optical isomers. Some examples of tetrahedral complexes of this type are *bis* (salicylaldehydato) boron (III) ion, $[B^{3+}(C_6H_4.O.CHO)_2]^+$ and *bis* (benzoyl-acetonato) beryllium (II), $[Be^{2+}(C_6H_5CO.CH.CH_3CO)_2]^0$ whose mirror-image isomers are given in Fig. 4.20 and 4.21 respectively.

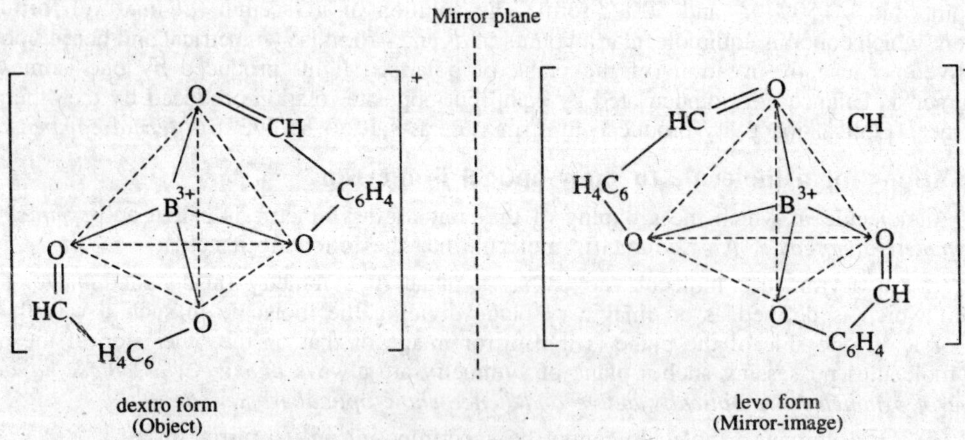

Fig. 4.20. Mirror-image isomers of tetrahedral *bis* (salicylaldehydato) boron (III) ion, $[B^{3+}(C_6H_4.O.CHO)_2]^+$.

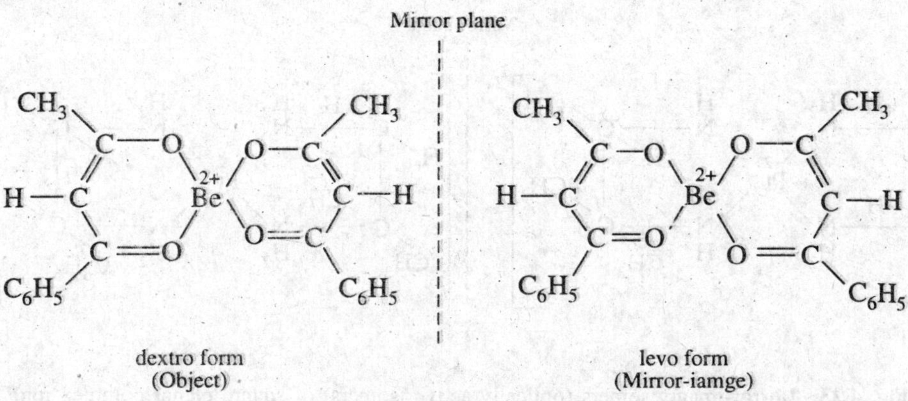

Fig. 4.21. Mirror-image isomers of tetrahedral *bis* (benzoylacetonato) beryllium (II) ion, $[Be^{2+}(C_6H_5CO.CH.CH_3CO)_2]^0$.

Optical isomerism in 4-coordinated complexes: Square planar complexes

Optical isomerism rarely occurs in square planar complexes, since they have all the four ligands and the central metal ion in the same plane and hence contain a plane or axis of symmetry. Thus these complexes are optically inactive, *i.e.*, they cannot show optical isomerism even if all the four ligands are different. Although square planar complexes seldom show optical isomerism, yet the following square planar complexes exhibit optical isomerism.

1. *Iso-butylene diamine*, $NH_2—CH_2—C(CH_3)_2—NH_2$ (bidentate ligand) and *meso-diphenyl ethylene diamine* (also called *meso-stilbene diamine*) (bidentate ligand), $NH_2—CH(C_6H_5)—CH(C_6H_5)—NH_2$ react with $[PtCl_4]^{2-}$ and forms a square planar complex which is called *iso-butylene diamine meso-stilbene diamine platinum* (II) *ion*, $[Pt^{2+} (NH_2—CH_2— C(CH_3)—NH_2)(NH_2—CH(C_6H_5)—CH(C_6H_5)—NH_2)]^{2+}$ (See Fig. 4.22).

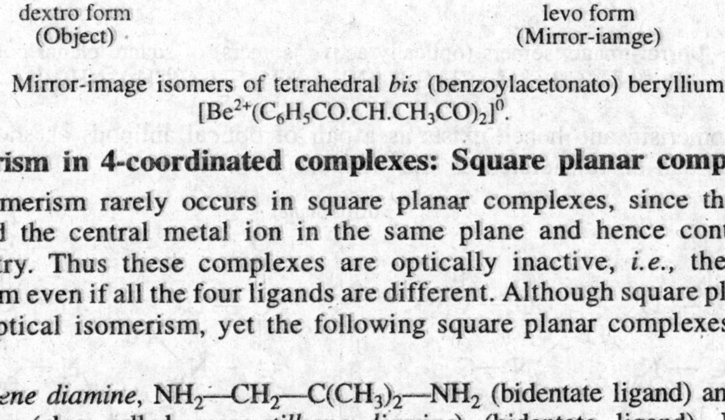

$H_2C—H_2N$ / $(CH_3)_2C—H_2N$ + $[PtCl_4]^{2-}$ + $NH_2—CH(C_6H_5)$ / $NH_2—CH(C_6H_5)$

Iso-butylene diamine | Meso-stilbene diamine

↓

$[(H_2C—NH_2)((CH_3)_2C—NH_2)Pt^{2+}(H_2N—CH(C_6H_5))(H_2N—CH(C_6H_5))]^{2+} + 4Cl^-$

Iso-butylene diamine meso-stilbene diamine paltinum (II) ion.

Fig. 4.22. Formation of iso-butylene diamine meso-stilbene diamine platinum (II) ion.

Mills and Quinbell (1935) have resolved the above complex into two optically active isomers which have been shown in Fig. 4.23.

2. Ethylene diamine tetraacetic acid forms a square planar complex ion with Pd(II). In this complex ion ethylene diamine tetraacetic acid acts as a tetradentate ligand. This complex ion

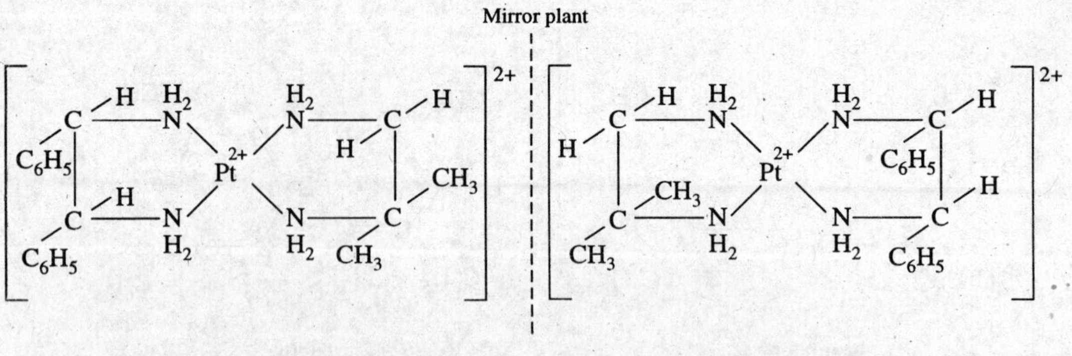

Fig. 4.23. Mirror-image isomers (optically active isomers) of square planar complex ion, $[Pt^{2+}(NH_2.CH(C_6H_5).CH(C_6H_5).NH_2)\ (NH_2.CH_2.C(CH_3)_2.NH_2)]^{2+}$

shows optical isomerism and hence exists as a pair of optical antipods as shown in Fig. 4.24. This complex ion can be formulated as $[Pd^{2+}(EDTA)]^{2-}$.

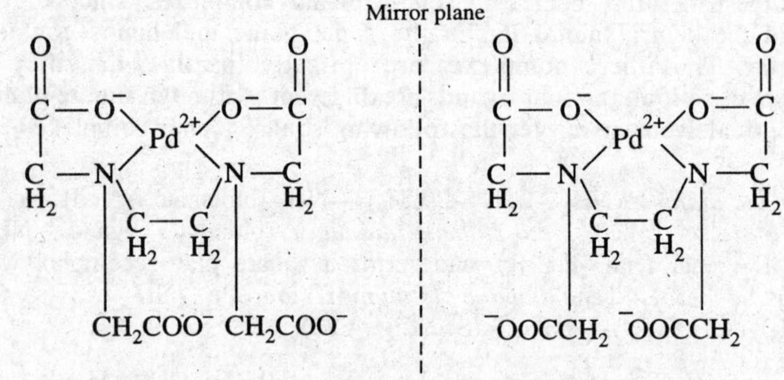

Fig. 4.24. Pair of optical antipods of $[Pd^{2+}(EDTA)]^{2-}$ ion.

3. The resolution of a complex of zinc with 8-hydroxy quinoline-5-sulphonic acid intc optical antipods has been described by Liu and Bailor (1951).

Optical isomerism in 6-coordinated complexes : Octahedral complexes

Here we shall discuss the optical isomerism in octahedral complexes of the followinǥ type:

1. $[Ma_4b_2]$ and $[Ma_3b_3]$ type complexes. (*a*) We have already seen that octahedral complexes of $[Ma_4b_2]$ type exist in *cis*- and *trans*-isomers. Both these isomers are optically inactive due to the presence of symmetry and hence donot show optical isomerism.

(*b*) We have already mentioned that octahedral complexes of $[Ma_3b_3]$ type exist in *cis* (or *facial*) and *trans* (or *meridional*) isomers. Both these isomers are optically inactive and hence donot show optical isomerism.

2. $[Ma_2b_2c_2]$ type complexes. $[Pt^{4+}(NH_3)_2(py)_2Cl_2]^{2+}$ ion is an important example of octahedral complex of $[Ma_2b_2c_2]$ type. We have already stated that $[Pt^{4+}(NH_3)_2(py)_2Cl_2]^{2+}$ ion can exist theoretically in five geometrical isomers, but only three have been isolated. In *cis*-isomer twc identical ligands occupy the adjacent positions of the octahedron. This *cis*-isomer exists in twc optical (mirror-image) isomers which are mirror-image to each other as shown in Fig. 4.25 *Trans*-form of this complex ion is symmetrical and hence is optically inactive form. Thus it is a *trans-meso form*.

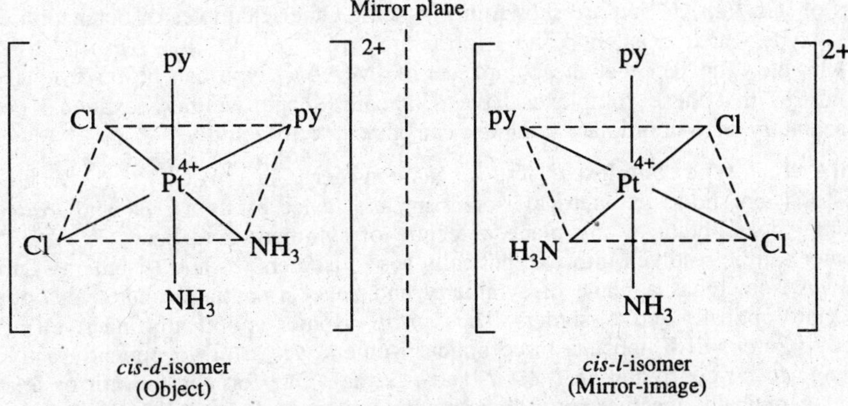

Fig. 4.25. Mirror-image isomers of $[Pt^{4+}(NH_3)_2(py)_2Cl_2]^{2+}$ ion.

3. [Mabcdef] type complexes. $[Pt^{4+}(py)(NH_3)(NO_2)ClBrI]^0$ is the only example of octahedral complex of this type. We have already stated that, theoretically, this complex can exist in 15 geometrical isomers. Each of these 15 geometrical isomers exists in optically active *d*-and *l*-foms, giving a total of 30 optically active isomers. *d*- and *l*-isomers for one of the 15 geometrical isomers are given in Fig. 4.26.

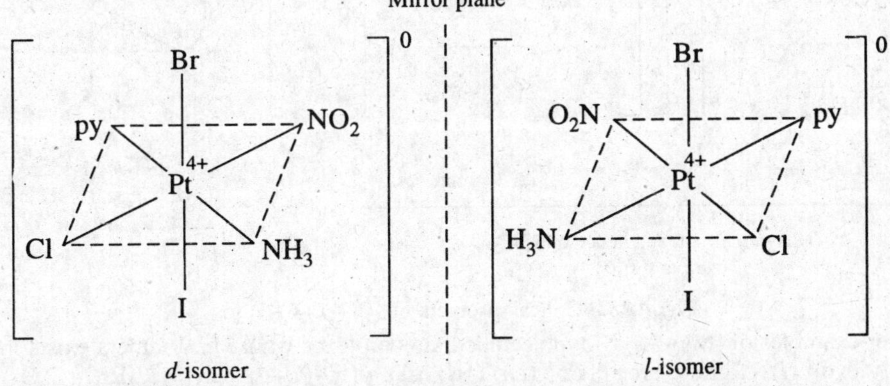

Fig. 4.26. Optical (*d*-and *l*-) forms of $[Pt^{4+}(py)(NH_3)(NO_2)(Cl)(Br)(I)]^0$.

4. [M(AA)$_3$] type complexes. Due to the absence of a plane or centre of symmetry, the octahedral complexes of this type are resolvable into *dextro* and *levo* forms. For example *d*-

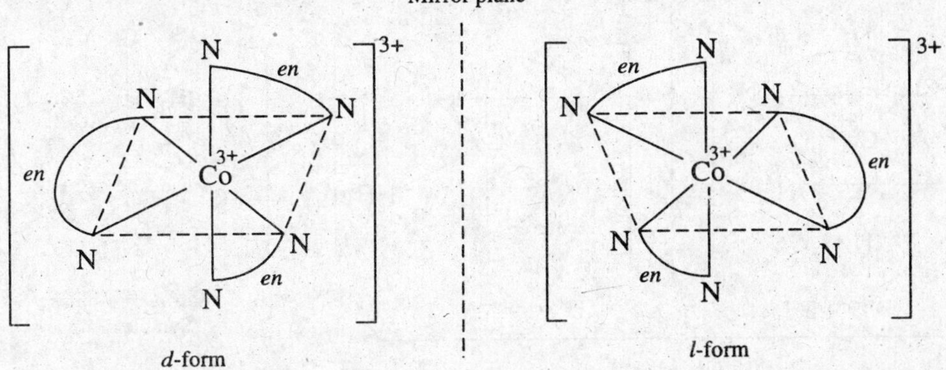

Fig. 4.27. Optical isomers of $[Co^{3+}(en)_3]^{3+}$ ion. These cannot be superimposed on each other.

and *l*-forms of $[Co^{3+}(en)_3]^{3+}$ ion are shown in Fig. 4.27. Other examples of octahedral complexes of $[M(AB)_3]$ type which exist in *d* and *l* forms are : $[Cr^{3+}(ox)_3]^{3-}$, $[Co^{3+}(pn)_3]^{3+}$, $[Pt^{4+}(en)_3]^{4+}$, $[Fe^{2+}(diph)_3]^{2+}$ etc. The fact that the complexes of $[M(AA)_3]$ type can be resolved into optical isomers confirms that these complexes have octahedral shape. Neither hexagonal nor trigonal prismatic geometry of 6-coordinate complex can give rise to activity.

5. $[M(AA)_2a_2]$ type complexes. Here we can consider $[Co^{3+}(en)_2Cl_2]^+$ ion. We have already stated that this ion shows geometrical isomerism and hence exists as *cis*- and *trans*- isomers. The *cis*-isomer does not have any plane or centre of symmetry and hence is *optically active*, *i.e.*, *cis*-isomer can be resolved into two optically active isomers (*d*- and *l*-isomers). On the other hand the *trans*-isomer has a plane of symmetry and hence is optically inactive, *i.e.*, it cannot be resolved into optically active isomers. Thus *trans*- isomer is optically inactive (meso-form). Consequently $[Co(en)_2Cl_2]^+$ ion has three optical isomers *viz.* (*a*) Two optically active isomers corresponding to *cis*-configuration (*i.e.*, *d*- and *l*-forms). One form is the mirror image of the other. (*b*) One optically inactive form corresponding to *trans*-form (meso-form). All the three forms have been shown in Fig. 4.28.

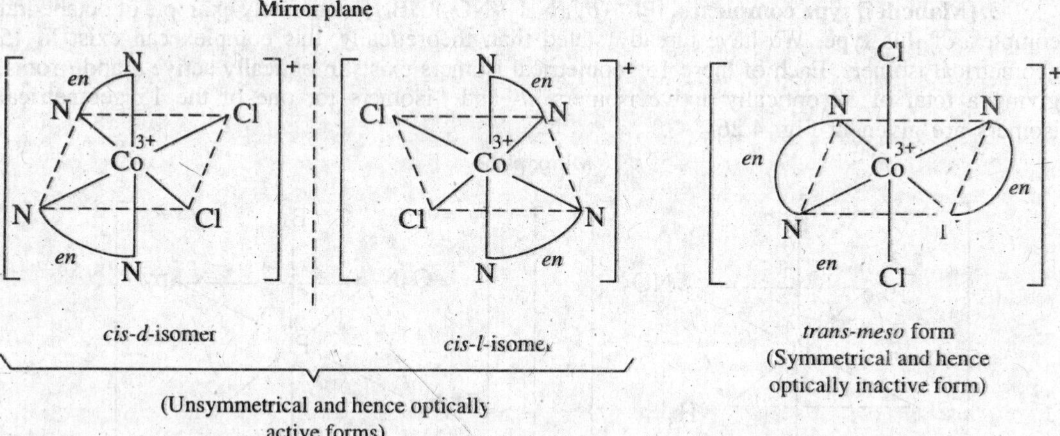

Fig. 4.28. Optical isomers of $[Co(en)_2Cl_2]^+$ ion.

Other examples of $[M(AA)_2a_2]$ type octahedral complexes whose *cis*-isomers exists in dextr (*d*) and levo (*l*) forms are: $[Co^{3+}(en)_2(NO_2)_2]^+$, $[Ir^{4+}(C_2O_4)_2Cl_2]^{2-}$, $[Rh^{4+}(C_2O_4)_2Cl_2]^{2-}$ $[Cr^{3+}(C_2O_4)_2(H_2O)_2]^-$, $[Ir^{3+}(C_2O_4)_2Cl_2]^{3-}$ etc.

6. $[M(AA)_2ab]$ type complexes. Here we shall consider $[Co^{3+}(en)_2(NH_3)Cl]^{2+}$ ion. We have already seen that this ion exists in *cis* and *trans*-isomers (geometrical isomers). Due to the

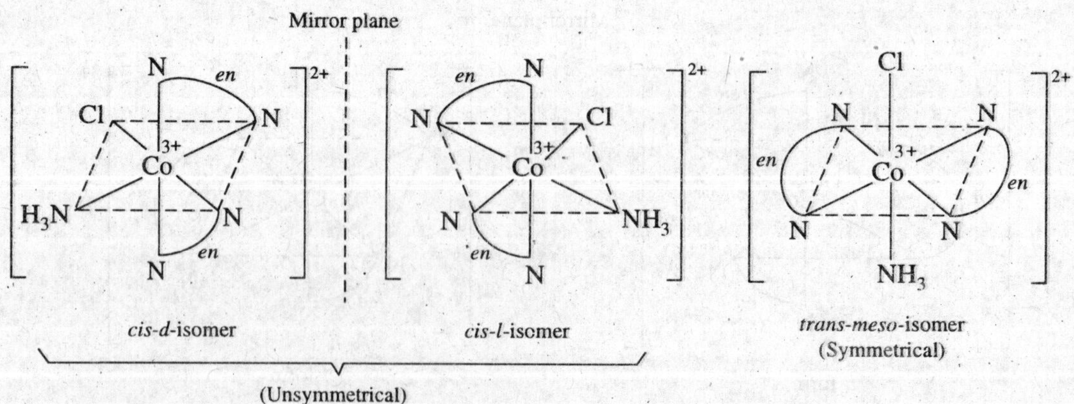

Fig. 4.29. Optical isomers of $[Co^{3+}(en)_2(NH_3)Cl]^{2+}$ ion.

absence of plane or centre of symmetry in *cis*-isomer, this isomer is optically active and hence can be resoved into *d* and *l*-forms. On the other hand the *trans*-isomer has a plane of symmetry and hence is optically inactive, *i.e.*, *trans*-isomer is *meso-trans* isomer (See Fig. 4.29).

Another example of $[M(AA)_2ab]$ type octahedral complexes whose *cis*-isomer exists in *d*- ɪd *l*-forms is $[Rh^{3+}(C_2O_4)_2(py)(NO_2)]^{2-}$.

7. $[M(AA)a_2b_2]$ type complexes. Here we shall consider $[Co^{3+}(en)(NH_3)_2Cl_2]^+$ ion. We have seen that this ion exists in *cis*- and *trans*-isomers (geometrical isomers). *Cis*-isomer is optically active and hence gives optically active *d*-and *l*-forms. On the other hand *trans*-isomer is optically inactive and hence does not give any optical active isomer. (See Fig. 4.30).

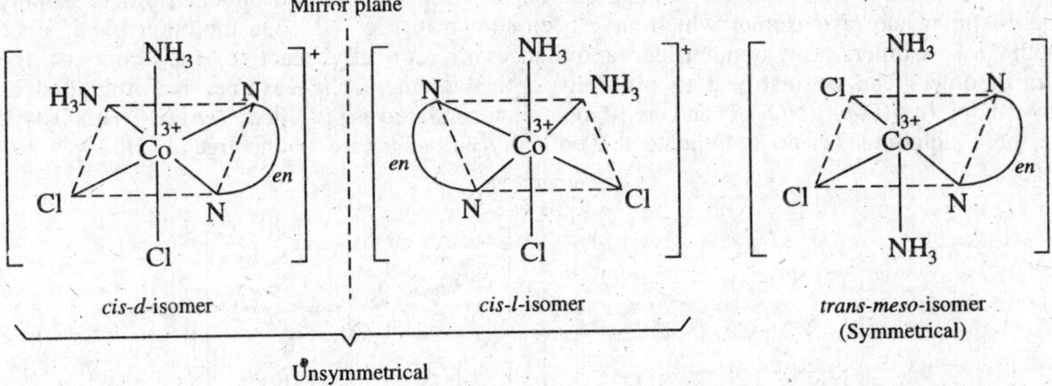

Fig. 4.30. Optical isomers of $[Co^{3+}(en)(NH_3)_2Cl_2]^+$ ion.

$[Co^{3+}(C_2O_4)(NH_3)_2(NO_2)_2]^-$ ion is another important example of $[M(AA)a_2b_2]$ type octahedral complexes whose *cis*-isomer gives *d*-and *l*-isomers.

8. $[M(AB)_3]$ type complexes. We have seen that an octahedral complex of this type exists in *cis*- and *trans*-isomers (geometrical isomers). Each of these forms is optically active and

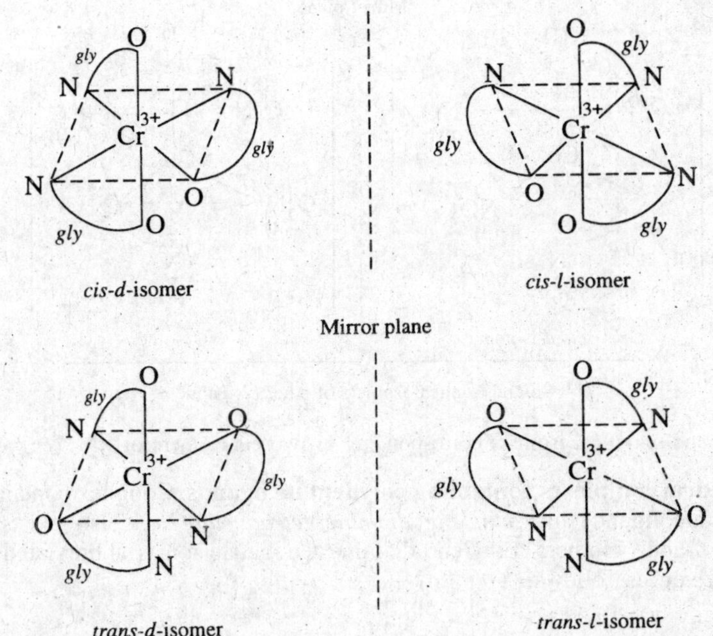

Fig. 4.31. Each of *cis*- and *trans*-isomers of $[Cr^{3+}(gly)_3]^0$ has *d* and *l* optical isomers.

hence each gives a pair of optical isomers (*d*-and *l*-forms) as shown in Fig. 4.31 for $[Cr^{3+}(gly)_3]$ complex compound.

9. Octahedral complexes containing optically active ligands. $[Co^{3+}(en)(pn)(NO_2)_2]^+$ ion is an important example of such type of octahedral complexes. Here en and *pn* are ethylene diamine and 1, 2-diamino propane respectively. Both these are bidentate ligands. *en* is a symmetrical and hence optically inactive ligand. On the other hand *pn* is unsymmetrical and is, therefore, optically active ligand. We have already seen that $[Co^{3+}(en)(pn)(NO_2)_2]^+$ ion exists in four geometrical isomers. Out of these four geometrical isomers, two are *cis*-isomers and two are *trans*-isomers. Since both *cis*-isomers are unsymmetrical and hence are optically active, *i.e.*, each of the two *cis*-isomers can be resovled to give two optical isomers namely *cis-d*-isomer and *cis-l*-isomer which have been shown in Fig. 4.32. On the other hand, since both *trans*-isomers are symmetrical and hence are optically inactive, *i.e.*, none of the *trans*-isomers can be resolved to give any optical isomers. *cis-d*-isomer is formulated as *cis-d*-$[Co^{3+}(en)(D\text{-}pn)(NO_2)_2]^+$ and *cis-l*-isomer is formulated as *cis-l*-$[Co^{3+}(en)(L\text{-}pn)(NO_2)_2]^+$. In these notations D and L indicate the optically active nature of the ligand, *pn*.

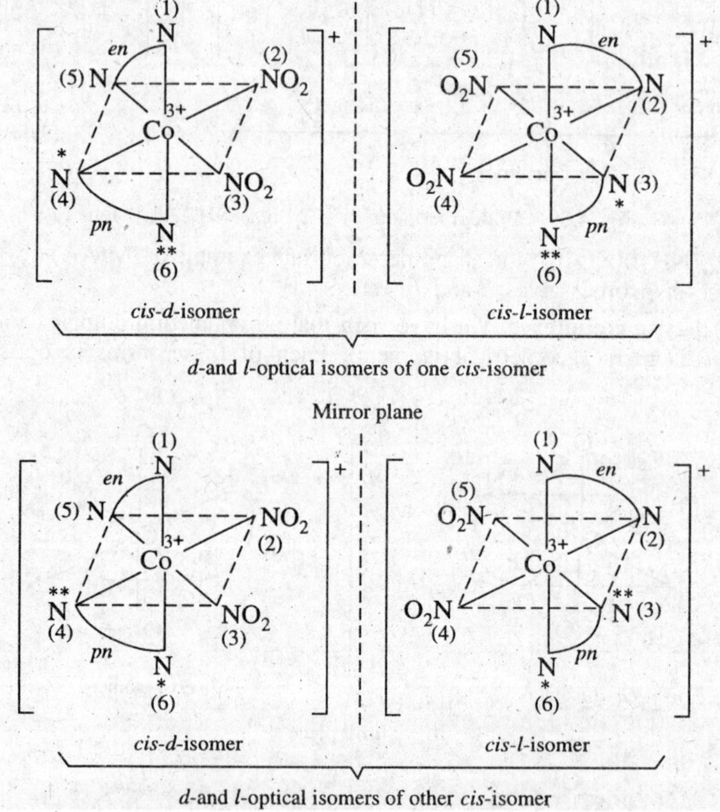

Fig. 4.32. *d*- and *l*-optical isomers corresponding to two cis-isomers of $[Co^{3+}(en)(pn)(NO_2)_2]^+$ ion.

10. Octahedral complexes containing polydentate ligands. Complex containing hexadentate ligand (polydentate ligand) like *ethylene diamine tetraacetate ion* also exists in two optical isomers namely *d*-and *l*-isomers. Ethylene diamine tetracetate ion is abbreviated as $EDTA^{4-}$ and its structure is given as follows:

$$\left[\begin{array}{l} :\ddot{\bar{O}}^{(5)}-\overset{O}{\overset{\|}{C}}-H_2C\diagdown^{(1)} \\ \qquad\qquad\qquad\quad \ddot{N}-CH_2-CH_2-\ddot{N}^{(2)} \\ :\ddot{\bar{O}}_{(6)}-\underset{O}{\underset{\|}{C}}-H_2C\diagup \end{array}\begin{array}{l} \diagup CH_2-\overset{O}{\overset{\|}{C}}-\ddot{\bar{O}}:^{(3)} \\ \\ \diagdown CH_2-\underset{O}{\underset{\|}{C}}-\ddot{\bar{O}}:_{(4)} \end{array}\right]^{4-}$$

$EDTA^{-4}$ has four donor oxygen atoms (numbered as 3, 4, 5 and 6) and two donor nitrogen atoms (numbered as 1 and 2). Thus it has six coordinating sites and four negative charges. In $[Co^{3+}(EDTA)]^{-}$ ion, Co^{3+} is attached with four oxygen of CH_2COO^{-} components and with two nitrogen atoms. $[Co^{3+}(EDTA)]^{-}$ ion has two optical isomers shown in Fig. 4.33.

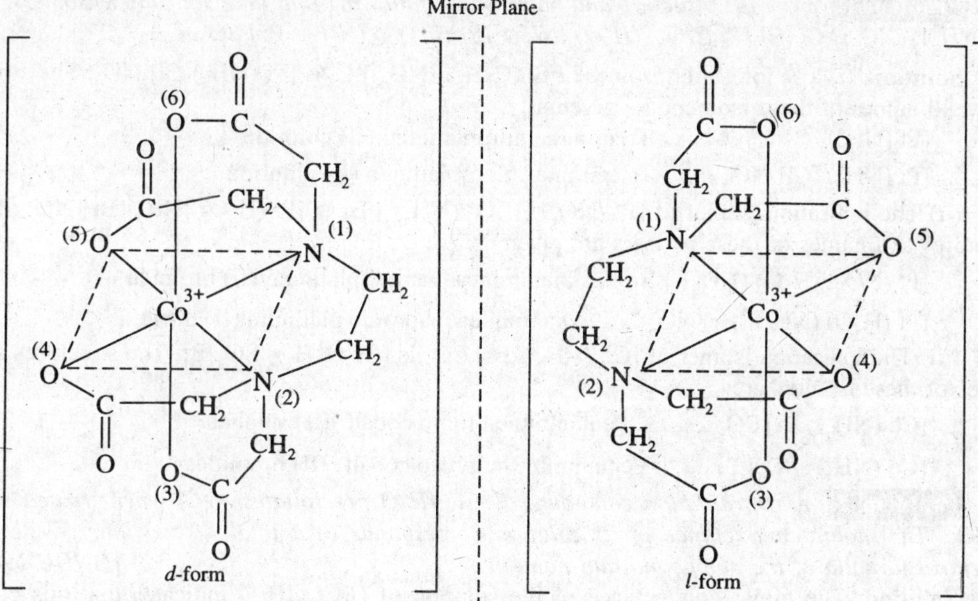

Fig. 4.33. Optical isomers of $[Co^{3+}(EDTA)]^{-}$ ion.

Two optical forms shown in Fig. 4.33 can also be shown in a simpler way as given in Fig. 4.34.

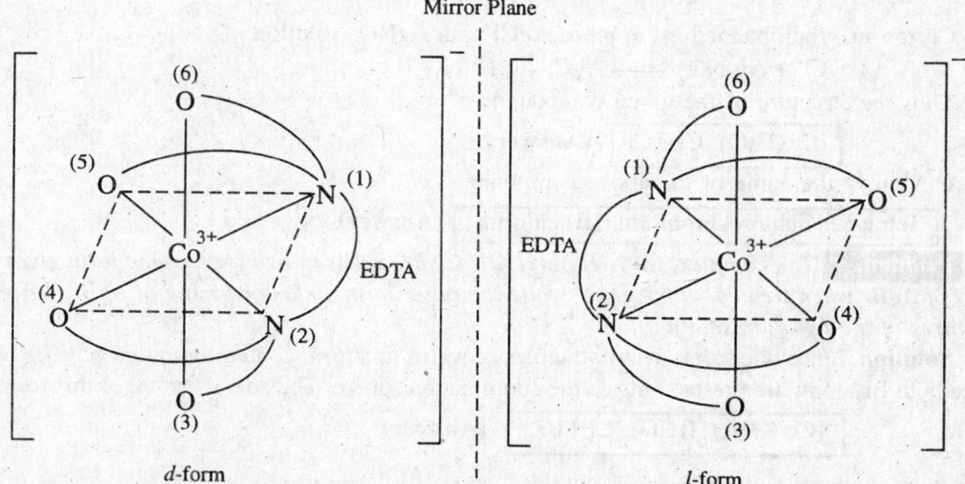

Fig. 4.34. *d*- and *l*-forms of $[Co^{3+}(EDTA)]^{-}$ ion shown in simpler way.

11. Bridged binuclear octahedral complex of Co (III). Co (III) forms a binuclear octahedral complex which contains NH_2^- and NO_2^- ions as bridging ligands. This complex is formulated as $[Co_2^{3+}(en)_4(NH_2)(NO_2)]^{4+}$ whose structure has already been shown at Fig. 1.45 given in chapter 1. This ion exists in *cis*- and *trans*-isomers whose structures are given in Fig. 1.46 in chapter 1. *Cis*-isomer is asymmetrical and hence is optically active, *i.e.*, *cis*-form gives two optical isomers namely *d-cis*-isomer and *l-cis*-isomer. On the other hand trans form, being symmetrical and optically inactive, does not give any isomer. The structures of *d-cis* and l-*cis* isomers of $[Co_2^{3+}(en)_2(NH_2)(NO_2)]^{4+}$ ion have been shown at Fig. 1.47 in chapter 1.

SOME TYPICAL PROBLEMS WITH SOLUTIONS

Problem 1. *Write the structure and name of the following and their ionisation isomers: (i) $[Pt(NH_3)_3(NO_2)]Cl$ (ii) $[Pt(NO_2)(H_2O)(NH_3)_2]Br$ (iii) $Co(NH_3)_5Br]SO_4$*

Solution. (*i*) The ionisation isomer of $[Pt(NH_3)_3(NO_2)]Cl$ is $[Pt(NH_3)_3Cl]NO_2$. The structures and names of these isomers are given as:

$[Pt(NH_3)_3(NO_2)]Cl$ ...... Triammine nitroplatinum (II) chloride

$[Pt(NH_3)_3Cl]NO_2$ .... Triammine chloroplatinum (II) chloride

(*ii*) The ionisation isomer of $[Pt(NO_2)(H_2O)(NH_3)_2]Br$ is $[Pt(H_2O)(NH_3)_2Br]NO_2$. The structures and names of these isomers are given as:

$[Pt(NO_2)(H_2O)(NH_3)_2]Br$ ...... Diammineaquanitro platinum (II) bromide

$[Pt(H_2O)(NH_3)_2Br]NO_2$ .... Diammineaquabromo platinum (II) nitrite

(*iii*) The ionisation isomer of $[Co(NH_3)_5Br]SO_4$ is $[Co(NH_3)_5SO_4]Br$. The structures and names of these isomers are:

$[Co(NH_3)_5Br]SO_4$ .... Pentammine bromo cobalt (III) sulphate

$[Co(NH_3)_5SO_4]Br$ .... Pentammine sulphato cobalt (III) bromide.

Problem 2. *A coordination compound $CrCl_3.4H_2O$ precipitates AgCl when treated with $AgNO_3$. The molar conductance of its solution corresponds to a total of two ions. Write the structural formula of the compound and name it.* *[GNDU 1982]*

Solution. The molar conductance of the solution of $CrCl_3.4H_2O$ indicates that this compound contains one Cl^- ion and one $[Cr(H_2O)_4Cl_2]^+$ ion.

$$[Cr(H_2O)_4Cl_2]Cl \rightleftharpoons \underbrace{[Cr(H_2O)_4Cl_2]^+ + Cl^-}_{\text{Total two ions}}$$

Cl^- ion may be obtained as a ppt. of AgCl with $AgNO_3$ solution.

$$Cl^- + AgNO_3 \longrightarrow AgCl\downarrow + NO_3^-$$

Thus the structure of the given compound is:

$[Cr(H_2O)_4Cl_2]Cl$ **(Answer)**

Obviously the name of the above compound is:

Tetraaquadichlorochromium (III) chloride. **(Answer)**

Problem 3. *The complex, $Co(NH_3)_2(H_2O)_2Cl_2Br_2$ exists in two forms. One form gives two moles of AgBr when treated with $AgNO_3$ while the other form yields one mole of AgBr. Write the structures of the two forms.*

Solution. Since one form gives two moles of AgBr, this form contains two ionisable Br^- ions. Hence both Br-atoms are present outside the coordination sphere. Thus the structure of this form is :

$[Co(NH_3)_2(H_2O)_2Cl_2]Br_2$. **(Answer)**

Since the other form gives only one mole of AgBr, only one Br^- ion is ionisable. Thus one Br^- ion is present outside and the other Br^- ion is present inside the coordination sphere. In order to keep C.N. equal to six, one H_2O molecule will be placed outside the coordination sphere. Thus the structure of the other form is:

$$\boxed{[Co(NH_3)_2(H_2O)Cl_2Br]Br.H_2O.}$$ **(Answer)**

Problem 4. *A compound with the empirical formula, $Co(NH_3)_5SO_4Br$ exists in two forms viz red and violet forms. Solution of red form gives a ppt. of AgBr with $AgNO_3$ solution. The violet form gives no ppt. with $AgNO_3$ but gives white ppt. with the aqueous solution of $BaCl_2$. From these observations give the structure of each form.*

Solution. Since red form gives a ppt. of AgBr with $AgNO_3$ solution, in this form Br atom is ionisable and hence is present outside the coordination sphere. Thus the structure of red form is $[Co(NH_3)_5SO_4]Br$.

Since violet form gives no ppt. with $AgNO_3$ solution, Br atom is not ionisable, *i.e.*, Br-atom is present inside the coordination sphere. Now since violet form gives a white ppt. with $BaCl_2$ solution, SO_4^{2-} ion is ionisable and hence is present outside the coordination sphere. Thus the structure of violet form is $[Co(NH_3)_5Br]SO_4$.

Problem 5. *A solution of $[Cu(NH_3)_4I_2]Cl$ when treated with $AgNO_3$ gives a white ppt. What must be the formula of the isomer of the dissolved compound that yields yellow ppt. What are the above isomers called?* *[Delhi 2000]*

Solution. The formation of white ppt (of AgCl) takes place, since Cl^- ion is present outside the coordination sphere of the given compound.

$$[Co(NH_3)_4I_2]Cl \longrightarrow [Co(NH_3)_4I_2]^+ + Cl^-$$

$$Cl^- + Ag^+ \text{ (From } AgNO_3 \text{ solution)} \longrightarrow AgCl \text{ (White ppt)}$$

Yellow ppt. is obtained when I^- ion is present outside the coordination sphere, *i.e.*, the structure of the new compound must be $[Cu(NH_3)_4ICl]I$. This compound has been obtained when one I^- ion present inside the coordination sphere is exchanged with Cl^- ion given outside the coordination sphere.

$$[Co(NH_3)_4ICl]I \longrightarrow [Co(NH_3)_4ICl]^+ + I^-$$

$$I^- + Ag^+ \text{ (From } AgNO_3 \text{ solution)} \longrightarrow AgI \text{ (Yellow ppt)}$$

$[Co(NH_3)_4I_2]Cl$ and $[Co(NH_3)_4ICl]I$ are ionisation isomers with each other.

Problem 6. *Complex compound with empirical formula $Co(NH_3)_3(H_2O)_2Br_2Cl$ exists in two isomeric forms A and B. Form A gives two moles of AgBr when treated with $AgNO_3$ solution whereas form B gives only one mole of AgBr. Write down the structural formula of both the forms. What are these isomers called?* *(Delhi 2001)*

Solution. The structure of the two isomers can be written as A = $[Co(NH_3)_3(H_2O)_2Cl]Br_2$, B = $[Co(NH_3)_3(H_2O)ClBr]Br.H_2O$. In both the structures C.N. of Co^{3+} ion is 6. The formation of two moles of AgBr from A and one mole of AgBr from B can be shown as:

$$\underset{\text{(A)}}{[Co(NH_3)_3(H_2O)_2Cl]Br_2} \longrightarrow [Co(NH_3)]_3(H_2O)_2Cl]^{2+} + 2Br^-$$

$$2Ag^+ \text{ (From } AgNO_3 \text{ solution)} + 2Br^- \longrightarrow 2AgBr \text{ (ppt) (Two moles)}$$

$$\underset{\text{(B)}}{[Co(NH_3)_3(H_2O)ClBr]Br.H_2O} \longrightarrow [Co(NH_3)_3(H_2O)ClBr]^+ + Br^- + H_2O$$

$$Ag^+ \text{ (From Ag } NO_3 \text{ solution)} + Br^- \longrightarrow AgBr \text{ (ppt) (One mole)}$$

Obviously $[Co(NH_3)_3(H_2O)_2Cl]Br_2$ (A) and $[Co(NH_3)_3(H_2O)ClBr]Br.H_2O$ (B) are hydrate isomers.

Problem 7. *A, B and C are three complexes of chromium (III) with empirical formula $H_{12}O_6Cl_3Cr$. All the three complexes have water and chloride ions as ligands. Complex A does* *not react with conc. H_2SO_4 whereas complexes B and C lose 6.75% and 13.5% of their original weight respectively on treatment with conc. H_2SO_4. Identify A, B and C.* (*IIT 1999*)

Solution. The given complex can also be written as Cr $(H_2O)_6Cl_3$. Since A does not react with conc. H_2SO_4, it does not lose any H_2O molecule, *i.e.*, A has no H_2O molecule outside the coordination sphere. In other words we say that all the six H_2O molecules are present inside the coordination sphere. Thus the composition of A can be given by:

$\boxed{A = [Cr\ (H_2O)_6]\ Cl_3}$ **(Answer)**

Molecular weight of $Cr(H_2O)_6Cl_3 = 52 + 6 \times 18 + 35.5 \times 3 = 266.5$ g

$$6.75\% \text{ of } 266.5 \text{ g} = \frac{266.5 \times 6.75}{100} = 17.98 \text{ g}$$
$$= 18.00 \text{ g}$$

Since the molecular weight of H_2O = 18, this weight of 18g shows that B has one H_2O molecule out side the coordination sphere and hence the composition of B can be given by:

$\boxed{B = [Cr\ (H_2O)_5\ Cl]\ Cl_2.H_2O}$ **(Answer)**

$$13.5\% \text{ of } 266.5 \text{ g} = \frac{266.5 \times 13.5}{100} = 35.97\text{g}$$
$$= 36 \text{ g}$$

This weight of 36 g shows that C has two molecules of H_2O ($2 \times H_2O = 2 \times 18 = 36$) outside the coordination sphere and hence the composition of C is given by:

$\boxed{C = [Cr\ (H_2O)_4\ Cl_2]\ Cl.2H_2O}$ **(Answer)**

Problem 8. *Addition of Ag NO_3 solution to aqueous solutions of each of the Pt (IV) amines viz. $PtCl_4.6NH_3$, $PtCl_4.5NH_3$, $PtCl_4.4NH_3$, $PtCl_4.3NH_3$ and $PtCl_4.2NH_3$ was found by Werner to give 4, 3, 2, 1 and o moles of AgCl per mole of the amines respectively. How did Werner explain these observations?*

Solution. (*i*) **$PtCl_4.6NH_3$** Since this amine gives 4 moles of AgCl, in this amine all the four Cl^- ions are ionisable and hence this amine has all the four Cl^- ions outside the coordination sphere. Thus Werner formulated this amine as [Pt $(NH_3)_6$] Cl_4.

(*ii*) **$PtCl_4.5NH_3$.** Since this amine gives 3 moles of AgCl, in this amine only three Cl^- ions are ionisable and hence this amine has only three Cl^- ions outisde the coordination sphere. Thus Werner formulated this amine as [Pt $(NH_3)_5Cl]Cl_3$.

(*iii*) **$PtCl_4.4NH_3$.** Since this amine gives 2 moles of AgCl, in this amine only two Cl^- ions are ionisable and hence this amine has only two Cl^- ions outside the coordination sphere. Thus Werner formulated this amine as [Pt $(NH_3)_4$ Cl_2] Cl_2.

(*iv*) **$PtCl_4.3NH_3$.** Since this amine gives 1 mole of AgCl, in this amine only one Cl^- ion is ionisable and hence this amine has only one Cl^- ions outside the coordination sphere. Thus Werner formulated this amine as [Pt $(NH_3)_3Cl_3$]Cl.

(*v*) **Pt $Cl_4.2NH_3$.** Since this amine gives no AgCl, in this amine no Cl^- ion is ionisable, *i.e.*, no Cl^- ion is present outside the coordination sphere or all the four Cl^- ions are present inside the coordination sphere. Thus Werner formulated this amine as [Pt $(NH_3)_2$ Cl_4] (Non-electrolyte).

Problem 9. *Three compounds A, B and C have empirical formula $CrCl_3.6H_2O$. When 1g of compound A was kept in a container with dehydrating agent, it lost water content and attained constant weight of 0.865 g. When 1g of B was kept in that vessel, it attained a constant weight of 0.932 g. Compound C showed no loss in water content.*

(*a*) *Find the composition of A, B and C.*

(*b*) *If an excess of aqueous $AgNO_3$ solution is added to 1g solution of A, B and C, what amount of AgCl will be precipitated in each case?*

Solution. (*a*) Molecular weight of $CrCl_3.6H_2O = 52 + 35.5 \times 3 + 6 \times 18 = 266.5$ g.
It is obvious from the question that:
1g of A gives (1 – 0.865) g = 0.135 g of H_2O on dehydration
$\therefore$ 266.5 g of A give 0.135 × 266.5 g = 36.0 g of H_2O
This weight of 36 g shows that on dehydration A gives two H_2O molecules (Molecular weight of $H_2O = 1 \times 2 + 16 = 18$ g), *i.e.*, two H_2O molecules are present outside the coordination sphere in the composition of A. Thus the composition of A is given by:

$$\boxed{A = [Cr(H_2O)_4Cl_2]Cl.2H_2O.}$$ **(Answer)**

On similar grounds it can be shown that:

266.5 g of B will give (1 – 0.932) × 266.5 or 18.122 g of H_2O

This weight (= 18.122 g) corresponds to one H_2O molecule. Thus B has one H_2O molecule outside the coordination sphere and hence the composition of B is given by:

$$\boxed{B = [Cr\ (H_2O)_5\ Cl]\ Cl_2.H_2O.}$$ **(Answer)**

Since C shows no loss in water content, no H_2O molecule is outside the coordination sphere, *i.e.*, all the six H_2O molecules are inside the coordination sphere and hence the composition of C is given by:

$$\boxed{C = [Cr\ (H_2O)_6]\ Cl_3.}$$

(*b*) Molecular weight of AgCl = 108 + 35.5
= 143.5 g

Since A, *i.e.*, $[Cr(H_2O)_4\ Cl_2]\ Cl.2H_2O$, on treatment with $AgNO_3$ solution, gives one AgCl molecule, 266.5 g of A gives 143.5 g of AgCl

$$\therefore \text{ 1g of A gives } \frac{143.5}{266.5} \text{ g} = 0.538 \text{ g of AgCl}$$ **(Answer)**

Since B, *i.e.*, $[Cr\ (H_2O)_5\ Cl]\ Cl_2.H_2O$ gives two molecules of AgCl, 266.5 g of B gives 2 × 143.5 g of AgCl

$$\therefore \text{ 1 g of B gives } \frac{2 \times 143.5}{266.5} \text{ g} = 1.076 \text{ g of AgCl}$$ **(Answer)**

Since C, *i.e.*, $[Cr\ (H_2O)_6]\ Cl_3$ gives three AgCl molecules, 266.5 g of C gives 3 × 143.5 g of AgCl

$$\therefore \text{ 1g of C gives } \frac{3 \times 143.5}{266.5} \text{ g} = 1.614 \text{ g of AgCl.}$$ **(Answer)**

Problem 10. *Write the structure and name of the following and their coordination isomers: (i) $[Co\ (en)_3]\ [Cr\ (CN)_6]$ (ii) $[Co\ (NH_3)_6]\ [Cr\ (C_2O_4)_3]$*

Solution. (*i*) The coordination isomer of $[Co\ (en)_3]\ [Cr(CN)_6]$ is obtained by interchanging the ligands between the two complex ions *viz* $[Co^{3+}\ (en)_3]^{3+}$ and $[Cr^{3+}\ (CN)_6]^{3-}$. Thus the coordination isomer of $[Co\ (en)_3]\ [Cr\ (CN)_6]$ is $[Co^{3+}\ (CN)_6]^{3-}\ [Cr^{3+}\ (en)_3]^{3+}$ which, in its correct form, should be written as $[Cr^{3+}\ (en)_3]^{3+}\ [Co^{3+}\ (CN)_6]^{3-}$ or simply as $[Cr\ (en)_3]\ [Co\ (CN)_6]$.

The structure and name of the above two isomers are given below :

$[Co\ (en)_3]\ [Cr\ (CN)_6]$ ...... *Tris* (ethylenediamine) cobalt (III) hexacyanochromate (III)

$[Cr\ (en)_3]\ [Co\ (CN)_6]$ ..... *Tris* (ethylenediamine) chromium (III) hexacyano cobaltate (III)

(*ii*) Coordination isomer of $[Co\ (NH_3)_6]\ [Cr\ (C_2O_4)_3]$ is $[Co\ (C_2O_4)_3]^{3-}\ [Cr\ (NH_3)_6]^{3+}$ or $[Cr\ (NH_3)_6]\ [Co\ (C_2O_4)_3]$. The structure and name of these isomers are given below:

$[Co(NH_3)_6][Cr(C_2O_4)_3]$ ...... Hexamine cobalt (III) tris (oxalato) chromate (III)

$[Cr(NH_3)_6][Co(C_2O_4)_3]$ ..... Hexamine chromium (III) tris (oxalato) cobaltate (III)

Problem 11. *Name the isomerism encountered in complexes containing ambidentate ligands.* (*Delhi 2000*)

Solution. Linkage isomerism.

Problem 12. *Draw the structures and write the names of all possible isomers of: (i) $[Pt(SCN)(NH_3)_3]SCN$ (ii) $[Co(SCN)(NH_3)_5]Cl_2$ (iii) $[Co(NH_3)_5(NO_2)]^{2+}$.*

Solution. (*i*) $[Pt(SCN)(NH_3)_3]SCN$ has one linkage isomer which is $[Pt(NCS)(NH_3)_3]SCN$. In $[Pt(SCN)(NH_3)_3]SCN$, SCN^- ion (ligand) is linked with Pt^{2+} ion (central metal ion) through negatively-charged *S*-atom while in $[Pt(NCS)(NH_3)_3]SCN$, SCN^- ion (ligand) is linked with Pt^{2+} ion through negatively-charged N-atom. Thus the structures and names of these linkage isomers can be written as given in Fig 4.35

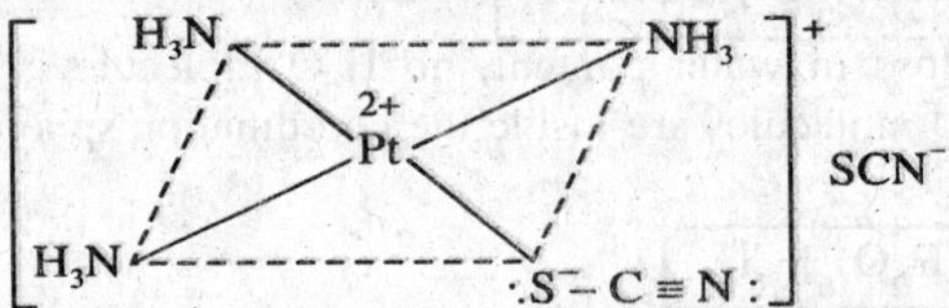

(*a*) Triamminethiocyanatoplatinum (II) thiocyanate, $[Pt(SCN)(NH_3)_3]SCN$

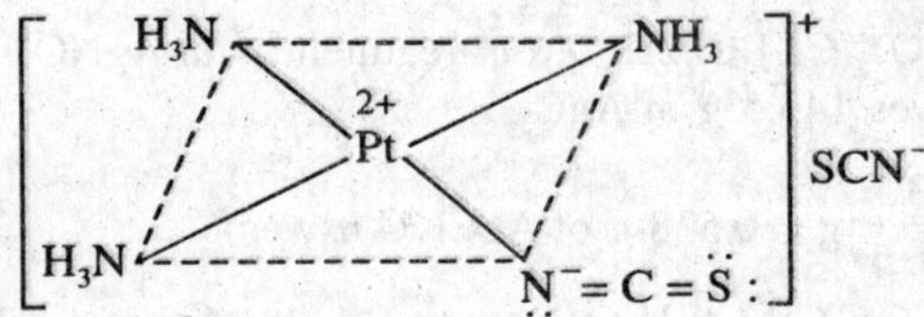

(*b*) Triammineisothiocyanatoplatinum (II) thiocyanate, $[Pt(NCS)(NH_3)_3]SCN$

Fig. 4.35 Structures and names of linkage isomers *viz* (*a*) $[Pt(SCN)(NH_3)_3]SCN$ (*b*) $[Pt(NCS)(NH_3)_3]SCN$.

(*ii*) $[Co(NCS)(NH_3)_5]Cl_2$ is the linkage isomer of $[Co(SCN)(NH_3)_5]Cl_2$

$[Co(NCS)(NH_3)_5]Cl_2$ .... Pentaammine isothiocyanato cobalt (III) chloride

$[Co(SCN)(NH_3)_5]Cl_2$ ...... Pentaammine thiocyanato cobalt (III) chloride

(*iii*) Since NO_2^- ion is an ambidentate ligand, the given complex ion has two linkage isomers. The structures of these isomers are shown in Fig. 4.36. In isomer (*a*), NO_2^- ion (ligand) is linked with the central Co^{3+} ion through its N-atom while in isomer (*b*), this ion is linked with Co^{3+} ion through its O-atom.

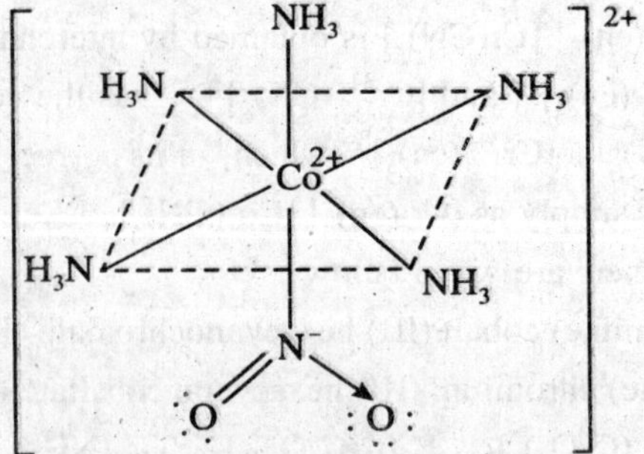

(*a*) Pentamminenitrocobalt (III) ion, $[Co(NH_3)_5(NO_2)]^{2+}$

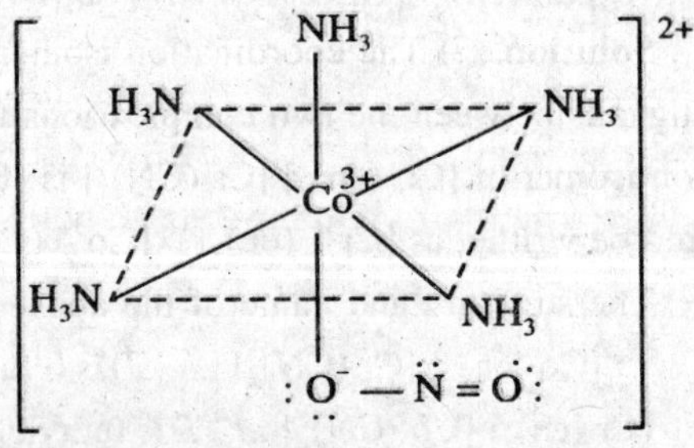

(*b*) Pentamminenitritocobalt (III) ion, $[Co(NH_3)_5(ONO)]^{2+}$.

Fig. 4.36 Structure of the two linkage isomers of $[Co(NH_3)_5NO_2]^{2+}$ ion.

Problem 13. *The complex ion, $[Co(NH_3)_5 Cl]^{2+}$ undergoes a substitution reaction when reacted with ($NaNO_2$ + HCl). An unstable scarlet red complex A is formed when dilute acid is used and a stable yellow complex B is formed when concentrated acid is used. Both A and B are isomeric pentammine complex ions. Give the structures of the ions and name the type of isomerism involved.*

Solution. Since the treatment of pentammine chlorocobalt (III) ion, $[Co^{3+}(NH_3)_5 Cl]^{2+}$ with ($NaNO_2$ + dil. HCl) gives an unstable red complex, this red complex should be pentaammine nitritocobalt (III) ion, $[Co^{3+}(NH_3)_5(ONO)]^{2+}$. Similarly, the ion obtained by treating $[Co^{3+}(NH_3)_5 Cl]^{2+}$ ion with ($NaNO_2$ + conc. HCl) should be pentaammine nitrocobalt (III) ion, $[Co^{3+}(NH_3)_5(NO_2)]^{2+}$.

Thus A= $[Co^{3+}(NH_3)_5(ONO)]^{2+}$ and B = $[Co^{3+}(NH_3)_5(NO_2)]^{2+}$. Obviously A and B are linkage isomers and the isomerism involved is linkage isomerism.

Problem 14. *What type of isomers are the following:*

(i) (a) $[(CO)_5 Mn(SCN)]$ and $[(CO)_5Mn(NCS)]$ (b) $[Co(en)_3][Cr(CN)_6]$ and $[Cr(en)_3][Co(CN)_6]$ (c) $[Co(NH_3)_5 NO_3]SO_4$ and $[Co(NH_3)_5 SO_4]NO_3$ (d) $[Co(py)_2(H_2O)_2Cl_2]Cl$ and $[Co(py)_2(H_2O)Cl_3].H_2O$

(ii) (a) $[Cr(NH_3)_6][Cr(CN)_6]$ and $[Cr(NH_3)_4(CN)_2][Cr(NH_3)_2(CN)_4]$ (b) $[Co(py)_2(H_2O)_2Cl_2]Cl$ and $[Co(py)_2(H_2O)Cl_3].H_2O$ (c) $[Pt(NH_3)_4Br_2]Cl_2$ and $[Pt(NH_3)_4Cl_2]Br_2$ (d) $[Co(NH_3)_5NO_2]Cl_2$ and $[Co(NH_3)_5ONO]Cl_2$

(iii) (a) $[Ru(NH_3)_4(CN)_2][Ru(NH_3)_2(CN)_4]$ and $[Ru(NH_3)_6][Ru(CN)_6]$ (b) $[Cr(NCS)_3(H_2O)_3]$ and $[Cr(SCN)_3(H_2O)_3]$ *(Delhi 2001)*

Solution. (*i*) (*a*) The given complexes are linkage isomers, since these complexes have ambidentate ligand *viz* SCN^- ion.

(*b*) The given complex compounds are coordination isomers, since each of them contains complex cation and complex anion as shown below and these complex compounds can be obtained by interchanging the ligands present in complex cation and complex anion.

Complex compound		*Complex cation*		*Complex anion*
$[Co(en)_3][Cr(CN)_6]$	$\rightleftharpoons$	$[Co(en)_3]^{3+}$	+	$[Cr(CN)_6]^{3-}$
$[Cr(en)_3][Co(CN)_6]$	$\rightleftharpoons$	$[Cr(en)_3]^{3+}$	+	$[Co(CN)_6]^{3-}$

(*a*) These are ionisation isomers, since they give different ions in aqueous solution as shown below:

$$\underset{\text{(I)}}{[Co(NH_3)_5NO_3]SO_4} \rightleftharpoons [Co(NH_3)_5NO_3]^{2+} + SO_4^{2-}$$

$$\underset{\text{(II)}}{[Co(NH_3)_5SO_4]NO_3} \rightleftharpoons [Co(NH_3)_5SO_4]^{+} + NO_3^{-}$$

(I) Gives SO_4^{2-} ions and (II) gives NO_3^- ions. Complex cations given by them are also different. Due to the production of SO_4^{2-} ions, (I) gives a white ppt. of $BaSO_4$ with $BaCl_2$ solution. On the other hand (II) gives the test of NO_3^- ions.

SO_4^{2-} [given by (I)] + Ba^{2+} (from $BaCl_2$ solution) $\longrightarrow BaSO_4\downarrow$ (white ppt.)

(*d*) These compounds are hydrate isomers, since they contain different number of H_2O molecules inside and outside coordination sphere

$$\underset{\text{Electrolyte}}{[Co(py)_2(H_2O)_2Cl_2]Cl} \rightleftharpoons [Co(py)_2(H_2O)_2Cl_2]^{+} + Cl^{-}$$

$$[Co(py)_2(H_2O)Cl_3].H_2O \rightleftharpoons \text{No ionisation}$$
Non-electrolyte

(*ii*) (*a*) The given complex compounds are coordination isomers, since each of them contains complex cation and complex anion as shown below and these isomers can be obtained by interchanging their ligands.

Complex compound		*Complex cation*		*Complex anion*
$[Cr(NH_3)_6]\,[Cr(CN)_6]$	$\rightleftharpoons$	$[Cr\,(NH_3)_6]^{3+}$	+	$[Cr\,(CN)_6]^{3-}$
$[Cr(NH_3)_4(CN)_2]\,[Cr(NH_3)_2(CN)_4]$	$\rightleftharpoons$	$[Cr\,(NH_3)_4\,(CN)_2]^+$	+	$[Cr\,(NH_3)_2\,(CN)_4]^-$

(*b*) The given complex compounds are hydrate isomers, since they contain different number of H_2O molecules inside and outside coordination sphere as shown below:

$$[Co(py)_2\,(H_2O)_2\,Cl_2]\,Cl \rightleftharpoons [Co(py)_2(H_2O)_2Cl_2]^+ + Cl^-$$
$$[Co(py)_2(H_2O)\,Cl_3]\,H_2O \longrightarrow \text{No ionisation.}$$
Non-electrolyte

(*c*) The given complex compounds are ionisation isomers, since they give different ions in aqueous solution as shown below:

$$\underset{\text{(I)}}{[Pt(NH_3)_4Br_2]Cl_2} \rightleftharpoons [Pt\,(NH_3)_4\,Br_2]^{2+} + 2Cl^-$$
$$\underset{\text{(II)}}{[Pt(NH_3)_4Cl_2]Br_2} \rightleftharpoons [Pt(NH_3)_4Cl_2]^{2+} + 2Br^-$$

(I) gives Cl^- ions and (II) gives Br^- ions. Complex cations given by them are also different. Due to the production of Cl^- ions, (I) gives white ppt. of AgCl with $AgNO_3$ solution. On the other hand, due to the production of Br^- ions, (II) gives yellow ppt. of AgBr with $AgNO_3$ solution

$$Cl^-\text{ [given by (I)]} + Ag^+\text{ (from }AgNO_3\text{ solution)} \longrightarrow AgCl\downarrow\text{ (white ppt)}$$
$$Br^-\text{ [given by (II)]} + Ag^+\text{ (from }AgNO_3\text{ solution)} \longrightarrow AgBr\downarrow\text{ (yellow ppt)}$$

(*d*) Since the given complex compounds contain ambidentate ligand *viz* NO_2^-, these are linkage isomers. In the first compound NO_2^- ion is called nitro group which is linked with the central Co^{3+} ion through N-atom while in the second compound NO_2^- ion is called nitrito group which is linked with Co^{3+} ion through O-atom.

(*iii*) (*a*) The given compounds are coordination isomers, since each of them has complex cation and complex anion as shown below and these isomers can be obtained by interchanging their ligands.

Complex compound		*Complex cation*		*Complex anion*
$[Ru^{4+}(NH_3)_4(CN)_2][Ru^{2+}(NH_3)_2(CN)_4]$	$\rightleftharpoons$	$[Ru^{4+}(NH_3)_4(CN)_2]^{2+}$	+	$[Ru^{2+}(NH_3)_2(CN)_4]^{2-}$
$[Ru^{4+}(NH_3)_6][Ru^{2+}(CN)_6]$	$\rightleftharpoons$	$[Ru^{4+}\,(NH_3)_6]^{4+}$	+	$[Ru^{2+}\,(CN)_6]^{4-}$

(*b*) The given complex compounds are linkage isomers, since these have ambidentate ligand *viz* SCN^- ion.

Problem 15. *Give the total number of geometrical and optical isomers given by: (i)* $[Co\,(NH_3)_4Cl_2]^+$ *(ii)* $[Pt\,(en)Br_2Cl_2]$ *(iii)* $CrCl_3(H_2O)_6$

Solution. (*i*) **$[Co(NH_3)_4Cl_2]^+$ ion.** This ion is an octahedral complex ion of $[Ma_4b_2]$ type and hence has *cis* and *trans* isomers (two geometrical isomers). Both these isomers are optically inactive and hence do not show optical isomerism.

(*ii*) **[Pt (*en*) Cl_2Br_2].** This complex compound is an octahedral compound of [M (AA) a_2b_2] type and hence has *cis* and *trans*-isomers (two geometrical isomers). *Cis* isomer is optically active and hence gives optically active *d*- and *l*-forms (two optical isomers). On the other hand *trans*-isomer is optically inactive and hence does not give any optical isomers. Thus [Pt (en) Cl_2Br_2] has three isomers *viz d*- and *l*-isomers (optical isomers) and *trans*-isomer (geometrical isomer).

(*iii*) **$CrCl_3(H_2O)_6$.** This compound has the structure, [Cr $(H_2O)_6$] Cl_3. [Cr $(H_2O)_6]^{3+}$ ion is an octahedral complex ion of [Ma_6] type. It does not show geometrical isomerism. It also does not show optical isomerism.

Problem 16. *How many geometrical isomers are there for: (a) [Co $(NH_3)_2$ $Cl_4]^-$ (b) [$AuCl_2Br_2]^-$ (square planar) (c) [$CuCl_2Br_2]^{2-}$ (tetrahedral)*

Solution. (*a*) Answer is 2. Since [Co $(NH_3)_2$ $Cl_4]^-$ is a [Ma_2b_4] type octahedral complex ion, it has two geometrical isomers which are *cis* and *trans* shown in Fig. 4.37

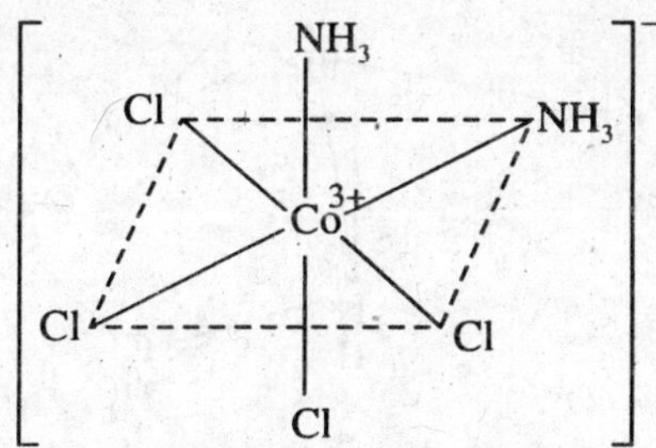

cis-isomer
(Two NH_3 ligands have *cis* position)

trans-isomer
(Two NH_3 ligands have *trans* position)

Fig. 4.37. *cis* and *trans* isomers (geometrical isomers) of $[Co(NH_3)_2Cl_4]^-$ ion (octahedral ion)

(*b*) Answer is 2. Since [Au Cl_2 $Br_2]^-$ ion is a [Ma_2 b_2] type square planar complex ion, it has two geometrical isomers *viz cis* and *trans* isomers (See fig. 4.38)

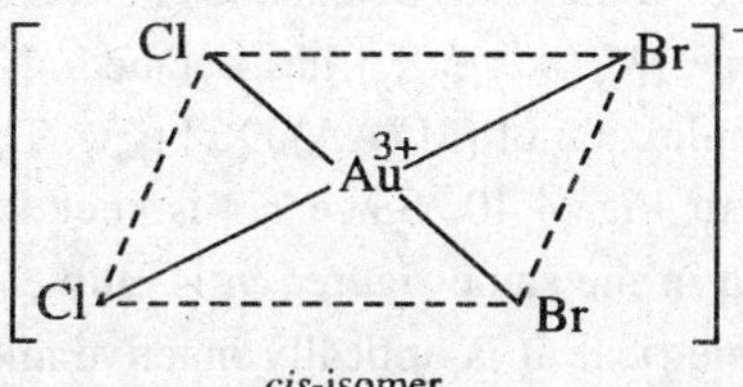

cis-isomer

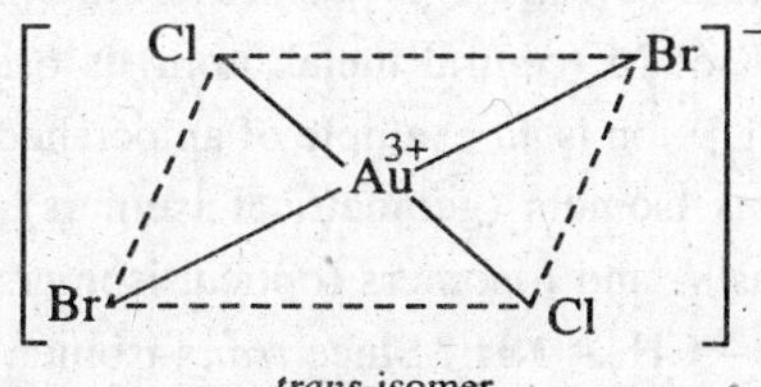

trans-isomer

Fig. 4.38 *cis* and *trams* isomers (geometrical isomers) of $[AuCl_2Br_2]^-$ ion (square planar)

(*c*) Since $[CoCl_2Br_2]^{2-}$ ion is a tetrahedral complex ion, it has no geometrical isomers.

Problem 17. *Which isomer of [$CoCl_2(en)_2]^+$ does not show optical isomerism.*

Solution. The given complex compound is an octahedral complex of [$Ma_2(AA)_2$] type and hence has two geometrical isomers *viz cis* and *trans*. *Cis*-isomer has no plane or centre of symmetry and hence is optically active *i.e., cis*-isomer can be resolved into two optically active isomers *viz d*- and *l*-isomers. In other words we say that *cis* isomer shows optical isomerism.

On the other hand, since *trans*-isomer has a plane of symmetry, it is optically inactive and hence is *meso* form, *i.e., trans* isomer does not show optical isomerism.

Problem 18. *A complex of the type [$M(AB)_2$] is known to be optically active. What does this indicate about the structure of the complex? Give one example of such complex.*

Solution. In [$M(AB)_2$], since AB is a bidentate (unsymmetrical) ligand, C.N. of M (central metal atom) is 4 and hence [$M(AB)_2$] can have either tetrahedral geometry or square planar geometry. Now since $[M(AB)_2]$ is known to be optically active, it has tetrahedral geometry. $[Be(gly)_2]^0$ is an example of such tetrahedral complex. Here gly^- ion is $NH_2.CH_2.COO^-$ ion.

Being tetrahedral in geometry, $[Be(gly)_2]^0$ does not have geometrical isomers. Now since $[Be(gly)_2]^0$ is optically active, it has *d* and *l*-isomers (optical isomers) shown in Fig 4.39

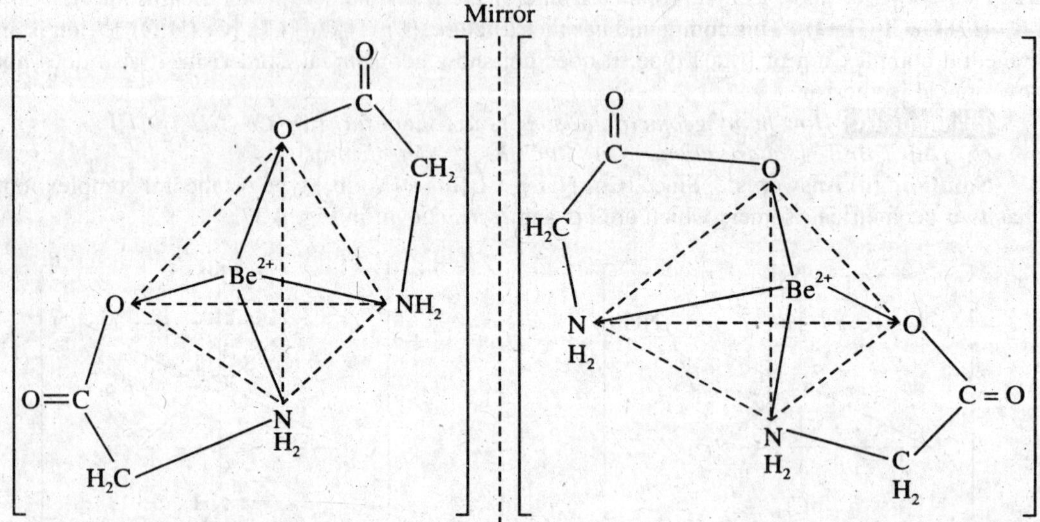

Fig. 4.39. Optical isomers of $[Be(gly)_2]^0$ $[gly^- = NH_2 . CH_2 . COO^-]$

Problem 19. *A complex of the type $[M(AA)_2X_2]$ is known to be optically active. What does this indicate about the structure of the complex? Give one example of such complex.*

Solution. In $[M(AA)_2X_2]$, since AA is a bidentate (symmetrical) ligand and X is a monodentate ligand, C.N. of M (central metal atom) is 6 and hence $[M(AA)_2X_2]$ has octahedral geometry. $[Co(en)_2Cl_2]^+$ ion is an example of an octahedral complex ion of $[M(AA)_2X_2]$ type. This ion has *cis* and *trans* isomers (geometrical isomers) shown in Fig. 4.40. Since *cis* isomer is optically active, it has *d*- and *l*-isomers (optical isomers) shown in the same figure. *en* is ethylenediamine, $NH_2 - CH_2 - CH_2 - NH_2$. Since *trans*-isomer is symmetrical, it is optically inactive and hence is *trans*-meso form .

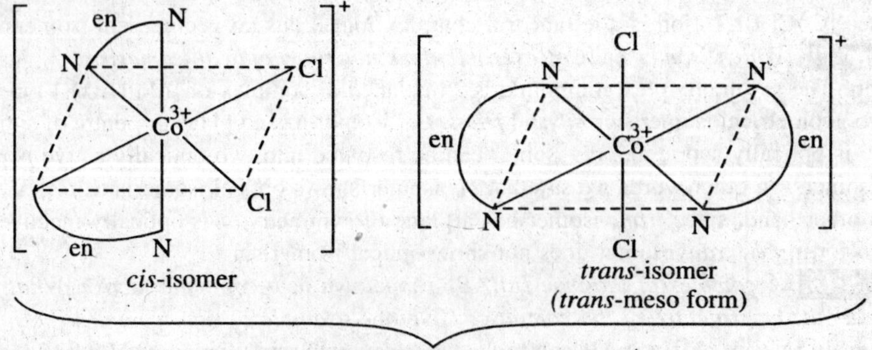

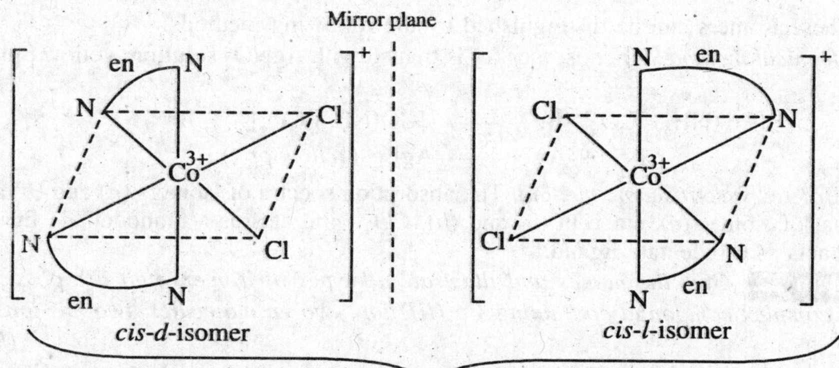

Fig. 4.40. Geometrical and optical isomers of $[Co(en)_2Cl_2]^+$ ion

Problem 20. *The complex ion, $[M(CN)(NO_2)(H_2O)(NH_3)]^+$ is optically active. What is the geometry of the complex ion ?*

Solution. Since each of the four ligands viz CN^-, NO_2^-, H_2O and NH_3 is a monodentate ligand, C.N. of M atom (central metal atom) is 4. Thus the given complex ion has either square planar or tetrahedral geometry. Since the given complex ion is optically active,it has tetrahedral geometry.

Problem 21. *The formula $Co(NH_3)_4CO_3Br$ represents three isomers. (i) Draw their structures (ii) How would you distinguish these isomers ?*

Solution. (*i*) The structures of the three isomers represented by the given formula are given in Fig 4.41. In isomers (*a*) and (*b*), CO_3^{2-} ion acts as a monodentate ligand but in isomer (*c*) this ion acts as a bidentate ligand.

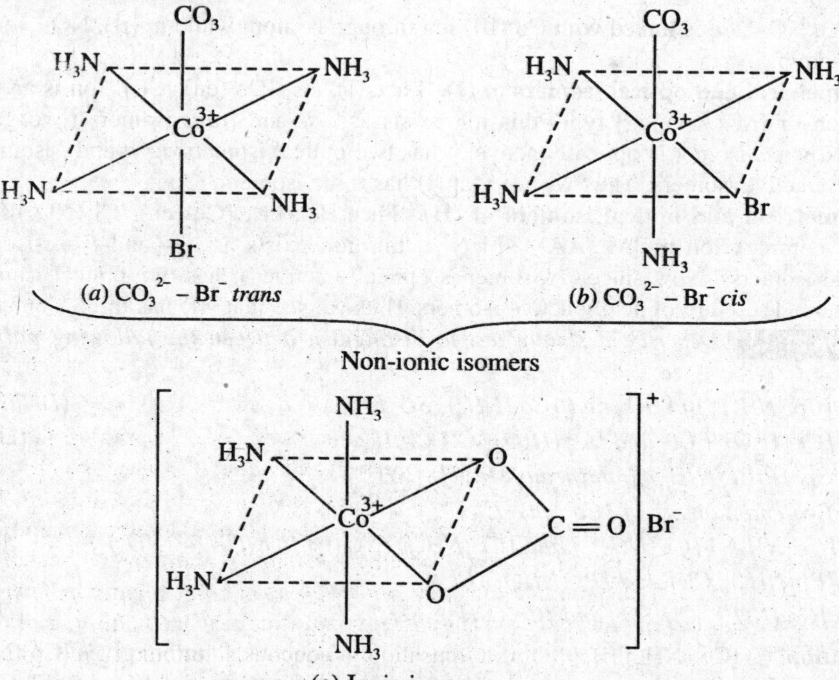

Fig. 4.41. Three isomers represented by $Co(NH_3)_4CO_3Br$

(*ii*) These isomers can be distinguished by the following methods:

(*a*) *Chemical method.* When isomer (*c*) is treated with $AgNO_3$ solution, yellow ppt. of AgBr is obtained.

$$[Co(NH_3)_4CO_3]^+Br^- \longrightarrow [CO(NH_3)_4CO_3]^+ + Br^-$$
$$Br^- + Ag^+ \longrightarrow AgBr \text{ (Yellow ppt.)}$$

(*b*) *Infra-red spectroscopic method.* The absorption spectra of isomers (*a*) and (*b*) are different from that of isomer (*c*), since in (*a*) and (*b*), CO_3^{2-} ion acts as a monodentate ligand but in (*c*) this ion acts as a bidentate ligand.

Problem 22. *Give the names and illustrate all types of isomers that are possible for an octahedral complex compound containing Co (III) ion, two en molecules, two Cl^- ions and one NO_2^- ion.* *(MD 1993)*

Solution. The complex compound can be represented by the formula, $Co(en)_2Cl_2NO_2$. Since the complex compound is octahedral, it may be represented by any of the following two structural formulae:

$[Co(en)_2Cl_2]NO_2$ (I) $\qquad$ $[Co(en)_2Cl(NO_2)]Cl$ (II)

(I) and (II) are ionisation isomers. On ionisation (I) gives $[Co(en)_2Cl_2]^+$ and NO_2^- ions and (II) gives $[Co(en)_2Cl(NO)_2]^+$ and Cl^- ions. Also (II) has been obtained from (I) by replacing one Cl^- ion by NO_2^- ion. Thus (I) and (II) are ionisation isomers.

(II) has linkage isomer. Since (II) has ambidentate ligand *viz* NO_2^- ion, this compound has two linkage isomers which are:

$[Co(en)_2Cl(NO_2)]Cl$
Bis (ethylene diamine) chloronitro cobalt (III) chloride (Nitro complex)
(A)

$[Co(en)_2Cl(ONO)]Cl$
Bis (ethylene diamine) chloronitrito cobalt (III) chloride (Nitrito complex)
(B)

In (A), NO_2^- ion is linked with Co (III) ion through N-atom while in (B), NO_2^- ion is linked through O-atom.

Geometrical and optical isomers of (I). Since (I), *i.e.*, $[Co(en)_2Cl_2]^+$ ion is an octahedral complex ion of $[M(AA)_2a_2]$ type, this ion exists as *cis-* and *trans*-isomers (two geometrical isomers). Now *cis*-isomer is optically active, it has two optical isomers *viz d*-and *l*-isomers. These are optically active isomers. Thus we see that (I) has three isomers.

Geometrical and optical isomers of (II). Since (II), *i.e.*, $[Co(en)_2Cl(NO_2)]^+$ ion is an octahedral complex ion of $[M(AA)_2ab]$ type, this ion exists as *cis-* and *trans*-isomers (two geometrical isomers). Now since *cis*-isomer is optically active, it has two optical isomers *viz d*- and *l*-isomers. These are optically active isomers Thus we see that (II) has three isomers.

Problem 23. *Describe a simple test to distinguish between the following pairs of compounds*

(*i*) $[Co(NH_3)_5Br]SO_4$ *and* $[Co(NH_3)_5SO_4]Br$ *(Delhi 1997, 99)*
(*ii*) $[Cr(H_2O)_6]Cl_3$ *and* $[Cr(H_2O)_5Cl]Cl_2.H_2O$ *(Delhi 1997)*
(*iii*) *Cis* $[PtCl_2(NH_3)_2]$ *and trans* $[PtCl_2(NH_3)_2]$
(*iv*) *Two enantiomers of* $[Co(en)_2Cl_2]^+$
(*v*) $[Cr(NH_3)_6][Cr(NO_2)_6]$ *and* $[Cr(NH_3)_4(NO_2)_2][Cr(NH_3)_2(NO_2)_4]$
(*vi*) $[Pt(NH_3)_2Cl_2]$ *and* $[Pt(NH_3)_6]Cl_2$
(*vii*) $[Co(NH_3)_5Cl]SO_4$ *and* $[Co(NH_3)_5SO_4]Cl$. *(Delhi 1999)*

Solution. (*i*) $[Co(NH_3)_5Br]SO_4$, on ionisation in aqueous solution, gives $[Co(NH_3)_5Br]^{2+}$ and SO_4^{2-} ions. This complex compound, therefore, gives white ppt. of $BaSO_4$ with $BaCl_2$ solution.

$$[Co(NH_3)_5Br]SO_4 \rightleftharpoons [Co(NH_3)_5Br]^{2+} + SO_4^{2-}$$

$$Ba^{2+} \text{ (From } BaCl_2) + SO_4^{2-} \longrightarrow BaSO_4 \text{ (\textit{White ppt})}$$

Since $[Co(NH_3)_5 SO_4]Br$ gives $[Co(NH_3)_5 SO_4]^+$ and Br^- ions, this compound gives light yellow ppt. of AgBr with $AgNO_3$ solution.

$$[Co(NH_3)_5 SO_4]Br \rightleftharpoons [Co(NH_3)_5 SO_4]^+ + Br^-$$

$$Ag^+ \text{ (From } AgNO_3) + Br^- \longrightarrow AgBr \text{ (\textit{Light yellow ppt.})}$$

(*ii*) When $[Cr(H_2O)_6]Cl_3$ is heated, no H_2O molecule is lost since all H_2O molecules are coordinated to the central metal atom. On the other hand when $[Cr(H_2O)_5 Cl]Cl_2.H_2O$ is heated, water molecule present outside the coordination sphere is easily lost, since it is not linked with the central metal atom.

(*iii*) *Cis*-$[Pt Cl_2 (NH_3)_2]$ has some value of dipole moment while the *trans*- $[Pt Cl_2 (NH_3)_2]$ has zero value for its dipole moment.

(*iv*) One enantiomer of $[Co(en)_2Cl_2]^+$ is dextro-rotory (*d*-form) and the other enantiomer is leavorotory (*l*-form).

(*v*) The given isomers can be distinguished by conductivity measurements. $[Cr(NH_3)_6][Cr(NO_2)_6]$ conducts as (3+, 3–) electrolyte while $[Cr(NH_3)_4(NO_2)_2][Cr(NH_3)_2(NO_2)_4]$ conducts as (1+, 1–) electrolyte.

$$\underset{\text{(I)}}{[Cr(NH_3)_6][Cr(NO_2)_6]} \longrightarrow \underset{\text{(3+, 3–) electrolyte}}{[Cr(NH_3)_6]^{3+} + [Cr(NO_2)_6]^{3-}}$$

$$\underset{\text{(II)}}{[Cr(NH_3)_4(NO_2)_2][Cr(NH_3)_2(NO_2)_4]} \longrightarrow \underset{\text{(1 +, 1 –) electrolyte.}}{[Cr(NH_3)_4(NO_2)_2]^+ + [Cr(NH_3)_2(NO_2)_4]^-}$$

Obviously (I) has higher value of molar conductance than (II).

(*vi*) The ionisation of the given complex compounds can be shown as:

$$\underset{\text{Non electrolyte}}{[Pt(NH_3)_2Cl_2]} \longrightarrow \text{No ions}$$

$$\underset{\text{Electrolyte}}{[Pt(NH_3)_6]Cl_2} \longrightarrow \underbrace{[Pt(NH_3)_6]^{2+} + 2Cl^-}_{\text{Three ions}}$$

We know that the total number of ions obtained as a result of ionisation of a complex compound is directly proportional to the magnitude of electrolytic conductance of the compound. Now since $[Pt(NH_3)_2 Cl_2]$ gives no ions, it has zero value of its electrolytic conductance. On the other hand, since $[Pt(NH_3)_6]Cl_2$ gives three ions, it has some value of electrolytic conductance.

(*vii*) Aqueous solution of $[Co(NH_3)_5 Cl]SO_4$ gives $[Co(NH_3)_5 Cl]^{2+}$ and SO_4^{2-} ions.

$$[Co(NH_3)_5 Cl]SO_4 \rightleftharpoons [Co(NH_3)_5 Cl]^{2+} + SO_4^{2-}$$

Thus the given complex gives white ppt. of $BaSO_4$ with $BaCl_2$ solution.

$$Ba^{2+} \text{ (From } BaCl_2) + SO_4^{2-} \longrightarrow BaSO_4 \text{ (\textit{White ppt})}$$

Since $[Co(NH_3)_5 SO_4]Cl$ gives Cl^- ions, this compound gives the ppt of AgCl (white) with $AgNO_3$ solution.

$$[Co(NH_3)_5SO_4]Cl \rightleftharpoons [Co(NH_3)_5 SO_4]^+ + Cl^-$$

$$Ag^+ \text{ (From } AgNO_3) + Cl^- \longrightarrow AgCl \text{ (\textit{White ppt.})}$$

Problem 24. *Explain the following:*

(i) SCN^- ion shows linkage isomerism in coordination compounds.

(ii) Tetrahedral complexes do not show geometrical isomerism while square planar complexes do show this kind of isomerism.

(iii) The square planar complex, $[Pt (NH_3) (H_2O) (py) (NO_2)]^+$ has three geometrical isomers.

Solution. (*i*) Since SCN^- is an ambidentate ligand, it can coordinate with metal atom either through *S*-atom (thiocyanato) or through N-atom (isothiocyanato) and gives two linkage isomers. Thus SCN^- ion shows linkage isomerism.

(*ii*) Tetrahedral complexes donot show geometrical isomerism because the relative positions of the ligands attached with the central metal atom are the same with respect to each other. The square planar complexes, on the other hand, show this kind of isomerism because, if same kind of ligands occupy positions adjascent to each other, we get *cis* geometrical isomer and if these ligands occupy positions opposite to each other, we get *trans* geometrical isomer.

(*iii*) Since the given complex compound is a aquare planar complex compound of [Mabcd] type, it shows geometrical isomerism and hence has three geometrical isomers shown in Fig. 4.42. These isomers arise because of the different arrangements of the ligands.

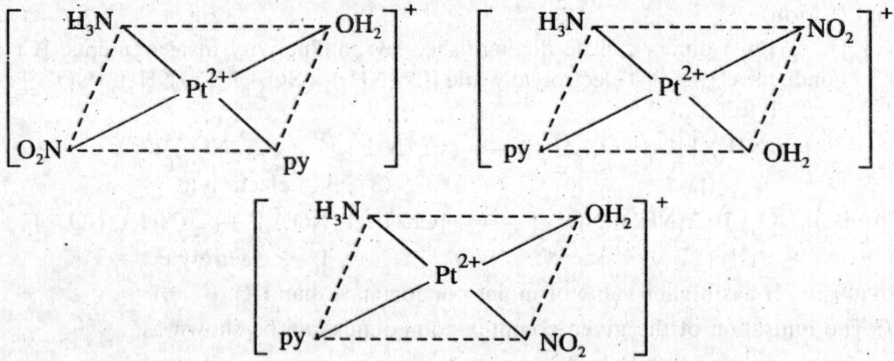

Fig. 4.42. Three geometrical isomers of $[Pt (NH_3)(H_2O)(py)(NO_2)]^+$ square planar complex ion

Problem 25. *Draw all the possible isomers of* : $[Pt (NH_3)_4 Cl_2]^{2+}$, $[Pt (gly)_2]^0$, $[Pt\{P(C_2H_5)_3\}Cl_2]_2$, $[Cr (NH_3) (OH)_2 Cl_3]^{2-}$, $[Pt (NH_3) (NH_2OH) (py) (NO_2)]^+$, $[Co (en) Cl_2 Br_2]^-$, $[Rh (py)_3 Cl_3]^0$, $[Co (en) Cl_3 Br]^-$

Solution. $[Pt (NH_3)_4 Cl_2]^{2+}$ ion. This ion is an octahedral ion of $[Ma_4b_2]$ type and hence has two geometrical isomers *viz cis* and *trans* isomers shown in Fig, 4.43 In *cis*-isomer two Cl^- ions are in *cis* position and in *trans*-isomer these ions have *trans* position.

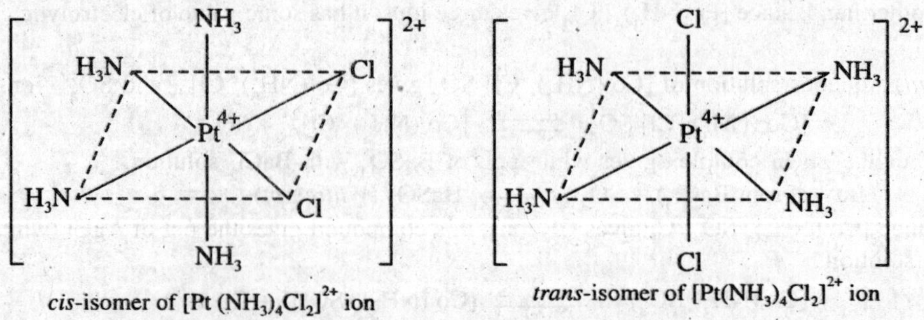

cis-isomer of $[Pt (NH_3)_4Cl_2]^{2+}$ ion

trans-isomer of $[Pt(NH_3)_4Cl_2]^{2+}$ ion

Fig. 4.43. *cis*-and *trans*-isomers of $[Pt(NH_3)_4Cl_2]^{2+}$ ion

$[Pt(gly)_2]^0$. It is a square planar complex of $[M(AB)_2]$ type in which gly^- ion ($NH_2.CH_2.COO^-$) is an unsymmetrical bidentate ligand. This complex has two geometrical isoners *viz cis* and *trans* shown in Fig 4.44

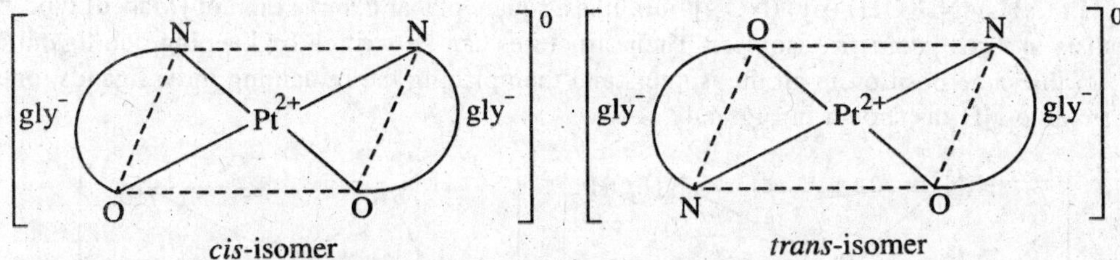

Fig. 4.44. *cis* and trans isomers of [Pt $(gly)_2$].

[Pt{P$(C_2H_5)_3$}Cl_2]$_2$. It is a bridged binuclear planar complex and exists in three isomeric forms viz *cis*, *trans* and *unsymmetrical* shown in Fig 4.45

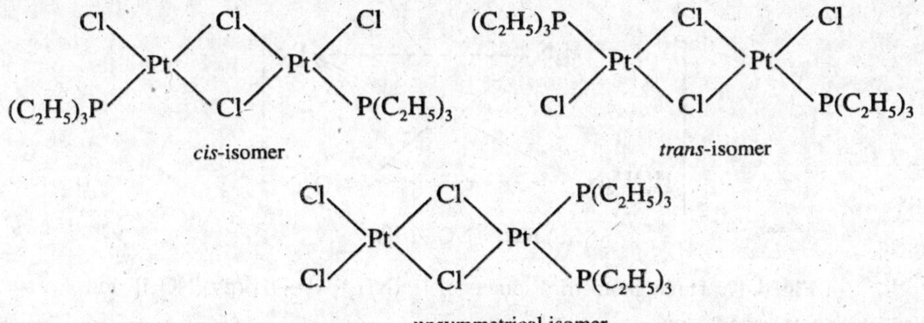

cis-isomer

trans-isomer

unsymmetrical isomer

Fig. 4.45 Three isomers of [Pt {P$(C_2H_5)_3$}Cl_2]$_2$

[Cr$(NH_3)(OH)_2Cl_3$]$^{2-}$ ion. This ion has three geometric isomers shown in Fig 4.46.

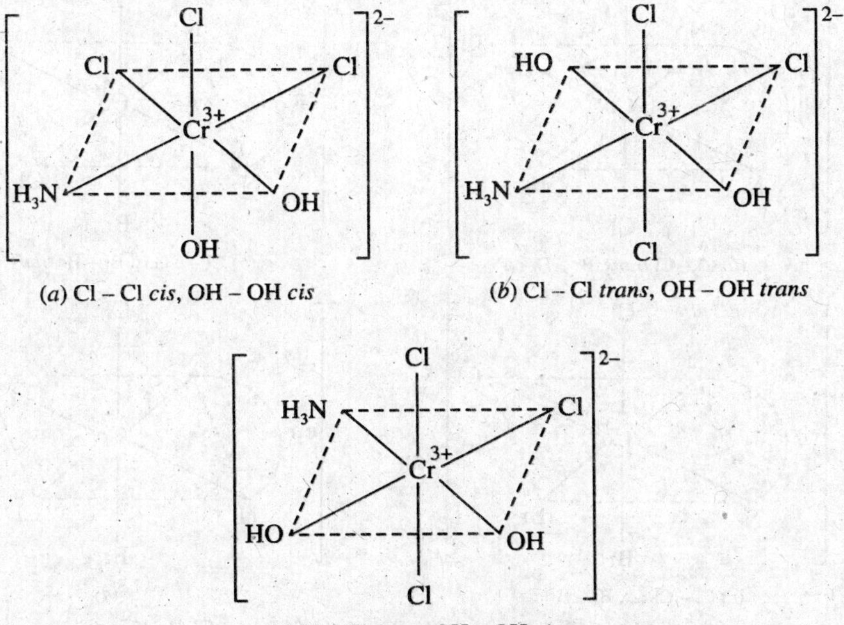

Fig. 4.46. Three geometric isomers of [Cr$(NH_3)(OH)_2Cl_3$]$^{2-}$ ion

$[Pt(NH_3)(NH_2OH)(py)(NO_2)]^+$ ion. It is a square planar complex ion of [Mabcd] type. This ion exists in three geometric isomers. Their structures can be written by keeping one ligand (say NH_3) in the same position in all the isomers and then placing the remaining three ligands, one by one, *trans* to NH_3 as shown in Fig. 4.47

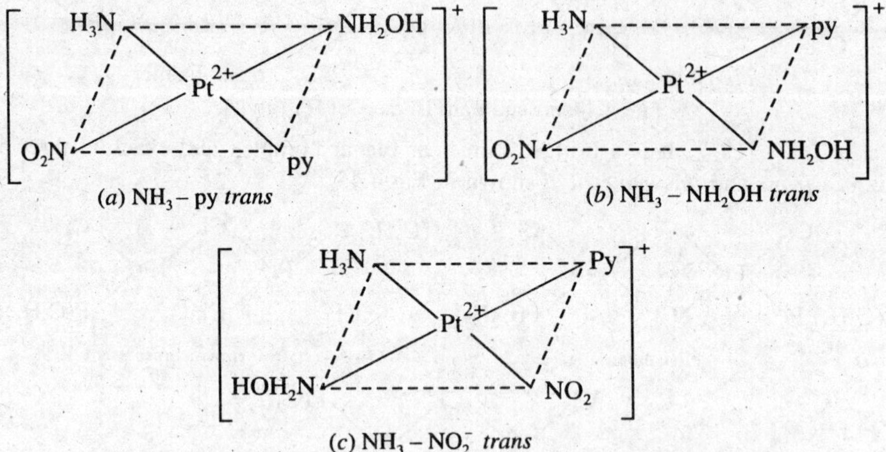

Fig. 4.47. Three geometric isomers of $[Pt(NH_3)(NH_2OH)(py)(NO_2)]^+$ ion

$[Co(en)Cl_2Br_2]^-$ ion. This ion has octahedral geometry. It has four geometric isomers shown in Fig. 4.48. While drawing their structures keep *en* group in the same position in all the structures.

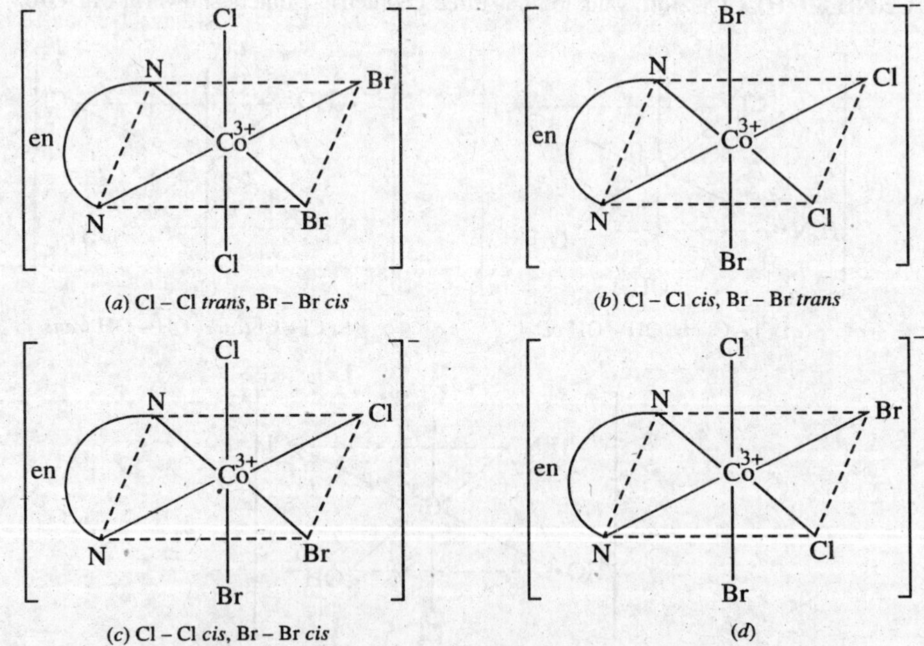

Fig. 4.48. Four geometric isomers of $[Co(en)Cl_2Br_2]^-$ ion

Since isomer (*c*) has both Cl^- ions and both Br^- ions *cis* to each other, this isomer is optically active and hence has two optical isomers (non superimposible mirror images) shown in Fig 4.49.

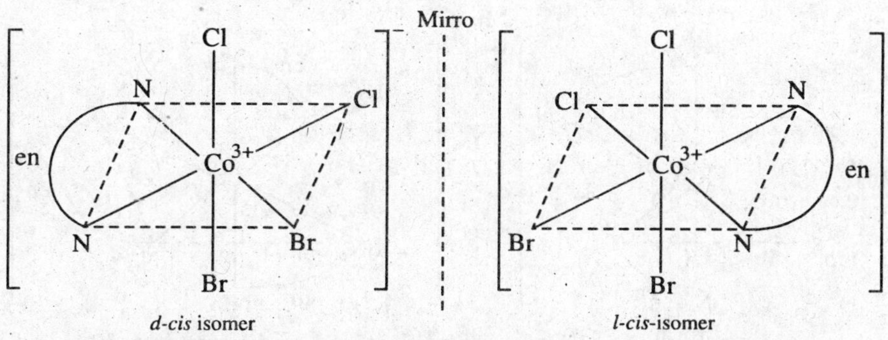

Fig. 4.49. Optical isomer corresponding to *cis* isomer (*c*) shown in Fig 4.48

$[Rh(py)_3Cl_3]^0$. This complex compound has octahedral geometry and is of $[Ma_3b_3]$ type. This compound has two geometric isomers shown in Fig. 4.50. In isomer (*a*) all the three Cl^- ions have *cis* positions and similarly all *py* groups have *cis* positions. In isomer (*b*) two Cl^- ions have *trans* positions and two *py* groups also have *trans* position.

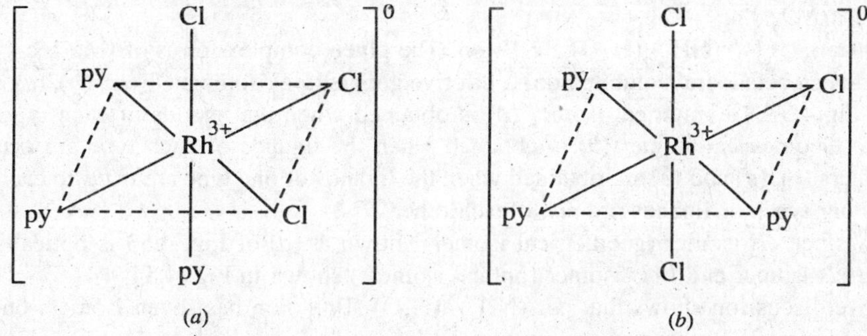

Fig. 4.50. Two geometric isomers of $[Rh(py)_3Cl_3]^0$

$[Co(en)Cl_3Br]^-$ ion. This ion does not have optical isomers but has two geometric isomers shown in Fig 4.51. In isomer (*a*) two Cl-atoms are *trans* to each other while in isomer (*b*) Cl atom is *trans* to Br atom.

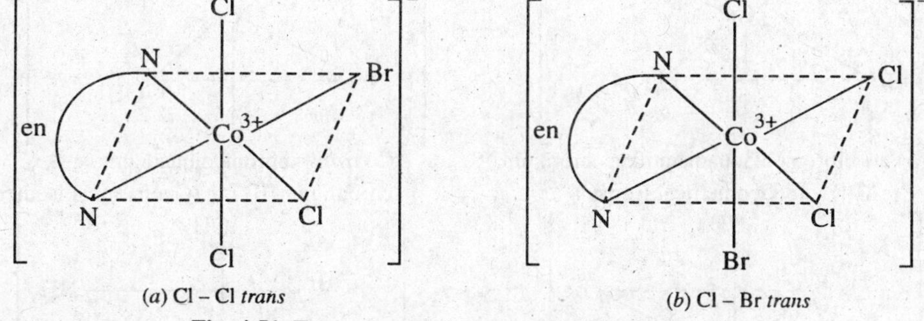

Fig. 4.51. Two geometric isomers of $[Co(en)Cl_3Br]^-$ ion.

Problem 26. *Draw the structure of all the possible isomers of $[Co(en)_3]^{3+}$ and $[Co(ox)_3]^{3-}$ ions.*

Solution. $[Co(en)_3]^{3+}$ ion. It is an octehedral complex ion of $[M(AA)_3]$ type. This ion does not have geometrical isomers, but, due to the absence of a plane or centre of symmetry, this ion can be resolved into dextro (*d*-form) and levo (*l*-form) forms. Thus $[Co(en)_3]^{3+}$ ion has two optical isomers *viz d*-isomer and *l*-isomer shown in Fig 4.52. These isomers are non-superimpossible mirror images, *i.e.,* these cannot be superimposed on each other.

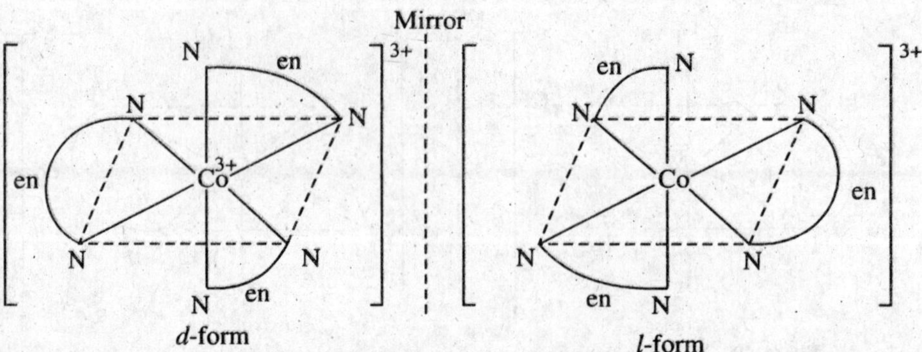

Fig. 4.52. Optical isomers of $[Co\ (en)_3]^{3+}$ ion

$[Co(ox)_3]^{3-}$ ion. Like $[Co\ (en)_3]^{3+}$, $[Co(ox)_3]^{3-}$ ion also does not have geometrical isomers. However, it has two optical isomers.

Problem 27. *Draw the structure of all the possible isomers of the following complex ions/compounds. (i) $[Cr\ (NH_3)_2\ (H_2O)_2\ Br_2]^+$ (ii) $[Cr\ (en)_2\ Cl_2]\ Cl$ (iii) $[M\ (AA)_2\ B_2]$ type complex (iv) $[Co\ (en)_2\ Cl\ (NO_2)]^+$*

Solution. (*i*) **$[Cr(NH_3)_2(H_2O)_2Br_2]^+$ ion.** The given complex ion is of $[Ma_2b_2c_2]$ type complex ion. It is an octahedral complex ion. It has five geometrical isomers *viz* (*a*), (*b*), (*c*), (*d*) and (*e*) shown in Fig 4.53. Geometrical isomer (*a*) is obtained when the ligands of each type are *cis* to each other. Geometrical isomer (*b*) is obtained when the ligands of each type are *trans* to each other. Isomers (*c*), (*d*) and (*e*) are obtained when the ligands of one type are *trans* to each other and the remaining types of ligands are *cis* to each other.

Now since *cis*-isomer (geometrical isomer) shown at (*d*) of Fig. 4.53 is optically active, it exists as *d-cis*-isomer and *l-cis*-isomer (optical isomers) shown in Fig. 4.54

Above discussion shows that $[Cr\ (NH_3)_2\ (H_2O)_2\ Br_2]^+$ ion has seven isomers out of which five are geometrical isomers and two are optical isomers.

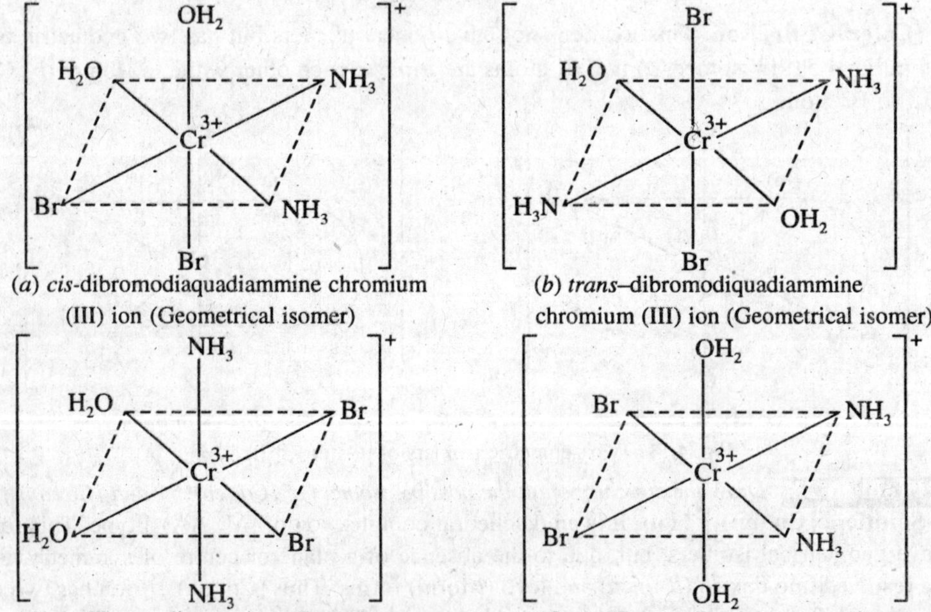

Fig. 4.53. Continued on the next page

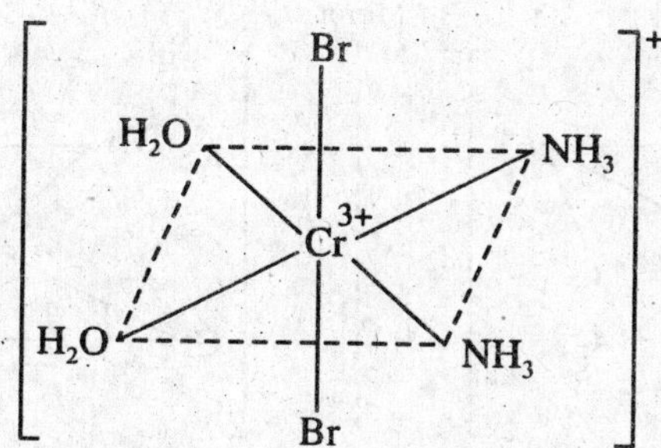

(e) *cis*-diamminediaqua *trans*-dibrome chromium (III) ion (Geometrical isomer)

Fig. 4.53. Five geometrical isomers of $[Cr(NH_3)_2(H_2O)_2Br_2)]^+$ ion

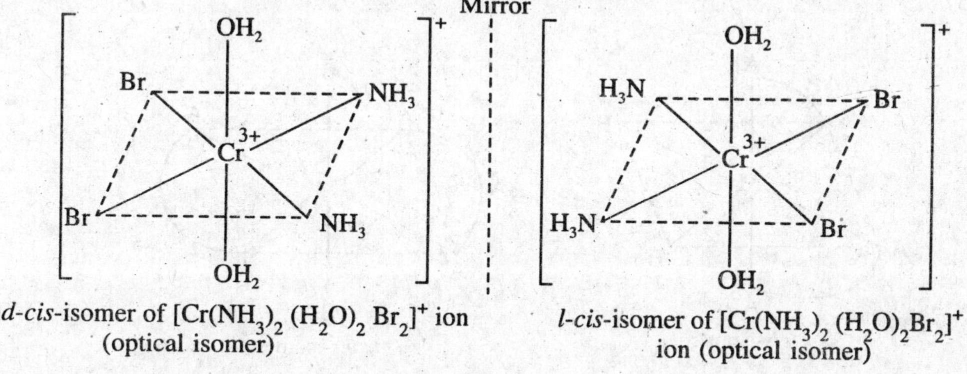

d-*cis*-isomer of $[Cr(NH_3)_2 (H_2O)_2 Br_2]^+$ ion (optical isomer)

l-*cis*-isomer of $[Cr(NH_3)_2 (H_2O)_2Br_2]^+$ ion (optical isomer)

Fig. 4.54. Optical isomers of *cis* isomer of $[Cr(NH_3)_2(H_2O)_2 Br_2]^+$ ion

(*ii*) $[Cr(en)_2Cl_2]Cl$**.** This compound contains $[Cr (en)_2 Cl_2]^+$ and Cl^- ions. Since $[Cr (en)_2 Cl_2]^+$ ion is an octahedral ion of $[M (AA)_2 a_2]$ type, it has two geometrical isomers viz *cis*-isomer and *trans*-isomer (See Fig. 4.55) Since *cis*-isomer has no plane of symmetry, it is unsymmetrical and hence is optically active. Being optically active, *cis*-isomer has two optical isomers namely *cis-d*-isomer and *cis-l*-isomer (See Fig. 4.56). Since *trans*-isomer has plane of symmetry, it is symmetrical and hence is optically inactive. Being optically inactive, *trans*-isomer has no optical isomers.

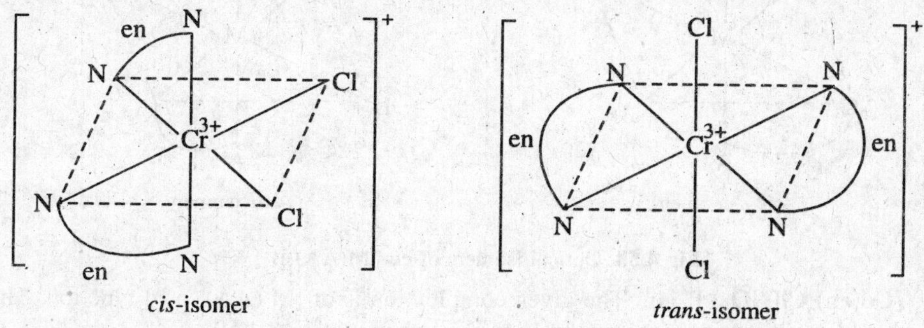

cis-isomer

trans-isomer

Fig. 4.55. *cis*- and *trans* isomers of $[Cr (en)_2Cl_2]^+$ ion (geometrical isomers)

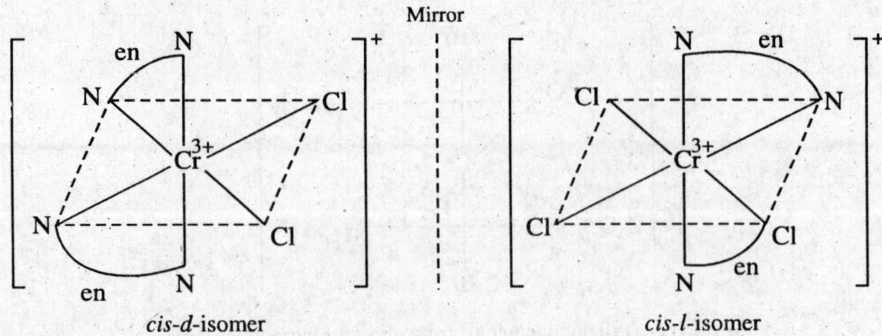

Fig. 4.56. Optical isomers of *cis-* isomer of $[Cr(en)_2Cl_2]^+$ ion.

(*iii*) **$[M(AA)_2B_2]$ type complex.** It has two geometrical isomers *viz cis* and *trans* shown in Fig. 4.57

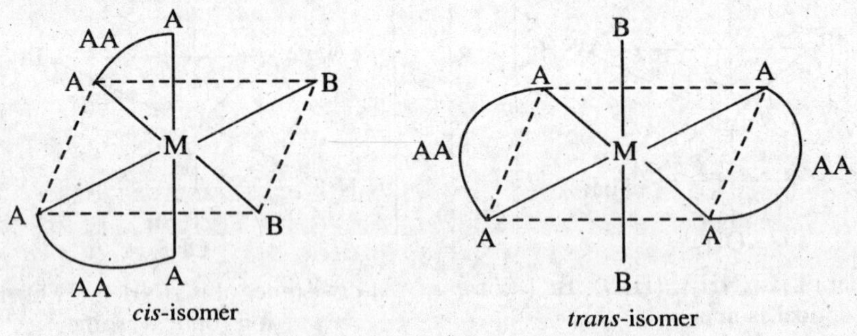

Fig. 4.57. *cis* and *trans* isomers of $[M(AA)_2B_2]$ (Geometrical isomers)

Cis-isomer does not have any plane or centre of symmetry and hence is optically active. Being optically active, *cis*-isomer can be resolved into two optical isomers which are *cis-d*-isomer and *cis-l*-isomer shown in Fig. 4.58.

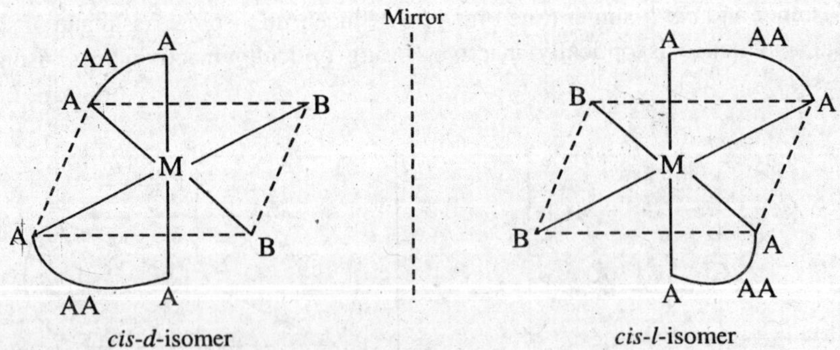

Fig 4.58. Optical isomers of *cis*-$[M(AA)_2B_2]$

(*iv*) **$[Co(en)_2Cl(NO_2)]^+$ ion.** The given complex ion is of $[M(AA)_2\ a\ b]$ type ion. This ion exists in *cis*-and *trans*-isomers (geometrical isomers) shown in Fig. 4.59.

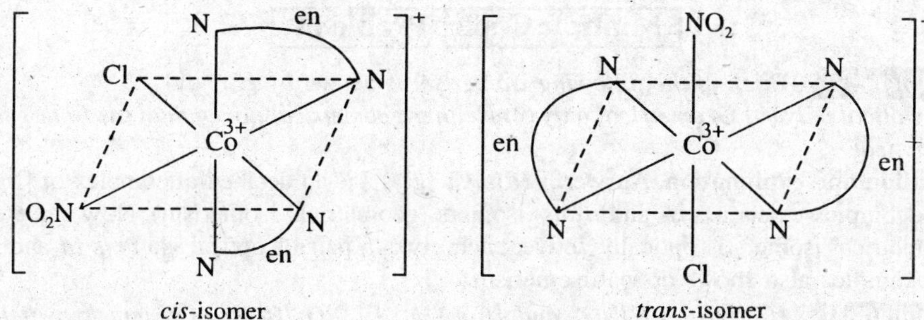

Fig. 4.59. Two geometrical isomers (*cis* and *trans*) of $[Co(en)_2Cl(NO_2)]^+$ ion.

Since *cis*-isomer does not have centre of symmetry, it is optically active and hence has *cis-d* and *cis-l* isomers (optical isomers) shown in Fig. 4.60

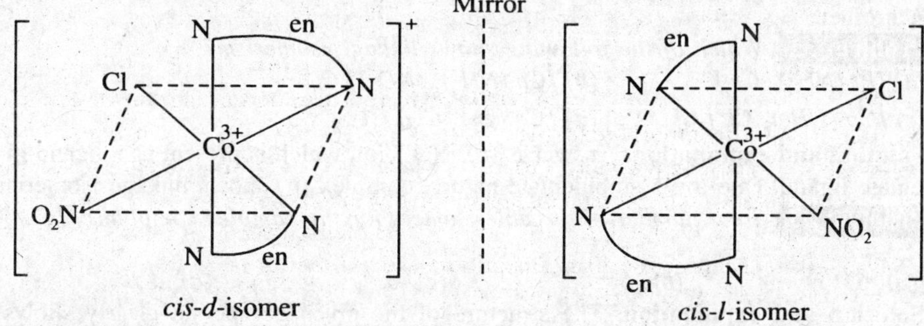

Fig 4.60 *cis-d* and *cis-l* isomers (optical isomers) of *cis*- $[Co(en)_2Cl(NO_2)]^+$ ion.

Problem 28. *Give the structure of all the possible isomers of the following complex species.*

(A) (i) $[Pt(gly)_2]$ *(ii)* $[Pt(NO_2)(C_5H_5N)(NH_3)(NH_2OH)]^+$

(iii) $[Cr(NH_3)_4Cl_2]^+$ *(Delhi 1998)*

(B) $[M(AA)_2B_2]$ *type complex.* *(Delhi 1999)*

(C) $[Co(en)_2Cl(NO_2)]^+$ *(Delhi 2000)*

(D) $[Co(ox)_3]^{3-}$ *ion (ox = oxalate ion,* $C_2O_4^{2-}$*)* *(Delhi 2001)*

Solution. Do these problems yourself

Problem 29. *The complex cation,* $[Pt(NH_3)(NH_2OH)(NO_2)(C_6H_5N)]^+$ *has been found to be optically active. What is the probable geometry of the complex cation ?*

Solution. In the given complex cation, since all the four ligands *viz* NH_3, NH_2OH, NO_2^-, C_6H_5N are monodentate ligands, the complex cation has either square planar (C.N. = 4) or tetrahedral (C.N. = 4) geometry. Now since the complex cation is optically active, it cannot have square planar geometry, because square planar complexes have plane of symmetry. However, it can have tetrahedral geometry (See Fig. 4.61), since it has four different ligands and hence the mirror-images of the complex cation are not superimposable, *i.e.*, it will show optical activity.

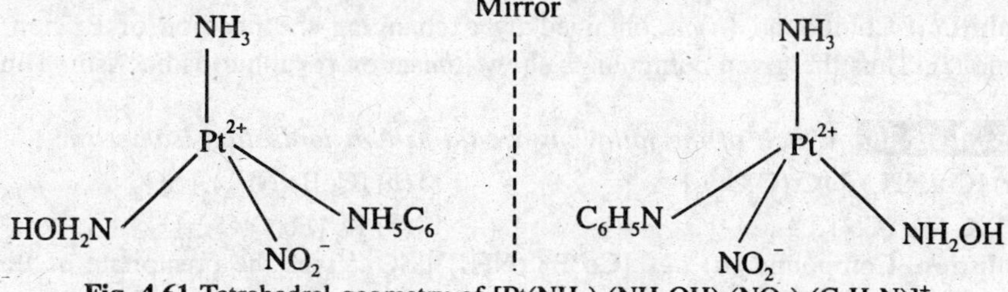

Fig. 4.61 Tetrahedral geometry of $[Pt(NH_3)(NH_2OH)(NO_2)(C_6H_5N)]^+$

Multiple Choice Problems

Problem 30. *Which of the following isomerism is shown by $[Cr(gly)_3]$*
(a) Optical only (b) Geometrical only (c) Neither geometrical nor optical (d) Both geometrical and optical

Solution and explanation. Answer is (*d*). $[Cr (gly)_3]$ is an octahedral complex of $[M(AB)_3]$ type. This complex exists in *cis*- and *trans*- isomers (geometrical isomerism). Now since each of these geometrical isomers is optically active, each gives a pair of optical isomers (*d*-and *l*-pair). Thus this complex also shows optical isomerism.

Problem 31. *$[Co(NH_3)_4Cl_2]NO_2$ and $[Co(NH_3)_4Cl\ NO_2]Cl$ are isomers of which type?*
(a) Geometrical (b) Optical (c) Linkage (d) Ionisation *[PMT 1993]*

Solution and explanation. Answer is (*d*). Since $[Co (NH_3)_4\ Cl\ NO_2]Cl$ can be obtained from $[Co (NH_3)_4\ Cl_2]\ NO_2$ by replacing Cl^- ion by NO_2^- ion, these compounds are ionisation isomers with each other.

Problem 32. *Which of the following shows linkage isomerism?*
(a) $[Pt (NH_3)_2\ Cl_2]$ *(b) $[Co (NH_3)_5 (NO_2)]\ Cl_2$*
(c) $[Co (NH_3)_4\ Cl_2]\ Cl$ *(d) $[Co (en)_2\ Cl_2]\ Cl$*

Solution and explanation. Answer is (*b*). NO_2^- ion which is present as a ligand in (*b*) is an ambidentate ligand. Due to its ambidentate nature, complex (*b*) shows linkage isomerism.

Problem 33. *Total number of possible isomers for the complex compound, $[Cu^{2+} (NH_3)_4]$ $[Pt^{2+} Cl_4]$ are:*
(a) 3 *(b) 6* *(c) 5* *(d) 4*

Solution and explanation. The structure of the given compound clearly shows that the compound will show coordination isomerism. Its coordination isomers are:

(*i*) $[Cu^{2+} (NH_3)_3Cl]^+ [Pt^{2+} Cl_3 (NH_3)]^-$
(*ii*) $[Cu^{2+} (NH_3)_2 Cl_2]^0 [Pt^{2+} Cl_2 (NH_3)_2]^0$
(*iii*) $[Cu^{2+} (NH_3) Cl_3]^- [Pt^{2+} Cl (NH_3)_3]^+$
(*iv*) $[Cu^{2+}Cl_4]^{2-} [Pt^{2+}(NH_3)_4]^{2+}$

Thus total number of isomers is 4. (*d*) is the correct answer.

Problem 34. *Out of the following which will not show geometrical isomerism.*
(a) $[Pt (NH_3)_2\ Cl_2]$ *(b) $[Co (NH_3)_5 (NO_2)]Cl_2$*
(c) $[Co (en)_2\ Cl_2]\ Cl$ *(d) $[Co (NH_3)_4\ Cl_2]\ Cl_2$*

Solution and explanation. (*a*) is a square planar compound and hence shows geometrical isomerism. Since (*c*) and (*d*) are octahedral complexes, both these compounds will show geometrical isomerism. (*b*) is also an octahedral complex but, since it is of $[M\ a_5b]$ type, it will not show geometrical isomerism. Thus the answer is (*b*)

Problem 35. *Which type of isomerism is shown by the complex compounds viz $[Co (NH_3)_5 Br]\ SO_4$ and $[Co (NH_3)_5\ SO_4]\ Br$?*
(a) Ionic *(b) Linkage* *(c) Coordination* *(d) Optical* *(PMT, MP 1998)*

Solution. Compound (*ii*) is obtained by exchanging the position of Br^- ion by SO_4^{2-} in compound (*i*). Thus the given compounds show ionisation (or ionic) isomerism. Thus the answer is (*a*)

Problem 36. *Which of the following would exhibit ionisation isomerism.*
(*a*) $[Co(NH_3)_6]\ [Cr(C_2O_4)_3]$ (*b*) $[CoBr(NH_3)_5]\ SO_4$
(*c*) $K_3\ [Fe (CN)_6]$ (*d*) $K_3\ [Fe(C_2O_4)_3]$

Solution. Compound (*b*) *i.e.*, $[Co\ Br\ (NH_3)_5]\ SO_4$ gives the precipitate of $BaSO_4$ with an aqueous solution of $BaCl_2$.

$$[Co\,Br(NH_3)_5]\,SO_4 \rightleftharpoons [CoBr(NH_3)_5]^{2+} + SO_4^{2-}$$
$$SO_4^{2-} + BaCl_2 \longrightarrow BaSO_4 + 2Cl^-$$

Thus compound (*b*) shows ionisation isomerism.

Problem 37. *$[Cr(NH_3)_5Br]SO_4$ and $[Co(NH_3)_5SO_4]Br$ are related to each other as :*

(a) Ionisation isomers *(b) Linkage isomers*
(c) Coordination isomers *(d) Optical isomers*

Solution. The given compounds can be obtained by exchanging the ligands given in coordination sphere and the ions given out of the coordination sphere. Thus the given compounds are related to each as ionisation isomers. The correct answer is, therefore, (*a*).

Problem 38. *Which of the following will show coordination isomerism*

(a) $[Cr(NH_3)_6][Co(CN)_6]$ *(b) $[Co(en)_2Cl_2]^+$*
(c) $[Cr(NH_3)_6]Cl_3$ *(d) $[Cr(en)_2Cl_2]^+$*

Solution and explanation. Answer is (*a*). It is only (*a*) which, on ionisation, gives complex cation and complex anion as shown below:

$$[Cr(NH_3)_6][Co(CN)_6] \rightleftharpoons [Cr(NH_3)_6]^{3+} + [Co(CN)_6]^{3-}$$

Thus only (*a*) can show coordination isomerism.

Problem 39. *Which one of the following complexes can exist as d and l isomers?*

(a) $[Co(NH_3)_4Cl_2]^+$ *(b) $[Co(NH_3)_3Cl_3]$*
(c) Trans $[Co(en)_2Cl_2]^+$ *(d) Cis $[Co(en)_2Cl_2]^+$*

Solution and explanation. Answer is (*d*). $[Co(en)_2Cl_2]^+$ ion exists as *cis* and *trans* isomers (geometrical isomers). Now since *cis*-isomer is unsymmetrical and hence optically active, it exists as *d* and *l* forms.

Problem 40. *Which one of the following square planar complexes exists in cis and trans isomeric forms?*

(a) Ma_4 *(b) Ma_3b* *(c) Ma_2b_2* *(d) Mabcd*

Solution and explanation. Answer is (*c*). Square planar complexes of Ma_4 and Ma_3b type do not show geometrical isomerism. Square planar complex of Mabcd type exists in three geometrical isomeric forms. Ma_2b_2 type complex exists as *cis* and *trans* isomer. Thus it is only Ma_2b_2 complex which exists in *cis* and *trans* isomeric forms.

Problem 41. *The planar complex, Mabcd may have :*

(a) Two optical isomers *(b) Two geometrical isomers*
(c) Three optical isomers *(d) Three geometrical isomers*

Solution and explanation. The answer is (*d*). Mabcd which is a square planar complex exists as three geometrical isomers which can be obtained by selecting one ligand and then placing the remaining three ligands, one by one, *trans* to the selected ligand. Example is $[Pt(NH_3)(py)(Cl)(Br)]^0$.

Problem 42. *Which of the following isomeric pairs show ionisation isomerism?*

(a) $[Co(NH_3)_6Cl][Cr(CN)_6]$ and $[Cr(NH_3)_6][Co(CN)_6]$

(b) $[Cr(H_2O)_6]Cl_3$ and $[Cr(H_2O)_5Cl]Cl_2.H_2O$

(c) $[Pt(NH_3)_2Cl_2]$ and $[Pt(NH_3)_4][PtCl_4]$

(d) $[Co(NH_3)_5Br]SO_4$ and $[Co(NH_3)_5SO_4]Br$ *(MP, PET 1993)*

Solution and explanation. Answer is (*d*). Since $[Co(NH_3)_5SO_4]Br$ can be oblained by replacing SO_4^{2-} ion by Br^- ion, these compounds are ionisation isomeric pairs to each other.

Problem 43. *$[Co(NH_3)_5(NO_2)]Cl_2$ and $[Co(NH_3)_5(ONO)]Cl_2$ are:*

(a) Ionisation isomers *(b) Linkage isomers*
(c) Hydrate isomers *(d) Polymerisation isomers.*

Solution with explanation: Since NO_2^- ion is an ambidentate ligand, the given compounds show linkage isomerism. Thus the given compounds are linkage isomers. Consequently answer is (*b*).

Problem 44. *Which of the following complexes is likely to show optical activity?* (*CBSE 1994*)

(*a*) *Trans* $[Co(NH_3)_4Cl_2]^+$ (*b*) $[Cr(H_2O)_6]^{3+}$
(*c*) *Cis* $[Co(NH_3)_2(en)_2]^{3+}$ (*d*) *Trans* $[Co(NH_3)_2(en)_2]^{3+}$

Solution with explanation. Answer is (c). Since (c), *i.e.*, cis $[Co(NH_3)_2(en)_2]^{3+}$ is unsymmetrical, it shows optical activity and hence exists as *d* and *l* isomers (optical isomers)

Problem 45. *The possible number of isomers for square planar complex* $[MCl_2Br_2]SO_4$ *is:*
(*a*) 1 (*b*) 2 (*c*) 4 (*d*) 5 (*AIIMS 1992*)

Solution with explanation. Answer is (*b*). Since square planar complexes of Ma_2b_2 type exist as *cis* and *trans* isomers, $[MCl_2Br_2]SO_4$ has two isomers.

Problem 46. *The number of isomers possible for square planar complex,* $[PtCl\,Br\,I(NH_3)]^-$ *would be:*
(*a*) 2 (*b*) 3 (*c*) 4 (*d*) 6 (*ISM Dhanbad 1995*)

Solution with explanation: Since the given square planar complex ion is of Mabcd type, it exists in three geometrical isomeric forms. Thus the answer is (*b*).

Problem 47. *Optical isomersim is not shown by the complex*
(*a*) $[Cr(ox)_3]^{3-}$ (*b*) *Cis* $[Co(en)_2Cl_2]^+$
(*c*) *Trans* $[Co(en)_2Cl_2]^+$ (*d*) $[Cr(en)_3]^{3+}$

Solution with explanation. Since (*c*) is symmetrical, it is optically inactive and hence does not show optical isomerism. Thus answer is (*c*).

Problem 48. *Which of the following will have three isomeric forms?*
(*a*) $[Co(en)_2ClBr]$ (*b*) $K_3[Co(C_2O_4)_2Cl_2]$
(*c*) $[Cr(NO_3)_3(NH_3)_3]$ (*d*) $K_3[Co(C_2O_4)_3]$ (*CBSE 1996*)

Solution with explanation. (*a*) and (*b*)are $[M(AA)_2ab]$ and $[M(AA)_2a_2]$ type octahedral complexes and hence have three isomeric forms namely two optically active isomers corresponding to cis configuration (*i.e.*, *d* and *l* forms) and one optically inactive form corresponding to trans configuration (meso form). Thus answer is (*a*) and (*b*).

Problem 49. *The number of geometrical isomers of the complex,* $[Co(NO_2)_3(NH_3)_3]$ *is:*
(*a*) 2 (*b*) 4 (*c*) 3 (*d*) 0 [*CBSE 1997*]

Solution with explanation. $[Co(NO_2)_3(NH_3)_3]$ is an octahedral complex of Ma_3b_3 type and hence exists as two geometrical forms (*d* and *l* forms). Thus answer is (*a*).

University Questions

1. Suggest a simple experiment to distinguish between the cis and trans isomers of the complex $[Pt(NH_3)_2Cl_2]^0$. Give reasons for your choice of experiment. [*Utkal 91*]
2. Name the individual isomers of each of the following:
   (*i*) $[Cr(NSC)_4(NH_3)_2]^-$ (*ii*) $[Co(en)_2Cl_2]^+$
   (*iii*) $[Pt(NH_3)_2Cl_2]$ (*iv*) $[Cr(H_2O)_6]Cl_3$ (*Lucknow 95*)
3. Write note on "Optical isomerism of octahedral complexes". (*Lucknow 95*)
4. What is isomerism? Explain stereo isomerism in metal complexes with coordination number 4 and 6. (*Osmania 93*)
5. Name different kinds of isomerism possible in coordination compounds, quoting one suitable example in each case. (*Meerut 96*)

6. Explain why tetrahedral complexes are unable to exhibit geometrical isomerism. (*Lucknow 2002*)

7. Write the structures of geometrical and optical isomers of $\left[Co(en)_2Cl_2\right]^+$ ion. (*Lucknow 2002*).

8. (*a*) What are different types of structural isomerism shown by complexes ? Name the isomerism shown by following pairs of complexes :

(*i*) $\left[Co(NH_3)_5NO_2\right]Cl_2$ and $\left[Co(NH_3)_5ONO\right]Cl_2$

(*ii*) $\left[Cr(NH_3)_6\left[Co(CN)_6\right]\right.$ and $\left[Cr(CN)_6\right]\left[Co(NH_3)_6\right]$

(*b*) Draw structures of all possible geometrical isomers of $\left[Co(en)_2Cl_2\right]^+$ ion. (*Nagpur 2002*).

9. (*a*) Predict and sketch all the stereo isomers of [Cr (glycine)$_3$] complex.
(*b*) Name the types of isomerism exhibited by the following pairs of complexes :

(*i*) $\left[(Co(en)_2(H_2O)Cl\right]Cl_2$ and $\left[(Co(en)_2Cl_2\right]Cl.H_2O$

(*ii*) $\left[Pt(NH_3)_4\right]^{2+}\left[PtCl_4\right]^{-2}$ and $\left[Pt(NH_3)_3Cl\right]^+\left[PtCl_3(NH_3)\right]^-$ (*Delhi 2002*).

Ans. (*a*) [Cr (glycine)$_3$]^0 exists in cis- and trans-isomers (geometrical isomers). Each of these forms is optically active and hence each gives a pair of optical isomers (*d* - and *l* - forms)

(*b*) (*i*) Hydrate isomerism (*ii*) Coordination isomerism.

10. (*a*) What type of isomerism is shown by the following pairs :

(*i*) $\left[Pt(NH_3)_4Cl_2\right]Br_2$ and $\left[Pt(NH_3)_4Br_2\right]Cl_2$

(*ii*) $\left[Co(NH_3)_5NO_2\right]Cl_2$ and $\left[Co(NH_3)_5ONO\right]Cl_2$

(*b*) Sketch the possible geometric isomers of $\left[Pt(NH_3)_2Cl_2\right]$. (*Rohilkhand 2003*).

11. (*a*) Explain and give one example each of hydrate and linkage isomers.
(*b*) Draw all possible isomers of dichlorobis (ethylene diamine) platinum (IV).

Which of these isomers will be optically inactive and why ? (*Delhi 2003*).

Ans. (*b*) $\left[Pt(en)_2Cl_2\right]^{2+}$ ion exists in cis- and trans-isomers. In cis- isomer the two monodentate ligands *viz.* Cl^- ions occupy the adjascent (*i.e., cis.*) positions while in trans isomer these ligands occupy opposite (*i.e. trans.*) postions. Since trans. isomer has a plane of symmetry, it is optically inactive.

12. Discuss the following types of isomerism shown by co-ordination compounds with suitable example :–
(*i*) Co-ordination isomerism
(*ii*) Ionisation isomerism
(*iii*) Linkage isomerism
(*iv*) Polymerisation isomerism. (*Nagpur 2003*).

13. (*a*) Explain the classification of isomerism in co-ordination compounds with the help of suitable examples. Explain hydrate and ligand isomerism.
(*b*) Give two examples of cis-trans isomerism in 4-coordination compounds.
(*c*) What is EAN rule and its limitations? (*Bhopal 2004*)

14. Draw the possible isomers of $\left[Co(en)^2Cl_2\right]^+$. (*GND Amritsar 2004*)

15. What type of isomerism is shown by the coordination compound, dichloro *bis* (ethylenediamine) cobalt (III) ion? (*GND Amritsar, 2004*)

5

Nature of Metal-Ligand Bonding In Complex Compounds

Introduction

We have already said that Werner was the first to explain the nature of bonding in complex compounds. However, with the advancement of theories of valence, modern theories have been proposed to explain the nature of metal-ligand bonding in complexes. These theories can also explain the colour, geometry and magnetic properties of the complex compounds. These modern theories are : (*i*) *Valence Bond Theory, VBT* (due to L. Pauling and J.L. Slater, 1935) (*ii*) *Crystal Field Theory, CFT* (due to H. Brethe, 1929 and Van Vleck, 1932) (*iii*) *Ligand Field Theory, LFT* or *Molecular Orbital Theory, MOT* (due to J. Van Vleck, 1935). Here we shall discuss only valence bond theory and crystal field theory.

Valence Bond Theory (VBT)

This theory is mainly due to Pauling. It deals with the electronic structure of the central metal ion in its ground state, kind of bonding, geometry and magnetic properties of the complexes.

Assumptions of valence bond theory

This theory is based on the following assumptions:

1. *The central metal atom or ion (as the case may be) makes available a number of empty s, p and d atomic orbitals equal to its coordination number. These vacant orbitals hybridise together to form hybrid orbitals which are the same in number as the atomic orbitals hybridising together. These hybrid orbitals are vacant, equivalent in energy and have definite geometry.* Important types of hybridisation occuring in the first row transition metal (3*d* metals) complexes and the geometry of the complexes are given in Table 5.1.
2. The ligands have at least one σ-orbital containing *a lone pair of electrons.*
3. Vacant hybrid orbitals of the metal atom or ion *overlap with the filled* (containing lone pair of electrons) σ-orbitals of the ligands to form ligand → metal σ-bond (represented as $L \xrightarrow{\sigma} M$). This bond which is generally known as coordinate bond is a special type of covalent bond and shows the characteristics of both the overlapping orbitals. However, it also possesses a considerable amount of polarity because of the mode of its formation.
4. The non-bonding electrons of the metal atom or ion are then rearranged in the metal orbitals (*viz.* pure *d, s* or *p* orbitals as the case may be) which do not participate in forming the hybrid orbitals. The rearrangement of non-bonding electrons takes place according to Hund's rule.

Table 5.1. *Important types of hybridisation found in the first row transition metal (3d-metal) complexes and the geometry of the complexes.*

Coordination number of the central metal atom/ion	*Type of hybridisation undergone by the central metal atom/ion*	*Geometry of the complex*	*Examples of complexes*
2	$sp(4s, 4p_x)$	Linear or diagonal	$[CuCl_2]^-$, $[Cu(NH_3)_2]^+$ etc.
3	$sp^2(4s, 4p_x, 4p_y)$	Trigonal planar or equilateral triangular	$\left[Cu^+\left(S{=}C\begin{matrix}NH{-}CH_2\\ \vert\\ NH{-}CH_2\end{matrix}\right)_3\right]^+$, $[Cu^+Cl(tu)_2]^0$ (distorted trigonal planar) etc.
4	$dsp^2(3d_{x^2-y^2}, 4s, 4p_x, 4p_y)$	Square planar	$[Ni(CN)_4]^{2-}$, $[PdCl_4]^{2-}$
4	$sp^2d(4s, 4p_x, 4p_y, 4d_{x^2-y^2})$	Square planar	$[Cu(NH_3)_4]^{2+}$ $[Pt(NH_3)_4]^{2+}$ etc.
4	$sp^3(4s, 4p_x, 4p_y, 4p_z)$	Tetrahedral	$[NiCl_4]^{2-}$, $[Cu(CN)_4]^{3-}$, $Ni(CO)_4$ etc.
5	$dsp^3(3d_{z^2}, 4s, 4p_x, 4p_y, 4p_z)$	Trigonal bipyramidal	$Fe(CO)_5$, $[CuCl_5]^{3-}$, $[Ni^{2+}(triars)\ Br_2]^0$
5	$dsp^3(3d_{x^2-y^2}, 4s, 4p_x, 4p_y, 4p_z)$	Square pyramidal	$[Co^{2+}(triars)\ I_2]^0$, $[Ni(CN)_5]^{3-}$ etc.
6	$d^2sp^3(3d_{x^2-y^2}, 3d_{z^2}, 4s, 4p_x, 4p_y, 4p_z)$	Inner-orbital octahedral	$[Ti(H_2O)_6]^{3+}$, $[Fe(CN)_6]^{3-}$ etc.
6	$sp^3d^2(4s, 4p_x, 4p_y, 4p_z, 4d_{x^2-y^2}, 4d_{z^2})$	Outer-orbital octahedral	$[Fe^+(NO^+)(H_2O)_5]^{2+}$, $[CoF_6]^{3-}$ etc.

Geometry of 6-coordinated complex ions

Examples of 6-coordinated complex ions formed by some transition metals are given in Table 5.2. In all these complex ions the coordination number of the central metal atom or ion is six and hence these complex ions have octahedral geometry. This octahedral geometry arises due to d^2sp^3 or sp^3d^2 hybridisation of the central metal atom or ion. What type of hybridisation (*i.e.*, whether sp^3d^2 or d^2sp^3) will occur depends on the number of unpaired or paired electrons present in the complex ion. d^2sp^3 or sp^3d^2 hybridisation is also called *octahedral hybridisation*

Octahedral complexes in which the central atom is d^2sp^3 hybridised are called *inner-orbital octahedral complexes* while the octahedral complexes in which the central atom is sp^3d^2 hybridised are called *outer-orbital octahedral complexes.*

Now let us see how d^2sp^3/sp^3d^2 hybridisation takes place in octahedral complexes.

d^2sp^3 Hybridisation in Inner Orbital Octahedral Complexes

This type of hybridisation takes place in those octahedral complexes which contain strong ligands. On the basis of the orientation of the lobes of *d*-orbitals in space, these orbitals have been classified into two sets *viz.* t_{2g} or d_ε and e_g or d_r. t_{2g} set consists of d_{xy}, d_{yz} and d_{zx}

orbitals while e_g set has d_{z^2} and $d_{x^2-y^2}$ orbitals. In the formation of six d^2sp^3 hybrid orbitals, *two* $(n - 1)$ *d*-orbitals of e_g set [*i.e.*, $(n - 1)d_{z^2}$ and $(n - 1)\ d_{x^2-y^2}$ orbitals], *one ns* and *three np* (np_x, np_y and np_z) orbitals combine together and form six d^2sp^3 hybrid orbitals. Thus we see that the two *d*-orbitals used in d^2sp^3 hybridisation are from penultimate shell [*i.e.* $(n - 1)$th shell] while *s* and three *p*-orbitals are from ultimate shell (*i.e.*, *n*th shell). This discussion shows that in case of octahedral complex ions of 3*d* transition series elements, two *d*-orbitals used in d^2sp^3 hybridisation are $3d_{z^2}$ and $3d_{x^2-y^2}$ orbitals (e_g set of orbitals) while *s*- and *p*-orbitals are 4*s* and 4*p* orbitals. Thus d^2sp^3 hybridisation taking place in such complexes can be represented as $3d_{x^2-y^2}.3d_{z^2}.4s.4p_x.4p_y\ 4p_z(d^2sp^3)$.

Since two *d*-orbitals used in d^2sp^3 hybridisation belong to inner shell [*i.e.*, $(n - 1)$th shell], the octahedral complex compounds resulted from d^2sp^3 hybridisation are called **inner orbital octahedral complexes.**

Since these complexes have comparatively lesser number of unpaired electrons than the outer-orbital octahedral complexes (see later on), these complexes are also called **low spin** or **spin paired octahedral complexes.** It is due to the presence of strong ligands in inner-orbital octahedral complexes of 3*d* transition series that the electrons present in $3d_{z^2}$ and $3d_{x^2-y^2}$ orbitals (e_g set) are forced to occupy $3d_{xy}$, $3d_{yz}$ and $3d_{xz}$ orbitals (t_{2g} set) and thus 3*d* orbitals of e_g set become vacant and hence can be used in $3d_{x^2-y^2}$. $3d_{z^2}$. 4*s*. $4p_x$. $4p_y$. $4p_z$ (d^2sp^3) hybridisation.

Now let us discuss the structure of some octahedral complex ions of 3*d* transition series elements which result from d^2sp^3 hybridisation.

1. Ferricyanide ion, $[Fe(CN)_6]^{3-}$ [Also called hexacynoferrate (III) ion]

In this ion the coordination number of Fe is six and hence the given complex ion is octahedral in shape. In this ion, Fe is present as Fe^{3+} ion whose valence-shell configuration is $3d^5$ or $t_{2g}^3\ e_g^2$ (Fe = $3d^64s^24p^0$, Fe^{3+} = $3d^5$ = $t_{2g}^3\ e_g^2$) as shown in Fig. 5.1. According to Hund's rule, each of the five electrons in 3*d* orbitals is unpaired in free Fe^{3+} ion (uncomplexed ion) and hence the number of unpaired electrons (*n*) is equal to 5 (see Fig. 5.1). However, magnetic study of $[Fe(CN)_6]^{3-}$ ion has shown that this ion has one unpaired electron (n = 1) and hence

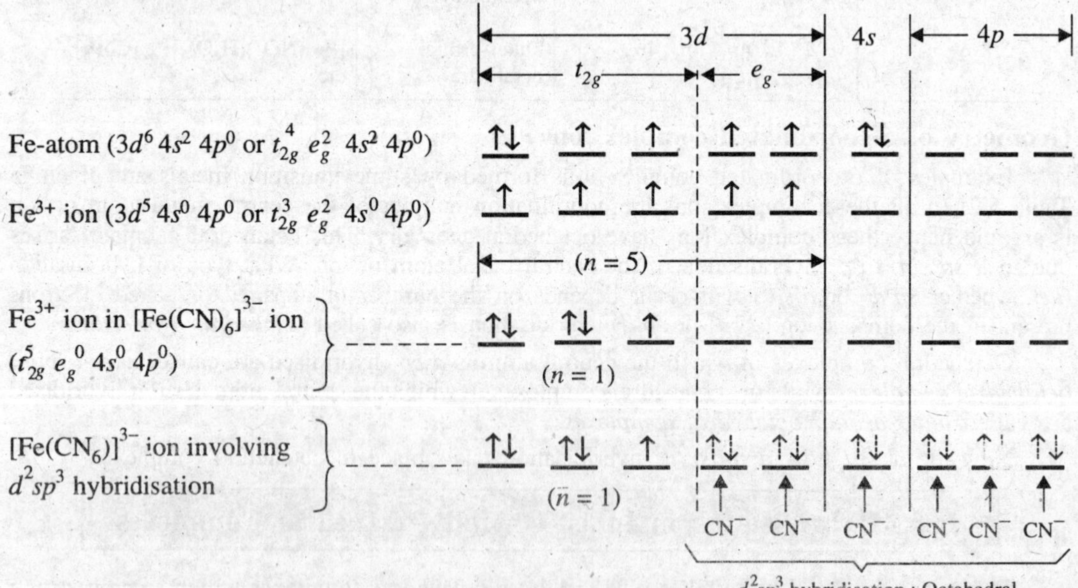

Fig. 5.1. Formation of $[Fe(CN)_6]^{3-}$ ion by d^2sp^3 hybridisation. ↑↓ indicates electron pair donated by each CN^- ion (ligand). (Inner-orbital octahedral complex ion).

is paramagnetic. Thus, in the formation of this ion, two electrons of e_g set of 3d orbitals (*i.e.*, $3d_{z^2}$ and $3d_{x^2-y^2}$ orbitals) pair up with the three electrons of t_{2g} set of orbitals (*i.e.*, $3d_{xy}$, $3d_{yz}$ and $3d_{zx}$ orbitals). This results in that e_g set of orbitals becomes vacant and is used in d^2sp^3 hybridisation. This also results in that the valence-shell configuration of Fe^{3+} ion gets changed from $t_{2g}^3\ e_g^2$ to $t_{2g}^5\ e_g^0$ and thus the number of unpaired electrons in 3d orbital now becomes equal to 1. Now $3d_{x^2-y^2}$, $3d_{z^2}$ (e_g set), 4s and three 4p ($4p_x$, $4p_y$ and $4p_z$) orbitals combine together and give rise to the formation of six $3d_{z^2} . 3d_{x^2-y^2} . 4s. 4p_x . 4p_y . 4p_z$ hybrid orbitals (d^2sp^3 hybridisation). Each of these hybrid orbitals is vacant. Each of the six CN^- ions (ligands) donates its lone pair of electrons to d^2sp^3 hybrid orbitals and six $N\bar{C}{\rightarrow}Fe^{3+}$ coordinate bonds are established (see Fig. 5.1). The above discussion shows that $[Fe(CN)_6]^{3-}$ ion has one unpaired electron and hence is paramagnetic. It is an inner orbital octahedral complex ion, since it is formed by d^2sp^3 hybridisation.

2. Ferrocyanide ion, $[Fe(CN)_6]^{4-}$ [Also called hexacyano ferrate (II) ion]

In this ion, since the coordination number of Fe is six, the given complex ion has octahedral geometry. In this ion, Fe is present as Fe^{2+} ion whose valence-shell configuration is $3d^6\ 4s^0\ 4p^0$ or $t_{2g}^4\ e_g^2\ 4s^0\ 4p^0$ which shows that Fe^{2+} ion has 4 unpaired electrons. Magnetic studies have, however, shown that the given complex ion is diamagnetic and hence it has no unpaired electrons (n = 0). Hence in order to get all the electrons in the paired state, two electrons of e_g orbitals are sent to t_{2g} orbitals so that n becomes equal to zero. Since CN^- ions (ligands) are strong ligands, they are capable of forcing the two electrons of e_g orbitals to occupy t_{2g} orbitals and thus make all the electrons paired. Now for the formation of $[Fe(CN)_6]^{4-}$ ion, two 3d orbitals of e_g set, 4s orbital (one orbital) and three 4p orbitals (all these six orbitals are vacant orbitals) undergo d^2sp^3 hybridisation (see Fig. 5.2). It is due to d^2sp^3 hybridisation that $[Fe(CN)_6]^{4-}$ ion is an inner orbital octahedral complex ion. The electron pair donated by CN^- ion (ligand) is accommodated in each of the six d^2sp^3 hybrid orbitals as shown in Fig. 5.2.

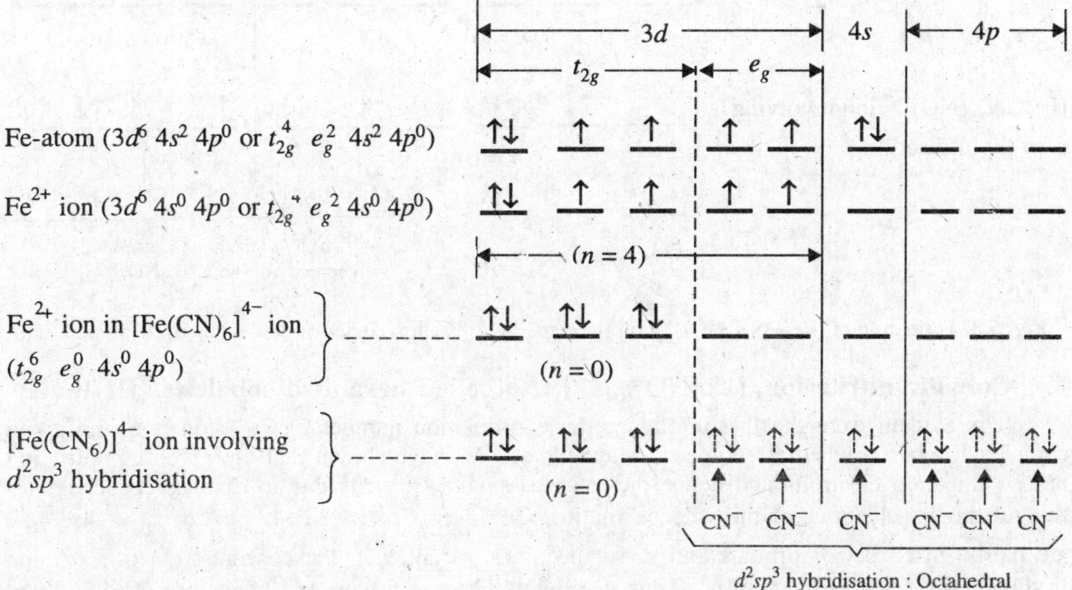

Fig. 5.2. Formation of $[Fe(CN)_6]^{4-}$ ion by d^2sp^3 hybridisation. ↑↓ indicates electron pair donated by each CN^- ion (ligand) (Inner-orbital octahedral complex ion).

3. Pentacyano nitrosyl ferrate (II) ion, $[Fe^{2+}(CN)_5(NO^+)]^{2-}$ (Also called nitroprusside ion)

Since the coordination number of Fe is six, the given ion has octahedral geometry. Pauling (1931) and Sidgwick (1935) have suggested that NO is present as NO^+ ion (nitrosyl ion) which is a two-electron donor. NO^+ ion is obtained when NO molecule (5 + 6 = 11 electrons) loses its unpaired electron residing in its π_y^* molecular orbital [Molecular orbital configuration of NO molecule is $(\sigma_{2s}^b)^2\ (\sigma_{2s}^*)^2\ (\sigma_x^b)^2\ (\pi_y^b)^2\ (\pi_z^b)^2\ (\pi_y^*)^1$. Thus molecular orbital configuration of NO^+ ion (5 + 6 – 1 = 10 electrons) is $(\sigma_{2s}^b)^2\ (\sigma_{2s}^*)^2\ (\sigma_x^b)^2\ (\pi_y^b)^2\ (\pi_z^b)^2\ (\pi_y^*)^0$. The presence of NO^+ ion in the given complex ion indicates that Fe is present as Fe^{2+} ion ($Fe^{2+} = 3d^6 4s^0 = t_{2g}^4\ e_g^2\ 4s^0$, having n = 0). Magnetic moment value of $[Fe(CN)_5(NO)]^{2-}$ ion has shown that the given ion has no unpaired electron (n = 0) and hence is diamagnetic. In order to account for its diamagnetic character, it is assumed that the two electrons of e_g orbitals are forced to occupy the t_{2g} orbitals so that we have all the electrons in the paired state and two 3d orbitals of e_g set become vacant and hence available for d^2sp^3 hybridisation. Thus the given ion results from d^2sp^3 hybridisation and hence is inner-orbital octahedral complex ion (see Fig. 5.3).

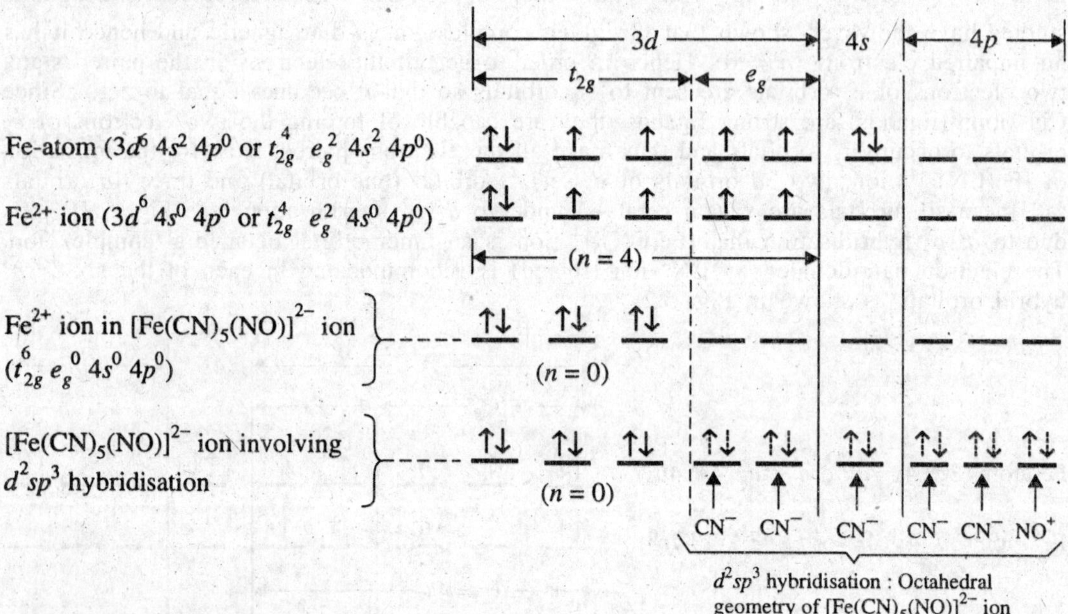

Fig. 5.3. Formation of $[Fe(CN)_5(NO)]^{2-}$ ion by d^2sp^3 hybridisation (Inner-orbital octahedral complex ion).

4. Cobaltic nitrite ion, $[Co(NO_2)_6]^{3-}$ [Also called hexanitro cobaltate (III) ion]

It is evident from the fact that since the coordination number of Co^{3+} ion is 6, the given complex ion has octahedral geometry. Magnetic studies have shown that the given complex ion has no unpaired electrons and hence the ion is diamagnetic. It is due to diamagnetic character that all the six electrons of the valence shell of Co^{3+} ion ($Co^{3+} = 3d^6\ 4s^0\ 4p^0$ or $t_{2g}^4\ e_g^2\ 4s^0\ 4p^0$) are paired in t_{2g} set of orbitals and e_g set becomes vacant, *i.e.*, the configuration of Co^{3+} ion in the complex ion is $t_{2g}^6 e_g^0 4s^0 4p^0$. Thus e_g orbitals become vacant and hence are used in d^2sp^3

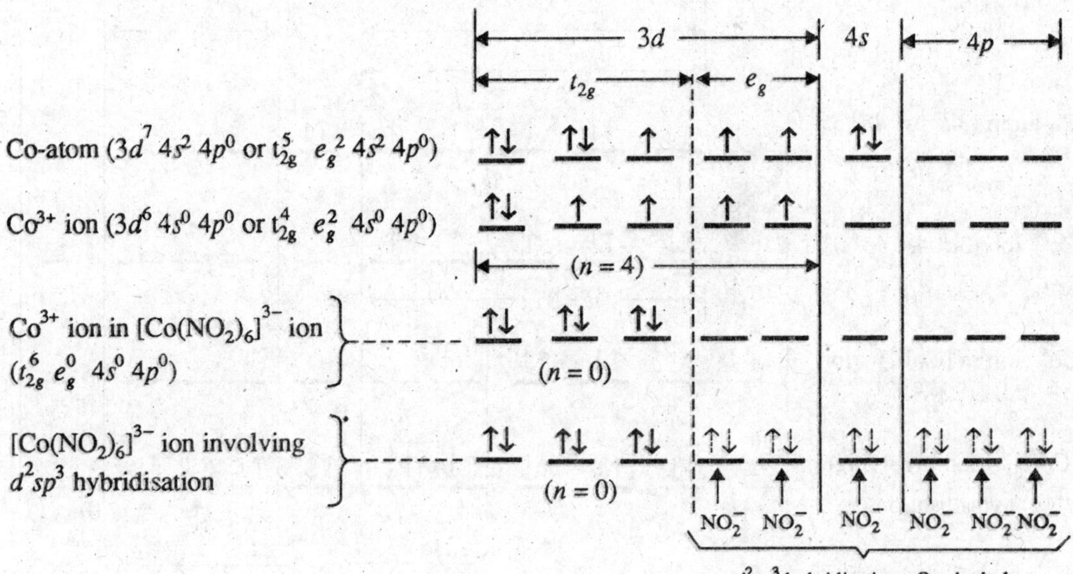

Fig. 5.4. Formation of $[Co(NO_2)_6]^{3-}$ ion by d^2sp^3 hybridisation (Inner-orbital octahedral complex ion).

hybridisation. It is due to strong character of NO_2^- ions (ligands) that the two electrons of e_g set are forced to occupy t_{2g} orbitals. (see Fig. 5.4).

5. Structure of hexacarbonyls of Cr, Mo and W

The hexacarbonyls of Cr, Mo and W can be represented as $M(CO)_6$ where M = Cr, Mo or W. In these molecules the central metallic atom is in zero oxidation state. All these molecules are diamagnetic ($n = 0$) and hence are formed by d^2sp^3 hybridisation. For their structures involving d^2sp^3 hybridisation see chapter 13 (*Transition Metal complexes with carbon monoxide : Metallic carbonyls*).

6. Inner orbital octahedral complexes of Co(+2)

Examples of such complexes are $[Co(NO_2)_6]^{4-}$ and $[Co(CN)_6]^{4-}$. Both these ions are paramagnetic corresponding to the presence of one unpaired electron. The valence-shell electronic configuration of Co^{2+} ion in its free state is $3d^7\ 4s^0\ 4p^0\ 5s^0$ or $t_{2g}^5\ e_g^2\ 4s^0\ 4p^0\ 5s^0$ which shows that this ion has 3 unpaired electrons. Pauling has suggested that, in order to have only one unpaired electron, one electron of e_g orbitals is forced to occupy t_{2g} orbitals and the other electron of e_g set is promoted to the higher energy vacant 5s orbital. Thus now the configuration of Co^{2+} ion becomes $t_{2g}^6\ e_g^0\ 4s^0\ 4p^0\ 5s^1$ in which the number of unpaired electrons is equal to one and both 3d orbitals of e_g set become empty and hence are available for d^2sp^3 hybridisation. The formation of $[CoL_6]^{4-}$ ion (L = NO_2^- or CN^-) by d^2sp^3 hybridisation has been shown in Fig. 5.5.

The presence of unpaired electron in 5s orbital is supported by the fact that, since 5s orbital is of very high energy, the single electron residing in it should be unstable and hence should be easily lost. Experimentally it has been found to be true. When $[CoL_6]^{4-}$ ion is acted upon by air or H_2O_2, the unpaired electron is lost and $[Co^{2+}L_6]^{4-}$ ion is oxidised to $[Co^{3+}L_6]^{3-}$ ion.

$$[Co^{2+}L_6]^{4-} \longrightarrow [Co^{3+}L_6]^{3-} + e^-$$

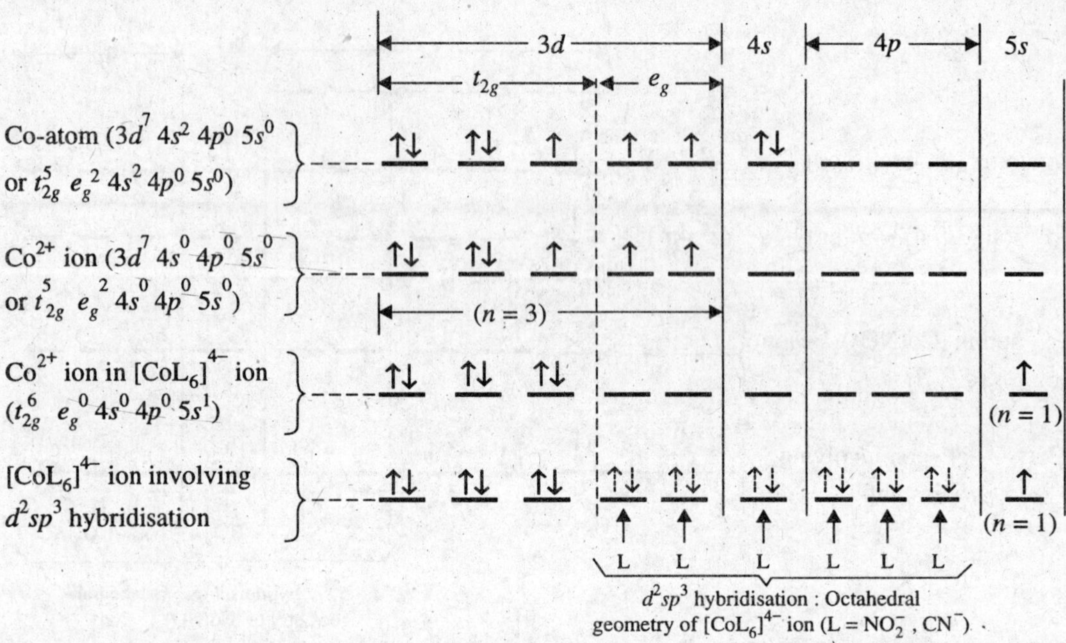

Fig. 5.5. Formation of $[CoL_6]^{4-}$ ion (L = NO_2^- or CN^-) by d^2sp^3 hybridisation (Inner-orbital octahedral complex ion).

Thus we find that Co(+2) low spin (inner orbital) octahedral complexes are unstable (are labile) and hence are easily oxidised to Co (+3) complexes by air or H_2O_2. Consequently such complexes should be prepared in an inert atmosphere.

7. Hexachloroplatinate (IV) ion, $[PtCl_6]^{2-}$

Since the coordination number of Pt^{4+} ion is six, the given complex ion is octahedral in

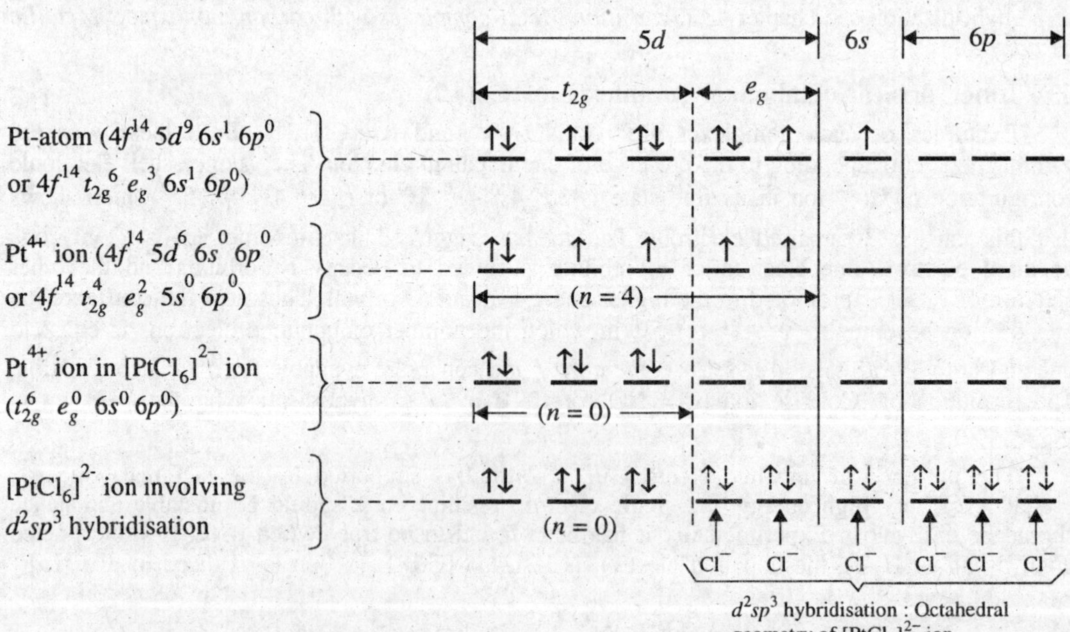

Fig. 5.6. d^2sp^3 hybridisation in $[PtCl_6]^{2-}$ ion (Inner-orbital octahedral complex ion).

shape. Valence-shell configuration of Pt^{4+} is $4f^{14}\ 5d^6\ 6s^0\ 6p^0$ or $4f^{14}\ t_{2g}^4\ e_g^2\ 6s^0\ 6p^0$ which has 4 unpaired electrons (n = 4). Magnetic study of $[PtCl_6]^{2-}$ ion has shown that this ion has no unpaired electron (n = 0) and hence is *diamagnetic*. Thus in the formation of this ion both electrons of e_g orbitals ($5d_{z^2}$ and $5d_{x^2-y^2}$ orbitals) are forced to occupy t_{2g} orbitals ($5d_{xy}$, $5d_{yz}$ and $5d_{zx}$ orbitals) and thus the number of unpaired electrons becomes equal to zero (n = 0) and also e_g orbitals become available for d^2sp^3 hybridisation. Thus we find that $[PtCl_6]^{2-}$ ion results from d^2sp^3 hybridisation of Pt^{4+} ion (see Fig. 5.6).

Examples of some inner-orbital octahedral complexes are given in Table 5.2.

sp^3d^2 Hybridisation in Outer Orbital Octahedral Complexes

This type of hybridisation takes place in those octahedral complex ions which contain weak ligands. Weak ligands are those which cannot force the electrons of d_{z^2} and $d_{x^2-y^2}$ orbitals (e_g set) of the inner shell to occupy d_{xy}, d_{yz} and d_{zx} orbitals (t_{2g} set) of the same shell. Thus in this hybridisation, $(n-1)\ d_{z^2}$ and $(n-1)\ d_{x^2-y^2}$ orbitals are not available for hybridisation. In place of these orbitals, we use nd_{z^2} and $nd_{x^2-y^2}$ orbitals (These d-orbitals belong to the outer shell) and hence sp^3d^2 hybridisation can be represented as $ns.\ np_x.\ np_y.\ np_z.\ nd_{z^2}.\ nd_{x^2-y^2}$. This hybridisation shows that all the six orbitals involved in hybridisation belong to the higher energy level (outer shell). This discussion shows that in case of octahedral ion of 3d transition series, d-orbitals used in hybridisation are $4d_{z^2}$ and $4d_{x^2-y^2}$ orbitals. Since two d-orbitals are from the outer shell (*i.e.*, nth shell), the octahedral complexes resulted from sp^3d^2 hybridisation are called **outer orbital octahedral complexes.** Since these complexes have comparatively greater number of unpaired electrons than the inner orbital octahedral complexes, these are also called **high spin** or **spin free octahedral complexes.**

Now let us discuss the structure of some octahedral complex ions of 3d transition series elements which are formed by sp^3d^2 hybridisation.

1. Hexafluroferrate (III) ion, $[FeF_6]^{3-}$

In this ion, the coordination number of Fe is six and hence the given complex ion has octahedral geometry. Here iron is present as Fe^{3+} whose valence shell electronic configuration is $3d^5 4s^0 4p^0$ or $t_{2g}^3\ e_g^2\ 4s^0\ 4p^0$. Each of the five electrons is unpaired and hence n = 5. Magnetic

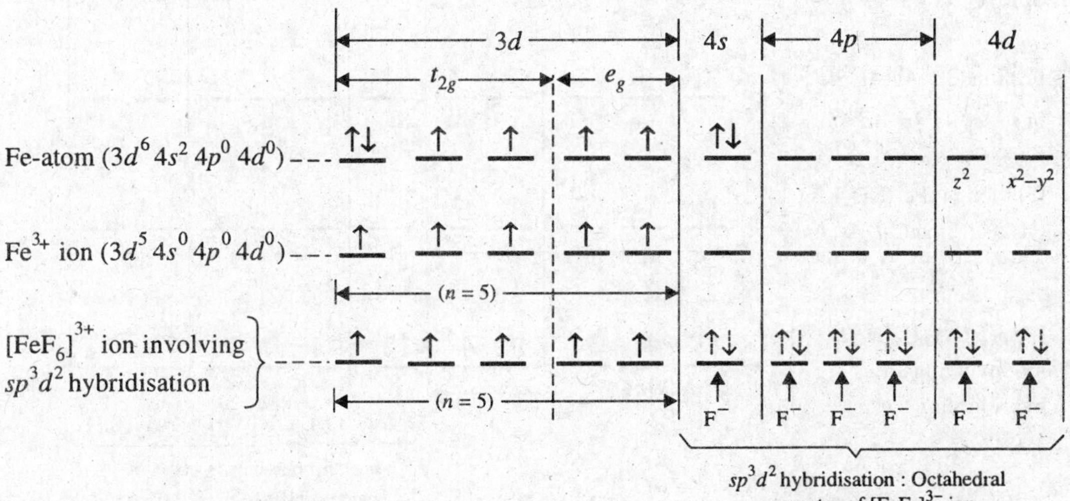

Fig. 5.7. Formation of $[FeF_6]^{3-}$ ion by sp^3d^2 hybridisation. (Outer-orbital octahedral complex ion).

properties of the given ion have also shown that the ion has five unpaired electrons and hence is paramagentic corresponding to the presence of five unpaired electrons. Thus two electrons residing in e_g orbitals cannot be forced to occupy t_{2g} orbitals as we have done in case of $[Fe(CN)_6]^{3-}$ ion, otherwise the number of electrons would become equal to one. Thus we find that in case of the given ion, the two *d*-orbitals used in hybridisation are $4d_{z^2}$ and $4d_{x^2-y^2}$ (and not $3d_{z^2}$ and $3d_{x^2-y^2}$ as in case of $[Fe(CN)_6]^{3-}$ ion] and *s* and *p* orbitals are 4*s* and 4*p*. Thus the given ion results from $(4s)(4p^3)(4d_{z^2})(4d_{x^2-y^2})$ hybridisation as shown in Fig. 5.7. This discussion shows that in the formation of $[FeF_6]^{3-}$ ion the original valence-shell configuration of Fe^{3+} ion is not disturbed.

2. Pentaquanitrosonium iron (I) ion, $[Fe^+(NO^+)(H_2O)_5]^{2+}$ ion

This ion is obtained when NO gas is passed through a freshly-prepared aqueous solution of $FeSO_4$.

$$FeSO_4 + NO + 5H_2O \longrightarrow [Fe(NO)(H_2O)_5]SO_4$$

or

$$Fe^{2+} + NO + 5H_2O \longrightarrow [Fe(NO)(H_2O)_5)]^{2+}$$

Since the coordination number of Fe atom is six, the given ion has octahedral geometry. The magnetic moment of this ion has been found to be equal to 3.89 B.M. which shows that the given ion has three unpaired electrons, *i.e.*, Fe is present as Fe^+ ion ($Fe^+ = 3d^6\,4s^1$ or $3d^7\,4s^0$ or $t_{2g}^5\,e_g^2\,4s^0$ having $n = 3$). In order to show + 2 charges on the complex ion, NO is assumed to be present as nitrosyl cation (NO^+) which is a two-electron donor. In order to show the presence of three unpaired electrons in the given ion, $t_{2g}^5\,e_g^2$ configuration of Fe^+ ion is not disturbed. This means that d^2sp^3 hybridisation scheme cannot be used to account for the presence of three unpaired electrons (d^2sp^3 hybridisation gives one unpaired electron which would not be present in t_{2g} set of orbitals but in some other orbital).

We assume that the given ion results from sp^3d^2 hybridisation and hence the given ion is an outer-orbital octahedral ion (see Fig. 5.8).

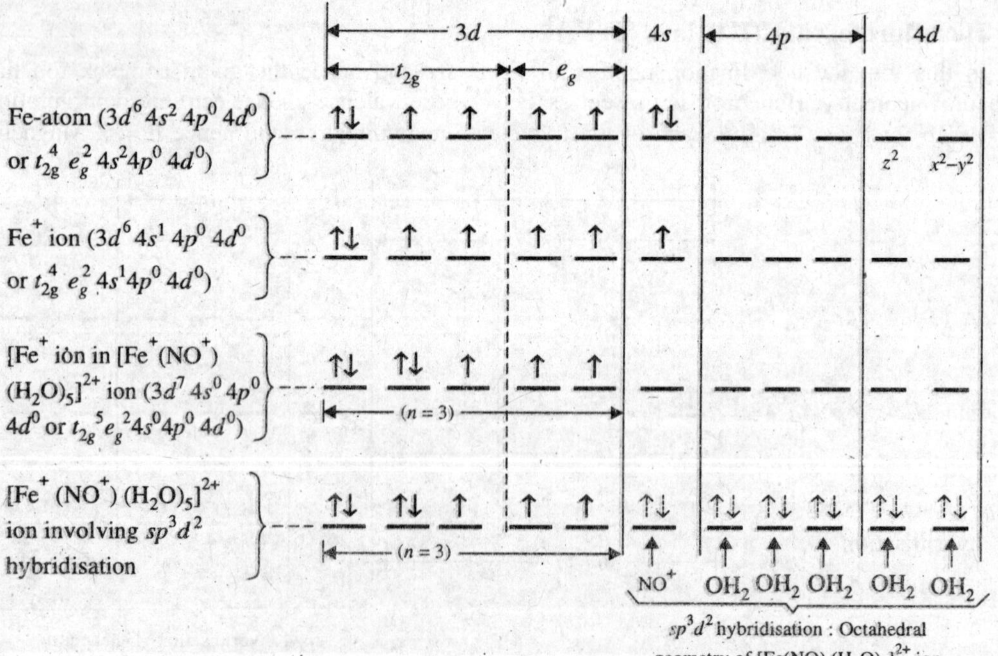

Fig. 5.8. Formation of $[Fe^+(NO^+)(H_2O)_5]^{2+}$ ion by sp^3d^2 hybridisation (Outer-orbital octahedral complex ion).

3. Outer-orbital octahedral complexes of Ni(+2)

Octahedral complexes of Ni^{2+} ion are outer-orbital octahedral complexes (sp^3d^2 hybridisation). The formation of inner-orbital octahedral complexes of Ni^{2+} ion ($Ni^{2+} = 3d^8 = t_{2g}^6\ e_g^2$) is not possible, since the two unpaired electrons present in e_g set of orbitals cannot be sent to t_{2g} orbitals which are already completely-filled. Thus e_g orbitals cannot be made empty for d^2sp^3 hybridisation. Outer orbital complexes of Ni^{2+} ion are paramagnetic corresponding to the presence of two unpaired electrons present in e_g orbitals.

As an example let us see how sp^3d^2 hybridisation takes place in $[Ni(NH_3)_6]^{2+}$ ion. (See Fig. 5.9).

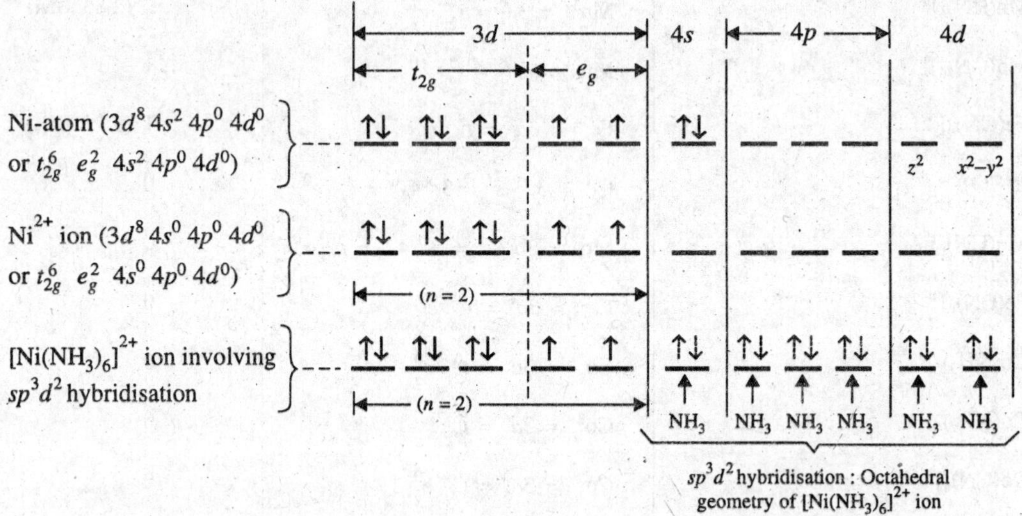

Fig. 5.9. Formation of $[Ni(NH_3)_6]^{2+}$ ion by sp^3d^2 hybridisation (Outer-orbital octahedral complex ion).

Exmaples of some outer-octahedral complexes are given in Table 5.2.

Table 5.2. *Examples of inner-orbital octahedral (d^2sp^3 hybridisation) and outer-orbital octahedral (sp^3d^2 hybridisation) complex (n = Number of unpaired electrons).*

Complex ion	*Configuration of the central atom / ion*	*n*
Inner-orbital octahedral complex ions (d^2sp^3 hybridisation)		
$[ScF_6]^{3-}$	$Sc^{3+} = 3d^0$	0
$[TiX_6]^{2-}$	$Ti^{4+} = 3d^0$	0
$[TiF_6]^{3-}$	$Ti^{3+} = 3d^1 = t_{2g}^1\ e_g^0$	1
$[Ti(H_2O_6]^{3+}$	$Ti^{3+} = 3d^1 = t_{2g}^1\ e_g^0$	1
$[V(H_2O)_6]^{3+}$	$V^{3+} = 3d^2 = t_{2g}^2\ e_g^0$	2
$[Cr(H_2O)_6]^{3+}$	$Cr^{3+} = 3d^3 = t_{2g}^3\ e_g^0$	3
$[Cr(NH_3)_6]^{3+}$	$Cr^{3+} = 3d^3 = t_{2g}^3\ e_g^0$	3

(Contd.)

$[Cr(CN)_6]^{3-}$	$Cr^{3+} = 3d^3 = t_{2g}^3\ e_g^0$	3
$[CrF_6]^{3-}$	$Cr^{3+} = 3d^3 = t_{2g}^3\ e_g^0$	3
$[Cr(NH_3)_4Cl_2]^+$	$Cr^{3+} = 3d^3 = t_{2g}^3\ e_g^0$	3
$[MnF_6]^{2-}$	$Mn^{4+} = 3d^3 = t_{2g}^3\ e_g^0$	3
$[Cr(CN)_6]^{4-}$	$Cr^{2+} = 3d^4 = t_{2g}^4\ e_g^0$	2
$[Mn(CN)_6]^{3-}$	$Mn^{3+} = 3d^4 = t_{2g}^4\ e_g^1$	2 (2.83 BM)
$[Mn(CN)_6]^{4-}$	$Mn^{2+} = 3d^5 = t_{2g}^5\ e_g^0$	1
$[Fe(CN)_6]^{3-}$	$Fe^{3+} = 3d^5 = t_{2g}^5\ e_g^0$	1 (1.73 BM)
$Cr(CO)_6$	$Cr^0 = 3d^54s^1 = 3d^64s^0 = t_{2g}^6\ e_g^0$	0
$[Mn(CN)_6]^{5-}$	$Mn^+ = 3d^54s^1 = 3d^64s^0 = t_{2g}^6\ e_g^0$	0
$[Fe(CN)_6]^{4-}$	$Fe^{2+} = 3d^6 = t_{2g}^6\ e_g^0$	0
$[Co(NH_3)_6]^{3+}$	$Co^{3+} = 3d^6 = t_{2g}^6\ e_g^0$	0
$[Co(NO_2)_6]^{3-}$	$Co^{3+} = 3d^6 = t_{2g}^6\ e_g^0$	0
$[Co(CN)_6]^{3-}$	$Co^{3+} = 3d^6 = t_{2g}^6\ e_g^0$	0
$[Fe(CN)_5(NO^+)]^{2-}$	$Fe^{2+} = 3d^6 = t_{2g}^6\ e_g^0$	0
$[PtCl_6]^{2-}$	$Pt^{4+} = 4f^{14}5d^6 = 4f^{14}\ t_{2g}^6\ e_g^0$	0
$[Co(NO_2)_6]^{4-}$	$Co^{2+} = 3d^7 = e_g^0\ 5s^1$	1 (in 5*s* orbital)
$[Co(CN)_6]^{4-}$	$Co^{2+} = 3d^7 = e_g^0\ 5s^1$	1 (in 5*s* orbital)
Outer-orbital octahedral complex ions (sp^3d^2 hybridisation)		
$[Cr(H_2O)_6]^{2+}$	$Cr^{2+} = 3d^4 = t_{2g}^3\ e_g^1$	4
$[Cr(NH_3)_6]^{2+}$	$Cr^{2+} = 3d^4 = t_{2g}^3\ e_g^1$	4
$[Mn(acac)_3]^0$	$Mn^{3+} = 3d^4 = t_{2g}^3\ e_g^1$	4
$[Mn(H_2O)_6]^{2+}$	$Mn^{2+} = 3d^5 = t_{2g}^3\ e_g^2$	5
$[FeF_6]^{3-}$	$Fe^{3+} = 3d^5 = t_{2g}^3\ e_g^2$	5
$[Fe(H_2O)_6]^{3+}$	$Fe^{3+} = 3d^5 = t_{2g}^3\ e_g^2$	5
$[Fe(NH_3)_6]^{2+}$	$Fe^{2+} = 3d^6 = t_{2g}^4\ e_g^2$	4

(Contd.)

$[CoF_6]^{3-}$	$Co^{3+} = 3d^6 = t_{2g}^4\ e_g^2$	4
$[Fe(NO^+)(H_2O)_5]^{2+}$	$Fe^+ = 3d^6 4s^1 = 3d^7 = t_{2g}^5\ e_g^2$	3
$[Co(NH_3)_6]^{2+}$	$Co^{2+} = 3d^7 = t_{2g}^5\ e_g^2$	3
$[MnF_6]^{3-}$	$Mn^{3+} = 3d^4 = t_{2g}^3\ e_g^1$	4
$[Ni(NH_3)_6]^{2+}$	$Ni^{2+} = 3d^8 = t_{2g}^6\ e_g^2$	2
$[Cu(NH_3)_6]^{2+}$	$Cu^{2+} = 3d^9 = t_{2g}^6\ e_g^3$	1
$[Zn(NH_3)_6]^{2+}$ (in crystals only)	$Zn^{2+} = 3d^{10} = t_{2g}^6\ e_g^4$	0

Table 5.3. *Differences between inner orbital and outer-orbital octahedral complexes.*

Inner orbital octahedral complexes (Low spin or spin paired octahedral complexes) (d^2sp^3 hybridisation)	Outer-orbital octahedral complexes (High spin or spin free octahedral complexes) (sp^3d^2) hybridisation
(*i*) These are formed by d^2sp^3 hybridisation *i.e.* in the formation of six d^2sp^3 hybrid orbitals, two $(n-1)\,d$ orbitals of e_g set [*i.e.*, $(n-1)\ d_{z^2}$ and $(n-1)\ d_{x^2-y^2}$ orbitals], one ns and three *np* orbitals are used. $(n-1)d$ orbitals belong to the inner (penultimate) shell while *ns* and *np* orbitals belong to the outer (*i.e.*, ultimate shell).	(*i*) These are formed by sp^3d^2 hybridisation, *i.e.*, in the formation of six sp^3d^2 hybrid orbitals, one *ns*, three *np* and two *nd* orbitals of e_g set (*i.e.*, nd_{z^2} and $nd_{x^2-y^2}$ orbitals) are used. Thus all the orbitals belong to the ultimate (outer) shell.
(*ii*) These complexes have comparatively lesser number of unpaired electrons and hence are also called low spin or spin paired octahedral complexes.	(*ii*) These complexes have comparatively greater number of unpaired electrons and hence are also called high spin or spin free octahedral complexes.
(*iii*) These are given by strong ligands.	(*iii*) These are given by weak ligands.

Geometry of 4-coordinated complex ions

Examples of 4-coordinated complex ions formed by some transition metals are given in Table 5.4. In these complex ions the coordination number of the central metal atom or ion is four. Such complex ions may have either *square planar* or *tetrahedral geometry,* depending on whether the central atom or ion is dsp^2 or sp^3 hybridised. What type of hybridisation (*i.e.*, whether dsp^2 or sp^3) the central metal atom or ion of a 4-coordinated complex ion undergoes depends on the number of unpaired or paired electrons present in the complex ion.

Square Planar Complex Ions

Here we shall consider the following complex ions.

1. $[Ni(CN)_4]^{2-}$ ion

In this complex ion, since the coordination number of Ni^{2+} ion is 4, the given complex ion may be square planar or tetrahedral in shape. In order to decide whether the given complex ion is square planar or tetrahedral, we take the help of magnetic property of the complex ion. Experimentally it has been shown that $[Ni(CN)_4]^{2-}$ ion has no unpaired electron ($n = 0$) and hence is diamagnetic. Two cases arise:

(*a*) **When $[Ni(CN)_4]^{2-}$ ion has tetrahedral geometry.** In this case Ni^{2+} ion is sp^3 hybridised as shown in Fig. 5.10. sp^3 hybridisation scheme given in Fig. 5.10 shows that $[Ni(CN)_4]^{2-}$ ion has two unpaired electrons (n = 2).

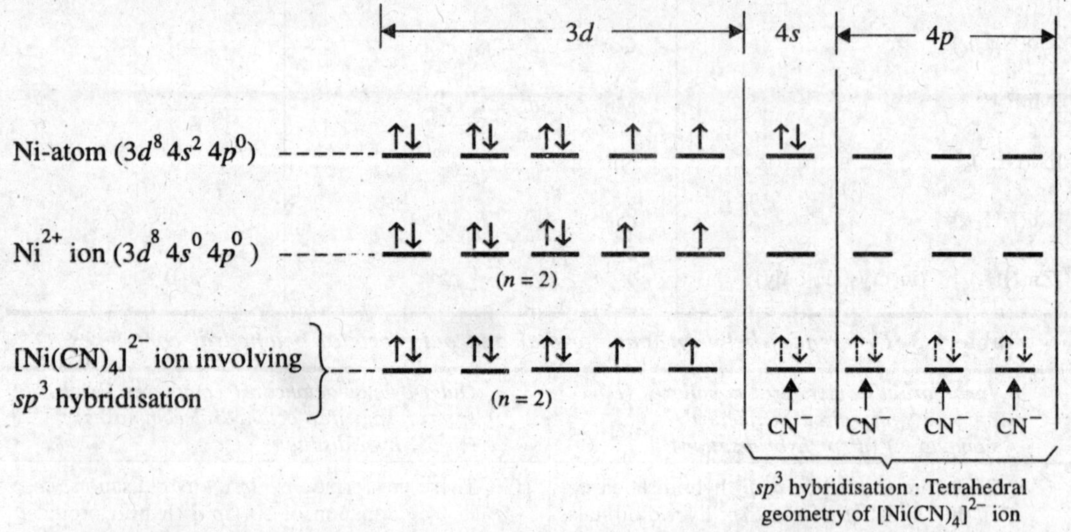

	3d					4s	4p		
Ni-atom ($3d^8 4s^2 4p^0$)	↑↓	↑↓	↑↓	↑	↑	↑↓	—	—	—
Ni^{2+} ion ($3d^8 4s^0 4p^0$)	↑↓	↑↓	↑↓	↑	↑	—	—	—	—
		(n = 2)							
$[Ni(CN)_4]^{2-}$ ion involving sp^3 hybridisation	↑↓	↑↓	↑↓	↑	↑	↑↓	↑↓	↑↓	↑↓
		(n = 2)				CN^-	CN^-	CN^-	CN^-

sp^3 hybridisation : Tetrahedral geometry of $[Ni(CN)_4]^{2-}$ ion

Fig. 5.10. Formation of $[Ni(CN)_4]^{2-}$ ion by sp^3 hybridisation (Tetrahedral complex ion with n = 2).

(*b*) **When $[Ni(CN)_4]^{2-}$ ion has square planar geometry.** For getting this geometry, Ni^{2+} ion should be dsp^2 hybridised. In this hybridisation, due to the energy made available by the approach of four CN^- ions (ligands), the two unpaired 3*d*-electrons are paired up, thereby, making one of the 3*d* orbitals empty. This empty 3*d* orbital (which is $3d_{x^2-y^2}$ orbital) is used in dsp^2 hybridisation. This hybridisation makes all the electrons paired (n = 0) as shown in Fig. 5.11.

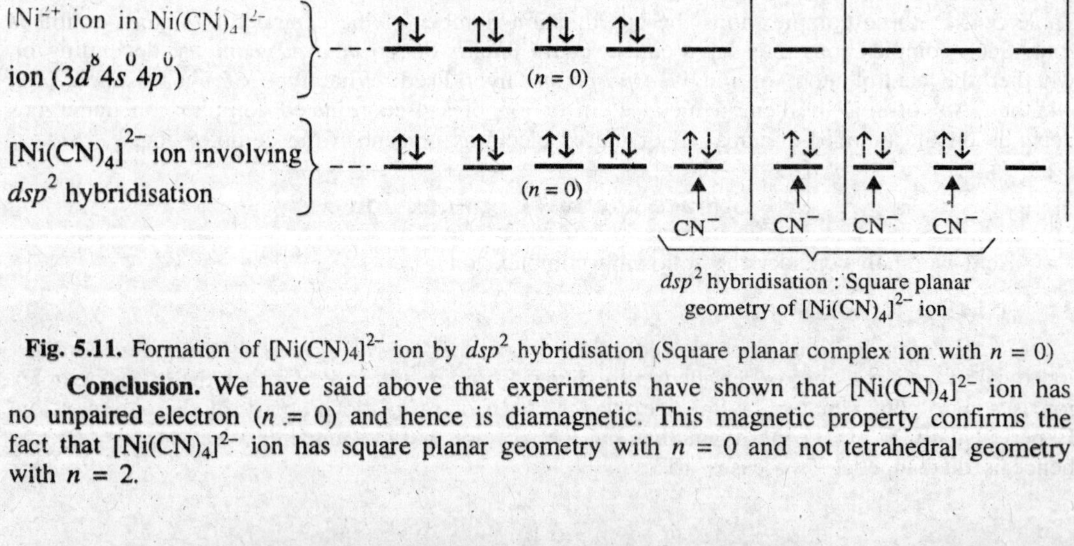

	3d					4s	4p		
Ni-atom ($3d^8 4s^2 4p^0$)	↑↓	↑↓	↑↓	↑	↑	↑↓	—	—	—
					x^2-y^2				
Ni^{2+} ion ($3d^8 4s^0 4p^0$)	↑↓	↑↓	↑↓	↑	↑	—	—	—	—
		(n = 2)							
[Ni^{2+} ion in $Ni(CN)_4]^{2-}$ ion ($3d^8 4s^0 4p^0$)	↑↓	↑↓	↑↓	↑↓	—	—	—	—	—
		(n = 0)							
$[Ni(CN)_4]^{2-}$ ion involving dsp^2 hybridisation	↑↓	↑↓	↑↓	↑↓	↑↓	↑↓	↑↓	↑↓	—
		(n = 0)			CN^-	CN^-	CN^-	CN^-	

dsp^2 hybridisation : Square planar geometry of $[Ni(CN)_4]^{2-}$ ion

Fig. 5.11. Formation of $[Ni(CN)_4]^{2-}$ ion by dsp^2 hybridisation (Square planar complex ion with n = 0)

Conclusion. We have said above that experiments have shown that $[Ni(CN)_4]^{2-}$ ion has no unpaired electron (n = 0) and hence is diamagnetic. This magnetic property confirms the fact that $[Ni(CN)_4]^{2-}$ ion has square planar geometry with n = 0 and not tetrahedral geometry with n = 2.

2. $[Cu(NH_3)_4]^{2+}$ ion

Since the coordination number of Cu^{2+} ion is 4, the given complex ion has either square planar or tetrahedral geometry. Square planar geometry arises due to dsp^2 hybridisation of Cu^{2+} ion as shown in Fig. 5.12 while tetrahedral geometry is due to sp^3 hybridisation of Cu^{2+} ion as shown in Fig. 5.13.

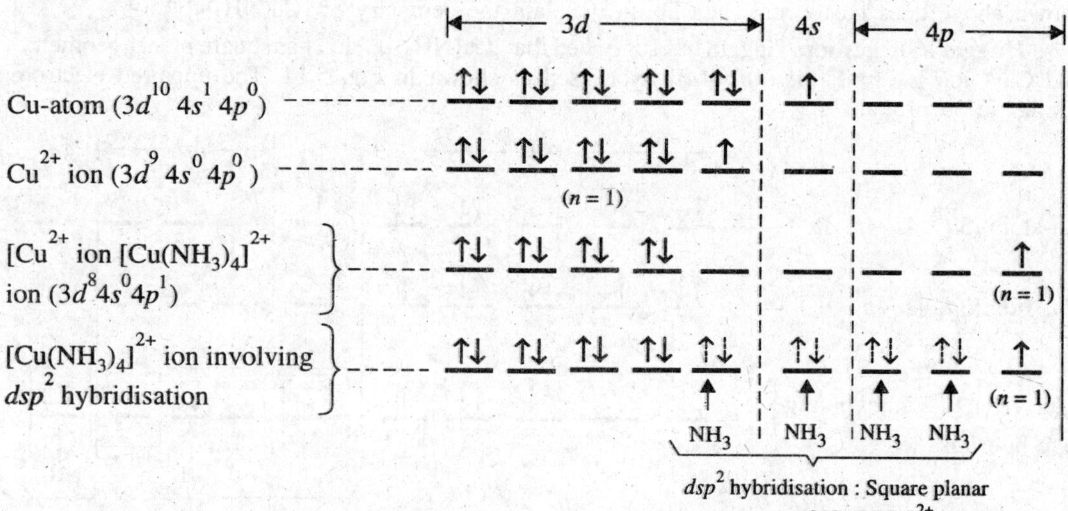

Fig. 5.12. dsp^2 hybridisation of Cu^{2+} ion in $[Cu(NH_3)_4]^{2+}$ ion which has square planar geometry with $n = 1$.

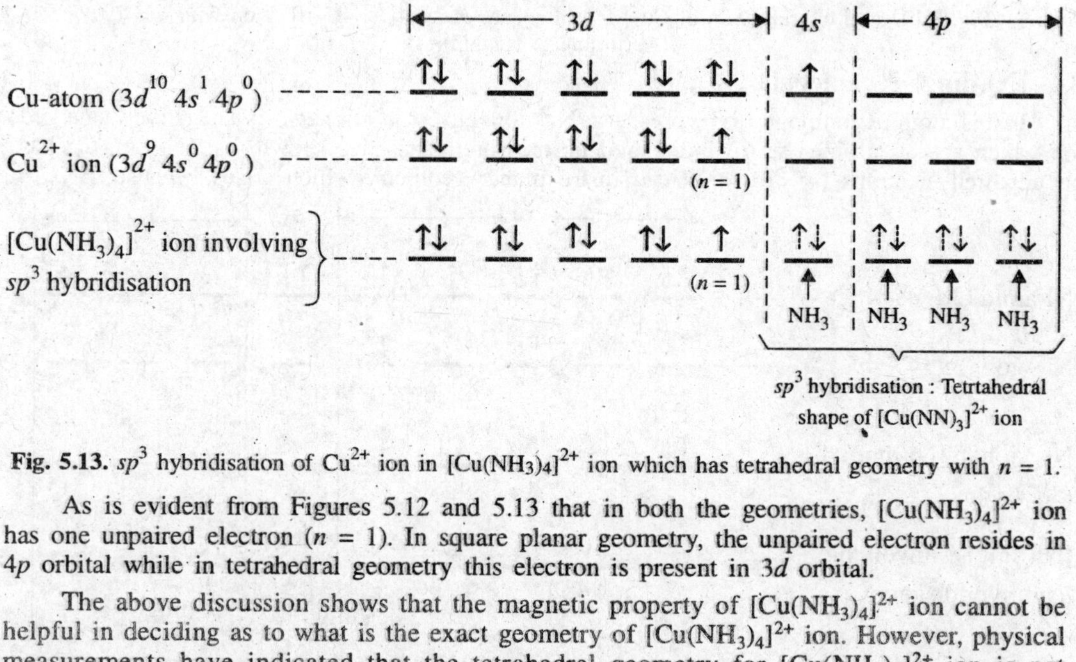

3d 4s 4p

Cu-atom ($3d^{10}$ $4s^1$ $4p^0$)

Cu^{2+} ion ($3d^9$ $4s^0$ $4p^0$) (n = 1)

$[Cu(NH_3)_4]^{2+}$ ion involving sp^3 hybridisation (n = 1)

NH_3 NH_3 NH_3 NH_3

sp^3 hybridisation : Tetrtahedral shape of $[Cu(NN)_3]^{2+}$ ion

Fig. 5.13. sp^3 hybridisation of Cu^{2+} ion in $[Cu(NH_3)_4]^{2+}$ ion which has tetrahedral geometry with $n = 1$.

As is evident from Figures 5.12 and 5.13 that in both the geometries, $[Cu(NH_3)_4]^{2+}$ ion has one unpaired electron ($n = 1$). In square planar geometry, the unpaired electron resides in $4p$ orbital while in tetrahedral geometry this electron is present in $3d$ orbital.

The above discussion shows that the magnetic property of $[Cu(NH_3)_4]^{2+}$ ion cannot be helpful in deciding as to what is the exact geometry of $[Cu(NH_3)_4]^{2+}$ ion. However, physical measurements have indicated that the tetrahedral geometry for $[Cu(NH_3)_4]^{2+}$ ion is not possible.

Now if the square planar geometry for $[Cu(NH_3)_4]^{2+}$ ion is supposed to be correct, the unpaired electron present in the higher energy $4p$ orbital (dsp^2 hybridisation) should be expected to be easily lost, *i.e.*, $[Cu(NH_3)_4]^{2+}$ ion (Cu = +2) should be easily oxidised to $[Cu(NH_3)_4]^{3+}$ ion (Cu = +3) by losing the unpaired electron residing in 4*p* orbital.

$$[Cu^{2+}(NH_3)_4]^{2+} \xrightarrow[(-e^-)]{\text{oxidation}} [Cu^{3+}(NH_3)_4]^{3+}$$

However, experiments have shown that the oxidation of $[Cu(NH_3)_4]^{2+}$ to $[Cu(NH_3)_4]^{3+}$ as shown above does not occur. Then how to explain the geometry of $[Cu(NH_3)_4]^{2+}$ ion.

Huggin's suggestion. Huggin has suggested that $[Cu(NH_3)_4]^{2+}$ ion has square planar geometry and Cu^{2+} ion is sp^2d $[(4s)(4p)^2(4d)]$ hybridised as shown in Fig. 5.14. The unpaired electron resides in 3*d* orbital.

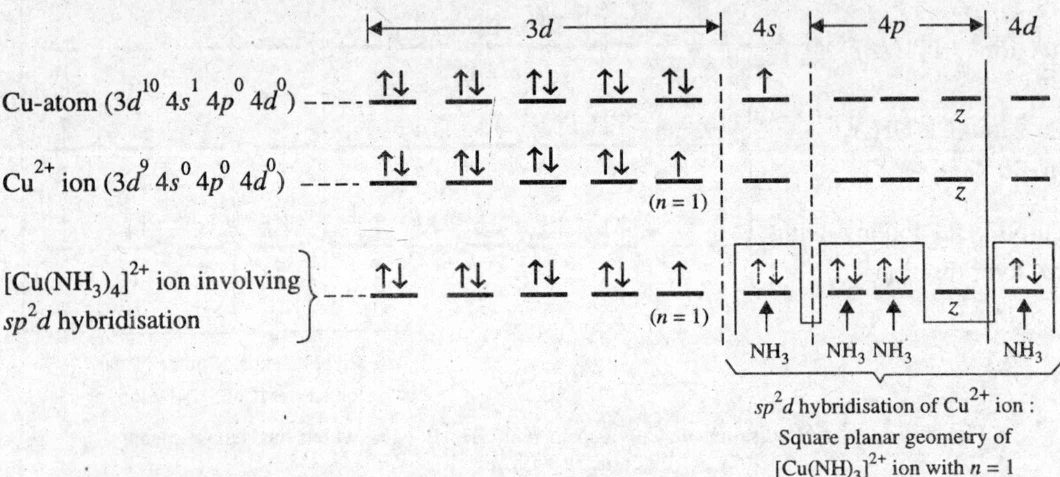

Fig. 5.14. sp^2d hybridisation of Cu^{2+} ion in square planar $[Cu(NH_3)_4]^{2+}$ ion with $n = 1$ (According to Huggin).

3. $[Ni(dmg)_2]^0$ molecule

In this complex compound, Ni is present as Ni^{2+} ion ($Ni^{2+} = 3d^8 4s^0$). *dmg* is a negatively-charged ion which acts as a bidentate ligand. Magnetic measurements have shown that $[Ni(dmg)_2]^0$ has no unpaired electrons ($n = 0$) and has square planar geometry which arises because of dsp^2

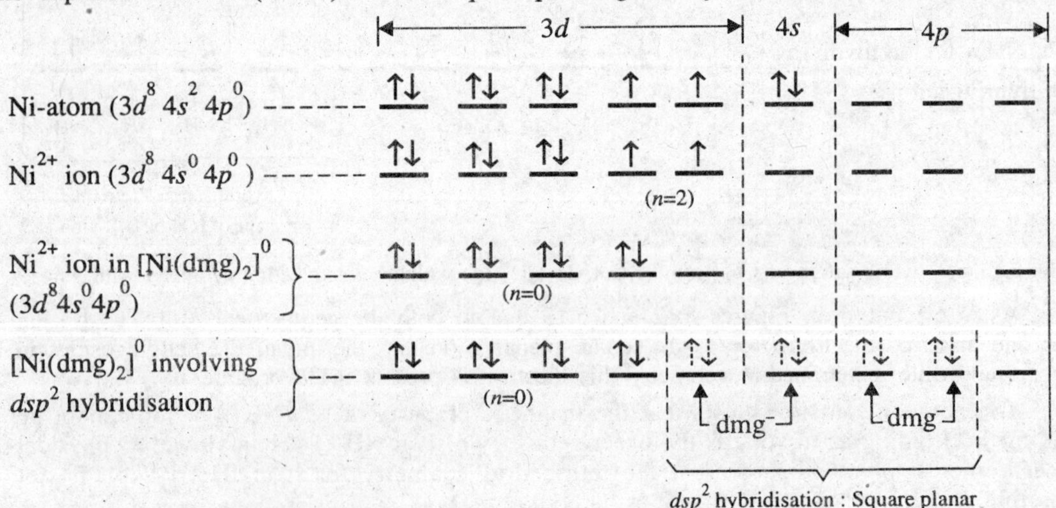

Fig. 5.15. dsp^2 hybridisation of Ni^{2+} ion in $[Ni(dmg)_2]^0$ which has square planar geometry. Note that dmg^- ion is a bidentate ligand and hence donates two electron pairs.

hybridisation of Ni^{2+} ion. For getting dsp^2 hybridisation, one $3d$ orbital should be vacant. In order to make $3d$ orbital vacant, two unpaired electrons present in $3d$ orbitals are forced to get paired (see Fig. 5.15).

Tetrahedral Complex Ions

Here we shall discuss the structure of the following complex compounds which have tetrahedral geometry.

1. $Ni(CO)_4$ molecule

In this complex compound Ni is in zero oxidation state and has its valence-shell configuration as $3d^8 4s^2$. This compound has tetrahedral geometry which arises due to sp^3 hybridisation of Ni atom.

The magnetic studies of $Ni(CO)_4$ have indicated that this molecule is diamagnetic ($n = 0$), showing that the two $4s$ electrons are forced to pair up with $3d$ orbitals. This results in sp^3 hybridisation and $Ni(CO)_4$ molecule has tetrahedral structure. (see Fig. 5.16).

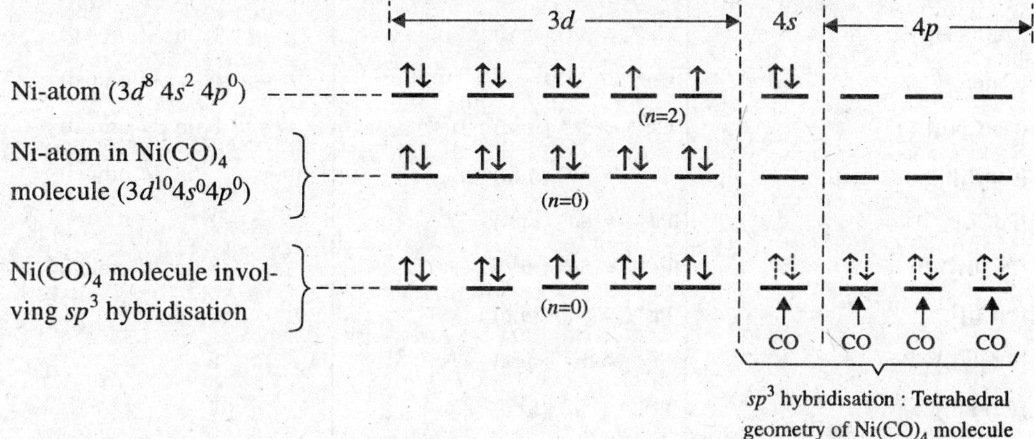

Fig. 5.16. sp^3 hybridisation of Ni-atom in $Ni(CO)_4$ molecule which has tetrahedral shape.

2. $[NiCl_4]^{2-}$ ion

This complex ion has Ni^{2+} ion whose valence-shell configuration as $3d^8 4s^0$. Magnetic measurements reveal that the given ion is paramagnetic and has two unpaired electrons ($n = 2$). This is possible only when this ion is formed by sp^3 hybridisation and has tetrahedral geometry (see Fig. 5.17).

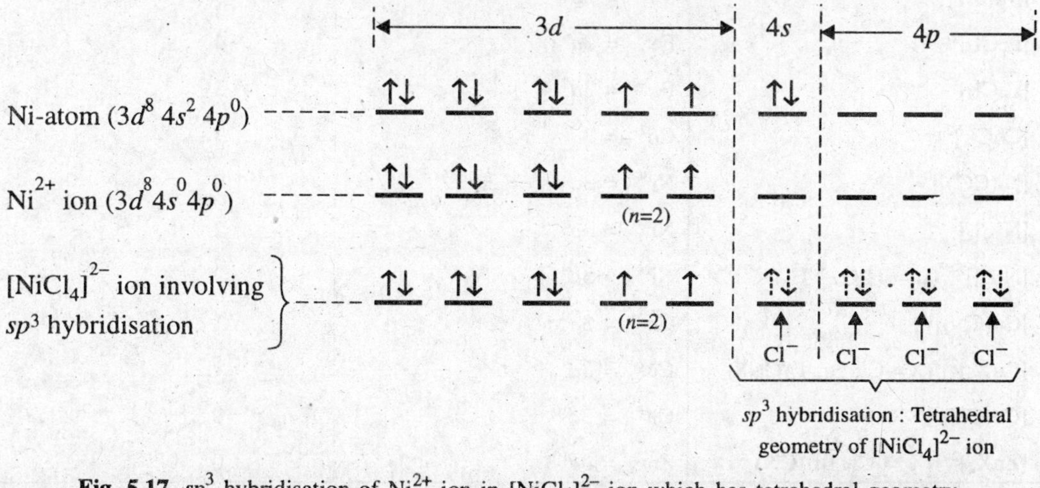

3d | 4s | 4p

Ni-atom ($3d^8\ 4s^2\ 4p^0$) ------ ↑↓ ↑↓ ↑↓ ↑ ↑ | ↑↓ | — — —

Ni^{2+} ion ($3d^8 4s^0 4p^0$) ------ ↑↓ ↑↓ ↑↓ ↑ ↑ (n=2) | — | — — —

$[NiCl_4]^{2-}$ ion involving sp^3 hybridisation ------ ↑↓ ↑↓ ↑↓ ↑ ↑ (n=2) | ↑↓ | ↑↓ ↑↓ ↑↓

Cl^- Cl^- Cl^- Cl^-

sp^3 hybridisation : Tetrahedral geometry of $[NiCl_4]^{2-}$ ion

Fig. 5.17. sp^3 hybridisation of Ni^{2+} ion in $[NiCl_4]^{2-}$ ion which has tetrahedral geometry.

Table 5.4. *Examples of 4-coordinated complex ions (square planar and tetrahedral complex ions)*

Complex ion	*Configuration of the central metal atom/ion*	*Number of unpaired electrons (n)*
Square planar complex ions (dsp^2 or sp^2d hybridisation)		
$[Ni(CN)_4]^{2-}$	Ni^{2+} = $3d^8$ (dsp^2)	0
$[Ni(NH_3)_4]^{2+}$	Ni^{2+} = $3d^8$ (dsp^2)	0
$[Ni(dmg)_2]^0$	Ni^{2+} = $3d^8$ (dsp^2)	0
$[Ni(bigH)_2]^{2+}$	Ni^{2+} = $3d^8$ (dsp^2)	0
$[Cu(NH_3)_4]^{2+}$	Cu^{2+} = $3d^9$ (sp^2d)	1 (in $3d$ orbital)
$[Cu(py)_4]^{2+}$	Cu^{2+} = $3d^9$ (sp^2d)	1 (in $3d$ orbital)
$[Cu(bigH)_2]^0$	Cu^{2+} = $3d^9$ (sp^2d)	1 (in $3d$ orbital)
$[Cu(en)_2]^{2+}$	Cu^{2+} = $3d^9$ (sp^2d)	1 (in $3d$ orbital)
$[Cu(CN)_4]^{2-}$	Cu^{2+} = $3d^9$ (sp^2d)	1 (in $3d$ orbital)
$[CuCl_4]^{2-}$	Cu^{2+} = $3d^9$ (sp^2d)	1 (in $3d$ orbital)
$[PdCl_4]^{2-}$	Pd^{2+} = $4d^8$ (dsp^2)	0
$[Pt(NH_3)_4]^{2+}$	Pt^{2+} = $5d^8$ (sp^2d)	1 (in $3d$ orbital)
$[PtCl_4]^{2-}$	Pt^{2+} = $5d^8$ (sp^2d)	1
$[Pt(gly)_2]^0$	Pt^{2+} = $5d^8$ (sp^2d)	1
$[Pt(PPh)_2Cl_2]^0$	Pt^{2+} = $5d^8$ (sp^2d)	1
Tetrahedral complex ions (sp^3 hybridisation)		
$[Zn(NH_3)_4]^{2+}$	Zn^{2+} = $3d^{10}$	0
$[MnCl_4]^{2-}$	Mn^{2+} = $3d^5$	5
$[MnBr_4]^{2-}$	Mn^{2+} = $3d^5$	5
$[FeCl_4]^{2-}$	Fe^{2+} = $3d^6$	4
$[FeCl_4]^-$	Fe^{3+} = $3d^5$	5
$[CoCl_4]^{2-}$	Co^{2+} = $3d^7$	3
$[Ni(CO)_4]^0$	Ni^0 = $3d^8 4s^2$ = $3d^{10}$	0
$[NiCl_4]^{2-}$	Ni^{2+} = $3d^8$	2
$[NiL_4]^{2+}$ (L = H_2O, NH_3)	Ni^{2+} = $3d^8$	2
$[Cu(CN)_4]^{2-}$	Cu^{2+} = $3d^9$	1
$[CuX_4]^{2-}$ (X= Cl, Br, I, CNS)	Cu^{2+} = $3d^9$	1
$[Cu(CN)_4]^{3-}$	Cu^+ = $3d^{10}$	0
$[ZnX_4]^{2-}$ [X = Cl, Br, CN)	Zn^{2+} = $3d^{10}$	0

To find out the number of unpaired electrons in a given complex ion

The number of unpaired electrons (n) present in a given complex ion depends on the geometry of the complex ion and the type of hybridisation undergone by the central metal atom or ion. This is evident from Table 5.5 in which we have shown a relation between the number of unpaired electrons and geometries of the complex ions.

Table 5.5. *Relation between the number of unpaired electrons and geometries of complex ions.*

Configuration of the central metal atom/ion	*Number of unpaired electrons*			
	Inner orbital octahedral complex ion (d^2sp^3 hybridisation)	*Outer orbital octahedral complex ion (sp^3d^2 hybridisation)*	*Tetrahedral complex ion (sp^3)*	*Square planar complex ion (dsp^2 or sp^2d)*
d^1	1	1	1	1
d^2	2	2	2	2
d^3	3	3	3	3
d^4	2	4	4	4
d^5	1	5	5	3
d^6	0	4	4	2
d^7	1	3	3	1
d^8	0	2	2	0
d^9	1	1	1	1

Limitations of VBT

(1) VBT cannot account for the relative stabilities for different shapes and different coordination numbers in metal complexes, *e.g.*, it cannot explain satisfactorily as to why Co(+2) (d^7 system) forms both octahedral and tetrahedral complexes while Ni(+2) (d^8 system) rarely forms tetrahedral complexes. Similarly, Fe(+3) (d^5 system) forms both tetrahedral and octahedral complexes while Cr(+3) (d^3 system) gives only octahedral complexes.

(2) VBT cannot explain as to why Cu(+2) forms only one distorted octahedral complexes even when all the six ligands are identical.

(3) This theory cannot account for the relative rates of reactions of analogous metal complexes, *e.g.* $[Mn(phen)_3]^{2+}$ dissociates instantaneusly in acidic aqueous solution while $[Fe(phen)_3]^{2+}$ dissociates at a *slow rate.*

(4) The classification of metal complexes on the basis of their magnetic behaviour into *covalent* (*inner-orbital*) and *ionic* (*outer-orbital*) complexes is not satisfactory and is often misleading, *e.g.*, many of the so-called ionic complexes like $[Fe^{3+}(acac)_3]^0$ have properties like high volatility, high solubility in organic solvents etc. which are characteristics of covalent compounds. Also it is known that the aquo ion, $[Cr(H_2O)_6]^{3+}$ (pKa = 3.8 at 25°C) is a much weaker acid than $[Fe(H_2O)_6]^{3+}$ (pKa = 2.2) and $[Co(H_2O)_6]^{3+}$ (pKa = 0.7) ions, despite the fact that only $[Fe(H_2O)_6]^{3+}$ is of outer-orbital type with ionic bonding while the other two aquo ions are of outer-orbital type with covalent bonding. Both $[Fe(H_2O)_6]^{3+}$ and $[Co(H_2O)_6]^{3+}$ ions exchange H_2O^{18} in solvent instantaneously while $[Cr(H_2O)_6]^{3+}$ ion exchanges H_2O^{18} very slowly. The acidity of an aquo metal ion is expected to increase with the increase of covalent character of the metal—oxygen bond. But neither the sequence of acidities nor the exchange rates of the above mentioned three aquo metal ions are in the order expected from their classification into ionic and covalent types on the basis of Pauling's VBT.

(5) VBT fails to explain the finer details of magnetic properties including the magnitude of the orbital contribution to the magnetic moments, *e.g.* although both tetrahedral (sp^3 hybridisation) and outer-orbital octahedral (sp^3d^2 hybridisation) complexes of Co(+2) (d^7 system) have three unpaired electrons and are, therefore, expected to have μ value equal to 3.87 B.M.; the tetrahedral complexes generally have μ value in the range of 4.4 – 4.8 B.M, while the octahedral complexes

have still higher value of μ in the range of 4.7 – 5.2 B.M. The increase in the value of μ is due to the orbital contribution. Similar is the case with tetrahedral and octahedral complexes of Ni(+2) (d^8 system). VBT cannot explain the increase in the value of μ.

(6) VBT cannot interpret the spectra (colour) of the complexes.

(7) This theory does not predict or explain the magnetic behaviours of complexes. This theory only predicts the number of unpaired electrons. Its prediction even for the number of unpaired electrons and their correlation with stereochemistry is misleading. For example it was assumed for long that all square planar complexes of Ni(+2) (d^8 system) formed by dsp^2 hybridisation were diamagnetic (n = 0), while all tetrahedral complexes of Ni(+2) formed by sp^3 hybridisation were paramagnetic due to the presence of two unpaired electrons (n = 2). X-ray study of 4-coordinate Ni(+2) complexes has shown that one and the same complex can be obtained in both a *paramagnetic form* (having blue colour) and a *diamagnetic form* (having yellow colour). VBT has no explanation for it. VBT cannot explain the temperature dependent paramagnetism of the complexes.

(8) VBT cannot give any explanation for the order of reactivities of the inner-orbital inert complexes of d^3, d^4, d^5 and d^6 ions and of the observed differences in the energies of activation in a series of similar complexes.

(9) The magnetic moment values of the complexes of certain ions (*e.g.*, Co^{2+}, Ni^{2+}, etc.) are much higher than those expected by spin-only formula. VBT can not explain the enhanced values of magnetic moment.

Crystal Field Theory (CFT)

This theory advanced by Brethe and Van Vleck was originally applied mainly to ionic crystals and is, therefore, called **Crystal Field Theory (CFT).** It was not until 1952 that Orgel popularised its use for inorganic chemists.

Important features of CFT

(*i*) The central metal cation is surrounded by ligands which contain one or more lone pairs of electrons.

(*ii*) The ionic ligands (*e.g.*, F^-, Cl^-, CN^- etc.) are regarded as negative point charges (also called point charges) and the neutral ligands (*e.g.*, H_2O, NH_3, etc.) are regarded as point dipoles or simply dipoles, *i.e.*, according to this theory neutral ligands are dipolar. If the ligand is neutral, the negative end of this ligand dipole is oriented towards the metal cation.

(*iii*) The CFT does not provide for electrons to enter the metal orbitals. Thus the metal ion and the ligands do not mix their orbitals or share electrons, *i.e.*, it does not consider any orbital overlap.

(*iv*) According to CFT, the bonding between the metal cation and ligand is not covalent but it is regarded as purely electrostatic or coulombic attraction between positively-charged (*i.e.*, cation) and negatively-charged (*i.e.*, anions or dipole molecules which act as ligands) species. Complexes are thus presumed to form when centrally situated cations electrically attract ligands which may be either anions or dipole molecules. The attraction between the cations and the ligands is because the cations are positively charged and the anions are negatively charged and the dipole molecules, as well, can offer their negatively incremented ends for such electrostatic attractions.

Grouping of five *d*-orbitals into t_{2g} and e_g sets of orbitals

On the basis of the orientation of the lobes of the five-*d*-orbitals with respect to coordinates these have been grouped into the following two sets.

1. e_g set of orbitals: d_{z^2} and $d_{x^2-y^2}$ orbitals. This set consists of the orbitals *which have their lobes along the axes* and hence are called *axial orbitals*. Quite obviously these are d_{z^2} and $d_{x^2-y^2}$ orbitals. Group theory calls these e_g orbitals in which *e* refers to doubly degenerate set.

2. t_{2g} set of orbitals: d_{xy}, d_{yz}, d_{zx} orbitals. This set includes the orbitals *whose lobes lie between the axes* and are called *non-axial orbitals*. Group theory calls these t_{2g} orbitals wherein *t* refers to triply degenerate set.

CFT as applied to octahedral complexes: Crystal field splitting of *d*-orbitals in octahedral complexes

In case of free metal ion all the five *d*-orbitals are degenerate, *i.e.*, these have the same energy. Now let us consider an octahedral complex, $[ML_6]^{n+}$ in which the central metal cation, M^{n+} is placed at the centre of the octahedron and is surrounded by six ligands which reside at the six corners of the octahedron as shown in Fig. 5.18. The three axes, *viz. x*-, *y*-, and *z*-axes which point along the corners have also been shown.

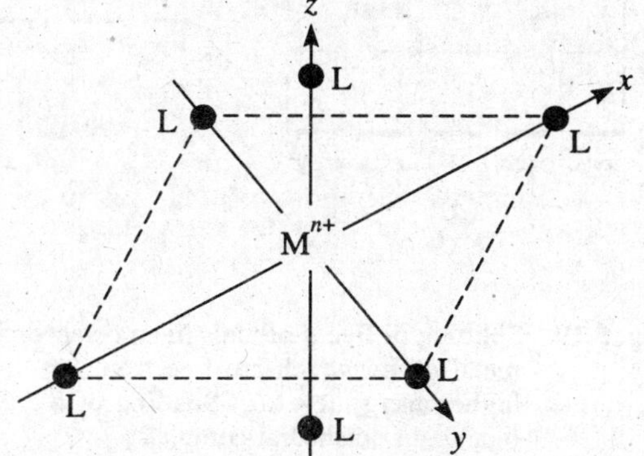

Fig. 5.18. Position of the central metal cation, M^{n+} and six ligands, L's in an octahedral complex, $[ML_6]^{n+}$.

Now suppose both the ligands on each of the three axes are allowed to approach towards the metal cation, M^{n+} from both the ends of the axes. In this process the electrons in *d*-orbitals of the metal cation are repelled by negative point charge or by the negative end of the dipole of the ligands. (Remember CFT regards the ionic ligands as negative point charges and neutral ligands as dipoles). This repulsion will raise the energy of all the five *d*-orbitals. If all the ligands approaching the central cation are at an equal distance from each of the *d*-orbitals (*i.e.*, the ligand field is spherically symmetrical), the energy of each of five *d*-orbitals will raise by the same amount, *i.e.*, all the *d*-orbitals will still remain degenerate, although they will have now higher energy than before. This is only a *hypothetical situation. Since the lobes of the two e_g orbitals* (*i.e.*, d_{z^2} and $d_{x^2-y^2}$ orbitals) *lie directly in the path of the approachinging ligands, the electrons in these orbitals experience greater force of repulsion than those in three t_{2g} orbitals* (*i.e.*, d_{xy}, d_{yz} and d_{zx} orbitals) *whose lobes are directed in space between the path of the approaching ligands i.e.*, the energy of e_g orbitals is increased while that of t_{2g} is decreased (Remember : *greater is the repulsion, greater the increase in energy*). Thus we find that under the influence of approaching ligands, the five *d*-orbitals which were originally degenerate in the free metallic cation are now split (or resolved) into two levels *viz.*, t_{2g} level which is triply degenerate and is of lower energy, and e_g level which is doubly degenerate and is of higher energy (see Fig. 5.19). In other words the degeneracy of the five *d*-orbitals is removed under the influence of the ligands. *The separation of five d-orbitals of the metal ion into two sets having different energies is called crystal field splitting* or *energy level splitting*. This concept of crystal field splitting makes the basis of CFT.

The energy gap between t_{2g} and e_g sets is denoted by Δ_0 or 10Dq where 0 in Δ_0 indicates an octahedral arrangement of the ligands round the central metal cation. This energy difference arises because of the difference in electrostatic field exerted by the ligands on t_{2g} and e_g sets of orbitals of the central metal cation. Δ_0 or 10Dq is called *crystal field splitting energy*. With the help of simple geometry it can be shown that the energy of t_{2g} orbitals is 0.4 Δ_0 (= 4Dq) less than that of hypothetical degenerate *d*-orbitals (No splitting state shown by dotted line in Fig. 5.19) and, hence, that of e_g orbitals is $0.6\Delta_0$ (= 6Dq) *above* that of the hypothetical degenerate *d*-orbitals. Thus, we find that t_{2g} set *loses* an energy equal to 0.4 Δ_0 (= 4Dq) while e_g set *gains* an energy equal to 0.6 Δ_0 (= 6Dq). In Fig 5.19 the loss and gain in energies of t_{2g} and e_g

orbitals is shown by negative (–) and positive (+) signs respectively. Δ_0 is generally measured in cm^{-1}.

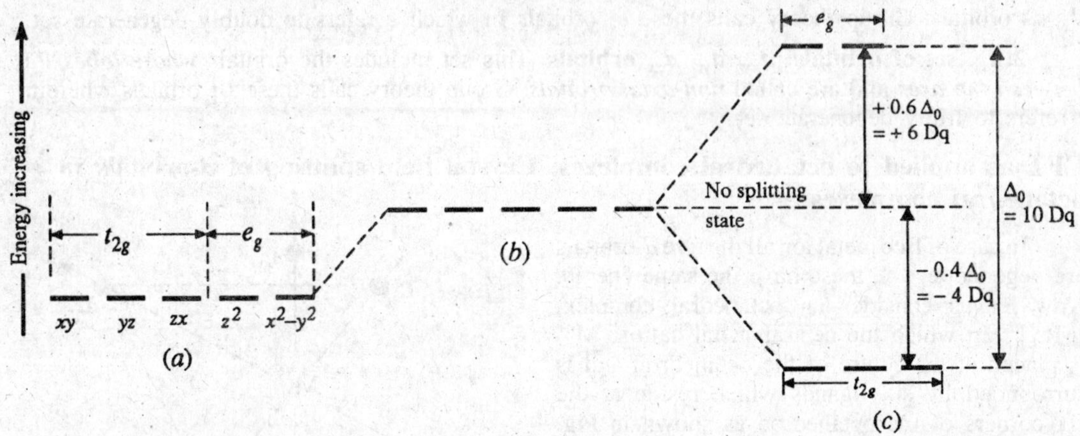

Fig. 5.19. Splitting of five *d*-orbitals in an octahedral complex. (*a*) Five degenegate *d*-orbitals on the central metal cation which are free from any ligand field. (*b*) Hypothetical degenerate *d*-orbitals at a higher energy level (*c*) Splitting of *d*-orbitals into t_{2g} and e_g orbitals under the influence of six ligands in octahedral complex.

Distribution of d^x electrons (x = 1 to 10) of the central metal cation in t_{2g} and e_g orbitals in high spin and low spin octahedral complexes

The distribution of d^x electrons of the central metal cation in t_{2g} and e_g orbitals in an octahedral complex depends on whether the six ligands are weak(er) or strong(er). Thus we have the following two cases:

1. When the ligands are weak(er). We have seen that under the influence of weak(er) ligands the energy difference, Δ_0 between t_{2g} and e_g sets is relatively small and hence all the five *d*-orbitals of these two sets may be supposed to be degenerate, *i.e.*, in presence of weak(er) ligands all the *d*-orbitals have the same energy and consequently the distribution of *d*-electrons in t_{2g} and e_g sets *takes place according to Hund's rule which states that electrons will pair up only when each of the five d-orbitals is at least singly filled.* Thus in weak(er) field the first three electrons enter t_{2g} level, *4th* and *5th* electrons (two electrons) go to e_g set, *6th*, *7th* and 8th electrons (three electrons) occupy t_{2g} set and *9th* and *10th* electrons (last two electrons) go to e_g set. (See Table 5.6). The octahedral complexes containing weak(er) ligands are called *weak-field* or *low-field complexes*.

2. When the ligands are strong (er). In octahedral complexes containing strong(er) ligands, the distribution of d-electrons in t_{2g} and e_g orbitals does not obey Hund's rule. Thus in strong(er) field the first six electrons go to t_{2g} orbitals and the remaining four electrons enter e_g orbitals (Table 5.7). The octahedral complexes having strong(er) ligands are called *strong-field* or *high-field complexes*.

The following points may be noted from tables 5.6 and 5.7.

(*i*) Whether the ligand field is strong(er) or weak(er), for d^1, d^2 and d^3 configurations, the electrons go to the lower energy t_{2g}-level (more stable), for d^8, d^9 and d^{10} configurations, the first six electrons go to the t_{2g}-level and the remaining two (in case of d^8 ion), three (in case of d^9 ion) and four (in case of d^{10} ion) electrons occupy the e_g-level. Thus, the distribution of electrons of d^1, d^2, d^3, d^8, d^9 and d^{10} configurations in t_{2g}- and e_g- levels for both strong(er) and weak(er) octahedral ligand field is the same.

(*ii*) For each of d^4, d^5, d^6 and d^7 configurations there is a difference in the arrangement of electrons in weak(er) and strong(er) ligand fields.

Table 5.6. *Distribution of d^x electrons ($x = 1$ to 10) in t_{2g} and e_g sets of orbitals in weak(er) field (high spin or spin free) octahedral complexes. (n = No. of unpaired electrons, S = Resultant spin = $n/2$, $p + q = x = 1, 2, \ldots\ldots 8, 9$ or 10). Here $\Delta_0 < P$.*

d^x ions	Distribution of d^x electrons in t_{2g} and e_g orbitals		$t_{2g}^p\ e_g^q$ configuration	n	$S = n/2$
	t_{2g} (Lower energy)	e_g (Higher energy)			
d^1	↑ — —	— —	$t_{2g}^1\ e_g^0$	1	1/2
d^2	↑ ↑ —	— —	$t_{2g}^2\ e_g^0$	2	1
d^3	↑ ↑ ↑	— —	$t_{2g}^3\ e_g^0$	3	3/2
d^4	↑ ↑ ↑	↑ —	$t_{2g}^3\ e_g^1$	4	2
d^5	↑ ↑ ↑	↑ ↑	$t_{2g}^3\ e_g^2$	5	5/2
d^6	↑↓ ↑ ↑	↑ ↑	$t_{2g}^4\ e_g^2$	4	2
d^7	↑↓ ↑↓ ↑	↑ ↑	$t_{2g}^5\ e_g^2$	3	3/2
d^8	↑↓ ↑↓ ↑↓	↑ ↑	$t_{2g}^6\ e_g^2$	2	1
d^9	↑↓ ↑↓ ↑↓	↑↓ ↑	$t_{2g}^6\ e_g^3$	1	1/2
d^{10}	↑↓ ↑↓ ↑↓	↑↓ ↑↓	$t_{2g}^6\ e_g^4$	0	0

Table. 5.7. *Distribution of d^x electrons ($x = 1$ to 10) in t_{2g} and e_g orbitals in strong(er) field (low spin or spin paired) octahedral complexes (n = No. of unpaired electrons, S = Resultant spin = $n/2$, $p + q = x = 1, 2, \ldots\ldots 8, 9$ or 10). Here $\Delta_0 > P$.*

d^x ions	Distribution of d^x electrons in t_{2g} and e_g orbitals		$t_{2g}^p\ e_g^q$ configuration	n	$S = n/2$
	t_{2g} (Lower energy)	e_g (Higher energy)			
d^1	↑ — —	— —	$t_{2g}^1\ e_g^0$	1	1/2
d^2	↑ ↑ —	— —	$t_{2g}^2\ e_g^0$	2	1
d^3	↑ ↑ ↑	— —	$t_{2g}^3\ e_g^0$	3	3/2
d^4	↑↓ ↑ ↑	— —	$t_{2g}^4\ e_g^0$	2	1
d^5	↑↓ ↑↓ ↑	— —	$t_{2g}^5\ e_g^0$	1	1/2
d^6	↑↓ ↑↓ ↑↓	— —	$t_{2g}^6\ e_g^0$	0	0
d^7	↑↓ ↑↓ ↑↓	↑ —	$t_{2g}^6\ e_g^1$	1	1/2
d^8	↑↓ ↑↓ ↑↓	↑ ↑	$t_{2g}^6\ e_g^2$	2	1
d^9	↑↓ ↑↓ ↑↓	↑↓ ↑	$t_{2g}^6\ e_g^3$	1	1/2
d^{10}	↑↓ ↑↓ ↑↓	↑↓ ↑↓	$t_{2g}^6\ e_g^4$	0	0

(iii) *Number of unpaired electrons (n) : High-spin and low-spin complexes.* Weak-field complexes of d^4, d^5, d^6 and d^7 ions have greater number of unpaired electrons (n) than those of (same ions) strong-field complexes and are thus with a higher value of resultant spin (S). It is for this reason that the weak-field and strong-field complexes are also called *spin-free* or *high-spin* (abbreviated as HS) and *spin-paired* or *low-spin* (LS) complexes respectively. Recall that VBT has called these complexes as *ionic* (*Pauling*) or *outer-orbital* (*Huggin*) and *covalent* (*Pauling*) or *inner-orbital* (*Huggin*) complexes respectively.

The number of unpaired electrons (*i.e.*, the value of the resultant spin, S) in the cases namely d^1 to d^3 and d^8 to d^{10} is the same in both the fields, and it is due to this reason that for these configurations the question of the formation of HS- and LS-complexes does not arise. The question does arise for the system d^4 to d^7.

The paramagnetism of HS-complexes is larger than that of LS-complexes, since, as is evident from Tables 5.6 and 5.7. HS-complexes have more unpaired electrons (*i.e.*, larger value of S) than the LS-complexes (*i.e.*, smaller value of S).

Solved examples

Example 1. *Give the number of unpaired electrons for the following complex ions (a) $[VF_6]^{3-}$, (b) $[Fe(CN)_6]^{4-}$ (c) $[Fe(CN)_6]^{3-}$ (d) $[RhCl_6]^{3-}$.*

Solution. (*a*) In $[VF_6]^{3-}$ vanadium is present as V^{3+} ion which is a $3d^2$ system. Thus, in this case the question of the formation of HS- or LS- complex does not arise and the distribution of two $3d$ electrons in t_{2g}- and e_g-sets is as V^{3+} $(3d^2) \longrightarrow t_{2g}^2\, e_g^0$, which gives the number of unpaired electrons, n, equal to two, *i.e.*, $n = 2$.

(*b*) In $[Fe(CN)_6]^{4-}$ ion, Fe is present as Fe^{2+} with $3d^6$ configuration. Since CN^- ion is a strong ligand, $[Fe(CN)_6]^{4-}$ ion is a LS-complex and hence the distribution of six $3d$-electrons in t_{2g} and e_g sets is as Fe^{2+} $(3d^6) \longrightarrow t_{2g}^6\, e_g^0$, giving $n = 0$.

(*c*) in this case Fe^{3+} $(3d^5) \longrightarrow t_{2g}^5\, e_g^0$, giving $n = 1$.

(*d*) Cl^- ion is a weak ligand. Thus $[RhCl_6]^{3-}$ is a HS-complex and consequently $Rh^{3+}(4d^6) \longrightarrow t_{2g}^4\, e_g^2$, giving $n = 4$.

Example 2. *Show, with a labelled diagram, the distribution of six d-electrons in Co^{3+} ion in t_{2g} and e_g orbitals of the complex ion pair viz. $[CoF_6]^{3-}$ and $[Co(NH_3)_6]^{3+}$.*

Solution. We know from the spectro-chemical series, that $6F^-$ ions are weaker ligands while $6NH_3$ molecules are stronger ligands. Thus the distribution of d^6 electrons of Co^{3+} ion

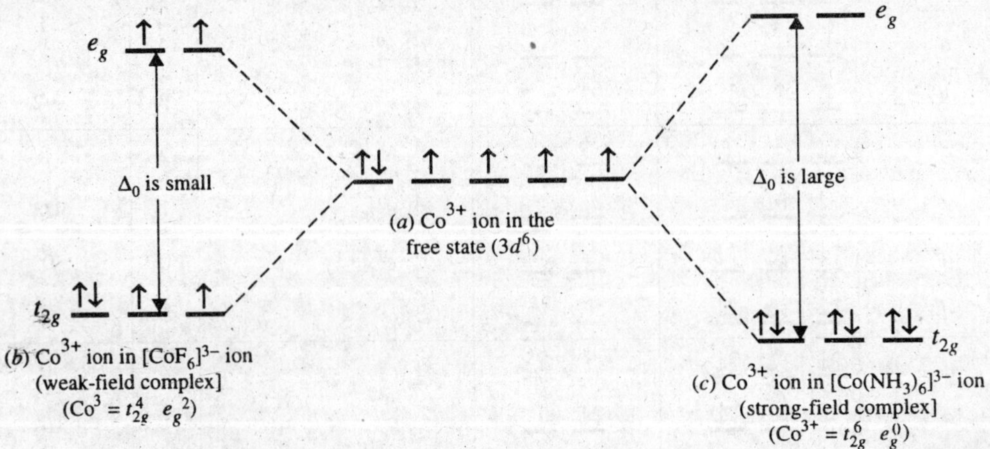

Fig. 5.20. Distribution of d^6 electrons of Co^{3+} ion in the weak-field complex, $[CoF_6]^{3-}$ and strong fieled complex, $[Co(NH_3)_6]^{3+}$.

in t_{2g} and e_g orbitals in $[CoF_6]^{3-}$ is $t_{2g}^4\ e_g^2$ and in $[Co(NH_3)_6]^{+3}$ ion is $t_{2g}^6\ e_g^0$. Both these distribution have been shown in Fig. 5.20.

Factors influencing the magnitude of Δ_0

A mass of experimental data show that the magnitude of Δ_0 depends on the following factors:

A. Nature of the metal cation

The influence of this factor can be studied under the following four heading:

1. Different charges on the cation of the same metal. The cations from atoms of the same transition series and having the same oxidation state have almost the same value of Δ_0 but the cation with a higher oxidation state has a larger value of Δ_0 than that with lower oxidation state, *e.g.*,

(*a*) Δ_0 for $[Fe^{2+}(H_2O)_6]^{2+}$ = 10,400 cm^{-1} ...... $3d^6$

Δ_0 for $[Fe^{3+}(H_2O)_6]^{3+}$ = 13,700 cm^{-1} ...... $3d^5$

(*b*) Δ_0 for $[Co^{2+}(H_2O)_6]^{2+}$ = 9,300 cm^{-1} ...... $3d^7$

Δ_0 for $[Co^{3+}(H_2O)_6]^{3+}$ = 18,200 cm^{-1} . ... $3d^6$

This effect is probably due to the fact that the central ion with higher oxidation state (*i.e.*, with higher charge) will polarise the ligands more effectively and thus the ligands would approach such a cation more closely than they can do the cation of lower oxidation state, resulting in larger splitting.

2. Different charges on the cation of different metals. Two different cations having the same number of *d*-electrons and the same geometry of the complex but with different charge can also be compared. The cation with a higher oxidation state has a larger value of Δ_0 than that with a lower oxidation state. For example, the behaviours towards the same ligand of V(II) and Cr(III), which are both d^3 ions can be compared. It is observed that the value of Δ_0 in $[V^{2+}(H_2O)_6]^{2+}$ is less than that in $[Cr^{3+}(H_2O)_6]^{3+}$ as is shown below:

Δ_0 for $[V^{2+}(H_2O)_6]^{2+}$ = 12,400 cm^{-1} ...... $3d^3$

and Δ_0 for $[Cr^{3+}(H_2O)_5]^{3+}$ = 17,400 cm^{-1} ...... $3d^3$

This fact can be explained in terms of the charge on the cation. The Cr^{3+} ion, which has greater positive charge than V^{2+} ion, exerts a greater attraction for water molecules (ligands) than does the V^{2+} ion. Hence the water molecules approach the Cr^{3+} ion more closely than they approach the V^{2+} ion and so exert a stronger crystal field effect on the *d*-electrons of Cr^{3+} ion.

3. Same charges on the cation but the number of *d*-electrons is different. In case of complexes having the cations with the same charges but with different number of *d*-electrons in the central metal cation the magnitude of Δ_0 decreases with the increase of the number of *d*-electrons, *e.g.*,

Δ_0 for $[Co^{2+}(H_2O)_6]^{2+}$ = 9,300 cm^{-1} ...... $3d^7$

Δ_0 for $[Ni^{2+}(H_2O)_6]^{2+}$ = 8,500 cm^{-1} ...... $3d^8$

From the combination of 1, 2 and 3 mentioned above it can be concluded that :

(*a*) For the complexes having the same geometry and the same ligands but having different number of *d*-electrons, the *magnitude of Δ_0 decreases with the increase of the number of d-electrons in the central metal cation* (No. of *d*-electrons $\propto \frac{1}{\Delta_0}$)

(*b*) In case of complexes having the same number of *d*-electrons the *magnitude of Δ_0 increases with the increase of the charges (i.e., oxidation state) on the central metal cation* (oxidation state $\propto \Delta_0$).

4. Quantum number (*n*) of the *d*-orbitals of the central metal ion. Δ_0 increases about 30% to 50% from $3d^n$ to $4d^n$ and by about the same amount again from $4d^n$ to $5d^n$ complexes, *e.g.*,

$$\Delta_0 \text{ for } [Co^{3+}(NH_3)_6]^{3+} = 23{,}000 \text{ cm}^{-1} \text{ ...... } 3d^6$$
$$\Delta_0 \text{ for } [Rh^{3+}(NH_3)_6]^{3+} = 34{,}000 \text{ cm}^{-1} \text{ ...... } 4d^6$$
$$\Delta_0 \text{ for } [Ir^{3+}(NH_3)_6]^{3+} = 41{,}000 \text{ cm}^{-1} \text{ ...... } 5d^6$$

Presumbaly the 5*d* and 4*d* valence orbitals of the central ion are better than the 3*d*-orbitals in σ-bonding with the ligands.

B. Strong(er) and weak(er) ligands and spectrochemical series

The magnitude of Δ_0 varies from strong(er) to weak(er) ligands. Strong(er) ligands are those which exert a strong(er) field on the central metal ion and hence have higher splitting power while the weak(er) ligands are those which have a weak(er) field on the central metal cation and consequently relatively lower splitting power. Thus strong(er) ligands (*e.g.* CN^-) give larger value of Δ_0 and weak(er) ligands (*e.g.* F^-) yield a smaller value of Δ_0 (See Fig. 5.21).

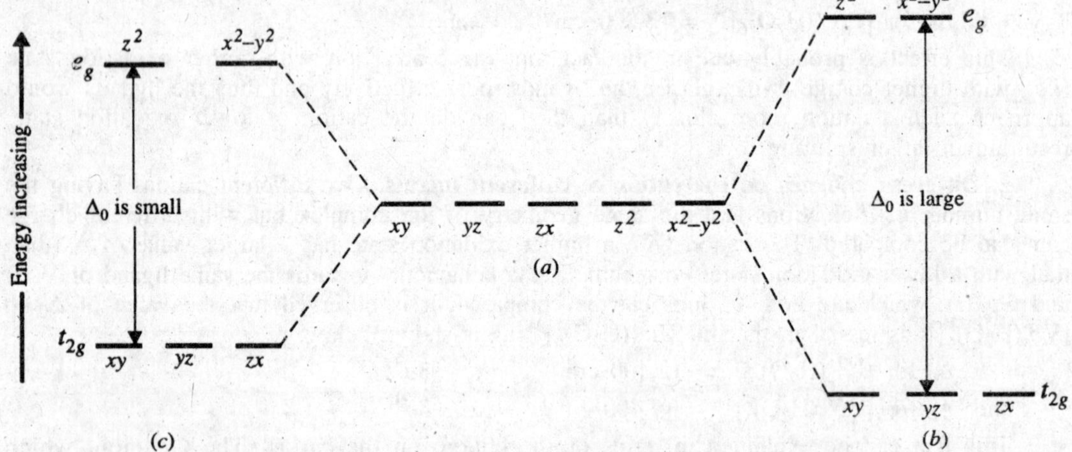

Fig. 5.21. Splitting of five *d*-orbitals in presence of strong(er) and weak(er) ligands in an octahecral complex. (*a*) Five *d*-orbitals in the free metal ion (*b*) Splitting of *d*-orbitals in presence of strong(er) ligands (*c*) Splitting of *d*-orbitals in presence of weak(er) ligands.

Fig. 5.21 shows that not only Δ_0, which represents the energy difference between the t_{2g} and e_g-sets of orbitals, is smaller in the weak(er) field complex than in the strong(er) field, but also that both the t_{2g} and e_g- levels of the weak(er) field are correspondingly closer to the level of the degenerate five *d*-orbitals of the free isolated metallic ion than are those, respectively, of the strong(er) field.

The common ligands can be arranged in the order of their increasing splitting power to cause *d*-orbitals splitting. This series is called *spectrochemical series* and is given below:

$I^- < Br^- < Cl^- \sim SCN^- \sim N_3^- < (C_2H_5O)_2\ PS_2^- < F^- < (NH_2)_2CO < OH^- < C_2O_4^{2-} \sim H_2O < NCS^- \sim H^- < CN^- < NH_2CH_2CO_2^- < NH_3 \sim C_5H_5N <$ en $\sim SO_3^{2-} < NH_2OH < NO_2^- <$ phen $< H^- < CH_3^- < CN^-$, CO.

This series shows that the value of Δ_0 in the series also increases from left to right.

The order of field strength of the common ligands shown above is, in fact, independent of the nature of the central metal ion and the geometry of the complex.

The increase in the value of Δ_0 on proceeding from left to right in the spectrochemical series is quite evident from the values of Δ_0 for some octahedral complexes given in Table 5.8 which clearly shows that since on proceeding from $6Br^- \rightarrow 3$ *en*, the field strength of the ligands increases, the value of Δ_0 also correspondingly increases.

Table 5.8. *Δ_0 values (i.e., energy difference between t_{2g} and e_g levels) in cm^{-1} for some octahedral complexes.*

Ligands →	6 Br^-	< 6Cl^-	< 6H_2O	< 6NH_3	< 3 en
Metal ion	Field strength increasing →				
Ni (II)	7,000 cm^{-1}	< 7,200 cm^{-1}	< 8500 cm^{-1}	< 10,800 cm^{-1}	< 11,500 cm^{-1}
Cr (III)	—	13,800	< 17,400	< 21,600	< 21,900
Co (III)	—	—	18,200	< 23,000	< 23,200
Rh (III)	19,000	< 20,300	< 27,000	< 34,100	< 34,600
	Δ_0 values (in cm^{-1}) also increasing →				

Mean pairing energy (P)

Mean pairing energy (P) is the energy which is required to pair two electrons against electron-electron repulsion in the same orbital. P is the pairing energy for one electron pair. P is generally expressed in cm^{-1}. Pairing energy depends on the principal energy level (*n*) of *d*-electrons.

Calculation of total pairing energy of d^x ion

If *m* is the total number of paired electrons in t_{2g} and e_g orbitals in d^x ion and P is the pairing energy for one electron, then

Total pairing energy for *m* electron pairs = *m*P cm^{-1}.

Solved examples

Example 1. *Calculate the total pairing energy of d^7 ion in high spin as well as in low spin octahedral complexes.*

Solution. We know that the configuration of d^7 ion in high spin state is $t_{2g}^5\ e_g^2$ which shows that $m = 2 + 0 = 2$.

∴ Total pairing energy for 2 paired electrons = 2 × P = 2P

The configuration of d^7 ion in low spin state is $t_{2g}^6\ e_g^1$ which gives $m = 3 + 0 = 3$.

∴ Total pairing energy for 3 paired electrons = 3 × P = 3P

Example 2. *Calculate the total pairing energy for $[Cr(H_2O)_6]^{2+}$ ion in high spin and low spin state. Given that mean pairing energy = 23,500 cm^{-1}.*

Solution. In $[Cr(H_2O)_6]^{2+}$ ion, Cr is present as Cr^{2+} which is a d^4 ion. Thus the configuration of d^4 ion in high spin state is $t_{2g}^3\ e_g^1$ which gives $m = 0$ and hence :

Total pairing energy of $[Cr(H_2O)_6]^{2+}$ ion in high spin state = 0 × P = 0

The configuration of d^4 ion in low spin state is $t_{2g}^4\ e_g^0$ which gives $m = 1$ and hence :

Total pairing energy of $[Cr(H_2O)_6]^{3+}$ ion in low spin state

$$= 1 \times P = 1 \times 23{,}500\ cm^{-1} = 23{,}500\ cm^{-1}$$

Example 3. *Which complex of the following pairs has the larger value of Δ_0*

(i) $[Co^{3+}(CN_6)]^{3-}$ and $[Co^{3+}(NH_3)_6]^{3+}$

(ii) $[Co^{3+}(NH_3)_6]^{3+}$ and $[Co^{3+}F_6]^{3-}$

(iii) $[Co^{3+}(H_2O)_6]^{3+}$ *and* $[Rh^{3+}(H_2O)_6]^{3+}$

(iv) $[Co^{2+}(H_2O)_6]^{2+}$ *and* $[Co^{3+}(H_2O)_6]^{3+}$*?*

Solution. *(i)* Since CN^- is stronger than NH_3 (*i.e.*, CN^- ion has greater field strength than NH_3), Δ_0 for $[Co(CN)_6]^{3-}$ is greater than Δ_0 for $[Co(NH_3)_6]^{3+}$.

(ii) $[Co(NH_3)_6]^{3+}$ ion has greater value of Δ_0 than that for $[CoF_6]^{3-}$.

(iii) Rh is a member of 4*d*-series while Co is a member of 3*d*-series. Thus principal quantum number, *n*, for Rh is greater than that of Co and consequently Δ_0 for $[Rh(H_2O_6)]^{3+}$ is greater.

(iv) The charge on Co in $[Co(H_2O)_6]^{3+}$ is higher than that on Co in $[Co(H_2O)_6]^{2+}$. Thus, Δ_0 for $[Co(H_2O)_6]^{3+}$ is higher.

How to predict the spin state of a given octahedral complex ion

By comparing the values of Δ_0 and P of a given metallic ion, the spin state of the octahedral complex ion formed by that metallic ion can be predicted. Δ_0 tends to force as many electrons to occupy t_{2g} orbitals while P tends to prevent the electrons to pair in t_{2g} orbitals. This discussion show that:

(i) When $\Delta_0 > P$, the electrons tend to pair and hence low spin (strong-field or spin paired) octahedral complex ions are obtained.

(ii) When $\Delta_0 < P$, the electrons tend to remain unpaired and hence high spin (weak-field or spin free) octahedral complex ions are obtained.

Examples of some HS- and LS-octahedral complexes are given in Table 5.9. In this table the value of P (in cm^{-1}) of the central metal ion of the corresponding complex determined from spectroscopic data and that of Δ_0 (in cm^{-1}) for the complexes are also listed. From this table it may be seen that the spin-state of the complexes predicted by CFT is the same as that observed experimentally. In every case where $\Delta_0 < P$, HS-complex is formed and in case where $\Delta_0 > P$, LS-complex is formed.

Table 5.9. *Examples of some LS-and HS-octahedral complexes.*

d^x configuration	*Examples of complexes*	*Value of P (cm^{-1})*	*Value of Δ_0 (cm^{-1})*	*spin-state*		*Relative magnitudes of Δ_0 and P*
				Predicted by CFT	*Observed experimentally*	
d^4	$[Cr(H_2O)_6]^{2+}$	23,500	13,900	HS	HS	$\Delta_0 < P$
	$[Mn(H_2O)_6]^{3+}$	28,800	21,000	HS	HS	$\Delta_0 < P$
	$[Mn(CN)_6]^{3-}$	28,800	38,500	LS	LS	$\Delta_0 > P$
d^5	$Mn(H_2O)_6]^{2+}$	25,500	7,800	HS	HS	$\Delta_0 < P$
	$[Fe(H_2O)_6]^{3+}$	30,000	13,700	HS	HS	$\Delta_0 < P$
d^6	$[Fe(H_2O)_6]^{2+}$	17,600	10,400	HS	HS	$\Delta_0 < P$
	$[Fe(CN)_6]^{4-}$	17,600	33,000	LS	LS	$\Delta_0 > P$
	$[Co(NH_3)_6]^{3+}$	21,000	23,000	LS	LS	$\Delta_0 > P$
	$[CoF_6]^{3-}$	21,000	13,000	HS	HS	$\Delta_0 < P$
d^7	$[Co(H_2O)_6]^{2+}$	22,500	9,300	HS	HS	$\Delta_0 < P$

Solved examples

Example 1. *For* Mn^{3+} *ion, the electron pairing energy, P is about 28,000* cm^{-1}*.* Δ_0 *values for the complexes* $[Mn(H_2O)_6]^{3+}$ *and* $[Mn(CN)_6]^{3-}$ *are 21,000* cm^{-1} *and 38,500* cm^{-1} *respectively. Do these complexes have high-spin or low-spin configurations? Also write down the configurations corresponding to these states.*

Solution. For $[Mn(H_2O)_6]^{3+}$ ion, since $P > \Delta_0$ ($P = 28,000\ cm^{-1}$, $\Delta_0 = 21,000\ cm^{-1}$), this ion is a high spin complex and the configuration of d^4 ion. (Mn^{3+} ion is a d^4 system) is therefore, $t_{2g}^3\ e_g^1$.

For $[Mn(CN)_6]^{3-}$ ion, since P < Δ_0 (P = 28,000 cm^{-1}, Δ_0 = 38,500 cm^{-1}), this ion is a low spin complex and the configuration of d^4 ion (Mn^{3+} ion is a d^4 system) is $t_{2g}^4\ e_g^0$.

Example 2. *For $[Cr(H_2O)_6]^{2+}$ ion, the mean pairing energy (P) is 23,500 cm^{-1} and the magnitude of Δ_0 is 13,900 cm^{-1}. State whether the given ion is stable in low spin state or in high spin state.*

Solution. We know that when Δ_0 < P, high spin state is more stable. Here since Δ_0 (= 13,900 cm^{-1}) < P (= 23,500 cm^{-1}), high spin state is more stable.

Crystal field stabilisation energies of d^x configuration (x = 0 to 10) of the central metal ion in octahedral complex

We have seen that according to CFT, under the influence of the six ligands approaching towards the central metal ion during the formaion of an octahedral complex, the *d*-orbitals of the central metal ion are split into two sets of orbitals *viz. lower energy triplet* (or *trio*) t_{2g} and *higher energy doublet* (or *pair*), e_g sets. t_{2g} set of orbitals is *triply degenerate* while e_g set is *doubly degenerate*. The energy gap between these two sets is equal to Δ_0 (or 10Dq). The energy of t_{2g} set is lowered by 2/5 Δ_0 (= 0.4 Δ_0) or 4Dq while hat of e_g set in raised by $\frac{3}{5}\Delta_0$ (= 0.6Δ_0) or 6Dq relative to the energy of hypotencial degenerate *d*-orbitals. Thus, each electron occupying t_{2g} orbitals decreases the energy of *d*-orbitals by – 0.4 Δ_0 (= –4Dq) while that going into e_g orbitals increases its energy by + 0.6 Δ_0 (= + 6Dq). – and + signs indicate respectively the decrease and increase in the energy of *d*-orbitals caused by their splitting under the influence of six ligands. Now let us consider a d^x ion containing $t_{2g}{}^p e_g{}^q$ configuration in which p is the number of electrons in t_{2g} set, q is the number of electrons in e_g set and $x = p + q$. Quite obviously :

Change in energy (in terms of Δ_0) for $t_{2g}{}^p e_g{}^q$ configuration

= Loss in energy due to p electrons in t_{2g} set
+ gain in energy due to q electrons in e_g set.

$$= -0.4\ \Delta_0 \times p + 0.6\ \Delta_0 \times q$$

$$= [-0.4\ p + 0.6q]\ \Delta_0 \qquad ...(i)$$

Now, since Δ_0 = 10Dq, the above expression can also be written as :

Change in energy (in terms of Dq) for $t_{2g}{}^p e_g{}^q$ configuration

$$= [-0.4p + 0.6q] \times 10\ \text{Dq}$$

$$= [-4p + 6q]\ \text{Dq} \qquad ...(ii)$$

Thus equations (*i*) and (*ii*) give the energies of d^x ion containing $t_{2g}{}^p e_g{}^q$ configuration. The change in energy for d^x ion containing $t_{2g}{}^p e_g{}^q$ configuration calculated as above is called *crystal field stabilisation energy* (CFSE) of d^x ion, since it stabilises *d*-orbitals by lowering their energy which results from their splitting into t_{2g} and e_g orbitals.

In the derivation of equation (*i*) and (*ii*) we have not considered the pairing energy, P, of d^x ion which is the energy required to pair two electrons against electron-electron repulsion in the same orbital. If the pairing energy of the ion is also involved in the $t_{2g}{}^p e_g{}^q$ configuration of a given d^x ion, then CFSE of the ion is given by the expression:

$$\text{CFSE} = [-0.4p + 0.6q]\ \Delta_0 + m\text{P} \qquad ...(iii)$$

$$= [-4p + 6q]\ Dq + m\text{P} \qquad (\because \Delta_0 = 10\ \text{Dq}) \qquad ...(iv)$$

Here m is the total number of paired electrons in t_{2g} and e_g sets of orbitals. Equations (*iii*) and (*iv*) have been used to calculate the CFSE values (in terms of Δ_0 and Dq respectively) for d^0 to d^{10} ions of HS and LS octahedral complexes. These values, as calculated from the above equations, are listed in Tables 5.10 and 5.11 respectively.

Table 5.10. *CFSE values (in the units of Δ_0 and Dq) for d^x configuration (x = 0 to 10) of the central metal ion in weak field (spin free or high spin) octahedral complexes. m = Total number of paired electrons in t_{2g} and e_g orbitals, P = Mean pairing energy, p + q = x = 0, 1, 2, ... 8, 0 or 10.*

d^x configuration	Distribution of d^x electrons in t_{2g} and e_g orbitals: t_{2g}	e_g	$t_{2g}^p\ e_g^q$ configuration	m	CFSE = $[-0.4p + 0.6q]\ \Delta_0 + mP$ = $[-0.4p + 0.6q] \times 10Dq + mP$ = $[-4p + 6q]\ Dq + mP$
d^0	— — —	— —	$t_{2g}^0\ e_g^0$	0	0.0 Δ_0 (0.0 Dq)
d^1	↑ — —	— —	$t_{2g}^1\ e_g^0$	0	– 0.4 Δ_0 (– 4Dq)
d^2	↑ ↑ —	— —	$t_{2g}^2\ e_g^0$	0	– 0.8 Δ_0 (– 8 Dq)
d^3	↑ ↑ ↑	— —	$t_{2g}^3\ e_g^0$	0	– 1.2 Δ_0 (– 12 Dq)
d^4	↑ ↑ ↑	↑ —	$t_{2g}^3\ e_g^1$	0	– 0.6 Δ_0 (– 6 Dq)
d^5	↑ ↑ ↑	↑ ↑	$t_{2g}^3\ e_g^2$	0	– 0.0 Δ_0 (0.0 Dq)
d^6	↑↓ ↑ ↑	↑ ↑	$t_{2g}^4\ e_g^2$	1	– 0.4 Δ_0 (– 4 Dq) + P
d^7	↑↓ ↑↓ ↑	↑ ↑	$t_{2g}^5\ e_g^2$	2	– 0.8 Δ_0 (– 8 Dq) + 2P
d^8	↑↓ ↑↓ ↑↓	↑ ↑	$t_{2g}^6\ e_g^2$	3	– 1.2 Δ_0 (– 12Dq) + 3P
d^9	↑↓ ↑↓ ↑↓	↑↓ ↑	$t_{2g}^6\ e_g^3$	4	– 0.6 Δ_0 (– 6 Dq) + 4P
d^{10}	↑↓ ↑↓ ↑↓	↑↓ ↑↓	$t_{2g}^6\ e_g^4$	5	0.0 Δ_0 (0.0 Dq) + 5P

Table 5.11. *CFSE values (in the units of Δ_0 and Dq) for d^x configuration (x = 0 to 10) of the central metal ion in strong field (spin paired or low spin) octahedral complexes. m = total number of paired electrons in t_{2g} and e_g orbitals, P = mean pairing energy, p + q = x = 0, 1, 2, ...8, 9 or 10.*

d^x configuration	Distribution of d^x electrons in t_{2g} and e_g orbitals: t_{2g}	e_g	$t_{2g}^p\ e_g^q$ configuration	m	CFSE = $[-0.4p + 0.6q]\ \Delta_0 + mP$ = $[0.4p + 0.6q] \times 10Dq + mP$ = $[-4p + 6q]\ Dq + mP$
d^0	— — —	— —	$t_{2g}^0\ e_g^0$	0	0.0 Δ_0 (0.0 Dq)
d^1	↑ — —	— —	$t_{2g}^1\ e_g^0$	0	– 0.4 Δ_0 (– 4Dq)
d^2	↑ ↑ —	— —	$t_{2g}^2\ e_g^0$	0	– 0.8 Δ_0 (– 8 Dq)
d^3	↑ ↑ ↑	— —	$t_{2g}^3\ e_g^0$	0	– 1.2 Δ_0 (– 12 Dq)
d^4	↑↓ ↑ ↑	— —	$t_{2g}^4\ e_g^0$	1	– 1.6 Δ_0 (– 16 Dq) + P

(Contd.)

d^5	↑↓ ↑↓ ↑	— —	$t_{2g}^5\ e_g^0$	2	$-2.0\ \Delta_0\ (-20\ Dq) + 2P$
d^6	↑↓ ↑↓ ↑↓	— —	$t_{2g}^6\ e_g^0$	3	$-2.4\ \Delta_0\ (-24\ Dq) + 3P$
d^7	↑↓ ↑↓ ↑↓	↑ —	$t_{2g}^6\ e_g^1$	3	$-1.8\ \Delta_0\ (-18\ Dq) + 3P$
d^8	↑↓ ↑↓ ↑↓	↑ ↑	$t_{2g}^6\ e_g^2$	3	$-1.2\ \Delta_0\ (-12Dq) + 3P$
d^9	↑↓ ↑↓ ↑↓	↑↓ ↑	$t_{2g}^6\ e_g^3$	4	$-0.6\ \Delta_0\ (-6\ Dq) + 4P$
d^{10}	↑↓ ↑↓ ↑↓	↑↓ ↑↓	$t_{2g}^6\ e_g^4$	5	$0.0\ \Delta_0\ (0.0\ Dq) + 5P$

Example. *For the $[Cr(H_2O)_6]^{2+}$ ion, the mean pairing energy, P, is found to be 23,500 cm^{-1}. The magnitude of Δ_0 is 13,900 cm^{-1}. Calculate the CFSE for this complex ion corresponding to high spin and low spin state.*

Solution. Quite evidently in the given complex ion Cr is present as Cr^{2+} ion which is a d^4 ion.

For d^4 ion in high spin state, ($m = 0$)

$$\text{CFSE} = -0.6\ \Delta_0 + mP$$
$$= -0.6 \times (13{,}900\ \text{cm}^{-1}) + 0 \times P = -8{,}340\ \text{cm}^{-1}$$

For d^4 ion in low spin state,

$$\text{CFSE} = -1.6\ \Delta_0 + 1 \times \text{P}$$
$$= -1.6 \times (13{,}900\ \text{cm}^{-1}) + 1 \times (23{,}500\ \text{cm}^{-1})$$
$$= +1{,}260\ \text{cm}^{-1}$$

CFT as applied to tetrahedral complexes : Crystal field splitting of *d*-orbitals in tetrahedral complexes

In case of a free metal ion (M^{n+}) all the *d*-orbitals are degenerate, *i.e.*, these have the same energy. Now let us consider a tetrahedral complex ion, $[ML_4]^{n+}$ in which the central metal ion (M^{n+}) is surrounded tetrahedrally by four ligands. A tetrahedron may be supposed to have been formed from a cube. The centre of the cube is the central of the tetrahedron at which is placed the central metal ion (M^{n+}). Four alternate corners of the cube are the four corners of the tetrahedron at which the four ligands, L, (which have been shown by black circles) are placed (see Fig. 5.22). It may be seen from the figure that the four ligands are lying between the three axes *viz*, *x*, *y* and *z* axes which pass through the centres of the six faces of the cube and thus go through the centre of the cube. Now since the lobes of t_2 orbitals (d_{xy}, d_{yz} and d_{zx}) are lying between the axes, *i.e.*, are lying directly in the path of the ligands, these orbitals will experience greater force of repulsion from the ligands than those of *e* orbitals (d_{z^2} and $d_{x^2-y^2}$) whose lobes are lying along the axes, *i.e.*, are lying in space between the ligands. Thus the energy of t_2 orbitals will be increased while that of *e* orbitals will be decreased. Consequently the *d* orbitals are again split into two sets as shown in Fig. 5.23 from which it may be

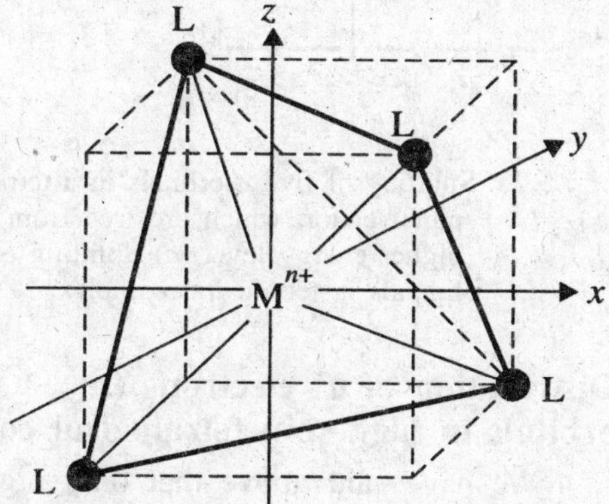

Fig. 5.22. Tetrahedral arrangement of four ligands (L) around the metal ion (M^{n+}) in tetrahedral complex ion, $[ML_4]^{n+}$.

seen that the order of energy of t_2 and e sets is the reverse of that seen for t_{2g} and e_g sets in octahedral complexes.

The energy difference between t_2 and e sets for tetrahedral complex is represented as Δt. In has been shown that $\Delta t < \Delta_0$, the cause of which is that t_2-orbitals, although now closest to the ligands, do not point directly at the ligands, *i.e.*, in an octahedral complex there is a ligand along each axis and in a tetrahedral complex no ligand lies directly along any axis. For this reason and also because there are only four ligands in the tetrahedral complex, while in an octahedral complex there are six ligands, the tetrahedral orbital splitting, Δt is less than Δ_0, for the same metal and ligands and the same internuclear distances. It has also been shown that $\Delta t = 0.45\ \Delta_0$. Thus the energy level of the t_2 set is raised by $0.4\ \Delta t = 0.18\ \Delta_0$ while that of e set is lowered by $0.6\ \Delta t = 0.27\ \Delta_0$. The relation namely $\Delta t = 0.45\ \Delta_0$ also shows that, other things being equal, the crystal field splitting in a tetrahedral complex will be about half the magnitude of that in an octahedral complex.

In case of tetrahedral complex, since Δt is generally less than P ($\Delta t <$ P), the electrons tend to remain unpaired and hence *only high spin tetrahedral complexes are known, i.e., low spin tetrahedral complexes are not known.*

Since $\Delta t < \Delta_0$, crystal field spliting of *d*-orbitals favours the formation of octahedral complexes.

Note. The subscript *g* is not used for the splitting of *d*-orbitals in tetrahedral complexes because a tetrahedron has no centre of symmetry. The symbol *g* is used for the ligand fields which have centre of symmetry.

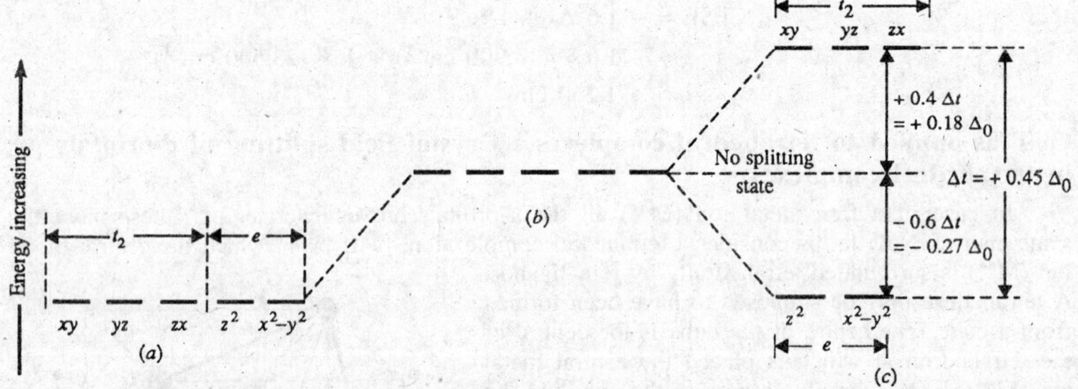

Fig. 5.23. Splitting of five *d*-orbitals in a tetrahedral complex (*a*) Five degenerate *d*-orbitals on the central metal cation which are free from any ligand field (*b*) Hypothetical degenerate *d*-orbitals at a higher energy level (*c*) Splitting of *d*-orbitals into *e* and t_2 orbitals under the influence of four ligands in tetrahedral complex.

Distribution of d^x electrons (x = 1 to 10) of the central metal ion in *e* and t_2 orbitals in high spin tetrahedral complexes

We have said above that only high spin tetrahedral complexes are known. In case of these complexes, the distribution of d^x electrons in *e* and t_2 orbitals takes place according to Hund's rule, *i.e.*, the electrons will pair up only when each of the five *d*-orbitals is at least singly-filled. The pairing of electrons will start from *e* orbitals, since these orbitals have less energy than t_2 orbitals. The distribution of d^x electrons in high spin tetrahedral complexes has been shown in Table 5.12.

Table 5.12. *Distribution of d^x electrons ($x = 1$ to 10) in e and t_2 orbitals in high spin tetrahedral complexes (n = Number of unpaired electrons, $p + q = x = 1, 2, ... 8, 9$ or 10). Here $\Delta t < P$.*

d^x ions	Distribution of d^x electrons in e and t_2 orbitals		$e^p t_2^q$ configuration	n	CFSE = $-0.27\ \Delta_0 \times p + 0.18\Delta_0 \times q$ = $[-0.27 \times p + 0.18 \times q]\Delta_0$
	e (Lower energy)	t_2 (Higher energy)			
d^1	↑ _	_ _ _	$e^1 t_2^0$	1	$[-0.27 \times 1 + 0.18 \times 0]\Delta_0 = -0.27\Delta_0$
d^2	↑ ↑	_ _ _	$e^2 t_2^0$	2	$[-0.27 \times 2 + 0]\Delta_0 = -0.54\ \Delta_0$
d^3	↑ ↑	↑ _ _	$e^2 t_2^1$	3	$[-0.27 \times 2 + 0.18 \times 1]\Delta_0 = -0.36\Delta_0$
d^4	↑ ↑	↑ ↑ _	$e^2 t_2^2$	4	$[-0.27 \times 2 + 0.18 \times 2]\Delta_0 = -0.18\ \Delta_0$
d^5	↑ ↑	↑ ↑ ↑	$e^2 t_2^3$	5	$[-0.27 \times 2 + 0.18 \times 3]\Delta_0 = 0.0\ \Delta_0$
d^6	↑↓ ↑	↑ ↑ ↑	$e^3 t_2^3$	4	$[-0.27 \times 3 + 0.18 \times 3]\Delta_0 = -0.27\Delta_0$
d^7	↑↓ ↑↓	↑ ↑ ↑	$e^4 t_2^3$	3	$[-0.27 \times 4 + 0.18 \times 3]\Delta_0 = -0.54\ \Delta_0$
d^8	↑↓ ↑↓	↑↓ ↑ ↑	$e^4 t_2^4$	2	$[-0.27 \times 4 + 0.18 \times 4]\Delta_0 = -0.36\ \Delta_0$
d^9	↑↓ ↑↓	↑↓ ↑↓ ↑	$e^4 t_2^5$	1	$[-0.27 \times 4 + 0.18 \times 5]\Delta_0 = -0.18\ \Delta_0$
d^{10}	↑↓ ↑↓	↑↓ ↑↓ ↑↓	$e^4 t_2^6$	0	$[0.27 \times 4 + 0.18 \times 6]\Delta_0 = 0.0\ \Delta_0$

Crystals field stabilisation energies of d^x configuration (x = 1 to 10) of the central metal ion in high spin tetrahedral complexes

We have seen that according to CFT, under the influence of four ligands approaching towardds the central metal ion during the formation of a high spin tetrahedral complex, the *d*-orbitals of the central metal ion are split into lower energy doublet *e* orbitals (d_{z^2} and $d_{x^2-y^2}$ orbitals) and higher energy triplet t_2 orbitals (d_{xy}, d_{yz} and d_{zx} orbitals). The energy gap between *e* and t_2 orbitals is denoted by Δt which is equal to 0.45 Δ_0. The energy of *e* orbitals is *lowered* by $0.6\ \Delta t = 0.6 \times 0.45\ \Delta_0 = 0.27\ \Delta_0$ and that of t_2 orbitals is *raised* by $0.4\Delta t = 0.4 \times 0.45\ \Delta_0 = 0.18\ \Delta_0$ relative to the energy of no spliting state. Thus each electron occupying *e* orbitals decreases the energy of *d*-orbitals by $-0.6\ \Delta t = -0.27\ \Delta_0$ while that going to t_2 orbitals increases its energy by $+0.4\ \Delta t = +0.18\ \Delta_0$. – and + signs indicate respectively the decrease and increase in the energy of *d*-orbitals caused by their splitting under the influence of four ligands. Now let us consider a d^x ion containing $e^p t_2^q$ configuration in which *p* is the number of electrons in *e* set of orbitals and *q* is the number of electrons in t_2 set of orbitals and $x = p + q$. Obviously:

CFSE for $e^p t_2^q$ configuration = Loss in energy due to *p* electrons in *e* set of orbitals + gain in energy due to *q* electrons in t_2 set of orbitals.

or $$\text{CFSE} = -0.27\ \Delta_0 \times p + 0.18\ \Delta_0 \times q$$
$$= [-0.27 \times p + 0.18 \times q]\Delta_0$$

In this equation the pairing energy has not been considered. With the help of this equation the value of CFSE for d^x ion in a high spin tetrahedral complex can be calculated. (see Table 5.12).

Crystal field splitting of *d*-orbitals in tetragonal (elongated distorted octahedral) and square planar complexes

Before considering the splitting of *d*-orbitals of the central metal cation in these complexes, we should understand how tetragonally distorted octahedral and square planar geometries are obtained from regular octahedral geometry. Consider a regular (symmetrical) octahedral complex, $[M(L_b)_4 (L_a)_2]$ in which M is the central metallic cation, L_a are two *trans*-ligands (*i.e.*, L_a are the ligands lying along the *z*-axis) and L_b are the basal equotorial ligands lying in *xy* plane. In this complex all the six bond distances (four M—L_b and two M—L_a distances) are equal [See Fig. 5.24 (*a*)]. Now if two L_a ligands are moved slightly longer from the central metal cation, M so that each of the two M—L_a distances becomes slightly longer than each of the four co-planar M—L_b distances, the symmetrical shape of octahedral complex gets distorted and

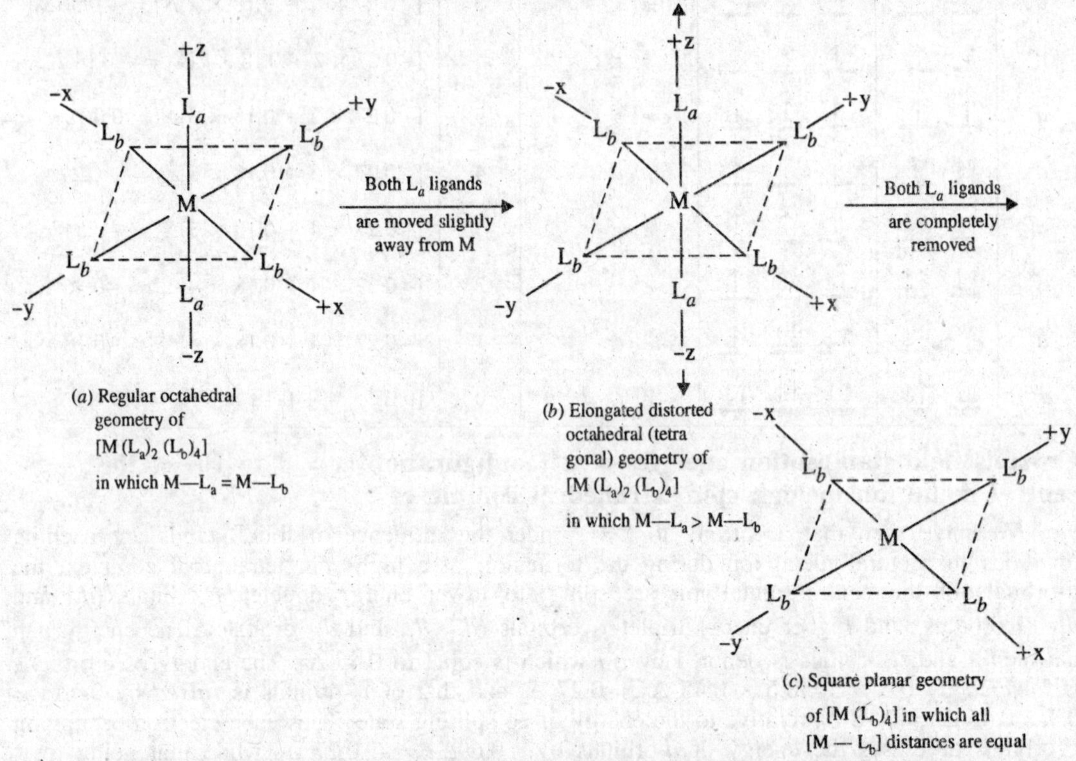

Fig. 5.24. To get tetragonal and square planar geometries from octahedral geometry.

becomes distorted octahedral shape [see Fig. 5.24(*b*)]. In this shape, since the two *trans* ligand have elongated, the distorted octahedral shape is also called *elongated distorted octahedral shape.* Obviously the elongation of two trans ligands takes place along $+z$ and $-z$ axes. Elangated distorted octahedral gemoetry is also called *tetragonally distorted octahedral shape* or simply *tetragonal shape.* Now if the two L_a ligands are completely removed away from the *z*-axis, the tetragonally distorted octahedral shape becomes square planar which is a four-coordinated complex [see Fig. 5.24(*c*)].

Now in order to consider the splitting of *d*-orbitals in elongated distorted octahedral and square planar complexes, we start with the splitting of *d*-orbitals in octahedral complexes. We have already seen that in octahedral complexes, the energy of d_{xy}, d_{yz} and d_{zx} orbitals (t_{2g}

orbitals) is decreased while that of d_{z^2} and $d_{x^2-y^2}$ orbitals (e_g orbitals) is increased [Fig. 5.25 (*b*)]. Now in elongated distorted octahedral complex, since the distance of the *trans*-ligands (L_a ligands) is increased from the central metal ion by removing them away along the *z*-axis, *d*-orbitals along the *z*-axis (*i.e.*, d_{z^2} orbital), *d*-orbital in *yz* plane (*i.e.* d_{yz} orbital) and *d*-orbital in *zx* plane (*i.e.* d_{zx} orbital) experience less repulsion from the ligands than they do in the octahedral complex while the *d*-orbitals in *xy* plane (*i.e.*, d_{xy} and $d_{x^2-y^2}$ orbitals) experience more repulsion than they do in the octahedral complex. Consequently the energy of d_{z^2}, d_{yz} and d_{zx} orbitals decrease while that of $d_{x^2-y^2}$ and d_{xy} orbitals increase [Fig. 5.25 (*d*)]. Note that d_{yz} and d_{zx} orbitals still remain degenerate as they are in the octahedral complex. In square planar geometry the energies of d_{z^2}, d_{yz} and d_{zx} orbitals again fall down while those of $d_{x^2-y^2}$ and d_{xy} orbitals rise up [see Fig. 5.25 (*c*)]. Thus the splitting of *d*-orbitals into various orbitals in square planar complexes takes place as shown at (*d*) of Fig. 5.25. The relative energy order between the various splitted *d*-orbitals in square planar complexes is uncertain but the order shown in Fig. 5.25(*d*) has been established for $5d^8$ configuration from spectroscopic data. The extent of splitting of *d*-orbitals in square planar complexes depends on the nature of the central metal atom and ligands. Semi-quantitative calculations for square planar complexes of Co^{2+} ($3d^7$), Ni^{2+} ($3d^8$) and Cu^{2+} ($3d^9$) have shown that $\Delta_1 = \Delta_0$, $\Delta_2 = \frac{2}{3}\Delta_0$ (or 0.66 Δ_0) and $\Delta_3 = \frac{1}{12}\Delta_0$ (or 0.08 Δ_0) and hence

$$\Delta_{sp} = \Delta_1 + \Delta_2 + \Delta_3 = \Delta_0 + \frac{2}{3}\Delta_0 + \frac{1}{12}\Delta_0$$

$$= \Delta_0 + 0.66\,\Delta_0 + 0.08\,\Delta_0 = 1.74\,\Delta_0$$

For the square planar complexes of Pd^{2+} ($4d^8$) and Pt^{2+} ($5d^8$) spectroscopic results have shown that:

$$\Delta_{sp} = \Delta_1 + \Delta_2 + \Delta_3 = 1.3\Delta_0$$

Applications of crystal field theory

The following properties of transition metal complexes can well be explained on the basis of CFT.

A. Colour of transition metal complexes. See chapter 6.

B. Uses of CFSE values. With the help of CFT, we have calculated the CFSE values for d^x configuration of the central metal ion in octahedral and in high spin tetrahedral complexes. These values have been given in Tables 5.10, 5.11 and 5.12.

With the help of CFSE values we can explain the following :

1. Crystal structure of spinels. Mixed oxides of the general formula, $A^{2+}B_2^{3+}O_4$ are called spinels after the name of the mineral spinel, $MgAl_2O_4$. Here A^{2+} = Mg^{2+}, Mn^{2+}, Fe^{2+}, Co^{2+}, Ni^{2+}, Cu^{2+}, Zn^{2+} etc., and B^{3+} = Al^{3+}, Cr^{3+}, Mn^{3+}, Co^{3+}, Fe^{3+} etc. A^{2+} and B^{3+} ions may be of different metals or of the same metal.

In spinels oxygen atoms are arranged in a *cubic close-packed lattice.* In such lattices each oxygen atom has 12 other oxygen atoms equidistant from it and the holes between oxygen atoms are of two types: (*a*) *octahedral holes* which are so called because these are surrounded by six oxygen atoms. There is one of such holes for each oxygen atom. (*b*) *tetrahedral holes* which are so called since these are surrounded by four oxygen atoms. There are two such holes for each oxygen atom. These are smaller than the octahedral holes. There are twice as many tetrahedral holes as there are octahedral holes. The cations occupy the octahedral and tetrahedral holes, since these are large enough to be filled by cations.

Spinels of $A^{2+}B_2^{3+}O_4$ type are classified as *normal* or *simple* and *inverse spinels.* In normal spinels all the A^{2+} cations occupy one of the eight available tetrahedral holes (positions where

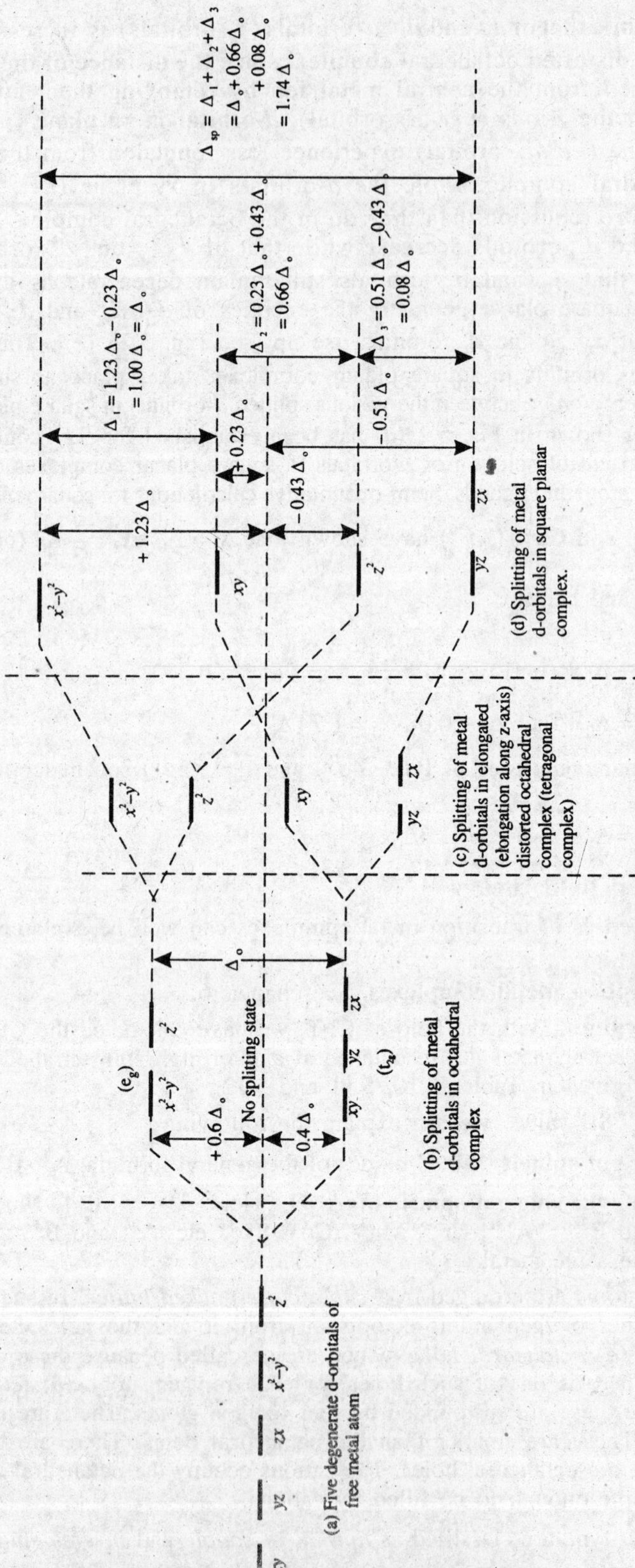

Fig 5.25 Splitting of metal d-orbitals in octahedral, tetragonal and square planar complexes.

a cation can be surrounded by four anions) and all B^{3+} cations occupy half of the available octahedral holes. Normal spinels are represented as A^{2+} $[B_2^{3+}]$ O_4. This representation shows that the cations outside the bracket (*i.e.*, A^{2+} cations) occupy the tetrahedral holes and the cations inside the bracket (*i.e.*, B^{3+} cations) occupy the octahedral holes.

Examples of normal spinels are $Mg^{2+}[Cr_2^{3+}]O_4$, $Ni^{2+}[Cr_2^{3+}]O_4$, Mn_3O_4 or $Mn^{2+}[Mn_2^{3+}]O_4$, Co_3O_4 or $Co^{2+}[Co_2^{3+}]O_4$ etc.

In inverse spinels all the A^{2+} and half of the B^{3+} cations are in octahedral and the other half of the B^{3+} cations are in tetrahedral holes. Inverse spinels are represented as $B^{3+}[A^{2+}B^{3+}]O_4$. This formulation shows that the tetrahedral holes are occupied by half of the B^{3+} ions and the octahedral holes are occupied by A^{2+} ions and the remaining half B^{3+} ions. Examples of inverse spinels are $CuFe_2O_4$ or $Fe^{3+}[Cu^{2+}Fe^{3+}]O_4$, $MgFe_2O_4$ or $Fe^{3+}[Mg^{2+}Fe^{3+}]O_4$, Fe_3O_4 or $Fe^{3+}[Fe^{2+}$ $Fe^{3+}]O_4$ etc.

Inverse spinels of $A^{4+}B_2^{2+}O_4$ type are also known. Examples are $TiZn_2O_4$ and $SnCo_2O_4$. These are represented as $Zn^{2+}[Ti^{4+}Zn^{2+}]O_4$ and $Co^{2+}[Sn^{4+}Co^{2+}]O_4$ respectively.

Now let us see how CFT helps in predicting the structure of spinels. For example with the help of CFT it can be shown why the oxide Mn_3O_4 or $Mn^{2+}Mn_2^{3+}O_4$ is a normal spinel while the oxide Fe_3O_4 or $Fe^{2+}Fe_2^{3+}O_4$ is an inverse spinel. CFSE values in octahedral and tetrahedral fields have been used for the interpretation. For this it is assumed that the oxide ions, O^{2-}, like water molecules, produce weak field. CFSE values (in terms of Δ_0) for $Mn^{3+}(d^4)$, Fe^{3+}, $Mn^{2+}(d^5)$ and $Fe^{2+}(d^6)$ ions in octahedral and tetrahedral weak ligand (*i.e.*, high spin) field are given below: (Negative sign has not been considered).

	Mn^{3+} (d^4)	Mn^{2+} (d^5)	Fe^{3+} (d^5)	Fe^{2+} (d^6)
CFSE (*octahedral weak field*):	0.60 Δ_0	0	0	0.40 Δ_0
CFSE (*tetrahedral weak field*):	0.18 Δ_0	0	0	0.27 Δ_0

It is obvious that for $Mn^{3+}(d^4)$ and $Fe^{2+}(d^6)$ ions the CFSE values are greater for octahedral than for tetrahedral sites. Thus Mn^{3+} and Fe^{2+} ions will preferentially occupy the octahedral sites, maximising the CFSE values of the system. Hence in Mn_3O_4 all the Mn^{3+} ions occupy octahedral sites and all Mn^{2+} ions are in the tetrahedral sites, *i.e.*, it is a normal spinel and its structure is, therefore, represented as $Mn^{2+}[Mn_2^{3+}]O_4$. In Fe_3O_4 all the Fe^{2+} ions and half of the Fe^{3+} ions are in the octahedral sites, while the remaining half of Fe^{3+} ions occupy tetrahedral sites. Thus it is an inverse spinel and is, therefore represented as $Fe^{3+}[Fe^{2+}Fe^{3+}]O_4$.

2. Stabilisation of oxidation states. CFSE values also explain why certain oxidation states are preferentially stabilised by coordinating with certain ligands. The following two examples illustrate this use:

(*a*) Although H_2O molecule which is a *weak ligand* should be expected to coordinate with Co^{2+} and Co^{3+} ions to form the high-spin octahedral complexes *viz.* $[Co(H_2O)_6]^{2+}$ and $[Co(H_2O)_6]^{3+}$ respectively, experiments show that H_2O stabilises Co^{2+} ion and not Co^{3+}, *i.e.*, $[Co(H_2O)_6]^{2+}$ is more stable than $[Co(H_2O)_6]^{3+}$. This is because of the fact that Co^{2+} (d^7) has a much higher value of CFSE in weak octahedral configuration (CFSE = 0.8 Δ_0) than Co^{3+} (d^6) in the same configuration (CFSE = 0.4 Δ_0).

(*b*) If we consider the coordination of NH_3 molecules with Co^{2+} and Co^{3+} ions, it may be seen that NH_3 which is a *strong ligand* stabilises Co^{3+} ion by forming $[Co(NH_3)_6]^{3+}$ rather than Co^{2+} ion. This is because of the fact that Co^{3+} ion (d^6 system) has much higher value of CFSE in strong octahedral configuration (CFSE = 2.4 Δ_0) than Co^{2+} ion (d^7 system) in the same configuration (CFSE = 1.8 Δ_0).

3. Stereochemistry of complexes. (*a*) CFSE values also predict why Cu^{2+} ion forms square planar complexes rather than tetrahedral or octahedral complexes in both the fields. This is because of the reason that Cu^{2+} ion (d^9 system) has a much higher CFSE value in a square planar configuration (CFSE = 1.22 Δ_0) than in octahedral (CFSE = 0.6Δ_0) or tetrahedral configuration (CFSF = 0.18 Δ_0).

(*b*) Most of the four coordinated complexes of Ni^{2+} ion (d^8 system) are square planar rather than tetrahedral [$(NiX_4)^{2-}$ is an exception, X = Cl^-, Br^-, I^-]. This is because CFSE values of d^8 ion are higher in square planar configuration (= 1.45 Δ_0) than those of the same ion in tetrahedral configuration (= 0.36 Δ_0).

C. Other applications. With the help of CFT, we can find out:

1. The number of unpaired electrons (*n*) in the central metal ion of a given complex ion and hence the value of magnetic moment (μ) of the ion. μ (in B.M.) is given by:

$$\mu = \sqrt{n(n+2)}$$

Thus, for $n = 0$, $\mu = 0.0$ (diamagnetic); $n = 1$, $\mu = 1.73$ B.M; $n = 2$, $\mu = 2.83$ B.M; $n = 3$, $\mu = 3.87$ B.M; $n = 4$, $\mu = 4.90$B.M; $n = 5$, $\mu = 5.92$ B.M.

2. Whether the given complex ion is HS or LS.

3. Whether the given complex ion is paramagnetic or diamagnetic.

All the above three applications can be understood by considering the following example.

Example. We are given two octahedral complex ions *viz.* $[CoF_6]^{3-}$ and $[Co(NH_3)_6]^{3+}$ ions. Answer the following: (*i*) Find out the number of unpaired electrons and the value of magnetic moment (μ). (*ii*) Which ion is HS and which is LS (*iii*) Which ion is paramagnetic and which is diamagnetic.

Solution. In both the complex ions Co is present as Co^{3+} ion which is a d^6 system. Now since F^- ions are weak ligands (see spectrochemical series), d^6 electrons of Co^{3+} ion in $[CoF_6]^{3-}$ ion remain unpaired ($d^6 = t_{2g}^4\, e_g^2$ with $n = 4$) and hence $[CoF_6]^{3-}$ ion is a HS (weak field) complex ion. Thus, $n = 2 + 2 = 4$ and hence:

$$\mu = \sqrt{n(n+2)} = \sqrt{4 \times 6} = 4.9 \text{ B.M.}$$

Since $(CoF_6)^{3-}$ ion has unpaired electrons ($n = 4$), it is a paramagnetic ion.

In case of $[Co(NH_3)_6]^{3+}$ ion, since NH_3 molecules are strong ligands (see spectrochemical series), d^6 electrons of Co^{3+} ion in the complex ion pair up in t_{2g} orbitals ($d^6 = t_{2g}^6\, e_g^0$ with $n = 0$) and $[Co(NH_3)_6]^{3+}$ ion is a LS (strong field) ion. Since this complex ion has all the electrons in the paired state, it is diamagnetic in nature.

Limitations of CFT

(*i*) CFT considers only the metal ion *d*-orbitals and gives no consideration at all to other metal orbitals such as *s*, p_x, p_y, and p_z orbitals and the ligand π-orbitals. Therefore, to explain all the properties of the complexes dependent on the π-ligand orbitals will be outside the scope of CFT. CFT does not consider the formation of π-bonding in complexes.

(*ii*) CFT is unable to account satisfactorily for the relative strengths of ligands, *e.g.*, it gives no explanation as to why H_2O appears in the spectrochemical series as a stronger ligand than OH^-.

(*iii*) According to CFT, the bond between the metal and ligand is purely ionic. It gives no account of the partly covalent nature of the metal–ligand bonds. Thus the effects directly dependent on covalency cannot be explained by CFT.

Experimental evidence for metal–ligand covalent bonding in complexes

The following evidences have been presented to show the metal–ligand covalent bonding in complexes.

1. Electron spin resonance (*esr*) spectra. Most direct evidence is obtained from esr spectrum of complexes, *e.g.*, *esr* spectrum of $[Ir^{IV}Cl_6]^{2-}$ ion shows that it has a complex pattern of sub-bands which is called the *hyperfine structure*. The hyperfine structure has been explained by assuming that certain of the iridium orbitals and certain orbitals of the surrounding Cl^- ions overlap to such as extent that the single unpaired *d*-electron is not localised entirely on the metal ion but instead is about 5% localised on each Cl^- ion. Such study of other complexes also gives similar results.

2. Nuclear magnetic resonance (*nmr*). nmr studies of complexes like $KMnF_3$ and $KNiF_3$ show that the metal t_{2g} and e_g electrons pass a fraction of time around the fluorine nuclei.

3. Nuclear quadrupole resonance (*nqr*). The *nqr* spectrum of some of the square planar complexes of Pt (II) and Pd (II) such as $[Pt^{II}X_4]^{2-}$ and $[Pd^{II}X_4]^{2-}$ suggest that there is considerable amount of covalency in the metal-ligand bonds (*i.e.*, Pt–X or Pd–X bonds).

4. The unusually large absorption band intensities observed for tetrahedral complexes like $[Co^{II}Cl_4]^{2-}$ have been explained by saying that the metal-ligand bonds have appreciable covalent character.

Comparison between VBT and CFT

The points showing the comparison between the two theories are given below.

1. Inner orbital octahedral complexes of VBT are the same as the spin-paired or low-spin octahedral complexes of CFT. Similarly outer-orbital complexes of VBT are the same as the spin-free or high-spin octahedral complexes of CFT.

2. In the formation of some inner-orbital octahedral complexes of VBT, the promotion of an electron from *d*-orbital to *s*-orbital is required, while in the formation of spin-paired octahedral complexes of CFT no such promotion is required.

3. According to VBT, the metal-ligand bonding in complexes is only covalent, since VBT assumes that ligand electrons are donated to the vacant *d*-orbitals on the central cation. On the other hand, CFT considers the bonding to be entirely electrostatic. Thus, CFT does not allow the ligand electrons to enter the metal *d*-orbitals.

SOME TYPICAL PROBLEMS WITH SOLUTIONS

Problem 1. *Using VBT for $[Ag(NH_3)_2]^+$ and $[Ag(CN)_2]^-$ ions, predict type of hybridisation involved, geometry of the ions and magnetic moment.*

Solution. In both the given complex ions, Ag has O.S. = + 1, *i.e.* the central metal ion is Ag^+ which is $4d^{10}$ ion. Since C.N. of Ag^+ is equal to 2, the given complex ions have linear geometry which arises due to *sp* hybridisation of Ag^+ ion as shown in Fig 5.26.

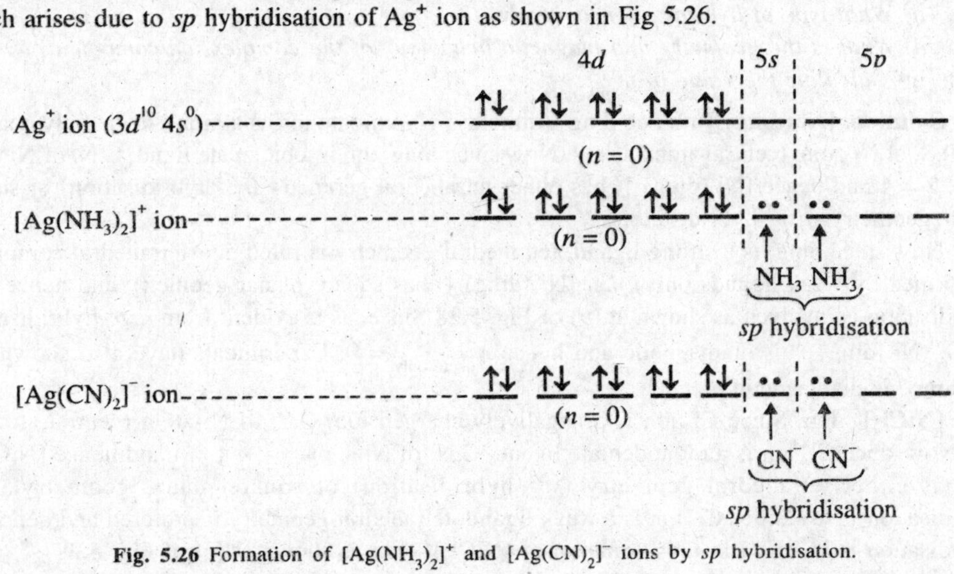

4*d* | 5*s* | 5*p*

Ag^+ ion $(3d^{10}\ 4s^0)$ ---- ↑↓ ↑↓ ↑↓ ↑↓ ↑↓ | __ | __ __ __
(*n* = 0)

$[Ag(NH_3)_2]^+$ ion ---- ↑↓ ↑↓ ↑↓ ↑↓ ↑↓ | •• | •• __ __
(*n* = 0)
NH_3 NH_3
sp hybridisation

$[Ag(CN)_2]^-$ ion ---- ↑↓ ↑↓ ↑↓ ↑↓ ↑↓ | •• | •• __ __
(*n* = 0)
CN^- CN^-
sp hybridisation

Fig. 5.26 Formation of $[Ag(NH_3)_2]^+$ and $[Ag(CN)_2]^-$ ions by *sp* hybridisation.

Since all the electrons in both the complex ions are in the paired state ($n = 0$), both the ions are diamagnetic and hence their magnetic value is zero.

Problem 2. *Discuss the geometry of $[HgI_3]^-$ ion on the basis of valence bond theory.*

Solution. $[HgI_3]^-$ ion is formed when HgI_2 reacts with KI solution

$$KI + HgI_2 \longrightarrow K[HgI_3]$$

or

$$I^- + HgI_2 \longrightarrow [HgI_3]$$

In $[HgI_3]^-$ ion, Hg is present as Hg^{2+} ion whose valence shell configuration is $5d^{10}6s^0$. Since C.N. of Hg^{2+} ion is equal to 3, $[HgI_3]^-$ ion has trigonal planar geometry which arises because of sp^2 hybridisation of Hg^{2+} ion as shown in Fig. 5.27. This figure shows that $[HgI_3]^-$ ion has no unpaired electron ($n = 0$) and hence is diamagnetic.

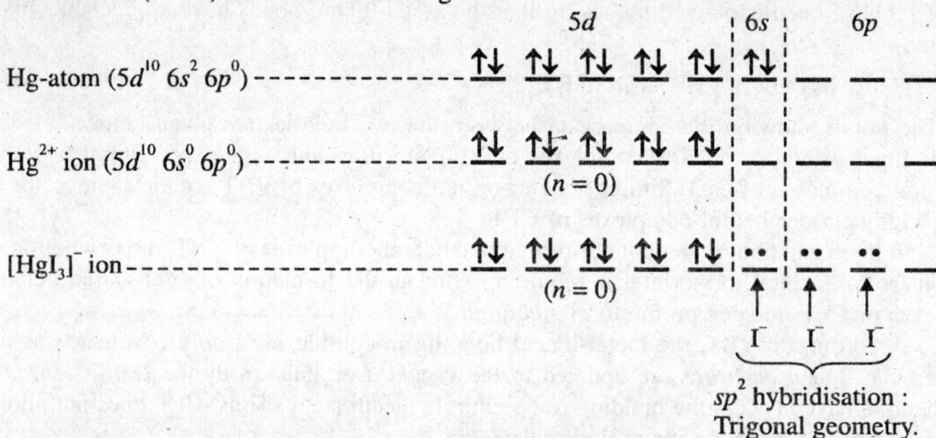

Fig. 5.27 Trigonal geometry of $[HgI_3]^-$ ion which arises becauses of sp^2 hybridasation of Hg^{2+} ion.

Problem 3. *On the basis of VBT, answer the following questions for the 4-coordinated complex compounds/ions: $[Ni\ (dmg)_2]^0$, $[Ni\ Cl_4]^{2-}$, $[Ni\ (CN)_4]^{2-}$, $Ni\ (CO)_4$, $[Cu\ (NH_3)_4]^{2+}$, $[Cu\ (CN)_4]^{2-}$, $[MX_4]^{2-}$, (X = halide ion), $[CoX_4]^{2-}$, (X = Cl, Br, I, NCS), $[Zn\ (CN)_4]^{2-}$, $[Zn\ (NH_3)_4]^{2+}$, $[Zn\ Cl_2\ (NH_3)_2]$.*

(i) What is the O.S. of the central metal atom/ion.

(ii) What type of hybridisation is involved?

(iii) What is the geometry and magnetic behaviour of the complex ion/compound?

(iv) Calculate the value of $m_{spin\ only}$.

Solution. $[Ni\ (dmg)_2]^0$. Since dmg^- (dimethyl glyoximate ion)is a mono negatively charged ion, O.S. of Ni atom (central atom) is + 2. Now since dmg^- ion is a bidentate lignd, C.N. of Ni^{2+} ion $= 2 \times 2 = 4$ and hence $[Ni\ (dmg)_2]^0$ has either tetrahedral geometry (sp^3 hybridisation) or square planar geometry (dsp^2 hybridisation).

Now since dmg^- is a strong ligand, tetrahedral geometry is ruled out (tetrahedral complexes are formed by weak ligands only), *i.e.* $[Ni\ (dmg)_2]^0$ has square planar geometry and hence dsp^2 hybridisation is involved as shown at (*b*) of Fig. 5.28 Since, as is evident from dsp^2 hybridisation, $n = 0$, $(Ni\ (dmg)_2)^0$ is diamagnetic and hence $\mu_{spin\ only} = 0$. Experiments have also shown that $Ni\ (dmg)_2$ is diamagnetic.

$[NiCl_4]^{2-}$ ion. Since Cl^- ion is a negatively charged ion, O.S. of Ni-atom (central atom) is + 2. Now since Cl^- ion is a monodentate ligand, C.N. of Ni^{2+} ion $= 1 \times 4 = 4$ and hence $[NiCl_4]^{2-}$ ion has either tetrahedral geometry (sp^3 hybridisation) or square planar geometry (dsp^2 hybridisation). Now since Cl^- ion is a weak ligand, tetrahedral geometry is prefered and hence sp^3 hybridisation is involved in the formation of $[NiCl_4]^{2-}$ ion as shown at (*c*) of Fig. 5.28

Since, as is evident from sp^3 hybridisation $n = 2$, $[NiCl_4]^{2-}$ ion is paramagnetic and hence:

$$\mu_{\text{spin only}} = \sqrt{n(n+2)} \text{ BM} = \sqrt{2\times(2+2)} \text{ BM} = 2.828 \text{ B.M.}$$

$[Ni(CN)_4]^{2-}$ ion. Since CN^- ion is a mono negatively charged ion, O.S. of Ni atom (central atom) is + 2. Now since CN^- ion is a monodentate ligand, C.N. of Ni^{2+} ion is = 1 × 4 = 4 and hence $[Ni(CN)_4]^{2-}$ ion has either tetrahedral geometry (sp^3 hybridisation) or square planar geometry (dsp^2 hybridisation). Now since CN^- ions are strong ligands, tetrahedral geometry is ruled out (tetrahedral complexes are given by weak ligands only), *i.e.* $[Ni\ (CN)_4]^{2-}$ ion has square planar geometry and hence dsp^2 hybridisation is involved as shown at (*d*) of Fig. 5.28 Since, as is evident from dsp^2 hybridisation, $n = 0$, $[Ni\ (CN)_4]^{2-}$ is diamagnetic and hence $\mu_{\text{spin only}} = 0$.

$Ni(CO)_4$. Since CO molecule is a neutral molecule, O.S. of Ni-atom (central atom) is zero. Now since CO molecule is a monodentate ligand, C.N. of Ni atom = 1 × 4 = 4 and hence Ni $(CO)_4$ has either tetrahedral shape (sp^3 hybridisation) or square planar geometry (dsp^2 hybridisation). Now since valence-shell configuration of Ni atom in Ni $(CO)_4$ is $3d^{10}4s^0$, Ni $(CO)_4$ is obtained by sp^3 hybridisation. For dsp^2 hybridisation (square planar geometry) one 3d orbital should be vacant. Thus in the formation of $Ni(CO)_4$ sp^3 hybridisation is involved and this molecule has tetrahedral shape. [See (*f*) of Fig. 5.28]. Since, as is evident from sp^3 hybridisation, $n = 0$, $Ni(CO)_4$ is diamagnetic and $\mu_{\text{spin only}}$ = Zero.

$[Cu(NH_3)_4]^{2+}$ and $[Cu(CN)_4]^{2-}$ ions. Obviously in both the given complex ions, O.S. of Cu (central metal atom) is + 2. Now since the ligands (NH_3 and CN^-) present in both the ions are strong ligands and C.N. of Cu^{2+} is 4, both the ions have the same geometry which may be either tetrahedral (sp^3 hybridisation) or square planar (dsp^2 or sp^2d hybridisation). As an example let us discuss the geometry of $[Cu\ (NH_3)_4]^{2+}$ ion. Since NH_3 molecules which act as ligands in $[Cu\ (NH_3)_4]^{2+}$ are strong ligands, tetrahedral geometry is ruled out (tetrahedral complexes are given by weak ligands only), *i.e.* $[Cu\ (NH_3)_4]^{2+}$ ion has square planar geometry which involves dsp^2 or sp^2d hybridisation of Cu^{2+} ion. Huggin has suggested that, since no $3d$ orbital in the valence shell of Cu^{2+} ion ($Cu^{2+} = 3d^9\ 4s^0\ 4p^0$) is vacant, $4d$ orbital is used for $(4s)\ (4p)^2\ (4d)$ or sp^2d hybridisation which gives square planar geometry to $[Cu\ (NH_3)_4]^{2+}$ ion as shown as (g) of Fig. 5.28 Since, as is evident from sp^2d hybridisation, $n = 1$ (present in $3d$ orbital), $[Cu\ (NH_3)_4]^{2+}$ is paramagnetic and hence :

$$\mu_{\text{spin only}} = \sqrt{n(n+2)} \text{ BM} = \sqrt{1\times(1+2)} \text{ BM}$$

$$= \sqrt{1\times 3} \text{ BM} = 1.73 \text{ BM}$$

$[MnX_4]^{2-}$ ions (X = halide ion). Since X^- ion (ligand) has one negative charge on it, O.S. of Mn-atom (central atom) in $[MnX_4]^{2-}$ ion is + 2. Now, since C.N. of Mn^{2+} ion (central metal ion) is 4, $[MnX_4]^{2-}$ ion may have either tetrahedral (sp^3 hybridisation) or square planar geometry (dsp^2 hybridisation). Since X^- ion (halide ion) is a weak ligand, $[MnX_4]^{2-}$ ion has tetrahedral geometry which involves sp^3 hybridisation as shown at (*h*) of Fig. 5.28 Since $n = 5$,

$$\mu_{\text{spin only}} = \sqrt{n(n+2)} \text{ BM} = \sqrt{5\times(5+2)} \text{ BM}$$

$$= 5.916 \text{ BM}$$

$[CoX_4]^{2-}$ ions (X = Cl, Br, I, NCS). In $[CoX_4]^{2-}$ ions since X^- ion is a mono negatively charged ion, O.S. of Co-atom (central metal atom) is + 2. Now since X^- ion is a monodentate ligand, C.N. of Co^{2+} ion (central metal ion) = 1 × 4 = 4. Thus $[CoX_4]^{2-}$ ion may have either tetrahedral geometry (sp^3 hybridisation) or square planar geometry (dsp^2 hybridisation). Now since X^- ion ($X^- = Cl^-, Br^-, I^-, NCS^-$) is a weak ligand, $[CoX_4]^{2-}$ ion has tetrahedral geometry which involves sp^3 hybridisation as shown at (*i*) of Fig. 5.28 Since $n = 3$,

$$\mu_{\text{spin only}} = \sqrt{3(3+2)} \text{ BM} = 3.872 \text{ BM.}$$

$[Zn[CN]_4]^{2-}$, $[Zn(NH_3)_4]^{2+}$ and $[Zn\,Cl_2(NH_3)_2]$. Obviously in all these species, O.S. of Zn atom (central atom) is + 2. Since C.N. of Zn^{2+} ion (central metal ion) is 4 in all the given species, each of the species has either square planar or tetrahedral geometry. Since the valence shell of Zn^{2+} ion has all the five 3*d*-orbitals completely-filled with electrons ($Zn^{2+} = 3d^{10}\,4s^0\,4p^0$), no *d*-orbital (3*d*-orbital) can be made available for dsp^2 hybridisation. Thus these species cannot have square planar geometry. Due to availability of vacant 4*s* (one) and 4*p* (three) orbitals, each of these species involves sp^3 hybridisation which gives tetrahedral geometry to the species with $n = 0$. Thus each of the given species has tetrahedral geometry and is diamagnetic with $\mu_{\text{spin only}} = 0$. Structure of $[Zn(NH_3)_4]^{2+}$ ion is given at (*j*) of Fig 5.28

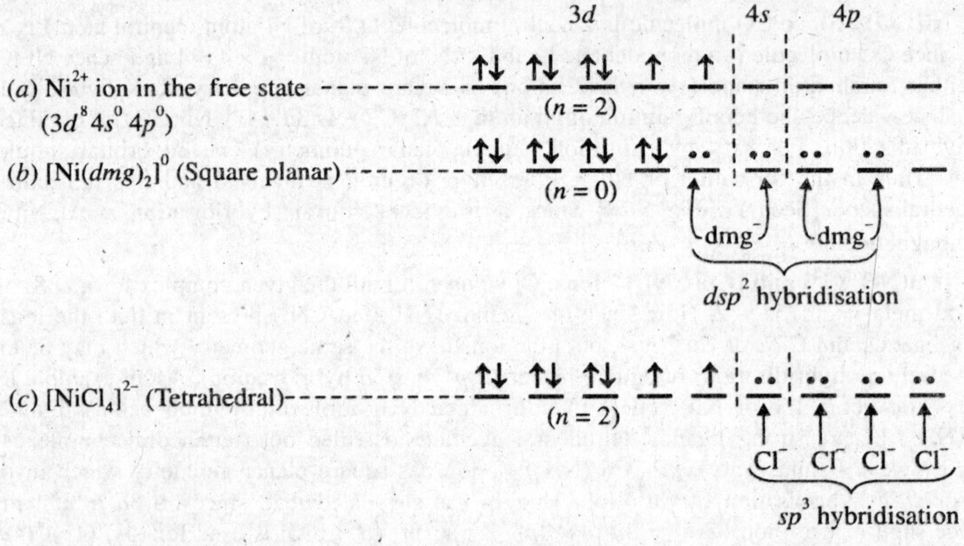

3*d* | 4*s* | 4*p*

(*a*) Ni^{2+} ion in the free state ($3d^8\,4s^0\,4p^0$) ---------- ↑↓ ↑↓ ↑↓ ↑ ↑ | — | — — — ($n = 2$)

(*b*) $[Ni(dmg)_2]^0$ (Square planar) ---------- ↑↓ ↑↓ ↑↓ ↑↓ •• | •• | •• •• — ($n = 0$)
dmg⁻ dmg⁻
dsp^2 hybridisation

(*c*) $[NiCl_4]^{2-}$ (Tetrahedral) ---------- ↑↓ ↑↓ ↑↓ ↑ ↑ | •• | •• •• •• ($n = 2$)
Cl^- Cl^- Cl^- Cl^-
sp^3 hybridisation

(*d*) $[Ni(CN)_4]^{2-}$ (Square planar) ---------- ↑↓ ↑↓ ↑↓ ↑↓ •• | •• | •• •• — ($n = 0$)
CN^- CN^- CN^- CN^-
dsp^2 hybridisation

(*e*) Ni-atom in $Ni(CO)_4$ ($3d^{10}\,4s^0\,4p^0$) ---------- ↑↓ ↑↓ ↑↓ ↑↓ ↑↓ | — | — — — ($n = 0$)

(*f*) $Ni(CO)_4$ (Tetrahedral) ---------- ↑↓ ↑↓ ↑↓ ↑↓ ↑↓ | •• | •• •• •• ($n = 0$)
CO CO CO CO
sp^3 hybridisation

Fig. 5.28. Continued on the next page

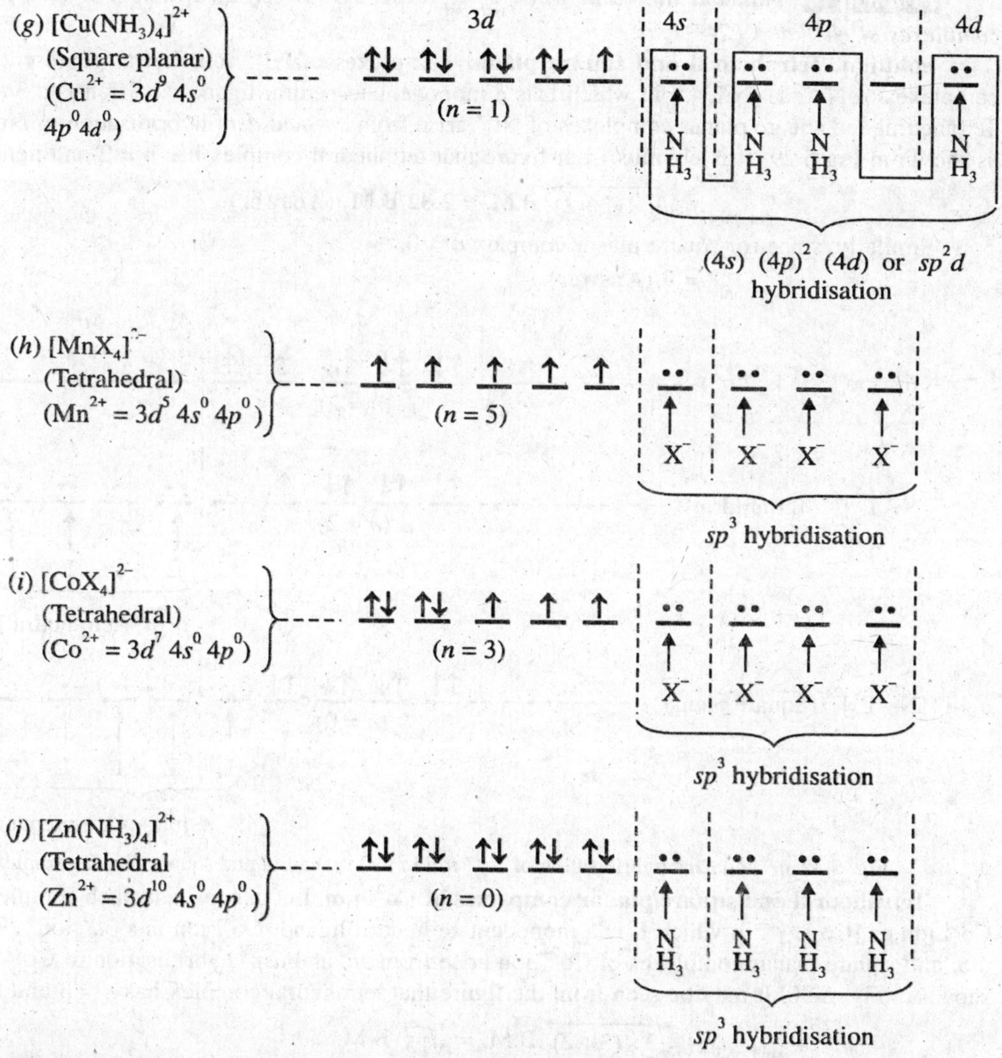

(g) $[Cu(NH_3)_4]^{2+}$ (Square planar) ($Cu^{2+} = 3d^9\ 4s^0\ 4p^0 4d^0$) — 3d: ↑↓ ↑↓ ↑↓ ↑↓ ↑ ($n = 1$); 4s, 4p, 4d: •• from NH_3 (4 ligands) — $(4s)\ (4p)^2\ (4d)$ or sp^2d hybridisation

(h) $[MnX_4]^{2-}$ (Tetrahedral) ($Mn^{2+} = 3d^5\ 4s^0\ 4p^0$) — ↑ ↑ ↑ ↑ ↑ ($n = 5$); •• from X^- X^- X^- X^- — sp^3 hybridisation

(i) $[CoX_4]^{2-}$ (Tetrahedral) ($Co^{2+} = 3d^7\ 4s^0\ 4p^0$) — ↑↓ ↑↓ ↑ ↑ ↑ ($n = 3$); •• from X^- X^- X^- X^- — sp^3 hybridisation

(j) $[Zn(NH_3)_4]^{2+}$ (Tetrahedral ($Zn^{2+} = 3d^{10}\ 4s^0\ 4p^0$) — ↑↓ ↑↓ ↑↓ ↑↓ ↑↓ ($n = 0$); •• from NH_3 NH_3 NH_3 NH_3 — sp^3 hybridisation

Fig. 5.28 Type of hybridisation involved and geometry of some 4-coordinated complex ions/compounds.

Problem 4. *On the basis of VBT, predict the number of unpaired electrons in $[NiCl_4]^{2-}$, $[Ni(CN)_4]^{2-}$ and $[Cu\ (NH_3)_4]^{2+}$ ions.* *(Delhi 1997, 98)*

Solution. The number of unpaired electrons in the given complex ions are as $[Ni\ Cl_4]^{2-} = 2$, $[Ni(CN)_4]^{2-} = 0$ and $[Cu\ (NH_3)_4]^{2+} = 1$. For detailed answer see problem number 3

Problem 5. *On the basis of VBT, explain the following:*

(i) $[NiCl_4]^{2-}$ and Ni $(CO)_4$ both are tetrahedral but differ in their magnetic moments.

(ii) A square planar complex of divalent metal having the outer configuration as $ns^1.(n-1)d^{10}$ is not oxidised. *(M.D. Rohtak 92)*

Solution. (*i*) See problem 3

(*ii*) $ns^1.(n-1)\ d^{10}$ is the outer configuration of an atom of Cu-group elements. X-ray studies have shown that $[Cu^{2+}\ (NH_3)_4]^{2+}$ ion has square planar geometry and cannot be oxidised to $[Cu^{3+}(NH_3)_4]^{3+}$. In order to explain the formation and square planar geometry of $[Cu\ (NH_3)_4]^{2+}$ ion see problem 3.

Problem 6. *Find out the value of $\mu_{spin\ only}$ expected for tetrahedral and square planar complexes of Ni^{2+} and Co^{2+} ions.*

Solution. Tetrahedral and square planar complexes of Ni^{2+} ion. Let us represent such complexes of Ni^{2+} as $[NiL_4]^{2+}$ in which L is a monodentate neutral ligand. Ni^{2+} ion is a $3d^8$ ion. Tetrahedral and square planar complexes of Ni^{2+} arise from sp^3 and dsp^2 hybridisation of Ni^{2+} ion as shown in Fig 5.29. It is obvious from figure that tetrahedral complex has $n = 2$ and hence:

$$\mu_{\text{spin only}} = \sqrt{2\times(2+2)}\ \text{B.M.} = \mathbf{2.82\ B.M.\ (Answer)}$$

Similarly, since for square planar complex $n = 0$,

$$\mu_{\text{spin only}} = \mathbf{0\ (Answer)}$$

	4d					4s	4p		
Ni^{2+} ion ($3d^8\,4s^0\,4p^0$)	↑↓	↑↓	↑↓	↑	↑	—	—	—	—
	$(n = 2)$								
$[Ni^{2+}L_4]^{2+}$ (Tetrahldral)	↑↓	↑↓	↑↓	↑	↑	•• L	•• L	•• L	•• L
	$(n = 2)$					sp^3 (tetrahedral)			
$[Ni^{2+}L_4]^{2+}$ (Square planar)	↑↓	↑↓	↑↓	↑↓	•• L	•• L	•• L	•• L	—
	$(n = 0)$				dsp^2 (square planar)				

Fig. 5.29 sp^3 and dsp^2 hybridisation of Ni^{2+} ion in its tetrahedral and square planar complexes

Tetrahedral and square planar complexes of Co^{2+} ion. Let us represent such complexes of Co^{2+} ion as $[Co\ L_4]^{2+}$ in which L is a monodentate neutral ligand. Co^{2+} ion is a $3d^7$ ion. Tetrahedral and square planar complexes of Co^{2+} ion arise from sp^3 and dsp^2 hybridisation of Co^{2+} ion as shown in Fig. 5.30. It may be seen from the figure that tetrahedral complex has $n = 3$ and hence:

$$\mu_{\text{spin only}} = \sqrt{3\times(3+2)}\ \text{B.M.} = \sqrt{15}\ \text{B.M.}$$
$$= \mathbf{3.87\ B.M\ (Answer)}$$

Similarly, since for square planar complex $n = 1$,

$$\mu_{\text{spin only}} = \sqrt{1\times(1+2)}\ \text{B.M.} = \mathbf{1.73\ B.M.\ (Answer)}$$

	3d					4s	4p		
Co^{2+} ion ($3d^7$)	↑↓	↑↓	↑	↑	↑	—	—	—	—
	$(n = 3)$								
$[Co^{2+}L_4]^{2+}$ (Tetrahedral)	↑↓	↑↓	↑	↑	↑	•• L	•• L	•• L	•• L
	$(n = 3)$					sp^3 (tetrehedral)			

Fig. 5.30 Continued on the next page

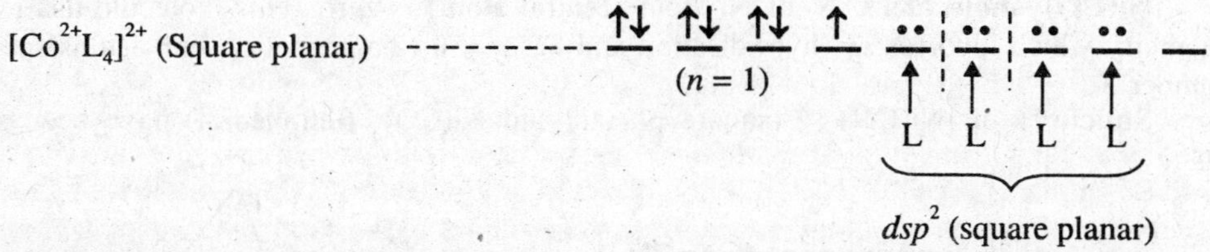

$[Co^{2+}L_4]^{2+}$ (Square planar) ---------- ↑↓ ↑↓ ↑↓ ↑ •• •• •• •• —
(n = 1)
L L L L
dsp^2 (square planar)

Fig. 5.30. sp^3 and dsp^2 hybridisation of Co^{2+} ion in its tetrahedral and square planar complexes

Problem 7. *Magnetic moment value of $[MnBr_4]^{2-}$ ion is 5.9 B.M. On the basis of VBT, predict the type of hybridisation and geometry of the ion.*

Solution. We know that:

$$\mu = \sqrt{n(n+2)} \text{ B.M.}$$

If we put $n = 5$ in the above equation, we get $\mu = \sqrt{35}$ B.M. = 5.91 B.M. Thus $[Mn\ Br_4]^{2-}$ ion has five unpaired electrons ($n = 5$). In $[Mn\ Br_4]^{2-}$ ion, the central atom is Mn^{2+} ion which is $3d^5$ ion. Now since C.N. of Mn^{2+} = 4, $[Mn\ Br_4]^{2-}$ ion may have either square planar (dsp^2 hybridisation) or tetrahedral (sp^3 hybridisation) geometry as shown in Fig. 5.31.

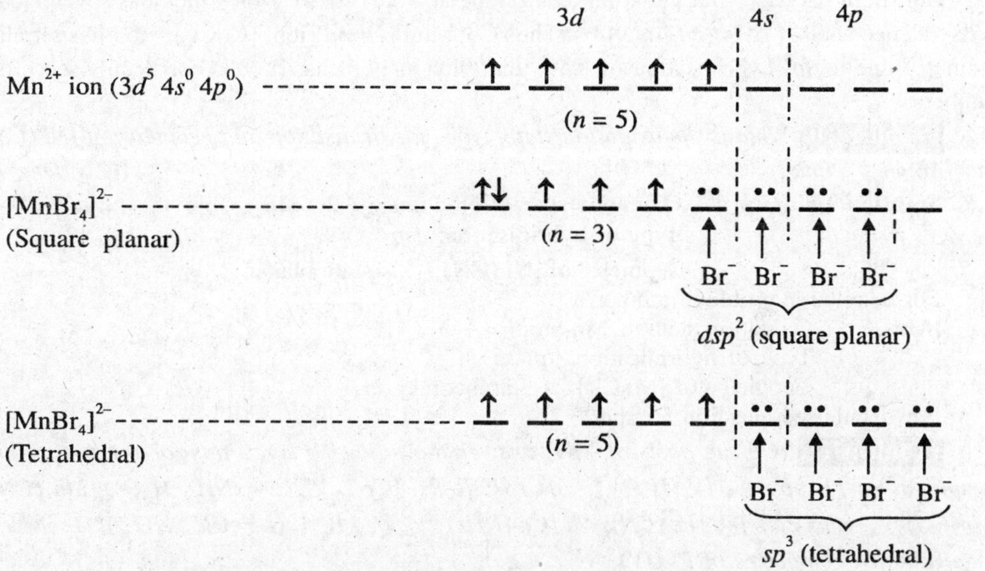

3d 4s 4p

Mn^{2+} ion ($3d^5\ 4s^0\ 4p^0$) ---------- ↑ ↑ ↑ ↑ ↑ | — | — — —
(n = 5)

$[MnBr_4]^{2-}$ (Square planar) ---------- ↑↓ ↑ ↑ ↑ •• | •• | •• •• | —
(n = 3)
Br^- Br^- Br^- Br^-
dsp^2 (square planar)

$[MnBr_4]^{2-}$ (Tetrahedral) ---------- ↑ ↑ ↑ ↑ ↑ | •• | •• •• ••
(n = 5)
Br^- Br^- Br^- Br^-
sp^3 (tetrahedral)

Fig. 5.31 Square planar and tetrahedral geometries of $[Mn\ Br_4]^{2-}$ ion.

Since dsp^2 hybridisation (square planar) gives n = 3 and sp^3 hybridisation (tetrahedral) gives $n = 5$, $[MnBr_4]^{2-}$ ion has tetrahedral geometry and not square planar. Alternatively, since Br^- ion is a weak field ligand, $[MnBr_4]^{2-}$ ion is tetrahedral in geometry. Tetrahedral complexes are given by weak field ligands, since these are HS complexes.

Problem 8. *Draw the structures of $[Ni\ (CN)_4]^{2-}$ and $Ni\ (CO)_4$. Write the hybridisation of atomic orbitals of transition metal in each case.* *(IIT 2000)*

Solution. $[Ni\ (CN)_4]^{2-}$ ion. O.S. of Ni-atom (central atom) is equal to +2. This ion has square planar geometry which involves dsp^2 hybridisation and has n = 0 (diamagnetic). For details see problem number 3

$Ni(CO)_4$ molecule. O.S. of Ni-atom (central atom) is zero. This molecule has tetrahedral geometry which involves sp^3 hybridisation and has $n = 0$ (diamagnetic). For details see problem number 3

Structures of $[Ni(CN)_4]^{2-}$ (square planar) and $Ni(CO)_4$ (tetrahedral) have been shown in Fig. 5.32.

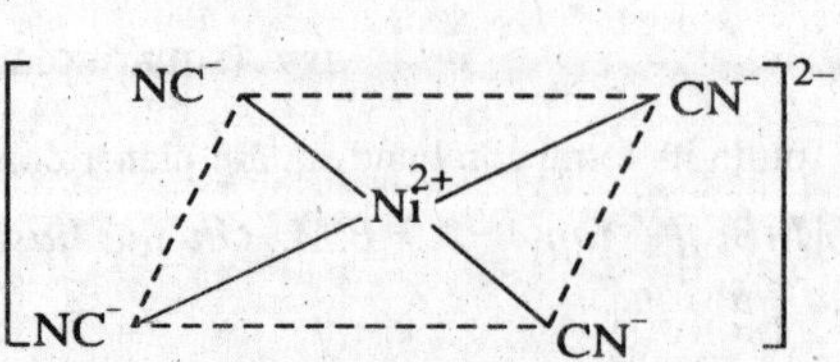

Square planar structure of $[Ni(CN)_4]^{2-}$ ion resulted from dsp^2 hybridisation of Ni^{2+} ion ($Ni^{2+} = 3d^8 4s^0 4p^0$)

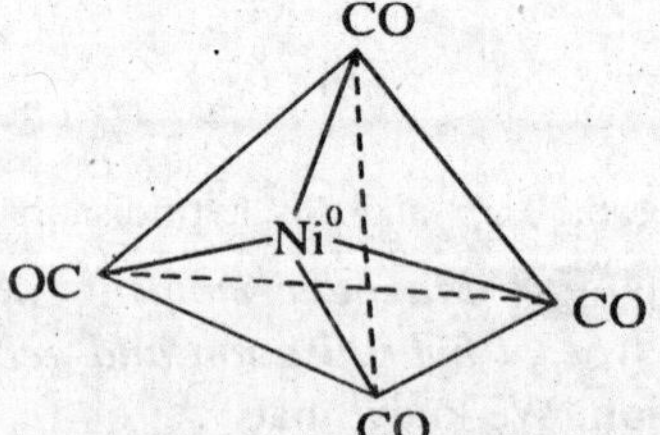

Tetrahedral geometry of $Ni(CO)_4$ resulted from sp^3 hybridisation of Ni-atom ($Ni = 3d^8\ 4s^2\ 4p^0 = 3d^{10}\ 4s^0 4p^0$)

Fig. 5.32 Structures of $[Ni(CN)_4]^{2-}$ and $Ni(CO)_4$.

Problem 9. *On the basis of VBT, determine the number of unpaired electrons in $[CoCl_4]^{2-}$ ion.* *(Kurukheshtra 97)*

Solution. $[CoCl_4]^{2-}$ has Co-atom (central metal atom) as Co^{2+} ion which has seven electrons in its valence-shell ($Co^{2+} = 3d^7\ 4s^0\ 4p^0$). Thus Co^{2+} ion is a $3d^7$ ion. $[CoCl_4]^{2-}$ ion has tetrahedral geometry due to sp^3 hybridisation of Co^{2+} ion. Obviously, it has $n = 3$. For details see problem number 3

Problem 10. *What is the oxidation state, type of hybridisation and geometry of $[Ni(CN)_4]^{2-}$ and $[MnCl_4]^{2-}$ ions?* *(G.N. Dev, 1996)*

Solution. $[Ni(CN)_4]^{2-}$. Oxidation state of Ni-atom = + 2
Type of hybridisation: dsp^2
Geometry of $[Ni(CN)_4]^{2-}$: square planar.

For details see problem number 3

$[MnCl_4]^{2-}$. Oxidation state of Mn-atom = + 2
Type of hybridisation: sp^3
Geometry of $[MnCl_4]^{2-}$: Tetrahedral

For details see question number 3

Problem 11. *On the basis of VBT, answer the following questions for the 6-coordinated complex ions: $[Ti(bpy)_3]^-$, $[V(H_2O)_6]^{3+}$, $[Cr(CN)_6]^{3-}$, $[CrL_6]^{3+}$ ($L = NH_3$, H_2O), $[Mn(CN)_6]^{4-}$, $[Mn(CN)_6]^{3-}$, $[Fe(CN)_6]^{4-}$, $[Fe(CN)_6]^{3-}$, $[Co(NH_3)_6]^{3+}$, $[CoL_6]^{3-}$ ($L = CN^-$, NO_2^-), $[Ir(NH_3)_6]^{3+}$, 6-coordinated complexes of Pt (IV).*

(a) What type of hybridisation is involved?

(b) Whether the given complex ion is inner orbital octahedral or outer orbital octahedral?

(c) What is the magnetic behaviour of the complex ion?

(d) Calculate the value of $\mu_{spin\ only}$.

[G.N. Dev 95, Raj. 96, Allahabad 96, Delhi 96, Guahati 96, Kurukshetra 97, Nagpur 97, Delhi 98, IIT 2000]

Solution. $[Ti(bpy)_3]^-$ ion. In this ion, Ti atom (central metal atom) is present as Ti^- ion which is $3d^3\ 4s^2 4p^0$ or $3d^5\ 4s^0\ 4p^0$ ion ($Ti = 3d^2\ 4s^2 4p^0$, $Ti^- = 3d^3\ 4s^2\ 4p^0 = 3d^5\ 4s^0\ 4p^0$). In the free state Ti^- ion has 5 unpaired electrons ($n = 5$). Now since *bpy* molecules are strong ligands, unpaired electrons present in $3d$ orbitals of Ti^- ion get paired so that n becomes equal to 1 ($n = 1$) and two $3d$ orbitals become vacant. These two vacant $3d$ orbitals are used for d^2sp^3 hybridisation (inner orbital octahedral hybridisation). Thus $[Ti(bpy)_3]^-$ ion is paramagnetic due to the presence of one unpaired electron ($n = 1$). Obviously,

$$\mu_{\text{spin only}} = \sqrt{n(n+2)}\ \text{BM} = \sqrt{1\times(1+2)}\ \text{BM}$$
$$= 1.732\ \text{B.M.}$$

d^2sp^3 hybridisation taking place in $[Ti\ (bpy)_3]^-$ ion has been shown in Fig. 5.33.

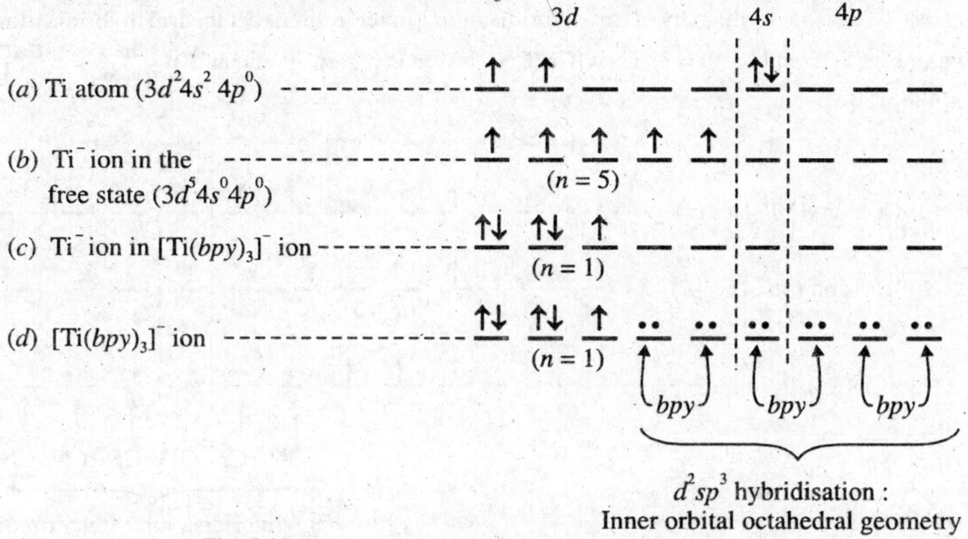

3d 4s 4p

(a) Ti atom ($3d^2 4s^2 4p^0$) ↑ ↑ _ _ _ | ↑↓ | _ _ _

(b) Ti^- ion in the free state ($3d^5 4s^0 4p^0$) ↑ ↑ ↑ ↑ ↑ | _ | _ _ _ ($n = 5$)

(c) Ti^- ion in $[Ti(bpy)_3]^-$ ion ↑↓ ↑↓ ↑ _ _ | _ | _ _ _ ($n = 1$)

(d) $[Ti(bpy)_3]^-$ ion ↑↓ ↑↓ ↑ •• •• | •• | •• •• •• ($n = 1$)

bpy bpy bpy

d^2sp^3 hybridisation : Inner orbital octahedral geometry

Fig. 5.33. Inner orbital octahedral geometry of $[Ti(bpy)_3]^-$ ion

$[V(H_2O)_6]^{3+}$ ion. In this complex ion, V-atom (central metal atom) is present as V^{3+} ion which is $3d^2$ ion ($V = 3d^3\ 4s^2\ 4p^0$, $V^{3+} = 3d^2\ 4s^0\ 4p^0$). In the free state, V^{3+} ion has two unpaired electrons ($n = 2$). In $[V\ (H_2O)_6]^{3+}$ ion, the arrangement of two unpaired electrons does not undergo any change, *i.e.*, even in $[V\ (H_2O)_6]^{3+}$ ion V^{3+} ion has the same number of unpaired electrons ($n = 2$). Due to the availability of two vacant $3d$ orbitals, $[V\ (H_2O)_6]^{2+}$ ion is formed by d^2sp^3 hybridisation (inner orbital octahedral hybridisation) with $n = 2$.

$$\mu_{\text{spin only}} = \sqrt{n(n+2)}\ \text{BM} = \sqrt{2\times(2+2)}\ \text{BM} = 2.828\ \text{BM}$$

d^2sp^3 hybridisation involved in the formation of $[V(H_2O)_6]^{3+}$ ion has been shown in Fig. 5.34.

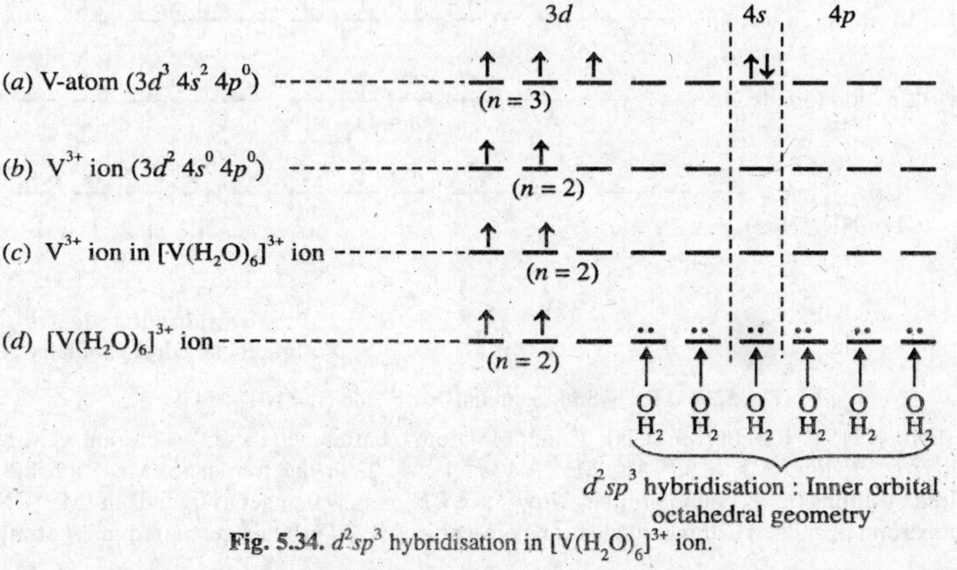

3d 4s 4p

(a) V-atom ($3d^3\ 4s^2\ 4p^0$) ↑ ↑ ↑ _ _ | ↑↓ | _ _ _ ($n = 3$)

(b) V^{3+} ion ($3d^2\ 4s^0\ 4p^0$) ↑ ↑ _ _ _ | _ | _ _ _ ($n = 2$)

(c) V^{3+} ion in $[V(H_2O)_6]^{3+}$ ion ↑ ↑ _ _ _ | _ | _ _ _ ($n = 2$)

(d) $[V(H_2O)_6]^{3+}$ ion ↑ ↑ _ •• •• | •• | •• •• •• ($n = 2$)

OH_2 OH_2 OH_2 OH_2 OH_2 OH_2

d^2sp^3 hybridisation : Inner orbital octahedral geometry

Fig. 5.34. d^2sp^3 hybridisation in $[V(H_2O)_6]^{3+}$ ion.

$[Cr(CN)_6]^{3-}$ ion. In this ion, Cr atom (central atom) is present as Cr^{3+} ion which is a $3d^3$ ion ($Cr = 3d^5 4s^1$, $Cr^{3+} = 3d^3\,4s^0\,4p^0$). Cr^{3+} ion in its free state has three unpaired electrons ($n = 3$) in $3d$ orbitals. Since two $3d$-orbitals are vacant, these orbitals are used for d^2sp^3 hybridisation. Thus $[Cr(CN)_6]^{3-}$ ion is formed by d^2sp^3 hybridisation (inner orbital octahedral hybridisation) as shown in Fig. 5.35 and has n = 3. Thus $[Cr(CN)_6]^{3-}$ ion is paramagnetic and $\mu_{\text{spin only}} = \sqrt{3(3+2)}$ BM = 3.872 BM.

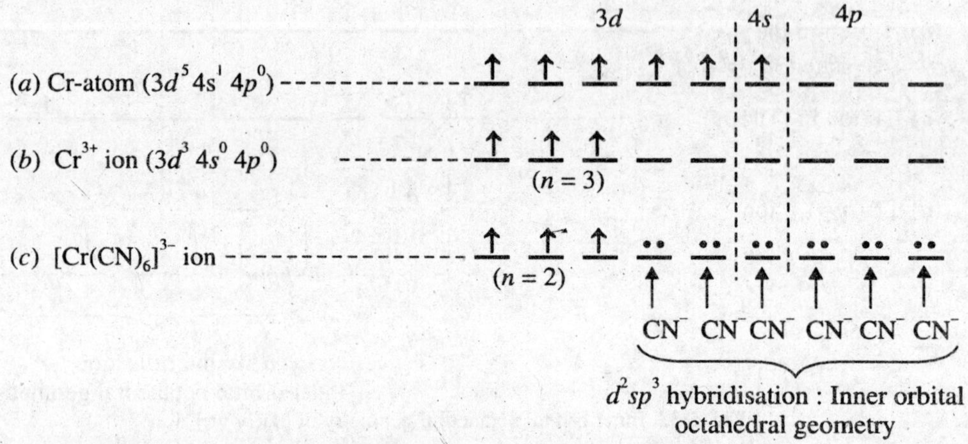

3d | 4s | 4p

(a) Cr-atom ($3d^5\,4s^1\,4p^0$) ↑ ↑ ↑ ↑ ↑ | ↑ | — — —

(b) Cr^{3+} ion ($3d^3\,4s^0\,4p^0$) ↑ ↑ ↑ — — | — | — — — ($n = 3$)

(c) $[Cr(CN)_6]^{3-}$ ion ↑ ↑ ↑ •• •• | •• | •• •• •• ($n = 2$)

CN^- CN^- CN^- CN^- CN^- CN^-

d^2sp^3 hybridisation : Inner orbital octahedral geometry

Fig. 5.35. d^2sp^3 hybridisation in $[Cr(CN)_6]^{3-}$ ion.

$[CrL_6]^{3+}$ ion ($L = NH_3, H_2O$). In $[CrL_6]^{3+}$ ion, Cr (central metal atom) is present as Cr^{3+} ion. This ion is a $3d^3$ ion which has three unpaired electrons in $3d$ orbitals ($Cr = 3d^5\,4s^1$, $Cr^{3+} = 3d^3\,4s^0\,4p^0$). Since two $3d$ orbitals are already vacant, these orbitals can be used for d^2sp^3 hybridisation. Thus $[CrL_6]^{3+}$ ion is formed by d^2sp^3 hybridisation (inner orbital octahedral hybridisation) (Fig. 5.36.) and has $n = 3$ (paramagnetic). Thus $\mu = \sqrt{3(3+2)}$ BM = 3.872 BM

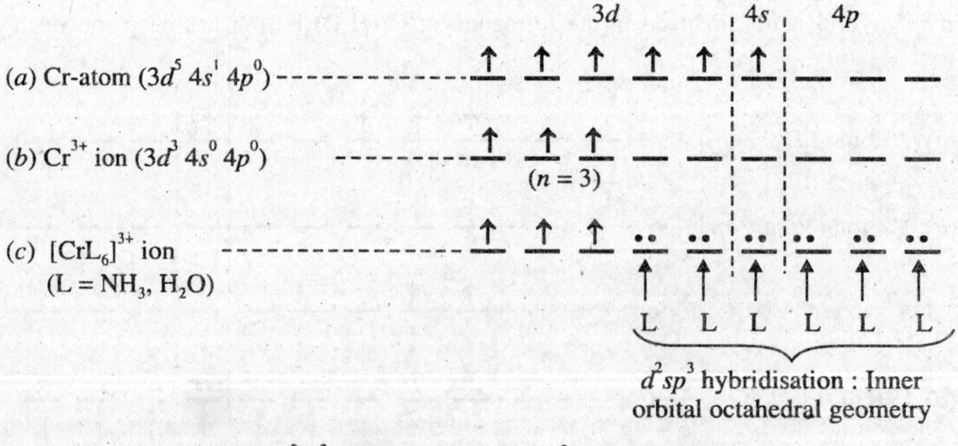

3d | 4s | 4p

(a) Cr-atom ($3d^5\,4s^1\,4p^0$) ↑ ↑ ↑ ↑ ↑ | ↑ | — — —

(b) Cr^{3+} ion ($3d^3\,4s^0\,4p^0$) ↑ ↑ ↑ — — | — | — — — ($n = 3$)

(c) $[CrL_6]^{3+}$ ion ($L = NH_3, H_2O$) ↑ ↑ ↑ •• •• | •• | •• •• ••

L L L L L L

d^2sp^3 hybridisation : Inner orbital octahedral geometry

Fig. 5.36. d^2sp^3 hybridisation in $[CrL_6]^{3+}$ ion ($L = NH_3, H_2O$)

$[Mn(CN)_6]^{4-}$ ion. In $[Mn(CN)_6]^{4-}$ ion, Mn-atom (central metal atom) is present as Mn^{2+} ion which is a $3d^5$ ion. ($Mn = 3d^5\,4s^2$, $Mn^{2+} = 3d^5\,4s^0\,4p^0$). In the free state, Mn^{2+} ion has five unpaired electrons ($n = 5$) in $3d$ orbitals. Now since CN^- ions which act as ligands in $[Mn(CN)_6]^{4-}$ ion are strong ligands, the unpaired electrons present in five $3d$ orbitals get paired in $3d$ orbitals so that n becomes equal to 1 (n =1) and two $3d$ orbitals become vavant. These vacant two $3d$ orbitals are used for d^2sp^3 hybridisation. Thus $[Mn(CN)_6]^{4-}$ ion results from d^2sp^3 hybridisation (inner orbital octahedral hybridisation) as shown in Fig. 5.37. Obviously, $[Mn(CN)_4]^{4-}$ ion has one unpaired electron (n = 1) and hence is paramagnetic. Thus $\mu_{\text{spin only}} = \sqrt{1(1+2)}$ BM = 1.732 BM.

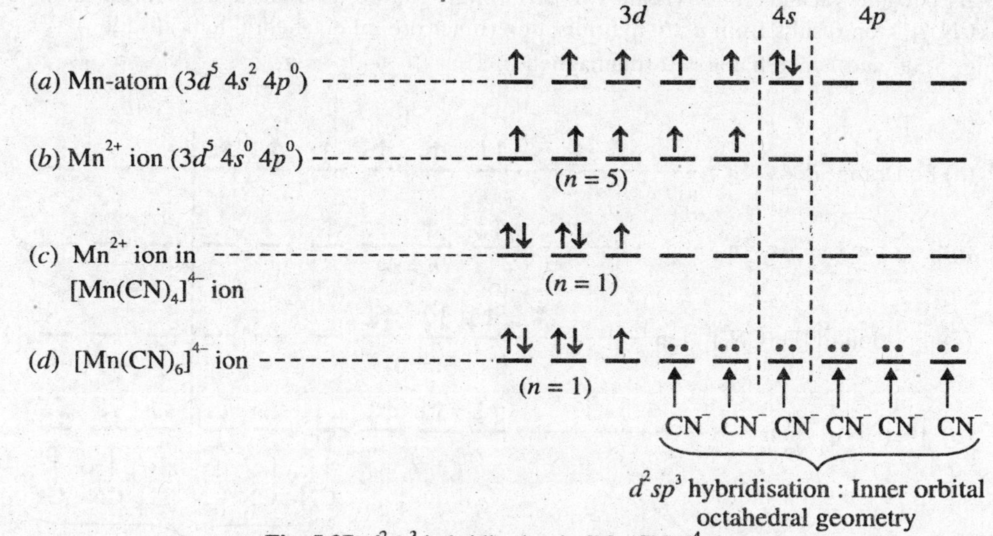

	3d					4s	4p		
(a) Mn-atom ($3d^5\,4s^2\,4p^0$)	↑	↑	↑	↑	↑	↑↓			
(b) Mn^{2+} ion ($3d^5\,4s^0\,4p^0$) (n = 5)	↑	↑	↑	↑	↑				
(c) Mn^{2+} ion in $[Mn(CN)_4]^{4-}$ ion (n = 1)	↑↓	↑↓	↑						
(d) $[Mn(CN)_6]^{4-}$ ion (n = 1)	↑↓	↑↓	↑	•• CN^-	•• CN^-	•• CN^-	•• CN^-	•• CN^-	•• CN^-

d^2sp^3 hybridisation : Inner orbital octahedral geometry

Fig. 5.37. d^2sp^3 hybridisation in $[Mn(CN)_6]^{4-}$ ion

$[Mn(CN)_6]^{3-}$ ion. In this ion, Mn (central metal atom) is present as Mn^{3+} ion which is a $3d^4$ ion (Mn = $3d^5\,4s^2$, Mn^{3+} = $3d^4\,4s^0\,4p^0$). In the free state, Mn^{3+} ion has four unpaired electrons (n = 4) in $3d$ orbitals. In the configuration of Mn^{3+} ion in the free state only one $3d$ orbital is vacant. Now since CN^- ions (ligands) are strong ligands, one unpaired electron gets shifted from one $3d$ orbital to another $3d$-orbital so that the number of unpaired electrons becomes equal to two (n = 2) and two $3d$ orbitals become vacant. These two vacant $3d$ orbitals are used for d^2sp^3 hybridisation. Thus $[Mn(CN)_6]^{3-}$ ion is inner orbital octahedral (d^2sp^3 hybridisation) complex ion with two unpaired electrons (n = 2). $\mu_{\text{spin only}} = \sqrt{n(n+2)}$ BM = $\sqrt{2\times(2+2)}$ BM = 2.828 BM. (Also See Fig. 5.38.)

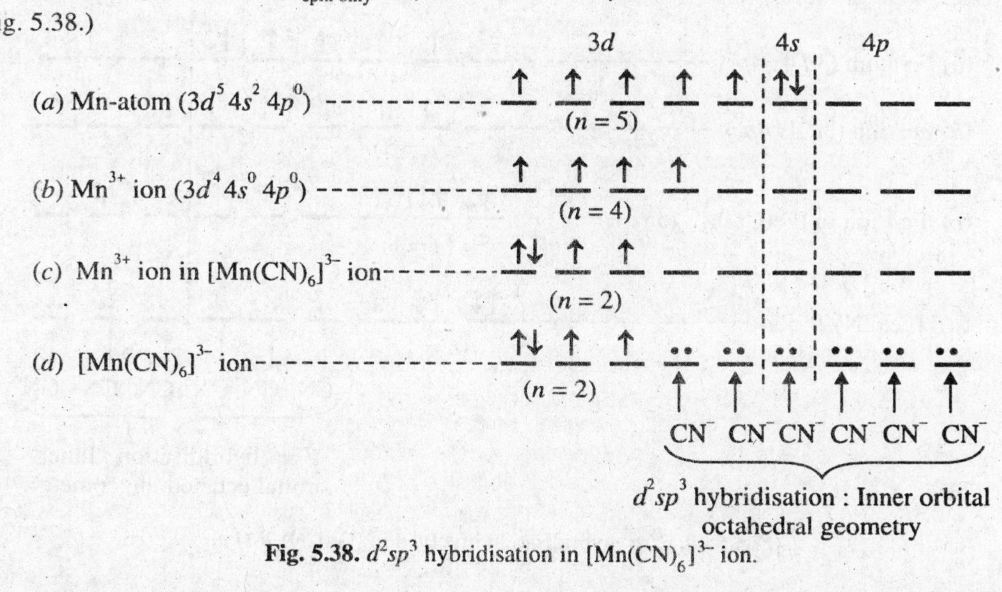

	3d					4s	4p		
(a) Mn-atom ($3d^5\,4s^2\,4p^0$) (n = 5)	↑	↑	↑	↑	↑	↑↓			
(b) Mn^{3+} ion ($3d^4\,4s^0\,4p^0$) (n = 4)	↑	↑	↑	↑					
(c) Mn^{3+} ion in $[Mn(CN)_6]^{3-}$ ion (n = 2)	↑↓	↑	↑						
(d) $[Mn(CN)_6]^{3-}$ ion (n = 2)	↑↓	↑	↑	•• CN^-	•• CN^-	•• CN^-	•• CN^-	•• CN^-	•• CN^-

d^2sp^3 hybridisation : Inner orbital octahedral geometry

Fig. 5.38. d^2sp^3 hybridisation in $[Mn(CN)_6]^{3-}$ ion.

$[Fe(CN)_6]^{4-}$ ion. In this ion, Fe is present as Fe^{2+} ion which is a $3d^6$ ion (Fe = $3d^6\ 4s^2\ 4p^0$, $Fe^{2+} = 3d^6\ 4s^0\ 4p^0$). In the free state Fe^{2+} ion has four unpaired electrons in 3*d* orbitals ($n = 4$). Now since CN^- ions which act as ligands in $[Fe(CN)_6]^{4-}$ ion are strong ligands, the unpaired electrons present in 3*d* orbitals get paired in three 3*d*-orbitals so that *n* becomes zero ($n = 0$) and two 3*d* orbitals become vacant. These vacant two 3*d* orbitals can be used for d^2sp^3 hybridisation. Thus $[Fe(CN)_6]^{4-}$ ion results from d^2sp^3 hybridisation (inner orbital octahedral hybridisation) as shown in Fig. 5.39. and hence has $n = 0$ (diamagnetic). Thus $\mu_{\text{spin only}}$ = zero.

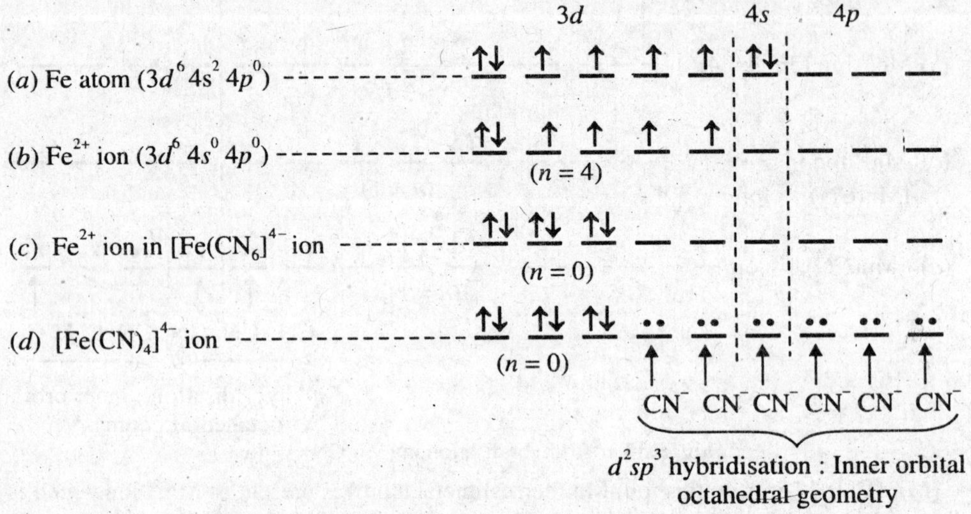

3*d* | 4*s* | 4*p*

(*a*) Fe atom ($3d^6\ 4s^2\ 4p^0$) ---------- ↑↓ ↑ ↑ ↑ ↑ | ↑↓ | — — —

(*b*) Fe^{2+} ion ($3d^6\ 4s^0\ 4p^0$) ---------- ↑↓ ↑ ↑ ↑ ↑ | — | — — — ($n = 4$)

(*c*) Fe^{2+} ion in $[Fe(CN_6]^{4-}$ ion ---------- ↑↓ ↑↓ ↑↓ — — | — | — — — ($n = 0$)

(*d*) $[Fe(CN)_4]^{4-}$ ion ---------- ↑↓ ↑↓ ↑↓ •• •• | •• | •• •• •• ($n = 0$)

CN^- CN^- CN^- CN^- CN^- CN^-

d^2sp^3 hybridisation : Inner orbital octahedral geometry

Fig. 5.39. d^2sp^3 hybridisatin involved in $[Fe(CN)_6]^{4-}$ ion

$[Fe(CN)_6]^{3-}$ ion. In this ion, Fe atom (central metal atom) is present as Fe^{3+} ion which is a $3d^5$ ion (Fe = $3d^6\ 4s^2\ 4p^0$, $Fe^{3+} = 3d^5\ 4s^0\ 4p^0$). In the free state, Fe^{3+} ion has five unpaired electrons in 3*d* orbitals ($n = 5$). Now since CN^- ions which act as ligands in $[Fe(CN)_6]^{3-}$ ion are strong ligands, the unpaired electrons present in five 3*d* orbitals get paired in three 3*d* orbitals so that *n*

3*d* | 4*s* | 4*p*

(*a*) Fe-atom ($3d^6 4s^2 4p^0$) ---------- ↑↓ ↑ ↑ ↑ ↑ | ↑↓ | — — —

(*b*) Fe^{3+} ion ($3d^5 4s^0 4p^0$) ---------- ↑ ↑ ↑ ↑ ↑ | — | — — — ($n = 5$)

(*c*) Fe^{3+} ion in $[Fe(CN)_6]^{3-}$ ion ---------- ↑↓ ↑↓ ↑ — — | — | — — — ($n = 1$)

(*d*) $[Fe(CN)_6]^{3-}$ ion ---------- ↑↓ ↑↓ ↑ •• •• | •• | •• •• •• ($n = 1$)

CN^- CN^- CN^- CN^- CN^- CN^-

d^2sp^3 hybridisation : Inner orbital octahedral geometry

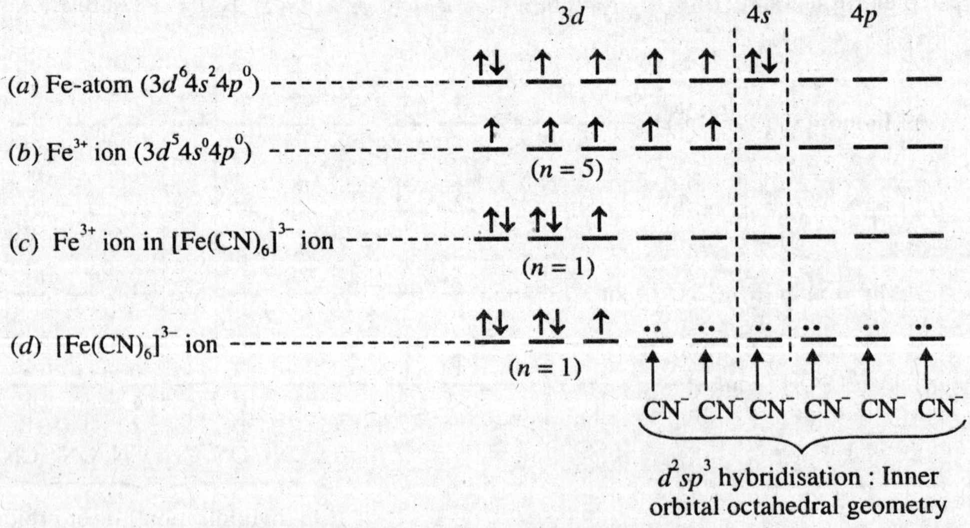

Fig. 5.40. d^2sp^3 hybridisation involved in $[Fe(CN)_6]^{3-}$ ion.

becomes equal to 1 ($n = 1$) and two $3d$ orbitals become vacant. These vacant two $3d$ orbitals can be used for d^2sp^3 hybridisation. Thus $[Fe(CN)_6]^{3-}$ ion results from d^2sp^3 hybridisation (inner orbital octahedral hybridisation) as shown in Fig. 5.40. and has $n = 1$ (paramagnetic). Thus $\mu_{\text{spin only}} = \sqrt{1(1+2)}$ BM = 1.732 BM.

$[Co(NH_3)_6]^{3+}$ ion. In $[Co(NH_3)_6]^{3+}$ ion, Co-atom (central metal atom) is present as Co^{3+} ion which is a $3d^6$ ion (Co = $3d^7 4s^2$, $Co^{3+} = 3d^6 4s^0 4p^0$). Since C.N. of Co^{3+} ion is 6, $[Co(NH_3)_6]^{3+}$ ion has octahedral geometry. In the free state Co^{3+} ion has four unpaired electrons in $3d$ orbitals ($n = 4$)(See Fig. 5.41). Now since NH_3 molecules which act as ligands in $[Co(NH_3)_6]^{3+}$ are strong ligands, the unpaired electrons present in $3d$ orbitals get paired in three $3d$ orbitals so that n becomes equal to zero ($n = 0$) and two $3d$ orbitals become vacant. These vacant two $3d$ orbitals are used for d^2sp^3 hybridisation. Thus $[Co(NH_3)_6]^{3+}$ ion results from d^2sp^3 hybridisation (inner orbital octahedral hybridisation) as shown in Fig. 5.41 and hence has $n = 0$ (diamagnetic). Experiments have also shown that $[Co(NH_3)_6]^{3+}$ is diamagnetic and $\mu_{\text{spin only}}$ = zero.

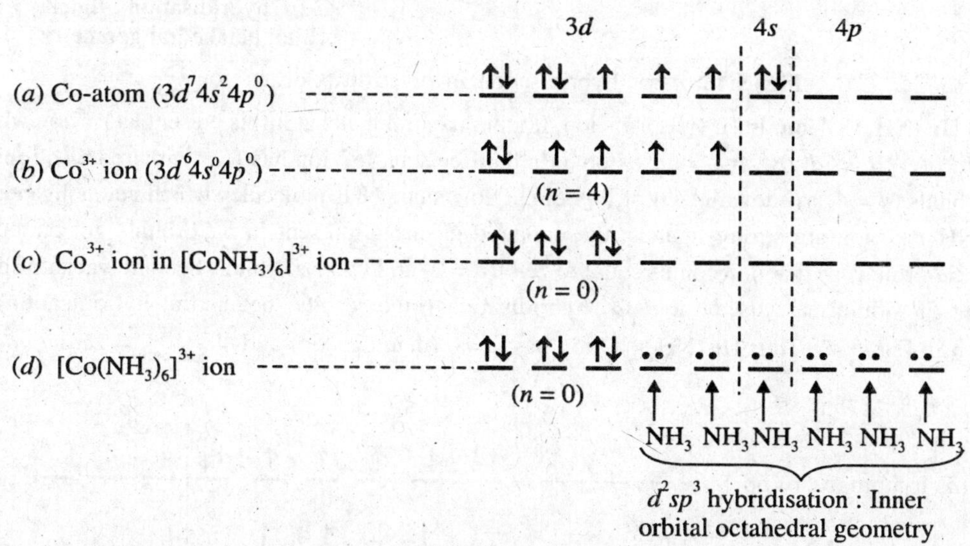

Fig. 5.41. d^2sp^3 hybridisation involved in $[Co(NH_3)]^{3+}$ ion.

$[CoL_6]^{3-}$ ions (L = CN^- or NO_2^-). Both these ions have Co-atom (central metal atom) as Co^{3+} ion which is a $3d^6$ ion (Co = $3d^7 4s^2 4p^0$, $Co^{3+} = 3d^6 4s^0 4p^0$). In the free state, Co^{3+} ion has 4 unpaired electrons in $3d$ orbitals ($n = 4$) as shown at (*b*) of Fig. 5.42 Now since CN^- and NO_2^- ions which act as ligands are strong ligands, the unpaired electrons become paired so that n becomes equal to zero ($n = 0$) and two $3d$ orbitals become vacant. These two vacant $3d$ orbitals are used for d^2sp^3 hybridisation. Thus $[Co(CN)_6]^{3-}$ and $[Co(NO_2)_6]^{3-}$ both result from d^2sp^3 hybridisation (inner orbital octahedral hybridisation) as shown at (*d*) of Fig 5.42 and hence $n = 0$ (diamagnetic).

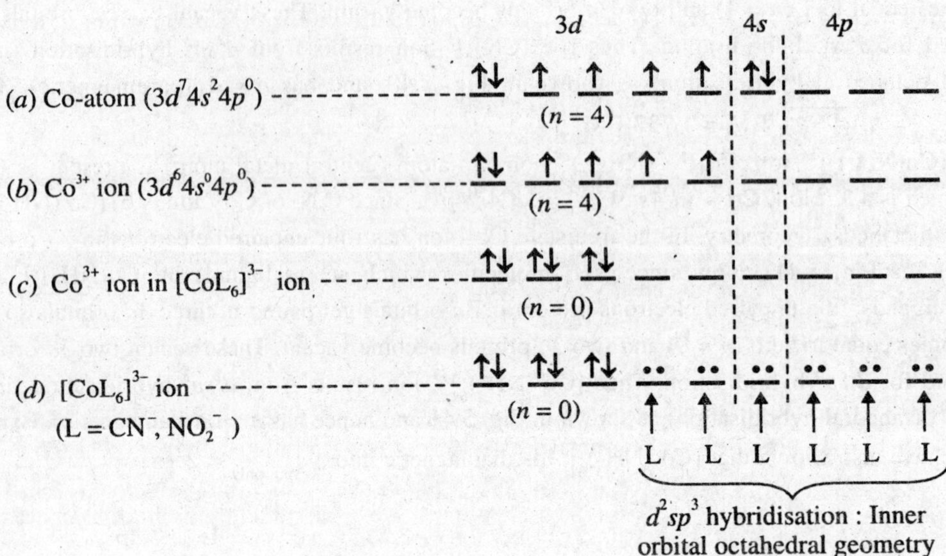

Fig. 5.42. d^2sp^3 hybridisation involved in $[CoL_6]^{3-}$ ion.

$[Ir(NH_3)_6]^{3+}$ ion. In $[Ir(NH_3)_6]^{3+}$ ion, Ir atom (central metal atom) is present as Ir^{3+} ion which is a $5d^6$ ion ($Ir = 5d^7\ 6s^2$, $Ir^{3+} = 5d^6 6s^0 6p^0$). In the free state Ir^{3+} ion has four unpaired electrons in $5d$ orbitals ($n = 4$) as shown at (b) of Fig. 5.43. Now since NH_3 molecules which act as ligands in $[Ir(NH_3)_6]^{3+}$ ion are strong ligands, the unpaired electrons present in $5d$ orbitals get paired in three $5d$ orbitals so that n becomes equal to zero ($n = 0$) and two $5d$ orbitals become vacant. These vacant $5d$ orbitals are used for d^2sp^3 hybridisation (inner orbital octahedral hybridisation) as shown in Fig. 5.43. Thus $[Ir(NH_3)_6]^{3+}$ ion has $n = 0$ (diamagnetic) and $\mu_{\text{spin only}}$ = zero.

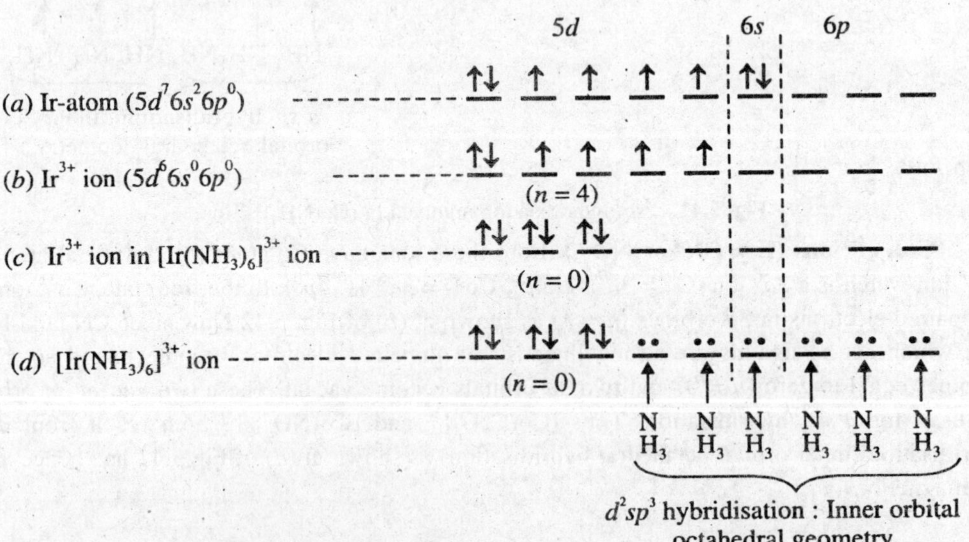

Fig. 5.43. d^2sp^3 hybridisation in $[Ir(NH_3)_6]^{3+}$ ion

6-coordinated complexes of Pt (IV). Hexamines of Pt (IV), $[Pt(Am)_6]X_4$ (Am = NH_3, CH_3NH_2, N_2H_4 etc, X = halide ion, CNS^-, OH^- or NO_2^-) and hexahaloplatinate (IV) salts, $M^{2+}[PtX_6]$ (X = F, Cl, Br, I) are all octahedral diamagnetic and involve d^2sp^3 hybridisation (inner orbital octahedral hybridisation). Out of these compounds $[Pt(NH_3)_6]^{4+}$ and $[PtCl_6]^{2-}$ are the most important. As an example let us discuss the geometry of $[PtCl_6]^{2-}$ ion. In this ion Pt is present as Pt^{4+} ion which is a $5d^6$ ion (Pt = $5d^9\,6s^1$, $Pt^{4+} = 5d^6\,6s^0\,6p^0$). In the free state Pt^{4+} ion has four unpaired electrons (n = 4) in $5d$ orbitals. In presence of Cl^- ions which act as ligands all the six $5d$ electrons get paired up so that n becomes equal to zero (n = 0) and two $5d$ orbitals become vacant and hence available for d^2sp^3 hybridisation. Thus $[PtCl_6]^{2-}$ ion has octahedral geometry and is diamagnetic (n = 0) and hence μ = 0. (Also see Fig. 5.44.)

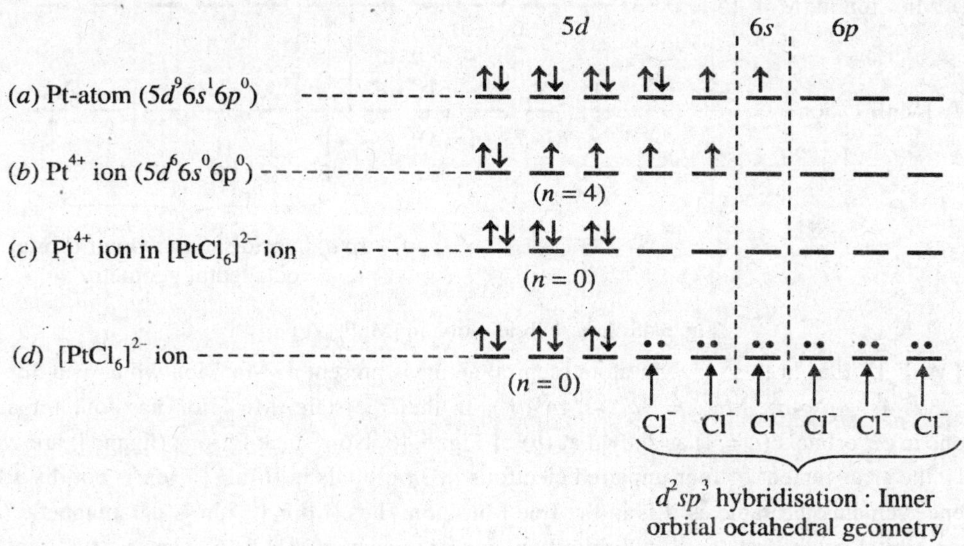

Fig. 5.44. d^2sp^3 hybridisation in $[PtCl_6]^{2-}$ ion

Problem 12. *On the basis of VBT, answer the following questions for 6-coordinated complex ions:* $[MnF_6]^{4-}$, $[MnF_6]^{3-}$, $[Fe(H_2O)_6]^{2+}$, $[Fe(H_2O)_6]^{3+}$, $[FeF_6]^{3-}$, $[CoF_6]^{3-}$, $[NiL_6]^{2+}$ (L = NH_3, H_2O), $[Fe(H_2O)_5(NO)]^{2+}$.

(a) What type of hybridisation is involved?

(b) Whether the given complex ion is inner orbital octahedral or outer orbital octahedral?

(c) What is the magnetic behaviour of the complex ion?

(d) Calculate the value of $\mu_{spin\ only}$.

(G.N. Dev 95, Raj 96, Allahabad 96, Delhi 96, Guahati 96, Kurukshetra 97, Nagpur 97, Delhi 98, IIT 2000)

Solution. $[MnF_6]^{4-}$ ion. In this ion, Mn atom (central metal atom) is present as Mn^{2+} ion which is a $3d^5$ ion (Mn = $3d^5\,4s^2$, $Mn^{2+} = 3d^5\,4s^0\,4p^0\,4d^0$). In the free state, Mn^{2+} ion has five unpaired electrons in $3d$ orbitals (n = 5) as shown at (b) of Fig. 5.45. Now since F^- ions which act as ligands in $[MnF_6]^{4-}$ ions are weak ligands, the arrangement of five unpaired electrons in $3d$ orbitals in $[MnF_6]^{4-}$ ion is not disturbed and hence remains the same as it is in the free Mn^{2+} ion. Thus $[MnF_6]^{4-}$ ion, like free Mn^{2+} ion, has five unpaired electrons (n = 5) and is formed by sp^3d^2 hybridisation (outer orbital octahedral hybridisation) as shown at (d) of Fig 5.45. Due to the presence of five unpaired electrons $[MnF_6]^{4-}$ ion is paramagnetic and theoretical value of $\mu_{\text{spin only}} = \sqrt{5(5+2)}$ BM = 5.916 BM.

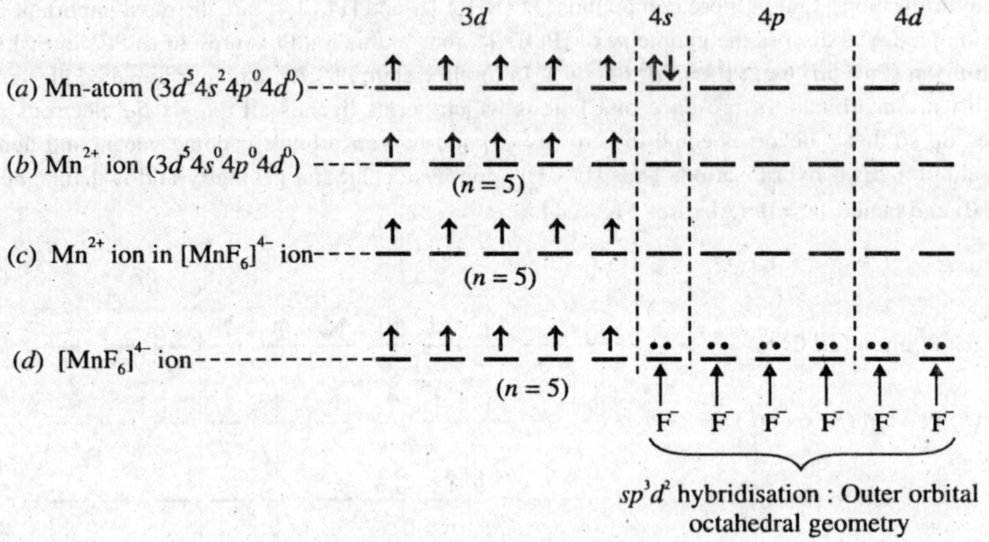

	3d					4s	4p			4d	
(a) Mn-atom ($3d^5 4s^2 4p^0 4d^0$)	↑	↑	↑	↑	↑	↑↓					
(b) Mn^{2+} ion ($3d^5 4s^0 4p^0 4d^0$) (n = 5)	↑	↑	↑	↑	↑						
(c) Mn^{2+} ion in $[MnF_6]^{4-}$ ion (n = 5)	↑	↑	↑	↑	↑						
(d) $[MnF_6]^{4-}$ ion (n = 5)	↑	↑	↑	↑	↑	•• (F^-)	•• (F^-)	•• (F^-)	•• (F^-)	•• (F^-)	•• (F^-)

sp^3d^2 hybridisation : Outer orbital octahedral geometry

Fig. 5.45. sp^3d^2 hybridisation in $[MnF_6]^{4-}$ ion

$[MnF_6]^{3-}$ ion. In this ion, Mn-atom (central atom) is present as Mn^{3+} ion which is a $3d^4$ ion (Mn = $3d^5$ $4s^2$ $4p^0$ $4d^0$, Mn^{3+} = $3d^4$ $4s^0 4p^0 4d^0$). In the free state, Mn^{3+} ion has four unpaired electrons in $3d$ orbitals (n = 4) as shown at (b) of Fig. 5.46. Now since F^- ions (ligands) are weak ligands, the arrangement of four unpaired electrons in $3d$ orbitals in $[MnF_6]^{3-}$ ion is not disturbed and hence remains the same as it is in the free Mn^{3+} ion.Thus $[MnF_6]^{3-}$ ion is paramagnetic (n = 4) outer orbital octahedral (sp^3d^2 hybridisation) ion as shown in Fig. 5.46. Thus :

$$\mu_{\text{spin only}} = \sqrt{n(n+2)} \text{ BM} = \sqrt{4\times(4+2)} \text{ BM} = 4.898 \text{ BM}.$$

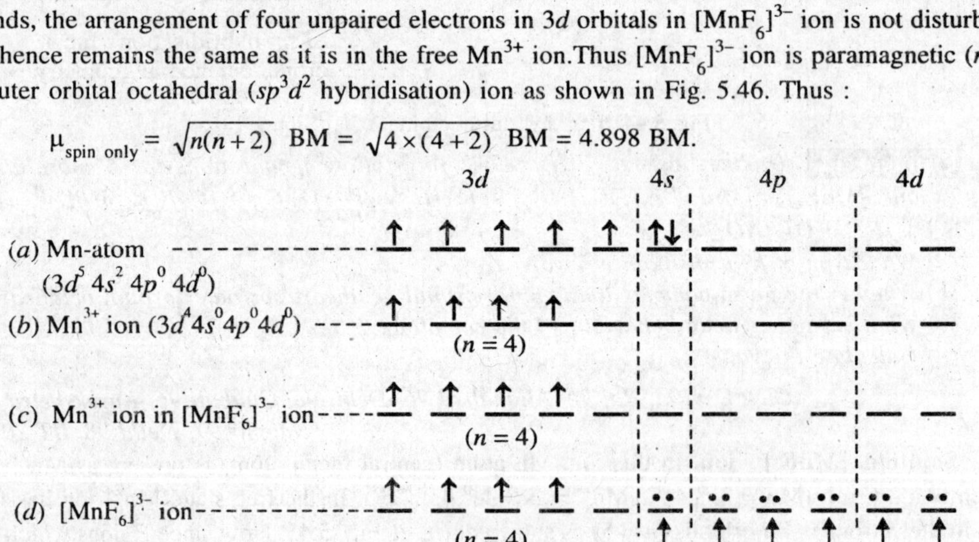

	3d					4s	4p			4d	
(a) Mn-atom ($3d^5$ $4s^2$ $4p^0$ $4d^0$)	↑	↑	↑	↑	↑	↑↓					
(b) Mn^{3+} ion ($3d^4 4s^0 4p^0 4d^0$) (n = 4)	↑	↑	↑	↑							
(c) Mn^{3+} ion in $[MnF_6]^{3-}$ ion (n = 4)	↑	↑	↑	↑							
(d) $[MnF_6]^{3-}$ ion (n = 4)	↑	↑	↑	↑		•• (F^-)	•• (F^-)	•• (F^-)	•• (F^-)	•• (F^-)	•• (F^-)

sp^3d^2 hybridisation : Outer orbital octahedral geometry

Fig. 5.46. sp^3d^2 hybridisation in $[MnF_6]^{3-}$ ion

$[Fe(H_2O)_6]^{2+}$ ion. In this ion, Fe (central metal atom) is present as Fe^{2+} ion which is a $3d^6$ ion (Fe = $3d^6\,4s^2$, Fe^{2+} = $3d^6\,4s^0\,4p^0\,4d^0$). In the free state Fe^{2+} ion has four unpaired electrons (n = 4) in $3d$ orbitals as shown at (*b*) of Fig. 5.47. Now since H_2O molecules which act as ligands in $[Fe(H_2O)_6]^{2+}$ ion are weak ligands, the arrangement of four unpaired electrons in $3d$ orbitals of Fe^{2+} ion in $[Fe(H_2O)_6]^{2+}$ is not disturbed and hence remains the same as it is in the free Fe^{2+} ion. Thus $[Fe(H_2O)_6]^{2+}$ ion has four unpaired electrons and is formed by sp^3d^2 hybridisation (outer orbital octahedral hybridisation) shown at (*d*) of Fig Fig. 5.47 Due to the presence of four unpaired electrons, $[Fe(H_2O)_6]^{2+}$ ion is paramagnetic and $\mu_{\text{spin only}} = \sqrt{4(4+2)}$ BM = 4.898 BM.

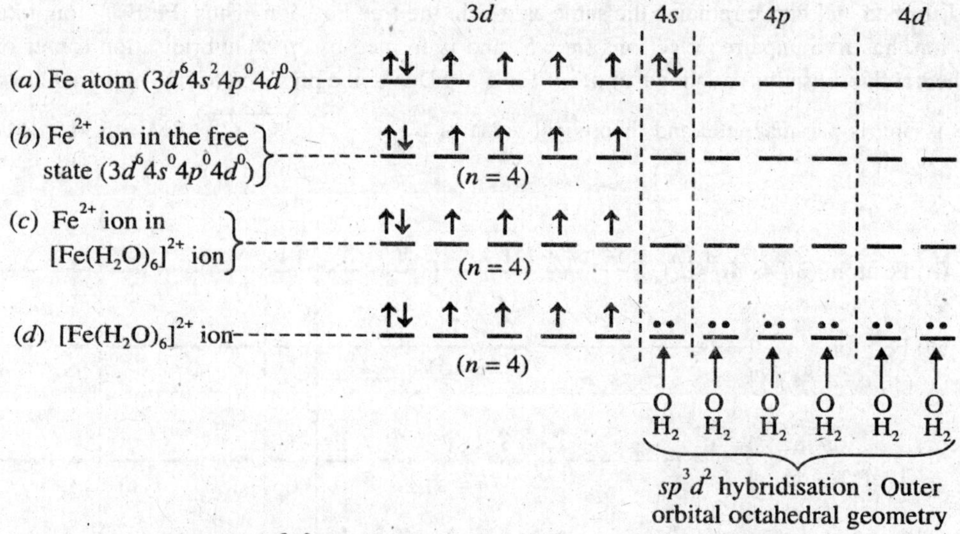

	3d					4s	4p			4d	
(*a*) Fe atom ($3d^6 4s^2 4p^0 4d^0$)	↑↓	↑	↑	↑	↑	↑↓	—	—	—	—	—
(*b*) Fe^{2+} ion in the free state ($3d^6 4s^0 4p^0 4d^0$)	↑↓	↑	↑	↑	↑	—	—	—	—	—	—
	(n = 4)										
(*c*) Fe^{2+} ion in $[Fe(H_2O)_6]^{2+}$ ion	↑↓	↑	↑	↑	↑	—	—	—	—	—	—
	(n = 4)										
(*d*) $[Fe(H_2O)_6]^{2+}$ ion	↑↓	↑	↑	↑	↑	••	••	••	••	••	••
	(n = 4)					↑ O H_2	↑ O H_2	↑ O H_2	↑ O H_2	↑ O H_2	↑ O H_2

sp^3d^2 hybridisation : Outer orbital octahedral geometry

Fig. 5.47. sp^3d^2 hybridisation of Fe^{2+} ion in $[Fe(H_2O)_6]^{2+}$ ion.

$[Fe(H_2O)_6]^{3+}$ ion. In this ion Fe (central metal atom) is present as Fe^{3+} which has five unpaired electrons in $3d$ orbitals [Fe = $3d^6 4s^2$, Fe^{3+} = $3d^5\,4s^0\,4p^0$]. Since H_2O molecules are weak

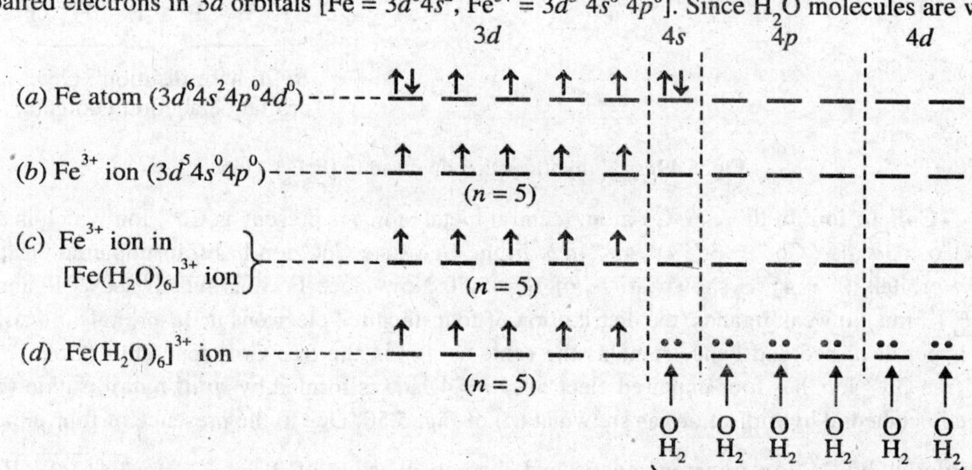

	3d					4s	4p			4d	
(*a*) Fe atom ($3d^6 4s^2 4p^0 4d^0$)	↑↓	↑	↑	↑	↑	↑↓	—	—	—	—	—
(*b*) Fe^{3+} ion ($3d^5 4s^0 4p^0$)	↑	↑	↑	↑	↑	—	—	—	—	—	—
	(n = 5)										
(*c*) Fe^{3+} ion in $[Fe(H_2O)_6]^{3+}$ ion	↑	↑	↑	↑	↑	—	—	—	—	—	—
	(n = 5)										
(*d*) $Fe(H_2O)_6]^{3+}$ ion	↑	↑	↑	↑	↑	••	••	••	••	••	••
	(n = 5)					↑ O H_2	↑ O H_2	↑ O H_2	↑ O H_2	↑ O H_2	↑ O H_2

sp^3d^2 hybridisation : Outer orbital octahedral geometry

Fig. 5.48. sp^3d^2 hybridisation involved in $[Fe(H_2O)_6]^{3+}$ ion

ligands, the distribution of five electrons in $3d$-orbitals in $[Fe(H_2O)_6]^{3+}$ ion remains the same as in free Fe^{3+} ion. Thus $[Fe(H_2O)_6]^{3+}$ ion has five unpaired electrons ($n = 5$) and is formed by sp^3d^2 hybridisation as shown in Fig 5.48. (outer orbital octahedral complex ion). Due to the presence of five unpaired electrons, $[Fe(H_2O)_6]^{3+}$ ion is paramagnetic. The value of $\mu_{\text{spin only}} = \sqrt{n(n+2)}$ BM $= \sqrt{5 \times (5+2)}$ BM = 5.916 BM.

$[FeF_6]^{3-}$ ion. In $[FeF_6]^{3-}$ ion, Fe atom (central metal atom) is present as Fe^{3+} ion which is a $3d^5$ ion ($Fe = 3d^6 4s^2$, $Fe^{3+} = 3d^5 4s^0 4p^0 4d^0$). In the free state Fe^{3+} ion has five unpaired electrons ($n = 5$) in $3d$ orbitals as shown at (*b*) of Fig. 5.49. Now since F^- ions which act as ligands in $[FeF_6]^{3-}$ ion are weak ligands, the distribution of five unpaired electrons in $3d$ orbitals in $[FeF_6]^{3-}$ ion is not disturbed and hence remains the same as it is in the free Fe^{3+} ion. Thus $[FeF_6]^{3-}$ ion, like free Fe^{3+} ion, has five unpaired electrons ($n = 5$) and is formed by sp^3d^2 hybridisation (outer orbital octahedral hybridisation) as shown at (*d*) of Fig 5.49. Due to the presence of five unpaired electrons, $[FeF_6]^{3-}$ ion is paramagnetic and theoretical value of $\mu_{\text{spin only}} = \sqrt{5(5+2)}$ BM = 5.916 BM

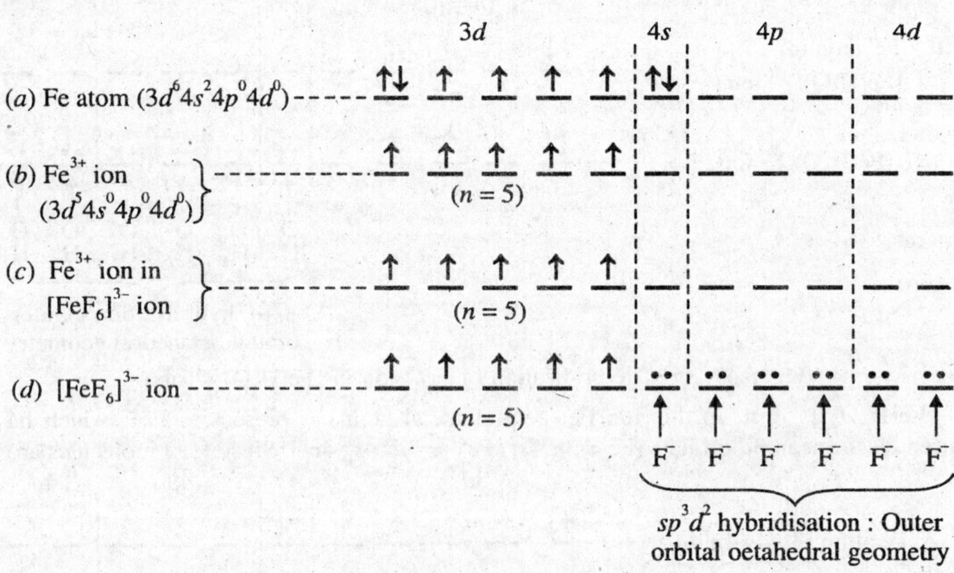

Fig. 5.49. sp^3d^2 hybridisation involved in $[FeF_6]^{3-}$ ion

$[CoF_6]^{3-}$ ion. In this ion, Co-atom (central metal atom) is present as Co^{3+} ion which is a $3d^6$ ion ($Co = 3d^7\ 4s^2$, $Co^{3+} = 3d^6\ 4s^0\ 4p^0\ 4d^0$). In the free state Co^{3+} ion has four unpaired electrons in $3d$ orbitals ($n = 4$) as shown at (*b*) of Fig. 5.50 Now since F^- ions which act as ligands in $[CoF_6]^{3-}$ ion are weak ligands, the distribution of four unpaired electrons in $3d$-orbitals in $[CoF_6]^{3-}$ ion is not disturbed and hence remains the same as it is in the free Co^{3+} ion. Thus $[CoF_6]^{3-}$ ion, like free Co^{3+} ion, has four unpaired electrons ($n = 4$) and is formed by sp^3d^2 hybridisation (outer orbital octahedral hybridisation) as shown at (*d*) of Fig. 5.50. Due to the presence of four unpaired electrons, $[CoF_6]^{3-}$ ion is paramagnetic and theoretical value of $\mu_{\text{spin only}} = \sqrt{4(4+2)}$ BM = 4.898 BM.

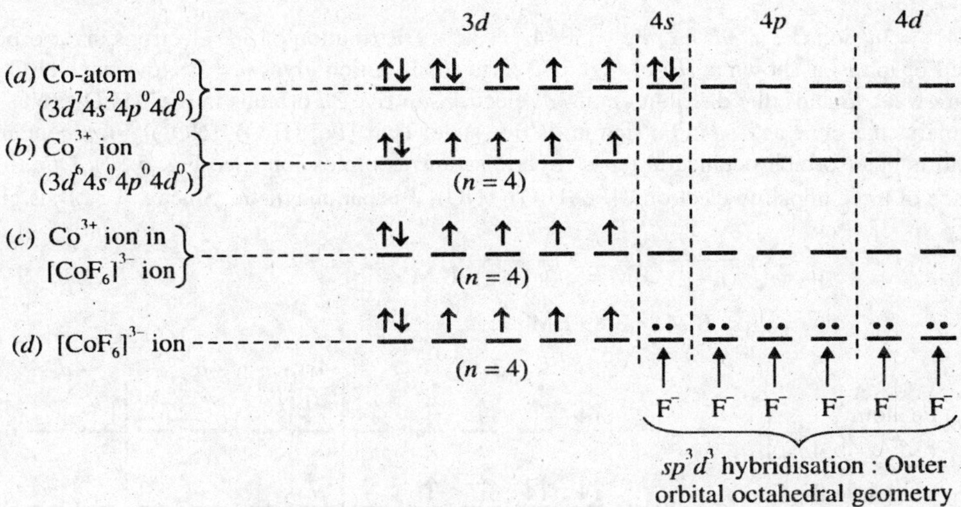

Fig. 5.50. sp^3d^2 hybridisation involved in $[CoF_6]^{3-}$ ion

$[NiL_6]^{2+}$ ion (L = NH_3 or H_2O). In this ion, Ni-atom (central metal atom) is present as Ni^{2+} ion which is a $3d^8$ ion (Ni = $3d^8\ 4s^2$, Ni^{2+} = $3d^8\ 4s^0\ 4p^0\ 4d^0$). In the free state, Ni^{2+} ion has two unpaired electrons in 3*d* orbitals (*n* = 2) as shown at (*b*) of Fig. 5.51. We know that for d^2sp^3 hybridisation,two vacant 3d orbitals are required. Now if, in the presence of six L ligands the pairing of two unpaired electrons takes place, we get only one vacant 3*d* orbital. Hence $[Ni\ L_6]^{2+}$ ion cannot be formed by d^2sp^3 hybridisation. In fact this ion results from sp^3d^2 hybridisation as shown in Fig. 5.51. This figure shows that *n* = 2 and hence $[NiL_6]^{2+}$ ion is paramagnetic with $\mu_{\text{spin only}} = \sqrt{2(2+2)}$ BM = 2.828 BM.

Examples of other paramagnetic (*n* = 2) outer orbital octahedral complexes (sp^3d^2 hybridisation) of Ni (II) are $[Ni(en)_3]^{2+}$, $[Ni(N_2H_4)_3]^{2+}$, $[Ni\ (dipy)_3]^{2+}$, $[Ni\ (NCS)_6]^{4-}$ etc.

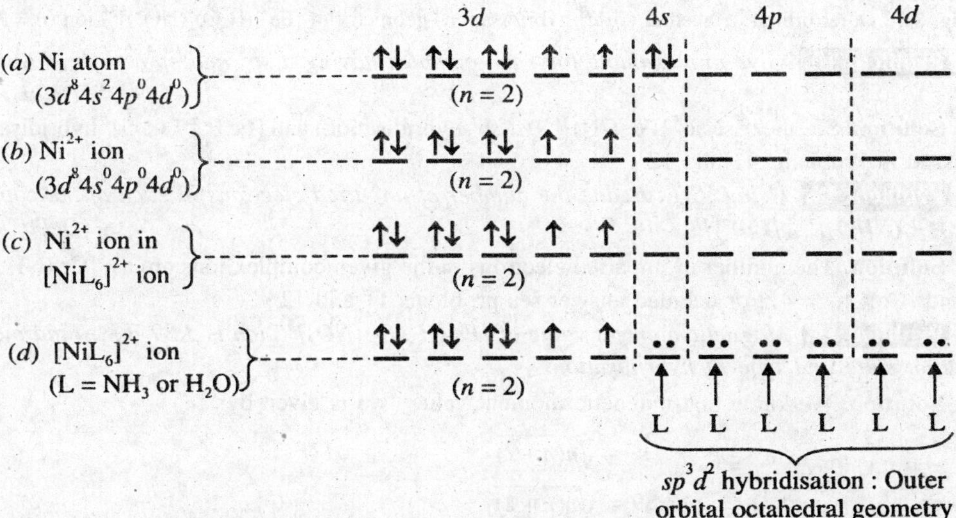

Fig. 5.51. sp^3d^2 hybridisation in $[NiL_6]^{2+}$ ion (L = NH_3 or H_2O)

$[Fe(H_2O)_5(NO)]^{2+}$ ion. In this ion, since NO which acts as ligand is present as NO^+ ion (nitrosyl ion), Fe (central metal atom) is present as Fe^+ ion. (Note that NO^+ ion is a two electron donor). Now since C.N. of Fe^+ ion = 6, $[Fe\ (H_2O)_5\ (NO)]^{2+}$ ion has octahedral geometry. Fe^+ ion is a $3d^6\ 4s^1 = 3d^7$ ion ($Fe = 3d^6\ 4s^2$, $Fe^+ = 3d^6 4s^1 = 3d^7$). Distribution of $3d^7$ electrons in $3d$ orbitals of free Fe^+ ion is as shown at (*b*) of Fig. 5.52 . This distribution gives $n = 3$. Now since H_2O and NO^+ are weak ligands, the distribution of $3d^7$ electrons in five $3d$ orbitals in $[Fe^+ (H_2O)_5 (NO^+)]^{+2}$ ion remains the same as it is in Fe^+ ion in its free state. Thus $[Fe^+ (H_2O)_5 (NO^+)]^{2+}$ ion contains $n = 3$ and is outer orbital octahedral (sp^3d^2 hybridisation) complex ion (See Fig. 5.52). Due to the presence of three unpaired electrons, $[Fe (H_2O)_5 (NO)]^{2+}$ is paramagnetic. Since $n = 3$, μ is given by :

$$\mu = \sqrt{n(n+2)} \text{ BM} = \sqrt{3(3+2)} \text{ B.M.}$$
$$= \sqrt{15} \text{ B.M.} = 3.89 \text{ BM}$$

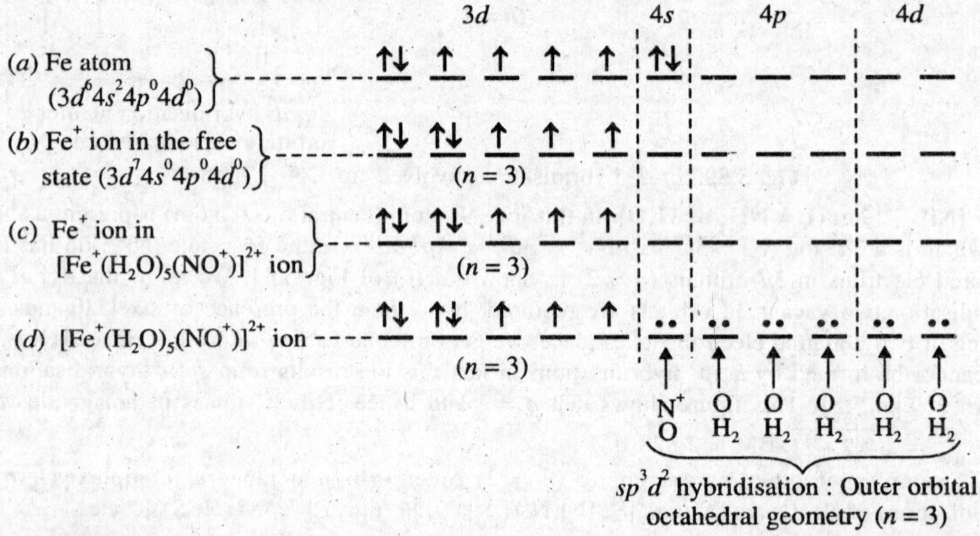

	3d	4s	4p	4d
(*a*) Fe atom ($3d^6 4s^2 4p^0 4d^0$)	↑↓ ↑ ↑ ↑ ↑	↑↓	— — —	— —
(*b*) Fe^+ ion in the free state ($3d^7 4s^0 4p^0 4d^0$)	↑↓ ↑↓ ↑ ↑ ↑ ($n = 3$)	—	— — —	— —
(*c*) Fe^+ ion in $[Fe^+(H_2O)_5(NO^+)]^{2+}$ ion	↑↓ ↑↓ ↑ ↑ ↑ ($n = 3$)	—	— — —	— —
(*d*) $[Fe^+(H_2O)_5(NO^+)]^{2+}$ ion	↑↓ ↑↓ ↑ ↑ ↑ ($n = 3$)	•• (N^+O)	•• •• •• (O H_2, O H_2, O H_2)	•• •• (O H_2, O H_2)

sp^3d^2 hybridisation : Outer orbital octahedral geometry ($n = 3$)

Fig. 5.52 Outer orbital octahedral (sp^3d^2 hybridisation) geometry of $[Fe^+ (H_2O)_5(NO^+)]^{2+}$ ion ($n = 3$)

Problem 13. *Give one example for each complex having d^2sp^3 and sp^3d^2 hybridisation.* (*Delhi 1999*)

Solution. See the cases of $[Fe (CN)_6]^{3-}$ (d^2sp^3 hybridisation) and $[Fe F_6]^{3-}$ (sp^3d^2 hybridisation) discussed in Problems 11 and 12.

Problem 14. *Using VBT, predict the number of unpaired electrons in each of the complex ions: $[Cr (NH_3)_6]^{3+}$, $[CoF_6]^{3-}$.* (*Delhi 1997*)

Solution. The number of unpaired electrons in the given complex ions are as: $[Cr (NH_3)_6]^{3+} = 3$ and $[CoF_6]^{3-} = 4$. For detailed answer see problems 11 and 12

Problem 15. *Magnetic moment value of $[Fe (H_2O)_5 (NO)]^{2+}$ ion is 3.89 B.M. Find out the oxidation state and type of hybridisation.*

Solution. We know that magnetic moment value (μ) is given by:

$$\mu = \sqrt{n(n+2)}$$

or $\quad 3.89 = \sqrt{n(n+2)}$

or $\quad (3.89)^2 = n\ (n + 2)$

or $\quad 15.132 = n^2 + 2n$

or $\quad n^2 + 2n - 15 = 0$

or $\quad (n + 5)\ (n - 3) = 0$

$\therefore \quad n = -5, +3$

Thus Fe-atom (central metal atom) in the given complex ion has three unpaired electrons (n = 3), Electronic configurations of Fe^0, Fe^+, Fe^{2+} and Fe^{3+} as given in Fig. 5.53. show that these species have 4, 3, 4 and 5 unpaired electrons respectively. Since it is only Fe^+ ion that contains n = 3, Fe is present as Fe^+ ion in $[Fe\ (H_2O)_5\ (NO)]^{2+}$ and hence NO is present as NO^+ (nitrosyl cation). Now since C.N. of Fe^+ = 6, $[Fe\ (H_2O)_5\ (NO)]^{2+}$ ion has octahedral geometry which arises because of sp^3d^2 hybridisation of Fe^+ ion as shown in Fig. 5.54.

	3d					4s
(a) Fe^0 atom ($3d^6 4s^2$)	↑↓	↑	↑	↑	↑	↑↓
	(n = 4)					
(b) Fe^+ ion ($3d^6 4s^1 = 3d^7$)	↑↓	↑↓	↑	↑	↑	—
	(n = 3)					
(c) Fe^{2+} ion ($3d^6 4s^0$)	↑↓	↑	↑	↑	↑	—
	(n = 4)					
(d) Fe^{3+} ion ($3d^5 4s^0$)	↑	↑	↑	↑	↑	—
	(n = 5)					

Fig. 5.53. Electronic configurations of Fe^0, Fe^+, Fe^{2+} and Fe^{3+}.

	3d					4s	4p			4d	
(a) Fe^+ ion in the free state ($3d^7 4s^0 4p^0 4d^0$)	↑↓	↑↓	↑	↑	↑	—	—	—	—	—	—
	(n = 3)										
(b) Fe^+ ion in $[Fe^+(H_2O)_5(NO^+)]^{2+}$ ion	↑↓	↑↓	↑	↑	↑	—	—	—	—	—	—
	(n = 3)										
(c) $[Fe^+(H_2O)_5(NO^+)]^{2+}$ ion	↑↓	↑↓	↑	↑	↑	••	••	••	••	••	••
	(n = 3)					↑ N^+O	↑ OH_2	↑ OH_2	↑ OH_2	↑ OH_2	↑ OH_2

sp^3d^2 hybridisation : Outer orbital octahedral geometry (n = 3)

Fig. 5.54. Outer orbital octahedral (sp^3d^2 hybridisation) geometry of $[Fe^+\ (H_2O)_5(NO^+)]^{2+}$ ion (n = 3)

Problem 16. *Derive the geometry of the complex compound corresponding to the brown ring in nitrate test. Predict the magnetic moment of the complex.*

Solution. Brown ring formed in the nitrate test is due to the formation of the complex ion, $[Fe\ (H_2O)_5\ (NO)]^{2+}$ in which Fe and NO both are present as Fe^+ and NO^+ (nitrosyl ion) ions respectively. Fe^+ ion acts as the central metal ion and NO^+ ion acts as a ligand. Five H_2O molecules also act as ligands. Since C.N. of Fe^+ ion is 5 + 1 = 6, $[Fe\ (H_2O)_5\ (NO)]^{2+}$ is an octahedral complex ion which is formed by sp^3d^2 hybridisation and has n = 3. Thus $[Fe\ (H_2O)_5\ (NO)]^{2+}$ ion is paramagnetic and its $\mu = \sqrt{3 \times (3+2)}$ BM = 3.89 B.M. For details see problem 15.

Problem 17. *Discuss the geometry of octahedral complexes of Co(II) on the basis of hybridisation concept.*

Solution. Octahedral complexes of Co(II) like $[Co(H_2O)_6]^{2+}$, $[Co\ (NH_3)_6]^{2+}$, $[Co(N_2H_5)_3]^{2+}$, $[Co\ (en)_3]^{2+}$ etc. are paramagnetic due to the presence of three unpaired electrons (n = 3). These complex ions involve sp^3d^2 hybridisation of Co^{2+} ion [$Co = 3d^7\ 4s^2$, $Co^{2+} = 3d^7$ ($n = 3$)] and hence are outer orbital octahedral complexes. As an example sp^3d^2 hybridisation in $[Co\ (NH_3)_6]^{2+}$ has been shown in Fig. 5.55. Due to sp^3d^2 hybridisation these are HS octahedral complexes.

3d | 4s | 4p | 4d

(a) Co-atom ($3d^7 4s^2 4p^0 4d^0$) ---- ↑↓ ↑↓ ↑ ↑ ↑ | ↑↓ | — — — | — — ($n = 3$)

(b) Co^{2+} ion ($3d^7 4s^0 4p^0 4d^0$) ---- ↑↓ ↑↓ ↑ ↑ ↑ | — | — — — | — — ($n = 3$)

(c) Co^{2+} ion in $[Co(NH_3)_6]^{2+}$ ion ---- ↑↓ ↑↓ ↑ ↑ ↑ | — | — — — | — — ($n = 3$)

(d) $[Co(NH_3)_6]^{2+}$ ---- ↑↓ ↑↓ ↑ ↑ ↑ | •• | •• •• •• | •• •• ($n = 3$)

NH_3 NH_3 NH_3 NH_3 NH_3 NH_3

sp^3d^2 hybridisation : Outer orbital octahedral geometry

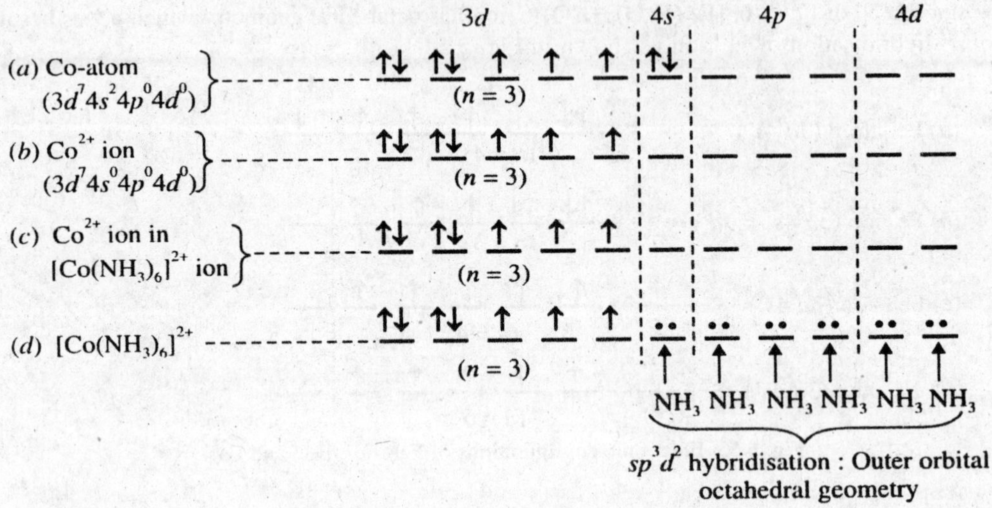

Fig. 5.55. sp^3d^2 hybridisation in $[Co(NH_3)_6]^{2+}$ ion

Co(II) also forms a few paramagnetic octahedral complexes with $n = 1$. These are obtained by d^2sp^3 hybridisation. Due to this type of hybridisation these complexes are LS octahedral complexes. Examples of such complexes are $[Co\ (NO_2)_6]^{4-}$, $[Co\ (diars)_3]^{2+}$, $[Co\ (triars)_2]^{2+}$ etc.

Problem 18. *Magnetic measurement on potassium hexacyanomanganate (III) indicates the presence of two unpaired electrons. Predict, on the basis of VBT, whether the given complex is inner orbital or outer orbital octahedral complex.*

Solution. The given complex is $K_3[Mn\ (CN)_6]$ in which Mn is present as Mn^{3+} ion which is a $3d^4$ ion. d^2sp^3 hybridisation (inner orbital) and sp^3d^2 hybridisation (outer orbital) of Mn^{3+} ion in

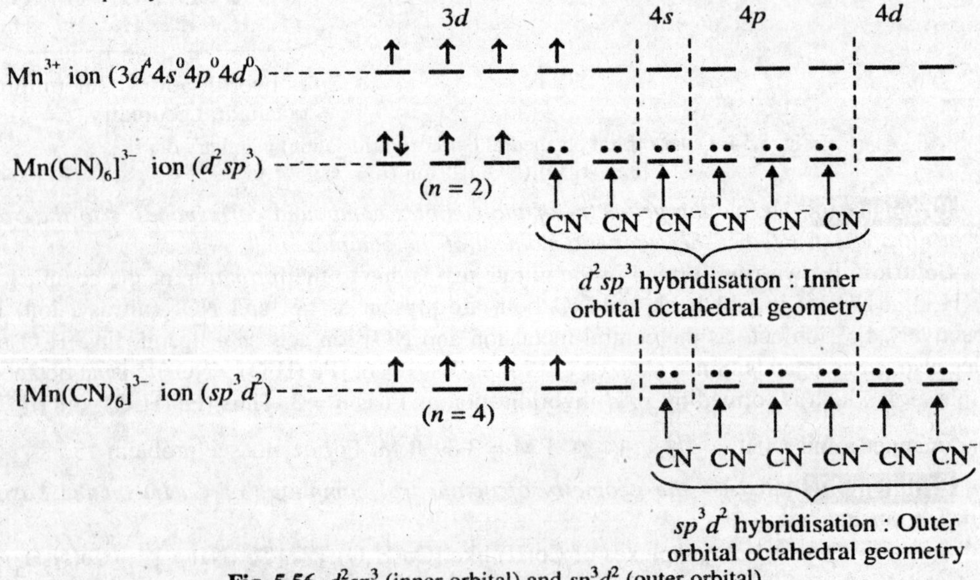

3d | 4s | 4p | 4d

Mn^{3+} ion ($3d^4 4s^0 4p^0 4d^0$) ---- ↑ ↑ ↑ ↑ — | — | — — — | — —

$Mn(CN)_6]^{3-}$ ion (d^2sp^3) ---- ↑↓ ↑ ↑ •• •• | •• | •• •• •• | — — ($n = 2$)

CN^- CN^- CN^- CN^- CN^- CN^-

d^2sp^3 hybridisation : Inner orbital octahedral geometry

$[Mn(CN)_6]^{3-}$ ion (sp^3d^2) ---- ↑ ↑ ↑ ↑ — | •• | •• •• •• | •• •• ($n = 4$)

CN^- CN^- CN^- CN^- CN^- CN^-

sp^3d^2 hybridisation : Outer orbital octahedral geometry

Fig. 5.56. d^2sp^3 (inner orbital) and sp^3d^2 (outer orbital) hybridisation of Mn^{3+} ion ($Mn^{3+} = 3d^4$) in $[Mn(CN)_6]^{3-}$ ion

the formation of $[Mn(CN)_6]^{3-}$ ion can be shown as depicted in Fig. 5.56. This figure shows that d^2sp^3 hybridisation gives $n = 2$, while sp^3d^2 hybridisation gives $n = 4$. Thus the given complex compound is an inner orbital octahedral complex.

Problem 19. *The magnetic moment value of $[Mn(CN)_6]^{3-}$ ion is 2.8 BM. Predict the type of hybridisation and geometry of the ion.*

Solution. We know that μ is given by:

$$\mu = \sqrt{n(n+2)}$$

or $$2.8 = \sqrt{n(n+2)}$$

or $$(2.8)^2 = n(n+2)$$

or $$n^2 + 2n - 7.84 = 0$$

or $$n^2 + 2n - 8 = 0$$

or $$(n+4)(n-2) = 0$$

$$\therefore \quad n = -4, +2$$

Thus $[Mn(CN)_6]^{3-}$ ion has two unpaired electrons ($n = 2$). In the given complex ion, Mn is present as Mn^{3+} ion which is a $3d^4$ ion ($n = 4$). Since C.N. of $Mn^{3+} = 6$, $[Mn(CN)_6]^{3-}$ ion has octahedral geometry which may arise either from d^2sp^3 hybridisation (inner orbital) or from sp^3d^2 hybridisation (outer orbital) as shown in Fig. 5.57. Now since d^2sp^3 hybridisation gives $n = 2$ and sp^3d^2 hybridisation gives $n = 4$, $[Mn(CN)_6]^{3-}$ ion has inner orbital octahedral geometry which results from d^2sp^3 hybridisation.

3d | 4s | 4p | 4d

Mn^{3+} ion ($3d^4 4s^0 4p^0 4d^0$) ↑ ↑ ↑ ↑ — | — | — — — | — —
($n = 4$)

$Mn(CN)_6]^{3-}$ ion (d^2sp^3) ↑↓ ↑ ↑ •• •• | •• | •• •• •• | — —
($n = 2$)

CN⁻ CN⁻ CN⁻ CN⁻ CN⁻ CN⁻

d^2sp^3 hybridisation : Inner orbital octahedral geometry

$[Mn(CN)_6]^{3-}$ ion (sp^3d^2) ↑ ↑ ↑ ↑ — | •• | •• •• •• | •• ••
($n = 4$)

CN⁻ CN⁻ CN⁻ CN⁻ CN⁻ CN⁻

sp^3d^2 hybridisation : Outer orbital octahedral geometry

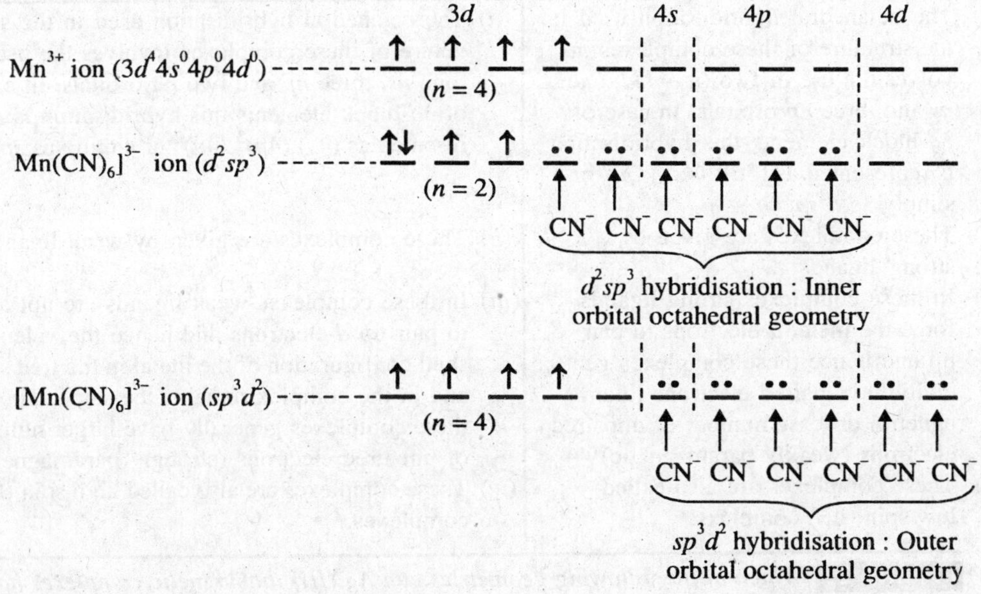

Fig. 5.57. d^2sp^3 and sp^3d^2 hybridisation of Mn^{3+} ion in $[Mn(CN)_6]^{3-}$ ion.

Problem 20. *On the basis of VBT explain the given magnetic moment (μ) values of the following complex compounds: $K_3[FeF_6] = 5.9$ BM, $[Co(NH_3)_6]Cl_3 = 0.0$ BM, $[Ni(H_2O)_6]Cl_2 = 2.9$ BM, $K_4[Mn(CN)_6] = 1.9$ BM, $K_4[MnF_6] = 5.9$ B.M.* *(Guru Nanak Dev 1995)*

Solution. We know that the complex ions obtained by the ionisation of the given compounds have octahedral geometry which arises because of d^2sp^3 or sp^3d^2 hybridisation as shown below. Number of unpaired electrons (n) and the value of μ are also given.

$[FeF_6]^{3-}$ ------------- sp^3d^2 (F^- ions are weak ligands),

$n = 5,\ \mu = \sqrt{5\times(5+2)}$ BM = 5.91 BM

$[Co(NH_3)_6]^{3+}$ -------- d^2sp^3 (NH_3 molecules are strong ligands),

$n = 0,\ \mu = 0.0$

$[Ni(H_2O)_6]^{2+}$ -------- sp^3d^2 (H_2O molecules are weak ligands),

$n = 2,\ \mu = \sqrt{2\times(2+2)}$ BM = 2.82 BM

$[Mn(CN)_6]^{4-}$ --------- d^2sp^3 (CN^- ions are strong ligands),

$n = 1,\ \mu = \sqrt{1\times(1+2)}$ BM = 1.732 BM

$[MnF_6]^{4-}$ ------------ sp^3d^2 (F^- ions are weak ligands),

$n = 5,\ \mu = \sqrt{5\times(2+5)}$ BM = 5.91 BM

Thus we see that values of μ calculated on the basis of VBT are almost the same as given in the question.

Problem 21. *In what respects do inner orbital octahedral complexes differ from outer orbital octahedral complexes.* *(Rajsthan 96; Delhi 97)*

Solution. The differences between inner orbital and outer orbital octahedral complexes have been given in the following table

Inner orbital octahedral complexes	*Outer orbital octahedral complexes*
(*i*) The octahedral hybridisation used in the structure of these complexes involves the use of two $(n-1)d$, one *ns* and three *np* orbitals. In case of 3*d*-block elements this hybridisation is represented as $(3d)^2\ (4s)\ (4p)^3$ or simply as d^2sp^3.	(*i*) The octahedral hybridisation used in the structure of these complexes involves the use of one *ns*, three *np* and two *nd* orbitals. In case of 3*d*-block elements this hybridisation is represented as $(4s)\ (4p)^3\ (4d)^2$ or simply as sp^3d^2.
(*ii*) These complexes are given by strong ligands	(*ii*) These complexes are given by weak ligands.
(*iii*) In these complexes strong ligands force the metal *d*-electrons to pair up and hence these complexes generally have paired electrons (diamagnetic) or lease number of unpaired electrons (weakly paramagnetic)	(*iii*) In these complexes weak ligands are not able to pair up *d*-electrons and hence the valence shell configuration of the metal in the free state and in the complex remains the same. Thus these complexes generally have larger number of unpaired electrons (strongly paramagnetic)
(*iv*) These complexes are also called low spin (LS) complexes.	(*iv*) These complexes are also called high spin (HS) complexes.

Problem 22. *Which of the following geometries can Ag (III) diamagnetic complexes have: inner orbital octahedral,outer orbital octahedral, tetrahedral or square planar?*

Solution. Ag (III) is a $4d^8$ ion. For diamagnetic complexes *n* must be equal to zero ($n = 0$). The type of hybridisation and the number of unpaired electrons (*n*) in different geometries have been shown in Fig. 5.58. This figure shows that since only inner orbital octahedral and square planar geometries give $n = 0$, Ag (III) diamagnetic complexes have only these geometries.

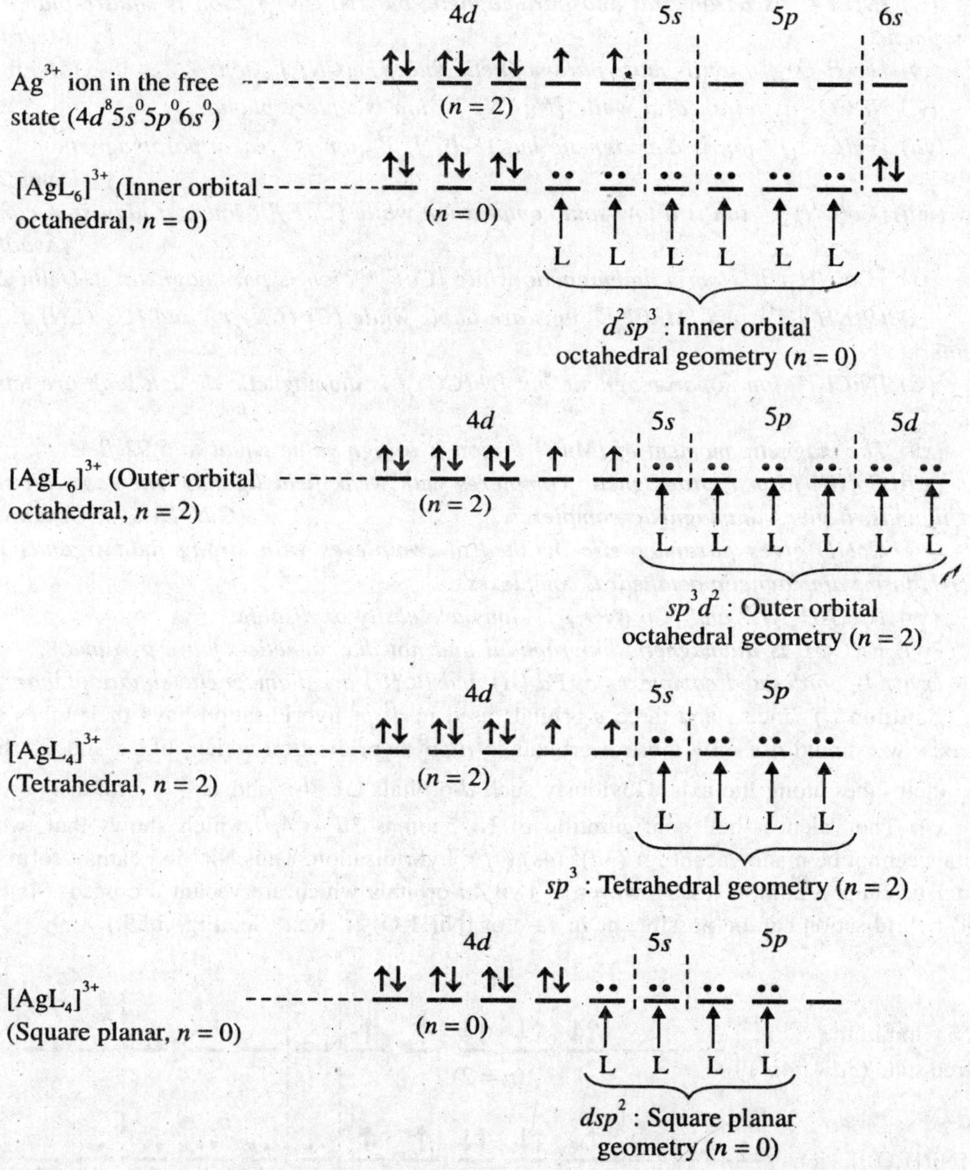

	4d					5s	5p			6s
Ag^{3+} ion in the free state ($4d^8 5s^0 5p^0 6s^0$)	↑↓	↑↓	↑↓	↑	↑	—	—	—	—	—
	($n = 2$)									
$[AgL_6]^{3+}$ (Inner orbital octahedral, $n = 0$)	↑↓	↑↓	↑↓	••	••	••	••	••	••	↑↓
	($n = 0$)			L	L	L	L	L	L	

d^2sp^3 : Inner orbital octahedral geometry ($n = 0$)

	4d					5s	5p			5d	
$[AgL_6]^{3+}$ (Outer orbital octahedral, $n = 2$)	↑↓	↑↓	↑↓	↑	↑	••	••	••	••	••	••
	($n = 2$)					L	L	L	L	L	L

sp^3d^2 : Outer orbital octahedral geometry ($n = 2$)

	4d					5s	5p		
$[AgL_4]^{3+}$ (Tetrahedral, $n = 2$)	↑↓	↑↓	↑↓	↑	↑	••	••	••	••
	($n = 2$)					L	L	L	L

sp^3 : Tetrahedral geometry ($n = 2$)

	4d					5s	5p		
$[AgL_4]^{3+}$ (Square planar, $n = 0$)	↑↓	↑↓	↑↓	↑↓	••	••	••	••	—
	($n = 0$)				L	L	L	L	

dsp^2 : Square planar geometry ($n = 0$)

Fig. 5.58. Type of hybridisation and the number of unpaired electrons in inner orbital octahedral, outer orbital octahedral, tetrahedral and square planar complexes of Ag (III)

Problem 23. *Explain the following, giving appropriate reasons.*

(i) The two d-orbitals used in d^2sp^3 hybridisation are d_{z^2} and $d_{x^2-y^2}$ orbitals of the central metal atom/ion.

(ii) Inner orbital octahedral complexes of Ni^{2+} ion are not formed

OR

All octahedral complexes of Ni^{2+} ion must be outer orbital octahedral complexes.

(iii) $[Fe(CN)_6]^{4-}$ ion is a low spin complex while $[Fe(H_2O)_6]^{3+}$ ion is a high spin complex.

(Lucknow 95)

(iv) $[NiCl_4]^{2-}$ is tetrahedral and paramagnetic but $[Ni(CN)_4]^{2-}$ ion is square planar and diamagnetic. *(Meerut 97)*

(v) $[Fe(H_2O)_6]^{3+}$ ion is more paramagnetic than $[Fe(CN)_6]^{3-}$ ion. *(Meerut 96)*

(vi) $Ni(CO)_4$ is tetrahedral while $[Ni(CN)_4]^{2-}$ ion is square planar.

(vii) $[Fe(CN)_6]^{4-}$ ion is diamagnetic but $[Fe(CN)_6]^{3-}$ ion is weakly paramagnetic. *(Delhi 2001)*

(viii) $[Co(CN)_6]^{3-}$ ion is a low spin complex ion while $[CoF_6]^{3-}$ ion is a high spin complex ion. *(Avadh 93)*

ix) $[Co(NH_3)_6]^{3+}$ ion is diamagnetic while $[CoF_6]^{3-}$ ion is paramagnetic. *(Delhi 2000)*

(x) $[V(NH_3)_6]^{3+}$ and $[MnCl_6]^{3-}$ ions are liable while $[Cr(CN)_6]^{3-}$ and $[Co(CN)_6]^{3-}$ ions are inert.

(xi) $[NiCl_4]^{2-}$ ion is paramagnetic but $[Ni(CO)_4]$ is diamagnetic though both are tetrahedral.

(xii) The magnetic moment of $[MnCl_4]^{2-}$ ion is a high value equal to 5.92 B.M.

(xiii) Co(III) forms paramagnetic complexes with weak field ligands whereas with strong field ligands, it gives diamagnetic complexes. *(G.N. Dev 94, Osmania 94)*

(xiv) Co(II) gives paramagnetic octahedral complexes with strong field ligands while Co (III) forms diamagnetic octahedral complexes.

(xv) $[Co(NH_3)_6]^{2+}$ and $[Co(NO_2)_6]^{4-}$ ions are easily oxidisable.

(xvi) $Fe(CO)_5$ is diamagnetic bipyramidal and not diamagnetic square pyramidal.

(xvii) 4-coordinated complexes of Pd (II) and Pt (II) are diamagnetic square planar.

Solution (*i*) Since *s*-and three *p*-orbitals used in d^2sp^3 hybridisation have their lobes along the axis, we should use only those *d*-orbitals in d^2sp^3 hybridisation which, like *s*-and *p*-orbitals, have their labes along the axis. Obviously such *d*-orbitals are d_{z^2} and $d_{x^2-y^2}$ orbitals.

(*ii*) The valence-shell configuration of Ni^{2+} ion is $3d^8 4s^0 4p^0$ which shows that two 3*d*-orbitals cannot be made vacant for $(3d)^2\ (4s)\ (4p)^3$ hybridisation. Thus Ni^{2+} ion cannot form inner orbital octahedral complexes. However, if two 4*d*-orbitals which are vacant are used, $(4s)\ (4p)^3\ (4d)^2$ hybridisation can be possible as in case of $[Ni(H_2O)_6]^{2+}$ ion. (See Fig. 5.59.)

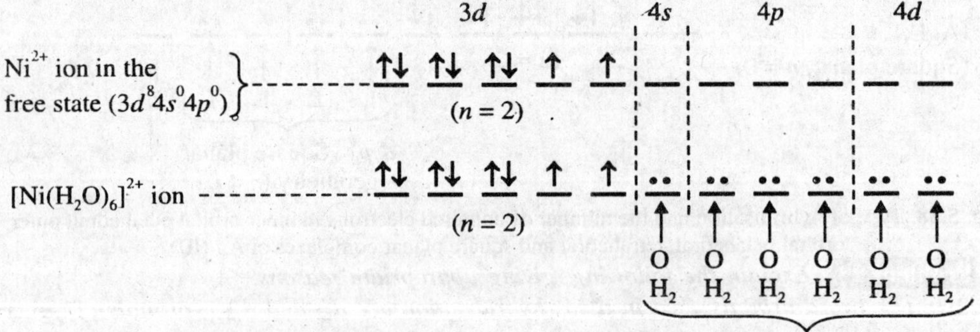

Fig. 5.59. Formation of $[Ni(H_2O)_6]^{2+}$ ion by sp^3d^2 hybridisation.

(*iii*) In $[Fe(CN)_6]^{4-}$ ion, Fe is present as Fe^{2+} ion whose valence-shell configuration is $3d^6 4s^0 4p^0$. Under the influence of six strong CN^- ligands, six electrons present in 3*d*-orbitals become paired and hence two 3d-orbitals *viz* $3d_{z^2}$ and $3d_{x^2-y^2}$ orbitals become available for d^2sp^3 hybridisation (inner orbital octahedral hybridisation). Thus, due to the presence of strong CN^- ligands, $[Fe(CN)_6]^{4-}$ ion has no unpaired electrons *i.e.* $[Fe(CN)_6]^{4-}$ ion is a low spin complex ion. (See Fig. 5.60.)

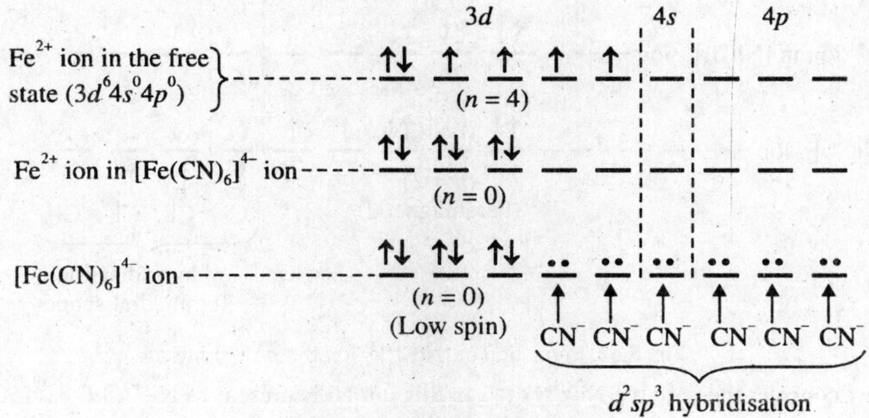

Fig. 5.60. Formation of $[Fe(CN)_6]^{4-}$ ion by d^2sp^3 hybridisation

Now consider the case of $[Fe(H_2O)_6]^{3+}$ ion. In this ion, the distribution of five electrons in 3d-orbitals of Fe^{3+} ion ($Fe^{3+} = 3d^54s^04p^0$) in the free state and in $[Fe(H_2O)_6]^{3+}$ ion remains the same, *i.e.*, these electrons are not paired. Thus $[Fe(H_2O)_6]^{3+}$ ion has five unpaired electrons and hence is high spin complex ion. It is formed by sp^3d^2 hybridisation (outer orbital octahedral hybridisation) as shown in Fig. 5.61.

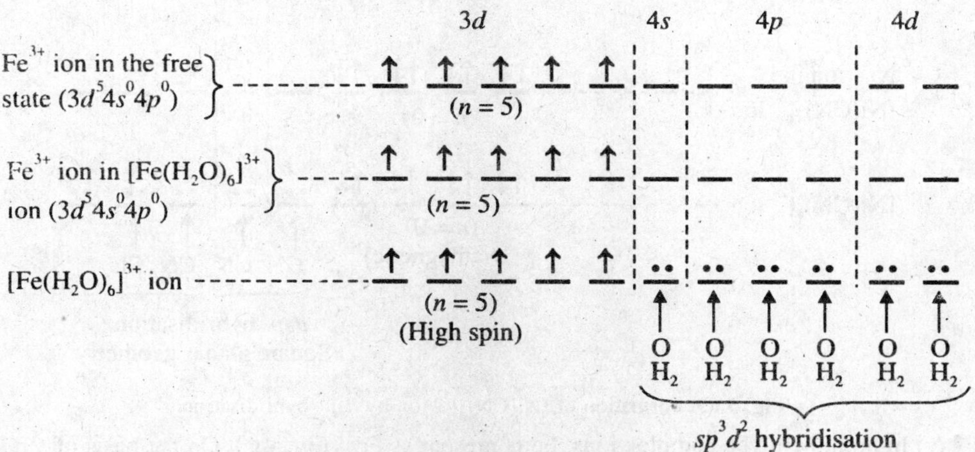

Fig. 5.61 Formation of $[Fe(H_2O)_6]^{3+}$ ion by sp^3d^2 hybridisation.

(*iv*) In $[NiCl_4]^{2-}$ ion, Ni is present as Ni^{2+} ion (central atom) whose valence-shell configuration in the free state is $3d^84s^04p^0$. Since Cl^- ions are weak ligands, the arrangement of eight 3d electrons in the free state of Ni^{2+} ion and in $[NiCl_4]^{2-}$ ion remains the same (n = 2). Thus $[NiCl_4]^{2-}$ ion is formed by sp^3 hybridisation as shown in Fig. 5.62. Due to sp^3 hybridisation, $[NiCl_4]^{2-}$ ion has tetrahedral geometry with n = 2. Thus $[NiCl_4]^{2-}$ is paramagnetic.

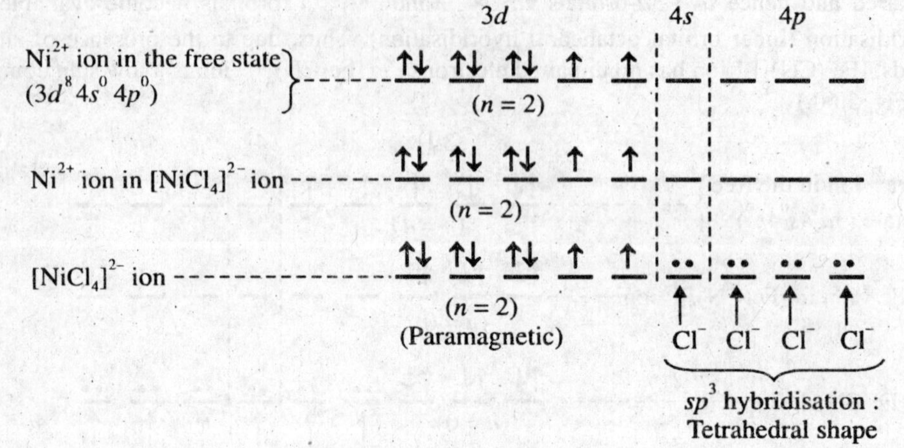

3d
4s
4p
Ni2+ ion in the free state (3d8 4s0 4p0)
(n = 2)
Ni2+ ion in [NiCl4]2− ion
(n = 2)
[NiCl4]2− ion
(n = 2)
(Paramagnetic)
Cl− Cl− Cl− Cl−
sp3 hybridisation :
Tetrahedral shape

Fig. 5.62. Formation of $[NiCl_4]^{2-}$ ion by sp^3 hybridisation.

Now consider the case of $[Ni(CN)_4]^{2-}$. In this ion Ni is present as Ni^{2+} ($3d^8\ 4s^0 4p^0$). Due to the presence of four strong CN^- ligands, one electron from $3d_{x^2-y^2}$ orbital gets paired with that in $3d_{z^2}$ orbital. Thus one *d*-orbital ($3d_{x^2-y^2}$ orbital) becomes available for dsp^2 hybridisation. dsp^2 hybridisation gives square planar geometry to $[Ni(CN)_4]^{2-}$ ion with $n = 0$. Thus $[Ni\ (CN)_4]^{2-}$ is diamagnetic. (See Fig. 5.63.)

3d
4s
4p
Ni2+ ion in the free state (3d8 4s0 4p0)
(n = 2)
z2 x2−y2
Ni2+ ion in [Ni(CN)4]2− ion
(n = 0)
z2 x2−y2
[Ni(CN)4]2− ion
(n = 0)
(Diamagnetic)
CN− CN− CN− CN−
dsp2 hybridisation :
Square planar geometry

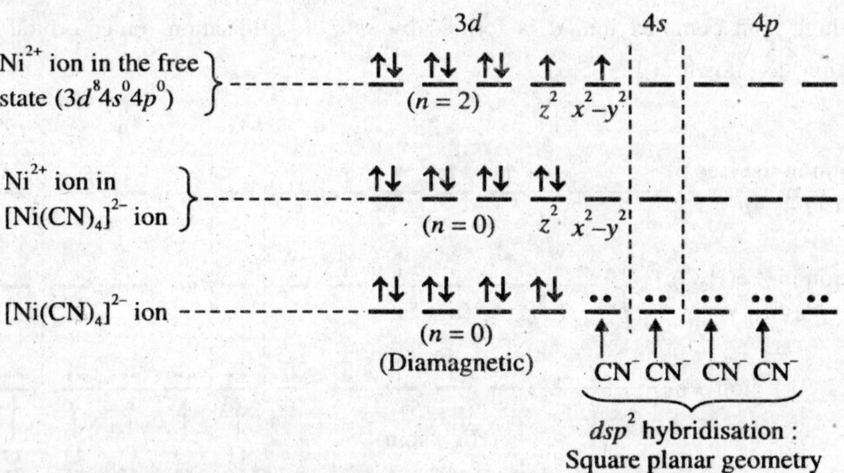

Fig. 5.63. Formation of $[Ni(CN)_4]^{2-}$ ion by dsp^2 hybridisation.

(*v*) In both the given complex ions, Fe is present as Fe^{3+} ($3d^5\ 4s^0$). On the basis of VBT, the formation of these ions can be explained as shown in Fig. 5.64.

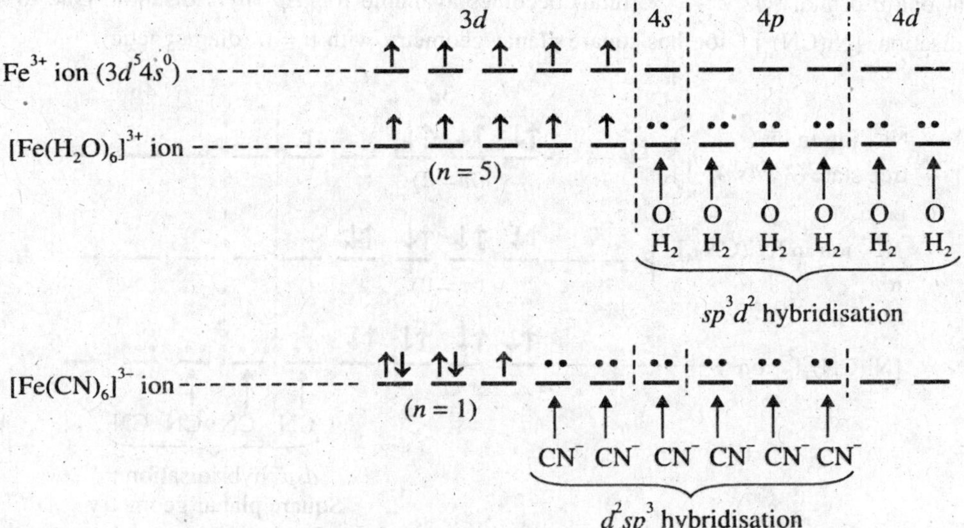

Fig. 5.64. Formation of $(Fe(H_2O)_6)^{3+}$ and $(Fe(CN)_6)^{3-}$ ions.

H_2O is much weaker ligand than CN^- ion and hence does not disturb the distribution of five electrons in 3*d* orbitals in the formation of $[Fe(H_2O)_6]^{3+}$ ion. Thus $[Fe(H_2O)_6]^{3+}$ ion has five unpaired electrons. On the other hand, since CN^- ions are strong ligands, they force the five 3*d* electrons to pair up in 3d orbitals. Thus $[Fe(CN)_6]^{3-}$ ion has only one unpaired electron. Due to the greater number of unpaired electrons in $[Fe(H_2O)_6]^{3+}$ ion ($n = 5$), this ion is more paramagnetic than $[Fe(CN)_6]^{3-}$ ion which has lesser number of unpaired electrons (n = 1).

(*vi*) In Ni $(CO)_4$, Ni is present as a neutral atom whose valence-shell configuration is $3d^8 4s^2 4p^0$. ($n = 2$). Under the influence of four CO ligands (which are strong ligands), two electrons present in 4*s* orbital get shifted to 3*d* orbitals to make all the electrons paired ($n = 0$) and to make 4*s* orbital available for sp^3 hybridisation. Thus, due to sp^3 hybridisation, $Ni(CO)_4$ molecule possesses tetrahedral shape (See Fig. 5.65.)

	3d	4s	4p
Ni-atom in free state $(3d^8 4s^2 4p^0)$	↑↓ ↑↓ ↑↓ ↑ ↑ ($n = 2$)	↑↓	— — —
Ni-atom in $Ni(CO)_4$ molecule	↑↓ ↑↓ ↑↓ ↑↓ ↑↓ ($n = 0$)	—	— — —
$Ni(CO)_4$ molecule	↑↓ ↑↓ ↑↓ ↑↓ ↑↓ ($n = 0$) (Diamagnetic)	•• ↑ CO	•• •• •• ↑ CO CO CO

sp^3 hybridisation : Tetrahedral geometry

Fig. 5.65. Formation of $Ni(CO)_4$ molecule by sp^3 hybridisation

In $[Ni(CN)_4]^{2-}$ ion, Ni is present as Ni^{2+} whose valence-shell configuration is $3d^8 4s^0 4p^0$ (n = 2). Under the influence of four strong CN^- ligands all the 8 electrons of 3*d*-orbitals become paired so that one *d*-orbital (*viz* $d_{x^2-y^2}$ orbital) becomes available for dsp^2 hybridisation. Due to dsp^2 hybridisation, $[Ni(CN)_4]^{2-}$ ion has square planar geometry with $n = 0$ (diamagnetic).

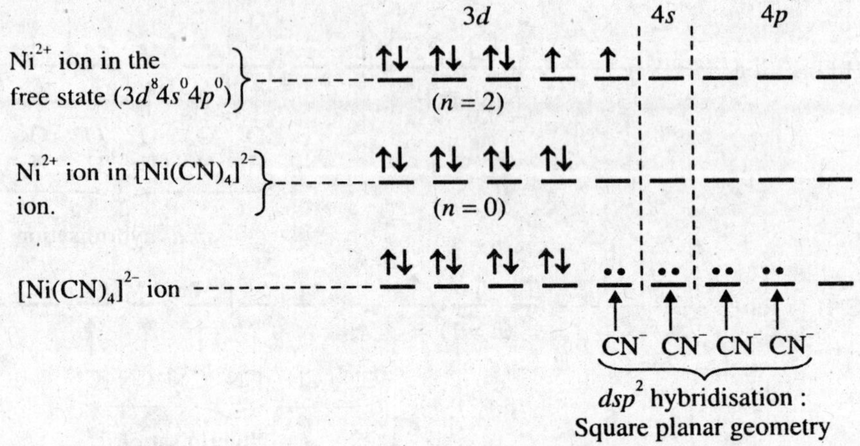

	3d					4s	4p		
Ni^{2+} ion in the free state ($3d^8 4s^0 4p^0$)	↑↓	↑↓	↑↓	↑	↑	—	—	—	—
	($n = 2$)								
Ni^{2+} ion in $[Ni(CN)_4]^{2-}$ ion.	↑↓	↑↓	↑↓	↑↓	—	—	—	—	—
	($n = 0$)								
$[Ni(CN)_4]^{2-}$ ion	↑↓	↑↓	↑↓	↑↓	••	••	••	••	—
					CN^-	CN^-	CN^-	CN^-	

dsp^2 hybridisation : Square planar geometry

Fig. 5.66. Formation of $[Ni(CN)_4]^{2-}$ ion by dsp^2 hybridisation.

(*vii*) In $[Fe(CN)_6]^{4-}$ ion, Fe is present as Fe^{2+} ion whose valence-shell configuration is $3d^6$ $4s^0$ $4p^0$. Under the influence of six CN^- ions (which are strong ligands), all the six 3*d* electrons become paired ($n = 0$) and two 3*d*-orbitals become available for d^2sp^3 hybridisation. Thus $[Fe(CN)_6]^{4-}$ ion is diamagnetic ($n = 0$). Due to the absence of any unpaired electron, $[Fe(CN)_6]^{4-}$ ion is diamagnetic.

	3d					4s	4p		
Fe^{2+} ion in the free state ($3d^6 4s^0 4p^0$)	↑↓	↑	↑	↑	↑	—	—	—	—
	($n = 4$)								
Fe^{2+} ion in $(Fe(CN)_6)^{4-}$ ion	↑↓	↑↓	↑↓	—	—	—	—	—	—
	($n = 0$)								
$[Fe(CN)_6]^{4-}$ ion	↑↓	↑↓	↑↓	••	••	••	••	••	••
($n = 0$) (Diamagnetic)				CN^-	CN^-	CN^-	CN^-	CN^-	CN^-

d^2sp^3 hybridisation : Octahedral geometry

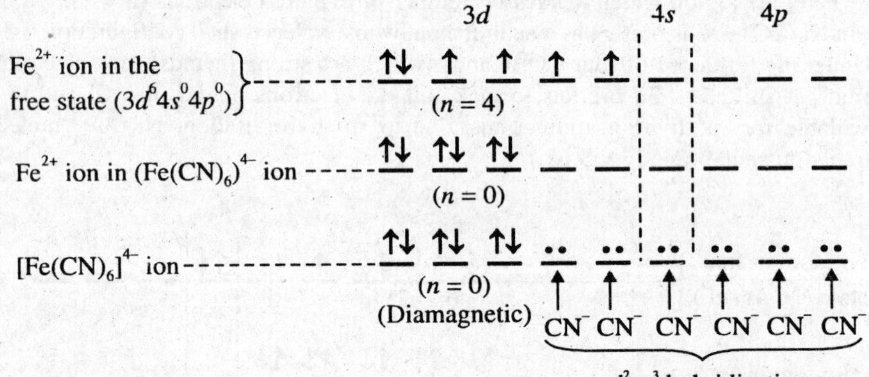

Fig. 5.67. Formation of $[Fe(CN)_6]^{4-}$ ion by d^2sp^3 hybridisation

In $[Fe(CN)_6]^{3-}$ ion, Fe is present as Fe^{3+} ion whose valence-shell configuration is $3d^5 4s^0 4p^0$. Under the influence of six CN^- ligands (strong ligands), all the five 3*d*-electrons occupy the three 3*d* orbitals and two 3*d* orbitals become available for d^2sp^3 hybridisation. This type of hybridisation gives octahedral shape to $[Fe(CN)_6]^{3-}$ ion with $n = 1$ (Fig. 5.68). Due to the presence of only one unpaired electron, $[Fe(CN)_6]^{3-}$ ion is weakly paremagnetic.

3d
4s
4p
Fe3+ ion in the free state (3d5 4s0 4p0)
(n = 5)
Fe3+ ion in [Fe(CN)6]3– ion
(n = 1)
[Fe(CN)6]3– ion
(n = 1)
(Paramagnetic)
CN– CN– CN– CN– CN– CN–
d2sp3 hybridisation :
Octahedral geometry

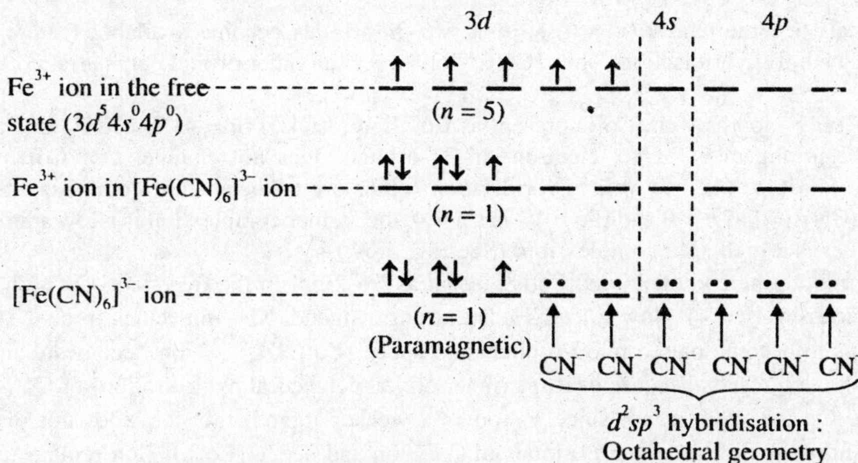

Fig. 5.68. Formation of $[Fe(CN)_6]^{3-}$ ion by d^2sp^3 hybridisation

(*viii*) In $[Co\ (CN)_6]^{3-}$ ion, Co (central metal atom) is present as Co^{3+} $(3d^6\ 4s^0)$ which has four unpaired electrons (n = 4). Under the influence of six strong ligands all the six electrons present

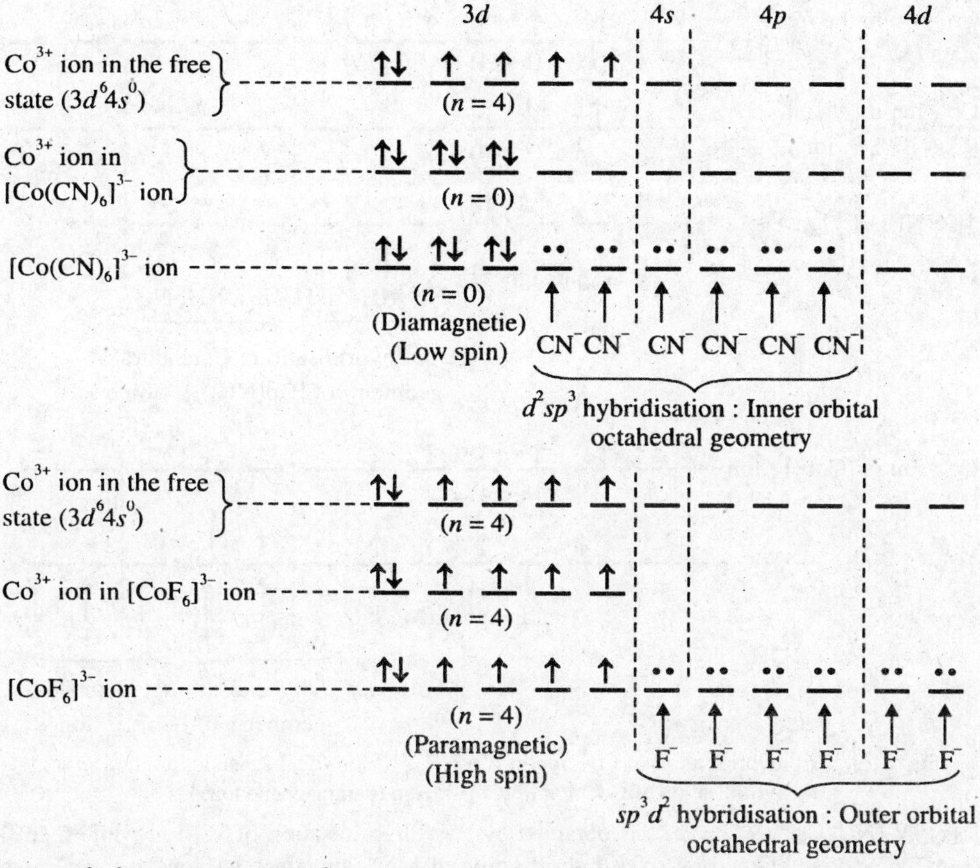

3d
4s
4p
4d
Co3+ ion in the free state (3d6 4s0)
(n = 4)
Co3+ ion in [Co(CN)6]3– ion
(n = 0)
[Co(CN)6]3– ion
(n = 0)
(Diamagnetie)
(Low spin)
CN– CN– CN– CN– CN– CN–
d2sp3 hybridisation : Inner orbital
octahedral geometry
Co3+ ion in the free state (3d6 4s0)
(n = 4)
Co3+ ion in [CoF6]3– ion
(n = 4)
[CoF6]3– ion
(n = 4)
(Paramagnetic)
(High spin)
F– F– F– F– F– F–
sp3d2 hybridisation : Outer orbital
octahedral geometry

Fig. 5.69. d^2sp^3 hybridisation of Co^{3+} ion in $[Co(CN)_6]^{3-}$ ion (low spin octahedral complex ion) and sp^3d^2 hybridisation of Co^{3+} ion in $[CoF_6]^{3-}$ ion (high spin octahedral complex ion)

in $3d$-orbitals become paired ($n = 0$) so that two $3d$-orbitals become available for d^2sp^3 (inner orbital octahedral) hybridisation. Thus $[Co(CN)_6]^{3-}$ ion is an inner orbital octahedral complex ion with n = 0.

In $[CoF_6]^{3-}$ ion as well, Co is present as Co^{3+} ion ($3d^6 4s^0$) ($n = 4$). Since F^- ions are weak ligands, the arrangement of six electrons in $3d$-orbitals does not change, *i.e.* $[CoF_6]^{3-}$ ion is formed by $(4s)$ $(4p)^3(4d)^2$ or sp^3d^2 hybridisation (outer orbital octahedral complex with $n = 4$) Since $[Fe (CN)_6]^{3-}$ has $n = 0$ and $[Fe F_6]^{3-}$ has $n = 4$, the former complex ion is a low spin complex and the latter is a high spin complex ion. (See Fig. 5.69.).

(*ix*) In both the complex ions, Co is present as Co^{3+} ion. In the free state, Co^{3+} ion has four unpaired electrons ($n = 4$). Now since NH_3 is a stronger ligand, NH_3 molecules force the unpaired $3d$-electrons to become paired in $3d$-orbitals and hence $[Co (NH_3)_6]^{3+}$ ion becomes diamagnetic. This complex ion results because of d^2sp^3 (inner orbital octahedral hybridisation) of Co^{3+} ion (See Fig. 5.70). On the other hand, since F^- ion is a weaker ligand, this ion does not disturb the arrangement of six electrons in $3d$ orbitals in Co^{3+} ion and hence $[CoF_6]^{3-}$ ion remains paramagnetic corresponding to the presence of four unpaired electrons. $[CoF_6]^{3-}$ ion results from sp^3d^2 hybridisation (outer orbital octahedral hybridisation) of Co^{3+} ion. (See Fig. 5.70.)

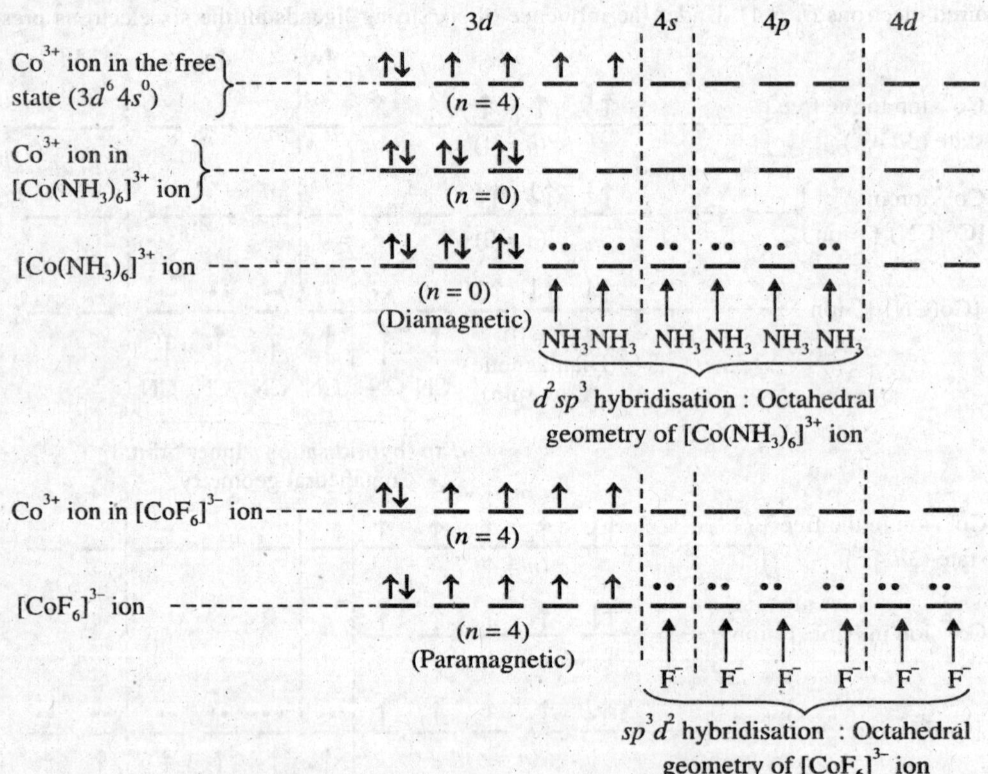

Fig. 5.70. d^2sp^3 hybridisation of Co^{3+} ion in $[Co(NH_3)_6]^{3+}$ ion (diamagenetic ion) and sp^3d^2 hybridisation of Co^{3+} ion in $[CoF_6]^{3-}$ ion (paramagnetic ion)

(*x*) $[V (NH_3)_6]^{3+}$ ($V^{3+} = 3d^2$) is obtained by d^2sp^3 hybridisation of V^{3+} ion while $[MnCl_6]^{3-}$ ion ($Mn^{3+} = 3d^4$) is obtained by sp^3d^2 hybridisation of Mn^{3+} ion, since Cl^- ions are weak ligands (See Fig. 5.71.). Fig 5.71 shows that $[V(NH_3)_6]^{3+}$ and $[MnCl_6]^{3-}$ both have one low lying unhybridised $3d$ orbital which is empty and hence is readily available to form bond with ligand. Thus both these complex ions are liable.

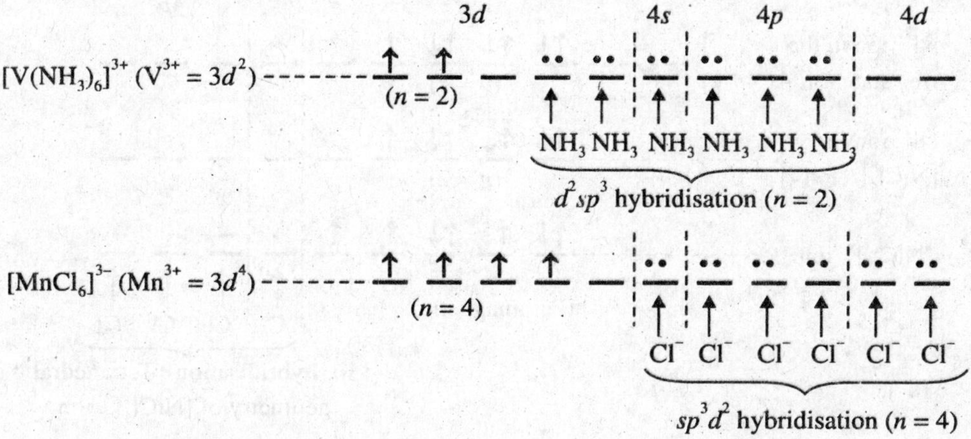

$[V(NH_3)_6]^{3+}$ ($V^{3+} = 3d^2$) ------ 3d: ↑ ↑ — •• •• | 4s: •• | 4p: •• •• •• | 4d: — —
(n = 2)
NH_3 NH_3 NH_3 NH_3 NH_3 NH_3
d^2sp^3 hybridisation (n = 2)

$[MnCl_6]^{3-}$ ($Mn^{3+} = 3d^4$) ------ ↑ ↑ ↑ ↑ — | •• | •• •• •• | •• ••
(n = 4)
Cl^- Cl^- Cl^- Cl^- Cl^- Cl^-
sp^3d^2 hybridisation (n = 4)

Fig. 5.71. d^2sp^3 and sp^3d^2 hybridisation in $[V(NH_3)_6]^{3+}$ and $[MnCl_6]^{3-}$ ions.

Since CN^- ions are strong ligands, $[Cr(CN)_6]^{3-}$ ($Cr^{3+} = 3d^3$) and $[Co(CN)_6]^{3-}$ ($Co^{3+} = 3d^6$) both involve d^2sp^3 hybridisation as shown in Fig 5.72. d^2sp^3 hybridisation scheme shows that none of these complex ions has any empty unhybridised 3*d* orbital which may be used to form bonds with ligands. Thus both these complex ions are inert.

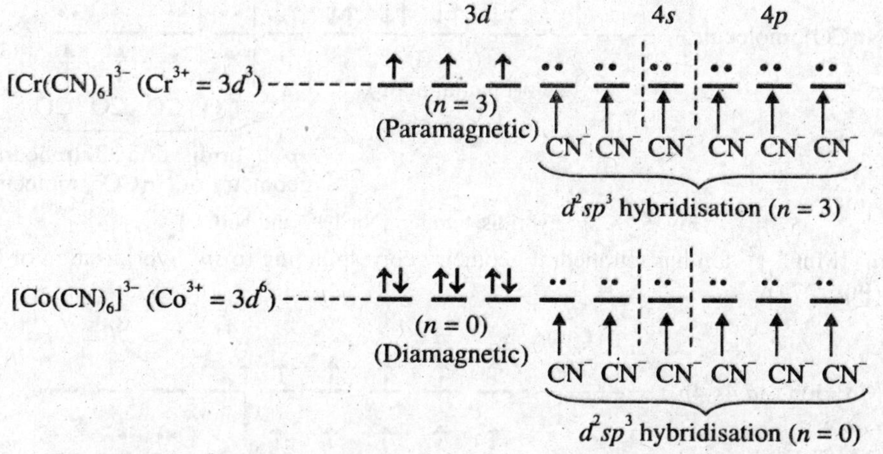

$[Cr(CN)_6]^{3-}$ ($Cr^{3+} = 3d^3$) ------ 3d: ↑ ↑ ↑ •• •• | 4s: •• | 4p: •• •• ••
(n = 3)
(Paramagnetic)
CN^- CN^- CN^- CN^- CN^- CN^-
d^2sp^3 hybridisation (n = 3)

$[Co(CN)_6]^{3-}$ ($Co^{3+} = 3d^6$) ------ ↑↓ ↑↓ ↑↓ •• •• | •• | •• •• ••
(n = 0)
(Diamagnetic)
CN^- CN^- CN^- CN^- CN^- CN^-
d^2sp^3 hybridisation (n = 0)

Fig. 5.72. d^2sp^3 hybridisation in $[Cr(CN)_6]^{3-}$ and $[Co(CN)_6]^{3-}$ ions.

(*xi*) In $[NiCl_4]^{2-}$ ion, Ni-atom is present as Ni^{2+} ion whose valence-shell configuration is $3d^8$ $4s^0$ ($Ni^{2+} = 3d^8\ 4s^0$ having $n = 2$). Since Cl^- ion is a weaker ligand, the arrangement of 8 electrons in 3*d*-orbitals in Ni^{2+} ion remains the same in $[NiCl_4]^{2-}$ ion. Thus $[NiCl_4]^{2-}$ ion results from sp^3 hybridisation (tetrahedral geometry) and is paramagnetic due to the presence of two unpaired electrons.

In case of $Ni(CO)_4$ molecule, Ni is present as a neutral atom whose valence-shell configuration is $3d^8\ 4s^2$. Since CO molecule is a stronger ligand, 4*s* electrons are forced to occupy 3*d* orbitals so that all the electrons become paired ($n = 0$). $[Ni(CO)_4]$ molecule results from sp^3 hybridisation (tetrahedral geometry) and is diamagnetic ($n = 0$). [See Fig. 5.73.]

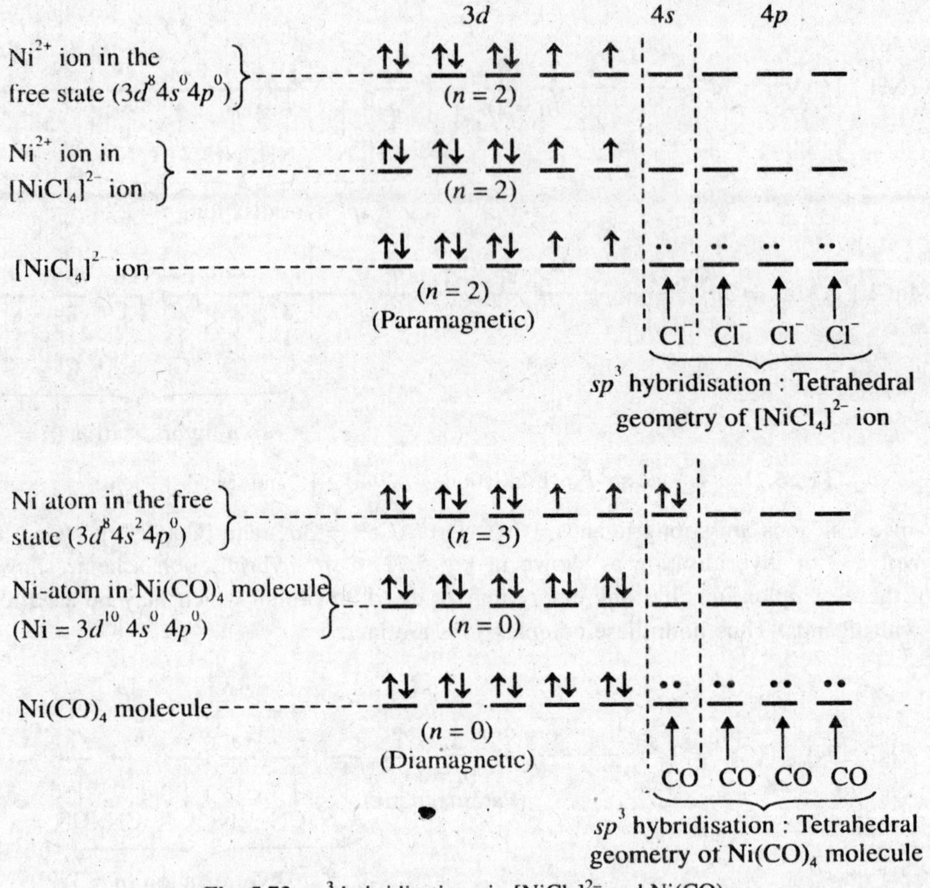

3d 4s 4p

Ni^{2+} ion in the free state ($3d^8 4s^0 4p^0$) ↑↓ ↑↓ ↑↓ ↑ ↑ (n = 2)

Ni^{2+} ion in $[NiCl_4]^{2-}$ ion ↑↓ ↑↓ ↑↓ ↑ ↑ (n = 2)

$[NiCl_4]^{2-}$ ion ↑↓ ↑↓ ↑↓ ↑ ↑ •• •• •• •• (n = 2) (Paramagnetic)

Cl^- Cl^- Cl^- Cl^-

sp^3 hybridisation : Tetrahedral geometry of $[NiCl_4]^{2-}$ ion

Ni atom in the free state ($3d^8 4s^2 4p^0$) ↑↓ ↑↓ ↑↓ ↑ ↑ ↑↓ (n = 3)

Ni-atom in $Ni(CO)_4$ molecule (Ni = $3d^{10} 4s^0 4p^0$) ↑↓ ↑↓ ↑↓ ↑↓ ↑↓ (n = 0)

$Ni(CO)_4$ molecule ↑↓ ↑↓ ↑↓ ↑↓ ↑↓ •• •• •• •• (n = 0) (Diamagnetic)

CO CO CO CO

sp^3 hybridisation : Tetrahedral geometry of $Ni(CO)_4$ molecule

Fig. 5.73. sp^3 hybridisatiuon in $[NiCl_4]^{2-}$ and $Ni(CO)_4$

(*xii*) $[MnCl_4]^{2-}$ ion has tetrahedral geometry corresponding to sp^3 hybridisation of Mn^{2+} ion ($n = 5$) (Fig. 5.74)

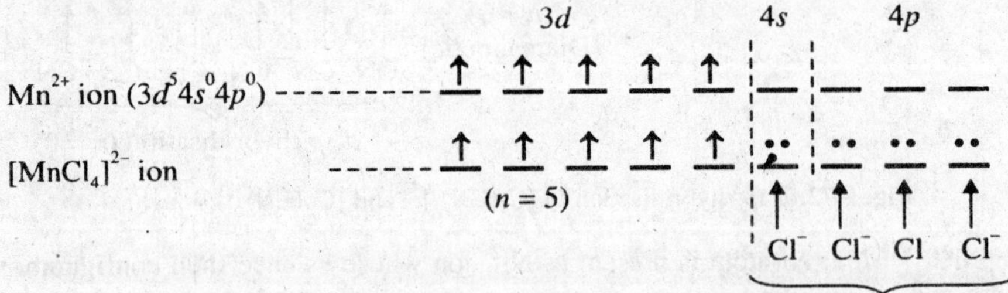

Fig. 5.74. sp^3 hybridisation in $[MnCl_4]^{2-}$ ion.

Due to the presence of five unpaired electrons ($n = 5$) magnetic moment value (μ) of $[MnCl_4]^{2-}$ ion is given by :

$$\mu = \sqrt{n(n+2)} = \sqrt{5 \times (5+2)}$$

$$= 5.916 \text{ BM}$$

Thus $[MnCl_4]^{2-}$ ion has high value for its magnetic moment.

(*xiii*) The valence-shell configuration of Co^{3+} ion in the free state is $3d^6 4s^0 4p^0 4d^0$ ($n = 4$) as shown at (*a*) of Fig. 5.75 This configuration has $n = 4$.

When weak field ligands like F^- ions get linked with Co^{3+} ion to form octahedral complexes, these ligands are not able to make the unpaired four 3*d* electrons to get paired in the same orbitals [Fig 5.75 (*b*)] and hence high spin (*i.e.*, paramagnetic) complexes are obtained. As an example see the formation of $[CoF_6]^{3-}$ ion given at (*c*) in Fig 5.75

On the other hand when strong field ligands like NH_3 molecules get attached with Co^{3+} ion, these ligands are able to pair up all the six 3*d* electrons [Fig (*d*)] so that two 3*d* orbitals become available for d^2sp^3 hybridisation. Thus low spin (*i.e.*, diamagnetic) complexes are obtained. See the formation of $[Co(NH_3)_6]^{3+}$ ion given at (*e*) of Fig. 5.75.

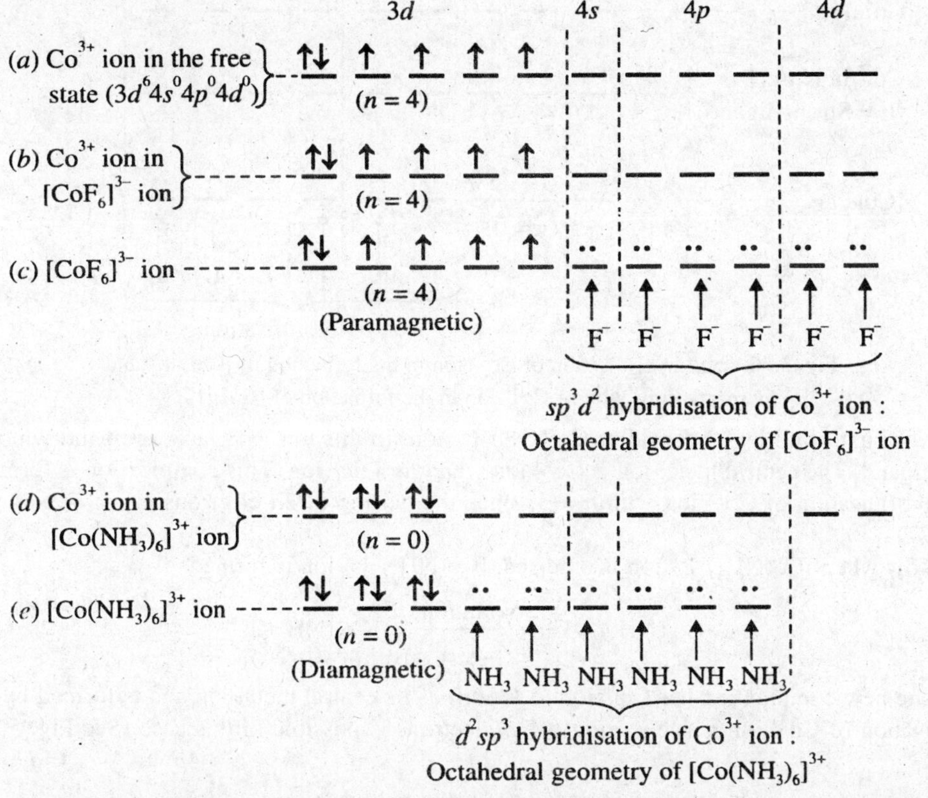

Fig. 5.75. sp^3d^2 hybridisation of Co^{3+} ion in $[CoF_6]^{3-}$ ion (high spin complex ion : paramagnetic) and d^2sp^3 hybridisation of Co^{3+} ion in $[Co(NH_3)_6]^{3+}$ ion (low spin complex ion : diamagnetic)

(*xiv*) For the formation of octahedral complexes of Co^{2+} ion ($Co^{2+} = 3d^7$), it is not possible to make two 3*d* orbitals vacant for d^2sp^3 hybridisation. Thus these complexes are obtained by sp^3d^2 hybridisation in which two *d*-orbitals used are 4*d* orbitals. sp^3d^2 hybridisation gives $n = 3$ which shows that these complexes are paramagnetic (See Fig. 5.76).

In presence of strong ligands, the electrons in two 3*d* orbitals of Co^{3+} ion ($Co^{3+} = 3d^6$) are forced to get paired and hence two 3*d* orbitals become empty for d^2sp^3 hybridisation. Thus these complexes are obtained by d^2sp^3 hybridisation which shows that $n = 0$ and hence these complexes are diamagnetic (See Fig. 5.76.)

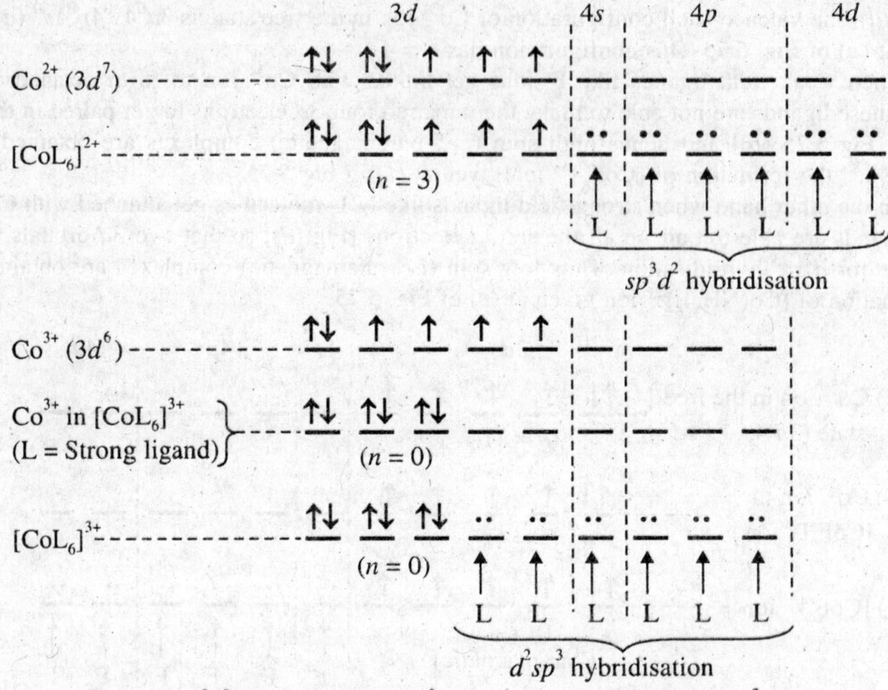

Fig. 5.76. sp^3d^2 hybridisation of Co^{2+} ion in the formation of $[CoL_6]^{2+}$ and d^2sp^3 hybridisation of Co^{3+} ion in the formation of $[CoL_6]^{3+}$.

(*xv*) First of all let us consider $[Co(NH_3)_6]^{2+}$ ion. In this ion, NH_3 acts as ligand which is a strong ligand. The central metal ion is Co^{2+} ion which is a $3d^7$ ion. This complex ion is formed by sp^3d^2 hybridisation of Co^{2+} ion with $n = 3$, since the pairing of $3d$ electrons is not possible (See Fig. 5.77)

Now when $[Co(NH_3)_6]^{2+}$ ion is oxidised, $[Co(NH_3)_6]^{3+}$ ion is formed.

$$\underset{sp^3d^2\ (n=3)}{[Co\ (NH_3)_6]^{2+}} \xrightarrow[(-e^-)]{\text{oxidation}} \underset{d^2sp^3\ (n=0)}{[Co\ (NH_3)_6]^{3+}}$$

The new complex ion has Co^{3+} ion ($3d^6$ ion) as its central metal ion and is formed by d^2sp^3 hybridisation of Co^{3+} ion, since pairing of $3d$ electrons is possible in this case (See Fig. 5.77.).

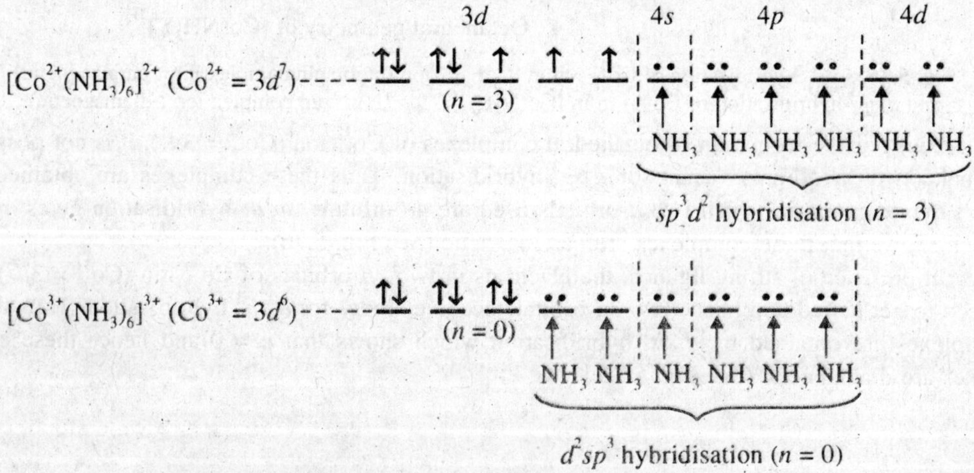

3d 4s 4p 4d

$[Co^{2+}(NH_3)_6]^{2+}$ ($Co^{2+} = 3d^7$) (n = 3)

NH_3 NH_3 NH_3 NH_3 NH_3 NH_3

sp^3d^2 hybridisation (n = 3)

$[Co^{3+}(NH_3)_6]^{3+}$ ($Co^{3+} = 3d^6$) (n = 0)

NH_3 NH_3 NH_3 NH_3 NH_3 NH_3

d^2sp^3 hybridisation (n = 0)

Fig. 5.77. Formation of $[Co(NH_3)_6]^{2+}$ and $[Co(NH_3)_6]^{3+}$ ions

Since in d^2sp^3 hybridisation, inner 3*d* orbitals are used, stronger bonds are formed in $[Co(NH_3)_6]^{3+}$ ion and hence this ion is more stable than $[Co(NH_3)_6]^{2+}$ ion. Thus it is due to the lesser stability of sp^3d^2 hybridisation in $[Co(NH_3)_6]^{2+}$ ion that this ion is easily oxidised to $[Co(NH_3)_6]^{3+}$ ion.

On the same lines of arguements, it can be shown that $[Co(NO_2)_6]^{4-}$ ion can easily be oxidised to $[Co(NO_2)_6]^{3-}$ ion.

(*xvi*) In Fe $(CO)_5$, CO molecules act as neutral monodentate ligands. Thus Fe has oxidation state equal to zero in Fe $(CO)_5$ molecule and has an oxidation number equal to 5. Thus, according to VBT, Fe $(CO)_5$ may have trigonal bipyramidal geometry (dsp^3 hybridisation) or square pyramidal geometry (d^2sp^2 hybridisation). Both the geometries have $n = 0$ and hence are diamagnetic as shown in Fig. 5.78 and 5.79 respectively.

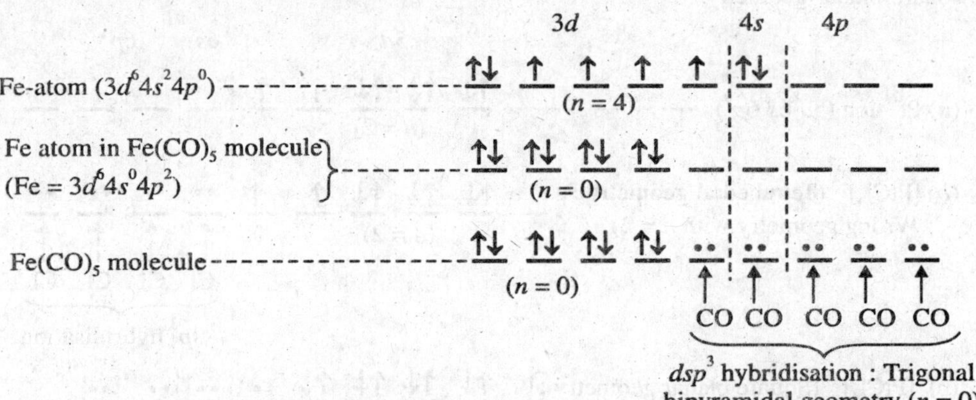

Fig. 5.78. Trigonal bipyramidal geometry ($n = 0$) of $Fe(CO)_5$ molecule resulted from dsp^3 hybridisation of Fe atom.

In d^2sp^2 hybridisation, since there are not enough 3*d* orbitals to accomodate eight electrons of Fe atom, two electrons from 3*d* orbitals have to be promoted to higher energy outer 4*p* orbitals. It is an unfavourable process, since the promotion of electrons to higher energy outer orbitals will make the electrons vulnerable to oxidation. Hence square pyramidal geometry is not a preferred geometry for Fe $(CO)_5$ molecule.

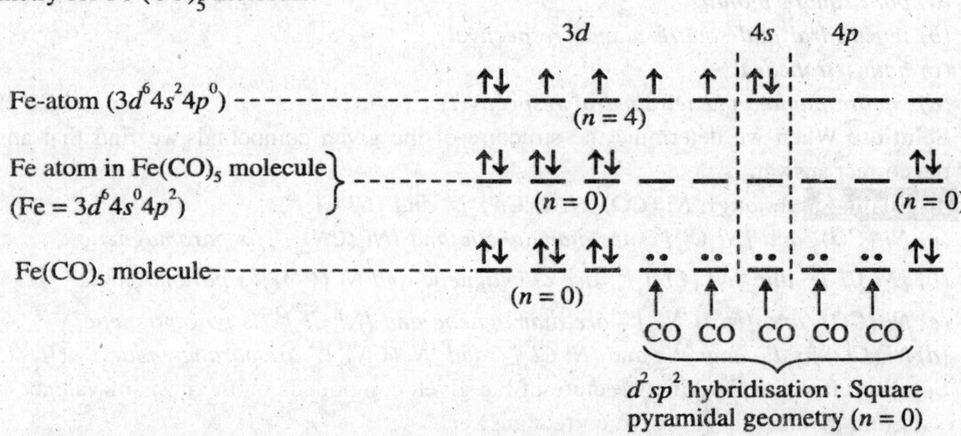

3*d* 4*s* 4*p*

Fe-atom ($3d^6 4s^2 4p^0$) ↑↓ ↑ ↑ ↑ ↑ | ↑↓ | — — —
($n = 4$)

Fe atom in $Fe(CO)_5$ molecule ($Fe = 3d^8 4s^0 4p^2$) ↑↓ ↑↓ ↑↓ — — | — | — — ↑↓
($n = 0$) ($n = 0$)

$Fe(CO)_5$ molecule ↑↓ ↑↓ ↑↓ ·· ·· | ·· | ·· ·· ↑↓
($n = 0$)

CO CO CO CO CO

d^2sp^2 hybridisation : Square pyramidal geometry ($n = 0$)

Fig. 5.79. Square pyramidal geometry with $n = 0$ (Wrong geometry) of $Fe(CO)_5$ molecule due to d^2sp^2 hybridisation of Fe atom.

(*xvii*) Examples of such 4-coordinated complexes are $[Pd(H_2O)_4]^{2+}$, $[PdCl_4]^{2-}$, $[Pd(en)_2]^{2+}$, $[PtCl_4]^{2-}$, $[Pt(NH_3)_4]^{2+}$, $[Pt(NH_3)_2Cl_2]^0$, K $[Pt(C_2H_4)Cl_3]$. H_2O (Zeiser's salt), $[Pt(CN)_4]^{2-}$ etc. All these complexes have square planar geometry. In all these complexes Pd and Pt are present as Pd^{2+} and Pt^{2+} (central metal ion). Both these ions are d^8 ions with $n = 2$. As an example let us discuss the structure of $[Pt\ Cl_4]^{2-}$ ion. Valence-shell configuration of Pt^{2+} ($5d^8\ 6s^0\ 6p^0$) contains two unpaired electrons ($n = 2$) as shown at (*a*) of Fig. 5.80. Since C.N. of Pt^{2+} ion $[Pt\ Cl_4]^{2-}$ is 4, $[Pt\ Cl_4]^{2-}$ ion has either tetrahedral geometry (sp^3 hybridisation) or square planar geometry (dsp^2 hybridisation) as shown at (*b*) and (*c*) of Fig. 5.80 respectively. From this figure it may be seen that sp^3 hybridisation gives $n = 2$ (paramagnetic) while dsp^2 hybridisation gives $n = 0$ (diamagnetic) Experiments have, however, shown that 4-coordinated complexes of Pt (II) are diamagnetic ($n = 0$). Thus it is dsp^2 hybridisation which is involved in the formation of $[PtCl_4]^{2-}$ ion, *i.e.*, 4-coordinated complexes of Pt (II) have square planar geometry with $n = 0$ (diamagnetic)

On the same lines of argument it can be shown that 4-coordinated complexes of Pd (II) also have square planar geometry.

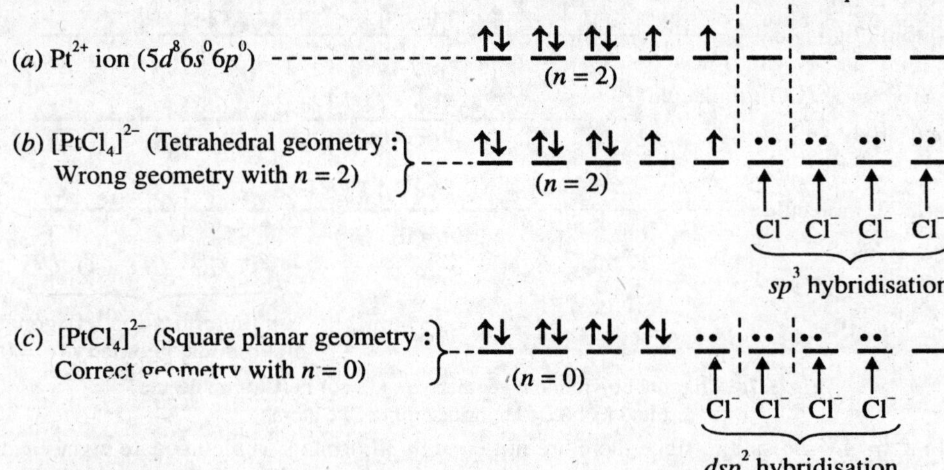

Fig. 5.80. Two possible geometries of $[PtCl_4]^{2-}$ ion.

Problem 24. *The geometry of [Ni (CO)$_4$] and [Ni (PPh$_3$)$_2$ Cl$_2$] are:*

(a) both square planar

(b) tetrahedral and square planar respectively

(c) both tetrahedral

(d) square planar and tetrahedral respectively. *(IIT 1999)*

Solution. When we determine the structure of the given compounds, we find that answer (*b*) is the correct answer.

Problem 25. *Amongst Ni (CO)$_4$, [Ni (CN)$_4$]$^{2-}$ and [Ni Cl$_4$]$^{2-}$*

(a) Ni (CO)$_4$ and [Ni Cl$_4$]$^{2-}$ are diamagnetic and [Ni (CN)$_4$]$^{2-}$ is paramagnetic.

(b) [Ni Cl$_4$]$^{2-}$ and [Ni (CN)$_4$]$^{2-}$ are diamagnetic and Ni (CO)$_4$ is paramagnetic.

(c) Ni (CO)$_4$ and [Ni (CN)$_4$]$^{2-}$ are diamagnetic and [Ni Cl$_4$]$^{2-}$ is paramagnetic.

(d) Ni (CO)$_4$ is diamagnetic and [Ni Cl$_4$]$^{2-}$ and [Ni (CN)$_4$]$^{2-}$ are paramagnetic. *(IIT 1991)*

Solution. If we explain the structure of the given compounds on the basis of valence-bond theory, we find that answer (*c*) is the correct answer.

Problem 26. *Based on CFT, draw the energy level diagram and write the electronic configuration of the central metal atom/ion in the following. Also determine the value of* μ.

$[Ni(NH_3)_6]^{2+}$, $[Cr(CN)_6]^{3-}$, $[CrL_6]^{3+}$ (L = H_2O) or NH_3), $[Mn(CN)_6]^{4-}$, $[Co(CN)_6]^{3-}$, $[Ir(NH_3)_6]^{3+}$, d^6 ion in octahedral low spin complex ion, $[PtCl_6]^{2-}$, $[Fe(CN)_6]^{4-}$, $[Fe(H_2O)_6]^{2+}$, $[IrCl_6]^{3-}$, $[Rh(NH_3)_6]^{3+}$, $[CoF_6]^{3-}$, $[Mn(CN)_6]^{3-}$, $[CoF_6]^{4-}$, $[Pt(NH_3)_6]^{4+}$, $[Co(H_2O)_6]^{2+}$, $[Co(NH_3)_6]^{2+}$ *(M.D. Rohtak 95, Raj 96, Punjab 94)*

Solution. $[Ni(NH_3)_6]^{2+}$ ion. In this ion, Ni (central metal atom) is present as Ni^{2+} which is a $3d^8$ ion. Distribution of $3d^8$ electrons of Ni^{2+} ion in t_{2g} and e_g orbitals in presence of octahedral ligand field due to six NH_3 molecules (ligands) is as shown in Fig. 5.81 This figure shows that:

Electronic configuration Ni^{2+} ion ($3d^8$ ion)

$$= t^6_{2g}\, e_g^{\,2}\ (n = 2)$$

$$\text{Since } n = 2,\ \mu = \sqrt{n(n+2)}\ \text{BM}$$

$$= \sqrt{2\times(2+2)}\ \text{BM}$$

$$= 2.82\ \text{BM}$$

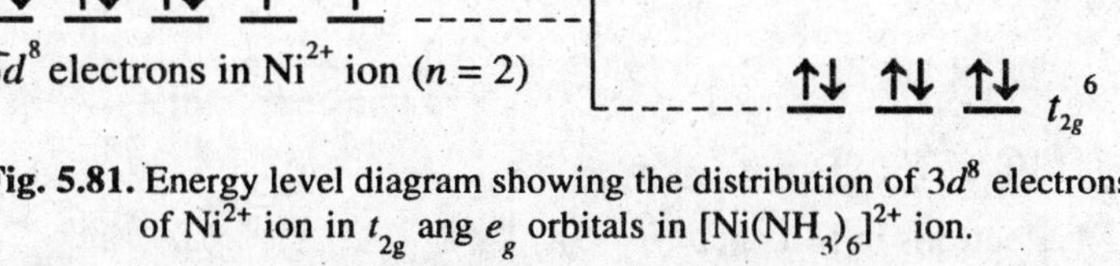

Fig. 5.81. Energy level diagram showing the distribution of $3d^8$ electrons of Ni^{2+} ion in t_{2g} ang e_g orbitals in $[Ni(NH_3)_6]^{2+}$ ion.

$[Cr(CN)_6]^{3-}$ ion. In this ion Cr is present as Cr^{3+} ion which is a $3d^3$ ion. Distribution of $3d^3$ electrons in t_{2g} and e_g orbitals in presence of octahedral ligand field due to six CN^- ions (ligands) is as shown in Fig. 5.82. This figure shows that :

Electronic configuration of Cr^{3+} ion ($3d^3$ ion)

$$= t^3_{2g}\, e_g^{\,0}\ (n = 3)$$

$$\text{Since}\quad n = 3,\quad \mu = \sqrt{n(n+2)}\ \text{BM}$$

$$= \sqrt{3\times(3+2)}\ \text{BM}$$

$$= 3.87\ \text{BM}$$

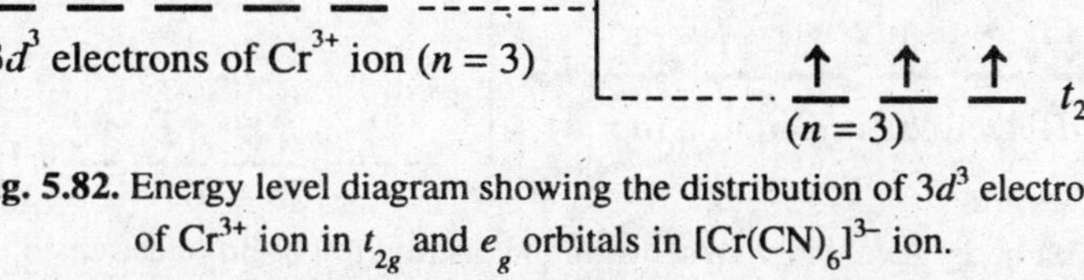

Fig. 5.82. Energy level diagram showing the distribution of $3d^3$ electrons of Cr^{3+} ion in t_{2g} and e_g orbitals in $[Cr(CN)_6]^{3-}$ ion.

$[CrL_6]^{3+}$ ion (L = H_2O or NH_3). In these complex ions, Cr is present as Cr^{3+} ion which is a $3d^3$ ion (Cr = $3d^5\ 4s^1$, Cr^{3+} = $3d^3$). The distribution of $3d^3$ electrons of Cr^{3+} ion in t_{2g} and e_g orbitals is as shown in energy level diagram given in Fig. 5.83. This figure shows that:

Electronic configuration of Cr^{3+} ion ($3d^3$ ion)

$$= t_{2g}^{\,3}\, e_g^{\,0}\ (n = 3)$$

Since $n = 3, \mu = \sqrt{n(n+2)}$ BM

$= \sqrt{3 \times (3+2)}$ BM = 3.87 BM

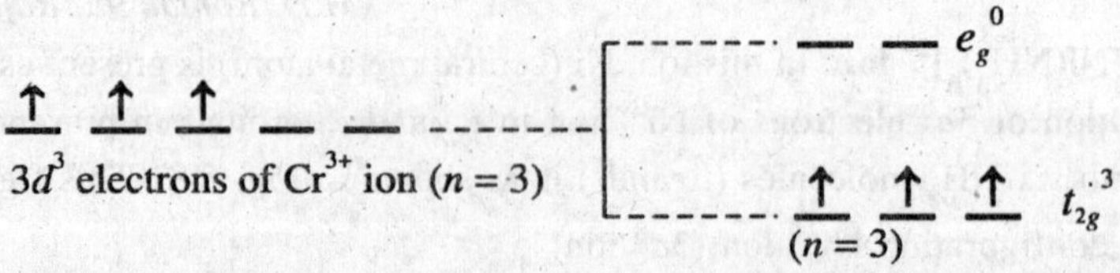

$3d^3$ electrons of Cr^{3+} ion ($n = 3$) — e_g^0, t_{2g}^3 ($n = 3$)

Fig. 5.83. Energy level diagram showing the distibution of $3d^3$ electrons of Cr^{3+} ion in t_{2g} and e_g orbitals in $[CrL_6]^{3+}$ ions (L = H_2O or NH_3)

$[Mn(CN)_6]^{4-}$ ion. This complex ion has Mn^{2+} ion as its central metal ion. This ion is a $3d^5$ ion. Now since CN^- ions are strong ligands, the distribution of $3d^5$ electrons of Mn^{2+} ion in t_{2g} and e_g orbitals is as shown in energy level diagram given in Fig. 5.84. This figure shows that:

Electronic configuration of Mn^{2+} ion ($3d^5$ ion) = $t^5_{2g}\ e_g^0$ ($n = 1$)

Since, as is evident from Fig. 5.84, $n = 1$,

$$\mu = \sqrt{n(n+2)}\ \text{BM} = \sqrt{1 \times (1+2)}\ \text{BM} = 1.73\ \text{BM}$$

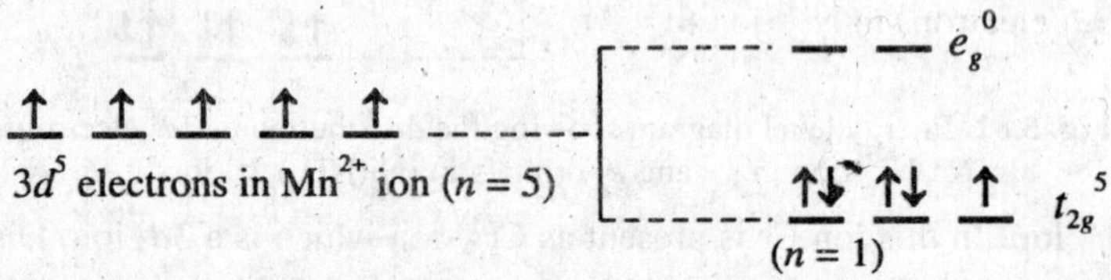

$3d^5$ electrons in Mn^{2+} ion ($n = 5$) — e_g^0, t_{2g}^5 ($n = 1$)

Fig. 5.84. Energy level diagram showing the distribution of $3d^5$ electrons of Mn^{2+} ion in t_{2g} and e_g orbitals in $[Mn(CN)_6]^{2-}$ ion.

$[Co(CN)_6]^{3-}$ ion. In this ion Co-atom (central metal atom) is present as Co^{3+} ion which is a $3d^6$ ion. Now since CN^- ions are strong ligands, the distribution of $3d^6$ electrons of Co^{3+} ion in t_{2g} and e_g orbitals in presence of six CN^- ligands is as shown in the energy level diagram given in Fig. 5.85 This figure shows that:

Electronic configuration of Co^{3+} ion ($3d^6$ ion)

$= t^6_{2g}\ e_g^0$ ($n = 0$)

Since $n = 0$, $[Co(CN)_3]^{3-}$ ion is diamagnetic and hence $\mu = 0$

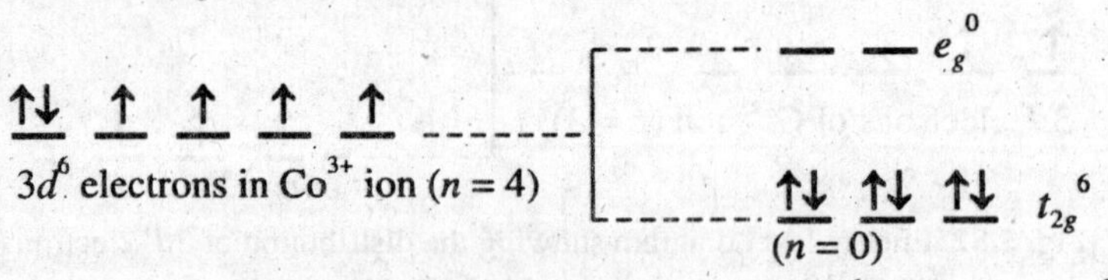

$3d^6$ electrons in Co^{3+} ion ($n = 4$) — e_g^0, t_{2g}^6 ($n = 0$)

Fig. 5.85. Energy level diagram showing the distribution of $3d^6$ electrons of Co^{3+} ion in t_{2g} and e_g orbitals in $[Co(CN)_6]^{3-}$ ion.

$[Ir(NH_3)_6]^{3+}$ ion. In this ion, Ir is present as Ir^{3+} ion which is a $5d^6$ ion (Ir = $5d^7\ 4s^2$, Ir^{3+} = $5d^6$). Since NH_3 molecules are strong ligands, the distribution of $5d^6$ electrons of Ir^{3+} ion in t_{2g} and e_g orbitals is as shown in Fig. 5.85 Thus:

Electronic configuration of Ir^{3+} ion ($5d^6$ ion)

$= t_{2g}^6\ e_g^0$ ($n = 0$)

This configuration gives $n = 0$ and hence $[Ir(NH_3)_6]^{3+}$ ion is diamagnetic with $\mu = 0$.

d^6 ion in octahedral low spin complex ion. According to CFT, under the influence of octahedral ligand field, *d*-orbitals of the central metal atom/ion of an octahedral complex ion are splitted into t_{2g} (lower energy) and e_g (higher energy) orbitals as shown in Fig. 5.86. Since we are given low spin state, d^6 electrons will get paired into lower energy t_{2g} orbitals (See Fig. 5.86) and hence:

Electronic configuration of d^6 ion = $t_{2g}^{\,6}\, e_g^{\,0}$ (n = 0)

Thus since $n = 0$, $\mu = 0$

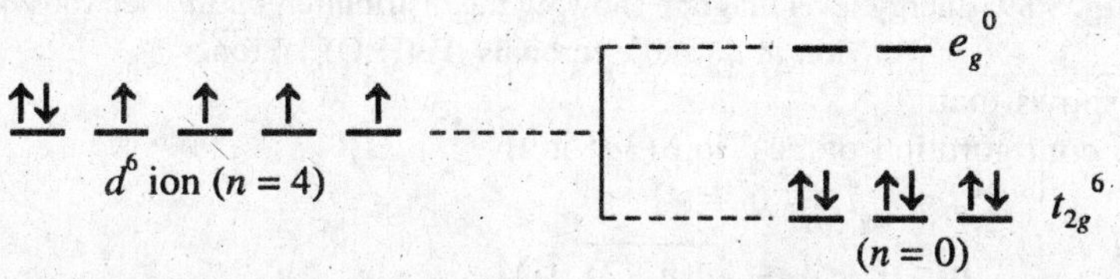

e_g^0

↑↓ ↑ ↑ ↑ ↑

d^6 ion ($n = 4$)

↑↓ ↑↓ ↑↓ t_{2g}^6

($n = 0$)

Fig. 5.86. Energy level diagram for d^6 ion in octahedral low spin complex ion

$[PtCl_6]^{2-}$ ion. Here Pt is present as Pt^{4+} ion which is a $5d^6$ ion (Pt = $5d^9\,6s^1$, $Pt^{4+} = 5d^6$). Under the influence of six Cl^- ions (ligands), all the six $5d$ electrons of Pt^{4+} ion get paired in t_{2g} orbitals and e_g orbitals remain vacant (See Fig. 5.87). Thus :

Electronic configuration of Pt^{4+} ion ($5d^6$ ion) = $t_{2g}^{\,6}\, e_g^{\,0}$ ($n = 0$)

Since $n = 0$, $[PtCl_6]^{2-}$ ion is diamagnetic having $\mu = 0$

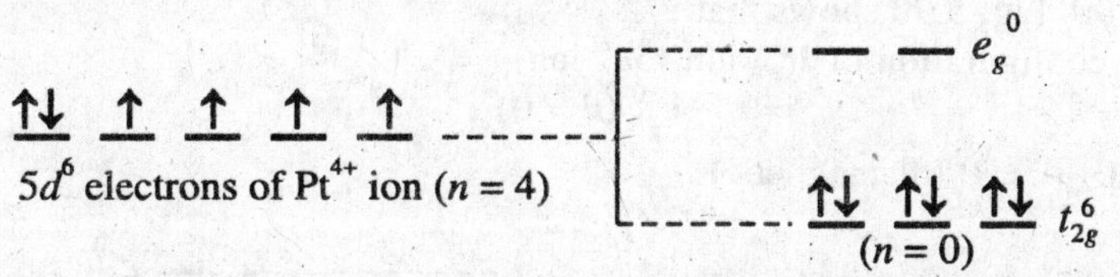

e_g^0

↑↓ ↑ ↑ ↑ ↑

$5d^6$ electrons of Pt^{4+} ion ($n = 4$)

↑↓ ↑↓ ↑↓ t_{2g}^6

($n = 0$)

Fig. 5.87. Energy level diagram for d^6 ion in Pt^{4+} ion in $[PtCl_6]^{2-}$ ion.

$[Fe(CN)_6]^{4-}$ ion. In this complex ion, the central metal ion is Fe^{2+} which is a $3d^6$ ion (Fe = $3d^6\,4s^2$, $Fe^{2+} = 3d^6$). Now since CN^- ions (ligands) are strong ligands, the distribution of $3d^6$ electrons of Fe^{2+} ion in t_{2g} and e_g orbitals is as shown in energy level diagram given in Fig. 5.88. This figure shows that:

Electronic configuration of Fe^{2+} ion ($3d^6$ ion)

$= t_{2g}^{\,6}\, e_g^{\,0}$ ($n = 0$)

Since $n = 0$, $\mu = 0$ (diamagnetic)

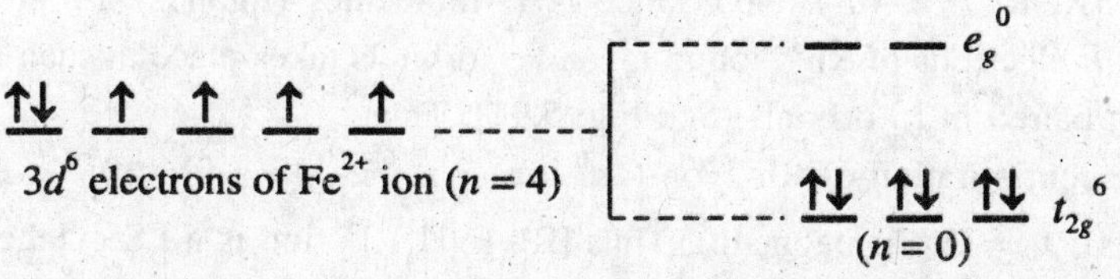

e_g^0

↑↓ ↑ ↑ ↑ ↑

$3d^6$ electrons of Fe^{2+} ion ($n = 4$)

↑↓ ↑↓ ↑↓ t_{2g}^6

($n = 0$)

Fig. 5.88. Energy level diagram showing the distribution of $3d^6$ electrons of Fe^{2+} ion in t_{2g} and e_g orbitals in $[Fe(CN_6)]^{4-}$ ion

$[Fe(H_2O)_6]^{2+}$ ion. In this ion Fe (central metal atom) is present as Fe^{2+} ion which is a $3d^6$ ion (Fe = $3d^6\,4s^2$, $Fe^{2+} = 3d^6$). Now since H_2O molecules (ligands) are weak ligands, the distribution of $3d^6$ electrons of Fe^{2+} ion in t_{2g} and e_g orbitals takes place as shown in energy level diagram given in Fig. 5.89

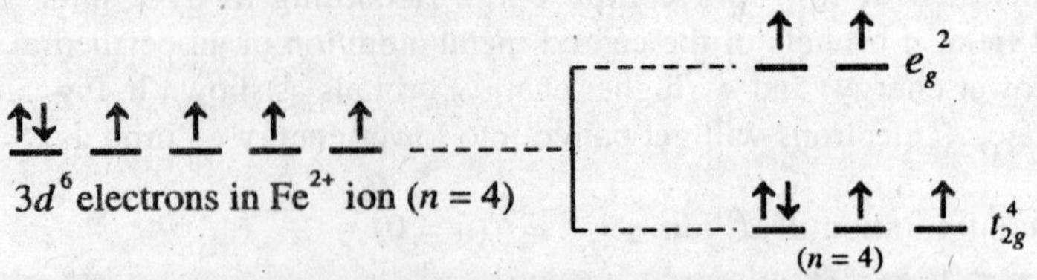

↑ ↑ e_g^2

↑↓ ↑ ↑ ↑ ↑

$3d^6$ electrons in Fe^{2+} ion ($n = 4$)

↑↓ ↑ ↑ t_{2g}^4

($n = 4$)

Fig. 5.89. Energy level diagram showing the distribution of $3d^6$ electrons of Fe^{2+} ion in t_{2g} and e_g orbitals $[Fe(H_2O)_6]^{2+}$ ion.

Fig. 5.89 shows that:

Electronic configuration of Fe^{2+} ion ($3d^6$ ion)

$$= t_{2g}^{\ 4}\ e_g^{\ 2}\ (n = 4)$$

Since $n = 4,\ \mu = \sqrt{n(n+2)}$ BM

$$= \sqrt{4 \times (4+2)} \text{ BM} = 4.898 \text{ BM.}$$

$[IrCl_6]^{3-}$ ion. In this complex ion, Ir is present as Ir^{3+} ion which is a $5d^6$ ion (Ir = $5d^7\ 6s^2$, Ir^{3+} = $5d^6$). Since in $[IrCl_6]^{3-}$ ion, Δ_0 has been found to be so large that it becomes greater than P ($P < \Delta_0$), $[Ir\ Cl_6]^{3-}$ is a LS octahedral ion, *i.e.*, all $5d^6$ electrons are paired up in t_{2g} orbitals as shown in Fig. 5.90. Fig. 5.90 shows that :

Electronic configuration of Ir^{3+} ion ($5d^6$ ion)

$$= t_{2g}^{\ 6}\ e_g^{\ 0}\ (n = 0)$$

Since $n = 0$, $\mu = 0$ (diamagnetic)

— — e_g^0

↑↓ ↑ ↑ ↑ ↑

$5d^6$ electrons of Ir^{3+} ion ($n = 4$)

↑↓ ↑↓ ↑↓ t_{2g}^6

($n = 0$)

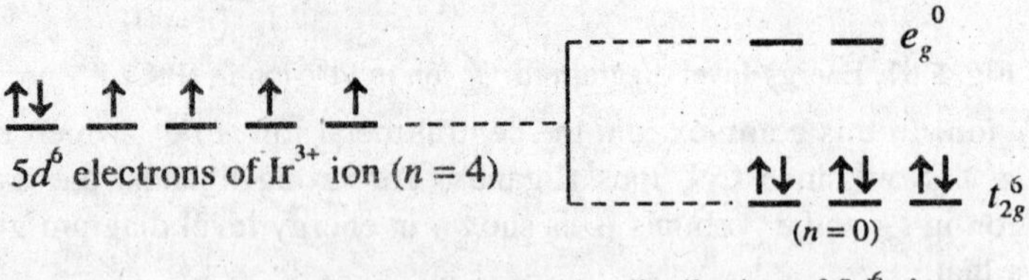

Fig. 5.90. Energy level diegram showing the distribution of $5d^6$ electrons of Ir^{3+} ion in $[IrCl_6]^{3-}$ ion in t_{2g} and e_g orbitals.

$[Rh[NH_3]_6]^{3+}$ ion. In this ion Rh (central metal atom) is present as Rh^{3+} ion which is a $4d^6$ ion (Rh = $4d^8\ 5s^1$, Rh^{3+} = $4d^6$). Now since NH_3 molecules (ligands) are strong ligands, the distribution of $4d^6$ electrons of Rh^{3+} ion in t_{2g} and e_g orbitals takes place in such a way that all the six electrons get paired in t_{2g} orbitals (See Fig. 5.91). Thus :

Electronic configuration of Rh^{3+} ion ($4d^6$ ion) = $t_{2g}^{\ 6}\ e_g^{\ 0}$ ($n = 0$)

Since $n = 0$, $\mu = 0$ (diamagnetic). Thus $[Rh\ (NH_3)_6]^{3+}$ ion is a LS octahedral diamagnetic complex ion.

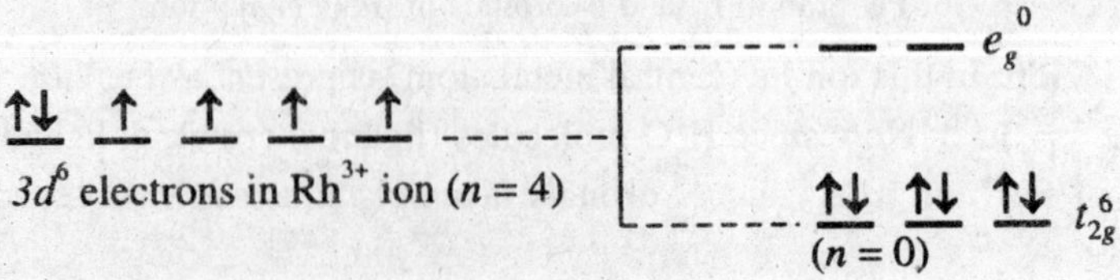

— — e_g^0

↑↓ ↑ ↑ ↑ ↑

$3d^6$ electrons in Rh^{3+} ion ($n = 4$)

↑↓ ↑↓ ↑↓ t_{2g}^6

($n = 0$)

Fig. 5.91. Energy level disagram showing the distribution of $4d^6$ electrons of Rh^{3+} ion in t_{2g} and e_g orbitals in $[Rh(NH_6)_3]^{3+}$ ion.

$[CoF_6]^{3-}$ ion. In this complex ion, Co-atom is present as Co^{3+} ion (central metal ion) which is a $3d^6$ ion (Co = $3d^7\ 4s^2$, $Co^{3+} = 3d^6$). Now since F^- ions (ligands) are weak ligands, the distribution of $3d^6$ electrons in t_{2g} and e_g orbitals takes place in such a way that we get maximum number of unpaired electrons (See Fig. 5.92) Thus $[CoF_6]^{3-}$ ion is a HS octahedral complex ion. Fig. 5.92 show that:

Electronic configuration of Co^{3+} ion ($3d^6$ ion)

$$= t_{2g}^{\ 4}\ e_g^{\ 2}\ (n = 4)$$

Since $n = 4$, $\mu = \sqrt{n(n+2)}$ BM $= \sqrt{4\times(4+2)}$ BM

$$= 4.898 \text{ BM}$$

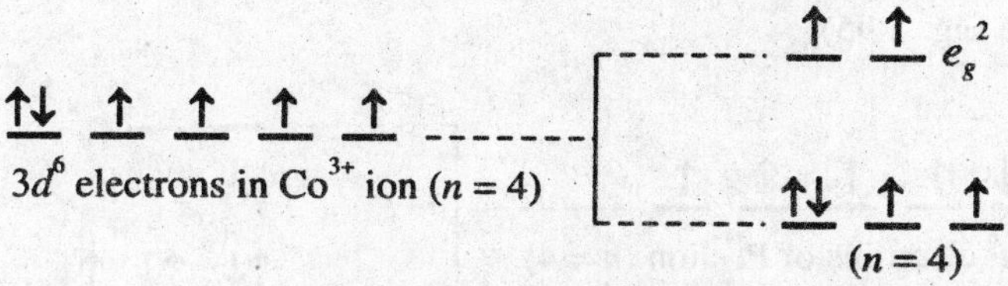

↑↓ ↑ ↑ ↑ ↑ ; ↑ ↑ e_g^2 ; ↑↓ ↑ ↑ t_{2g}^4

$3d^6$ electrons in Co^{3+} ion ($n = 4$) ; ($n = 4$)

Fig. 5.92. Energy level diagram showing the distribution of $3d^6$ electrons of Co^{3+} ion in t_{2g} and e_g orbitals in $[CoF_6]^{3-}$ ion.

$[Mn(CN)_6]^{3-}$ ion. In this ion Mn is present as Mn^{3+} ion (central metal ion) which is a $3d^4$ ion (Mn = $3d^5\ 4s^2$, $Mn^{3+} = 3d^4$). Now since CN^- ions are strong ligands, the distribution of $3d^4$ electrons in t_{2g} and e_g orbitals takes place in such a way that we get minimum number of unpaired electrons as shown in Fig. 5.93

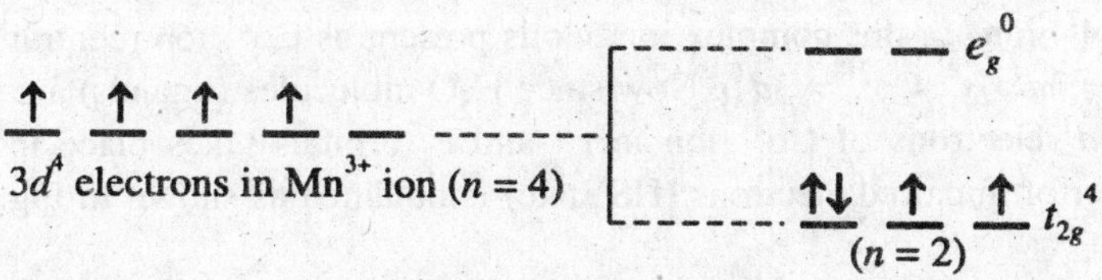

↑ ↑ ↑ ↑ — ; — — e_g^0 ; ↑↓ ↑ ↑ t_{2g}^4

$3d^4$ electrons in Mn^{3+} ion ($n = 4$) ; ($n = 2$)

Fig. 5.93. Energy level diagram showing the distribution of $3d^4$ electrons of Mn^{3+} ion in t_{2g} and e_g orbitals in $[Mn(CN)_6]^{3-}$ ion.

Fig. 5.93 show that:

Electronic configuration of Mn^{3+} ion ($3d^4$ ion)

$$= t^4_{2g}\ e_g^{\ 0}\ (n = 2)$$

Since $n = 2$, $\mu = \sqrt{n(n+2)}$ BM $= \sqrt{2\times(2+2)}$ BM

$$= 2.828 \text{ BM}$$

$[CoF_6]^{4-}$ ion. In this complex ion Co is present as Co^{2+} ion which is $3d^7$ ion (Co = $3d^7$. $4s^2$, $Co^{2+} = 3d^7$). Now since F^- ions (ligands) are weak ligands, the distribution of $3d^7$ electrons in t_{2g} and e_g orbitals takes place in such a way that we get maximum number of unpaired electrons as shown in Fig. 5.94.

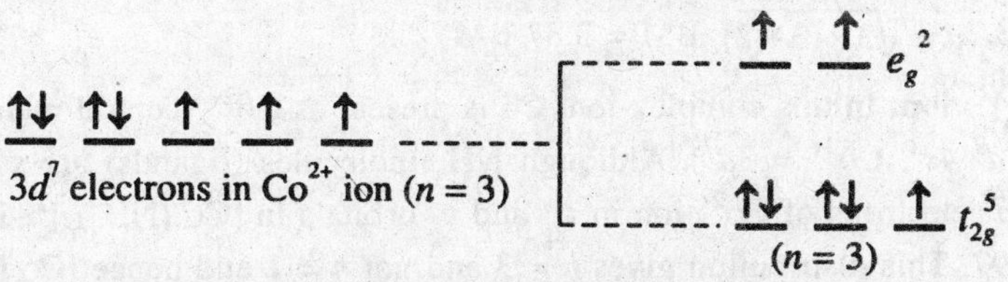

↑↓ ↑↓ ↑ ↑ ↑ ; ↑ ↑ e_g^2 ; ↑↓ ↑↓ ↑ t_{2g}^5

$3d^7$ electrons in Co^{2+} ion ($n = 3$) ; ($n = 3$)

Fig. 5.94. Energy level diagram showing the distribution of $3d^7$ electrons of Co^{2+} ion in t_{2g} and e_g orbitals in $[CoF_6]^{4-}$ ion.

Fig. 5.94 shows that:
Electronic configuration of Co^{2+} ion ($3d^7$ ion)

$$= t_{2g}^5\, e_g^2 \quad (n = 3)$$

Since $n = 3$, $\mu = \sqrt{n(n+2)}$ BM $= \sqrt{3 \times (3+2)}$ BM

$$= 3.872 \text{ BM}$$

$[Pt(NH_3)_6]^{4+}$ ion. This ion has Pt as Pt^{4+} ion (central metal ion) which is a $5d^6$ ion (Pt = $5d^9$ $6s^1$, $Pt^{4+} = 5d^6$). In the free state Pt^{4+} ion has four unpaired electrons ($n = 4$) in $5d$ orbitals. Under the influence of six NH_3 molecules (ligands) all the six $5d$ electrons become paired in t_{2g} orbitals and e_g orbitals remain vacant. Thus n becomes equal to zero ($n = 0$) and hence $[Pt(NH_3)_6]^{4+}$ ion is diamagnetic. (See Fig. 5.95)

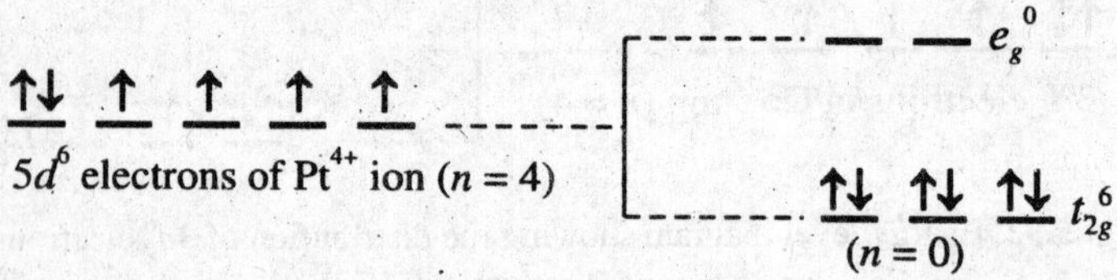

e_g^0

↑↓ ↑ ↑ ↑ ↑

$5d^6$ electrons of Pt^{4+} ion ($n = 4$)

↑↓ ↑↓ ↑↓ t_{2g}^6

($n = 0$)

Fig. 5.95. Energy level diagram for d^6 ion in Pt^{4+} ion in $[Pt(NH_3)_6]^{4+}$ ion

Fig. 5.95 show that:
Electronic configuration of Pt^{4+} ion ($5d^6$ ion)

$$= t_{2g}^6\, e_g^0 \quad (n = 0)$$

Since $n = 0$, $[Pt(NH_3)_6]^{4+}$ ion is diamagnetic and $\mu = 0$.

$[Co(H_2O)_6]^{2+}$ ion. In this complex ion, Co is present as Co^{2+} ion (central metal ion) which is a $3d^7$ ion (Co = $3d^7$ $4s^2$, $Co^{2+} = 3d^7$). Now since H_2O molecules (ligands) are weak ligands, the distribution of $3d^7$ electrons of Co^{2+} ion in t_{2g} and e_g orbitals takes place in such a way that maximum number of unpaired electrons (HS state) is obtained as shown in Fig. 5.96

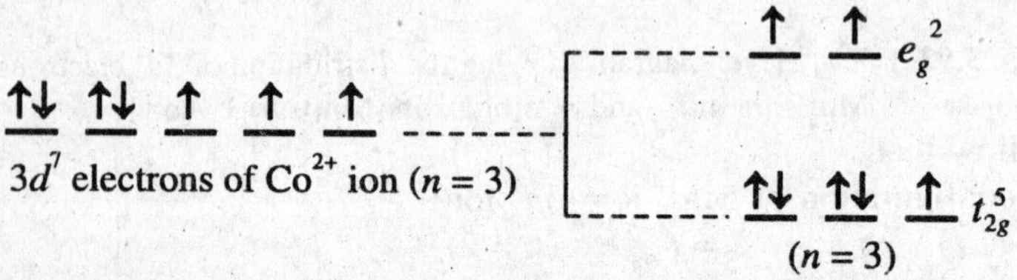

↑ ↑ e_g^2

↑↓ ↑↓ ↑ ↑ ↑

$3d^7$ electrons of Co^{2+} ion ($n = 3$)

↑↓ ↑↓ ↑ t_{2g}^5

($n = 3$)

Fig. 5.96. Energy level diagram showing the distribution of $3d^7$ electrons of Co^{2+} ion in t_{2g} and e_g orbitals in $[Co(H_2O)_6]^{2+}$ ion.

Fig. 5.96 show that :
Electronic configuration of Co^{2+} ion ($3d^7$ ion)

$$= t_{2g}^5\, e_g^2 \quad (n = 3)$$

Since $n = 3$, $\mu = \sqrt{3 \times (3+2)}$ BM = 3.87 BM

$[Co(NH_3)_6]^{2+}$ ion. In this complex ion, Co is present as Co^{2+} (central metal ion) which is a $3d^7$ ion (Co = $3d^7$ $4s^2$, $Co^{2+} = 3d^7$). Although NH_3 molecules (ligands) are strong ligands, the distribution of $3d^7$ electrons of Co^{2+} ion in t_{2g} and e_g orbitals in $[Co(NH_3)_6]^{2+}$ ion takes place as shown in Fig. 5.97. This distribution gives $n = 3$ and not $n = 1$ and hence $[Co(NH_3)_6]^{2+}$ ion is a HS complex ion. Fig. 5.97 shows that :

Electronic configuration of Co^{2+} ion ($3d^7$ ion)

$= t_{2g}^{5} e_{g}^{2}$ (and not $t_{2g}^{6} e_{g}^{1}$)

This configuration gives $n = 3$ and hence $\mu = \sqrt{3(3+2)}$ BM = 3.87 BM.

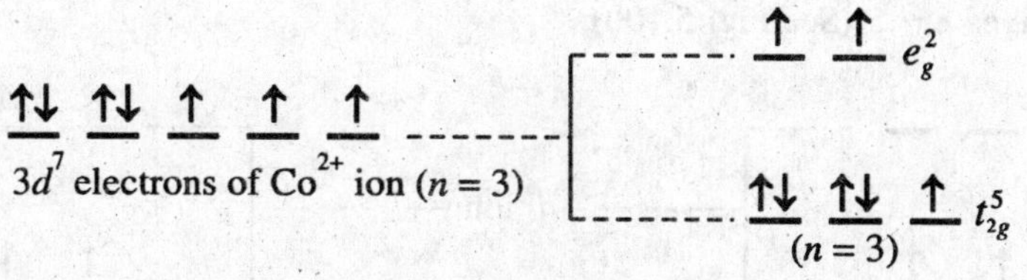

↑↓ ↑↓ ↑ ↑ ↑ — $3d^7$ electrons of Co^{2+} ion ($n = 3$); ↑ ↑ e_g^2; ↑↓ ↑↓ ↑ t_{2g}^5 ($n = 3$)

Fig. 5.97. Energy level diagram showing the distribution of $3d^7$ electrons of Co^{2+} ion in t_{2g} and e_g orbitals in $[Co(NH_3)_6]^{2+}$ ion.

Problem 28. *Determine the number of unpaired electrons in octahedral complexes of d^9, d^3, d^4 and d^5 ions when $\Delta_0 > P$ and $\Delta_0 < P$.*

Solution. We know that in case of the octahedral complexes of d^4 and d^5 ions, when $\Delta_0 > P$, LS complexes are formed and when $\Delta_0 < P$, HS complexes are formed. In case of octahedral complexes of d^9 and d^3 ions, the question of LS and HS complexes does not arise, since in both cases the number of unpaired electrons is the same for a given d^9 or d^3 ion.

***d^9* and *d^3* ions.** The distribution of 9 electrons of d^9 ion and 3 electrons of d^3 ion in t_{2g} and e_g orbitals is as shown in Fig. 5.98 This figure shows that for d^9 ion, $n = 1$ and for d^3 ion, $n = 3$.

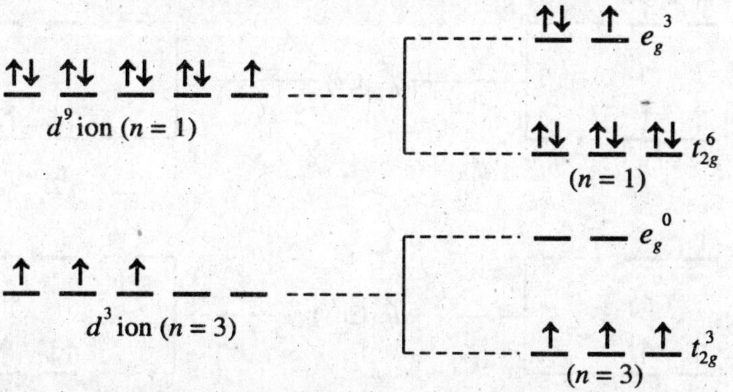

↑↓ ↑↓ ↑↓ ↑↓ ↑ — d^9 ion ($n = 1$); ↑↓ ↑ e_g^3; ↑↓ ↑↓ ↑↓ t_{2g}^6 ($n = 1$)

↑ ↑ ↑ — — — d^3 ion ($n = 3$); — — e_g^0; ↑ ↑ ↑ t_{2g}^3 ($n = 3$)

Fig. 5.98. Distribution of 9 electrons of d^9 ion and 3 electrons of d^3 ion in t_{2g} and e_g orbitals

***d^4* ion.** (*a*) When $\Delta_0 > P$, LS complex is obtained and hence the distribution of d^4 electrons in t_{2g} and e_g orbitals gives the electronic configuration $t_{2g}^4 e_g^0$ in which $n = 2$. (See Fig. 5.99)

(*b*) When $\Delta_0 < P$, HS complex is formed and hence electronic configuration for d^4 ion is $t_{2g}^3 e_g^1$ which gives $n = 4$ (See Fig. 5.99)

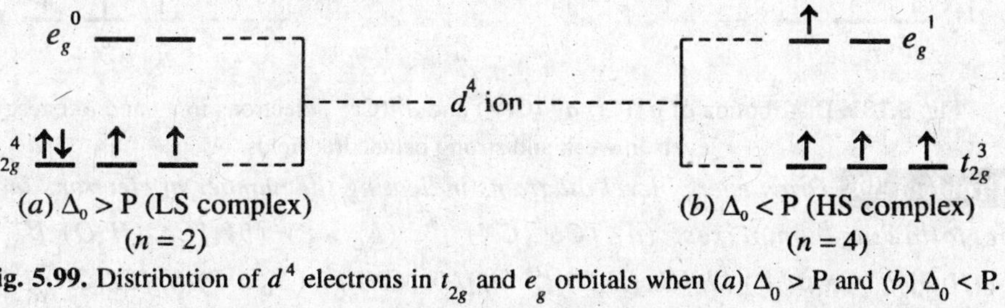

e_g^0 — — ; t_{2g}^4 ↑↓ ↑ ↑ — d^4 ion — ↑ — e_g^1 ; ↑ ↑ ↑ t_{2g}^3

(*a*) $\Delta_0 > P$ (LS complex) ($n = 2$) (*b*) $\Delta_0 < P$ (HS complex) ($n = 4$)

Fig. 5.99. Distribution of d^4 electrons in t_{2g} and e_g orbitals when (*a*) $\Delta_0 > P$ and (*b*) $\Delta_0 < P$.

d^5 ion. (*a*) When $\Delta_0 > P$, LS complex is formed and hence electronic configuration for d^5 ion is $t_{2g}^5\ e_g^0$ which gives n = 1. (See Fig. 5.100)

(*b*) When $\Delta_0 < P$, HS complex is formed and hence electronic configuration for d^5 ion is $t_{2g}^3\ e_g^2$ which gives n = 5 (See Fig. 5.100)

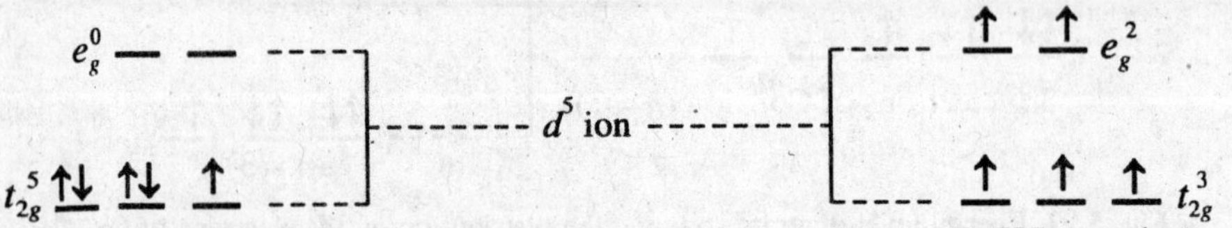

e_g^0 — —

d^5 ion

t_{2g}^5 ↑↓ ↑↓ ↑

(*a*) $\Delta_0 > P$ (LS complex)
(n = 1)

↑ ↑ e_g^2

↑ ↑ ↑ t_{2g}^3

(*b*) $\Delta_0 < P$ (HS complex)
(n = 5)

Fig. 5.100. Distribution of d^5 electrons in t_{2g} and e_g orbitals when (*a*) $\Delta_0 > P$ and (*b*) $\Delta_0 < P$

Problem 29. *Consider the species viz Cr, Cr^{2+} and Cr^{3+} in an octahedral field. Draw the energy level diagram depicting the occupation of energy levels in weak and strong fields around the metal atom or ion.*

Solution. Cr, Cr^{2+} and Cr^{3+} are $3d^5\ 4s^1$ or $3d^6$, $3d^4$ and $3d^3$ systems respectively. The distributionof $3d^6$, $3d^4$ and $3d^3$ electrons in t_{2g} and e_g energy levels in weak and strong octahedral fields has been shown in Fig. 5.101

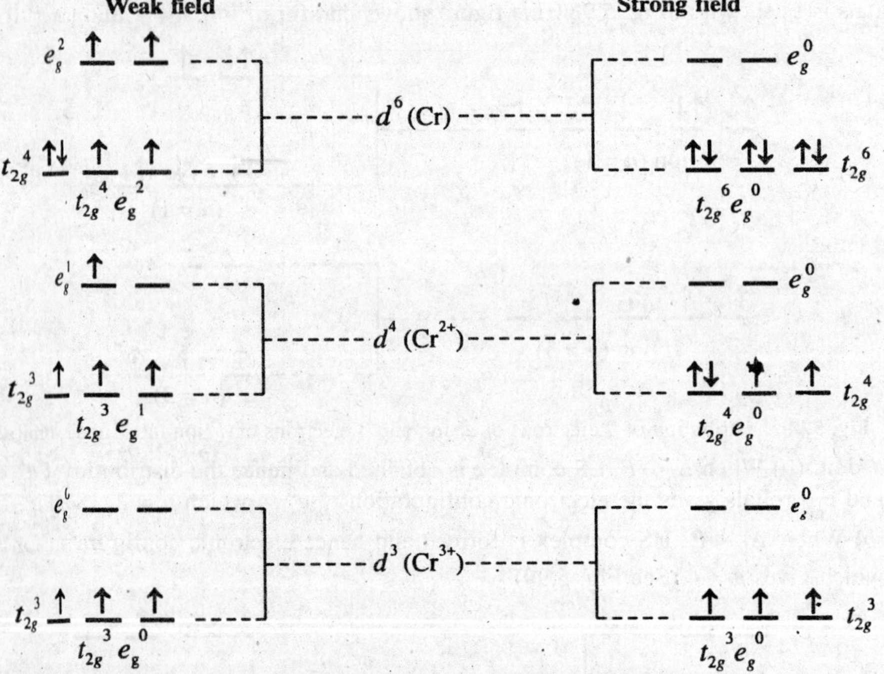

Fig. 5.101. Distribution of d^6(Cr) d^4 (Cr^{2+}) and d^3 (Cr^{3+}) electrons in t_{2g} and e_g energy levels in weak and strong octahedral fields.

Problem 30. *Draw energy level diagrams indicating the number of electrons in each level for the following complexes: (a) $[Co\ (CN)_6]^{3-}$ ($\Delta_0 > P$) (b) $[Co\ (H_2O)_6]^{3+}$ ($\Delta_0 < P$) (c) $[Mn\ (H_2O)_6]^{2+}$ ($\Delta_0 < P$) (d) $[Cu\ (H_2O)_6]^{2+}$ (e) $[Mn\ (CN)_6]^{3-}$ ($\Delta_0 > P$) (f) $[Fe\ (CN)_6]^{3-}$ ($\Delta_0 > P$).*

Solution. (*a*) **$[Co(CN)_6]^{3-}$ ion ($\Delta_0 > P$).** Here Co is present as Co^{3+} ion which is a $3d^6$ ion. Since $P < \Delta_0$, the given complex ion is a LS octahedral complex ion and hence energy level diagram for d^6 ion in LS state is as shown at (*a*) of Fig. 5.102. Thus the configuration of d^6 ion in LS state is $t_{2g}^6 e_g^0$ ($n = 0$).

(*b*) **$[Co(H_2O)_6]^{3+}$ ion ($\Delta_0 < P$).** Here Co is present as Co^{3+} ion which is a $3d^6$ ion. Since $P > \Delta_0$, the given complex ion is a HS octahedral complex ion and hence the energy level diagram for d^6 ion in HS state is as shown at (*b*) of Fig. 5.102. Thus the configuration of d^6 ion in HS state is $t_{2g}^4 e_g^2$ ($n = 4$).

(*c*) **$[Mn(H_2O)_6]^{2+}$ ion ($\Delta_0 < P$).** Here Mn is present as Mn^{2+} ion which is a $3d^5$ ion. Since $P > \Delta_0$, the given complex ion is a HS octahedral complex ion and hence the energy level diagram for d^5 ion in HS state is as shown at (c) of Fig. 5.102 Thus the configuration of d^5 ion in HS state is $t_{2g}^3 e_g^2$ ($n = 5$).

(*d*) **$[Cu(H_2O)_6]^{2+}$ ion.** Cu^{2+} ion is a $3d^9$ ion. Whether P is larger or smaller than Δ_0, the energy level diagram for d^9 ion is as shown at (*d*) of Fig. 5.102. This figure shows that the configuration of d^9 ion is $t_{2g}^6 e_g^3$ ($n = 1$).

(*e*) **$[Mn(CN)_6]^{3-}$ ion ($\Delta_0 > P$).** Mn is present as Mn^{3+} ion which is a $3d^4$ ion. Since $\Delta_0 > P$, tire given complex ion is a LS octahedral complex ion and hence energy level diagram of d^4 ion in LS state is as shown at (*e*) of Fig. 5.102. This figure shows that the configuration for d^4 ion is $t_{2g}^4 e_g^0$ ($n = 2$).

(*f*) **$[Fe(CN)_6]^{3-}$ ion ($\Delta_0 > P$).** Fe^{3+} ion is $3d^5$ ion. Since $\Delta_0 > P$, the given complex ion is a LS octahedral complex ion and hence energy level diagram for d^5 ion in LS state is as shown at (*f*) of Fig. 5.102. Thus the configuration for d^5 ion in LS state is $t_{2g}^5 e_g^0$ ($n = 1$).

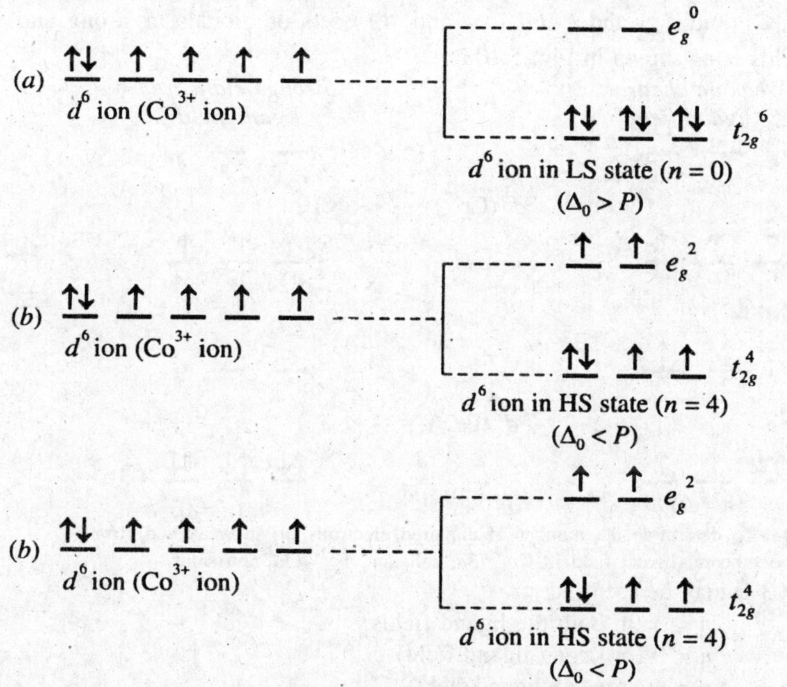

(*a*) ↑↓ ↑ ↑ ↑ ↑

d^6 ion (Co^{3+} ion)

e_g^0

↑↓ ↑↓ ↑↓ t_{2g}^6

d^6 ion in LS state ($n = 0$)
($\Delta_0 > P$)

(*b*) ↑↓ ↑ ↑ ↑ ↑

d^6 ion (Co^{3+} ion)

↑ ↑ e_g^2

↑↓ ↑ ↑ t_{2g}^4

d^6 ion in HS state ($n = 4$)
($\Delta_0 < P$)

(*b*) ↑↓ ↑ ↑ ↑ ↑

d^6 ion (Co^{3+} ion)

↑ ↑ e_g^2

↑↓ ↑ ↑ t_{2g}^4

d^6 ion in HS state ($n = 4$)
($\Delta_0 < P$)

Fig. 5.102 Continued on the next page

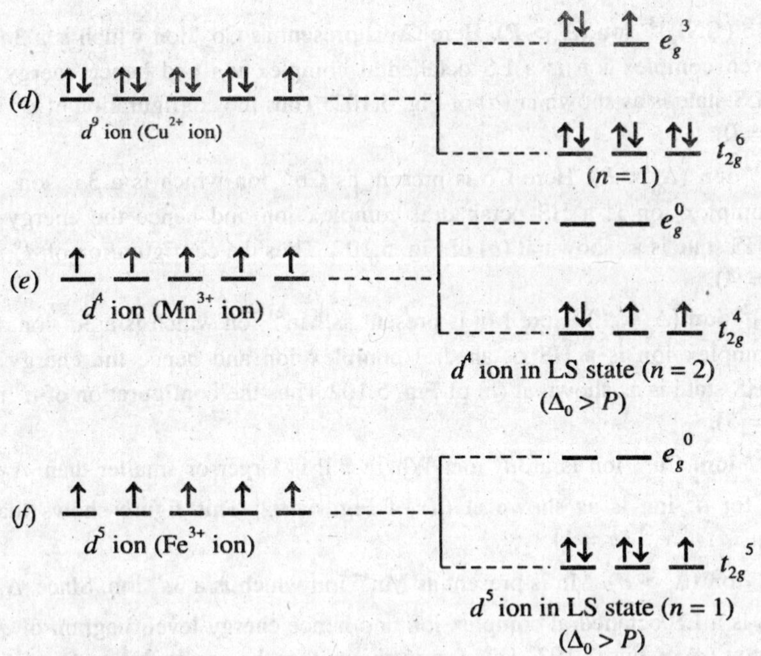

Fig. 5.102. Energy level diagrams for d^6(LS), d^6(HS), d^5 (HS), d^9, d^4 (LS) and d^5 (LS) ions.

Problem 31. *Find out the number of unpaired electrons in strong and weak octahedral field for Cr^{3+} and Fe^{2+} ions.* *(G.N. Dev 1996)*

Solution. Cr^{3+} and Fe^{2+} ions have $3d^3$ and $3d^6$ as their valence-shell electronic configurations respectively. Distribution of three 3*d* electrons of Cr^{3+} ion ($3d^3$) and six 3*d* electrons of Fe^{2+} ion ($3d^6$) in t_{2g} (d_{xy}, d_{yz} and d_{zx}) and e_g ($d_{x^2-y^2}$ and d_{z^2}) sets of orbitals in strong and weak octahedral ligand fields is as shown in Fig. 5.103

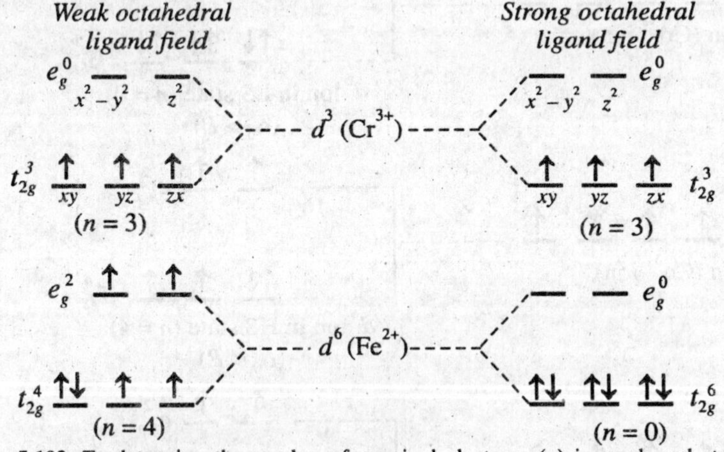

Fig. 5.103. To determine the number of unpaired electrons (*n*) in weak and strong octahedral ligand field in Cr^{3+} ($3d^3$ ion) and Fe^{2+} ($3d^6$ ion) ions.

From Fig. 5.103 it may be seen that :

For Cr^{3+} ion, $n = 3$ (in both the ligand fields)

For Fe^{2+} ion, $n = 0$ (in strong ligand field)

$n = 4$ (in weak ligand field)

Problem 32. *On the basis of CFT, predict the geometry of the compound, $K_3[Mn(CN)_6]$. Also calculate the value of $\mu_{spin\ only}$ of the compound.*

Solution. The given compound contains $[Mn(CN)_6]^{3-}$ ion as its complex ion. This ion has Mn^{3+} ion ($3d^4$ ion) as its central metal ion. Since C.N. of Mn^{3+} ion in $[Mn(CN)_6]^{3-}$ is 6, $[Mn(CN)_6]^{3-}$ ion has octahedral geometry. Now since CN^- ions are strong ligands, $[Mn(CN)_6]^{3-}$ ion is a LS octahedral ion and the distribution of d^4 electrons of Mn^{3+} ion in t_{2g} (lower energy orbitals) and e_g (higher energy orbitals) is as $t^4_{2g}\,e_g^0$ which shows that $n = 2$ and hence:

$$\mu_{spin\ only} = \sqrt{n(n+2)}\ \text{BM}$$
$$= \sqrt{2\times(2+2)}\ \text{BM}$$
$$= 2.82\ \text{BM Answer}$$

Problem 33. *Which of the following complex ions is LS and which is HS? $[CoF_6]^{3-}$ and $[Rh(NH_3)_6]^{3+}$.*

Solution. In both the given complex ions, the central metal atom is a d^6 ion ($Co^{3+} = 3d^6$, $Rh^{3+} = 4d^6$).

Now since F^- ion is a weak field ligand, the distribution of d^6 electrons of Co^{3+} ion in $[CoF_6]^{3-}$ ion in t_{2g} and e_g orbitals is $t^4_{2g}\,e^2_g$ which gives $n = 4$.

Since NH_3 is a strong ligand, the distribution of d^6 electrons of Rh^{3+} ion in $[Rh(NH_3)_6]^{3+}$ ion in t_{2g} and e_g orbitals is $t_{2g}^6\,e_g^0$ which gives $n = 0$.

Now since n for $[CoF_6]^{3-}$ ion ($n = 4$) is higher than that for $[Rh(NH_3)_6]^{3+}$ ion ($n = 0$), $[CoF_6]^{3-}$ ion is HS complex and $[Rh(NH_3)_6]^{3+}$ is LS complex.

Problem 34. *On the basis of CFT explain the given values of magnetic moment of the following complex compounds: $K_3[FeF_6] = 5.9$ BM, $[Co(NH_3)_6]Cl_3 = 0.0$ BM, $[Ni(H_2O)_6]Cl_2 = 2.9$ BM, $K_4[Mn(CN)_6] = 1.9$ BM, $K_4[MnF_6] = 5.9$ BM* *(Guru Nanak Dev 1995)*

Solution. The given complex compounds contain $[FeF_6]^{3-}$, $[Co(NH_3)_6]^{3+}$, $[Ni(H_2O)_6]^{2+}$, $[Mn(CN)_6]^{4-}$ and $[MF_6]^{4-}$ complex ions respectively. These complex ions have Fe^{3+} ($3d^5$), Co^{3+} ($3d^6$), Ni^{2+} ($3d^8$), Mn^{2+} ($3d^5$) and Mn^{2+} ($3d^5$) ions as their central metal ions. On the basis of CFT, the distribution of 3d-electrons in t_{2g} and e_g sets of 3d orbitals in Fe^{3+} ($3d^5$), Co^{3+} ($3d^6$), Ni^{2+} ($3d^8$) and Mn^{2+} ($3d^5$) ions is as given below :

$3d^5$ electrons of Fe^{3+} ion in $[FeF_6]^{3-}$ ion (F^- ions are weak ligands) = $t_{2g}^3\,e_g^2$ ($n = 5$)

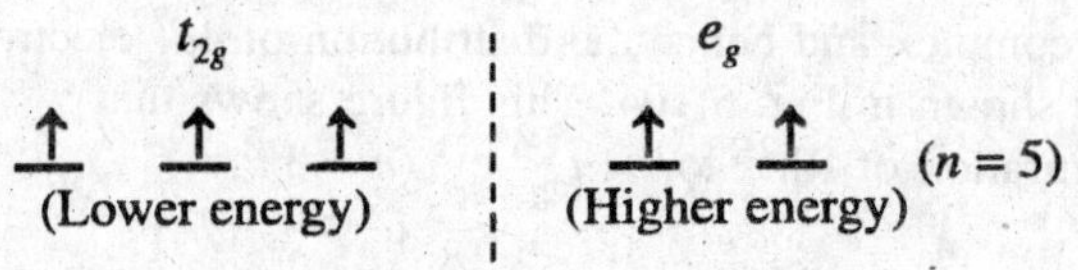

t_{2g}	e_g	
↑ ↑ ↑	↑ ↑	($n = 5$)
(Lower energy)	(Higher energy)	

$3d^6$ electrons of Co^{3+} ion in $[Co(NH_3)_6]^{3+}$ ion (NH_3 molecules are strong ligands) = $t_{2g}^6\,e_g^0$ ($n = 0$).

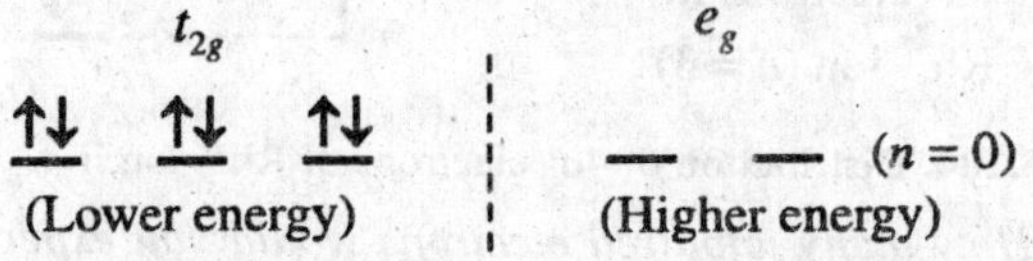

t_{2g}	e_g	
↑↓ ↑↓ ↑↓	— —	($n = 0$)
(Lower energy)	(Higher energy)	

$3d^8$ electrons of Ni^{2+} ion in $[Ni(H_2O)_6]^{2+}$ ion (H_2O molecules are weak ligands) = $t_{2g}^6\,e_g^2$ ($n = 2$)

t_{2g}		e_g	
↑↓ ↑↓ ↑↓		↑ ↑	$(n = 2)$
(Lower energy)		(Higher energy)	

$3d^5$ electrons of Mn^{2+} ion in $[Mn(CN)_6]^{4-}$ ion (CN^- ions are strong ligands) = $t_{2g}^5 e_g^0$ $(n = 1)$

t_{2g}		e_g	
↑↓ ↑↓ ↑		— —	$(n = 1)$
(Lower energy)		(Higher energy)	

$3d^5$ electrons of Mn^{2+} ion in $[MnF_6]^{4-}$ ion (F^- ions are weak ligands) = $t_{2g}^3 e_g^2$ $(n = 5)$

t_{2g}		e_g	
↑ ↑ ↑		↑ ↑	$(n = 5)$
(Lower energy)		(Higher energy)	

It is clear from the above discussion that :

μ of $[FeF_6]^{3-}$ ion $(n = 5) = \sqrt{n(n+2)} = \sqrt{5\times(5+2)} = \sqrt{35}$

$= 5.9$ BM

μ of $[Co(NH_3)_6]^{3+}$ ion $(n = 0) = \sqrt{0\times(0+2)} = 0.0$ BM

μ of $[Ni(H_2O)_6]^{2+}$ ion $(n = 2) = \sqrt{2\times(2+2)} = 2.9$ BM

μ of $[Mn(CN)_6]^{4-}$ ion $(n = 1) = \sqrt{1\times(2+1)} = \sqrt{3} = 1.98$ BM

μ of $[MnF_6]^{4-}$ ion $(n = 5) = \sqrt{5\times(2+5)} = 5.9$ BM

Thus we see that the values of μ calculated as above on the basis of CFT are the same as given in the problem.

Problem 35. *On the basis of CFT, give the electronic configuration of Rh^{2+} ion in an octahedral complex for which $\Delta_o > P$.*

Solution. Rh^{2+} ion is a $4d^7$ ion ($Rh = 4d^8, 5s^1$, $Rh^{2+} = 4d^7$). Since $\Delta_o > P$, the octahedral complex would be LS complex and hence the distribution of $4d^7$ electrons of Rh^{2+} ion in t_{2g} and e_g orbitals would be as shown in Fig. 5.104. This figure shows that :

Electronic configuration of Rh^{2+} ion = $t_{2g}^6 e_g^1$ $(n = 1)$.

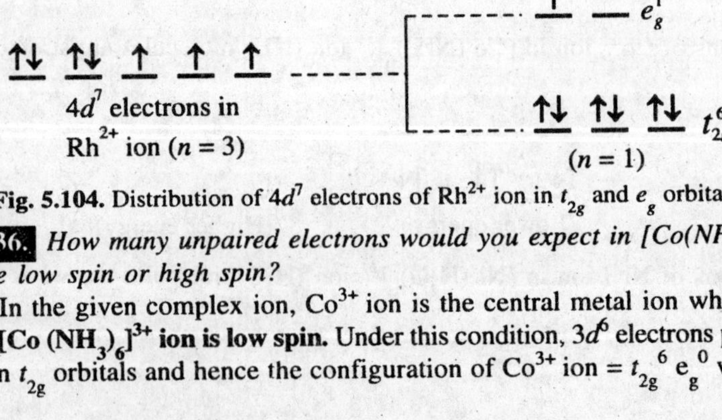

	↑ —	e_g^1
↑↓ ↑↓ ↑ ↑ ↑ $4d^7$ electrons in Rh^{2+} ion $(n = 3)$		
	↑↓ ↑↓ ↑↓ $(n = 1)$	t_{2g}^6

Fig. 5.104. Distribution of $4d^7$ electrons of Rh^{2+} ion in t_{2g} and e_g orbitals.

Problem 36. *How many unpaired electrons would you expect in $[Co(NH_3)_6]^{3+}$ ion, if it is considered to be low spin or high spin?*

Solution. In the given complex ion, Co^{3+} ion is the central metal ion which is a $3d^6$ ion.

(a) **When $[Co(NH_3)_6]^{3+}$ ion is low spin.** Under this condition, $3d^6$ electrons present in Co^{3+} ion will get paired in t_{2g} orbitals and hence the configuration of Co^{3+} ion = $t_{2g}^6 e_g^0$ which gives $n = 0$.

(*b*) **When $[Co(NH_3)_6]^{3+}$ ion is high spin.** Under this condition, the distribution of $3d^6$ electrons would be as $t_{2g}^4\, e_g^2$ which gives $n = 2 + 2 = 4$.

Problem 37. *On the basis of CFT, explain the following, giving appropriate reasons for your answer.*

(*i*) *The magnetic moment of $[Fe(H_2O)_6]^{3+}$ ion is 5.92 B.M. and that of $[Fe(CN)_6]^{3-}$ ion is 1.73 B.M.* (*Guwahati 95, Kurukshetra 95; M.D. Rohtak 98*)

(*ii*) *$[CoF_6]^{3-}$ ion is paramagnetic while $[Co(CN)_6]^{3-}$ ion is diamagnetic.* (*Kurukshetra 94; Guwahati 96; Bangalore 95; Meerut 96; Allahabad 95*)

(*iii*) *$[Fe(CN)_6]^{3-}$ ion is weakly paramagnetic while $[Fe(H_2O)_6]^{3+}$ ion strongly paramagnetic.* (*Allahabad 96*)

(*iv*) *Complexes of Co(III) like $[Co(NH_3)_6]^{3+}$, $[Co(en)_3]^{3+}$ and $[Co(NO_2)_6]^{3-}$ are diamagnetic while $[CoF_6]^{3-}$ and $[Co(H_2O)_6]^{3+}$ are paramagnetic.*

(*v*) *$[Fe(CN)_6]^{4-}$ ion is diamagnetic while $[Fe(CN)_6]^{3-}$ ion is paramagnetic, although both the ions have strong ligands*

(*vi*) *$[Co(CN)_6]^{4-}$ ion is paramagnetic while $[Co(CN)_6]^{3-}$ ion is diamagnetic, although both the ions have strong ligands.*

Solution. (*i*) In $[Fe(H_2O)_6]^{3+}$ ion, H_2O molecules are weak ligands and central metal ion is Fe^{3+} which is $3d^5$ system ($Fe^{3+} = 3d^5$). Thus the distribution of five 3*d* electrons ($Fe^{3+} = 3d^5$) in t_{2g} and e_g sets of orbitals in octahedral weak ligand field is as $t_{2g}^3\, e_g^2$ which gives $n = 5$.

The magnetic moment (μ) of $[Fe(H_2O)_6]^{3+}$ ion is, therefore, given by:

$$\mu = \sqrt{n(n+2)} = \sqrt{5 \times (5+2)} = \sqrt{35}$$
$$= 5.91 \text{ BM}$$

In $[Fe(CN)_6]^{3-}$ ion, CN^- ions are strong ligands and the central metal ion is Fe^{3+} which is a $3d^5$ system ($Fe^{3+} = 3d^5$). Thus the distribution of five 3*d* electrons ($Fe^{3+} = 3d^5$) in t_{2g} and e_g orbitals in octahedral strong ligand field is as $t_{2g}^5\, e_g^0$ ($n = 1$).

The magnetic moment (μ) of $[Fe(CN)_6]^{3-}$ ion is, therefore, given by:

$$\mu = \sqrt{n(n+2)} = \sqrt{1 \times (1+2)} = 1.732 \text{ BM}$$

(*ii*) Both the given complex ions are octahedral and have the same central metal ion *viz* Co^{3+} ion which is a $3d^6$ ion ($Co^{3+} = 3d^6$). In $[CoF_6]^{3-}$ ion F^- ions are weak ligands. Thus, according to CFT, the distribution of six 3*d* electrons of Co^{3+} ion in t_{2g} and e_g orbitals in octahedral weak ligand field is as $t_{2g}^4\, e_g^2$ ($n = 4$). Due to the presence of 4 unpaired electrons, $[CoF_6]^{3-}$ ion is paramagnetic.

In $[Co(CN)_6]^{3-}$ ion, CN^- ions are strong ligands and hence the distribution of six 3*d* electrons ($Co^{3+} = 3d^6$) in t_{2g} and e_g orbitals in octahedral strong ligand field is as $t_{2g}^6\, e_g^0$ ($n = 0$). Due to the absence of unpaired electrons, $[Co(CN)_6]^{3-}$ ion is diamagnetic.

(*iii*) Both the complex ions are octahedral and have Fe^{3+} ion as their central metal ion. Fe^{3+} ion is a $3d^5$ ion ($Fe^{3+} = 3d^5 4s^0$). In $[Fe(CN)_6]^{3-}$ ion, since CN^- ions are strong ligands, the distribution of five 3*d* electrons of Fe^{3+} ion in t_{2g} and e_g orbitals in octahedral strong ligand field produced by CN^- ions is as $t_{2g}^5\, e_g^0$ ($n = 1$). The presence of one unpaired electron shows that $[Fe(CN)_6]^{3-}$ ion is paramagnetic to extent of only 1 unpaired electron.

In $[Fe(H_2O)_6]^{3+}$ ion, since H_2O molecules are weak ligands, the distribution of five 3*d* electrons in t_{2g} and e_g orbitals in octahedral weak ligand field created by H_2O molecules is $t_{2g}^3\, e_g^2$ ($n = 5$). The presence of 5 unpaired electrons shows that $[Fe(H_2O)_6]^{3+}$ ion paramagnetic.

The above discussion ion shows that since $[Fe(CN)_6]^{3-}$ ion has lesser number of unpaired electrons ($n = 1$) than $[Fe(H_2O)_6]^{3+}$ ion ($n = 5$), the former ion is weakly paramagnetic while the later ion is strongly paramagnetic.

(*iv*) The complex ions *viz* $[Co(NH_3)_6]^{3+}$, $[Co(en)_3]^{3+}$ and $[Co(NO_2)_3]^{3-}$ have Co^{3+} ion ($3d^6$ ion) as their central metal ion and have NH_3, en and NO_2^- as ligands. Since these ligands are strong field ligands, the distribution of $3d^6$ electrons of Co^{3+} ion (central metal ion) in t_{2g} and e_g orbitals takes place in such a way that all $3d^6$ electrons become paired in t_{2g} orbitals and e_g orbitals remain vacant. Thus the electronic configuration of Co^{3+} ion ($3d^6$ ion) is $t_{2g}^6 e_g^0$. This configuration gives $n = 0$ and hence each of the three complex ions is diamagnetic.

The complex ions *viz* $[CoF_6]^{3-}$ and $[Co(H_2O)_6]^{3+}$ also have Co^{3+} ion ($3d^6$ ion) as their central metal ion and have F^- and H_2O as ligands. Since these ligands are weak field ligands, the distribution of $3d^6$ electrons of Co^{3+} ion in t_{2g} and e_g orbitals takes place in accordance with Hund's rule of maximum multiplicity. Thus the electronic configuration of Co^{3+} ion ($3d^6$ ion) is $t_{2g}^4 e_g^2$. This configuration gives $n = 4$. Due to the pressence of four unpaired electrons each of these two complex ions is paramagnetic.

(*v*) In $[Fe(CN)_6]^{4-}$ ion, Fe is present as Fe^{2+} ion (central metal ion) which is a $3d^6$ ion ($Fe = 3d^6 4s^2$, $Fe^{2+} = 3d^6$). Since CN^- ions (lignads) are strong ligands, all $3d^6$ electrons of Fe^{2+} ion occupy the lower energy t_{2g} orbitals and hence electronic configuration of Fe^{2+} ion in $[Fe(CN)_6]^{4-}$ ion is $t_{2g}^6 e_g^0$ which gives $n = 0$. $n = 0$ makes $[Fe(CN)_6]^{4-}$ ion diamagnetic.

In $[Fe(CN)_6]^{3-}$ ion, Fe is present as Fe^{3+} ion which is a $3d^5$ ion. Since CN^- ions (ligands) are strong ligands, all $3d^5$ electrons of Fe^{3+} ion occupy the lower energy t_{2g} orbitals and hence electronic configuration of Fe^{3+} ion in $[Fe(CN)_6]^{3-}$ ion is $t_{2g}^5 e_g^0$ which gives $n = 1$. $n = 1$ makes $[Fe(CN)_6]^{3-}$ ion paramagnetic.

(*vi*) In $[Co(CN)_6]^{4-}$ ion, Co is present as Co^{2+} ion (central metal ion) which is a $3d^7$ ion [$Co = 3d^7 4s^2$, $Co^{2+} = 3d^7$]. Since CN^- ions (ligands) are strong ligands, out of $3d^7$ electrons, six electrons occupy t_{2g} orbitals and one electron goes to e_g level. Thus the electronic configuration of Co^{2+} ion in $[Co(CN)_6]^{4-}$ ion is as $t_{2g}^6 e_g^1$ which gives $n = 1$. Due to the presence of one unpaired electron, $[Co(CN)_6]^{4-}$ ion is paramagnetic.

In $[Co(CN)_6]^{3-}$ ion, Co is present as Co^{3+} ion which is a $3d^6$ ion. Since CN^- ions are strong ligands, all $3d^6$ electrons of Co^{3+} ion occupy the lower energy t_{2g} orbitals and e_g orbitals remain vacant. Thus the electronic configuration of Co^{3+} ion in $[Co(CN)_6]^{3-}$ ion is $t_{2g}^6 e_g^0$ which gives $n = 0$. Thus $[Co(CN)_6]^{3-}$ ion diamagnetic.

Problem 38. *How is the crystal field splitting energy (Δ) of a complex compound obtained from its absorption spectrum?* *(Gujarat 95; Punjab 95)*

Solution. Splitting energy (Δ) of a complex ion is usually measured from the absorption spectrum of the complex ion. Suppose we have to calculate the value of Δ_0 for $[Ti(H_2O)_6]^{3+}$ ion (Octahedral complex ion). When white light is allowed to fall on this ion, the ion absorbs the light of frequency of 20,000 cm^{-1} (wave number) as is evident from the absorption spectrum of $[Ti(H_2O)_6]^{3+}$ ion shown in Fig 5.106. This wave number (= 20,000 cm^{-1}) lies in visible region of the spectrum. This spectrum has been obtained by plotting a graph between wave number (in cm^{-1}) of the absorbed light and the absorbance (absorbance is the amount of absorbed light). 20,000 cm^{-1} is the frequency (wave number) of the band (peak) of maximum absorption (*i.e.*, peak or maxima). This wave number of 20,000 cm^{-1} is associated with energy which is equal to 239.23 KJ mol^{-1} as shown below. We know that since

$$\therefore \quad 83.6\ cm^{-1} = 1\ KJ\ mol^{-1}$$

$$1\ cm^{-1} = \frac{1}{83.6}\ KJ\ mol^{-1}$$

$$\therefore \quad 20000\ cm^{-1} = \frac{20,000}{83.6}\ KJ\ mol^{-1} = 239.23\ KJ\ mol^{-1}$$

This energy is the energy difference betweent t_{2g} and e_g sets of orbitals (See Fig. 5.105) and is called the crystal field splitting energy (Δ_0) of $[Ti\ (H_2O)_6]^{3+}$ ion. Thus crystal field splitting energy (Δ_0) for $[Ti\ (H_2O)_6]^{3+}$ ion = 20,000 cm^{-1} = 239.23 KJ mol^{-1}

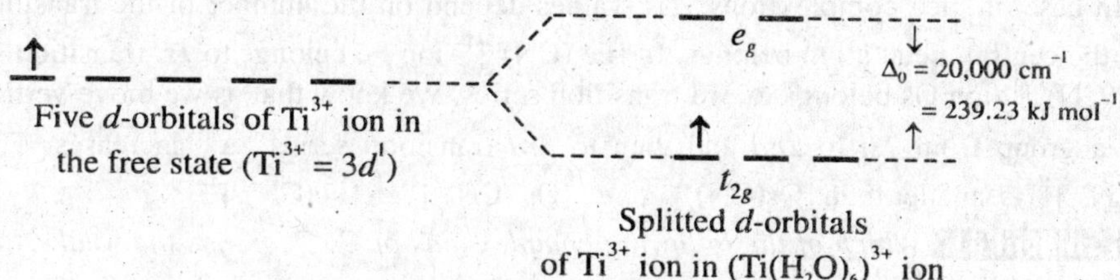

Fig. 5.105. Crystal field splitting energy (Δ_0) of $[Ti(H_2O)_6]^{3+}$ ion.

It may be seen from the above discussion that higher is the value of the frequency in cm^{-1} (wave number) of the peak of maximum absorption in the absorption spectrum of a given complex ion, greater is the magnitude of Δ_0. Thus in routine work, Δ_0 is expressed in cm^{-1} and not in KJ mol^{-1}].

Problem 39. *Which of the following pairs of complex ions has higher value of Δ_0 and why? (a) $[Co\ (NH_3)_6]^{3+}$ and $[Rh\ (NH_3)_6]^{3+}$ (b) $[Fe\ (C_2O_4)_3]^{4-}$ and $[Fe\ (C_2O_4)_3]^{3-}$ (c) $[Cr\ (en)_3]^{3+}$ and $[Cr(C_2O_4)_3]^{3+}$.*

Solution. (*a*) Since in $[Co\ (NH_3)_6]^{3+}$, Co belongs to *1st* transition series while in $[Rh\ (NH_3)_6]^{3+}$ Rh belongs to *2nd* transition series, Δ_0 for $[Rh\ (NH_3)_6]^{3+}$ is greater than that for $[Co\ (NH_3)_6]^{3+}$.

Explanation. Since Rh^{3+} ion is larger in size than Co^{3+} ion (Rh^{3+} = 80.5 pm, Co^{3+} = 63 pm), interaction of ligands with the orbitals of the larger Rh^{3+} ion is greater than that of the same ligands with the orbitals of smaller Co^{3+} ion. Greater interaction increases the value of Δ_0.

(*b*) In $[Fe\ (C_2O_4)_3]^{4-}$ oxidation state of Fe is + 2 while in $[Fe\ (C_2O_4)_3]^{3-}$ oxidation state of Fe is + 3. Due to higher oxidation state, Δ_0 value for $[Fe(C_2O_4)_3]^{3-}$ is higher than that for $[Fe(C_2O_4)_3]^{4-}$.

Explanation. The cation having higher oxidation state or higher positive charge has greater polarising power. Greater polarising power of the cation results in larger splitting of five d-orbitals into t_{2g} and e_g sets.

(*c*) In both the given complex ions the central metal atom is the same and has the same oxidation state equal to + 3. Ligands in both the ions are bidentate ligands. In such complex ions the magnitude of Δ_0 depends on the nature of ligands. Stronger is the ligand, higher the value of Δ_0. Thus since *en* (ethylenediamine) is a stronger ligand than $C_2O_4^{2-}$ (oxalate ion), $[Cr(en)_3]^{3+}$ ion has higher value of Δ_0 than $[Cr\ (C_2O_4)_3]^{3-}$ ion.

Problem 40. *Arrange the following pairs of complex ions on the basis of their crystal field splitting parameters (Δ_0): (a)* $[V\ (H_2O)_6]^{2+}$ and $[V\ (H_2O)_6]^{3+}$ *(b) $[Fe\ (CN)_6]^{4-}$ and $[Os\ (CN)_6]^{4-}$*

Solution. (*a*) Both the given complex ions have the same ligands, same geometry and same central metal atom, but the oxidation states of the central metal atoms are different. In case of such complexes, Δ_0 values depend on the oxidation state of the central metal atom. Δ_0 increases with the increase in oxidation state of the central metal atom. Thus since in $[V(H_2O)_6]^{2+}$ vanadium is in +2 oxidation state and in $[V(H_2O)_6]^{3+}$ the same metal atom (i.e., vanadium) is in + 3 oxidation state, Δ_0 for $[V(H_2O)_6]^{3+}$ is greater than that for $[V(H_2O)_6]^{2+}$, *i.e.*, $[V(H_2O)_6]^{3+} > [V(H_2O)_6]^{2+}$.

(*b*) Both the given complex ions have the same ligands and the same geometry but the central metal atom is different. Note oxidation state of both the central metal atoms is the same (= + 2). In case of such complex ions, Δ_0 values depend on the number of the transition series to which the central metal atom belongs. In $[Fe(CN)_6]^{4-}$ ion Fe belongs to *1st* transition series while in $[Os(CN)_6]^{4-}$ ion Os belongs to 3rd transition series. We know that as we move vertically downwards a group from *1st* to *2nd* and then to *3rd* transition series, Δ_0 increases. Thus Δ_0 for $[Fe(CN)_6]^{4-}$ is smaller than $[Os(CN)_6]^{4-}$, *i.e.*, $[Os(CN)_6]^{4-} > [Fe(CN)_6]^{4-}$.

Problem 41. *Which of the following complex ions of Co^{2+} ion has the weakest crystal field splitting (Δ_0): $[CoCl_6]^{4-}$, $[Co(CN)_6]^{4-}$, $[Co(NH_3)_6]^{2+}$, $[Co(en)_3]^{2+}$ and $[Co(H_2O)_6]^{2+}$.*

Solution. Since all the given complex ions have the same central metal ion which is Co^{2+} ion and the same geometry (octahedral geometry) but different ligands, the magnitude of Δ_0 depends on the position of ligands in the spectrochemical series. Now since, as is evident from spectrochemical series (--- < Cl^- < ... < H_2O ... < NH_3 < en < ---- < CN^-), Cl^- ion has the lowest crystal field splitting power, its complex *viz* $[CoCl_6]^{4-}$ has the minimum value of Δ_0, *i.e.*, $[CoCl_6]^{4-}$ has the weakest crystal field splitting.

Problem 42. *The energy difference (Δ_0) between d-levels (i.e., t_{2g} and e_g levels) for $[Co(NH_3)_6]^{3+}$ and $[Cr(H_2O)_6]^{3+}$ ions is 22,900 cm^{-1} and 17,400 cm^{-1} respectively. Express these wave numbers in terms of wavelength in Å.*

Solution. We know that:

$$\text{Wavelength } (\lambda) = \frac{1}{\text{Wave number } (\bar{\nu})}$$

Thus for $[Co(NH_3)_6]^{3+}$ ion:

$$\lambda = \frac{1}{22{,}900\,\text{cm}^{-1}} = \frac{1}{22{,}900}\ \text{cm}$$

$$= \frac{1}{22{,}900} \times 10^8\ \text{Å} \qquad \left[\begin{array}{l} \because\ 10^2\,\text{cm} = 10^{10}\,\text{Å} \\ \text{or}\ \ 1\,\text{cm} = 10^8\,\text{Å} \end{array}\right]$$

$$= \frac{10^6}{229}\ \text{Å} = 4366.8\ \text{Å}$$

= **4367 Å (Answer).**

Similarly, for $[Cr(H_2O)_6]^{3+}$ ion:

$$\lambda = \frac{1}{17{,}400\,\text{cm}^{-1}} = \frac{1}{17{,}400}\ \text{cm}$$

$$= \frac{1}{17{,}400} \times 10^8\ \text{Å}$$

$$= \frac{10^6}{174} \text{ Å} = 5747.12 \text{ Å}$$

= **5747 Å (Answer)**

Problem 43. *$[Co(NH_3)_6]^{3+}$ ion has a maximum absorption at a wavelength of 437 nm. Express this value of Δ for this ion in cm^{-1}.*

Solution. We know that :

$$\text{Wave number} = \frac{1}{\text{Wavelength}}$$

$$= \frac{1}{437 \text{ nm}}$$

$$= \frac{1}{437 \times 10^{-7} \text{ cm}} \qquad \left[\begin{array}{ll} \because & 10^9 \text{ nm} = 10^2 \text{ cm} \\ \text{or} & 1 \text{ nm} = 10^{-7} \text{ cm} \end{array}\right]$$

$$= \frac{10^7}{437} \text{ cm}^{-1} = 22883.2 \text{ cm}^{-1}$$

Problem 44. *(a) The crystal field splitting parameter (Δ_0) of $[IrCl_6]^{3-}$ is 27,600 cm^{-1}. Convert this wave number to nm. (b) What is your prediction about the magnetic properties of $[IrCl_6]^{3-}$ ion.*

Solution. (*a*) We know that:

$$\text{Wavelength} = \text{Inverse of wave number}$$

$$= \text{Inverse of } 27{,}600 \text{ cm}^{-1}$$

$$= \frac{1}{27{,}600 \text{ cm}^{-1}}$$

$$= \frac{1}{27{,}600} \text{ cm}$$

$$= \frac{1}{27{,}600} \times 10^7 \text{ nm} \qquad (\because 10^2 \text{ cm} = 10^9 \text{ nm})$$

$$= \frac{10^5}{276} \text{ nm} = \mathbf{362.3 \text{ nm (Answer)}}$$

(*b*) $[IrCl_6]^{3-}$ ion is an octahedral complex ion. Ir is present as Ir^{3+} ion. This ion is a $5d^6$ ion which is *3rd* transition series ion. We know that in octahedral complexes having the same central metal atom in the same oxidation state, the value of Δ_0 increases as we proceed from $3d^n$ (1st transition series ion) to $4d^n$ (*2nd* transition series ion) to $5d^n$ (*3rd* transition series ion). Thus Ir^{3+} ion which is $5d^6$ ion has the highest value of Δ_0 and hence lowest value of pairing energy (P). Value of P is so low that Δ_0 exceeds P ($\Delta_0 > P$) and hence low spin diamagnetic complex is formed. Thus $[IrCl_6]^{3-}$ ion is a low spin octahedral complex (inner orbital octahedral complex involving d^2sp^3 hybridisation) with diamagnetic character ($n = 0$).

Problem 45. *Soluble compounds of the complex ion, $[Co(NH_3)_6]^{3+}$ have a maximum absorption of visible light at 437 nm. What is the colour of the ion ?*

Solution. We know that:

$$10^9 \text{ nm} = 10^{10} \text{ Å}$$

$$\therefore \qquad 437 \text{ nm} = \frac{10^{10}}{10^9} \times 437 \text{ Å}$$
$$= 4370 \text{ Å}$$

Wavelength equal to 4370 Å shows that $[Co(NH_3)_6]^{3+}$ ion absorbs the light in the violet-blue region and hence transmits the light in yellow green-yellow region. Thus $[Co(NH_3)_6]^{3+}$ ion has yellow colour.

Problem 46. *If the frequency of the radiation absorbed by a complex ion in the visible region is 5.30×10^{14} s^{-1}, what is the colour of the complex ion?*

Solution. Wavelength of the absorbed radiation (λ)

$$= \frac{c}{\nu}$$
$$= \frac{3.0 \times 10^{10} \text{cm.s}^{-1}}{5.30 \times 10^{14} \text{s}^{-1}}$$
$$= 0.566 \times 10^{-4} \text{ cm}$$
$$= 0.566 \times 10^{-4} \times \frac{10^{10}}{10^2} \text{ Å} \qquad [\because 10^2 \text{cm} = 10^{10} \text{ Å}]$$
$$= 0.566 \times 10^4 \text{ Å}$$
$$= 5660 \text{ Å}$$

Table (page 217) shows that when the complex ion absorbs the radiation of $\lambda = 5660$ A^0, the complex ion absorbs the radiation in violet-blue region and transmits the radiation in yellow green-yellow region. Thus the complex ion has yellow colour.

Problem 47. *Explain the following, giving appropriate reason for your answer.*

(i) $[Ti(H_2O)_6]^{3+}$ ion has purple colour. *(Gujarat 95, Punjab 96)*

(ii) Both $[Fe(CN)_6]^{4-}$ and $[Fe(H_2O)_6]^{2+}$ appear colourless in dilute solutions.

(iii) $[Fe(H_2O)_6]^{3+}$ ion is practically colourless, but its solution becomes red when NCS^- ions are added to it.

(iv) A solution of $[Ni(H_2O)_6]^{2+}$ is green but that of $[Ni(CN)_4]^{2-}$ is colourless.

(v) A solution of $[Co(CN)_6]^{3-}$ ion colourless but that of $[CoF_6]^{3-}$ has reddish colour.

(vi) Complexes of Co (III) like $[Co(NH_3)_6]^{3+}$, $[Co(en)_3]^{3+}$ and $[Co(NO_2)_6]^{3-}$ are orange-yellow while $[CoF_6]^{3-}$ and $[Co(H_2O)_6]^{3+}$ are blue in colour.

Solution. (*i*) We know that white light is composed of many different colours. When white light is allowed to fall on $[Ti(H_2O)_6]^{3+}$ ion, this ion absorbs the light of green colour and transmits the light whose colour is white minus green (green is the colour of the absorbed light). White minus green colour (absorbed colour) is almost purple colour. Thus $[Ti(H_2O)_6]^{3+}$ ion transmits the light of purple colour and hence this ion has purple colour. In other words, we say that the colour of the transmitted light is actually the colour of $[Ti(H_2O)_6]^{3+}$ ion. The purple colour of the transmitted light is called complimentary colour of that of absorbed light. Thus purple colour is the complimentary colour of green colour.

In Table (page 217) we have shown the colour of the absorbed light (*1st* column), range of wavelength in Å (*2nd* column) and frequency in cm^{-1} (wave number) (*3rd* column) of the absorbed light. In *4th* column we have shown the colour of the transmitted light corresponding to the colour of the absorbed light.

The light absorbed by $[Ti(H_2O)_6]^{3+}$ ion has frequency of 20,000 cm^{-1} (wave number) or wavelength of 5000 Å, since the only one absorption peak (maxima) present in the absorption spectrum of $[Ti(H_2O)_6]^{3+}$ ion shown in Fig 5.106. lies at a frequency of 20,000 cm^{-1}. This

Table. Colours of the absorbed and transmitted light along with the range of wavelength in Å and frequency in cm^{-1} (wave number) of the absorbed light.

Colour of the absorbed light	*Range of wavelength (in Å) of the absorbed light*	*Range of frequency in cm^{-1} (wave number) of the absorbed light*	*Colour of the transmitted light. The colour of the transmitted light is the colour of the complex ion or compound. Colour of the transmitted light is called complimentary colour of the absorbed light.*
Energy (Δ_0) drereases →	Wave length increases →	Frequency decreases → Energy (Δ_0) drereases →	
Violet (Highest energy)	4000 – 4500	25,000 – 22,222.2	Yellow green
Blue	4500 – 4800	22,222.2 – 20,833.3	Yellow
Green blue	4800 – 4900	20,833.3 – 20,408.1	Orange
Blue green	4900 – 5000	20,408.1 – 20,000	Red
Green	5000 – 5600	20,000 – 17,857.1	Violet (almost purple)
Yellow green	5600 – 5750	17,857.1 – 17391.3	Violet
Yellow	5750 – 5900	17391.3 – 16,949.1	Blue
Orange	5900 – 6250	16,949.1 – 16,000	Blue green
Red (Lowest energy)	6250 – 7500	16,000 – 13,333.3	Blue green

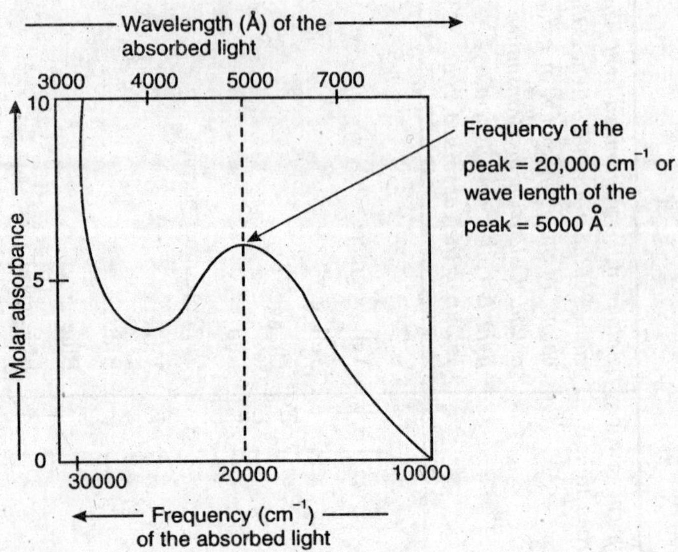

Fig. 5.106. Absorption spectrum of $[Ti\ H_2O)_6]^{3+}$ ion which has only one peak at a frequency of 20,000 cm^{-1} (wave number) or wavelength of 5000 Å. Note the relation :

$$\text{Wave length} = \frac{10^8}{\text{Wave number (cm}^{-1})} \text{ Å} = \frac{10^8}{20,000} \text{ Å} = 5000 \text{ Å}$$

frequency (= 20,000 cm^{-1}) lies in the visible region. The absorption spectrum has been obtained by plotting a graph between frequency in cm^{-1} (wave number) of the absorbed light and the molar absorbance (absorbance is the amount of absorbed light).

The absorbed light of frequency of 20,000 cm^{-1} is associated with energy equal to 239.23 KJ mol^{-1} as shown below:

$$\therefore \quad 83.6 \text{ cm}^{-1} \equiv 1 \text{ KJ mol}^{-1}$$

$$1 \text{ cm}^{-1} = \frac{1}{83.6} \text{ KJ mol}^{-1}.$$

$$\therefore \quad 20,000 \text{ cm}^{-1} \equiv \frac{20,000}{83.6} \text{ KJ mol}^{-1}$$

$$= 239.23 \text{ KJ mol}^{-1}.$$

Explanation of the purple colour of $[Ti\ (H_2O)_6]^{3+}$ ion. In $[Ti\ (H_2O)_6]^{3+}$ ion, Ti is present as Ti^{3+} ion (central metal ion) which is a $3d^1$ ion.Under the ligand field strength of six H_2O molecules (H_2O molecules act as ligands), the five *d*-orbitals are split into t_{2g} (lower energy) and e_g (higher energy) sets of orbitals and hence ground state configuration of d^1 ion is $t_{2g}^{\ 1}\ e_g^{\ 0}$. The energy of 239.23 KJ mol^{-1} which is associated with the frequency of 20,000 cm^{-1} of the absorbed light excites the single electron residing in lower energy t_{2g} set to higher energy e_g set. Thus the excited state configuration of d^1 ion becomes $t_{2g}^{\ 0}\ e_g^{\ 1}$. (See Fig. 5.107) Thus we see that the purple colour of $[Ti\ (H_2O)_6]^{3+}$ ion is due to the excitation of an electron from t_{2g} level (lower energy level) to e_g level (higher energy level). This type of transition of an electron which takes place within the same *d*-orbitals is called *d–d* electronic transition.

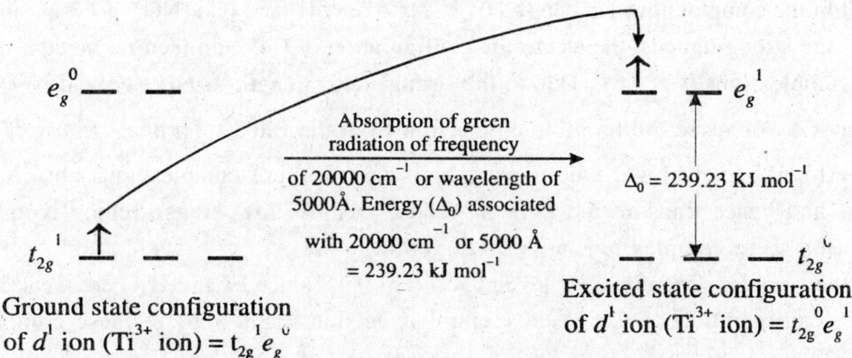

Fig. 5.107. When $[Ti(H_2O)_6]^{3+}$ ion absorbs green radiation (frequency = 20000 cm^{-1} and wave length = 5000Å), single electron of Ti^{3+} ion ($Ti^{3+} = 3d^1$) is promoted from t_{2g} orbitals to e_g orbitals.

Note. It may be noted that the energy associated with absorbed light is directly proportional to its frequency in cm^{-1} and inversely proportional to its wavelength given in Å! Thus since the frequency decreases from the absorbed violet light to the absorbed red light, the absorbed violet light has the highest energy and the absorbed red light has the lowest energy.

(*ii*) We know that CN^- ions are strong ligands but H_2O molecules are weak ligands. Due to the presence of strong ligands in $[Fe(CN)_6]^{4-}$ ion, this ion has so large value of Δ_0 (in cm^{-1}) that the wavelength of the absorption peak of this complex ion lies in the ultraviolet region. On the other hand, due to the presence of weak ligands in $[Fe(H_2O)_6]^{2+}$ ion, this complex ion has such a value of Δ_0 (in cm^{-1}) that the wavelength of the absorption peak of this complex ion lies in the infrared region. Thus we find that none of the given complex ions absorbs light in the visible region and hence each of them appears colourless.

(*iii*) When NCS^- ions are added to $[Fe(H_2O)_6]^{3+}$ ion, $[Fe(NCS)_6]^{3-}$ ion is obtained.

$$\underset{\text{Colourless}}{[Fe(H_2O)_6]^{3+}} + 6NCS^- \longrightarrow \underset{\text{Red}}{[Fe(NCS)_6]^{3-}} + 6H_2O$$

Since H_2O is a weak ligand, $[Fe(H_2O)_6]^{3+}$ ion has so small value of Δ_0 (in cm^{-1}) that the wavelength of the absorption peak of this ion lies in the infrared region and hence this ion appears colourless. Now since NCS^- ions are strong ligands, the presence of these ligands in $[Fe(NCS)_6]^{3-}$ ion makes the value of Δ_0 high so that the wavelength of the absorption peak lies in the visible region (actually in the blue-green region).

(*iv*) Since H_2O is a weak ligand, the value of Δ_0 (in cm^{-1}) for $[Ni(H_2O)_6]^{2+}$ ion is small and hence this value lies in the visible region. Thus $[Ni(H_2O)_6]^{2+}$ ion has colour (green colour). On the other hand, since CN^- ion is a strong ligand, the value of Δ_0 for $[Ni(CN)_4]^{2-}$ ion is large and hence lies in the ultraviolet region. Thus $[Ni(CN)_4]^{2-}$ ion has no colour (*i.e.*, colourless).

(*v*) We know that CN^- ions are strong ligands but F^- ions are weak ligands. Due to the presence of strong ligands in $[Co(CN)_6]^{3-}$ ion, this ion has so large value of Δ_0 that the wavelength of the absorption peak of this complex ion lies in the ultraviolet region. On the other hand, due to the presence of weak ligands in $[CoF_6]^{3-}$ ion, this complex ion has such a value of Δ_0 that the wavelength of the absorption peak of this complex ion lies in the blue-green portion of the visible region and hence transmits reddish light. Thus $[CoF_6]^{3-}$ ion is reddish in colour.

(*vi*) In the complex ions *viz* $[Co(NH_3)_6]^{3+}$, $[Co(en)_3]^{3+}$ and $[Co(NO_2)_6]^{3-}$ ions since NH_3, *en* and NO_2^- are strong ligands, the electronic configuration of Co^{3+} ion (central metal ion = $3d^6$ ion) in these complex ions is $t_{2g}^6 e_g^0$. Due to the strong field strength of NH_3, *en* and NO_2^- (ligands), the energy (Δ_0) required to promote an electron from the filled t_{2g} orbitals to empty e_g orbitals (*d–d* transition) is very high. Due to very high energy, these complex ions absorb violet/blue radiations and hence transmit the light in yellow green-yellow orange region (complementary colour). Thus these complex ions are orange yellow.

In the complex ions *viz* $[CoF_6]^{3-}$ and $[Co(H_2O)_6]^{3+}$ since F^- and H_2O are weak ligands, the electronic configuration of Co^{3+}, ion (central metal ion = $3d^6$ ion) in these complex ions is $t_{2g}^4 e_g^2$. Due to the weak field strength of F^- and H_2O (ligands), the energy (Δ_0) required to promote an electron from t_{2g}^4 orbitals to e_g^2 orbitals (*d–d* transition) is comparatively small. Due to small energy, these complex ions absorb orange/yellow light and hence transmit blue green/ blue radiations (complementary radiations). Thus these complex ions are blue in colour.

Problem 48. *On the basis of CFT, predict whether the following ions are square planar or tetrahedral. Give energy level diagram, showing the distribution of d-electrons of the central metal atom in the splitted d-orbitals. Also find out the spin state of each of the complex ions.* $[Pt(NH_3)_4]^{2+}$, $[Cu(NH_3)_4]^{2+}$ $[Pt(en)_2]^{2+}$, $[NiCl_4]^{2-}$, $[CoCl_4]^{2-}$.

Solution. $[Pt(NH_3)_4]^{2+}$ ion. This ion has Pt^{2+} ion as its central metal ion. Pt^{2+} ion is a $5d^8$ ion. Since NH_3 molecule (ligand) is a strong ligand, $[Pt(NH_3)_4]^{2+}$ ion in which C.N. of Pt^{2+} ion = 4 does not have tetrahedral geometry; rather it has square planar geometry. The distribution of $5d^8$ electrons in Pt^{2+} ion in different splitted *d*-orbitals in square planar field can be shown as given in Fig. 5.108. Obviously, since $n = 0$, $[Pt(NH_3)_4]^{2+}$ ion is diamagnetic.

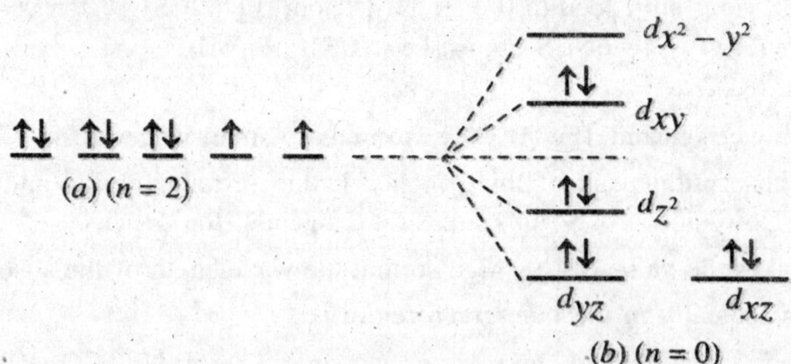

Fig. 5.108. (*a*) Distribution of $5d^8$ electrons of Pt^{2+} ion in five degenerate *d*-orbitals. (*b*) Distribution of $5d^8$ electrons of Pt^{2+} ion in $[Pt(NH_3)_4]^{2+}$ ion in the splitted *d*-orbitals in square planar field.

$[Cu(NH_3)_4]^{2+}$ ion. This ion has Cu^{2+} ion as its central metal ion. Cu^{2+} ion is a $3d^9$ ion. Since NH_3 molecule (ligand) is a strong ligand, $[Cu(NH_3)_4]^{2+}$ ion in which C.N. of Cu^{2+} ion = 4 does not have tetrahedral geometry; rather it has square planar geometry. The distribution of $3d^9$ electrons in Cu^{2+} ion in different splitted *d*-orbitals in square planar field can be shown as given in Fig. 5.109 Obviously, since $n = 1$, $[Cu(NH_3)_4]^{2+}$ ion is paramagnetic.

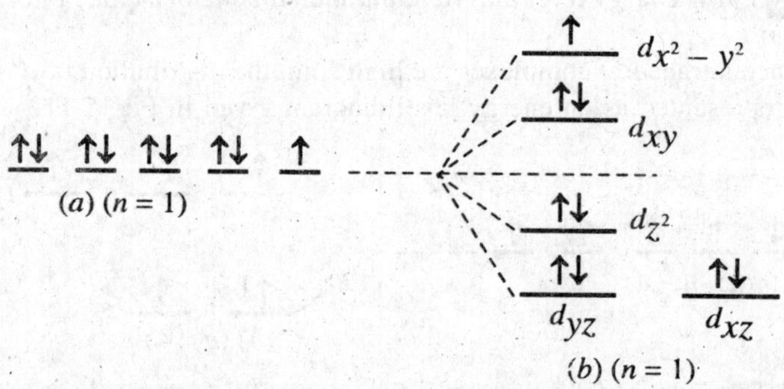

Fig. 5.109. (*a*) Distribution of $3d^9$ electrons of Cu^{2+} ion in five degenerate *d*-orbitals. (*b*) Distribution of $3d^9$ electrons of Cu^{2+} ion in $[Cu(NH_3)_4]^{2+}$ ion in the splitted *d*-orbitals in square plannar field.

$[Pt(en)_2]^{2+}$ ion. In this ion, Pt (central metal atom) is present as Pt^{2+} ion which is a $5d^8$ ion. *en* acts as ligand which is a strong ligand. Now since C.N. of Pt^{2+} ion is 4 and *en* acts as strong ligand, $[Pt(en)_2]^{2+}$ ion does not have tetrahedral geometry; rather it has square planar geometry. The distribution of $5d^8$ electrons in different splitted *d*-orbitals in square planar ligand field generated by *en* ligands is as shown in Fig. 5.108. Since $n = 0$, $[Pt\ (en)_2]^{2+}$ ion is diamagnetic.

$[NiCl_4]^{2-}$ ion. This ion has Ni^{2+} ion has its central metal ion. Ni^{2+} ion is a $3d^8$ ion. Since Cl^- ion is a weak ligand, $[NiCl_4]^{2-}$ ion in which C.N. of Ni^{2+} ion = 4 has tetrahedral geometry (4-coordinated complex ions having weak ligands have tetrahedral geometry). The distribution of $3d^8$ electrons of Ni^{2+} ion in *e* (lower energy orbitals) and t_2 (higher energy orbitals) orbitals in tetrahedral ligand field generated by four Cl^- ligands is as $e^4t_2^{\ 4}$ as shown in Fig. 5.110. This distribution shows that $n = 2$. Thus $[NiCl_4]^{2-}$ ion is paramagnetic.

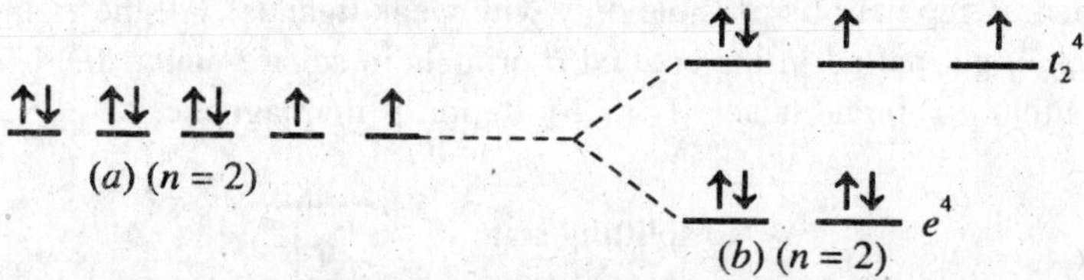

Fig. 5.110. (*a*) Distribution of $3d^8$ electrons of Ni^{2+} ion in five degenerate *d*-orbitals (*b*) Distribution of $3d^8$ electrons of Ni^{2+} ion in $[NiCl_4]^{2-}$ ion in *e* and t_2 orbitals in tetrahedral ligand field.

$[CoCl_4]^{2-}$ ion. This ion has Co^{2+} ion as its central metal ion. Co^{2+} ion is a $3d^7$ ion. Since Cl^- ion is a weak ligand, $[Co\ Cl_4]^{2-}$ ion in which C.N. of Co^{2+} ion = 4 has tetrahedral geometry. The distribution of $3d^7$ electrons of Ni^{2+} ion in *e* (lower energy orbitals) and t_2 (higher energy orbitals) orbitals in tetrahedral ligand field generated by four Cl^- ligands is as $e^4t_2^{\ 3}$ (See Fig. 5.111) which gives $n = 3$. Thus $[CoCl_4]^{2-}$ ion is paramagnetic.

$t_2^{\ 3}$

(*a*) ($n = 3$)

e^4

(*b*) ($n = 3$)

Fig. 5.111 (*a*) Distribution of $3d^7$ electrons of Co^{2+} ion in five degenerate *d*-orbitals (*b*) Distribution of $3d^7$ electrons of Co^{2+} ion in $[CoCl_4]^{2-}$ ion in *e* and t_2 orbitals in tetrahedral ligand field.

Problem 49. *Draw energy level diagram and indicate the occupacy of orbitals (splitted) in d^6 (tetrahedral) ion.* *(Punjabi 94)*

Solution. Since tetrahedral complexes are high spin, the distribution of d^6 electrons in e and t_2 orbitals can be represented as an energy level diagram given in Fig. 5.112

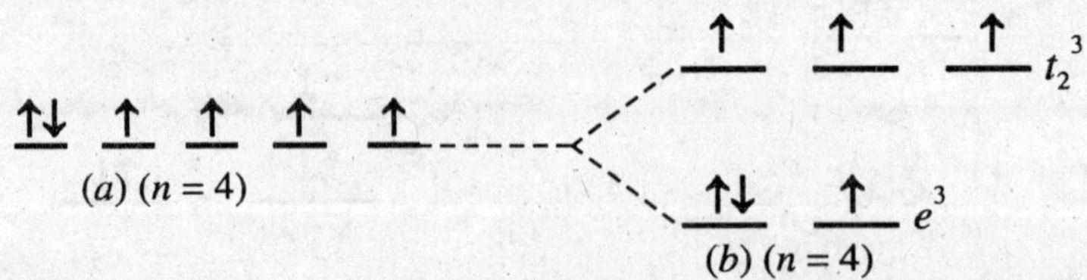

Fig. 5.112. (*a*) Distribution of $3d^6$ electrons in five degererate *d*-orbitals (*b*) Distribution of $3d^6$ electrons in *e* and t_2 orbitals in tetrahedral ligand Field.

Problem 50. *On the basis of CFT, explain the following, giving appropriate reasons for your answer.*

(i) Tetrahedral complexes are generally high spin. *(Kurukshetra 95, G.N. Dev 95)*

(ii) $[Ni(CN)_4]^{2-}$ is diamagnetic but $[NiCl_4]^{2-}$ is paramagnetic. *(Punjabi 94, Patna 94)*

Solution. (*i*) In case of tetrahedral complexes, since Δ_t [Δ_t is the energy difference between t_2 (d_{xy}, d_{yz} and d_{zx} orbitals) and *e* ($d_{x^2-y^2}$ and d_{z^2} orbitals) sets of orbitals] is generally less than P (P is the energy which is required to pair two electrons against electron-electron repulsion in the same orbital. P is the pairing energy for one electron pair), the electrons tend to remain unpaired and hence tetrahedral complexes are high spin complexes even when strong field ligands are involved. In high spin complexes electrons occupy the five *d*-orbitals in accordance with Hund's rule of maximum multiplicity.

(*ii*) Since CN^- ions are strong ligands, $[Ni(CN)_4]^{2-}$ ion has square planar geometry but not tetrahedral (tetrahedral geometry is possible only with weak ligands). All the eight 3*d* electrons of Ni^{2+} ion ($Ni^{2+} = 3d^8$) get paired in the splitted *d*-orbitals in square planar field (see Fig. 5.113). This figure shows that $n = 0$ and hence $[Ni\ (CN)_4]^{2-}$ ion is diamagnetic.

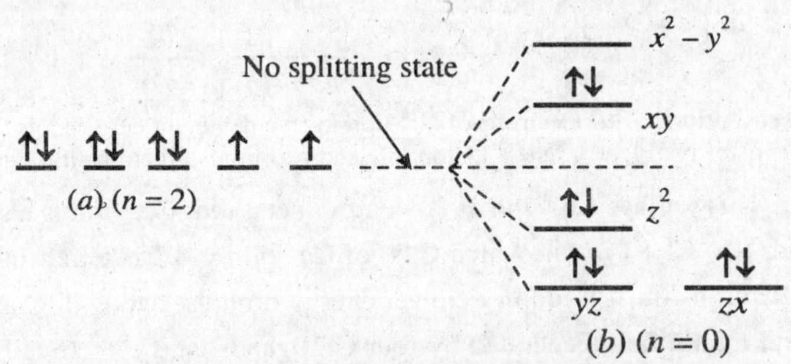

Fig. 5.113. (*a*) Five degenerate *d*-orbitals of Ni^{2+} ion ($Ni^{2+} = 3d^8$)
(*b*) Splitted *d*-orbitals in square planar $[Ni(CN)_4]^{2-}$ ion.

Since Cl^- ions are weak ligands, $[NiCl_4]^{2-}$ ion has tetrahedral geometry. Ni-atom is present as Ni^{2+} ion which is a $3d^8$ ion. The distribution of eight 3*d* electrons of Ni^{2+} ion in tetrahedral field takes place as shown in Fig. 5.114 This figure shows that $n = 2$ and hence $[NiCl_4]^{2-}$ ion is paramagnetic.

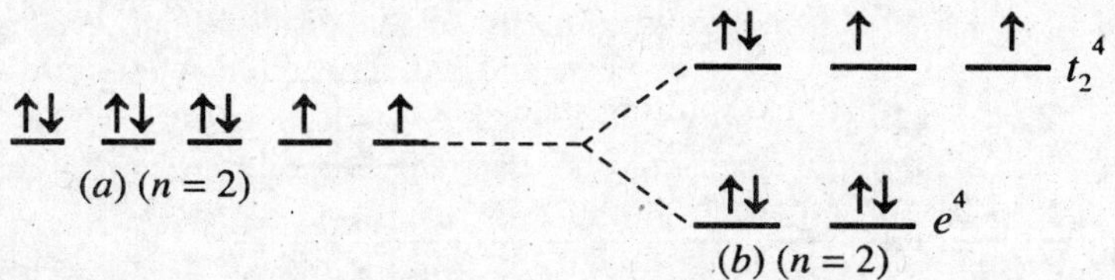

Fig. 5.114. (*a*) Five degenerate *d*-orbitals of Ni^{2+} ion ($Ni^{2+} = 3d^8$)
(*b*) Splitted *d*-orbitals in tetrahedral $[NiCl_4]^{2-}$

Problem 51. *Bis (dimethylglyoximate) nickel (II) is experimentally known to be diamagnetic. Predict, on the basis of VBT and CFT, whether the complex is square planar, tetrahedral or octahedral.*

Solution. VBT. The given complex has the formula $[Ni(dmg)_2]^0$ in which Ni^{2+} ion is the central metal ion which is a $3d^8$ ion. Dimethylglyoxime ion (dmg^-) is a mono-negatively charged ion and acts as a bidentate ligand. Due to the bidentate nature of dmg^-, the C.N. of Ni^{2+} ion is 4 and hence the given complex has either square planar (dsp^2 hybridisation) or tetrahedral (sp^3 hybridisation) geometry but not octahedral geometry as shown in Fig. 5.115

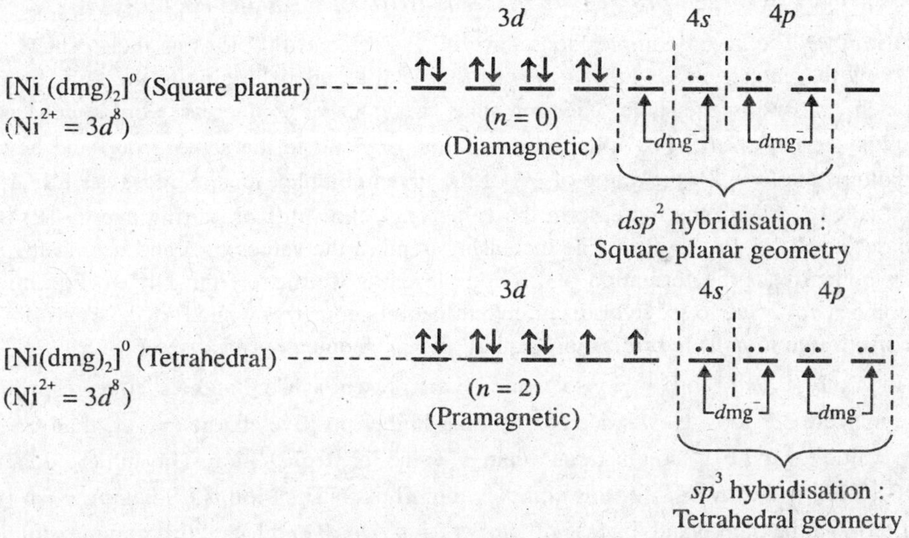

3*d* 4*s* 4*p*

$[Ni(dmg)_2]^0$ (Square planar) ↑↓ ↑↓ ↑↓ ↑↓ | •• | •• | •• | •• |
($Ni^{2+} = 3d^8$) ($n = 0$) (Diamagnetic) dmg^- dmg^-

dsp^2 hybridisation : Square planar geometry

3*d* 4*s* 4*p*

$[Ni(dmg)_2]^0$ (Tetrahedral) ↑↓ ↑↓ ↑↓ ↑ ↑ | •• | •• | •• | •• |
($Ni^{2+} = 3d^8$) ($n = 2$) (Pramagnetic) dmg^- dmg^-

sp^3 hybridisation : Tetrahedral geometry

Fig. 5.115. Two possible geometries *viz* square planar and tetrahedral of $[Ni(dmg)_2]^0$.

Fig. 5.115 shows that it is only dsp^2 hybridisation (square planar geometry) which gives $n = 0$ or which makes $[Ni\ (dmg)_2]^0$ diamagnetic. Thus $[Ni\ (dmg)_2]^0$ has square planar geometry.

CFT. Since dmg^- ion is a strong ligand, according to CFT, tetrahedral geometry is not possible (tetrahedral geometry is given by weak ligands), *i.e.*, the given complex has square planar geometry. Now we know that the relative order of energy of the splitted *d*-orbitals of Ni^{2+} ion in square planar field is as shown in Fig. 5.116. Since there is no unpaired electron ($n = 0$), the complex compound is diamagnetic.

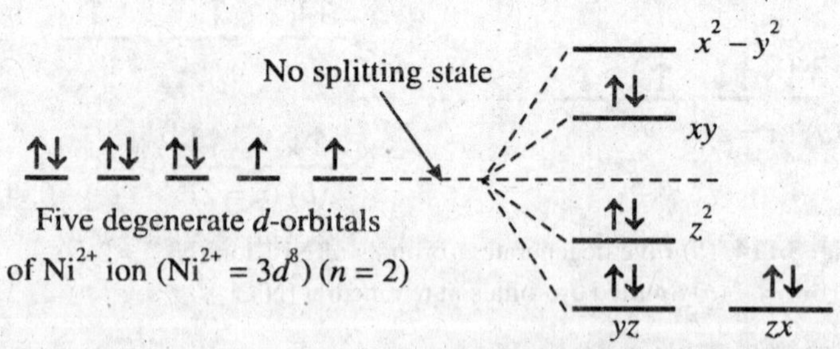

Fig. 5.116. According to CFT, $[Ni(dmg)_2]^0$ has square planar geometry and is diamagnetic ($n = 0$)

Problem 52. *Explain the following:*

(i) $[NiCl_4]^{2-}$ *ion is paramagnetic tetrahedral but* $[PdCl_4]^{2-}$ *and* $[PtCl_4]^{2-}$ *ions are diamagnetic square planar.*

(ii) $[CoF_6]^{3-}$ *and* $[NiF_6]^{2-}$ *ions both have* F^- *ions as ligands which are weak ligands, yet* $[CoF_6]^{3-}$ *ion is paramagnetic* ($\mu \approx 4.89$ *BM) and* $[NiF_6]^{2-}$ *ion is diamagnetic* ($\mu = 0$)

(iii) Δ_0 *increases in the order:* $[CrCl_6]^{3-} < [Cr(NH_3)_6]^{3+} < [Cr(CN)_6]^{3-}$.

Solution. (*i*) The given complex ions have Ni^{2+}, Pd^{2+} and Pt^{2+} ions as their central metal ions. We know that in complexes having the central metal atoms belonging to the same group of transition elements and in the same oxidation state, the value of Δ increases on descending the group. Now since Ni^{2+} ($3d^8$), Pd^{2+} ($4d^8$) and Pt^{2+} ($5d^8$) belong to the same group and have the same oxidation state (= + 2), the value of Δ of the given complex ions increases as $[NiCl_4]^{2-} < [PdCl_4]^{2-} < [PtCl_4]^{2-}$. With the increase in the value of Δ, the value of pairing energy (P) of the ions decreases as $Ni^{2+} > Pd^{2+} > Pt^{2+}$. The increasing trend in the values of Δ and decreasing trend in P values both fovour the formation of LS complexes as we move from Ni^{2+} to Pd^{2+} to Pt^{2+}. $[NiCl_4]^{2-}$ ion has $n = 2$ due to sp^3 hybridisation (tetrahedral geometry) while $[PdCl_4]^{2-}$ and $[PtCl_4]^{2-}$ both have $n = 0$ due to dsp^2 hybridisation (square planar geometry) (Also see Fig. 5.117)

(*ii*) In $[Co\ F_6]^{3-}$ and $[Ni\ F_6]^{2-}$ ions Co and Ni are present as Co^{3+} and Ni^{4+} ions respectively. Both these ions are $3d^6$ ions. Now since Ni^{4+} ion has higher positive charge (= + 4) than Co^{3+} ion (= + 3), Δ_0 value for $[Ni\ F_6]^{2-}$ ion is greater than Δ_0 value for $[CoF_6]^{3-}$ ion. The enhanced value of Δ_0 for $[NiF_6]^{2-}$ ion makes $\Delta_0 > P$ for this complex ion. Thus $[NiF_6]^{2-}$ ion is LS having $n = 0$ (d^2sp^3 hybridisation). On the other hand, in case of $[CoF_6]^{3-}$ ion $\Delta_0 < P$ and hence this complex ion is HS (sp^3d^2 hybridisation) with $n = 4$ and $\mu = \sqrt{4 \times (4+2)} = 4.898$ BM.

(*iii*) Since the given complex ions have the same metal ion and the same geometry but different ligands, Δ_0 values of these complex ions increase with the strength of the ligands. The spectrochemical series shows that the strength of Cl^-, NH_3 and CN^- ligands is in the order: $Cl^- < NH_3 < CN^-$. Thus the magnitude of Δ_0 of the given complex ions increases as:

$$[CrCl_6]^{3-} < [Cr\ (NH_3)_6]^{3+} < [Cr\ (CN)_6]^{3-}.$$

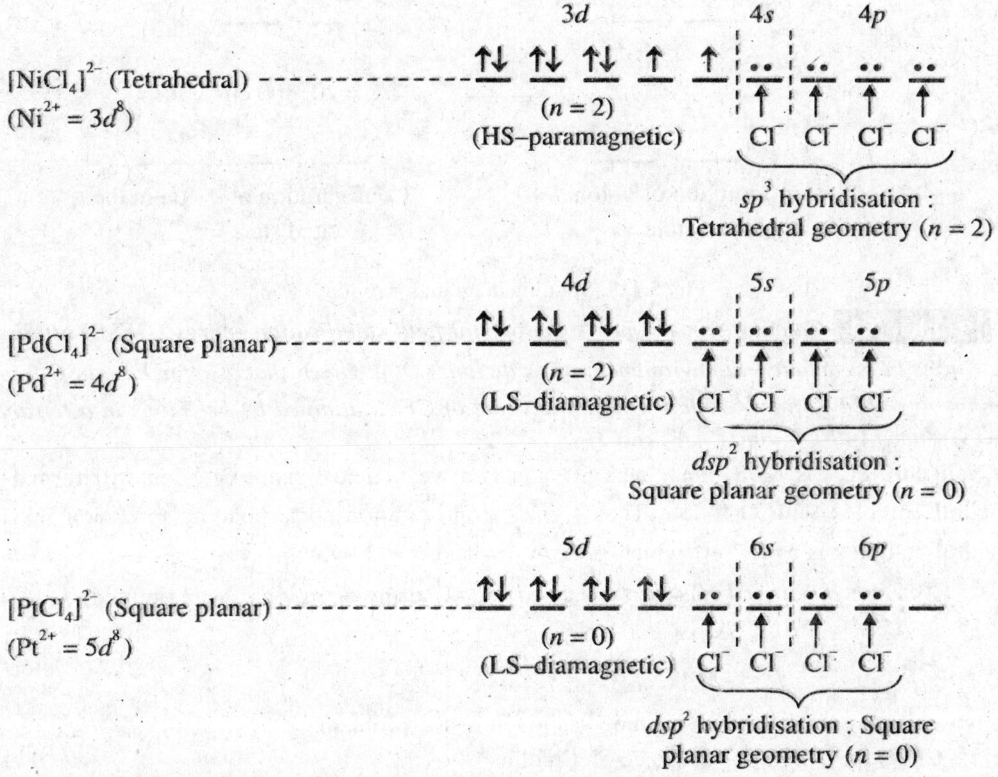

$[NiCl_4]^{2-}$ (Tetrahedral) — 3d: ↑↓ ↑↓ ↑↓ ↑ ↑ ; 4s: •• ; 4p: •• •• ••
($Ni^{2+} = 3d^8$)
(n = 2) (HS–paramagnetic) Cl⁻ Cl⁻ Cl⁻ Cl⁻
sp^3 hybridisation : Tetrahedral geometry (n = 2)

$[PdCl_4]^{2-}$ (Square planar) — 4d: ↑↓ ↑↓ ↑↓ ↑↓ •• ; 5s: •• ; 5p: •• •• —
($Pd^{2+} = 4d^8$)
(n = 2) (LS–diamagnetic) Cl⁻ Cl⁻ Cl⁻ Cl⁻
dsp^2 hybridisation : Square planar geometry (n = 0)

$[PtCl_4]^{2-}$ (Square planar) — 5d: ↑↓ ↑↓ ↑↓ ↑↓ •• ; 6s: •• ; 6p: •• •• —
($Pt^{2+} = 5d^8$)
(n = 0) (LS–diamagnetic) Cl⁻ Cl⁻ Cl⁻ Cl⁻
dsp^2 hybridisation : Square planar geometry (n = 0)

Fig. 5.117. Formation $[NiCl_4]^{2-}$, $[PdCl_4]^{2-}$ and $[PtCl_4]^{2-}$ ions by sp^3 (tetrahedral), dsp^2 (square planar) and dsp^2 (square planar) hybridisation.

Problem 53. *Calculate CFSE in KJ mol⁻¹ for $[Ti(H_2O)_6]^{3+}$ ion for which d–d transition is single broad absorption peak with a maximum at 20,300 cm⁻¹. Draw the splitting diagram also.*

Solution. In the given complex ion, Ti is present as Ti^{3+} which is a $3d^1$ ion. We know that:

$$\text{CFSE for } d^1 \text{ ion} = -0.4\ \Delta_0$$

Now since $\Delta_0 = 20{,}300\ cm^{-1}$ as given in the problem,

$$\text{CFSE for } d^1 \text{ ion} = -0.4 \times 20{,}300\ cm^{-1}$$

$$= -8120\ cm^{-1}$$

$$= \frac{-8120}{83.7}\ \text{KJ mol}^{-1} \qquad (\because 83.7\ cm^{-1} = 1\ \text{KJ mol}^{-1})$$

$$\mathbf{= 97.0\ KJ\ mol^{-1}\ (Answer)}$$

The splitting daigram can be shown as given in fig. 5.118

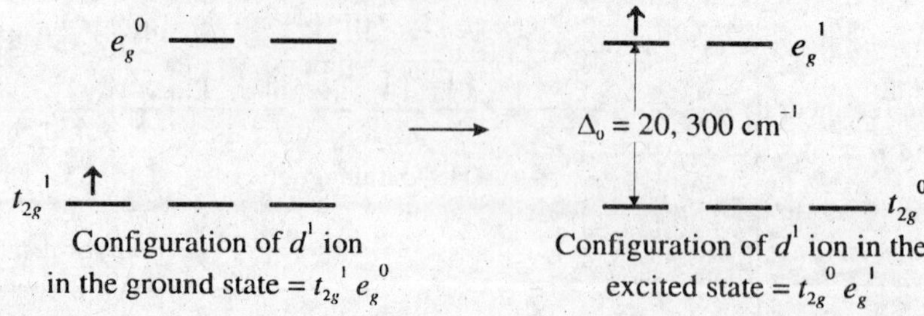

Configuration of d^1 ion in the ground state = $t_{2g}^1 e_g^0$

Configuration of d^1 ion in the excited state = $t_{2g}^0 e_g^1$

Fig. 5.118. Splitting diagram for d^1 ion

Problem 54. *Calculate, in KJ mol^{-1}, the crystal field stabilisation energy (CFSE) attained by Fe^{2+} ions in an octahedral environment of oxide ions (O^{2-}). Given that Δ_0 for Fe^{2+} in O^{2-} ion environment = 124 KJ mol^{-1}. What will be the value of CFSE attained by Fe^{2+} ions in a tetrahedral environment of O^{2-} ions?*

Solution. Fe^{2+} is a $3d^6$ ion. Since O^{2-} ion is a weak field ligand, Fe^{2+} ion will form HS octahedral complex with O_2^{2-} ion. Thus $t_{2g}^p e_g^q$ configuration corresponding to d^6 ion in HS octahedral complex is as $t_{2g}^4 e_g^2$ which gives $p = 4$ and $q = 2$. Obviously :

$$\text{CFSE for } d^6 \text{ ion} = (-0.4 \times p + 0.6 \times q)\ \Delta_0 \quad \text{(Pairing energy has been omitted)}$$
$$= (-0.4 \times 4 + 0.6 \times 2)\ \Delta_0$$
$$= -0.4\ \Delta_0$$

Now since $\Delta_0 = 124$ KJ mol^{-1} as given in the problem,

$$\text{CFSE for } d^6 \text{ ion} = -0.4 \times 124 \text{ KJ mol}^{-1}$$
$$= \mathbf{-49.6\ KJ\ mol^{-1}}$$ **(Answer)**

Tetrahedral complexes are always high spin. Thus $e^p t_2^q$ configuration corresponding to d^6 ion in HS tetrahedral complex is as $e^3 t_2^3$ giving $p = 3$ and $q = 3$. Obviously :

$$\text{CFSE for } d^6 \text{ ion} = (-0.6 \times p + 0.4 \times q)\ \Delta t$$
$$= (-0.6 \times 3 + 0.4 \times 3) \times 0.45\ \Delta_0 \qquad (\because \Delta t = 0.45\ \Delta_0)$$
$$= -0.6 \times 0.45\ \Delta_0$$
$$= -0.27\ \Delta_0$$
$$= -0.27 \times 124 \text{ KJ mol}^{-1}$$
$$= \mathbf{-33.48\ KJ\ mol^{-1}}$$ **(Answer)**

Problem 55. *For Mn^{3+} ion, the electron pairing energy is about 28,000 cm^{-1}. Δ_0 value for $[Mn(CN)_6]^{3-}$ ion is 38,500 cm^{-1}. Does this complex has high spin or low spin configuration? Also write down the configuration corresponding to this state.*

Solution. The answer can be given in the following two ways:

(*i*) Since P for Mn^{3+} ion is lower than Δ_0 for $[Mn(CN)_6]^{3-}$ ion (P = 28,000 cm^{-1}, Δ_0 = 38,500 cm^{-1}), $[Mn(CN)_6]^{3-}$ ion is a LS octahedral complex ion.

(*ii*) In $[Mn(CN)_6]^{3-}$ ion Mn is present as Mn^{3+} ion which is a $3d^4$ ion. Let us calculate CFSE value for d^4 ion (Mn^{3+} ion) in its HS and LS octahedral complexes as shown below :

$$\text{CFSE for } d^4 \text{ ion } (Mn^{3+} \text{ ion}) \text{ in HS octahedral complex} = -0.6\ \Delta_0$$
$$= -0.6 \times 38{,}500 \text{ cm}^{-1}$$

$= -23{,}100\ cm^{-1}$

CFSE for d^4 ion (Mn^{3+} ion) in LS octahedral complex $= -1.6\ \Delta_0 + P$

$= (-1.6 \times 38{,}500 + 28000)\ cm^{-1}$

$= -33{,}600\ cm^{-1}$

Now since CFSE value for d^4 ion (Mn^{3+} ion) in LS octahedral complex is more negative ($= -33{,}600\ cm^{-1}$) than that for this ion in HS octahedral complex ($= -23{,}100\ cm^{-1}$), $[Mn\ (CN)_6]^{3-}$ ion a LS octahedral complex ion.

$t_{2g}^{\ p}\ e_g^{\ q}$ configuration of d^4 ion in LS octahedral complex is $t_{2g}^{\ 4}\ e_g^{\ 0}$.

Problem 56. *10 Dq of $[Mn\ (H_2O)_6]^{3+}$ is known from electronic spectrum to be 21,000 cm^{-1}. The pairing energy of Mn (III) is 28,800 cm^{-1}. Predict whether the given complex ion is high spin or low spin.*

Solution. The answer can be found out by the following two methods:

(*i*) Since pairing energy (P) is larger than 10Dq, the given complex ion is high spin octahedral complex ion.

(*ii*) The given complex ion is an octahedral complex ion which has Mn^{3+} ion as its central metal ion. Mn^{3+} ion is $3d^4$ ion. Let us calculate CFSE value for d^4 ion (Mn^{3+} ion) in its HS and LS octahedral complexes as shown below:

CFSE value for d^4 ion (Mn^{3+} ion) in HS octahedral complex

$$= -6\ Dq$$

$$= -6 \times \frac{21{,}000}{10} \qquad (\because\ 10\ Dq = 21{,}000\ cm^{-1})$$

$$= -12{,}600\ cm^{-1}$$

CFSE value for d^4 ion (Mn^{3+} ion) in LS octahedral complex

$$= -16\ Dq + P$$

$$= -16 \times \frac{21{,}000}{10} + 28{,}800$$

$$= -4800\ cm^{-1}$$

Now since CFSE value for d^4 ion (Mn^{3+} ion) in HS octahedral complexes is more negative than that in LS octahedral complexes, $[Mn\ (H_2O)_6]^{3+}$ ion is a HS octahedral complex ion.

Problem 57. *Calculate CFSE values in terms of Δ_0 and P for HS and LS octahedral complexes of Fe (II) and Co (II). Predict whether the complexes are paramagnetic or diamagnetic.*

Solution. HS octahedral complexes of Fe (II) (P > Δ_0). Fe (II) is a $3d^6$ ion. The distribution of $3d^6$ electrons in t_{2g} and e_g sets of orbitals in HS octahedral complexes of Fe (II) is as $t_{2g}^{\ 4}\ e_g^{\ 2}$. This distribution gives $m = 1$, $n = 4$ and hence:

$$\text{CFSE} = (-0.4 \times t_{2g} + 0.6 \times e_g)\ \Delta_0 + mP$$

$$= (-0.4 \times 4 + 0.6 \times 2)\ \Delta_0 + 1 \times P$$

$$= \mathbf{-0.4\ \Delta o\ (Answer)}$$

Due to the presence of 4 unpaired electrons ($n = 4$), the complexes are paramagnetic.

LS octahedral complexes of Fe (II) (P < Δ_0). The distribution of $3d^6$ electrons in t_{2g} and e_g sets of orbitals in LS octahedral complexes of Fe (II) is as $t_{2g}^{\ 6}\ e_g^{\ 0}$. This distribution gives $m = 3$, $n = 0$ and hence:

$$\text{CFSE} = (-0.4 \times t_{2g} + 0.6 \times e_g)\ \Delta_0 + mP$$

$$= (-0.4 \times 6 + 0.6 \times 0)\ \Delta_0 + 3P$$

$$= \mathbf{-2.4\ \Delta_0 + 3P\ (Answer)}$$

Due to the absence of unpaired electron ($n = 0$), the complexes are diamagnetic.

HS octahedral complexes of Co (II) (P > Δ_0) Co (II) is a $3d^7$ ion. The distribution $3d^7$ electrons in t_{2g} and e_g sets of orbitals in these complexes is as $t_{2g}^5 e_g^2$. This distribution gives $m = 2$, $n = 3$ and hence:

$$\text{CFSE} = (-0.4 \times t_{2g} + 0.6 \times e_g)\ \Delta_0 + mP$$
$$= (-0.4 \times 5 + 0.6 \times 2)\ \Delta_0 + 2P$$
$$= \mathbf{-0.8\ \Delta + 2P\ (Answer)}$$

Due to presence of 3 unpaired electrons ($n = 3$), the complexes are paramagnetic.

LS octahedral complexes of Co (II) ($P < \Delta_0$). The distribution of $3d^7$ electrons in t_{2g} and e_g sets of orbitals in these complexes is as $t_{2g}^6 e_g^1$. This distribution gives $m = 3$, $n = 1$ and hence:

$$\text{CFSE} = (-0.4 \times t_{2g} + 0.6 \times e_g)\ \Delta_0 + mP$$
$$= (-0.4 \times 6 + 0.6 \times 1)\ \Delta_0 + 3P$$
$$= \mathbf{-1.8\ \Delta_0 + 3P\ (Answer)}$$

Due to the presence of one unpaired electron ($n = 1$), the complexes are paramagnetic.

Problem 58. *Calculate CFSE values for the following systems: (i) d^1-octahedral (ii) d^1–tetrahedral (iii) d^5 - low spin octahedral (iv) d^5 - high spin octahedral.*

(M.D. Rohtak 94; Delhi 96)

Solution. (*i*) d^1-octahedral system. Single d^1 electron will occupy t_{2g} orbitals and e_g orbitals will remain vacant. Thus $t_{2g} = 1$ and $e_g = 0$, and hence:

$$\text{CFSE} = (-0.4 \times t_{2g} + 0.6 \times e_g)\ \Delta_0$$
$$= (-0.4 \times 1 + 0.6 \times 0)\ \Delta_0 = \mathbf{-0.4\ \Delta_0\ (Answer)}$$

(*ii*) **d^1-tetrahedral system.** Single d^1 electron will occupy e set of orbital and t_2 set of orbitals will remain vacant. Thus $e = 1$ and $t_2 = 0$, and hence:

$$\text{CFSE} = (-0.6 \times e + 0.4 \times t_2)\ \Delta t$$
$$= -(0.6 \times 1 + 0.4 \times 0)\ \Delta t$$
$$= \mathbf{-0.6\ \Delta t\ (Answers)}$$

(*iii*) **d^5-low spin octahedral system.** In this case, the distribution of d^5 electrons in t_{2g} and e_g sets of orbitals is as $t_{2g}^5 e_g^0$. Thus $t_{2g} = 5$ and $e_g = 0$, and hence:

$$\text{CFSE} = (-0.4 \times t_{2g} + 0.6 \times e_g)\ \Delta_0$$
$$= (-0.4 \times 5 + 0.6 \times 0)\ \Delta_0$$
$$= \mathbf{-2.0\ \Delta_0\ (Answers)}$$

(*iv*) **d^5-high spin octahedral system.** In this case, the distribution of d^5 electrons in t_{2g} and e_g sets of orbitals is as $t_{2g}^3 e_g^2$. Thus $t_{2g} = 3$ and $e_g = 2$ and hence:

$$\text{CFSE} = (-0.4 \times t_{2g} + 0.6 \times e_g)\ \Delta_0$$
$$= (-0.4 \times 3 + 0.6 \times 2)\ \Delta_0$$
$$= \mathbf{0.0\ (Answer)}$$

Problem 59. *Calculate CFSE of the following complexes: (i) $[CoF_6]^{3-}$ (ii) $[Fe(CN)_6]^{4-}$ (iii) $[Cu(NH_3)_6]^{2+}$*

(M.D. Rohtak 95, Punjabi 96)

Solution. (*i*) **$[CoF_6]^{3-}$ ion.** This ion is octahedral and contains Co-atom as Co^{3+} ion which is a d^6 ion ($Co^{3+} = 3d^6\ 4s^0$). Since F^- ions are weak ligands, the distribution of six *d*-electrons in Co^{3+} ion in t_{2g} and e_g orbitals is as $t_{2g}^{\ 4} e_g^{\ 2}$. Thus $t_{2g} = 4$ and $e_g = 2$ and hence :

$$\text{CFSE of } d^6 \text{ ion} = (-0.4 \times 4 + 0.6 \times 2)\ \Delta_0$$
$$= \mathbf{-0.4\ \Delta_0\ (Answer)}$$

(*ii*) **$[Fe(CN)_6]^{4-}$ ion.** This ion is octahedral and contains Fe-atom as Fe^{2+} ion which is d^6 ion. Since CN^- ions are strong ligands, the distribution of six *d*-electrons in Fe^{2+} ion in t_{2g} and e_g orbitals is as $t_{2g}^6\ e_g^{\ 0}$. Thus $t_{2g} = 6$ and $e_g = 0$, and hence :

$$\text{CFSE of } d^6 \text{ ion} = (-0.4 \times 6 + 0.6 \times 0)\ \Delta_0$$
$$= \mathbf{-2.4\ \Delta_0\ Ans.}$$

(*iii*) **$[Cu(NH_3)_6]^{2+}$ ion.** This ion has octahedral geometry. Since Cu^{2+} ion is a $3d^9$ ion, the distribution of nine $3d$ electrons in t_{2g} and e_g sets of orbitals is $t_{2g}^{\ 6}\ e_g^{\ 3}$. Thus $t_{2g} = 6$ and $e_g = 3$ and hence.

$$\text{CFSE of } d^9 \text{ ion} = (-0.4 \times 6 + 0.6 \times 3)\ \Delta_0$$
$$= \mathbf{-0.6\ \Delta_0}.$$

Problem 60. *Compute CFSE (in terms of Δ_0) values for d^7 (octahedral) and d^7 (tetrahedral) ions in strong field.* *(Punjabi 1994)*

Solution. (*i*) **d^7 ion (octahedral) in strong field.** The distribution of d^7 electrons in strong field in t_{2g} and e_g sets is as $t_{2g}^{\ 6}\ e_g^{\ 1}$ which gives $t_{2g} = 6$ and $e_g = 1$ and hence:

$$\text{CFSE for } d^7 \text{ ion} = (-0.4 \times 6 + 0.6 \times 1)\ \Delta_0$$
$$= \mathbf{-1.8\ \Delta_0\ (Answer)}$$

(*ii*) **d^7 ion (tetrahedral) in strong field.** Since tetrahedral complexes are always high spin, the distribution of d^7 electrons in *e* and t_2 sets of orbits is as $e^4\ t_2^{\ 3}$. Now since $e = 4$ and $t_2 = 3$,

$$\text{CFSE of } d^7 \text{ ion} = (-0.27 \times e + 0.18 \times t_2)\ \Delta_0$$
$$= (-0.27 \times 4 + 0.18 \times 3)\ \Delta_0$$
$$= \mathbf{-0.54\ \Delta_0\ (Answer)}$$

Problem 61. *Calculate the number of unpaired electrons (n) and CFSE for the following complex ions:* (*i*) $[Fe(H_2O)_6]^{3+}$ (*ii*) $[Cr(NH_3)_6]^{3+}$ (*iii*) $[CoCl_4]^{2-}$. *(Punjabi 95, Kurukshetra 97)*

Solution. (*i*) **$[Fe(H_2O)_6]^{3+}$ ion.** Since H_2O molecules are weak ligands, the distribution of five $3d$ electrons in Fe^{3+} ion ($Fe^{3+} = 3d^5$) in t_{2g} and e_g sets of orbitals is as : $t_{2g}^{\ 3}\ e_g^{\ 2}$.

Thus the number of unpaired electrons (*n*) in $[Fe(H_2O)_6]^{3+}$ ion = **5 (Answer)**

Now since $t_{2g} = 3$ and $e_g = 2$,

$$\text{CFSE of } d^5 \text{ ion} = (-0.4\ t_{2g} + 0.6\ e_g)\ \Delta_0$$
$$= (-0.4 \times 3 + 0.6 \times 2)\ \Delta_0$$
$$= \mathbf{0.0\ (Answer).}$$

(*ii*) **$[Cr(NH_3)_6]^{3+}$ ion.** Here Cr is present as Cr^{3+} ion which is a $3d^3$ ion ($Cr^{3+} = 3d^3$). Thus the distribution of three $3d$ electrons of Cr^{3+} ion in t_{2g} and e_g orbitals is as $t_{2g}^{\ 3}\ e_g^{\ 0}$.

Thus the number of unpaired electrons (*n*) in $[Cr(NH_3)_6]^{3+}$ ion = **3 (Answer)**

Now since $t_{2g} = 3$ and $e_g = 0$,

$$\text{CFSE of } d^3 \text{ ion} = (-0.4 \times t_{2g} + 0.6 \times e_g)\ \Delta_0$$

$$= (-0.4 \times 3 + 0.6 \times 0)\ \Delta_0$$
$$= \mathbf{-1.2\ \Delta_0\ (Answer)}$$

(*iii*) **$[CoCl_4]^{2-}$ ion.** This complex ion has tetrahedral geometry. Here Co is present as Co^{2+} ion which is a $3d^7$ ion. The distribution of d^7 electrons in t_2 (d_{xy}, d_{yz}, d_{zx}) and e (d_{z^2} and $d_{x^2-y^2}$) orbitals is as $e^4t_2^3$. Thus the number of unpaired electrons (n) in $[Co\ Cl_4]^{2-}$ ion = 3. Now since $e = 4$ and $t_2 = 3$,

$$\text{CFSE of } d^7 \text{ ion in } [CoCl_4]^{2-} = (-0.27\ e + 0.18\ t_2)\ \Delta_0$$
$$= (-0.27 \times 4 + 0.18 \times 3)\ \Delta_0$$
$$= \mathbf{-0.54\ \Delta_0\ (Answer)}$$

Problem 62. *Calculate the value of n, μ and CFSE in terms of Δ_0 / Δ_t of the following systems: (i) d^4 system in weak field octahedral complex (ii) d^7 system in tetrahedral complex.*

Solution. (*i*) **d^4 system in weak field octahedral complex.** The distribution of four electrons of d^4 system in t_{2g} and e_g sets of orbitals is as $t_{2g}^3\ e_g^1$ which gives n = **4. (Answer)**

Hence,
$$\mu = \sqrt{n(n+1)}\ \text{BM} = \sqrt{4 \times (4+2)}\ \text{BM}$$
$$= \mathbf{4.898\ BM\ (Answer)}$$
$$\text{CFSE} = (-0.4 \times t_{2g} + 0.6 \times e_g)\ \Delta_0$$
$$= (-0.4 \times 3 + 0.6 \times 1)\ \Delta_0$$
$$= \mathbf{-0.6\ \Delta_0\ (Answer)}$$

(*ii*) **d^7 system in tetrahedral complex.** We know that since all tetrahedral complexes are HS, the distribution of d^7 electrons in e and t_2 sets of orbitals is as $e^4\ t_2^3$ which gives n = **3 (Answer)**

Hence,
$$\mu = \sqrt{n(n+2)}\ \text{B.M.} = \sqrt{3 \times (3+2)}\ \text{B.M.}$$
$$= \mathbf{3.872\ BM\ (Answer)}$$
$$\text{CFSE} = (-0.27 \times e + 0.18 \times t_2)\ \Delta_0$$
$$= (-0.27 \times 4 + 0.18 \times 3)\ \Delta_0$$
$$= -0.54\ \Delta_0 = -0.54 \times \frac{\Delta t}{0.45} \quad (\because\ \Delta_t = 0.45\ \Delta_0)$$
$$= -1.2\ \Delta t$$

Thus
$$\text{CFSE} = \mathbf{-0.54\ \Delta_0} \text{ or } \mathbf{-1.2\ \Delta t\ (Answer)}$$

Problem 63. *Explain the following, giving appropriate reasons for your answer.*

(i) Almost all octahedral complexes of Co^{3+} ion are low spin diamagnetic octahedral complexes.

(ii) Tetrahedral Co(II) complexes are stable while Ni (II) prefer to form square planar complexes.

(iii) Co(III) is stabilised in presence of strong field ligands while Co(II) is stabilised in presence of weak field ligands.

(iv) $[Co^{3+}(NH_3)_6]^{3+}$ ion is more stable than $[Co^{2+}(NH_3)_6]^{2+}$ ion but $[Co^{3+}(H_2O)_6]^{3+}$ ion is less stable than $[Co^{2+}(H_2O)_6]^{2+}$ ion.

(v) Cr (III), under forcing conditions, forms $[CrCl_3]^{3-}$ but the corresponding complexes are never formed by Mn (III) and Fe (III).

Solution. (*i*) Excepting $[CoF_6]^{3-}$, all other octahedral complexes of Co^{3+} ion ($Co^{3+} = 3d^6$) are LS diamagnetic ($n = 0$) octahedral complexes. Their diamagnetic nature is because t_{2g} set has three pairs of electrons and e_g set is vacant ($t_{2g}^6 e_g^0$). These complexes are LS, since the value of CFSE for d^6 ion in LS state is the highest or most negative. (= – 2.4 Δ_0 + 3P) . $[CoF_6]^{3-}$ ion, as said above, is not LS diamagnetic octahedral ion, rather it is HS paramagnetic octahedral ion with $n = 4$. Since F^- ion is a weak field ligand, Δ_0 cannot exceed the pairing energy (P) of Co^{3+} ion, *i.e.*, $\Delta_0 < P$. Higher value of P as compared to that of Δ_0 makes $[CoF_6]^{3-}$ HS paramagnetic octahedral.

(*ii*) We know that Co^{2+} and Ni^{2+} ions are $3d^7$ and $3d^8$ ions respectively. Now we know that:

CFSE of d^7 ion (Co^{2+} ion) in tetrahedral complexes = – 0.54 Δ_0

CFSE of d^8 ion (Ni^{2+} ion) in tetrahedral complexes = – 0.36 Δ_0

CFSE of d^7 ion (Co^{2+} ion) in HS square planar complexes = + 1.028 Δ_0

CFSE of d^8 ion (Ni^{2+} ion) in HS square planar complexes = + 1.456 Δ_0

Now since CFSE of Co^{2+} ion (d^7 ion) in tetrahedral complexes is more negative than that of Ni^{2+} ion (d^8 ion) (Co^{2+} = – 0.54 Δ_0, Ni^{2+} = – 0.36 Δ_0), Co^{2+} ion prefers to form tetrahedral complexes, *i.e.*, tetrahedral complexes of Co (II) are stable.

Similarly, since CFSE of Ni^{2+} ion (d^8 ion) in HS square planar complexes is greater than that of Co^{2+} ion (d^7 ion) (Ni^{2+} = + 1.456 Δ_0, Co^{2+} = + 1.028 Δ_0), Ni^{2+} ion prefers to form square planar complexes.

(*iii*) This question can be answered by compairing CFSE values of Co^{2+} (d^7 ion) and Co^{3+} (d^6 ion) ions in octahedral strong and weak field ligands as given below:

CFSE value of Co^{2+} ion (d^7 ion) in octahedral strong field ligands = – 1.8 Δ_0

CFSE value of Co^{3+} ion (d^6 ion) in octahedral strong field ligands = – 2.4 Δ_0

CFSE value of Co^{2+} ion (d^7 ion) in octahedral weak field ligands = – 0.8 Δ_0

CFSE value of Co^{3+} ion (d^6 ion) in octahedral weak field ligands = – 0.4 Δ_0

Now since CFSE of Co^{3+} ion in octahedral strong field ligands (= – 2.4 Δ_0) is more negative than that of Co^{2+} ion in the same field ligands (= –1.8 Δ_0), Co^{3+} ion is stabilised in the presence of strong field ligands.

Similarly, since CFSE value of Co^{2+} ion in octahedral weak field ligands (= – 0.8 Δ_0) is more negative than that of Co^{3+} ion in the same field ligands (= – 0.4 Δ_0), Co^{2+} ion is stabilised in the presence of weak field ligands.

(*iv*) We know that NH_3 is a strong ligand and H_2O is a weak ligand. Now since CFSE Co^{3+} ion (d^6 ion) in octahedral strong ligand field (= – 2.4 Δ_0) is more negative than that of Co^{2+} ion (d^7 ion) (= – 1.8 Δ_0), $[Co^{3+}(NH_3)_6]^{3+}$ ion is more stable than $[Co^{2+}(NH_3)_6]^{2+}$ ion.

Similarly, since CFSE of Co^{3+} ion (d^6 ion) in octahedral weak ligand field (= – 0.4 Δ_0) is less negative than that of Co^{2+} ion (d^7 ion) (= – 0.8 Δ_0), $[Co^{3+}(H_2O)_6]^{3+}$ ion is less stable than $[Co^{2+}(H_2O)_6]^{2+}$ ion.

(*v*) Cr (III), Mn (III) and Fe (III) are d^3, d^4 and d^5 systems respectively. Since Cl^- ion is a weak field ligand, all the complexes, if formed at all, must be weak field (HS) octahedral complexes.

The CFSE values of d^3, d^4 and d^5 ions are $-1.20\ \Delta_0$, $-0.6\ \Delta_0$ and zero respectively. Now since d^3 ion has the most negative value of CFSE ($= -1.2\ \Delta_0$), $[Cr\ Cl_6]^{3-}$ ion is the most stable, *i.e.*, Cr (III) has the maximum chances to form $[Cr\ Cl_6]^{3-}$ ion.

Problem 64. *Determine CFSE of a d^6 octahedral complex having $\Delta_0 = 25{,}000\ cm^{-1}$ and $P = 15{,}000\ cm^{-1}$.*

Solution. Since $P < \Delta_0$ ($P = 15{,}000\ cm^{-1}$, $\Delta_0 = 25{,}000\ cm^{-1}$), d^6 octahedral complex is a LS octahedral complex and hence:

CFSE for d^6 ion in LS state $= -2.4\ \Delta_0 + 3P$

$= (-2.4 \times 25{,}000 + 3 \times 15{,}000)\ cm^{-1}$

$= (-60{,}000 + 45{,}000)\ cm^{-1}$

$= \mathbf{-15{,}000\ cm^{-1}}$ **(Answer)**

Problem 65. *For $[Cr(H_2O)_6]^{2+}$ ion. the mean pairing energy (P) is found to be equal to 23,500 cm^{-1}. The magnitude of Δ_0 is 13,900 cm^{-1}. (a) Calculate the CFSE for this complex ion corresponding to HS and LS state. (b) Which state is more stable ?*

Solution. (*a*) Quite obviously, in the given complex ion Cr is present as Cr^{2+} ion which is a d^4 ion.

For d^4 ion (Cr^{2+} ion) in HS state,

CFSE $= -0.6\ \Delta_0$

$= -0.6 \times 13{,}900\ cm^{-1}$

$= \mathbf{-8340\ cm^{-1}}$ **(Answer)**

For d^4 ion (Cr^{2+} ion) in LS state,

CFSE $= -1.6\ \Delta_0 + P$

$= (-1.6 \times 13{,}900 + 23{,}500)\ cm^{-1}$

$= \mathbf{+1260\ cm^{-1}}$ **(Answer)**

(*b*) Which of the two states (HS state and LS state) of $[Cr\ (H_2O)_6]^{2+}$ ion is more stable can be determined with the help of any of the following two concepts.

(*i*) We have shown above that since CFSE of d^4 ion in HS state is negative ($= -83{,}40\ cm^{-1}$) and that of the same ion in LS state is positive ($= +12{,}60\ cm^{-1}$), $[Cr(H_2O)_6]^{2+}$ ion in HS state is more stable than in LS state.

(*ii*) Since P ($= 23{,}500\ cm^{-1}$) $> \Delta_0$ ($= 13{,}900\ cm^{-1}$), $[Cr(H_2O)_6)]^{2+}$ is more stable in HS state.

Problem 66. *(a) Using CFT, give the electronic configuration of rhodium (II) ion (Rh^{2+}) in an octahedral field for which the crystal field splitting parameter (Δ_0) is greater than the pairing energy (P). (b) Calculate CFSE for this configuration in terms of Δ_0 and P.*

Solution. (a) Rh^{2+} ion is a $4d^7$ ion. Since $\Delta_0 > P$, the octahedral complex of Rh^{2+} ion is a LS octahedral complex. Thus electronic configuration corresponding to d^7 ion in LS octahedral field (strong field) is $t_{2g}^6 e_g^1$ which gives m (number of paired electrons) = 3. Thus answer is :

$= \boldsymbol{t_{2g}^6 e_g^1}$

(*b*) CFSE of d^7 ion in LS octahedral complex

$= (-0.4 \times t_{2g} + 0.6 \times e_g) + mP$

$= (-0.4 \times 6 + 0.6 \times 1) \times 3P$

$= \mathbf{-1.8 + 3P}$ **(Answer)**

Problem 67. *Observed (experimental) lattice energies (in KJ/mol) of octahedrally coordinated crystals of VO and FeO are – 3917 and – 3923 respectively. The lattice energies of these* *crystals in the absence of CFSE are – 3691 and – 3856 KJ/mol respectively. Assuming that oxide ion (O^{2-}) is a weak field ligand, calculate CFSE values of V^{2+} and Fe^{2+} ion.*

Solution. Since O^{2-} ion is a weak field ligand (HS),

CFSE for V^{2+} ion (d^3 ion) in HS state = $U_{exp} - U_{cal}$

$$-1.20\ \Delta_0 = -3917 - (-3691)$$

$$-1.20\ \Delta_0 = -226 \text{ KJ/mol}$$

$$\therefore \quad \Delta_0 = \frac{226}{1.2} \text{ KJ/mol}$$

$$= 188.33 \text{ KJ/mol}$$

$$= 188.33 \times 83.6 \text{ cm}^{-1} \qquad (\because 1 \text{ KJ/mol} = 83.6 \text{ cm}^{-1})$$

$$= \mathbf{15744.38\ cm^{-1}} \textbf{ (Answer)}$$

CFSE for Fe^{2+} ion (d^6 ion) in HS state

$$= U_{exp} - U_{cal}$$

$$-0.4\ \Delta_0 = -3923 - (-3856)$$

$$-0.4\ \Delta_0 = -67 \text{ KJ/mol}$$

$$\therefore \quad \Delta_0 = \frac{67}{0.4} \text{ KJ/mol}$$

$$= 167.5 \text{ KJ/mol}$$

$$= 167.5 \times 83.6 \text{ cm}^{-1}$$

$$= \mathbf{14003\ cm^{-1}} \textbf{ (Answer)}$$

Problem 68. *The enthalpy of hydration of Cr^{2+} ion is – 460 Kcal/mol. In the absence of crystal field stabilisation energy, the value of enthalpy of hydration is – 435 Kcal/mol. Calculate the value of Δ_0 for $[Cr(H_2O)_6]^{2+}$ ion.*

Solution. We know that :

$$\text{CFSE} = \begin{bmatrix}\text{Experimental} \\ \text{value of enthalpy} \\ \text{of hydration}\end{bmatrix} - \begin{bmatrix}\text{Value of enthalpy} \\ \text{of hydration in the} \\ \text{absence of CFSE}\end{bmatrix} \qquad \text{... (i)}$$

Now it is given that :

Experimental value of enthalpy of hydration of Cr^{2+} ion = – 460 Kcal/mol

Value of enthalpy of hydration of Cr^{2+} ion in the absence of CFSE = – 435 Kcal/mol.

Now since H_2O molecules are weak ligands, $[Cr(H_2O)_6]^{2+}$ ion is a HS complex ion. Cr^{2+} ion (central metal ion) is a d^4 ion whose CFSE is in HS state is equal to $-0.6\ \Delta_0$. Thus equation (*i*) can be written as:.

$$-0.6\ \Delta_0 = [-460 - (-435)] \text{ Kcal/mol}$$

or

$$-0.6\ \Delta_0 = -25 \text{ Kcal/mol}$$

$$\therefore \quad \Delta_0 = \frac{-25}{-0.6} \text{ Kcal/mol}$$

$$= 41.66 \text{ Kcal/mol}$$

$$= 41.66 \times 350 \text{ cm}^{-1} \qquad (\because 1 \text{ Kcal/mol} = 350 \text{ cm}^{-1})$$

$$= \mathbf{14583.3\ cm^{-1}} \textbf{ (Answer)}$$

Problem 69. *The enthalpy of hydration of Fe^{2+} ion is – 11.4 Kcal/mol which is higher than that expected in the absence of CFSE. Assuming that $[Fe(H_2O)_6]^{2+}$ ion is HS ion, estimate the value of Δ_0 for this ion.*

Solution. We know that CFSE of Fe^{2+} ion in $[Fe(H_2O)_6]^{2+}$ ion is given by:

$$CFSE = (\Delta H_{hyd})_{exp} - (\Delta H_{hyd})_{cal}$$

or $$CFSE = -11.4 \qquad ...(i)$$

Now since $[Fe(H_2O)_6]^{2+}$ ion is a HS complex ion, CFSE of Fe^{2+} ion in HS state = – 0.4 Δ_0

Thus equation (i) becomes:

$$-0.4\ \Delta_0 = -11.4 \text{ Kcal/mol}$$

or $$\Delta_0 = \frac{11.4}{0.4} = 28.5 \text{ Kcal/mol}$$

$$= 28.5 \times 350 \text{ cm}^{-1} \qquad (\because 1 \text{ Kcal/mol} = 350 \text{ cm}^{-1})$$

$$= \mathbf{9975\ cm^{-1}} \textbf{ Ans (Answer)}$$

Problem 70. *Write a note on spinels, mentioning the definition, structure and classification.*

Solution. What is spinel? A spinel is a combination of two metallic oxides. One of these oxides is an oxide of a divalent metallic cation (*e.g.* Co^{2+}, Mn^{2+}, Mg^{2+}, Ni^{2+}, Fe^{2+}, Cu^{2+} etc.) and the other oxide is an oxide of a trivalent metalic cation (*e.g.* Co^{3}, Cr^{3+}, Mn^{3+}, Al^{3+}, Fe^{3+} etc.) These cations may be of the same metal or of different metals. Thus a spinel can be represented by any of the following fomulae :

(*a*) $\overset{2+}{\mathbf{A}}\mathbf{O}.\overset{3+}{\mathbf{A}}_2\mathbf{O}_3$ **or** $\mathbf{A_3O_4}$. In this formula divalent (A^{2+}) and trivalent (A^{3+}) cations are of the same metal

(*b*) $\overset{2+}{\mathbf{A}}\mathbf{O}.\ \overset{3+}{\mathbf{B}}_2\mathbf{O}_3$ **or** $\overset{2+}{\mathbf{A}}\ \overset{3+}{\mathbf{B}}_2\mathbf{O}_4$. In this formula divalent (A^{2+}) and trivalent (A^{3+}) cations are of different metals.

Both the formulae given above show that in spinels the bivalent cations (A^{2+}) and trivalent cations (A^{3+} or B^{3+}) are present in the ratio of 1 : 2.

Examples of spinels are given below :

$$Co_3O_4 \equiv \overset{2+}{Co}O.\overset{3+}{Co}_2O_3 \equiv \overset{2+}{Co}\overset{3+}{Co}_2O_4$$

$$Mn_3O_4 \equiv \overset{2+}{Mn}O.\overset{3+}{Mn}_2O_3 \equiv \overset{2+}{Mn}\overset{3+}{Mn}_2O_4$$

$$MgCr_2O_4 \equiv \overset{2+}{Mg}O.\overset{3+}{Cr}_2O_3 \equiv \overset{2+}{Mg}\overset{3+}{Cr}_2O_4$$

$$NiCr_2O_4 \equiv \overset{2+}{Ni}O.\overset{3+}{Cr}_2O_3 \equiv \overset{2+}{Ni}\overset{3+}{Cr}_2O_4$$

$$MnAl_2O_4 \equiv \overset{2+}{Mn}O.\overset{3+}{Al}_2O_3 \equiv \overset{2+}{Mn}\overset{3+}{Al}_2O_4$$

$$MgAl_2O_4 \text{ (Spinel mineral)} \equiv \overset{2+}{Mg}O.\overset{3+}{Al}_2O_3$$

$$\equiv \overset{2+}{Mg}.\overset{3+}{Al}_2O_4$$

$$Fe_3O_4 \text{ (Magnetite)} \equiv \overset{2+}{Fe}O.\overset{3+}{Fe_2}O_3 \equiv \overset{2+}{Fe}\overset{3+}{Fe_2}O_4$$

$$Ni\,Al_2\,O_4 \equiv \overset{2+}{Ni}O.\overset{3+}{Al_2}\,O_3 \equiv \overset{2+}{Ni}\overset{3+}{Al_2}\,O_4$$

$$Cu\,Fe_2\,O_4 \equiv \overset{2+}{Cu}\,O.\overset{3+}{Fe_2}\,O_3 \equiv \overset{2+}{Cu}\overset{3+}{Fe}O_4$$

$$Mg\,Fe_2\,O_4 \text{ (Ferrite)} \equiv \overset{2+}{Mg}O.\overset{3+}{Fe_2}\,O_3 \equiv \overset{2+}{Mg}\,\overset{3+}{Fe_2}\,O_4$$

Structure of spinels. In spinels, oxide ions (O^{2-}) have cubic close packing (ccp) arrangement. This arrangement of O^{2-} ions produces octahedral and tetrahedral holes in the crystal structure of spinels. In a close packed arrangement of O^{2-} ions, each O^{2-} ion has one octahedral and two tetrahedral holes. Thus if the number of O^{2-} ions in unit cell of the lattice = x, then,

No. of octahedral holes = $1 \times x = x$

and, No. of tetrahedral holes = $2 \times x = 2x$

Thus :

No. of O_2^{2-} ions per unit cell = No. of octahedral holes

$$= \frac{1}{2} \times \text{No. of tetrahedral holes}$$

Octahedral and tetrahedral holes are occupied by divalent (A^{2+}) and trivalent (A^{3+} or B^{3+}) cations.

Classification of spinels. On the basis whether A^{2+} and A^{3+} or B^{3+} cations occupy octahedral or tetrahedral holes, spinels have been classified into the following two categories.

(*i*) **Simple or normal spinels.** In these spinels, A^{2+} cations (divalent cations) occupy one-eight of the tetrahedral holes and A^{3+} or B^{3+} cations (trivalent cations) occupy one-half of the octahedral holes. Thus we see that in normal spinels A^{2+} cations are present in tetrahedral holes and A^{3+} or B^{3+} cations are present in octahedral holes. Examples of normal spinels are:

$$Co_3O_4 \equiv Co^{2+}\,O.\,Co_2^{3+}\,O_3 \equiv Co^{2+}\,Co_2^{3+}\,O_4$$

$$Mn_3\,O_4 \equiv Mn^{2+}O.\,Mn_2^{3+}\,O_3 \equiv Mn^{2+}\,Mn_2^{3+}\,O_4$$

$$MgCr_2O_4 \equiv Mg^{2+}O.Cr_2^{3+}O_3 \equiv Mg^{2+}\,Cr_2^{3+}\,O_4$$

$$NiCr_2O_4 \equiv Ni^{2+}O\,.Cr_2^{3+}O_3 \equiv Ni^{2+}\,Cr_2^{3+}\,O_4$$

$$MgAl_2O_4 \text{ (\textit{Spinel mineral})} \equiv Mg^{2+}.Al_2^{3+}O_3$$

$$\equiv Mg^{2+}\,Al_2^{3+}\,O_4.$$

(*ii*) **Inverted spinels.** In these spinels, all A^{2+} cations (divalent cations)occupy octahedral holes while 50% of A^{3+} or B^{3+} (trivalent cations) cations occupy octahedral holes and the remaining 50% of A^{3+} or B^{3+} cations occupy tetrahedral holes. Examples of inverted spinels are:

$$Fe_3O_4 \text{ (Magnetite)} \equiv Fe^{2+}O.Fe_2^{3+}O_3 \equiv Fe^{2+}Fe_2^{3+}O_4$$

$$NiAl_2O_4 \equiv Ni^{2+}O.Al_2^{3+}O_3 \equiv Ni^{2+}\,Al_2^{3+}\,O_4$$

$$MFe_2O_4 \text{ (Ferrites)} \equiv M^{2+}O.Fe_2^{3+}O_3 \equiv M^{2+}Fe_2^{3+}\,O_4$$

($M = Cu^{2+}, Mg^{2+}, Zn^{2+}$)

The classification of spinels into simple and inverted spinels is shown in Fig 5.119

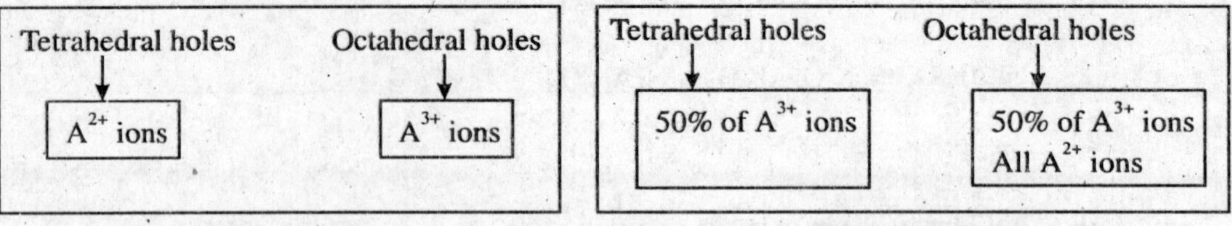

Simple spinel **Inverted spinel**

Fig. 5.119. Classification of spinels into simple and inerted spinels

Problem 71. *On the basis of CFSE values predict which of the oxides viz Mn_3O_4 and Fe_3O_4 is a simple spinel and which one is an inverted spinel?*

Solution. Mn_3O_4 oxide. Mn_3O_4 is a combination of $Mn^{2+}O$ and $Mn_2^{3+}O_3$ oxides ($Mn_3O_4 \equiv Mn^{2+}O.\,Mn_2^{3+}O_3$). Thus Mn_3O_4 contains Mn^{2+} ($3d^5$ ion) and Mn^{3+} ($3d^4$ ion) cations. O^{2-} ions are weak field ligands. CFSE values (in terms of Δ_0) of Mn^{2+} ($3d^5$ ion) and Mn^{3+} ($3d^4$ ion) cations in octahedral and tetrahedral weak ligand fields are given below:

Cation	*CFSE value in octahedral weak ligand field*	*CFSE value in tetrahedral weak ligand field*
Mn^{2+} ($3d^5$) ion	0	0
Mn^{3+} ($3d^4$) ion	$-0.60\,\Delta_0$	$-0.18\,\Delta_0$

Zero values of CFSE will not contribute to the stability of Mn_3O_4. Now since CFSE value of Mn^{3+} ion in octrahedral weak ligand field is more negative ($= -0.60\ \Delta_0$) than that of the same ion (*i.e.*, Mn^{3+} ion) in tetrahedral weak ligand field ($= -0.18\ \Delta_0$), Mn^{3+} ions (trivalent cations) prefer to occupy octahedral holes and Mn^{2+} ions (divalent cations) occupy tetrahedral sites. Thus:

Mn_3O_4 is a simple spinel.

Fe_3O_4 oxide. Fe_3O_4 is a combination of $Fe^{2+}O$ and $Fe_2^{3+}O_3$ oxides ($Fe_3O_4 \equiv Fe^{2+}O.Fe_2^{3+}O_3$). Thus Fe_3O_4 contains Fe^{2+} ($3d^6$ ion) and Fe^{3+} ($3d^5$ ion) cations. O^{2-} ions are weak field ligands. CFSE values (in terms of Δ_0) of Fe^{2+} ($3d^6$ ion) and Fe^{3+} ($3d^5$ ion) cations in octahedral and tetrahedral weak ligand fields are given below:

Cation	*CFSE value in octahedral weak ligand field*	*CFSE value in tetrahedral weak ligand field*
Fe^{2+} ($3d^6$) ion	$-0.40\,\Delta_0$	$-0.27\,\Delta_0$
Fe^{3+} ($3d^5$) ion	0	0

Zero values of CFSE will not contribute to the stability of Fe_3O_4. Now since CFSE value of Fe^{2+} ion in octahedral weak ligand field is more negative ($= -0.40$ V) than that of the same ion (*i.e.* Fe^{2+} ion) in tetrahedral weak ($= -0.27\ \Delta_0$), all Fe^{2+} ions (divalent cations) prefer to occupy ocathedral holes while 50% of Fe^{3+} ions (trivalent cations) occupy octahedral holes and the remaining 50% of Fe^{3+} ions occupy tetrahedral holes. Thus :

Fe_3O_4 is an inverted spinel.

University Questions

1. Explain the following:
   (*i*) Inner orbital octahedral complexes of Co^{2+} ion are unstable.
   (*ii*) The formation of inner orbital octahedral complexes of Ni^{2+} ion is not possible.
   (*iii*) Low spin tetrahedral complexes are not known.
   OR
   Only high spin tetrahedral complexes are known.
   (*iv*) Mn_3O_4 is represented as Mn^{2+} $[Mn_2{}^{3+}]O_4$ while Fe_3O_4 is represented as Fe^{3+} $[Fe^{2+}\ Fe^{3+}]O_4$.
   (v) $[Co(H_2O)_6]^{2+}$ ion is more stable than $[Co(H_2O)_6]^{3+}$ ion.
   (*vi*) $[Co\ (NH_3)_6]^{3+}$ ion is more stable than $[Co(NH_3)_6]^{2+}$ ion.
   (*vii*) Cu^{2+} ion forms square planar complexes rather than tetrahedral or octahedral complexes in both the fields.
   (*viii*) Most of the 4-coordinated complexes of Ni^{2+} ion are square planar rather than tetrahedral.
   (*ix*) $[Fe(CN)_6]^{4-}$ is a low spin complex, while $[Fe(H_2O)_6]^{3+}$ is a high spin complex. [*Lucknow 95*]
   (*x*) $[NiCl_4]^{2-}$ is tetrahedral and paramagnetic but $[Ni(CN)_4]^{2-}$ is square planar and diamagnetic. [*Lucknow 95*]
   (*xi*) $[Ni(CN)_4]^{2-}$ ion is diamagnetic but $[NiCl_4]^{2-}$ ion is paramagnetic with two unpaired electrons. (*Meerut 97*)
   (*xii*) $[Fe(H_2O)_6]^{3+}$ ion is more paramagnetic than $[Fe(CN)_6]^{3-}$ ion. (*Meerut 96*)
   (*xiii*) $Ni(CO)_4$ is tetrahedral while $[Ni(CN_4)]^{2-}$ ion is square planar.
   (*xiv*) $[Fe(CN)_6]^{4-}$ ion is diamagnetic but $[Fe(CN_6)]^{3-}$ ionis paramagnetic.
   (*xv*) $[Sc(H_2O)_6]^{3+}$ ion is colourless while $[Ti(H_2O)_6]^{3+}$ ion is coloured. (*Meerut 96*)
   (*xvi*) $[Co(NH_3)_6]^{3+}$ ion is diamagnetic but $[CoF_6]^{3-}$ ion is paramagnetic. (*Utkal 1991, Rohilkhand 91*)
   (*xvii*) $[CoF_6]^{3-}$ ion is a high spin complex ion whereas $[Co(CN)_6]^{3-}$ ion is a low spin complex ion. (*Avadh 93*)
   (*xviii*) Most of the coordination compounds are coloured. (*Rohilkhand 91*)
   (*xix*) The *d–d* transition band in $[Ti(H_2O_6)]^{3+}$ ion is very broad and unsymmetrical. (*Berhangam 91*)
   (*xx*) Low spin complexes are usually not encountered for tetrahedral complexes. (*Berhangum 91*)
   (*xxi*) CN^- forms 6-coordinate complex with Cr(III) and Co(III) but, with Ni(II) the most stable complex formed is four coordinate, $[Ni(CN)_4]^{2-}$. (*Berhangam 91*)
   (*xxii*) The halide ions (F^-, Cl^-, Br^-, I^-)are arranged as $I^- < Br^- < Cl^- < F^-$ in the spectrochemical series but in nephelauxetic series these are arranged as $F^- < Cl^- < Br^- < I^-$. (*Berhangam 91*)
2. Explain the following:
   (*a*) Spin-forbidden transitions are allowed
   (*b*) *d-d* transition in transition metal complexes is Laporte forbidden.
   (*c*) Octahedral NI(II) complexes possess magnetic moments 3.2 B.M.
   (*d*) A light pink coloured cobalt (II) chloride becomes deep blue on addition of excess of HCl.
   (*e*) Crystal field splitting ability is the in the order: $CN^- > NH_3 > H_2O > F^-$.
   (*f*) The stereochemistry of Ni (II) complexes is reflected by their magnetic properties.
   (*g*) Cu^{2+} complexes are coloured but Zn^{2+} complexes are colourless. (*Berhangam 91*)
   (*h*) Square planar complexes are more stable than octahedral complexes. (*Raj. 82*)
   (*i*) Fe^{3+} ion usually forms octahedral complexes that are low spin. (*Delhi Hons. 87*)
   (*j*) Ni^{2+} gives low spin square planar complex with CN– only whereas Pt^{2+}, Pd^{2+} and Au^{2+} form low spin square planar complexes with NH_3 and even Cl^-.
   (*k*) In octahedral geometry $d_{x^2-y^2}$ and d_{z^2} orbitals of a metal ion should be vacant. (*Delhi Hons.81*)
   (*l*) The value of the observedmagnetic moments (in BM) for the following complexes are:
      (*i*) $[Fe(CN)_6]^{3-} = 1.86$, $[FeI_6]^{3-} = 5.89$, $[Ni(CN)_4]^{3-}$ = diamagnetic. (*Madras 86*)
      (*ii*) $K_3[FeF_6] = 5.9$, $[Co(NH_3)_6]Cl_3 = 0.0$, $[Ni(H_2O)_6]Cl_2 = 2.9$,
      $K_4[Mn(CN)_6] = 1.9$, $K_4[MnF_6] = 1.0$ (*Raj. 85*)
      (*iii*) $Co(NH_3)_6]^{2+} = 5.0$, $[Co(NO_2)_6]^{4-} = 1.9$, $[Fe(CN)_6]^{3-} = 2.3$,
      $[Fe(H_2O)_6]^{2+} = 5.3$ (*Raj. 86*)
      (*iv*) $[Fe(H_2O)_6]^{2+} = 4.9$, $[(FeCN)_6]^{4-} = 0.0$ (*Poona 86*)
   (*m*) $[CoF_6]^{3-}$ is paramagnetic while $[Co(NH_3)_6]^{3+}$ is not. (*M.D. Rohtak 86*)

4. Explain the following :
   (*i*) Structure of $Ni(CO)_4$. *(Lucknow 95)*
   (*ii*) (*a*) Hybridisationo of Cu in $[Cu(NH_3)_4]SO_4$ and Fe in $[Fe(H_2O)_6]Cl_3$
   (*b*) Complex compound behaviour of $CoCl_3.6NH_3$.
   (*c*) Sidgwick's electronic interpretation of $[Fe(CN)_6]^{4-}$ ion. *(Avadh 93)*
5. Define and explain "Inner and outer orbital complexes". *(Avadh 93)*
6. (*i*) What do you understand by crystal field stabilisation energy (CFSE)?
   (*ii*) Calculate the value of CFSE (in Dq units) for $[Fe(CN)_6]^{3-}$ and $[FeF_6]^{3-}$ ions. *(Avadh 93)*
7. Represent schematically the splitting of *d*-orbitals in octahedral and tetrahedral fields. *(Avadh 93)*
8. What is CFT? How does it differ form VBT? *(Rohilkhand 91)*
9. What in crystal field splitting? Mention its salient features. Write the tetrahedral crystal field splitting diagram. *(Osmania 93)*
10. (a) What is the origin of paramagnetism in inorganic compounds? Calculate the magnetic moment of a d^3 system in an octahedral field.
    (*b*) Write explanatory note on crystal field stabilisation energy and its significance. *(Berhangam 91)*
11. (*a*) Explain the theoretical failure of an ionic model of CFT.
    (*b*) Discuss in brief the adjusted CFT. *(Berhangum 91)*
12. (*a*) What are the limitations of VBT of coordinate bonding and how far have these been rectified in the CFT.
    (*b*) What is meant by CFSE and explain how this affects the stabilityof complexes. *(Berhangam 91)*
13. Select the correct answer:
    (*i*) Amongst $Ni(CO)_4$, $[Ni(CN)_4]^{2-}$ and $[NiCl_4]^{2-}$
14. (*i*) Mention the solient features of crystal field theory and explain what you understand by crystal field splitting and crystal field stabilisation energy.
    (*ii*) Complex ion, $\left[Ti(H_2O)_6\right]^{3+}$ records an absorption peak at 20, 300 cm^{-1}. Calculate the value of 10 Dq in KJ mol^{-1} for the complex ion. *(Lucknow 2002).*
15. (*a*) On the basis of V. B. T. discuss the following :–
    (*i*) $\left[Fe(CN)_6\right]^{4-}$ ion is diamagnetic but $\left[Fe(CN)_6\right]^{3-}$ ion is paramagnetic in nature.
    (*ii*) $\left[Ni(CN)_4\right]^{2-}$ ion is square planar but $\left[Ni(Cl)_4\right]^{2-}$ ion is tetrahedral.

    (*b*) Explain the following :
    (*i*) Ti^{4+} is more stable than Ti^{3+}.
    (*ii*) Cu (I) is diamagnetic and Cu (II) is paramagnetic
    (*iii*) Cu salts are generally coloured whereas Zn salts are colourless. *(Nagpur 2002).*
16. Give reasons for the following :
    (*a*) $\left[Pt(NH_3)_2 Cl_2\right]$ complex has square planar geometry but Ni $(CO)_4$ has tetrahedral geometry.
    (*b*) Predrict which of the following are inner or outer orbital and diamagnetic or paramagnetic complex :
    (*i*) $\left[Cr(NH_3)_6\right]^{3+}$ (*ii*) $\left[Cr(H_2O)_6\right]^{3+}$ *(Delhi 2002).*

Ans. (*a*) $\left[Pt(NH_3)_4 Cl_2\right]$ complex has a square planar geometry due to dsp^2 hybridisation while $\left[Ni(CO)_4\right]$ due to sp^3 hybridisation, has tetrahedral geometry.
(*b*) (*i*)and (*ii*) both are inner orbital complexes due to d^2sp^3 hybridisation and paramagnetic.

17. Explain the following :
    (*i*) Ni $(CO)_4$ is diamagnetic
    (*ii*) Cu^{2+} complexes are coloured and Zn^{2+} complexes are colourless. *(VBS Purvanchal 2003).*
18. What do you mean by crystal field splitting ? Discuss the splitting of d-orbitals in weak and strong field with reference to octahedral complexes.
    (*b*) What are the facters on which the magnitude of CFSE depends. *(VBS Purvanchal 2003).*

19. (*a*) Discuss the electronic spectrum of $[Cu\ H_2O)_6]^{2+}$ Give its diagram also.
(*b*) Calculate CFSE (in Dq units) for $[Fe\ (CN)_6]^{4-}$ and $[FeF_6]^{3-}$
(*c*) What is spectrochemical series ? Discuss its applications for predicting the spectra of complexes. *(Rohilkhand 2003).*

20. (*a*) How does an "inner orbital complex" differ from "outer orbital complex"? 2+5
(*b*) Indicate the geometry and type of hybrid orbitals involved in any two of the following complex ions :
(*i*) $[Fe(CN)_6]^{4-}$ (*ii*) $[Zn\ (NH_3)_4]^{2+}$ (*iii*) $[Cu(NH_3)_4]^{2+}$
(*c*) Sketch the crystal field splitting pattern of d-orbitals and indicate the occupancy of the orbitals in the following complexes :
(*i*) d^6 octahedral (low spin) (*ii*) d^8 tetrahedral
(*d*) Which complex ion in each of the following pair will be more stable ? Give reasons :
(*i*) $[Ni\ (NH_3)_6]^{2+}$ or $[Ni\ (en)_3]^{2+}$ (*ii*) $[Ni\ (en)_2\ (H_2O)_2]^{2+}$ or $[Ni\ (tren)\ (H_2O)_2]^{2+}$
(en = ethylenediamine ; tren = triaminotriethylamine) *(Rohilkhand 2003).*

21. (*a*) What are essential elements of a crystal field theory ? How do the d orbitals split in (*i*) tetrahedral field (*ii*) square planar field ? Show the energies of different levels.
(*b*) Assign the ground state terms for Cr^{3+} and Mn^{2+} ions.
(*c*) CrO_4^{2-} is a d^0 complex and yet it is coloured.
(*d*) The brown ring compound in the nitrate test shows μ_{eff} = 3.8 BM. why ? *(Rohilkhand 2003).*

22. Explain with illustration labile and inert complexes. *(Gujrat 2003).*

23. Predict the shape and magnetic moments of the following on the basis of VB theory : (*i*) $[NiCl_4]^{2-}$ (*ii*) $[Co(CN)_6]^{3-}$ *(Delhi 2003).*

24. (*a*) What are the postulates of Pauling's Valence Bond Theory (V. B. T.) ? Discuss the geometry of $[Co(NH_3)_6]^{3+}$ on the basis of V. B. T. 4
(*b*) Explain how magnetic criteria is useful to explain geometry of $[Mn\ (Br)_4]^{-2}$ complex ion. 2
(*c*) What are the limitations of V. B. T. ? *(Nagpur 2003).*

25. (*a*) Discuss about the colour, number of unpaired electrons and magnetic properties of the following ions :
(*i*) Fe^{2+} (*ii*) Cu^{2+} (*iii*) Mn^{2+}
(*b*) Explain the following : (*i*) Transition metal exhibit variable oxidation states (*ii*) Cu (II) ion is coloured while Zn (II) ion is colourless.
(*c*) What do you mean by *d-d* transition.
(*d*) Explain magnetic properties and spectral properties of $[Ti(H_2O)_6]^{3+}$ complex ion.
(*e*) Discuss the differences between high spin and low spin complexes.
(*f*) Write short notes on inner complexes. *(Bhopal 2004)*

26. Explain the following on the basis of VBT :
(*i*) $[Ni(CN)_4]^{2-}$ is diamagnetic and square planar
(*ii*) $[NiCl_4]^{2-}$ is paramagnetic and tetrahedral
(*iii*) $[Ni(CO)_4]$ is diamagnetic and tetrahedral. *(Bhopal 2004)*

27. Explain CFT of metal ligand bond. How *d*-orbital splitting in octahedral complexes takes place? *(Bhopal 2004)*

28. (*a*) Explain the following :
(*i*) Transition elements form coloured compounds
(*ii*) $Cu^{2+}(3d^9)$ is more stable than $Cu^{+}(3d^{10})$
(*b*) How is magnetic moment (μ) related to the number of unpaired electrons, if the magnetism is due solely to the electron spins. *(GND Amritsar, 2004)*

29. (*a*) Define crystal field theory.
(*b*) Define crystal field stabilisation energy. Calculate its value for d^5 low and high spin octahedral complexes.
(*c*) Explain crystal field splitting in tetrahedral complexes. Account for the smaller value of crystal field splitting in tetrahedral than in octahedral complexes. *(GND Amritsar, 2004)*

Part II : Transition Metals (*d*-Block Elements) and Their Compounds

6

General Characteristics of Transition Metals : *d*-Block Elements

Introduction

The elements from Sc_{21} to Zn_{30} (ten elements), Y_{39} to Cd_{48} (10 elements) La_{57}, Hf_{72} to Hg_{80} (ten elements), Ac_{89}, Ku_{104}, Ha_{105}, Unh_{106} etc. are called transition metals. Since the last electron in the atoms of these elements enters *d*-sub-shell belonging to penultimate shell, [*i.e.*, $(n - 1)$th shell], these elements are also called *d*-block elements. Thus in these elements the last electron goes to $(n - 1)d$ sub-shell.

Position of *d*-block elements in the periodic table

The position of *d*-block elements in the periodic table has been shown in Fig. 6.1. From this figure the following points may be noted:

(*i*) *d*-block elements lie in between *s*-and *p*-block elements, *i.e.*, these elements are located in the middle of the periodic table.

(*ii*) *d*-block elements are present in 4th (Sc_{21} to Zn_{30} = 10 elements), 5th (Y_{39} to Cd_{48} = 10 elements), 6th (La_{57}, Hf_{72} to Hg_{80} = 10 elements) and 7th (incomplete) period which contains 8 elements *viz.* Ac_{89}, Ku_{104} to Uun_{110}

(*iii*) *d*-block elements are present in IIIB (3), IVB (4), VB (5), VIB (6), VIIB (7), VIII (8, 9, 10), IB (11) and IIB (12) groups.

Electronic configurations and definition

The electronic configurations of the atoms of *d*-block elements are given in Table 6.1. In this table $[Ar]_{18}$, $[Kr]_{36}$, $[Xe]_{54}$ and $[Rn]_{86}$ indicate the electronic configurations of Ar, Kr, Xe and Rn respectively, which are given below:

$[Ar]_{18}$ = 2, 8, 8 (Three-shells)

$[Kr]_{36}$ = 2, 8, 18, 8 (Four shells)

$[Xe]_{54}$ = 2, 8, 18, 18, 8 (Five shells)

$[Rn]_{86}$ = 2, 8, 18, 32, 18, 8 (Six shells)

The configurations given in Table 6.1 show that:

(*a*) *d*-block elements can be defined as those elements in which the last electron (differentiating electron) enters $(n - 1)d$ orbitals (*i.e.*, *d*-orbitals of the penultimate shell) or in which $(n - 1)d$ orbitals are progressively filled up with electrons.

(*b*) *d*-block elements are also defined as those elements whose two outer-most shells are incomplete (*i.e.*, partially-filled).

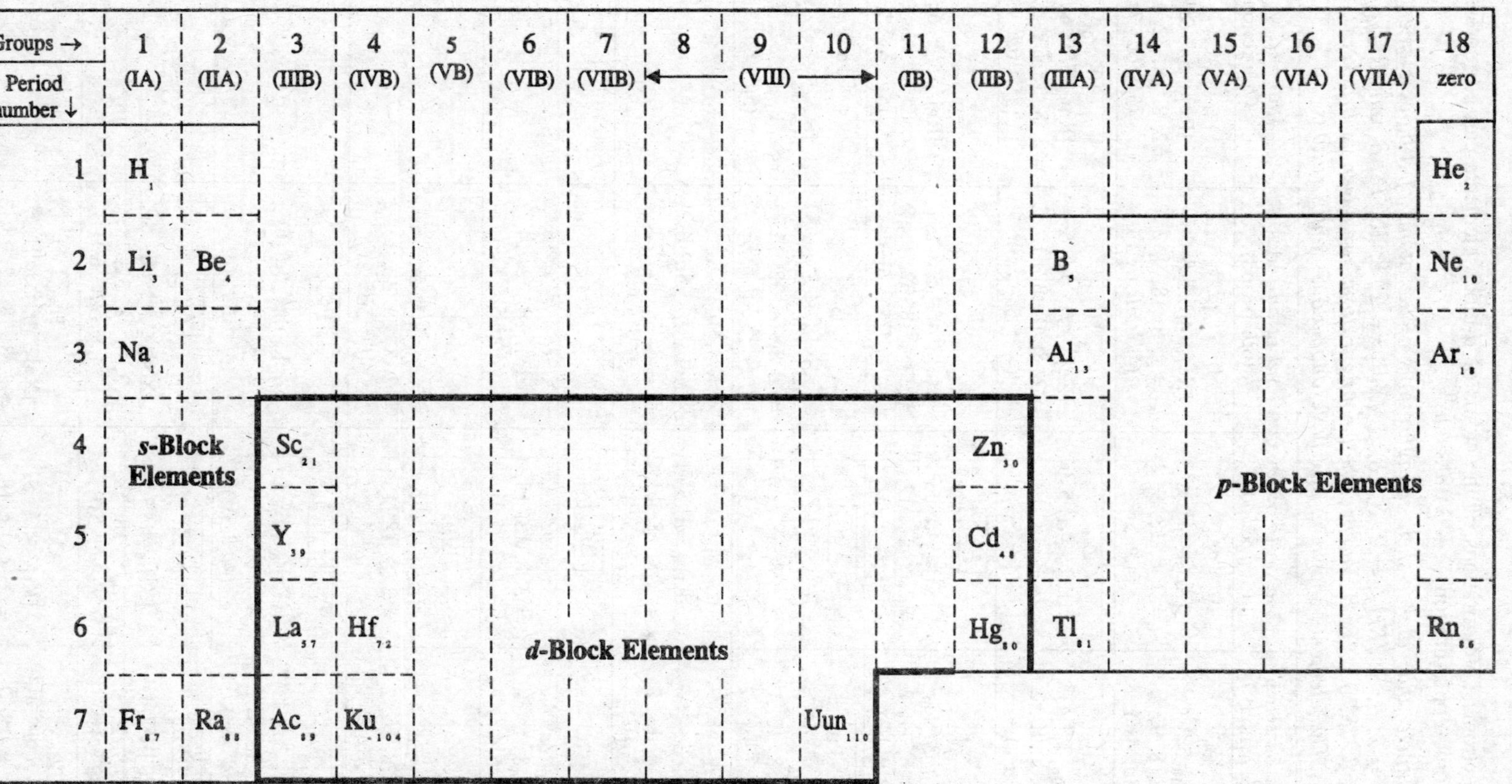

Fig. 6.1. Position of d-block elements in the periodic table

(*c*) The electronic configurations of the atoms of *d*-block elements can be represented by a general electronic configuration $(n - 1)\ d^{1-10}\ ns^{0-2}$.

Table 6.1. *Complete and valence-shell configurations of the atoms of d-block elements. Here* $[Ar]_{18}$ *= 2, 8, 8;* $[Kr]_{36}$ *= 2, 8, 18, 8;* $[Xe]_{54}$ *= 2, 8, 18, 18, 8 and* $[Rn]_{86}$ *= 2, 8, 18, 32, 18, 8. Valence-shell configurations are given in bracket. The elements shown in box have anomalous configurations.*

Group ↓	1st or 3d series (4th period)	2nd or 4d series (5th period)	3rd or 5d series (6th period)	4th or 6d series (7th period)
IIIB (3)	Sc_{21} $[Ar]_{18}\ 3d^1\ 4s^2$ = 2, 8, 9, 2 $(3d^1\ 4s^2)$	Y_{39} $[Kr]_{36}\ 4d^1\ 5s^2$ = 2, 8, 18, 9, 2 $(4d^1\ 5s^2)$	La_{57} $[Xe]_{54}\ 4f^0\ 5d^1\ 6s^2$ = 2, 8, 18, 9, 2 $(4f^0\ 5d^1\ 6s^2)$	Ac_{89} $[Rn]_{89}\ 5f^0\ 6d^1\ 7s^2$ = 2, 8, 18, 32, 18, 9, 2 $(5f^0\ 6d^1\ 7s^2)$
IVB (4)	Ti_{22} $[Ar]_{18}\ 3d^2\ 4s^2$ = 2, 8, 10, 2 $(3d^2\ 4s^2)$	Zr_{40} $[Kr]_{36}\ 4d^2\ 5s^2$ = 2, 8, 18, 10, 2 $(4d^2\ 5s^2)$	Hf_{72} $[Xe]_{54}\ 4f^{14}\ 5d^2\ 6s^2$ = 2, 8, 18, 32, 10, 2 $(4f^{14}\ 5d^2\ 6s^2)$	Ku_{104} $[Rn]_{86}\ 5f^{14}\ 6d^2\ 7s^2$ = 2, 8, 18, 32, 32, 10, 2 $(5f^{14}\ 6d^2\ 7s^2)$
VB (5)	V_{23} $[Ar]_{18}\ 3d^3\ 4s^2$ = 2, 8, 11, 2 $(3d^3\ 4s^2)$	**Nb_{41}** $[Kr]_{36}\ 4d^4\ 5s^1$ = 2, 8, 18, 12, 1 $(4d^4\ 5s^1)$	Ta_{73} $[Xe]_{54}\ 4f^{14}\ 5d^3\ 6s^2$ = 2, 8, 18, 32, 11, 2 $(4f^{14}\ 5d^3\ 6s^2)$	Ha_{105} $[Rn]_{86}\ 5f^{14}\ 6d^3\ 7s^2$ = 2, 8, 18, 32, 32, 11, 2 $(5f^{14}\ 6d^3\ 7s^2)$
VIB (6)	**Cr_{24}** $[Ar]_{18}\ 3d^5\ 4s^1$ = 2, 8, 13, 1 $(3d^5\ 4s^1)$	**Mo_{42}** $[Kr]_{36}\ 4d^5\ 5s^1$ = 2, 8, 18, 13, 1 $(4d^5\ 5s^1)$	W_{74} $[Xe]_{54}\ 4f^{14}\ 5d^4\ 6s^2$ = 2, 8, 18, 32, 12, 2 $(4f^{14}\ 5d^4\ 6s^2)$	Unh_{106} $[Rn]_{86}\ 5f^{14}\ 6d^4\ 7s^2$ = 2, 8, 18, 32, 32, 12, 2 $(5f^{14}\ 6d^4\ 7s^2)$
VIIB (7)	Mn_{25} $[Ar]_{18}\ 3d^5\ 4s^2$ = 2, 8, 13, 2 $(3d^5\ 4s^2)$	Tc_{43} $[Kr]_{36}\ 4d^5\ 5s^2$ = 2, 8, 18, 13, 2 $(4d^5\ 5s^2)$	Re_{75} $[Xe]_{54}\ 4f^{14}\ 5d^5\ 6s^2$ = 2, 8, 18, 32, 13, 2 $(4f^{14}\ 5d^5\ 6s^2)$	
VIII (8)	Fe_{26} $[Ar]_{18}\ 3d^6\ 4s^2$ = 2, 8, 14, 2 $(3d^6\ 4s^2)$	**Ru_{44}** $[Kr]_{36}\ 4d^7\ 5s^1$ = 2, 8, 18, 15, 1 $(4d^7\ 5s^1)$	Os_{76} $[Xe]_{54}\ 4f^{14}\ 5d^6\ 6s^2$ = 2, 8, 18, 32, 14, 2 $(4f^{14}\ 5d^6\ 6s^2)$	
VIII (9)	Co_{27} $[Ar]_{18}\ 3d^7\ 4s^2$ = 2, 8, 15, 2 $(3d^7\ 4s^2)$	**Rh_{45}** $[Kr]_{36}\ 4d^8\ 5s^1$ = 2, 8, 18, 16, 1 $(4d^8\ 5s^1)$	Ir_{77} $[Xe]_{54}\ 4f^{14}\ 5d^7\ 6s^2$ = 2, 8, 18, 32, 15, 2 $(4f^{14}\ 5d^7\ 6s^2)$	
VIII (10)	Ni_{28} $[Ar]_{18}\ 3d^8\ 4s^2$ = 2, 8, 16, 2 $(3d^8\ 4s^2)$	**Pd_{46}** $[Kr]_{36}\ 4d^{10}\ 5s^0$ = 2, 8, 18, 18, 0 $(4d^{10}\ 5s^0)$	**Pt_{78}** $[Xe]_{54}\ 4f^{14}\ 5d^9\ 6s^1$ = 2, 8, 18, 32, 17, 1 $(4f^{14}\ 5d^9\ 6s^1)$	
IB (11)	**Cu_{29}** $[Ar]_{18}\ 3d^{10}\ 4s^1$ = 2, 8, 18, 1 $(3d^{10}\ 4s^1)$	**Ag_{47}** $[Kr]_{36}\ 4d^{10}\ 5s^1$ = 2, 8, 18, 18, 1 $(4d^{10}\ 5s^2)$	**Au_{79}** $[Xe]_{54}\ 4f^{14}\ 5d^{10}\ 6s^1$ = 2, 8, 18, 32, 18, 1 $(4f^{14}\ 5d^{10}\ 6s^1)$	
IIB (12)	Zn_{30} $[Ar]_{18}\ 3d^{10}\ 4s^2$ = 2, 8, 18, 2 $(3d^{10}\ 4s^2)$	Cd_{48} $[Kr]_{36}\ 4d^{10}\ 5s^2$ = 2, 8, 18, 18, 2 $(4d^{10}\ 5s^2)$	Hg_{80} $[Xe]_{54}\ 4f^{14}\ 5d^{10}\ 6s^2$ = 2, 8, 18, 32, 18, 2 $(4f^{14}\ 5d^{10}\ 6s^2)$	

Classifications of *d*-block elements in 3*d*, 4*d*, 5*d* and 6*d* series (four series)

The electronic configurations of *d*-block elements as given in Table 6.1 show that depending on whether the last electron enters 3*d*, 4*d*, 5*d* or 6*d* orbitals, these elements can be grouped into the following four series:

1. 3*d*-series (1st series : 4th period). This series contains ten elements *viz.* Sc_{21} to Zn_{30}. These elements are present in 4th period. In the atoms of these elements the last electron goes to 3*d*-orbitals, *i.e.*, in this series 3*d* orbitals are progressively filled up with electrons as we move from Sc_{21} to Zn_{30}. It may be noted that the configurations of Cr_{24} and Cu_{29} (two elements) are anomalous, since Cr_{24} has 5 electrons (instead of 4) in 3*d* orbitals and Cu_{29} has 10 electrons (instead of 9) in these orbitals. Thus the correct electronic configurations of Cr_{24} and Cu_{29} are $[Ar]_{18}\ 3d^5\ 4s^1$ [instead of $[Ar]_{18}\ 3d^4\ 4s^2$] and $[Ar]_{18}\ 3d^{10}\ 4s^1$ [instead of $[Ar]_{18}\ 3d^9\ 4s^2$] respectively. Complete and valence-shell electronic configurations of the atoms of 3*d*-series elements can be written as follows. Here $[Ar]_{18}$ = 2, 8, 8 (*three shells*).

Complete configuration = $[Ar]_{18}\ 3d^{1-10}\ 4s^{1-2}$
= 2, 8, (8 + 1 to 10), 1 or 2 (four shells)
= 2, 8, (9 to 18), 1 or 2
= 2, 8, $3s^2p^6d^{1-10}$, $4s^{1,2}$

Valence-shell configuration = $3d^{1-10}\ 4s^{1,2}$

2. 4*d*-series (2nd series : 5th period). Like 3*d*-series, this series also has ten elements namely Y_{39} to Cd_{48}. These elements are present in 5th period. In this series, the differentiating electron occupies 4*d* orbitals, *i.e.*, the elements of this series involve the progressive filling of 4*d* orbitals as we proceed from Y_{39} to Cd_{48}. In this series there are more elements which have anomalous configurations. The elements having anomolous configurations are Nb_{41}, Mo_{42}, Ru_{44}, Rh_{45}, Pd_{46} and Ag_{47} (*six elements*). These anamolous configurations are explained on the basis of nuclear-electron and electron-electron forces existing in these atoms. Complete and valence-shell electronic configurations of the atoms of 4*d*-series elements can be written as follows. Here $[Kr]_{36}$ = 2, 8, 18, 8 (*four shells*).

Complete configuration = $[Kr]_{36}\ 4d^{1-10}\ 5s^{0-2}$
= 2, 8, 18, (8 + 1 to 10), 0 to 2 (*five shells*)
= 2, 8, 18, (9 to 18), 0 to 2
= 2, 8, 18, $4s^2p^6d^{1-10}$, $5s^{0-2}$

Valence-shell configuration = $4d^{1-10}\ 5s^{0-2}$

3. 5*d*-series (3rd series: 6th period). Like 3*d* and 4*d* series, this series also consists of *ten* elements which are La_{57} (one element) and Hf_{72} to Hg_{80} (nine elements). The elements of this series involve the gradual filling of 5*d* orbitals. In between La_{57} and Hf_{72}, there are 14 elements *viz.* Ce_{58} to Lu_{71} which are called *lanthanides* or *lanthanones*. These 14 elements involve the progressive filling of 4*f* orbitals and hence donot belong to 5*d* series. Thus at Lu_{71}, 4*f* orbitals are completely-filled. Consequently at La_{57}, 4*f* orbitals are vacant ($4f^0$ configuration) while in the remaining nine elements (Hf_{72} to Hg_{80}) 4*f* orbitals are completely-filled ($4f^{14}$ configuration). The elements namely Pt_{78} and Au_{79} (*two elements*) have anamolous configurations. Complete and valence-shell electronic configurations of the atoms of 5*d*-series elements can be written as follows. Here $[Xe]_{54}$ = 2, 8, 18, 18, 8 (*five shells*)

Complete configuration = $[Xe]_{54}\ 4f^{0,14}\ 5d^{1-10}\ 6s^2$
= 2, 8, 18, $4s^24p^6d^{10}f^{0,14}\ 5s^2p^6d^{1-10}$, $6s^2$ (*six shells*)

Valence-shell configuration = $4f^{0,14}\ 5d^{1-10}\ 6s^2$

4. 6*d*-series (4th series: 7th period-incomplete period). The elements of this series are present in 7th period which is an incomplete period. At present this series consists of Ac_{89}, Ku_{104}, Ha_{105} and Unh_{106} (*four elements*). These elements involve the gradual filling of 6*d* orbitals.

In between Ac_{89} and Ku_{104} there are 14 elements *viz.* Th_{90} to Lw_{103} which are called *actinides* or *actinones*. These 14 elements involve the progressive filling of 5*f* orbitals and hence donot belong to 6*d* series. Thus at Lw_{103} 5*f* orbitals are completely-filled. Consequently at Ac_{89}, 5*f* orbitals are vacant ($5f^0$ configuration) while in the remaining elements *viz.* Ku_{104}, Ha_{105} and Unh_{106} (three elements), 5*f* orbitals are completely-filled ($5f^{14}$ configuration). Complete and valence-shell configurations of the atoms of 6*d*-series elements can be written as follows. Here $[Rn]_{86}$ – 2, 8, 18, 32, 18, 8 (*six shells*).

$$\textit{Complete configuration} = [Rn]_{86}\ 5f^{0,14}\ 6d^{1-4}\ 7s^2$$

$$= 2, 8, 18, 32, 5s^2p^6d^{10}f^{0,14}\ 6s^2p^6d^{1-4}, 7s^2 \text{ (7 shells)}$$

$$\textit{Valence-shell configuration} = 5f^{0,14}, 6d^{1-4}\ 7s^2$$

Physico-chemical Properties

The transition elements show several properties. Some of these properties and their trends of variation are discussed below:

1. Atomic radii

The atomic radii of the atoms of *d*-block elements are given in Table 6.2. The following trends may be observed :

(*i*) **Variation of atomic radii in a given series (period).** The atomic radii of the elements of a particular series decrease gradually upto the midway element and then these values remain almost constant upto the element of group IB (Cu, Ag and Au). The last element of each series (Zn, Cd and Hg) shows an increase in its atomic radius.

For example, for the elements of 1st transition series the atomic radii decrease gradually from Sc to Mn but from Fe to Cu these values remain practically constant. The atomic radius of Zn is higher than that of Cu. Similar behaviour has been observed for the elements of 2nd and 3rd transition series.

Table 6.2. *Atomic radii (pm) of d-block elements*

	IIIB		IVB	VB	VIB	VIIB	←	VIII	→	IB	IIB
	(3)		(4)	(5)	(6)	(7)	(8)	(9)	(10)	(11)	(12)
1st series	Sc_{21} 162		Ti 147	V 134	Cr 127	Mn 126	Fe 126	Co 125	Ni 124	Cu 128	Zn_{30} 138
2nd series	Y_{39} 180		Zr 160	Nb 146	Mo 139	Tc 136	Ru 134	Rh 134	Pd 137	Ag 144	Cd_{48} 154
3rd series	La_{57} 187	Ce_{58} — Lu_{71} 165 156 ↑ 14 lanthanides	Hf_{72} 158	Ta 146	W 139	Re 137	Os 135	Ir 136	Pt 138	Au 144	Hg_{80} 157

Explanation. For the elements from Sc to Mn the atomic radii decrease. This decrease is because of the gradual increase in nuclear charge with the increase in atomic number. The increased nuclear charge makes the atom to shrink in size and hence the size of the atom decreases. However, since the electrons added to 3*d* orbitals screen the 4*s* electron(s), the attraction between the nucleus and the 4*s* electron(s) decreases, *i.e.*, due to the screening effect caused by 3*d* electrons the magnitude of nuclear charge decreases and consequently the atomic radii for the elements from Fe to Cu remain almost uncharged. Towards the end of each series, there is an increase in electron-electron repulsion between the electrons being added to 3*d* orbitals. This increase in repulsion becomes greater than that of the attraction between the nucleus and the 4*s* electron(s). Because of the greater magnitude of electron-electron repulsion, the electron cloud of Zn expands and hence it size also becomes greater than that of Cu.

Note. Upto Mn ($3d^5\ 4s^2$) electron-electron repulsions donot come into picture because there is no pairing of electrons in 3*d* orbitals. However, in case of the elements from Fe to Zn, the pairing of electrons in 3*d* orbitals takes place and hence electron-electron repulsions are produced. Now the increase in electron-electron repulsion makes the atom to *expand* in size while the nuclear charge makes the atom to *shrink* in size. In the elements from Fe to Cu, these opposing effects almost counterbalance each other and hence there is no major change in size on going from Fe to Cu (similarly from Ru to Ag in 2nd series and from Os to Au in 3rd series). In the last elements of each period *viz.* Zn, Cd and Hg, the electron-electron repulsion becomes greater than the attraction between the nucleus and outermost *s* electrons (*i.e.*, nuclear charge) and, therefore, there is an increase in size.

(*ii*) **Variation of atomic radii in a given group.** The covalent radii of the atoms of the elements of group III B (Sc, Y and La) increase (as expected) from top to bottom as the atomic number increases. When we compare these values for the elements of other groups of transition elements, it is found that these values for the elements of 2nd transition series are higher (as expected) than those for the elements of 1st transition series, but these values for the elements of 3rd transition series are almost equal to those for the elements of 2nd transition series. The similarity in the covalent radii for the elements of 2nd and 3rd series is due to *lanthanide contraction* (discussed in the chapter in "*Lanthanides*") which takes place among lanthanides (Ce_{58} to Lu_{71} = 14 elements). These 14 lanthanides are present between La_{57} (group III B) and Hf_{72} (group IV B) in 3rd transition series.

Due to the similarity in the covalent radii of the elements of 2nd and 3rd transition series, these elements show close similarity in their properties while the elements of 1st and 2nd transition series donot exhibit so close similarity. For example the solubility of the salts of the elements of 2nd and 3rd transition series are very close and hence the separation of these elements by precipitation method is difficult. Examples of the pairs of elements which show similar properties and hence are difficult to separate are Zr—Hf, Nb—Ta, Mo—W, Ru—Os, Rh—Ir, Pd—Pt, Ag—Au. The above discussion shows that due to lanthanide contraction the elements of 2nd and 3rd transition series resemble each other much more than the elements of 1st and 2nd transition series.

2. Ionic radii

The ionic radii of the common cations derived from transition metals have been given in Table 6.3. The following trends in their values may be noted:

(*i*) Since the transition metals show many oxidation states (variable oxidation states), the radii of the ions in different oxidation states are also different.

(*ii*) In general the ionic radii of different cations of the same element in different oxidation states decrease with the increase in oxidation state. For example, $Ti^{2+} > Ti^{3+} > Ti^{4+}$; $Cr^{2+} > Cr^{3+} > Cr^{4+} > Cr^{5+} > Cr^{6+}$; $Cu^{+} > Cu^{2+}$ etc.

(*iii*) Ionic radii of the cations of different elements in the same oxidation state generally decrease with the increase in nuclear charge or atomic number. Thus the ionic radii of the bivalent cations derived from the elements of 1st transition series decrease as we move from left to right in the series, *i.e.*, $Ti^{2+} > V^{2+} > Cr^{2+} > Mn^{2+} > Fe^{2+} > Co^{2+} > Ni^{2+} > Cu^{2+}$.

(*iv*) These values for the cations of the elements of group III B (*i.e.*, Sc^{3+}, Y^{3+} and La^{3+}) increase as $Sc^{3+} < Y^{3+} < La^{3+}$, but these values for the cations derived from most of the elements of 3rd transition series are almost the same as those of the corresponding cations derived from the elements of 2nd transition series. The similarity in their ionic radii is due to the presence of 14 lanthanides (Ce_{58} to Lu_{71}) between La_{57} (group III B) and Hf_{72} (group IV B).

Table 6.3. *Ionic radii (pm) of the common cations derived from d-block elements.*

	III B	IV B	V B	VI B	VII B	VIII			I B	II B
	(3)	(4)	(5)	(6)	(7)	(8)	(9)	(10)	(11)	(12)
1st series	Sc_{21}	Ti	V	Cr	Mn	Fe	Co	Ni	Cu	Zn_{30}
	+2 = 81 +3 = 88.5	+2 = 91 +3 = 76 +4 = 74.5	+2 = 88 +3 = 74 +4 = 72 +5 = 68	+2 = 84 +3 = 69 +4 = 68 +5 = 63 +6 = 58	+2 = 80 +3 = 66 +7 = 60	+2 = 76 +3 = 64	+2 = 76 +3 = 63	+2 = 72	+1 = 91 +2 = 69	+2 = 74
2nd series	Y_{39}	Zr	Nb	Mo	Tc	Ru	Rh	Pd	Ag	Cd_{48}
	+3 = 104	+4 = 86	+3 = 86 +4 = 82 +5 = 78	+3 = 83 +4 = 79 +5 = 75 +6 = 73	+4 = 78.5 +5 = 74 +7 = 70	+3 = 82 +4 = 76 +5 = 70.5	+3 = 80.5 +4 = 74 +5 = 69	+2 = 100 +3 = 90 +4 = 75.5	+1 = 129 +2 = 108 +3 = 89	+2 = 109
3rd series	La_{57}	Hf_{72}	Ta	W	Re	Os	Ir	Pt	Au	Hg_{80}
	+3 = 117.2	+4 = 85	+3 = 86 +4 = 82 +5 = 78	+4 = 80 +5 = 76 +6 = 74	+4 = 77 +5 = 72 +6 = 69 +7 = 67	+4 = 77 +5 = 71.5	+3 = 82 +4 = 76.5 +5 = 71	+2 = 94 +4 = 76.5 +5 = 71	+1 = 51 +3 = 99 +5 = 71	+1 = 133 +2 = 116

3. Metallic character and related properties

All the transition elements show metallic character, *i.e.*, all these elements are metals.

Explanation. Since the transition elements have low values for their ionisation energies, *ns* electrons can easily be lost by them to form M^+ or M^{2+} cations and thus these elements show metallic character.

As usual, the metallic character of these elements increases on descending the group, *e.g.* basic character of the pentoxides of the elements of group V B (V_2O_5, Nb_2O_5 and Ta_2O_5) increases as $V_2O_5 < Nb_2O_5 < Ta_2O_5$. Thus V_2O_5 is amphoteric (but more basic than acidic) while Nb_2O_5 and Ta_2O_5 are more basic than acidic.

The metallic character of the transition elements is evident from the following properties of these elements.

(*i*) **Electrical and thermal conductivity.** These elements are good conductors of electricity and heat because of the existance of metallic bonding (*i.e.*, presence of free mobile electrons) among the atoms of these elements. Cu, Ag and Au show exceptionally high thermal and electrical conductivity.

(*ii*) **Hardness.** These elements are hard (*i.e.*, cannot be cut with a knife) and brittle. Their hardness is due to the presence of strong metallic bonding among the atoms of these elements. The metallic bonding arises due to the overlap between the singly-filled orbitals of different atoms of the elements. Greater is the number of unpaired electrons, greater is the number of metallic bonds and, therefore, greater is the strength of these bonds or hardness of the element. Since Cr, Mo and W have maximum number of unpaired electrons, these are very hard metals. Zn, Cd and Hg contain no unpaired electrons, and hence are not very hard. As a matter of fact Hg is a liquid and is soft like alkali metals.

(*iii*) **Crystal structure.** These elements have simple hexagonal close-packed (*hcp*), cubic close-packed (*ccp*) or body-centred cubic (*bcc*) lattices which are characteristic of metals.

4. Atomic volumes and densities

The atomic volumes (in ml) and density (in g/cm³) of transition elements are given below in Table 6.4. The values given against *ml* are atomic volumes while those given against *g/cm³* represent densities.

Table 6.4. *Atomic volumes (ml) and densities (g/cm³) of d-block elements*

	III B	IV B	V B	VI B	VII B	←	VIII	→	I B	II B
	Sc_{21}	Ti_{22}	V	Cr	Mn	Fe	Co	Ni	Cu	Zn_{30}
ml	15.02	10.60	8.35	7.23	7.39	7.10	6.70	6.60	7.10	9.20
g/cm³	3.01	4.51	6.10	7.19	7.43	7.86	8.90	8.90	8.96	7.14
	Y_{39}	Zr_{40}	Nb	Mo	Tc	Ru	Rh	Pd	Ag	Cd_{48}
ml	19.8	14.1	10.8	9.4	—	8.3	8.3	8.9	10.4	13.1
g/cm³	4.47	6.49	8.40	10.21	11.50	12.20	12.40	12.00	10.51	8.65
	La_{71}	Hf_{72}	Ta	W	Re	Os	Ir	Pt	Au	Hg_{80}
ml	20.50	13.60	10.91	9.53	8.85	8.43	8.54	9.10	10.22	14.80
g/cm³	6.17	13.10	16.60	19.30	21.00	22.60	22.50	22.40	19.30	13.60

Atomic volumes of *d*-block elements are much lower than those of the elements of *s*- and *p*-blocks.

We know that density and atomic volume are inversely proportional to each other. All the transition metals have low values for their atomic volumes and hence their densities are high. The atomic volumes are low because the electrons are added to $(n - 1)$ *d* orbitals and not to *ns* orbital. The increased nuclear charge is partly screened by $(n - 1)$ *d* electrons and *ns* electrons are strongly attracted by the nucleus. Consequently the densities of transition metals are high.

Variation of atomic volumes in a given series. In a given transition series, the atomic volumes of the elements first decrease on proceeding from left to right and attain a maximum value for the elements of group VIII. They then start increasing as we proceed further to the elements of group IB and IIB.

Variation of densities in a given series. In a given transition series, the density of the elements increases on moving from left to right and attain a maximum value for the elements of group VIII. It then starts decreasing as we move further to the elements of groups IB and II B. This variation is due to the small radius and close packed structure of the atoms of these elements.

Variation of densities in a given sub-group. Densities of elements belonging to the same sub-group increases on moving down the sub-group. The densities of the elements of 2nd transition series are only slightly higher than those of the corresponding elements of 1st transition series while these values for the transition elements from Hf_{72} to Hg_{80} (elements of 3rd transition series) are almost double these values for the elements from Zr_{40} to Cd_{48} respectively (elements of 2nd transition series). Note that the density of La_{57} (= 6.17) is not double that of the value for Y_{39} (= 4.47).

The variation of densities of transition series elements in a given sub-group, as discused above, can be explained as follows:

(*a*) The atomic weights of the elements of 3rd transition series are almost double the atomic weights of the corresponding elements of 2nd transition series. This makes the densities of the elements of 3rd transition series almost double those of the corresponding elements of 2nd transition series.

(*b*) Because of lanthanide contraction occuring in lanthanides, the atomic sizes of the elements of 3rd transition series coming after Lu_{71} (*i.e.*, elements from Hf_{72} to Hg_{80}) become very small. Consequently the packing of the atoms in their metallic crystals become so much compact that their densities become very high. Thus the densities of the elements of 3rd transition series become almost double the densities of the corresponding elements of 2nd transition series.

5. Melting and boiling points

The melting and boiling points of transition metals are given in Table 6.5. These values suggest that these are very high values. The high values are due to the fact that transition metals have strong metallic bonding between their atoms.

Table 6.5. *Melting and boiling points (°C) of d-block elements.*

3d-series elements:	Sc	Ti	V	Cr	Mn	Fe	Co	Ni	Cu	Zn
M.P. (°C)	1539	1668	1900	1875	1245	1536	1495	1453	1083	419.5
B.P. (°C)	2730	3260	3450	2665	2150	3000	2900	2730	2595	906.00
4d-series elements:	Y	Zr	Nb	Mo	Tc	Ru	Rh	Pd	Ag	Cd
M.P. (°C)	1509	1852	2415	2610	2200	2500	1966	1552	960.8	320.9
B.P. (°C)	2927	1500	3300	5560	—	4900	4500	3980	2210	765
5d-series elements:	La	Hf	Ta	W	Re	Os	Ir	Pt	Au	Hg
M.P. (°C)	920	2222	2996	3410	3180	2700	2454	1769	1063	−38.4
B.P. (°C)	3470	5400	5425	5930	5900	5500	5300	4530	2920	357

As we proceed along a particular series, the melting points increase from Sc to Cr in 1st series (V and Cr have almost equal melting points), from Y to Mo in 2nd series and from La to W in 3rd series, and then decrease. This variation can be explained as follows:

We know that the strength of metallic bonding is roughly related with the number of unpaired electrons present in the valence-shell of the element. In general, greater in the number of unpaired electrons, stronger is the metallic bonding and hence greater is the melting point of the element. As we move along a particular series, since the number of unpaired electrons increases upto Cr (in 1st series), Mo (in 2nd series) and W (in 3rd series), the melting points of the elements also increase upto these elements and then with decrease in the number of unpaired electrons, the melting points also decrease. Since Zn, Cd and Hg donot have any unpaired electrons, their melting points are the lowest in their respective periods. Their melting points are so low that these elements can be volatilised off easily. In fact, Hg is a liquid at room temperature.

Although Mn, Tc and Re also have the maximum number of unpaired electrons, these elements have lower values of their melting points than Cr, Mo and W respectively. Their low values are due to the fact that the metallic bonding in these elements are relatively weak.

6. Ionisation energies

The values of first ionisation energy (in kJ/mole) of 1st, 2nd and 3rd transition series elements are given in Table 6.6. From this table the following points may be noted:

(*i*) The values of ionisation energy increase as we move from left to right in each series, although the increase is not quite regular, *i.e.*, the increase in ionisation energies is quite gradual or slow. The reason for the gradual increase is given below.

Explanation. We know that when we move from left to right in a given series, the nuclear charge increases. The increased nuclear charge would attract the ns^2 electron cloud with greater force and hence the ionisation energies are expected to increase at each step. However, as the electron is added to $(n-1)d$ sub-shell at each next element, the ns^2 electrons are shielded more and more. The effect of increasing nuclear charge is opposed by the additional screening effect of the nucleus and consequently, the ionisation energies increase but quite slowly along a given series.

Table 6.6. *Ionisation energies (kJ mol^{-1}) of d-block elements*

These values have been taken from Table 2.2A, page 40-42, "*Inorganic Chemistry: Principles of Structure and Reactivity*" (2nd edition) by James E. Huheey.

Groups ⟶	IIIB (3)		IVB (4)	VB (5)	VIB (6)	VIIB (7)	⟵VIII⟶ (8)	(9)	(10)	IB (11)	IIB (12)
1*st* or 3*d* series (4*th* period) ⟶	Sc_{21} 631		Ti_{22} 658	V_{23} 650	Cr_{24} 652.8	Mn_{25} 717.4	Fe_{26} 759.4	Co_{27} 758	Ni_{28} 736.6	Cu_{29} 745.5	Zn_{30} 906.4
2*nd* or 4*d* series (5*th* period) ⟶	Y_{39} 616		Zr_{40} 660	Nb_{41} 664	Mo_{42} 685	Tc_{43} 702	Ru_{44} 711	Rh_{45} 720	Pd_{46} 805	Ag_{47} 731	Cd_{48} 867.7
3*rd* or 5*d* series (6*th* period) ⟶	La_{57} 538.1	(Ce_{58}—Lu_{71})	Hf_{72} 654	Ta_{73} 761	W_{74} 770	Re_{75} 760	Os_{76} 840	Ir_{77} 880	Pt_{78} 870	Au_{79} 890.1	Hg_{80} 1007

(*ii*) When we proceed from top to bottom in group III B, ionisation energy values decrease as $Sc_{21} > Y_{39} > La_{57}$. However, in most of the remaining groups the ionisation energies increase in the same direction, *i.e.*, the ionisation energies of the elements lying after La_{57} in 3rd series are greater than those of their corresponding elements lying in 1st and 2nd series.

Explanation. In the atoms of 14 elements following La_{57} (Ce_{58} to Lu_{71}), 4*f* orbitals are completely filled and hence cannot shield the nucleus effectively. Due to ineffective (*i.e.*, poor) shielding of the nucleus by the 4*f* electrons, the elements lying after La_{57} in 3rd series have greater effective nuclear charge acting on the ns^2 elctrons. Therefore, the ionisation energies of the elements following La_{57} in 3rd series are higher than those of their corresponding elements lying in 1st and 2nd series.

(*iii*) In each series the elements of group II B (Zn, Cd and Hg) have the highest values for their ionisation energy. This is due to the extra stability associated with their completely-filled $(n-1)d$ and ns orbitals ($Zn_{30} = 3d^{10}\ 4s^2$, $Cd_{48} = 4d^{10}\ 5s^2$ and $Hg_{80} = 5d^{10}\ 6s^2$).

(*iv*) When we compare these values of *d*-block elements with those of *s*-and *p*-block elements, we find that these values are higher than those of *s*-block elements but lower than those of *p*-block elements. It is also seen that in any particular series, the difference in ionisation energies between any two successive *d*-block elements is much less as compared to the difference in case of successive *s*-block or *p*-block elements.

Relation between the ionisation energy and stability of a metal in a given oxidation state

With the help of ionisation energy values we can predict which of the two metals in a given oxidation state is thermodynamically more stable. As an example let us compare the relative stability of Ni (II)—Pt(II) and Ni(IV)—Pt(IV) pairs. When a metal, M(0) is converted into M(II), energy required for this conversion is equal to the sum of 1st (I_1) and 2nd (I_2) ionisation energies of the metal. Similarly when a metal, M(0) is converted into M(IV), energy required is equal to the sum of I_1, I_2, I_3 and I_4 of the metal. Thus :

$$\text{Ni(0)} \xrightarrow{-2e^-} \text{Ni (II), Energy required} = I_1 + I_2 = 2.49 \times 10^3 \text{ KJ mol}^{-1}$$

$$\text{Pt(0)} \xrightarrow{-2e^-} \text{Pt (II), Energy required} = I_1 + I_2 = 2.66 \times 10^3 \text{ KJ mol}^{-1}$$

$$\text{Ni(0)} \xrightarrow{-4e^-} \text{Ni (IV), Energy required} = I_1 + I_2 + I_3 + I_4 = 11.29 \times 10^3 \text{ KJ mol}^{-1}$$

$$\text{Pt(0)} \xrightarrow{-4e^-} \text{Pt (IV), Energy required} = I_1 + I_2 + I_3 + I_4 = 9.36 \times 10^3 \text{ KJ mol}^{-1}$$

The above values show that since ($I_1 + I_2$) for Ni(II) is less than ($I_1 + I_2$) for Pt(II), Ni(II) compounds are more stable than Pt(II) compounds. Similarly since the value of ($I_1 + I_2 + I_3 + I_4$) for Pt (IV) is less than that for Ni(IV), Pt(IV) compounds are more stable than Ni(IV) compounds. This fact becomes evident when we find that $K_2[PtCl_6]$ is a well known compound of Pt(IV) but the corresponding compound of Ni (IV) (*i.e.*, K_2NiCl_6) does not exist at all.

7. Standard reduction potential values

These values (in volts) for the elements of 1st transition series are given below:

Elements	V	Cr	Mn	Fe	Co	Ni	Cu
E^o_{red} *(volts)*	– 1.18	– 0.91	– 1.18	– 0.44	– 0.28	– 0.25	+ 0.34

These values are for the reduction reaction, $M^{2+}(aq) + 2e^- \rightleftharpoons M(s)$.

It may be seen that these values, except for Cu, are negative and there is no regular trend in these values, since the ionisation energies ($I_1 + I_2$) and sublimation energies of these metals donot show any regular trend. These values have been used to explain the reducing property and reactivity of these metals.

8. Variable oxidation states

With the exception of a few elements, most of the *d*-block elements show more than one oxidation state (*i.e.*, variable or several oxidation states). The following two reasons have been given to explain why *d*-block elements show several oxidation states:

(*i*) $(n-1)d$ and ns orbitals in the atoms of *d*-block elements have almost the same energies and hence electrons can be removed from $(n-1)d$ orbitals as easily as they can be removed from ns orbital.

(*ii*) After removing ns electron(s) the remainder is called *core* or *kernel* (penultimate shell).

In case of most of the *d*-block elements, the core is unstable and hence can lose one or more electrons, resulting in the formation of cations having different oxidation states, *i.e.*, the removal of one or more electrons from the unstable core gives several oxidation states to transition metals.

Various oxidation states shown by transition elements are given in Table 6.7. Oxidation states shown in bracket are unstable oxidation states.

Table 6.7. *Various oxidation states of transition metals. Unstable oxidation states are shown in bracket.*

Elements of 1st transition series (3*d*-series) (n = No. of unpaired electrons)

Sc	$3d^14s^2$ (n = 1)	+2, +3
Ti	$3d^24s^2$ (n = 2)	(0, –1), +2, +3, +4
V	$3d^34s^2$ (n = 3)	(–1, 0, +1), +2, +3, +4, +5
Cr	$3d^54s^1$ (n = 6)	(0, +1), +2, +3, (+4, +5), +6
Mn	$3d^54s^2$ (n = 5)	(–1, 0, +1),+2, +3, +4, (+5), +6, +7
Fe	$3d^64s^2$ (n = 4)	(–2, 0, +1), +2, +3, (+4, +5, +6)
Co	$3d^74s^2$ (n = 3)	(–1, 0, +1), +2, +3, (+4)
Ni	$3d^84s^2$ (n = 2)	(0, +1), +2, +3, +4
Cu	$3d^{10}4s^1$ (n = 1)	+1, +2, (+3)
Zn	$3d^{10}4s^2$ (n = 0)	+2

Elements of 2nd transition series (4*d*-series)

Y	+3
Zr	(0, + 3), +4
Nb	(–1, 0, +1, +2, +3, +4), +5
Mo	(0, +1), +2, +3, +4, +5, +6
Tc	(0), +2, (+3, +4, +5, +6, +7)
Ru	(0), +2, +3, +4, (+5, +6, +7, +8)
Rh	(0, + 1), +2, +3, +4, (+6)
Pd	(0), +2, (+3), +4
Ag	+1, (+2, +3)
Cd	+2

Elements of 3rd transition series (5*d*- series)

La	+3
Hf	(+3), +4
Ta	(–1, +1, +2, +3, +4), +5
W	(0, +1), +2, (+3), +4, +5, +6
Re	(–1, 0, +1, +2), +3, +4, +5, (+6), +7
Os	(0), +2, +3, +4, (+5), +6, (+7), +8
Ir	(0, + 1), +2, +3, +4, (+6)
Pt	(0), +2, (+3), +4, (+5, +6)
Au	+1, +3
Hg	+1, +2

Oxidation states shown by *d*-block elements of 1st transition series

In connection with the oxidation states exhibited by the *d*-block elements of 1st transition series the following points may be noted:

***(i)* +1 and +2 oxidation states.** The oxidation states shown in bracket are unstable oxidation states. These are low oxidation states. The remaining oxidation states are stable. Out of these stable oxidation states, +1 is the lowest oxidation state for Cu while for other elements the lowest oxidation state is +2. Cr and Cu both show +2 oxidation state when one $4s$ and one $3d$ electrons are lost while the remaining elements exhibit this oxidation state (*i.e.* +2 state) when both ns electrons are involved in bonding.

***(ii)* Stabilisation of unstable (low) oxidation states.** It may be seen that the low oxidation states like –1, 0, +1 are unstable. These states are stabilised by forming complexes with π-acid ligands like CO, NO, N_2, PR_3, AsR_3, CNR (isocyanide), CN^-, C_6H_6, toluene, α, α′-dipyridyl, o-phenanthroline, ethylene etc. These ligands are electron-deficient, since they have vacant π-orbitals, called acceptor orbitals. In addition to vacant π-orbitals, they also have filled orbitals, called donor orbitals. Thus these ligands have vacant π-acceptor orbitals as well as filled donor orbitals. These ligands can stabilise transition metal atoms in their low oxidation states (*e.g.* +1, 0, –1) by forming complexes with them. The complexes formed have double bonding which can be represented as $L \rightleftharpoons M$. This double bonding indicates that in these complexes both the metal and the ligand act both as donors and acceptors. How $L \rightarrow M\sigma$ and $L \leftarrow M\pi$ bonds are formed has been explained below.

The donor orbital on the ligand overlaps with the vacant orbital on transition metal atom in low oxidation state and forms a $L \rightarrow M\sigma$ bond. The formation of $L \rightarrow M\sigma$ bond accumulates negative charge on the transition metal atom. The accumulation of negative charge leads to substantial repulsion between the negative charges of the metal and of the ligands. In order to decrease the magnitude of the negative charge accumulated on the metal atom, suitable filled *d*-orbital on the metal atom overlaps with the vacant π-orbital (acceptor orbital) on the ligand. This type of overlap transfers some negative charge from the metal to the ligand and $M \rightarrow L\pi$ bond is formed. Thus the vacant π-orbital of the ligand receives the back-donated electrons from the filled *d*-orbitals on the transition metal and thus stabilises the metal atom in lower oxidation state. Thus we have double bonding between the ligand and the metal atom. One is a $L \rightarrow M\sigma$ bond and the other is a $M \rightarrow L\pi$ bond. The complexes formed are called *π-acid complexes*, since $M \rightarrow L\pi$ bond is formed in these complexes. In the formation of $M \rightarrow L\pi$ bond, since the ligand accepts the back-donated electrons from the metal into its vacant π-orbitals, the ligand is said to behave as a π-acid and hence is called *π-acid ligand* or *π-bonding ligand.* Examples of some complex compounds formed by transition metal atoms in their low oxidation states (+1, 0, –1) with π-acid ligands like CO, CNR, C_6H_6, PCl_3 etc. are $[Cr^0(CO)_6]^0$, $[Cr(CNR)_6]^0$, $[Cr^0(C_6H_6)_2]^0$, $[Cr^{+1}(C_6H_6)_2]^+$, $[Fe^0(CO)_5]^0$, $[Co_2^0(CO)_8]^0$, $Na^+[Co^{-1}(CO)_4]$, $[Ni^0(Co)_4]^0$, $[Ni^0(CNR)_4]^0$, $[Ni^0(PCl_3)_4]^0$, $Na^+[V^{-1}(CO)_6]$, $[V^0(dipy)_3]^0$, $[V^{+1}(dipy)_3]I$.

It may easily be understood that in order to get $L \underset{\pi}{\overset{\sigma}{\rightleftharpoons}} M$ bond, transition metal in its lower oxidation state should have both vacant as well as filled *d*-orbitals.

***(iii)* High oxidation state.** The highest (maximum) oxidation state for the first five elements (Sc to Mn) is equal to the sum of the electrons in $4s$ and $3d$ orbitals. On the other hand, the highest oxidation state for the remaining five elements (Fe to Zn) is not equal to the sum of the electrons in $4s$ and $3d$ orbitals.

The high oxidation states of many transition metals like V(V), Cr(VI), Mn(IV), Mn(VII), Fe(III), Fe(VI), Co(III), Ni(III), Cu(III), Ag(II), Ag(III) etc. are oxidising (*i.e.*, unstable). These unstable oxidation states which are high oxidation states get stabilised by forming complexes with small highly electronegative ligands like O^{2-}, F^-, Cl^-, per iodate (IO_6^{5-}), tellurate (TeO_6^{6-}) etc. Examples are given below:

(a) O^{2-} ion stabilises Fe(VI), V(V), Cr(VI) and Mn(VII) by forming FeO_4^{2-} (ferrate),

VO_4^{3-} (vanadate), CrO_4^{2-} (chromate) and MnO_4^{2-} (permanganate) ions respectively. FeO_4^{2-} is obtained by the anodic oxidation of $Fe(OH)_3$ in conc. alkali.

(*b*) V(V) is stabilised by F^- ion by forming VF_5 (but not VCl_5 and VBr_5).

(*c*) IO_6^{5-} and TeO_6^{6-} ions stabilise Ag (III) by forming the complex compounds namely $M_7^+[Ag^{3+}(IO_6)_2].xH_2O$ and $M_9^+[Ag^{3+}(TeO_6)_2].xH_2O$ respectively. TeO_6^{6-} also stabilises Cu(III) by forming $M_9^+[Cu^{3+}(TeO_6)_2].xH_2O$. All these complexes are obtained in alkaline medium. IO_6^{5-} and TeO_6^{6-} also stablise Mn(IV) by forming the red-coloured complex compounds namely $Na_7H_4[Mn^{+4}(IO_6)_3].7H_2O$ and $K_7H_4[Mn^{+4}(IO_6)_3].8H_2O$. These complexes are obtained by the alkaline oxidation of $MnCl_2$ in presence of sodium (or potassium) periodate, $Na_2H_3IO_6$ (or $K_2H_3IO_6$).Alkaline oxidation of $MnCl_2$ in presence of sodium (or potassium) tellurate gives dark red $Na_7H_7[Mn^{+4}(TeO_6)_3].3H_2O$ or $K_6H_8[Mn^{+4}(TeO_6)_3].5H_2O$ (μ = 3.30 B.M.).

(*d*) Ag(II) has been stabilised by forming complexes with *pic*$^-$, *py, dipy* and *o-phen*. The composition of the complexes formed are $[Ag^{2+}(pic)_2]^0$, $[Ag^{2+}(py)_4]^{2+}$, $[Ag^{2+}(dipy)_2]^{2+}$ and $[Ag^{2+}(o\text{-}phen)]^{2+}$ respectively. Ag(III) has been stabilised by forming the complex with *big*H. The composition of the complex compound is $[Ag^{3+}(bigH)_2](OH)(SO_4)$ (silky crystals).

(*e*) In aqueous solution high oxidation states are favoured in alkaline medium. Thus in order to stabilise the lower oxidation states in aqueous solution non-oxidising acids should be used. Transition metal hydroxides are usually oxidised by atmospheric oxygen, *e.g.*, pale green $Fe(OH)_2$ gets oxidised to brown $Fe(OH)_3$ on standing in air.

(*f*) F^- and O^{2-} ions can form compounds in which Fe and Co achieve oxidation states greater than +3 or Ni and Cu acquire oxidation state greater than +2.

(*iv*) Ionic/covalent character of compounds of a given transition metal in various oxidation states. With the increase in the oxidation state of a given transition metal, the covalent character of its compounds increases. Thus the compounds of a given transition metal in lower oxidation states are ionic while those of the same metal in high oxidation states are covalent, *e.g.* $\overset{+2}{V}Cl_2$ is ionic, $\overset{+3}{V}Cl_3$ is less ionic while $\overset{+4}{V}Cl_4$ is covalent.

Explanation. As the oxidation state of the transition metal increases, the charge density on the metal also increases. This results in the increase of the polarisation of the anion charge cloud by the metal and hence covalent character increases (see *Fajan's Rules*).

(*v*) Acidic/basic character of the compounds of a given transition metal in various oxidation states. We have said earlier that with the increase in the oxidation state of a given transition metal, the covalent character of its compounds increases. The increase in the covalent character also increases the acidic character of the compound. Thus the compounds of a given transition metal becomes more and more acidic as the oxidation state of the transition metal increases. This fact becomes evident when we study the following :

(*a*) The variation of the basic, amphoteric and acidic character of the oxides of V, Cr and Mn is as shown below:

(*i*)	*Oxides of vanadium* :	VO	V_2O_3	VO_2	V_2O_5
	Oxidation state of vanadium :	+2	+3	+4	+5
	Nature of the oxides :	Basic	Basic	Amphoteric	Acidic
(*ii*)	*Oxides of Cr* :	CrO	Cr_2O_3	CrO_2	CrO_3
	Oxidation state of Cr :	+2	+3	+4	+6
	Nature of the oxides :	Basic	Amphoteric	Amphoteric	Acidic

(*ii*) *Oxides of Cr* :

Oxides of Cr :	CrO	Cr_2O_3	CrO_2	CrO_3
Oxidation state of Cr :	+2	+3	+4	+6
Nature of the oxides :	Basic	Amphoteric	Amphoteric	Acidic

(*iii*) *Oxides of Mn* :

Oxides of Mn :	MnO	Mn_3O_4	Mn_2O_3	MnO_2	MnO_3 (Unstable)	Mn_2O_7
Oxidation state of Mn :	+2	+8/3	+3	+4	+6	+7
Nature of the oxides :	Basic	Amphoteric	Amphoteric	Amphoteric	Weakly acidic	Strongly Acidic

The variation of acidic, basic and amphoteric character of the oxides as shown above indicates that the oxides of a given transition metal in lower oxidation states are basic, those of the same metal in intermediate oxidation states are amphoteric while those of the metal in higher oxidation states are acidic.

(*b*) The acidic character of $TiCl_2$, $TiCl_3$ and $TiCl_4$ increases in the order : $\overset{2+}{TiCl_2} < \overset{3+}{TiCl_3} < \overset{4+}{TiCl_4}$. Similarly the acidic character of VCl_2, VCl_3 and $VOCl_3$ is in the order : $\overset{2+}{VCl_2} < \overset{3+}{VCl_3} < \overset{5+}{VOCl_3}$.

(*c*) Acidic character of $Mn(OH)_2$, $Mn(OH)_3$, $Mn(OH)_4$, H_2MnO_4 and $HMnO_4$ is in the order : $\overset{2+}{Mn}(OH)_2$ (base of medium strength) < $\overset{3+}{Mn}(OH)_3$ (weak base) < $\overset{4+}{Mn}(OH)_4$ (amphoteric hydroxide) < $H_2\overset{6+}{Mn}O_4$ (strong acid) < $H\overset{7+}{Mn}O_4$ (very strong acid).

(*vi*) No. of various oxidation states shown by elements. When we proceed from Sc (Sc = $3d^1\ 4s^2$, n = 1) to Mn (Mn = $3d^5 4s^2$, n = 5) in the series, the number of unpaired electrons (n) increases from 1 (in Sc) to 5 (in Mn) (donot consider Cr which has 6 unpaired electrons) and hence the number of various oxidation states shown by the elements increases from Sc to Mn. On proceeding further from Mn to Zn (Zn = $3d^{10}\ 4s^2$, n = 0) the number of unpaired electrons decreases from 5 in Mn to zero in Zn and hence the number of various oxidation states shown by these elements decreases from Mn to Zn. Thus we find that the number of oxidation states shown by Mn is maximum while that shown by the elements lying at the right and left of Mn goes on decreasing when we proceed in either direction of Mn.

(*vii*) Oxidation state of Zn. $3d^{10}$ orbitals in Zn^{2+} ion ($Zn^{2+} = 3d^{10}4s^0$), being completely-filled, are stable and hence donot lose any more electrons. Thus Zn atom cannot show any oxidation state higher than +2.

(*viii*) Relative stability of various oxidation states. (*a*) The relative stability of various oxidation states of a given element can be explained on the basis of the stability of d^0, d^5 and d^{10} orbitals. For example: Ti^{4+} ($3d^0.4s^0$) is more stable than Ti^{3+} ($3d^1.4s^0$) because of the presence of $3d^0$ orbital in Ti^{4+} ion. Mn^{2+}($3d^5.4s^0$) is more stable than Mn^{3+} ($3d^4.4s^0$), since Mn^{2+} ion has $3d^5$ orbital.

(*b*) The stability of +2 oxidation state increases as we proceed from Sc to Zn in the series, *i.e.*, the reducing power of M^{2+} ions decreases from left to right. For example the salts of V^{2+} (vonadous salts) and Cr^{2+} (chromous salts) are powerful reducing agents while the salts of other M^{2+} ions are not. Mn^{2+} salts are not reducing agents because of the presence of half-filled $3d$ orbital ($Mn^{2+} = 3d^5$) Ni^{2+} and Cu^{2+} donot have reducing property.

(*c*) The stability of +3 oxidation state decreases on proceeding from Sc to Cu (Zn does not show +3 oxidation state) in the series, *i.e.*, the oxidising power of M^{3+} ions increases along the series. +3 is a stable oxidation stable for Sc. Ti^{3+} ion acts as a reducing agent, since it is readily oxidised to Ti^{4+} ion. V^{3+} and Cr^{3+} ions donot act as oxidising agents (*i.e.*, these ions are stable), since they donot get reduced to cations with lower oxidation states. Mn^{3+} ion acts as an oxidising agent (*i.e.*, this ion is unstable), since it is reduced to Mn^{2+} ion. Fe^{3+} ion ($3d^5\ 4s^0$) is stable because of the presence of stable half-filled $3d$-orbitals in it.

(*e*) The stability of a given lower oxidation state of different transition metals of a given sub-group decreases as we descend the sub-group.

9. Colour of transition metal complex ions

The study of the colour of transition metal complex ions can be made under the following two heads (A) and (B).

(A) Colour of the complex ions whose central metal atom contains partially-filled *d*-orbitals

The transition metal complex ions and complex compounds whose central atom contains partially-filled *d*-orbitals are usually coloured in the solid form or in solution form.

When white light which is composed of many different colours falls on a coloured complex ion or complex compound of a transition metal, the ion or the compound absorbs some portion of the white light and the remaining portion of the incident light in transmitted or reflected by the complex compound or ion. The absorbed light has the following characteristics :

(*i*) Since white light is composed of many different colours, the colours of the absorbed light and transmitted light are different from each other. The colour of the transmitted light is called complementary colour of the absorbed light (see Table 6.8). As a matter of fact, the colour of a given complex ion or compound is the colour of the transmitted light. For example:

(*a*) Since hydrated Cu^{2+} ion, *i.e.*, Cu^{2+} ion present in $[Cu(H_2O)_5]SO_4$ or $CuSO_4.5H_2O$ absorbs *yellow* radiation, it transmits *blue* radiation and hence this ion looks blue to our eyes.

(*b*) Hydrated Ti^{3+} ion, *i.e.*, Ti^{3+} ion present in $[Ti(H_2O)_6]^{3+}$ absorbs *green* radiation and hence transmits the radiation of *purple* colour. Hydrated Ti^{3+} ion, therefore, looks violet (almost purple).

(*c*) Anhydrous Co^{2+} compounds absorb the radiation of *red* colour and therefore appear *blue green*. On the other hand the hydrated Co^{2+} ion, *i.e.*, Co^{2+} ion in $[Co(H_2O)_6]^{2+}$ ion, absorbs *blue green* radiation and therefore, appears *red*.

Table 6.8. *Colours of the absorbed and transmitted radiations. Range of the wavelength (in Å) of the absorbed light is given in the central column.*

Colour of the absorbed radiation	*Range of the wavelength (Å) of the absorbed light.*				*Colour of the transmitted light (This colour is called complimentary colour of that of the absorbed light)*
Violet	4000 – 4500	Low	High	High	Yellow green
Blue	4500 – 4800	↓	↓	↓	Yellow
Green blue	4800 – 4900	Wavelength increasing	Frequency (wave number) decreasing	Energy (Δ_0) decreasing	Orange
Blue green	4900 – 5000				Red
Green	5000 – 5600				Violet (almost purple)
Yellow green	5600 – 5750				Violet
Yellow	5750 – 5900				Blue
Orange	5900 – 6250	↓	↓	↓	Green blue
Red	6250 – 7500	High	Low	Low	Blue green

The colours of some common hexa hydrated transition metal complex ions, $[M(H_2O)_6]^{n+}$ are given in Table 6.9.

(*ii*) The colour of the absorbed light and hence that of the complex ion or compound depends on the nature of the central metal atom/ion. For example Ti^{3+} ion in $[Ti(H_2O)_6]^{3+}$ absorbs green radiation and hence $[Ti(H_2O)_6]^{3+}$ ion looks violet (almost purple). On the other hand Co^{2+} ion in $[Co(H_2O)_6]^{2+}$ ion absorbs blue green radiation and, therefore, appears red. Thus we see that although Ti^{3+} and Co^{2+} ions are attached with the same ligands (*viz.* $6H_2O$ molecules), they absorb the radiations of different colour and hence $[Ti(H_2O)_6]^{3+}$ and $[Co(H_2O)_6]^{2+}$ ions appear different in their colour.

Table 6.9. *Colours of some hexahydrated transition metal complex ions, $[M(H_2O)_6]^{n+}$. For the sake of convenience, $6H_2O$ molecules have not been shown (n = No. of unpaired electrons).*

Ions	Outer electronic configuration	*Colour of the ion*
Ti^{3+}	$3d^1$ $(n = 1)$	Violet (almost purple)
V^{3+}	$3d^2$ $(n = 2)$	Green
V^{2+}, Cr^{3+}	$3d^3$ $(n = 3)$	Violet
Cr^{2+}	$3d^4$ $(n = 4)$	Blue
Mn^{3+}	$3d^4$ $(n = 4)$	Violet
Mn^{2+}	$3d^5$ $(n = 5)$	Pink
Fe^{3+}	$3d^5$ $(n = 5)$	Yellow
Fe^{2+}	$3d^6$ $(n = 4)$	Green
Co^{3+}	$3d^6$ $(n = 4)$	Pink
Co^{2+}	$3d^7$ $(n = 3)$	Pink (red)
Ni^{2+}	$3d^8$ $(n = 2)$	Blue-green
Cu^{2+}	$3d^9$ $(n = 1)$	Blue

(*iii*) The colour of the absorbed light also depends on the nature of the ligands. For example Ni^{2+} ion in $[Ni(H_2O)_6]^{2+}$ ion absorbs red radiation and hence has blue green colour. On the other hand Ni^{2+} ion in $[Ni(NH_3)_6]^{2+}$ ion absorbs yellow colour and is, therefore, blue in colour. Thus, we see that although the central metal ion is the same in both the complex ions, these ions absorb radiations of different colours and, therefore, have different colours.

(*iv*) The wavelength of the absorbed light lies in the visible region of the electromagnetic radiation (4000 – 8000 Å), *i.e*, the transition metal ions absorb only that portion of the incident light whose wavelength lies in between 4000Å and 8000Å (visible region).

(*v*) The energy associated with the wavelength of the radiation absorbed by the complex ion can also be calculated. Suppose $[M(H_2O)_6]^{n+}$ ion which is a complex ion absorbs the radiation whose wavelength is equal to λÅ. Obviously the frequency in wave number (in cm^{-1}) corresponding to this wavelength is given by:

Frequency in wave number (in cm^{-1}) of the absorbed radiation of wave length of λÅ

$$= \frac{1}{\lambda \text{ (in Å)}} = \frac{1}{\lambda \times 10^{-8} \text{ cm}}$$

$$= \frac{10^8}{\lambda} \text{ cm}^{-1} \qquad (\because 1\text{Å} = 10^{-8} \text{ cm})$$

This frequency is associated with energy whose magnitude can be calculated in kcal mol^{-1} or kJ mol^{-1} by using the fact that a frequency of 350 cm^{-1} is associated with energy equal to 1 kcal mol^{-1} and a frequency of 83.7 cm^{-1} is associated with energy equal to 1 kJ mol^{-1}, *i.e.*,

$$350 \text{ cm}^{-1} = 1 \text{ kcal mol}^{-1}$$

and $$83.7 \text{ cm}^{-1} = 1 \text{ kJ mol}^{-1}$$

Thus since $$350 \text{ cm}^{-1} = 1 \text{ kcal mol}^{-1}$$

$$\therefore \quad \frac{10^8}{\lambda}\ \text{cm}^{-1} = \frac{10^8}{350 \times \lambda\ (\text{in Å})}\ \text{kcal mol}^{-1} \qquad ...(i)$$

Equation (*i*) shows that :

Energy associated with a wavelength of λÅ

$$= \frac{10^8}{350 \times \lambda\ (\text{in Å})}\ \text{kcal mol}^{-1} \qquad ...(ii)$$

Similarly, since $83.7\ \text{cm}^{-1} = 1\ \text{kJ mol}^{-1}$

$$\therefore \quad \frac{10^8}{\lambda}\ \text{cm}^{-1} = \frac{10^8}{83.7 \times \lambda\ (\text{in Å})}\ \text{kJ mol}^{-1} \qquad ...(iii)$$

Equation (*iii*) shows that:

Energy associated with a wavelength of λÅ

$$= \frac{10^8}{83.7 \times \lambda\ (\text{in Å})}\ \text{kJ mol}^{-1} \qquad ...(iv)$$

Equation (*ii*) and (*iv*) both show that the energy associated with the absorbed radiation is inversely proportional to the wavelength of the radiation but directly proportional to the frequency of the radiation.

To explain the purple colour of octahedral $[Ti(H_2O)_6]^{3+}$ ion by *d-d* electron transition

$[Ti(H_2O)_6]^{3+}$ is an octahedral complex ion in which Ti-atom is present as Ti^{3+} ion whose valence-shell electronic configuration is $3d^1$. According to crystal field theory when six water molecules (which act as ligands) approach the central Ti^{3+} ion to form the octahedral complex ion, $[Ti\ (H_2O)_6]^{3+}$, the five *d*-orbitals of Ti^{3+} ion split into t_{2g} (d_{xy}, d_{yz} and d_{zx}) and e_g ($d_{x^2-y^2}$ and d_{z^2}) sets of orbitals. This phenomenon is called **crystal field splitting.** t_{2g} set of orbitals has lower energy than e_g set of orbitals and hence d^1 electron of Ti^{3+} ion resides in t_{2g} set and e_g set remains vacant. Thus the ground state electronic configuration of Ti^{3+} ion can be written as $t_{2g}^1 e_g^0$.

Now when white light is allowed to fall on $[Ti(H_2O)_6]^{3+}$ ion, this ion absorbs green radiation. Since the ion absorbs green radiation, the colour of the transmitted radiation would be white light minus green light (colour of the absorbed light) which is purple. Thus $[Ti(H_2O)_6]^{3+}$ ion looks purple to our eyes. The absorption of green light by $[Ti(H_2O)_6]^{3+}$ ion takes place at a

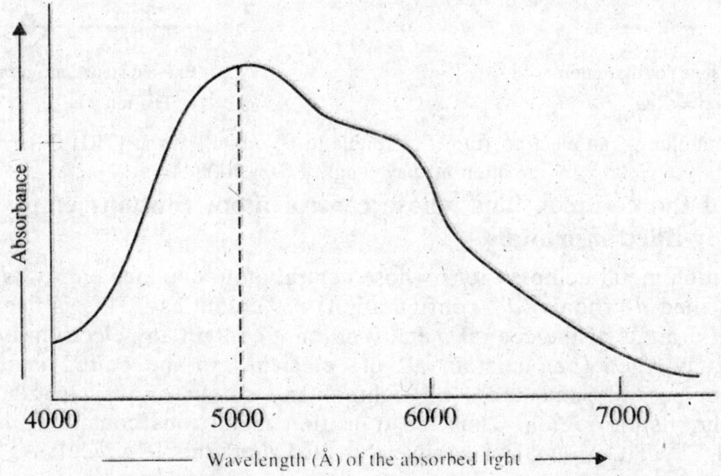

Fig. 6.2. Visible absorption spectrum of $[Ti(H_2O)_6]^{3+}$ ion.

wavelength of about 5000 Å as is evident from the visible absorption spectrum which has been obtained by plotting a graph between the wavelength (in Å) of the absorbed light and the amount of absorbed light (absorbance) (see Fig. 6.2). 5000 Å is the wavelength of the band of maximum absorption. Obviously the energy (Δ_0) associated with the wavelength of 5000 Å.

$$= \frac{10^8}{83.7 \times 5000} \text{ kJ mol}^{-1} = 239 \text{ kJ mol}^{-1}$$

This energy (= 239 kJ mol^{-1}) is close to the *crystal field spliting energy*, Δ_0 (also called *excitation energy*) (Excitation energy is the energy difference between t_{2g} and e_g orbitals) and, therefore, the electron present in t_{2g} orbitals absorbs 239 kJ mol^{-1} and is excited to occupy the vacant e_g orbitals ($t_{2g}^1 \longrightarrow e_g^0$ electron transition) so that the electronic configuration of Ti^{3+} ion in the excited state becomes $t_{2g}^0\ e_g^1$. (see Fig. 6.3). Thus we see that the purple colour of $[Ti(H_2O)_6]^{3+}$ ion is due to the excitation of an electron from the *d*-orbitals of lower energy (t_{2g} orbitals) to the *d*-orbitals of higher energy (e_g orbitals).

We have already said that the magnitude of Δ_0 (Δ_0 is the energy difference between t_{2g} and e_g orbitals) is inversely proportional to that of the wavelength of the absorbed radiation. This means that a complex ion having higher value of Δ_0 will absorb the radiation of lower wavelength and *vice versa*. This fact is evident from the following examples of octahedral complex ions of Co^{3+} and Ni^{2+} ions.

(*a*) Since the value of Δ_0 for $[Co(CN)_6]^{3-}$ ion is higher than that of $[Co(NH_3)_6]^{3+}$, $[Co(CN)_6]^{3-}$ ion absorbs violet radiation which has lower wavelength. On the other hand $[Co(NH_3)_6]^{3+}$ ion absorbs blue green radiation which has higher wavelength. Thus $[Co(CN)_6]^{3-}$ ion appears yellow green in solution and $[Co(NH_3)_6]^{3+}$ ion appears orange or pink in solution.

(*b*) Since $[Ni(H_2O)_6]^{2+}$ ion has higher value of Δ_0 than $[Ni(NH_3)_6]^{2+}$ ion, $[Ni(H_2O)_6]^{2+}$ ion absorbs yellow colour which has lower wavelength while $[Ni(NH_3)_6]^{2+}$ ion absorbs red colour which has higher wavelength. Thus $[Ni(H_2O)_6]^{2+}$ ion in solution has blue colour but $[Ni(NH_3)_6]^{2+}$ ion in solution has blue green colour.

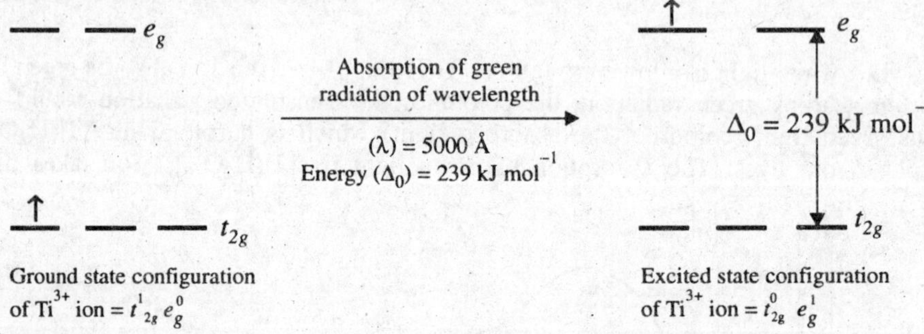

Fig. 6.3. Promotion of an electron from t_{2g} orbitals to e_g orbitals when $[Ti(H_2O_6]^{3+}$ absorbs green radiation at wavelength (λ) = 5000 Å.

(B) Colour of the complex ions whose central atom contains empty or completely-filled *d*-orbitals

The transition metal complex ions whose central atom contains empty (d^0 configuration) or completely-filled *d*-orbitals (d^{10} configuration) are colourless. This is because of the fact that since the *d*-orbitals of the central metal ion donot contain any electron (empty *d*-orbitals) or are completely-filled (*i.e.*, contain all the electrons in the paired state), *d*-*d* electron transition is not possible and hence no light of any colour or wavelength is absorbed by such ions in the visible region. Thus the transition metal ions containing empty *d*-orbitals [*e.g.*, Sc^{3+} and Ti^{4+}(TiO_2) ions)] or completely-filled *d*-orbitals [*e.g.*, Cu^+, Ag^+, Zn^{2+} ($ZnSO_4$), Cd^{2+}, Hg^{2+} etc.] are colourless.

10. Magnetic properties of transition metal ions and their complexes

When the transition metal ions or their complexes are placed in a magnetic field, they show different behaviour. On the basis of this behaviour these substances have been classified as *paramagnetic substances, diamagnetic substances, ferromagnetic substances, anti ferromagnetic substances* and *ferrimagnetic substances.*

The transition metal complexes whose central atom/ion contains one or more unpaired electrons are paramagnetic while those whose central atom/ion has no unpaired electrons (*i.e.*, all the electrons are paired) behave as diamagnetic substances. The paramagnetic character increases with the increase in the number of unpaired electrons. The paramagnetism of a substance is expressed in terms of its magnetic moment (μ). The larger is the number of unpaired electrons in the central atom/ion of a given complex compound, the greater is its paramagnetic character and hence larger is the value of the magnetic moment of that complex compound. The magnetic moment is expressed in *Bohr Magnetons* (B.M.). Paramagnetic substnaces have some value of magnetic moment while diamagnetic substances have zero value of their magnetic moment.

To deduce a formula for calculating the magnetic moment of transition metal complexes

Let us consider a transition metal complex of $3d$-elements whose central metal ion contains only unpaired electrons in its $3d$ orbitals. The magnetic moment (μ) of one unpaired electron is equal to the sum of its magnetic moment due to its orbital motion, called *orbital magnetic moment* (μ_l) and that due to its spin motion, called *spin magnetic moment* (μ_s). Thus:

$$\mu = \mu_l + \mu_s$$

Now since $$\mu_l = \sqrt{l\,(l+1)}\,\frac{eh}{4\pi mc}$$

and $$\mu_s = 2\sqrt{s\,(s+1)}\,\frac{eh}{4\pi mc}$$

The value of μ is given by :

$$\mu = [\sqrt{l\,(l+1)} + 2\sqrt{s\,(s+1)}]\,\frac{eh}{4\pi mc}$$

$$= [\sqrt{l\,(l+1)} + 2\sqrt{s\,(s+1)}] \text{ B.M.}$$

where B.M. is called Bohr Magneton which is equal to $\frac{eh}{4\pi mc}$. B.M. is the unit of magnetic moment. One B.M. is equal to 9.27×10^{-21} ergs/gauss. Here e = charge on the electron, h = Plank's quantum number, m = mass of the electron and c = velocity of light.

In case of transition metal ions of $3d$-series, the crystal fields (*i.e.*, the ligands surrounding the metal ions) restrict the movement of the electron round the nucleus in orbitals and hence the magnetic moment due to the orbital motion of the electron (μ_l) is quenched and hence $\sqrt{l\,(l+1)} = 0$. Thus the magnetic moment of an unpaired electron residing in a $3d$-orbital of the ions of $3d$-series elements is given by:

$$\mu = 2\sqrt{s\,(s+1)} \text{ B.M.}$$

Since the value of s for one electron = $\frac{1}{2}$, μ is given by:

$$\mu = 2\sqrt{\frac{1}{2}\left(\frac{1}{2}+1\right)} \text{ B.M.}$$

If a metal ion of 3*d*-series elements has two, three, four...... *n* unpaired electrons in its 3*d* orbitals, the value of magnetic moment for *n* unpaired electrons (called *effective magnetic moment*, μ_{eff}) will be obtained by putting the sum of the spin quantum numbers of *n* unpaired electrons. Since $s = \frac{1}{2}$ for one electron, the sum of *s* for two, three, four, ... *n* unpaired electrons is equal to $2 \times \frac{1}{2}$, $3 \times \frac{1}{2}$, $4 \times \frac{1}{2}$ ...... $n \times \frac{1}{2}$ respectively. Thus the value of μ_{eff} for *n* unpaired electrons is given by:

$$\mu_{eff} = 2\sqrt{\frac{n}{2}\left(\frac{n}{2} + 1\right)} \text{ B.M.}$$

$$= \sqrt{n(n+2)} \text{ B.M.}$$

Since the value of μ_{eff} is only due to the spin motion of *n* unpaired electrons, it is also called *spin-only magnetic moment* and is represented as μ_S. Thus:

$$\mu_{eff} = \mu_S = \sqrt{n(n+2)} \text{ B.M.} \qquad ...(i)$$

Equation (*i*) shows that the magnitude of μ_{eff} depends on the number of unpaired electrons (*n*) present in a given species. The value of μ_{eff} for different number of unpaired electrons can be calculated with the help of equation (*i*) as shown below:

For $n = 0$ $\quad \mu_{eff} = \sqrt{0(0+2)} = 0.0$ B.M.

For $n = 1$ $\quad \mu_{eff} = \sqrt{1(1+2)} = \sqrt{3} = 1.73$ B.M.

For $n = 2$ $\quad \mu_{eff} = \sqrt{2(2+2)} = \sqrt{8} = 2.83$ B.M.

For $n = 3$ $\quad \mu_{eff} = 3.87$ B.M.

For $n = 4$ $\quad \mu_{eff} = 4.90$ B.M.

For $n = 5$ $\quad \mu_{eff} = 5.92$ B.M.

Thus we see that greater is the number of unpaired electrons, greater is the value of μ_{eff}. The value of μ_{eff} of a given complex ion whose central metal atom is from 3*d*-transition series depends on the number of unpaired electrons present in 3*d* orbitals of the central metal atom. Greater is the number of unpaired electrons present in 3*d* orbitals of central metal atom, greater is the value of μ_{eff} (see Table 6.10). The complex ions which have a definite value of μ_{eff} are paramagnetic.

Now let us consider a complex ion of 3*d* series elements whose central metal atom/ion contains all the electrons in the paired state in its 3*d* orbitals (*i.e.*, the central atom/ion contains no unpaired electrons in its 3*d* orbitals) (*e.g.*, Cu^+, Zn^{2+} = $3d^{10}$, $n = 0$). Since the two electrons residing in the same orbital have opposite spins, the magnetic field produced by one electron is neutralised by that produced by the other electron which has opposite spin and consequently the resultant magnetic moment is zero ($\mu_{eff} = 0$). The substances having $\mu_{eff} = 0$ are diamagnetic. The value of μ_{eff} for the complex compounds whose central ion contains no electron at all in its 3*d*-orbitals or has vacant 3*d* orbitals (*e.g.* Sc^{+3}, Ti^{+4} = $3d^0$, $n = 0$) is also equal to zero and hence such complex compounds are also diamagnetic.

Table 6.10 contains the values of μ_{eff} [as calculated from equation (*i*)] for some complex compounds of 3*d*-elements. The observed (experimental) values are also given in the same Table. It may be seen that the complex compounds whose central metal ion contains one or more unpaired electrons in its 3*d*-orbitals have a definite value of μ_{eff} and hence are paramagnetic. On the other hand, the complex compounds whose central metal ion has all the paired electrons or no unpaired electrons ($n = 0$) in 3*d* orbitals has zero value for μ_{eff} and hence are diamagnetic.

Table 6.10. *Calculated and observed (experimental) values of effective magnetic moment (in B.M.) of some complex compounds of 3d-series elements. n is the number of unpaired electrons present in 3d orbitals of the central metal ion. The numbers given in bracket in the column of n (3rd column) indicate the number of unpaired electrons predicted on the basis of the experimental value of μ given in the last column.*

Ion	*Valence-shell configuration*	*n*	*Complex compounds*	*Shape and hybridisation*	*Calculated value of μ (μ_{cal})*	*Experimental value of μ (μ_{exp})*
Ti^{3+}	$3d^1$	1	$K_3[Ti^{3+}F_6]$	Octahedral (d^2sp^3)	1.73	1.70
V^{4+}	$3d^1$	1	$[V^{+4}O(acac)_2]$	Square pyramidal (d^2sp^2)	1.73	1.70
V^{3+}	$3d^2$	2	$NH_4[V^{3+}(SO_4)_2].12H_2O$	Octahedral (d^2sp^3)	2.84	2.80
Cr^{3+}	$3d^3$	3	$[Cr^{3+}(NH_3)_6]Br_3$	Octahedral (d^2sp^3)	3.87	3.77
Cr^{2+}	$3d^4$	4	$[Cr^{2+}(H_2O)_6]SO_4$	Octahedral (sp^3d^2)	4.90	4.80
Fe^{3+}	$3d^5$	5	$[Na_3[Fe^{3+}F_6]$	Octahedral (sp^3d^2)	5.92	5.85
Fe^{2+}	$3d^6$	4	$(Et_4N)_2\ [Fe^{2+}Cl_4]$	Tetrahedral (sp^3)	4.90	5.40
Co^{2+}	$3d^7$	3	$Cs_2[Co^{2+}F_4]$	Tetrahedral (sp^3)	3.87	4.60
Ni^{2+}	$3d^8$	2	$[Ni^{2+}(NH_3)_6]Cl_2$	Octahedral (sp^3d^2)	2.84	3.32
Cu^{2+}	$3d^9$	1	$[Cu^{2+}(bigH)_2^0]Cl_2$	Square planar (dsp^2)	1.73	1.79
Zn^{2+}	$3d^{10}$	0	$[Zn^{2+}(NH_3)_4]^{2+}$	Tetrahedral (sp^3)	0.00	0.00
Sc^{3+}	$3d^0$	0	$[Sc^{3+}F_6]^{3-}$	Octahedral (d^2sp^3)	0.00	0.00
Fe^{3+}	$3d^5$	5 (1)	$K_3[Fe^{3+}(CN)_6]$	Octahedral (d^2sp^3)	5.92	1.73
Co^{3+}	$3d^6$	4 (0)	$[Co^{3+}(NH_3)_6]Cl_3$	Octahedral (d^2sp^3)	4.90	0.00
Co^{2+}	$3d^7$	3 (1) presumably in 5*s* orbital	$K_2Ba[Co^{2+}(NO_2)_6]$	Octahedral (d^2sp^3)	3.87	1.88
Ni^{2+}	$3d^8$	2 (0)	$[Ni^{2+}(bigH)_2^0\ ]Cl_2$	Square planar (dsp^2)	2.84	0.00

Although the experimental values of the magnetic moment of most of the 3*d*-metal complexes given in Table 6.10 are almost equal to those calculated on the basis of the number of unpaired electrons in 3*d* orbitals, yet there are many complexes like $K_3[Fe(CN)_6]$, $[Co(NH_3)_6]Cl_3$, $K_2Ba[Co(NO_2)_6]$, $[Ni(bigH)_2]Cl_2$ which donot obey this rule, *i.e.*, the actual number of unpaired electrons (n) in $3d$ orbitals (given in bracket) is different from that expected from the valence-shell configuration of the central metal ion and hence the experimental value of μ is different from the calculated value of μ. The diserepancy between the experimental and calculated values of μ, is due to the fact that the number of unpaired electrons in a given complex compound depends on the experimental value of μ and not on the valence-shell configuration of the central metal cation. Thus it is experimental value of μ which determines the actual number of unpaired electrons present in the complex compound. This can be illustrated by considering $[FeF_6]^{3-}$ and $[Fe(CN)_6]^{3-}$ ions (see "*Uses of magnetic moment values*").

Uses of magnetic moment values

The experimental magnetic moment value of a given complex compound can be used to determine the number of unpaired electrons occupying the d-orbitals of the central metal ion of a given complex compound. The number of unpaired electrons determined in this way can be used to predict:

(*i*) Whether a given 4-coordinated complex compound is square planar (dsp^2 hybridisation) or tetrahedral (sp^3 hybridisation).

(*ii*) Whether a given 6-coordinated complex compound (octahedral) results from d^2sp^3 hybridisation (inner orbital octahedral complex) or sp^3d^2 hybridisation (outer orbital octahedral complex).

This point would be clear when we consider the geometries of $[Ni(CN)_4]^{2-}$, $[NiCl_4]^{2-}$, $[Fe(CN)_6]^{3-}$ and $[FeF_6]^{3-}$ ions as discussed earlier in this chapter. $[Ni(CN)_4]^{2-}$ ion has $n = 0$ (diamagnetic) and is square planar (dsp^2 hybridisation) while $[NiCl_4]^{2-}$ ion has $n = 2$ (paramagnetic) and is tetrahedral (sp^3 hybridisation). $[Fe(CN)_6]^{3-}$ ion has $n = 1$ (paramagnetic) and is inner-orbital octahedral in shape (d^2sp^3 hybridisation) but $[FeF_6]^{3-}$ ion has $n = 5$ (paramagnetic) and is outer-orbital octahedral in shape (sp^3d^2 hybridisation).

How to determine the paramagnetic and diamagnetic nature of a given substance?

The paramagnetic/diamagnetic nature of a given substance can be determined by weighing the substance in a *Gouy mangetic balance* in the absence and presence of magnetic field as shown in Fig. 6.4. The substance is taken in a small test tube and weighed in air (*i.e.*, in the absence of magnetic field) as shown at (*a*) of Fig. 6.4 and then weighed between the poles of a magnet. If the substance shows an increase in its weight, this implies that the substance is attracted by the magnetic field and hence is paramagnetic [See (*b*) of Fig. 6.4]. On the other hand, if the substance shows a decrease in its weight, it implies that the substance is repelled by the magnetic field and hence is diamagnetic. [See (*c*) of Fig. 6.4]

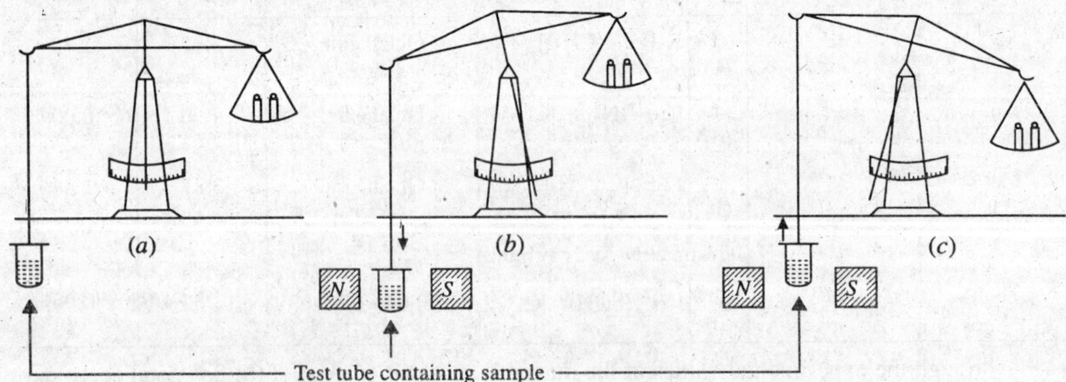

Fig. 6.4. Gouy magnetic balance used to determine the paramagnetic and diamagnetic nature of a given substance. (*a*) weight of the substance in the absence of magnetic field (*b*) weight of the substance increases when placed between the poles of a magnet–paramagnetic substance (*c*) weight of the substance decreases when placed between the poles of a magnet–diamagnetic substance.

The change in weight (*i.e.*, increase or decrease in weight) is used for calculating the magnetic moment of the substance.

11. Tendency of transition metals to form complex compounds

The transition metal atoms or the cations derived from them have an ability to form complex compounds with certain molecules (*e.g.*, CO, NO, NH_3 etc.) or ions (*e.g.*, F^-, Cl^-, CN^- etc.). These molecules or ions are called ligands. These ligands contain one or more lone pairs of electrons which they can easily donate to the transition metal atom or cation and thus complete the vacant orbitals of those atoms or cations through the formation of L $\rightarrow$ M coordinate bonds. The tendency of transition metal atom or cation to form the complex compounds is due to its following properties:

(*a*) Due to the small size of the transition metal atoms or their cations and high effective nuclear charge, they have a high positive charge density on them. This high positive change density makes the atoms or cations to attract (accept) the lone pairs of electrons from the ligands.

(*b*) The transition metal cations or atoms have vacant *d*-orbitals [$(n - 1)$ *d* orbitals] in which they can accommadate the lone pairs of electrons donated by the ligands and thus can form L $\rightarrow$ M coordinate bonds.

Stability of complex compounds

(*i*) Stability of complex compounds decreases with the increase in atomic number of the central transition metal.

(*ii*) Transition metals in high oxidation state form stable complexes with the small highly electronegative and basic ligands like F^-, Cl^-, NH_3, IO_6^{5-}, TeO_6^{6-} etc. while these metals in low oxidation states (*e.g.*, +1, 0, –1) give stable complexes with π-acid ligands like CO, NO, CN^-, PCl_3, C_6H_6 etc. The complex compounds formed by transition metal atoms in low oxidation states contain L $\rightleftharpoons$ M coordinate bond. Also see "*Variable Oxidation States*".

(*iii*) When a given transition metal forms complexes with the same ligand in different oxidation states, the complex compound having the metal in higher oxidation state is more stable, *e.g.* $[Co^{3+}(NH_3)_6]^{3+}$ is more stable than $[Co^{2+}(NH_3)_6]^{2+}$. Greater stability is due to the fact that the metal cation having higher charge has greater power to attract the lone pair of electrons on the ligand. In other words we can say that since the metal cation with high charge is smaller in size, it can attract the lone pair of electrons of the ligand more strongly and hence can give more stable complex ion.

12. Formation of interstitial compounds

We know that transition metals have defects in their structures. These defects create vacant interstitial spaces (also called interstitial holes or sites) between the atoms of these metals. The vacant spaces can easily accommadate small atoms like those of H, B, C, N etc. (non-metals) and give rise to the formation of interstitial compounds. It is due to the filling of the vacant spaces by the non-metallic atoms that the transition metals become rigid and hard. For example steel and cast iron are hard because of the formation of interstitial compounds with carbon.

According to Nagg, the stable interstitial compounds are obtained when the non-metallic to metallic atom radius ratio is less than 0.59. These compounds cannot be represented by a definite composition and hence are non-stoichiometric in nature, *i.e.*, these compounds have variable composition and hence cannot be represented by a definite composition. Non-stoichiometry is produced due to variable valency of transition metals and due to the defects in solid structures.

These compounds have the same chemical properties as the parent metal but different physical properties such as density, hardness, electrical conductivity, thermal conductivity, metallic lusture etc. These have extremely high melting points, showing that the metal-non-metal bonds in these compounds are stronger than the metal-metal bonds in metals.

Interstitial hydrides, carbides, nitrides and oxides are important examples of interstitial compounds. Now we shall discuss these compounds one by one in the following sections.

A. Metallic or interstitial hydrides

H-atoms, being small in size, fit into the interstitial sites of many transition and inner-transition metal lattices at elevated temperatures and give rise to the formation of interstitial hydrides. Since these hydrides resemble their parent metal in chemical properties, these are also called *metallic* or *alloy-type hydrides*. Examples of such hydrides are $TiH_{1.73}$, $VH_{0.6}$, $ZrH_{1.92}$, $PdH_{0.8}$, $TbH_{3.07}$, $LaH_{2.76}$, $CeH_{2.69}$ etc. These are interstitial hydrides because H-atoms occupy the interstitial sites in the metallic lattices. Some chemists regard them as solid solutions.

Preparation. (*i*) The hydrides may be prepared by heating the metal in hydrogen under varying pressures and temperatures. Due to the catalytic action of the metals, H_2 molecules dissociate into H-atoms which then pass into the interstial sites between metal atoms.

(*ii*) On heating, V, Nb and Ta combine with H_2 and form the interstitial hydrides. The amount of hydrogen combined with the metal changes with the change of temperature and pressure.

(*iii*) These hydrides can also be prepared by electrodeposition of hydrogen on the metal used as cathode. In this process hydrogen is absorbed by the metal. The temperature at which the absorption of hydrogen takes place on the surface of the metal depends on the energy of activation involved in the reactions, $H_2 \rightleftharpoons 2H$ occuring at the metal surface. Palladium hydride is obtained by electrolysing $2N$ H_2SO_4 solution, using Pd as cathode. As H-atoms are obtained by the electrolysis of H_2SO_4, they are absorbed by metal and then react with it to form the interstitial hydride.

The absorption of hydrogen by certain metals such as Cu, Ag,Cr, Mo, W, Fe, Co, Ni and Pt is endothermic while that by other metals like Ti, Th, Zr, V, Nb, Ta and Pd is exothermic.

Properties. They show similarities with the parent metal, are non-volatile and have metallic properties such as hardness, lustre, electrical conductivity etc. Their densities are usually less than those of parent metal, showing that the metal lattice expands during the absorption of hydrogen. In these compounds hydrogen acts as a metal and hence these hydrides behave as alloys.

These compounds give up hydrogen when heated strongly in vacuum, but small amount of hydrogen is still retained in the metal lattices. Hydrogen gas thus obtained has been found to be more reactive than the molecular hydrogen; this suggests that hydrogen present in these hydrides is in the atomic state. It is due to the presence of atomic hydrogen that these compounds are used as powerful reducing agents.The case with which these compounds lose hydrogen on heating depends on the nature of the metal. These hydrides are non-volatile, brittle and semi-conductors.

The hydrides of Ti, Zr, and Hf are brittle, metal-like, black or grey powders with varying composition. Maximum hydrogen content is found in the hydrides namely $TiH_{1.73}$, $ZrH_{1.92}$. These hydrides are intermediate between solid solutions and salt-like (ionic) hydrides of CaH_2 type. These hydrides are chemically stable and donot react with water or with dil. acids.

V, Nb and Ta also form interstitial hydrides. In these hydrides, the maximum hydrogen content is seen in the formula $VH_{0.71}$, $NbH_{0.88}$ and $TaH_{0.76}$. The hydrides of vanadium are brittle, grey and black metal-like powders having varying composition. Hydrides of V, Nb and Ta are stable and are acted upon by water or dil. acids.

Uses. These are used as powerful reducing agents. Some of these hydrides are used in powder metallurgy and in the manufacture of vacuum tubes. The decomposition of the interstitial hydrides of Ti, Zr and Hf is used to obtain the metals in produced form and to coat with them the surfaces of various articles.

Structure. Mostly these hydrides are non-stoichiometric solids as is evident from their formulae $TiH_{1.73}$, $ZrH_{1.92}$, $VH_{0.71}$, $NbH_{0.88}$, $TaH_{0.76}$, $PdH_{0.6}$ etc. It has not yet been confirmed whether these compounds are *true compounds* or *solid solutions.*

Some of the hydrides of transition metals have stoichiometric composition (*e.g.*, CuH, NiH_2, CoH_2, FeH_2, CrH_3) but these hydrides donot resemble ionic hydrides and are, therefore, called *borderline hydrides.* NiH_2, CoH_2 and FeH_2 are obtained as black precipitate, when phenol magnesium bromide (C_6H_5MgBr) solution is mixed with metal (II) chloride and H_2 is passed through it.

$$MCl_2\ (M = Ni, Co, Fe) + 2C_6H_5MgBr \longrightarrow M(C_6H_5)_2 + MgBr_2 + MgCl_2$$

$$M(C_6H_5)_2 + 2H_2 \longrightarrow 2C_6H_6 + MH_2$$

B. Interstitial or metallic carbides

These carbides are also called *refractory carbides.* These carbides are formed by those transition metals whose atomic radii (metallic radii for coordination number 12) are generally greater than 1.3 Å ($R_M > 1.3$ Å), since the radius ratio namely R_C/R_M must be either 1 : 0.41 or 1 : 0.59 to permit C-atoms to enter into the octahedral holes (also called cavities, interstices or interstitial positions) of the cubic close packing of the metallic lattice without distorting it. The radius of the metals Ti, Zr, Hf (Group IV B); V, Nb, Ta (Group V B); Mo and W (Group VI B) is greater than 1.3 Å and hence these metals give the interstitial carbides which are of two types *viz.* MC type (M = Ti, Zr, Hf, V, Nb, Ta, Mo, W) and MC_2 type (M = V, Mo, W).

Preparation. Interstitial carbides are prepared:

(*i*) *by direct combination of the metal with carbon.* In this method powdered metal is heated with powdered carbon at high temperature. Carbides obtained by this method are purified and made compact by sintering them in vacuum or in an atmosphere of inert gases.

(*ii*) *by reducing the metallic oxide with carbon in an electric furnace at about 2000°C.* This method has been used for the preparation of TiC, ZrC, HfC, VC, NbC, TaC, WC.

$$TiO_2 + 3C \longrightarrow TiC + 2CO$$

$$WO_3 + 4C \longrightarrow WC + 3CO$$

$$MoO_3 + 4C \longrightarrow MoC + 3CO$$

(*iii*) *by heating the metal in an atmosphere of CH_4 at different temperatures.* HfC, MoC, TaC, WC etc. carbides have been prepared by this method, *e.g.*,

$$Hf + CH_4 \longrightarrow HfC + 2H_2$$

$$Mo + CH_4 \longrightarrow MoC + 2H_2$$

Properties and structure.

(*i*) The melting points of these carbides are very high. For example the melting points of TaC and ZrC are 3900°C and 3800°C respectively.

(*ii*) These are chemically inert and extremely hard like diamond. For example WC is used for cutting tools on account of its hardness. The hardness of Moh's scale is between 9 and 10.

(*iii*) The crystal lattice of these carbides is a close-packed type with C-atoms occupying the octahedral holes in the metal lattice. Presence of C-atoms, therefore, does not affect the electrical conductivity of the metal. Thus these carbides possess metallic lustre and high electrical conductivity which increases at lower temperature and may become infinite at absolute zero.

(*iv*) These are weakly paramagnetic which is characteristic of the presence of metal lattice in their crystal structure.

(*v*) They are attacked by strong oxidising agents at red heat.

(*vi*) Carbides of Mo and W are somewhat more reactive than the metals from which they are derived but other carbides are less reactive than the parent metals.

(*vii*) MC type carbides (M = V, Nb, Ta) are refractory and hard while those of MC_2 type are ionic and react with H_2O to liberate acetylene.

(*viii*) Carbides of MC type (M = Ti, Zr, Hf) are crystalline solids, very hard, refractory (*i.e.*, can withstand high temperature like 3000 to 4000°C), good conductors of electricity and chemically inert.

(*ix*) TiC, ZnC and HfC have a structure of NaCl type and form solid solutions with one another.

C. Interstitial nitrides

Transition metals like Ti, Zr, Hf (elements of group IVB), V, Nb, Ta (elements of group VB) absorb nitrogen to give the interstitial nitrides of MN type (M = Ti, Zr, V, Nb, Ta). Nitrides of M_2N (M = Mo, W) and M_4N (M = Fe, Mn) type are also known.

Preparation. These nitrides are obtained :

(*i*) by heating the powdered metal in N_2 or NH_3 at 1100-1300°C.

(*ii*) by heating the metal filament in N_2, *e.g.*, TiN and nitrides of V, Nb and Ta are obtained by this method.

(*iii*) by heating a carrier filament (*e.g.*, Pt, W, C) in a mixture of N_2 and metal halide vapour and subsequently removing the carrier by high temperature volatilisation.

Properties. Being deficient in nitrogen, these compounds are non-stoichiometric in nature. These, like metals, have considerable hardness and electrical conductivity. These are stable and very inert (*i.e.*, chemically unreactive). They donot hydrolyse. They have high m.pts. (*e.g.*, VN, NbV = 2300°C, WN = 2570°C, TaV = 3087°C). Some nitrides are isomorphous to oxides and carbides, *e.g.*, TiN is isomorphous to TiO and TiC.

MN type nitrides of Ti, Zr and Hf (elements of group IVB) are crystalline solids, very hard, refractory (*i.e.*, can withstand high temp. like 3000 to 4000°C), good conductors of electricity and chemically inert. Interstitial nitrides of MN type of V, Nb and Ta (elements of group VB), Nb_2N and Ta_2N have corrosion resistance power.

Structure. In the nitrides of MN type, all the octahedral interstices in transition metal lattices are filled up with N-atoms. Thus we find that these nitrides have NaCl type structure. In the nitrides of M_2N type only half of the octahedral interstices have been filled up with N-atoms while in the nitrides of M_4N type only one-quarter of the octahedral interstices are occupied by N-atoms.

Fulminating metals. The nitrides of Ag, Cu and Hg (Ag_3N, AuN and Hg_3N_2) are called fulminating metals. These metals should not be confused with fulminic acid, C═N—OH. These metals are obtained by the action of NH_3 on Ag_2O, Au_2O_3 and HgO.

Fulminating metals

$$2NH_3 + 3Ag_2O \longrightarrow 2Ag_3N + 3H_2O$$
$$2NH_3 + Au_2O_3 \longrightarrow 2AuN + 3H_2O$$
$$2NH_3 + 3HgO \longrightarrow Hg_3N_2 + 3H_2O$$

D. Interstitial oxides

Examples of interstitial oxides formed by transition metals are given below:

(*i*) The lowest oxide of titanium (TiO) is a non-stoichiometric oxide. Its actual composition varies between $TiO_{1.9}$ and $TiO_{2.0}$. In order to show the interstitial nature of TiO, a bar is placed over its formula, *i.e.*, TiO is represented as $\overline{TiO}$ or as $TiO_{1.9} \longrightarrow TiO_{2.0}$ or as $Ti \longrightarrow O_{1.9-2.0}$.

(*ii*) The analysis of FeO has shown that the actual composition of this oxide varies from $Fe_{0.84}O$ to $Fe_{0.94}O$, *i.e.*, FeO is represented as $Fe_{0.84-0.94}O$.

(*iii*) The monoxides of Nb, NbO (grey) has varying composition ($NbO_{0.94-1.04}$), a metallic lustre and metallic conductivity. The dioxide, NbO_2 (black) also has a variable composition ($NbO_{1.9-2.09}$) but is a semi conductor.

13. Catalytic activity

Many transition metals (*e.g.*, Fe, Ni, Pt etc.), their alloys (*e.g.*, iron-molybdenum alloy) and their compounds (*e.g.*, V_2O_3, V_2O_5, MnO_2, Co^{2+} salts etc.) act as catalysts in many chemical reactions.

Explanation. The catalytic activity of transition metals or their compounds is due to the following reasons:

(*i*) Since the transition metals have vacant orbitals and show variable oxidation states, these metals can form unstable intermediate compounds with suitable reactants. These intermediate compounds give reaction paths for lower activation energy and, therefore, increase the rate of the reaction. These intermediate compounds readily decompose and give the final products and original substance. This original substance is the transition metal or its compounds which has been used as a catalyst. Thus:

Reactants + Transition metal or its compound which acts as a catalyst $\longrightarrow$ Unstable intermediate compound $\xrightarrow{\text{Decomposition}}$ Products + Transition metal or its compound (catalyst).

(*ii*) In some cases the transition metals and their compounds provide a large surface area on which the reactants may be absorbed and, therefore, come closer to each other for the reaction (*Adsorption theory*).

Example. Solid V_2O_5 is used as a catalyst for the oxidation of SO_2 to SO_3 required in the contact process which is used for the manufacture of H_2SO_4. During the conversion of SO_2 to SO_3, solid V_2O_5 absorbs a molecule of SO_2 on its surface to form V_2O_4 and O_2. O_2 is taken by SO_2 and V_2O_4 to form SO_3 and V_2O_5 (*catalyst*) respectively.

$$2V_2O_5 \longrightarrow 2V_2O_4 + O_2$$

$$SO_2 + \frac{1}{2}O_2 \longrightarrow SO_3$$

$$V_2O_4 + \frac{1}{2}O_2 \longrightarrow V_2O_5 \ (\textit{catalyst})$$

Examples of catalysts. (*i*) The *finely-divided nickel* can adsorb fairly large quantity of hydrogen especially at high temperatures. Due to this property, finely-divided Ni is used as a catalyst in various reactions like those shown below:

(*a*) Reduction of unsaturated ethylenic compounds (*Sabatier process*).

$$CH_2{=}CH_2 + H_2 \xrightarrow{\text{Ni (410 K)}} CH_3{-}CH_3$$

(*b*) *In the hydrogenation of oils.* In this process the unsaturated oils are converted into saturated fats (hardening of oils) in presence of Ni.

(*c*) The following hydrogenation reactions are catalysed by heated nickel.

$$CO + 3H_2 \longrightarrow CH_4 + H_2O$$

$$2NO + 5H_2 \longrightarrow 2NH_3 + 2H_2O$$

(*ii*) *Platinum black* is used as catalyst in the preparation of HCHO from CH_3OH and in the decomposition of H_2O_2. Copper is also used as a catalyst for converting CH_3OH to HCHO.

$$2CH_3OH + O_2 \ (\textit{From air}) \xrightarrow{\text{Pt black}} 2HCHO + 2H_2O$$

$$2H_2O_2 \xrightarrow{\text{Pt black}} 2H_2O + O_2$$

$$2CH_3OH + O_2 \ (\textit{From air}) \xrightarrow{\text{Cu (770 K)}} 2HCHO + 2H_2O$$

(*iii*) *Platinised asbestos and V_2O_5* are used as catalyst in the oxidation of SO_2 to SO_3 required for the manufacture of H_2SO_4 by contact process.

(*iv*) *Platinised gauge* is used as a catalyst in the oxidation of NH_3 to NO required for the manufacture of HNO_3 by Ostwald's process.

$$4NH_3 + 5O_2 \xrightarrow[800°C]{\text{Pt gauge}} 4NO + 6H_2O$$

(*v*) Finely divided Fe mixed with Mo (as promotor) is used for catalysing the combination of N_2 and H_2 to form NH_3 by Haber's process.

$$N_2 + 3H_2 \underset{770\text{ K}}{\overset{\text{Finely divided Fe + Mo as promotor}}{\rightleftharpoons}} 2NH_3$$

(*vi*) Fenton's reagent ($FeSO_4 + H_2O_2$) is used for oxidising alcohols to aldehydes.

(*vii*) Pd is used for hydrogenation of phenol to cyclohexanone.

(*viii*) Ziegler-Natta catalyst ($TiCl_4$) is used for the manufacture of polythene.

(*ix*) V_2O_3 is used as a catalyst for converting ethylene to ethyl alcohol.

(*x*) $VOCl_3$ is used as a catalyst in olefine polymerisation including ethylene propylene rubber.

(*xi*) NiO catalyses the following reaction which is used for the manufacture of sulphur from coal gas.

$$Fe_2O_3 + 3H_2S \xrightarrow{\text{NiO}} 2FeS + 3H_2O + S$$

(*xii*) MnO_2 is used to catalyse the decomposition of H_2O_2 solution.

(*xiii*) Co(II) salts catalyse the decomposition of bleaching powder (ClO^-) as Co (II) salts can easily be oxidised to Co (III) salts.

$$Co^{2+} + OCl^- + H_2O \longrightarrow Co^{3+} + Cl^- + 2OH^-$$

$$2Co^{3+} + 2OH^- \longrightarrow 2Co^{2+} + H_2O + \frac{1}{2}O_2$$

14. Alloy formation

We have seen above that the atoms of many transition metals have almost the same size and hence the atoms of one metal can be replaced by the atoms of other metal. This replacement gives rise to the formation of alloys. Thus when Mn is dissolved in molten iron and the solution is cooled, manganese-iron alloy is obtained. Such alloys are hard, have high melting points and are more resistant to corrosion than the parent metals.

General study of the compounds formed by the metals of 1st transition series

The metals of 1st transition series react with a number of non-metals like C,N,P,O,S and halogens to form binary compounds. Here we shall discuss oxides, halides and sulphides.

A. Oxides

The oxides of the metals of 1st transition series are generally obtained by heating the metals with O_2 at high temperatures. Important oxides given by these metals are given below. The acidic, basic and amphoteric character of these oxides has been indicated by writing (*a*), (*b*) and (*c*) respectively after their formulae.

Sc : Sc_2O_3(*b*)

Ti : TiO(*b*), Ti_2O_3(*b*), TiO_2(*c*)

V : VO(*b*), V_2O_3(*b*), VO_2(*c*), V_2O_5 (*a*)

Cr : CrO(*b*), Cr_2O_3(*c*), CrO_2(*c*), CrO_3 (*a*)

Mn : MnO(*b*), Mn_3O_4(*c*), Mn_2O_3(*c*), MnO_2(*c*), MnO_3(unstable)(*c*), Mn_2O_7(*a*)

Fe : $FeO(b)$, $Fe_2O_3(b)$, $Fe_3O_4(b)$,

Co : $CoO(b)$

Ni : $NiO(b)$

Cu : $Cu_2O(b)$, $CuO(c)$

Zn : $ZnO(c)$

Properties. (*i*) *Variation in acidic, basic and amphoteric nature of oxides.* It has already been discussed under "*Variable Oxidation States of Transition Metals*".

(*ii*) *Solubility of oxides.* The basic and amphoteric oxides dissolve in non-oxidising acids and form hexaaquo ions, $[M(H_2O)_6]^{n+}$. The acidic oxides like CrO_3 and Mn_2O_7 dissolve in water to form the oxy acids. These oxides also reacts with bases (*e.g.*, KOH) to form their oxo salts.

$$CrO_3 + H_2O \longrightarrow H_2CrO_4 \text{ (Chromic acid)}$$
$$CrO_3 + 2KOH \longrightarrow K_2CrO_4 + H_2O$$
$$2CrO_3 + H_2O \longrightarrow H_2Cr_2O_7 \text{ (Dichromic acid)}$$
$$2CrO_3 + 2KOH \longrightarrow K_2Cr_2O_7 + H_2O$$
$$Mn_2O_7 + H_2O \longrightarrow 2HMnO_4 \text{ (Permanganic acid)}$$
$$Mn_2O_7 + 2KOH \longrightarrow 2KMnO_4 + H_2O$$

(*iii*) *Reducing properties of oxides.* As we move along a given transition series, +2 oxidation state of the metals becomes more and more stable and hence the tendency of these metals to lose electrons becomes more. Thus we can say that the reducing property of the metals increases as we move along a given transition series. VO and CrO are strong reducing agents. As we move down a group of transition metals, the higher oxidation state becomes more stable. For example OsO_4 (Os = +8) and RuO_4 (Ru = +8) both have been isolated (RuO_4 is unstable and decomposes on standing) while FeO_4 (Fe = +8) has not been isolated.

B. Halides

Transition metals react with halogens at high temperatures and form halides. The reactions of the metals with halogens have high activation energies and hence require high temperatures. The order of the tendency of the halogens to react with a given metal decreases from F to I (F > Cl > Br > I). Transition metals in higher oxidation states give fluorides.

Properties. (*i*) Transition metal halides have low volatility and high susceptibility to hydrolysis.

(*ii*) Halides of metals in high oxidation state have greater tendency to undergo hydrolysis, *e.g.*,

$$TiCl_4 + 2H_2O \longrightarrow TiO_2 + 4HCl$$

(*iii*) The elements of 1st transition series tend to form halides in low oxidation states, *e.g.*, Cr to Zn form dihalides.

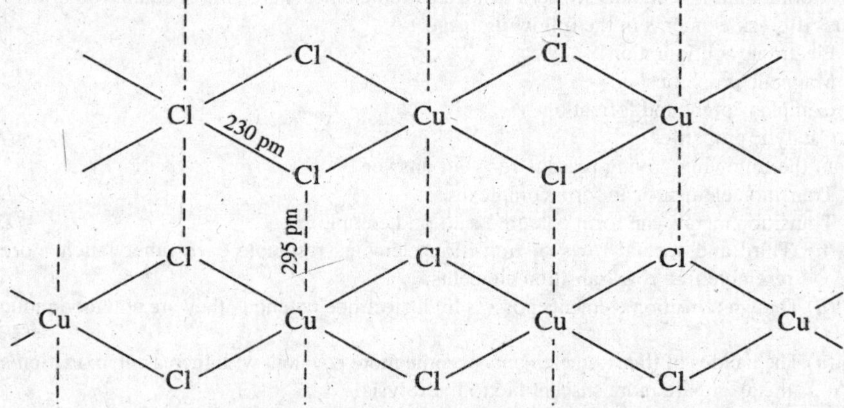

Fig. 6.5. Structure of copper (II) chloride.

(*iv*) Fluorides are mainly ionic while chlorides, bromides and iodides have ionic as well as covalent character. The ionic character of halides decreases with increasing atomic mass, e.g., CuF_2 is ionic while $CuCl_2$ and $CuBr_2$ are covalent and have infinite chains. The structure of Cu (II) chloride is shown in Fig. 6.5.

C. Sulphides

Sulphides are obtained when the metal combines directly with sulphur or when the compounds like H_2S, Na_2S etc. react with a solution of the metallic salts. Since sulphur is not a powerful oxidising agent, sulphides are formed by the metals in lower oxidation states.

Properties. (*i*) Transition metal sulphides are highly coloured or black.

(*ii*) They are insoluble in water.

(*iii*) In some cases, when a sulphide is treated with an oxidising agent, sulphur gets precipitated and sulphide is oxidised to sulphate.

University Questions

1. Give the reasons for the following:
   (*i*) Transition elements have high enthalpies of atomisation.
   (*ii*) Most of the compounds formed by transition metals are coloured.
   (*iii*) Transition metals and their compounds show paramagnetic behaviour. *(Meerut 97)*
   (*iv*) Transition metals form complex compounds.
   (*v*) Zn and Cd are normally not considered as transition metals.
   (*vi*) Transition metals show variable oxidation states. *(Meerut 97)*
   (*vii*) The paramagnetic character of 3*d*-transition series elements increases upto Mn and then decreases.
   (*viii*) Mn(II) ion shows maximum magnetic character among the bivalent ions of first transition series.
   (*ix*) $K_2[PtCl_6]$ is a well known compound whereas corresponding Ni compound is not known.
   (*x*) Ni^{2+} compounds are thermodynamically more stable than Pt^{2+} compounds while Pt^{4+} compounds are relatively more stable than Ni^{4+} compounds.
   (*xi*) Ionisation energies of 5d-elements are greater than those of 3*d*-and 4*d*-elements. *(Meerut 97)*
   (*xii*) Most of the transition metals do not displace hydrogen from dilute acids.
   (*xiii*) The ionisation energies of the elements of 1st transition series do not vary much with increasing atomic number.
   (*xiv*) Transition metals have high enthalpy of hydration.
   (*xv*) Cu(I) is diamagnetic while Cu(II) is paramagnetic. *(Merat 97)*
   (*xvi*) Zn^{2+} salts are white while Cu^{2+} salts are blue. *(Meerut 97)*
   (*xvii*) [Ti(H2O)6]3+ is coloured while $[Sc(H_2O)_6]^{3+}$ is colourless.
   (*xviii*) Though Cu, Ag and Au have completely-filled sets of *d*-orbitals, yet they are considered as transition metals.
   (*xix*) Variations in the radii of transitiion elements are not as pronounced as those of representative elements. *(ISM Dhanbad 90)*
   (*xx*) Atomic radii of 2nd and 3rd series of transition elements are almost equal. *(Meerut 97)*
2. Discuss d-block elements in the following respect:
   (*a*) Electronic configuration
   (*b*) Magnetic properties
   (*c*) Complex compound formation
   (*d*) Catalytic properties. *(Meerut 98)*
3. Explain the following, giving reasons for your answer.
   (*i*) Transition elements can form complexes. *(Delhi 1997)*
   (*ii*) Transition metals can form coloured salts and complexes. *(Delhi 1998)*
   (*iii*) (*a*) Third and second rows of transition elements resemble each other much more than they resemble first row transition elements.
   (*b*) Though transition elements possess high electrode potential, they are not good reducing agents. *(Delhi 1999)*
   (*iv*) (*a*) The halides of transiton elements become more covalent with increasing oxidation states of the metal and are more susceptible to hydrolysis.

(b) Transition elements show more oxidation states than other elements. *(Delhi 2000)*

(v) (a) Transition elements are less reactive than alkali metals.

(b) Transition elements and their compounds are generally good catalysists.

(c) The radii of the elements of third transition series are very similar to those of second transition series. *(Delhi 2001)*

4. Discuss the properties of first transition series with referencve to :

(i) Oxidation states and (ii) Complex forming tendency. *(Nagpur 2002)*

5. In moving down a transition series, the atomic sizes increases from 3d to 4d but there is hardly any increase from 4d to 5d. Explain. *(Delhi 2002)*

6. (a) What do you understand by magnetic moment and magnetic susceptibility ?

(b) How will you show that magnetic moment can be related directly with the number of unpaired electrons ? *(VBS, Purvanchal 2003)*

7. Write the general electronic configuration of transition elements. *(VBS, Purvanchal 2003)*

8. How the magnetic properties of transition metal complexes can be explained on the basis of CFT ? Discuss with at least two examples. *(VBS Purvanchal 2003)*

9. (a) List the oxidation states of transition elements in a tabular form. How do the stability of various oxidation states vary in the transition elements ?

(b) Compare the magnetic properties of 3d elements with 4d and 5d elements. *(Rohilkhand 2003)*

10. Explain the following :

(a) In the first transition series, the atomic radii decreases from Sc to Cr but from Cr to Cu, it is nearly the same. 2½ + 2½ + 2

(b) Sc^{3+} is colourless but Ti^{3+} is coloured.

(c) Transition elements have a strong tendency to form complex compounds. *(Rohilkhand 2003)*

11. Explain the following properties of transition elements : (i) Magnetic properties (ii) Variable oxidation states. *(Delhi 2003)*

12. (a) Discuss the properties of 3d-series elements with respect to their :

(i) Electronic configuration

(ii) Oxidation states.

(b) Discuss in brief the catalytic activities shown by the first transition series elements and their compounds.

(c) Why 3*d*-elements are called as 'Transition Elements' ?

(d) Discuss magnetic properties of 3d-elements. *(Nagpur 2003)*

13. Although Cd is a transition metal, its salts are colourless. Explain. *(Delhi 2003)*

Ans. Cd salts have Cd^{2+} ion which has $4d^{10}\ 5s^0$ configuration. Since 4d orbitals in this configuration are completely-filled, it is not possible to promote the electrons within 4d orbitals and hence Cd^{2+} salts do not show any colour.

14. (a) How do the size of atoms and ions of transition elements vary?

(b) Give a brief comparative account of the properties of 1st transition series with 2nd and 3rd transition series.

(c) Describe Gouy's method for measuring magnetic susceptibilies. *(GND Amritsar, 2004)*

15. (a) Give the name and electronic configuration of transition elements of first transition series. Why are they called transition elements. Explain.

(b) What do you mean by catalytic properties of transition elements.

(c) Describe the following properties of 4*d* transition elements.

(i) Colour (ii) Oxidation state (iii) Atomic size *(Bhopal 2004)*

16. Discuss the electronic configuration, colour, magnetic properties of first transition series of elements. *(Bhopal 2004)*

7

Group-wise Study of the General Properties of Transition Elements (*d*-Block Elements)

Introduction

Transition elements have been grouped into ten vertical columns called groups. These groups are IIIB (3), IVB (4), VB (5), VIB (6), VIIB (7), VIII (8, 9, 10), IB (11) and IIB (12). These elements are placed between *s*-and *p*-block elements in the long form of periodic table as shown in Fig. 7.1. Here we shall make a group-wise study of the general properties of these elements and the variation of these properties in moving down each of the groups.

←— *d*-Block Elements (Transtion Elements) —→

	III B (3)	IV B (4)	V B (5)	VI B (6)	VII B (7)	VIII (8)	VIII (9)	VIII (10)	I B (11)	II B (12)	
s-Block Elements	Sc_{21}	Ti_{22}	V_{23}	Cr_{24}	Mn_{25}	Fe_{26}	Co_{27}	Ni_{28}	Cu_{29}	Zn_{30}	*p*-Block Elements
	Y_{39}	Zr_{40}	Nb_{41}	Mo_{42}	Tc_{43}	Ru_{44}	Rh_{45}	Pd_{46}	Ag_{47}	Cd_{48}	
	La_{57}	Hf_{72}	Ta_{73}	W_{74}	Re_{75}	Os_{76}	Ir_{77}	Pt_{78}	Au_{79}	Hg_{80}	
	Ac_{89}	Ku_{104}	Ha_{105}								

Fig. 7.1. Location of *d*-block elements (transition elements) in the periodic table.

Group IIIB (3) : Sc_{21}, Y_{39}, La_{57}, Ac_{89}

Group IIIB of the periodic table consists of four elements which are *Scandium* (Sc), *Yttrium* (Y), *Lanthanum* (La) and *Actinium* (Ac). These elements are generally called *scandium group elements.*

Electronic configurations

The electronic configurations of the atoms of the elements of group IIIB are given in Table 7.1. These configurations show that the valence-shell configurations of these elements can be represented by $(n-1)d^1.ns^2$, indicating that the extra electron goes in $(n-1)d$ orbital. Thus, these are *d*-block (transition) elements. All the elements have 2 electrons in the outermost shell.

General properties and gradation in them

Some of the properties are given in Table 7.1. and are discussed below.

1. Oxidation states. All the elements show +3 oxidation state in solution, *i.e.*, in solution all the elements exist as M^{3+} ions. These ions are obtained by removing both *ns* and one $(n-1)d$ electrons.

Table 7.1. *Important physical properties of scandium group elements: Sc, Y, La, Ac [Group IIIB (3)]. Here* $[Ar]_{18}$ = 2, 8, 8; $[Kr]_{36}$ = 2, 8, 18, 8 *and* $[Xe]_{54}$ = 2, 8, 18, 18, 8.

Property	*Scandium (Sc)*	*Yttrium (Y)*	*Lanthanum (La)*	*Actinium (Ac)*
Atomic number	21	39	57	89
Atomic weight	44.96	88.92	138.92	227
Electronic configuration	$[Ar]_{18}\ 3d^1\ 4s^2$ = 2, 8, 9, 2	$[Kr]_{36}\ 4d^1\ 5s^2$ =2, 8, 18, 9, 2	$[Xe]_{54}\ 4f^0 5d^1 6s^2$ = 2, 8, 18, 9, 2	$[Rn]_{89}\ 5f^0 6d^1 7s^2$ = 2, 8, 18, 32, 18, 9, 2
Atomic radius (pm)	162	180	187	—
Ionic radius (pm)	81 (+2) 88.5 (+3)	— 104 (+3)	— 117.2 (+3)	— 118 (+3)
Atomic volume (ml)	15.02	19.8	20.50	—
Density (g/cm³)	3.01	4.47	6.17	—
Melting point (°C)	1539	1509	920	—
Boiling point (°C)	2730	2927	3470	—
Ionisation energy (kJ/mol)	631	616	538.1	—
Oxidation states	+2, +3	+3	+3	+3
Colour of M^{3+} ions	colourless	colourless	colourless	colourless
Heat of fusion (kcal/mol)	3.85	4.1	2.7	—
Heat of vaporisation (kcal/mol)	72.85	94.0	95.5	—
Reduction potential (volts $M^{3+} + 3e^- \rightleftharpoons M$	–2.1	–2.37	–2.52	—
Electronegativity	1.3	1.2	1.1	1.1

2. Nature of salts of M^{3+} ions. The salts of M^{3+} ions have *paired electrons* and hence are *colourless* and *diamagnetic*. The metals have high melting points.

3. Action of water on metals. Due to their high electropositive character, these metals are highly reactive. For example they react with H_2O to form basic oxides, MO.OH and liberate H_2 gas. These metals react slowly with cold water and more rapidly with hot water.

4. Oxides and hydroxides. The oxides of these metals can be prepared by heating hydroxides or salts such as carbonate, nitrate or sulphate, while hydroxides can be prepared by treating these metals with H_2O. Oxides and hydroxides have basic charcter. The *basic character of the hydroxides increases with the increase in size of M^{3+} ion.* [$Sc(OH)_3 < Y(OH)_3 < La(OH)_3$]. Thus $La(OH)_3$ is the *strongest base*. Because of their basic character the oxides and hydroxides react with acids to form salts. Their oxides are very stable and hence cannot be reduced by thermite reaction.

5. Atomic and ionic radii. The size of the atoms (atomic radii) and M^{3+} ions (ionic radii) increase as we proceed from Sc to Ac.

6. Complex formation. *The tendency of M^{+3} ions to form complexes decreases with the increase of their size.* The Sc^{3+} ion which is the *smallest* shows some tendency to form the complexes like $[ScF_4]^-$, $[ScF_5]^{2-}$, $[ScF_6]^{3-}$ etc., while the other members do not have a strong tendency to form complexes.

7. Formation of halides. All these elements form trichlorides (MCl_3). These halides are non-volatile and are not hydrolysed by water.

8. Formation of double sulphates. These elements form double sulphates such as $La_2(SO_4)_3.3K_2SO_4$ but these are not alums.

Similarities between scandium and aluminium

1. Formation of M^{3+} ions. Outer electronic configuration of scandium is $3d^1\,4s^2$ and that of aluminium is $3s^2\,3p^1$. Both of them lose three electrons to give trivalent ions Sc^{3+} and Al^{3+}. Both these ions have a noble gas configuration as shown below:

$Sc_{21} = 2, 8, 9, 2$ $\qquad$ $Sc^{3+} = 2, 8, 8$

$= [Ar]_{18}\,3d^1\,4s^2$ $\qquad$ $= [Ar]_{18}$

$Al_{13} = 2, 8, 3$ $\qquad$ $Al^{3+} = 2, 8$

$= [Ne]_{10}\,3s^2\,3p^1$ $\qquad$ $= [Ne]_{10}$

2. Formation of colourless and diamagnetic compounds. Both form colourless and diamagnetic compounds.

3. Reaction with water. Both react with water vigorously liberating hydrogen.

$$2\,Sc + 3\,H_2O \longrightarrow Sc_2O_3 + 3H_2$$

$$2\,Al + 3\,H_2O \longrightarrow Al_2O_3 + 3\,H_2O$$

4. Nature of oxides. Both Sc_2O_3 and Al_2O_3 are insoluble in water. However, Al_2O_3 is amphoteric whereas Sc_2O_3 is basic.

5. Formation of double salts. Scandium and aluminium both from double sulphates given below:

$$K_2SO_4\,.\,Sc_2(SO_4)_3\,.\,xH_2O \text{ and } K_2SO_4\,.\,Al_2(SO_4)_3.24H_2O$$

Sulphates differ in the size of their M^{3+} ions. Sc^{3+} ion is much larger in size than Al^{3+} ion.

Similarities between scandium and lanthanides

Scandium resembles lanthanides in the formation of insoluble salts like oxalates, phosphates, carbonates and fluorides.

Discimilarities between scandium and lanthanides

1. Size of M^{3+} ions. Scandium differs from lanthanides in the size of their M^{3+} ions. Ionic radius of Sc^{3+} is 0.18 Å whereas ionic radii of lanthanides are in the range 0.85—1.06 Å.

2. Nature of their oxides. Scandium oxide is less basic than oxide of lanthanides.

3. Nature of chloride, nitrate and sulphate. Scandium chloride, $ScCl_3$ is volatile and sublimes at a much lower temperature than lanthanide chlorides, scandium nitrate readily decomposes and sulphate is readily soluble in cold as well as hot water. These differ from corresponding salts of lanthanides.

4. Formation of double sulphate. Scandium forms a double sulphate $K_2SO_4\,.\,Sc_2(SO_4)_3\,.\,24\,H_2O$ and no such double sulphate is given by lanthanides.

5. Formation of thiocyanate. Scandium salts on treatment with KSCN give a yellow precipitate of $Sc(SCN)_3$. Salts of the lanthanides do not give any precipitate of sulphocyanide.

6. Hydrolysis of M^{3+} ions. Sc^{3+} ions being smaller in size than the corresponding lanthanide ions, have a greater tendency to hydrolyse.

7. Formation of complexes. Sc^{3+} ion on account of its smaller size forms complex ions more readily than the lanthanide ion. The fluoride is readily soluble in excess HF and NH_4F to form definite fluoro complexes, *e.g.*, ScF_6^{3-}.

8. Polarizing power of M^{3+} ions. On account of its smaller size Sc^{3+} has greater polarizing power and greater tendency to form covalent linkage is compounds.

Group IVB (4) : Ti_{22}, Zr_{40}, Hf_{72}, Ku_{104}

Group IVB of the periodic table consists of four elements which are *Titanium* (Ti), *Zirconium* (Zr), *Hafnium* (Hf) and *Kurchatovium* (Ku). These elements are generally called *titanium group elements.*

Electronic configurations

The electronic configurations of the atoms of these elements as given in Table 7.2. indicate that the valence-shell configurations can be represented by $(n-1)d^2.ns^2$ showing that the extra electron goes to $(n-1)d$ orbitals. Thus these are *d*-block (transition) elements. All the elements have two electrons in the outer-most shell (ns^2 configuration). The penultimate shell has the arrangement, $(n-1)s^2p^6d^2$ (unsaturated). This is the first group in which the elements show definitely transitional character both in their isolated atoms and in their compounds.

General properties and variation in them

Some properties of the elements of this group are given in Table 7.2 and have been discussed below:

1. Metallic character. All are typical metals. They are very hard and good conductors of heat and electricity.

2. Atomic volume. Atomic volume increases with the increase in atomic number from titanium to zirconium but shows a little decrease in hafnium as a result of lanthanide contraction, since hafnium comes immediately after the lanthanide series of elements.

3. Atomic radius. Like atomic volume, it increases from titanium to zirconium and then falls in hafnium.

As a result of *lanthanide contraction* Hf has almost the same atomic (covalent) ionic radius as Zr. The similarity in size of the atoms and ions of these two elements is responsible for similarity in their properties. The chemical properties of the two elements are so similar that till recently their separation from each other was considered difficult. Now these are separated by ion-exchange technique.

4. Melting and boiling points. They posses very high melting and boiling points.

5. Ionisation potential. These values are high and have a decreasing trend.

6. Electronegativity. The metallic nature of the elements increases with the rise in atomic number and hence electronegativity decreases slightly from titanium to hafnium.

7. Valency. These metals are generally quadrivalent in their compounds.

8. Oxidation states. The different oxidation states shown by these elements are +2, +3, and +4. The +2 state arises from the removal of two *ns*-electrons and +3 and +4 states are obtained when one or both $(n-1)$ *d*-electrons are successively removed. The +4 state is the most common oxidation state for each element.

Table 7.2. *Important physical properties of Ti, Zr and Hf [Group IVB (4)]. Here $[Ar]_{18}$ = 2, 8, 8; $[Kr]_{36}$ = 2, 8, 18, 8 and $[Xe]_{54}$ = 2, 8, 18, 18, 8.*

Property	*Titanium (Ti)*	*Zirconium (Zr)*	*Hafnium (Hf)*
Atomic number	22	40	72
Electronic configuration	$[Ar]_{18}\ 3d^2\ 4s^2$ = 2, 8, 10, 2	$[Kr]_{36}\ 4d^2\ 5s^2$ = 2, 8, 18, 10, 2	$[Xe]_{54}\ 4f^{14}\ 5d^2\ 6s^2$ = 2, 8, 18, 32, 10, 2
Atomic weight	47.90	91.22	178.49
Atomic volume (ml)	10.60	14.1	13.6
Density (g/cm^3)	4.51	6.49	13.1
Atomic radius (pm)	147	160	158
Ionic radii (pm)	91 (+2) 74.5 (+4)	86 (+4)	85 (+4)
Melting point (°C)	1668	1852	2222
Boiling point (°C)	3260	1500	5400
Ionisation potential (kJ/mol)	658	660	654
Electronegativity	1.5	1.4	1.3
Stable oxidation state	+4	+4	+4
Specific heat	0.113	0.068	0.0341
Heat of fusion (kcal/g.atom)	3.7	4.0	5.2
Abundance in earth's crust	0.63	0.022	0.00045
Heat of vaporisation (kcal/g.atom)	106.5	120.0	155.0

9. Nature of bonds in compounds. As considerable amount of energy is required to remove all the four $(n-1)\ d^2.ns^2$ electrons to realize +4 oxidation state, most of the compounds of these elements in +4 oxidation state are the covalent. Thus $TiCl_4$ is a volatile liquid which boils at 136°C. The existence of M^{+4} ions even in strongly acidic solution is doubtful. The bivalent and trivalent compounds of these elements behave as a reducing agent, since they have a marked tendency to disproportionate.

$$2Ti^{2+} \xrightarrow{H_2O} Ti^0 + Ti^{4+}$$

$$2Ti^{3+} \xrightarrow{Heat} Ti^{2+} + Ti^{4+}$$

The tetravalent compounds of these elements are colourless and diamagnetic whereas those from the lower valency states are coloured and paramagnetic.

10. Formation of interstitial compounds. These elements yield interstitial hydrides, nitrides, carbides and borides of the compositions, MH_2, MN, MC and MB with hydrogen, nitrogen, carbon and boron respectively.

11. Formation of oxides and hydroxides. No definite hydroxides of the type, $M(OH)_4$ are known. The hydroxides are really hydrated oxides. These metals, on heating with O_2 above

600°C, give dioxides (MO_2) whicn are stable, non-volatile and insoluble and become refractory on heating. TiO_2 is amphoteric (though more acidic than basic). It dissolves in alkalies to form titanates like Na_2TiO_3 and $Na_2Ti_2O_5$. ZrO_2 and HfO_2 are more basic and less acidic. The metals also form peroxides and proxy compounds when reacted with H_2O_2.

12. Formation of halides. Ti and Zr form a complete series of tetrahalides. Hf gives only HfF_4 and $HfCl_4$. Tetrafluorides are largely ionic while other tetrahalides are predominantly covalent volatile substances. Except tetrafluorides all other tetrahalides are readily hydrolysed by H_2O. The extent of hydrolysis decreases from Ti to Hf. TiX_4 compounds are hydrolysed to TiO_2 while ZrX_4 and HfX_4 give $ZrOX_2$ and $HfOX_2$.

$$TiCl_4 + 2H_2O \longrightarrow TiO_2 + 4HCl$$

$$\underset{(M = Zr, Hf)}{MCl_4} + H_2O \longrightarrow MOCl_2 + 2HCl$$

13. Formation of salts with oxy-acids. Ti^{4+} does not form any normal salt with oxy-acids. However the normal salts of Zr^{4+} ion like $Zr(SO_4)_2$ have been obtained.

14. Formation of complexes. The tetrahalides react with alkali metal halides to form the *halocomplexes* like $M_2^+[TiX_6]$, $M_2^+[ZrF_6]$, $M_3^+[ZrF_7]$ etc.

$$TiX_4 + 2M^+X \longrightarrow M_2^+[TiX_6]$$

$$ZrX_4 + 2M^+X \longrightarrow M_2^+[ZrF_6]$$

$$ZrF_4 + 3M^+X \longrightarrow M_3^+[ZrX_7]$$

The tetrahalides also act as *Lewis acids* and hence from 1 : 2 addition compounds with nitrogen and oxygen donors, *e.g.*,

$$TiCl_4 + 2\text{ diars} \longrightarrow \underset{1\,:\,2\text{ addition compound}}{TiCl_4.2\text{diars}}$$

15. Action of acids and alkalies. These metals are not attacked by acids in cold. Alkali solutions, cold or hot, have no action on the metals. They are best dissolved by hydrofluoric acid when hexafluoro complex is formed.

$$M + 6HF \longrightarrow H_2MF_6 + 2H_2$$

In this respect they resemble silicon.

Resemblance between titanium and tin

There is a good deal of resemblance between titanium and tin. For example:

1. Tetravalency. Both of them are tetravalent in most of their compounds.

2. Ionic radii. Ionid radii of Ti^{4+} (= 0.68Å) and Sn^{4+} (0.71 Å) are quite close to each other.

3. Oxides. These oxides TiO_2 and SnO_2 are amphoteric in character and are identical in structure.

4. Chlorides. Both from tetrachlorides $TiCl_4$ and $SnCl_4$ which are colourless, volatile liquids which fume in air. They are largely covalent compounds and behave as Lewis acids.

Group VB (5) : V_{23}, Nb_{41}, Ta_{73}

Group VB of the periodic table consists of three elements which are *Vanadium* (V), *Niobium* (Nb) and *Tentalum* (Ta).

Electronic configurations

The electronic configurations of the atoms of the elements as given in Table 7.3 show that, excepting Nb, the valence-shell configuration of the remaining two elements can be represented as $(n-1)d^3.ns^2$. In case of Nb one electron from $5s$ orbital gets shifted to $4d$ orbital and hence the valence-shell configuration of Nb becomes $4d^4.5s^1$ instead of the expected configuration namely $4d^3.5s^2$. The valence-shell configurations also show that the extra electron goes to $(n-1)d$ orbital and hence these are *d*-block elements. It is also evident that these elements have five electrons in their valence-shell.

General properties and gradation in them

Some important physical properties are given in Table 7.3.

Table 7.3. *Important properties of V, Nb and Ta [Group VB (5)]. Here $[Ar]_{18}$ = 2, 8, 8; $[Kr]_{36}$ = 2, 8, 18, 8 and $[Xe]_{54}$ = 2, 8, 18, 18, 8.*

Property	*Vanadium (V)*	*Niobium (Nb)*	*Tantalum (Ta)*
Atomic number	23	41	73
Atomic weight	50.42	92.908	180.948
Electronic configuration	$[Ar]_{18}\ 3d^3\ 4s^2$ = 2, 8, 18, 2	$[Kr]_{36}\ 4d^4\ 5s^1$ = 2, 8, 18, 12, 1	$[Xe]_{54}\ 4f^{14}\ 5d^3\ 6s^2$ = 2, 8, 18, 32, 11, 2
Atomic volume (ml)	8.35	10.8	10.91
Density (g/cm^3)	6.10	8.40	16.60
Covalent radius (pm)	122.4	134.2	134.3
Atomic radius (pm)	134	146	146
Ionic radius (pm)	74 (+3) 72 (+4) 68 (+5)	86 (+3) 82 (+4) 78 (+5)	86 (+3) 82 (+4) 78 (+5)
Melting point (°C)	1900	2415	2996
Boiling point (°C)	3450	3300	5425
Ionisation energy (kJ/mol)	650	664	761
Electronegativity	1.7	1.6	1.5
Most common oxidation state	+5	+5	+5
Specific heat	0.1203	0.0645	0.0333
Heat of fusion (kcal/g.atom)	4.2	6.4	7.5
Heat of vaporisation (kcal/g.atom)	106	—	180
Abundence in earth's crust (%)	0.017	0.0024	0.00021

1. Metallic character. All are typical, silver-white metals. They are very hard and good conductors of heat and electricity. The elements of this group are penta positive. The electrical conductivity of these elements increases with increase in atomic number.

2. Atomic volume. Atomic volume increases with increase in atomic number.

3. Atomic radius. As was observed in the case of zirconium and hafnium—the members of group IV B, the two heavier members of this sub-group, niobium and tantalum are almost equal in atomic radius (because of the lanthanide contraction) and, therefore, very unequal in density.

4. Ionic and covalent radii. Due to *lanthanide contraction*, the ionic and covalent radii of Nb and Ta are almost equal. Thus these two elements are similar in many properties.

5. Melting and boiling points. The high melting and boiling points for these metals indicate strong metallic bonding in the liquid state, which is due to the availability of five valence electrons and appreciably low energy orbitals.

6. Valency. The metals of this group being transitional elements, show variable valency. Since the energy difference between the *d*-level of penultimate shell and *s*-level of valency shell is small, whereby transition of electron is possible from one to the other sub-levels.

7. Catalytic behaviour. All the elements possess good catalytic property. Vanadium is widely used as a catalyst in many industrial preparations. In the manufacture of sulphuric acid by contact process, V_2O_5 has been found to be more superior catalyst than platinum as it is not easily poisoned by arsenic trioxide besides being much cheaper than platinum.

8. Oxidation states. They show a variety of oxidation states. The stability of +5 state increases from V to Ta. Thus V(+5) compounds are reduced all the way to V(+2) compounds and Nb (+5) and Ta (+5) compounds remain stable towards reduction. Consequently V (+5) compounds are good oxidising agents while those of Nb (+5) and Ta(+5), being stable, are not. Lower oxidation states become less and less stable on moving down from V to Ta. The ionic character of compounds of each element decreases with the decrease of cationic size (*Fajan's rule*). Thus, since V(+2) > V(+3) > V(4), V(+2) compounds are more ionic than those of V(+3) and V(+4). Consequently V(+4) compounds (*e.g.*, VCl_4) are covalent.

9. Formation of oxides. Vanadium, niobium and tantalum form the pentaoxides namely V_2O_5 (orange red), Nb_2O_5 (white), Ta_2O_5 (white). They are decidedly acidic and are known as *acid earths.* Basic character of these oxides increases from V to Ta. Thus, V_2O_5 is amphoteric but more acidic than basic. Nb_2O_5 and Ta_2O_5 are less acidic than basic. Besides pentoxides, vanadium also forms the dioxide VO_2 (deep blue), sesquioxide V_2O_3 (black) and monoxide VO (black). Niobium also forms the dioxide NbO_2 (black), and the monoxide NbO (black). Tantalum forms the dioxide TaO_2 (greyish brown). The sesquioxides are not formed by niobrium and tantalum and the latter also does not form monoxide.

10. Formation of hydrides, nitrides and carbides. All these elements absorb hydrogen with the evolution of heat when the metal lattice swells. These are considered as alloy type hydrides. Their composition is $VH_{0.71}$, $NbH_{0.86}$ and $TaH_{0.76}$.

All these elements form the nitrides VN (m.p. 2300°C), NbN (m.p. 2300°C), and TaN (m.p. 3087°C). These are interstitial in nature and possess good electrical conductivity like the metals.

These metals form the carbides V_2C, Nb_2C, Ta_2C and VC. These are extremely hard materials and resistant to chemical reagents.

11. Formation of halides. All the three metals form pentafluorides. Niobium and tantalum also form pentahalides with chlorine, bromine and iodine whereas the highest halides of vandium with the remaining halogens are VCl_4, VBr_3 and VI_3. Att these halides are volatile covalent compounds and hydrolysed by water.

12. Action of acids. These elements donot form simple salts even with the strongest acids.

13. Formation of complexes. The tendency of these elements to form the complexes increases with the decrease of atomic size. Thus vanadium which has the smallest size forms the largest number of complexes while Ta gives the smallest number.

$$\underset{(M = V, Nb)}{MF_5} + 2F^- \longrightarrow [MaF_7]^{2-}$$

$$TaF_5 + 3F^- \longrightarrow [TaF_8]^{3-}$$

14. Colour of the compounds. Since the compounds with lower oxidation states have *incomplete d-orbitals,* these are *coloured,* while those with higher oxidation states which have *completely filled d-orbitals* are *colourless.*

Group VIB (6) : Cr_{24}, Mo_{42}, W_{74}

Group VIB of the periodic table consists of the three elements which are *Chromium* (Cr). *Molybdenum* (Mo) and *Tungsten* (W). These are called *chromium group elements.*

Electronic configurations

Electronic configurations of the atoms of these elements as given in Table 7.4 show that excepting W, the valence-shell configuration of the remaining two elements can be represented as $(n-1)d^5.ns^1$. In case of these two elements one electron from ns^2 orbital gets shifted to $(n-1)d$ orbitals and hence the observed configuration becomes $(n-1)d^5.ns^1$ instead of the expected configuration, $(n-1)d^4.ns^2$. The observed valence-shell configuration of W is the same as its expected configuration. The valence-shell configuration of all the three elements show that the extra electron goes to $(n-1)d$ orbital and hence these are *d*-block (transition) elements. It is also evident that these elements have six electrons in their valence-shell.

General characteristic of Cr-group elements

Some of the properties of the elements of Cr-group are given in Table 7.4. These properties have been discussed below:

Table 7.4. *Important physical properties of the elements of croup VIB (6). Here* $[Ar]_{18}$ = 2, 8, 8; $[Kr]_{36}$ = 2, 8, 18, 8 *and* $[Xe]_{54}$ = 2, 8, 18, 18, 8.

Property	*Chromium (Cr)*	*Molybdenum (Mo)*	*Tungsten (W)*
Atomic number	24	42	74
Electronic configuration	$[Ar]_{18}\ 3d^5\ 4s^1$ = 2, 8, 13, 1	$[Kr]_{36}\ 4d^5\ 5s^1$ = 2, 8, 18, 13, 1	$[Xe]_{54}\ 4f^{14}\ 5d^4\ 6s^2$ = 2, 8, 18, 32, 12, 2
Atomic weight	52.00	95.94	183.85
Atomic radius (pm)	127	139	139
Ionic radius (pm)	84 (+2) 69 (+3) 58 (+6)	83 (+3) 73 (+6)	74 (+6)
Atomic volume (ml)	7.23	9.4	9.3
Density (g/cm^3)	7.19	10.21	19.30
Melting point (°C)	1875	2610	3410
Boiling point (°C)	2665	5560	5930
Ionisation energy (kJ/mol)	652.8	685	770
Electronegativity	1.6	1.8	1.7
Stable oxidation state	+3	+6	+6
Heat of vaporisation (kcal/g.atom)	72.97	128	185
Heat of fusion (kcal/g.atom)	3.3	6.6	8.5

1. Physical state. All are hard, silvery white metals.

2. Density, melting points and boiling points. Their density, melting points and boiling point gradually increase with the increase as we move down the group.

3. Atomic radius and atomic volume. The atomic radius and the atomic volume gradually increase with the increase in atomic number. Mo and W possess nearly the same values due to lanthanide contraction.

4. Ionisation potential. As is evident from the above table the value of ionisation potential gradually increases as we move down the group. The inicrease is due to the addition of electrons in the penultimate *d*-orbital which causes tightening of the binding in the atom.

5. Electronegativity. All have nearly the same electronegativity.

6. Variable valency and coloured ions. All are transition elements with incomplete *d*-orbitals and hence exhibit variable valency. The maximum valency 6 of these elements corresponds to the number of the group in which they are placed. Due to variable valency they possess excellent catalytic character. Their ions are coloured because due to incomplete *d*-orbital transition of electron from one energy level to other is possible.

7. Paramagnetic character. All these elements are paramagnetic *i.e.*, attract magnetic lines of force. This is due to the presence of unpaired electrons in the *d*-orbital.

8. Oxides and oxy-acids. All form trioxides of the general formula MO_3. *e.g.*, CrO_3, MoO_3 and WO_3. They are acidic as shown by their ability to dissolve in alkali to give chromates (CrO_4^{2-}), molybdates (MoO_4^{2-}) and tungstates (WO_4^{2-}) respectively.

$$CrO_3 + 2OH^- \longrightarrow CrO_4^{2-} + H_2O$$

The acidic character of the trioxide decreases from CrO_3 to WO_3.

CrO_3	MoO_2	WO_3
(strongly acidic)	(acidic)	(amphoteric)

They form oxy-acids of the type H_2MO_4 *e.g*, H_2CrO_4 (chromic acid), H_2MoO_4 (molybdic acid) and H_2WO_4 (tungstic acid). The strength of these oxy-acids decreases from H_2CrO_4 (a very strong acid) to H_2WO_4 (a weak acid). Moreover, their oxidising property also decreases as we move down the group. Thus chromic acid is a powerful oxidising agent, molybdic acid only a weak oxidizing agent while tungstic acid is a stable acid and is devoid of oxidising character.

9. Acid chlorides. All form similar acid chlorides of the type MO_2Cl_2 *e.g.*, CrO_2Cl_2 (chromyl chloride), MoO_2Cl_2(molybdenyl chloride) and WO_2Cl_2 (tungstenyl chloride).

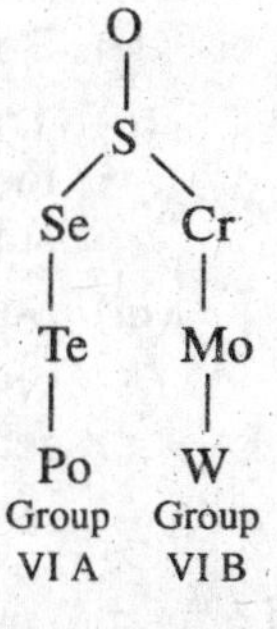

10. Carbonyls. With carbon monoxide, all these elements form carbonyls of the type $M(CO)_6$ *e.g.*, $Cr(CO)_6$, $Mo(CO)_6$ and $W(CO)_6$. These carbonyls are insoluble in water but soluble in organic solvents such as benzene. They sublime without decomposition under reduced pressure.

11. Combination with elements. All these elements combine with a large number of metallic and non-metallic elements. Thus they directly form oxides with oxygen, sulphides with sulphur and nitrides with nitrogen.

O
S
Se Cr
Te Mo
Po W
Group VI A Group VI B

12. Formation of alloys. They form excellent alloys. All these elements are used in making different varieties of steel.

Similarities between Cr and Mn

Since Mn (group VII B) is the right hand neighbour of Cr (VIB) (See Fig. 7.1), these two elements show many similar properties as shown below:

(*i*) Both are hard lustrous metals of high melting point and boiling point.

(*ii*) Both are transition elements and hence exhibit variable valency, form coloured salts, possess complex forming tendency and paramagnetic character.

(*iii*) Both exhibit valency of 2 and 3 and form oxides of the type MO (*ous*) and M_2O_3 (*ic*).

(*iv*) Both exhibit hexavalency in compounds such as trioxides (CrO_3 and MnO_3), chromates (*e.g.*, Na_2CrO_4), dichromates (*e.g.*, $K_2Cr_2O_7$) and manganates (*e.g.*, K_2MnO_4).

(*v*) The trioxides of both *i.e.*, CrO_3 and MoO_3 are acidic and react with water to form respectively chromic (H_2CrO_4) and manganic (H_2MnO_4) acids. These acids are analogous and form salts—chromates and manganates which are isomorphous and strong oxidising agents.

Similarities between Cr And Fe

Since Fe (group VIII) is also right hand neighbour of Cr (group VIB) (See Fig. 7.1), these two elements also show many similarities as shown below :

(*i*) Both are hard lustrous metals of high melting point and boiling point.

(*ii*) Both are transition elements and hence exhibit variable valency, form coloured salts, possess complex forming tendency and paramagnetic character.

(*iii*) Both exhibit valency of 2 and 3 and form oxides of the type MO (*ous*) and M_2O_3 (*ic*). The oxides CrO and FeO give rise to similar chromous and ferrous salts. Similarly chromic salts derived from Cr_2O_3 resemble ferric salts derived from Fe_2O_3. The stable salts are chromic and ferric.

(*iv*) Both become passive with concentrated nitric acid.

(*v*) Both from alums which are isomorphous *e.g.*,

$K_2SO_4.Cr_2(SO_4)_3.24H_2O$ .... *Chrome alum*

$K_2SO_4.Fe_2(SO_4)_3.24H_2O$ .... *Potassium iron alum*

(*vi*) The chromates (*e.g.*, K_2CrO_4) and ferrates (*e.g.*, K_2FeO_4) are isomorphous and possess strong oxidising properties.

(*vii*) Chromium forms complex cyanides like iron.

Similarities between Cr and S

Sulphur acts as a bridge element between the elements of sub-groups VIA (O, S, Se, Te, Po) and VIB (Cr, Mo, W) as shown in the margin and hence shows similarities with Cr. Some points of similarity are given below :

(*i*) Both exhibit hexavalency.

(*ii*) Trioxides of both (*i.e.*, CrO_3 and SO_3 respectively) are acidic and react with water to form chromic acid and sulphuric acid. The salts of these acids—chromates and sulphates are isomorphous.

(*iii*) Sulphates and chromates of both barium and leads are insouble in water.

(*iv*) Oxychlorides of both *i.e.*, CrO_2Cl_2 (chromyl chloride) and SO_2Cl_2 (sulphuryl chloride) react with water to give chromic acid and sulphuric acid.

$$CrO_2Cl_2 + 2H_2O \longrightarrow H_2CrO_4 + 2HCl$$

$$SO_2Cl_2 + 2H_2O \longrightarrow H_2SO_4 + 2HCl$$

Comparison between Cr and S

The comparison between Cr and S consists of making a discussion of the similarities and dissimilarities between Cr and S.

(*a*) **Similarities.** These have already been discussed above.

(*b*) **Dissimilarities.** (*i*) Sulphur is a *typical element* with *six* electrons in its outermost orbit ($3s^2p^4$) while chromium is a *transition element* with one electron in its outermost orbit ($3d^54s^1$).

(*ii*) Sulphur is a *typical non-metal* while chromium shows *metallic behaviour.* It is hard, lustrous and has high density, melting and boiling point.

(*iii*) Sulphur readily burns in air and forms SO_2 while chromium is not acted upon by air.

(*iv*) Sulphur dissolves in alkalies to form thiosulphate while chromium remains unaffected by alkalies.

(*v*) Dilute acids have no action on sulphur but chromium dissolves in them to form salts. Concentrated HNO_3 oxidises sulphur to form H_2SO_4 but chromium remains passive.

(vi) Sulphur combines with metals and non-metals to give sulphides. Chromium forms alloys with metals.

Position of Cr in the periodic table

A look at the position of Cr in the periodic table (See Fig. 7.1) shows that (1) Cr lies with Mo and W in the same sub-group (2) Mn and Fe lie to its right hand side (3) S is acting as a bridge element between the elements of sub-groups VIA and VIB. Thus the position of Cr in the periodic table is justified by discussing the similarities of Cr with the above said elements.

1. Similarities of Cr with Mo and W. The similarities of Cr with Mo and W have already been discussed under "*General characteristics of Cr group elements*". (See page 280)

2. Similarities betwen Cr rand Mn. See above.

3. Similarities between Cr and Fe. See above.

4. Similarities between Cr and S. See above.

Group VIIB (7) : Mn_{25}, Tc_{43}, Re_{74}

Group VIIB of the periodic table consists of three eements which are *Manganese* (Mn), *Technitium* (Tc) and *Rhenium* (Re). Of these the last two are rare metals. The elements of this group are called *manganese group elements*. Out of these metals Tc does not occur in nature and was the first man-made element. It is radioactive and its chemistry is not fully known.

Electronic configurations

The electronic configurations of the atoms of these elements as given in Table 7.5, show that the valence-shell configurations of these elements can be represented by $(n - 1)d^5.ns^2$ showing that the extra electron goes to $(n - 1)d$ orbital and hence these are *d*-block (transition elements.

General properties of Mn, Tc and Re

Important physical properties of these metals are given in Table 7.5. Some of the physical and chemical properties of these elements are discussed below:

1. Typical properties of transition elements. All these elements have typical properties of transition elements. For example:

(a) These elements exhibit variable oxidation states. In accordance with their electronic configuration, these elements, like halogens, exhibit a maximum oxidation state of +7 in compounds like $KMnO_4$, Mn_2O_7, Re_2O_7 etc. In addition to this oxidation state, these elements also show lower oxidation states. The stability of higher oxidation states increases gradually from Mn to Re, *e.g.* the permanganates, MnO_4^-, decompose readily in acidic, alkaline and neutral solutions evolving O_2 and thus act as strong oxidising agents, while pertechnetates, TcO_4^- and perrhenates, ReO_4^- are more stable and hence have less oxidising power. The stability of lower oxidation states decreases from Mn to Re. All these elements show *zero* oxidation states in carbonyls like $Mn_2(CO)_{10}$, $Tc_2(CO)_{10}$ and $Re(CO)_{10}$.

(b) All the elements have high melting and boiling points.

(c) They form coloured ions.

(d) They have a tendency to form complex ions.

(e) They have paramagnetic character.

Table 7.5. *Important physical properties of Mn, Tc and Re [Group VIIB (7)].Here $[Ar]_{18}$ = 2, 8, 8; $[Kr]_{36}$ = 2, 8, 18, 8 and $[Xe]_{54}$ = 2, 8, 18, 18, 8.*

Property	*Manganese (Mn)*	*Technetium (Tc)*	*Rhenium (Re)*
Atomic number	25	43	75
Electronic configuration	$[Ar]_{18}\ 3d^5\ 4s^2$ = 2, 8, 13, 2	$[Kr]_{36}\ 4d^5\ 5s^2$ = 2, 8, 18, 13, 2	$[Xe]_{54}\ 4f^{14}\ 5d^5\ 6s^2$ = 2, 8, 18, 32, 13, 2
Atomic weight	54.94	99	186.23
Atomic radius (pm)	126	136	137
Ionic radius (pm)	80 (+2) 66 (+3) 66 (+7)	78.5 (+4) 74 (+5) 70 (+7)	77 (74) 72 (+5) 67 (+7)
Atomic volume (ml)	7.39	—	8.85
Density (g/cm³)	7.43	11.50	21.00
Melting point (°C)	1245	2200	3180
Boiling point (°C)	2150	—	5900
Specific heat	0.1214	—	0.0327
Ionisation potential (kJ/mol)	717.4	702	760
Electronegativity	1.5	1.9	1.9
Stable oxidation state	+2	+4, +7	+3, +4, +7
Electrode potential (volts) $M^{2+}(aq) + 2e^- \rightleftharpoons M(s)$	1.18	—	—

2. Formation of oxides. Both manganese and rhenium form a number of oxides. Important oxides are:

MnO_2 MnO_3 Mn_2O_7

ReO_2 ReO_3 Re_2O_7

The most important oxide of technetium is the heptoxide, Tc_2O_7. The basic character of the oxides decreases gradually with the increase of oxidation state of the element. For example, the oxides *viz.* MnO and Mn_2O_3 (in which Mn is in +2 and +3 state respectiveiy) are *basic* and *ionic*, MnO_2 (in which Mn is in +4) is *amphoteric* and *does not exist as* Mn^{4+} ion, MnO_3 (in which Mn is in +6 state) is an *unstable acidic oxide* and Mn_2O_7 (in which Mn is in +7 state) is *strongly acidic*. Thus we see that *with the increase of oxidation state of* Mn *atom in its oxides, the character of these oxides changes from basic to amphoteric and finally to acidic.*

3. Formation of oxy-acids. All form oxy-acids of the type HRO_4, *e.g.*, $HMnO_4$ (permanganic acid), $HTcO_4$ (pertechnetic acid) and $HReO_4$ (perrhenic acid). The salts of these acid—permanganates, pertechnetates and perrhenates respectively are analoguus.

Similarities between Mn and Cr

Since Cr (group VIB) is the left hand neighbour of Mn (group VIIIB) (See Fig. 7.1), these two elements show many similarities with each other. These similarities have already been given in the discussion of group VIB elements (Cr,Mo,W).

Similarities between Mn and Fe

Fe (group VIII) is also right hand neighbour of Mn (group VIIIB) (See Fig. 7.1), these two elements have many similarities as shown below:

(*i*) Both are hard lustrous metals of high melting point and boiling point.

(*ii*) Both are transition elements and hence exhibit variable valency, form coloured salts, possess complex forming tendency etc.

(*iii*) Both exhibit valency 2 and 3 and form oxides of the type MO (*ous*) and M_2O_3 (*ic*). The oxide Mn_3O_4 is a mixed oxide ($MnO + Mn_2O_3$) like Fe_3O_4 ($FeO + Fe_2O_3$).

(*iv*) Manganous and ferrous salts (derived from manganous and ferrous oxides) form double salts with alkali metal salts *e.g.*, $MnSO_4.K_2SO_4.6H_2O$ is isomorphous with $FeSO_4$ $(NH_4)_2SO_4.6H_2O$. Similarly, manganic and ferric salts (derived from manganic and ferric oxides) form ismorphous double salts called alum, *e.g.*, $K_2SO_4.Mn_2(SO_4)_3.24H_2O$ (manganic alum) and $K_2SO_4.Fe_2(SO_4)_3.24H_2O$ (potassium iron alum).

(*v*) The manganates, MnO_4^{2-} and ferrates, FeO_4^{2-} are similar.

Similarities between Mn and Cl

Chlorine acts as a bridge element between the elements of sub-groups VIIA (F, Cl, Br, I, At) and VIIB (Mn, Tc, Re) as shown in the margin and hence shows similarities with Mn. Some points of similarity between Mn and Cl are given below. These similarities are very slight and are shown in their highest oxidation states.

F
|
Cl
Br — Mn
I — Tc
At — Re
Group VIIA — Group VIIB

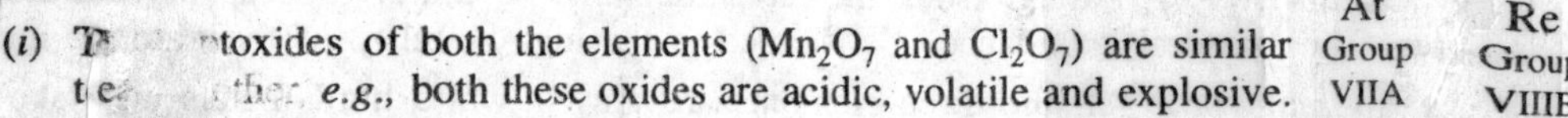

(*i*) The heptoxides of both the elements (Mn_2O_7 and Cl_2O_7) are similar to each other, *e.g.*, both these oxides are acidic, volatile and explosive.

(*ii*) Permanganates (MnO_4^-) and perchlorates (ClO_4^-) are isomorphous salts and have similar solubilities, *e.g.*, both $AgMnO_4$ and $AgClO_4$ are sparingly soluble in water.

Position of Mn in the periodic table

A look at the position of Mn in the periodic table as shown in Fig. 7.1 shows that (1) Mn lies with Tc and Re in the same sub-group (2) Cr lies to its left hand side (3) Fe lies to its right hand side (4) Cl is acting as a bridge element between the elements of sub-groups VIIA and VIIB. Thus the position of Mn in the periodic table is justified by discussing the similarities of Mn with the above said elements. These similarities are discussed below.

1. Similarities between Mn, Tc and Re. See "*General properties of Mn, Tc and Re* discussed in this chapter on page 283.

2. Similarities between Mn and Cr. See page 284.

3. Similarities between Mn and Fe. See page 284.

4. Similarities between Mn and Cl . See page 285.

Group VIII (8, 9, 10) : Fe_{26}, Co_{27}, Ni_{28}, Ru_{44}, Rh_{45}, Pd_{46}, Os_{76}, Ir_{77}, Pt_{78}

Group VIII of the periodic table consists of nine elements which are grouped into three triads namely : (*i*) First triad which consists of iron (Fe), cobalt (Co) and nickel (Ni) (*ii*) second triad, consisting of ruthenium (Rh), rhodium (Rh) and palladium (Pd) (*iii*) third triad, consisting of osmium (Os), iridium (Ir) and platinum (Pt). The elements of the first triad belong to the first transition series and are called *ferrous metals* or *iron family metals* while those of second and third triads are the members of the second and third transition series respectively and are collectively known as the *platinum metals*. Ru, Rh and Pd are called *light platinum metals*, while Os, Ir and Pt are known as *heavy platinum metals* on the basis of their density as compared to platinum.

The above discussion can be summarised as follows:

1st triad : Fe_{26}, Co_{27}, Ni_{28} .... *Ferrous metals*
2nd triad : Ru_{44}, Rh_{45}, Pd_{46} .... *Light platinum metals*
3rd triad : Os_{76}, Ir_{77}, Pt_{78} .... *Heavy platinum metals.*

Ferrous Metals or Iron Family Metals: Fe_{26}, Co_{27}, Ni_{28}

Electronic configurations

The electronic configurations of the atoms of these metals as given in Table 7.6 show that Fe, Co and Ni have *two* electrons in their outer-most shell (*i.e.*, *4th shell*) but their *kernels* are incomplete. As the atomic number increases from Fe to Ni, the number of electrons in the d-orbitals of the penultimate shell (*i.e.*, *3rd shell*) increases by one in each succeeding elements.

General properties of ferrous metals and gradation in them: Horizontal relationship between ferrous metals

A. Physical properties. Important physical properties of ferrous metals are given in Table 7.6 and have been discussed below.

1. Occurrence. All the ferrous metals particularly Fe and Ni occur in nature in large quantities. Their natural abundance is attributed to the high stability of their atomic nuclei.

2. Physical state. All the three elements are greyish-white hard metals and take a high polish. Their hardness increaes as Fe < Co < Ni. All the three elements are very tenacious and ductile metals.

3. Atomic and ionic radii. Their atomic and ionic radii of all the three metals are very similar to each other.

4. Density. All the elements have high values of density which varies as Fe (= 7.86) < Co (= 8.90) = Ni (= 8.90).

5. Ionisation energies. Their ionisation energies are about the same. These elements have, therefore, the same tendency to form ions.

6. Melting point. These have fairly close melting points which are uniformly high. Very similar melting points seem to indicate that the average number of electrons per atom contributing to the metallic bonding is nearly the same for each of them.

7. Standard reduction potentials. Standard reduction potential values for Co and Ni corresponding to the reduction reaction,

$$M^{2+}\ (aq) + 2e^- \rightleftharpoons M(s)$$

are very close to each other and are negative. Negative values indicate that these metals can liberate H_2 from acids.

8. Magnetic property. All the three metals are *ferromagnetic, i.e.*, these are strongly attracted by a magnet and remain magnetised even after the removal of the magnetic field. The ferromagnetic property *decreases* from Fe to Ni. At elevated temperatures called their Curie points, the metals lose ferromagnetic property and become paramagnetic.

9. Catalytic properties. All the three metals occlude hydrogen, nickel being superior to other two in this respect. As a result of this property, these metals are used as catalysts in certain reactions involving the addition of hydrogen.

10. Oxidation states. Fe shows both +2 (in *ferrous salts*) and +3 (in *ferric salts*) oxidation states, cobalt mostly exhibits +2 oxidation state but also forms alums and complexes [*e.g.*, $CoCl_6^{3-}$, $Co(CN)_6^{3-}$] in which it shows +3 state while nickel usually shows +2 state. In certain complexes nickel also shows +1 state, *e.g.*, $K_2[Ni(CN)_3]$, NiCN etc. Nickel does not form

Table 7.6. *Important physical properties of ferrous metals: Fe, Co, Ni [Group VIII (8, 9, 10)]. Here* $[Ar]_{18}$ = 2, 8, 8.

Property	*Iron (Fe)*	*Cobalt (Co)*	*Nickel (Ni)*
Atomic number	26	27	28
Atomic weight	55.85	58.93	58.71
Electronic configuration	$[Ar]_{18}\ 3d^6\ 4s^2$ = 2, 8, 14, 2	$[Ar]_{18}\ 3d^7\ 4s^2$ = 2, 8, 15, 2	$[Ar]_{18}\ 3d^8\ 4s^2$ = 2, 8, 16, 2
Atomic radius (pm)	126	125	128
Ionic radii (pm), M^{2+}, M^{3+}	76, 64	74, 63	72, —
Atomic volume (ml)	7.10	6.70	6.60
Density (g/cm^3)	7.86	8.90	8.90
Melting point (°C)	1536	1495	1453
Boiling point (°C)	3000	2900	2730
Ionisation potential ($kJ\ mol^{-1}$)	759.4	758	736.6
Standard reduction potential (volts) $M^{2+}(aq) + 2e^- \rightleftharpoons M(s)$	– 0.44	– 0.28	– 0.25
Common oxidation states	+2, +3	+2, +3	+2
Specific heat	0.10983	0.103	0.10842
Magnetic property	Ferromagnetic Curie point = 766°C	Ferromagnetic Curie point = 1100°C	Ferromagnetic Curie point = 360°C
Allotropy	Exists in two solid phases	Exists in two solid phases	Exists in two solid phases

Ni^{3+} ion. The removal of a 3*d*-electron from M^{2+} cations to oxidise them into M^{3+} cations becomes more and more difficult as we proceed from Fe^{2+} to Ni^{2+} *i.e.*, the stability of M^{2+} cations is in the order : $Fe^{2+} < Co^{2+} < Ni^{2+}$ and hence the reducing properties of these ions is in the order: $Fe^{2+} > Co^{2+} > Ni^{2+}$. The increase in stability from Fe^{2+} to Ni^{2+} ions is evident from the behaviour of their M(II) oxide on heating, *e.g.*, FeO is oxidised to Fe_2O_3 simply on heating, CoO changes to Co_3O_4 but does not undergo further oxidation to Co_2O_3 while NiO is stable. *i.e.*, it remains unaffected. The tendency of these elements to show +3, state *decreases* sharply from Fe to Ni, *i.e.*, the stability of M^{3+} cations is in the order : $Fe^{3+} > Co^{3+} > Ni^{3+}$, indicating that the oxidising properties of Fe^{3+}, Co^{3+} and Ni^{3+} cations are in the order: $Fe^{3+} < Co^{3+} < Ni^{3+}$. The reason for this difference in the +3 state is that, while Fe^{3+} ion has a half-filled 3*d*-orbital ($Fe^{3+} = 3d^5 4s^0$) and therefore stable, Co^{3+} and Ni^{3+} have relatively unstable $3d^6$ and $3d^7$ configurations respectively ($Co^{3+} = 3d^6 4s^0$, $Ni^{3+} = 3d^7 4s^0$). The +4 state is relatively uncommon for these elements. While +6 oxidation state is reported for iron in ferrate ion, $[FeO_4]^{2-}$ which is unstable and oxidising agent, this state has not been reported for Co and Ni. Lower oxidation states such as *zero* are found in complexes having π-bonding ligands such as CO. Thus in the carbonyls, $Fe(CO)_5$, $Co_2(CO)_8$ and $Ni(CO)_4$ these metals show zero oxidation state.

B. Chemical properties. Important chemical properties of ferrous metals are discussed below.

1. Reactivity. All these metals are reactive and their reactivity decreases as Fe > Co > Ni.

2. Action of acids. Since the standard reduction potentials of these metals are negative, they can liberate H_2 from acids like HCl and H_2SO_4. Fe reacts readily while Co and Ni react only slowly.

3. Action of alkalies. None of these metals reacts with alkalies.

4. Formation of halides. (*a*) All the three metals form dihalides, *e.g.*, $FeCl_2$, $CoCl_2$, $NiCl_2$.

(*b*) All the trihalides of Fe are known but Co gives only CoF_3. Ni does not form any trihalide.

5. Action of strong oxidising agents: Passivity. Strong oxidising agents like conc. HNO_3 and dichromates render these metals passive due to the formation of a protective film of their oxide on their surface. This tendency increases as : Fe < Co < Ni.

6. Formation of complex ions. All these elements form stable complex ions. More common complex ions are those in which the metal is in +2 or +3 oxidation state. For example:

Oxidation state	*Complex ions*
+2	$[Fe(CN)_6]^{4-}$, $[Fe(H_2O)_6]^{2+}$, $[Co(CN)_6]^{4-}$, $[Co(H_2O)_6]^{2+}$ $[Ni(CN)_6]^{4-}$, $[Ni(H_2O)_6]^{2+}$
+3	$[Fe(CN)_6]^{3-}$, $[Fe(H_2O\}_6]^{3+}$, $[Co(CN)_6]^{3-}$,$[Co(H_2O)_6]^{3+}$ $[Ni(H_2O)_6]^{2+}$.

All these complex ions have octahedral geometry.

7. Formation of sulphates. All the three elements from *isomorphous vitriol type sulphates* $MSO_4.7H_2O$, where M = Fe(II), Co(II) or Ni(II). For example $FeSO_4.7H_2O$ is isomorphous with $CoSO_4.7H_2O$. All these sulphates form *schonite type double sulphates,* $M_2SO_4.MSO_4.6H_2O$. Of these Mohr's salt, $(NH_4)_2SO_4.FeSO_4.6H_2O$ is very important. These schonites are also isomorphous with each other, *e.g.*, Mohr's salt is isomorphous with nickel ammonium sulphate, $(NH_4)_2SO_4.NiSO_4.6H_2O$. Since Fe(III) and Co(III) [but not Ni(III)] are known as aquo ions, these form alum type sulphates: $M_2SO_4.M_2'(SO_4)_3.24H_2O$, M = Na, K, NH_4^+ etc. and M' = Fe(III) or Co(III).

8. Nature of sulphides. The sulphides of all the three metals are black and soluble in HCl.

9. Formation of oxides. These metals form the following oxides:

Oxidation state of the metal	*Oxides*		
+2	FeO	CoO	NiO
+3	Fe_2O_3	Co_2O_3	Ni_2O_3
+2, +3	Fe_3O_4 or $FeO.Fe_2O_3$	Co_3O_4 or $CoO.Co_2O_3$	—

Thermal stability of FeO, CoO and NiO increases as FeO < CoO < NiO.

10. Formation of carbides. Only Fe and Co form carbide of M_3C type (M = Fe, Co) Fe_3C is called *cementite*. NiC is much less stable than Fe_3C. Ni does not form the carbide, Ni_3C. All the three metals are present in *cohenite* mineral (FeCoNiC) which is a carbide.

11. Formation of carbonyls. All the three metals form carbonyls, *e.g.*, $Fe(CO)_5$, $Fe_2(CO)_9$, $Fe_3(Co)_{12}$, $Co_2(Co)_8$, $Ni(CO)_4$.

Important differences between ferrous metals

(*i*) Fe is present in the 3rd group of qualitative analysis while Co and Ni are in 4th group.

(*ii*) Fe and Co exhibit a coordinate number of 6 only while Ni has coordination number equal to 4 and 6.

(*iii*) Ni generally does not form compound in +3 oxidation state.

(*iv*) Fe does not form complex ammine like $[Fe(NH_3)_6]^{3+}$ while Co and Ni both can form the ammines like $[Co(NH_3)_6]^{3+}$ (*blue*) and $[Ni(NH_3)_4]^{2+}$ (*blue*).

Silarities between Fe and Mn

Since Mn (group VIIB) is the left hand neighbour of Fe (group VIII), these two elements show many similarities with each other. These similarities have already been discussed in group VIIB (Mn, Tc,Re) in this chapter on page 284.

Similarities between Fe and Cr

Since Cr (group VIB) is also left hand neighbour of Fe (group VIII), these two elements have many similarities with each other.The points showing similarities between Fe and Cr have already been discussed in group VIB (Cr, Mo, W) in this chapter on page 282.

Similarities between Ni and Cu

Since Cu (group IB) is the right hand neighbour of Ni (group VIII) (See Fig. 7.1), these two elements show many similar properties. Some of these properties are given below :

(*i*) Both exhibit a co-ordination number of 4.

(*ii*) Their ions have similar ionic radii.

(*iii*) Their hydrated salts in solution react with acid due to hydrolysis.

(*iv*) Basic carbonates with the formula $XCO_3.X(OH)_2$ are precipitated on treating their salt solutions with sodium carbonate.

(*v*) Both of them show catalytic properties, *e.g.*,

$$2CH_3OH + O_2 \xrightarrow[500°C]{Cu} 2HCHO + 2H_2O$$

$$CH_2{=}CH_2 + H_2 \xrightarrow[140°C]{Ni} CH_3{-}CH_3$$

(*vi*) Both exhibit bivalency.

(*vii*) Both form coloured salts. Copper salts are bluish-green while the salts of nickel are green.

(*viii*) Both form two series of salts—cuprous and cupric; nickelous and nickelic.

(*ix*) Both form black sulphides.

(*x*) Both form complex compounds, *e.g.*

$[Cu(NH_3)_4]SO_4$ $[Zn(NH_3)_4]SO_4$

$K_3[Cu(CN)_4]$ $K_2[Ni(CN)_4]$

(*xi*) Both form oxides on heating their carbonates or nitrates.

(*xii*) Both form double salts of the following composition.

$MSO_4.(NH_4)_2SO_4.6H_2O$ where M = Cu or Ni

Platinum Metals: Ru_{44}, Rh_{45}, Pd_{46}, Os_{76}, Ir_{77}, Pt_{78}

Platinum metals occur together in nature and hence show many similar properties. Some of their physical properties are given in Table 7.7 and discussed below.

1. Electronic configurations. Pd has completely-filled $(n-1)\,d$ orbitals while other elements have partially-filled $(n-1)d$ orbitals.

2. Atomic and ionic radii. It is due to almost identical values of atomic and ionic radii tnat the six elements have similar properties.

3. Physical states. All are rare metals and are grey-white (Ru, Os, Pd, Pt) or silver-white (Rh, Ir). They are hard and inert towards acids. Hardness and ductility increases with the increase of atomic number in each triad.

4. Melting and boiling points. All the platinum metals have high melting and boiling points and high heat of vaporisation. This indicates that metallic lattice is very stable which is responsible for the inert nature of these elements.

5. Magnetic property. All the six metals are paramagnetic.

6. All the metals form coloured compounds and the cations are coloured.

7. Catalytic property. All the six metals have a remarkable property of occluding hydrogen and they can also activate the gas, and hence they possess important catalytic properties.

8. Specific gravity. It is interesting to note that the three light platinum metals (Ru, Rh, Pd) have almost the same specific gravity of about 12 and the three heavy platinum metals (Os, Ir, Pt) have nearly the same specific gravity of about 22.

9. Affinity for oxygen. The metals are difficult to burn in oxygen. The affinity for oxygen decreases in general from left to right. They react slowly at very high temperatures to form oxides of the type MO, M_2O_3, MO_2,MO_3 and MO_4. The oxides are sparingly soluble in water and dilute acids. The mono, sesqui and di-oxides are weekly basic, whereas, the tri-oxides are acidic and tetra-oxides are neutural.

10. Hydroxides. The metallic hydroxides in some cases are weakly basic, in others weakly acidic, and in some cases even amphoteric.

11. Sulphides. The platinum metals have a high affinity for sulphur. They form simple sulphides and complex compounds with organic sulphides as well.

12. Action with acids. Even strong acids hardly dissolve the metals. Powdered osmium and palladium react with hot and concentrated nitric acid to form OsO_4 and $Pd(NO_3)_2$ respectively. It is difficult to dissolve Ru, Rh and Os even in aqua-regia, these, however, dissolve after being alloyed with other metals. Pd and Pt dissolve in aqua-regia on digestion to yield H_2PdCl_6 and H_2PtCl_6 respectively.

13. Formation of halides: Powdered metals react with halogens forming halides. The reactivity of halogens decreases from fluorine to iodine.Iridium combines with fluorine readily while other metals combine with the halogens on heating.

14. Action of alkalies. All the platinum metals are attacked by fused alkalies in the presence of the oxidising agents, *e.g.*, KNO_2 or $KClO_4$.

15. Oxidation states. The oxidation state of higher and lower are most important in the chemistry of platinum metals. The highest oxidation state is not of much importance in these elements. Ruthenium and osmium alone display the maximum valency of +8 in forming their tetroxides *viz.* RuO_4 and OsO_4. Thus the tendancy to show higher oxidation state decreases from left to right.

16. Complex formation All the platinum metals have a characteristic property of passing from one valency state to the other and form a number of complex compounds like complex amines, complex cyanides, nitrosyls and carbonyls.

17. Formation of hydrated cations. Hydrated cations are rarely formed by light platinum metals and never by heavy platinum metals.

Similarities between Fe, Ru and Os : Vertical relationship between Fe, Ru and Os

(*i*) All the three metals form isomorphous complex cyanides.

(*ii*) Ferrites (FeO_3^{2-}) and ruthenites (RuO_3^{2-}) are similar.

(*iii*) Ferrates (FeO_4^{2-}), ruthenates (RuO_4^{2-}) and osmates (OsO_4^{2-}) are similar.

Similarities between Co, Rh and Ir : Vertical relationship between Co, Rh and Ir

(*i*) All the three metals form similar complex cyanides, ammines and nitrites.

(*ii*) The sulphates, $M_2(SO_4)_3$ [M = Co(III), Rh (III), Ir(III)] form true alums.

Similarities between Ni, Pd and Pt : Vertical relationship between Ni, Pd and Pt

(*i*) They form similar complex ammines, cyanides and nitrites.

(*ii*) They are relatively inert in their respective triads.

Table 7.7. *Important physical properties of paltinum metals : Ru, Rh, Pd, Os, Ir, Pt [Group VIII (8, 9, 10)]. Here $[Kr]_{36}$ = 2, 8, 18, 8 and $[Xe]_{54}$ = 2, 8, 18, 18, 8.*

Property	*Ruthenium (Ru)*	*Rhodium (Rh)*	*Palladium (Pd)*	*Osmium (Os)*	*Iridium (Ir)*	*Platinum (Pt)*
Atomic number	44	45	46	76	77	78
Atomic weight	101.1	102.9	106.4	109.2	192.2	195.0
Electronic configuration	$[Kr]_{36}\ 4d^7\ 5s^1$ = 2, 8, 18, 15, 1	$[Kr]_{36}\ 4d^8\ 5s^1$ = 2, 8, 18, 16, 1	$[Kr]_{36}\ 4d^{10}\ 5s^0$ = 2, 8, 18, 18, 0	$[Xe]_{54}\ 4f^{14}\ 5d^6\ 6s^2$ = 2, 8, 18, 32, 14, 2	$[Xe]_{54}\ 4f^{14}\ 5d^7\ 6s^2$ = 2, 8, 18, 32, 15, 2	$[Xe]_{54}\ 4f^{14}\ 5d^9\ 6s^1$ = 2, 8, 18, 32, 17, 1
Atomic volume (ml)	8.3	8.2	8.9	8.43	8.54	9.1
Density (g/cm^3)	12.20	12.40	12.0	22.6	22.5	22.4
Atomic radius (pm)	134	134	137	135	136	138
Ionic radius (pm)	82 (+3) 76 (+4) 70.5 (+5)	80.5 (+3) 74 (+4) 69 (+5)	100 (+2) 90 (+3) 75.5 (+7)	77 (+4) 71.5 (+5)	82 (+3) 76.5 (+4) 71 (+5)	94 (+2) 76.5 (+4) 71 (+5)
Melting point (°C)	2500	1966	1552	2700	2454	1769
Boiling point (°C)	4900	4500	3980	5500	5300	4530
Ionisation potential (kJ/mol)	711	720	805	840	880	870
Standard reduction potential (volts) $M^{2+}(aq) + 2e^- \rightarrow M(s)$	+0.5	+0.6	+ 1.2	+ 0.9	+ 1.0	+ 1.2
Stable oxidation states	+3, +4	+3	+2	+3, +4	+3, +4	+2, +4
Electronegativity	2.2	2.2	2.2	2.2	2.2	2.2
Coordination number	4,6	6	4, 6	4, 6	6	4, 6
Heat of vaporisation or sublimation (kcal/mol)	144	133	91	165	155	135
Heat of fusion (kcal/mol)	6.1	5.2	4.0	6.4	6.6	5.2

Similarities between Pd and Ag

Since Ag (IB) is the right hand neighbour of Pd (group VIII) (See Fig. 7.1), these two elements show many similarities with each other. Some similar properties are given below:

(*i*) Both are white in colour.

(*ii*) Both are noble metals.

(*iii*) Both act as catalyst in their finely divided state.

(*iv*) Both form complex compounds.

Similarities between Pt and Au

Since Au (group IB) is the right hand neighbour of Pt (group VIII) (See Fig. 7.1), these two elements are similar in many properties as shown below:

(*i*) Both occur in the native state in nature.

(*ii*) Both are not affected by air or oxygen.

(*iii*) Both are used in jewellery.

(*iv*) Both are highly malleable and ductile metals possessing bright metallic lustre. Both have high specific gravity and metling point.

(*v*) Both exhibit variable valency. Oxidation states of gold are +1 and +3 while of platinum are +2 and +4.

(*vi*) Both are noble metals, unreacted by acids but are dissolved in aqua regia.

(*vii*) Both the metals are obtained in the colloidal form by Bredig's arc method or by the reduction of their chlorides.

(*viii*) Both are easily attacked by fused cyanides and nitrates.

(*ix*) Both form —*ous* and —*ic* oxides and halides *e.g.*,

aurous oxide, Au_2O	and	auric oxide, Au_2O_3
platinous oxide, PtO	and	platinic oxide, PtO_2
aurous chloride, $AuCl$	and	auric chloride, $AuCl_3$
platinous chloride, $PtCl_2$	and	platinic chloride,$PtCl_4$

(*x*) Both form complex halogen acids *e.g.*, $HAuCl_4$ (chloro auric acid) and H_2PtCl_4 (chloro platinic acid). Both these acids form double salts with amines.

(*xi*) Both form complex cyanides *e.g.*, $K[Au(CN)_2]$ and $K_2[Pt(CN)_6]$

Comparison between iron family metals (Fe, Co, Ni) and platinum metals (Ru, Rh, Pd, Os, Ir, Pt)

1. The platinum metals are much more noble than the metals of the iron family and are little affected by acids.
2. The reactivity of the metals of the iron family decreases from iron to nickel but in platinum metals the reactivity increases from Ru to Pd and from Os to Pt.
3. The halogens react with platinum metals only at higher temperatures and bring out the higher valencies *e.g.*, OsF_6, IrF_6 and PtF_4.
4. The lower valency states in the metals of iron family exist as simple ions whereas in platinum metals they are unstable (except in complexes) and few simple ions exist.
5. The triads Ru, Rh, Pd and Os, Ir, Pt are similar because of lanthanide contradiction.
6. The platinum metals are rare while those of iron family are relatively abundant.
7. All the metals are transition metals. They show variable valency, are paramagnetic and produce coloured ions. They form complex compounds like ammines and cyanides. They also show catalytic properties.

How to justify the inclusion of ferrous metals (Fe, Co, Ni) in the same group (group VIII)

We know that ferrous metals have been placed in group VIII. The inclusion of these metals in this group is justified by discussing the following similarities.

1. Similarities between Fe, Co and Ni. Since all these three metals are lying in the same row (4th period) of the periodic table, these show similarities with each other. The similarities between Fe, Co and Ni have already been discussed under "*General properties of ferrous metals and gradation in them: Horizontal relationship between ferrous metals*" given in this chapter on page 286.

2. Similarities between Fe and Cr. Cr (group VIB) is the left hand neighbour of Fe (group VIII) and hence these elements show many similarities. These similarities have already been given on page 282.

3. Similarities between Fe and Mn. Mn (group VIIB) is also left hand neighbour of Fe (group VIII) and hence these two elements show many similar properties with each other. These properties have already been discussed on page 284.

4. Similarities between Ni and Cu. Cu (group IB) is the right hand neighbour of Ni and hence shows similarities with it. The similarities between these two elements have already been given in this chapter on page 289.

5. Similarities of Fe with Ru and Os. Since all these metals are lying in the same column (8th column) of the periodic table, Fe shows many similarities with Ru and Os. These similarities have already been discussed under "*Vertical relationship between Fe, Ru and Os*" given in this chapter on page 290.

6. Similarities of Co with Rh and Ir. Since all the three metals are lying in the same column (9th column) of the periodic table, Co shows similarities with Ru and Ir. For these similarities see "*Vertical relationship between Co, Rh and Ir*" discussed in this chapter on page 290.

7. Similarities of Ni with Pd and Pt. Since all these elements are present in the same column (10th column) of the periodic table, Ni shows similarities with Pd and Pt. For these similarities see "*Vertical relationship between Ni, Pd and Pt*" discussed in this chapter on page 290.

Position of cobalt in the periodic table

Cobalt is present in group VIII of the periodic table. To its left and right are lying Fe and Ni respectively (See Fig. 7.1) which belong to the same group and period. Below it are lying Rh and Ir (See Fig. 7.1). Thus, the position of Co in the periodic table is justified by discussing the following similarities.

1. Similarities between Fe, Co and Ni. The similarities between Fe, Co and Ni have already been discussed under "*Ferrous metals*" in this chapter on page 286.

2. Similarities between Co, Rh and Ir. The similarities between Co, Rh and Ir have already been discussed in this chpater under "*Vertical relationship between Co, Rh and Ir*" on page 290.

Position of nickel in the periodic table

Nickel is present in group VIII of the periodic table. To its left are lying Fe and Co (same group) while to its right is lying Cu (group IB) (See Fig. 7.1). Below it are Pd and Pt which are in the same sub-group. Thus, the position of nickel in the periodic table is justified by discussing the following similarities.

1. Similarities between Fe, Co and Ni. These similarities have already been discussed in this chapter on page 286 under "*Ferrous metals*".

2. Similarities between Ni, Pd and Pt. These have been discussed in this chapter under "*Vertical relationship between Ni, Pd and Pt*" on page 290.

3. Similarities between Ni and Cu. The similarities between these elements have been discussed in this very chapter on page 289.

Position of platinum in the periodic table

We know that Pt is present in group VIII of the periodic Table. To its right is lying Au (Group IB) and above it are lying Ni and Pd (same group) (See Fig. 7.1). Thus the position of Pt in the periodic table is justified by discussing the following similarities.

1. Similarities between Pt and Au. The similarities between these two elements have been discussed in this chapter on page 292.

2. Similarities between Ni, Pd and Pt. These similarities have been discussed in this chapter under " *Vertical relationship between Ni, Pd and Pt*" on page 290.

Group IB (11) : Cu_{29}, Ag_{47}, Au_{79}

Group IB of the long form of periodic table consists of three elements which are: *Copper* (Cu), *Silver* (Ag) and *Gold* (Au). These are collectively known as *coinage metals* or *currency metals* because they have been used in the past for the manufacture of coins for currency. However, now their main uses are quite different. These are also called *copper group metals.*

Electronic configurations

The electronic configurations of the atoms of coinage metals as given in Table 7.8 show that the ultimate shell of these elements has *one* electron (s^1 electron) while the penultimate shell contains 18 ($s^2\ p^6\ d^{10}$) electrons.

Similarities between Cu, Ag and Au

Some of the similarities existing between Cu, Ag and Au are given below.

1. Occurrence. On account of their inert character, all these metals are found in the native state in nature.

2. Physical state. All are hard, malleable and ductile metals. The malleability and ductility increase from copper to gold. They possess characteristic metallic lustre and take high polish. They have high density, melting point and boiling point while low atomic volume.

3. Conductivity. All are good conductors of heat and electricity. Silver is the best conductor of heat and electricity.

4. Electronic configuration. All have the same electronic configuration (Table 7.8).

5. Electropositive character. They are weakly electropositive and hence almost unreactive.

6. Variable valency. All these metals show variable valency. The oxidation states of all these elements are as follows :

For copper + 1 and + 2
For silver + 1 and + 2
For gold + 1 and + 3

The +1 oxidation state is very important in silver and not so important in copper and gold. The +2 oxidation state is most important in copper. This state occurs only in a few compounds of silver *e.g.*, argentic fluoride, AgF_2, silver peroxide, Ag_2O_2 etc. as well as a certain complexes with organic ligands like pyridine etc. *e.g.*, tetrapyridine argentic persulphate $[Ag.4py]S_2O_8$, pyridine argentic nitrate $[Ag.4py](NO_3)_2$ etc. while it is unknown in gold. The +3 state is the common, stable and important state of gold and is not at all important in copper and silver.

In aqueous solution, the number of oxidation states is further restricted. In aqueous solution copper exists only in +2 state as Cu^{2+} ion, silver only in +1 state as Ag^+ ion while gold only in +3 state as Au^{3+} ion. The univalent Cu^+ and Au^+ ions decompose almost completely in their aqueous solutions as follows :

$$2Cu^+ \rightleftharpoons Cu^{2+} + Cu^0$$

$$3Au^+ \rightleftharpoons Au^{3+} + 2Au^0$$

Therefore, a readily soluble curprous salt *e.g.*, Cu_2SO_4, $CuNO_3$ etc. is unstable in solution and decomposes into the corresponding cupric salt and metallic copper. Soluble aurous salts also behave similarly.

$$Cu_2SO_4 \longrightarrow CuSO_4 + Cu$$
$$3AuCl \longrightarrow AuCl_3 + 2Au$$

Hence, the only stable cuprous and aurous compounds are those which are insoluble in water or are present in complexes.

7. Displacement of the hydrogen from acids. The coinage metals do not displace hydrogen from acids because they are less electropositive than hydrogen.

8. Displacement of the metals from their salt solutions. Zinc, iron and magnesium displace coinage metals from their salt solutions.

9. Nature of monovalent salts. The monovalent salts of coinage metals are generally colourless and insoluble in water.

10. Complex salts. Being the members of the transition series, the coinage metals readily form complex salts in which the metal forms a part of the complex cation or anion *e.g.*,

$[Cu(NH_3)_4]^{2+}$	$[Ag(NH_3)_2]^+$	$[AuCl_4]^-$
Tetrammine cupric ion	Diammine silver ion	Auric chloride ion
$[Cu(CN)_4]^{3-}$	$[Ag(CN)_2]^-$	$[Au(CN)_2]^-$
Cupro cyanide ion	Argento cyanide ion	Auro cyanide ion

Table 7.8. *Important physical properties of copper group elements: Cu, Ag, Au [Group IB (11)]. Here* $[Ar]_{18}$ = 2, 8, 8; $[Kr]_{36}$ = 2, 8, 18, 8 *and* $[Xe]_{54}$ = 2, 8, 18, 18, 8.

Property	*Copper (Cu)*	*Silver (Ag)*	*Gold (Au)*
Atomic number	29	47	79
Atomic weight	63.54	107.87	196.97
Electronic configuration	$[Ar]_{18}\ 3d^{10}\ 4s^1$ = 2, 8, 18, 1	$[Kr]_{36}\ 4d^{10}\ 5s^1$ = 2, 8, 18, 18, 1	$[Xe]_{54}\ 4f^{14}\ 5d^{10}\ 6s^1$ = 2, 8, 18, 32, 18, 1
Atomic radius (pm)	128	144	144
Ionic radius (pm)	91 (+1) 69 (+2)	129 (+1)	51 (+1) 99 (+3)
Atomic volume (ml)	7.1	10.4	10.22
Density (g/cm^3)	8.96	10.51	19.30
Melting point (°C)	1083	960.8	1063
Boiling point (°C)	2595	2210	2920
Specific heat	0.092	0.056	0.031
Ionisation energy (kJ/mol)	745.5	731	890.1
Heat of sublimation (kcal/g.atom)	81.1	68.4	87.3
Heat of fusion (kcal/g.atom)	3.11	2.7	3.03

Gradation in properties

All the three elements show a regular gradation in their physical and chemical properties as discused below :

1. Atomic radius, density, malleability and ductility increase while specific heat, ionization potential decrease as we pass from copper to gold.

   [**Note**: Melting point, boiling point, heat of sublimation and heat of fusion do not show any order].

2. **Nobility.** This property increases as the atomic weight increases *e.g.*, copper is acted upon by all dilute and concentrated mineral acids; silver is acted upon by dilute and concentrated nitric acid hot concentrated sulphuric acid while gold is dissolved by aqua regia or nascent chlorine.

3. **Number of salts.** This decreases as we pass from copper to gold, *e.g.*, copper forms a large number of salts, the number is much less for silver while gold salts are very few.

4. **Affinity for oxygen.** They have a weak affinity for oxygen and it decreases as the atomic weight increases. Copper combines with oxygen on heating to form cupric oxide; the latter on strong heating gives cuprous oxide. Silver and gold are not oxidised even when heated to redness in air. Molten silver dissolves oxygen and rejects some of it on cooling. This reaction is often violent enough to spurt some of the metal slightly away from the main bulk, hence it is termed as **spitting of silver.**

5. **Reduction of their salt solutions.** The ease of reduction of their salt solutions increases from copper to gold. Ferrous sulphate readily precipitates gold from its salts but less readily silver while copper is not precipitated from its salts.

$$AuCl_3 + 3FeSO_4 \longrightarrow Au + Fe_2(SO_4)_3 + FeCl_3$$

   Glucose reduces alkaline cupric salt solutions to cuprous oxide but precipitates silver and gold from their salt solutions.

Similarities between the elements of group IB (Cu, Ag, Au : coinage metals) with those of group IIB (Zn, Cd, Hg)

Since the elements of group IIB (Zn, Cd, Hg) are the right hand neighbours of the elements of group IB (Cu, Ag, Au) as shown in Fig. 7.1, Cu shows similarities with Zn, Ag with Cd and Au with Hg. The similarities between these pairs of elements (Cu—Zn, Ag—Zn and Au—Hg pairs) have been shown below.

A. Similarities between Cu and Zn

(*i*) Both exhibit bivalency.

(*ii*) Sulphides, phosphates and carbonates of both are insoluble in water.

(*iii*) Chlorides and sulphates of both are soluble in water.

(*iv*) Both form stable oxides on heating their carbonates or nitrates.

(*v*) Both form complex compounds. For example:

$[Cu(NH_3)_4]SO_4$; $[Zn(NH_3)_4]SO_4$, $K_2[Cu(CN)_4]$, $K_2[Zn(CN)_4]$

(*vi*) Both form double salts of the following composition.

$MSO_4.(NH_4)_2SO_4.6H_2O$ where M = Cu or Zn

B. Similarities between Ag and Cd

(*i*) Both are white metals.

(*ii*) Both form complex compounds. For example:

$[Ag(NH_3)_2]Cl$, $[Cd(NH_3)_4]SO_4$, $K[Ag(CN)_2]$, $K_2[Cd(CN)_4]$

(*iii*) The oxides of both are coloured and are easily reduced when heated with carbon.

C. Similarities between Au and Hg

(*i*) Both are not attacked by air, water and alkalies.

(*ii*) Both form two series of salts—aurous and auric; mercurous and mercuric.

(*iii*) The oxides of both are coloured and are easily reduced when heated with carbon.

Similarities between the elements of group IB (Cu, Ag, Au) with those of group VIII (Ni, Pd, Pt)

Since the elements of group VIII *viz.* Ni, Pd and Pt are the left hand neighbours of the elements of group IB (Cu, Ag, Au) (See Fig. 7.1), Cu shows similarities with Ni, Ag with Pd and Au with Pt. The similarities between Cu and Ni, Ag and Pd and Au and Pt have already been discussed on pages 182, 185 and 185 respectively.

Similarities between Ag and alkali metals

In the following properties, silver not only differs from the members of its own group but resembles alkali metals.

(*i*) *Colour.* Its silver-white lustre is similar to the lustre of a freshly cut surface of an alkali metal.

(*ii*) *Valency.* The most common and important oxidation state of silver is +1. Moreover, Ag^+ ion is stable in aqueous solution. In this respect it resembles alkali metals.

(*iii*) *Nature of oxides.* Silver oxide, Ag_2O like the oxides of alkali metals, is basic.

(*iv*) *Nature of chlorides.* The chlorides of both silver and alkali metals are isomorphous.

(*v*) *Formation of alums.* Silver sulphate like sulphates of alkali metals forms alums *e.g.*,

$$Ag_2SO_4.Al_2(SO_4)_3.24H_2O; \quad K_2SO_4.Al_2(SO_4)_3.24H_2O$$

How to justify the inclusion of coinage metals (Cu, Ag, Au) in the same sub-group (IB) of the periodic table

A look at the periodic table given in Fig. 7.1 shows that (*A*) coinage metals (Cu, Ag, Au) are present in the same sub-group *viz* group IB (*B*) Zn, Cd and Hg (group IIB) are the right hand neighbours of coinage metals (*C*) Ni, Pd and Pt (group VIII) are the left hand neighbours of coinage metals. Thus the inclusion of coinage metals in group IB can be justified by discussing the following similarities.

A. Similarities between Cu, Ag and Au. See page 294.

B. Similarities between the coinage metals (Cu, Ag and Au) and the metals of group IIB (Zn, Cd and Hg). See page 296.

C. Similarities between coinage metals and the metals *viz.* Ni, Pd and Pt (group VIII). See the similarities between Cu and Ni (page 289), between Ag and Pd (page 292) and Au and Pt (page 292) pairs.

Group IIB (12) : Zn_{30}, Cd_{48}, Hg_{80}

Group IIB of the long form of periodic table consists of three elements namely *Zinc* (Zn), *Cadmium* (Cd) and *Mercury* (Hg). These elements are collectively known as *zinc group elements.*

Electronic configurations

The electronic configurations of the atoms of these elements as given in Table 7.9 show that the ultimate shell has *two* electrons (s^2 electrons) while the penultimate shell contains 18 electrons ($s^2p^6d^{10}$ electrons). Thus the valence-shell configuration can be represented as $(n - 1)\ d^{10}\ ns^2$. Thus, these are *transitional elements*, but they do not show many peroperties which are shown by transition elements, *e.g.*, (*i*) They do not show variable valency (*ii*) Many of their compounds are white. (*iii*) Their melting and boiling points are very low.

Similarities in physical properties and gradation in them

1. Occurrence. None of these elements occurs in the free state in nature. Important ore of these elements is sulphide.

2. Oxidation state. We have seen that the valence-shell configuration of these elements can be represented as $(n - 1)d^{10}.ns^2$. When s^2 electrons are lost by these elements, M^{2+} ion produced possesses $(n - 1)d^{10}$ configuration. This configuration is stable and hence cannot lose any more electrons. Thus all the elements show +2 oxidation state. It may be noted that Hg forms two series of salts—*mercurous* salts containing Hg_2^{2+} ion and *mercuric* salts containing Hg^{2+} ion.

Table 7.9. *Important physical properties of zinc group elements: Zn, Cd, Hg [Group II B(12)]. Here* $[Ar]_{18}$ = 2, 8, 8; $[Kr]_{36}$ = 2, 8, 18, 8 *and* $[Xe]_{54}$ = 2, 8, 18, 18, 8.

Property	*Zinc (Zn)*	*Codmium (Cd)*	*Mercury (Hg)*
Atomic number	30	48	80
Atomic weight	65.38	112.41	200.06
Electronic configuration	$[Ar]_{18}\ 3d^{10}\ 4s^2$ = 2, 8, 18, 2	$[Kr]_{36}\ 4d^{10}\ 5s^2$ = 2, 8, 18, 18, 2	$[Xe]_{54}\ 4f^{14}\ 5d^{10}\ 6s^2$ = 2, 8, 18, 32, 18, 2
Atomic radius (pm)	138	154	157
Ionic radius, M^{2+} or M^{2+} (pm)	74 (Zn^{2+})	109 (Cd^{2+})	133 (Hg^+) 116 (Hg^{2+})
Atomic volume (ml)	9.20	13.1	14.80
Density (g/cm^3)	7.14	8.65	13.60
Melting point (°C)	419.5	320.9	– 38.4
Boiling point (°C)	906.0	765.0	357.0
Specific heat	0.093	0.053	0.033
Ionisation energy (kJ/mol)	906.4	867.4	1007.0
Electronegativity	1.66	1.46	1.44
Oxidation states	+2	+2	+1, +2
Heat of sublimation (kJ/mol)	114.66	100.02	58.17

3. Physical state. They are greyish white lustrous metals, malleable and ductile. The malleability and ductility gradually decrease as the atomic number increases. Except mercury which is a liquid, the other two are moderately hard; cadmium being harder than zinc.

4. Conductivity. All are good conductors of electricity.

5. Atomic radii. The atomic radii of these elements are fairly large and increase as the atomic number increases.

6. Ionic radii. The ionic radii are much smaller as compared to atomic radii and increase from zinc to mercury. This accounts for the covalent character in them.

7. Atomic volume. The atomic volume increases from zinc to mercury due to an increase in the atomic radii with an increase in the atomic number.

8. Density. All these elements possess high density and this property gradually increases as the atomic number increases.

9. Melting and boiling points. These elements possess very low melting and boiling points as compared to other elements except the alkali metals. Moreover it is striking to note that they decrease from zinc to mercury.

10. Ionization potential. The metals of zinc group have fairly high ionization potentials which indicate that these elements should be less reactive. Infact, they are even more reactive than coinage metals. This is due to their high electrode potentials. It can also be noted here that ionization potentials do not show regular trends.

11. Electronegativity. As can be seen from Table 7.9 the metals of zinc group have low electronegativity values and these values decrease from zinc to mercury.

Similarities in chemical properties and gradation in them

1. Reactivity. They are not very reactive and the reactivity decreases in moving down from zinc to mercury.

2. Action of air. Zn is stable in air. In moist air, it is covered with a protective film of basic zinc carbonate. This film protects Zn from the further action of air on the metal.

Cadmium is slowly oxidised to CdO in moist air. This oxide forms a coating on the surface of the metal.

Mercury is not attacked by air either dry or moist at ordinary temperature.

3. Action of water. The metals of zinc group do not decompose cold water. Zinc and cadmium decompose steam at red heat while mercury does not react with water at any temperature.

$$\text{Zn } (\textit{impure metal}) + 2H_2O\ (\textit{steam}) \xrightarrow[\text{heat}]{\text{Red}} Zn(OH)_2 + H_2$$

4. Action of acids. Very pure Zn reacts very slowly with acids while commercial Zn reacts readily with all acids. The nature of the products formed depends on the strength of the acid used as shown below :

$$Zn + 2HCl\ (\textit{dil. or conc.}) \longrightarrow ZnCl_2 + H_2$$

$$Zn\ (\textit{commercial}) + 2H_2SO_4\ (\textit{dil.}) \longrightarrow ZnSO_4 + 2H_2$$

$$Zn + 2H_2SO_4\ (\textit{hot and conc.}) \longrightarrow ZnSO_4 + SO_2 + 2H_2O$$

$$7Zn + 9H_2SO_4\ (\textit{moderately conc. 20\%}) \longrightarrow 7ZnSO_4 + H_2S + 8H_2O + S\downarrow$$

$$4Zn + 10HNO_3\ (\textit{cold and very dil. 6\%}) \longrightarrow 4Zn(NO_3)_2 + NH_4NO_3 + 3H_2O$$

or $$4Zn + NO_3^- + 10H^+ \longrightarrow 4Zn^{2+} + NH_4^+ + 3H_2O$$

$$4Zn + 10HNO_3\ (\textit{cold and dil. 20\%}) \longrightarrow 4Zn(NO_3)_2 + N_2O + 5H_2O$$

or $$4Zn + 2NO_3^- + 10H^+ \longrightarrow 4Zn^{2+} + N_2O + 5H_2O$$

$$Zn + 4HNO_3\ (\textit{cold and conc.}) \longrightarrow ZnNO_3 + 2NO_2 + 2H_2O$$

or $$Zn + 2NO_3^- + 4H^+ \longrightarrow Zn^{2+} + 2NO_2 + 2H_2O$$

$$3Zn + 8HNO_3\ (\textit{cold and moderately conc. 50\%}) \longrightarrow 3Zn(NO_3)_2 + 4H_2O + 2NO$$

or $$3Zn + 2NO_3^- + 8H^+ \longrightarrow 3Zn^{2+} + 2NO + 4H_2O$$

Cadmium dissolves in dilute or concentrated hydrochloric acid and dilute sulphuric acid giving off hydrogen. It dissolves in dilute nitric acid liberating nitric oxide.

$$3Cd + 8HNO_3 \longrightarrow 3Cd(NO_3)_2 + 2NO + 4H_2O$$

Dilute hydrochloric acid and dilute sulphuric acid have no action on mercury. Cold and dilute nitric acid gives mercurous nitrate and nitric oxide.

$$6Hg + 8HNO_3 \longrightarrow 3Hg_2(NO_3)_2 + 2NO + 4H_2O$$

Concentrated nitric acid gives mercuric nitrate.

$$3Hg + 8HNO_3 \longrightarrow 3Hg(NO_3)_2 + 2NO + 4H_2O$$

Mercury is attacked by concentrated sulphuric acid only on heating when sulphur dioxide is evolved.

$$Hg + 2H_2SO_4 \longrightarrow HgSO_4 + 2H_2O + SO_2$$

Mercury is attacked by hydriodic acid on account of the formation of a complex ion, $[HgI_4]^{2-}$.

5. Action of caustic alkalies on metals and their salts. Metallic zinc dissolves in caustic alkalies forming zincates with the evolution of hydrogen, while cadmium and mercury remain unaffected by alkalies. Moreover, when caustic alkali is added to solutions of zinc and cadmium salts, hydroxides of both are precipitated while mercuric salts give a precipitate of mercuric oxide when treated with alkalies (the hydroxide of mercury is unknown).

$$Zn + 2NaOH \ (\textit{hot and conc.}) \longrightarrow \underset{\text{Sod. zincate}}{Na_2ZnO_2} + H_2$$

$$\text{Cd or Hg} + NaOH \longrightarrow \text{No reaction}$$

$$\underset{(M^{2+} = Zn^{2+} \text{ or } Cd^{2+})}{M^{2+}} + 2OH^- \longrightarrow \underset{\text{ppt.}}{M(OH)_2\downarrow}$$

$$Hg^{2+} + 2OH^- \longrightarrow \underset{\text{Yellow ppt.}}{HgO\downarrow} + H_2O$$

6. Action of ammonia on metallic salts. On adding ammonia solution to the solutions of zinc or cadmium salts, their hydroxides are precipitated which form soluble complexes in an excess of the reagent *e.g*., $[Zn(NH_3)_4](OH)_2$ and $[Cd(NH_3)_4](OH)_2$. Mercury salts produce complex mercuric amido complex like $[Hg(NH_2)]Cl$.

$$\underset{(M^{2+} = Zn^{2+},\ Cd^{2+})}{M^{2+}(aq)} + 2NH_3 + 2H_2O \longrightarrow \underset{\text{ppt}}{M(OH)_2\downarrow} + 2NH_4^+$$

$$M(OH)_2 + 4NH_3 \ (\textit{excess}) \longrightarrow \underset{\text{Soluble complex}}{[M(NH_3)_4](OH)_2}$$

When Hg_2Cl_2 or $HgCl_2$ is treated with NH_3, it forms a bulky ppt. of mercuric amido chloride, $[Hg(NH_2)]Cl$ (complex compound).

$$Hg_2Cl_2 + 2NH_3 \longrightarrow Hg + [Hg(NH_2)]Cl + NH_4Cl$$

$$HgCl_2 + 2NH_3 \longrightarrow [Hg(NH_2)]Cl + NH_4Cl$$

7. Nature of oxides and hydroxides. Zinc oxide is amphoteric while the oxides of the other two are basic in character. The amphoteric nature of ZnO is evident from the following reactions.

$$ZnO + 2HCl \longrightarrow ZnCl_2 + H_2O$$

$$ZnO + 2NaOH \longrightarrow Na_2ZnO_2 + H_2O$$

The stability of oxides towards heat decreases from zinc to mercury. The oxides of zinc and cadmium sublime on heating while mercuric oxide decomposes on heating.

$$2HgO \xrightarrow{\text{High temp.}} 2Hg + O_2$$

The hydroxide of mercury is unknown while the hydroxides of zinc and cadmium are obtained as crystalline precipitates and dissolve in ammonia forming soluble complexes. (See above). Zinc hydroxide also dissolves in caustic alkalies forming zincates.

$$Zn(OH)_2 + 2NaOH \longrightarrow Na_2ZnO_2 + 2H_2O$$

8. Formation of peroxides, hydrides and nitrides. Zinc and cadmium form peroxides while mercury does not form peroxide. Moreover, these peroxide are less stable than the peroxides of alkaline earth metals and alkali metals.

None of the zinc group elements forms any hydride or nitride directly.

9. Sulphides. All these elements form sulphides. Their solubility decreases from zinc to mercury. Zinc sulphide is soluble in dilute acids, cadmium sulphide is soluble in concentrated acids while mercuric sulphide dissolves in aqua-regia only.

10. Halides. All these elements directly form halides which have very low melting points and are easily volatile. The nature of the bond in the halides depends upon the polarising power of the cation and the polarisability of the anion. It may be noted that polarising power increases from zinc to mercury while polarisability amongst the halide ions increases as $F^- < Cl^- < Br^- < I^-$. Therefore, $CdCl_2$ is much less ionic than CdF_2. Similarly HgF_2 is ionic while $HgCl_2$ is covalent.

11. Thermal decomposition of salts. The ease with which their various salts such as carbonates, nitrates, sulphates etc. decompose on heating increases from zinc to mercury. In some cases, mercury salts on decomposition give even metallic mercury.

12. Flame test. The salts of the metals of zinc group do not impart any characteristic coloration to the flame.

Differences of mercury from zinc and cadmium

From the above account it is clear that mercury and its compounds differ from zinc and cadmium, the members of the same group, in the following respects :

1. Mercury is a liquid at ordinary temperature, whereas the other two are solids.
2. Mercury also occurs in the native state in nature whereas the other two donot occur in the native state.
3. Mercury is less electropositive than hydrogen and, therefore, does not displace hydrogen from hydrochloric or dilute sulphuric acids whereas the other two being more electropositive than hydrogen, displace hydrogen from hydrochloric acid or dilute sulphuric acid.
4. Metallic mercury remains unaffected by air and dilute acids; even concentrated hydrochloric and sulphuric acids have no action in cold. Moreover, like cadmium but unlike zinc, it remains unaffected by caustic alkalies.
5. Mercury forms two series of salts—mercurous and mercuric.
6. Mercuric chloride is a covalent compound and does not ionize in solution. It does not give HCl when heated with concentrated sulphuric acid. Moreover, it does not respond to chromyl chloride test. The chlorides of zinc and cadmium are ionic.
7. Mercury does not form its hydroxide or peroxide.
8. The oxide of mercury is reduced to mercury by heat alone while the oxides of zinc and cadmium are more stable towards heat and are reduced to the metals when heated with carbon.
9. Mercury does not react with water at any temperature while zinc and cadmium decompose steam at red heat.
10. The sulphides and iodides of mercury show polymorphism.
11. Mercuric sulphide is insoluble in concentrated acids while ZnS and CdS dissolve in acids forming the corresponding salts.
12. With ammonia, mercuric chloride gives a white precipitate of $Hg(NH_2)Cl$ while zinc and cadmium salts form complex ions, $[Zn(NH_3)_4]^{2+}$ and $[Cd(NH_3)_4]^{2+}$ respectively.
13. Mercuric compounds are reduced to mercurous compounds by reducing agents while no such reduction takes place in zinc and cadmium compounds.
14. Zinc and cadmium form alloys with other metals while mercury with other metals forms amalgams.

Comparison between the elements of group IB (Cu, Ag, Au) and those of group IIB (Zn, Cd, Hg)

1. Melting and boiling points. The melting and boiling points of Zn-group elements are lower than those of Cu-group elements. The difference in these values has been explained as follows :

(*a*) Due to greater shielding effect of $(n-1)$ d electrons in Zn-group elements as compared to that in Cu-group elements, the atoms of Zn-group elements are larger in size than those of Cu-group elements (Cu = 1.17 Å, Zn = 1.25 Å, Ag = 1.34 Å, Cd = 1.48 Å, Au = 1.34 Å, Hg = 1.49 Å). With the increase in size of the atoms of Zn-group elements, the atoms get farther away from one another and hence the inter-atomic forces of attraction decreases, *i.e.*, the inter-atomic forces of attraction become very weak. Thus lesser amount of energy is required to melt or boil such elements whose atoms are large in size. Consequently, the m.pts. and b.pts. of Zn-group elements are lower than those of Cu-group elements.

(*b*) Since the atoms of Zn, Cd and Hg have completely-filled $(n-1)d$ orbitals, the electrons in these orbitals are not in a position to form covalent bonds resulting from *d–d* overlap among the atoms of these elements. Thus in the atoms of these elements the inter-atomic forces are very weak and hence lesser amount of energy is required to melt or boil these elements. Consequently these elements have lower m.pts. and b.pts.

The inter-atomic forces in Hg are so weak that its m.pt. (= – 38.4°C) is even below room temperature. As a matter of fact Zn-group elements have the lowest values of their m.pts. and b.pts in their respective periods.

2. Electronic configurations. Cu-group elements have only *one* electron in their *ns* orbital (Cu = $3d^{10}.4s^1$, Ag = $4d^{10}.5s^1$, Au = $5d^{10}.6s^1$) while Zn-group elements have *two* electrons in their *ns* orbital (Zn = $3d^{10}.4s^2$, Cd = $4d^{10}.5s^2$, Hg = $5d^{10}.6s^2$).

3. Oxidation states. With the exception of Ag, Cu and Au both can lose one or two electrons from the completely-filled $(n-1)d$ orbitals to form M^{2+} (Cu^{2+}) or M^{3+} (Au^{3+}) cations [$Cu^0(3d^{10}.4s^1) \xrightarrow{-e^-} Cu^+ (3d^{10}.4s^0) \xrightarrow{-e^-} Cu^{2+} (3d^9.4s^0)$; $Ag^0 (4d^{10}.5s^1) \xrightarrow{-e^-} Ag^+ (4d^{10}.5s^0)$; $Au^0 (5d^{10}.6s^1) \xrightarrow{-e^-} Au^+ (5d^{10}.6s^0) \xrightarrow{-2e^-} Au^{3+} (5d^8.6s^0)$], while Zn, Cd and Hg can lose only two *ns* electrons to form M^{2+} ion [$Zn^0 (3d^{10}.4s^2) \xrightarrow{-2e^-} Zn^{2+} (3d^{10}.4s^0)$; $Cd (5d^{10}.5s^2) \xrightarrow{-2e^-} Cd^{2+} (4d^{10}.5s^0)$; $Hg(5d^{10}.6s^2) \xrightarrow{-2e^-} Hg^{2+} (5d^{10}.6s^0)$]. Thus we find that Cu, Ag and Au show +1 (Cu, Ag, Au), +2 (only Cu) and +3 (only Au) oxidation states while Zn-group elements show only +2 oxidation states. In addition to +2 oxidation state, Hg also shows +1 oxidation state in compounds, called *mercurous compounds* which exist as dimers, *e.g.*, mercurous chloride exists as $(HgCl)_2$ or Hg_2Cl_2 or Cl—Hg—Hg—Cl and not as HgCl or Cl—Hg. Consequently mercurous ion exists as Hg_2^{2+} and not as Hg^+.

4. Reactivity of the metals. We know that the reactivity of a metal is favoured when its heat of sublimation (also called heat of vaporisation) [$M(s)$ + heat of sublimation $\longrightarrow M(g)$] and ionisation energy [$M(g)$ + ionisation energy $\xrightarrow{-e^-} M^+(g)$] both are *low* and heat of hydration [$M^+(g) + aq \longrightarrow M^+(aq)$ + heat of hydration] is *high.* Since heats of hydration of the metals of both the groups are almost the same, their effect on the reactivity can be ignored. Thus the metals having low value of (heat of sublimation + ionisation energy) will be more reactive. Now, as shown below, since the sum of the heat of sublimation and ionisation for Cu, Ag and Au is comparatively high as compared to that for Zn, Cd and Hg, Cu-group metals are less reactive than Zn-group metals.

	Cu	Ag	Au	Zn	Cd	Hg
Heat of sublimation (in kJ/mole)	338.98	286.25	365.35	114.66	100.02	58.17
Ionisation energy (in kJ/mole)	745.50	731.00	890.10	906.40	867.70	1007.00
Sum	1084.48	1017.25	1255.45	1021.06	967.72	1065.17

It is due to the less reactivity shown by Cu-group metals that these metals are not attacked by non-oxidising agents. Impure Zn and Cd readily displace H_2.

5. Transition and non-transition metals. Some chemists regard Zn, Cd and Hg as non-transition metals, since many of their properties are different from those of transition metals, *e.g.*, these metals are diamagnetic, donot show variable oxidation states, most of their compounds are white, M^{2+} ions contain completely-filled $(n-1)$ *d* orbitals etc. On the other hand, although Cu, Ag and Au have completely-filled $(n-1)d$ orbitals, these are considered to be transition metals because these metals show many properties which are similar to those of other transition metals, *e.g.*, (*a*) Cu^{2+} and Au^{3+} have partially-filled $(n-1)d$ orbitals (*b*) Cu^+, Cu^{2+}, Ag^+, Au^+, and Au^{3+} form many complexes (*c*) Cu and Au show variable oxidation states.

6. Displacement of hydrogen from acids. We know that the metals having positive values of their standard electrode potential can replace hydrogen from the acids while those having negative values cannot do so. Now an examination of the standard oxidation potential values of Cu-group and Zn-group elements as given in bracket [$E^o_{Cu/Cu^{2+}} = -0.52V$, $E^o_{Ag/Ag^+} = -0.80V$, $E^o_{Au/Au^+} = -1.7V$, $E^o_{Zn/Zn^{2+}} = +0.76$ V, $E^o_{Cd/Cd^{2+}} = +0.40V$, $E^o_{Hg/Hg^{2+}} = -0.85V$] show that Cu-group elements cannot displace hydrogen from acids while Zn-group elements (excepting Hg) can do so.

University Questions

Group IIIB: Sc, Y, La and Ac

1. Name the elements included in the scandium group.Give their electronic configurations. *(Andhra 85)*
2. Give reasons for the following:
   (*i*) The salts of scandium group elements are colourless and diamagnetic.
   (*ii*) Scandium group elements undergo hydrolysis readily.
   (*iii*) Scandium forms complexes easily.
   (*iv*) Sc^{3+} ion has greater polarising power.
3. Write the electronic configuration of yttrium. *(Andhra 85)*
4. How do the following vary in IIIB group elements:
   (*i*) valency (*ii*) magnetic property (*iii*) reacitivity (*iv*) complexing ability. *(Delhi 87)*
5. What happens when scandium is treated with NaOH? *(Poona 86)*
6. Compare the properties of Sc and Al. *(Punjab 83)*
7. Give the use of lanthanum nitrate in qualitative analysis. *(Delhi 85)*
8. Taking La and Ac as examples, compare and contrast the properties of *d*-block elements with *f*-block elements. *(Delhi Hons. 85)*
9. Lanthanium belongs to *f*-block although it has no electron in 5*f* orbitals. Comments. *(Bombay 84)*

Group IVB: Ti, Zr and Hf

10. Discuss briefly the general characteristics of the elements of titanium group under the following heads: (*i*) oxidation states (*ii*) reactivity and passive behaviour (*iii*) nature of oxides (*iv*) formation of complexes. *(Bharathiar 86)*
11. Name the elements of titanium group and give their electronic configuration. *(Delhi 87)*
12. Compare the properties of Zr and Hf. *(Raj. 85)*
13. How do Zr and Hf show similarities with each other? How are they separated from each other in a pure state? *(Raj. 85, 87)*
14. Write a note on "Chemistry of Zr and Hf in a different oxidation states". *(Raj. 82)*

Group VB : V, Nb and Ta

15. Discuss the position of V, Nb and Ta in the periodic table.
16. How do the following change when we move from V to Ta *via* Nb?
    (*i*) Stability of various oxidation states shown by V, Nb and Ta.
    (*ii*) Ionic character of the compounds of V, Nb and Ta with change in their oxidation states.
    (*iii*) Basic character of the oxides *viz.* V_2O_5, Nb_2O_5 and Ta_2O_5. *(Delhi Hons. 86)*
17. Give the reason for the following :
    (*i*) Vanadium has a greater tendency to form complexes than do niobium and tantalum.
    (*ii*) The compounds of VB group elements in their lower oxidation states are coloured while those in higher oxidation states are colourless. *(Banaras 84; Lucknow 86)*

Group VIB : Cr, Mo and W

18. Give a comparative account of the general chemistry of Cr, Mo and W. Also discuss the position of these elements in the periodic table. *(Allahabad 88)*
19. Why do Mo and W show similarities with each other ? How are they separated from each other in a pure state? *(Raj. 85, 87)*
20. How many oxidation states are shown by chromium in its compounds? Give one example of each. *(Raj. 87 S; Meerut 82; Garhwal 88)*

Group VIIB : Mn, Tc and Re

21. Discuss the position of Mn in the periodic table. *(Madras 86; Bharathiar 86; M.D. Rohtak 84)*

22. Indicate the resemblance in properties between (*i*) Mn and Cr (*ii*) Mn and Fe. *(Madras 86)*

23. Give reason for the following :

(*i*) The aqueous solution of Mn^{2+} has only a pale-pink colour. *(Madras 85)*

(*ii*) Mn with five unpaired electrons is less ferromagnetic than Fe with four unpaired electrons. *(Delhi Hons. 87)*

(*iii*) Mn is an example of an element exhibiting a high degree of paramagnetic character. *(Andhra 85)*

24. How many oxidation states are exhibited by Mn in its compounds? Give one example of each. *(Raj. 87; Meerut 86)*

25. Write a note on quadrivalent manganese. *(Utkal 86)*

Ferrous metals : Fe, Co and Ni

26. Make a comparative study of Fe, Co and Ni and their compounds. *(Madras 86; Bharathiar 86; Madurai 85)*

27. Discuss the resemblances in properties of : (*i*) Co and Ni (*ii*) Co and Fe. *(Madras 86)*

28. Explain why Fe (II) and Fe (III) form complexes with CN^- ions but not with NH_3. *(Delhi 87)*

29. Discuss the position of nickel in the periodic table. *(Garhwal 88)*

30. Discuss the important oxidation states shown by Ni in its compounds. *(Garhwal 88)*

Platinum metals : Ru, Rh, Pd, Os, Ir and Pt

31. Discuss the position of Pt in the periodic table. *(Raj. 81; Allahabad 80)*

32. Why do Pd and Pt show similarities with each other? How would you separate them from each other in a pure state. *(Raj. 85; 87)*

33. Discuss the following : (*i*) Electronic configuration and position of Pd in the periodic table (*ii*) Resembalance between Pt and Pd (*iii*) Various oxidation states of Pd (*iv*) Similarities between Au and Pt. *(Vikram 82; Raj. 86)*

Group IB : Cu, Ag and Au

34. Give the reasons for the following :

(*i*) Coinage metals show variable valency. *(Madras 85)*

(*ii*) The first ionisation potential of copper is higher than those of alkali metals but second and third ionisation potentials are lower. *(Punjab 85)*

35. Discuss the position of Cu, Ag and Au in the periodic table. *(Delhi 85)*

36. Cu, Ag and Au are related on the one hand to Ni, Pd and Pt and on the other hand to Zn, Cd and Hg. Justify. *(Meerut 84; Punjab 86)*

37. Name three compounds containing copper in the anionic portion. *(Allahabad 84)*

38. Compare and contrast the properties of (*i*) Au and Ag (*ii*) Au and Pt. *(Madras 86; Lucknow 85)*

Group IIB : Zn, Cd, Hg

39. Write the electronic configuration and give a comparative study of group IIB elements (Zn, Cd and Hg) in the periodic table and their compounds. *(Sambalpur 86; Utkal 85; Bharathiar 86; Bharathidasan 86; Madras 85; Agra 84)*

40. Discuss the position of 3d elements in periodic table. *(Nagpur 2002)*

41. What are the important oxidation states of copper ? Discuss the chemistry of copper in +1 and +2 oxidation states. *(VBS, Purvanchal 2003)*

42. Taking the example of chromium discuss the important characteristics of transition elements. *(VBS, Purvanchal 2003)*

43. Compare the following properties of Fe, Co and Ni. (*i*) Position in the periodic table (*ii*) Ionic radii (*iii*) Oxidation states (*iv*) Magnetic properties (*v*) Colour of the compounds. *(VBS, Purvanchal 2003)*

8

Occurrence, Extraction, Properties and Uses of Some Transition Metals

We have already said that the elements lying in groups IIIB (3), IVB (4), ... IB (11), IIB (12) are called transition elements. In this chapter we shall discuss the various methods of extraction, properties and uses of those metals which have been shown below :

IIIB	IVB	VB	VIB	VIIB	←	VIII	→	IB	IIB
(3)	(4)	(5)	(6)	(7)	(8)	(9)	(10)	(11)	(12)
21	Ti_{22}	V_{23}	Cr_{24}	Mn_{25}	Fe_{26}	Co_{27}	Ni_{28}	29	Zn_{30}
39	40	41	Mo_{42}	43	44	45	Pd_{46}	47	48
57	72	73	W_{74}	75	Os_{76}	77	Pt_{78}	Au_{79}	80
89	104	105	106						

Titanium

Occurrence

Titanium is the ninth most abundant element in the earth's crust. Its main ores are : (*i*) *Rutile,* TiO_2 which is one of the three forms of TiO_2, and (*ii*) *Ilmetite,* $FeTiO_3$.

Extraction of titanium metal from rutile ore, TiO_2

Since titanium has high melting point and it reacts quite readily with air, O_2,N_2, H_2 and carbon at elevated temperatures, the extraction of the metal is a difficult task. The extraction of the metal from the rutile mineral consists of the following steps :

1. Conversion of rutile ore into $TiCl_4$. Rutile (TiO_2) mineral is heated with carbon in a current of Cl_2.

$$\underset{\text{Rutile ore}}{TiO_2} + C + 2Cl_2 \longrightarrow TiCl_4 + CO_2$$

2. Reduction of $TiCl_2$ to Ti metal. $TiCl_4$ obtained as above is reduced to Ti metal by any of the following methods:

(*a*) *Kroll process.* In this process vapours of $TiCl_4$ are passed over molten magnesium, when $TiCl_4$ is reduced to spongy Ti metal.

$$TiCl_4 + 2Mg \xrightarrow{800°C} \underset{\text{Spongy}}{Ti} + 2MgCl_2$$

The spongy metal is heated in an inert atmosphere of Ar to get a compact mass.

(b) *I.C.I. process.* In this process $TiCl_4$ is reduced by heating it with Na in an atmosphere of Ar.

$$TiCl_4 + 4Na \xrightarrow{\text{Heat}} Ti + 4NaCl$$

(c) $TiCl_4$ can also be reduced to Ti metal by heating it with NaH.

$$TiCl_4 + 4NaH \xrightarrow{\Delta} Ti + 4NaCl + 2H_2$$

Extraction of titanium metal from ilmenite ore, $FeTiO_3$

The extraction of titanium metal from ilmenite ore consists of the following steps:

1. Conversion of ilmenite ore, $FeTiO_3$ into TiO_2. TiO_2 can be obtained from the ilmenite ore by the following processes :

(a) When ilmenite ore is heated with HCl in a current of Cl_2, TiO_2 is obtained.

$$\underset{\text{Ilmenite}}{2FeTiO_3} + 4HCl + Cl_2 \xrightarrow{\Delta} 2FeCl_3 + 2TiO_2 + 2H_2O$$

(b) The powdered ilmenite ore is fused with Na_2S. The product obtained consists of FeS, TiO_2 and Na_2O.

$$\underset{\text{Ilmenite}}{FeTiO_3} + Na_2S \xrightarrow{\text{Fuse}} \underbrace{FeS + TiO_2}_{\text{Insoluble}} + \underset{\text{Soluble}}{2Na_2O}$$

Now when the product consisting of FeS, TiO_2 and Na_2O is extracted with H_2O, FeS and TiO_2 remain insoluble and Na_2O goes into solution. The insoluble portion containing FeS and TiO_2 is treated with H_2SO_3 in presence of atmospheric O_2 when FeS is converted into soluble FeS_2O_3 but TiO_2 remains unaffected.

$$FeS + H_2SO_3 + \tfrac{1}{2}O_2 \ (\textit{From air}) \longrightarrow \underset{\text{Soluble}}{FeS_2O_3} + H_2O$$

$$TiO_2 + H_2SO_3 + \tfrac{1}{2}O_2 \ (\textit{From air}) \longrightarrow \text{No reaction}$$

(c) When ilmenite ore is mixed with carbon and heated in a current of Cl_2 at 360°, a mass is obtained which consists of $FeCl_3$ and $TiCl_4$. Now when this mass is heated at 550°C, volatile $TiCl_4$ distils over. $TiCl_4$ thus obtained is hydrolysed by H_2O to get TiO_2.

$$\underset{\text{Ilmenite}}{2FeTiO_3} + 2C + 7Cl_2 \xrightarrow{350°C} \underbrace{2FeCl_3 + 2TiCl_4}_{\downarrow\, 550°C} + 3CO_2\uparrow$$

$$TiCl_4 \text{ distils over}$$

$$TiCl_4 + 2H_2O \xrightarrow{\text{Hydrolysis}} TiO_2 + 4HCl$$

2. Reduction of TiO_2 to Ti metal. TiO_2 obtained as above is reduced to titanium metal by any of the following methods:

(a) *By alumino-thermic process.* In this method TiO_2 is mixed with Al and the mixture is ignited with a suitable fuse.

$$3TiO_2 + 4Al \longrightarrow 2Al_2O_3 + 3Ti$$

(b) TiO_2 can also be reduced to free Ti metal by heating it with carbon.

$$TiO_2 + C \longrightarrow Ti + CO_2\uparrow$$

Purification of titanium metal

Titanium metal obtained either from rutile or from ilmenite contains many impurities.

Impure metal canbe purified by *Van Arkel process* in which impure titanium is heated with a limited amount of I_2 in an evacuated glass apparatus. TiI_4 vapours formed are allowed to difuse on to a tungsten filament at a very high temperature. The vapurs of TiI_4 decompose and the pure Ti metal gets deposited on the W-filament.

$$\text{Impure Ti metal} + 2I_2 \xrightarrow{150\text{–}250°} TiI_4(g) \xrightarrow[\text{filament}]{1800°\text{ Tungsten}} \text{Pure Ti metal} + 2I_2(g)\uparrow$$

(gets deposited on W-filament)

Physical properties

Metallic titanium has a silvery lustre and when polished resembles steel appearance. Amorphous form is a dark grey powder resembling iron (reduced) in appearance. It is brittle at ordinary temperature but at low red heat it is readily malleable. The pure metal is very ductile. Its m.pt. and b.pt are 1668°C and 3260°C. The fused metal is very hard and crystalline and scratches quartz.

Chemical properties

1. Action of air. On exposure to air at ordinary temperature, titanium metal reacts with O_2 and N_2 of the air and forms a thin film (coating) of TiO_2 and TiN on the surface of the metal. This coating prevents the metal from corrosion. This coating remains stable in air upto 400°C.

2. Action of halogens. The metal readily combines with Cl_2 (at 350°C), Br_2 and I_2 to form the tetrahalides, TiX_4.

3. Combination with non-metals. On being heated, the metal combines with H_2, B, C and N_2 to produce TiH_2, TiB, TiC and TiN. These are interstitial compounds and hard. These have refractory properties.

4. Action of acids and alkalies. The metal remains passive towards the action of acids and alkalies. This behaviour is due to the formation of a protective film of TiO_2 on the surface of the metal. The metal dissolves in HF to give $[TiF_6]^{2-}$ ion.

$$Ti + 4HF \longrightarrow TiF_4 + 2H_2$$

$$TiF_4 + 2F^- \longrightarrow [TiF_6]^{2-}$$

The metal also dissolves in HCl and H_2SO_4 to give soluble salts.

$$2Ti + 6HCl \longrightarrow 2TiCl_3 + 3H_2$$

$$Ti + 4H_2SO_4\ (conc.) \longrightarrow Ti(SO_4)_2 + 2SO_2 + 4H_2O$$

With HNO_3, metatitanic acid (H_2TiO_3) is obtained.

$$Ti + 4HNO_3 \longrightarrow H_2TiO_3 + 4NO_2 + H_2O$$

5. Action on steam. Steam is decomposed by the metal.

$$Ti + 2H_2O\ (steam) \longrightarrow TiO_2 + 2H_2O$$

6. Action on $TiCl_4$. $TiCl_4$ is reduced to $TiCl_2$ when heated with Ti metal.

$$TiCl_4 + Ti \xrightarrow{\Delta} 2TiCl_2$$

Uses of titanium metal, its compounds and its alloys

(*i*) Due to the strength, lightness, corrosion, resistance and low thermal expansion of Ti metal, it is used in high speed air-craft, naval and military projects and also in chemical industry.

(*ii*) It is used to remove the last traces of gases like O_2, H_2 and N_2 from evacuated vessels.

(*iii*) A titanium cathode is used in special photoelectric cell.

(*iv*) Pure Ti metal is used in making incandescent filament.

(*v*) A number of titanium alloys have been found to be very useful in industry. For example: (*a*) Ferrotitanium alloy is used in steel industry as a scavanger to remove adsorbed O_2, H_2 and N_2, and as a cleansing and deoxidising agent in the metallurgical operations. Ferrotitanium is obtained in the electric furnace by reducing ilmenite ore, $FeTiO_3$ with carbon or Al metal.

$$\underset{\text{Ilmenite}}{FeTiO_3} + 3C \text{ (or 2Al)} \longrightarrow \underbrace{Fe + Ti}_{\text{Ferrotitanium alloy}} + 3CO \text{ (or } Al_2O_3)$$

(*b*) The alloy of Ti with Cr, Fe and Mo are used for making rails, gears, propellers, car wheels etc. (*c*) The alloys of Ti with Co and Ni are used as filaments in vacuum tubes in place of Pt—Ir alloys which are very costly.

(*vi*) Among its compounds, its carbide is used as refractory material, its nitride in high speed tools and its oxide (TiO_2) is used as a white pigment in paints under the name of *titan white*.

Vanadium

Occurrence

Vanadium does not occur in the free state in nature but is found in may rocks. This element occurs in sun, stars and meteorites. It also occurs in the ash or flue dust of numerous lignites and coal. It is also found in clays and crude oil. Important minerals of this element are :

1. Vanadinite, $3Pb_3(VO_4)_2.PbCl_2$ or $Pb_{10}(VO_4)_6Cl_2$ or $Pb_5(VO_4)_3Cl$ which is lead chloro orthovanadate.

2. Carnotite, $K_2O.2UO_3.V_2O_5.3H_2O$ or $2K(UO_2)VO_4.3H_2O$ which is potassium uranyl orthovanadate. This ore contains mainly vanadium, uranium and traces of Ba, Fe, Al, Ra etc. It is used for the extraction of *uranium, vanadium* and *radium*. It is found in Colorado and Utah in U.S.A. and in South Australia. In India it is found in Joduguna mines of Bihar.

3. Patronite, V_2S_5 (Vandium sulphide). This ore contains pyrites and sulphur as impurities. It is found in North. Rhodesia.

Extraction

The extraction of vanadium metal from its ores consists of the following steps :

1. Preparation of ammonium metavanadate, NH_4VO_3. Ammonium metavanadate can be obtained either from carnotite or vanadinite ore as shown below :

(*a*) **From carnotite ore $2K(UO_2)VO_4.3H_2O$.** Carnotite ore is boiled with Na_2CO_3 solution and concentrated. On cooling, the crystals of sodium uranyl carbonate, $Na_4(UO_2)(CO_3)_2$ are obtained and sodium orthovanadate, Na_3VO_4 is left in solution. The solution containing Na_3VO_4 is treated with NH_4Cl when ammonium metavanadate, NH_4VO_3 is precipitated.

$$\underset{\text{Carnotite ore}}{2K(UO_2)VO_4} + 7Na_2CO_3 \longrightarrow \underset{\text{crystals}}{2Na_4(UO_2)(CO_3)_3} + \underset{\text{In solution}}{2Na_3VO_4} + 2K_2CO_3$$

$$\underset{\text{In solution}}{Na_3VO_4} + 3NH_4Cl \longrightarrow \underset{\text{ppt}}{NH_4VO_3} + 3NaCl + H_2O + 2NH_3\uparrow$$

(*b*) **From vanadinite ore, $Pb_5(VO_4)_3Cl$.** Powdered vanadinite ore is treated with conc. HCl when $PbCl_2$ is obtained as precipitate and a complex salt, $[V(OH)_4]Cl$ goes into solution. When the solution containing $[V(OH)_4]Cl$ is boiled with NH_4Cl, ammonium metavanadate, NH_4VO_3 is obtained as precipitate.

$$\underset{\text{Vanadinite ore}}{Pb_5(VO_4)_3Cl} + 12HCl \longrightarrow \underset{\text{ppt}}{5PbCl_2} + \underset{\text{In solution}}{3[V(OH)_4]Cl}$$

$$\underset{\text{In solution}}{[V(OH)_4]Cl} + NH_4Cl \longrightarrow \underset{\text{ppt.}}{NH_4VO_3} + 2HCl + H_2O$$

(*c*) **From patronite, V_2S_5.** The ore is roasted in the excess of air to convert it into V_2O_5.

$$\underset{\text{Patronite ore}}{2V_2S_5} + 15O_2 \longrightarrow 2V_2O_5 + 10SO_2$$

V_2O_5 thus obtained contains a number of impurities. In order to remove the impurities, V_2O_5 in fused (melted) with Na_2CO_3 solution so that V_2O_5 is converted into soluble sodium orthovanadate, Na_3VO_4 while the impurities remain as residue.

$$V_2O_5 + 3Na_2CO_3\ (solution) \longrightarrow 2Na_3VO_4\ (solution) + 3CO_2\uparrow$$

Now to the solution containing Na_3VO_4 is added NH_4Cl so that NH_4VO_3 gets precipitated.

$$Na_3VO_4 + 3NH_4Cl \longrightarrow \underset{\text{ppt}}{NH_4VO_3} + 3NaCl + H_2O + 2NH_3\uparrow$$

2. Conversion of NH_4VO_3 to V_2O_5. When NH_4VO_3 obtained as above is ignited, vanadium pentoxide (V_2O_5) is obtained.

$$2NH_4VO_3 \xrightarrow{\text{Ignite}} V_2O_5 + 2NH_3 + H_2O$$

3. Reduction of V_2O_5 to vanadium metal. Vanadium metal can be obtained by reducing V_2O_5.

(*a*) With Ca and $CaCl_2$ (used as a flux) in a steel bomb at 900-950°C (*Marden and Rich process*).

$$V_2O_5 + 5Ca + 5CaCl_2 \longrightarrow 2V + 5(CaO.CaCl_2)$$

(*b*) With Al (*Alumino-thermic process*)

$$3V_2O_5 + 10Al \longrightarrow 5Al_2O_3 + 6V$$

4. Purification of vanadium metal. Vanadium metal obtained as above is impure, since it contains some impurities associated with it. The impure metal is purified by *Van Arkel method* in which the impure vanadium metal is heated with a limited amount of I_2 in a evacuated glass apparatus. VI_4 vapours so formed are allowed to diffuse on a tungsten filament at a very high temperature. The vapours of VI_4 decompose and the pure vanadium metal gets deposited on the filament.

$$\text{Impure V metal} + 2I_2 \longrightarrow VI_4(g) \xrightarrow{\text{W-filament}} \underset{\text{Pure metal}}{V(s)} + 2I_2(g)\uparrow$$

Preparation of vanadium metal

Vanadium metal can be obtained by the following methods :

1. By reducing the halides of vanadium. When ferro-vanadium alloy is heated in a current of Cl_2, VCl_4 and $FeCl_3$ are obtained. These chlorides are separated by the distillation process. VCl_4 thus obtained is reduced by heating in current of hydrogen or with Mg metal to get vanadium.

$$VCl_4 + 4H_2 \xrightarrow{600°C} 2V + 8HCl\ (\textit{Jantesh process})$$

$$VCl_4 + 2Mg \longrightarrow V + 2MgCl_2$$

VCl_3 on being reduced by H_2, Na or Mg also gives vanadium.

$$VCl_3 + 3Na \longrightarrow V + 3NaCl$$

$$2VCl_3 + 3H_2 \longrightarrow 2V + 6HCl$$

$$2VCl_3 + 3Mg \longrightarrow 2V + 3MgCl_2$$

2. By reducing V_2O_5. See the extraction of vanadium.

Properties of vanadium metal

It is a steel-grey hard metal which melts above 1700°C and boils at 3450°C. When pure, it is malleable as well as ductile. When heated in air, the metal is converted into V_2O_5. When heated upon 1400°C, it forms VO_2. When treated with excess of dry Cl_2, VCl_4 is formed. Bromine water has no action on vanadium metal. HCl and H_2SO_4 both have no action on the metal. However, with hot and conc.H_2SO_4 it gives SO_2. With HF it gives H_2. It is easily attacked by HNO_3 and forms $V(NO_3)_3$. Aqua regia attacks the metal very vigorously. Fused alkalies give orthovanadates (Na_3VO_4), with the liberation of H_2.

$$2V + 6NaOH + 2H_2O \longrightarrow 2Na_3VO_4 + 5H_2$$

Oxidising agents like chloric acid, perchloric acid, bromic acid, $KClO_3$, bromate, iodate and perchlorate convert vanadium into V_2O_5. Vanadium metal reduces the solution of $AgNO_3$, $AuCl_3$, $PtCl_4$ and $IrCl_4$. It combines with non-metals like N_2, C, S and Si to form nitride, carbide, sulphide and silicide respectively. Vanadium combines with a number of metals like Fe, Cu, Co, Ni and Al forms alloys. For example:

(*a*) *Ferro-vanadium alloy* is obtained by heating V_2O_5 with iron ore (Fe_2O_3) and coke in presence of a flux (lime or fluorspar) in an electric furnace.

$$V_2O_5 + 5C \longrightarrow 2V + 5CO \uparrow$$
$$Fe_2O_3 + 3C \longrightarrow 2Fe + 3CO \uparrow$$

This alloy contains 40-80% vanadium.

(*b*) *Nickel-vanadium alloy* is obtained by heating a mixture of V_2O_5 and NiO.

(*c*) *Alumino-vanadium alloy* is prepared by adding Al and then thermite mixture of V_2O_5 and aluminium soot in the electric furnace having melting fluorspar.

Uses of vanadium

The most important use of vanadium is in the form of its alloys. For example, ferro-vanadium alloy is used as a scavanger by removing efficiently N_2 and O_2 in the manufacture of special steels. The presence of a small amount of vanadium (0.1 to 0.3%) increases the toughness, tensile strength and resistance to shock which are valuable characteristics for the manufacture of machinery parts subjected to vibration. Thus, ferro-vanadium is used in the construction of motor parts. Alumino-vanadium alloy is used in the manufacture of aircraft parts. The addition of 0.2% vanadium in glass manufacture is recommended for cutting of all the wavelengths below 3580Å which are harmful for the eyes and fade fabrics.

Chromium

Occurrence

Metallic chromium does not occur in the native state. The ores of chromium are : (*i*) *chromite* or *chrome ore*, $FeO.Cr_2O_3$. (*ii*) *chrome-ochre*, Cr_2O_3 (*iii*) *crocosite*, $PbCrO_4$. Chromite ore is the most common and is the industrial source of the metal. In India the chromite ore occurs in Singhbhum district of Bihar,Mysore, Orissa, Madras and Bombay.

Extraction of chromium metal from chromite ore, $FeO.Cr_2O_3$

The extraction of Cr metal from chromite ore consistts of the following steps:

1. Concentration of the chromite ore. The ore is first crushed and then concentrated by *Wilfley table method* (*Gravity separation method*). In this method the crushed ore is collected in the form of a heap on a sloping floor. A stream of water is allowed to flow through the heap. By this operation the lighter impurities of earthy matter go with water flowing down the slope while the heavier impurities remain with the ore in the heap.

2. Conversion of the concentrated chromite ore into Na_2CrO_4: Roasting of the concentrated ore. The concentrated ore is mixed with excess of Na_2CO_3 and roasted in a reverberatory furnace at 900 – 1000°C in the presence of free supply of air. By this operation the chromite ore, $FeO.Cr_2O_3$ is converted into soluble sodium chromate, Na_2CrO_4.

$$\underset{\text{Chromite ore}}{4(FeO.Cr_2O_3)} + 8Na_2CO_3 + 7O_2\ (\textit{From air}) \xrightarrow{900\text{–}1000°C} \underbrace{\underset{\text{Soluble}}{8Na_2CrO_4} + \underset{\text{Insoluble}}{2Fe_2O_3}}_{\text{Roasted mass (Yellow Solution)}} + 8CO_2\uparrow$$

The roasted mass containing Na_2CrO_4 and Fe_2O_3 is extracted with water when Na_2CrO_4 goes into solution while Fe_2O_3 is left behind as a residue.

In the process of roasting, a small amount of lime (CaO) is also added with the mixture of the ore and Na_2CO_3. The addition of CaO makes the mass porous so that the process of roasting is facilitated.

3. Conversion of Na_2CrO_4 into $Na_2Cr_2O_7$. The solution containing Na_2CrO_4 is treated with a calculated quantity of H_2SO_4 whereby Na_2CrO_4 is converted into sodium dichromate, $Na_2Cr_2O_7$ and less soluble Na_2SO_4 is also obtained.

$$2Na_2CrO_4 + H_2SO_4 \longrightarrow \underset{\substack{\text{More soluble}\\ \text{(orange solution)}}}{Na_2Cr_2O_7} + \underset{\text{Less soluble}}{Na_2SO_4} + H_2O$$

The solution containing $Na_2Cr_2O_7$ (more soluble) and Na_2SO_4 (less soluble) is concentrated when the less soluble Na_2SO_4 crystallises out leaving behind more soluble $Na_2Cr_2O_7$ in solution. The solution containing $Na_2Cr_2O_7$ is further concentrated when crystals of $Na_2Cr_2O_7$ are obtained.

4. Conversion of $Na_2Cr_2O_7$ into Cr_2O_3. $Na_2Cr_2O_7$ obtained as above can be converted into Cr_2O_3 by the following methods :

(*a*) $Na_2Cr_2O_7$ is heated with carbon to get sodium chromite, $NaCrO_2$ which, on being treated with H_2O, gives the precipitate of chromic oxide, Cr_2O_3.

$$Na_2Cr_2O_7 + 3C \xrightarrow{\Delta} \underset{\text{Sod. chromite}}{2NaCrO_2} + 3CO\uparrow$$

$$\underset{\text{Sod. chromite}}{2NaCrO_2} + H_2O \longrightarrow \underset{\text{ppt}}{Cr_2O_3} + 2NaOH$$

(*b*) $Na_2Cr_2O_7$ is heated with NH_4Cl when the precipitate of Cr_2O_3 is obtained.

$$Na_2Cr_2O_7 + 2NH_4Cl \xrightarrow{\Delta} \underset{\text{ppt}}{Cr_2O_3} + 2NaCl + 4H_2O + N_2\uparrow$$

5. Reduction of Cr_2O_3 to chromium metal. Cr_2O_3 obtained as above is reduced to Cr-metal:

(*a*) by heating Cr_2O_3 with Al-powder (*alumino-thermite process*).

$$Cr_2O_3 + 2Al \longrightarrow Al_2O_3 + 2Cr;\ \Delta H = -\ 468.6\ \text{kJ}.$$

(*b*) by heating Cr_2O_3 with a calculated quantity of Si in presence of CaO which forms a slag of $CaSiO_3$ with SiO_2.

$$2Cr_2O_3 + 3Si \longrightarrow 4Cr + 3SiO_2$$

$$3SiO_2 \xrightarrow{+3CaO} \underset{\text{Slag}}{3CaSiO_3}$$

(*c*) by passing H_2 over Cr_2O_3 (*Industrial preparation*)

$$Cr_2O_3 + 3H_2 \rightleftharpoons 2Cr + 3H_2O\ (\textit{water steam})$$

As is obvious, the reduction of Cr_2O_3 by H_2 is reversible and hence water formed in the reaction is to be removed. It is removed by getting it absorbed by Na metal kept near the crucible in which the reduction reaction is taking place. Cr obtained by this process is highly pure.

(*d*) by heating Cr_2O_3 with sugar charcoal (*Daville's process*).

Physical properties

Chromium is a bluish-white lustrous metal and crystalline in nature. It is capable of taking good polish on its surface and this polish produces a brilliant surface and lasts long because of the formation of an invisible and then self-protective oxide coating. This coating is not tarnished even after long exposure to the atmosphere. The sp. gravity is 7.19, m.pt. = 1875°C and b.pt. = 2665°C. The metal is malleable, ductile and paramagnetic.

The pure metal is soft but becomes hard when it contains small amounts of carbon as impurity. It is resistant to corrosive agents.

Chemical properties

1. Action of air. Cr is extremely stable and is not affected by moist air, rain, sea-water, H_2S, SO_2 etc. The inertness is probably due to the formation of a protective thin oxide film on the surface of the metal. In oxy-hydrogen flame, the metal burns at about 2000°C brilliantly and forms Cr_2O_3.

$$2Cr + 3H_2O \xrightarrow{2000°C} Cr_2O_3 + 3H_2O$$

2. Action of steam. Cr has no action on water at ordinary temperature but at red heat steam is decomposed by the metal.

$$2Cr + 3H_2O\ (\textit{steam}) \xrightarrow{\text{Red heat}} Cr_2O_3 + 3H_2$$

3. Action of halogens. F_2 and dry Cl_2 combines with heated metal and give Cr (III) halides.

$$2Cr + 3X_2 \xrightarrow{\Delta} 2CrX_3$$
$$(X = F, Cl)$$

4. Action of alkalies. Alkalies have no action on the metal at ordinary temperature.

5. Action of NH_3 or N_2. When heated in NH_3 or N_2, CrN is formed.

$$2Cr + 2NH_3 \longrightarrow 2CrN + 3H_2$$

6. Reducing properties. In the finely-divided state, Cr acts as a good reducing agent, because in this state Cr readily reacts with O_2.

7. Action of acids. (*a*) In the absence of air, dil. HCl and dil. H_2SO_4 both dissolve the metal due to the formation of blue solutions containing $CrCl_2$ and $CrSO_4$ respectively. H_2 is evolved in both the cases.

$$Cr + 2HCl\ (\textit{dil.}) \longrightarrow \underset{\text{Blue solution}}{CrCl_2} + H_2$$

$$Cr + H_2SO_4\ (\textit{dil.}) \longrightarrow \underset{\text{Blue solution}}{CrSO_4} + H_2$$

In the presence of air or O_2, $CrCl_2$ (blue solution) is very easily oxidised to green $CrCl_3$.

$$\underset{\text{Blue}}{4CrCl_2} + 4HCl + O_2 \longrightarrow \underset{\text{Green}}{4CrCl_3} + 2H_2O$$

(*b*) Conc. H_2SO_4 reacts with metal on heating and forms $Cr_2(SO_4)_3$ and evolves SO_2.

$$2Cr + 6H_2SO_4\ (\textit{conc.}) \longrightarrow Cr_2(SO_4)_3 + 3SO_2 + 6H_2O$$

(*c*) Dil. HNO_3 has no action on the metal while conc. HNO_3 renders the metal passive due to the formation of a protective oxide film on the surface of the metal.

Uses of chromium

1. Preparation of chrome steels : Alloys of chromium. Chromium is used in the preparation of various types of chrome steels and alloys. Examples of such chrome steels are *stainless steel, chrome steel* (Cr = 2.5 – 4.0%, C = 1.0 – 1.5%), *chrome-vanadium steel, chrome-nickel steel* etc. The composition, properties and uses of some important chrome steels are given in Table 8.1. The presence of chromium in its alloys imparts greater strength, resistance towards corrosion, hardness and brightness to the chrome steels. Ferrochrome is an important alloy of chromium which is used for making special steels. For this purpose ferrochrome is added to the steel. Special steel thus obtained has special properties. Alloys to Cr with Ni containing about 20% of Cr are used in making electrical resistances for heaters, electrical furnaces etc., since these alloys resist oxidation effectively even at red heat. Cr also forms alloys which are not acted up by acids.

Table 8.1. *Alloys of chromium*

Name of the alloy	*Composition*	*Properties and uses*
(*i*) Ferrochrome	Cr = 65%, Fe = 35%. It is prepared by heating the chromite ore, $FeO.Cr_2O_3$ with carbon in an electric furnace. $$\underbrace{FeO.Cr_2O_3 + 4C \xrightarrow{\Delta} Fe + 2Cr}_{\text{Ferrochrome}} + 4CO \uparrow$$	It is used in the manufacture of chrome-steel. Chrome-steel is drill proof and is, therefore, used in the manufacture of burglar-proof safes.
(*ii*) Stainless steel	Cr = 18%, Fe = 73%, Ni = 8%, carbon = remaining part	It is used for cutlery and house hold wares.
(*iii*) Chromium-vanadium steel		It is very hard and strong and is used for gears, locomotive wheels, oxide and bearings etc.
(*iv*) Nichrome	Cr = 15%, Ni = 60%, Fe = 25%	It is resistant to atomspheric corrosion and can withstand high temperature. It is used in resistance wires for electrical heating.
(*v*) Stellite	Cr = 20 – 35%, Co = 40 – 80%, Ni = 0.25%, C = 0.75 – 2.5%, W = remaining part	It is extremely hard, bright and very resistant to corrosion and is used in cutlery, surgical instruments, cutting and high speed tools etc.
(*vi*) Chrome-nickel steel		It is used for armour plates.

2. Chrome plating. Cr is widely used for plating metals (*chrome plating*), since it is resistant to corrosion. Even very thin layer of Cr on an article imparts a bright bluish-white appearance to the article to be plated and makes the article resistant to corrosion. The layer of Cr on the article is very hard.

Molybdenum

Occurrence

Molybdenum is a rare metal which constitutes 7.5×10^{-4}% of the earth's crust. It does not occur in the native state. Its important minerals are *molybdenite* (MoS_2) found in Australia and Austria and *wulfenite* ($PbMoO_4$) found in Australia and U.S.A. U.S.A. produces about 90% of the total world production. In India it occurs in Travancore,, Rajasthan, Mysore Madhya Pradesh etc.

Extraction of molybdenum from molybdenite ore, MoS_2

The extraction of molybdenum from molybdenite ore consists of the following steps :

1. Concentration of the ore. Molybdenite ore is crushed in jaw-crushes and powdered in ball mills. Now the powdered ore is concentrated by forth floatation process.

2. Conversion of molybdenite ore, MoS_2 into MoO_3: Roasting. Concentrated molybdenite ore is roasted in a reverberatory furnace in a plenty supply of air to get the crude MoO_3.

$$\underset{\text{Molybdenite}}{2MoS_2} + 7O_2 \xrightarrow{\text{Roasting}} 2MoO_2\ (\textit{Impure}) + 4SO_2$$

The crude MoO_3 obtained as above is extracted with hot conc. NH_4OH to get the crystals of ammonium molybdate, $(NH_4)_2MoO_4$.

$$MoO_3\ (\textit{Impure}) + 2NH_4OH \longrightarrow (NH_4)_2MoO_4 + H_2O$$

Now the crystals of $(NH_4)_2MoO_4$, on heating, give relatively pure MoO_3.

$$(NH_4)_2MoO_4 \xrightarrow{\Delta} MoO_3 + 2NH_3 + H_2O$$

3. Reduction of MoO_3 into Mo metal. MoO_3 obtained as above is reduced to Mo metal either by H_2, carbon or Al (*Alumino-thermic process*). Mo metal obtained by the second method is contaminated with some carbide of molybdenum.

$$MoO_3 + 3H_2 \longrightarrow Mo + 3H_2O$$

$$MoO_3 + 3C \longrightarrow Mo + 3CO$$

$$MoO_3 + 2Al \longrightarrow Mo + Al_2O_3$$

4. Purification of Mo metal. Mo metal obtained as above is contaminated with impurities. In order to remove impurities, crude Mo metal obtained as above is heated with CO at 200°C under pressure of about 200 atmospheres when $Mo(CO)_6$ is formed.

$$Mo\ (\textit{Impure}) + 6CO \longrightarrow Mo(CO)_6$$

$Mo(CO)_6$, on being heated at 360–410°C, gets decomposed to form pure Mo metal.

$$Mo(CO)_6 \xrightarrow{\Delta} Mo\ (\textit{pure}) + 6CO \uparrow$$

Preparation of pure molybdenum metal

Thermal decomposition of molybdenum pentachloride on an electrically heated molybdenum filament gives pure metal.

$$2MoCl_5 \xrightarrow[\text{filament (Heat)}]{\text{Molybdenum}} 2Mo\ (\textit{pure}) + 5Cl_2$$

Properties

Massive Mo has typical silver white appearance and metallic properties. Mo has density 10.21 and high m.pt. (= 2610°C). The metal does not combine at ordinary temperature with non-metals but at elevated temperature it readily combines with O_2, S, C, N_2 and halogens to form MoO_3, MoS_2, MoC, MoN, MoF_6, $MoCl_5$ etc.

NH_3 has no action on Mo metal even when it is hot. H_2S attacks the metal at 1200°C to form the sulphide. It is oxidised to MoO_3 by CO_2.

$$Mo + 3CO_2 \longrightarrow MoO_3 + 3CO$$

It is only slowly attacked by acids. With hot and conc. HNO_3 it is oxidised to MoO_3.

$$Mo + 6HNO_3 \longrightarrow MoO_3 + 6NO_2 + 3H_2O$$

Fused caustic potash oxidises the metal to MoO_3. The metal dissolves in $AgNO_3$ with separation of Ag.

Uses of molybdenum and its compounds

(*i*) Molybdenum finds most important use in the production of alloy steels and high speed tools.

(*ii*) Molybdenum is used in thermo-couples for high temperature measurement.

(*iii*) Both in radio valves and bulbs molybdenum is used as filament support.

(*iv*) Mo is used in the manufacture of standard weights.

(*v*) In potteries, molybdenum compounds are used as blue pigments.

(*vi*) Molybdenum compounds are used for dyeing silk and wool and for colouring leather and rubber.

(*vii*) $(NH_4)_2MoO_4$ is used to detect and estimate phosphate and arsenic in chemical analysis.

(*viii*) MoO_3 is used as a catalyst in many chemical reactions.

Tungsten

Occurrence

Tungsten is a rare element, found combined in minerals which are usually associated with tin ores. The chief minerals are *wolframite* (Fe, Mn)WO_4; *scheelite*, $CaWO_4$ and *hubnerite*, $MnWO_3$, all the three containing 70 to 80 per cent of WO_3.

Extraction of tungsten from wolframite ore, (Fe, Mn)WO_4

The extraction of tungsten metal from wolframite involves the following steps :

1. Concentration of the ore. The powdered wolframite ore is concentrated by electromagnetic separation method to remove the non-magnetic impurities.

2. Conversion of wolframite ore into WO_3. The concentrated wolframite ore is fused with Na_2CO_3 in air. This operation gives a fused mass consisting of sodium tungstate, Na_2WO_4 (soluble) and a mixture of Fe_2O_3 and MnO_3 (insoluble).

$$\underbrace{4[(Fe, Mn)\ WO_4] \text{ or } (4FeWO_4 + 4MnWO_4)}_{\text{Wolframite ore}} + 8Na_2CO_3$$

$$+ 3O_2 \ (\textit{from air}) \xrightarrow{\text{Fusion}} \underbrace{\underset{\text{Soluble}}{8Na_2WO_4} + \underbrace{2Fe_2O_3 + 4MnO_2}_{\text{Insoluble}}}_{\text{Fused mass}} + 8CO_2 \uparrow$$

Now the fused mass is leached with water so that Na_2WO_4 goes into solution and the mixture of Fe_2O_3 and MnO_2 is left as residue. The solution containing Na_2WO_4, on being acidified with HCl, gives a precipitate of tungstic acid, H_2WO_4 which on ignition gives WO_3.

$$Na_2WO_4 + 2HCl \longrightarrow \underset{\text{ppt}}{H_2WO_4 \downarrow} + 2NaCl$$

$$H_2WO_4 \xrightarrow{\text{Ignition}} WO_3 + H_2O$$

3. Reduction of WO_3 into W metal. WO_3 obtained as above is reduced to W metal by either of the following methods.

$$WO_3 + 3H_2 \xrightarrow{1200°C} W + 3H_2O \ ; \qquad WO_3 + 3C \xrightarrow{\text{White heat}} W + 3CO$$

$$WO_3 + 2Al \longrightarrow Al_2O_3 + W$$

Preparation of pure tungsten metal

Pure tungsten metal can be obtained :

(*i*) by heating the tungstic acid, H_2WO_4 in a current of H_2.

$$H_2WO_4 + 3H_2 \longrightarrow W\ (pure) + 4H_2O$$

(*ii*) by thermal decomposition of WCl_6 vapour on a tungsten electric arc.

$$WCl_6 \longrightarrow W\ (pure) + 3Cl_2$$

Properties

Massive tungsten possesses typical white appearance and the metallic properties. Its density and m.pt. are 19.3 and 3410°C respectively. The metal is not acted upon by dil. and conc. acids and solutions of alkalies. It is, however, readily attacked by fused alkalies in presence of oxidising agents like KNO_3. The metal does not combine with elements at ordinary temperature but at high temperatures it interacts with O_2, S, N_2, halogens, C and H_2.

$2W + 3O_2 \longrightarrow 2WO_3$; $W + 2S \longrightarrow WS_2$; $W + C \longrightarrow WC$

$2W + C \longrightarrow W_2C$; $4W + N_2 \longrightarrow 2W_2N$; $2W + N_2 \longrightarrow 2WN$

$W + 3X_2 \longrightarrow WX_6$ (X = F, Cl, Br); $W + I_2 \longrightarrow WI_2$

Uses of tungsten and its compounds

The metal is used in making high speed tool steels, alloy steels, filaments of electric lamps, pen-points, strings of musical instruments, contact points, telegraphic keys, gramophone needles, voltage regulators etc. Surgical instruments are manufactured from its alloy namely *stellite* which contains W along with Co and Cr.

Manganese

Occurrence

Manganese does not ocur in the free state in nature. The chief ores of manganese are oxide ores which are : (*i*) *Pyrolusite*, MnO_2. This ore contains 60–63% manganese and is black in colour. It is found mainly in India (Panch Mahal, Karnatak, Singbhum, Madhya Pradesh, Orissa, Mysore and Rajsthan), Russia, South Africa, U.S.A. and Germany (*ii*) *Braunite*, Mn_2O_3. This ore is brownish-black in colour. (*iii*) *Hausmannite*, Mn_3O_4 which is brownish-black. (*iv*) *Manganite*, $Mn_2O_3.H_2O$ or 2MnO(OH) which is grey or black. Other less important ores of manganese are *manganese blende* (MnS), *rhodonite* ($MnSiO_3$), and *rhodochrosite* ($MnCO_3$). Manganese is commonly associated with iron ores. Its estimated abundance is 0.01% of the earth's crust. Traces of Mn occur in water, plants, animals and in many amethyst coloured minerals.

Extraction of Mn from oxide ores

Manganese may be extracted from its oxide ores like *pyrolusite* (MnO_2) and *hausmannite* (Mn_3O_4) by any of the following methods:

1. Carbon-reduction process. Mn is obtained by heating the mixture of the oxide ore *viz.* MnO_2 or Mn_3O_4 and carbon to white heat. In this process carbon reduces the oxides to free Mn-metal.

$$MnO_2 + 2C \longrightarrow Mn + 2CO \ ; \qquad Mn_3O_4 + 4C \longrightarrow 3Mn + 4CO$$

Mn obtained by this method is not pure, since it is contaminated with unreacted carbon. For obtaining pure metal less than the theoretical amount of carbon is used in this method.

2. Alumino-thermic process. The reaction between the pyrolusite ore (MnO_2) and aluminium is an explosive reaction. Hence MnO_2 is first converted into Mn_3O_4 by heating it to redness and then Mn_3O_4 thus obtained is reduced to Mn by metallic aluminium.

$$3MnO_2 \xrightarrow{\text{Heat to redness}} Mn_3O_4 + O_2$$

$$3Mn_3O_4 + 8Al \longrightarrow 9Mn + 4Al_2O_3 + \text{Heat evolved}$$

The heat produced in the reaction makes the reaction proceed very rapidly and vigorously and Mn melts and collects as a liquid at the bottom of the crucible and alumina (Al_2O_3) remains on the surface as a slag which solidifies to form artificial corundum. Mn in the molten state is prevented from being oxidised by the layer of Al_2O_3. Molten Mn may be allowed to flow out through the tapping plug. Molten manganese obtained by this method is fairly pure.

3. Electrolytic method : Production of pure manganese. In this method the oxide ores namely MnO_3 and Mn_3O_4 are converted into $MnSO_4$ by the methods represented by the following chemical equations :

(*i*) MnO_2 (*pyrolusite*) + $CO \xrightarrow{\text{Reduction}} MnO + CO_2 \uparrow$

$$MnO + H_2SO_4 \longrightarrow MnSO_4 \text{ (solution)} + H_2O$$

(*ii*) Mn_3O_4 (*hausmannite*) + $H_2SO_4 \longrightarrow 2MnSO_4$ (*solution*) + $MnO_2.2H_2O$ (*residue*)

Now when the solution of $MnSO_4$ obtained as above is electrolysed in a cell (cathode = mercury, anode = carbon), Mn metal gets liberated at Hg cathode with which it forms Mn-Hg amalgam. When this amalgam is distilled in vacuum at 250°C, Hg is removed and Mn is obtained.

Properties of manganese metal

Pure manganese is silvery white but is usually very pink tinge metal in industrial samples. It exists in three forms which are : (*a*) α-form which is stable below 742°C and has complex cubic structure (*b*) β-form which is stable between 742°C and 1190°C and like α-form, has complex cubic structure (*c*) γ-form which is stable between 1190°C and 1244°C and forms face-centred cubic tetragonal crystals. Finely divided Mn, when heated in air, Mn_3O_4 is obtained

$$3Mn + 2O_2 \longrightarrow Mn_3O_4$$

Dil.HCl and H_2SO_4 react with Mn to form Mn (+2) salts and H_2 gas is liberated

$$Mn + 2H^+ \longrightarrow Mn^{2+} + H_2\uparrow$$

With slightly more conc. HNO_3, NO is obtained

$$3Mn + 8HNO_3 \longrightarrow 3Mn(NO_3)_2 + 4H_2O + 2NO$$

With moderately conc. HNO_3, NO_2 is obtained

$$Mn + 4HNO_3 \longrightarrow Mn(NO_3)_2 + 2H_2O + 2NO_2$$

Conc. H_2SO_4 gives $MnSO_4$ and SO_2 is evolved

$$Mn + 2H_2SO_4 \text{ (conc.)} \longrightarrow MnSO_4 + 2H_2O + SO_2$$

Hot water gets decomposed by the action of Mn

$$Mn + 2H_2O \text{ (hot)} \longrightarrow Mn(OH)_2 + H_2$$

When heated with halogens, sulphur, nitrogen and carbon, halides, sulphide, nitrides (Mn_5N_2 and Mn_3N_2) and carbide are formed.

Uses of manganese

(*i*) Manganese forms a number of alloys with other metals. These alloys are very useful, *e.g.*, *spiegeleisen*, an alloy of Mn (Mn = 20–30%, C = 5%, rest is Fe) is used in the manufacture of steel.

(*ii*) Manganese compounds are very useful, *e.g.*, (*a*) MnO_2 is used as an oxidising agent in the production of hydroquinone and also used in dry cells (*b*) Manganese lactate, $Mn(C_3H_5O_3).3H_2O$ and manganese pyro phosphate $Mn(H_2PO_2)_2.H_2O$ are used in medicines.

Iron : Cast iron, Wrought iron and Steel

Occurrence of iron

Iron is the fourth most abundant element and second most abundant metal—the first three elements being oxygen, silicon and aluminium. Iron rarely occurs in the free state, since it is highly reactive.

The most important ores of iron are : (*i*) *Magnetite,* Fe_3O_4 which contains 70% or even more of Fe (*ii*) *Haematite*, Fe_2O_3. (*iii*) *Limonite or hydrated oxide*, $Fe_2O_3 . 3H_2O$. (*iv*) *Copper-iron pyrites,* $CuFeS_2$ or CuS.FeS (also called *chalco pyrites*). It is used for the extraction of Cu (*v*) *Sinderite or spathic iron,* $FeCO_3$. It is also called *clay iron stone* due to the presence of the excess of clay in it (*vi*) *Iron pyrites,* FeS_2. It is mainly used for the preparation of SO_2 (sulphuric acid) and not iron. (*vii*) *Arsenical pyrites,* FeAsS. India possesses about one fourth of the total world reserve of iron ore, compaired to U.S.A. (14%) and U.S.S.R. (24.7%). Huge deposits of iron are found in Singhbhum and Bonai. The iron ores of these places are of highest quality. Iron ores are widely distributed in Madhya Pradesh (Durg, Bastar,Katni, Bijori). Important deposits have been found in Maharashtra (Chanda, Loharu, Akola, Ratanpur, Ratnagiri), Mysore (Hospet, Bellary districts), Orissa (Cuttack, Koraput, Sambalpur, Mayurbhanj, Brrybil-Koira Valley). Small deposits of iron ore have also been located in Kerala, Himachal Pradesh (Mandi district), Assam (Kashi Hills), Punjab (Mahendragarh), Rajasthan (Alwar, Bundi, Udaipur and Jaipur), Uttar Pradesh and West Bengal.

Physical properties of iron

Pure iron possesses a grey colour and is malleable ductile and a good conductor of heat and electricity. It is the most magnetic of all metals but loses this property above 766°C. It melts at 1536°C, boils at 3000°C and density 7.86.

Chemical properties

1. Action of air or oxygen. When iron is strongly heated in air or oxygen, ferrosoferric oxide (tri ferric tetroxide) is formed.

$$3Fe + 2O_2 \longrightarrow Fe_3O_4$$

Iron wire burns when heated in oxygen or in oxy-coal gas flame.

2. Action of moisture : Rusting. Iron when exposed to moisture or water, or allowed to remain damp, is covered with a reddish yellow film or compound. This is known as *rust* and the phenomenon is known as *rusting*. Rust is a hydrated iron oxide with an approximate chemical composition $2Fe_2O_3.3H_2O$. Pure iron does not rust in pure air or pure water.

3. Action of water (steam). Iron decomposes steam.

$$3Fe + 4H_2O \rightleftharpoons Fe_3O_4 + 4H_2$$

4. Action of acids. Iron dissolves in dilute acids forming hydrogen and ferrous salts.

$$Fe + H_2SO_4 \longrightarrow FeSO_4 + H_2$$

$$Fe + 2HCl \longrightarrow FeCl_2 + H_2$$

Dilute nitric acid gives a mixture of ferrous nitrate and ammonium nitrate.

$$[Fe + 2HNO_3 \longrightarrow Fe(NO_3)_2 + 2H] \times 4$$

$$2HNO_3 + 8H \longrightarrow NH_4NO_3 + 3H_2O$$

On adding: $4Fe + 10HNO_3 \longrightarrow 4Fe(NO_3)_2 + NH_4NO_3 + 3H_2O$

Concentrated H_2SO_4 gives a mixture of ferrous and ferric sulphates.

$$Fe + 2H_2SO_4 \longrightarrow FeSO_4 + SO_2 + 2H_2O$$

$$2FeSO_4 + 2H_2SO_4 \longrightarrow Fe_2(SO_4)_3 + 2H_2O + SO_2$$

Fairly strong nitric acid gives ferric nitrate and a mixture of the oxides of nitrogen. Highly concentrated and pure nitric acid makes it *passive.*

5. Action of alkalies, halogens and sulphur. Alkalies have no action of iron. Halogens and sulphur combine with the heated metal forming halides and sulphide respectively.

$$2Fe + 3Cl_2 \longrightarrow 2FeCl_3$$

$$Fe + S \longrightarrow FeS$$

6. Displacement of less electropositive metals. For example, it displaces copper from a solution of copper sulphate.

$$Fe + Cu^{2+} \longrightarrow Fe^{2+} + Cu$$

7. Passivity. Concentrated nitric acid renders iron *passive.* When dipped in strong nitric acid, chromic acid, chloric acid or hydrogen peroxide iron becomes chemically inert and does not react further, *i.e.*, it does not liberate hydrogen with dilute acids. When scratched or hammered the activity is restored. The passivity of iron is due to the formation of very thin invisible coating of oxides on the metal surface.

Passivity of Iron

When a piece of iron is brought in contact with dilute sulphuric acid, a vigorous reaction begins and hydrogen gas starts coming out. The same piece of iron will displace copper from a solution of copper sulphate. It is noticed, however, that if the iron piece be kept for sometime in conc. HNO_3, then washed free from nitric acid and allowed to react in the above manner, it can neither liberate hydrogen from H_2SO_4 nor precipitate copper from copper sulphate solution. In other words, by treatment with nitric acid iron has so changed that it would not behave in the same manner as any other untreated piece of iron would do. *Iron in this form is said to be passive, i.e., it shows lesser reactivity than it would normally show.* Iron can also be made passive (*i.e.*, can be passivated) (*i*) by dipping it in a solution of other oxidising agents like H_2CrO_4, acidified $KMnO_4$, H_2O_2 (*ii*) by treating it with dry NO_2 (*iii*) by subjecting it to anode oxidation, *i.e.*, in electroplating or electrolysis, the anodes sometimes become passive, if a high current density is employed.

Degree of passivity

Passivity is not an absolute property and a metal exhibits variable degree of passivity depending on the chemical conditions. For example: (*i*) Oxidising conditions tend to promote passivity while reducing conditions tend to decrease it. (*ii*) Fe coupled with Pt is more easily passivated than iron alone while Fe coupled with Zn cannot be passivated.

Theories of passivity

1. Smith and Russel's theory. According to Smith and Russel (1926) the passivity is due to the change in electron configuration of atoms on the surface of the metal.

2. Oxide film theory. Faraday suggested that the passivity of iron is due to the formation of an extremely thin and impervious film on the surface of iron. This film is usually of ferroso-ferric oxide, Fe_3O_4. This theory is confirmed by the fact that if the oxide film on the surface is removed by scratching or by heating it in a reducing atmosphere of H_2 or CO or by dissolving it in iodine solution, iron again becomes active and begins to give its usual reactions. Evans strongly supported this theory in 1927 by his researches on the subject of passivity.

Rusting or Corrosion of Iron

When a piece of iron is left exposed to ordinary moist air, it is found covered with a reddish brown coating which can easily be detached and is called *rust*. The *process of formation of rust is called rusting.* Partington suggests that rust is chiefly *hydrated ferric oxide,* $2Fe_2O_3.3H_2O$. Some authorities believe that rust is an indefinite mixture which varies in its composition according to its age. Analysis shows that it is probably a mixture of *ferric oxide, hydrated ferrous* and *ferric oxides* and *basic ferrous* and *ferric carbonates.* If the rust is long exposed to the air, the amount of ferrous compounds becomes very small and ultimately hydrated ferric oxide predominates.

Conditions required for rusting

Conditions essentially required for the rusting of iron are; (*i*) *Presence of moisture,* (*ii*) *Presence of a weakly acidic atmosphere,* (*iii*) *Presence of impurity in the iron.*

Theories of rusting

A number of theories have been put forward to explain the phenomena of rusting. But the following theories are mentioned here.

1. Acid theory (Carbonate formation theory). According to this theory put foward by Calvert and Crum Brown in 1888, iron reacts with H_2CO_3 formed by the dissolution of CO_2 in H_2O to form $FeCO_3$ which is then hydrolysed in presence of dissolved oxygen to give a precipitate of $Fe(OH)_3$ and an acidic solution of H_2CO_3 which continues to attack iron.

$$CO_2 + H_2O \longrightarrow H_2CO_3$$

$$H_2CO_3 + Fe \longrightarrow FeCO_3 + H_2$$

$$FeCO_3 + 10H_2O + O_2 \longrightarrow \underset{\text{ppt}}{4Fe(OH)_3} + 4H_2CO_3$$

This theory fails to explain rusting in the absence of CO_2 and even in alkaline solutions. Alkalies prevent rusting, since they neutralise the CO_2 present.

2. Electrochemical theory. According to this theory rusting is an electrochemical phenomenon and takes place due to local action. Iron contains a number of impurities of lower oxidation potential and hence a kind of tiny voltaic cell (or an *electronic couple* or a *corrosion couple*) is set up between the impurities and iron itself in contact with H_2O containing dissolved O_2 and CO_2 in which iron acts as an anode.

Reaction. The surface of the moist iron is covered with water film which contains dissolved O_2 and CO_2. CO_2 reacts with H_2O to give H_2CO_3. Thus the surface of Fe is covered with a solution of an electrolyte *viz.* H_2CO_3 and a depolariser *viz.* dissolved O_2. H_2O and H_2CO_3 give H^+, HCO_3^- and OH^- ions as follows :

$$CO_2 + H_2O \longrightarrow H_2CO_3$$

At cathode : $$H_2CO_3 \rightleftharpoons H^+ + HCO_3^-$$

At anode : $$H_2O \rightleftharpoons H^+ + OH^-$$

H^+ ions originated from H_2CO_3 and H_2O go to the cathode where these are converted into H_2 molecules. These molecules of hydrogen are oxidized to H_2O by the dissolved O_2 which acts as a depolariser.

$$2H^+ + 2e^- \longrightarrow H_2 \;; \qquad 2H_2 + O_2 \longrightarrow 2H_2O$$

HCO_3^- and OH^- ions obtained from H_2CO_3 and H_2O respectively migrate to the anode where OH^- ions combine with Fe^{3+} ions (formed at the anode from Fe^{2+} ions—see below) to form $Fe(OH)_3$—a powdery deposit of rust.

$$Fe^{3+} + 3OH^- \longrightarrow \underset{\text{deposit of rust}}{Fe(OH)_3}$$

Fe^{3+} ions are formed at the anode as follows : Fe atoms at the anode are oxidised to Fe^{2+} ions by losing two electrons and these Fe^{2+} ions pass into solution and then are oxidised to Fe^{3+} ions by the dissolved O_2.

$Fe \longrightarrow Fe^{2+} + 2e^-$ (*Oxidation of Fe atoms to Fe^{2+} ions*)

$Fe^{2+} + H_2O + O_2$ (*Dissolved oxygen*) $\rightarrow 4Fe^{3+} + 4OH^-$ (*Oxidation of Fe^{2+} ions to Fe^{3+} ions*).

Protection of iron from rusting

The protection of iron and steel from rusting may be brought about :

(i) By coating iron with an impervious film of oils, paints (specially aluminium paints, red oxide), *coal tar, grease, pitch* etc. which do not allow iron to come in contact with air and moisture and thus protect iron from rusting.

(ii) By blowing steam over red-hot iron so that an adherent film of Fe_3O_4 gets deposited on the surface of iron (*Russian iron*) and thus iron is protected from rusting (*Barf process*).

(iii) By alloying iron water with Ni, Cr, Si etc. The alloys obtained by alloying iron water with NI, Cr Si etc. have excellent resistance towards rusting due to homogeneous make up and formation of a film of the oxide on the surface.

(iv) By metallic coatings. In this method the surface of iron is covered with a layer of a metal more active than iron, *e.g.*, Zn, Sn, Al, Cr etc. The zinc coating is usually obtained by galvanising which is carried out by dipping iron into a bath of molten zinc. The zinc coating of the galvanised iron does not allow iron to come in contact with air and moisture and thus protects it from rusting. Even when the zinc coating is partially removed from the surface of iron, iron does not get rusted. In this case both the metals *viz.* Fe and Zn come in contact with air. But, because Zn has higher oxidation potential, it gets oxidised to Zn^{2+} in preference to Fe, *i.e.*, reaction (*a*) takes place while the reaction (*b*) which is necessary for rusting does not take place.

(*a*) $Zn \longrightarrow Zn^{2+} + 2e^-$ (*Takes place*)

(*b*) $Fe \longrightarrow Fe^{2+} + 2e^-$ (*Does not take place*)

In other words it is the zinc and not iron that goes into solution.

(v) By other coatings. Some other types of coating like phosphate coatings and concrete coatings are also effective in protecting iron from rusting.

(vi) By using corrosion inhibitors. The substances that bring about chemical alteration of the metal surface so as to make the same rustproof are called corrosion inhibitors. The most commonly used inhibitor of this class is K_2CrO_4. In a solution of this salt iron will remain free form corrosion indefinitely.

Commercial varieties of iron

Depending on the carbon content present in iron, there are three commercial varieties of iron. These are *wrought iron or mellable iron, steel and cast iron or pig iron.*

The amount of carbon in steel is midway between that in wrought iron and in cast iron. First cast iron is manufactured from iron ores and then both wrought iron and steel are obtained from cast iron.

Cast Iron or Pig Iron

Composition of cast iron

This is the *least pure* form of commercial iron and contains the *highest percentage* of carbon *viz.*, 2.5 to 4.5% and traces of impurities like S, P, Mn and Si. The average composition of cast iron is : Fe = 93 – 95%, C = 2.5 – 4.5%, Si = 0.6 – 2.8%, P = 0.4 – 1.0%, S = 0.1 – 0.3%, Mn = 0.3 – 1.2%.

Varieties of cast iron

There are two varieties of cast iron.

1. White cast iron. This variety contains carbon mostly as *iron carbide*, Fe_3C and very small amount of graphite. It is obtained by cooling the molten pig iron quickly.

2. Grey cast iron. This variety contains carbon mostly as *graphite* and very small amount of Fe_3C. It is obtained by cooling the molten pig iron slowly.

Manufacture of cast iron from haematite ore, Fe_2O_3

The cast iron or pig iron is generally extracted from haematite ore, Fe_2O_3. The process involves the following steps:

1. Concentration of the ore. The oxide ore is first crushed in jaw crushers and is broken with small pieces of 3 to 5 cm in size. These are then washed with water to remove earthy and siliceous impurities.

2. Roasting in a reverberatory furnace. The concentrated ore is mixed with a little of coal and roasted (*i.e.*, heated strongly with excess of air) in a reverberatory furnace. During the process of roasting, the following changes occur :

(*a*) Moisture gets removed.

(*b*) The impurities of S, As, P etc. are removed as their volatile oxides

$$S + O_2 \longrightarrow SO_2\uparrow, \quad 4As + 3O_2 \longrightarrow 2As_2O_3\uparrow, \quad P_4 + 5O_2 \longrightarrow P_4O_{10}\uparrow$$

(*c*) If $FeCO_3$ is present as impurity, it gets decomposed into FeO which is oxidised to Fe_2O_3.

$$4FeCO_3 \longrightarrow FeO + CO_2$$

$$4FeO + O_2 \longrightarrow 2Fe_2O_3$$

Thus the formation of the slag of $FeSiO_3$ by the combination of SiO_2 and FeO is prevented during melting, *i.e.*, the following reaction does not occur.

$$SiO_2 + FeO \longrightarrow FeSiO_3 \ (slag)$$

(*d*) The entire mass becomes porous and hence the reduction of Fe_2O_3 to spongy iron becomes easy at later slage.

3. Reduction or smelting in blast furnace. The roasted ore (8 parts) is mixed with coke (4 parts) and lime stone (1 part). The mixture, known as *charge* is introduced from the top of a furnace called *blast furnace* with the help of cup and cone arrangement. The furnace is made up of high quality steel and is lined with fire proof bricks. At the base of the furnace, it is provided with (*i*) tuyers arrangement for the introduction of preheated air (*ii*) a tapping hole for withdrawing molten iron and (*iii*) an outlet through which slag can be flown out. (Construction and working of blast furnace have already been discussed in chapter 20 of *Advanced Inorganic Chemistry: Vol. I* by the same authors).

Various reactions taking place in the furnace

(*a*) A blast of dry preheated air is blown into the furnace from near the bottom of the furnace. Near the bottom (*combustion zone*), the preheated air comes in contact with the falling coke and combustion of coke into CO_2 takes place.

$$C + O_2 \longrightarrow CO_2 \qquad \Delta H = -393.5 \text{ kJ}$$

Heat produced (exothermic reaction) in the above reaction raises the temperature of the furnace to about 1600°C. This part of the furnace where the combustion of coke to CO_2 takes place is called *combustion zone*. In this part of the furnace, the oxide like P_4O_{10}, SiO_2 and MnO_2 which are present as impurities are reduced to the free elements by red-hot coke.

$$P_4O_{10} + 10C \longrightarrow 4P + 10CO\uparrow$$

$$SiO_2 + 2C \longrightarrow Si + 2CO\uparrow$$

$$MnO_2 + 2C \longrightarrow Mn + 2CO\uparrow$$

Si and Mn get dissolved into the molten iron and phosphorus combines with Fe to form Fe_3P.

(*b*) CO_2 produced in the combustion zone rises up and meets with more coke and gets reduced to CO.

$$CO_2 + C \longrightarrow 2CO \qquad \Delta H = +163.0 \text{ kJ}$$

Since the reduction of CO_2 to CO is an endothermic reaction (heat is required), temperature is decreased to about 1500°C. CO produced as above reduces the oxide ore, Fe_2O_3 to spongy Fe which might not have been reduced in the reduction zone.

$$Fe_2O_3 + 3CO \longrightarrow 2Fe + 3CO_2$$

This part of the furnace in which Fe_2O_3 is reduced to Fe is called *fusion zone.*

(*c*) In the middle of the furnace (*slag formation zone*) where the temperature varies from 800–1000°C, lime stone ($CaCO_3$) present in the charge decomposes into CaO and CO_2.

$$CaCO_3 \xrightarrow{800\text{–}1000°C} CaO + CO_2$$

CaO combines with the impurity of SiO_2 and forms a fusible slag of $CaSiO_3$. Thus CaO acts as a basic flux.

$$CaO\ (\textit{basic flux}) + SiO_2\ (\textit{acidic impurity}) \longrightarrow CaSiO_3\ (\textit{slag})$$

Slag is lighter than the molten iron. It moves down and floats over molten iron. This region where slag is formed is called *slag formation zone.*

(*d*) Near the *top* of the furnace (*reduction zone*), the temperature varies from 250–700°C. Here the oxide ore (Fe_2O_3) is reduced to spongy iron with the help of uprising vapours of CO produced in the *zone of fusion.*

$$Fe_2O_3 + 3CO \xrightarrow{400\text{–}700°C} 2Fe + 3CO_2$$

This part of the furnace where Fe_2O_3 is reduced to Fe is called *reduction zone.* The spongy iron produced as above in the reduction zone moves down slowly and melts in the fusion zone. Here the metal dissolves some C, Si, Mn and P (as Fe_3P) and forms the lower layer at the base of the furnace. The slag, being lighter, floats over the molten iron and prevents the oxidation of the molten iron. The slag and molten are removed from their respective outlets. The molten iron obtained in this way is solidified into block called *pigs.* This form of iron is, therefore, called pig iron.

Properties and uses of cast iron

It is the most impure form of commercial iron and contains the highest percentage of carbon and traces of impurities of S, P, Mn and Si.

The impurities present in cast iron have a remarkable influence on its properties. Cast iron melts at about 1250°C whereas pure iron melts at 1536°C. The presence of carbon and silicon increases the hardness and strength of the material. Phosphorus increases the fluidity of iron when molten, while sulphur makes it brittle when cold. Such an iron is called **cold short.** Sulphur increases the brittleness when red hot. Such a specimen is called **red short**.

Cast iron is very hard and can not be welded. It is so brittle that it cannot be used for making parts of machinery that are subjected to great stress. It expands on solidification and is used for casting various articles such as stoves, pipes, radiators, railway sleepers, gutter pipes, toys etc. A large quantity of cast iron is used in the manufacture of wrought iron and steel.

Wrought Iron

Composition of wrought iron

It is the purest form of commercial iron and contains the lowest percentage of carbon *viz.* 0.12 to 0.25% and 0.3% of impurities like S, P, Si and Mn.

Manufacture of wrought iron

Wrought iron is manufactured from *puddling process* in which the cast iron is heated in a special type of reverberatory furnace called *puddling furnace.* This furnace has a low roof to deflect the hot gases and flames downwards and to melt cast iron. The hearth of furnace is lined with haematite (Fe_2O_3).

The cast iron is piled on the hearth, melted by the hot gases and stirred or puddled with long iron rods called *rabbles* through the door to bring it in thorough contact with the lining of the hearth.

The haematite (Fe_2O_3) lining supplies oxygen necessary for the oxidation of carbon, sulphur, silicon, manganese and phosphorus present in the cast iron. Oxides of carbon and sulphur, being volatile, escape while those of manganese, silicon and phosphorus form slags. Thus :

$$3C + \underset{\text{Lining}}{Fe_2O_3} \longrightarrow 2Fe + 3CO\uparrow$$

$$3S + 2Fe_2O_3 \longrightarrow 4Fe + 3SO_2\uparrow$$

$$3Mn + Fe_2O_3 \longrightarrow 2Fe + 3MnO$$

$$3SiO_2 + 2Fe_2O_3 \longrightarrow 4Fe + 3SiO_2$$

$$6P + 5Fe_2O_3 \longrightarrow 10Fe + 2P_2O_5$$

$$MnO + SiO_2 \longrightarrow MnSiO_3 \ (\textit{slag})$$

$$P_2O_5 + Fe_2O_3 \longrightarrow 2FePO_4 \ (\textit{slag})$$

With the removal of impurities, the melting point of the metal rises and it becomes a semi-solid mass. As this stage, it is taken out in the form of balls or *blooms* on the ends of rabbles. While still hot, these balls are subjected to hammering to squeeze out as much of slag as possible. The product so obtained is known as wrought iron.

Properties

(*i*) It is extremely tough, highly malleable, ductile and shows resistance towards rusting and corrosion. This nature of wrought iron is due to the presence of slag in it.

(*ii*) It is brittle and can be easily welded and worked under hammer so as to assume any desired shape. Its melting point is 1500°C.

Uses

(*i*) Due to its tough nature, it is used to make articles which are to stand severe and sudden stress such as anchors, wires, bolts, chains and agricultural implements.

(*ii*) It is also used to make magnets and dynamos for electric motors.

Steel

Composition of steel

Steel is an alloy of iron (as ferrite) containing 0.25 to 2.5% of carbon (as cementite, Fe_3C) and traces of S, P, Si and Mn. Thus, we see that the percentage of carbon in steel is intermediate between that in wrought iron and cast iron. S and P are objectionable impurities. The percentage of S and P are usually below 0.05%, depending on the method used for the manufacture of steel. Si is usually between 0.15 to 0.35% and is present as iron silicide, FeSi which is soluble in ferrite. High content of Si gives a fibrous structure to steel. Mn is added to steel during its manufacture to serve as a deoxidiser and to neutralise the harmful effect of sulphur by forming MnS which is inert, while FeS makes steel brittle in hot working.

Varieties of steel

There are many varieties of steel depending on the amount of carbon present in it.

1. Mild and quenched steels. Mild steel is a variety which contains *lower percentage* of carbon. It possesses the properties of wrought iron along with elasticity and hardness of steel.

If mild steel is heated to a high temperature (*i.e.,* to bright redness) and then suddenly cooled by plunging in oil or water, it becomes as hard and brittle as glass. This is called *quenched steel* and process is called *quenching* or *hardening*.

2. Hard steel. This variety contains *higher percentage* of carbon. It is hard and brittle like glass.

3. Special steels or alloy steels. Addition of small amounts of nickel, cobalt, chromium, tungsten, molybdenum, manganese and silicon confer special properties on steel considerably altering its hardness, tenacity, resistance to corrosion and coefficient of expansion. Such products are called special steels or alloy steels and find extensive use in industry.

Some of the alloy steels along with their composition, properties and uses are listed in Table 8.2. given below :

Table 8.2. *Some important alloy steels.*

Name	*Percentage composition*	*Properties*	*Uses*
1. Stainless steel	Fe = 73 Cr = 18 Ni = 8 and carbon	Hard and rust-proof	Utensils, cycle and automobile parts, curtlery.
2. Nickel steel	Fe = 98 – 96 Ni = 2 – 4	Hard, elastic and rust-proof	Cables, automobile and aeroplane parts, armour plates, gears
3. Invar	Fe = 64, C = 0.3 Ni = 36	Low expansion on heating	Metre scales, measuring instruments, clock pendulums.
4. Tungsten steel	Fe = 83 – 72 W = 14 – 20 Cr = 3 – 8	Hard, resistant to corrosion	High speed cutting tools, springs
5. Silicon steel	Fe = 84 Si = 15	Hard, and resistant to acid	Pumps and pipes for carrying acids.
6. Manganese steel	Fe = 88 – 85 Mn = 12 – 15	Extremely hard, high m.pt.	Rock crushers, burglar proof safes, rail tracks.
7. Perma alloy	Fe = 21 Ni = 78 and carbon	Strongly magnetised by electric current and demagnetised upon cutting off current.	Electromagnets, cables for oceans.
8. Alnico	Fe = 63, Ni = 20 Al = 12, Co = 5	Highly magnetic	Permanent magnets.

(*iv*) **Steel containing S and P.** Iron or steel containing excess of S is brittle when hot (*hot short*) and that containing excess of P is also brittle when cold (*cold short*).

Manufacture of steel

We have already said that the amount of carbon in steel is mid-way between that of cast iron and in wrought iron so that steel can be obtained either by removing a part of carbon from cast iron or by adding some carbon to wrought iron. Thus steel is prepared either from cast iron or from wrought iron. Following methods are used:

1. Bessemer's process

This process was discovered by Sir Henry Bessemer of England in 1856.

Principle of the process. This process is based on the fact that the impurities present in cast iron are completely oxidised in presence of hot air blast, *i.e.*, virtually wrought iron is obtained. This is then mixed with known amount of *spiegeleisen*, an alloy of Fe, Mn and C to obtain steel.

In this process molten cast iron is taken directly from the blast furnace and is put into Bessemer's convertor.

Construction of Bessemer's convertor. Bessemer's convertor is a pear-shaped furnace, about 20m high and 10m in diameter. It is made of steel plates and has a number of holes (called tuyers) at its base through which a blast of hot air can be admitted into the convertor. The convertor is rotated on its horizontal central axis (Fig. 8.1).

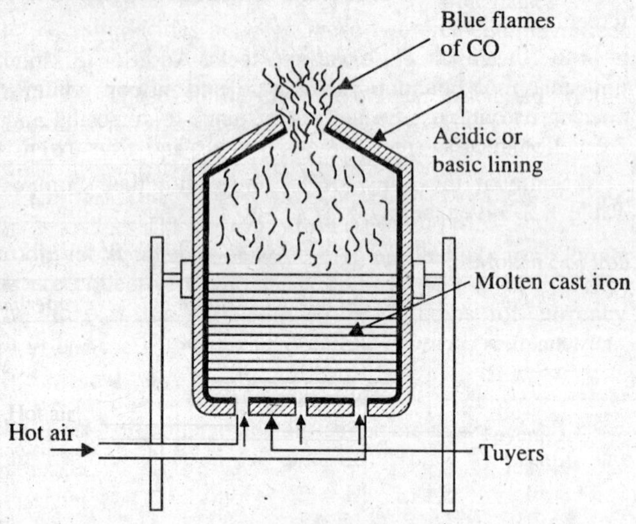

Fig. 8.1. Bessemer's convertor used for the manufacture of steel from cast iron.

Depending on the nature of the impurities present in the cast iron, the convertor is lined on its inside with silica, SiO_2 (acidic lining) or lime (CaO)/magnesia (MgO) (basic lining). If the cast iron contains C, S, Si and Mn as impurities silica lining is used and the process is called *acid Bessemer's process*. On the other hand, if the cast iron contains P as impurity, CaO/MgO lining is used and the process is known as *basic Bessemer's process.* In actual practice, basic lining is obtained by using a mixture of CaO and MgO mixed with tar to bind them. The materials (SiO_2, CaO and MgO) used for lining the convertor act as flux as will be evident from the following discussion.

Working. To start with, the convertor is brought into the horizontal position, molten cast iron is put into it and then a blast of hot air or O_2 is blown into the convertor through its tuyers. The convertor is then rotated so that its mouth comes in vertically upwards position. The blowing of hot air is continued.

Reactions taking place in Bessemer's convertor

(*i*) In the beginning Si, Mn and P (impurities) get oxidised by O_2 to their respective oxides.

$$Si + O_2 \longrightarrow SiO_2$$
$$2Mn + O_2 \longrightarrow 2MnO$$
$$4P + 5O_2 \longrightarrow 2P_2O_5$$

MnO produced as above combines with SiO_2 lining (acidic lining) of the Bessemer's convertor to form the slag of $MnSiO_3$ and is thus removed.

MnO (*basic impurity*) + SiO_2 (*acidic lining acting as an acidic flux*) $\longrightarrow$ $MnSiO_3$ (*slag*)

P_2O_5 and SiO_2 produced react with CaO lining (*basic lining*) of the Bessemer's convertor to form the slags of $Ca_3(PO_4)_2$ and $CaSiO_3$ and are thus removed.

P_2O_5 (*acidic impurity*) + 3CaO (*basic lining acting as a basic flux*) $\longrightarrow$ $Ca_3(PO_4)_2$ (*slag*)

SiO_2 (*acidic impurity*) + CaO (*basic lining acting as a basic flux*) $\longrightarrow$ $CaSiO_3$ (*slag*)

The mixture of the three slags *viz.* $MnSiO_3$, $Ca_3(PO_4)_2$ and $CaSiO_3$ is called basic slag which is used as a fertiliser. The present discussion makes it evident that SiO_2, CaO and MgO which are used for lining the convertor act as fluxes.

(*ii*) Later on C and S are oxidised to CO and SO_2 respectively by the blast of air. SO_2 escapes from the mouth of the convertor and CO burns with a blue flame at the mouth of the

Bessmer convertor (*Bessemer's candles*). When the whole of carbon present in the cast iron has been oxidised to CO, the burning of CO at the mouth of the convertor stops. At this stage, the blowing of O_2 into the convertor is stopped. Here it should be noted that the oxidation of impurities into their oxides is exothermic and hence no external heat is required for the formation of slags. After all the impurities have been removed, the calculated quantity of *spiegeleisen* alloy (an alloy of iron and manganese having Mn = 20–30%, C = 5%, Fe = remaining quantity) is added to the molten cast iron into the convertor and blast of hot O_2 is again allowed to enter into the convertor so that carbon and Mn present in spiegeleisen alloy get mixed with cast iron and steel is obtained which is taken out of the convertor in the molten state. Steel obtained in this way is called *manganese steel.* Spiegeleisen alloy being added does the following functions:

(*a*) It acts as a scavenger (*i.e.*, cleaning agent), since Mn present in the alloy removes O_2, N_2, P and S, if dissolved in the molten steel, in the form of its oxide (MnO), nitride (Mn_3N_2), phosphide (Mn_3P_2) and sulphide (MnS). If O_2 and N_2 are not removed, these gases, on cooling, become less soluble and form bubbles (called blow holes) in steel. These bubbles weaken the strength of the steel. Thus Mn prevents the formation of blow holes. Some Al or ferro-silicon may also be added to remove blow holes. Due to the formation of MnS, the harmful effects of sulphur are neutralised.

(*b*) It acts as a *de-oxidiser*, since Mn present in the alloy reduces oxidised iron, if present.

(*c*) The alloy supplies the required quantity of carbon, Mn and Fe to get the steel of the required composition. Note that carbon is an essential constituent of all types of steel. Mn makes the steel hard and increases its tensile strength. If more quantity of Mn (upto 14%) is added in the form of spiegeleisen alloy, manganese steel obtained is very hard, tough for making steel helmets, safes, cross overs etc.

Requisite quantity of other metals like Cr, Ni etc. along with carbon are also added to get the steel of a particular composition.

2. Open-hearth process or Siemens-Martin process

This is the modern process and the furnace used in this process is called *open-hearth furnace*. Since this furnace works on the regenerative principle of heat economy, it is also called *regenerative furnace*. In this process steel is prepared from cast iron. Open-hearth furnace consists of an open hearth. The hearth is lined with silica or calcined dolomite (CaO.MgO) depending upon the nature of impurities present in pig or cast iron. Silica lining is used if the impurities are manganese, silicon, etc., and calcined dolomite lining is used if much of phosphorus is present. A high temperature of about 1500°C is generated by burning producer gas which works on regenerative system of heat economy (Fig. 8.2).

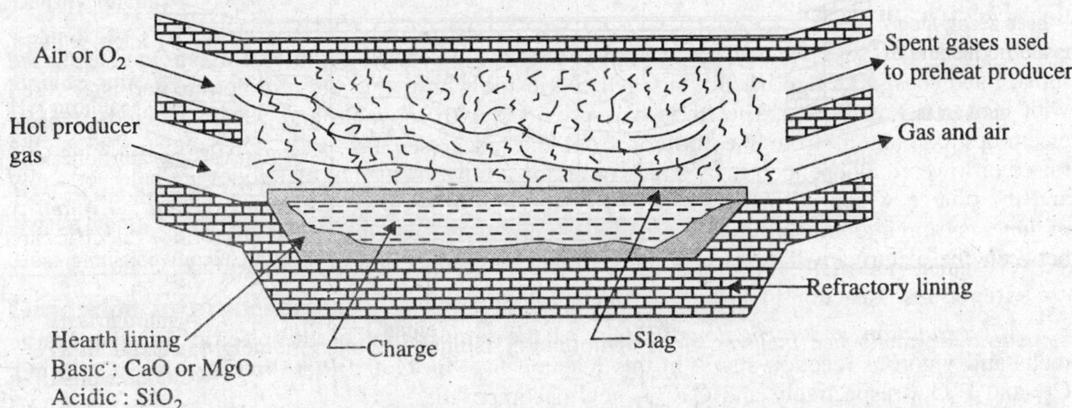

Fig. 8.2. Open hearth furnace for the manufacture of steel from cast iron.

The charge consists of pig or cast iron, iron scraps, iron ore (haematite) and lime. The charge is heated on the hearth of the furnace. The impurities are oxidised by iron ore.

$$3Si + 2Fe_2O_3 \longrightarrow 4Fe + 3SiO_2$$
$$3Mn + Fe_2O_3 \longrightarrow 2Fe + 3MnO$$
$$MnO + SiO_2 \longrightarrow MnSiO_3 \ (slag)$$
$$3C + Fe_2O_3 \longrightarrow 2Fe + 3CO$$
$$12P + 10Fe_2O_3 \longrightarrow 3P_4O_{10} + 20Fe$$
$$6CaO + P_4O_{10} \longrightarrow 2Ca_3(PO_4)_2 \ (slag)$$
$$CaO + SiO_2 \longrightarrow CaSiO_3 \ (slag)$$
$$2Fe_2O_3 + 3S \longrightarrow 3SO_2 + 4Fe$$

Samples of steel are drawn from time to time and tested for carbon content. Finally *spiegleisen* (an alloy of iron, manganese and carbon) is added to the molten mass and obtain desired steel. The process takes about 8 to 10 hours for completion. The process takes longer time than Bessemer process.

Advantages of open-hearth process over Bessemer's process. Open-hearth process has the following advantages over the Bessemer's process.

(*i*) The temperature can be controlled as the heating is done externally.

(*ii*) As it is a slower process, it can be controlled in better way. The composition and qualtity can be well controlled.

(*iii*) The loss of iron in this process is only 4% while the loss is about 15% in Bessemer's process.

(*iv*) In this process scrap iron is re-used.

(*v*) This yields better quality of steel.

(*vi*) A considerable economy of the fuel is achieved by using the regenerative system.

3. Duplex process

Now-a-days Duplex process is being used for the manufacture of large quantities of steel. This process is a combination of Bessemer's process and open-hearth process. In this process the molten pig iron is first treated in an acid Bessemer's convertor to remove Si, Mn and a part of C. The molten pig iron is then transferred to open-hearth furnace with a basic lining to remove P and remaining C and is finished off as usual. Tata Iron and Steel Works are using Duplex process for the manufacture of steel.

4. Electric process

In this process steel is prepared in arc furnace (also called Heroult's furnace). This is a crucible-shaped furnace and consists of steel shell which is lined inside with magnesite or dolomite and is covered with heat resisting bricks. The furnace is provided with movable and water jacketed graphite electrodes coming in from the roof or from the sides (see Fig. 8.3). Each electrode can be raised or lowered independently by a rack or pinion arrangement. The electrodes are held vertically and the charge which consists of *Bessemer's steel, iron ore haematite* and *calculated quantity of lime* is run into the arc furnace and the current is switched on the produce electric arc between the electrodes. This electric arc produces a temperature of 3000–3500°C.

Reactions. Reactions taking place in this process are :

(*a*) *Oxidation and formation of slag.* At the temperature of the electric arc, the charge melts and vigorous reaction starts. In this reaction Si, Mn, C and P are oxidised to SiO_2, MnO, CO and P_4O_{10} respectively and $CaCO_3$ gets decomposed.

$$3Si + \underset{\text{(Iron ore)}}{2Fe_2O_3} \longrightarrow 4Fe + 3SiO_2$$

$$3Mn + Fe_2O_3 \longrightarrow 2Fe + 3MnO$$
$$C + O_2 \longrightarrow CO_2$$
$$CO_2 + C \longrightarrow 2CO$$
$$P_4 + 5O_2 \longrightarrow P_4O_{10}$$
$$CaCO_3 \longrightarrow CaO + CO_2$$

The oxides *viz.* SiO_2, MnO, CO and P_4O_{10} are removed as slags.

$$SiO_2 + CaO \longrightarrow \underset{\text{Slag}}{CaSiO_3}$$
$$MnO + SiO_2 \longrightarrow \underset{\text{Slag}}{MnSiO_3}$$
$$Fe_2O_3 + CO \longrightarrow CO_2 + 2FeO$$
$$FeO + SiO_2 \longrightarrow \underset{\text{Slag}}{FeSiO_3}$$
$$P_4O_{10} + 6CaO \longrightarrow \underset{\text{Slag}}{2Ca_3(PO_4)_2}$$

The slag being lighter floats over the metallic portion. As soon as the oxidation and formation of slag is complete, the furnace is carefully tilted and the slag is poured out of it.

(*b*) *Deoxidation and desulphurization.* After the slag is poured out of the furnace, a charge consisting of *lime, sand, fluorspar* and some *coke* and *ferro-silicon* is added to the furnace. Any FeO present in the steel is removed as Fe by reducing it with the added coke.

$$FeO + C \longrightarrow Fe + CO \quad (\textit{Removal of FeO})$$

Sulphur present as FeS is removed as CaS by treating it with CaO

$$FeS + CaO \longrightarrow FeO + CaS \quad (\textit{Removal of S})$$

FeO thus produced is at once reduced to Fe as before while CaS goes into the slag. When desulphurization (*i.e.*, removal of sulphur) is complete, the steel is poured out into moulds.

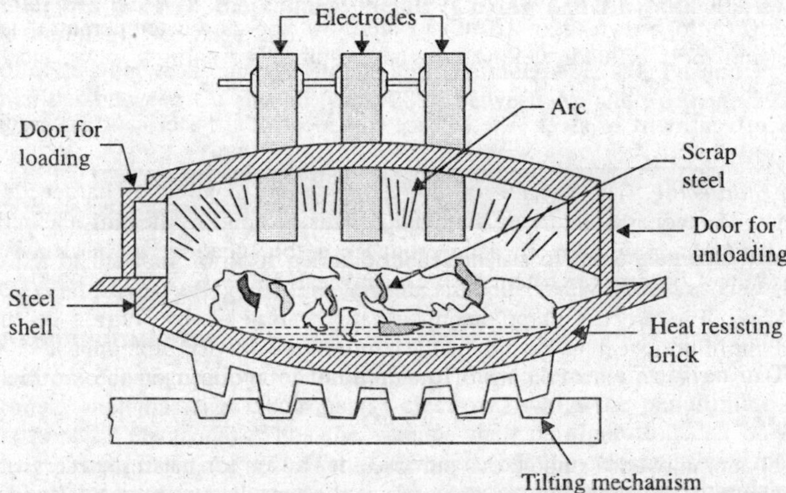

Fig. 8.3. Arc funrnace (Heroult's furnace) for the manufacture of steel.

This method is also used for the manufacture of alloy steels, tool steels, stainless steels and special quality steels.

5. L.D. process

This process has recently been developed in Lintz and Dusenverfahren in Austria and is in use at the *Rourkela steel plant*. The main advantages of the process are :

(*i*) Less capital expenditure and more productivity.

(*ii*) It takes only 45 minutes to treat one lot.

(*iii*) Even scrap iron can be used.

(*iv*) Less operating cost and better quality of steel.

The process is carried out in a convertor lined with magnesite which differs from the Bessemer convertor in not having the perforated base but a solid bottom. Pure O_2 is blown from the top through a copper lance (Fig. 8.4).

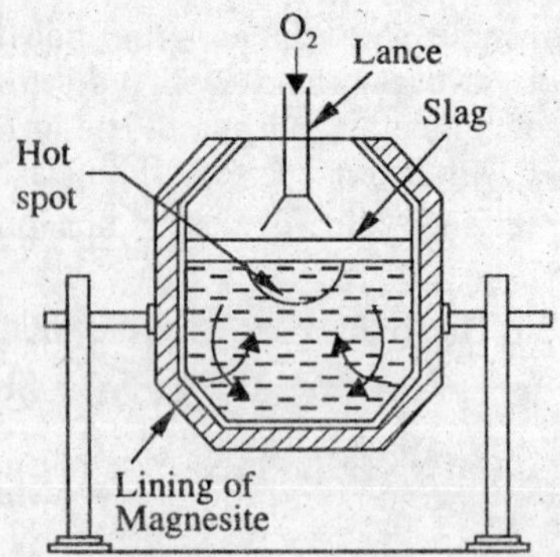

Fig. 8.4. Converter for L.D. process.

Properties of steel

1. Mechanical properties. Steel combines the useful properties of cast iron and wrought iron, being both *hard* and *elastic* and provided the proportion of carbon is not high, it can be *forged* and *welded.*

2. Action of Heat : Heat treatment of steel. A characteristic behaviour of steel which distinguishes it from other commercial forms of iron is that its hardness and elasticity can be varied by proper heat treatment of steel. For example :

(*a*) *Anealing.* When hard steel is heated to bright redness and then allowed to cool slowly, hardness of steel is considerably decreased and it gets softened. This process is called *anealing*.

(*b*) *Quenching or hardening.* When mild steel is heated to a high temperature (*i.e.*, to bright redness) and then suddenly cooled by plunging it in oil or water, it becomes as hard and brittle as glass. This process is called *quenching* or *hardening* and the steel obtained by this method is called *quenched steel.*

(*c*) *Tempering.* The quenched steel can be made to lose its brittleness by heating it to varying temperatures for varying periods. The process is called *tempering* and is employed for bringing the steel into a suitable state of hardness and elasticity. The temperature is generally judged from the colour of a thin film of oxide which is formed on the surface and varies from yellow (200–230°C) to brown (225–270°C) to blue (300°C) as the temperature rises. Steel for cutting blades and tools is made yellow but that required for springs, saws, etc., is tampered blue.

3. Surface treatment of steel. The surface treatment of the steel is done by the following proceses :

(*a*) *Case hardening.* The process of producing a thin coating of hardened steel on the surface of the mild steel is called case hardening. This is done by heating the mild steel with charcoal and then plunging into oil. This produces a thin coatinig of hardened steel on the surface. Such a steel becomes resistant to wear and tear.

(*b*) *Nitriding.* The process of producing a hard coating of iron nitride on the surface of steel is called nitriding. Steel is heated in the atmosphere of dry ammonia at 500–600°C for about 3 to 4 days when a hard coating of iron nitride is produced on the surface.

Uses of steel

(*i*) Steel is employed for numerous purposes. It is used for making girders, armour plates, machinery, knives, razors, swords, tools, springs, dies, etc.

(*ii*) It is also used for making bar magnets and magnetic needles.

Comparison between cast iron, wrought iron and steel

The comparative study between the three varities of iron is given in Table 8.3.

Table 8.3. *Comparison between cast iron, wrought iron and steel.*

	Property	*Cast iron*	*Wrought iron*	*Steel*
1.	Chemical composition	Iron 93–95%, Carbon 2.5–5% Impurities about 2%	Iron 99.5–99.8%, Carbon 0.1–0.25% Impurities about 0.3%	Iron 99.5–98.0%, Carbon 0.25 to 2.0%
2.	Melting point	Lowest about 1200°C	Highest about 1500°C	Between 1300–1400°C
3.	Hardness	Very hard	Soft	Medium hardness
4.	Malleability	Brittle	Malleable	Malleable and brittle
5.	Welding	Can not be welded	Can be welded easily	Can be welded with difficulty
6.	Tempering	Can not be tempered	Can not be tempered	Can be tempered
7.	Magnetisation	Can not be permanently magnetised	Magnetisation is not permanent but easy	Can be permanently magnetised
8.	Structure	Crystalline	Fibrous	Granular

Osmium

Occurrence

Osmium mainly occurs in nature associated with platinum metals. Minerals like *laurite* and *osmiridium* also contain osmium 3% and 0.8–20% respectively. Osmiridium is found in Russia, California, Spain, Canada etc. The mineral *irite* which is a black magnetic substance contains Ir, Os, Fe_2O_3 and chromite ($FeO.Cr_2O_3$). Osmiridium is an alloy of Os and Ir. This alloy also contains some quantity of ruthenium (Ru)

Extraction by fusion process

We have said above that osmiridium is an alloy of Os and Ir and also contains some amount of ruthenium (Ru). This alloy is fused with alkaline oxidising mixture of KOH and KNO_3 when Ru is converted into soluble K_2RuO_4 (pot. ruthenate). After dissolving and acidifying, the mixture is boiled when Os distils off as OsO_4 which is reduced to osmium metal with the help of hydrogen or some other suitable reducing agent in acidic medium.

$$OsO_4 + 8H^+ + 8e^- \longrightarrow Os(s) + O + 4H_2O$$

OsO_4 can be reduced to osmium metal *easily* in acidic medium, since the reduction potential for the above reduction reaction is a positive quantity which is equal to + 0.85 V.

Physical properties

The compact metal is bluish-grey, and in the finely divided state it is bluish-black. Osmium is a hard and brittle metal. It is the heaviest element known, its sp. gr. being 22.6. It has the highest m.pt. equal to 2700°C and very high boiling point, 5500°C. It is very difficult to prepare very pure metal.

Chemical properties

Compact metal is not affected by air or oxygen even at 100°C, but is oxidised at higher temperatures. Finely divided osmium undergoes slow oxidation in air even at room temperature which is indicated by the smell of its oxidation product, the tetroxide OsO_4. The metal burns on strong heating in air forming OsO_4, in some cases the spongy osmium even glows.

The metal powder reacts with fluorine at 100°C or above with incandescence, forming tetra, hexa and octa fluorides. With chlorine it forms a dark green dichloride, $OsCl_2$ and a red mass of the tetrachloride, $OsCl_4$ at 650°C. Bromine and iodine do not react with the metal.

It is not attacked by hydrochloric acid in the absence of oxygen, but is oxidised by HNO_3 and H_2SO_4. The metal is dissolved readily in a concentrated solution of sodium or potassium hypo-chlorite.

Osmium burns brightly in sulphur vapour forming its sulphide. It combines with phosphorus vapour with incandescence forming its phosphite. It forms alloys with metals such as copper, silver, gold, zinc, mercury, iron, cobalt, nickel etc.

When osmium is fused with potassium hydroxide in oxygen or air, potassium osmate, K_2OsO_4, is formed.

$$Os + 2KOH + 3O \longrightarrow K_2OsO_4 + H_2O$$

Uses

Osmium is used in making hard alloys for fountain pens, phonograph needles and for sparking points, it is also used in lamp mantles. A solution of OsO_4 known as osmic acid blacknes organic materials due to reduction to metallic state and is, therefore, used in differentiating tissues under microscope.

Cobalt

Occurrence

Cobalt does not occur free in nature. It constitutes about 0.001% of the earth's crust. Cobalt occurs mainly as *arsenide* and *sulphide* ores with which other metals like Fe, Ni and Ag are also associated. Its principal ores are : *smaltite* or *speiss cobalt,* (Ni, Co, Fe)As_2; *cobaltite* or *cobalt glance,* (Co, Fe, As) S; *crythrite* or *cobalt bloom*, $Co_3(AsO_4)_2.8H_2O$ and *linnacite*, $(Fe,Co, Ni)_3S_4$. All these ores are associated with Fe, Ni, Cu and sometimes with traces of Ag. Many cobalt ores also contain profitable amounts of Ag. Small amounts of Co are also present in the ores of some other metals and are obtained from them as by-product, *e.g.*, Co is also obtained as by-product during the refining of Cu, Ni, Fe and Pb. Cobalt is also present in vitamin-12 and its presence in this vitamin is of physiological interest. Cobalt ores are found in Canada, Northern Rhodesia and Belgian Congo. In India cobalt ores are found in Rajasthan.

Extraction from cobaltite ore, (Co, Fe, As)S

Due to the presence of several other metals in the ores of cobalt, the extraction of cobalt is a very complex process. Cobalt ores are generally worked for the manufacture of cobalt salts and not for the free metal. The extraction of cobalt from cobaltite ore, (Co, Fe, As)S involves the following steps:

1. Concentration. The ore is crushed and finely powdered and then concentrated by the froth-floation process.

2. Roasting. The concentrated ore contains Fe, Co, Ni, Ag, As, Cu and S as the chief constituents. It is roasted in a free supply of air which removes S as SO_2 and As as As_2O_3 and FeS is oxidised to FeO.

$$S + O_2 \longrightarrow SO_2$$

$$4As + 3O_2 \longrightarrow 2As_2O_3\uparrow$$

$$2FeS + O_2 \longrightarrow 2FeO + 2S$$

As_2O_3 is condensed and used as a source of As.

3. Smelting. The roasted mass is smelted in a small furnace with a fuel *e.g.*, coke and fluxing material (*e.g.*, sand, SiO_2 and lime stone, $CaCO_3$). Here FeO produced during roasting and CaO produced from $CaCO_3$ combine with SiO_2 to form a fusible slag at the top of the molten mass.

$$CaCO_3 \longrightarrow CaO + CO_2$$

$$CaO + SiO_2 \longrightarrow \underset{\text{slag}}{CaSiO_3} \ ; \quad FeO + SiO_2 \longrightarrow \underset{\text{slag}}{FeSiO_3}$$

The molten mass obtained at the end of smelting separates into the following three layers which contain different materials :

(*i*) *Slag.* This is the top layer and consists of earthy matter present in the ore and slag of $FeSiO_3$ and $CaSiO_3$ formed during smelting. This is rejected.

(*ii*) *Speiss.* This is the middle layer and consists of fused arsenides of Co, Ni, Fe and Cu and some As and S.

(*iii*) *Bullion.* This is the bottom layer and consists of easily reducible and impure metals, mainly Ag and Cu. It is used for the recovery of 99% pure Ag.

4. Roasting the speiss with NaCl. The speiss is finely ground and roasted in air in a reverberatory furnace to convert As and S, if any, into As_2O_3 and SO_2 which are driven off. It is then mixed with NaCl and then roasted whereby the metals *viz.* Co, Ni, Fe and Cu are converted into their chlorides.

5. Separation of Co, Ni, Fe and Cu. The roasted speiss is extracted with H_2O to get the aqueous solution of the chlorides. This aqueous solution is then treated with lime stone ($CaCO_3$) to precipitate Fe as $FeCO_3$ and the filtrate (after filtration) with Na_2CO_3 to precipitate Cu as $CuCO_3$. By this treatment the chlorides of Co and Ni remain unaffected and hence remain in solution. Ag (in the form of AgCl), if present, is extracted by using a dil. solution of NaCN (*Cyanide process*).

6. Separation of Ni from Co. The solution containing the chlorides of Ni and Co is cautiously treated with lime, $Ca(OH)_2$ and bleaching powder, $CaOCl_2$ whereby $CoCl_2$ is converted into a precipitate of $Co(OH)_3$ and $NiCl_2$ remains in solution.

$$CoCl_2 + \underset{\text{Lime}}{Ca(OH)_2} \longrightarrow Co(OH)_2 + CaCl_2$$

$$2Co(OH)_2 + H_2O + \underset{\substack{\text{Bleaching}\\\text{powder}}}{CaOCl_2} \longrightarrow \underset{\text{ppt}}{2Co(OH)_3} + CaCl_2$$

$NiCl_2$ that remains in solution can be converted into basic carbonate, if necessary.

7. Formation of Co_3O_4. The precipitate of $Co(OH)_3$ obtained as above is dried and ignited, when cobalto-cobaltic oxide Co_3O_4 is obtained.

$$12Co(OH)_3 \xrightarrow{\text{Ignite}} 4Co_3O_4 + 18H_2O + O_2$$

8. Reduction of Co_3O_4 to Co. Generally Co_3O_4 is used as such for the manufacture of cobalt compounds. In case cobalt metal is needed, it may be obtained by reducing Co_3O_4 by heating it with coke in an oil-fired furnace or with aluminium (*Alumino thermic process*).

$$Co_3O_4 + 4C \longrightarrow 3Co + 4CO$$

$$3Co_3O_4 + 4Al \longrightarrow 9Co + 4Al_2O_3$$

Preparation of pure cobalt

Pure cobalt can be prepared by electrolysing the solution of $CoSO_4$ containing $(NH_4)_2SO_4$ and NH_3 using platinum electrodes.

Physical properties

(*i*) Cobalt is a silver-white lustrous metal, slightly harder than iron. It is malleable and ductile.

(*ii*) It is more magntic than all other metals except Fe.

(*iii*) It has a density equal to 8.9. It melts at 1495°C and boils at 2900°C. It can take up a good polish.

(*iv*) The finely-divided Co absorbs 59–153 volumes of H_2 and thus resembles Ni and Pd. The finely-divided Co is pyrophoric.

Chemical properties

1. Action of air. Dry air or moist air has no action on cobalt but the metal reacts with O_2 to form Co_3O_4 when red hot.

$$3Co + 2O_2 \xrightarrow{\text{Red hot}} Co_3O_4$$

2. Action of steam. Co metal decomposes steam at red heat into H_2.

$$Co + H_2O\ (steam) \xrightarrow{\text{Red hot}} CaO + H_2$$

3. Action of acids. (*a*) Dil.HCl and H_2SO_4 form the salts and H_2 is evolved, *e.g.*,

$$Co + 2HCl \longrightarrow CoCl_2 + H_2$$

(*b*) Dil. HNO_3 dissolves the metal and renders it passive due to the formation of a protective thin layer of the metallic oxide on the surface of the metal.

4. Action of alkalies. Even conc. alkali has no action on the metal. The metal is dissolved in fused caustic potash at 550°C.

5. Action of CO. Finely-divided Co combines with CO at 150°C and under 30 atmospheric pressure to form orange-red $Co_2(CO)_8$.

6. Action of NH_3. Co_4N_2 is formed.

$$4Co + 2NH_3 \xrightarrow{\Delta} Co_4N_2 + 3H_2$$

7. Action of other elements. Cobalt readily combines with H_2, S, P and B but N_2 has no action on it.

Uses of cobalt and its compounds

(*i*) Cobalt is used in electroplating.

(*ii*) Cobalt is used for the preparation of various types of alloys which have been put to many uses. For example :

(*a*) *Cobalt steel.* This alloy contains 35% of Co and is used for making permanent magnets.

(*b*) *Stellite.* It is an alloy of Co, Cr and W. It is very hard and is resistant towards corrosion. It is used for making surgical instruments.

(*c*) *Co-chrome or festel metal.* It is an alloy of Co, Cr and Fe and is used for making resistance wire like ni-chrome.

(*iii*) Cobalt oxides and silicates are used for colouring glass, pottery and glazes. Thenard's blue ($CoO.Al_2O_3$), Rinmann's blue (CoO.ZnO) and smelt ($CoO.K_2O.3SiO_2$) are used as pigments.

Nickel

Occurrence

Nickel occurs in the form of sulphide, arsenide and silicate ores. Its important ores are:

(*i*) *Pentlandite,* (Ni, Cu, Fe) S which is found in the mines of Sudbury district of Canada, South Africa, U.S.S.R., Finland and U.S.A. This ore is also called *Sudbury ore* and also contains platinum metals (Ru, Rh, Pd, Os, Ir, Pt) along with other metals like Cu, Ag, Au etc. This ore also contains FeS and free sulphur as impurity.

(*ii*) *Nickel glance,* NiAsS.

(*iii*) *Garnierite,* (Ni, Mg)$SiO_3.xH_2O$, found in Australia, Cuba, Brazil, Russia.

(*iv*) *Kupfer nickel or nicolite,* NiAs.

(*v*) *White nickel ore,* $NiAs_2$.

(*vi*) Milerite, NiS.

Extraction from sulphide ore or Sudbury ore viz. pentlandite, (Ni, Cu, Fe)S

The extraction of nickel from Sudbury ore consists of the following steps:

1. Removal of impurities : Concentration of the ore. The ore is crushed, finely powdred and then concentrated by the froth flotation process. The concentrated ore consists of FeS, NiS and CuS, and some impurities like sulphur etc.

2. Roasting. The concentrated ore is roasted in the presence of excess of air in a reverberatory furnace so that large amount of S and FeS is converted to SO_2 and ($FeO + Fe_2O_3$) respectively while NiS and CuS remain unaffected.

$$S\ (\textit{large amount}) + O_2 \longrightarrow SO_2\uparrow$$
$$3FeS\ (\textit{large amount}) + 5O_2 \longrightarrow FeO + Fe_2O_3 + 3SO_2\uparrow$$

Thus at the end of roasting we are left with roasted mass which consists of CuS, NiS, FeO, Fe_2O_3, some amount of S and FeS which could not be converted into oxides during roasting.

3. Smelting. The roasted mass is mixed with coke and SiO_2 (being used as an acidic flux) and then smelted in a blast furnace so that Fe_2O_3 formed during roasting is reduced by coke to FeO.

$$Fe_2O_3 + 3C\ (\textit{coke}) \longrightarrow 2FeO + CO$$

FeO produced during roasting and smelting combines with SiO_2 (flux) and forms a slag of $FeSiO_3$ which, being lighter, floats over the molten mass which consists of NiS,CuS and small amounts of S and FeS.

$$FeO + SiO_2\ (\textit{flux}) \longrightarrow FeSiO_3\ (\textit{slag})$$

4. Bessemerisation : Formation of bessemerised matte. The molten mass obtained as above is taken in a Bessemer's convertor with the lining of SiO_2 (acidic lining) and fitted with tuyers to admit into the convertor a blast of hot air under pressure. When a blast of hot air is admitted into the convertor, the whole of the remaining S and FeS are converted into SO_2 and FeO respectively.

$$S + O_2 \longrightarrow SO_2\uparrow$$
$$2FeS + 3O_2 \longrightarrow 2FeO + 3SO_2\uparrow$$

SO_2 escapes but FeO combines with the acidic lining of SiO_2 (flux) and forms the slag of $FeSiO_3$. Thus S and FeS both are removed and we are left with a mass which consists of (NiS + CuS) mixture. This mixture is called *bessemerised matte*.

5. To get NiO from the bessemerised matte. The bessemerised matte which consists of (NiS + CuS) mixture is roasted in a free supply of O_2 so that NiS and CuS are converted into NiO and CuO respectively.

$$2NiS + 3O_2 \longrightarrow 2NiO + 2SO_2$$
$$2CuS + 3O_2 \longrightarrow 2CuO + 2SO_2$$

The residue consisting of (NiO + CuO) mixture is leached with dil. H_2SO_4 at 80°C which converts CuO into $CuSO_4$ (soluble) while NiO remains unaffected (*i.e.*, as residue).

$$CuO + H_2SO_4 \longrightarrow CuSO_4\ (\textit{solution}) + H_2O$$
$$NiO + H_2SO_4 \longrightarrow \text{No reaction.}$$

6. To get crude Ni metal from NiO. The residue of NiO thus obtained can be reduced to crude nickel metal by the following methods:

(*a*) NiO is reduced by CO, carbon or H_2.

$$NiO + CO \longrightarrow Ni\ (\textit{Crude metal}) + CO_2\uparrow$$
$$NiO + C \longrightarrow Ni\ (\textit{Crude metal}) + CO$$

$$NiO + H_2 \longrightarrow Ni\ (\textit{Crude metal}) + H_2O$$

(*b*) The residue of NiO is heated in a *reducing tower* made of iron at 300–350°C in a current of water gas, ($CO + H_2$) when NiO is reduced to crude Ni-metal which collects as a black powder.

$$2NiO + \underset{\text{water gas}}{(H_2O + CO)} \longrightarrow 2Ni\ (\textit{crude metal}) + H_2O + CO_2$$

Crude nickel thus obtained contains platinum metals (Ru, Rh, Pd, Os, Ir, Pt), coinage metals (Cu, Ag, Au) etc. as impurities.

7. Purification of crude Ni-metal by Mond's process. Crude Ni-metal obtained above is purified by Mond's process which was discovered by German chemist, Ludwig Mond in 1893. In this process crude Ni-metal is treated with CO at 60–80°C in a tower called *volatiliser* which is fixed close to the reducing tower. Nickel from the crude matter forms volatile $Ni(CO)_4$ while platinum metals (Ru, Rh, Pd, Os, Ir, Pt) and coinage metals (Cu, Ag, Au) are left as non-volatile residue. This residue is called *concentrate* and is used to extract platinum metal from it.

$$Ni(s) + 4CO\ (g) \xrightarrow{60-80°C} Ni(CO)_4\ (g) \qquad ...(i)$$

Vapours of $Ni(CO)_4$ formed as above are pumped to the bottom of another tower called *decomposer* which is made of cast iron, filled with granules of nickel and kept at 180°C. In this decomposer, the vapours of $Ni(CO)_4$ get decomposed to pure nickel.

$$Ni(CO)_4 \xrightarrow{180°C} Ni\ (\textit{pure metal}) + 4CO \qquad ...(ii)$$

Both reactions namely (*i*) and (*ii*) can be represented as :

$$Ni(s) + 4CO(g) \underset{180°C}{\overset{60–80°C}{\rightleftharpoons}} Ni(CO)_4\ (g)$$

Nickel obtained gets deposited on the granules of nickel filled in the decomposer. CO produced is sent back to the volatiliser for the formation of $Ni(CO)_4$. This process should be operated carefully, since CO and $Ni(CO)_4$ both are poisonous substances.

8. Further purification of nickel by electrolytic process. Nickel obtained by Mond's process may further be purified by electrotytic process in which nickel (II) ammonium sulphate solution is electrolysed in a cell (cathode = pure nickel, anode = impure nickel) when pure nickel gets deposited on cathode and inpurities are collected as *anode mud* below anode.

Physical properties

Nickel is a soft silvery white metal, quite malleable and ductile. It is slightly magnetic and possesses a high melting and boiling point (m.pt. = 1453°C, b.pt. = 2730°C). Like iron and cobalt, it absorbs a large volume of hydrogen gas (one volume of the metal absorbing 17 volumes of the gas).

Chemical properties

The action of air on the metal is slow at the ordinary temperature, but it vigorously burns in oxygen. It is attacked by chlorine, when nickel chloride, $NiCl_2$ is formed. Nickel decomposes steam at a red heat, forming NiO and H_2. Dilute hydrochloric and sulphuric acid solutions have a very slow action on this metal. But it readily dissolves in nitric acid and aqua regia.

$$3Ni + 8HNO_3 \longrightarrow 3Ni(NO_3)_2 + 2NO + 4H_2O$$

When Ni is kept for some time under the surface of fuming HNO_3, the metal becomes passive.

Ni decomposes steam at red heat

$$Ni + H_2O\ (\textit{steam}) \overset{\text{Red heat}}{\rightleftharpoons} NiO + H_2$$

The metal remains unaffected by caustic alkalies, even by fused alkalies. Nickel crucibles are, therefore, commonly used for alkali fusion in qualitative experiments. When CO gas is passed over nickel kept at about 50°C, a volatile colourless liquid, nickel tetra-carbonyl, $Ni(CO)_4$ is formed. The formation of $Ni(CO)_4$ finds an industrial application in the purification (or extraction) of nickel by Mond's process.

Uses of nickel

(*i*) Due to high melting point, great resistance to corrosion, acids and alkalies, nickel is largely used in making crucibles, dishes and other chemical wares and also in electrical industry.

(*ii*) It is used for nickel plating in which $NiSO_4.7H_2O$ or $NiSO_4.(NH_4)_2SO_4.6H_2O$ solution is used as an electrolyte.

(*iii*) Finely divided nickel is used as a catalyst in the hydrogenation reactions like,

$$R.CH = CH.R + H_2 \longrightarrow R.CH_2 - CH_2.R$$
$$CO + 3H_2 \longrightarrow CH_4 + H_2O$$
$$2NO + 5H_2 \longrightarrow 2NH_3 + 2H_2O$$

(*iv*) $Ni(CO)_4$ is used to get pure nickel (99.9% pure) by heating it at 180°C.

(*v*) Nickel metal combines with many other metals and forms alloys which find a number of applications. Some important alloys of Ni and their uses are given in Table 8.4.

Table 8.4. *Some important alloys of nickel and their uses.*

Alloy	*Approx. composition*	*Properties and uses*
1. Platinite	Ni = 46%, Fe = 54%	Same coefficient of expansion as that of glass and platinum, used in electric bulbs and radio-valves.
2. Cupro nickel	Ni = 25%, Cu = 75%	Coinage
3. Constantan	Ni = 40%, Cu = 60%	Electrical resistance
4. Platinoid	Ni = 14%, Cu = 60% Zn = 24%, W = 2%	Used for thermocouples and resistance coils.
5. Permalloy	Ni = 21.5%, Fe = 78.5%	Has unusual magnetic properties, used in telephony and telegraphy.

Palladium

Occurrence

Palladium occurs in Brazil and Urals. Its common minerals are:

1. *Braggite,* (Pt,Pd, Ni)S which contains 20% of Pd.
2. *Stibiopallinite,* Pb_3Sb.
3. *Potarite,* PdHg.
4. *Palladium gold or propezite* which containis 85–90% gold and 6–7% Pd.

Pd is also present in *Sudbury nickel ores* of Canada as PdSe which contains Pt, Rh and Ir. It also occurs associated with Au and Ag.

Extraction

Palladium may be extracted by the following two methods.

1. From Sudbury nickel ore. Sudbury nickel ore is concentrated and then digested with aqua regia under pressure so that we get a solution which contains $PdCl_2$ and $PtCl_4$. This solution is treated with NH_4Cl when a yellow ppt. of *ammonium chloroplatinate* (IV), $(NH_4)_2[PtCl_6]$ is obtained.

$$PtCl_4 + 2NH_4Cl \longrightarrow (NH_4)_2[PtCl_6]$$

$PdCl_2$ remains unaffected by NH_4Cl and is present in solution. The solution which contains mostly $PdCl_2$ is mixed with NH_3 and concentrated by heating. On cooling, the crystals of *tetrammine palladium (II) chloride monohydrate*, $[Pd(NH_3)_4]Cl_2.H_2O$ are obtained.

$$PdCl_2 + 4NH_3 + H_2O \longrightarrow [Pd(NH_3)_4]Cl_2.H_2O$$

$[Pd(NH_3)_4]Cl_2.H_2O$, on being heated strongly, gives palladium in the form of a spongy or a fine powder.

2. From alloys. Pd can also be extracted from its natural alloys with other metals of the family. When the alloy is fused with KCN and the product is extracted with water, only $Pd(CN)_2$ separates out as a precipitate. On ignition, $Pd(CN)_2$ gets decomposed to give pure palladium metal.

Forms of palladium

1. Metallic palladium. It is obtained by igniting palladous dichloro-diammine, $[Pd(NH_3)_2Cl_2]$ or ammonium chloropalladite, $(NH_4)_2PdCl_4$ in a current of hydrogen.

2. Crystalline palladium. It is obtained by electrically heating palladium ribbon dusted over with finely divided topaz to redness.

3. Spongy palladium. It is conveniently prepared by the reduction of palladous salt solutions with sodium formate.

$$HCOONa + PdCl_2 \longrightarrow Pd + HCl + NaCl + CO_2$$

It has a great capacity to absorb hydrogen. After absorption it swells up and falls to a powder. It is used as a catalyst in hydrogenation and in various organic reactions.

4. Palladium black. It is the finely divided form of palladium and is obtained by adding zinc or iron strips to the acidic solution of palladous salts. It has a greater capacity to absorb hydrogen than spongy palladium. It is widely used as a catalyst.

5. Coloidal palladium. It is obtained by sparking palladium electrodes under ice cold, water, containing trace of alkali, or by reducing palladous chloride with hydrazine hydrate containing gum arabic.

Physical properties

Palladium is a lustrous silvery-white metal which is malleable and ductile. Its density is 12.0 and is lighter than platinum and gold. It is as hard as paltinum and takes a good polish. Its sp. heat is 0.05928. It melts at 1552°C and boils at 3980°C. This metal has the characteristic property of absorbing gases, more specially hydrogen. Freshly prepared palladium absorbs 370 volumes of hydrogen at room temperature and 650 volumes at 100°C. Spongy palladium absorbs the gas upto 980 volumes. When palladium metal is heated in oxy-hydrogen flame, a greenish vapour of palladium is evolved. Palladium black, dry or suspended in water or alcohol, absorbs 4000 volumes, and colloidal palladium about 5000 volumes of acetylene per volume of palladium.

Chemical properties

Palladium is not tranished by air and on heating a blue oxide layer is formed on the metallic surface. When heated to dull redness in oxygen, palladium sponge is oxidised to palladous oxide which decomposes on heating above 807°C. On heating with sodium peroxide, a soluble palladite is formed.

Palladium combines with fluorine at a dull red heat, forming PdF_3. Spongy palladium at dull redness forms palladous chloride. Palladium is also attacked by chlorine or bromine water. Hydrochloric acid attacks the metal only in presence of air or free chlorine. When palladium foil is evaporated to dryness with an alcoholic solution of iodine, palladous iodide is formed.

The metal combines with sulphur with incandascence. On heating the metal in a current of sulphur dioxide, palladium sulphide is formed. The metal also dissolves in the fused potassium hydrogen sulphate forming palladium sulphate.

Palladium also dissolves in nitric acid at ordinary temperature more easily when alloyed with copper or silver.It also dissolves in boiling concentrated sulphuric acid but easily soluble in aqua-regia.

$$Pd + 4HNO_3(conc.) \longrightarrow Pd(NO_3)_2 + 2H_2O + 2NO_2$$

$$Pd + 2H_2SO_4 \longrightarrow PdSO_4 + SO_2 + 2H_2O$$

Palladium combines on heating with phosphorus to form its phosphite, with antimony it forms antimonides, PdSb, Pd_3Sb and $PdSb_2$.

On fusion with potassium hydrogen sulphate, palladous sulphate is formed; with potassium pyrosulphate, $K_2S_2O_7$, palladous sulphate and SO_2 are formed. With ammonium sulphate or nitrate, the metal forms a diamine. Palladium is not oxidised on fusion with oxidising fusion mixture (Na_2CO_3 and KNO_3).

Uses

1. Palladium and its compounds are used as catalyst in different types of reactions such as oxidation, reduction, chlorination, isomerisation, polymerisation, synthetic reactions, hydrogenations of oils etc. Usually, palladous oxide, palladium black and colloidal platinum are used for this purpose. The use of Pd-metal as a catalyst in hydrogenation reactions is due to the fact that in presence of Pd, molecular hydrogen gets dissociated into atomic hydrogen. Since atomic hydrogen is more reactive, the reaction is brought about very speedily and very efficiently.

Palladium catalysts also find wide application in gas analysis, manufacture of synthetic petrolium, manufacture of HCN, acetylene and alkaline earth formates.

2. It is used for the purification of hydrogen, in dentistry and jewellery as a substitute for gold and platinum.

3. Palladium is also used in astronomical and other scientific instruments.

4. It is used as a coating for mirrors for search lights. Its salts are used in photography.

Occurrence

Platinum is found in the free state as well as in the combined state.

1. In the free state. (*a*) Platinum metal is found in nature in the free state in *platini-ferrous sands* and gravels in the river beds along with other metals like Ru, Rh, Pd, Ir (platinum metals), Ag, Au, Ni etc. Sand particles are present as impurity.

(*b*) The non-volatile residue obtained at the end of Mond's process used in the extraction of Ni metal from Sudbury ore contains Ru, Rh, Pd, Os, Ir, Pt, Cu, Ag, Au, etc.

2. In the combined state. Platinum is found in the form of the following minerals:

(*a*) *Sperrylite*, $PtAs_2$

(*b*) *Cooperite*, $Pt(AsS)_2$

(*c*) *Braggite*, (Pt, Pd, Ni)S

Extraction of platinum

The extraction of platinum metal consists of the following steps :

1. Preparation of concentrate which consists of Ru, Rh, Pd, Os, Ir, Pt, Ag, Au etc. metals. The concentrate can be obtained from any of the following two sources :

(*i*) As already said, the platini-ferrous sands and gravels present in the river beds contain Ru, Rh, Pd, Os, Ir, Pt, Ag, Au, Ni metals along with sand particles as impurity. Sand particles are removed by Welfley table method in which a stream of water is made to flow accross the table so that heavier metals are retained behind the riffles (residue) while the sand particles (gangue particles) are washed away.

The residue thus obtained contains Ru, Rh, Pd, Os, Ir, Pt (50–85%), Ag, Au, Ni etc. Now this residue is treated with CO gas so that Ni forms volatile nickel tetracarbonyl, $Ni(CO)_4$ and is thus removed from the residue. The non-volatile residue left consists of Ru, Rh, Pd, Os, Ir, Pt, Ag,Au etc. and is called *concentrate*.

(*ii*) The non-volatile residue obtained at the end of Mond's process used in the extraction of Ni metal from Sudbury ore contains Ru, Rh, Pd, Os, Ir, Pt, Cu, Ag, Au, etc. This residue is roasted in air to convert Cu into CuO. The roasted mass obtained consists of Ru, Rh, Pd, Os, Ir, Pt, Ag,Au, and CuO. This mass is treated with H_2SO_4 when CuO gets dissolved due to the formation of soluble $CuSO_4$ while the metals remain undissolved (residue). This undissolved portion contains Ru, Rh, Pd, Os, Ir, Pt, Ag and Au.

2. Separation of Pt-metal from the concentrate. The concentrate obtained by any of the above methods is used to extract platinum metal. It is treated with hot aqua regia so that Pd, Pt and Au go into solution as their soluble chloro complex compounds *viz.* $H_2[Pd^{+4}Cl_6]$, $H_2[Pt^{+4}Cl_6]$ and $H[Au^{3+}Cl_4]$ respectively while the insoluble portion consists of $RuCl_4$, Rh, Os, Ir, AgCl.

$$2HNO_3 + 8HCl + M(M = Pd \text{ or } Pt) \longrightarrow 2NOCl + 4H_2O + H_2[M^{+4}Cl_6]$$

$$Au + 3HNO_3 + 4HCl \longrightarrow H[Au^{3+}Cl_4] + 3NO_2 + 3H_2O$$

The soluble portion is treated with $FeSO_4$ solution so that $H_2[Pd^{+4}Cl_6]$ and $H[Au^{3+}Cl_4]$ get reduced to $H_2[Pd^{2+}Cl_4]$ (soluble portion) and free metallic gold, Au^0 (insoluble portion) while $H_2[PtCl_6]$ (soluble portion) remains unaffected. Thus the soluble portion contains $H_2[PdCl_4]$ and $H_2[PtCl_6]$ and insoluble fraction consists of metallic gold.

$$[Pt^{+4}Cl_6]^{2-} + Fe^{2+} \longrightarrow \text{No reaction}$$

$$[Pd^{+4}Cl_6]^{2-} + Fe^{2+} + e^- \longrightarrow [Pd^{+2}Cl_4]^{2-} + Fe^{3+} + 2Cl^-$$

$$H[Au^{3+}Cl_4] + 3Fe^{2+}SO_4 \longrightarrow Au^0\downarrow + Fe^{3+}Cl_3 + Fe_2^{+3}(SO_4)_2 + HCl$$

or $$[Au^{3+}Cl_4]^- + 3Fe^{2+} \longrightarrow Au^0 + 3Fe^{3+} + 4Cl^-$$

The soluble portion containing $H_2[PdCl_4]$ and $H_2[PtCl_6]$ is treated with NH_4Cl whereby $H_2[PdCl_4]$ remaiins unaffected and hence remains in solution but $H_2[PtCl_6]$ is converted into ammonium hexachloroplatinate (IV), $(NH_4)_2[PtCl_6]$ (insoluble portion) which, on being ignited, gives spongy platinum as a greyish black porous mass.

$$H_2[PtCl_6] + 2NH_4Cl \longrightarrow (NH_4)_2[PtCl_6] + 2HCl$$

$$(NH_4)_2[PtCl_6] \xrightarrow{\text{Ignite}} 2NH_4Cl + 2Cl_2 + Pt \text{ (spongy platinum)}$$

Various forms of platinum

Various forms of platinum are prepared and used for different purposes. Following are important forms of platinum.

1. Spongy platinum. *Preparation.* This form of platinum is obtained by igniting $(NH_4)_2[PtCl_6]$

$$(NH_4)_2[PtCl_6] \xrightarrow{\text{Ignite}} Pt \text{ (spongy platinum)} + 2Cl_2 + 2NH_4Cl$$

Properties. It is soft and spongy greyish-black porous mass. On fusing it in oxy-hydrogen flame in an electric furnace, it forms a compact lustrous mass when Cl_2 is passed through a solution of spongy platinum in conc. HCl, $H_2[PtCl_6]$ is obtained.

$$Pt\ (\textit{spongy platinum}) + 2HCl + Cl_2 \longrightarrow H_2[PtCl_6]$$

Uses. Spongy platinum has great surface activity and hence is used for absorbing gases. It causes a mixture of H_2 and O_2 to unite with explosion. Coal gas can be ignited by spongy platinum and is, therefore, used in self lightinig Bunsen burners. Spongy platinum absorbs large quantity of O_2 and is then used as an oxidising agent.

2. Platinum black. *Preparation.* It is obtained as a black precipitate by adding Zn-pieces or formic acid (HCOOH) to the solution of platinum in aqua regia (HNO_3 + 3HCl). First of all chloroplatinic acid, $H_2[PtCl_6]$ is produced by the action of aqua regia on platinum metal.

$$[HNO_3 + 3HCl \longrightarrow NOCl + 2H_2O + 2Cl] \times 2$$
$$Pt + 4Cl \longrightarrow PtCl_4$$
$$PtCl_4 + 2HCl \longrightarrow H_2[PtCl_6]$$

On adding : $$3HNO_3 + 8HCl + Pt \longrightarrow 2NOCl + 4H_2O + H_2[PtCl_6]$$

Now $H_2[PtCl_6]$ is reduced to black paltinum by nascent hydrogen produced by the action of Zn-pieces on HCl or by HCOOH.

$$Zn + 2HCl \longrightarrow 2H + ZnCl_2$$
$$H_2[PtCl_6] + 4H \longrightarrow Pt\ (\textit{black platinum}) + 6HCl$$
$$H_2[PtCl_6] + 2HCOOH \longrightarrow Pt\ (\textit{black platinum}) + 6HCl + 2CO_2$$

Uses. Platinum black is used as an absorbent of gases like H_2, O_2 etc. which can be liberated again. Thus platinum black is used as an oxidising agent. It is also used as a catalyst in the preparation of HCHO and in the decomposition of H_2O_2.

$$2CH_3OH + O_2\ (\textit{from air}) \xrightarrow{\text{Pt–black}} 2HCHO + 2H_2O$$
$$2H_2O_2 \xrightarrow{\text{Pt–black}} 2H_2O + O_2\downarrow$$

Platinum black is used in automatic gas lightning. It catalyses oxidation of oil gas by air. Heat thus produced makes fine platinum wire glow which lights the gas.

3. Platinised asbestos. *Preparation.* It is prepared as follows. Pt is converted into $H_2[PtCl_6]$ by the action of aqua regia on it.

$$2HNO_3 + 8HCl + Pt \longrightarrow 2NOCl + 4H_2O + H_2[PtCl_6]$$

Now the asbestos is soaked in 10% solution of H_2PtCl_6, dried and heated to 500°C. By this operation H_2PtCl_6 decomposes to give finely-divided platinum which is distributed throughout the asbestots fibre. The asbestos fibre having finely-divided platinum in it is called *platinised asbestos.*

Uses. It is used as a catalyst in the combination of SO_2 and O_2 to form SO_3 required in the manufacture of H_2SO_4 by contact process. Platinised asbestos absorbs large volume of O_2 and then acts as oxidising agent.

4. Colloidal platinum. *Preparation.* It can be obtained either by *Bredig's arc method* in which an electric arc is passed between two platinum electrodes dipped in ice-cold water so that platinum particles are scattered through the solution to give a brown solution which is called colloidal platinum or by converting platinum into H_2PtCl_6 by the action of aqua regia on it and then reducing H_2PtCl_6 by hydrazine ($NH_2.NH_2$) in presence of some protective colloid to get colloidal platinum.

$$H_2PtCl_6 + NH_2.NH_2 \longrightarrow Pt\ (\textit{colloidal platinum}) + N_2 + 6HCl$$

Uses. Colloidal platinum has been found to be an effective catalyst, *e.g.*, it decomposes H_2O_2 easily.

Physical properties of platinum

Platinum is a soft and silver-white metal. It has very high density equal to 22.4. Its m.pt. and b.pt. are 1769°C and 4530°C respectively. Excepting Ag and Au, it is the most malleable and ductile of all the metals. It can be rolled into foils and drawn into wires. It softens before melting. Due to this property, this metal can be welded. Its coefficient of expansion (= 0.000,009 per °C) is almost the same as that of glass. It is due to this property that Pt wires and foils can be sealed into glass without causing any cracks on cooling. The hardness of the metal increases by the presence of Ir metal. The metal is a good conductor of electricity. Molten Pt metal absorbs O_2 which is given out on cooling. This phenomenon is called *spitting*.

Chemical properties of platinum

1. Action of air or O_2. Platinum is a noble metal and hence is not affected by air or O_2 even on heating the metal.

2. Action of acids. Pure metal does not react with any of the acids. However, pure metal reacts with aqua regia and forms $H_2[PtCl_6]$.

$$2HNO_3 + 8HCl + Pt \longrightarrow 2NOCl + 4H_2O + H_2[PtCl_6]$$

$$HNO_3 + 18HCl + 3Pt \longrightarrow 4NO + 8H_2O + 3H_2[PtCl_6]$$

HF has no action on platinum. The impure metal is attacked by H_2SO_4.

3. Action of elements. Platinum metal combines with many elements like F_2, Cl_2, P, S, As etc. and forms the compounds like PtF_4, $PtCl_4$, Pt_3P_5, PtS, Pt_2As_3 etc. respectively.

With carbon, Pt forms platinum carbide which is brittle. Hence Pt-crucible should not be heated in a smoky flame. When Cl_2 is passed through a solution of spongy platinum in conc. HCl, $H_2[PtCl_6]$ is obtained.

$$\underbrace{\text{Pt } (\textit{spongy platinum}) + 2\text{HCl } (\textit{conc.}) + \text{Cl}_2}_{\text{Solution}} \longrightarrow \text{H}_2[\text{PtCl}_6]$$

4. Action of metals. Pt combines directly with many metals like Sn, Pb, Ag, Zn etc. but not with Hg and form alloys. Hence the above said metals should not be heated in platinum crucibles.

5. Action of compounds. Pt metal is attacked by various compounds like fused alkalies, Na_2O_2, alkali cyanides, alkali sulphides and nitrates. Such compounds should, therefore, not be heated in Pt crucible. With CS_2 vapour the metal forms $PtCS_2$. At 100°C, the metal forms $PtCl_4.2NOCl$ with NOCl. Na_2CO_3 has no action on the metal.

Uses of platinum

(*i*) *In scientific instruments and laboratory apparatus.* Due to its remarkable properties like high melting point and corrosion resistance, platinum is largely used to manufacture scientific instruments and laboratory apparatus such as crucibles, dishes, basins etc. Pt is also used for contacts in electrical apparatus and for making electrodes.

(*ii*) *As catalyst.* Different forms of platinum are used as catalyst in many manufactural and laboratory processes. Examples are given below:

(*a*) The use of platinum black, platinised asbestos and colloidal platinum as catalyst has already been discused.

(*b*) *Platinum gauge* is used as a catalyst in the oxidation of NH_3 to NO required for the manufacture of HNO_3 by Ostwald's process.

$$4NH_3 + 5O_2 \xrightarrow[800°C]{\text{Pt–gauge}} 4NO + 6H_2O$$

(*c*) When mixture of H_2 and Br_2 or I_2 is passed over red hot *finely-divided platinum*, **HBr** and HI is obtained.

$$H_2 + X_2 \ (X = Br \text{ or } I) \rightleftharpoons 2HX$$

Since *finely-divided platinum* can absorb several hundred times it own volume of hydrogen gas, it is used in many hydrogenation or dehydrogenation reactions. Large quantity of platinum metal is also used in the petroleum, synthetic fibre and glass industry.

(*iii*) *In jewellery.* Pt is also used in jewellery as a setting for diamonds. For this purpose the metal is alloyed with metals like Ir, Ru or Cu. Pt—Ir alloy is also used in pen points, standard weights, coins etc.

(*iv*) *In dentistry.* The metal is used for making artificial teeth, pins, dental filling and support for dentures.

(*v*) Pt-salts are used in printing and toning in photography. Pt wire is also used in the qualitative analysis in chemical. In the medical fields many instruments like X-ray tubes, radium applicators, lypodermic needles require platinum.

Occurrence

Gold is found in the native (free) as well as in the combined state. In the native state this metal is found in quartz rocks which are called *auriferous quartz rocks* or simply *auriferous quartz.* Gold present in auriferous quartz is called *vein gold.* Gold is also found in sandy deposits found in the beds of certain rivers. These sandy deposits are called *allovial deposits* and gold present in these deposits is called *alluvial gold* or *placer gold. In the combined state*, the metal occurs as *bismuth aurite* (BiAu), *sylvanite* [(Au, Ag) Te_2], *calaverite* ($AuTe_2$) and *petzite* [$(Au, Ag)_2$ Te]

India produces about 2% of the world's gold production and Mysore is the richest centre, the *Kolar gold fields* there producing 99.97% of our gold output. Alluvial gold washing is carried out in small scale in Bihar, Assam and Orissa.

Extraction of gold from auriferrous quartz rocks : Extraction of vein gold

The following methods are used :

1. Amalgamation process. The auriferrous quartz rocks are broken to smaller pieces by means of rock crushers. The smaller pieces thus obtained are finely powdered in big stamp mills. This is called *pulverisation* of the quartz. The pulverised ore is carried off by a current of H_2O through a screen at the other end of the stamp mill and thus the *slurry* (*i.e.*, mixture of finer particles of quartz and H_2O) is obtained. The coarse particles, if any, are retained back and are again pulverised.

Now the slurry is allowed to flow over Cu-plates coated with Hg and placed in a sloping position. The particles of Au-metal form Au-Hg amalgam with Hg which is retained on the plates. The amalgam is scrapped off from plates and distilled in iron retorts when Hg, being more volatile, distills off and Au-metal is left behind in the retort. Nearly, 75–80% of total gold originally present in the quartz rocks is thus recovered and the remaining 25–20% gold adheres to the rock particles (called *tailings*) from which gold can be recovered by *Mac-Arthur Forest cyanide process* discussed below.

2. Mac-Arthur Forest cyanide process. This method is used to extract gold metal from the tailings obtained from amalgamation process and also from the auriferous quartz rocks. The tailings or quartz rocks are filled in large vats made of cement or iron and consists of false wooden bottoms with holes. A dilute solution of NaCN (0.3%) is added in vats and the whole mass is thoroughly agitated with a powerful current of air for 12–24 hours. Au present in the tailings or quartz rocks reacts with NaCN solution in presence of O_2 (from air) and forms soluble complex compound, $Na[Au(CN)_2]$ (sodium aurocyanide) which, on being treated with Zn shavings, gives the ppt. of Au.

$$4Au + 8NaCN + 2H_2O + O_2 \longrightarrow \underset{\text{Solution}}{4Na[Au(CN)_2]} + 4NaOH$$

$$2Na[Au(CN)_2] + Zn \longrightarrow Na_2[Zn(CN)_4] + 2Au\downarrow$$

3. Plattner's chlorination process. The powdered auriferous quartz is roasted in order to remove S and As as their oxides and is then placed in wooden vats with false perforated bottoms moistened with H_2O and is treated with Cl_2. This treatment gives $AuCl_3$.

$$2Au\ (\textit{from auriferous quartz}) + 3Cl_2\ (\textit{moist}) \longrightarrow 2AuCl_3$$

$AuCl_3$ obtained as above is dissolved in water and the solution is treated with $FeSO_4$ so that Au-metal gets precipitated.

$$AuCl_3 + 3FeSO_4 \longrightarrow Fe_2(SO_4)_3 + FeCl_3 + Au\downarrow$$

Extraction of gold from alluvial deposits: Extraction of alluvial or placer gold by placer mining method

Gold is extracted from alluvial deposits by placer mining method which is one of the washing methods. In this method gold-bearing sand is introduced in a sluice way system and powerful streams of water passed over it. A *sluice way system* is an inclined channel provided with a series of sluices *i.e.*, troughs and cross wire strips called *riffles*. The heavy gold particles are retained in the sluices while the light sand particles are carried away by the water and are collected at the end of the system.

Refining of gold

Gold obtained by any of the above methods usually contains Ag, Cu, Pb etc. as impurities. These are removed by the following methods :

1. Removal of Pb by cupellation process. Pb is removed by cupellation process in which impure gold is heated in the presence of air so that Pb is oxidised to PbO which volatilises.

2. Removal of Ag and Cu by parting process. The removal of Ag and Cu is called parting which can be done with dry Cl_2 or conc. H_2SO_4.

(*i*) *Parting with dry Cl_2.* The impure gold is melted in a fire clay crucible and dry Cl_2 is bubbled through the melt. Ag is converted into AgCl and Au remains unaffected. AgCl forms a scum on the surface of molten gold metal and this scum can easily be removed.

(*ii*) *Parting with conc. H_2SO_4.* If the impure gold sample contains pure gold more than 25%, some Ag is added to the impure sample so that the pure gold content in the impure sample becomes 25%. Now the impure gold sample containing 25% pure gold is treated with boiling conc. H_2SO_4 when Ag and Cu get dissolved and Au is left behind. To reduce the gold content in impure gold sample to 25% or one quarter by adding Ag is called *quartation*.

3. Electrolytic refining. Gold can also be purified by electrolytic process using 4% gold chloride solution acidified with 10% hydrochloric acid as the electrolyte. Impure gold is made the anode while the cathode is a plate of pure gold.

Physical properties

Gold is a lustrous yellow metal which melts at 1063°C to a green liquid and boils at 2920°C. Its specific gravity is 19.3. It is the most malleable and ductile of all metals and is a good conductor of heat and electricity. With mercury, it forms an amalgam.

Chemical properties

1. Action of air and water. Neither dry or moist air nor water has any action on gold at any temperature. It is due to this property that gold is called *noble metal*.

2. Action of acids. (*a*) Selenic acid (H_2SeO_4) reacts with Au at 225°C, forming auric selenate, $Au_2(SeO_4)_3$.

$$2Au + 6H_2SeO_4 \longrightarrow Au_2(SeO_4)_3 + 3SeO_2 + 6H_2O$$

(*b*) Metal dissolves in aqua regia forming $H[AuCl_4]$.

$$[HNO_3 + 3HCl \longrightarrow 2H_2O + NOCl + 2Cl] \times 3$$
$$[Au + HCl + 3Cl \longrightarrow H[AuCl_4]] \times 2$$
$$\overline{3HNO_3 + 11HCl + 2Au \longrightarrow 2H[AuCl_4] + 6H_2O + 3NOCl}$$

(*c*) HF even mixed with HNO_3 has no action on the metal.

(*d*) Very conc. and hot HNO_3 dissolves only a few milligrams of the metal per litre.

3. Action of halogens. Cl_2, Br_2 and I_2 form AuX_3 when Au is heated with these halogens.

$$2Au + 3X_2 \xrightarrow{\Delta} 2AuX_3$$

4. Action of alkali cyanide solutions. In presence of O_2, complex compound, $M[Au(CN)_2]$ (M = alkali metal) is formed.

$$4Au + 8KCN + 2H_2O + O_2 \longrightarrow \underset{\text{soluble}}{4K[Au(CN)_2]} + 4KOH$$

5. Action of alkaline sulphides and thiosulphates. These compounds dissolve the metal at high temperature.

6. Action of sulphur. Sulphur has no action on the metal.

7. Action of oxidising agents. Oxidising agents like H_2O_2, alkali metal, nitrates, chlorates, perchlorates, permanganates and bleaching powder etc. along with HCl dissolve the metal.

Uses

(*i*) In coinage, jewellery and ornamental vessels.

(*ii*) In gold plating.

(*iii*) Gold leaf is used for decorations, lettering and in medicine as tonic.

(*iv*) Purple of Cassium which is a combination of gold with colloidal stannic acid is used for making ruby red glass and high class pottery.

(*v*) Gold is used in making white gold which is a gold-nickel alloy. This alloy resembles platinum in appearance and is used in jewellery.

Purity or fineness of gold : Concept of carat

We have already said that gold is used in making ornaments, jewellery and watches. Since pure gold is very soft, to make it harder, it is mixed with small amounts of Cu or Ag for making jewellery. The purity or fineness of gold in a given ornament is generally expressed as *carats*. Pure gold is 24 carats. These days ornaments are generally made of 22 carats.

A 22 carat gold means that the sample of gold contains 22 parts of gold and $24 - 22 = 2$ parts of Cu or Ag. Thus the percentage of gold in 22 carats $= \frac{22}{24} \times 100 = 91.6$ and % of Cu or Ag $= 100 - 91.6 = 8.4$.

Preparation of colloidal gold

Colloidal gold is obtained by two methods.

1. Bredig's arc method. It is a dispersion process in which an electric arc is struck between two gold electrodes dipped under cold water. As a result of sparking the gold particles distintegrate forming a colloidal solution. A trace of KOH is added to cold water and helps to stabilise the colloidal solutions.

2. Reduction process. When $AuCl_3$ is reduced by means of $SnCl_2$, colloidal gold is obtained.

This colloidal gold is called *purple of Cassius* after the name of Cassius who discovered it. For details see "*Reduction properties of $AuCl_3$.*" discussed in chapter 9.

Faraday's gold

It is obtained by reducing $AuCl_3$ solution with formaldehyde. It is red, brown or purple in colour.

Zinc

Occurrence

Zinc is not found in the free (native) state in nature. In the combined state it widely occurs in nature in association with Cu, Ag, Pb etc. It occurs as :

(*i*) *Oxide, e.g.*, zincite or red zinc, ZnO and franklinite, $ZnO.Fe_2O_3$.

(*ii*) *Sulphide, e.g.*, sphalerite or zinc blende, ZnS.

(*iii*) *Carbonate, e.g.*, calamine or zinc spar, $ZnCO_3$.

(*iv*) *Silicate, e.g.*, willemite, Zn_2SiO_4 or $2ZnO.SiO_2$.

The most abundant ore is *zinc blende* (ZnS) and next to it is *calamine* ($ZnCO_3$). ZnS is found in U.S.A., Mexico, Australia, Germany, England, Canada, and Burma. In India, ZnS is found in Udaipur (Rajsthan). Zinc is also found in animal cells.

Extraction of zinc from zinc blende/calamine

There are two main processes for the extraction of zinc from its important ores viz. *zinc blende* (ZnS) and *calamine* ($ZnCO_3$). These processes are *carbon reduction process* and *electrolytic process.*

A. Carbon reduction process

This process consists of the following steps :

1. Concentration. *Zinc blende ore* (ZnS) is first dressed to remove the superfluous rocky materials and then is finely powdered. From the finely powdered ore, the impurity of iron is removed by electromagnetic separation process and the earthy and silicious impurities are removed by froath floatation process.

Calamine ore ($ZnCO_3$) does not require concentration since it does not contain the impurity of *galena* (PbS).

2. Calcination and roasting. (*a*) If the ore is calmine ($ZnCO_3$), the concentrated ore is calcined to remove moisture and CO_2.

$$ZnCO_3 \longrightarrow ZnO + CO_2$$

(*b*) If the ore is zinc blende (ZnS), it is roasted by heating it in a reverberatory furnace at 950°C. During roasting, ZnS is converted into ZnO and some part of ZnS is also converted into $ZnSO_4$ which is decomposed into ZnO at 950°C.

$$2ZnS + 3O_2 \longrightarrow 2ZnO + 2SO_2$$

$$ZnS + 2O_2 \longrightarrow ZnSO_4$$

$$2ZnSO_4 \xrightarrow{950°C} 2ZnO + 2SO_2 + O_2$$

SO_2 formed as above is used for the manufacture of H_2SO_4.

3. Reduction of ZnO to Zn metal. ZnO obtained as above is mixed with powdered coke or anthracite and heated in retorts to 1400–1450°C. ZnO is reduced to Zn metal.

$$ZnO + C \longrightarrow Zn + CO$$

$$ZnO + CO \longrightarrow Zn + CO_2$$

$$CO_2 + C \rightleftharpoons 2CO$$

Reduction is carried out by any of the following processes.

(*a*) *Belgian process.* The roasted ore is heated with coke to about 1100°C in small fire clay retorts. The heating is often done by producer gas. Each retort is provided with a clay pipe which serves as a condenser and an earthenware pot in which is collected the condensed metal. Zinc, being fairly volatile, is distilled in the retort and collected in the pot. The chemical reactions involved in this process are :

$$ZnO + C \longrightarrow Zn + CO$$
$$ZnO + CO \longrightarrow Zn + CO_2$$
$$CO_2 + C \rightleftharpoons 2CO$$

There are a number of such retorts which are arranged in three of four rows one above the other. The lowermost is supported on the hearth.

(*b*) *New Jersy on vertical retort process.* It is a modern process and is much more economical than the Belgian process. It is a continuous process and yields metal of much greater purity (95–99%). Powdered ZnO and coal (called charge) is made into small brickets (small blocks) with the help of molasses which bends the particles together. These brickets are fed into the tall vertical retort from the top. The retort is lined on the inside with fire caly bricks. The retort is heated externally by means of a heating jacket in which a mixture of producer gas and air is burnt. The vapours of zinc metal obtained by the reduction of zinc oxide with carbon rise to the top and are condensed in the condenser. The crude metal so obtained is called *spelter.* (see Fig. 8.5).

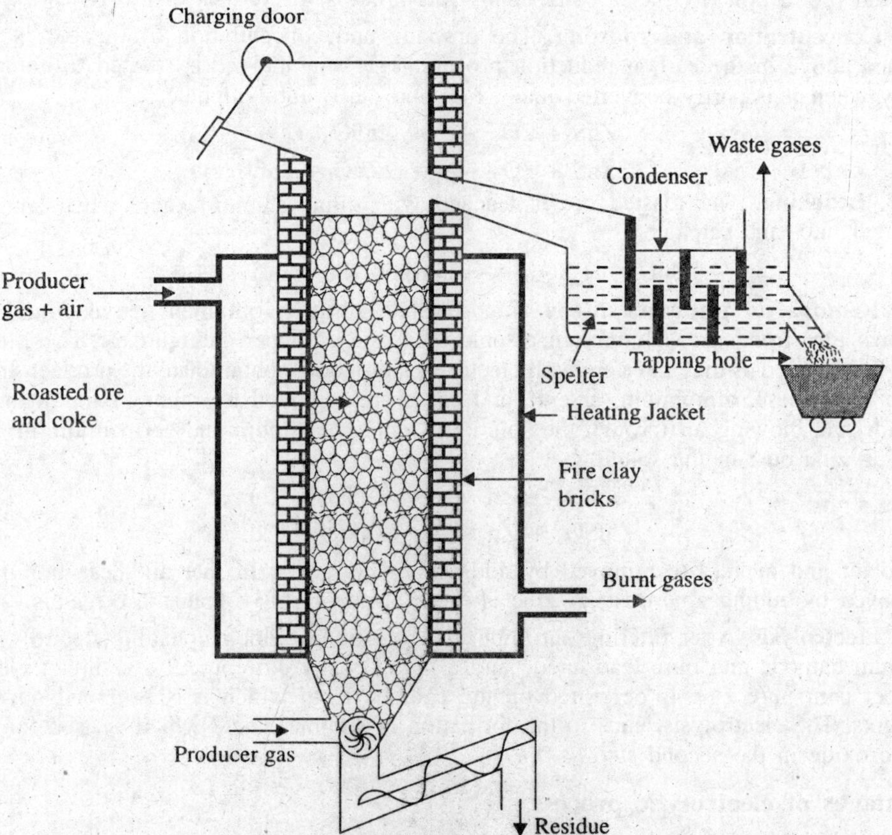

Fig. 8.5. Vertical retort process for the reduction of ZnO to Zn-metal.

4. Purification. The crude zinc metal or the spelter, as it is called, contains non-fusible impurities and the impurities of Pb, Fe, Cd and As. These impurities are removed by the following methods.

(*a*) *Liquation.* The crude zinc is melted on the sloping hearth of a reverberatory furnace. Molten zinc flows down while non-fusible impurities are left on the hearth.

(*b*) *Fractional distillation.* Spelter or commercial zinc obtained by any of the processes contains about 97–98 percent of zinc; the impurities being mostly lead, iron, cadmium and arsenic. These are easily separated by utilizing the differences in their boiling points which are iron = 3000°C, Zn = 906°C, Pb = 1725°C and Cd = 765°C. The spelter is distilled between temperature of 950°C to 1000°C. First arsenic volatilizes and soon after cadmium distils over and iron and lead remain behind. The mixed vapours of zinc and cadmium are fed through a column maintained at 800°C, when zinc condenses while vapours of cadmium pass over. On repeating the process nearly 99.9 per cent pure zinc is obtained.

(*c*) *Electrolysis.* Very pure zinc can be obtained by electrolysis. Impure zinc is made the anode and pure aluminium sheet as the cathode in an electrolytic cell containing pure zinc sulphate solution acidified with dilute sulphuric acid. On passing electric current pure zinc from the electrolyte is deposited on the cathode which grows in size. An equal amount of zinc passes into the solution from the anode.

B. Electrolytic process

This method is employed where cheap electricity is available. About 50 per cent of the total world production of zinc is obtained by this process. It consists of the following steps :

1. Concentration and roasting. The dressing and concentration of the ore is done as mentioned above in the carbon reduction process. The sulphide ore is roasted to about 700°C carefully when it is partly converted into oxide and partly into sulphate.

$$ZnS + 2O_2 \longrightarrow ZnSO_4$$

$$2ZnS + 3O_2 \longrightarrow 2ZnO + 2SO_2$$

2. Leaching. The roasted ore is leached with dilute sulphuric acid when zinc oxide i converted into zinc sulphate.

$$ZnO + H_2SO_4 \longrightarrow ZnSO_4 \text{ (Solution)} + H_2O$$

3. Removal of impurities. The zinc sulphate solution obtained above contains many impurities, *e.g.*, iron, silica, aluminium, arsenic, manganese, copper, cadmium etc. These impurities have to be removed as they interfere with electrolysis or would contaminate the product. Impurities of iron, manganese, aluminium, arsenic and silica are removed by precipitation with calcium hydroxide and blowing air through the solution. Copper, cadmium and germanium are removed by adding zinc dust in the solution.

$$CuSO_4 + Zn \longrightarrow ZnSO_4 + Cu$$

$$CdSO_4 + Zn \longrightarrow ZnSO_4 + Cd$$

Cobalt and nickel are removed by adding organic acids. In fact all these impurities can be removed by adding zinc dust, as zinc is more electro-positive than these metals.

4. Electrolysis. After filtering out impurities, solution of zinc sulphate is electrolysed using aluminium cathode and pure lead anode and a high density current. As a result of electrolysis 99.98 per cent pure zinc is deposited on the cathode from which it is scrapped out and cast into ingots. The electrolysis leads to the formation of sulphuric acid which is used for leaching out zinc oxide in the second step.

Advantages of electrolytic process

The advantages of this process are :

(*i*) The metal obtained is highly pure and possesses greater corrosion property.

(*ii*) Poor ores can also be satisfactorily employed for the extraction of the metal.

(*iii*) Zinc is obtained more efficiently and completely.

(*iv*) Other metals associated with zinc in the ores are profitably recovered.

Preparation of granulated zinc

This zinc is used in laboratory and is made by melting the metal in a clay crucible and pouring the melt in the form of a thin stream into the bucket of cold water.

Preparation of zinc foil or sheet

It is prepared by heating the metal to 100–150°C when it becomes soft and rolled into sheets.

Preparation of zinc dust

It is obtained by atomising zinc with a blast of air. It always contains oxide; good varities may contain over 90 per cent of metallic zinc. Zinc oxide may be removed by washing with very dilute hydrochloric acid, water and alcohol and drying.

Physical properties

Zinc is a bluish white metal, brittle at the room temperature. It loses its brittleness at about 120°C but regains it at 200°C when it can be easily powdered. Its specific gravity is 7.14, melting point 419.5°C and boiling point 906°C. It is a good conductor of heat and electricity. At 100°–150°C it becomes ductile and malleable.

Chemical properties

1. Action of air. Zinc is stable in air. In moist air a protective covering of basic zinc carbonate is formed. When heated to 500°C, it catches fire with a bluish white flame forming zinc oxide which is very light and is called *philosopher's wool*.

2. Action of water. Pure zinc is not acted upon by water. Impure zinc decomposes boiling water slowly and steam rapidly.

$$Zn + 2H_2O \longrightarrow Zn(OH)_2 + H_2$$

Zn-Cu couple decomposes water at ordinary temperature evolving hydrogen.

3. Action of acids. Very pure zinc reacts very slowly with acids whereas commercial zinc reacts readily with all acids. The reason for this has not been definitely established. Most probably a layer of condensed hydrogen forms on the surface of pure zinc and stops the action of the acid.

Impure zinc contains particles of other metals as impurity which are electronegative to zinc and set up a galvanic action and form a number of tiny voltaic cells similar to that of copper and zinc.

(*a*) *Hydrochloric acid :* Dilute or concentration hydrochloric acid forms zinc chloride with the evolution of hydrogen.

$$HCl + Zn \longrightarrow ZnCl_2 + H_2$$

(*b*) *Sulphuric acid:* Dilute acid reacts with commercial zinc evolving hydrogen. Hot and concentrated acid attacks zinc giving off sulphur dioxide.

$$Zn + H_2SO_4 \ (\textit{Dilute}) \longrightarrow ZnSO_4 + H_2$$

$$Zn + 2H_2SO_4 \ (\textit{Hot and concentrated}) \longrightarrow ZnSO_4 + SO_2 + 2H_2O$$

Moderately concentrated acid (20 per cent) when heated with the metal evolves hydrogen sulphide and precipitates a little sulphur.

$$4Zn + 5H_2SO_4 \ (\textit{Moderately concentrated}) \xrightarrow{\Delta} 4ZnSO_4 + H_2S + 4H_2O$$

$$3Zn + 4H_2SO_4 \ (\textit{Moderately concentrated}) \xrightarrow{\Delta} 3ZnSO_4 + S + 4H_2O$$

(*c*) *Nitric acid.* Zn reacts with very dil. HNO_3 (6%) and is oxidised to $Zn(NO_3)_2$.HNO_3 is reduced to NH_4NO_3.

$$4Zn + 10HNO_3\ (\textit{very dilute}) \longrightarrow 4Zn(NO_3)_2 + NH_4NO_3 + 3H_2O$$

or
$$4Zn + NO_3^- + 10H^+ \longrightarrow 4Zn^{2+} + NH_4^+ + 3H_2O$$

Zn reacts with cold and dil. HNO_3 (20%) and is oxidised to $Zn(NO_3)_2$. HNO_3 is reduced to N_2O.

$$4Zn + 10HNO_3\ (\textit{cold and dilute}) \longrightarrow 4Zn(NO_3)_2 + 2N_2O + 5H_2O$$

or
$$Zn + 2NO_3^- + 10H^+ \longrightarrow 4Zn^{2+} + N_2O + 5H_2O$$

Zn reacts with cold and conc. HNO_3 (70%) and is oxidised to $Zn(NO_3)_2$. HNO_3 is reduced to NO_2.

$$Zn + 4HNO_3\ (\textit{cold and conc.}) \longrightarrow Zn(NO_3)_2 + 2NO_2 + 2H_2O$$

or
$$Zn + 2NO_3^- + 4H^+ \longrightarrow Zn^{2+} + 2NO_2 + 2H_2O$$

Zn also reacts with cold and moderately conc. HNO_3 (50%). In this reaction Zn is oxidised to $Zn(NO_3)_2$ and NO gas is evolved.

$$3Zn + 8HNO_3 \longrightarrow 3Zn(NO_3)_2 + 4H_2O + 2NO$$

or
$$3Zn + 2NO_3^- + 8H^+ \longrightarrow 3Zn^{2+} + 2NO + 4H_2O$$

4. Action of alkalies. Zinc dissolves in hot and concentrated caustic alkalies forming alkali zincates with the evolution of hydrogen.

$$Zn + 2NaOH \longrightarrow Na_2ZnO_2 + H_2$$

5. Displacement of metals. Most metals being less electropositive than zinc are precipitated from their salt solutions. The reaction is utilized in the precipitation of silver and gold from the complex cyanide solution in the *cyanide process*.

$$Zn + CuSO_4 \longrightarrow ZnSO_4 + Cu$$

$$Zn + 2NaAu(CN)_2 \longrightarrow Na_2Zn(CN)_4 + 2Au\downarrow$$

6. Action of non-metals. Zinc reacts with halogens and sulphur directly on heating.

$$Zn + Cl_2 \longrightarrow ZnCl_2$$

$$Zn + S \longrightarrow ZnS$$

7. Action of ammonia. Zinc nitride is formed.

$$3Zn + 2NH_3 \longrightarrow Zn_3N_2 + 3H_2$$

8. Reducing action. It acts as a powerful reducing agent *e.g.*, it reduces ferric salts to ferrous salts and nitrate to ammonia.

$$Fe_2(SO_4)_3 + Zn \longrightarrow 2FeSO_4 + ZnSO_4$$

$$NaNO_3 + 4Zn + 7NaOH \longrightarrow 4Na_2ZnO_2 + 2H_2O + NH_3$$

Zinc in presence of an acid, is an excellent reducing agent; the general reaction being :

$$Zn + 2HCl + XO \longrightarrow ZnCl_2 + H_2O + X$$

It has reducing power much greater than that of ordinary hydrogen. The greater chemical activity is explained on the fact that the whole of the liberated energy in the reaction does not appear as heat, but a part is associated with the hydrogen molecules increasing its chemical energy. Zinc amalgam is an effective reducing agent.

9. Action of alkyl halides. Zn, like Mg, forms addition compounds with alkyl halides.

$$Zn + C_2H_5I \longrightarrow ZnC_2H_5I$$

Uses of zinc

(*i*) Zinc is extensively employed for galvanizing iron *i.e.*, for the manufacture of iron sheets for roofing, iron gutters, utensils, buckets, pipes etc.

(*ii*) For sherardizing iron for the manufacture of conduit pipes, laying telephone cables, electric wire etc.

(*iii*) Zinc is used in the laboratory for the preparation of hydrogen gas and as a powerful reducing agent with dilute hydrochloric or sulphuric acid or concentrated solution of a caustic alkali and also as zinc dust or zinc-copper couple.

(*iv*) It is also used as cathode plate in dry cells and electric batteries and in producing smoke screens in war.

(*v*) For desilverizing lead in Parke's process and for the precipitation of silver and gold in cyanide process.

(*vi*) In the form of zinc dust it is used for the manufacture of dye-stuffs, paints, drugs and face powders, snow, creams, etc.

(*vii*) Many zinc compounds find extensive uses, *e.g.*, the oxide in the manufacture of paints, filling rubber and in medicine; the sulphate, in preserving hides, calico printing, in medicine as eye lotion and paints etc.

(*viii*) For the preparation of useful alloys like *brass, muntz metal, german silver* and *die casting metal.*

(*a*) *Brass :* It is yellow, malleable and ductile alloy containing 60 to 80 per cent copper and 20 to 40 per cent zinc.

(*b*) *Muntz metal :* It is an alloy containing 60 per cent copper and 40 per cent zinc.

(*c*) *German silver :* It is a shining white alloy containing 25–50 per cent copper, 25–35 per cent zinc and 10 to 35 per cent nickel. It is used for preparing cheap quality ornaments and utensils.

(*d*) *Die casting metal:* Zinc alloyed with 4 per cent aluminium, upto 3.5 percent copper and 0.05 per cent manganese is used for the production of die castings.

Galvanisation

What is galvanisation?

It is the *process of depositing a thin layer of zinc over the surface of iron.* The zinc coating protects the iron from corrosion because zinc is more electropositive and would be attacked first. The protection is, therefore, *sacrificial.*

Steps of galvanisation process

The process of galvanisation consists of the following steps :

1. Cleaning of iron sheets : Pickling. The iron articles to be galvanised are first cleaned by means of a sand blast. The sheets are then dipped in dilute sulphuric acid solution to remove any oxide film and washed with water. This is known as *pickling*.

2. Galvanising. In this step zinc metal is deposited on the cleaned iron sheets. Following methods are employed for this purpose.

(*a*) *Sherardizing.* The articles made of iron are dipped into zinc dust in a revolving drum at a temperature just below the melting point of zinc. After keeping there for some time it is taken out and heated to 800°C when an alloy of iron and zinc is formed at this temperature at the surface.

(*b*) *Metalizing.* Molten zinc is sprayed on the cleaned iron articles and an adherent coat of zinc metal is obtained. This method is especially suitable for articles too large for dipping, as for example, bridges, ships etc.

(*c*) *Electro-galvanising.* Iron articles are continuously passed through a zinc plating bath of sodium zinc tetracyanide. The article is made the cathode when metallic zinc is deposited on it. The deposit so obtained on the article is very smooth, shining and durable.

Typical Problem Metallic ore (A) when heated with K_2CO_3 in air gives a yellow solid (B) which is soluble in water. On acidification of (B), an orange solution of (C) is obtained which was crystallised to solid (C). On heating the solid (C) with NH_4Cl, water soluble green coloured compound (D) is formed. Identify A to D and also explain the chemical reactions involved in it. *(Delhi 2000)*

Solution. Reactions can be written as :

(*i*) $\underset{\text{Chromite are (A)}}{4\,(FeO.Cr_2O_3)} + 8K_2CO_3 + 7O_2 \text{ (From air)} \longrightarrow \underset{\text{Yellow solution (B)}}{8Na_2CrO_4} + 2Fe_2O_3 + 8CO_2$

(*ii*) $\underset{\text{(B)}}{2Na_2CrO_4} + H_2SO_4 \longrightarrow \underset{\text{Orange solution (C)}}{Na_2Cr_2O_7} + Na_2SO_4 + H_2O$

(*iii*) $\underset{\text{(C)}}{Na_2Cr_2O_7} + 2NH_4Cl \longrightarrow \underset{\text{Green soluble (D)}}{2Cr(OH)_3} + 2NaCl + H_2O + N_2$

Above reactions show that A = FeO. Cr_2O_3 (chromite ore), B = Na_2CrO_4 (sodium chromate), C = $Na_2Cr_2O_7$ (sodium dichromate), D = $Cr(OH)_3$ (chromium hydroxide)

University Questions

Titanium

1. Write a note on "oxidation states of titanium". *(Jodhpur 84)*
2. What are the different oxidation states of titanium? In which of these states the compound will be (*a*) paramagnetic (*b*) diamagnetic (*c*) coloured (*d*) colourless. Give reasons. *(Raj. 86)*
3. What are the important ores of titanium? *(Meerut 82; Jodhpur 85; Punjab 85; Bharathiar 86; Allahabad 88)*
4. Write formulae of rutile and illmenite, whichmetalis extracted from them? *(Andhra 85)*
5. How is titanium metal extracted in a pure form from illmeniteand rutile ores. *(Meerut 82, 88;Jodhpur 85; Punjab 85; Bharathiar 86; Gorakhpur 81, 86; Allahabad 88; Bundelkhand 88)*
6. What are Kroll processand I.C.I. processes in the extraction of titanium metal? *(Madras 85)*
7. What is Van Arkel method for purificationof titanium? *(Delhi 86)*
8. Explain why direct reduction with carbon may not be appropriate for the extraction of titanium from its oxide. *(Bharathiar 86)*
9. Give reasons for the following:
   (*i*) Nearly all compounds of titanium in + 2 oxidation state are purple. *(Poona 86)*
   (*ii*) Titanium is unreactive at low temperature. *(Calcuta 85)*
   (*iii*) Titanium dissolves in HF readily.
   (*iv*) Ti (III) compounds are strong reducing agents. *(M.D. Rohtak 85)*
10. Write notes on: (*i*) uses of titanium metal (*ii*) titanium alloys. *(Jodhpur 85; Punjab 85; Bharathiar 86)*

Vanadium

11. Vanadiumis a transitionmetal and phosphorus is a non-metal, yet there is a similarity between them. Illustrate with examples. *(Meerut 84)*
12. Write a note on the important oxidation states of vanadium. Give suitable examples. *(Gorakhpur 85; Madras 86)*
13. How is vanadium extracted from its ores? *(Gorakhpur 81)*
14. Discuss the chemistry of vanadium on the basis of its electronic configuration and important oxidation states. *(Raj. 87)*
15. How does vanadium occur innature? Discuss the extraction of vanadium from either vanadinite or carnotite. *(Raj. 87; Meerut 82; Indore 86)*
16. What happens when vanadium (i) is heated in air, in N_2 and in Cl_2 (*ii*) is dissolved in perchloric acid, conc. H_2SO_4 and hot fused alkali. *(Delhi 85; Punjab 85)*

Chromium

17. What are important ores of chromium? Describe the extraction of the metal fromits chief ores. Give the chemical principles involved in the extraction process. *(Meerut 95; Garhwal 88; Allahabad 89; Gorakhpur 86; Sambalpur 86; Agra 85)*
18. Discuss in detail the occurrence extraction, properties and uses of Cr-metal andits compounds. *(Sambalpur 86; Agra 85; meerut 97)*
19. Write the electronic configuration of the atom and the ion in oxidation state + 2 of Cr-metal. *(Raj. 85)*
20. Discuss the position of Cr in the periodic table. *(Garhwal 88; Meerut 81, 82)*

Molybdenum

21. Name the important ores of molybdenum. How is the metal extracted? What are its uses? *(Meerut M. Sc. 82, 85; Bharathiar 85)*
22. Comment on various oxidation states shown by Mo. Give one example of each. *(Meerut M.Sc. 84)*
23. Write a note on "Pentavalent molybdenum". *(Utkal 86)*

Tungsten

24. What are the important ores of tungsten? How is the metal obtained from its ores? *(Madras 85, 86; Madurai 85)*
25. Write a note on "Quadrivalent tungsten" *(Utkal 86)*

Manganese

26. How does Mn occur in nature? *(Allahabad 87)*

    OR

    What are chief ores of Mn? *(Meerut 86; Bundelkhand 88)*
27. Outline the method for the extraction of Mn-metalfromits ores. Give all the equations of reactions involved. *(Allahabad 87; Punjab 96; Meerut 86; Bundelkhand 88; Madras 85;Lucknow 86)*

Iron

28. Give thenames of important ores of iron. *(Raj. 86)*
29. How is pig iron produced from iron ore? *(Agra 86; Raj.86)*
30. How is pig iron converted into wrought iron and steel? *(Agra 86)*
31. What is meantby cast iron? *(Meerut 89)*
32. Give equations of reactions that occur in the blast furnace for the productionof pig iron. *(M.D. Rohtak 84)*
33. What is meant by wrought iron? *(Meerut 89)*
34. What are alloy steels? Mention threeelements that are used inpreparing them? What specialproperties are producedin these metals? *(Madras 86; Bharathiar 86)*
35. Describe the manufactureof steel by L.D. process. State themerits of the process. *(Bharathiar 85, 86; Bharathidasan 85; Madras 86)*
36. Write a note on alloy steels. Name their uses. *(Bharathiar 86; Punjab 84; Delhi 85; Madurai 85)*
37. What is meant by steel? *(Meerut 89)*
38. Write a note on steel industry. *(Jiwaji 83)*
39. Compare the composition, properties and uses of pig iron, cast iron, wrought iron and steel. *(Meerut 82, 89; Raj. 86; Punjab 86)*
40. Write notes on: (*i*) Tempering of steel, (*ii*) Hardening of steel (*iii*) Case hardening. *(Bharathiar 85, 86; Bharathidasan 85; Madras 86; Meerut 89)*

Osmium

41. How is osmium extracted from osmiridium by fusion process?
42. Write some uses of osmium metal.

Cobalt

43. Menion two important ores of cobalt. Explain with equations, how is the metal is extracted from its ores? *(Madras 86; Punjab 85)*
44. What happens when cobalt salt is treated with excess of KCN? *(Agra 84)*

Nickel

45. What are the chief ores of nickel? Describe the extraction of nickel (*i*) by Mond's process (*ii*) from Sudbery ores. What by-products are obtained in (*ii*) process. Give the utility of themetal in industry. *(Jiwaji 84; Allahabad 88; Meerut 82,84; Garhwal 88; Gorakhpur 85; Usmania 93)*
46. Name important alloys of nickel and mention their uses. *(Jiwaji 84)*

47. (*a*) How willyou extract nickel from pentaldite ore? *(Bundelkhand 83; Meerut 88)*

Palladium

48. Write the names of important ores of palladium.

49. Write the names of different forms of palladium.

50. How does palladium react with HNO_3 and H_2SO_4 ?

Platinum

51. Name the important ores of Pt. *(Meerut 96; Jodhpur 84; Bharathiar 86)*

52. Discuss the method and give the chemical principles involved in the extraction of Pt from native ore. *(Allahabad 85, 88; Jodhpur 84; Agra 82, 84, 87, 88; Bharathiar 86)*

53. Write notes on: (*i*) Different from of Pt (*ii*) Preparation of Pt black and its uses. *(Jodhpur 84; Agra 84; Bundeldhand 86)*

54. Starting form platinum how would you prepare platinum black, platinised asbestos and spongy platinum ? *(Indore 86; Punjab 82)*

55. (*a*) Write notes on : (*i*) Important uses of platinum (*ii*) Platinum catalysis. *(Allahabad 80, Agra 85)*

(*b*) How will you extract and isolate Pt-metal ? *(Meerut 98)*

(c) How will you prepare platinum black ? *(Meerut 96)*

Gold

56. How is gold extracted from its ores ?Discuss the chemical reactions involved in method employed. *(Allahabad 89)*

57. How in fineness of gold expressed ? *(Meerut 82)*

58. Explain why gold dissolves in aqua regia but neither in conc. HCl nor in conc. HNO_3 alone. *(R.E.E. 85; Delhi Hons. 87)*

59. What happens whin gold is dissolved in aqua regia. Explain with equations. *(Jiwaji 84; I.IT, 87 madras 85)*

60. Describe the Mac Arthur Forest cyanide process for the extraction of gold from its ores. *(Meerut 82)*

61. Give the names of two important minerals of gold. Give the method of extraction of gold from its important minerals. *(Agra 88)*

Zinc

62. What are important ores of molybdenum ? How is this metal extracted from its ores ? Mention important uses of this metal. *(Lucknow 2002)*

63. How is titanium extracted from ilmenite are ? How is the impurity of iron removed from it? *(VBS, Purvanchal 2003)*

64. (*a*) Name the principle ores of copper. Explain electrolytic refining process for purification of copper.

(*b*) How does copper react with air below 1370 K ? *(Nagpur 2003)*

65. Write the formula of rutile are . *(VBS, Purvanchal 2003)*

66. Write the formula of chromite are. *(VBS, Purvanchal 2003)*

67. (i) Explain the following terms :

(a) Cementite (b) Ferrite.

(ii) Write a note on annealing of steel

(iii) How is ferro manganese manufactured ?

(iv) How does the following elements influence the properties of steel ?

(a) Manganese (b) Chromium

(v) Account of the following

(a) Ferro alloys but not pure elements are used in the production of alloy steels.

(b) Sodium cyanide is used in the extraction of gold by Mac-Arthur cyanide process

(c) Case hardened steel is used for making automobile axles

(vi) Describe the extraction of Gold by hydrometallurgy

(vii) Describe the extraction of tungsten powder from wolframite.

(viii) (a) How is ferromanganese prepared ? Mention its use.

(b) Discuss the influence of carbon on the properties of steel.

(ix) Give one use of Manganese steel. *(Banglore 2004)*

68. Discuss the metallurgy of Ni on the following points :

(*i*) Ores (*ii*) Extraction (*iii*) Recovery of Ni. *(Bhopal 2004)*

9

Important Compounds of Transition Metals

Compounds of Lanthanum

Lanthanum forms stable compounds in +3 oxidation state. Its important compounds are discussed below :

1. Lanthanum hydrides, LaH_2 and LaH_3

Lanthanum reacts with H_2, forming dihydride (LaH_2) and trihydride (LaH_3). At the initial stage of the reaction between La and H_2, heating to 300–400°C is required, but later on heat is evolved, *i.e.*, the reaction is exothermic. Even upto 900°C both the hydrides are stable.

2. Lanthanum oxide, La_2O_3

It is obtained by heating lanthanum carbonate or oxalate. It is snow-white powder. When put into water, it is converted into $La(OH)_3$ and a large amount of heat is evolved. This behaviour of La_2O_3 is similar to that shown by CaO. It is sparingly soluble in water. It shows basic character and dissolves in dil. acids. On heating a mixture of La_2O_3 and carbon in a current of Cl_2, anhydrous $LaCl_3$ is obtained. La_2O_3, on being treated with dil. H_2SO_4, gives $La_2(SO_4)_3$.La_2O_3 reacts with dil. HNO_3 to form $La(NO_3)_3$.

La_2O_3 is used (*i*) in the manufacture of crookes glass, since it can absorb ultraviolet rays to some extent (*ii*) in the preparation of $LaCl_3$, $La_2(SO_4)_3$, $La(NO_3)_3$ etc.

3. Lanthanum halides, LaX_3

(*a*) LaF_3 is obtained as a white gelatinous precipitate when HF reacts with a lanthanum salt. If the solution containing HF and lanthanum salt is allowed to stand for sometime, crystals of $LaF_3.3H_2O$ are obtained. It is insoluble in dil. but soluble in conc. minerals acids.

(*b*) When a mixture of La_2O_3 and carbon is heated in a current of Cl_2, anhydrous $LaCl_3$ is obtained. It is a white deliquescent powder and is soluble in water. Crystals of heptahydrate ($LaCl_3.7H_2O$) can be obtained from the aqueous solution of $LaCl_3$.

(*c*) $LaBr_3$ and LaI_3 are also known and are prepared by similar methods.

4. Lanthanum sulphate, $La_2(SO_4)_3$

Anhydrous sulphate is obtained by reacting La_2O_3 with dil. H_2SO_4. It is a white hygroscopic powder with a density of 3.6. It is soluble in ice-cold water. On adding K_2SO_4 to the solution of $La_2(SO_4)_3$, a double sulphate, $La_2(SO_4)_3.3K_2SO_4$ is obtained. When a concentrated solution of $La_2(SO_4)_3$ is warmed, sparingly soluble *nona-hydrate* $La_2(SO_4)_3.9H_2O$ separates out.

5. Lanthanum nitrate, $La(NO_3)_3.6H_2O$

It is obtained by treating La_2O_3 with dil.HNO_3. It gets separated from water as *hexahydrate,* $La(NO_3)_3.6H_2O$. When concentrated solutions of $La(NO_3)_3$ and NH_4NO_3 are mixed, crystals of $La(NO_3)_3. 2NH_4NO_3.4H_2O$ (double salt) are obtained.

6. Lanthanum carbonate, $La_2(CO_3)_3$

This salt occurs in nature as *lanthanite*, $La_2(CO_3)_3.8H_2O$. When $(NH_4)_2CO_3$ is added to a solution of $La(NO_3)_3$, $La_2(CO_3)_3$ is obtained as a white ppt. When the solution containing the two salts is allowed to stand, crystals of $La_2(CO_3)_3.(NH_4)_2CO_3$ (double salt) separate out.

Compounds of Titanium

Titanium forms compounds in 0, +2, +3, +4 and –1 oxidation states.

Compounds of Ti(0)

Ti in zero oxidation state has its electronic configuration as 2, 8, $3s^2p^6d^2$, $4s^2$. $[Ti(dipy)_3]^0$ is an important example of Ti(0) complex compound. Compounds of Ti(0) are rare.

Compounds of Ti (+2)

The configuration of Ti^{2+} ion is 2, 8, $3s^2p^6d^2$, $4s^0$. Ti (+2) compounds are not very stable. They donot have aqueous chemistry, since they are readily oxidised by H_2O liberating H_2. TiO, $TiCl_2$, $TiBr_2$, TiI_2 etc. are the examples of Ti(+2) compounds.

Compounds of Ti (+3) : Titanous compounds

The configuration of Ti^{3+} ion is 2, 8, $3s^2p^6d^1$, $4s^0$, showing that Ti^{3+} ion is a d^1 system. Compounds of Ti(+3) are called *titanous compounds.*

These compounds are prepared when the corresponding Ti(+4) compounds are reduced with Zn or dil. HCl or by electrolytic method. Ti(+3) compounds are mainly ionic. They undergo disproportionation.

$$2Ti^{3+} \xrightarrow{\Delta} Ti^{4+} + Ti^{2+}$$

Examples of Ti(+3) compounds are Ti_2O_3, $Ti(OH)_3$, TiX_3 (X = F,Cl), $TiBr_3.6H_2O$, $[Ti(H_2O)_6]Cl_3$, $M_3[TiF_6]$, $M_2[TiCl_5(H_2O)]$ (M = alkali metal) etc.

Compounds of Ti (+4) : Titanic compounds

The configuration of Ti^{+4} ion is 2, 8, $3s^2p^6d^0$, $4s^0$. Ti (+4) compounds are the most stable. These are generally covalent and generally resemble, in certain properties, with the corresponding compounds of group IVA elements. Ti (+4) compounds are called *titanic compounds*. Examples are TiO_2, TiX_4 (X = F, Cl, Br, I), $Ti(SO_4)_2$, $TiOCl_2$, $TiOSO_4$ etc.

Compounds of Ti (–1)

Electronic configuration of Ti^{-1} ion is 2, 8, $3s^2p^6d^3$, $4s^2$. Compounds of Ti(–1) are rare. $[Ti(dipy)_3]^-$ is an example of Ti(–1) compound.

In this chapter we shall make a study of oxides and halides formed by titanium metal in different oxidation states.

Oxides of Titanium

Titanium forms three oxides *viz.* TiO (Ti = +2), Ti_2O_3 (Ti = +3) and TiO_2 (Ti = + 4).

1 Titanium oxide, TiO

It is formed by heating TiO_2 with Ti metal or carbon.

$$TiO_2 + Ti \longrightarrow 2TiO$$
$$TiO_2 + C \longrightarrow 2TiO + CO$$

It is a golden-yellow crystalline substance. It is a basic oxide and dissolves in acids, evolving H_2.TiO reduces H_2O to H_2.

$$TiO + H_2O \longrightarrow TiO_2 + H_2$$

It has ionic character.

2. Titanium sesquioxide, Ti_2O_3

It is obtained by heating TiO_2 with H_2 at 1000°C.

$$2TiO_2 + O_2 \longrightarrow Ti_2O_3 + H_2O$$

It dissolves in H_2SO_4 forming titanous sulphate. It has *corundum structure.*

3. Titanium dioxide or titanic dioxide or titania, TiO_2

This oxide occurs in nature as rutile which is an important ore of titanium.

Preparation. (*i*) *From $Ti(OH)_4$.* TiO_2 can be obtained by calcining the ppt. of $Ti(OH)_4$ formed by adding NH_4OH to a solution of $TiCl_4$.

$$TiCl_4 + 4NH_4OH \longrightarrow \underset{\text{ppt}}{Ti(OH)_4} + 4NH_4Cl$$

$$Ti(OH)_4 \xrightarrow{\Delta} TiO_2 + 2H_2O$$

(*ii*) *From ilmenite ore, $FeTiO_3$.* See *"Extraction of Titanium"*.

(*iii*) *From $TiCl_4$.* The spraying of liq. $TiCl_4$ into moist air produces white smokes of TiO_2.

$$TiCl_4 + 4H_2O \longrightarrow Ti(OH)_4 + 4HCl$$
$$Ti(OH)_4 \xrightarrow{\Delta} TiO_2 + 2H_2O$$
$$\overline{TiCl_4 + 2H_2O \longrightarrow TiO_2 + 4HCl}$$

Properties. (*i*) TiO_2 occurs in nature in three crystalline forms which are *rutile, anatase* and *brookite.* Rutile is the commonest form and forms tetragonal transparent to opaque, red or yellowish crystals. This form is stable above 1040°C. The anatase modification also crystallises in the tetragonal system but with the different axial ratio. This form is stable upto 860°C. Brookite form crystallises in rhombic form and is stable between 860–1040°C. The rutile form of TiO_2 is considered to be the most stable. The rutile or brookite forms are metastable below 860°C and thus are converted into anatase by suitable heat treatment.

$$\textit{Anatase} \underset{}{\overset{860°C}{\rightleftharpoons}} \textit{Brookite} \overset{1040°C}{\rightleftharpoons} \textit{Rutile}$$

(*ii*) Natural TiO_2 is impure due to the presence of iron oxide as an impurity contaminated with it. Pure TiO_2 is colourless (amorphous white powder) when cold, and yellowish when hot.

(*iii*) TiO_2 is insoluble in water.

(*iv*) *Reduction* (*a*) When TiO_2 is heated with Ti-metal or in a carbon crucible, TiO_2 is reduced to TiO.

$$TiO_2 + Ti \longrightarrow 2TiO$$
$$TiO_2 + C \longrightarrow 2TiO + CO$$

(*b*) When reduced by H_2 at 1000°C, Ti_2O_3 is obtained.

$$2TiO_2 + H_2 \longrightarrow Ti_2O_3 + H_2O$$

(*c*) TiO_2 melts at about 1800°C and loses O_2 to form Ti_2O_3.

$$4TiO_2 \xrightarrow{1800°C} 2Ti_2O_3 + O_2$$

(*d*) When heated with carbon, Ca or Al, TiO_2 gets reduced to Ti-metal.

$$TiO_2 + C \xrightarrow{2500°C} Ti + CO_2$$
$$3TiO_2 + 4Al \longrightarrow 3Ti + 2Al_2O_3$$
$$TiO_2 + 2Ca \longrightarrow Ti + 2CaO$$

(*v*) *Action of Cl_2 on a mixture of TiO_2 and carbon.* When TiO_2 is heated with carbon in a stream of Cl_2, $TiCl_4$ is obtained.

$$TiO_2 + C + 2Cl_2 \xrightarrow{\Delta} TiCl_4 + CO_2$$

(*vi*) *Action of Cl_2.* On heating TiO_2 in a current of dry Cl_2, it forms $TiCl_4$ with the evolution of O_2.

$$TiO_2 + 2Cl_2\ (dry) \xrightarrow{\Delta} TiCl_4 + O_2$$

(*vii*) *Amphoteric nature.* TiO_2 is an amphoteric oxide and hence reacts with acids to form *titanyl salts* (TiO^{2+}) and with alkalies to form *titanates* (TiO_3^{2-}).

$$TiO_2 + H_2SO_4 \longrightarrow TiOSO_4 + H_2O$$
$$TiO_2 + 2KOH \longrightarrow K_2TiO_2 + H_2O$$

(*viii*) *Action of vapours of CCl_4.* On passing the vapours of CCl_4 over TiO_2 at high temperature, $TiCl_4$ is formed.

$$TiO_2 + CCl_4\ (vapours) \longrightarrow TiCl_4 + CO_2$$

(*ix*) *Action of HF.* TiO_2 dissolves in HF to form hydro fluotitanic acid, $H_2[TiF_6]$.

$$TiO_2 + 6HF \longrightarrow H_2[TiF_6] + 2H_2O$$

Uses. (*i*) Because of its high opacity,inertness towards H_2S, high covering power, light weight, non-poisonous nature, ease of wetting and disperssion in oil as well as resistance to high temperature, it is used as a white pigment under the name *titan white.* Mixed with $BaSO_4$ it is sold under the trade name *titanox.* (*ii*) It can be also be used for interior protection purpose. When applied externally the film gets chalked out. (*iii*) It is also used (*a*) in making non-greasable paper (*b*) in the manufacture of arc lamp electrodes (*c*) in gas mantles (*d*) for the preparation of glazes (*e*) in making components for use in radio and television sets (*f*) as an additive in the manufacture of fibre glass (*g*) in the manufacture of artificial gems (*h*) in enamelling of porcelain goods (*i*) in the surface of white leather (*j*) in rubber industry to soften rubber (*k*) as a whitener in cosmetics (*e.g.* in face powder, cold creams etc.) (*l*) in preparing printing and decorating inks.

Halides of Titanium

The well known halides formed by titanium are TiF_4, TiF_3, $TiCl_4$, $TiCl_3$, $TiCl_2$, $TiBr_4$, $TiBr_3 \cdot 6H_2O$, $TiBr_2$, TiI_2, TiI_4 etc. Out of these chlorides are important.

1. Titanium dichloride, $TiCl_2$

Preparation. It is prepared:

(*i*) by reducing $TiCl_4$ with Ti-metal

$$TiCl_4 + Ti \longrightarrow 2TiCl_2$$

(*ii*) by the disproportionation of $TiCl_3$

$$\underset{(Ti\,=\,+3)}{2TiCl_3} \longrightarrow \underset{(Ti\,=\,+2)}{TiCl_2} + \underset{(Ti\,=\,+4)}{TiCl_4}$$

Properties. (*i*) It is a black powder and is volatile.

(*ii*) When heated in air, it catches fire.

(*iii*) With dry NH_3 at red heat, Ti_3N_4 is formed.

$$3TiCl_2 + 4NH_3 \longrightarrow Ti_3N_4 + 6HCl + 3H_2O$$

2. Titanium trichloride or titanous chloride, $TiCl_3$

Preparation. It is the most important compound of Ti(+3). It is prepared by reducing $TiCl_4$ with H_2 in an electrically heated spiral in a water cooled tube or with Ag at 200°C.

$$2TiCl_4 + H_2 \longrightarrow 2TiCl_3 + 2HCl$$

$$TiCl_4 + Ag \longrightarrow TiCl_3 + AgCl$$

$TiCl_3$ can also be obtained by dissolving Ti metal in HCl.

$$2Ti + 6HCl \longrightarrow 2TiCl_3 + 3H_2$$

Properties and uses. Anhydrous $TiCl_3$ is a violet powder. It disproportionates into $TiCl_4$ and $TiCl_2$.

$$2TiCl_3 \longrightarrow TiCl_4 + TiCl_2$$

It is a powerful reducing agent since it reduces chlorates and perchlorates to chlorides, Fe^{3+} to Fe^{2+}, Cu^{2+} to Cu, and persulphate to SO_4^{2-}. Due to this property $TiCl_3$ is used in volumetric analysis of the above mentioned compounds. An aqueous solution of $TiCl_3$, when added to a dilute solution of a gold salt, produces an intense violet colour, due to the reduction of the gold salt followed by the absorption of colloidal gold on hydrous TiO_2. This reaction has been used as a very delicate test for gold. $TiCl_3$ is also used for removing certain dyes from the goods so that they may be redyed. It is also used for removing strains.

3. Titanium tetrachloride or titanic chloride, $TiCl_4$

Preparation. It is prepared : (*i*) by passing Cl_2 over heated Ti-metal or over a mixture of TiO_2 and carbon.

$$Ti + 2Cl_2 \xrightarrow{350°C} TiCl_4$$

$$TiO_2 + C + 2Cl_2 \longrightarrow TiCl_4 + CO_2$$

(*ii*) by heating TiO_2 in a current of dry Cl_2.

$$TiO_2 + 2Cl_2\ (dry) \xrightarrow{\Delta} TiCl_4 + O_2$$

(*iii*) by passing the vapours of CCl_4 over TiO_2 at high temperature.

$$TiO_2 + CCl_4\ (vapours) \longrightarrow TiCl_4 + CO_2$$

(*iv*) from ilmenite ore, $FeTiO_3$. See "*Extraction of Ti-metal*".

Properties. (*i*) It is a colourless strongly fuming liquid and has m.pt. = – 30°C and b.pt. = 136.4°C.

(*ii*) *Hydrolysis.* (*a*) The spraying of liq. $TiCl_4$ into moist air results in its hydrolysis, forming dense white smokes of TiO_2. These smokes have irritating nature.

$$TiCl_4 + 2H_2O \longrightarrow TiO_2 + 4HCl$$

It is due to this property that $TiCl_4$ is used for producing white smoke screens and for making writings in the sky in army operations.

(*b*) In presence of a little amount of water, $TiCl_4$ is hydrolysed to $TiOCl_2$.

$$TiCl_4 + H_2O \longrightarrow TiOCl_2 + 2HCl$$

(*iii*) *Reduction.* (*a*) It can be reduced to $TiCl_2$ by Ti metal.

$$TiCl_4 + Ti \longrightarrow 2TiCl_2$$

(*b*) When reduced by H_2 or Ag, $TiCl_3$ is obtained.

$$2TiCl_4 + H_2 \longrightarrow 2TiCl_3 + 2HCl$$

$$TiCl_4 + Ag \longrightarrow TiCl_3 + AgCl$$

(*c*) When reduced by Mg(*Kroll process*), Na (*I.C.I process*) or NaH, Ti metal is obtained.

$$TiCl_4 + 2Mg \longrightarrow Ti + 2MgCl_2$$
$$TiCl_4 + 4Na \longrightarrow Ti + 4NaCl$$
$$TiCl_4 + 4NaH \longrightarrow Ti + 4NaCl + 2H_2$$

(*iv*) *Action of NH_4Cl.* Yellow crystals of *ammonium chloro titanate*, $(NH_4)_2[TiCl_6].2H_2O$ is obtained.

$$TiCl_4 + 2NH_4Cl + 2H_2O \longrightarrow (NH_4)_2[TiCl_6].2H_2O$$

(*v*) *Action of acids.* (*a*) Conc. HCl dissolves $TiCl_4$, forming chloro complexes like $H[TiCl_5(H_2O)]$, $H_2[TiCl_6]$ etc.

$$TiCl_4 + HCl + H_2O \longrightarrow H[TiCl_5(H_2O)]$$
$$TiCl_4 + 2HCl \longrightarrow H_2[TiCl_6]$$

(*b*) Anhydrous HF gives TiF_4.

$$TiCl_4 + 4HF \longrightarrow TiF_4 + 4HCl$$

(*c*) The action of HI on $TiCl_4$ produces TiI_4.

(*d*) When $TiCl_4$ is treated with H_2SO_4 and the solution is evaporated in vacuum, $TiSO_4$ is obtained.

(*vi*) *Lewis acid character.* It is a strong Lewis acid (*i.e.*, electron pair acceptor) and hence can accept electron pairs from donor molecules like NH_3 and PCl_5, forming the addition compounds like $TiCl_4.6NH_3$, $TiCl_4.8NH_3$ and $TiCl_4.PCl_5$ etc. $TiCl_4$ also forms addition compounds with alcohols and ethers of the type $(R_2O)TiCl_4$, $(R_2O)_2TiCl_4$ etc.

Uses. It is used : (*i*) for the production of dense white irritating smoke screens and for making writings in the sky in army operations (*ii*) as a catalyst in the manufacture of polyethylene (*iii*) in the preparation of $TiCl_2$ and $TiCl_3$ (*iv*) in the production of Ti metal by reducing it with Mg, Na or NaH (*v*) in the manufacture of rutile gems and organic titanium compounds.

Uses of titanium compounds

(*i*) The uses of TiO_2 and $TiCl_4$ have already been described.

(*ii*) Double salts of Ti like $TiOC_2O_4.K_2C_2O_4.H_2O$, $Na_2SO_4.Ti_2(SO_4)_3.5H_2O$ etc. are used as mordants in the dyeing or textile and leather.

(*iii*) $Ti_2(SO_4)_3$ is used for removing certain dyes from the goods so that they may be redyed. It is also used for removing stains.

(*iv*) TiC has been used as a catalyst during the preparation of nitrogen-hydrogen and nitrogen-oxygen compounds. TiC is also used for the preparation of hard metals or cutting alloys.

(*v*) Titanium boride is one of the hardest substances known and hence is used as an abrasive.

Compounds of Vanadium

Vanadium forms compounds in –1, 0, +1, +2, +3, +4 and +5 oxidation states. Compounds in lower oxidation states are very unstable and act as powerful reducing agents. The most stable oxidation states are +4 and +5.

V (–1) compounds

Important examples of compounds in which vanadium is in –1 oxidation state are $[V(CN)_5(NO^+)]^{5-}$, $[V(CO)_6]^-$ etc.

V (0) compounds

Important compounds in which vanadium is in zero oxidation state are $V(CO)_6$, $V(dipy)_3$.

V (+1) compounds

V_2O, $[V(dipy)_3]^+$ etc. are the compounds having vanadium in +1 oxidation state.

V (+2) compounds

Examples of compounds having vanadium in +2 oxidation state are VO or V_2O_2,VX_2 (X = Cl, Br, I), VS or V_2S_2, $VSO_4.7H_2O$, $[V(H_2O)_6]^{2+}$, $[V(CN)_6]^{4-}$ etc.

V(+3) compounds

Examples of simple compounds having vanadium in +3 state are V_2O_3, $V_2(SO_4)_3$, V_2S_3, $V(OH)_3$, VX_3 (X = F, Cl, Br, I) ,VOX etc. Examples of complex ions of V(+3) are $[V(H_2O)_6]^{3+}$, $[V(C_2O_4)_3]^{3-}$ $[V(CN)_6]^{3-}$, $[V(NCS)_6]^{3-}$, $[VCl_4]^-$, $[VF_6]^{3-}$ etc.

V(+4) compounds

VO_2 or V_2O_4, $VOSO_4$, VX_4 (X = F, Cl), VOX_2 (X = F, Cl, Br), $[VCl_6]^{2-}$ etc. are the examples of compounds having vanadium in +4 oxidation state. Compounds of V(+4) are the most stable compounds. VO_2 or V_2O_4 and tetrachlorides are *covalent* while oxyhalides (*e.g.*, VOF_2, $VOCl_2$, $VOBr_2$) are *ionic* in character and contain the vanadyl cation, VO^{2+} which is blue in colour. Compounds containing vanadyl cation are called *vanadyl compounds*. Here we shall discuss $VOCl_2$ and $VOSO_4$.

1. Vanadyl chloride, $VOCl_2$. It is obtained by the hydrolysis of VCl_4.

$$VCl_4 + 2H_2O \rightleftharpoons VOCl_2 + 2HCl$$

Pure $VOCl_2$ is obtained as grassy-green crystals by heating $VOCl_3$ with Zn in a sealed tube.

$$2VOCl_3 + Zn \longrightarrow 2VOCl_2 + ZnCl_2$$

$VOCl_2$ is a strong reducing agent and is used in dyeing and calico pringing.

2. Vanadyl sulphate, $VOSO_4$. This salt exists in two modifications; one is soluble while the other is insoluble in water. When a solution of V_2O_5 in H_2SO_4 is reduced by means of SO_2 and the resulting solution is evaporated, bright blue crystals are obtained. These crystals have the composition, $2VOSO_4.H_2SO_4.xH_2O$. The pure substance is greyish-green crystalline powder which changes to blue soluble modification on heating to 130°C with a little water.

With K_2SO_4, it forms a blue double salt, $K_2SO_4.VOSO_4.xH_2O$. It is used for the titration of weak oxidising agents and for removing traces of O_2 from gases.

V(+5) compounds

Important compounds having vanadium in +5 oxidation state are V_2O_5, VOX_3 (X = F, Cl, Br), VX_5 (X = F, Cl), V_2S_5, vanadic acids, their salts and complex compounds. Here we shall discuss V_2O_5, vanadic acids, their salts and some complex compounds.

Vanadium Pentoxide, V_2O_5

This oxide is also called *vanadic anhydride,* since it can be supposed to be anhydride of orthovanadic acid, H_3VO_4.

$$2H_3VO_4 \xrightarrow{-3H_2O} V_2O_5 + 3H_2O$$

Preparation

V_2O_5 is obtained : (*i*) by heating ammonium meta vanadate, (NH_4VO_3) or by adding dil.H_2SO_4 to its aqueous solution.

$$2NH_4VO_3 \xrightarrow{\Delta} V_2O_5 + 2NH_3 + H_2O$$

$$2NH_4VO_3 + H_2SO_4 \longrightarrow V_2O_5 + (NH_4)_2SO_4 + H_2O$$

NH_4VO_3 used in the above reactions can be obtained from any of the ores of vanadium like carnotite, vanadinite and patronite.

(*ii*) by the hydrolysis of vanadium oxytrichloride ($VOCl_3$).

$$2VOCl_3 + 3H_2O \longrightarrow V_2O_5 + 6HCl$$

(*iii*) by heating vanadium in an atmosphere of O_2 or in air.

$$4V + 5O_2 \longrightarrow 2V_2O_5$$

Along with V_2O_5, little amount of lower oxides is also formed.

Properties

1. Physical state and dissociation. V_2O_5 is a yellowish-red poisonous powder having m.pt. = 670°C. It is slightly soluble in H_2O and the resulting pale yellow solution is acidic in nature. It undergoes reversible dissociation in the temperature range 700–1125°C.

$$2V_2O_5 \rightleftharpoons 2V_2O_4 + O_2$$

2. Action of Cl_2. When V_2O_5 is heated in Cl_2 at elevated temperature or when Cl_2 is passed over heated mixture of V_2O_5 and carbon, $VOCl_3$ is obtained.

$$2V_2O_5 + 6Cl_2 \xrightarrow{\Delta} 4VOCl_3 + 3O_2$$

$$V_2O_5 + 3C + 3Cl_2 \longrightarrow 2VOCl_3 + 3CO$$

3. Action of acids. (*a*) V_2O_5 dissolves in HCl to give salts like vanadium oxydichloride ($VOCl_2$) and VCl_5.

$$V_2O_5 + 6HCl \longrightarrow 2VOCl_2 + 3H_2O + Cl_2$$

The liberation of Cl_2 indicates that V_2O_5 has oxidised HCl to Cl_2, *i.e.*, V_2O_5 can be used as an oxidising agent. V_2O_5 also reacts with HCl in the following manner.

$$V_2O_5 + 10HCl \longrightarrow 2VCl_5 + 2H_2O$$

(*b*) With HNO_3, V_2O_5 gives metavanadic acid, HVO_3.

$$V_2O_5 + 2HNO_3 \longrightarrow 2HVO_3 + N_2O_5$$

4. Action of H_2O_2 in alkaline medium. When H_2O_2 reacts with V_2O_5 in alkaline medium, peroxy vanadate $[V(O_2)_4]^{3-}$ is obtained (blue-violet) is obtained.

$$V_2O_5 + 8H_2O_2 + 6OH^- \longrightarrow 2[V(O_2)_4]^{3-} + 11H_2O$$

5. Action of carbon. When heated with carbon at high temperature, vanadium carbide (VC) is obtained.

$$2V_2O_5 + 4C \xrightarrow{\Delta} 4VC + 5O_2$$

6. Action of Na_2CO_3. When V_2O_5 is melted with Na_2CO_3, sodium orthovanadate (Na_3VO_4) is obtained,

$$V_2O_5 + 3Na_2CO_3 \longrightarrow 2Na_3VO_4 + 3CO_2\uparrow$$

7. Reduction. (*a*) Potassium reduces V_2O_5 to hypovanadium oxide or vanadium monoxide (V_2O_2 or VO).

$$V_2O_5 + 6K \longrightarrow V_2O_2 + 3K_2O$$

(*b*) H_2, carbon and S reduce V_2O_5 to vanadium trioxide, V_2O_3.

$$V_2O_5 + 2H_2 \xrightarrow{\Delta} V_2O_3 + 2H_2O$$

$$V_2O_5 + S \longrightarrow V_2O_3 + SO_2$$

$$V_2O_5 + 2C \xrightarrow{\Delta} V_2O_3 + 2CO$$

(*c*) SO_2, H_2S and $H_2C_2O_4$ reduce V_2O_5 to vanadium tetroxide, V_2O_4 or VO_2.

$$V_2O_5 + SO_2 \longrightarrow V_2O_4 + SO_3$$
$$V_2O_5 + H_2S \longrightarrow V_2O_4 + H_2O + S$$
$$V_2O_5 + H_2C_2O_4 \longrightarrow V_2O_4 + H_2O + 2CO_2$$

(*d*) (*i*) Al-power reduces V_2O_5 to V-metal.

$$3V_2O_5 + 10Al \longrightarrow 5Al_2O_3 + 6V$$

(*ii*) When V_2O_5 is heated with excess of Ca in presence of $CaCl_2$ in a steel bomb at 900–950°C (*Marden and Rich, 1927*), pure vanadium metal is obtained.

$$V_2O_5 + 5Ca \xrightarrow[\text{of } CaCl_2]{\text{Heat in presence}} 2V + 5CaO$$

(*e*) When a solution of V_2O_5 in H_2SO_4 is allowed to undergo electrolytic reduction, a green solution is obtained. This solution, on evaporation, deposits green lustrous needles of $V_2(SO_4)_3.H_2SO_4.12H_2O$. These needles, on being heated to 180°C, loses $13H_2O$ and SO_3, leaving a yellow powder of $V_2(SO_4)_3$.

$$V_2O_5 + 4H_2SO_4 + 9H_2O \xrightarrow[\text{reduction}]{\text{Electrolytic}} V_2(SO_4)_3.H_2SO_4.12H_2O + O_2$$

$$\downarrow 180°C$$

$$\underset{\text{yellow powder}}{V_2(SO_4)_3} + 3H_2O + SO_3$$

(*f*) When a solution of V_2O_5 in H_2SO_4 is reduced by means of SO_2 and the resulting solution is evaporated, bright blue crystals are obtained. Probably these crystals have the composition, $2VOSO_4.H_2SO_4.xH_2O$.

8. Formation of various types of vanadates at different pH values. V_2O_5 dissolves in strong alkalies (pH = 14) to give colourless ortho vanadate (VO_4^{3-}).

$$V_2O_5 + 6NaOH \longrightarrow 2Na_3VO_3 + 3H_2O$$

or

$$V_2O_5 + 6OH^- \longrightarrow 2VO_4^{3-} + 3H_2O$$

When this solution containing VO_4^{3-} ion is progressively acidified, *i.e.*, when the pH value of the solution is decreased from 14, VO_4^{3-} ion gets polymerised to give more complicated vanadate ions and in acidic solution (pH = 6.8) pentavanadate, $[V_5O_{14}]^{3-}$ can be isolated. With further decrease in pH value (pH < 6.8) hydrated vanadium pentoxide gets precipitated. Two definite hydrates namely $V_2O_5.H_2O$ or $H_2V_2O_6$ or HVO_3 (meta vanadic acid) and $2V_2O_5.H_2O$ or $H_2V_4O_{11}$ (tetravanadic acid) have been identified. In strongly acidic solution (pH < 2), VO_2^+ is formed. A typical scheme is shown below :

$$V_2O_5 \xrightarrow{+OH^-\,(pH = 14)} \underset{\substack{\text{Ortho vanadate} \\ \text{(colourless)}}}{[VO_4]^{3-}} \xrightarrow{pH = 12} [VO_3(OH)]^{2-} \xrightarrow{pH = 10}$$

$$\underset{\text{colourless}}{[V_2O_6(OH)]^{3-}} \xrightarrow{pH = 9} \underset{\text{colourless}}{[V_3O_9]^{3-}} \xrightarrow{pH = 6.8} \underset{\text{pentavanadate}}{[V_5O_{14}]^{3-}} \xrightarrow{pH < 6.8} \underset{\text{hydrated pentaoxide}}{mV_2O_5.nH_2O}$$

$$\xrightarrow{pH = 2.2} [V_{10}O_{28}]^{6-} \xrightarrow{pH < 2} [VO_2]^+$$

It may be seen that vanadium in all the above mentioned ions is in +5 oxidation state.

9. Formation of ferro-vanadium alloy. When a mixture of V_2O_5 and haematite ore (Fe_2O_3) is reduced by carbon or Al-powder in an electric furnace (using carbon electrodes) in the presence of a flux viz. lime or flourspar, ferro-vanadium alloy is obtained in the molten state.

Reduction by carbon $\begin{cases} V_2O_5 + 5C \longrightarrow 2V + 5CO\uparrow \\ Fe_2O_3 + 3C \longrightarrow 2Fe + 3CO\uparrow \end{cases}$

Reduction by Al-powder $\begin{cases} 3V_2O_5 + 10Al \longrightarrow 6V + 5Al_2O_3 \\ Fe_2O_3 + 2Al \longrightarrow 2Fe + Al_2O_3 \end{cases}$

Formation of ferro-vanadium alloy: $V + Fe + C \longrightarrow$ Ferro-vanadium alloy

The moleten alloy and the slag are removed by separate pipes. The alloy formed by Al-powder may contain small amounts of C, Si and other metals as impurities.

Ferro-vanadium alloy is used as a scavenger (*i.e.*, clearing agent) in steel industry, since vanadium present in the alloy removes O_2 and N_2, if dissolved in the molten steel in the form of its oxide (V_2O_5) and nitride (VN). Also at the same time, vanadium dissolves in the molten steel and thus (*a*) it increases the resistance of the steel towards shock, strain, vibratain etc. (*b*) it imparts a higher tensile strength to the steel (*c*) it improves the machinery qualities of the steel (*d*) it improves the ductility.

Uses of V_2O_5

V_2O_5 is used: (*i*) As a catalyst in a number of reactions. Some such reactions are : (*a*) Oxidation of SO_2 to SO_3 used in the manufacture of H_2SO_4 by contact process. (*b*) Oxidation of anthrancene to anthraquinone. (*c*) Oxidation of phenanthrene to phenanthraquinone (*d*) Oxidation of tolune to benzaldehyde to benzoic acid. (*e*) Oxidation of ethyl alcohol to acetaldehyde. (*ii*) In the preparation of vanadium glass. This glass is prepared by adding 0.02% vanadium as V_2O_5 to ordinary glass. This glass cuts out completely the harmful actinic rays which injure the eye and brings about fadings of organic compounds. Such glass is used particularly in textile store windows. Such glasses protect the eyes against day-light glare. (*iii*) In the preparation of ferro-vanadium alloy which is used as a scavengener for removing O_2 dissolved in molten cast iron used for the manufacture of special steels.

Structure of V_2O_5 molecule

In V_2O_5, the unit cell is VO_6 octahedron which is strongly distroted, almost to a trigonal bipyramid by an additional O-atom.

Vanadic Acids and Their Salts

These acids may be regarded as having been formed by adding one, two or three H_2O molecules to one molecule of V_2O_5 as shown below.

$$V_2O_5 + H_2O \longrightarrow 2HVO_3 \text{ (Metavanadic acid)}$$
$$V_2O_5 + 2H_2O \longrightarrow H_4V_2O_7 \text{ (Pyrovanadic acid)}$$
$$V_2O_5 + 3H_2O \longrightarrow 2H_3VO_4 \text{ (Orthovanadic acid)}$$

The salts of these acids are important. Some important salts are discussed below :

1. Ammonium metavanadate, NH_4VO_3

It is obtained : (*i*) by adding an ammonium salt like NH_4Cl to a solution of $NaVO_3$, (*ii*) by treating metavanadic acid, HVO_3 with NH_3.

$$NaVO_3 + NH_4Cl \longrightarrow NH_4VO_3 + NaCl$$
$$HVO_3 + NH_3 \longrightarrow NH_4VO_3$$

On large scale, NH_4NO_3 can be obtained from any of the ores of vanadium like carnotite $[2K(UO_2)VO_4.3H_2O]$, vanadite $[Pb_5(VO_4)_3Cl]$ and patronite (V_2S_5) (*See the extraction of vanadium*)

2. Sodium pyrovanadate, $Na_4V_2O_7$

In weaker alkaline solutions, sodium vanadate (Na_3VO_4) changes to sodium pyrovanadate ($Na_4V_2O_7$).

$$2Na_3VO_4 \xrightarrow[\text{solution}]{\text{Alkaline}} Na_4V_2O_7 + Na_2O$$

or $$2VO_4^{3-} \longrightarrow V_2O_7^{4-} + O^{2-}$$

On heating with NaOH, $Na_2V_2O_7$ is changed into Na_3VO_4.

$$Na_2V_2O_7 + 2NaOH \longrightarrow 2Na_3VO_4 + H_2O$$

or $$V_2O_7^{4-} + 2OH^- \longrightarrow 2VO_4^{3-} + H_2O$$

3. Sodium orthovanadate, $Na_3VO_4.12H_2O$

It is prepared : (*i*) by dissolving V_2O_5 in NaOH, followed by crystallisation or by treating sodium pyrovanadate, $Na_4V_2O_7$ with NaOH (*ii*) by melting V_2O_5 with Na_2CO_3.

$$V_2O_5 + 6NaOH \longrightarrow 2Na_3VO_4 + 3H_2O$$

$$Na_4V_2O_7 + 2NaOH \longrightarrow 2Na_3VO_4 + H_2O$$

$$V_2O_5 + 3Na_2CO_3 \longrightarrow 2Na_3VO_4 + 3CO_2\uparrow$$

It is soluble in H_2O. It is isomorphous with sodium orthophosphate dodecahydrate, $Na_3PO_4.12H_2O$. This compound is used: (*i*) as a catalyst in the dyeing of leather to black colour. (*ii*) in making good quality, fire-flowing blue-black ink and marking inks. (*iii*) in volumetric analysis for certain oxidation-reduction reactions. The estimations by means of this reagent include metals such as Fe, Mo, Sn and Cu, and organic compounds such as citirc acid, lactic acid, tartaric acid, aldehydes, ketones and sugars.

Complex Compounds

Examples are $K_2[VF_7]$, $Na_3[VOF_6]$, $M[VOF_4]$, $M_2[VOF_5]$, $M_2[VO_2F_3]$, $[VF_6]^-$, etc. (M = an alkali metal such as Na, K etc.).

$K_2[VF_7]$ and $Na_3[VOF_6]$ are obtained by the following reactions :

$$2KF + VF_5 \longrightarrow K_2[VF_7]$$

$$3NaF + VOF_3 \longrightarrow Na_3[VOF_6]$$

Fluorovanadate complexes like $M[VOF_4]$, $M_2[VOF_5]$ and $M_2[VO_2F_3]$ are obtained by the action of V_2O_5 on lakali metal fluorides in HF solution.

Compounds of Chromium

Chromium forms compounds in 0, +1, +2, +3 and +6 oxidation states. All compounds of Cr are coloured.

Compounds of Cr (0)

Electronic configuration of Cr-atom is 2, 8, $3s^2p^6d^5$, $4s^1$. $Cr(CO)_6$ is an example of a compound in which Cr is in zero oxidation state.

Compounds of Cr (+1)

Electronic configuration of Cr^+ ion is 2, 8, $3s^2p^6d^5$, $4s^0$. $[Cr(dipyridyl)_3]^+ ClO_4^-$ is an important compound of Cr(+1).

Compounds of Cr (+2) : Chromous compounds

Cr^{2+} ion is called chromous ion and hence its compounds are called chromous compounds. Electronic configuration of Cr^{2+} ion is 2, 8, $3s^2p^6d^4$, $4s^0$. Examples of compounds of Cr(+2) are CrF_2, $CrCl_2.4H_2O$, $CrSO_4.5H_2O$, $(CH_3COO)_2Cr.2H_2O$, CrO, $Cr(OH)_2$.Cr(+2) compounds are not very stable. These compounds have sky blue colour and are very good reducing agents as they are readily oxidised to Cr(+3) compounds in acid solution and on exposure to O_2 or air, *e.g.*,

$$4CrCl_2 + 4HCl + O_2 \longrightarrow 4CrCl_3 + 2H_2O$$

Cr(+3) compounds : Chromic compounds

Cr^{3+} ion is called chromic ion and hence its compounds are called chromic compounds. The electronic configuration of Cr^{3+} ion is 2, 8, $3s^2p^6d^3$, $4s^0$. Important compounds formed by Cr^{3+} ion are $CrCl_3$, $CrBr_3$, CrI_3, $Cr_2(SO_4)_3.18H_2O$ (violet), $Cr_2(SO_4)_3.6H_2O$ (green), $K_2SO_4.Cr_2(SO_4)_3.24H_2O$ (chrome alum), Cr_2O_3, $Cr(OH)_3$ etc. The compounds of Cr^{3+} ion are the most stable and are obtained by the reduction of Cr (+6) compounds such as chromates (CrO_4^{2-}) and dichromates ($Cr_2O_7^{2-}$) in acid solution.

$$CrO_4^{2-} + 8H^+ + 3e^- \longrightarrow Cr^{3+} + 4H_2O$$

$$Cr_2O_7^{2-} + 14H^+ + 6e^- \longrightarrow 2Cr^{3+} + 7H_2O$$

Properties. (*i*) Cr(+3) compounds dissolve in water to give violet solution because of the formation of hydrated Cr(+3) ion, $[Cr(H_2O)_6]^{3+}$.

$$Cr^{3+} + 6H_2O \longrightarrow [Cr(H_2O)_6]^{3+}$$

(*ii*) The solution of Cr(+3) ion in water is slightly acidic because of the formation of H^+ ions.

$$Cr^{3+} + H_2O \longrightarrow [Cr(OH)]^{2+} + H^+$$

(*iii*) When a base is added to the aqueous solution of a Cr(+3) salt, a dirty ppt. of $Cr(OH)_3.xH_2O$ is obtained.

$$Cr^{3+} + 3OH^- + xH_2O \longrightarrow Cr(OH)_3.xH_2O$$

The ppt. of $Cr(OH)_3.xH_2O$ dissolves in the excess of the base and forms a deep green solution which contains chromite ion, CrO_2^- (Cr = +3).

$$Cr(OH)_3.xH_2O + OH^- \longrightarrow CrO_2^- + (x + 2)\ H_2O$$

Cr^{3+} ion also gives many complex ions. Examples are $[Cr(CN)_6]^{3-}$, $[Cr(NH_3)_6]^{3+}$, $[Cr(H_2O)_6]^{3+}$, $[CrCl_6]^{3-}$ etc.

Compounds of Cr(+6)

Electronic configuration of Cr^{6+} ion is 2, 8, $3s^2p^6d^0$, $4s^0$. Examples of Cr(+6) compounds are chromic trioxide (CrO_3), chromates (CrO_4^{2-}), dichromates ($Cr_2O_7^{2-}$), chromyl chloride (CrO_2Cl_2), CrF_6 etc. It may be noted that most of Cr(+6) compounds are oxo compounds and are powerful oxidising agents.

Oxides of Chromium

Well known oxides formed by chromium are *chromous oxide*, CrO (Cr = +2), *chromic oxide*, Cr_2O_3(Cr = +3) (also called *chromic sesquioxide*) and *chromium trioxide*, CrO_3 (Cr = +6) (also called *chromic acid anhydride*). Here we shall discuss Cr_2O_3 and CrO_3 only.

1. Chromic oxide or chromium sesquioxide, Cr_2O_3

Preparation. This oxide is prepared by the reactions given below:

(*a*) $$2Cr(OH)_3 \xrightarrow{\Delta} Cr_2O_3 + 3H_2O$$

(*b*) $$(NH_4)_2Cr_2O_7 \xrightarrow{\Delta} Cr_2O_3 + 4H_2O + N_2\uparrow$$

(*c*) $$K_2Cr_2O_7 + 2NH_4Cl \xrightarrow{\Delta} Cr_2O_3 + 2KCl + 4H_2O + N_2\uparrow$$

(*d*) $$Na_2Cr_2O_7 + S \xrightarrow{\text{Heat in an iron pot}} Cr_2O_3 + Na_2SO_4$$

(*e*) $$4CrO_2Cl_2 \ (vapours) \xrightarrow{\text{Pass vapours in red hot tube}} Cr_2O_3 + O_2 + 4Cl_2$$

(*f*) $$4Hg_2CrO_4 \xrightarrow{\Delta} 8Hg + 2Cr_2O_3 + 5O_2$$

(*g*) $$4CrO_3 \xrightarrow{250° - 400°C} 2Cr_2O_3 + 3O_2$$

(*h*) $$4K_2Cr_2O_7 \xrightarrow{\text{Heat strongly}} 4K_2CrO_4 + 2Cr_2O_3 + 3O_2$$

$Cr(OH)_3$ required in reaction (*a*) is obtained by adding an alkali to a chromic salt. In reactions (*c*) and (*d*) the residue of Cr_2O_3 is washed with water to removed KCl and Na_2SO_4 respectively and then dried.

Properties. Cr_2O_3 is a green powder with very high melting point and is very refractory. It dissolves in fused borax or glass and gives a green colour and hence is used in tining glass and painting procelain. It is not decomposed by heat nor reduced by H_2. It is reduced to metal when heated with carbon (better with Al powder) at a high temperature. It is an amphoteric oxide because it forms salts with acids and chromites when fused with alkalies.

$$Cr_2O_3 + 3H_2SO_4 \longrightarrow Cr_2(SO_4)_3 + 3H_2O$$

$$Cr_2O_3 + 2NaOH \longrightarrow 2NaCrO_2 + H_2O$$

When it is fused with an alkali in presence of air or an oxidising agent, a chromate is formed.

$$2Cr_2O_3 + 8NaOH + 3O_2 \longrightarrow 4Na_2CrO_4 + 4H_2O$$

If ignited too strongly, it becomes passive towards both acid and base. When fused with an oxide of MO type, it gives a crystalline compound of the composition, $MO.Cr_2O_3$.

When Cl_2 is pased over a mixture of Cr_2O_3 and carbon and heated strongly, anhydrous $CrCl_3$ is obtained.

$$CrO_3 + 3C + 3Cl_2 \xrightarrow{\Delta} 2CrCl_3 + 3CO$$

When S_2Cl_2 vapour is passed over hot Cr_2O_3, anhydrous $CrCl_3$ is obtained.

$$2Cr_2O_3 + 6S_2Cl_2 \longrightarrow 4CrCl_3 + 3SO_2 + 9S$$

Uses. Cr_2O_3 is used as a green pigment for oil paints and in lining glass and painting porcelain.

2. Hydrated chromic oxide, $Cr_2O_3.2H_2O$: Guigret's green

It is a powder which has brilliant green colour and hence is used as a green pigment. It is prepared by fusing equimolecular amounts of $K_2Cr_2O_7$ and boric acid and treating the product with water. It also contains some amount of boric acid.

3. Chromium trioxide or chromic acid anhydride, CrO_3

This oxide is the anhydride of unknown chromic acid, H_2CrO_4 and hence is also called *chromic acid anhydride.*

Preparation. It is obtained as deep rud needle-like crystals when a well-cooled solution of $K_2Cr_2O_7$ is treated with conc. H_2SO_4.

$$K_2Cr_2O_7 + H_2SO_4 \longrightarrow 2CrO_3 + K_2SO_4 + H_2O$$

or $$C_2O_7^{2-} + H_2SO_4 \longrightarrow 2CrO_3 + SO_4^{2-} + H_2O$$

The crystals of CrO_3 are filtered by means of a Buchner funnel containing asbestos, washed with pure HNO_3 and dried in a current of warm air to remove the adhering HNO_3. This compound forms deep red needle like crystals which are deliquescent.

Properties. It is highly poisonous and damages kidneys. When heated to 198°C it melts and gives a dark liquid and then, on further heating, O_2 gas is evolved.

$$4CrO_3\ (solid) \xrightarrow{198°C} 4CrO_3\ (liquid) \xrightarrow{>198°C} 2Cr_2O_3 + 3O_2$$

It is soluble in water, giving a red solution. This red solution is strongly acidic and contains several chromic acids. This simplest of these acids is *chromic acid*, H_2CrO_4.

$$CrO_3 + H_2O \rightleftharpoons H_2CrO_4$$

It is a powerful oxidising agent and its oxidising action is based on the following equation.

$$2CrO_3 \longrightarrow Cr_2O_3 + O$$

When CrO_3 oxidises any substance, the oxide itself is reduced to Cr(+3) salt. Thus it oxidises SO_2, acidified H_2S, $FeSO_4$, $SnCl_2$ as shown by the following equations:

$$2CrO_3 + 3SO_2 \longrightarrow Cr_2(SO_4)_3$$

$$2CrO_3 + 3H_2S + 3H_2SO_4 \longrightarrow Cr_2(SO_4)_3 + 6H_2O + S$$

$$2CrO_3 + 6FeSO_4 + 6H_2SO_4 \longrightarrow Cr_2(SO_4)_3 + 3Fe_2(SO_4)_3 + 6H_2O$$

$$2CrO_3 + 3SnCl_2 + 12HCl \longrightarrow 2CrCl_3 + 3SnCl_4 + 6H_2O$$

When CrO_3 is dissolved in calculated amount of NH_3, ammonium dichromate, $(NH_4)_2Cr_2O_7$ is formed.

$$2CrO_3 + 2NH_3 + H_2O \longrightarrow (NH_4)_2Cr_2O_7$$

CrO_3 reacts with gaseous HCl to give chromyl chloride, CrO_2Cl_2.

$$CrO_3 + 2HCl\ (gas) \longrightarrow CrO_2Cl_2 + H_2O$$

It dissolves in alkalies to give chromates (CrO_4^{2-}).

$$CrO_3 + 2NaOH \longrightarrow Na_2CrO_4 + H_2O$$

Uses. The mixture of $K_2Cr_2O_7$ and conc. H_2SO_4 is called *chromic acid mixture.* This mixture is used as an oxidising agent in volumetric analysis and for cleaning glass apparatus. CrO_3 dissolved in glacial acetic acid is used as an oxidising agent and as a disinfectant to wash foul ulcers and ores.

Halides of Chromium

Well known halides given by Cr are CrF_2, $CrCl_2.4H_2O$, $CrCl_3$, $CrBr_3$, CrI_3 etc.

1. Chromous chloride, $CrCl_2.4H_2O$

Preparation. Anhydrous $CrCl_2$ is prepared: (*a*) by heating chromic chloride, $CrCl_3$ in a current of H_2 or (*b*) by heating Cr metal in a current of dry HCl gas.

$$2CrCl_3 + H_2 \longrightarrow 2CrCl_2 + 2HCl$$

$$Cr + 2HCl \longrightarrow CrCl_2 + H_2$$

A solution of $CrCl_2$ is obtained by reducing $K_2Cr_2O_7$ solution with Zn and HCl in the absence of air. When $K_2Cr_2O_7$ is heated with conc. HCl, Cl_2 is given off and $CrCl_3$ is obtained.

$$K_2Cr_2O_7 + 14HCl \longrightarrow 2KCl + 2CrCl_3 + 7H_2O + 3Cl_2\uparrow$$

Now when Zn metal is added to the above solution, nascent hydrogen is evolved which reduces $CrCl_3$ to blue chromous chloride, $CrCl_2$.

$$CrCl_3 + H\ (From\ Zn/HCl) \longrightarrow CrCl_2 + HCl$$

Properties. $CrCl_2$ forms white deliquescent silky needles which dissolve in H_2O to give a blue solution. It is a powerful reducing agent and is readily oxidised in air to $CrCl_3$. Its vapour density, even above 1500°C, indicates appreciable association [$2CrCl_2 \rightleftharpoons Cr_2Cl_4$]. When treated with HF, it gives CrF_2.

2. Chromic chloride, $CrCl_3$

Preparation. Anhydrous $CrCl_3$ is obtained:

(*i*) by passing Cl_2 over heated Cr.

$$2Cr + 3Cl_2 \longrightarrow 2CrCl_3$$

(*ii*) by passing Cl_2 over a mixture of Cr_2O_3 and carbon heated to redness.

$$Cr_2O_3 + 3C + 3Cl_2 \xrightarrow{\Delta} 2CrCl_3 + 3CO$$

(*iii*) by passing S_2Cl_2 vapour over heated Cr_2O_3

$$2Cr_2O_3 + 6S_2Cl_2 \longrightarrow 4CrCl_3 + 3SO_2 + 9S$$

(*iv*) by heating $K_2Cr_2O_7$ with conc. HCl.

$$K_2Cr_2O_7 + 14HCl \xrightarrow{\Delta} 2KCl + 2CrCl_3 + 7H_2O + 3Cl_2$$

Properties. (*i*) It is a pinkish-violet solid which is almost insoluble in cold water but readily dissolves in water in presence of traces of $CrCl_2$ or a reducing agent like CuCl or $SnCl_2$ and gives a green solution.

(*ii*) It volatalises at 1065°C and dissociates when heated above 1300°C.

$$2CrCl_3 \rightleftharpoons 2CrCl_2 + Cl_2$$

(*iii*) *Hydrate isomerism*. Hexahydrate, $CrCl_3.6H_2O$ exists in three crystalline forms. Each of them is soluble in water. They have different colours. All of them react with $AgNO_3$ solution to give the white ppt. of AgCl. Coordination number of Cr^{3+} ion in each form is six and hence each has octahedral geometry. These three forms are discussed below :

(*a*) *Violet form, $[Cr(H_2O)_6]Cl_3$.* This form does not lose water when treated with conc. H_2SO_4 and all the three Cl^- ions can be precipitated as AgCl by means of $AgNO_3$ solution. The above properties show that in this form all the six H_2O molecules are present in the coordination sphere and all the three Cl-atoms are present outside the coordination sphere. Thus this form can be represented by the structure, $[Cr(H_2O)_6]Cl_3$. All the three Cl-atoms are ionisable.

$$[Cr(H_2O)_6]Cl_3 \rightleftharpoons [Cr(H_2O)_6]^{3+} + 3Cl^-$$
$$3Cl^- + 3Ag^+ \longrightarrow 3AgCl\downarrow$$

All the six H_2O molecules act as ligands and this form has no water of crystallisation.

(*b*) *Blue-green form, $[Cr(H_2O)_5Cl]Cl_2.H_2O$.* This form loses only one H_2O molecule when treated with conc.H_2SO_4 and two Cl^- ions can be precipitated as AgCl by means of $AgNO_3$ solution. These properties show that in this form five H_2O molecules and one Cl-atom are palced in the bracket and two Cl-atoms are placed outside the bracket. These two Cl-atoms are ionisable. Thus this form can be represented as $[Cr(H_2O)_5Cl]Cl_2.H_2O$.

$$[Cr(H_2O)_5Cl]Cl_2.H_2O \rightleftharpoons [Cr(H_2O)_5Cl]^{2+} + 2Cl^- + H_2O$$
$$2Ag^+ + 2Cl^- \longrightarrow 2AgCl\downarrow$$

Five H_2O molecules act as ligands and one H_2O molecule behaves as water of crystallisation.

(*c*) *Green form, $[Cr(H_2O)_4Cl_2]Cl.2H_2O$.* This form loses two H_2O molecules on treatment with conc.H_2SO_4 and only one Cl^- ion can precipitated as AgCl by $AgNO_3$ solution. These properties show that in this form four H_2O molecules and two Cl-atoms are present in the bracket and only one Cl-atom is present outside the bracket. This one Cl-atom is ionisable. Thus this form has its structure as $[Cr(H_2O)_4Cl_2]Cl.2H_2O$.

$$[Cr(H_2O)_4Cl_2]Cl.2H_2O \rightleftharpoons [Cr(H_2O)_4Cl_2]^+ + Cl^- + 2H_2O$$
$$Cl^- + Ag^+ \longrightarrow AgCl\downarrow$$

Four H_2O molecules act as ligands and two H_2O molecules behave as water of crystallisation.

The above discussion shows that since in $CrCl_3.6H_2O$ water molecules can be placed either inside the coordination sphere as ligands or outside the sphere as water of hydration, $CrCl_3.6H_2O$ shows *hydrate isomerism*. All the three forms are called *hydrate isomers*.

In solution, these isomers exist in equilibrium as shown below :

$$[Cr(H_2O)_4Cl_2]^+ \underset{-Cl^-}{\overset{+H_2O}{\rightleftharpoons}} [Cr(H_2O)_5Cl]^{2+} \underset{-Cl^-}{\overset{+H_2O}{\rightleftharpoons}} [Cr(H_2O)_6]^{3+}$$

Chromic Acids and Their Salts

CrO_3 dissolves readily in water to give a red solution which is strongly acidic. The acidic nature of the aqueous solution is due to the fact that the solution has several chromic acids.

$$H_2O \xrightarrow{+CrO_3} \underset{\text{Chromic acid}}{H_2CrO_4} \xrightarrow{+CrO_3} \underset{\text{Dichromic acid}}{H_2Cr_2O_7} \xrightarrow{+CrO_3} \underset{\text{Trichromic acid}}{H_2Cr_3O_{10}} \xrightarrow{+CrO_3}$$

$$\underset{\text{Tetrachromic acid}}{H_2Cr_4O_{13}} \xrightarrow{+CrO_3} \ldots\ldots\ldots\ldots$$

Higher chromic acids can be supposed to have been formed by the condensation of 2, 3 or more H_2CrO_4 molecules with the elimination of water molecules. Thus:

$$2H_2CrO_4 \xrightarrow{-H_2O} H_2Cr_2O_7 + H_2O$$

$$3H_2CrO_4 \xrightarrow{-2H_2O} H_2Cr_3O_{10} + 2H_2O$$

$$3H_2CrO_4 \xrightarrow{-3H_2O} H_2Cr_4O_{13} + 3H_2O$$

Chromates (CrO_4^{2-}) and Dichromates ($Cr_2O_7^{2-}$)

Chromic acid (H_2CrO_4) forms only normal salts of $M_2^+CrO_4$ and $M^{2+}CrO_4$ type, and no acid salts of $MHCrO_4$ type. These salts are called *chromates*. Chromates are soluble in water and give brilliant yellow solution. When acidified, the yellow colour of the solution changes to orange due to the conversion of CrO_4^{2-} ion into $Cr_2O_7^{2-}$ ion.

$$\underset{\text{Yellow}}{2CrO_4^{2-}} + 2H^+ \longrightarrow \underset{\text{Orange}}{Cr_2O_7^{2-}} + H_2O$$

Examples of important chromates are K_2CrO_4, Na_2CrO_4, $PbCrO_4$, basic lead chromate $[Pb(OH)_2.PbCrO_4]$, basic zinc chromate $[Zn(OH)_2.ZnCrO_4.H_2O]$ etc.

The salts of dichromic acid, $H_2Cr_2O_7$ are called dichromates. In presnce of dil.H_2SO_4, $Cr_2O_7^{2-}$ ion acts as a strong oxidising agent as it gets reduced to Cr^{3+} ion.

$$Cr_2O_7^{2-} + 14H^+ + 6e^- \longrightarrow 2Cr^{3+} + 7H_2O$$

$Cr_2O_7^{2-}$ ion reacts with conc. H_2SO_4 to give a solution of red CrO_3.

$$Cr_2O_7^{2-} + H_2SO_4\ (conc.) \longrightarrow 2CrO_3 + H_2O + SO_4^{2-}$$

$(NH_4)_2Cr_2O_7$, $Na_2Cr_2O_7$ and $K_2Cr_2O_7$ are the examples of important dichromates.

In this section we shall discuss the preparation, properties and uses of some important chromates and $K_2Cr_2O_7$.

1. Potassium chromate, K_2CrO_4

Commercial preparation. (*i*) On commercial scale it is prepared from chromite ore, $FeO.Cr_2O_3$. The ore is first concentrated by Wilfley table method (gravity separation method).

The concentrated ore is mixed with excess of K_2CO_3 and roasted in a reverberatory furnace at a high temperature in the presence of free supply of air. By this operation the chromite ore, $FeO.Cr_2O_3$ is converted into soluble potassium chromate, K_2CrO_4.

$$\underset{\text{Chromite ore}}{4[FeO.Cr_2O_3]} + 8K_2CO_2 + 7O_2 \text{ (From air)} \xrightarrow{\Delta} \underbrace{\underset{\text{Soluble}}{8K_2CrO_4} + \underset{\text{Insoluble}}{2Fe_2O_3}}_{\text{Roasted mass}} + 8CO_2\uparrow$$

The roasted mass containing K_2CrO_4 and Fe_2O_3 is extracted with water when K_2CrO_4 goes into solution and Fe_2O_3 is left behind as a residue. In the process of roasting some quantity of CaO is also added so that the mixture of chromite ore and K_2CO_3 may become porous.

(*ii*) K_2CrO_4 can also be prepared from $K_2Cr_2O_7$. To the water solution of $K_2Cr_2O_7$ calculated amount of KOH or K_2CO_3 is added. On evaporating the resulting solution lemon-yellow crystals of anhydrous potassium chromate are obtained.

$$K_2Cr_2O_7 + 2KOH \longrightarrow 2K_2CrO_4 + H_2O$$
$$K_2Cr_2O_7 + K_2CO_3 \longrightarrow 2K_2CrO_4 + CO_2\uparrow$$

Properties. (*i*) K_2CrO_4 forms lemon yellow crystals having m.pt. = 968°C. K_2CrO_4 is isomorphous with K_2SO_4.

(*ii*) It is highly soluble in water and the resulting solution is yellow in colour.

(*iii*) *Action of H_2SO_4.* When H_2SO_4 is added to the yellow solution of K_2CrO_4, yellow colour changes to orange-red due to the formation of $K_2Cr_2O_7$.

$$2K_2CrO_4 + H_2SO_4 \longrightarrow K_2Cr_2O_7 + K_2SO_4 + H_2O$$

(*iv*) *Action of heat.* When heated above 670°C, it changes to a red modification which melts at 970°C.

(*v*) *Oxidising properties.* Acidified solution of K_2CrO_4 acts as an oxidising agent.

$$2K_2CrO_4 + 5H_2SO_4 \longrightarrow 2K_2SO_4 + Cr_2(SO_4)_3 + 5H_2O + 3O$$

Thus acidified solution of K_2CrO_4 liberates I_2 from KI solution and oxidises H_2S to S, SO_2 to H_2SO_4, HCl to Cl_2 etc.

$$2K_2CrO_4 + 5H_2SO_4 + 3H_2O \longrightarrow 2K_2SO_4 + Cr_2(SO_4)_3 + 8H_2O + 3S$$

(*vi*) *Precipitation reactions.* K_2CrO_4 precipitates chromates from different salts, *e.g.*, Ag_2CrO_4 (*brick-red*) from Ag salt solution, *yellow* $BaCrO_4$ from Ba salt and *yellow* $PbCrO_4$ from Pb salt solutions etc.

$$CrO_4^{2-} + 2Ag^+ \longrightarrow \underset{\text{Brick-red ppt}}{Ag_2CrO_4\downarrow}$$

$$\underset{(M = Ba, Pb)}{CrO_4^{2-} + M^{2+}} \longrightarrow \underset{\text{Yellow ppt}}{MCrO_4\downarrow}$$

(*vii*) *Action on neutral solution of zinc salt.* When K_2CrO_4 is added to the neutral solution of zinc salt, basic zinc chromate, $Zn(OH)_2.ZnCrO_4.H_2O$ is obtained.

$$K_2CrO_4 + 2Zn^{2+} + H_2O \longrightarrow Zn(OH)_2.ZnCrO_4.H_2O + 2K^+$$

(*viii*) *Action of H_2O_2.* When H_2O_2 is added to a solution of K_2CrO_4 containing dil.H_2SO_4 and a small amount of ether, an intense blue colour is produced which persists in etharal solution. This blue colour is due to the formation of chromium diperoxide, $Cr(O_2)_2O$ or CrO_5.

$$K_2CrO_4 + H_2SO_4 + 2H_2O_2 \xrightarrow{\text{Ether}} K_2SO_4 + CrO_5 + 3H_2O$$

or $$CrO_4^{2-} + 2H^+ + 2H_2O_2 \xrightarrow{\text{Ether}} CrO_5 + 3H_2O$$

Uses. K_2CrO_4 is used (*i*) as a reagent in the laboratory (*ii*) as an indicator in Mohr's volumetric method of silver estimation (*iii*) as an inhibitor for corrosion (*iv*) for the preparation of basic zinc chromate, $Zn(OH)_2.ZnCrO_4.H_2O$ which is used as a yellow pigment.

2. Lead chromate, $PbCrO_4$ (also called chrome yellow)

In nature of occurs as *crocosite.*

Preparation. It is obtained as a yellow precipitate :

(*a*) by mixing an aqueous solution of K_2CrO_4 with that of $Pb(CH_3COO)_2$

$$(CH_3COO)_2Pb + K_2CrO_4 \longrightarrow PbCrO_4 + 2CH_3COOK$$

(*b*) by adding a solution of $Na_2Cr_2O_7$ or $K_2Cr_2O_7$ to a solution of $Pb(CH_3COO)_2$ or $Pb(NO_3)_2$.

$$Na_2Cr_2O_7 + 2(CH_3COO)_2Pb + H_2O \longrightarrow 2PbCrO_4 + 2CH_3COONa + 2CH_3COOH$$

$$Na_2Cr_2O_7 + 2Pb(NO_3)_2 + H_2O \longrightarrow 2PbCrO_4 + 2KNO_3 + 2HNO_3$$

Properties. (*i*) It is a yellow solid and is least soluble of all the lead salts.

(*ii*) It is insoluble in dilute but dissolves in conc. HNO_3.

(*iii*) *Action of NaOH solution.* (*a*) Conc. solution of NaOH gives sod. plumbite (Na_2PbO_2) and sod. chromate (Na_2CrO_4).

$$PbCrO_4 + 4NaOH \text{ (conc. solution)} \longrightarrow Na_2PbO_2 + Na_2CrO_4 + 2H_2O$$

(*b*) When $PbCrO_4$ is boiled with cold and dil. solution of NaOH, basic lead chromate, $Pb(OH)_2$, $PbCrO_4$ (chrome red) is formed as a red ppt.

$$2PbCrO_4 + 2NaOH \text{ (cold and dil. solution)} \longrightarrow Pb(OH)_2.PbCrO_4 + Na_2CrO_4$$

(*iv*) *Action on prussian blue.* When chrome yellow is mixed intimately with prussian blue, chrome green is obtained. Chrome green is used as a green pigment.

Uses. Lead chromate is used as a yellow pigment under the name of *chrome yellow, cologne yellow* or *lemon chrome*. It is used for the preparation of chrome green which is an important green pigment.

3. Basic lead chromate (also called chrome red), $Pb(OH)_2.PbCrO_4$

When $PbCrO_4$ is boiled with cold dil. solution of NaOH, basic lead chromate is formed as a red ppt.

$$2PbCrO_4 + 2NaOH \text{ (cold and dil. solution)} \longrightarrow \underset{\text{red ppt}}{Pb(OH)_2.PbCrO_4} + Na_2CrO_4$$

$PbCrO_4$ required in the above reaction is prepared by the reaction between the solutions of $K_2Cr_2O_7$ and $Pb(NO_3)_2$ or $(CH_3COO)_2Pb$.

$$2Pb(NO_3)_2 + K_2Cr_2O_7 + H_2O \longrightarrow 2PbCrO_4 + 2KNO_3 + 2HNO_3$$

$$2Pb(CH_3COO)_2 + K_2Cr_2O_7 + H_2O \longrightarrow 2PbCrO_4 + 2CH_3COONa + 2CH_3COOH$$

Basic lead chromate is used as a red pigment under the name of *chrome red.*

4. Basic zinc chromate, $Zn(OH)_2.ZnCrO_4.H_2O$

It is prepared by adding K_2CrO_4 solution to a neutral zinc salt solution.

$$K_2CrO_4 + 2Zn^{2+} + 2H_2O \longrightarrow Zn(OH)_2.ZnCrO_4.H_2O + 2K^+ + 2H^+$$

K_2CrO_4 required in the above reaction is obtained from $K_2Cr_2O_7$ by the following reactions.

$$K_2Cr_2O_7 + 2KOH \longrightarrow 2K_2CrO_4 + H_2O$$

$$K_2Cr_2O_7 + K_2CO_3 \longrightarrow 2K_2CrO_4 + CO_2$$

Basic zinc chromate is used as a yellow pigment.

5. Potassium dichromate, $K_2Cr_2O_7$

Laboratory preparation. $K_2Cr_2O_7$ is prepared in the laboratory by adding the calculated quantity of H_2SO_4 to a saturated solution of K_2CrO_4.

$$2K_2CrO_4 + H_2SO_4 \longrightarrow K_2Cr_2O_7 + K_2SO_4 + H_2O$$

Manufacture. $K_2Cr_2O_7$ is manufactured from chromite ore, $FeO.Cr_2O_3$. The method consists of the following steps :

(*a*) *Concentration of the chromite ore.* See "*Extraction of Cr*".

(*b*) *Conversion of the concentrated chromite ore ($FeO.Cr_2O_3$) into Na_2CrO_4 : Roasting of the concentrated ore.*

$$\underset{\text{Chromite ore}}{4(FeO.Cr_2O_3)} + 8Na_2CO_3 + 7O_2\ (\textit{From air}) \xrightarrow{900-1000°C} \underbrace{\underset{\text{Soluble}}{8Na_2CrO_4} + \underset{\text{Insoluble}}{2Fe_2O_3}}_{\text{Roasted mass}} + 8CO_2\uparrow$$

For the details of this step see "*Extraction of Cr*".

(*c*) *Conversion of Na_2CrO_4 into $Na_2Cr_2O_7$.*

$$2Na_2CrO_4 + H_2SO_4 \longrightarrow \underset{\text{more soluble}}{Na_2Cr_2O_7} + \underset{\text{less soluble}}{Na_2SO_4} + H_2O$$

For the details of this step see "*Extraction of Cr*".

(*d*) *Conversion of $Na_2Cr_2O_7$ into $K_2Cr_2O_7$.* Hot solution of $Na_2Cr_2O_7$ obtained as above in step (*c*) is treated with the calculated quantity of KCl whereby less soluble $K_2Cr_2O_7$ separates out from the solution on crystallisation.

$$Na_2Cr_2O_7 + 2KCl \longrightarrow \underset{\text{More soluble}}{2NaCl} + \underset{\text{Less soluble}}{K_2Cr_2O_7}$$

Physical properties. (*i*) $K_2Cr_2O_7$ forms orange-red prismatic crystals. Its sp. gra. = 2.676 and m.pt. = 396°C.

(*ii*) *Solubility.* It is moderately soluble in cold water but freely so in hot water. It is insoluble in alcohol.

Chemical properties. (*i*) *Action of heat.* When heated to a white heat, it decomposes to give potassium chromate, K_2CrO_4 and O_2 gas is evolved.

$$4K_2Cr_2O_7 \xrightarrow{\Delta} 4K_2CrO_4 + 2Cr_2O_3 + 3O_2\uparrow$$

(*ii*) *Action of alkalies and acids.* When an alkali is added to an orange-red solution of $K_2Cr_2O_7$; a yellow solution results. The yellow colour of the solution is due to the formation of K_2CrO_4.

$$\underset{\text{Orange-red}}{K_2Cr_2O_7} + 2KOH \longrightarrow \underset{\text{Yellow}}{2K_2CrO_4} + H_2O$$

$$\underset{\text{Orange-red}}{K_2Cr_2O_7} + 2NaOH \longrightarrow \underset{\text{Yellow}}{K_2CrO_4} + H_2O + Na_2CrO_4$$

or $$Cr_2O_7^{2-} + 2OH^- \longrightarrow 2CrO_4^{2-} + H_2O$$

or $$2Cr_2O_7^{2-} + 4OH^- \longrightarrow 4CrO_4^{2-} + 2H_2O \quad ...(a)$$

On acidifying, the yellow colour of K_2CrO_4 again changes to orange-red due to the reformation of $K_2Cr_2O_7$.

$$\underset{\text{Yellow}}{2K_2CrO_4} + H_2SO_4 \longrightarrow \underset{\text{Orange-red}}{K_2Cr_2O_7} + K_2SO_4 + H_2O$$

or $$2CrO_4^{2-} + 2H^+ \longrightarrow Cr_2O_7^{2-} + H_2O$$

or $$4CrO_4^{2-} + 4H^+ \longrightarrow 2Cr_2O_7^{2-} + 2H_2O \quad ...(b)$$

Equations (*a*) and (*b*) show that $Cr_2O_7^{2-}$ and CrO_4^{2-} ions exist in equilibrium which are inter-convertible by changing the pH of the solution. Thus the reactions shown by (*a*) and (*b*) can be combined and can be written as

$$\underset{\text{Orange-red solution}}{2Cr_2O_7^{2-}} \underset{+4H^+ \ (-2H_2O)}{\overset{+4OH^- \ (-2H_2O)}{\rightleftharpoons}} \underset{\text{Yellow solution}}{4CrO_4^{2-}}$$

Thus we see that the acidic solution contains $Cr_2O_7^{2-}$ ions while alkaline solution has CrO_4^{2-} ions.

On adding an alkali, the concentration of H^+ ions decreases and hence the above reaction proceeds in the forward direction producing yellow solution containing CrO_4^{2-} ions. On the other hand when an acid is added, the concentration of H^+ ions increases and hence the above reaction proceeds in the backward direction producing the orange-red solution containing $Cr_2O_7^{2-}$ ions.

(*iii*) *Precipitation reactions.* The above discussion shows that the orange red solution of $K_2Cr_2O_7$ actually contains both $Cr_2O_7^{2-}$ and CrO_4^{2-} ions. Thus when the orange-red solution of $K_2Cr_2O_7$ is treated with a metallic salt solution [*e.g.* $Pb(NO_3)_2$ solution], theoretically $PbCrO_4$ or $PbCr_2O_7$ may be precipitated.

$$2Pb(NO_3)_2 + K_2Cr_2O_7 + H_2O \longrightarrow 2PbCrO_4 + 2KNO_3 + 2HNO_3$$

$$Pb(NO_3)_2 + K_2Cr_2O_7 \longrightarrow PbCr_2O_7 + 2KNO_3$$

Since $PbCrO_4$ has less solubility product than $PbCr_2O_7$, it is only $PbCrO_4$ (chrome yellow) which actually gets precipitated when $K_2Cr_2O_7$ solution is added to the solution of $Pb(NO_3)_2$, *i.e.*, $PbCr_2O_7$ is not precipitated.

(*iv*) *Action of conc. H_2SO_4 and HCl.* (*a*) When a well-cooled solution of $K_2Cr_2O_7$ is treated with conc. H_2SO_4, red crystals of CrO_3 are obtained.

$$K_2Cr_2O_7 + H_2SO_4 \longrightarrow 2CrO_3 + K_2SO_4 + H_2O$$

or $$Cr_2O_7^{2-} + H_2SO_4 \longrightarrow 2CrO_3 + SO_4^{2-} + H_2O$$

(*b*) On heating the mixture of $K_2Cr_2O_7$ and conc. H_2SO_4, O_2 is evolved.

$$2K_2Cr_2O_7 + 8H_2SO_4 \xrightarrow{\Delta} 2K_2SO_4 + 2Cr_2(SO_4)_3 + 8H_2O + 3O_2$$

(*c*) $K_2Cr_2O_7$ reacts with conc. HCl to evolve Cl_2.

$$K_2Cr_2O_7 + 14HCl \xrightarrow{\Delta} 2KCl + 2CrCl_3 + 7H_2O + 3Cl_2$$

(*d*) When a solution containing 3 parts of powdered $K_2Cr_2O_7$, 4 parts of warm conc.HCl and a little water is cooled, red crystals of pot. chlorochromate, $KCrO_3Cl$ are formed. This compound is called *Petigot's salt.*

$$K_2Cr_2O_7 + 2HCl \longrightarrow KCrO_3Cl + H_2O$$

(*v*) *Chromyl chloride test.* When solid $K_2Cr_2O_7$ is heated with conc. H_2SO_4 and a soluble chloride like NaCl, deep red vapours of chromyl chloride, CrO_2Cl_2 are produced.

$$K_2Cr_2O_7 + 4NaCl + 3H_2SO_4\ (concentrated) \xrightarrow{\Delta} \underset{\text{Red vapours}}{2CrO_2Cl_2} + K_2SO_4 + 2Na_2SO_4 + 3H_2O$$

The above reaction is used for the identification of Cl^- ion. It is called chromyl chloride test for Cl^- ion. This reaction has also been used for the preparation of CrO_2Cl_2.

(*vi*) *Oxidising properties.* $K_2Cr_2O_7$ is a powerful oxidising agent. In presence of dil. H_2SO_4, one molecule of this compound gives three atoms of oxygen as shown below :

$$K_2Cr_2O_7 + 4H_2SO_4\ (dil) \longrightarrow K_2SO_4 + Cr_2(SO_4)_3 + 4H_2O + 3O$$

In terms of electronic concept, $Cr_2O_7^{2-}$ ion takes electrons in the acidic medium and hence is reduced to Cr^{3+} ion. Thus $Cr_2O_7^{2-}$ acts as an oxidising agent in acidic medium.

$$Cr_2O_7^{2-} + 14H^+ + 6e^- \longrightarrow 2Cr^{3+} + 7H_2O$$

Note. Both $Na_2Cr_2O_7$ and $K_2Cr_2O_7$ are oxidising agents, but $K_2Cr_2O_7$ is preferred because it is not hygroscopic and can be used as a primary standard.

Thus in presence of dil. H_2SO_4, $K_2Cr_2O_7$ liberates I_2 from KI solution, oxidises Fe(+2) salts to Fe(+3) salts, oxidises H_2S to S, oxidises SO_2 to H_2SO_4, sulphites to sulphates, halogen acids to halogens, arsenites to arsenates.

(*a*) $K_2Cr_2O_7 + 7H_2SO_4 + 6KI \longrightarrow 4K_2SO_4 + Cr_2(SO_4)_3 + 7H_2O + 3I_2$

or $Cr_2O_7^{2-} + 14H^+ + 6I^- \longrightarrow 2Cr^{3+} + 7H_2O + 3I_2$

(*b*) $K_2Cr_2O_7 + 7H_2SO_4 + 6FeSO_4 \longrightarrow K_2SO_4 + Cr_2(SO_4)_3 + 7H_2O + 3Fe_2(SO_4)_3$

or $Cr_2O_7^{2-} + 14H^+ + 6Fe^{2+} \longrightarrow 2Cr^{3+} + 7H_2O + 6Fe^{3+}$.

(*c*) $K_2Cr_2O_7 + 4H_2SO_4 + 3H_2S \longrightarrow K_2SO_4 + Cr_2(SO_4)_3 + 7H_2O + 3S$

or $Cr_2O_7^{2-} + 8H^+ + 3H_2S \longrightarrow 2Cr^{3+} + 7H_2O + S$

(*d*) $K_2Cr_2O_7 + 4H_2SO_4 \longrightarrow K_2SO_4 + Cr_2(SO_4)_3 + 3O$

$[SO_2 + O + H_2O \longrightarrow H_2SO_4] \times 3$

$3SO_2 + K_2Cr_2O_7 + H_2SO_4 \longrightarrow K_2SO_4 + Cr_2(SO_4)_3 + H_2$

or $3SO_2 + Cr_2O_7^{2-} + 2H^+ \longrightarrow 2Cr^{3+} + 3SO_4^{2-} + H_2O$

The above reaction has been used to prepare chrome alum, K_2SO_4. $Cr_2(SO_4)_3.24H_2O$.

(*e*) $3Na_2SO_3 + K_2Cr_2O_7 + 4H_2SO_4 \longrightarrow 3Na_2SO_4 + K_2SO_4 + Cr_2(SO_4)_3 + 4H_2O$

or $3SO_3^{2-} + Cr_2O_7^{2-} + 8H^+ \longrightarrow 3SO_4^{2-} + 2Cr^{3+} + 4H_2O$

(*f*) $K_2Cr_2O_7 + 14HCl \longrightarrow 2KCl + 2CrCl_3 + 7H_2O + 3Cl_2$

or $Cr_2O_7^{2-} + 14H^+ + 6Cl^- \longrightarrow 2Cr^{3+} + 7H_2O + 3Cl_2$

Acidified solution of $K_2Cr_2O_7$ also oxidises primary alcohols to aldehydes and then to acids.

(*g*) $K_2Cr_2O_7 + 4H_2SO_4 \longrightarrow K_2SO_4 + Cr_2(SO_4)_3 + 4H_2O + 3O$

$C_2H_5OH + O \longrightarrow CH_3CHO + H_2O$

$\xrightarrow{+O} CH_3OOH$

Acidified solution of $K_2Cr_2O_7$ oxidises chloroform ($CHCl_3$) to carbonyl chloride ($COCl_2$) and aniline to *p*-benzoquinone.

(*h*) $2CHCl_3 + 3O \longrightarrow 2COCl_2 + Cl_2 + H_2O$

(*i*) Aniline ($C_6H_5NH_2$) $\longrightarrow$ *p*-benzoquinone (O=C_6H_4=O)

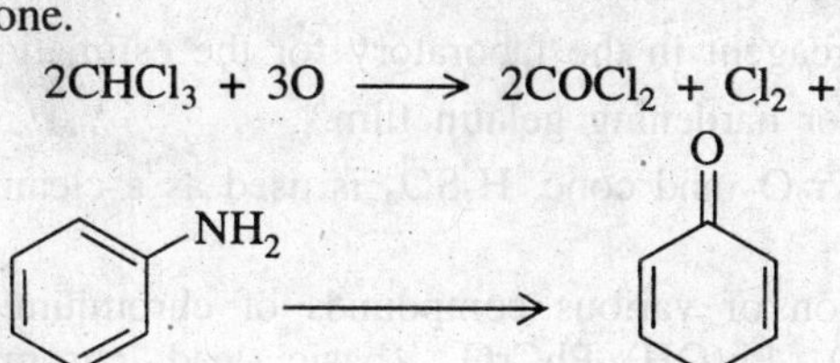

Aniline

p-benzoquinone

(*viii*) *Action of H_2O_2* (*a*) When a solution of $K_2Cr_2O_7$ containing dil.H_2SO_4 and some ether, $(C_2H_5)_2O$ is treated with H_2O_2 and the solution mixture is shaken, the layer of ether becomes blue due to the formation of chromic diperoxide, CrO_5 or $CrO(O_2)_2$.

$$K_2Cr_2O_7 + H_2SO_4 + 4H_2O_2 \xrightarrow{\text{Ether}} K_2SO_4 + 2CrO_5 + 5H_2O$$

or $Cr_2O_7^{2-} + 2H^+ + 4H_2O_2 \xrightarrow{\text{Ether}} 2CrO_5 + 5H_2O$

(*b*) When neutral or slightly acidic solution of $K_2Cr_2O_7$ is treated with H_2O_2, *violet diperoxy chromate*, $K^+[CrO(O_2)_2(OH)]^-$ is obtained.

$$K_2Cr_2O_7 + 6H_2O_2 \longrightarrow 2K[CrO(O_2)_2(OH)] + 5H_2O + O_2$$

or
$$Cr_2O_7^{2-} + 6H_2O_2 \longrightarrow 2[CrO(O_2)_2(OH)]^- + 5H_2O + O_2$$

(*viii*) *Action of boric acid.* When a mixture having equimolecular amounts of $K_2Cr_2O_7$ and boric acid is fused and the product so formed is treated with water, hydrated chromic oxide, $Cr_2O_3.2H_2O$ (called *Guigret's green*) is obtained. This oxide has brilliant colour and is used as a green pigment.

(*ix*) *Action of K_2CO_3.* When $K_2Cr_2O_7$ solution is treated with the calculated amount o K_2CO_3 and the resulting solution is evaporated, lemon-yellow crystals of anhydrous K_2CrO_4 are obtained.

$$K_2Cr_2O_7 + K_2CO_3 \longrightarrow 2K_2CrO_4 + CO_2$$

This reaction has been used for the preparation of K_2CrO_4.

(*x*) *Action of NH_4Cl.* When $K_2Cr_2O_7$ is heated with NH_4Cl, chromic oxide (Cr_2O_3) is obtained.

$$K_2Cr_2O_7 + NH_4Cl \xrightarrow{\Delta} Cr_2O_3 + 2KCl + 4H_2O + N_2\uparrow$$

(*xi*) *Action of carbon.* When the solution containing $K_2Cr_2O_7$ and carbon is heated, Cr_2O_3 is obtained.

$$K_2Cr_2O_7 + 3C + H_2O \xrightarrow{\Delta} Cr_2O_3 + 2KOH + 3CO\uparrow$$

(*xii*) *Action of $K_4[Fe(CN)_6]$.* When $K_2Cr_2O_7$ is fused with dry $K_4[Fe(CN)_6]$, pot. cyanate, KCNO is obtained.

$$2K_2Cr_2O_7 + 2K_4[Fe^{2+}(CN)_6] \longrightarrow 2K_2O + 2Cr_2O_3 + 6KCNO + 2K_2Fe^+[Fe^{3+}(CN)_6]$$

(*xiii*) *Action of H_2SO_4.* (*a*) With cold H_2SO_4, red crystals of CrO_3 are obtained.

$$K_2Cr_2O_7 + H_2SO_4\text{ (cold)} \longrightarrow 2CrO_3 + K_2SO_4 + H_2O$$

(*b*) When $K_2Cr_2O_7$ is heated with H_2SO_4, O_2 gas is evolved.

$$K_2Cr_2O_7 + 8H_2SO_4 \xrightarrow{\Delta} 2K_2SO_4 + 2Cr_2(SO_4)_3 + 8H_2O + 3O_2\uparrow$$

Uses. $K_2Cr_2O_7$ is used :

(*i*) in dyeing, calico printing and chrome tanning.

(*ii*) as an oxidising agent in laboratory.

(*iii*) as a volumetric reagent in the laboratory for the estimation of Fe^{2+}, I^- etc. ions.

(*iv*) in photography for hardening gelatin film.

(*v*) a mixture of $K_2Cr_2O_7$ and conc. H_2SO_4 is used as a cleaning agent in the laboratory for glass wages.

(*vi*) for the preparation of various compounds of chromium such as K_2CrO_4, $PbCrO_4$ (chrome yellow), $Pb(OH)_2.PbCrO_4$ (basic lead chromate, called chrome red), $Zn(OH)_2.ZnCrO_4$ (basic zinc chromate), chrome alum $[K_2SO_4.Cr_2(SO_4)_3.24H_2O]$, chromyl chloride ($CrO_2Cl_2$), $Cr_2O_3.2H_2O$ (hydrated chromic oxide) etc. These compounds have industrial applications.

Uses of $K_2Cr_2O_7$ in analytical chemistry

The uses of $K_2Cr_2O_7$ in analytical chemistry can be studied under the following heading.

A. Uses in qualitative analysis. $K_2Cr_2O_7$ is used in the detection of SO_3^{2-} and Cl^- radicals.

(*a*) *Detection of SO_3^{2-} radical.* We have seen that acidified solution of $K_2Cr_2O_7$ oxidises SO_2 to K_2SO_4. This property of $K_2Cr_2O_7$ is used to detect SO_3^{2-} ion in a given inorganic salt.

The procedure is that the sulphite is heated with dil.H_2SO_4 in a test tube when SO_2 gas which has suffocating smell comes out of the test tube.

$$Na_2SO_3 + H_2SO_4 \longrightarrow Na_2SO_4 + H_2O + SO_2$$

Now when a piece of filter paper moistened with acidified $K_2Cr_2O_7$ is brought on the mouth of test tube, paper becomes green due to formation of green $Cr_2(SO_4)_3$ which is produced by the reduction of $K_2Cr_2O_7$ to $Cr_2(SO_4)_3$ (green) by SO_2 and SO_2 is oxidised by acidified $K_2Cr_2O_7$ to K_2SO_4.

$$K_2Cr_2O_7 + H_2SO_4 + 3SO_2 \longrightarrow \underset{\text{green}}{Cr_2(SO_4)_3} + K_2SO_4 + H_2O$$

or $$Cr_2O_7^{2-} + 2H^+ + 3SO_2 \longrightarrow 2Cr^{3+} + 3SO_4^{2-} + H_2O$$

(*b*) *Detection of Cl^- radical by chromyl chloride test.* We have seen that when solid $K_2Cr_2O_7$ mixed with a soluble chloride (*e.g.*, NaCl) is heated with conc. H_2SO_4, deep red vapours of chromyl chloride (CrO_2Cl_2) are produced.

$$K_2Cr_2O_7 + 4NaCl + 3H_2SO_4 \xrightarrow{\Delta} \underset{\text{red vapours}}{2CrO_2Cl_2} + K_2SO_4 + Na_2SO_4 + 3H_2O$$

On passing these vapours in NaOH solution, yellow solution, due to the formation of sodium chromate (Na_2CrO_4), is obtained. Now this yellow solution is acidified with CH_3COOH and is added lead acetate, $(CH_3COO)_2Pb$ to it. This treatment gives yellow ppt. of lead chromate ($PbCrO_4$).

$$\underset{\text{red vapours}}{CrO_2Cl_2} + 4NaOH \longrightarrow \underset{\text{yellow solution}}{Na_2CrO_4} + 2NaCl + H_2O$$

$$Na_2CrO_4 + (CH_3COO)_2Pb \longrightarrow \underset{\text{yellow ppt}}{PbCrO_4} + CH_3COOH$$

B. Uses in quantitative analysis. We have seen that the acidified solution of $K_2Cr_2O_7$ oxidises ferrous salts to ferric salts.

$$2FeSO_4 + K_2Cr_2O_7 + 7H_2SO_4 \longrightarrow 3F_2(SO_4)_3 + Cr_2(SO_4)_3 + 7H_2O + K_2SO_4$$

or $$6Fe^{2+} + Cr_2O_7^{2-} + 14H^+ \longrightarrow 6Fe^{3+} + 2Cr^{3+} + 7H_2O$$

This property of $K_2Cr_2O_7$ has been used for the determination of the following :

(*a*) Determination of the amount of ferrous ammonium sulphate (FAS) (also called Mohr's salt) present per litre of its given solution. The procedure is that a known volume (say 20 ml) of the given FAS solution containing some dil. H_2SO_4 is taken in a conical flask and this solution is titrated with a standard solution of $K_2Cr_2O_7$ taken in a burette (intermediate solution). Potassium ferricyanide, $K_3[Fe(CN)_6]$ placed on a tile is used as an external indicator.

Before the end point, the reaction mixture gives deep blue colour with $K_3[Fe(CN)_6]$ solution drop placed on the tile, since Fe^{2+} ions (present in the reaction mixture) which have not been oxidised to Fe^{3+} ions by the acidified $Cr_2O_7^{2-}$ ions react with $[Fe(CN)_6]^{3-}$ ion and produce the blue coloured ferro-ferricyanide, $Fe_3^{2+}[Fe^{3+}(CN)_6]_2$ (I) which is called Turnbull's blue

$$3Fe^{2+} + 2[Fe(CN)_6]^{3-} \rightleftharpoons \underset{\text{(I)}}{Fe_3^{2+}[Fe^{3+}(CN)_6]}$$

(*b*) Determination of percentage of ferric ammonium sulphate, $Fe_2(SO_4)_3.(NH_4)_2SO_4.24H_2O$ (ferric alum).

(*c*) Determination of the percentage purity of a given sample of ferrous ammonium sulphate.

(*d*) Determination of the percentage of ferrous iron present in a given sample of an iron ore like haematite (Fe_3O_4) or spanthic iron ($FeCO_3$).

In all the above determinations $K_3[Fe(CN)_6]$ (external indicator) or diphenyl amine (internal indicator) is used as an indicator.

6. Ammonium dichromate, $(NH_4)_2Cr_2O_7$

When CrO_3 is dissolved in calculated amount of ammonia, $(NH_4)_2Cr_2O_7$ is obtained as orange-red crystals.

$$2CrO_3 + 2NH_3 + H_2O \longrightarrow (NH_4)_2Cr_2O_7$$

When red crystals of $(NH_4)_2Cr_2O_7$ are heated, a violent action takes place accompanied by flaskes of light and N_2 is liberated leaving behind a dark green residue of Cr_2O_3.

$$(NH_4)_2Cr_2O_7 \xrightarrow{\Delta} N_2 + 4H_2O + Cr_2O_3$$

Structures of CrO_4^{2-} and $Cr_2O_7^{2-}$ ions

X-ray study of $(NH_4)_2Cr_2O_7$ crystal has shown that $Cr_2O_7^{2-}$ ion is made up of two CrO_4^{2-} tetrahedral ions which are linked together through O-atom. Thus this O-atom acts as a bridge between two CrO_4^{2-} tetrahedral ions and hence is called *bridging O-atom.* Since CrO_4^{2-} ion is tetrahedral and results from sp^3 hybridisation of Cr^{6+} ion, each Cr-atom in $Cr_2O_7^{2-}$ is also sp^3 hybridised and hence is linked with four O-atoms which are placed at the four corners of the tetrahedron. Excepting the bridging O-atom, all the remaining seven O-atoms are called terminal O-atoms. Cr—O (terminal) and Cr—O (bridging) distances have been found to be equal to 1.6Å and 1.19Å respectively. CrOCr bond angle = 115°. Structures of CrO_4^{2-} and $Cr_2O_7^{2-}$ ions have been shown in Fig. 9.1.

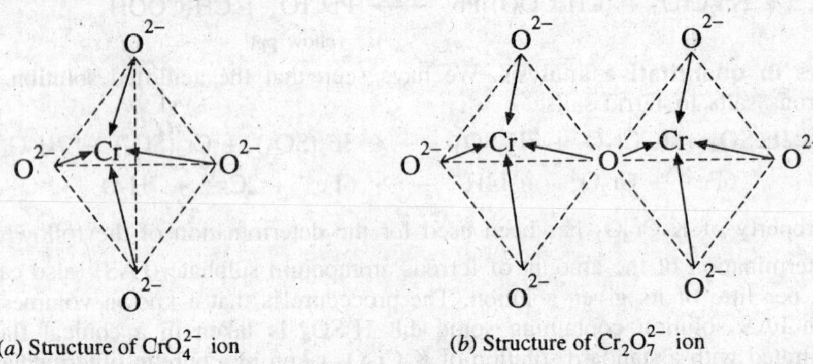

(*a*) Structure of CrO_4^{2-} ion (*b*) Structure of $Cr_2O_7^{2-}$ ion

Fig. 9.1. Structures of CrO_4^{2-} and $Cr_2O_7^{2-}$ ions. In both these ions Cr is in +6 oxidation state.

Potassium chromic sulphate or chrome alum, $K_2SO_4.Cr_2(SO_4)_3.24H_2O$ or $KCr(SO_4)_2.12H_2O$

It is a double sulphate of K^+ and Cr^{3+} cations.

Preparation

(*a*) In laboratory, it is prepared by reducing a solution of $K_2Cr_2O_7$ acidified with H_2SO_4 by SO_2. In this reaction SO_2 is oxidised to K_2SO_4.

$$K_2Cr_2O_7 + 3SO_2 + H_2SO_4 \longrightarrow K_2SO_4 + \underset{\text{green}}{Cr_2(SO_4)_3} + H_2O$$

or $$Cr_2O_7^{2-} + 2H^+ + 3SO_2 \longrightarrow 2Cr^{3+} + 3SO_4^{2-} + H_2O$$

On standing, beautiful dark purple octahedral crystals of chrome alum are separated.

(*b*) It is obtained as a by-product when anthracene is oxidised to anthraquinone by means of $K_2Cr_2O_7$ solution acidified with H_2SO_4.

Properties

It forms dark purple octahedral crystals which are isomorphous with other alums. It is fairly soluble in water and forms a bluish-red solution. When this solution is heated to 90°C, it turns green due to the conversion of violet $Cr_2(SO_4)_3.18H_2O$ into the green form, $Cr_2(SO_4)_3.6H_2O$.

$$Cr_2(SO_4)_3.18H_2O \text{ (violet form)} \xrightarrow[-12H_2O]{90°C} Cr_2(SO_4)_3.6H_2O \text{ (green form)} + 12H_2O$$

Uses

It is used: (*i*) as a mordant in dyening and calico printing, (*ii*) for tanning leather, (*iii*) in photography (during fixing) for hardening of the negatives.

Chromium oxy chloride or chromyl chloride, CrO_2Cl_2

Preparation

CrO_2Cl_2 is prepared by the following methods:

(*i*) It is obtained as a dark red liquid when a dry mixture of $K_2Cr_2O_7$ and NaCl is distilled with conc. H_2SO_4 and the resulted vapours of CrO_2Cl_2 are condensed.

$$K_2Cr_2O_7 + 4NaCl + 3H_2SO_4 \text{ (conc.)} \xrightarrow{\Delta} 2CrO_2Cl_2 + K_2SO_4 + 2Na_2SO_4 + 3H_2O$$

(*ii*) It is also prepared by adding small quantity of conc. H_2SO_4 to a cold solution of CrO_3 in conc. HCl.

$$CrO_3 + 2HCl \text{ (conc.)} \longrightarrow CrO_2Cl_2 + H_2O$$

The solution mixture is separated into two layes. CrO_2Cl_2 is present in the lower layer from which CrO_2Cl_2 can be separated by a separating funnel.

(*iii*) CrO_2Cl_2 can also be prepared by treating chromic acid, H_2CrO_4 or $CrO_2(OH)_2$ with HCl gas.

$$O_2Cr(OH)_2 + 2HCl \xrightarrow{-2H_2O} O_2CrCl_2 + 2H_2O$$

or $$CrO_2(OH)_2 + 2HCl \longrightarrow CrO_2Cl_2 + 2H_2O$$

Physical properties

(*i*) It is a deep red liquid having b.pt. = 116.7°C and m.pt. = –96.5°C.

(*ii*) It is miscible with organic solvents like CCl_4 and CS_2.

Chemical properties

(*i*) *Hydrolysis*. It is hydrolysed by water, forming chromic acid (H_2CrO_4) and HCl.

$$CrO_2Cl_2 + 2H_2O \longrightarrow H_2CrO_4 + 2HCl$$

(*ii*) *Stability*. It is photosensitive but otherwise it is stable. When heated, Cr_2O_3 is obtained.

$$4CrO_2Cl_2 \xrightarrow{\Delta} 2Cr_2O_3 + 4Cl_2 + O_2$$

(*iii*) *Action of $(CH_3COO)_2$ Pb*. The solution of CrO_2Cl_2 gives a yellow precipitate of $PbCrO_4$ with $(CH_3COO)_2Pb$.

$$CrO_2Cl_2 + 2H_2O + (CH_3COO)_2Pb \longrightarrow \underset{\text{Yellow ppt}}{PbCrO_4} + 2HCl + 2CH_3COOH$$

(*iv*) *Oxidising properties.* It is a mild oxidising agent. A solution of CrO_2Cl_2 in CS_2 at 25–40°C oxidises toluene ($C_6H_5CH_3$) to benzaldehyde (C_6H_5CHO) (*Etard's reaction*).

$$CrO_2Cl_2 \longrightarrow CrCl_2 + 2O$$

$$C_6H_5CH_3 + 2O \longrightarrow C_6H_5CHO + H_2O$$

(*v*) *Action of NaCl or KCl solution.* When the vapours of CrO_2Cl_2 are passed into the saturated solution of NaCl or KCl, red crystals of the *metal chloro chromate* ($MCrO_3Cl$) are obtained. $MCrO_3Cl$ is called *petigot's salt.*

$$\underset{(M\,=\,Na,\ K)}{CrO_2Cl_2 + MCl + H_2O} \longrightarrow \underset{\text{Petigot's salt}}{MCrO_3Cl} + 2HCl$$

(*vi*) *Action of NaOH.* Yellow solution of Na_2CrO_4 is obtained.

$$CrO_2Cl_2 + 4NaOH \longrightarrow \underset{\text{Yellow solution}}{Na_2CrO_4} + 2NaCl + 2H_2O$$

(*vii*) *Action of F_2 or HF.* In both cases chromyl fluoride (CrO_2F_2) is formed.

$$CrO_2Cl_2 + F_2 \longrightarrow CrO_2F_2 + Cl_2$$

$$CrO_2Cl_2 + 2HF \longrightarrow CrO_2F_2 + 2HCl$$

Uses

It is used: (*i*) as an intermediate product in the identification of Cl^- ions. (*ii*) as an excellent oxidising agent in organic chemistry, *e.g.*, its solution of CS_2 is used to oxidise toluene to benzaldehyde.

Structure

CrO_2Cl_2 molecule can be regarded as having been obtained by removing two H_2O molecules from the combination of one molecule of chromic acid, H_2CrO_4 or $CrO_2(OH)_2$ and two molecules of HCl. Thus :

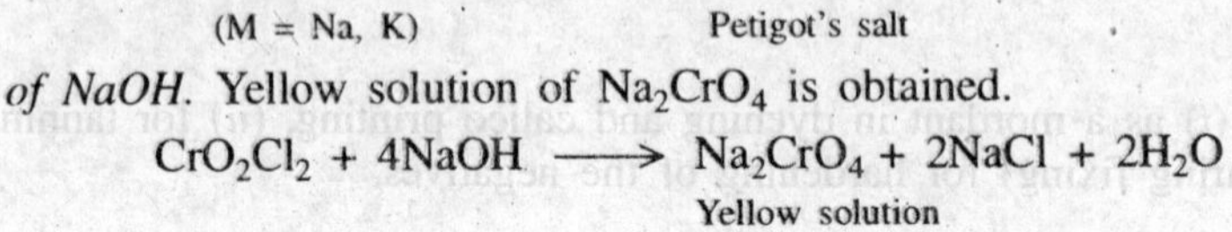

$$O_2Cr(OH)_2 + 2HCl \longrightarrow O_2CrCl_2 + 2H_2O$$

It may be concluded from the above discussion that CrO_2Cl_2 is a dichloro derivative of H_2CrO_4 or $CrO_2(OH)_2$

Peroxy Compounds of Chromium: Peroxy Chromates

Transition metals like Ti, V, Cr etc. form well definied per oxy compounds. Peroxy compounds of Ti and V form the basis of colorimetric estimation of the elements. In these compounds each peroxy group, O_2^{2-} assumes two coordination positions round the transition metal atom. A three membered strained ring is formed at the metal atom.

Chromium in its higher oxidation state (*e.g.*, +6, +5, +4) form well defined peroxy compounds which are called *peroxy chromates.* These compounds are more or less unstable both in and out of solution, decompose slowly with evolution of O_2. Some are explosive and flammable in air. Examples of peroxy chromates which we shall discuss here are given below :

Peroxy Compounds of Cr (+6)

Chromium in +6 oxidation state forms the following peroxy compounds:

1. Chromic peroxide or chromic diperoxide or blue peroxy chromate, CrO_5

Preparation. When a solution of K_2CrO_4 or $K_2Cr_2O_7$ containing dil. H_2SO_4 and some ether, $(C_2H_5)_2O$ is treated with H_2O_2 and the solution mixture is shaken, the layer of ether becomes blue due to the formation of chromic diperoxide, CrO_5.

$$K_2CrO_4 + H_2SO_4 + 2H_2O_2 \longrightarrow K_2SO_4 + CrO_5 + 3H_2O$$

or
$$CrO_4^{2-} + 2H^+ + 2H_2O_2 \longrightarrow \underset{\text{(blue)}}{CrO_5} + 3H_2O$$

$$K_2Cr_2O_7 + H_2SO_4 + 4H_2O_2 \longrightarrow K_2SO_4 + 2CrO_5 + 5H_2O$$

or
$$Cr_2O_7^{2-} + 2H^+ + 4H_2O_2 \longrightarrow 2CrO_5 + 5H_2O$$

Properties. (*i*) *Decomposition of CrO_5.* Blue colour of ether solution of CrO_5 is unstable and hence fades rapidly. The fading of the colour is due to the fact that CrO_5, in presence of H_2SO_4, decomposes to give $Cr_2(SO_4)_3$ with evolution of O_2.

$$4CrO_5 + 6H_2SO_4 \longrightarrow 2Cr_2(SO_4)_3 + 6H_2O + 7O_2$$

CrO_5 has two peroxide groups (O_2^{2-}) and one oxide group (O^{2-}) and hence is formulated as $CrO(O_2)_2$.

(*ii*) *Action of H_2O_2.* When an alkaline solution of CrO_5 is treated with H_2O_2, *blue pot. peroxy chromate*, $K_2Cr_2O_{12}.2H_2O$ and *red pot. peroxychromate*, $K_2Cr_2O_{16}$ are obtained.

$$2CrO_5 + H_2O_2 + 2KOH \longrightarrow K_2Cr_2O_{12}.2H_2O\ (blue)$$

$$2CrO_5 + 3H_2O_2\ (excess) + 6KOH \longrightarrow K_2Cr_2O_{16}\ (red) + 6H_2O$$

(*iii*) *Action of organic bases.* When the blue coloured solution containing $CrO(O_2)_2$ is shaken with ether, $(C_2H_5)_2O$, an addition compound (adduct), $(C_2H_5)_2O \rightarrow CrO(O_2)_2$ is obtained.

$$(C_2H_5)_2O + CrO(O_2)_2 \longrightarrow \underset{\text{Ether adduct}}{[(C_2H_5)_2O \rightarrow CrO(O_2)_2]}$$

When an organic base like pyridine (C_5H_5N), 2, 2′-dipyridine, piperidine, aniline, trimethyl aniline, quinoline etc., is added to the ether solution of $CrO(O_2)_2$, crystalline violet solid adducts are obtained. In the formation of these adducts, $(C_2H_5)_2O$ molecule of $(C_2H_5)_2O \rightarrow CrO(O_2)_2$ adduct is replaced by the organic base molecule and hence these adducts are formulated as $L \rightarrow CrO(O_2)_2$ (L = organic base).

$$\underset{\text{Ether adduct}}{(C_2H_5)_2O \rightarrow CrO(O_2)_2} + L \longrightarrow L \rightarrow CrO(O_2)_2 + (C_2H_5)_2O$$

These adducts are non-electrolytes and insoluble in water but soluble in organic solvents like C_6H_6.

The pyridine adduct, $C_5H_5N \rightarrow CrO(O_2)_2$ is monomeric, soluble in benzene and is diamagnetic in nature.

Structure of $C_5H_5N \rightarrow CrO(O_2)_2$ adduct. This adduct has pentagonal pyramidal structure in which oxide ion (O^{2-}) is placed at the apex while the two peroxy groups (O_2^{2-}) and N-atom of C_5H_5N molecule are placed at the corners of the base of the pentagon. Cr-atom is located at the centre of the pentagonal base. (see Fig. 9.2). Thus we see that Cr-atom is coordinated with two peroxide groups (O_2^{2-}), one NC_5H_5 molecule and one oxide (O^{2-}) oxygen atom. The O—O distance in the peroxide groups is equal to 1.41 Å. Here Cr is in +6 oxidation state.

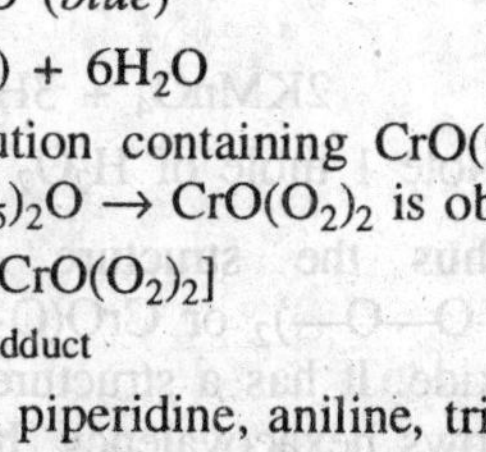

Fig. 9.2. Pentagonal pyramidal structure of $C_5H_5N \rightarrow CrO(O_2)_2$ adduct.

(*iv*) *Action of aqueous ammonia.* When an excess of aqueous ammonia is added to the ether solution of CrO_5, *red-brown diperoxy chromate*, $[(NH_3)_3Cr^{+4}(O_2)_2]^0$ is obtained.

$$Cr^{+6}(O_2)_2O + 3NH_4OH \longrightarrow [(NH_3)_3Cr^{4+}(O_2)_2]^0 + 3H_2O + \frac{1}{2}O_2$$

Structure of CrO_5 molecule. Formerly the blue compound (CrO_5) was assumed to be perchromic acid ($HCrO_5$) and hence its structure was represented as $HO{-}O{-}\overset{\overset{\displaystyle O}{\|}}{\underset{\underset{\displaystyle O}{\|}}{Cr}}{=}O$ which shows that the oxidation state of Cr-atom is 7 which is not possible. Schwarz and Giese in 1932 showed that the blue-coloured compound is a peroxide of chromium (VI) which contains two peroxide groups, O_2^{2-} and one oxide ion, O^{2-} and hence is formulated as $CrO(O_2)_2$. The presence of two peroxide groups is evident from the following experimental observations:

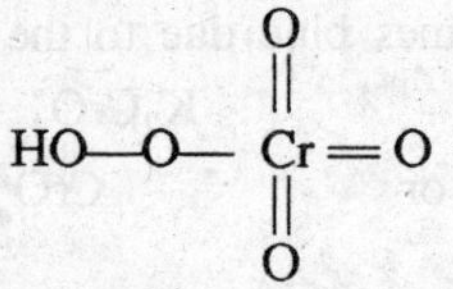

(*a*) Dilute acids liberate 7 atoms of active oxygen per two moles of the peroxy compound rather than 6 atoms of active oxygen as expected from H_2CrO_5 formulation.

$$2CrO_5 \longrightarrow Cr_2O_3 + 7O$$
$$2HCrO_5 \longrightarrow Cr_2O_3 + H_2O + 6O$$

(*b*) Titration of the peroxy compound with $KMnO_4$ in acidic medium has shown that the compound consumes four equivalents of $KMnO_4$. This observation shows that the peroxy compound has two peroxy groups $(-O-O-)^{2-}$ or O_2^{2-}.

$$CrO_5 + 2H_2O \longrightarrow CrO_3 + 2H_2O_2$$
$$2KMnO_4 + 3H_2SO_4 + 5H_2O_2 \longrightarrow K_2SO_4 + 2MnSO_4 + 8H_2O + 5O_2$$

(Note 1 mole of $H_2O_2 \equiv 2$ equivalents of $KMnO_4$).

Thus the structure of CrO_5 can be written as $CrO(-O-O-)_2$ or $CrO(O_2)_2$ and should be called chromic diperoxide. It has a structure in which Cr-atom is hexavalent and shows hexacovalency. It may be seen from Fig. 9.3 that each peroxy group assumes two coordination positions around Cr-atom and thus a three-membered strained ring is formed at Cr-atom.

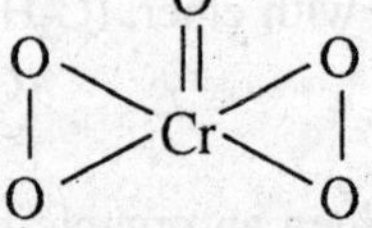

Fig. 9.3. Structure of chromic diperoxide molecule, CrO_5 or $CrO(O_2)_2$.

2. Violet diperoxy chromates, $M^+[CrO(O_2)_2(OH)]^-$ ($M^+ = K^+, NH_4^+$ or Tl^+)

These compounds are obtained by treating neutral or slightly acidic solutions of K^+, NH_4^+ or Tl^+ dichromates with H_2O_2, *e.g.*,

$$K_2Cr_2O_7 + 6H_2O_2 \longrightarrow 2K[CrO(O_2)_2(OH)] + 5H_2O + O_2$$

or

$$Cr_2O_7^{2-} + 6H_2O_2 \longrightarrow 2[CrO(O_2)_2(OH)]^- + 5H_2O + O_2$$

These are violet compounds, are diamagnetic and violently explosive. The presence of two peroxy groups in these compounds has been established analytically and the magnetic properties show that Cr is in +6 oxidation state. Analysis shows the presence of one H-atom in the molecule. Since the molecular conductivity of the ammonium salt in aqueous solution is about 100 ohm^{-1} cm^2, it is a uni-univalent electrolyte. Like $CrO(O_2)_2.NC_5H_5$ adduct, $[CrO(O_2)_2(OH)]^-$ ion also has pentagonal pyramidal structure with oxide oxygen at the apex. Two peroxy groups and OH group are placed at the corners of the pentagonal pyramidal base while Cr-atom is located at the centre (See Fig. 9.4).

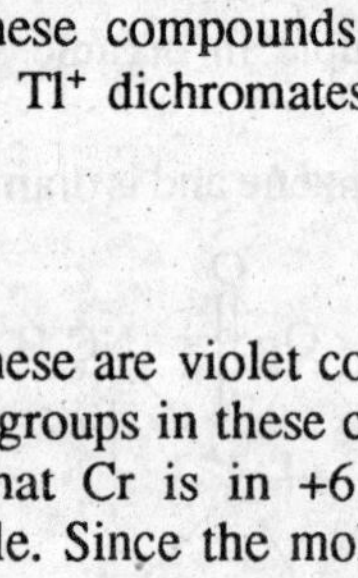

Fig. 9.4. Structure of $[CrO(O_2)_2(OH)]^-$ ion.

This structure shows that Cr-atom is coordinated with two peroxide groups (O_2^{2-}), one OH group and one oxide oxygen atom (Cr$=$O bond).

3. Blue potassium peroxy chromate, $K_2Cr_2O_{12}.2H_2O$

Preparation. It is obtained by the action of H_2O_2 on alkaline solution of CrO_5.

$$2CrO_5 + H_2O_2 + 2KOH \longrightarrow K_2Cr_2O_{12}.2H_2O \ (Blue)$$

Properties. This salt decomposes at room temperature to form the anhydrous $K_2Cr_2O_{12}$. It is diamagnetic in nature. When this salt is treated with an alkali or dil. acid, it decomposes to give CrO_4^{2-} or Cr^{3+}.

$$2K_2Cr_2O_{12} + 4KOH \longrightarrow 4K_2CrO_4 + 2H_2O + 5O_2$$

or $$2Cr_2O_{12}^{2-} + 4OH^- \longrightarrow 4CrO_4^{2-} + 2H_2O + 5O_2$$

$$K_2Cr_2O_{12} + 4H_2SO_4 \longrightarrow Cr_2(SO_4)_3 + K_2SO_4 + 4H_2O + 4O_2$$

or $$Cr_2O_{12}^{2-} + 8H^+ \longrightarrow 2Cr^{3+} + 4H_2O + 4O_2$$

Structure of $K_2Cr_2O_{12}$ molecule. Schwarz and Giese have shown that in $K_2Cr_2O_{12}$ molecule each Cr-atom is in +6 oxidation state and shows *hexacovalency*. It is linked with $2\frac{1}{2}$ peroxo groups and one oxide ion (O^{2-}). Potassium is attached with O^{2-} ion. Thus the structure of $K_2Cr_2O_{12}$ molecule is as shown in Fig. 9.5. This structure clearly shows that the given molecule can be represented as $K^+O^-(O_2)_2Cr—O—O—Cr(O_2)_2O^-K^+$.

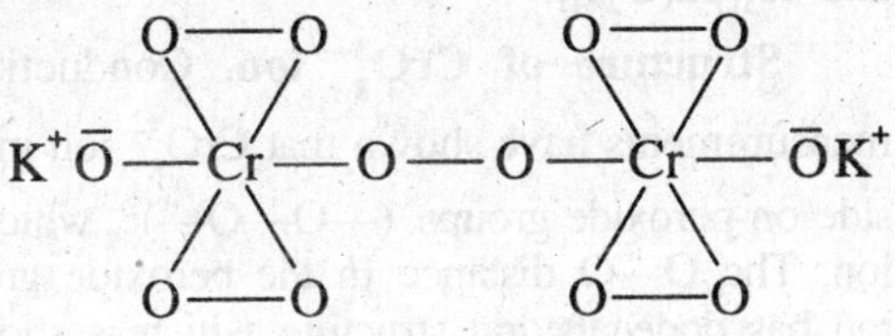

$$K^+\bar{O}—Cr(O_2)_2—O—O—Cr(O_2)_2—\bar{O}K^+$$

Fig. 9.5. Structure of $K_2Cr_2O_{12}$ molecule.

4. Red potassium peroxy chromate, $K_6Cr_2O_6$

Preparation. It is prepared by the action of H_2O_2 (excess) on an alkaline solution of CrO_5.

$$2CrO_5 + 3H_2O_2 \ (excess) + 6KOH \longrightarrow K_6Cr_2O_{16} \ (red) + 6H_2O$$

Structure. In $K_2Cr_2O_{16}$ molecule each Cr-atom is in +6 oxidation state and exhibits *hexacovalency.* It is linked with $3\frac{1}{2}$ peroxo groups and one neutral O-atom (Cr═O bond). Thus the structure of this molecule is as shown in Fig. 9.6.

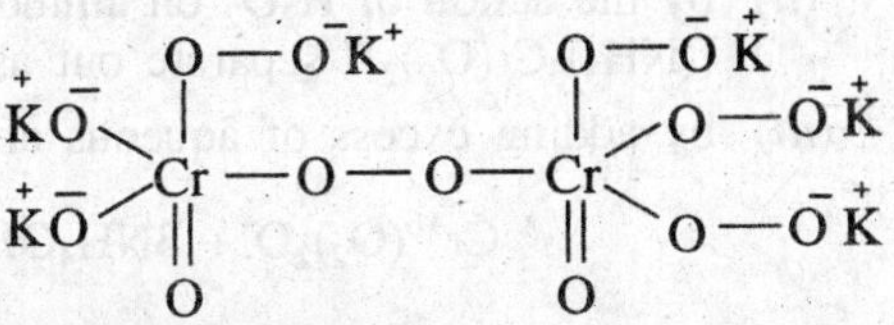

$$(K^+\bar{O})_2(K^+\bar{O}—O)Cr(=O)—O—O—Cr(=O)(O—\bar{O}K^+)_3$$

Fig. 9.6. Structure of $K_6Cr_2O_{16}$ molecule.

Peroxy Compounds of Cr (+5)

Chromium in +5 oxidation state forms *red tetraperoxy chromates* which can be represented by the general formula, $M_3^+[CrO_8]^{3-}$ or $M_3^+[Cr(O_2)_4]^{3-}$ (M = K, NH_4).

Preparation. These compounds are obtained by treating alkaline solution of a chromate with 30% H_2O_2 solution below 0°C. For example:

$$2K_2CrO_4 + 2KOH + 7H_2O_2 \longrightarrow 2K_3[Cr(O_2)_4] + 8H_2O$$

or $$2CrO_4^{2-} + 2OH^- + 7H_2O_2 \longrightarrow 2[Cr(O_2)_4]^{3-} + 8H_2O$$

Properties. (*i*) These compounds are paramagnetic due to the presence of one unpaired electron per Cr-atom. The presence of one unpaired electron per Cr-atom indicates that Cr is present in +5 oxidation state in these compounds (Cr = $3d^54s^1$, Cr^{5+} = $3d^14s^0$)

(*ii*) These compounds are stable at room temperature and also for some days in aqueous solution.

(*iii*) Under neutral or alkaline conditions these compounds decompose to give chromate (CrO_4^{2-}) and O_2, but in acidic solution they produce Cr(III) salts and O_2.

$$4K_3[Cr(O_2)_4] + 2H_2O \longrightarrow 4K_2CrO_4 + 4KOH + 7O_2$$

or $$4[Cr(O_2)_4]^{3-} + 2H_2O \longrightarrow 4CrO_4^{2-} + 4OH^- + 7O_2$$

$$2[K_3Cr(O_2)_4] + 6H_2SO_4 \longrightarrow Cr_2(SO_4)_3 + 3K_2SO_4 + 6H_2O + 5O_2$$

or $$[Cr(O_2)_4]^{3-} + 12H^+ \longrightarrow 2Cr^{3+} + 6H_2O + 5O_2$$

(*iv*) When solution of $(NH_4)_3[Cr(O_2)_4]$ is heated to 50° or when it is treated with 10% NH_3 solution at 313K, *red brown diperoxy chromate,* $[(NH_3)_3Cr^{+4}(O_2)_2]^0$ is obtained.

$$(NH_4)_3[Cr^{+5}(O_2)_4] \xrightarrow{50°C} \underset{\text{Red-brown diperoxy chromate}}{[(NH_3)_3Cr^{4+}(O_2)_2]^0}$$

$$(NH_4)_3[Cr(O_2)_4] + NH_3(10\%) \longrightarrow [(NH_3)_3Cr(O_2)_2]^0$$

(*v*) $K_3[Cr(O_2)_4]$ is isomorphous with $K_3[V(O_2)_4]$, $K_3[Nb(O_2)_4]$ and $K_3[Ta(O_2)_4]$.

Structure of CrO_8^{3-} ion. Conductivity and freezing point measurements have shown that CrO_8^{3-} ion in $K_3[CrO_8]$ has four broad side-on-peroxide groups, $(—O—O—)^{2-}$ which are arranged round Cr^{5+} ion. The O—O distance in the peroxide group is 1.41Å. $[Cr(O_4)_2]^{3-}$ ion has dodecahedral structure which is shown in Fig. 9.7.

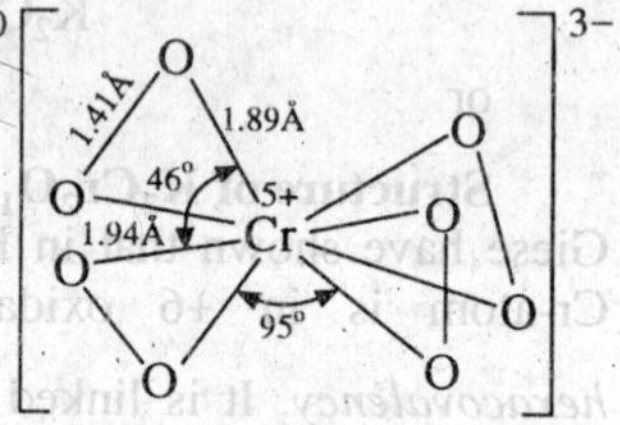

Fig. 9.7. Dodecahedral structure of $[Cr(O_2)_4]^{3-}$ ion.

Peroxy Compounds of Cr(+4)

The well known peroxy compound in which Cr shows +4 oxidation state is *red-brown diperoxy chromate* $CrO_4.3NH_3$ or $[(NH_3)_3Cr(O_2)_2]^0$.

Preparation. It is obtained as brown crystals :

(*i*) by heating the solution of $(NH_4)_3[Cr(O_2)_4]$ to 50°C or with 10% NH_3 solution at 313K.

(*ii*) by the action of H_2O_2 on ammoniacal solution of $(NH_4)_2CrO_2$ at 273°C. On cooling, $[(NH_3)_3Cr(O_2)_2]^0$ separate out as crystals.

(*iii*) by adding excess of aqueous ammonia to an ether solution of $Cr(O_2)_2O$.

$$Cr^{+6}(O_2)_2O + 3NH_4OH \longrightarrow [(NH_3)_3Cr^{4+}(O_2)_2]^0 + 3H_2O + \frac{1}{2}O_2$$

Properties. The compound is stable at room temperature but explodes on heating. Its paramagnetic value is equal to 2.8 B.M. which indicates that the compound has two unpaired electrons on Cr-atom and hence the metal is in +4 oxidation state.It is slightly soluble in H_2O. On gentle warning with KCN solution, $K_3[Cr(O_2)_2(CN)_3]$ is obtained.

$$[(NH_3)_3Cr^{4+}(O_2)_2] + 3KCN \longrightarrow K_3[Cr^{4+}(O_2)_2(CN)_3] + 3NH_3$$

Structure. $[(NH_3)_3Cr(O_2)_2]$ has roughly pentagonal bipyramidal structure with two NH_3 molecules in the apical (axial) positions. Two broad side-on peroxy groups and third NH_3 molecule are roughly in the equitorial positions. The O—O distance in the peroxide groups is equal to 1.41Å. Cr-atom is in +4 oxidation state and is coordinated with two peroxide ions and three NH_3 molecules as shown in Fig. 9.8 given in the margin.

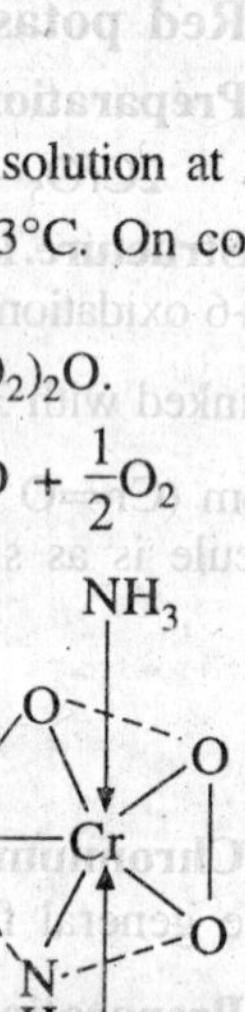

Fig. 9.8. Pentagonal bi-pyramidal structure of $[(NH_3)_3Cr(O_2)_2]^0$.

Compounds of Molybdenum

Introduction

Molybdenum forms compounds in −2, 0, +2, +3, +4, +5 and +6 oxidation states. Examples are given below :

Mo(zero) compounds : $Mo(CO)_6$, $[Mo(py)_3(CO)_3]$etc.
Mo(+2) compounds : $MoCl_2$, $[Mo(diars)_2X_2]^0$, $[Mo(CN)_6]^{4-}$
Mo(+3) compounds : Mo_2O_3, Mo_2S_3, $[Mo(H_2O)_3Cl_3]^0$, $[MoCl_6]^{3-}$, $MoCl_3$
Mo(+4) compounds : MoO_2, $MoCl_4$, $[Mo(NCS)_6]^{2-}$, MoS_2
Mo(+5) compounds : Mo_2S_5, MoF_5, $MoOCl_3$, $MCl_5(g)$, $[Mo(CN)_8]^{3-}$
Mo(+6) compounds : MoO_3, MoS_3, MoF_6, $MoOF_2$, $[MoO_4]^{2-}$, $[MoF_8]^{2-}$, MoO_2Cl_2 etc.

Let us discuss the following compounds.

1. Molybdenum trioxide, MoO_3

Preparation. It is prepared : (*i*) by heating Mo metal or MoS_2 in air or O_2. (*ii*) by heating MoO_2 with HNO_3, (*iii*) by heating normal ammonium molybdate, $(NH_4)_2MoO_4$ alone or by heating ammonium paramolybdate, $(NH_4)_6Mo_7O_{24}.4H_2O$ alone or with HNO_3.

$$2Mo + 3O_2 \longrightarrow 2MoO_3$$
$$2MoS_2 + 7O_2 \longrightarrow 2MoO_3 + 4SO_2$$
$$MoO_2 + 2HNO_3 \longrightarrow MoO_3 + H_2O + 2NO_2$$
$$(NH_4)_2MoO_4 \xrightarrow{\Delta} 2NH_3 + H_2O + MoO_3$$
$$(NH_4)_6Mo_7O_{24}.4H_2O \xrightarrow{\Delta} 7H_2O + 6NH_3 + 7MoO_3$$
$$(NH_4)_6Mo_7O_{24}.4H_2O + 6HNO_3 \xrightarrow{\Delta} 7H_2O + 6NH_4NO_3 + 7MoO_3$$

Properties. At room temperature MoO_3 is colourless but on heating it becomes yellow. Although MoO_3 is regarded as the anhydride of molybdic acid, H_2MoO_4, it does not form this acid direclty. It is more soluble in cold water than in hot water. MoO_3 dissolves in alkalies to form *normal molybdates* of $M_2[MoO_4]$ type.

$$MoO_3 + 2MOH \longrightarrow M_2[MoO_4]$$

A solution of MoO_3 in HNO_3, when evaporated, gives a white crystalline powder of molybdic acid, H_2MoO_4. When heated with dry HCl at about 150–200°C, it gives a pale yellow water-soluble sublimate of the composition, $MoO_3.2HCl$ or $MoO(OH)_2Cl_2$. When MoO_3 is dissolved in an alkali sulphide (*e.g.*, Na_2S) and then the solution so obtained is acidified, MoS_3 is formed.

$$MoO_3 \xrightarrow{+Na_2S} Na_2MoS_4 \xrightarrow{+H^+} M_2S_3$$

On being reduced by H_2, it gives MoO_2 and then Mo metal.

$$MoO_3 + H_2 \xrightarrow{500-470°C} MoO_2 + H_2O \;; \quad MoO_2 + 2H_2 \xrightarrow{600°C} Mo + 2H_2O$$

With hot conc. NH_4OH, normal ammonium molybdate, $(NH_4)_2MoO_4$ is formed.

$$MoO_3 + 2NH_4OH \longrightarrow (NH_4)_2MoO_4 + H_2O$$

Now when the solution containing $(NH_4)_2MoO_4$ is evaporated, crystals of *ammonium paramolybdate*, $(NH_4)_6[Mo_7O_{24}]$ or $6NH_3.7MoO_3.3H_2O$ are obtained.

$$7MoO_3 + 6NH_4OH \longrightarrow (NH_4)_6[Mo_7O_{24}] + 3H_2O$$

2. Molybdic acid, $H_2MoO_4.H_2O$

It is obtained by treating ammonium molydate with HNO_3 and allowing the solution to crystallise in *cold*. It is a yellow crystalline powder slightly soluble in water, giving an acidic solution. This acid is much weaker than H_2CrO_4. It dissolves in alkalies, forming molybdates.

3. Ammonium molybdate, $(NH_4)_2MoO_4$

This compound is obtained by the treatment of MoO_3 with NH_4OH and evaporating the solution to crystallisation. It is a white crystalline powder, soluble in water and insoluble in alcohol. $(NH_4)_2MoO_4$ easily loses NH_3 while $(NH_4)_6[Mo_7O_{24}]$ is quite stable in air. The solution of ammonium molybdate in dil. HNO_3 slowly deposits yellow α-molybdic acid, H_2MoO_4. H_2O which, on warming at 70°C with H_2O forms white β-molybdic acid, H_2MoO_4. The two acids have different X-ray spectra and vapour pressure. When a solution of ammonium molybdate and HCl is dialysed, colloidal molybdic acid is obtained. On heating the salt alone or with HNO_3, MoO_3, is obtained (see the preparation of MoO_3). When a phosphate or an arsenate is heated with conc. HNO_3 and $(NH_4)_2MoO_4$ solution, yellow precipitate of ammonium phosphomolybdate, $(NH_4)_3PO_4.12MoO_3$ or ammonium arsenomolybdate, $(NH_4)_3AsO_4.12MoO_3$ is obtained.

$$Na_3PO_4 + 12(NH_4)_2MoO_4 + 24HNO_3 \longrightarrow (NH_4)_3PO_4.12MoO_3 + 12H_2O + 3NaNO_3 + 21NH_4NO_3$$

or $$Na_3PO_4 + 12(NH_4)_2MoO_4 + 24HNO_3 \longrightarrow (NH_4)_3[PMo_{12}O_{40}] + 12H_2O + 3NaNO_3 + 2NH_4NO_3$$

or $$PO_4^{3-} + 12MoO_4^{2-} + 24NO_3^- \longrightarrow \underset{\text{Phospho-12 molybdate}}{[PMo_{12}O_{40}]^{3-}} + 12H_2O$$

$$AsO_4^{3-} + 12MoO_4^{2-} + 24NO_3^- \longrightarrow \underset{\text{Phospho-12 arsenate}}{[PAs_{12}O_{40}]^{3-}}$$

On the basis of the above reactions, $(NH_4)_2MoO_4$ is used in the laboratory for the detection and estimation PO_4^{3-} and AsO_4^{3-} ions. When As_2S_3 or As_2S_5 is boiled with conc. HNO_3, arsenic acid, H_3AsO_4 is produced which gives a yellow precipitate of ammonium arsenomolybdate, $(NH_4)_3AsO_4.12MoO_3$ with $(NH_4)_2MoO_4$.

$$As_2S_3 + 10HNO_3 \longrightarrow 2H_3AsO_4 + 10NO_2 + 2H_2O + 3S$$

$$As_2S_5 + 10HNO_3 \longrightarrow 2H_3AsO_4 + 10NO_2 + 2H_2O + 5S$$

$$H_3AsO_4 + 12(NH_4)_2MoO_4 + 21HNO_3 \longrightarrow \underset{\text{yellow ppt.}}{(NH_4)_3AsO_4.12MoO_3} + 21NH_4NO_3 + 12H_2O$$

$(NH_4)_2MoO_4$ is used: (*i*) as an analytical reagent for the detection and estimation of PO_4^{3-} and AsO_4^{3-} ions (*ii*) as a catalyst for dehydrogenation in petroleum and coal technology.

4. Molybdenum blue

It is obtained: (*i*) by treating molybdic acid or an acidic solution of a molybdate with a reducing agent like SO_2, H_2S, HI, hydrazine, glucose, zinc or molybdenum metal or (*ii*) by reducing MoO_3 in 25–75% H_2SO_4 with Cu^+ ions or with Sn^{2+} ions in presence of sunlight. Molybdenum blue is a deep blue colloidal solution. The actual colour of the solution depends on the concentration of molybdenum, pH of the medium, nature of the reducing agent and temperature of the reaction. The solution resembles indigo blue in appearance. It is very soluble in water and also in some of the organic solvents. The exact structure of molybdenum blue is not known. It is supposed to be a mixture of hydrated oxides of molybdenum with hydroxyl groups. The mean oxidation state (0.5) of Mo atom is between 5 and 6. Some molybdenum blues are crystalline while others are amorphous. Examples are : (*i*) *Crystalline molybdenum blues:* $MoO_{2.0}(OH)$ (Mean O.S. of Mo = 5), $MoO_{2.6}.nH_2O$ (O.S. = 5.20), $MoO_{2.5}(OH)_{0.25}$ (O.S = 5.50), $MoO_{2.38}.H_2O$(O.S. = 5.76), (*ii*) *Amorphous molybdenum blues:* $MoO_{2.75}.nH_2O$ (O.S. = 5.50), $MoO_{2.88}.nH_2O$ (O.S. = 5.76). On heating molybdenum blue above 350°C, only H_2O molecules are lost while the OH groups remain attached. This may suggest that OH groups are involved in bridging of different groups. Formation of molybdenum blue is utilised in the detection and calorimetric determination of Mo. Phosphorus is indirectly estimated by this method. Because of the colloidal nature of molybdenum blue, it is easily absorbed on vegetable and animal fibres and hence is used for dyeing those fibres.

Compounds of Tungsten

Introduction

Tungsten forms compounds in 0 *e.g.* [$(W(CO)_6$,$[W(CO)_5X]^-$ etc.], + 2(*e.g.* WX_2, [W(diars)

$(CO)_4I]^+$ etc.), +3 (*e.g.*, $[W(diars)(CO)_3\ Br_2]^+$), +4(*e.g.*, WO_2, WCl_4, $[W(OH)Cl_5]^{2-}$) + 5(*e.g.* WF_6^-) and +6 (*e.g.* WO_3, WX_6, WO_2Cl_2, WO_4^{2-} etc.) oxidation states. Compounds containing tungsten in +6 oxidation state are the most stable. Important compounds are discussed below :

1. Tungsten trioxide or tungstic oxide, WO_3

Preparation. (*i*) It is obtained by the following methods :

$$\underset{\text{Tungstic acid}}{H_2WO_4} \xrightarrow{\text{Dehydration}} WO_3 + H_2O$$

$$2W + 3O_2 \xrightarrow{\Delta\ (500°C)} 2WO_3$$

$$2WS_2 + 7O_2\ (\textit{from air}) \xrightarrow{\text{Roasting}} 2WO_3 + 4SO_2$$

$$\underset{\text{Amm. tungstate}}{(NH_4)_2WO_4} \xrightarrow{\Delta} WO_3 + 2NH_3 + H_2O$$

$$Na_2WO_4 + 2HNO_3 \longrightarrow WO_3 + 2NaNO_3 + H_2O$$

(*ii*) WO_3 can also be obtained from wolframite ore, (Fe, Mn)WO_4.

Properties. (*i*) It is a soft lemon-yellow coloured powder, but turns orange on heating. It melts at 1470° to a green liquid. It is insoluble in water and acids (excepting HCl).

$$WO_3 + 4HF \longrightarrow \underset{\text{Hydrofluo-tungstic acid}}{H_2WO_2F_4} + H_2O$$

(*ii*) It dissolves in alkalies to form tungstate.

$$WO_3 + 2NaOH \longrightarrow Na_2WO_4 + H_2O$$

(*iii*) When heated with H_2, it is reduced to W metal *via* W_2O_5 and WO_2.

$$\underset{\text{Yellow}}{WO_3} \xrightarrow{H_2} \underset{\text{Blue}}{W_2O_5} \xrightarrow{H_2} \underset{\text{Brown}}{WO_2} \xrightarrow{H_2} W$$

When heated with C, Mg etc., W metal is obtained directly. When heated with Cl_2 it gives volatile oxychloride, WO_2Cl_2 while with H_2S it gives WS_3.

$$WO_3 + 3C \longrightarrow 3CO + W$$

$$WO_3 + 3Mg \longrightarrow 3MgO + W$$

$$2WO_3 + 2Cl_2 \longrightarrow 2WO_2Cl_2 + O_2$$

$$WO_3 + 3H_2S \longrightarrow WS_3 + 3H_2O$$

Uses. It is used (*i*) as a mordant in dyeing and ceramic industry (*ii*) as a pigment in oil and water colours.

2. Tungstic acid, H_2WO_4

A precipitate of this acid may be obtained by treating a solution of any soluble tungstate with acid. The precipitate of the acid may be white ($H_2WO_4.H_2O$) or yellow (H_2WO_4) according as the precipitation has been made from cold solution or hot solution.

$$(NH_4)_2WO_4 + H_2SO_4 \longrightarrow \underset{\text{Yellow ppt}}{H_2WO_4} + (NH_4)_2SO_4$$

The hydrated acid ($H_2WO_4.H_2O$) is somewhat soluble in water but the anhydrous acid (H_2WO_4) is neither soluble in H_2O nor in any acid, excepting HF. With HF hydrofluotungstic acid, $H_2WO_3F_4$ is formed. With alkalies, H_2WO_4 forms tungstates. On heating WO_3 is obtained.

$$H_2WO_4 + 4HF \longrightarrow H_2WO_2F_4 + 2H_2O$$

$$H_2WO_4 + 2NaOH \longrightarrow Na_2WO_4 + 2H_2O$$

$$H_2WO_4 \xrightarrow{\Delta} WO_3 + H_2$$

The composition of various tungstic acids is as : *meta tungstic acid*—$WO_3.H_2O$ or H_2WO_4; *para tungstic acid*—$WO_3.2H_2O$ or $H_2WO_4.H_2O$ or H_2WO_5; *ortho tungstic acid*—$WO_3.3H_2O$ or $H_2WO_4.2H_2O$ or H_6WO_6. Tungstic acid is used in ceramic industry and in making pigments.

3. Tungsten bronzes, M_xWO_3

These are the *non-stoichiometric* compounds having a general formula, M_xWO_3 (M = Li, Na, K etc. and $x < 1$). These are also called *metal-tungstate bronzes*. They are obtained by reducing alkali metal or alkaline earth metal tungstates by H_2, metallic Na, metallic W or electrically. These are insoluble in water, chemically inert, have metallic lustre and are conductors of electricity. These are highly coloured. Their colour depends on the value of x as shown below :

x =	0.93	0.64	0.46	0.32
Colour =	*Gold yellow*	*Orange*	*Red-violet*	*Blue-violet*

These are resistant to the action of acids excepting HF. In the presence of an alkali, these are oxidised by O_2. Tungsten bronzes are used as pigments.

4. Tungsten halides

The known halides given by tungsten are : (*i*) *Fluorides:* WF_6. (*ii*) *Chlorides:* WCl_6, WCl_5, WCl_4, WCl_2. (*iii*) *Bromides*: WBr_6 (*very unstable*), WBr_5, WBr_2 (*iv*) *Iodides* : WI_4, WI_2. Of these WF_6 and WCl_6 are the most important.

WF_6 can be prepared by any of the following methods:

$$W + 3F_2 \longrightarrow WF_6$$

$$WCl_6 + 6HF\ (anhydrous) \longrightarrow WF_6 + 6HCl$$

$$WCl_6 + 2AsF_3 \longrightarrow WF_6 + 2AsCl_3$$

$$WCl_6 + 3SbF_5 \longrightarrow WF_6 + 2SbF_3Cl_2$$

It is colourless or slightly yellow coloured gas. It is hydrolysed by H_2O, forming WO_3.

$$WF_6 + 3H_2O \longrightarrow WO_3 + 6HF$$

It is extremely volatile diamagnetic compound. It attacks all metals except Pt and Au. In presence of traces of moisture, it attacks glass because of the production of HF (due to hydrolysis). It behaves a Lewis acid and forms addition compounds with NH_3, pyridine (py) and methylamine ($CH_3.NH_2$) (*e.g.*, $WF_6.4NH_3$, $WF_6.3py$, $WF_6.3CH_2NH_2$). With NaF it gives crystalline solid $WF_6.2NaF$ in which W is 8-coordinated. With $SbCl_6$ it gives WCl_6.

WCl_6 can be prepared by the following reactions :

$$W + 3Cl_2 \xrightarrow{300°C} WCl_6$$

$$WS_2 + 4Cl_2 \xrightarrow{\Delta} WCl_6 + S_2Cl_2$$

$$2WO_3 + 6Cl_2 \xrightarrow{\text{In } CCl_4 \text{ in a sealed tube at } 280°} 2WCl_6 + 3O_2$$

Pure WCl_6 is steel blue coloured compound, slowly attacked by cold water, but rapidly by hot H_2O.

$$WCl_6 + 4H_2O \longrightarrow H_2WO_4 + 6HCl$$

Other reactions of WCl_6 are given below :

$$WCl_6 + 5HBr \longrightarrow WBr_5 + HCl + \frac{1}{2}Cl_2$$

$$WCl_6 + 2NH_3 \longrightarrow WN_2 + 6HCl$$

$$WCl_6 \xrightarrow{H_2} WCl_5 \longrightarrow WCl_4 \longrightarrow WCl_2$$

Compounds of Manganese

Introduction

Manganese forms compounds in –1 to 7 oxidation states. Compounds having Mn in +2 oxidation state are the most stable. Electronic configuration of Mn in different oxidation states and the compounds formed by it are given in Table 9.1.

Table 9.1. *Different oxidation states of Mn and examples of compounds formed by it.*

Mn atom or Mn^{n+} ion	*Configuration (n = No. of unpaired electrons)*	*Examples of compounds*
Mn^0	2, 8, $3s^2p^6d^5$, $4s^2$ (n = 5)	$Mn_2(CO)_{10}$
Mn^-	2, 8, $3s^2p^6d^6$, $4s^2$ (n = 4)	$[Mn(CO)_5]^-$
Mn^+	2, 8, $3s^2p^6d^5$, $4s^1$ (n = 6)	$[Mn(CO)_3]Cl$, $[Mn(CN)_6]^{5-}$
Mn^{2+}	2, 8, $3s^2p^6d^5$, $4s^0$ (n = 5)	$MnCl_2$, MnO, MnS, $Mn(OH)_2$, $Mn(NO_3)_2$, $MnSO_4.5H_2O$, $MnCO_3$, $Mn(ClO_4)_2.6H_2O$ or $[Mn(H_2O)_6](ClO_4)_2$, $[Mn(CN)_5(NO)]^{3-}$, $[MnX_4]^{2-}$, $MnSO_4.4H_2O$ or $[Mn(H_2O)_4]SO_4$, $[Mn(NH_3)_6]^{2+}$, $[Mn(CN)_6]^{4-}$,
Mn^{3+}	2, 8, $3s^2p^6d^4$, $4s^0$ (n = 4)	Mn_2O_3, $Mn(CH_3COO)_3.2H_2O$, $[Mn(C_2O_4)_3]^{3-}$, MnF_3, $MnO(OH)$, $MnPO_4$, $[Mn(CN)_6]^{3-}$, $Mn_2(SO_4)_3$
Mn^{4+}	2, 8, $3s^2p^6d^3$, $4s^0$ (n = 3)	MnO_2, MnF_4, $Mn(SO_4)_2$, $[MnCl_6]^{2-}$, K_2MnO_3
Mn^{5+}	2, 8, $3s^2p^6d^2$, $4s^0$ (n = 2)	$Na_3MnO_4.10H_2O$
Mn^{6+}	2, 8, $3s^2p^6d^1$, $4s^0$ (n = 1)	K_2MnO_4, $Na_2MnO4.nH_2O$, HMnO
Mn^{7+}	2, 8, $3s^2p^6d^0$, $4s^0$ (n = 0)	Mn_2O_7, $HMnO_4$, $KMnO_4$, MnO_3F

Compounds of Mn (0)

Dimanganese decacarbonyl, $Mn_2(CO)_{10}$ is an example of the compound in which Mn exhibits zero oxidation state. Preparation, properties and structure of this compound will be discussed in chapter 13.

Compounds of Mn(+2) : Manganous compounds

These compounds are the most stable. Since the potential for the reaction, $Mn^{2+} \rightarrow Mn^{3+} + e^-$ is negative (= – 1.51 Volts), Mn^{2+} ion is not oxidised to Mn^{3+} ion. Thus Mn^{2+}, in neutral or acidic medium, remains unaffected even on prolonged exposure to O_2. Mn^{2+} ion, therefore, cannot act as a reducing agent. We know that the valence-shell configurations of Cr^{2+}, Mn^{2+} and Fe^{2+} ions are $Cr^{2+} = 3d^4$, $Mn^{2+} = 3d^5$ and $Fe^{2+} = 3d^6$. Since it is only Mn^{2+} ion which has half-filled $3d$ sub-shell, this ion is more stable than Cr^{2+} ion as well as Fe^{2+} ion. In neutral or acidic aqueous solution, Mn^{2+} ion exists as $[Mn(H_2O)_6]^{2+}$ (pale pink).

Out of various compounds of Mn(+2) given in Table 9.1 we shall discuss here the preparation, properties and uses of mangeneous oxide (MnO) only.

Manganeous oxide, MnO

Preparation. This oxide is prepared by the following reactions :

$$MnO_2 + H_2 \xrightarrow{\Delta} MnO + H_2O$$

$$Mn(COO)_2 \xrightarrow{\Delta} MnO + CO + O_2$$

Properties. This is a greyish-green powder and insoluble in water. It is strongly basic and forms manganous salts [*i.e.*, Mn(II) salts] with acids. It is quickly oxidised by air in the cold to Mn_2O_3.

Uses. It has a NaCl lattice. It is used as an intermediate for the preparation of Mn(II) salts, as a nutrient in agriculture, and in welding.

Compounds of Mn (+3) : Manganic Compounds

Mn^{3+} ion is not stable in aqueous solution, since it is easily reduced to Mn^{2+} ion. The unstability of Mn^{2+} ion is evident from the fact that even in the absence of a reducing agent, this ion undergoes disproportionation in aqueous solution.

$$\underset{(Mn = +3)}{2Mn^{3+}} + 2H_2O \longrightarrow \underset{(Mn = +2)}{Mn^{2+}} + \underset{(Mn = +4)}{MnO_2} + 4H^+$$

Since Mn^{3+} ion is easily reduced to Mn^{2+} ion, this ion acts as a strong oxidising agent. It is so strong that it can oxidise water and can evolve O_2 from it. Mn^{3+} ion becomes stable when it forms complex compounds like $[Mn(C_2O_4)_3]^{3-}$, $[Mn(CN)_6]^{3-}$ etc.

The above discussion shows that Mn^{3+} ion is stable either in the solid state or in its complex compounds.

1. Manganic oxide, Mn_2O_3

This compound is also called *manganese sesquioxide*. It can be obtained by the following reactions :

$$4MnO + O_2 \xrightarrow{\Delta} 2Mn_2O_3$$

$$4MnO_2 \xrightarrow{\Delta} 2Mn_2O_3 + O_2$$

The *hydrous oxide*, $Mn_2O_3.H_2O$ is obtained as :

$$2Mn(OH)_2 + \frac{1}{2} O_2 \longrightarrow Mn_2O_3.H_2O$$

It is a black amorphous powder. Above 1000°C it decomposes into Mn_3O_4, liberating O_2. This oxide is weakly basic and dissolves in hot dil. acids to form Mn(II) salts and MnO_2.

$$6Mn_2O_3 \xrightarrow{> 1000°C} 4Mn_3O_4 + O_2$$

$$Mn_2O_3 + H_2SO_4 \longrightarrow MnSO_4 + MnO_2 + H_2O$$

$$Mn_2O_3 + 2HNO_3 \longrightarrow Mn(NO_3)_2 + MnO_2 + H_2O$$

Mn_2O_3 dissolves in conc. HCl, forming $MnCl_2$ with the evolution of Cl_2.

$$Mn_2O_3 + 6HCl \longrightarrow 2MnCl_2 + Cl_2 + 3H_2O$$

Mn_2O_3 has O═Mn—O—Mn═O structure which is supported by magnetic susceptibility. In this oxide Mn is in +3 oxidation state.

2. Manganese tetroxide, Mn_3O_4

It is also called *mangano-manganic oxide* or *red oxide of manganese.* It may be prepared by heating any oxide of Mn.

$$6MnO + O_2 \xrightarrow{\Delta} 2Mn_3O_4$$

$$3MnO_2 \xrightarrow[\text{absence of air}]{\text{Heat in the}} Mn_3O_4 + O_2$$

$$6Mn_2O_3 \xrightarrow[\text{(in air)}]{>95°C} 4Mn_3O_4 + O_2$$

It is a brownish-red powder. It slowly dissolves in conc. cold H_2SO_4, forming a red solution containing $MnSO_4$ and $Mn_2(SO_4)_3$. With dil. HNO_3 it gives $Mn(NO_3)_2$ and MnO_2 while with HCl, $MnCl_2$ is formed and Cl_2 is evolved.

Structure of Mn_3O_4

$$Mn_3O_4 + 4H_2SO_4 \longrightarrow MnSO_4 + Mn_2(SO_4)_3 + 4H_2O$$
$$Mn_3O_4 + 4HNO_3 \longrightarrow 2Mn(NO_3)_2 + MnO_2 + 2H_2O$$
$$Mn_3O_4 + 8HCl \longrightarrow 3MnCl_2 + Cl_2 + 4H_2O$$

Mn_3O_4 has the structure shown in the margin. This oxide is represented as $MnO.Mn_2O_3$.

Compounds of Mn(+4)

Compounds of Mn(+4) are limited in number. The most familiar compound of Mn(+4) is MnO_2 which is discussed below :

Manganese dioxide, MnO_2

Preparation. The pure MnO_2 is prepared by heating $Mn(NO_3)_2$ while the hydrated form, $MnO_2.H_2O$ (called manganous acid, H_2MnO_3) is prepared by the alkaline oxidation of Mn(II) salt by sodium hypochlorite or bleaching powder.

$$Mn(NO_3)_2 \xrightarrow{\Delta} MnO_2 + 2NO_2$$
$$Mn^{2+} + 3OH^- + ClO^- \longrightarrow MnO_2.H_2O + Cl^-$$

Properties. It is a black powder, insoluble in water and a good conductor of electricity. When heated above 530°C in air, it loses O_2 to form Mn_3O_4. It acts as a basic oxide towards cold acids like HCl, forming salts containing Mn in +4 oxidation state. With conc. H_2SO_4, $MnSO_4$ is formed and O_2 is evolved. HNO_3 has no action on MnO_2.

$$3MnO_2 \xrightarrow{>530°C} Mn_3O_4 + O_2$$
$$MnO_2 + 4HCl \textit{ (cold and concentrated)} \longrightarrow MnCl_4 + 2H_2O$$
$$2MnO_2 + 2H_2SO_4 \textit{ (concentrated)} \longrightarrow 2MnSO_4 + 2H_2O + O_2$$

In the presence of excess of H^+ ions, MnO_2 acts as a strong oxidising agent. In showing this behaviour Mn^{4+} changes to Mn^{2+} ion.

$$\underset{(Mn\,=\,+4)}{MnO_2} + 4H^+ + 4e^- \longrightarrow \underset{(Mn\,=\,+2)}{Mn^{2+}} + 2H_2O$$

Thus in presence of hot dil. H_2SO_4, MnO_2 oxidises oxalic acid to CO_2, Fe(II) salt to Fe(III) salt and chloride (Cl^-) to Cl_2

$$C_2O_4^{2-} + MnO_2 + 4H^+ \longrightarrow Mn^{2+} + 2H_2O + 2CO_2$$
$$2Fe^{2+} + MnO_2 + 4H^+ \longrightarrow 2Fe^{3+} + Mn^{2+} + 2H_2O$$
$$MnO_2 + 4H^+ + 2Cl^- \longrightarrow Mn^{2+} + 2H_2O + Cl_2$$

In presence of air or oxidising agents (*e.g.*, KNO_3, $KClO_3$) MnO_2 reacts with fused KOH to give potassium manganate, K_2MnO_4.

$$2MnO_2 + 4KOH + O_2 \longrightarrow 2K_2MnO_4 + 2H_2O$$

Since in MnO_2, Mn is in +4 state, the structure of MnO_2 is $O{=}Mn{=}O$ which has also been supported by magnetic susceptibility.

Uses. MnO_2 is used : (*i*) as a depolariser in Lechanehe cell and dry cells.In these cells MnO_2 oxidises H_2 liberated at the cathode and thus counteracts polarisation. (*ii*) as a decolouriser in glass. (*iii*) in forming with ferric oxide, a dark brown glaze which is used in pottery. (*iv*) as a catalyst in the preparation of O_2 from $KClO_3$. (*v*) as a dier in paints and varnishes. (*vi*) in the preparation of Cl_2 in the laboratory. (*vii*) in dyestuff industry and in varnish manufacture. (*viii*) in the manufacture of metallic manganese, manganates, permanganates and a number of its salts.

Compounds of Mn (+6): Manganates (MnO_4^{2-})

Mn in +6 oxidation state forms manganates (MnO_4^{2-}) which are the salts of hypothetical manganic acid, H_2MnO_4. The acid is unknown even in solution.

Preparation of manganates (MnO_4^{2-})

Manganates can be obtained by the following methods :

1. From pyrolusite mineral, MnO_2 : Large scale preparation. When finely powdered mineral (MnO_2) is fused with KOH or K_2CO_3 in presence of air or an oxidising agent like KNO_3, $KClO_3$ etc., green coloured mass containing K_2MnO_4 is obtained.

(*a*) $$2MnO_2 + 4KOH + O_2 \longrightarrow 2K_2MnO_4 + 2H_2O$$

or $$2MnO_2 + 4OH^- + O_2 \longrightarrow 2MnO_4^{2-} + 2H_2O$$

(*b*) $$2MnO_2 + 2K_2CO_3 + O_2 \longrightarrow 2K_2MnO_4 + 2CO_2$$

or $$2MnO_2 + 2CO_3^{2-} + O_2 \longrightarrow 2MnO_4^- + 2CO_2$$

(*c*) $$MnO_2 + 2KOH + KNO_3 \longrightarrow K_2MnO_4 + KNO_2 + H_2O$$

or $$MnO_2 + 2OH^- + NO_3^- \longrightarrow MnO_4^{2-} + NO_2^- + H_2O$$

(*d*) $$3MnO_2 + 6KOH + KClO_3 \longrightarrow 3K_2MnO_4 + KCl + 3H_2O$$

or $$3MnO_2 + 6OH^- + ClO_3^- \longrightarrow 3MnO_4^{2-} + Cl^- + 3H_2O$$

The fused mass is dissolved in small quantity of water and the green solution thus obtained is allowed to evaporate in vacuum when dark green crystals of K_2MnO_4 are obtained.

2. From permanganate (MnO_4^-). When a permanganate (MnO_4^-) is boiled with an alkali hydroxide solution, a solution containing manganate (MnO_4^{2-}) is obtained.

$$4KMnO_4 + 4KOH \longrightarrow 4K_2MnO_4 + 2H_2O + O_2$$

or $$4MnO_4^- + 4OH^- \longrightarrow 4MnO_4^{2-} + 2H_2O + O_2$$

General properties of manganates

1. Stability. Manganates are stable only in highly alkaline solutions which have green colour. On adding large quantity of H_2O or on acidifying the solution, the green MnO_4^{2-} ion (Mn = +6) undergoes disproportionation to $MnO_2(s)$ (Mn = +4) (*brown ppt.*) and permanganate ion, MnO_4^- (Mn = +7).

$$3K_2MnO_4 + 2H_2O \longrightarrow MnO_2(s) + 2KMnO_4 + 4KOH$$

or $$3MnO_4^{2-} + 2H_2O \longrightarrow MnO_2(s) + 2MnO_4^- + 4OH^-$$

$$3K_2MnO_4 + 2H_2SO_4 \longrightarrow MnO_2(s) + 2KMnO_4 + 2K_2SO_4 + 2H_2O$$

or $$3MnO_4^{2-} + 4H^+ \longrightarrow MnO_2(s) + 2MnO_4^- + 2H_2O$$

MnO_4^{2-} ion is changed to MnO_2 and MnO_4^- ion even on passing CO_2 in its solution.

$$3K_2MnO_4 + 2H_2O + 4CO_2 \longrightarrow MnO_2 + 2KMnO_4 + 4KHCO_3$$

or $$3MnO_4^{2-} + 2H_2O + 4CO_2 \longrightarrow MnO_2 + 2MnO_4^- + 4HCO_3^-$$

(2) Conversion of K_2MnO_4 to $KMnO_4$. Fused mass (green coloured) of K_2MnO_4 obtained as above is extracted with H_2O and the solution, after filtration, is converted into $KMnO_4$:

(*a*) by the disproportionation of K_2MnO_4 (Mn = +6) to MnO_2 (Mn = +4) and $KMnO_4$ (Mn = +7) by acidifying the solution of K_2MnO_4 with H_2SO_4 or by passing CO_2 gas through the solution.

$$3K_2MnO_4 + 2H_2SO_4 \longrightarrow K_2SO_4 + 2KMnO_4 + 2H_2O + MnO_2\downarrow$$

or $$3MnO_4^{2-} + 4H^+ \longrightarrow 2MnO_4^- + 2H_2O + MnO_2\downarrow$$

$$3K_2MnO_4 + 4CO_2 + 2H_2O \longrightarrow 2KMnO_4 + MnO_2 + 4KHCO_3$$

or $$3MnO_4^{2-} + 4CO_2 + 2H_2O \longrightarrow 2MnO_4^- + MnO_2 + 4HCO_3^-$$

(*b*) by oxidising K_2MnO_4 (Mn = +6) to $KMnO_4$ (Mn = +7). This can be done:

(*i*) by passing Cl_2 or O_3 through the solution of K_2MnO_4.

$$2K_2MnO_4 + Cl_2 \longrightarrow 2KMnO_4 + 2KCl$$

or $$2MnO_4^{2-} + Cl_2 \longrightarrow 2MnO_4^- + 2Cl^-$$

$$2K_2MnO_4 + H_2O + O_3 \longrightarrow 2KMnO_4 + 2KOH + O_2$$

or $$2MnO_4^{2-} + H_2O + O_3 \longrightarrow 2MnO_4^- + 2OH^- + O_2$$

(*ii*) by electrolysing the solution of K_2MnO_4 in a cell (Ni : anode, Fe : cathode). On passing the electric current the following reactions take place in the cell.

Ionisation : $$[K_2MnO_4 \rightleftharpoons K^+ + KMnO_4^-] \times 2$$

At anode : $$[KMnO_4^- \longrightarrow KMnO_4 + e^-] \times 2$$

At cathode : $$[K^+ + e^- \longrightarrow K] \times 2$$

$$2K + 2H_2O \longrightarrow 2KOH + H_2\uparrow$$

On adding all the above equations, we get the final equation as :

$$2K_2MnO_4 + 2H_2O \longrightarrow \underset{\text{Solution}}{2KMnO_4} + 2KOH + H_2\uparrow$$

The solution of $KMnO_4$ obtained as above is concentrated by evaporation so that crystal of $KMnO_4$ are obtained.

Properties. Potassium permanganate forms dark purple lustrous crystals giving deep pink colour in solution. It is isomorphous with potassium perchlorate. Its solubility in water at 20° is only about 7% whilst it dissolves more at higher temperatures (25% at 63°C). It gives off O_2 at 200°C.

$$2KMnO_4 \longrightarrow K_2MnO_4 + MnO_2 + O_2$$

At red heat K_2MnO_4 formed as above is decomposed into permanganite, K_2MnO_3 and O_2.

$$2K_2MnO_4 \longrightarrow 2K_2MnO_3 + O_2$$

When an alkaline solution of $KMnO_4$ is heated, O_2 is evolved.

$$4KMnO_4 + 4KOH \xrightarrow{\Delta} 4K_2MnO_4 + 2H_2O + O_2$$

or $$4MnO_4^- + 4OH^- \longrightarrow 4MnO_4^- + 2H_2O + O_2$$

With *well cooled* conc. H_2SO_4, it gives Mn_2O_7 which decomposes on warming.

$$2KMnO_4 + 2H_2SO_4 \longrightarrow Mn_2O_7 + 2KHSO_4 + H_2O$$

$$2Mn_2O_7 \longrightarrow 4MnO_2 + 3O_2$$

With *warmed* conc. H_2SO_4 oxygen gas is evolved.

$$4KMnO_4 + 6H_2SO_4 \longrightarrow 2K_2SO_4 + 4MnSO_4 + 6H_2O + 5O_2$$

When heated in a current of H_2, solid $KMnO_4$ gives KOH, MnO and steam.

$$2KMnO_4 + 5H_2 \longrightarrow 2KOH + 2MnO + 4H_2O$$

A mixture of $KMnO_4$ with sulphur and charcoal burns when heated, and explodes like a gun powder when rubbed. A well-powdered mixture of $KMnO_4$ and oxalic acid catches fire spontaneously in a few seconds. When a drop of glycerine is poured over powdered $KMnO_4$, the same thing happens.

Oxidising properties. $KMnO_4$ is a powerful oxidising agent. The course of reaction depends on whether the oxidation takes place in *alkaline, neutral* or *acid* solution.

(*a*) **In alkaline solution.** In alkaline solution, MnO_4^- ion (Mn = +7) is first reduced to MnO_4^{2-} ion (Mn = +6) and then to insoluble MnO_2 (Mn = +4). The solution first changes from purple (due to MnO_4^-) to green (due to MnO_4^{2-}) and finally it becomes colourless (MnO_2).

$$\underset{\substack{\text{Purple}\\(\text{Mn}=+7)}}{2KMnO_4} + 2KOH \longrightarrow \underset{\substack{\text{Green}\\(\text{Mn}=+6)}}{2K_2MnO_4} + H_2O + O \quad ...(i)$$

$$\underset{\substack{\text{Green}\\(\text{Mn}=+6)}}{2K_2MnO_4} + 2H_2O \longrightarrow \underset{\substack{\text{Colourless}\\(\text{Mn}=+4)}}{2MnO_2} + 4KOH + 2O \quad ...(ii)$$

On adding :
$$\underset{\substack{\text{Purple}\\(\text{Mn}=+7)}}{2KMnO_4} + H_2O \xrightarrow{\text{Alkali}} \underset{\substack{\text{Colourless}\\(\text{Mn}=+4)}}{2MnO_2} + 2KOH + 3O \quad ...(iii)$$

This is the complete equation which represents the oxidising action of $KMnO_4$ in alkaline medium.

From equation (*iii*) it is evident that two molecules of $KMnO_4$ in alkaline solution give three oxygen atoms (*i.e, six equivalents*) and therefore, *the equivalent weight of $KMnO_4$ as an oxidising agent in alkaline solution is one-third of its molecular weight.*

Equations (*i*), (*ii*) and (*iii*) can also be written in the ionic forms as follows :

$$\underset{\substack{\text{Purple}\\(\text{Mn}=+7)}}{2MnO_4^-} + 2OH^- \longrightarrow \underset{\substack{\text{Green}\\(\text{Mn}=+6)}}{2MnO_4^{2-}} + H_2O + O$$

$$\underset{\substack{\text{Green}\\(\text{Mn}=+6)}}{2MnO_4^{2-}} + 2H_2O \longrightarrow \underset{\substack{\text{Colourless}\\(\text{Mn}=+4)}}{2MnO_2} + 4OH^- + 2O$$

On adding :
$$\underset{\substack{\text{Purple}\\(\text{Mn}=+7)}}{2MnO_4^-} + H_2O \xrightarrow{\text{Alkali}} \underset{\substack{\text{Colourless}\\(\text{Mn}=+4)}}{2MnO_2} + 2OH^- + 3O$$

or
$$MnO_4^- + 2H_2O + 3e^- \longrightarrow MnO_2 + 4OH^-$$

Thus in alkaline medium, $KMnO_4$ oxidises :

(*i*) *Iodides (I^-) to iodates (IO_3^-)*

$$2KMnO_4 + H_2O \longrightarrow 2MnO_2 + 2KOH + 3O \text{ (\textit{Reduction reaction})}$$

$$KI + 3O \longrightarrow KIO_3 \text{ (\textit{Oxidation reaction})}$$

$$2KMnO_4 + H_2O + KI \longrightarrow 2MnO_2 + 2KOH + KIO_3$$

OR

$$2\times [MnO_4^- + 2H_2O + 3e^- \longrightarrow MnO_2 + 4OH^-] \text{ (\textit{Reduction half-reaction})}$$

$$I^- + 6OH^- \longrightarrow IO_3^- + 3H_2O + 6e^- \text{ (\textit{Oxidation half-reaction})}$$

$$2MnO_4^- + H_2O + I^- \longrightarrow 2MnO_2 + 2OH^- + IO_3^-$$

(ii) NH_3 *(N = –3) to* N_2 *(N = 0)*

$$2KMnO_4 + H_2O \longrightarrow 2MnO_2 + 2KOH + 3O \text{ (Reduction reaction)}$$
$$2NH_3 + 3O \longrightarrow N_2 + 3H_2O \text{ (Oxidation reaction)}$$
$$2KMnO_4 + 2NH_3 \longrightarrow 2MnO_2 + 2KOH + N_2 + 2H_2O$$

OR

$$2\times [MnO_4^- + 2H_2O + 3e^- \longrightarrow MnO_2 + 4OH^-] \text{ (Reduction half-reaction)}$$
$$2NH_3 + 6OH^- \longrightarrow N_2 + 6H_2O + 6e^- \text{ (Oxidation half-reaction)}$$
$$2MnO_4^- + 2NH_3 \longrightarrow 2H_2O + 2OH^- + N_2 + 2MnO_2$$

(iii) Nitrotoluene to nitrobenzoic acid

$$2KMnO_4 + H_2O \longrightarrow 2MnO_2 + 2KOH + 3O \text{ (Reduction reaction)}$$
$$C_6H_4\begin{cases}CH_3\\NO_2\end{cases} + 3O \longrightarrow C_6H_4\begin{cases}COOH\\NO_2\end{cases} + H_2O \text{ (Oxidation reaction)}$$
$$2KMnO_4 + \underset{\text{Nitrotoluene}}{C_6H_4\begin{cases}CH_3\\NO_2\end{cases}} \longrightarrow 2MnO_2 + 2KOH + \underset{\text{Nitrobenzoic acid}}{C_6H_4\begin{cases}COOH\\NO_2\end{cases}}$$

OR

$$2 \times [MnO_4^- + 2H_2O + 3e^- \longrightarrow MnO_2 + 4OH^-] \text{ (Reduction reaction)}$$
$$C_6H_4\begin{cases}CH_3\\NO_2\end{cases} + 6OH^- \longrightarrow C_6H_4\begin{cases}COOH\\NO_2\end{cases} + 4H_2O + 6e^- \text{ (Oxidation reaction)}$$
$$2MnO_4^- + C_6H_4\begin{cases}CH_3\\NO_2\end{cases} \longrightarrow MnO_2 + C_6H_4\begin{cases}COOH\\NO_2\end{cases} + 2OH^-$$

(iv) Ethylene $(H_2C{=}CH_2)$ *to ethylene glycol* $(HO{-}CH_2{-}CH_2{-}OH)$

$$2KMnO_4 + H_2O \longrightarrow 2MnO_2 + 2KOH + 3O \text{ (Reduction reaction)}$$
$$3 \times [H_2C{=}CH_2 + O + H_2O \longrightarrow HO{-}CH_2{-}CH_2{-}OH] \text{ (Oxidation reaction)}$$
$$2KMnO_4 + 4H_2O + 3H_2C{=}CH_2 \longrightarrow 2MnO_2 + 2KOH + 3HO{-}CH_2{-}CH_2{-}OH$$

OR

$$2\times [MnO_4^- + 2H_2O + 3e^- \longrightarrow MnO_2 + 4OH^-] \text{ (Reduction reaction)}$$
$$3 \times [H_2C{=}CH_2 + 2OH^- \longrightarrow HO{-}CH_2{-}CH_2{-}OH + 2e^-] \text{ (Oxidation reaction)}$$
$$2MnO_4^- + 4H_2O + 3H_2C{=}CH_2 \longrightarrow 2MnO_2 + 2OH^- + 3HO{-}CH_2{-}CH_2{-}OH$$

(b) **In neutral solution.** In neutral solution, $KMnO_4$ is directly reduced to MnO_2.

$$2KMnO_4 + H_2O \longrightarrow 2KOH + 2MnO_2 + 3O$$

or $$2MnO_4^- + H_2O \longrightarrow 2OH^- + 2MnO_2 + 3O$$

or $$MnO_4^- + 2H_2O + 3e^- \longrightarrow MnO_2 + 4OH^-$$

This equation is the same as for alkaline medium. Evidently, *in neutral medium as well, the equivalent weight of* $KMnO_4$ *is one-third of its molecular weight.* It is obvious from the equation representing the oxidising action of $KMnO_4$ in neutral medium that an alkali is generated in the reaction. This alkali makes the medium alkaline even if we start with the neutral solution. It is for this reason that the equations representing the oxidising property of $KMnO_4$ in alkaline and neutral media are the same. Thus in neutral medium, $KMnO_4$ oxidises.

(*i*) *Manganous salt (e.g. $MnSO_4$) to insoluble MnO_2* ($Mn^{2+} \longrightarrow Mn^{4+}O_2$)

$$2KMnO_4 + H_2O \longrightarrow 2KOH + 2MnO_2 + 3O$$
$$[MnSO_4 + O + H_2O \longrightarrow MnO_2 + H_2SO_4] \times 3$$
$$2KOH + H_2SO_4 \longrightarrow K_2SO_4 + 2H_2O$$
$$2KMnO_4 + 4H_2O + 3MnSO_4 + H_2SO_4 \longrightarrow 5MnO_2 + 3H_2SO_4 + K_2SO_4 + 2H_2O$$
or
$$2KMnO_4 + 2H_2O + 3MnSO_4 \longrightarrow 5MnO_2 + 2H_2SO_4 + K_2SO_4$$

OR

$$[MnO_4^- + 2H_2O + 3e^- \longrightarrow MnO_2 + 4OH^-] \times 2 \text{ (Reduction reaction)}$$
$$[Mn^{2+} + 2H_2O \longrightarrow MnO_2 + 4H^+ + 2e^-] \times 3 \text{ (Oxidation reaction)}$$
$$2MnO_4^- + 10H_2O + 3Mn^{2+} \longrightarrow 5MnO_2 + 8OH^- + 12H^+$$
or
$$2MnO_4^- + 10H_2O + 3Mn^{2+} \longrightarrow 5MnO_2 + 8H_2O + 4H^+$$
or
$$2MnO_4^- + 2H_2O + 3Mn^{2+} \longrightarrow MnO_2 + 4H^+$$

(*ii*) *Sodium thiosulphate, $Na_2S_2O_3$ (S = +2) to Na_2SO_4 (S = +6)* [$S_2O_3^{2-} \rightarrow SO_4^{2-}$]

$$[2KMnO_4 + H_2O \longrightarrow 2KOH + 2MnO_2 + 3O] \times 4 \text{ (Reduction reaction)}$$
$$[Na_2S_2O_3 + 2KOH + 4O \longrightarrow Na_2SO_4 + K_2SO_4 + H_2O] \times 3 \text{ (Oxidation reaction)}$$
$$8KMnO_4 + H_2O + 3Na_2S_2O_3 \longrightarrow 8MnO_2 + 2KOH + 3Na_2SO_4 + 3K_2SO_4$$

OR

$$[MnO_4^- + 2H_2O + 3e^- \longrightarrow MnO_2 + 4OH^-] \times 8$$
$$[5H_2O + S_2O_3^{2-} \longrightarrow 2SO_4^{2-} + 10H^+ + 8e^-] \times 3$$
$$8MnO_4^- + 16H_2O + 15H_2O + 3S_2O_3^{2-} \longrightarrow 8MnO_2 + 32OH^- + 6SO_4^{2-} + 30H^+$$
or
$$8MnO_4^- + 31H_2O + 3S_2O_3^{2-} \longrightarrow 8MnO_2 + 30H_2O + 2OH^- + 6SO_4^{2-}$$
or
$$8MnO_4^- + H_2O + 3S_2O_3^{2-} \longrightarrow 8MnO_2 + 2OH^- + 6SO_4^{2-}$$

(*iii*) *Nitrogen dioxide, NO_2 (N = +4) to HNO_3 (N = +5)* [$NO_2 \rightarrow NO_3^-$]

$$2KMnO_4 + H_2O \longrightarrow 2KOH + 2MnO_2 + 3O \text{ (Reduction reaction)}$$
$$2NO_2 + H_2O + O \longrightarrow 2HNO_3] \times 3 \text{ (Oxidation reaction)}$$
$$2KMnO_4 + 4H_2O + 6NO_2 \longrightarrow 2KOH + 2MnO_2 + 2MnO_2 + 6HNO_3$$

OR

$$MnO_4^- + 2H_2O + 3e^- \longrightarrow MnO_2 + 4OH^- \text{ (Reduction reaction)}$$
$$[NO_2 + H_2O \longrightarrow NO_3^- + 2H^+ + e^-] \times 3 \text{ (Oxidation reaction)}$$
$$MnO_4^- + 5H_2O + 3NO_2 \longrightarrow MnO_2 + 4OH^- + 3NO_3^- + 6H^+$$
or
$$MnO_4^- + 5H_2O + 3NO_2 \longrightarrow MnO_2 + 4H_2O + 3NO_3^- + 2H^+$$
or
$$MnO_4^- + H_2O + 3NO_2 \longrightarrow MnO_2 + 3NO_3^- + 2H^+$$

(*c*) **In acidic medium.** In acidic medium (dil. H_2SO_4), MnO_4^- ion is reduced to Mn^{2+} ion and five O-atoms are rendered available from two MnO_4^- ions.

$$2KMnO_4 + 3H_2SO_4\ (Dilute) \longrightarrow K_2SO_4 + 2MnSO_4 + 3H_2O + 5O$$

or

$$\underset{\substack{\text{Purple} \\ (Mn\ =\ +7)}}{2MnO_4^-} + 6H^+ \longrightarrow \underset{\substack{\text{Colourless} \\ (Mn\ =\ +2)}}{2Mn^{2+}} + 3H_2O + 5O$$

The reduction of MnO_4^- ion into Mn^{2+} ion as shown above can also be represented by the following ionic equation :

$$\underset{\substack{\text{Purple} \\ (Mn\ =\ +7)}}{MnO_4^-} + 8H^+ + 5e^- \longrightarrow \underset{\substack{\text{Colourless} \\ (Mn\ =\ +2)}}{Mn^{2+}} + 4H_2O$$

It is obvious that *in acid solution the equivalent weight of* $KMnO_4$ *is one-fifth of its molecular weight.*

In acidic medium $KMnO_4$ oxidises :

(*i*) *KI to* I_2 ($I^- \rightarrow I_2$)

$$2KMnO_4 + 3H_2SO_4 \longrightarrow K_2SO_4 + 2MnSO_4 + 3H_2O + 5O$$
$$[2KI + H_2SO_4 + O \longrightarrow K_2SO_4 + H_2O + I_2] \times 5$$
$$\overline{\underset{O.A.}{2KMnO_4} + \underset{R.A.}{10KI} + 8H_2SO_4 \longrightarrow 2MnSO_4 + \underset{R.A.}{5I_2} + \underset{O.A.}{8H_2O} + 6K_2SO_4}$$

OR

$$2 \times [MnO_4^- + 8H^+ + 5e^- \longrightarrow Mn^{2+} + 4H_2O]\ (Reduction\ half\text{-}reaction)$$
$$5 \times [2I^- \longrightarrow I_2 + 2e^-]\ (Oxidation\ half\text{-}reaction)$$
$$\overline{\underset{O.A.}{2MnO_4^-} + 16H^+ + \underset{R.A.}{10I^-} \longrightarrow \underset{R.A.}{2Mn^{2+}} + 8H_2O + \underset{O.A.}{5I_2}}$$

(*ii*) *Ferrous salts to ferric salts* ($Fe^{2+} \rightarrow Fe^{3+}$)

$$2KMnO_4 + 3H_2SO_4 \longrightarrow K_2SO_4 + 2MnSO_4 + 3H_2O + 5O$$
$$[2FeSO_4 + H_2SO_4 + O \longrightarrow Fe_2(SO_4)_3 + H_2O] \times 5$$
$$\overline{\underset{O.A.}{2KMnO_4} + 8H_2SO_4 + \underset{R.A.}{10FeSO_4} \longrightarrow K_2SO_4 + \underset{R.A.}{2MnSO_4} + \underset{O.A.}{5Fe_2(SO_4)_3} + 8H_2O}$$

OR

$$MnO_4^- + 8H^+ + 5e^- \longrightarrow Mn^{2+} + 4H_2O\ (Reduction\ half\text{-}reaction)$$
$$5 \times [Fe^{2+} \longrightarrow Fe^{3+} + e^-]\ (Oxidation\ half\text{-}reaction)$$
$$\overline{\underset{\substack{\text{Purple} \\ O.A.}}{MnO_4^-} + 8H^+ + \underset{R.A.}{5Fe^{2+}} \longrightarrow \underset{\substack{\text{Colourless} \\ R.A.}}{Mn^{2+}} + \underset{O.A.}{5Fe^{3+}} + 4H_2O}$$

The above reaction has been utilised to determine the strength of a given ferrous salt solution (*e.g.*, ferrous ammonium sulphate solution) by titrating its known volume with pre-standardised $KMnO_4$ solution. In this titration $KMnO_4$ solution acts both as an intermediate solution and as a self indicator. When all MnO_4^- ions (purple) have been reduced to Mn^{2+} ions (colourless) by Fe^{2+} ions, the mixture solution becomes colourless due to the production of colourless Mn^{2+} ions. Now if more quantity of $KMnO_4$ solution is added to Fe^{2+} ions solution, the colour of the reaction mixture becomes pink due to the pink colour of the added $KMnO_4$ solution. This stage indicates end point, *i.e.*, end point is indicated by the production of pink colour.

(iii) Oxalic acid ($H_2C_2O_4$) or oxalate ($C_2O_4^{2-}$) to CO_2

$$2KMnO_4 + 3H_2SO_4 \longrightarrow K_2SO_4 + 2MnSO_4 + 3H_2O + 5O$$
$$[H_2C_2O_4 + O \longrightarrow 2CO_2 + H_2O] \times 5$$
$$\underset{O.A.}{2KMnO_4} + 3H_2SO_4 + \underset{R.A.\ (C = +3)}{5H_2C_2O_4} \longrightarrow K_2SO_4 + \underset{R.A.}{2MnSO_4} + 8H_2O + \underset{O.A.\ (C = +4)}{10CO_2}$$

OR

$$2 \times [MnO_4^- + 8H^+ + 5e^- \longrightarrow Mn^{2+} + 4H_2O] \text{ (Reduction half-reaction)}$$
$$5 \times [C_2O_4^{2-} \longrightarrow 2CO_2 + 2e^-] \text{ (Oxidation half-reaction)}$$
$$\underset{\substack{O.A.\\ Purple}}{2MnO_4^-} + 16H^+ + \underset{R.A.}{5C_2O_4^{2-}} \longrightarrow \underset{\substack{R.A.\\ Colourless}}{2Mn^{2+}} + 8H_2O + \underset{O.A.}{10CO_2}$$

The above reaction has been used to determine the strength of a given solution of oxalic acid or an oxalate by titrating its known volume with pre-standardised $KMnO_4$ solution. $KMnO_4$ acts both as an intermediate solution and as a self indicator.

The above reaction has also been used to detect $C_2O_4^{2-}$ ion when present alone or present along with F^- ion in a given inorganic mixture. The procedure is that Na_2CO_3 extract is acidified with CH_3COOH and then $CaCl_2$ solution is added so that a mixture of CaC_2O_4 and CaF_2 (both are white precipitates) is obtained.

$$Na_2C_2O_4 + CaCl_2 \longrightarrow \underset{\text{white ppt}}{CaC_2O_4} + 2NaCl$$
$$NaF + CaCl_2 \longrightarrow \underset{\text{white ppt}}{CaF_2} + 2NaCl$$

Both CaC_2O_4 and CaF_2 are sparingly soluble in CH_3COOH. This mixture is put into dil. H_2SO_4 when CaC_2O_4 is converted into oxalic acid ($H_2C_2O_4$), but CaF_2 remains as such. Now when a few drops of $KMnO_4$ solution (purple) are added to the solution containing $H_2C_2O_4$, its purple colour gets discharged due to the formation of colourless Mn^{2+} ions. These ions are formed by the reduction of MnO_4^- ions (purple) to Mn^{2+} ions (colourless) by $C_2O_4^{2-}$ ions as shown above.

(iv) H_2S to S ($S^{2-} \rightarrow S^0$)

$$2KMnO_4 + 3H_2SO_4 \longrightarrow K_2SO_4 + 2MnSO_4 + 3H_2O + 5O$$
$$[H_2S + O \longrightarrow H_2O + S] \times 5$$
$$\underset{O.A.}{2KMnO_4} + 3H_2SO_4 + \underset{R.A.}{5H_2S} \longrightarrow 2MnSO_4 + 8H_2O + 5S + K_2SO_4$$

OR

$$2 \times [MnO_4^- + 8H^+ + 5e^- \longrightarrow Mn^{2+} + 4H_2O] \text{ (Reduction half-reaction)}$$
$$5 \times [S^{2-} \longrightarrow S^0 + 2e^-] \text{ (Oxidation half-reaction)}$$
$$\underset{O.A.}{2MnO_4^-} + 16H^+ + \underset{R.A.}{5S^{2-}} \longrightarrow 2Mn^{2+} + 8H_2O + 5S^0\downarrow$$

(v) *Nitrite to nitrate* ($NO_2^- \rightarrow NO_3^-$)

$$2KMnO_4 + 3H_2SO_4 \longrightarrow K_2SO_4 + 2MnSO_4 + 3H_2O + 5O$$
$$[KNO_2 + O \longrightarrow KNO_3] \times 5$$

$$\underset{O.A.}{2KMnO_4} + 3H_2SO_4 + \underset{R.A.}{5KNO_2} \longrightarrow K_2SO_4 + 2MnSO_4 + 3H_2O + 5KNO_3$$

OR

$$2 \times [MnO_4^- + 8H^+ + 5e^- \longrightarrow Mn^{2+} + 4H_2O] \text{ (Reduction reaction)}$$
$$5 \times [NO_2^- + H_2O \longrightarrow NO_3^- + 2H^+ + 2e^-] \text{ (Oxidation reaction)}$$

$$\underset{O.A.}{2MnO_4^-} + 6H^+ + \underset{R.A.}{5NO_2^-} \longrightarrow 2Mn^{2+} + 3H_2O + 5NO_3^-$$

(vi) *Arsenite,* AsO_3^{3-} *(As = +3) to arsenate,* AsO_4^{3-} *(As = +5)*

$$2KMnO_4 + 3H_2SO_4 \longrightarrow K_2SO_4 + 2MnSO_4 + 3H_2O + 5O$$
$$[Na_3AsO_3 + O \longrightarrow Na_3AsO_4] \times 5$$

$$2KMnO_4 + 3H_2SO_4 + 5Na_3AsO_3 \longrightarrow K_2SO_4 + 2MnSO_4 + 3H_2O + 5Na_3AsO_4$$

OR

$$[MnO_4^- + 8H^+ + 5e^- \longrightarrow Mn^{2+} + 4H_2O] \times 2 \text{ (Reduction reaction)}$$
$$[AsO_3^{3-} + H_2O \longrightarrow AsO_4^{3-} + 2H^+ + 2e^-] \times 5 \text{ (Oxidation reaction)}$$

$$2MnO_4^- + 6H^+ + 5AsO_3^{3-} \longrightarrow 2Mn^{2+} + 3H_2O + 5AsO_4^{3-}$$

(vii) *Sulphite,* SO_3^{2-} *(S = +4) to sulphate,* SO_4^{2-} *(S = +6)*

$$2KMnO_4 + 3H_2SO_4 \longrightarrow K_2SO_4 + 2MnSO_4 + 3H_2O + 5O$$
$$[Na_2SO_3 + O \longrightarrow Na_2SO_4] \times 5$$

$$\underset{O.A.}{2KMnO_4} + 3H_2SO_4 + \underset{R.A.(S = +4)}{5Na_2SO_3} \longrightarrow K_2SO_4 + 2MnSO_4 + 3H_2O + \underset{(S = +6)}{5Na_2SO_4}$$

OR

$$[MnO_4^- + 8H^+ + 5e^- \longrightarrow Mn^{2+} + 4H_2O] \times 2 \text{ (Reduction reaction)}$$
$$[SO_3^{2-} + H_2O \longrightarrow SO_4^{2-} + 2H^+ + 2e^-] \times 5 \text{ (Oxidation reaction)}$$

$$2MnO_4^- + 6H^+ + 5SO_3^{2-} \longrightarrow 2Mn^{2+} + 3H_2O + 5SO_4^{2-}$$

(viii) *Halogen acid, HX (X = –1) to the corresponding halogen,* X_2 *(X = 0)* ($X^- \rightarrow X_2$)

$$2KMnO_4 + 3H_2SO_4 \longrightarrow K_2SO_4 + 2MnSO_4 + 3H_2O + 5O$$
$$[2HX + O \longrightarrow H_2O + X_2] \times 5$$

$$\underset{O.A.}{2KMnO_4} + 3H_2SO_4 + \underset{\substack{R.A.\\(X = -1)}}{10HX} \longrightarrow K_2SO_4 + 2MnSO_4 + 8H_2O + \underset{(X = 0)}{5X_2}$$

OR

$$2 \times [MnO_4^- + 8H^+ + 5e^- \longrightarrow Mn^{2+} + 4H_2O] \text{ (Reduction half-reaction)}$$
$$5 \times [2X^- \longrightarrow X_2 + 2e^-] \text{ (Oxidation reaction)}$$

$$2MnO_4^- + 16H^+ + 10X^- \longrightarrow 2Mn^{2+} + 8H_2O + 5X_2$$

The oxidation of HCl by $KMnO_4$ has been utilised for the preparation of Cl_2 gas. $KMnO_4$ alone can also oxidise conc. HCl to Cl_2.

$$2KMnO_4 + 16HCl \longrightarrow 2MnCl_2 + 2KCl + 8H_2O + 5Cl_2$$

(ix) H_2O_2 (O = –1) to O_2 (O = 0)

$$2KMnO_4 + 3H_2SO_4 \longrightarrow K_2SO_4 + 2MnSO_4 + 3H_2O + 5O$$
$$[H_2O_2 + O \longrightarrow H_2O + O_2] \times 5$$

$$\underset{\text{O.A.}}{2KMnO_4} + 3H_2SO_4 + \underset{\substack{\text{R.A.}\\(\text{O.} = -1)}}{5H_2O_2} \longrightarrow K_2SO_4 + 2MnSO_4 + 8H_2O + \underset{(\text{O} = 0)}{5O_2}$$

OR

$$2 \times [MnO_4^- + 8H^+ + 5e^- \longrightarrow Mn^{2+} + 4H_2O] \times 2 \text{ (Reduction half-reaction)}$$
$$5 \times [H_2O_2 \longrightarrow O_2 + 2H^+ + 2e^-] \text{ (Oxidation half-reaction)}$$

$$2MnO_4^- + 6H^+ + 5H_2O_2 \longrightarrow 2Mn^{2+} + 8H_2O + 5O_2$$

(x) SO_2 (S = +4) to H_2SO_4 (S = +6) ($SO_2 \rightarrow SO_4^{2-}$)

$$KMnO_4 + 3H_2SO_4 \longrightarrow K_2SO_4 + 2MnSO_4 + 3H_2O + 5O$$
$$[SO_2 + H_2O + O \longrightarrow H_2SO_4] \times 5$$

$$2KMnO_4 + 5SO_2 + 2H_2O \longrightarrow K_2SO_4 + 2MnSO_4 + 2H_2SO_4$$

OR

$$2 \times [MnO_4^- + 8H^+ + 5e^- \longrightarrow Mn^{2+} + 4H_2O] \text{ (Reduction reaction)}$$
$$5 \times [SO_2 + 2H_2O \longrightarrow SO_4^{2-} + 4H^+ + 2e^-] \text{ (Oxidation reaction)}$$

$$2MnO_4^- + 5SO_2 + 2H_2O \longrightarrow 2Mn^{2+} + 5SO_4^{2-} + 4H^+$$

(xi) Hydrazine, N_2H_4 (N = –2) to N_2 (N = 0)

$$[2KMnO_4 + 3H_2SO_4 \longrightarrow K_2SO_4 + 2MnSO_4 + 3H_2O + 5O] \times 2$$
$$[N_2H_4 + 2O \longrightarrow N_2 + 2H_2O] \times 5$$

$$4KMnO_4 + 6H_2SO_4 + 5N_2H_4 \longrightarrow 2K_2SO_4 + 4MnSO_4 + 16H_2O + 5N_2$$

OR

$$4 \times [MnO_4^- + 8H^+ + 5e^- \longrightarrow Mn^{2+} + 4H_2O] \text{ (Reduction reaction)}$$
$$5 \times [N_2H_4 \longrightarrow N_2 + 4H^+ + 4e^-] \text{ (Oxidation reaction)}$$

$$4MnO_4^- + 12H^+ + 5N_2H_2 \longrightarrow 4Mn^{2+} + 16H_2O + 5N_2$$

(xii) Hydrazoic acid, HN_3 (N = –1/3) to N_2 (N = 0)

$$2KMnO_4 + 3H_2SO_4 \longrightarrow K_2SO_4 + 2MnSO_4 + 3H_2O + 5O$$
$$2HN_3 + O \longrightarrow 3N_2 + H_2O] \times 5$$

$$2KMnO_4 + 3H_2SO_4 + 10HN_3 \longrightarrow K_2SO_4 + 2MnSO_4 + 8H_2O + 15N_2$$

OR

$$2 \times [MnO_4^- + 8H^+ + 5e^- \longrightarrow Mn^{2+} + 4H_2O] \text{ (Reduction reaction)}$$
$$5 \times [2HN_3 \longrightarrow 3N_2 + 2H^+ + 2e^-] \text{ (Oxidation reaction)}$$

$$2MnO_4^- + 6H^+ + 10HN_3 \longrightarrow 2Mn^{2+} + 8H_2O + 15N_2$$

(*xiii*) *Nitric oxide, NO (N = +2) to HNO_3 (N = +5)*

$$[2KMnO_4 + 3H_2SO_4 \longrightarrow K_2SO_4 + 2MnSO_4 + 3H_2O + 5O] \times 3$$
$$[2NO + 3O + H_2O \longrightarrow 2HNO_3] \times 5$$
$$6KMnO_4 + 9H_2SO_4 + 10NO \longrightarrow 4H_2O + 3K_2SO_4 + 6MnSO_4 + 10HNO_3$$

OR

$$3 \times [MnO_4^- + 8H^+ + 5e^- \longrightarrow Mn^{2+} + 4H_2O] \text{ (Reduction reaction)}$$
$$5 \times [NO + 2H_2O \longrightarrow HNO_3 + 3H^+ + 3e^-] \text{ (Oxidation reaction)}$$
$$3MnO_4^- + 9H^+ + 5NO \longrightarrow 2H_2O + 3Mn^{2+} + 5HNO_3$$

(*xiv*) *Potassium ferrocyanide, $K_4[Fe^{2+}(CN)_6]$ to potassium ferricyanide, $K_3[Fe^{3+}(CN)_6]$*

$$2KMnO_4 + 3H_2SO_4 \longrightarrow K_2SO_4 + 2MnSO_4 + 3H_2O + 5O$$
$$[2K_4[Fe^{2+}(CN)_6] + O + H_2O \longrightarrow 2K_3[Fe^{3+}(CN)_6] + 2KOH] \times 5$$
$$2KMnO_4 + 3H_2SO_4 + 10K_4[Fe^{2+}(CN)_6] \longrightarrow 2H_2O + K_2SO_4 + 2MnSO_4 + 10K_3[Fe^{3+}(CN)_6] + 10KOH$$

OR

$$[MnO_4^- + 8H^+ + 5e^- \longrightarrow Mn^{2+} + 4H_2O] \text{ (Reduction reaction)}$$
$$5 \times [(Fe^{2+}(CN)_6]^{4-} \longrightarrow [Fe^{3+}(CN)_6]^{3-} + e^-] \text{ (Oxidation reaction)}$$
$$MnO_4^- + 8H^+ + 5[Fe^{2+}(CN)_6]^{4-} \longrightarrow Mn^{2+} + 4H_2O + 5[Fe^{3+}(CN)_6]^{3-}$$

(*xv*) *Sod. thiosulphate, $Na_2S_2O_3$ (S = +2) to sod. dithionate, $Na_2S_2O_6$ (S = +5)*

$$[2KMnO_4 + 3H_2SO_4 \longrightarrow K_2SO_4 + 2MnSO_4 + 3H_2O + 5O] \times 3$$
$$[Na_2S_2O_3 + 3O \longrightarrow Na_2S_2O_6] \times 5$$
$$6KMnO_4 + 6H_2SO_4 + 5Na_2S_2O_3 \longrightarrow 3K_2SO_4 + 6MnSO_4 + 9H_2O + 5Na_2S_2O_6$$

OR

$$6 \times [MnO_4^- + 8H^+ + 5e^- \longrightarrow Mn^{2+} + 4H_2O] \text{ (Reduction reaction)}$$
$$5 \times [S_2O_3^{2-} + 3H_2O \longrightarrow S_2O_6^{2-} + 6H^+ + 6e^-] \text{ (Oxidation reaction)}$$
$$6MnO_4^- + 18H^+ + 5S_2O_3^{2-} \longrightarrow 9H_2O + 6Mn^{2+} + 5S_2O_6^{2-}$$

$KMnO_4$ reacts with H_2SiF_6 and gives $HMnO_4$

$$2KMnO_4 + H_2SiF_6 \longrightarrow K_2SiF_6 + 2HMnO_4$$

Uses. $KMnO_4$ is used: (*i*) as an oxidising agent in the laboratory as well as in industry; (*ii*) in volumetric estimations of ferrous salts, oxalic acid, H_2O_2 etc.; (*iii*) as a disinfectant and a germicide on account of its oxidising property; (*iv*) in qualitative analysis for detecting halides and oxalates; (*v*) in the manufacture of saccharin, benzoic acid, acetaldelyde etc.; (*vi*) for bleaching waxes; (*vii*) for the detection of double bond in organic compounds. For this purpose alkaline $KMnO_4$ solution under the name *Bayer's reagent* is used. (*viii*) as a source of reactive form of MnO_2 which is used in place of pyrolusite in the manufacture of dry cells.

Compounds of Iron

Introduction

Iron forms compounds in –2, 0, +1, +2, +3, +4, +5 and +6 oxidation states, but the compounds having iron in +2 (*ferrous*) and +3 (*ferric*) oxidation states are the most stable. Examples of compounds formed by iron in different oxidation states are given below :

−2 ⟶ $[Fe(CO)_4]^{2-}$

Zero ⟶ $Fe(CO)_5$, $Fe_2(CO)_9$, $Fe_3(CO)_{12}$

+1 ⟶ $[Fe(H_2O)_5(NO)]SO_4$

+2 ⟶ FeO, $Fe(OH)_2$, $FeCl_2$, FeS, $FeSO_4.7H_2O$, $(NH_4)_2SO_4.FeSO_4.6H_2O$, $[Fe(NH_3)_6]Cl_2$, $K_4[Fe(CN)_6].3H_2O$, $Na_2[Fe(CN)_5(NO)].2H_2O$, etc.

+3 ⟶ Fe_2O_3, $Fe(OH)_3$, FeF_3, $FeCl_3$, Fe_2S_3, $Fe_2(SO_4)_3$, $(NH_4)_2SO_4.Fe_2(SO_4)_3.24H_2O$, $K_2SO_4.Fe_2(SO_4)_3.24H_2O$, $[FeCl_4]^-$, $K_3[Fe(CN)_6]$, $H_3[Fe(CN)_6]$, $Fe(CNS)_3$, $[Fe(CNS)_6]^{3-}$

+4 ⟶ FeS_2, $[Fe(diars)_2Cl_2]^{2+}$

+5 ⟶ $[FeO_4]^{3-}$

+6 ⟶ $[FeO_4]^{2-}$

Compounds containing both Fe^{2+} and Fe^{3+} ions are also known. Examples are *ferroso-ferric oxide*, Fe_3O_4 or $FeO.Fe_2O_3$ (also called *magnetic oxide of iron*), *prussian blue and turnbull's blue.*

Compounds of Fe (+1)

Nitrosonium pentaquo iron (I) sulphate, $[Fe^+(NO^+)(H_2O)_5]SO_4$ (also called nitroso ferrous sulphate) is an important example of the complex compound containing iron in +1 oxidtion state. Five H_2O molecules and one NO^+ ion act as ligands. It is an octahedral complex. Its preparation and properties have been given in chapter 14. The structure of $[Fe(NO)(H_2O)_5]^{2+}$ ion has been discussed in chapter 5.

Compounds of Fe(+2) : Ferrous Compounds

Compounds of Fe (+2) give Fe^{2+} ions (ferrous ions) in aqueous solution. This ion is pale green. All Fe(+2) salts are, therefore, pale green in colour. In aqueous solution, Fe^{2+} ion can readily be oxidised to Fe^{3+} ion even by the atmospheric O_2.

$$4Fe^{2+} + O_2 + 4H^+ \longrightarrow 4Fe^{3+} + 2H_2O$$

It is due to this property that Fe^{2+} ion is a strong reducing agent. Thus this ion reduces Hg^{2+} to Hg_2^{2+}, Sn^{4+} to Sn^{2+}, Au^{3+} to Au, acidified solution of MnO_4^- to Mn^{2+}, $Cr_2O_7^{2-}$ to Cr^{3+}, HNO_3 to NO etc.

$$2Hg^{2+} + 2Fe^{2+} \longrightarrow Hg_2^{2+} + 2Fe^{3+}$$

$$Sn^{4+} + 2Fe^{2+} \longrightarrow Sn^{2+} + 2Fe^{3+}$$

$$Au^{3+} + 3Fe^{2+} \longrightarrow Au + 3Fe^{3+}$$

$$MnO_4^- + 8H^+ + 5Fe^{2+} \longrightarrow Mn^{2+} + 7H_2O + 6Fe^{3+}$$

$$Cr_2O_7^{2-} + 14H^+ + 6Fe^{2+} \longrightarrow 2Cr^{3+} + 7H_2O + 3Fe^{3+}$$

$$NO_3^- + 4H^+ + 3Fe^{2+} \longrightarrow NO + 2H_2O + 3Fe^{3+}$$

Examples of salts containing Fe in +2 oxidation state are FeO, $Fe(OH)_2$, $FeCl_2$, FeS, $FeSO_4.7H_2O$ (green vitriol), $(NH_4)_2SO_4.FeSO_4.6H_2O$ (Mohr's salt) etc.

Complex Compounds of Fe (+2)

Iron in +2 oxidation state forms several complex compounds. Some examples are

$[Fe(NH_3)_6]Cl_2$, $Na_2[Fe(CN)_5(NO)]$, $K_4[Fe(CN)_6]$, $Ag_4[Fe(CN)_6]$, $Cu_2[Fe(CN)_6]$, $Zn_2[Fe(CN)_6]$, $Ca_2[Fe(CN)_6]$, $K_2Fe^{2+}[Fe^{2+}(CN)_6]$ etc. Out of these $[Fe(NH_3)_6]Cl_2$, $Na_2[Fe(CN)_5(NO)]$, $K_4[Fe(CN)_6$ and $K_2Fe[Fe(CN)_6]$ are the most important and are discussed below :

1. Hexammine ferrate (II) chloride, $[Fe(NH_3)_6]Cl_2$

$FeCl_2$ combines with NH_3, giving $[Fe(NH_3)_6]Cl_2$.

$$FeCl_2 + 6NH_3 \longrightarrow [Fe(NH_3)_6]Cl_2$$

$[Fe(NH_3)_6]^{2-}$ ion has octahedral geometry.

2. Sodium pentacyano nitrosyl ferrate (II), $Na_2[Fe^{2+}(CN)_5(NO^+)].2H_2O$ (also called sod. nitroprusside)

In this complex compound five CN^- ions and one nitrosyl ion (NO^+) act as ligands. It is also an octahedral complex. Preparation and properties of this compound are given in chapter 14 while the structure of $[Fe(CN)_5(NO)]^{2-}$ ion has been explained by "*Valence Bond Theory*" of complex compounds in chapter 5.

3. Potassium ferrocyanide or potassium hexacyano ferrate (II), $K_4[Fe(CN)_6]$

Preparation. (*i*) *Laboratory preparation.* In laboratory $K_4[Fe(CN)_6]$ is obtained by adding KCN to $FeSO_4$ (green vitriol) solution when a precipitate of $Fe(CN)_2$ is first formed and then gets dissolved in the excess of KCN solution and forms $K_4Fe(CN)_6$.

$$FeSO_4 + 2KCN \longrightarrow Fe(CN)_2\downarrow + K_2SO_4$$

$$Fe(CN)_2 + 4KCN \longrightarrow F_4Fe(CN)_6$$

The solution on concentration gives the lemon yellow crystals of $K_4[Fe(CN)_6]$.

(*ii*) When $K_3[Fe(CN)_6]$ is reduced by H_2O_2 in alkaline medium, $K_4[Fe(CN)_6]$ is obtained. In this reaction H_2O_2 is oxidised to O_2.

$$2K_3[Fe(CN)_6] + 2KOH + H_2O_2 \longrightarrow 2K_4[Fe(CN)_6] + 2H_2O + O_2$$

or $$2[Fe(CN)_6]^{3-} + 2OH^- + H_2O_2 \longrightarrow 2[Fe(CN)_6]^{4-} + 2H_2O + O_2$$

(*iii*) *Commercial method: From crude coal gas.* On large scale, $K_4[Fe(CN)_6]$ is obtained from crude gas. This gas contains 0.3% of HCN as impurity. The following two methods are used:

(*a*) Crude coal gas is passed over hydrated iron oxide [mixture of $Fe(OH)_2$ and $Fe(OH)_3$], when prussian blue, $Fe_4^{3+}[Fe^{2+}(CN)_6]_3$ is obtained.

$$Fe(OH)_2 + 2HCN \longrightarrow Fe(CN)_2 + 2H_2O$$

$$Fe(OH)_3 + 3HCN \longrightarrow Fe(CN)_3 + 3H_2O$$

$$3Fe(CN)_2 + 4Fe(CN)_3 \longrightarrow \underset{\text{Prussian blue}}{Fe_4^{3+}[Fe^{2+}(CN)_6]_3}$$

$Fe_4[Fe(CN)_6]_3$, when heated with $Ca(OH)_2$, gives calcium ferrocyanide, $Ca_2[Fe^{2+}(CN)_6]$

$$Fe_4^{3+}[Fe^{2+}(CN)_6]_3 + 6Ca(OH)_2 \longrightarrow 3Ca_2[Fe^{2+}(CN)_6] + \underset{\text{ppt}}{4Fe(OH)_3}$$

$Ca_2[Fe(CN)_6]$ is extracted with H_2O and then treated with K_2CO_3 solution when $K_4[Fe(CN)_6]$ is obtained.

$$Ca[Fe(CN)_6] + 2K_2CO_3 \longrightarrow K_4[Fe(CN)_6] + \underset{\text{ppt}}{2CaCO_3}$$

$CaCO_3$ is removed by filtration and the filtrate is evaporated to get the crystals of $K_4[Fe(CN)_6]$.

(*b*) Coal gas, when passed through a washer containing KOH and $FeSO_4$, gives $K_4Fe(CN)_6$.

$$[HCN + KOH \longrightarrow KCN + H_2O] \times 6$$
$$6KCN + FeSO_4 \longrightarrow K_4[Fe(CN)_6] + K_2SO_4$$
$$6HCN + 6KOH + FeSO_4 \longrightarrow K_4[Fe(CN)_6] + K_2SO_4 + 6H_2O$$

Properties. 1. Potassium ferrocyanide forms lemon yellow crystal which are soluble in water. This compound contains $3H_2O$ as water of crystalisation.

2. *Action of heat.* When heated to 42°C, all the three H_2O molecules are lost. On further heating, it gives iron carbide (FeC_2), KCN, N_2. etc.

$$K_4Fe(CN)_6 \xrightarrow{\Delta} FeC_2 + 4KCN + N_2$$
$$K_4Fe(CN)_6 \xrightarrow{\Delta} Fe + 4KCN + (CN)_2$$

3. *Action of H_2SO_4.* (*a*) When $K_4Fe(CN)_6$ is heated with dil.H_2SO_4, hydrocyanic acid (HCN) is produced.

$$K_4Fe(CN)_6 + 3H_2SO_4\ (dil.) \longrightarrow 2K_2SO_4 + FeSO_4 + 6HCN$$

(*b*) On heating with conc. H_2SO_4, CO gas is evolved.

$$K_4Fe(CN)_6 + 6H_2SO_4\ (conc.) + 6H_2O \longrightarrow 2K_2SO_4 + 3(NH_4)_2SO_4 + FeSO_4 + 6CO$$

4. *Action of Zn(II) salts.* With Zn (II) salts [*e.g.* $(CH_3COO)_2Zn$, $ZnSO_4$ etc.] bluish-white ppt. of zinc ferro-cyanide, $Zn_2[Fe^{2+}(CN)_6]$ is obtained, *e.g.*,

$$2(CH_3COO)\ Zn + K_4[Fe(CN)_6] \longrightarrow Zn_2[Fe(CN)_6] + 4CH_3COOK$$
$$2ZnSO_4 + K_4[Fe(CN)_6] \longrightarrow Zn_2[Fe(CN)_6] + 2K_2SO_4$$

or
$$Zn^{2+} + [Fe(CN)_6]^{4-} \longrightarrow \underset{\text{Bluish white ppt}}{Zn_2[Fe^{2+}(CN)_6]}$$

The above reaction has been used in the identification of Zn^{2+} ion present in a given inorganic salt.

5. *Action of Fe(III) salts: Formation of blue precipitate of prussian blue.* See Prussian Blue discussed on page 305 of this chapter.

6. *Action of Fe(II) salts.* When $K_4[Fe(CN)_6]$ solution is added to the solution of a Fe (II) salt, bluish-white ppt. of pot. ferrous ferro hexacyanide, $K_2Fe^{2+}[Fe^{2+}(CN)_6]$ is obtained.

$$K_4Fe^{2+}[(CN)_6] + Fe^{2+}Cl_2 \longrightarrow \underset{\text{Bluish-white ppt.}}{K_2Fe^{2+}[Fe^{2+}(CN)_6]} + 2KCl$$

or
$$[Fe(CN)_6]^{4-} + Fe^{2+} \longrightarrow [Fe\{Fe(CN)_6\}]^{2-}$$

The bluish-white ppt. of $K_2Fe[Fe(CN)_6]$ slowly becomes blue due to its oxidation by air to prussian blue, $KFe^{3+}[Fe^{2+}(CN)_6]$

$$K_2Fe^{2+}[Fe^{2+}(CN)_6] \xrightarrow{\text{Air oxidation}} \underset{\text{Prussian blue}}{KFe^{3+}[Fe^{2+}(CN)_6]} + K$$

7. *Action of Cu(II) salts.* Cu (II) salts (*e.g.*, $CuSO_4$) give a red-brown ppt. of cupric ferro hexacyanide, $Cu_2^{2+}[Fe^{2+}(CN)_6]$.

$$2CuSO_4 + K_4[Fe(CN)_6] \longrightarrow Cu_2[Fe(CN)_6] + 2K_2SO_4$$

or
$$2Cu^{2+} + [Fe(CN)_6]^{4-} \longrightarrow \underset{\text{Red-brown ppt}}{Cu_2^{2+}\ [Fe(CN)_6]}$$

This reaction is used for the detection of Cu^{2+} ions present in an inorganic salt.

8. *Action of Ag (I) salts.* White ppt. of silver ferro cyanide, $Ag_4[Fe(CN)_6]$ is obtained.

$$K_4[Fe(CN)_6] + 4AgNO_3 \longrightarrow Ag_4[Fe(CN)_6] + 4KNO_3$$

or
$$[Fe(CN)_6]^{4-} + 4Ag^+ \longrightarrow \underset{\text{White ppt}}{Ag_4[Fe(CN)_6]}$$

9. *Action of Ca(II) salt + NH_4Cl.* When a Ca (II) salt (*e.g.*, $CaCl_2$) solution containing some NH_4Cl is treated with $K_4[Fe(CN)_6]$, a white ppt. of cal. ammonium pot. ferro-cyanide, $Ca(NH_4)K[Fe(CN)_6]$ is obtained.

$$CaCl_2 + NH_4Cl + K_4[Fe(CN)_6] \longrightarrow 3KCl + \underset{\text{White ppt}}{Ca(NH_4)K[Fe(CN)_6]}$$

or $$Ca^{2+} + NH_4^+ + K^+ + [Fe(CN)_6]^{4-} \longrightarrow Ca(NH_4)K[Fe(CN)_6]$$

This reaction is used in the detection of Ca^{2+} ions present in an inorganic salt.

10. *Action of fused $K_2Cr_2O_7$.* When fused $K_2Cr_2O_7$ reacts with dry $K_4Fe(CN)_6$, potassium cyanate (KCNO) is produced.

$$2K_2Cr_2O_7 + 2K_4[Fe(CN)_6] \longrightarrow 6KCNO + 2K_2O + 2Cr_2O_3 + K_2Fe^+[Fe^{3+}(CN)_6]$$

11. *Action of K_2CO_3 and sulphur.* When $K_4Fe(CN)_6$ is heated with a mixture of (K_2CO_3 + S), potassium thiocyanate (KCNS) is obtained.

$$K_4Fe(CN)_6 + K_2CO_3 + 6S \xrightarrow{\Delta} 6KCNS + CO_2 + FeO$$

12. *Reducing properties: Oxidation of $K_4[Fe^{2+}(CN)_6]$ to $K_3[Fe^{3+}(CN)_6]$.* It is a reducing agent and hence reduces many oxidising agents like Cl_2, O_3, acidified solution of H_2O_2 and $KMnO_4$ to chloride (Cl^-), O_2, and $MnSO_4$ respectively. In all these reactions $K_4[Fe^{2+}(CN)_6]$ is oxidised to $K_3[Fe^{3+}(CN)_6]$.

(*a*) $$2K_4[Fe^{2+}(CN)_6] + Cl_2 \longrightarrow 2KCl + 2K_3[Fe^{3+}(CN)_6]$$

or $$2[Fe^{2+}(CN)_6]^{4-} + Cl_2^0 \longrightarrow 2Cl^- + 2[Fe^{3+}(CN)_6]^{3-}$$

(*b*) $$2K_4[Fe(CN)_6] + O_3 + H_2O \longrightarrow 2K_3[Fe(CN)_6] + 2KOH + O_2$$

or $$2[Fe(CN)_6]^{4-} + O_3 + H_2O \longrightarrow 2[Fe(CN)_6]^{3-} + 2OH^- + O_2$$

(*c*) $$2K_4[Fe(CN)_6] + H_2SO_4 + H_2O_2 \longrightarrow 2K_3[Fe(CN)_6] + K_2SO_4 + 2H_2O$$

or $$2[Fe(CN)_6]^{4-} + 2H^+ + H_2O_2 \longrightarrow 2[Fe(CN)_6]^{3-} + 2H_2O$$

(*d*) $$5K_4[Fe(CN)_6] + 4H_2SO_4 + KMnO_4 \longrightarrow 5K_3[Fe(CN)_6] + MnSO_4 + 4H_2O$$

or $$5[Fe(CN)_6]^{4-} + 8H^+ + MnO_4^- \longrightarrow 5[Fe(CN)_6]^{3-} + Mn^{2+} + 4H_2O$$

(*e*) Electrolytic oxidation of slightly alkaline solution of $K_4Fe(CN)_6$ gives $K_3Fe(CN)_6$

$$2K_4[Fe(CN)_6] + 2H_2O \longrightarrow 2K_3[Fe(CN)_6] + 2KOH + H_2$$

or $$2[Fe(CN)_6]^{4-} + 2H_2O \longrightarrow 2[Fe(CN)_6]^{3-} + 2OH^- + H_2$$

(*f*) When a concentrated solution of $K_4Fe(CN)_6$ is heated with 50% HNO_3, $K_4Fe(CN)_6$ gets oxidised to $K_3Fe(CN)_6$ and HNO_3 is reducted to NO.

$$3K_4[Fe(CN)_6] + 4HNO_3 \longrightarrow 3K_3[Fe(CN)_6] + NO + 3KNO_3 + 2H_2O$$

or $$3[Fe(CN)_6]^{4-} + 4H^+ + NO_3^- \longrightarrow 3[Fe(CN)_6]^{3-} + NO + 2H_2O$$

(*g*) When a solution of $K_4[Fe(CN)_6]$ and KNO_2 is added to dil. CH_3COOH, it is oxidised to $K_3[Fe(CN)_6]$ and KNO_2 is reduced to NO. This reaction is used for the preparation of NO.

$$2K_4[Fe(CN)_6] + KNO_2 + 2CH_3COOH \longrightarrow 2K_3[Fe(CN)_6] + 2CH_3COOK + H_2O + NO$$

or $$[Fe(CN)_6]^{4-} + NO_2^- + 2H^+ \longrightarrow [Fe(CN)_6]^{3-} + NO + H_2O$$

Uses. $K_4Fe(CN)_6$ is used :

(*i*) for the preparation of CO gas

(*ii*) for the preparation of prussian blue pigments

(*iii*) for the manufacture of $K_3[Fe(CN)_6]$ as shown under its reducing properties.

(*iv*) as a reagent in the laboratory for the detection of Cu^{2+}, Fe^{3+}, Zn^{2+} and Ca^{2+} ions present in an inorganic salt.

(v) to distinguish between ferrous and ferric salts.

(vi) for making blue points

Structure of $[Fe(CN)_6]^{4-}$ ion. See "*Valence Bond Theory*" of complex compounds (Chapter 5).

4. Potassium ferrous ferro hexacyanide $K_2Fe^{2+}[Fe^{2+}(CN)_6]$

It is obtained as a bluish-white ppt. by adding Fe(II) salt (*e.g.*, $FeCl_2$) to $K_4[Fe(CN)_6]$

$$FeCl_2 + K_4[Fe(CN)_6] \longrightarrow K_2Fe[Fe(CN)_6] + 2KCl$$

$K_2Fe[Fe(CN)_6]$ is oxidised by air to prussian blue, $KFe^{3+}[Fe^{2+}(CN)_6]$

$$K_2Fe[Fe(CN)_6] \xrightarrow{\text{Air oxidation}} KFe[Fe(CN)_6] + K$$

Compounds of Fe(+3) : Ferric Compounds

Preparation. These compounds are obtained when the corresponding Fe(+2) compounds are oxidised.

Properties. (*i*) *Hydrolysis.* Due to its small size and high charge density, Fe^{3+} ion undergoes hydrolysis to give H^+ ions. The production of H^+ ions makes the aqueous solution of Fe^{3+} ions acidic.

$$Fe^{3+} + H_2O \rightleftharpoons [Fe(OH)]^{2+} + H^+$$

The value of equilibrium constant of the above reaction at ordinary temperature has been found to be equal to 9×10^{-4}.

(*ii*) *Ionisation.* Fe(+3) compounds are much less ionised than Fe(+2) compounds, *e.g.*,

(*a*) $FeCl_3$ is so little ionised in aqueous solution that this solution does not give any test for Fe^{3+} and F^- ions.

(*b*) It is also due to little ionisation of $Fe(OH)_3$ that this compound is less basic than $Fe(OH)_2$.

(*iii*) *Reduction.* Fe^{3+} ion can be reduced to Fe^{2+} ion by reducing agents like H, H_2S, Sn^{2+} etc.

$$Fe^{3+} + H \longrightarrow Fe^{2+} + H^+$$

$$2Fe^{3+} + H_2S \longrightarrow 2Fe^{2+} + 2H^+ + S$$

$$2Fe^{3+} + Sn^{2+} \longrightarrow 2Fe^{2+} + Sn^{4+}$$

Examples of salts containing Fe^{3+} ion are Fe_2O_3, $Fe(OH)_3$, FeF_3, $FeCl_3$, Fe_2S_3, $Fe_2(SO_4)_3$, $(NH_4)_2SO_4.Fe_2(SO_4)_3.24H_2O$ (ammonium iron alum), $K_2SO_4.Fe_2(SO_4)_3.24H_2O$ (potassium iron alum) etc.

Complexes of Fe(+3)

Fe^{3+} forms several complex compounds. Examples are $[FeCl_4]^-$, $[Fe(C_2O_4)_3]^{3-}$, $[Fe(H_2O)_6]^{3+}$, $[FeCl_6]^{3-}$, $[Fe(CN)_6]^{3-}$ etc. Coordination number (C.N.) of Fe^{3+} ion in $[FeCl_4]^-$ ion is 4 and this complex ion has tetrahedral geometry. C.N. of Fe^{3+} ion in other complex compounds is 6 and these ions have octahedral shape with one unpaired electron (d^2sp^3 hybridisation). Thus these are low spin octahedral complex ions. Here we shall discuss $H_3[Fe(CN)_6]$ and $K_3[Fe(CN)_6]$.

1. Hydroferricyanic acid, $H_3[Fe(CN)_6]$

It is obtained by decomprising $Pb_3[Fe(CN)_6]_2$ with dil. H_2SO_4, filtering and evaporating.

$$Pb_3[Fe(CN)_6]_2 + 3H_2SO_4 \longrightarrow 2H_3[Fe(CN)_6] + 3PbSO_4$$

$H_3[Fe(CN)_6]$ forms brown needles.

2. Potassium ferricyanide or potassium hexacyano ferrate (III), $K_3[Fe(CN)_6]$

Preparation. $K_3[Fe(CN)_6]$ may be prepared by oxidising $K_4[Fe(CN)_6]$ to $K_3[Fe(CN)_6]$. The oxidation can be done by the following methods :

(*i*) by the oxidation of $K_4[Fe(CN)_6]$ by Cl_2, O_3 acidified solution of H_2O_2 or $KMnO_4$.

(*a*) $$2K_4[Fe(CN)_6] + Cl_2 \longrightarrow 2KCl + 2K_3[Fe(CN)_6]$$

or $$2[Fe(CN)_6]^{4-} + Cl_2 \longrightarrow 2Cl^- + 2[Fe(CN)_6]^{3-}$$

(*b*) $$2K_4[Fe(CN)_6] + O_3 + H_2O \longrightarrow 2K_3[Fe(CN)_6] + 2KOH + O_2$$

or $$[Fe(CN)_6]^{4-} + O_3 + H_2O \longrightarrow 2[Fe(CN)_6]^{3-} + 2OH^- + O_2$$

(*c*) $$2K_4[Fe(CN)_6] + H_2SO_4 + H_2O_2 \longrightarrow 2K_3[Fe(CN)_6] + K_2SO_4 + 2H_2O$$

or $$2[Fe(CN)_6]^{4-} + 2H^+ + H_2O_2 \longrightarrow 2[Fe(CN)_6]^{3-} + 2H_2O$$

(*d*) $$5K_4[Fe(CN)_6] + 4H_2SO_4 + KMnO_4 \longrightarrow 5K_3[Fe(CN)_6] + MnSO_4 + 4H_2O$$

or $$5[Fe(CN)_6]^{4-} + 8H^+ + MnO_4^- \longrightarrow 5[Fe(CN)_6]^{3-} + Mn^{2+} + 4H_2O$$

(*ii*) by the electrolytic oxidation of slightly alkaline solution of $K_4[Fe(CN)_6]$

$$2K_4[Fe(CN)_6] + 2H_2O \longrightarrow 2K_3[Fe(CN)_6] + H_2 + 2KOH$$

or $$2[Fe(CN)_6]^{4-} + 2H_2O \longrightarrow 2[Fe(CN)_6]^{3-} + H_2 + 2OH^-$$

(*iii*) by heating a concentrated solution of $K_4[Fe(CN)_6]$ with 50% HNO_3.

$$3K_4[Fe(CN)_6] + 4HNO_3 \xrightarrow{\Delta} 3K_3[Fe(CN)_6] + NO + 3KNO_3 + 2H_2O$$

or $$3[Fe(CN)_6]^{4-} + 4H^+ + NO_3^- \longrightarrow 3[Fe(CN)_6]^{3-} + NO + 2H_2O$$

Properties. 1. It is a dark red crystalline solid. It is readily soluble in water and gives a yellow solution. In solution, it ionises as :

$$K_3[Fe(CN)_6] \rightleftharpoons 3K^+ + \underset{\text{Complex ion}}{[Fe^{3+}(CN)_6]^{3-}}$$

The complex ion is very stable. This ion does not give the usual tests of Fe^{3+}, since six CN^- ions are strongly linked with Fe^{3+} in an octahedral manner.

2. *Oxidising properties: Reduction of $K_3[Fe^{3+}(CN)_6]$ to $K_4[Fe^{2+}(CN)_6]$.* $K_3[Fe(CN)_6]$ oxidises a number of substances in alkaline as well as in acidic medium and is itself reduced to $K_4[Fe(CN)_6]$.

(*i*) *In alkaline medium.* In this medium, $K_3[Fe(CN)_6]$ is reduced to $K_4[Fe(CN)_6]$, liberating nascent oxygen which oxidises the substances.

$$2K_3[Fe^{3+}(CN)_6] + 2KOH \longrightarrow 2K_4[Fe^{2+}(CN)_6] + H_2O + O$$

or $$2[Fe^{3+}(CN)_6]^{3-} + 2OH^- \longrightarrow 2[Fe^{2+}(CN)_6]^{4-} + H_2O + O$$

or $$[Fe^{3+}(CN)_6]^{3-} + e^- \longrightarrow [Fe^{2+}(CN)_6]^{4-}$$

Thus, $K_3[Fe(CN)_6]$ in alkaline medium oxidises H_2S to S, PbO (lithrage) to PbO_2, MnO to MnO_2 ($Mn^{2+}O + O \rightarrow Mn^{4+}O_2$), Cr_2O_3 to K_2CrO_4 ($Cr_2^{3+}O_3 + 4KOH + O \rightarrow 2K_2Cr^{+6}O_4 + 2H_2O$), $CrCl_3$ to K_2CrO_4, SO_2 to H_2SO_4 ($S^{4+}O_2 + H_2O \rightarrow H_2S^{6+}O_4$), H_2O_2 to O_2($H_2O_2^{1-} + O \rightarrow H_2O + O_2^0$), N_2H_4 (hydrazine) to N_2 ($N_2^{2-}H_4 + 2O \rightarrow N_2^0 + 2H_2O$). As an example, let us see how the oxidation of H_2S to S can be represented by various chemical equations given below :

$$2K_3[Fe^{3+}(CN)_6] + 2KOH \longrightarrow 2K_4[Fe^{2+}(CN)_6] + H_2O + O \text{ (\textit{Reduction reaction})}$$

$$H_2S^{-2} + O \longrightarrow H_2O + S^0 \text{ (\textit{Oxidation reaction})}$$

$$2K_3[Fe^{3+}(CN)_6] + 2KOH + H_2S^{-2} \longrightarrow 2K_4[Fe^{2+}(CN)_6] + 2H_2O + S^0 \text{ (\textit{Complete equation})}$$

OR

$$2[Fe^{3+}(CN)_6]^{3-} + 2OH^- \longrightarrow 2[Fe^{2+}(CN)_6]^{4-} + H_2O + O$$

$$H_2S^{-2} + O \longrightarrow H_2O + S^0$$

$$2[Fe^{3+}(CN)_6]^{3-} + 2OH^- + H_2S^{-2} \longrightarrow 2[Fe^{2+}(CN)_6]^{4-} + 2H_2O + S^0$$

OR

$$[[Fe^{3+}(CN)_6]^{3-} + e^- \longrightarrow [Fe^{2+}(CN)_6]^{4-}] \times 2$$
$$S^{-2} \longrightarrow S^0 + 2e^-$$
$$2[Fe^{3+}(CN)_6]^{3-} + S^{-2} \longrightarrow 2[Fe^{2+}(CN)_6]^{4-} + S^0$$

Similar equations can be written for other oxidation reactions.

(*a*) $2K_3[Fe(CN)_6] + 2KOH + M^{2+}O \longrightarrow 2K_4[Fe(CN)_6] + H_2O + M^{4+}O_2$
(M = Pb, Mn)

or $2[Fe^{2+}(CN)_6]^{3-} + 2OH^- + M^{2+}O \longrightarrow 2[Fe(CN)_6]^{4-} + H_2O + M^{4+}O_2$

(*b*) $6K_3[Fe(CN)_6] + 10KOH + Cr_2^{3+}O_3 \longrightarrow 6K_4[Fe(CN)_6] + 5H_2O + 2K_2Cr^{+6}O_4$

or $6[Fe^{3+}(CN)_6]^{3-} + 10OH^- + Cr_2^{3+}O_3 \longrightarrow 6[Fe^{2+}(CN)_6]^{4-} + 5H_2O + 2[Cr^{+6}O_4]^{2-}$

(*c*) $3K_3[Fe(CN)_6] + 8KOH + CrCl_3 \longrightarrow 3K_4[Fe(CN)_6] + 4H_2O + K_2CrO_4 + 3KCl$

or $3[Fe^{3+}(CN)_6]^{3-} + 8OH^- + Cr^{3+}Cl_3 \longrightarrow 3[Fe^{2+}(CN)_6]^{4-} + 4H_2O + [Cr^{+6}O_4]^{2-} + 3Cl^-$

(*d*) $2K_3[Fe(CN)_6] + 2KOH + SO_2 \longrightarrow 2K_4[Fe(CN)_6] + H_2SO_4$

or $2[Fe^{3+}(CN)_6]^{3-} + 2OH^- + S^{4+}O_2 \longrightarrow 2[Fe^{2+}(CN)_6]^{4-} + H_2S^{6+}O_4$

(*e*) $2K_3[Fe(CN)_6] + 2KOH + H_2O_2 \longrightarrow 2K_4[Fe(CN)_6] + 2H_2O + O_2$

or $2[Fe^{3+}(CN)_6]^{3-} + 2OH^- + H_2O_2^{-1} \longrightarrow 2[Fe^{2+}(CN)_6]^{4-} + 2H_2O + O_2^0$

(*f*) $4K_3[Fe(CN)_6] + 4KOH + N_2H_4 \longrightarrow 4K_4[Fe(CN)_6] + 4H_2O + N_2$

or $4[Fe^{3+}(CN)_6]^{3-} + 4OH^- + N_2^{2-}H_4 \longrightarrow 4[Fe(CN)_6]^{4-} + 4H_2O + N_2^0$

(*ii*) *In acidic medium.* In acidic medium, $K_3[Fe(CN)_6]$ oxidises HCl to Cl_2 and H_2S to S.

(*a*) $2K_3[Fe(CN)_6] + 8HCl \longrightarrow 2H_4[Fe(CN)_6] + 6KCl + Cl_2$

or $2[Fe(CN)_6]^{3-} + 2Cl^- \longrightarrow 2[Fe(CN)_6]^{4-} + Cl_2$

(*b*) $6K_3[Fe^{3+}(CN)_6] + 3H_2S \longrightarrow 4K_4[Fe^{2+}(CN)_6] + K_2Fe^{2+}[Fe^{2+}(CN)_6] + 6HCN + 3S$

3. *Action of Fe(II) salts : Formation of blue precipitate of Turnbull's blue.* See "*Turnbull's Blue*" discussed on page 305 of this chapter.

4. *Action of Fe(III) salts.* Redish-brown coloured ferric ferri cyanide, $Fe^{3+}[Fe^{3+}(CN)_6]$ is obtained.

$$K_3[Fe(CN)_6] + FeCl_3 \longrightarrow Fe[Fe(CN)_6] + 3KCl$$

or $[Fe(CN)_6]^{3-} + Fe^{3+} \longrightarrow \underset{\text{Redish-brown}}{Fe[Fe(CN)_6]}$

5. *Action on $CoCl_3$.* $K_3[Fe(CN)_6]_3$ gives a reddish-brown ppt. of $Co_3[Fe(CN)_6]_3$ with $CoCl_2$ solution.

$$3CoCl_3 + K_3[Fe(CN)_6] \longrightarrow Co_3^{3+}[Fe^{3+}(CN)_6]_3$$

Cobaltic ferric cyanide complex (reddish-brown ppt.)

6. *Action of Na_2CO_3 in presence of NO.* When $K_3[Fe(CN)_6]$ is treated with Na_2CO_3 and NO gas is passed through it and the solution is evaporated to crystallisation, ruby-red crystals of sodium nitroprusside, $Na_2[Fe^{2+}(CN)_5(NO^+)]$ are obtained.

$$2K_3[Fe(CN)_6] + 3Na_2CO_3 + 2NO \longrightarrow 2Na_2[Fe(CN)_5(NO)] + 2NaCN + 3K_2CO_3$$

Uses. $K_3[Fe(CN)_6]$ is used :

(*i*) to detect Fe^{2+} ion present in an inorganic salt.

(*ii*) for distinguishing ferrous and ferric salts.

(*iii*) as an external indicator in the determination of the strength of ferrous salt (ferrous ammonium sulphate) solution by titrating it with standard solution of $K_2Cr_2O_7$.

(*iv*) in making blue prints.

(*v*) in making sodium nitroprusside, $Na_2[Fe(CN)_5(NO)]$ which is used to detect S^{2-} ion.

Structure of $[Fe(CN)_6]^{3-}$ ion. This ion has octahedral geometry corresponding to d^2sp^3 hybridisation of Fe^{3+} ion. Due to the presence of one unpaired electron, the ion is paramagnetic. For the detailed discussion of its structure see chapter 5 (*Nature of metal-ligand bonding in complex compounds*).

Complex Compounds Having Iron in +2 as well as in +3 Oxidation State

Prussian blue and turnbull's blue are the examples of complex compounds which contain Fe^{2+} as well as Fe^{3+} ions. These are discussed below :

Prussian blue and Turnbull's blue

Formerly it was proposed that these two substances have different composition and have different formula, but now it has been established by analysis that both these substances are the names of the same substance and hence have the same chemical formula. Some chemists say that each of them is *potassium ferric hexacyano ferrate (II)* or *potassium ferric ferro hexacyanide*, $KFe^{3+}[Fe^{2+}(CN)_6]$ while others are of the opinion that each of them is *ferric hexacyano ferrate* (II) or *ferric ferro hexacyanide*, $Fe_4^{3+}[Fe^{2+}(CN)_6]_3$. Both have blue colour but the blue colour of Turnbull's blue is less intense than the blue colour of Prussian blue.

Preparation of Prussian blue

It is obtained as a blue ppt. by adding the excess of Fe(III) salt solution to that of $K_4[Fe^{2+}(CN)_6]$.

$$Fe^{3+}Cl_3 + K_4[Fe^{2+}(CN)_6] \longrightarrow \underset{\text{Blue ppt. of prussian blue}}{KFe^{3+}[Fe^{2+}(CN)_6]} + 3KCl$$

or

$$Fe^{3+} + [Fe^{2+}(CN)_6]^{4-} \longrightarrow [Fe^{3+}\{Fe^{2+}(CN)_6\}]^-$$

$$4FeCl_3 + 3K_4[Fe(CN)_6] \longrightarrow \underset{\text{Blue ppt. of prussian blue}}{Fe_4^{3+}[Fe^{2+}(CN)_6]_3} + 12KCl$$

or

$$4Fe^{3+} + 3[Fe^{2+}(CN)_6]^{4-} \longrightarrow Fe_4^{3+}[Fe^{2+}(CN)_6]_3$$

If the amount of Fe(III) salt solution being adding to $K_4[Fe(CN)_6]$ solution is less than the equivalent amount, then the prussian blue is obtained in its colloidal solution.

The formation of prussian blue by the reaction between Fe(III) salt solution and $K_4[Fe^{2+}(CN)_6]$ has been utilised for the identification of Fe^{3+} ion in inorganic analysis.

Preparation of Turnbull's blue

It is obtained as a blue ppt. by adding Fe(II) salt solution (*e.g.*, $FeCl_2$, $FeSO_4$ etc.) to the solution of $K_3[Fe^{3+}(CN)_6]$. The formation of Turnbull's blue having the formula, $KFe[Fe^{2+}(CN)_6]$ can be shown as follows :

$$Fe^{2+}Cl_2 + K_3[Fe^{3+}(CN)_6] \longrightarrow KFe^{3+}[Fe^{2+}(CN)_6] + 2KCl$$

$$FeSO_4 + K_3[Fe(CN)_6] \longrightarrow \underset{\text{Blue ppt. of Turnbull's blue}}{KFe[Fe(CN)_6]} + K_2SO_4$$

Both the above equations can be written in ionic form as :

$$Fe^{2+} + [Fe^{3+}(CN)_6]^{3-} \longrightarrow [Fe^{3+}\{Fe^{2+}(CN)_6\}]^-$$

Actually the formation of $KFe[Fe(CN)_6]$ or $[Fe\{Fe(CN)_6\}]^-$ ion takes place through the following steps :

(*a*) Fe^{2+} ion present in Fe(II) salt is oxidised by $[Fe(CN)_6]^{3-}$ ion present in $K_3[Fe(CN)_6]$

$$FeCl_2 \longrightarrow Fe^{2+} + 2Cl^-$$

$$K_3[Fe(CN)_6] \longrightarrow 3K^+ + [Fe(CN)_6]^{3-}$$

$$Fe^{2+} + [Fe(CN)_6]^{3-} \longrightarrow Fe^{3+} + [Fe(CN)_6]^{4-}$$

(*b*) Now Fe^{3+} and $[Fe(CN)_6]^{4-}$ ions formed as above in step (*a*) unite together to form $[Fe\{Fe(CN)_6\}]^-$ ion which combines with K^+ ion to give the blue ppt. of Turnbull's blue, $KFe[Fe(CN)_6]$

$$Fe^{3+} + [Fe(CN)_6]^{4-} \longrightarrow [Fe\{Fe(CN)_6\}]^- \xrightarrow{+K^+} \underset{\text{Turnbull's blue}}{KFe[Fe(CN)_6]}$$

It has been observed that $[Fe(CN)_6]^{4-}$ ions formed at (*a*) also combine with Fe^{2+} ions present initially in Fe(II) salt and form bluish-white ppt. of ferrous ferro hexacyanide, $[Fe^{2+}\{Fe^{2+}(CN)_6\}]^{2-}$.

$$[Fe(CN)_6]^{4-} + Fe^{2+} \longrightarrow \underset{\text{Bluish-white ppt}}{[Fe^{2+}\{Fe^{2+}(CN)_6\}]^{2-}}$$

or

$$K_4[Fe(CN)_6] + FeCl_2 \longrightarrow \underset{\text{Bluish-white ppt}}{K_2Fe^{2+}[Fe^{2+}(CN)_6]} + 2KCl$$

Thus we see that $K_2Fe[Fe(CN)_6]$ (bluish-white ppt.) is also formed along with $KFe[Fe(CN)_6]$ (blue ppt.). It is due to the formation of bluish-white ppt. of $K_2Fe[Fe(CN)_6]$ that the blue colour of $KFe[Fe(CN)_6]$ (Turnbull's blue) is diluted.

The formation of Turnbull's blue having the formula $Fe_4^{3+}[Fe^{2+}(CN)_6]_3$ can be shown as :

$$4FeCl_2 + 3K_3[Fe(CN)_6] \longrightarrow Fe_4[Fe(CN)_6]_3 + 8KCl + K$$

or

$$4Fe^{2+} + 3[Fe(CN)_6]^{3-} + K^+ \longrightarrow \underset{\text{Blue ppt. of Turnbull's blue}}{Fe_4[Fe(CN)_6]_3} + K$$

The formation of Turnbull's blue by the action of $K_3[Fe(CN)_6]$ on Fe(II) salt is utilised to detect the presence of Fe^{2+} ion in a salt. The use of $K_3[Fe(CN)_6]$ as an external indicator in the determination of the strength of FAS solution (ferrous ammonium sulphate solution) by titrating it (*i.e.*, FAS solution) with standard solution of $K_2Cr_2O_7$ is based on the reaction between Fe(II) salt solution and $K_3[Fe(CN)_6]$.

Properties of $KFe^{3+}[Fe^{2+}(CN)_6]$

(*i*) It is a deep blue crystalline solid. It is insoluble in water and gives colloidal dispersion in pure water but is coagulated by the addition of salt. It is also insoluble in dil. acids. It is soluble in oxalic acid and gives a deep blue colour which is used in the manufacture of ink.

(*ii*) *Oxidation.* When $KFe[Fe(CN)_6]$ is oxidised by means of Cl_2 or HNO_3, it gives a green pigment which is called *Berlin green*, $Fe^{3+}[Fe^{3+}(CN)_6]$ (Ferri ferri cyanide).

$$2KFe^{3+}[Fe^{2+}(CN)_6] + Cl_2 \longrightarrow 2K + \underset{\text{Berlin green}}{Fe^{3+}[Fe^{3+}(CN)_6]}$$

(*iii*) *Decomposition.* When treated with caustic alkalies, it gets decomposed to form potassium ferro cyanide, $K_4[Fe(CN)_6]$ and a brown precipitate of $Fe(OH)_3$.

$$KFe[Fe(CN)_6] + 3KOH \longrightarrow K_4[Fe(CN)_6] + \underset{\text{Brown ppt}}{Fe(OH)_3\downarrow}$$

Uses. It is used as a dye for blue print paper, for ink and as a bluing agent in laundry industry.

Properties of $Fe_4^{3+}[Fe^{2+}(CN)_6]_3$

(*i*) It is blue powder and has metallic lusture.

(*ii*) *Action of conc. H_2SO_4.* When boiled with conc. H_2SO_4, it gets decomposed to form CO and HCN.

(*iii*) *Action of an alkali or $Ca(OH)_2$ solution.* When boiled with an alkali or $Ca(OH)_2$

solution, it decomposes to give the precipitate of $Fe(OH)_3$.

$$Fe_4[Fe(CN)_6]_3 + 12KOH \longrightarrow 3K_4[Fe(CN)_6] + 4Fe(OH)_3\downarrow$$

$$Fe_4[Fe(CN)_6]_3 + 6Ca(OH)_2 \longrightarrow 3Ca_2[Fe(CN)_6] + 4Fe(OH)_3\downarrow$$

Uses. It is employed as a permanent blue pigment which has great covering power.

Comparison between prussian blue and turnbull's blue

(*i*) Both are the same substances and hence have identical chemical formula.

(*ii*) Both have blue colour, but the blue colour of turnbull's blue is less intense than the blue colour of prussian blue. The explanation of this difference has already been given.

(*iii*) Both are obtained by different methods. Prussian blue is obtained by adding Fe(III) salt solution to that of $K_4[Fe^{2+}(CN)_6]$ while turnbull's blue is prepared by adding Fe(II) salt solution to that of $K_3[Fe^{3+}(CN)_6]$.

Differences between ferrous and ferric salts

(*i*) *Colours of the anhydrous and hydrated salts.* Anhydrous ferrous and ferric salts are white. Hydrated ferrous salts are light green while hydrated ferric salts are yellow or reddish-brown.

(*ii*) *Colour of the solution in HCl.* The solution of ferrous salts in HCl is light green while that of the solution of ferric salts in this acid is yellow.

(*iii*) *Nature of the hydroxides.* $Fe(OH)_2$ is a weak base, light green in colour and can easily be oxidised. $Fe(OH)_3$ is a much weaker base and shows amphoteric character.

(*iv*) *Oxiding and reducing property.* Ferrous salts are strong reducing agents and hence can reduce acidified $KMnO_4$, $K_2Cr_2O_7$ etc. On the other hand ferric salts are mild oxidising agents and hence can oxidise KI, SO_2 etc.

(*v*) *Action of nitric oxide, NO.* When aqueous solution of a ferrous salt (*e.g.*, $FeSO_4$) reacts with NO, brown-black solution is obtained. This solution is due to the formation of nitroso ferrous cation, $[Fe^+(NO^+)]^{2+}$; *e.g.*,

$$FeSO_4 + NO \longrightarrow [Fe^+(NO^+)]^{2+}SO_4^{2-}$$

or $$Fe^{2+} + NO \longrightarrow [Fe^+(NO^+)]^{2+}$$

Ferric salt has no action on NO.

(*vi*) *Action of NaOH solution.* In the absence of air ferrous salts give a white ppt. of $Fe(OH)_2$. This ppt. is oxidised by air to reddish-brown ppt. of $Fe(OH)_3$.

$$Fe^{2+} + 2NaOH \longrightarrow \underset{\text{White ppt}}{Fe(OH)_2\downarrow} + 2Na^+$$

$$4Fe(OH)_2 + 2H_2O + O_2 \ (\textit{form air}) \longrightarrow \underset{\text{Reddish-brown ppt}}{4Fe(OH)_3\downarrow}$$

Ferric salts give reddish-brown ppt. of $Fe(OH)_3$.

$$Fe^{3+} + 3NaOH \longrightarrow Fe(OH)_3 + 3Na^+$$

(*vii*) *Action of ammonium thiocyanate solution, NH_4CNS.* This reagent has no action on ferrous salts. When solution of NH_4CNS is added to the solution of a ferric salt, blood red colour is produced. This colour is supposed to be due to the formation of $[Fe^{3+}(CNS)]^{2+}$ or $[Fe^{3+}(CNS)_6]^{3-}$ or $Fe^{3+}(SCN)_3$ or $Fe^{3+}[Fe^{3+}(CNS)_6]$.

$$Fe^{3+} + CNS^- \longrightarrow [Fe^{3+}(CNS)]^{2+} \ (\textit{blood red colour})$$

$$Fe^{3+} + 6CNS^- \longrightarrow [Fe^{3+}(CNS)_6]^{3-} \ (\textit{blood red colour})$$

$$Fe^{3+} + 3SCN^- \longrightarrow Fe^{3+}(SCN)_3 \ (\textit{blood red colour})$$

$$2Fe^{3+} + 6SCN^- \longrightarrow Fe^{3+}[Fe^{3+}(CNS)_6] \ (\textit{blood red colour})$$

(*viii*) *Action of $K_3[Fe(CN)_6$.* (*a*) When $K_3[Fe(CN)_6]$ solution is added to the solution of a

Fe (II) salt, blue ppt. of Turnbull's blue is obtained. Turnbull's blue can be represented by the formula, $KFe^{3+}[Fe^{2+}(CN)_6]$ or $Fe_4^{3+}[Fe^{2+}(CN)_6]_3$.

$$FeCl_2 + K_3[Fe(CN)_6] \longrightarrow KFe^{3+}[Fe^{2+}(CN)_6] + 2KCl$$

or $$Fe^{2+} + [Fe(CN)_6]^{3-} \longrightarrow [Fe^{3+}\{Fe^{2+}(CN)_6\}]^-$$

$$4FeCl_2 + 3K_3[Fe(CN)_6] \longrightarrow Fe_4[Fe(CN)_6]_3 + 8KCl + K$$

or $$4Fe^{2+} + K^+ + 3[Fe(CN)_6]^{3-} \longrightarrow Fe_4[Fe(CN)_6]_3 + K$$

(*b*) When $K_3[Fe(CN)_6]$ solution is added to the solution of a Fe(III) salt, redish-brown coloured ferric-ferri cyanide, $Fe^{3+}[Fe^{3+}(CN)_6]$ is obtained.

$$FeCl_3 + K_3[Fe(CN)_6] \longrightarrow Fe^{3+}[Fe^{3+}(CN)_6] + 3KCl$$

or $$Fe^{3+} + [Fe(CN)_6]^{3-} \longrightarrow \underset{\text{Reddish-brown}}{Fe[Fe(CN)_6]}$$

(*ix*) *Action of $K_4[Fe(CN)_6]$.* (*a*) When $K_4[Fe(CN)_6]$ solution is added to that of a Fe(II) salt, bluish-white ppt. of pot. ferrous ferro hexacyanide, $K_2Fe^{2+}[Fe^{2+}(CN)_6]$ is obtained.

$$K_4[Fe(CN)_6] + FeCl_2 \longrightarrow K_2Fe[Fe(CN)_6] + 2KCl$$

or $$[Fe^{2+}(CN)_6]^{4-} + Fe^{2+} \longrightarrow [Fe^{2+}\{Fe^{2+}(CN)_6\}]^{2-}$$

Bluish-white ppt. of $K_2Fe[Fe(CN)_6]$ slowly becomes blue due to its oxidation by air to prussian blue, $KFe^{3+}[Fe^{2+}(CN)_6]$.

$$K_2Fe^{2+}[Fe^{2+}(CN)_6] \xrightarrow{\text{Air oxidation}} KFe^{3+}[Fe^{2+}(CN)_6] + K$$

(*b*) When $K_4[Fe(CN)_6]$ solution is added to that a Fe(III) salt, blue ppt. of prussian blue is obtained. Prussian blue may have the formula $KFe^{3+}[Fe^{2+}(CN)_6]$ or $Fe_4^{3+}[Fe^{2+}(CN)_6]_3$.

$$FeCl_3 + K_4[Fe(CN)_6] \longrightarrow KFe^{3+}[Fe^{2+}(CN)_6] + 3KCl$$

or $$Fe^{3+} + [Fe(CN)_6]^{4-} \longrightarrow [Fe^{3+}\{Fe^{2+}(CN)_6\}]^-$$

$$4FeCl_3 + 3K_4[Fe(CN)_6] \longrightarrow [Fe_4^{3+}[Fe^{2+}(CN)_6]_3 + 12KCl$$

or $$4Fe^{3+} + 3[Fe(CN)_6]^{4-} \longrightarrow Fe_4[Fe(CN)_6]_3$$

Compounds of Osmium

Osmium forms compounds in +2, +3, +4, +6 and +8 oxidation states. Most important compounds are those in which Os is in +4 and +6 oxidation state. Examples of some compounds formed by osmium are oxides (OsO, Os_2O_3, OsO_2, OsO_3, OsO_4), halides (OsF_4, OsF_6, OsF_8, $OsCl_2$, $OsCl_3$, $OsCl_4$, $Os_2Br_9.6H_2O$, OsI_4, OsI_3, OsI_2,OsI). Important compounds are OsO_2, OsO_4, OsF_6, OsF_8 and $OsCl_4$. These are discussed below :

1. Osmium dioxide, OsO_2

It is obtained by the controlled reduction of OsO_4 or by fusing $K_2[OsCl_6]$ with sodium carbonate. It is also obtained when salts of tetravalent osmium are heated with sodium carbonate in an atmosphere of CO_2. The dioxide is a greyish-black powder, forming copper-red octahedral crystals. It decomposes when heated above 450°C :

$$2OsO_2 \rightleftharpoons 2Os + 2O_2$$

It is insoluble in water and acids. When heated in air, it forms OsO_4.

2. Osmium tetroxide, OsO_4

It is the volatile oxide of osmium and has a characteristic smell and the vapours are poisonous and attack eyes and nose. It is formed by heating osmium in oxygen. The finely divided osmium inflames in air at 400°C forming OsO_4. It is pale-yellow solid with chlorine like irritating odour. The tetroxide is commonly known as **osmic acid,** whereas, actually it is the anhydride of perosmic acid. Its aqueous solution is weakly-acidic. It forms unstable addition compounds with strong alkalies.

Aqueous solutions of OsO_4 are used in histology for staining the tissues. It hardens the tissues without distorting them.

3. Osmium hexafluoride, OsF_6

It is formed as one of the products when fluorine reacts with osmium. It is a yellow, volatile solid which melts at 32.2°C. It dissolves in water to form clear solution which undergoes hydrolysis to form OsO_4.

4. Osmium octafluoride, OsF_8

It is formed as vapours when the metal is heated in fluorine at 250°C. The vapours are cooled by solid CO_2. It is a lemon yellow solid which melts at 34.5°C and boils at 47.5°C. It is hydrolysed by water and alkaline solutions and reacts readily with organic substances.

5. Osmium tetrachloride, $OsCl_4$

It may be prepared by the action of chlorine on osmium at 700°C. It is insoluble in all common solvents but is slowly attacked by water and oxidising acids. With water it forms hydrated osmium dioxide, $OsO_2.nH_2O$.It is a black crusty solid and vaporises without melting.

Compounds of Cobalt

Introduction

Important compounds formed by cobalt are those in which this element has +2 (cobaltous) or +3 (cobaltic) oxidation state.

Compounds of Co(+2) : Cobaltous Compounds

Co^{2+} ion ($3d^7 4s^0$) is more stable than Fe^{2+} ion ($3d^6 4s^0$), since this ion cannot easily be oxidised to Co^{3+} ion. However, in basic solutions, Co(+2) compounds are not stable, since in this medium Co^{2+} ion can easily be oxidised to Co^{3+} ion. In aqueous solution, Co^{2+} ion exists as $[Co(H_2O)_6]^{2+}$ ion and has pink colour. When heated, it turns blue due to dehydration. Thus $CoCl_2$ is blue when anhydrous and turns pink when hydrated ($CoCl_2.6H_2O$). Most of Co(+2) compounds are soluble in water. When a base is added to a solution having Co^{2+} ion, $Co(OH)_2$ is obtained as a blue ppt. Important examples of Co(+2) compounds are CoO, $Co(OH)_2$, $CoCl_2.6H_2O$, $Co(NO_3)_2.6H_2O$, $CoSO_4.7H_2O$.Co_3O_4 is a combination of CoO and Co_2O_3. Here we shall discuss CoO and Co_3O_4 only.

1. Cobaltous oxide, CoO

Preparation. It is prepared :

(*i*) by heating $Co(OH)_2$

$$Co(OH)_2 \xrightarrow{\Delta} CoO + H_2O$$

(*ii*) by passing steam over red hot cobalt.

$$Co + H_2O\ (steam) \rightleftharpoons CoO + H_2O$$

Properties. (*i*) It is an olive-green powder.

(*ii*) *Action of heat.* When heated at 600–700°C in air, it gives Co_3O_4

$$3CoO + \frac{1}{2}O_2 \longrightarrow Co_3O_4$$

(*iii*) *Basic nature.* CoO is a basic oxide and hence gives cobaltous salts (Co^{2+}) with acids.

(*iv*) *Combination with oxides.* CoO combines with various oxides to form coloured compounds. For example CoO combines with Al_2O_3 and forms *cobalt meta aluminate*, $CoO.Al_2O_3$ or $Co(AlO_2)_2$ (also called *Thenard's blue*). Similarly, CoO gives *cobalt zincate*, CoO.ZnO or $CoZnO_2$ (also called *Rinman's green*) with ZnO. *Smelt*, $CoO.K_2O.3SiO_2$ can be obtained by mixing the three oxides. These coloured substances are used as pigments. The formation of these coloured double oxides makes the basis of *cobalt nitrate test.*

Uses. CoO is used (*i*) in making enamels and also for desorating china ware blue (*ii*) in the preparation of pigments like Thenard's blue, Rinman's green etc.

2. Cobalto-cobaltic oxide or Co(+2, +3) oxide, Co_3O_4 or $CoO.Co_2O_3$

Preparation. It is prepared :

(*i*) by strongly heating (below 1000°C) cobaltous nitrate.

(*ii*) by heating CoO at 600–700°C in air

$$3CoO + \frac{1}{2}O_2 \longrightarrow Co_3O_4$$

(*iii*) by heating cobaltic oxide, Co_2O_3.

Properties. It is black powder. Unlike Fe_3O_4, it is non-magnetic. It dissolves in acids to form cobaltous salts, *e.g.*,

$$Co_3O_4 + 8HCl \longrightarrow 3CoCl_2 + 4H_2O + Cl_2$$

Structure. Co_3O_4 has spinel structure and is represented as $Co^{2+}O.Co_2^{3+}O_3$

Compounds of Co(+3): Cobaltic Compounds

Co^{3+} ion is a strong oxidising agent, since it readily gets reduced to Co^{2+} ion in aqueous solution.

$$4Co^{3+} + 2H_2O \longrightarrow 4Co^{2+} + O_2(g) + 4H^+$$

The only examples of simple cobaltic compounds are $Co(OH)_3$, CoF_3 and $Co_2(SO_4)_3.18H_2O$.

Complex compounds of Co^{3+} ion are quite stable, since they are not easily reduced to Co^{2+} complex compounds. Thus complex compounds of Co^{3+} ion are not good oxidising agents.

Important cobaltic complex compounds are potassium cobaltic cyanide, $K_3[Co(CN)_6]$; sod. cobalticyanide, $Na_3[Co(CN)_6]$; pot. cobaltinitrite (Fischcher's salt), $K_3[Co(NO_2)_6]$; sod. cobaltinitrite, $Na_3[Co(NO_2)_6]$ and cobaltammines like $[Co(NH_3)_6]Cl_3$, $[Co(NH_3)_5(H_2O)]Cl_3$, $[Co(NH_3)_5Cl]Cl_2$, $[Co(NH_3)_4Cl_2]Cl$, $[Co(NH_3)_4(NO_2)_2]Cl$ etc. Here we shall discuss only $Na_3[Co(NO_2)_6]$, $K_3[Co(NO_2)_6]$ and $[Co(NH_3)_6]Cl_3$.

1. Sodium hexanitro cobaltate (III) or sodium cobalti nitrite, $Na_3[Co(NO_2)_6]$

Preparation. This compound is obtained as a yellow precipitate by the action of Co(II) salt (*e.g.*, $CoCl_2$) on $NaNO_2$ in presence of CH_3COOH.

$$CoCl_2 + 7NaNO_2 + 2CH_3COOH \longrightarrow 2CH_3COONa + 2NaCl + NO + H_2O + \underset{\text{Yellow ppt}}{Na_3[Co(NO_2)_6]} \quad ...(A)$$

or $$Co^{2+} + 3Na^+ + 7NO_2^- + 2H^+ \longrightarrow Na_3[Co(NO_2)_6] + H_2O + NO$$

Actually the above reaction takes place through the following steps :

(*a*) $$CoCl_2 + 2NaNO_2 \longrightarrow Co(NO_2)_2 + 2NaCl$$

(*b*) $$CH_3COOH + NaNO_2 \longrightarrow CH_3COONa + HNO_2$$

HNO_2 produced in reaction (*b*) oxidises $Co(NO_2)_2$ formed in reaction (*a*) to $Co(NO_2)_3$ and is itself reduced to NO.

(*c*) $$\underset{(N = +3)}{2HNO_2} + \underset{(Co = +2)}{Co(NO_2)_2} \longrightarrow \underset{(Co = +3)}{Co(NO_2)_3} + \underset{(N = +2)}{NO} + H_2O$$

$Co(NO_2)_3$ formed in reaction (*c*) combines with $NaNO_2$ to form yellow ppt. of $Na_3[Co(NO_2)_6]$.

(*d*) $$Co(NO_2)_3 + 3NaNO_2 \longrightarrow \underset{\text{Yellow ppt}}{Na_3[Co(NO_2)_6]}$$

Now multiply equation (*b*) by 2 and then add all the equations to get equation (*A*).

On treating $Na_3[Co(NO_2)_6]$ with alcohol, yellow crystals of $Na_3[Co(NO_2)_6]$ are obtained.

Properties. (*i*) It forms yellow crystals which are soluble in water. Its aqueous solution ionises as :

$$Na_3[Co(NO_2)_6] \rightleftharpoons 3Na^+ + [Co(NO_2)_6]^{3-}$$

(*ii*) *Action of ammonium and potassium salts.* With aqueous solution of ammonium and potassium salts, this compound gives a yellow ppt. of ammonium cobaltinitrite, $(NH_4)_3[Co(NO_2)_6]$ and potassium cobaltinitrite (called *Fischer's salt*), $K_3[Co(NO_2)_6]$ respectively. These complexes are slightly soluble in H_2O.

$$\underset{(M\,=\,NH_4,\ K)}{Na_3[Co(NO_2)_6] + 3MCl} \longrightarrow \underset{\text{Yellow ppt}}{M_3[Co(NO_2)_6]} + 3NaCl$$

or $$[Co(NO_2)_6]^{3-} + 3M^+ \longrightarrow \underset{\text{Yellow ppt}}{M_3[Co(NO_2)_6]}$$

The above reaction is used for the detection and estimation of K^+ and NH_4^+ ions.

Uses. $Na_3[Co(NO_2)_6]$ is used (*i*) for the detection and estimation of K^+ and NH_4^+ ions. (*ii*) for the preparation of $K_3[Co(NO_2)_6]$ (*Fischer's salt*) which is an important compound.

Structure of $[Co(NO_2)_6]^{3-}$ ion. See "*Valence Bond Theory*" of complex compounds. (Chapter 5).

2. Potassium hexanitro cobaltate (III) or potassium cobaltinitrite, $K_3[Co(NO_2)_6]$

$K_3[Co(NO_2)_6]$ is also called *Fischer's salt.*

Preparation. It is prepared : (*i*) by the action of Co(II) salt solution [*e.g.*, $Co(NO_3)_2$, $CoCl_2$ etc.] on KNO_2 in presence of CH_3COOH.

(*a*) $$2CH_3COOH + Co(NO_3)_2 + 7KNO_2 \longrightarrow 2CH_3COOK + 2KNO_3 + H_2O + NO + K_3[Co(NO_2)_6]$$

or $$2H^+ + Co^{2+} + 3K^+ + 7NO_2^- \longrightarrow H_2O + NO + K_3[Co(NO_2)_6] \text{ (Yellow ppt)}$$

(*b*) $$2CH_3COOH + CoCl_2 + 7KNO_2 \longrightarrow 2CH_3COOK + 2KCl + H_2O + NO + K_3[Co(NO_2)_6]$$

or $$2H^+ + Co^{2+} + 3K^+ + 7NO_2^- \longrightarrow H_2O + NO + K_3[Co(NO_2)_6] \text{ (Yellow ppt)}$$

The formation of $K_3[Co(NO_2)_6]$ by the reactions shown above is utilised in the detection of Co^{2+} ion in a given inorganic salt.

(*ii*) by the action aqueous solution of potassium salt on $Na_3[Co(NO_2)_6]$.

$$Na_3[Co(NO_2)_6] + 3KCl \longrightarrow K_3[Co(NO_2)_6] + 3NaCl$$

Properties. (*i*) $K_3[Co(NO_2)_6]$ is a yellow crystalline solid and is insoluble in water.

(*ii*) $[Co(NO_2)_6]^{3-}$ ion is decomposed by alkalies to form $Co(OH)_3$ while strong acids decompose it into oxides of nitrogen.

Uses. This compound is used (*i*) as a yellow pigment under the name of *Indian yellow* or *cobalt yellow* (*ii*) in printing porcelain blue.

3. Hexammine cobalt (III) chloride or hexammine cobaltic chloride, $[Co(NH_3)_6]Cl_3$

When a vigorous stream of air is drawn for several hours through a solution of a cobalt (II) chloride, $CoCl_2$, containing ammonia, ammonium chloride and some activated charcoal, good yield of hexammine cobalt (III) chloride is obtained.

$$4CoCl_2 + 4NH_4Cl + 20NH_3 + O_2 \longrightarrow 4[Co(NH_3)_6]Cl_3 + 2H_2O$$

In the absence of charcoal, replacement usually occurs to give, for example, $[Co(NH_3)_5Cl]^{2+}$ and $[Co(NH_3)_4(CO_3)]^+$.This forms orange-yellow crystals. Conductance measurement indicates the presence of four ions in its solution, namely one $[Co(NH_3)_6]^{3+}$ ion and three Cl^- ions as shown below:

$$[Co(NH_3)_6]Cl_3 \rightleftharpoons [Co(NH_3)_6]^{3+} + 3Cl^-$$

Compounds of Nickel

Introduction

Nickel forms compounds in 0, +2, +3 and +4 oxidation states. Examples are given below:

Ni (0) compounds : $Ni(CO)_4$, $Ni(PF_3)_4$, $[Ni(CN)_4]^{4-}$

Ni (+2) compounds (called nickelous compounds) : NiO, $Ni(OH)_2$, $NiCl_2.6H_2O$, NiS, $NiSO_4.7H_2O$, $(NH_4)_2SO_4.NiSO_4.6H_2O$, $Ni(NO_3)_2.6H_2O$, $[Ni(NH_3)_6]Cl_2$, $[Ni(CN)_4]^{2-}$, $[Ni(H_2O)_6]^{2+}$, $Ni(dmg)_2$, (where $dmg = C_4H_7N_2O_2^-$), $[NiCl_4]^{2-}$, $[NiL_4]^{2+}$ ($L = NH_3$, H_2O)

Ni(+3) compounds : Ni_2O_3, $[NiBr_3(PEt_3)_2]^0$

Ni (+4) compounds : NiO_2, $K_2[NiF_6]$

Nickel in +3 oxidation state forms limited number of compounds. These compounds are very strong oxidising agents, since these are readily reduced to Ni(+2) compounds. Ni in +4 oxidation state forms certain oxide systems and a few complexes. Ni in zero oxidation state forms $Ni(CO)_4$ which is an important compound of nickel. Nickel forms most stable compounds in +2 oxidation state.

Ni(+2) Compounds : Nickelous Compounds

In aqueous sulution, Ni^{2+} ion exists as $[Ni(H_2O)_6]^{2+}$ which has green colour. Thus the salts of Ni^{2+} ion are green or blue.

Ni^{2+} ion gets precipitated as $Ni(OH)_2$ which is apple-green, when treated with NaOH or KOH.

$$Ni^{2+} + 2OH^- \longrightarrow Ni(OH)_2$$

This apple-green ppt. dissolves in aqueous ammonia or ammonium salts to give violet-blue solutions which contain $[Ni(NH_3)_6]^{2+}$.

$Ni(OH)_2$ is not oxidised by atmospheric O_2 or H_2O_2. When treated with a mild oxidising agent like Br_2 in alkaline medium, it is converted into black powder to $N_2O_3.H_2O$. When treated with strong oxidising agent like Cl_2 or hypochlorite (ClO^-), a black ppt. of $NiO_2.xH_2O$ is obtained.

$$Ni(OH)_2 + OCl^- \longrightarrow NiO_2 + H_2O + Cl^-$$

Here we shall discuss in some detail the following compounds in which Ni exhibits +2 oxidation state.

1. Nickel sulphate, $NiSO_4.7H_2O$

This compound forms heptahydrate rhombic crystals, isomorphous with cobalt and ferrous sulphates. Nickel sulphate is prepared by dissolving its hydroxide or carbonate in dilute sulphuric acid or by, heating nickel nitrate, $Ni(NO_3)_2.6H_2O$ with concentrated sulphuric acid. If a saturated solution of nickel sulphate is heated to 54°C, monoclinic crystals of $NiSO_4.6H_2O$ are obtained. The crystals are green. When heated above 280°C, yellow anhydrous $NiSO_4$ is formed. If anhydrous nickel sulphate is dissolved in a concentrated solution of ammonia, an unstable dark blue salt, tetrammine nickel sulphate, $Ni(NH_3)_4SO_4.2H_2O$ is formed similar to cuprammonium salts. Besides this complex salt, nickel sulphate also forms a double salt with ammonium sulphate, known as nickel ammonium sulphate $NiSO_4.(NH_4)_2SO_4.6H_2O$, which is extensively used in nickel plating.

2. Nickel ammonium sulphate, $NiSO_4.(NH_4)_2SO_4.6H_2O$

This compound is obtained by mixing, the solution of $NiSO_4.7H_2O$ with that of $(NH_4)_2SO_4$ and then evaporating the solution to crystallisation. The compound forms bluish green crystals. This compound is used for nickel electroplating.

3. Bis (dimethyl glyoximato) nickel (II), $[Ni(C_4H_7N_2O_2)_2]^0$ or $[Ni(dmg)_2]^0$

Dimethyl glyoxime has the molecular formula, $C_4H_8N_2O_2$ or $(C_4H_7N_2O_2)H$ or dmgH and its structural formula is as shown below:

When dimethyl glyoxime is added to a nickel (II) salt (*e.g.*, $NiCl_2$) containing NH_4OH, red ppt. of *bis* (dimethyl glyoximato) nickel (II), $[Ni(C_4H_7N_2O_2)_2]^0$ or $[Ni(dmg)_2]$ is obtained. The formation of this complex compound takes place through the following steps :

(*a*) $$NiCl_2 \rightleftharpoons Ni^{2+} + 2Cl^-$$

(*b*) Dimethyl glyoxime, $C_4H_8N_2O_2$ is a dibendate ligand and has one OH group attached to each of the two N-atoms of two oxime groups $\left(>N{-}OH \right)$. One OH group of each molecule ionises to give H^+ and glyoxime ion, $C_4H_7N_2O_2^-$.

$$C_4H_8N_2O_2 \rightleftharpoons C_4H_7N_2O_2^- + H^+$$

or

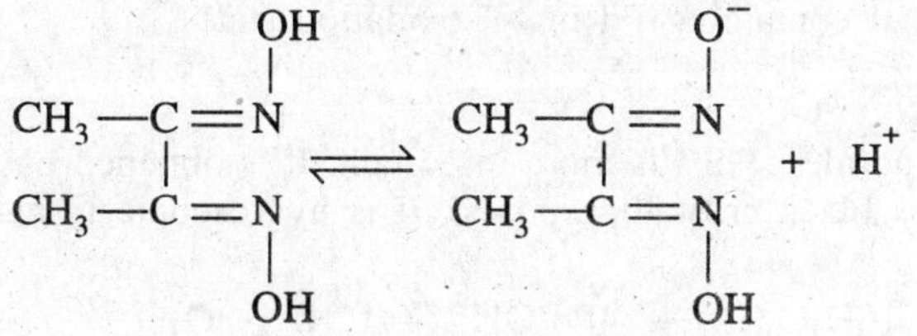

$$\begin{array}{c} OH \\ | \\ CH_3{-}C{=}N \\ | \\ CH_3{-}C{=}N \\ | \\ OH \end{array} \rightleftharpoons \begin{array}{c} O^- \\ | \\ CH_3{-}C{=}N \\ | \\ CH_3{-}C{=}N \\ | \\ OH \end{array} + H^+$$

or

$$dmgH \rightleftharpoons dmg^- + H^+$$

Now two dmg^- ions combine with one Ni^{2+} ion to form electrically neutral *bis* (dimethyl glyoximato) nickel (II), $[Ni(C_4H_7N_2O_2)_2]^0$ or $[Ni(dmg)_2]^0$.

$$\begin{array}{c} O^- \\ | \\ CH_3{-}C{=}N \\ | \\ CH_3{-}C{=}N \\ | \\ O{-}H \end{array} + Ni^{2+} + \begin{array}{c} H{-}O \\ | \\ N{=}C{-}CH_3 \\ | \\ N{=}C{-}CH_3 \\ | \\ O^- \end{array} \longrightarrow$$

$$\begin{array}{ccc} & H & \\ {}^-O & & O \\ | & & | \\ CH_3{-}C{=}N & \rightarrow Ni^{2+} \leftarrow & N{=}C{-}CH_3 \\ CH_3{-}C{=}N & \rightarrow \quad \leftarrow & N{=}C{-}CH_3 \\ | & & | \\ O & & O^- \\ & H & \end{array}$$

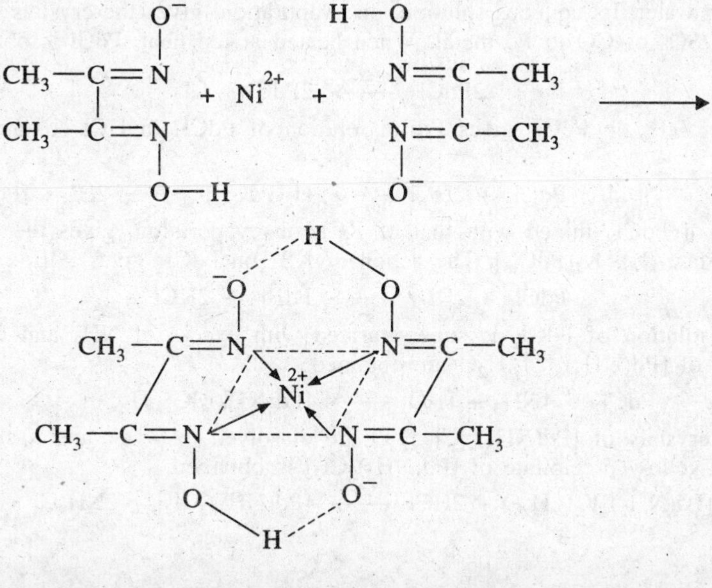

or $C_4H_7N_2O_2^- + Ni^{2+} + C_4H_7N_2O_2^- \longrightarrow [Ni(C_4H_7N_2O_2)_2]^0$

or $dmg^- + Ni^{2+} + dmg^- \longrightarrow [Ni(dmg)_2]^0$

Bis-(dimethylgly oximato) nickel (II) (red ppt)

Bis (dimethyl glyoximato) nickel (II), $[Ni(C_4H_7N_2O_2)_2]^0$ or $[Ni(dmg)_2]^0$ is a chelate complex. It is electrically neutral and covalent in character. Its covalent character accounts for its low solubility in water. It may be seen from the structure that the co-ordination of dmg^- ion (ligand) to Ni^{2+} ion occurs through both its N-atoms making two 5-membered rings. Further two 6-membered rings are also formed by H-bonding between NOH and ON. H-bonding stabilises the structure of the complex.

Structure of $[Ni(dmg)_2]^0$. This complex compound has square planar geometry corresponding to dsp^2 hybridisation of Ni^{2+} ion. In this complex compound Ni^{2+} ion has its valence-shell configuration as $3d^8 = t_{2g}^6\ 3d_{x^2-y^2}^2\ 3d_{z^2}^0$. The compound has no unpaired electrons, *i.e.*, all the eight electrons are present as paired electrons and hence the complex compound is diamagnetic.

Uses of $[Ni(dmg)_2]^0$. The formation of this complex compound is used in detecting as well as in estimating nickel and also in distinguishing Ni from Co.

Compounds of Palladium

In most of its compounds, palladium exists in +2 oxidation state. $PdCl_2$ is the most important compound of Pd(+2). A few compounds of Pd in +3 oxidation state are also known. PdF_3 is the most important compound containing Pd in +3 oxidation state.

Fluorides

PdF_2 and PdF_3 are important fluorides. PdF_3 is more important. It is obtained by the action of F_2 on $PdCl_2$ at 200–250°C. It is a black crystalline power. It is hygroscopic in nature. With cold or hot water, it forms $Pd(OH)_2$.

$$4PdF_3 + 10H_2O \longrightarrow 4Pd(OH)_2 + 12HF + O_2$$

Chlorides

Its common chlorides are PdCl, $PdCl_2$, $PdCl_3$ and $PdCl_4$. $PdCl_2$ is the most important chloride. It is obtained by the action of Cl_2 on Pd at red heat. It is a dark red solid which is highly soluble in water. Its aqueous solution, on evaporation, gives the crystals of $PdCl_2.2H_2O$. It is reduced by SO_2 or CO to Pd metal. When heated at red heat, PdCl is obtained.

$$2PdCl_2 \xrightarrow{\Delta} 2PdCl + Cl_2$$

The action of F_2 on $PdCl_2$ gives PdF_3. Solution of $PdCl_2$ in HCl contains the complex ion, $[PdCl_4]^{2-}$.

$$PdCl_2 + 2HCl \longrightarrow H_2[PdCl_4] \rightleftharpoons 2H^+ + [PdCl_4]^{2-}$$

A solution of $PdCl_2$ mixed with that of KCl, on evaporation, gives the crystals of pot. tetrachloro palladate (II), $K_2[PdCl_4]$. The action of KBr on $PdCl_2$ gives $PdBr_2$.

$$PdCl_2 + 2KBr \longrightarrow PdBr_2 + 2KCl$$

When the solution of $PdCl_2$ is supersaturated with excess of NH_3 and concentrated by heating, crystals of $[Pd(NH_3)_4]Cl_2.H_2O$ are obtained.

$$PdCl_2 + 4NH_3 + H_2O \longrightarrow [Pd(NH_3)_4]Cl_2.H_2O$$

When the crystals of $[Pd(NH_3)_4]Cl_2.H_2O$ are dissolved in water and a small amount of HCl is added, a yellow precipitate of $[Pd(NH_3)_2Cl_2]$ is obtained.

$$[Pd(NH_3)_4]Cl_2.H_2O + 2HCl \longrightarrow [Pd(NH_3)_2Cl_2] + 2NH_4Cl + H_2O$$

When KCN is added to a solution of $PdCl_2$, PdCN is obtained.

$$PdCl_2 + 2KCN \longrightarrow Pd(CN)_2 + 2KCl$$

$PdCl_2$ exists as a linear polymer in the solid state.

$$>Pd\begin{smallmatrix}Cl\\ \\Cl\end{smallmatrix}Pd\begin{smallmatrix}Cl\\ \\Cl\end{smallmatrix}Pd\begin{smallmatrix}Cl\\ \\Cl\end{smallmatrix}Pd<$$

Bromide

$PdBr_2$ is the most important bromide which is obtained by the action of KBr on $PdCl_2$. It is a brown solid, insoluble in water but soluble in HBr.

Iodide

PdI_2 is prepared as a black precipitate by the action of KI on a solution of palladous salt. It is insoluble in water, alcohol and ether but is soluble in a solution of KI and forms $K_2[PdI_4]$.

Oxides

The known oxides are Pd_2O, PdO and PdO_2. Palladous oxide (PdO) is the most important and is obtained by heating Pd-metal in O_2 or by heating the solution of palladous salts with NaOH solution. It is a black powder insoluble in water, acids and aqua regia. It decomposes on heating to 87°C.

Compounds of Platinum

Platinum chiefly forms two series of compounds—*platinous* and *platinic* corresponding to +2 and +4 oxidation states of platinum respectively. The compounds in +1 (*e.g.*, PtCl), +3($PtCl_3$), +5(PtF_5) and +6(PtO_3) oxidtion states are also known but these are rare.

Pt(+2) Compounds : Platinous Compounds

These compounds may be of the following types :

(*i*) *Simple compounds, e.g.*, PtO, $PtCl_2$, PtI_2.

(*ii*) *Complex compounds*. These are of the following types:

(*a*) *Platinous ammines, e.g.*, $[Pt(NH_3)_4]Cl_2$, $[Pt(NH_3)_3Cl]Cl$, $[Pt(NH_3)_2Cl_2]$, $K[Pt(NH_3)Cl_3]$.

(*b*) *Tetrachloro complexes, e.g.*, $H_2[PtCl_4]$, $K_2[PtCl_4]$

(*c*) *Tetracyano complexes, e.g.*, $K_2[Pt(CN)_4].3H_2O$, $Ba[Pt(CN)_4].4H_2O$.

Most of the Pt(+2) complexes are four co-ordinated. They are diamagnetic with dsp^2 hybridisation and have square planar geometry. These are quite inert.

Here we shall discuss $H_2[PtCl_4]$, $K_2[Pt(CN)_4].3H_2O$ and $Ba[Pt(CN)_4].4H_2O$ in which Pt is in +2 oxidation state.

1. Hydrogen tetrachloro platinate (II) or tetra chloroplatinous acid, $H_2[PtCl_4]$

In this complex compound Pt is in +2 oxidation state. Here the coordination number of Pt is 4. It is obtained by dissolving PCl_2 in HCl or by the reduction of H_2PtCl_6 with SO_2.

$$PtCl_2 + 2HCl \longrightarrow H_2PtCl_4$$

$$H_2PtCl_6 + 2H_2O + SO_2 \longrightarrow H_2PtCl_4 + H_2SO_4 + 2HCl$$

The acid does not crystallise but forms a dark-brown amorphous solid on evaporation. Its salts are called *chloroplatinites*. Potassium chloro platinite, K_2PtCl_4 may be better prepared by reducing chloroplatinate, K_2PtCl_6 by cuprous chloride.

$$K_2PtCl_6 + Cu_2Cl_2 \longrightarrow 2CuCl_2 + K_2PtCl_4$$

The salt forms dark red crystals and is used in platinum printing in photography. The corresponding ammonium, silver or lead salts are also known.

2. Potassium tetracyano platinate (II) or potassium platinocyanide, $K_2[Pt(CN)_4].3H_2O$

It is also a complex compound of Pt in +2 oxidation state. It is obtained as yellow needles by mixing solutions of $PtCl_2$ and KCN in proper proportions and evaporating the solution to crystallisation.

$$PtCl_2 + 4KCN \longrightarrow K_2[Pt(CN)_4] + 2KCl$$

3. Barium tetracyano platinate (II) or barium platinocyanide, $Ba[Pt(CN)_4].4H_2O$

In this complex compound Pt is in +2 oxidation state. It is prepared by warming $H_2[PtCl_6]$ solution with baryta water, $Ba(OH)_2$ and HCN and passing in SO_2 until the solution becomes colourless.

$$H_2[PtCl_6] + 5Ba(OH)_2 + 4HCN + SO_2 \longrightarrow Ba[Pt(CN)_4] + \underset{ppt}{BaSO_4} + 3BaCl_2 + 8H_2O$$

$BaSO_4$ is filtered off and the filtrate is crystallised to get $Ba[Pt(CN)_4].4H_2O$. It is a lemon-yellow powder which is soluble in water. It is used for fluorescent screens in X-ray work.

Pt(+4) Compounds : Platinic Compounds

These compounds may be of the following types :

(*i*) *Simple compounds, e.g.*, PtO_2, $PtCl_4$.

(*ii*) *Complex compounds.* These may be :

(*a*) *Platinic ammines, e.g.*, $[Pt(NH_3)_6]Cl_4$, $[Pt(NH_3)_5Cl]Cl_3$, $[Pt(NH_3)_4Cl_2]Cl_2$, $[Pt(NH_3)_3Cl_3]Cl$, $[Pt(NH_3)_2Cl_4]^0$, $K[Pt(NH_3)Cl_5]$.

(*b*) *Hexachloro complexes. e.g.*, $H_2[PtCl_6].6H_2O$, $(NH_4)_2[PtCl_6]$, $Li_2[PtCl_6].6H_2O$, $Na_2[PtI_6]$ $6H_2O$, $K_2[PtCl_6]$.

Complexes of Pt(+4) are remarkably stable and kinetically inert. All these are octahedral resulting from d^2sp^3 hybridisation (inner orbital). Since d^6 electrons of Pt^{4+} ion are distributed as $t_{2g}^6\ e_g^0$ (d^2sp^3 hybridisation), Pt(+4) complexes have all d^6 electrons in the paired state and hence these complexes are diamagnetic.

Now let us discuss some important compounds of platinum (+4).

1. Platinum (IV) chloride or platinic chloride, $PtCl_4$

In this compound Pt is in +4 oxidation state.

Preparation. $PtCl_4$ is obtained :

(*i*) by heating $H_2[PtCl_6]$ at 370°C in a current of Cl_2 or at 165°C in H_2.

$$H_2[PtCl_6] \xrightarrow{\Delta} PtCl_4 + 2HCl$$

(*ii*) by the action of Cl_2 on Pt-metal at 200–300°C.

$$Pt + 2Cl_2 \longrightarrow PtCl_4$$

Properties. (*i*) It forms reddish-brown slightly hygroscopic crystalline mass.

(*ii*) *Hydrolysis*. It dissolves in water, giving yellowish-red solution which is strongly acidic. The acidic nature of the solution is due to the formation of a complex acid, $H_2[PtCl_4(OH)_2]$.

$$PtCl_4 + 2H.OH \longrightarrow H_2[PtCl_4(OH)_2]$$

On electrolysis, this acid gives the complex anion on the anode showing platinum is discharged at the anode as a part of $[PtCl_4(OH)_2]^{2-}$ ion.

$$H_2[PtCl_4(OH)_2] \longrightarrow 2H^+(\textit{At cathode}) + [PtCl_4(OH)_2]^{2-}(\textit{At anode})$$

(*iii*) *Oxidising property : Reduction of $PtCl_4$.* (*a*) $PtCl_4$ oxidises KI to iodine and is itself reduced to $PtCl_2$. Thus $PtCl_4$ liberates I_2 from KI solution.

$$2KI + PtCl_4 \longrightarrow I_2 + 2KCl + PtCl_2$$

or
$$2I^- + Pt^{+4} \longrightarrow I_2^0 + Pt^{2+}$$

(*b*) When $PtCl_4$ reacts with hydroxyl amine (NH_2OH) in acidic medium, it gets reduced to Pt-metal. Hydrazine (N_2H_4) also reduces $PtCl_4$ to Pt-metal. NH_2OH (N = –1) gets oxidised to N_2O (N = +1) while N_2H_4 (N = –2) gets oxidised to N_2 (N = 0).

$$PtCl_4 + 2NH_2OH \longrightarrow Pt + N_2O + 4HCl + H_2O$$

$$PtCl_4 + N_2H_4 \longrightarrow Pt + N_2 + 4HCl$$

(*iv*) *Action of HCl and alkali metal halides.* Complexes of $[PtX_6]^{2-}$ type (X = Cl, Br, I) are obtained.

$$PtCl_4 + 2HCl \longrightarrow H_2[PtCl_6]$$

$$PtCl_4 + 2NaCl \longrightarrow Na_2[PtCl_6]$$

$$PtCl_4 + 2KCl \longrightarrow K_2[PtCl_6]$$

$$PtCl_4 + 2NH_4Cl \longrightarrow (NH_4)_2[PtCl_6]$$

(*v*) *Action of heat.* When heated at 853 K, $PtCl_2$ is obtained.

$$PtCl_4 \xrightarrow{\Delta} PtCl_2 + Cl_2$$

(*vi*) *Action of KOH.* $Pt(OH)_2$ gets precipitated when $PtCl_4$ reacts with KOH.

$$PtCl_4 + 4KOH \longrightarrow Pt(OH)_4\downarrow + 4KCl$$

(*vii*) *Action of KCN.* Potassium tetra cyano platinate (II), $K_2[Pt(CN)_4]$ along with cynogen, $(CN)_2$ are produced.

$$PtCl_4 + 6KCN \longrightarrow K_2[Pt(CN)_4] + 4KCl + (CN)_2\downarrow$$

Uses. $PtCl_4$ is used for the preparation of $H_2[PtCl_6]$ and $K_2[PtCl_6]$. Both these compounds are very useful compounds.

2. Hydrogen hexachloro platinate (IV) or hexachloro platinic acid, $H_2[PtCl_6]$

Preparation. It is prepared:

(*i*) by the action of HCl on $PtCl_4$ or $Pt(OH)_4$.

$$2HCl + PtCl_4 \longrightarrow H_2[PtCl_6]$$

$$6HCl + Pt(OH)_4 \longrightarrow H_2[PtCl_6] + 4H_2O$$

(*ii*) by the action of aqua regia (HNO_3 + 3HCl) on pure platinum metal.

$$[HNO_3 + 3HCl \longrightarrow NOCl + 2H_2O + 2Cl] \times 2$$

$$Pt + 4Cl \longrightarrow PtCl_4$$

$$PtCl_4 + 2HCl \longrightarrow H_2[PtCl_6]$$

On adding:
$$2HNO_3 + 8HCl + Pt \longrightarrow 2NOCl + 4H_2O + H_2[PtCl_6]$$

OR

$$[HNO_3 + 3HCl \longrightarrow NO + 2H_2O + 3Cl] \times 4$$

$$[Pt + 4Cl \longrightarrow PtCl_4] \times 3$$

$$[PtCl_4 + 2HCl \longrightarrow H_2PtCl_6] \times 3$$

On adding:
$$4HNO_3 + 18HCl + 3Pt \longrightarrow 4NO + 8H_2O + 3H_2[PtCl_6]$$

The solution is evaporated with HCl till HNO_3 is completely removed from the reaction mixture. On cooling, the red-brown crystals of $H_2PtCl_6.6H_2O$ are left behind.

(*iii*) by passing Cl_2 through a solution of spongy platinum in conc. HCl.

$$Pt(\textit{spongy platinum}) + 2HCl + Cl_2 \longrightarrow H_2[PtCl_6]$$

Properties. (*i*) H_2PtCl_6 is a red-brown crystalline solid that deliquesces in air and is readily soluble in H_2O, alcohol and ether. It is associated with six H_2O molecules, $H_2[PtCl_6].6H_2O$.

(*ii*) *Electrolysis.* When the aqueous solution of H_2PtCl_6 is electrolysed, hydrogen is liberated. This hydrogen reduces H_2PtCl_6 to liberate Pt-metal at cathode.

$$H_2PtCl_6 \rightleftharpoons 2H^+ + [PtCl_6]^{2-}$$

$$2H^+ + 2e^- \longrightarrow H_2$$

$$H_2[PtCl_6] + 2H_2 \xrightarrow{\text{Reduction}} Pt\ (\textit{At cathode}) + 6HCl.$$

(*iii*) *Action of heat.* (*a*) When heated at 165°C in H_2 or at 370°C in Cl_2, $PtCl_4$ is formed. This reaction has been used to prepare $PtCl_4$.

$$H_2[PtCl_6] \xrightarrow[Cl_2]{370°C} PtCl_4 + 2HCl$$

(*b*) When heated to 390°C in Cl_2, $PtCl_3$ is formed.

$$2H_2[PtCl_6] \xrightarrow[Cl_2]{390°C} 2PtCl_3 + 4HCl + Cl_2$$

(*c*) The heating of $H_2[PtCl_6]$ in Cl_2 at 580°C gives $PtCl_2$.

$$H_2[PtCl_6] \xrightarrow[Cl_2]{580°C} PtCl_2 + 2HCl + Cl_2$$

(*iv*) *Reduction: Oxidising properties.* $H_2[PtCl_6]$ can be reduced by various substances. The nature of the products obtained by reduction depends on the nature of the reducing agent used. For example:

(*a*) Nascent hydrogen obtained by the action of Zn on HCl or formic acid (HCOOH) reduces the acid to platinum black.

$$2Zn + 4HCl \longrightarrow 2ZnCl_2 + 4H$$

$$H_2[PtCl_6] + 4H \longrightarrow Pt\ (\textit{platinum black}) + 6HCl$$

$$H_2[PtCl_6] + 2HCOOH \longrightarrow Pt\ (\textit{platinum black}) + 2CO_2\uparrow$$

(*b*) Hydrazine (NH_2—NH_2) reduces the acid to colloidal platinum in presence of some protective colloid. NH_2—NH_2 (N = –2) is oxidised to N_2 (N = O).

$$H_2[PtCl_6] + NH_2—NH_2 \longrightarrow Pt\ (\textit{colloidal platinum}) + N_2 + 6HCl$$

(*c*) When asbestos is soaked in 10% solution of $H_2[PtCl_6]$, dried and heated to 500°C, H_2PtCl_6 gets reduced to finely-divided platinum which is distributed throughout the asbestos fibre. The asbestos fibre having finely-divided platinum in it is called asbestos.

(*d*) KI solution reduces H_2PtCl_6 to PtI_2 and is itself oxidised to I_2.

$$H_2[PtCl_6] + 4KI \longrightarrow I_2 + PtI_2 + 4KCl + 2HCl$$

(*e*) When H_2PtCl_6 is treated with $Ba(OH)_2$ and HCN and SO_2 gas is passed through the reaction mixture till the solution becomes colourless, $Ba[Pt(CN)_4].3H_2O$ is obtained in solution.

$$H_2[PtCl_6] + 5Ba(OH)_2 + 4HCN + SO_2 \longrightarrow Ba[Pt(CN)_4]\ (\textit{solution}) + BaSO_4\ (\textit{ppt}) + 3BaCl_2$$

$BaSO_4$ is filtered off and the filtrate containing $Ba[Pt(CN)_4]$ is crystallised to get the crystals of $Ba[Pt(CN)_4]$. In the above reaction $H_2[PtCl_6]$ (Pt = +4) is reduced to $Ba[Pt(CN)_4]$ (Pt = +2) and SO_2 (S = +4) is oxidised to $BaSO_4$ (S = +6). Thus here SO_2 acts as a reducing agent.

(*f*) The aqueous solution of SO_2 reduces $H_2[PtCl_6]$ to $H_2[PtCl_4]$ and is itself oxidised to H_2SO_4.

$$H_2[PtCl_6] + SO_2 + H_2O \longrightarrow H_2[PtCl_4] + H_2SO_4 + 2HCl$$

(*v*) *Formation of salts of the acid–called hexachloro platinates.* $H_2[PtCl_6]$ is a strong dibasic acid and ionises as :

$$H_2[PtCl_6] \rightleftharpoons 2H^+ + [PtCl_6]^{2-}$$

The mode of ionisation shown above indicates that $H_2[PtCl_6]$ forms salts of the type $M_2^+[PtCl_6]^{2-}$ ($M^+ = Li^+, Na^+, K^+, NH_4^+, Ag^+$ etc.) . These salts are called hexachloro platinates which contain hexachloro platinate ion, $[PtCl_6]^{2-}$.

(*vi*) *Action of H_2S.* When H_2S is passed into a hot solution of H_2PtCl_6, platinic sulphide (PtS_2) is obtained as a black precipitate.

$$H_2[PtCl_6] + 2H_2S \longrightarrow PtS_2 \ (\textit{black ppt.}) + 6HCl$$

(*vii*) *Formation of addition compound with organic bases.* The organic bases like $C_6H_5NH_2$ form a stable crystalline addition compound with $H_2[PtCl_6]$. The formation of such compounds is used in the gravimetric estimation of these bases.

$$2C_6H_5NH_2 + H_2[PtCl_6] \longrightarrow \underset{\text{Addition compound}}{2C_6H_5NH_2.H_2[PtCl_6]}$$

Uses. H_2PtCl_6 is used:

(*i*) for the preparation of a number of compounds of platinum (*e.g.* $PtCl_4$)

(*ii*) for the determination of the molecular weight of bases.

(*iii*) for the gravimetric estimation of organic bases.

(*iv*) for the preparation of hexachloro platinates.

(*v*) for preparing platinum black, colloidal platinum, platinised asbestos.

(*vi*) as a tanning agent in photography.

Structure of hexachloro platinate (IV) ion, $[PtCl_6]^{2-}$. See "*Valence Bond Theory of complex compounds*" discussed in chapter 5.

Salts of Hexachloro Platinic Acid : Hexachloroplatinates

Hexachloroplatinic acid, $H_2[PtCl_6]$ is a strong dibasic acid and ionises as :

$$H_2[PtCl_6] \rightleftharpoons 2H^+ + [PtCl_6]^{2-}$$

Thus this acid gives the stable salts of the type $M_2^+[PtCl_6]$ ($M^+ = Li^+, Na^+, K^+, NH_4^+, Ag^+$ etc.) which are called *hexachloro platinates* (IV) of M^+. Examples of such salts are $Li_2[PtCl_6].6H_2O$, $Na_2[PtCl_6].6H_2O$, $K_2[PtCl_6]$, $(NH_4)_2[PtCl_6]$, $Ag_2[PtCl_6]$etc.

$Li_2[PtCl_6]$, and $Na_2[PtCl_6]$ are obtained by the action of LiOH or NaOH on $H_2[PtCl_4]$.

$$2LiOH + H_2[PtCl_6] \longrightarrow Li_2[PtCl_6] + 2H_2O$$

$$2NaOH + H_2PtCl_6] \longrightarrow Na_2[PtCl_6] + 2H_2O$$

$Na_2[PtCl_6]$ can also be obtained by this action of $PtCl_4$ on NaCl.

$$PtCl_4 + 2NaCl \longrightarrow Na_2[PtCl_6]$$

$Ag_2[PtCl_6]$ is obtained as a yellow precipitate when $H_2[PtCl_6]$ is treated with $AgNO_3$ solution.

$$H_2[PtCl_6] + 2AgNO_3 \longrightarrow Ag_2[PtCl_6] + 2HNO_3$$

$Li_2[PtCl_6].6H_2O$ and $Na_2[PtCl_6].6H_2O$ are soluble in water.

Here we shall discuss in some detail $K_2[PtCl_6]$ and $(NH_4)_2[PtCl_6]$.

1. Potassium hexachloro platinate (IV), $K_2[PtCl_6]$

Preparation. It is obtained as a yellow precipitate by the action of KCl on $PtCl_4$ or on $H_2[PtCl_6]$.

$$2KCl + PtCl_4 \longrightarrow K_2[PtCl_6]$$
$$2KCl + H_2[PtCl_6] \longrightarrow K_2[PtCl_6] + 2HCl$$

Properties. (*i*) This compound is poorly soluble in H_2O. It is yellow in colour.

(*ii*) *Action of an alkali.* When $K_2[PtCl_6]$ is treated with KOH, $K_2[Pt(OH)_6]$ is obtained.

$$K_2[PtCl_6] + 6KOH \longrightarrow K_2[Pt(OH)_6] + 6KCl$$

or
$$[PtCl_6]^{2-} + 6OH^- \longrightarrow [Pt(OH)_6]^{2-} + 6Cl^-$$

(*iii*) *Action of heat.* When heated, it gives spongy platinum.

$$K_2[PtCl_6] \xrightarrow{\Delta} 2KCl + 2Cl_2 + Pt \text{ (spongy platinum)}$$

(*iv*) *Reduction.* It gets reduced to potassium tetrachloro platinate (II), $K_2[PtCl_4]$ by CuCl.

$$K_2[PtCl_6] + 2CuCl \longrightarrow 2CuCl_2 + K_2[PtCl_4]$$

Uses. $K_2[PtCl_6]$ is used in photography.

2. Ammonium hexachloro platinate (IV), $(NH_4)_2[PtCl_6]$ (Also called ammonium chloroplatinate)

Preparation. When $H_2[PtCl_6]$ is treated with neutral solution of NH_4Cl or NH_4OH, yellow crystalline ppt. of $(NH_4)_2[PtCl_6]$ is obtained.

$$H_2[PtCl_6] + 2NH_4Cl \longrightarrow \underset{\text{Yellow ppt}}{(NH_4)_2[PtCl_6]} + 2HCl$$

$$H_2[PtCl_6] + 2NH_4OH \longrightarrow \underset{\text{Yellow ppt}}{(NH_4)_2[PtCl_6]} + 2H_2O$$

$(NH_4)_2[PtCl_6]$ can also be obtained from the platini-ferrous sands and gravels present in the river beds and also from the non-volatile residue obtained at the end of Mond's process. (*See the extraction of platinum*).

Properties. (*i*) It is a yellow substance and is poorly soluble in H_2O.

(*ii*) *Action of heat.* When heated to redness, spongy platinum is obtained.

$$(NH_4)_2[PtCl_6] \xrightarrow{\Delta} 2NH_4Cl + 2Cl_2 + Pt \text{ (spongy platinum)}$$

(*iii*) *Action of liquid NH_3.* Hexammine platinum (IV) chloride, $[Pt(NH_3)_6]Cl_4$ is obtained.

$$(NH_4)_2[PtCl_6] + 6NH_3 \longrightarrow [Pt(NH_3)_6]Cl_4 + 2NH_4Cl$$

(*iv*) *Reduction.* Hydrazine (N_2H_4) reduces $(NH_4)_2[PtCl_6]$ to platinum metal and is itself oxidised to N_2. Some of the platinum produced gets deposited on the inside of the tube as a platinum mirror.

$$(NH_4)_2[PtCl_6] + N_2H_4 \longrightarrow Pt + N_2 + 4HCl + 2NH_4Cl$$

Uses. $(NH_4)_2[PtCl_6]$ is used (*i*) in the extraction of platinum metal, (*ii*) for the detection and estimation of potassium and ammonium and for the equivalent weight determination of organic bases.

Compounds of Copper

Introduction

Copper forms two types of compounds: (*i*) *Copper* (+1) or *cuprous compounds* in which copper is *monovalent*, and (*ii*) *Copper* (+2) or *cupric compounds* in which copper is *divalent*. When cupric or cuprous is not stated in the case of a salt of copper, cupric is implied *e.g.*, copper sulphate implies cupric sulphate. Most of the *cuprous compounds are usually colourless and diamagnatic* as 3*d* orbital is completely filled ($Cu^+ = 3d^{10}$). Cu_2O, Cu_2S etc. which are red and black are exceptions. These are generally insoluble in water. The soluble compounds of Cu(+1) are *unstable in aqueous solution, since they disproportionate* to Cu^{2+} and Cu.

$$2Cu^+ \longrightarrow Cu + Cu^{2+}$$

The above reaction is supported by the fact that when Cu_2O is dissolved in dil.H_2SO_4, the products formed are ($CuSO_4$ + Cu) and not (Cu_2SO_4 + Cu).

$$Cu_2O + H_2SO_4 \longrightarrow CuSO_4 + Cu(s) + H_2O$$

or

$$Cu_2O + 2H^+ \longrightarrow Cu^+ + Cu(s) + H_2O$$

Actually the above reaction takes place through the following two steps :

$$Cu_2O + H_2SO_4 \longrightarrow Cu_2SO_4 + H_2O$$

$$Cu_2SO_4 \longrightarrow CuSO_4 + Cu(s)$$

$$Cu_2O + H_2SO_4 \longrightarrow CuSO_4 + Cu(s) + H_2O$$

Most of the *anhydrous cupric compounds* are colourless while the *hydrated cupric compounds are blue* due to the formation of blue hydrated ion, $[Cu(H_2O)_4]^{2+}$ or $[Cu(H_2O)_6]^{2+}$. Solutions of Cu(+2) salts are acidic in nature due to hydrolysis.

$$Cu^{+2} + H_2O \longrightarrow [Cu(OH)]^+ + H^+$$

Since the outer most shell of Cu^{2+} ion has *one unpaired electron* (d^9 configuration), Cu(+2) compounds are paramagnetic.

Let us discuss the following compounds of Cu.

Oxides

Well known oxides of copper are *cuprous oxide* (CuO) and *cupric oxide* (CuO).

1. Cuprous oxide (also called red oxide of copper), Cu_2O

It occurs in nature as *cuprite* or *red copper* ore.

Preparation. It is prepared:

(*i*) by heating CuO alone or with Cu powder.

$$4CuO \xrightarrow{1000°C} 2Cu_2O + O_2.$$

$$CuO + Cu \xrightarrow{\Delta} Cu_2O$$

(*ii*) by boiling a solution containing $CuSO_4$, Na_2CO_3 and Na_2SO_3.

$$2CuSO_4 + 2Na_2CO_3 + Na_2SO_3 \longrightarrow Cu_2O + 3Na_2SO_4 + 2CO_2$$

(*iii*) by fusing CuCl with Na_2CO_3 and dissolving NaCl in water.

$$2CuCl + 2Na_2CO_3 \longrightarrow Cu_2O + 2NaCl + CO_2$$

(*iv*) by boiling Fehling solution (Fehling solution is an alkaline solution of $CuSO_4$) with glucose ($C_5H_{11}O_5.CHO$) or aldehyde (*e.g.*, CH_3CHO).

$$\underbrace{CuSO_4 + 2NaOH}_{\text{Fehling solution}} \longrightarrow CuO + H_2O + Na_2SO_4$$

$$2CuO + C_5H_{11}O_5.CHO \longrightarrow Cu_2O + C_5H_{11}O_6.COOH$$

$$2CuO + CH_3COOH \longrightarrow Cu_2O + CH_3COOH$$

Properties. (*i*) Cu_2O is a red powder, insoluble in water and alkalies. It is basic in nature.

(*ii*) *Action of heat.* When heated in air, Cu_2O (Cu = +1) is oxidised to CuO (Cu = +2).

$$Cu_2O + \frac{1}{2}O_2 \longrightarrow 2CuO$$

(*iii*) *Action of acids.* (*a*) H_2SO_4. With dil. acid, a solution of $CuSO_4$ is formed and Cu-metal gets separated.

$$Cu_2O + H_2SO_4\ (dil.) \longrightarrow \underset{\text{Solution}}{CuSO_4} + Cu\downarrow + H_2O$$

or $$Cu_2O + 2H^+ \longrightarrow Cu^{2+} + Cu\downarrow + H_2O$$

In boiling conc. H_2SO_4, $CuSO_4$ is formed (in solution) and SO_2 is evolved.

$$Cu_2O + 3H_2SO_4\ (boiling\ and\ conc.) \longrightarrow 2CuSO_4 + 3H_2O + SO_2$$

(*b*) *HNO_3*. The reaction of Cu_2O with HNO_3 takes place as shown below :

$$Cu_2O + 2HNO_3\ (dil.\ and\ cold) \longrightarrow \underset{\text{Solution}}{Cu(NO_3)_2} + H_2O + Cu\downarrow$$

$$3Cu_2O + 14HNO_3\ (boiling\ dilute\ or\ cold\ conc.) \longrightarrow \underset{\text{Solution}}{6Cu(NO_3)_2} + 7H_2O + 2NO$$

(*c*) *HCl*. Conc. HCl dissolves Cu_2O, forming CuCl which dissolves in the excess of HCl to produce hydro cuprochloric acid, $H[CuCl_2]$

$$Cu_2O + 2HCl \longrightarrow 2CuCl + H_2O$$
$$[CuCl + HCl \longrightarrow H[CuCl_2]] \times 2$$
$$Cu_2O + 4HCl \longrightarrow 2H[CuCl] + H_2O$$

(*iv*) *Action of NH_4OH*. In the absence of air, Cu_2O gives colourless soluble diammine cuprous hydroxide, $[Cu^+(NH_3)_2]OH$ (complex compound).

$$Cu_2O + 4NH_4OH \longrightarrow 2[Cu^+(NH_3)_2]OH + 3H_2O$$

In the presence of air the above cuprous complex compound is oxidised to tetrammine cupric hydroxide, $[Cu^{2+}(NH_3)_4](OH)_2$.

(*v*) *Action of alkali cyanide*. It dissolves in alkali cyanide (*e.g.*, KCN) solution, forming a colourless complex, $K_2[Cu(CN)_3]$.

$$Cu_2O + 6KCN + H_2O \longrightarrow 2K_2[Cu(CN)_3] + 3KOH$$

Uses. It is used. (*i*) for making ruby glass and enamel. (*ii*) as pigment. (*iii*) in the manufacture of antirust paints. (*iv*) as a pesticide.

2. Cupric oxide, CuO (Also called black oxide of copper)

It occurs in nature as *tenorite*.

Preparation. CuO is prepared:

(*i*) by prolonged heating Cu-turnings in air or O_2.

(*ii*) by thermal decomposition of $Cu(NO_3)_2$, $Cu(OH)_2$ or $CuCO_3$.

$$2Cu(NO_3)_2 \xrightarrow{\Delta} 2CuO + 4NO_2 + O_2$$

This method gives *pure* CuO.

(*iii*) by heating malachite, $Cu(OH)_2.CuCO_3$ and azurite, $Cu(OH)_2.2Cu(CO_3)_2$. Both these compounds are basic cupric carbonates.

$$Cu(OH)_2.CuCO_3 \xrightarrow{200°C} 2CuO + H_2O + CO_2$$

$$Cu(OH)_2.2CuCO_3 \xrightarrow{220°C} 3CuO + H_2O + 2CO_2$$

The above methods are used to get CuO on commercial scale.

Properties. (*i*) CuO is a black powder having its m.pt. = 1000°C. It is insoluble in water.

(*ii*) It is a basic oxide and forms salts with acids.

$$CuO + H_2SO_4 \longrightarrow CuSO_4 + H_2O$$

(*iii*) *Action of heat.* On heating to 1000°C, CuO gets decomposed into Cu_2O.

$$4CuO \xrightarrow{1000°C} 2Cu_2O + O_2$$

(*iv*) *Reduction*. (*a*) CuO (Cu = +2) may be reduced to Cu_2O (Cu = +1) by heating it to 1000°C

$$4CuO \xrightarrow{1000°C} 2Cu_2O + O_2$$

(*b*) CuO may also be reduced to Cu-metal by H_2, carbon, CO or hydrocarbons.

$$CuO + C \longrightarrow Cu + CO$$

(*v*) It is rapidly attacked by halogens and many halogen compounds at 300–400°C.

Uses. CuO is used (*i*) in organic analysis for the estimation of carbon and H_2 (*ii*) in petroleum refining (*iii*) in glass and glaze industries.

Basic Cupric Carbonates

Normal cuprous carbonate (Cu_2CO_3) and cupric carbonate ($CuCO_3$) have not been isolated but basic cupric carbonates viz. **malachite,** $Cu(OH)_2.CuCO_3$ or $CuCO_3.CuO.H_2O$ (green) and **azurite**, $Cu(OH)_2.2CuCO_3$(blue) are known. Azurite is also known as *chessylite*, since it is found at chesy near Lyons. Both are found native in nature.

Preparation. Malachite is obtained:

(*i*) by passing CO_2 gas through an aqueous suspension of $Cu(OH)_2$.

$$2Cu(OH)_2 + CO_2 \longrightarrow Cu(OH)_2.CuCO_3 + H_2O$$

(*ii*) by heating a mixture of $CuSO_4$ and lime stone, $CaCO_3$ in a sealed tube at a temperature of 150-170°C.

$$2CuSO_4 + 2CaCO_3 + H_2O \xrightarrow{150–170°C} Cu(OH)_2.CuCO_3 + 2CaSO_4 + CO_2$$

If the temperature of the above reaction is less than 150–170°C, azurite is obtained.

Properties. (*i*) In presence of moisture and CO_2, malachite is transformed into azurite while the latter is transformed into the former under the conditions favouring the loss of CO_2 and gain of water.

(*ii*) Copper carbonate of laboratory resembles malachite in composition and is prepared as a fine green powder by boiling the mixture of the solutions of $CuSO_4$ and Na_2CO_3.

$$2CuSO_4 + 2Na_2CO_3 + H_2O \longrightarrow CuCO_3.Cu(OH)_2 + 2Na_2SO_4 + CO_2\uparrow$$

(*iii*) *Action of heat.* Both the basic carbonates decompose into black CuO, H_2O and CO_2 on heating.

$$\underset{\text{Malachite}}{Cu(OH)_2.CuCO_3} \xrightarrow{200°C} 2CuO + H_2O + CO_2$$

$$\underset{\text{Azurite}}{Cu(OH)_2.2CuCO_3} \xrightarrow{220°C} 3CuO + H_2O + 2CO_2$$

(*iv*) These basic carbonates dissolve in mineral acids to evolve CO_2. They also dissolve in aqueous NH_3 and NH_4^+ salt solutions.

Uses. Azurite is used as a paint pigment and also for blue flare mixtures in pyrotechniques.

Basic Cupric Acetate or Verdigris, $Cu(CH_3COO)_2.Cu(OH)_2$

Preparation. It is prepared by exposing thin sheets of copper to the action of vinegar, CH_3COOH and air.

$$2Cu + 2CH_3COOH + O_2 \longrightarrow (CH_3COOH)_2Cu.Cu(OH)_2$$

Properties. It is greenish blue powder. It is sparingly soluble in cold water but dissolves in dilute acids and ammonia.

Uses. (*i*) Because of its fairly good covering power, it is used as green pigment, (*ii*) used in dyeing and calico-printing, (*iii*) used in making insecticides and fungicides, (*iv*) when verdigris is mixed with As_2O_3, cupric aceto arsenite, (called *paris green*) $Cu_3(AsO_3)_2.Cu(CH_3COO)_2$ which is poisonous is obtained.

Halides

Excepting CuF, copper forms all the halides in +1 as well as in +2 oxidation state. These halides are CuCl, CuBr, CuI, CuF_2, $CuCl_2.2H_2O$, $CuBr_2$ and CuI_2.

1. Cuprous chloride, CuCl

Preparation. It is obtained by the following reactions:

$$CuCl_2 + Cu \xrightarrow[\text{of conc. HCl}]{\text{Heat in presence}} 2CuCl$$

$$2CuCl_2 + 2H \xrightarrow{\text{Reduction}} 2CuCl + 2HCl$$

$$2CuCl_2 + SO_2 + 2H_2O \xrightarrow{\text{Reduction}} 2CuCl + H_2SO_4 + 2HCl$$

$$2CuCl_2 \xrightarrow{450°C} 2CuCl + Cl_2$$

Properties. (*i*) Cuprous chloride is a colourless crystalline substance, insoluble in water but soluble in HCl and NH_4OH. When exposed to air, it is slowly oxidised to green basic cupric chloride, $3CuO.CuCl_2.3H_2O$.

(*ii*) *Action of oxidising agents.* With oxidising agents, it is oxidised into cupric chloride.

$$2CuCl + 2HCl + O \longrightarrow 2CuCl_2 + H_2O$$

(*iii*) An acidified solution of cuprous chloride absorbs oxygen and carbon monoxide while an ammoniacal solution absorbs acetylene.

$$4CuCl + 4HCl + O_2 \longrightarrow 4CuCl_2 + 2H_2O$$

This reaction is utilised in the manufacture of chloride by *Deacon's process.*

$$CuCl + 2CO \longrightarrow \underset{\text{Addition compound}}{CuCl.2CO}$$

This addition compound decomposes on heating, regenerating carbon monoxide. This reaction is, therefore, employed in gas analysis for separation of carbon monoxide.

$$2CuCl + \underset{\text{Acetylene}}{C_2H_2} \longrightarrow \underset{\text{Copper acetylide}}{C_2Cu_2} + 2HCl$$

Acetylene can be regenerated by treating the acetylide (C_2Cu_2) with strong HCl. The reaction is, therefore, used in the *purification* and *separation of acetylene.*

(*iv*) *Action of alkalies.* Cuprous chloride solution gives with alkalies a yellow ppt. of cuprous oxide which gradually changes to a red colour.

$$2CuCl + 2NaOH \longrightarrow \underset{\substack{\text{Yellow ppt}\\\text{changing to red}}}{Cu_2O} + 2NaCl + H_2O$$

(*v*) *Action of H_2S.* With H_2S, a black precipitate of cuprous sulphide Cu_2S is produced.

$$CuCl + H_2O \longrightarrow \underset{\text{Black ppt}}{Cu_2S} + 2HCl$$

Uses. Ammoniacal solution of cuprous chloride is used in gas analysis for absorbing acetylene while HCl solution of it is used for absorbing CO and oxygen.

2. Cuprous bromide, CuBr

Preparation. It is formed by heating metallic copper with bromine or by dissolving metallic copper in hot concentrated HBr and on pouring the solution into water or by passing SO_2 gas through a solution of copper sulphate and sodium bromide in water.

$$2CuSO_4 + 2NaBr + SO_2 + 2H_2O \longrightarrow 2CuBr + Na_2SO_4 + 2H_2SO_4$$

Properties. It is a greenish-yellow crystalline salt, insoluble in water, but forms soluble complex like cuprous chloride.

3. Cuprous iodide, CuI

Preparation. (*i*) It is obtained as white crystalline powder when KI is added to a solution of $CuSO_4$.

$$2CuSO_4 + 4KI \longrightarrow 2K_2SO_4 + \underset{\text{Green ppt (Unstable)}}{2CuI_2}$$

$$2CuI_2 \longrightarrow 2CuI + I_2$$

$$2CuSO_4 + 4KI \longrightarrow 2K_2SO_4 + \underset{\text{ppt}}{2CuI} + I_2$$

or

$$2Cu^{2+} + 4I^- \longrightarrow \underset{\text{ppt}}{2CuI} + I_2$$

In this reaction, $CuSO_4$ acts as an oxidising agent, since I_2 is liberated from KI.

The liberated of I_2 does not take place, if the above reaction is carried out in presence of a reducing agent like $FeSO_4$ or aqueous SO_2.

$$2CuSO_4 + 2KI + 2FeSO_4 \longrightarrow 2CuI + Fe_2(SO_4)_3 + K_2SO_4$$

$$2CuSO_4 + 2KI + 2H_2O + SO_2 \longrightarrow 2CuI + 2H_2SO_4 + K_2SO_3$$

Insoluble ppt. of CuI is washed and dried by means of alcohol and ether.

(*ii*) CuI is also obtained as a white precipitate:

(*a*) by dissolving metallic copper in hot conc. HI and pouring the resulting solution in water or (*b*) by heating metallic copper in iodine vapour.

Properties. CuI is a white powder which turns red and then black on heating but again becomes white on cooling. It is insoluble in water but dissolves readily in NH_3 and in the solutions of KCN and $Na_2S_2O_3$.

4. Cupric chloride, $CuCl_2.2H_2O$

Preparation. When the metal, its oxide, hydroxide or carbonate is dissolved in concentrated hydrochloric acid and the resulting solution is evaporated to crystallization, emerald green crystals of cupric chloride dihydrate, $CuCl_2.2H_2O$ are separated. The crystals are blue in colour when free from moisture. Anhydrous cupric chloride is obtained as a dark brown mass either by burning copper metal in the excess of chlorine gas or by heating the hydrate, $CuCl_2.2H_2O$ in hydrogen chloride gas at 150°C.

Properties. (*i*) The hydrated salt is emerald green while the anhydrous salt is brown powder (m.p. 498°C).

(*ii*) It is very deliquescent and is readily soluble in water. Its *dilute* solution in blue due to the formation of $[Cu(H_2O)_4]^{2+}$ ion and its *concentrated* aqueous solution is green due to the formation of $[CuCl_4]^{2-}$ ion (*yellow*) together with $[Cu(H_2O)_4]^{2+}$ ion (*blue*).

$$CuCl_2 \rightleftharpoons Cu^{2+} + 2Cl^-$$

$$Cu^{2+} + 4H_2O \longrightarrow [Cu(H_2O)_4]^{2+} \ (\textit{blue})$$

$$Cu^{2+} + 4Cl^- \longrightarrow [CuCl_4]^{2-} \ (\textit{yellow})$$

(*iii*) *Action of conc. HCl.* In conc. HCl, $CuCl_2$ forms a yellow solution due to the formation of $[CuCl_4]^{2-}$ ion.

$$2CuCl_2 \rightleftharpoons Cu[CuCl_4] \rightleftharpoons \underset{\text{yellow}}{Cu^{2+} + [CuCl_4]^{2-}}$$

(*iv*) *Acidic nature.* The aqueous solution of $CuCl_2$ is acidic towards litmus due to hydrolysis.

$$CuCl_2 + 2H_2O \longrightarrow Cu(OH)_2 + 2HCl$$

(*v*) It dissolves easily in many organic solvents such as alcohol, acetone and pyridine. The alcoholic solution burns with a fine green flame.

(*vi*) *Action of heat.* The anhydrous salt on heating gives cuprous chloride and chlorine.

$$2CuCl_2 \longrightarrow 2CuCl + Cl_2$$

The hydrated salt gives cuprous chloride, hydrochloric acid and chlorine.

$$CuCl_2.2H_2O \rightleftharpoons CuCl_2 + 2H_2O$$
$$CuCl_2 + H_2O \rightleftharpoons CuO + 2HCl$$
$$2CuCl_2 \rightleftharpoons 2CuCl + Cl_2$$

(*vii*) *Action of NH_4OH.* $Cu(OH)_2$ is first precipitated, which dissolves in the excess of ammonium hydroxide giving a deep blue solution of the complex tetrammine cupric ion, $[Cu(NH_3)_4]^{2+}$.

$$CuCl_2 + 2NH_4OH \longrightarrow Cu(OH)_2\downarrow + 2NH_4Cl$$
$$Cu(OH)_2 + 4NH_4OH\ (excess) \longrightarrow [Cu(NH_3)_4](OH)_2 + 4H_2O$$

or

$$Cu^{2+} + 4NH_4OH\ (excess) \longrightarrow [Cu(NH_3)_4]^{2+} + 4H_2O$$

On evaporating the solution containing $[Cu(NH_3)_4]^{2+}$ ions, deep blue crystals of tetrammine cupric chloride, $[Cu(NH_3)_4]Cl_2.H_2O$ are obtained.

$$[Cu(NH_3)_4](OH)_2 + 2HCl \longrightarrow \underset{\text{Blue crystals}}{[Cu(NH_3)_4]Cl_2.H_2O} + H_2O$$

(*viii*) *Action of NaOH.* A pale blue precipitate of basic cupric chloride, $CuCl_2.3Cu(OH)_2$ or $[Cu.3Cu(OH)_2]Cl_2$ is obtained.

(*ix*) *Reduction.* It is readily reduced to cuprous chloride.

(*a*) by boiling with Cu turnings.

$$CuCl_2 + Cu \longrightarrow 2CuCl$$

(*b*) by passing SO_2 gas.

$$2CuCl_2 + 2H_2O + SO_2 \longrightarrow 2CuCl + H_2SO_4 + 2HCl$$

(*c*) by the action of zinc dust and hydrochloric acid.

$$2CuCl_2 + 2H \longrightarrow 2CuCl + 2HCl$$

(*d*) by the action of stannous chloride.

$$2CuCl_2 + SnCl_2 \longrightarrow 2CuCl + SnCl_4$$

Uses. It is used : (*i*) as a oxygen carrier in the preparation of organic dyestuffs, (*ii*) in pyrotechnics to produce green flames. (*iii*) in medicine, (*iv*) as a catalyst in the manufacture of chlorine by Deacon's process.

5. Cupric bromide, $CuBr_2$

Preparation. The hydrated salt containing two or four molecules of water of crystallization has been reported. It forms green crystals which deliquesee in air and is, therefore, not easily obtained. The *anhydrous* salt is obtained by:

(*i*) by evaporating a solution of cupric oxide or carbonate in hydrobromic acid in a vacuum desiccator over quick lime.

(*ii*) by treating metallic copper with bromine water and d ying the product in vacuum.

Properties. The anhydrous salt forms black crystals which are very soluble in water and in aqueous solution of mineral acids. It is also soluble in alcohol and acetone. On heating in a dry tube, it is decomposed into cuprous bromide and bromine.

Complex Compounds of Cu^+ ion

Cu^+ ion forms various types of complex compounds. In some compounds Cu^+ ion is present in the cationic portion while in others it is present in anionic portion. Examples are $[Cu(NH_3)_2]Cl$, $H[CuCl_2]$, $K[CuCl_2]$, $[Cu(NH_3)_2]OH$, $K_2[Cu(CN)_3]$, $K_3[Cu(CN)_4]$, $Na[Cu(S_2O_3)]$ etc. These compounds are prepared as follows :

(*i*) Conc. HCl dissolves Cu_2O, forming CuCl which dissolves in the excess of HCl to produce *hydrocuprochloric acid*, $H[CuCl_2]$.

$$Cu_2O + 2HCl \longrightarrow 2CuCl + H_2O$$
$$[CuCl + HCl \longrightarrow H[CuCl_2]] \times 2$$
$$\overline{Cu_2O + 4HCl \longrightarrow 2H[CuCl] + H_2O}$$

(*ii*) In the absence of air, Cu_2O gives colourless soluble *diammine cuprous hydroxide*, $[Cu(NH_3)_2]OH$.

$$Cu_2O + 4NH_4OH \longrightarrow 2[Cu(NH_3)_2]OH + 3H_2O$$

In the presence of air, the above cuprous complex compound is oxidised to *tetrammine cupric hydroxide*, $[Cu(NH_3)_4](OH)_2$.

(*iii*) Cu_2O dissolves in alkali cyanide (*e.g.* KCN) solution, forming a colourless complex compound, $K_2[Cu(CN)_3]$.

$$Cu_2O + 6KCN + H_2O \longrightarrow 2K_2[Cu(CN)_3] + 3KOH$$

(*iv*) CuCl reacts with KCl and forms $K[CuCl_2]$.

$$CuCl + KCl \longrightarrow K[CuCl_2]$$

(*v*) CuCl dissolves readily in NH_4OH forming a colourless solution which contains $[Cu(NH_3)_2]Cl.H_2O$ in the absence of any trace of O_2.

$$CuCl + 2NH_4OH \longrightarrow \underset{\text{Solution}}{[Cu(NH_3)_2]Cl} + 2H_2O$$

The solution of $[Cu(NH_3)_2]Cl$ turns deep blue when exposed to air due to the formation of $[Cu(NH_3)_4]Cl_2$.

(*vi*) $CuSO_4$ reacts with KCN to form $Cu(CN)_2$ which decomposes to give $Cu_2(CN)_2$ and $(CN)_2$. $Cu_2(CN)_2$ dissolves in the excess of KCN and forms *pot. cuprous cyanide*, $K_3[Cu(CN)_4]$.

$$[CuSO_4 + 2KCN \longrightarrow Cu(CN)_2 + K_2SO_4] \times 2$$
$$2Cu(CN)_2 \longrightarrow Cu_2(CN)_2 + (CN)_2$$
$$Cu_2(CN)_2 + 6KCN \longrightarrow 2K_3[Cu(CN)_4]$$
$$\overline{2CuSO_4 + 10KCN \longrightarrow 2K_3[Cu(CN)_4] + 2K_2SO_4 + (CN)_2}$$

(*vii*) When $CuSO_4$ reacts with $Na_2S_2O_3$ (hypo), blue colour of $CuSO_4$ gets discharged and *sodium coprous thiosulphate*, $Na[Cu(S_2O_3)]$ is formed.

$$CuSO_4 + Na_2S_2O_3 \longrightarrow CuS_2O_3 + Na_2SO_4$$
$$2CuS_2O_3 + 2Na_2S_2O_3 \longrightarrow 2Na[Cu(S_2O_3)] + Na_2S_4O_6$$

$Na[Cu(S_2O_3)]$, on boiling or on acidification gives a ppt. of Cu_2S.

$$2Na[Cu(S_2O_3)] + H_2O \longrightarrow Cu_2S + Na_2S_2O_3 + H_2SO_4$$

Complex Compounds of Cu^{2+} ion

Cu^{2+} ion forms a number of complex compounds. Some important complexes are $K_2[Cu(CN)_4]$, $[Cu(py)_2](CN)_2$, $[Cu(NH_3)_4](OH)_2$, $[Cu(NH_3)_4]SO_4$ etc. These compounds are prepared as follows:

(*i*) $Cu(CN)_2$ dissolves in the excess of KCN, forming *pot. cupricyanide*, $K_2[Cu(CN)_4]$.

$$Cu(CN)_2 + 2KCN \longrightarrow K_2[Cu(CN)_4]$$

(*ii*) $Cu(CN)_2$ reacts with pyridine (*py*) and produces the complex compound, $[Cu(py)_2](CN)_2$.

$$Cu(CN)_2 + 2py \longrightarrow [Cu(py)_2](CN)_2$$

(*iii*) When $Cu(OH)_2$ dissolves in the excess of NH_4OH, a blue soluble *cupric ammine hydroxide*, $[Cu(NH_3)_4](OH)_2$ is obtained.

$$Cu(OH)_2 + 4NH_4OH \longrightarrow [Cu(NH_3)_4](OH)_2 + 4H_2O$$

(*iv*) With NH_4OH, $CuSO_4$ forms soluble *tetrammine cupric sulphate*, $[Cu(NH_3)_4]SO_4$.

$$CuSO_4 + 2NH_4OH \longrightarrow \underbrace{Cu(OH)_2}_{ppt} + (NH_4)_2SO_4$$

$$Cu(OH)_2 \xrightarrow{+2NH_4OH} [Cu(NH_3)_4]SO_4 + 4H_2O$$

$[Cu(NH_3)_4]SO_4$ is known as *Schwitzer's reagent* which is used for dissolving cellulose in the manufacture of artificial silk.

Compounds of Gold

Introduction

Gold forms compounds in +1 and +3 oxidation state. Compounds having Au in +1 oxidation state are called *aurous compounds* while those containing Au in +3 oxidation state are known as *auric compounds*. In general auric compounds are more stable and more common than aurous compounds. All gold compounds are easily decomposed. They are easily reduced to free gold metal either by weak reducing agents or simply by heating.

Au(+1) Compounds: Aurous Compounds

Au^+ ion exists in aqueous solution in very minute concentration, since this ion undergoes disproportionation in aqueous solution.

$$3Au^+ \rightleftharpoons Au^{3+} + 2Au$$

Thus the presence of Au^+ ions cannot be detected in aqueous solution. Au^+ ion becomes stable when it forms a complex compound. Thus $K[Au(CN)_2]$ is highly stable in solution.

Examples of compounds containing Au in +1 oxidation state are given below :

(*a*) *Simple compounds.* Au_2O, $Au(OH)$, AuX (X = Cl, Br, I), Au_2S, AuCN etc.

(*b*) *Complex compounds.* $K[Au(CN)_2]$, $Na_3[Au(S_2O_3)_2].\frac{1}{2}H_2O$ etc. 2-coordinate complex compounds given by Au(+1) are *linear* involving *sp* hybridisation whereas 4-coordinate complexes are probably *tetrahedral* resulting from sp^3 hybridistion. Both the types of complexes are *diamagnetic*. Here we shall discuss the following compunds of Au(+1)

1. Aurous chloride, AuCl

Preparation. It is prepared by heating auric chloride ($AuCl_3$) to 175°C.

$$AuCl_3 \xrightarrow{175°C} AuCl + Cl_2$$

Properties. It is a light-yellow powder which is insoluble in water. When heated above 175°C, it gets decomposed into metallic gold and free Cl_2.

$$2AuCl \xrightarrow{>175°C} 2Au + Cl_2$$

When treated with H_2O at ordinary temperature, it is decomposed into metallic Au and $AuCl_3$.

$$3AuCl \longrightarrow AuCl_3 + 2Au$$

AuCl combines with alkali metal chlorides (*e.g.*, KCl) to form chloroaurates (I), $M[AuCl_2]$. When treated with Cl_2, $AuCl_3$ is formed.

$$AuCl + KCl \longrightarrow K[AuCl_2] \text{ [Pot. dichloroaurate (I)]}$$
$$AuCl + Cl_2 \longrightarrow AuCl_3$$

2. Potassium dicyano aurate (I) or potassium aurocyanide, $K[Au(CN)_2]$

Preparation. It is obtained by the following methods:

(*a*) Gold is dissolved in aqua regia and precipitated with ammonia forming fulminating gold. It is washed and dissolved in potassium cyanide and crystallised.

(*b*) Fine gold is treated with potassium cyanide solution in contact with atmosphere oxygen:

$$4Au + 8KCN + O_2 + 2H_2O \longrightarrow 4KAu(CN)_2 + 4KOH$$

Properties. It is a colourless substance and soluble in water. When heated with HCl at 50°C, yellow crystals of AuCN are produced.

$$KAu(CN)_2 + HCl \longrightarrow KCl + HCN + AuCN$$

When Zn is added, gold is precipitated.

$$KAu(CN)_2 + Zn \longrightarrow K_2[Zn(CN)_2] + \underset{ppt}{2Au}$$

Uses. It is used in gold plating.

3. Sodium dithiolsulphato aurate (I) or sodium auro thiosulphate, $Na_3[Au(S_2O_3)_2]$

Just as hypo forms a complex with silver chloride, a similar complex is obtained by adding auric chloride to hypo solution.

$$4Na_2S_2O_3 + AuCl_3 \longrightarrow Na_3[Au(S_2O_3)_2] + Na_2S_4O_6 + 3NaCl$$

Hypo is side by side oxidised to sodium tetrathionate as in iodine reaction.

It crystallises in colourless needles, fairly soluble in water. It is not reduced by $FeSO_4$, oxalic acid or $SnCl_2$. It is used in medicine as *sanochrysin* (Greek *chrusos*).

Au(+3) Compounds: Auric Compounds

These compounds are covalent in nature and donot have Au^{3+} ions in pure aqueous solutions. Auric compounds are powerful oxidising agents. Examples of auric compounds are given below:

(*a*) *Simple compounds.* Au_2O_3, $Au(OH)_3$, AuX_3(X = Cl, Br, I), Au_2S_3, $Au(CN)_3$, $Au(NH_2)(NH)$ etc.

(*b*) *Complex compounds.* $K[Au(CN)_4]$, $Na[AuO_2]$, $H[AuX_4]$ (X = Cl, Br, I), $[Au(NH_3)_4](NO_3)_2$ $[Au(NH_3)_4](C_2O_4)(NO_3)$. These are 4-coordinated complex compounds. These are diamagnetic and square planar resulting from dsp^2 hybridisation. 5-coordinated complexes are *trigonal bipyramidal* while 6-coordinated complexes are *octahedral*. Here we shall discuss the following compounds in some detail.

1. Auric chloride, $AuCl_3$

Preparation. (*i*) $AuCl_3$ is prepared by treating AuCl with water or with Cl_2.

$$3AuCl \longrightarrow AuCl_3 + 2Au$$
$$AuCl + Cl_2 \longrightarrow AuCl_3$$

(*ii*) When metallic gold is dissolved in aqua regia, a bright yellow solution of $HAuCl_4$ is obtained.

$$\underbrace{2Au + 3HNO_3 + 11HCl}_{\text{Aqua regia}} \longrightarrow \underset{\text{Solution}}{2HAuCl_4} + 6H_2O + 3NOCl$$

On evaporating the solution, yellow deliquescent crystals of tetrahydrate chloroauric acid, $HAuCl_4.4H_2O$ deposit. This is called *gold chloride* of commerce. When these yellow crystals are heated at 200°C, deep red crystals of anhydrous auric chloride, $AuCl_3$ are obtained.

$$HAuCl_4.4H_2O \xrightarrow{200°C} AuCl_3 + HCl + 4H_2O$$

(*iii*) The preparation of *anhydrous auric chloride* can also be done as follows: Gold is first reacted with chlorine to give auric chloride which is dissolved in a little water. This is gently heated, in order to convert any contaminated aurous chloride into metallic gold which is thrown down as a precipitate and is filtered out, and then the solution evaporated and heated to 150°C. This gives anhydrous auric chloride. It is soluble in ether.

(*iv*) $AuCl_3$ can also be obtained by the action of moist Cl_2 on gold.

$$2Au + 3Cl_2\ (moist) \longrightarrow 2AuCl_3$$

Properties. (1) Anhydrous $AuCl_3$ is a deep red crystalline substance which is freely soluble in water. When crystallised from water solution, orange crystals of dihydrate, $AuCl_3.2H_2O$ are obtained.

(2) Auric chloride is non-polar (covalent) and does not ionise in aqueous solution. The non-polar character is further shown by its low melting point (290°), volatility and its solubility in organic solvents as alcohol and ether.

(3) *Action of sunlight or heat.* On exposure to sunlight or on heating at 450K, $AuCl_3$ decomposes into AuCl (yellow powder) and Cl_2.

$$AuCl_3 \longrightarrow AuCl + Cl_2$$

(4) *Action of conc. NH_3.* When concentrated ammonia is added to $AuCl_3$, fulminating gold, AuN_2H_3 or HN=Au—NH_2 is obtained.

$$AuCl_3 + 2NH_3 \longrightarrow HN{=}Au - NH_2 + 3HCl$$

(5) *Decomposition of $AuCl_3$ solution.* Solution of $AuCl_3$ is not very stable, since it decomposes to form AuCl and H_2O_2.

$$AuCl_3 + 2H_2O \longrightarrow AuCl + 2HCl + H_2O_2$$

(6) *Reduction.* (*i*) $AuCl_3$ can be reduced to free gold by various reducing agents. For example:

(*a*) ***Reduction by $SnCl_2$ and preparation of colloidal gold or purple of Cassius.*** When $AuCl_3$ solution is treated with $SnCl_2$, $AuCl_3$ is reduced to colloidal particles of Au and $SnCl_2$ is oxidised to $SnCl_4$.

$$2AuCl_3 + 3SnCl_2 \longrightarrow 2Au + 3SnCl_4 \qquad ...(i)$$

$SnCl_4$ formed undergoes hydrolysis to form stannic acid, H_4SnO_4 or $Sn(OH)_4$ which absorbs colloidal particles of Au and purple of Cassius is obtained.

$$SnCl_4 + 4H_2O \longrightarrow H_4SnO_4 \text{ or } Sn(OH)_4 + 4HCl$$

or
$$SnCl_4 + 4H_2O \longrightarrow SnO_2.2H_2O + 4HCl$$

or
$$3SnCl_4 + 12H_2O \longrightarrow 3SnO_2.6H_2O + 12HCl \qquad ...(ii)$$

On adding equations (*i*) and (*ii*) we get

$$2AuCl_3 + 3SnCl_2 + 6H_2O \longrightarrow \underbrace{2Au + 2SnO_2}_{\text{Purple of Cassius}} + 12HCl$$

This equation shows that purple of Cassius may be considered as the colloidal form of SnO_2 containing absorbed colloidal gold.

Purple of Cassius is generally used for making ruby glass and high glass pottery.

(*b*) *Reduction by hydrazine (N_2H_4) and hydroxylamine (NH_2OH).* N_2H_4 and NH_2OH both reduce $AuCl_3$ to free Au metal in acidic medium.

$$4AuCl_3 + 2N_2H_4 \longrightarrow 4Au + 3N_2 + 12HCl$$

$$4AuCl_3 + 6NH_2OH \longrightarrow 4Au + 3N_2O + 12HCl + 3H_2O$$

(*c*) *Reduction by $FeSO_4$.* When $AuCl_3$ solution is treated with $FeSO_4$, gold is precipitated and $FeSO_4$ is oxidised to $Fe_2(SO_4)_3$.

$$AuCl_3 + 3FeSO_4 \longrightarrow Au\downarrow + Fe_2(SO_4)_3 + FeCl_3$$

This reaction has been utilised in the extraction of gold from auriferrous quarts by chlorination process.

(*d*) *Reduction by Ag metal.* When $AuCl_3$ solution is treated with Ag, Au is replaced by Ag and Au is precipitated.

$$AuCl_3 + 3Ag \longrightarrow 3AgCl + Au\downarrow$$

The above reaction is used for toning process used in photography. In this process the photographic plate obtained after fixing is dropped in $AuCl_3$ solution. In this process silver grains of the image are replaced by those of gold (purple), as is evident from the above chemical equation.

(*e*) *Reduction by formaldehyde.* When $AuCl_3$ is reduced by formaldehyde, Faraday's gold is obtained. It is red, brown or purple in colour.

(*f*) *Reduction by KI solution.* $AuCl_3$ (Au = +3) is reduced to AuI (Au = +1) by KI solution (I = –1). In this reaction KI is oxidised to I_2 (I = 0)

$$AuCl_3 + 3KI \longrightarrow AuI + 3KCl + I_2$$

(*g*) *Reduction by H_2S gas.* $AuCl_3$ (Au = +3) is reduced to AuS (Au = +2) by H_2S gas (S = –2). In this reaction H_2S is oxidised to H_2SO_4 (S = +6).

$$8AuCl_3 + 9H_2O + 4H_2O \longrightarrow 8AuS + H_2SO_4 + 24HCl$$

(7) *Formation of complex compounds.* $AuCl_3$ reacts with various substances and forms complex compounds. For example :

(*a*) With excess of HCl, $H[AuCl_4]$ is obtained. With alkali chloride (*e.g.*, NaCl), well-defined crystalline, $Na[AuCl_4]$ is obtained. $Na[AuCl_4]$ is a salt of $H[AuCl_4]$.

$$AuCl_3 + HCl\ (excess) \longrightarrow H[AuCl_4]$$

$$AuCl_3 + NaCl \longrightarrow Na[AuCl_4]$$

(*b*) When conc. solution of $AuCl_3$ is treated with KCN, solution which contains $K[Au(CN)_4]$ is obtained. On evaporating and cooling the solution, colourless crystals of $K[Au(CN)_4].\frac{3}{2}H_2O$ are obtained.

$$AuCl_3 + 4KCN \longrightarrow K[Au(CN)_4] + 3KCl$$

(*c*) When a solution of $AuCl_3$ is treated with $Na_2S_2O_3$ (hypo), $Na_3[Au(S_2O_3)_2]$ is obtained.

$$AuCl_3 + 4Na_2S_2O_3 \longrightarrow Na_3[Au(S_2O_3)_2] + 2Na_2S_4O_6 + 3NaCl$$

(*d*) When KOH or NaOH is added to a solution of $AuCl_3$, $Au(OH)_3$ is precipitated which dissolves in the excess of the alkali forming potassium or sodium aurate (III), $M[AuO_2]$ (M = K or Na).

$$AuCl_3 + 3NaOH \longrightarrow \underset{ppt}{Au(OH)_3} + 3NaCl$$

$$Au(OH)_3 + NaOH \longrightarrow Na[AuO_2] + 2H_2O$$

Structure. The vapour density of auric chloride shows that it is a dimeric molecule, Au_2Cl_6.

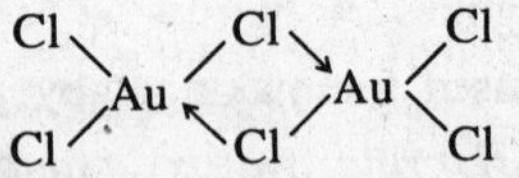

Cl Cl Cl
Au Au
Cl Cl Cl

Uses. Hydrochloroauric acid, $H[AuCl_4].4H_2O$ under the name of *gold chloride* of commerce is used in photography as a toning agent.

2. Fulminating gold or gold fulminate, $Au(NH_2)(NH)$

Preparation. It is prepared :

(*i*) by treating auric chloride solution with ammonia :

$$AuCl_3 + 2NH_3 \longrightarrow HN{=}Au{-}NH_2 + HCl$$

(*ii*) by treating auric hydroxide with ammonia.

$$Au(OH)_3 + 2NH_3 \longrightarrow HN{=}Au{-}NH_2 + 3H_2O$$

Properties. (*i*) It is an olive green powder of variable composition.

(*ii*) The dry mass explodes violently on heating or percussion, forming Au, N_2, NH_3 and steam.

$$2Au(NH_2)(NH) \longrightarrow 2Au + N_2 + 2NH_3$$

(*iii*) It dissolves in HCl to form $H[AuCl_4]$ and NH_4Cl

$$Au(NH_2)(NH) + 6HCl \longrightarrow H[AuCl_4] + 2NH_4Cl$$

(*iv*) It readily dissolves in alkali cyanide solutions. It is converted to non-explosive compounds by the action of H_2S and by careful heating with S. It is reduced to Au with explosion by $SnCl_2$.

Structure. Raschig (1986) formulated it as $HN{=}Au{-}NH_2.\frac{3}{2}H_2O$. According to Weitz (1915) the fulminating gold produced by the action of NH_3 on $AuCl_3$ solution contains some chlorine and, therefore, he suggested the following structure to this compound.

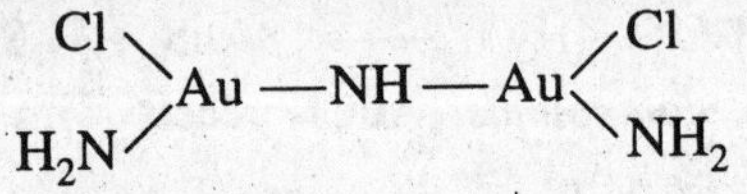

Cl Cl
Au —NH— Au
H_2N NH_2

3. Tetrachloroauric acid or hydrogen tetrachloro aurate (III), $H[AuCl_4]$

Preparation. (*i*) When metallic gold is dissolved in aqua regia, a bright yellow solution of $HAuCl_4$ is obtained

$$2Au + 3HNO_3 + 11HCl \longrightarrow \underset{Solution}{2HAuCl_4} + 6H_2O + 3NOCl$$

Actually the above reaction takes place through the following steps :

$$[HNO_3 + 3HCl \longrightarrow 2H_2O + NOCl + 2Cl] \times 3$$
$$[Au + 3Cl \longrightarrow AuCl_3] \times 2$$
$$[AuCl_3 + HCl \longrightarrow H[AuCl_4]] \times 2$$

$$2Au + 3HNO_3 + 11HCl \longrightarrow 2H[AuCl_4] + 6H_2O + 3NOCl$$

On evaporating the solution, yellow deliquescent crystals of tetrahydrate chloroauric acid, $HAuCl_4.4H_2O$ deposit. This is called *gold chloride* of commerce.

(*ii*) When $AuCl_3$ reacts with excess of HCl, $H[AuCl_4]$ is obtained.

$$HCl\ (excess) + AuCl_3 \longrightarrow H[AuCl_4]$$

(*iii*) When fulminating gold, $Au(NH_2)(NH)$ reacts with HCl, $H[AuCl_4]$ is produced.

$$Au(NH_2)(NH) + 6HCl \longrightarrow H[AuCl_4] + 2NH_4Cl$$

Properties. (*i*) It is soluble in water and donor organic solvents.

(*ii*) *Action of heat.* When the crystals of $H[AuCl_4]$ are heated in a current of Cl_2, HCl and $AuCl_3$ (deep red crystals) are obtained.

$$H[AuCl_4] \xrightarrow[\text{of } Cl_2]{\text{Heat in a current}} AuCl_3 + HCl$$

(*iii*) *Reduction.* It is easily reduced to metallic gold by reducing agents like $FeSO_4$, $SnCl_2$ and even by H_2.

$$HAuCl_4 + 3FeSO_4 \longrightarrow Fe_2(SO_4)_2 + FeCl_3 + HCl + Au$$

$$2HAuCl_4 + 3SnCl_2 \longrightarrow 3SnCl_4 + 3HCl + 2Au$$

$$2HAuCl_4 + 3H_2 \longrightarrow 8HCl + 2Au$$

(*iv*) *Action of KI.* Reaction with KI precipitates AuI.

$$HAuCl_4 + 3KI \longrightarrow \underset{ppt}{AuI} + 3KCl + HCl + I_2$$

(*v*) *Reaction with alkalies.* Alkalies precipitate $Au(OH)_3$ which dissolves in excess of the alkali solution to form soluble complex, $Na[Au(OH)_4]$.

$$HAuCl_4 + 4NaOH \longrightarrow \underset{ppt}{Au(OH)_3} + 4NaCl + H_2O$$

$$Au(OH)_3 + NaOH \longrightarrow \underset{\text{Soluble complex}}{Na[Au(OH)_4]}$$

(*vi*) *Reaction with KCN.* First AuCN is obtained as precipitate which in excess of KCN dissolves due to the formation of soluble complex, $K[Au(CN)_2]$.

$$HAuCl_4 + 3KCN \longrightarrow \underset{ppt}{AuCN} + 3KCl + HCl + (CN)_2$$

$$AuCN + KCN \longrightarrow \underset{\text{Soluble complex}}{K[Au(CN)_2]}$$

(*vii*) *Reaction with H_2S.* With H_2S the acid yields a precipitate which is probably a mixture of Au_2S and Au_2S_3.

(*viii*) *Formation of tetrachloroaurates.* $H[AuCl_4]$ is a monobasic acid and gives salts of $M[AuCl_4]$ type (M = alkali metal). These salts are called *tetrachloroaurates* which contain tetra chloro aurate (III) ion, $[AuCl_4]^-$. Alkali tetrachloro aurates can be obtained by the action of an alkali chloride on $H[AuCl_4]$ or on $AuCl_3$.

$$\underset{(M = Na,\ K)}{MCl} + H[AuCl_4] \longrightarrow M[AuCl_4] + HCl$$

$$MCl + AuCl_3 \longrightarrow M[AuCl_4]$$

$[AuCl_4]^-$ ion is diamagnetic and has square planar geometry arising from dsp^2 hybridisation of Au atom.

Uses. The acid and its sodium and potassium salts are used in photography for toning and printing.

4. Potassium tetracyano aurate (III) or potassium auricyanide, $K[Au(CN)_4]$

Preparation. A hot concentrated solution of $AuCl_3$ when mixed with KCN gives $K[Au(CN)_4]$ in solution.

$$AuCl_3 + 4KCN \longrightarrow \underset{\text{Solution}}{K[Au(CN)_4]} + 3KCl$$

When the solution is evaporated and cooled, colourless crystals of $K[Au(CN)_4].\frac{3}{2}H_2O$ are obtained.

Properties. The crystals of $K[Au(CN)_4]$ are colourless and give a ppt. of $Ag[Au(CN)_4]$ when treated with a solution of $AgNO_3$.

$$K[Au(CN)_4] + AgNO_3 \longrightarrow Ag[Au(CN)_4] + KNO_3$$

Uses. $K[Au(CN)_4]$ is used in gold plating.

Compounds of Zinc

1. Zinc oxide or zinc white or Chinese white, ZnO

This oxide is also called *philosopher's wool.* In nature it occurs as *zincite* or *red zinc ore.*

Preparation. (*i*) *Commercial preparation.* On the large scale, ZnO is obtained by burning Zn-vapours in air. ZnO is collected as white wooly flocks and hence is called philosopher's wool.

(*ii*) It also obtained by roasting zinc blende (ZnS) or by the calcination of calamine ($ZnCO_3$), $Zn(NO_3)_2$, $Zn(OH)_2$ etc.

$$\underset{\text{Zinc blende}}{2ZnS} + 3O_2 \xrightarrow{\Delta} 2ZnO + 2SO_2$$

$$ZnCO_3 \longrightarrow ZnO + CO_2$$

$$2Zn(NO_3)_2 \longrightarrow 2ZnO + 4NO_2 + O_2$$

$$Zn(OH)_2 \longrightarrow ZnO + H_2O$$

Properties. It is a white amorphous powder which becomes sulphur-yellow on heating but again white on cooling. It is insoluble in water. It is an amphoteric oxide since, it dissolves readily in acids forming the zinc salt and in alkalies forming zincates.

$$ZnO + 2HCl \longrightarrow ZnCl_2 + H_2O$$

$$ZnO + NaOH \longrightarrow \underset{\text{Sod. zincates}}{Na_2ZnO_2} + H_2O$$

When heated in dry hydrogen above 450° or with carbon, it is reduced to metallic zinc. When it is heated with a solutionn of $Ca(NO_3)_2$, a green mass is formed which is due to the formation of *cobalt zincate*, $CoZnO_2$ (double oxide, ZnO.CoO) which is known as *Rinmann's green.*

$$ZnO + H_2 \longrightarrow H_2O + Zn$$

$$ZnO + C \longrightarrow Zn + CO\uparrow$$

$$2Co(NO_3)_2 \longrightarrow 2CoO + 4NO_2 + O_2$$

$$ZnO + CoO \longrightarrow ZnO.CoO \text{ or } CoZnO_2$$

Uses. Zinc oxide is used : (*i*) as a white paint under the name of *zinc white* or *chinese white* although it does not possess as good a covering powder as white lead. It is, however, sometimes preferred because it is not blackened by H_2S. (*ii*) as a catalyst when mixed with Cr_2O_3 in the manufacture of methyl alcohol from water gas and H_2. (*iii*) as a filler for soft rubber (*iv*) as an absorbent in surgical dressing. (*v*) for the preparation of cement and hard resins. (*vi*) for the preparation of other zinc compounds. (*vii*) the green mass obtained by heating ZnO with $Co(NO_3)_2$ is used as a pigment under the name of *Rinmann's green.*

2. Lithopone, ($ZnS + BaSO_4$)

It is a white pigment and is a mixture of ZnS and $BaSO_4$.

Preparation. It is obtained as a precipitate when solutions of BaS and $ZnSO_4$ are mixed together.

$$BaS + ZnSO_4 \longrightarrow \underbrace{ZnS + BaSO_4}_{\text{Lithopone}}$$

The precipitate of ($ZnS + BaSO_4$) thus obtainedis dried, heated to redness in a muffle furnace, quenched with H_2O and finally ground to fine powder. The process of quenching with H_2O increases the whitness of lithopone.

Properties and uses. It is a valuable white pigment and is used in wall paints and cheap enamel paints. It does not get blackened when exposed to H_2S but slowly darkens in light due to the liberation of Zn-metal from ZnS. In dark the liberated Zn-metal is oxidised to ZnO due to which the paint regains its original whiteness. Lithopone is much cheaper than white lead, $2PbCO_3.Pb(OH)_2$ and is non-poisonous. It has a good covering power. Lithopone is also used in rubber industry in the manufacture of white goods.

SOME TYPICAL PROBLEMS WITH SOLUTIONS

Problem 1. *How is $K_2Cr_2O_7$ obtained from $Cr_2(SO_4)_3$?*

Solution. $$Cr_2(SO_4)_3 + 3H_2O \longrightarrow \boxed{Cr_2O_3} + 3H_2SO_4$$

$$2Cr_2O_3 + 8NaOH + 3O_2 \longrightarrow \boxed{4\,Na_2CrO_4} + 4H_2O$$

$$2Na_2CrO_4 + H_2SO_4 \longrightarrow \boxed{Na_2Cr_2O_7} + Na_2SO_4 + H_2O$$

$$Na_2Cr_2O_7 + 2KCl \longrightarrow K_2Cr_2O_7 + 2NaCl$$

Problem 2. *What happens when 30% solution of H_2O_2 is added to an acidic solution of K_2CrO_4.* *(Delhi 1999)*

Solution. When 30% solution of H_2O_2 is added to an acidic solution of K_2CrO_4 containing a small amount of ether, an intense blue colour [due to the formation of chromium diperoxide, Cr $(O_2)_2O$ or CrO_5] is produced in the ether solution.

$$K_2CrO_4 + H_2SO_4 + 2H_2O_2 \xrightarrow{\text{Ether}} K_2SO_4 + 3H_2O + \underbrace{CrO_5 \text{ or } Cr\,(O_2)_2\,O}_{\text{Blue colour}}$$

or $$CrO_4^{2-} + 2H^+ + 2H_2O_2 \xrightarrow{\text{Ether}} \underbrace{CrO_5 \text{ or } Cr\,(O_2)_2\,O}_{\text{Blue colour}} + 3H_2O$$

Problem 3. *What happens when CH_3COOH is added to a mixture containing $NaNO_2$ and $Co\,(NO_3)_2$.* *(Delhi 1999)*

Solution. When CH_3COOH is added to a mixture containing $NaNO_2$ and $Co\,(NO_3)_2$, yellow ppt. of sodium cobaltinitrite, $Na_3\,[Co\,(NO_2)_6]$ is obtained.

$$2CH_3\,COOH + 7Na\,NO_2 + Co\,(NO_3)_2 \longrightarrow 2CH_3\,COO\,Na + 2NaNO_3 + H_2O + NO + \underset{\text{yellow ppt.}}{Na_3\,[Co\,(NO_2)_6]}$$

or $$Co^{2+} + 2H^+ + 3Na^+ + 7NO_2^- \longrightarrow \underset{\text{yellow ppt.}}{Na_3\,[Co\,(No_2)_6]} + H_2O + NO$$

Problem 4. *What happens when:*

(i) H_2S is passed through the acidified solution of K_3 [Fe (CN)$_6$]. *(Delhi 1997)*

(ii) Hydrazine is added into alkaline solution of K_3 [Fe (CN)$_6$] *(Delhi 2000)*

Solution. (*i*) H_2S (S = – 2) is oxidised to S (S = 0) and $K_3[Fe(CN)_6]$ is reduced to $K_4[Fe(CN)_6]$

$$6K_3[Fe^{3+}(CN)_6] + 3H_2S \longrightarrow 4K_4[Fe^{2+}(CN)_6] + 3S + K_2Fe^{2+}[Fe^{2+}(CN)_6] + 6HCN.$$
(Fe = + 3) (S = – 2) (Fe = +2) (S = O)

(*ii*) Hydrazine (N_2H_4) is oxidised to N_2 and $K_3[Fe(CN)_6]$ is reduced to $K_4[Fe(CN)_6]$

$$4K_3[Fe(CN)_6] + 4KOH + N_2H_4 \longrightarrow 4K_4[Fe(CN)_6] + N_2 + 4H_2O$$
(Fe = + 3) (N = – 2) (Fe = + 2) (N = 0)

or $4[Fe(CN)_6]^{3-} + 4OH^- + N_2H_4 \longrightarrow 4[Fe(CN)_6]^{4-} + 4H_2O + N_2$

Problem 5. *What happens when:*

(i) Potassium ferrocyanide is heated with conc. H_2SO_4 *(Delhi 1999, 2001)*

(ii) Potassium ferrocyanide is treated with dil. HNO_3 *followed by neutralisation with* Na_2CO_3. *(Delhi 2001)*

Solution. (*i*) CO gas is evolved.

$$K_4[Fe(CN)_6] + 6H_2SO_4\,(conc.) + 6H_2O \longrightarrow 2K_2SO_4 + 3(NH_4)_2SO_4 + FeSO_4 + 6CO.$$

(*ii*) When $K_4[Fe(CN)_6]$ is treated with HNO_3, it gets oxidised to $K_3[Fe(CN)_6]$ and HNO_3 is reduced to NO

$$3K_4[Fe(CN)_6] + 4HNO_3 \longrightarrow 3K_3[Fe(CN)_6] + 3KNO_3 + 2H_2O + NO$$

$K_3[Fe(CN)_6]$ obtained as above reacts with Na_2CO_3 in presence of NO and ruby-red crystals of sodium nitroprusside, $Na_2[Fe^{2+}(CN)_5(NO^+)]$ are obtained.

$$2K_3[Fe^{2+}(CN)_6] + 3Na_2CO_3 + 2NO \longrightarrow 2Na_2[Fe^{2+}(CN)_5(NO^+)] + 2NaCN + 3K_2CO_3.$$

Problem 6. *An orange coloured compound (A) when treated with NaCl and conc.* H_2SO_4 *gives red brown vapour (B). This vapour when passed through NaOH gives yellow coloured solution (C). On addition of* CH_3COOH *and* $(CH_3COO)_2Pb$ *to this yellow solution, yellow coloured ppt. (D) is obtained. Identify A to D and also give the chemical equations involved.*

(Delhi 2001)

Solution. Reactions are:

(*i*) $K_2Cr_2O_7 + 4NaCl + 3H_2SO_4\,(conc.) \longrightarrow 2CrO_2Cl_2 + K_2SO_4 + Na_2SO_4 + 3H_2O$
Orange (A) — Red vapour (B)

(*ii*) $CrO_2Cl_2 + 4NaOH \longrightarrow Na_2CrO_4 + 2NaCl + H_2O$
(B) — Yellow solution (C)

(*iii*) $Na_2CrO_4 + (CH_3COO)_2Pb \longrightarrow PbCrO_4 + CH_3COOH$
(C) — yellow ppt (D)

Thus A = $K_2Cr_2O_7$, B = CrO_2Cl_2, C = Na_2CrO_4 and D = $PbCrO_4$.

Problem 7. *Write the ionic equation for the convers ion of: (i) Dichromateion to chromate ion (ii) Chromate ion to dichromate ion.*

Solution. (*i*) $Cr_2O_7^{2-} + 2OH^- \longrightarrow 2CrO_4^{2-} + H_2O$
Orange — Yellow

(*ii*) $2CrO_4^{2-} + 2H^+ \longrightarrow Cr_2O_7^{2-} + H_2O$
Yellow — Orange

Problem 8. *Explain why orange solution of* $K_2Cr_2O_7$ *turns yellow on adding NaOH to it.*

Solution. When NaOH is added to a solution of $K_2Cr_2O_7$ (orange solution), it is converted into K_2CrO_4 which has yellow colour. Thus the change from orange colour to yellow colour is due to the formation of K_2CrO_4

$$K_2Cr_2O_7 + 2NaOH \longrightarrow 2K_2CrO_4 + H_2O$$
Orange — Yellow

or $\underset{\text{Orange}}{Cr_2O_7^{2-}} + 2OH^- \longrightarrow \underset{\text{Yellow}}{2CrO_4^{2-}} + H_2O$

Multiple Choice Problems

Problem 9. *The equilibrium, $Cr_2O_7^{2-} \rightleftharpoons 2CrO_4^{2-}$ is shifted to the right in:*

(a) an acidic medium *(b) a basic medium*

(c) a neutral medium *(d) it does not exist.*

Solution. Answer is (*b*). We know that the given equilibrium is actually written as:

$$Cr_2O_7^{2-} + OH^- \underset{+\text{ Acid}}{\overset{+\text{ Alkali}}{\rightleftharpoons}} 2\,CrO_4^{2-} + H^+$$

This equilibrium shows that the equilibrium is shifted to the right only in presence of an alkali (i.e., in basic medium).

Problem 10. *$KMnO_4$ acts as an oxidant in neutral, alkaline as well as acidic media. The final products obtained from it in the three conditions are respectively.*

(a) MnO_2, MnO_2, Mn^{2+} *(b) MnO_4^-, Mn^{3+}, Mn^{2+}*

(c) MnO_2, MnO_4^{2-}, Mn^{3+} *(d) MnO, MnO_2, Mn^{2+}* *(PMT, MP 1997)*

Solution with explanation. Answer is (*a*). We know that the reduction of $KMnO_4$ or MnO_4 (purple) in acidic, alkaline and neutral media can be represented by the following ionic equations:

$$\underset{\substack{\text{purple}\\(Mn = +7)}}{2MnO_4} + 8H^+ + 5e^- \xrightarrow[\text{acidic medium}]{\text{Reduction in}} \underset{\substack{\text{colourless}\\(Mn = +2)}}{2Mn^{2+}} + 4H_2O$$

$$\underset{\substack{\text{purple}\\(Mn = +7)}}{MnO_4^-} + 2H_2O + 3e^- \xrightarrow[\text{alkaline / neutral medium}]{\text{Reduction in}} \underset{\substack{\text{colourless}\\(Mn = +4)}}{MnO_2} + 4OH^-$$

Thus answer (*a*) is the correct answer.

Problem 11. *The number of moles of $KMnO_4$ that will be needed to react with one mole of SO_3^{2-} ion in acidic solution is:*

(a) $\frac{2}{5}$ *(b)* $\frac{3}{5}$ *(c)* $\frac{4}{5}$ *(d) 1* *(IIT 1997)*

Solution. When $KMnO_4$ reacts with SO_3^{2-} ion in acidic solution, it gets reduced to Mn^{2+} ion and SO_3^{2-} ion is oxidised to SO_4^{2-} ion. Thus:

$$[\underset{(Mn = +7)}{MnO_4^-} + 8H^+ + 5e^- \xrightarrow{\text{Reduction}} \underset{(Mn = +2)}{Mn^{2+}} + 4H_2O] \times 2$$

$$[\underset{(S = +4)}{SO_3^{2-}} + H_2O \xrightarrow{\text{Oxidation}} \underset{(S = +6)}{SO_4^{2-}} + 2H^+ + 2e^-] \times 5$$

On adding : $2MnO_4^- + 5SO_3^{2-} + 6H^+ \longrightarrow 2Mn^{2+} + 5SO_4^{2-} + 3H_2O$

This equation shows that:

No. of moles of MnO_4^- needed to react with 5 moles of $SO_3^{2-} = 2$

$\therefore$ No. of moles of MnO_4^- needed to react with 1 mole of $SO_3^{2-} = \frac{2}{5}$.

Thus answer is (*a*).

Question 12. *The number of moles of $KMnO_4$ that will be needed to react completely with one mole of ferrous oxalate in acidic solution is:*

(a) 3/5 *(b) 2/5* *(c) 4/5* *(d) 1* *(IIT 1997)*

Solution. When $KMnO_4$ reacts with Fe (C_2O_4) in acidic medium, $Fe(C_2O_4)$ is converted into $Fe_2(C_2O_4)_3$ and CO_2 is evolved.

$$2KMnO_4 + 3H_2SO_4 \longrightarrow K_2SO_4 + 2MnSO_4 + 3H_2O + 5O] \times 3$$

$$\underset{(Fe = +2)}{2Fe(C_2O_4)} + 3H_2SO_4 + 3O \longrightarrow \underset{(Fe = +3)}{Fe_2(SO_4)_3} + 4CO_2 + 3H_2O] \times 5$$

On adding: $\underset{(Mn = +7)}{6KMnO_4} + 24H_2SO_4 + \underset{(Fe = +2)}{10\,Fe(C_2O_4)} \longrightarrow 3K_2SO_4 + 20CO_2$

$$+ \underset{(Mn = +2)}{6\,MnSO_4} + 24H_2O + \underset{(Fe = +3)}{5Fe_2(SO_4)_3}$$

This equation show that :

10 moles of Fe $(C_2O_4) \equiv$ 6 moles of $KMnO_4$

$$1 \text{ mole of Fe } (C_2O_4) \equiv \frac{6}{10} \text{ moles of } kMnO_4$$

$$\equiv \frac{3}{5} \text{ moles of } KMnO_4.$$

The above equations can also be written in the ionic forms as:

$$\underset{(Mn = +7)}{MnO_4^-} + 8H^+ + 5e^- \xrightarrow{\text{Reduction}} Mn^{2+} + 4H_2O] \times 3$$

$$\underset{(Fe = +2)}{Fe(C_2O_4)} \xrightarrow{\text{Oxidation}} Fe^{3+} + 2CO_2 + 3e^-] \times 5$$

On adding: $\underset{(Mn = +7)}{3MnO_4^-} + 24H^+ + \underset{(Fe = +2)}{5Fe(C_2O_4)} \longrightarrow \underset{(Mn = +2)}{3Mn^{2+}} + 12H_2O + \underset{(Fe = +3)}{5Fe^{3+}} + 10CO_2$

This equation shows that:

5 moles of Fe $(C_2O_4) \equiv$ 3 moles of MnO_4^-

$$\therefore \quad 1 \text{ mole of Fe } (C_2O_4) \equiv \frac{3}{5} \text{ moles of } MnO_4^-$$

Thus the correct answer is *(a)*

Question 13. *For the redox reaction:*

$MnO_4^- + C_2O_4^{2-} + H^+ \longrightarrow Mn^{2+} + CO_2 + H_2O,$

the correct coefficients of the reactants, for the balanced reaction are:

	MnO_4^-	$C_2O_4^{2-}$	H^+
(a)	2	5	16
(b)	16	5	2
(c)	5	16	2
(d)	2	16	5

(IIT 1992)

Solution. On balancing the given redox reaction, we find that the balanced equation is:

$5C_2O_4^{2-} + 2MnO_4^- + 16H^+ \longrightarrow 10CO_2\ 2Mn^{2+} + 8H_2O$

Thus the correct answer is *(a)*

University Questions

1. Describe the preparation, properties and uses of titanium oxide. *(Meerut 96)*
2. Give the preparation of $KMnO_4$. *(Meerut 90, 95, 97)*
3. What happen when metavanadate is heated. *(Meerut 88)*
4. Explain the following :
   (*i*) V_2O_5 is an important compound of vanadium. *(Raj. 79)*
   (*ii*) In V_2O_5, vanadium is in + 5 oxidation state and hence should be colourless, but it is actually orange, in colour. *(Delhi Hans.86)*
5. How would you get vanadium anhydride from vanadinite. *(Meerut 90)*
6. Give an account of the preparation, properties anduses of V_2O_5. *(Meerut 89)*
7. Compare vanadium (III) with vanadium (V) with respect to redox property and tendency to form anionic complexes. *(Meerut, M.Sc. 1989)*
8. Give the preparation, properties and uses of:
   (*i*) Sodium nitroprusside. (*ii*) Prussian blue *(Meerut 95, 96)*
9. Give the reasons for the following:
   (*i*) Hexacyanoferrate (III) is orange while hexacyanoferrate (II) is yellow. *(Madurai 1985)*
   (*ii*) Fe (II) and Fe (III) form complexes with CN^- ions but not with NH_3. *(Delhi 1987)*
   (*iii*) Ferric salts are more stable than ferrous salts. *(Osmania 93)*
10. How will you get $K_2[PtCl_6]$ from platinum? *(Meerut 1990)*
11. What happens when:
    (*i*) $H_2[PtCl_6]$ reacts with SO_2. (*ii*) NH_4Cl is added to $H_2[PtCl_6]$. *(Agra 85)*
12. Write a note on "Platinum ammines". *(Lucknow 95)*
13. Give the preparation, properties and uses of chloroplatinic acid. *(Meerut 95, 96)*
14. Explain why Pt(IV) complexes are generally octahedral while Pt(II) ocmplexes are square planar. *(Meerut 1983)*
15. What happens when:
    (*i*) Gold chloride solution reacts with NH_2OH? *(Agra 84, Delhi 87)*
    (*ii*) Gold chloride is treated with stannous chloride. *(Meerut 1989 S)*
16. Discuss the structure of hydrogen tetrachloro aurate. *(Bundelkhand 88)*
17. Give preparation, properties and uses of gold fulminate. *(Vikram 86)*
18. How is $K_2Cr_2O_7$ obtained from chromite ore *(Meerut 89)*
19. Discuss the electronic structure of $K_2Cr_2O_7$ molecule. *(Meerut 89)*
20. Give the preparation, properties and uses of $Na_2Cr_2O_7$. *(Meerut 84)*
21. What happens when $(NH_4)_2Cr_2O_7$ is heated? *(Agra 85)*
22. How would you get CrO_2Cl_2 from chromite ore? Give necessary equations. *(Meerut 89)*
23. How is K_2CrO_4 prepared? Give its uses. *(Utkal 86)*
24. How will you prepare CrO_2Cl_2 from Cr-metal? *(Meerut 95)*
25. Discuss the chemistry of the peroxy compounds of chromium. *(Meerut M.Sc. 85)*
26. What happens when:
    (*i*) A soluble of K_2CrO_4 containing H_2O_2 is acidified with H_2SO_4 in presence of ether or pyridine? *(Delhi B.Sc. Hans. 87)*
    (*ii*) A solution of $K_2Cr_2O_7$ containing H_2O_2 is acidified in presence of ether or pyridine.
    (*iii*) H_2O_2 is added to an aqueous solution of $K_2Cr_2O_7$ and the solution is acidified with H_2SO_4. *(Sambalpur 86)*

27. Starting from chrome iron, how would you obtain a pure specimen of $K_2Cr_2O_7$, $(NH_4)_2Cr_2O_7$, Cr_2O_3, CrO_2Cl_2 and metallic chromium? *(Meerut 89)*

28. Give the preparation, properties and uses of (*i*) $K_2Cr_2O_7$, (*ii*) perchromic acid. *(Meerut 95)*

29. Explain why $Cr(CH_3COO)_3$ is diamagnetic. *(Berhampur 91)*

30. An aqueous solution of $CoCl_2$ has pink colour. When a little conc.HCl is added to it and heated, the colour of the solution changes to deep blue. Explainwhy. *(Utkal 91)*

31. Describe the preparation, properties and uses of sodium cobaltinitrite. *(Meerut 96)*

32. Give the preparation properties and uses of:

(*a*) $K_2Cr_2O_7$, pot. ferrocyanide. (*b*) $K_2Cr_2O_7$, $TiCl_4$ *(Meerut 97, 98)*

33. How will you prepare?

(*i*) CrO_2Cl_2 from Cr-metal (*ii*) $Ni(CO)_4$ from Ni-metal

(*iii*) $PbCrO_4$ from $Cr(OH)_3$ (*iv*) K_2MnO_4 from $KMnO_4$

(*v*) K_2CrO_4 from $K_2Cr_2O_7$ (*vi*) Cr-metal from Cr_2O_3 *(Meerut 97, 98)*

34. What happens when ?

(*i*) NaCl is heated with solid $K_2Cr_2O_7$ and conc. H_2SO_4 and the evolved gas is passed into NaOH solution.

(*ii*) Pt-metal is dissolved in aqua regia.

(*iii*) Chromic chloride solution is treated with NaOH and then with H_2O_2.

(*iv*) $KMnO_4$ is heated with HCl. *(Meerut 97)*

35. Write short notes on : (*i*) Prussian blue (*ii*) Pot. ferri cyanide *(Meerut 98)*

36. Write methods for the preparation, properties and uses of $HAuCl_4$. *(Lucknow 2002).*

37. (*a*) Giving chemical equations to explain what happens when :

(*i*) Potassium dichromate is heated with sodium sulphite in dilute sulfuric acid in aqueous medium.

(*ii*) Ferric chloride in excess is treated with potassium ferrocyanide.

(*iii*) Cuprous chloride is heated with acetylene.

(*b*) A blue solution of copper sulfate turns brown on addition of Kl. This brown colour is discharged on treatment with excess of sodium thiosulfate and a white solid settles down. How can these observations be explained ? *(Delhi 2002)*

Ans. (*a*) (*i*) $K_2Cr_2O_7 + 3Na_2SO_3 + 3H_2SO_4 \longrightarrow K_2SO_4 + Cr_2(SO_4)_3 + 3Na_2SO_4 + 3H_2O$

(*ii*) $4FeCl_3 + 3K_4[Fe(CN)_6] \longrightarrow Fe_4[Fe(CN)_6]_3 + 12KCl$

(*iii*) $CuCl + CH \equiv CH \longrightarrow CuC \equiv CCu + 2HCl$

(*b*) $2CuSO_4 + 4KI \longrightarrow 2CuI + K_2SO_4 + I_2$

$2Na_2S_2O_3 + I_2 \longrightarrow Na_2S_4O_6 + 2NaI$

white ppt.

38. (*i*) Give preparation, properties and uses of the following compounds : (*a*) Sodium tungstate (*b*) Molybdenum blue (*c*) Vanadyl sulphate

(*ii*) Write a short note on : Tunsten bronzes. *(Lucknow 2002)*

39. Describe the important complexes of Zn^{2+} ion. *(VBS, Purvanchal 2003)*

40. Discuss the preparation, properties and application of $K_2Cr_2O_7$ and $KMnO_4$. *(VBS Purvanchal 2003)*

41. (*a*) What happens when NH_4VO_3 is heated ?

(*b*) Complete and balance the following reaction :

$K_2Cr_2O_7 + KCl + H_2SO_4 \longrightarrow ......$ *(VBS Purvanchal 2003)*

42. (*a*) Explain giving equations what happens when –

(*i*) Alkali is added to a solution of $KMnO_4$

(*ii*) H_2O_2 is added to the solution of a dichromate in dil H_2SO_4

(*iii*) $K_4[Fe(CN)_6]$ is heated with conc H_2SO_4

(*b*) Give one method of preparation and structure of $K_2Cr_2O_7$. *(Delhi 2003)*

43. (*a*) Draw the structure of TiO_2.

(*b*) How is prussian blue formed? Give its properties. *(GND Amritsar, 2004)*

Part III: Inner-transition Elements and Their Compounds

10

General Characteristics of Inner-transition Elements : f-Block Elements (Lanthanides and Actinides)

Introduction

The elements from Ce_{58} to La_{71} (14 elements) and from Th_{90} to Lw_{103} (14 elements) are called **inner-transition elements.** These elements have been shown at a separate place at the bottom of the periodic table. The first series of 14 elements lies between La_{57} and Hf_{72} in 6th period while the other series of 14 elements lies between Ac_{89} and Ku_{104} in 7th period (incomplete period). Both the series are present in IIIB (3) group.

The elements of 1st series are called **lanthanides** or **lanthanones** (rare earth elements) while those of the 2nd series are called **actinides** or **actinones.** The names lanthanides and actinides have been given, since these elements follow La_{57} and Ac_{89} respectively with which these elements show close similarities.

Since the last electron (differentiating electron) in the atoms of these elements enters the *f*-sub-shell belonging to anti-penultimate shell [*i.e.*, $(n - 2)$th shell], these elements are also called ***f*-block elements.** Thus the last electron enters $(n - 2)f$ sub-shell in inner-transition elements.

Since lanthanides are present in 6th period ($n = 6$) of the periodic table, the last electron in these elements goes to $(6 - 2)f$ or 4*f* sub-shell. Thus lanthanides are also called 4*f* **block elements,** *i.e., lanthanides can be defined as those elements in which the last electron goes to 4f sub-shell.*

Similarly, since actinides are present in 7th period ($n = 7$) of the periodic table, the last electron in these elements goes to $(7 - 2)f$ or 5*f* sub-shell. Thus actinides are also called **5*f*-block elements**, *i.e., actinides are those elements in which the last electron enters 5f sub-shell.*

Valence-shell electronic configuration of the atoms of *f*-block elements can be represented as $(n - 2)f^{0-14}$. $(n - 1)d^{0-2}$. ns^2 which shows that in these elements the outermost three shells are partially-filled while the remaining inner shells are completely-filled.

Now we shall discuss lanthanides (4*f*-block elements) and actinides (5*f*-block elements) under separate sections.

Lanthanides: 4*f*-Block Elements

As already said, the 14 elements *viz.* Ce_{58} to Lu_{71} are called lanthanides.

General characteristics of lanthanides

The following are important properteis of lanthanides.

1. Electronic configurations

We know that the electronic configuration of La_{57} which is followed by 14 lanthanides is $[Xe]_{54}\,4f^0\,5d^1\,6s^2$ in which $5d$ sub-shell is singly-filled and $4f$ sub-shell is vacant. This configuration shows that as we move in the series of 14 lanthanides (Ce_{58} to Lu_{71}), the additional electron should occupy the vacant $4f$ orbitals and $5d$ orbitals should remain singly-filled, *i.e.*, the expected configurations of the atoms of lanthanides should be those in which $5d$ orbitals are singly-filled and $4f$ orbitals are progressively filled up with electrons. In other words, we can say that the expected configuration of the atoms of lanthanides should be $[Xe]_{54}\,4f^{1-14}\,5d^1\,6s^2$. But, since the energies of $5d$ and $4f$ orbitals are closely similar, in all the lanthanides, excepting Gd_{64} and Lu_{71}, $5d^1$ electron gets shifted to 4f orbitals and hence $5d$ orbitals remain vacant. In Gd_{64}, the shifting of $5d^1$ electron to $4f$ orbitals does not take place, since this type of shifting gives unstable configuration *viz.* $[Xe]_{54}$ $5f^8 5d^0 6s^2$ to Gd_{64}. In Lu_{71} the shifting of $5d^1$ electron to $4f$ orbitals is also not possible because $4f$ orbitals are already filled to their maximum capacity of 14 electrons.

Observed as well as expected configurations are given in Table 10.1. Observed configurations show that the complete and valence configurations can be written as :

Complete configuration : 2, 8, 18, 20 to 32, 8 or 9, 2

or 2, 8, 18, $4s^2p^6d^{10}f^{2-14}$, $5s^2p^6d^{0-1}$, $6s^2$

or $[Xe]_{54}\,4f^{2-14}\,5d^{0-1}6s^2$

Valence-shell configuration : $4f^{2-14}\,5d^{0-1}\,6s^2$.

2. Oxidation states

Main oxidation states exhibited by lanthanides are shown in Table 10.2. This table shows that +3 is the principal oxidation state which is shown by all the lanthanides. Although a few lanthanides also show +2 and +4 oxidation states, +3 is the most stable oxidation state for all the lanthanides since some of the M^{2+} and M^{4+} cations are converted into M^{3+} ions, *e.g.*, Sm^{2+} is a good reducing agent while Ce^{4+} is a good oxidising agent, since both these ions are converted into M^{3+} ions which are the most stable ions. This is evident from the following reactions.

$$\underset{}{2Sm^{2+}} + \underset{(H\,=\,+1)}{2H_2O} \longrightarrow 2Sm^{3+} + 2OH^- + \underset{(H\,=\,0)}{H_2}$$

$$Ce^{4+} + Fe^{2+} \longrightarrow Ce^{3+} + Fe^{3+}$$

For some lanthanides, +2, +3 and +4 oxidation states are explained on the basis of the fact that M^{2+}, M^{3+} and M^{4+} ions attain $4f^0$, $4f^7$ and $4f^{14}$ configurations respectively which are very stable configurations. For example:

(*i*) La and Ce attain $4f^0$ configuration when they are in +3 and +4 oxidation states respectively.

$La^{3+} = [Xe]_{54}\,4f^0$; $Ce^{4+} = [Xe]_{54}\,4f^0$

(*ii*) Eu, Gd and Tb get $4f^7$ configuration in +2, +3 and +4 oxidation states respectively.

$Eu^{2+} = [Xe]_{54}\,4f^7$; $Gd^{3+} = [Xe]_{54}\,4f^7$; $Tb^{4+} = [Xe]_{54}\,4f^7$

(*iii*) Yb and Lu show +2 and +3 oxidation states, since these oxidation states have $4f^{14}$ configuration

$Yb^{2+} = [Xe]_{54}\,4f^{14}$; $Lu^{3+} = [Xe]_{54}\,4f^{14}$

Exceptions. Many of the lanthanides show +2, +3 and +4 oxidation states but the elements in these oxidation states *donot* have $4f^0$, $4f^7$ and $4f^{14}$ configurations. For example :

$Ce^{2+} = [Xe]_{54}\,4f^2$ $Sm^{2+} = [Xe]_{54}\,4f^6$

$Tm^{2+} = [Xe]_{54}\,4f^{13}$ $Ce^{3+} = [Xe]_{54}\,4f^1$

$Pr^{3+} = [Xe]_{54}\,4f^2$ $Pr^{4+} = [Xe]_{54}\,4f^1$

$Nd^{4+} = [Xe]_{54}\,4f^2$ $Dy^{2+} = [Xe]_{54}\,4f^8$

The above oxidation states can be explained on the basis of thermodynamic and kinetic factors.

Table 10.1. *Expected and observed (actual) electronic configurations of lanthanide atoms. The configuration of La_{57} is also given. $[Xe]_{54}$ = 2, 8, 18, 18, 8*

Lanthanide element	*Expected configuration*	*Observed (actual) configuration*
Lanthanum, La_{57}	$[Xe]_{54}\ 4f^1 5d^0 6s^2$ = 2, 8, 18, (18 + 1), 8, 2 = 2, 8, 18, 19, 8, 2	$[Xe]_{54}\ 4f^0\ 5d^1\ 6s^2$ = 2, 8, 18, (18 + 0), (8 + 1), 2 = 2, 8, 18, 18, 9, 2
Cerium, Ce_{58}	$[Xe]_{54}\ 4f^1 5d^1 6s^2$ = 2, 8, 18, (18 + 1), (8 + 1), 2 = 2, 8, 18, 19, 8, 2	$[Xe]_{54}\ 4f^2 5d^0 6s^2$ = 2, 8, 18, (18 + 2), 8, 2 = 2, 8, 18, 20, 8, 2
Praseodymium, Pr_{59}	$[Xe]_{54}\ 4f^2 5d^1 6s^2$ = 2, 8, 18, (18 + 2), 8, 2 = 2, 8, 18, 20, 9, 2	$[Xe]_{54}\ 4f^3 5d^0 6s^2$ = 2, 8, 18, (18 + 3), 8, 2 = 2, 8, 18, 21, 8, 2
Neodymium, Nd_{60}	$[Xe]_{54}\ 4f^3 5d^1 6s^2$ = 2, 8, 18, (18 + 3), 9, 2 = 2, 8, 18, 21, 9, 2	$[Xe]_{54}\ 4f^4\ 5d^0\ 6s^2$ = 2, 8, 18, (18 + 4), 8, 2 = 2, 8, 18, 22, 8, 2
Promethium, Pm_{61}	$[Xe]_{54}\ 4f^4 5d^1 6s^2$ = 2, 8, 18, (18 + 4), 9, 2 = 2, 8, 18, 22, 9, 2	$[Xe]_{54}\ 4f^5\ 5d^0\ 6s^2$ = 2, 8, 18, (18 + 5), 8, 2 = 2, 8, 18, 23, 8, 2
Samarium, Sm_{62}	$[Xe]_{54}\ 4f^5 5d^1 6s^2$ = 2, 8, 18, (18 + 5), 9, 2 = 2, 8, 18, 23, 9, 2	$[Xe]_{54}\ 4f^6\ 5d^0\ 6s^2$ = 2, 8, 18, (18 + 6), 8, 2 = 2, 8, 18, 24, 8, 2
Europium, Eu_{63}	$[Xe]_{54}\ 4f^6 5d^1 6s^2$ = 2, 8, 18, (18 + 6), 9, 2 = 2, 8, 18, 24, 9, 2	$[Xe]_{54}\ 4f^7\ 5d^0\ 6s^2$ = 2, 8, 18, (18 + 7), 8, 2 = 2, 8, 18, 25, 8, 2
Gadolinium, Gd_{64}	$[Xe]_{54}\ 4f^7 5d^1 6s^2$ = 2, 8, 18, (18 + 7), 9, 2 = 2, 8, 18, 25, 9, 2	$[Xe]_{54}\ 4f^7\ 5d^1\ 6s^2$ = 2, 8, 18, (18 + 7), 9, 2 = 2, 8, 18, 25, 8, 2
Terbium, Tb_{65}	$[Xe]_{54}\ 4f^8 5d^1 6s^2$ = 2, 8, 18, (18 + 8), 9, 2 = 2, 8, 18, 26, 9, 2	$[Xe]_{54}\ 4f^9\ 5d^0\ 6s^2$ = 2, 8, 18, (18 + 9), 8, 2 = 2, 8, 18, 27, 8, 2
Dysprosium, Dy_{66}	$[Xe]_{54}\ 4f^9 5d^1 6s^2$ = 2, 8, 18, (18 + 9), 9, 2 = 2, 8, 18, 27, 9, 2	$[Xe]_{54}\ 4f^{10}\ 5d^0\ 6s^2$ = 2, 8, 18, (18 + 10), 8, 2 = 2, 8, 18, 28, 8, 2
Holmium, Ho_{67}	$[Xe]_{54}\ 4f^{10} 5d^1 6s^2$ = 2, 8, 18, (18 + 10), 9, 2 = 2, 8, 18, 28, 9, 2	$[Xe]_{54}\ 4f^{11}\ 5d^0\ 6s^2$ = 2, 8, 18, (18 + 11), 8, 2 = 2, 8, 18, 29, 8, 2
Erbium, Er_{68}	$[Xe]_{54}\ 4f^{11} 5d^1 6s^2$ = 2, 8, 18, (18 + 11), 9, 2 = 2, 8, 18, 29, 9, 2	$[Xe]_{54}\ 4f^{12}\ 5d^0\ 6s^2$ = 2, 8, 18, (18 + 12), 8, 2 = 2, 8, 18, 30, 8, 2
Thulium, Tm_{69}	$[Xe]_{54}\ 4f^{12} 5d^1 6s^2$ = 2, 8, 18, (18 + 12), 9, 2 = 2, 8, 18, 30, 9, 2	$[Xe]_{54}\ 4f^{13}\ 5d^0\ 6s^2$ = 2, 8, 18, (18 + 13), 8, 2 = 2, 8, 18, 31, 8, 2
Ytterbium, Yb_{70}	$[Xe]_{54}\ 4f^{13} 5d^1 6s^2$ = 2, 8, 18, (18 + 13), 9, 2 = 2, 8, 18, 31, 9, 2	$[Xe]_{54}\ 4f^{14}\ 5d^0\ 6s^2$ = 2, 8, 18, (18 + 14), 8, 2 = 2, 8, 18, 32, 8, 2
Lutetium, Lu_{71}	$[Xe]_{54}\ 4f^{14} 5d^1 6s^2$ = 2, 8, 18, (18 + 14), 9, 2 = 2, 8, 18, 32, 9, 2	$[Xe]_{54}\ 4f^{14}\ 5d^1\ 6s^2$ = 2, 8, 18, (18 + 14), 9, 2 = 2, 8, 18, 32, 9, 2

3. Atomic and ionic radii of M^{3+} ions: Lanthanide contraction

Atomic and ionic radii of M^{3+} ions have already been given in Table 10.2. A look at these values shows that there in a *steady* decrease in these values as we move from Ce to Lu and from Ce^{3+} to Lu^{3+}.

Table 10.2. *Oxidation states, atomic radii and ionic radii (of M^{3+} ions) of lanthanides.*

Lanthanides	*Oxidation states*	*Atomic radii (pm)*	*Ionic radii (M^{3+} ions) (pm)*
La	+3	169	106
Ce	+3, +4	165	103
Pr	+3, +4	165	101
Nd	+2, +3, +4	164	100
Pm	+3	—	98
Sm	+2, +3	166	96
Eu	+2, +3	185	95
Gd	+3	161	94
Tb	+3, +4	159	92
Dy	+3, +4	159	91
Ho	+3	158	89
Er	+3	157	88
Tm	+2, +3	156	87
Yb	+2, +3	170	86
Lu	+3	156	85

The steady decrease in the atomic and ionic radii (M^{3+} ions) of lanthanide elements with increasing atomic number is called **lanthanide contraction.**

When we say that there is a steady decrease in the values of atomic and ionic radii, this means that the decrease is *very small.* For example on moving from Ce to Lu the atomic radii decrease from 165 pm to 156 pm and the decrease is only equal to 165 – 156 = 9 pm. Similarly, the ionic radii decrease from 103 to 85 on moving from Ce^{3+} to Lu^{3+} and the decrease in only equal to 103 – 85 = 18 pm. Thus we see that although the atomic number or nuclear charge increases by +14 units which is a large number, the decrease in atomic and ionic radii is only equal to 9 pm and 18 pm respectively. This decrease in atomic and ionic radii is very small as compared to that found in the elements of other groups and periods.

It may be seen from the table that although the atomic radii show some irregularities (*e.g.*, the atomic radii of Eu and Yb are abnormally high), the ionic radii of M^{3+} ions contract steadily.

Cause of lanthanide contraction

We know that as we proceed from one element to the next one in lanthamide series, the nuclear charge (*i.e.*, atomic number) increases by +1 at each next element. Thus as we move from Ce to Lu, the attraction between the nucleus and the outermost shell electron increases gradually at each step. It is also known that as we move from Ce to Lu, the addition of extra electron takes place to 4*f* orbitals. Since 4*f* orbitals have very diffused shape, the electrons in these orbitals are not able to shield (decrease) effectively the attraction of the nucleus for the electrons in the outer-most shell as the atomic number of lanthanides increases. Thus it is only due to the gradual increase in the nuclear charge (*i.e.*, increase in the attraction between the nucleus and the outer-most shell electrons) that the size of the lanthanide atoms and M^{3+} ions decreases gradually with atomic number. The above discussion shows that it is due to the poor shielding effect of 4*f* electrons and gradual increase in the nuclear charge that the lanthanide contraction takes place among lanthanides.

Consequences of lanthanide contraction

Lanthanide contraction plays an important role in determining the chemistry of lanthanides and heavier transition series elements. Some, important consequences of lanthanide contraction are discussed below :

(*a*) **Basic character of lanthanide hydroxides, $M(OH)_3$.** Due to lanthanide contraction, the size of +3 lanthanide ions (M^{3+} ions) decreases regularly with increase in atomic number. As a result of this decrease in size, the covalent character between M^{3+} ion and OH^- ions increases from $La(OH)_3$ to $Lu(OH)_3$. (*Fajan's rules*). Therefore, the basic character of the hydroxides decreases with increase in atomic number. Consequently, $La(OH)_3$ is most basic while $Lu(OH)_3$ is the least basic.

(*b*) **Similarities among lanthanides.** Because of very small change in the radii of lanthanides, their chemical properties are quite similar. It is due to their similar properties that the lanthanides cannot be separated from each other in pure state. Recently, methods based on repeated fractional crystallisation or ion exchange techniques, which take the advantage of slight differences in their properties like solubility, complex ion formation, hydration etc. arising from very slight size differences of their trivalent ions, have been used.

(*c*) **Comparison between the densities of the elements of 1st, 2nd and 3rd transition series.** See "*Atomic Volumes and Densities*" discussed in the chapter of "*General Characteristics of a d-Block Elements*" (Chapter 6).

(*d*) **Resemblance between the atomic radii of the elements of 2nd and 3rd transition series**. See "*Atomic Radii*" discussed in the chapter of "*General characteristics of d-blcok Elements*". (Chapter 6).

4. Colour of M^{3+} ions

Most of the trivalent cations of lanthanide elements are coloured in the solid as well as in aqueous solution while only a few ions are colourless. (see Table 10.3).

Table 10.3. *Colour of trivalent cations (M^{3+}) of lanthanide ions.*

M^{3+} ions	*Colour*	*Valence-shell configuration and the number of 4f-electrons = n*	*M^{3+} ions*	*Colour*	*Valence-shell configuration and the number of 4f-electrons = x = (14 – n)*
La^{3+}	Colourless	$4f^0$ $(n = 0)$	Lu^{3+}	Colourless	$4f^{14}$ $(x = 14 - 0 = 14)$
Ce^{3+}	Colourless	$4f^1$ $(n = 1)$	Yb^{3+}	Colourless	$4f^{13}$ $(x = 14 - 1 = 13)$
Pr^{3+}	Green	$4f^2$ $(n = 2)$	Tm^{3+}	Pale green	$4f^{12}$ $(x = 14 - 2 = 12)$
Nd^{3+}	Bright pink	$4f^3$ $(n = 3)$	Er^{3+}	Pink	$4f^{11}$ $(x = 14 - 3 = 11)$
Pm^{3+}	Pink yellow	$4f^4$ $(n = 4)$	Ho^{3+}	Pale yellow	$4f^{10}$ $(x = 14 - 4 = 10)$
Sm^{3+}	Yellow	$4f^5$ $(n = 5)$	Dy^{3+}	Yellow	$4f^9$ $(x = 14 - 5 = 9)$
Eu^{3+}	Pale pink	$4f^6$ $(n = 6)$	Tb^{3+}	Pale pink	$4f^8$ $(x = 14 - 6 = 8)$
Gd^{3+}	Colourless	$4f^7$ $(n = 7)$			

It may be seen from the table that the colour depends on the number of electrons present in 4*f* orbitals. The ion having *n* electrons in 4*f* orbitals has the same colour as the ion which has (14 – *n*) electrons in 4*f* orbitals. For example La^{3+} ($4f^0$) which has no electron in its 4*f* orbitals (*n* = 0) is colourless and Lu^{3+} ion ($4f^{14}$) which has (14 – 0) = 14 electrons in 4*f* orbitals is also colourless. Similarly Pr^{3+} ion ($4f^2$) which has two electrons in its 4*f* orbitals (*n* = 2) and Tm^{3+} ion ($4f^{12}$) which has (14 – 2) = 12 electrons in its 4*f* orbitals have the same colour (green). (See Fig. 10.1)

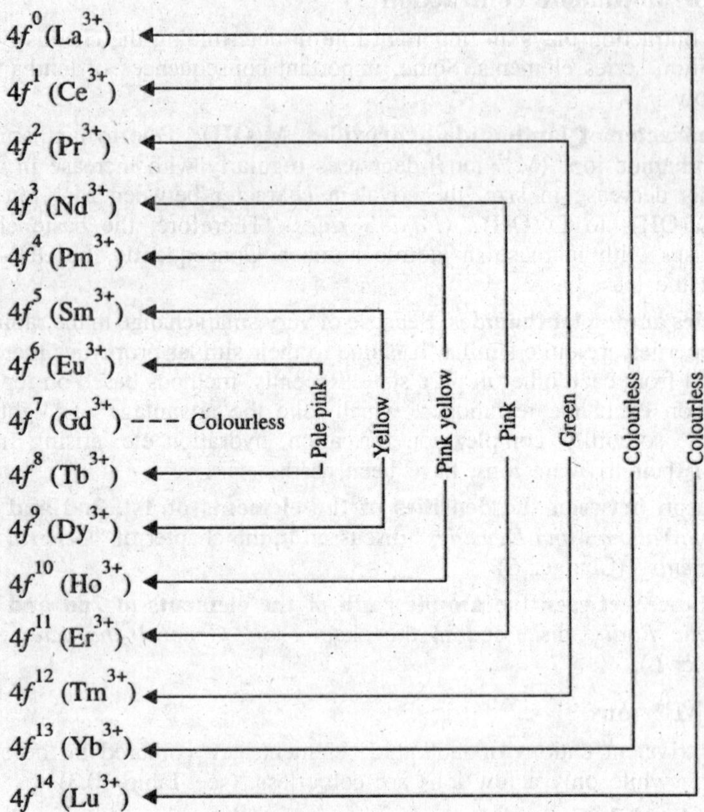

Fig. 10.1. Colour of tripositive lanthanide cations (M^{3+} ions).

5. Magnetic properties of M^{3+} ions

We know that ions having unpaired electrons are paramagnetic while those having all the electrons paired are diamagnetic. Thus, since La^{3+} ($4f^0.5d^0.6s^0$) and Lu^{3+} ($4f^{14}$ $5d^0$ $6s^0$) ions have all the electrons paired, these ions are diamagnetic. On the other hand, the remaining tripositive lanthanide ions (M^{3+} ions) are paramagnetic. since they contain 1, 2, ...7 unpaired electrons.

Cause of paramagnetism

In case of M^{3+} cations, since 4*f* orbitals are quite deeply seated, these are well shielded from the ligands surrounding the M^{3+} ions by the overlapping 5*s* and 5*p* orbitals and hence the electrons residing in 4*f* orbitals are not able to interact with the surrounding ligands. Thus the orbital motion of electrons in 4*f* orbitals is not quenched as in the case of *d*-block elements, and hence the observed paramagnetism in 4*f*-block elements is due to both factors namely orbital motion of the electron and its spinning round its axis.

Calculated an experimental values of magnetic moment (in BM) of the sulphates of triposotive lanthanide ions (M^{3+} ions) are given in Table 10.4 In this table *n* indicates the number of unpaired electrons and J is the total angular momentum. This table shows that the calculated value increases upto Nd^{3+} (= 3.62) and then drops to 0.84 for Sm^{3+}. It starts rising again, becoming maximum for Dy^{3+} (= 10.65). It again starts decreasing becoming zero for Lu^{3+} which is diamagnetic.

Table 10.4. *Calculated and experimental values of magnetic moment (in BM) of the sulphates of M^{3+} ions (n = No. of unpaired electrons in 4f-orbitals)*

M^{3+} ion	$4f^x$ electrons	n	Calculated value, $\mu_{cal} = g\sqrt{J(J+1)}$ BM (Theoretical values)	Experimental value, μ_{exp} in (BM)
La^{3+}	$4f^0$ (x = 0)	0	0.00	Diamagnetic
Ce^{3+}	$4f^1$	1	2.54	2.3–2.5
Pr^{3+}	$4f^2$	2	3.58	3.4–3.6
Nd^{3+}	$4f^3$	3	3.62	3.5–3.6
Pm^{3+}	$4f^4$	4	2.70	—
Sm^{3+}	$4f^5$	5	0.84	1.5–1.6
Eu^{3+}	$4f^6$	6	0.00	3.4–3.6
Gd^{3+}	$4f^7$	7	7.94	7.8–8.0
Tb^{3+}	$4f^8$	6	9.70	9.4–9.6
Dy^{3+}	$4f^9$	5	10.65	10.4–10.5
Ho^{3+}	$4f^{10}$	4	10.60	10.3–10.5
Er^{3+}	$4f^{11}$	3	9.60	9.4–9.6
Tm^{3+}	$4f^{12}$	2	7.60	7.1–7.4
Yb^{3+}	$4f^{13}$	1	4.50	4.4–4.9
Lu^{3+}	$4f^{14}$ (x = 14)	0	0.00	Diamagnetic

6. Physical properties

All the lanthanides are soft, malleable and ductile, and have low tensile strength. They are not good conductors of heat and electricity. In general, the atomic volumes and densities of these elements increase with the increase in atomic number.

7. Chemical reactivity of lanthanides

All the lanthanides are almost equally chemical reactive. Their similar chemical reactivity is due to the fact that since 4*f* electrons in lanthanides are very effectively shielded from the interaction with other elements by the overlapping 5*s*, 5*p* and 6*s* electrons (5*d* orbitals donot contain any electron), these elements have very little difference in their chemical reactivity. It is because of similarity in their chemical reactivity that lanthanides occur together in nature and hence it is difficult to separate these elements from each other.

Lanthanides are generally more reactive them *d*-block elements.

8. Formation of compounds

(*a*) Many lanthanides react with carbon to form salt-like carbides and with H_2 to give salt- like hydrides. Lanthanides react with air or O_2 and S to form sesquioxides (M_2O_3) and sulphides (M_2S_3) respectively. Cerium gives CeO_2. Ytterbium, however, resists the action of air even at 1000°C due to the formation of a protective coating of its oxide. M_2O_3 oxides react with H_2O to form in soluble hydroxides, $M(OH)_3$. Oxides react with CO_2 to give carbonates, $M_2(CO_3)_3$.

All the lanthanides react with H_2O slowly in cold but rapidly on heating and liberate H_2 and form $M(OH)_3$.

(*b*) Lanthanide compounds are generally ionic, coloured and show paramagnetism. Fluorides, oxides, hydroxides, carbonates, phosphates, chromates and oxalates of lanthanides are largely soluble in water. On the other hand, the halides other than fluorides, nitrates, acetates, perchlorates and salts of oxy acids of lanthanides follow the pattern of solubility of the salts of group 2 elements. However, lanthanide sulphates are soluble in water.

All hydroxides, $M(OH)_3$ are ionic and basic in nature. These are stronger bases than $Al(OH)_3$, but weaker than $Ca(OH)_2$. Due to lanthanide contraction, the basicity of the hydroxides decreases from $La(OH)_3$ to $Lu(OH)_3$. Thus $La(OH)_3$ is the strongest base and $Lu(OH)_3$ is the weakest base.

Hydrides react with CO_2 to form $M_2(CO_3)_3$.

9. Formation of complexes

Although tripositive lanthanide cations have a high charge equal to +3 on them, yet their size is so large that their charge-to-radius ratio becomes so small that these ions have very poor tendency to form complexes. Common ligands with which M^{3+} ions form stable complexes are: (*i*) *chelating oxygen containing ligands* like EDTA citric acid, oxalic acid, acetylacetone, (*ii*) *nitrogen containing ligands* like ethylenediamine, NCS etc.

10. Properties dependent on standard reduction potential values

The values of standard reduction potential (E^0 values) for the reduction reaction:

$$M^{3+}\ (aq) + 3e^- \rightleftharpoons M(s)$$

increase from La to Lu, *i.e.*, E^0 values becomes less negative as we move from La to Lu as shown below (values are given in volts):

——————— *E*° *values become less negative* ———————→

La = – 2.52, Ce = –2.48, Pr = – 2.46, Nd = – 2.43, Pm = – 2.42, Sm = – 2.41, Eu = – 2.40, Gd = –2.39, Tb = – 2.39, Dy = – 2.35, Ho = – 2.32, Er = – 2.30, Tm = – 2.28, Yb = – 2.27, Lu = – 2.25.

These E° values explain the following properties of lanthanides:

(*a*) **Reducing property.** Since lanthanides have high negative E° values, these elements have a strong tendency to lose their three electrons to undergo oxidation and hence are strong reducing agents. With the increase of E^0 values from La to Lu, *i.e.*, as E° values become less negative from La to Lu, the reducing power of these elements decreases from La to Lu.

(*b*) **Electropositive character.** Due to negative E^0 values, the lanthanides readily lose their electrons and hence show strong electropositive (or metallic) character. With the increase of E° values from La to Lu, the electropositive character of these elements decreases from La to Lu.

(*c*) **Liberation of H_2 from H_2O.** Due to highly electropositive character of lanthanides, these elements slowly react with cold water and rapidly with hot water to liberate H_2 from it.

Extraction of lanthanides from monazite sand

The method used for the extraction of lanthanides from monazite sand consists of the steps which have been shown in flow sheet 1 given on page 346. The solution obtained after removing cerium as basic ceric nitrate, $Ce(OH)(NO_3)_3.3H_2O$ contains the nitrates of La, Pr, Nd and Sm. The individual lanthanides are separated from this solution by suitable methods.

Separation of lanthanide elements

All the lanthanide elements form tripositive ions (M^{3+} ions) which are almost identical in size. The chemical properties of lanthanides which depend on the size and charge of M^{3+} ions are almost the same. Due to the similarity in chemical properties, it is difficult to separate the lanthanides from each other. However, the following methods have been used to separate them.

1. Fractional crystallisation method. Simple salts of lanthanides like nitrates, sulphates, oxalates, bromates, perchlorates and double salts such as $2M(NO_3)_3.\ 3Mg(NO_3)_2.24H_2O$ crystallise well and form well-defined crystals. Since the solubility of these simple and double salts decreases from La to Lu, these lanthanides can be separated from each other by repeating the fractional crystallisation a number of times. In the separation of $Nd(NO_3)_3$ from $Pr(NO_3)_3$ the use of non-aqueous solvent like ether has been made.

2. Fractional precipitation method. When NaOH is added to a solution of lanthanide nitrates, $Lu(OH)_3$ which is the *weakest base* and has the *lowest solubility* product is precipitated *first* while $La(OH)_3$ which is the *strongest base* and has the *highest solubility product* remains dissolved and precipitates out *last* of all. By dissolving and precipitating the hydroxides for a number of times it is possible to get the complete separation of lanthanides.

3. Change in oxidation state method. As already mentioned, some of the lanthanides show +2 and +4 oxidation states in addition to the usual +3 state which is the common state for all the lanthanides. The properties of M^{2+} and M^{4+} ions are different from those of M^{3+} ions. A change in oxidation state results in a remarkable change in properties. By taking advantages of different properties of M^{2+}, M^{3+} and M^{4+} ions, it has been possible to separate some of the lanthanides. The following examples illustrate this procedure.

(*i*) Suppose, we have a mixture containing Ce^{3+} and other tripositive lanthanide ions. This mixture is treated with a strong oxidising agent like alkaline solution of $KMnO_4$ so that Ce^{3+} ion gets oxidised to Ce^{4+} ion while other M^{3+} ions remain as such in solution. Now the solution containing Ce^{4+} and other M^{3+} ions is treated with an alkali whereby Ce^{4+} ion gets precipitated as $Ce(OH)_4$ while other M^{3+} ions remain unaffected and hence remain as such in solution.

(*ii*) Eu^{3+} can almost quantitatively be separated from a solution containing Eu^{3+} and other M^{3+} ions by reducing it to Eu^{2+} by means of Zn-amalgam and then precipitating it as $EuSO_4$ which is insoluble in water and hence can be separated. The sulphates of all other tripositive lanthanides, on the other hand, are soluble in water.

4. Ion exchange method—A modern method. This is the most rapid and most effective method. When an aqueous solution containing the mixture of usual trivalent lanthanide ions, $M^{3+}(aq)$ is passed through a column having synthetic cation-exchange resin [abbreviated as HR (*solid*)], the $M^{3+}(aq)$ ions replace H^+ ion of the resin and thus get fixed on it.

$$M^{3+}(aq) + 3HR\ (solid) \longrightarrow MR_3(solid) + 3H^+\ (aq)$$

Since $Lu^{3+}(aq)$ is largest in size and $La^{3+}(aq)$ is smallest, $La^{3+}(aq)$ is attached to the column with maximum and $Lu^{3+}(aq)$ with minimum firmness.

In order to recover the M^{3+} ions fixed on the resin, the column is *eluted* (*i.e.*, leached) with a citric acid—ammonium citrate solution (called **eluant** or **eluante**). During elution process NH_4^+ ions replace M^{3+} ions and M-citrate complexes are formed.

$$MR_3 + 3NH_4^+ \longrightarrow 3NH_4R + M^{3+}$$

$$M^{3+} + \text{citrate ions} \longrightarrow \text{M-citrate complex}$$

We have seen that since La^{3+} (*aq*) is attached to the resin with maximum and $Lu^{3+}(aq)$ with minimum firmness, *Lu-citrate complex comes out of the column first* and *La-citrate complex comes out last*. In actual practice the process has to be repeated several times by careful control of concentration of citric acid-ammonium citrate solution.

5. Solvent extraction method. This method makes the use of the difference in the value of partition coefficient of lanthanides between two solvents. $La(NO_3)_3$ and $Gd(NO_3)_3$ have been separated from each other by this method. The partition coefficient of $Gd(NO_3)_3$ between water and tributyl phosphate in kerosene is different from that of $La(NO_3)_3$ between the same solvents. This means that $Gd(NO_3)_3$ can be separated from $La(NO_3)_3$ by continuous extraction with water from a solution of these salts in tributyl phosphate in kerosene.

Flow sheet 1. *Extraction of lanthanides from monazite sand.*

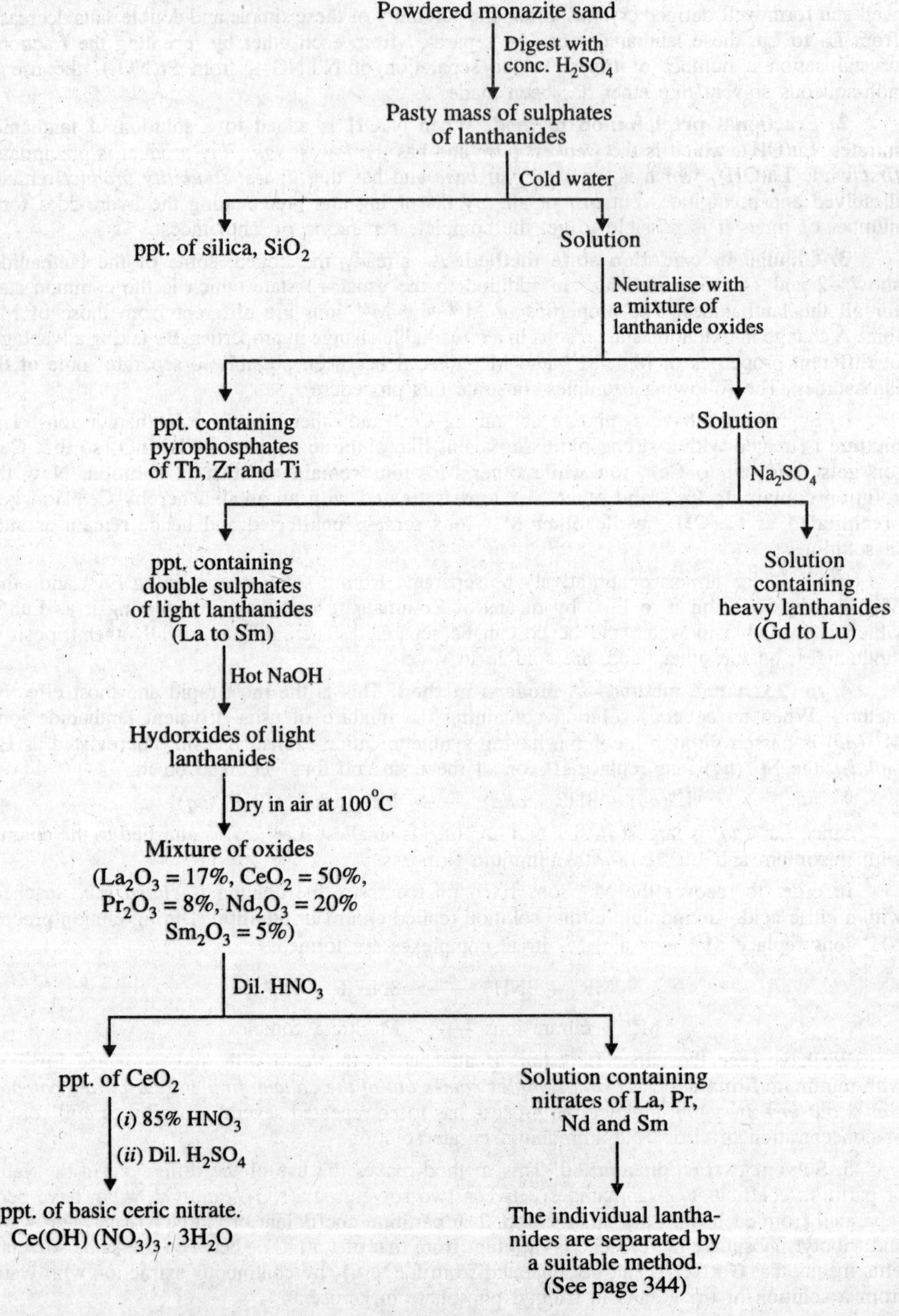

Production of lanthanide metals

The following methods may be used for this purpose:

1. Electrolysis of fused chlorides. This method is similar to that used in the metallurgy of Ca by the electrolysis of $CaCl_2$.

2. Reduction of anhydrous chlorides with Na. Lighter lanthanides such as La, Ce and Gd can easily be prepared by the reduction of their anhydrous chlorides with Na at 100°C.

$$LaCl_3 + 3Na \xrightarrow{100°C} La + 3NaCl$$

3. Reduction of anhydrous fluorides and chlorides with Mg or Ca. Heavier lanthanides such as Lu are prepared when anhydrous fluorides and chlorides are reduced by Ca or Mg metals at a temperature above 1000°C, since the fluorides are less volatile than the chlorides and consequently the loss caused by evaporation in case of fluorides is small.

Uses of lanthanides and their compounds

(*i*) Certain alloys of the lanthanide elements, known as *misch metals* containing predominantly 30–35% of Ce together with smaller quantities of other light lanthanides are used as reducing agents in metalllurgical operations. Mg-alloys containing about 30% misch metal and 1% Zr are used for making parts of jet engine.

(*ii*) La, Ce, Pr, Nd mixed with steel are used in cigarette lighters, toys, flame throwing tanks.

(*iii*) Nd_2O_3 and Pr_2O_3 are used as colouring agents for glass and in the production of standard filters. Lanthanum oxides are used for polishing glass. Ceria (CeO_2) is used in gas mantles.

(*iv*) Cerium phosphate is used as a catalyst in petroleum cracking.

(*v*) Cerium salts are used in analysis, dyeing cotton, lead accumulators, medicines etc.

(*vi*) Lanthanide compounds are used as good catalysts in a number of reactions like hydrogenation, dehydrogenation, oxidation and cracking of petroleum.

(*vii*) Lanthanide compounds are filled into arc carbon electrodes to give more brilliant light.

Comparison between *d*- and *f*-Block elements *(particularly between 3d and 4f-block elements)*

Some of the differences are given below:

1. Entry of extra electron. In case of *d*-block elements the extra electron enters $(n - 1)$ *d*-orbitals while in case of *f*-block elements the extra electron occupies $(n - 2)$ *f*-orbitals.

2. Energy of orbitals. $(n - 1)$ *d*-orbitals have higher energy than $(n - 2)$ *f*-orbitals.

3. Cause of paramagnetism. The maximum value of magnetic moment (μ value) for *d*-block elements is due to the presence of 5 unpaired electrons while the maximum value of μ for *f*-block elements is due to the presence of 7 unpaired electrons. Further, in case of *d*-block elements, since *d*-orbitals project out, these are not well shielded from the ligands surrounding the transition metal atoms or ions, *i.e.*, the electrons residing in *d*-orbitals are able to interact with the surrounding ligands. Thus in case of transition metals, the orbital motion of the electron is quenched and hence the observed paramagnetism in these elements is only due to the spin motion of the electron. Due to the quenching of orbital motion, the value of μ is calculated from the spin-only formula, $\mu = \sqrt{n(n+2)}$.

Now consider the case of tripositive lanthanide ions. In case of these ions since 4*f*-orbitals are quite deeply seated, these are well shielded from the ligands surrounding these ions by the overlapping 5*s* and 5*p* orbitals, *i.e.*, the electrons residing in 4*f* orbitals are not able to interact with the surrounding ligands. Thus in case of lanthanides, the orbital motion of the electron is not quenched as in case of *d*-block elements and hence the observed paramagnetism in these elements is due to both the motions of the electron *viz.* orbital motion and spin motion. In other words, the value of μ for M^{3+} ion, cannot be calculated by using spin-only formula. Rather μ is calculated from : $\mu = g\sqrt{J(J+1)}$.

4. Nature of absorption bands. *d*-block elements give broad bands in the absorption spectra in the visible region while, excepting La^{3+} and Lu^{3+}, all the remaining tripositive lanthanide ions give sharp almost line-like bands in the absorption spectra.

5. Oxidation states. *d*-block elements show much greater number of oxidation states while *f*-block elements, particularly lanthanides, show a limited number of oxidation states.

6. Formation of complex compounds. *d*-block elements form complexes more easily while *f*-block elements, particularly lanthanides, do so with great difficulty.

7. Coordination geometry. Because of their larger size, lanthanide ions generally have higher coordination numbers than transitional metal ions. Coordination numbers greater than 6 for transition metal ions are only formed with difficulty because of the strong ligand repulsion in the coordination spheres.

Actinides: 5*f*-Block Elements

As already said the 14 elements *viz.* Th_{90} to Lw_{103} are called actinides.

Trans-uranic elements

The actinide elements lying beyond U_{92}, *i.e.,* the actinide elements with atomic numbers 93 to 103 (Np_{93} to Lw_{103}) are called *trans-uranic* or *trans-uranium elements.*

Position of actinides in the periodic table

Prior to the discovery of trans-uranium elements (trans-uranium elements are the elements beyond uranium in the periodic table), the naturally occurring heaviest known elements namely Pa_{91}, and U_{92}, were placed below Hf_{72}, Ta_{73} and W_{74} in IVB, VB and VIB groups of the dic table, because these elements showed +4, +5 and +6 oxidation states and resembled Ta and W respectively in many of their properties. The then undiscovered trans-uranium ments with atomic numbers 93 to 100 were thus expected to occupy the positions in the riodic table below Re_{75}, Os_{76}, Ir_{77}, Pt_{78}, Au_{79}, Hg_{80}, Tl_{81} and Pb_{82} respectively as shown below:

IVB	*VB*	*VIB*	*VIIB*		*VIII*		*IB*	*IIB*	*IIIA*	*IVA*
—	—	—	—	—	—	—	—	—	—	—
Hf_{72}	Ta_{73}	W_{74}	Re_{75}	Os_{76}	Ir_{77}	Pt_{78}	Au_{79}	Hg_{80}	Tl_{81}	Pb_{82}
Th_{90}	Pa_{91}	U_{92}	93	94	95	96	97	98	99	100

The discovery of the element neptunium (Np_{93}) came in 1940 and this discovery was followed shortly by the discovery of plutonium (Pu_{94}) in 1941. The tracer chemical experiments with Np_{93} and Pu_{96} showed that the chemical properties of these two elements very much resemble those of U_{92} and not at all those of Re_{75} and Os_{76}. On this basis in 1944, all the three elements namely U_{92}, Np_{93} and Pu_{94} in their chemical properties, and hence the elements with atomic numbers 95 and 96 were also placed along with U_{92}, Np_{93} and Pu_{94} below W_{74} in group VI B as shown below :

IVB	*VB*	*VIB*
—	—	—
Hf_{72}	Ta_{73}	W_{74}
Th_{90}	Pa_{91}	U_{92}, Np_{93}, Pu_{94}, 95, 96

This assumption, however, proved to be wrong, since the experiments directed towards the discovery of elements with atomic numbers 95 and 96 on the pattern of discovery of Np_{93} and Pu_{94} failed. Later on, in the same year (1944) Seaborg suggested that all the elements having atomic number greater than 89 (Ac_{89}) constitute a second series of inner-transition elements. This series is similar to lanthanide series. The elements of this new series which have atomic number greater than 89 were called *actinides*. The elements of this series were placed below lanthanides out of the main body of the periodic table as shown below :

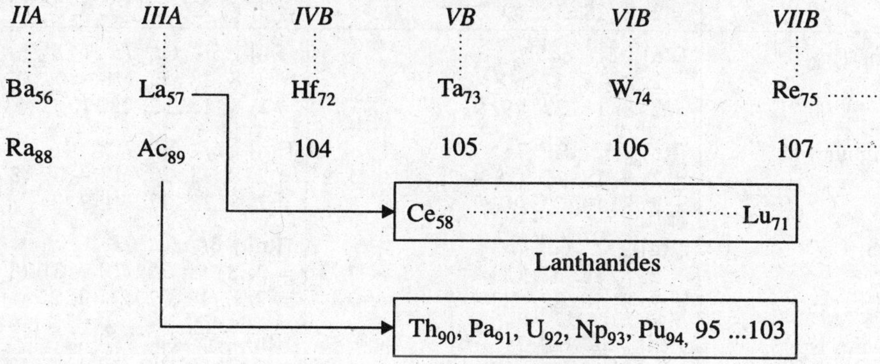

Actinides

The position of actinides shown above was confirmed by the discovery of the elements from atomic numbers 95 to 103 in 1961 and by the electronic configuration of these elements.

General characteristics of actinides

The following are important properties :

1. Occurrence

Excepting Ac, Th, Pa and U which occur in nature in uranium minerals, all the remaining actinides are unstable and are made artificially.

2. Raioactive nature

All the actinides are radioactive.

3. Electronic configurations

We known that the electronic configuration of Ac_{89} which is followed by 14 actinides (Th_{90}—Lw_{103}) is $[Rn]_{86}\ 5f^0 6d^1 7s^2$ in which 6*d* sub-shell is singly filled and 5*f* sub-shell is vacant. This configuration shows that as we move in the series of 14 actinides (Th_{90} to Lw_{103}), the additional electron should occupy the vacant 5*f* orbitals and 6*d* orbitals should remain singly-filled, *i.e.*, the expected configuration of the atoms of actinides should be those in which 6*d* orbitals are singly-filled and 5*f* orbitals are progressively filled up with electrons. This means that the expected configuration should be $[Rn]_{86}\ 5f^{1-14}\ 6d^1\ 7s^2$. But, since the energies of 6*d* and 5*f* orbitals are almost the same, in Th_{90} $5f^1$ electron gets shifted to 6*d* orbitals and in Pu_{94}, Am_{95}, Bk_{97}, Cf_{98}, Es_{99}, Fm_{100}, Md_{101} and No_{102}, $6d^1$ electron gets shifted to 5*f* orbitals. In Cm_{96} the shifting of $6d^1$ electron to 5*f* orbitals does not take place, since this type of shifting gives unstable configuration *viz*. $[Rn]_{86}\ 5f^8\ 6d^0\ 7s^2$ to Cm_{96}. In Lw_{103}, the shifting of $6d^1$ electron to 5*f* orbitals is also not possible, since 5*f* orbitals are already having 14 electrons.

Observed (actual) and expected configurations are given in Table 10.5. Observed configurations show that the complete and valence-shell configurations can be written as :

Complete configuration : 2, 8, 18, 32, 18 to 32, 8 to 10, 2

or 2, 8, 18, 32, $5s^2p^6d^{10}f^{0-14}$, $6s^2p^6d^{0-2}$, $7s^2$

or $[Rn]_{86}\ 5f^{0-14}\ 6d^{0-2}\ 7s^2$

Valence-shell configuration : $5f^{0-14}\ 6d^{0-2}\ 7s^2$

Table 10.5. *Expected and observed (actual) electronic configuration of the atoms of actinides. The configuration of Ac_{89} is also given. $[Rn]_{86}$ = 2, 8, 18, 32, 18, 8.*

Actinide element	*Expected configuration*	*Observed configuration*
Actinium, Ac_{89}	$[Rn]_{86}\ 5f^1\ 6d^0\ 7s^2$ = 2, 8, 18, 32, (18 + 1), (8 + 0), 2 = 2, 8, 18, 32, 19, 8, 2	$[Rn]_{86}\ 5f^0\ 6d^1\ 7s^2$ = 2, 8, 18, 32, (18 + 0), (8 + 1), 2 = 2, 8, 18, 32, 19, 9, 2
Thorium, Th_{90}	$[Rn]_{86}\ 5f^1\ 6d^1\ 7s^2$ = 2, 8, 18, 32, (18 + 1), (8 + 1), 2 = 2, 8, 18, 32, 19, 9, 2	$[Rn]_{86}\ 5f^0\ 6d^2\ 7s^2$ = 2, 8, 18, 32, (18 + 0), (8 + 2), 2 = 2, 8, 18, 32, 18, 10, 2
Protoactinium, Pa_{91}	$[Rn]_{86}\ 5f^2\ 6d^1\ 7s^2$ = 2, 8, 18, 32, (18 + 2), (8 + 1), 2 = 2, 8, 18, 32, 20, 9, 2	$[Rn]_{86}\ 5f^2\ 6d^1\ 7s^2$ = 2, 8, 18, 32, (18 + 2), (8 + 1), 2 = 2, 8, 18, 32, 20, 9, 2
Uranium, U_{92}	$[Rn]_{86}\ 5f^3\ 6d^1\ 7s^2$ = 2, 8, 18, 32, (18 + 3), (8 + 1), 2 = 2, 8, 18, 32, 21, 9, 2	$[Rn]_{86}\ 5f^3\ 6d^1\ 7s^2$ = 2, 8, 18, 32, (18 + 3), (8 + 1), 2 = 2, 8, 18, 32, 21, 9, 2
Neptunium, Np_{93}	$[Rn]_{86}\ 5f^4\ 6d^1\ 7s^2$ = 2, 8, 18, 32, (18 + 4), (8 + 1), 2 = 2, 8, 18, 32, 22, 9, 2	$[Rn]_{86}\ 5f^4\ 6d^1\ 7s^2$ = 2, 8, 18, 32, (18 + 4), (8 + 1), 2 = 2, 8, 18, 32, 22, 9, 2
Plutonium, Pu_{94}	$[Rn]_{86}\ 5f^5\ 6d^1\ 7s^2$ = 2, 8, 18, 32, (18 + 5), (8 + 1), 2 = 2, 8, 18, 32, 23, 9, 2	$[Rn]_{86}\ 5f^6\ 6d^0\ 7s^2$ = 2, 8, 18, 32, (18 + 6), (8 + 0), 2 = 2, 8, 18, 32, 24, 8, 2
Americium, Am_{95}	$[Rn]_{86}\ 5f^6\ 6d^1\ 7s^2$ = 2, 8, 18, 32, (18 + 6), (8 + 1), 2 = 2, 8, 18, 32, 24, 9, 2	$[Rn]_{86}\ 5f^7\ 6d^0\ 7s^2$ = 2, 8, 18, 32, (18 + 7), 8, 2 = 2, 8, 18, 32, 25, 8, 2
Curium, Cm_{96}	$[Rn]_{86}\ 5f^7\ 6d^1\ 7s^2$ = 2, 8, 18, 32, (18 + 7), (8 + 1), 2 = 2, 8, 18, 32, 25, 9, 2	$[Rn]_{86}\ 5f^7\ 6d^1\ 7s^2$ = 2, 8, 18, 32, (18 + 7), (8 + 1), 2 = 2, 8, 18, 32, 25, 9, 2
Beakelium, Bk_{97}	$[Rn]_{86}\ 5f^8\ 6d^1\ 7s^2$ = 2, 8, 18, 32, (18 + 8), (8 + 1), 2 = 2, 8, 18, 32, 26, 9, 2	$[Rn]_{86}\ 5f^9\ 6d^0\ 7s^2$ = 2, 8, 18, 32, (18 + 9), 8, 2 = 2, 8, 18, 32, 27, 8, 2
Californium, Cf_{98}	$[Rn]_{86}\ 5f^9\ 6d^1\ 7s^2$ = 2, 8, 18, 32, (18 + 9), (8 + 1). 2 = 2, 8, 18, 32, 27, 9, 2	$[Rn]_{86}\ 5f^{10}\ 6d^0\ 7s^2$ = 2, 8, 18, 32, (18 + 10), 8, 2 = 2, 8, 18, 32, 28, 8, 2
Einsteinium, Es_{99}	$[Rn]_{86}\ 5f^{10}\ 6d^1\ 7s^2$ = 2, 8, 18, 32, (18 + 10), (8 + 1), 2 = 2, 8, 18, 32, 28, 9, 2	$[Rn]_{86}\ 5f^{11}\ 6d^0\ 7s^2$ = 2, 8, 18, 32, (18 + 11), 8, 2 = 2, 8, 18, 32, 29, 8, 2
Fermium, Fm_{100}	$[Rn]_{86}\ 5f^{11}\ 6d^1\ 7s^2$ = 2, 8, 18, 32, (18 + 11), (8 + 1), 2 = 2, 8, 18, 32, 29, 9, 2	$[Rn]_{86}\ 5f^{12}\ 6d^0\ 7s^2$ = 2, 8, 18, 32, (18 + 12), 8, 2 = 2, 8, 18, 32, 30, 8, 2
Mendelevium, Md_{101}	$[Rn]_{86}\ 5f^{12}\ 6d^1\ 7s^2$ = 2, 8, 18, 32, (18 + 12), (8 + 1), 2 = 2, 8, 18, 32, 30, 9, 2	$[Rn]_{86}\ 5f^{13}\ 6d^0\ 7s^2$ = 2, 8, 18, 32, (18 + 13), 8, 2 = 2, 8, 18, 32, 31, 8, 2
Nobelium, No_{102}	$[Rn]_{86}\ 5f^{13}\ 6d^1\ 7s^2$ = 2, 8, 18, 32, (18 + 13), (8 + 1), 2 = 2, 8, 18, 32, 31, 9, 2	$[Rn]_{86}\ 5f^{14}\ 6d^0\ 7s^2$ = 2, 8, 18, 32, (18 + 14), 8, 2 = 2, 8, 18, 32, 32, 8, 2
Lawrencium, Lw_{103} or Lr_{103}	$[Rn]_{86}\ 5f^{14}\ 6d^1\ 7s^2$ = 2, 8, 18, 32, (18 + 14), (8 + 1), 2 = 2, 8, 18, 32, 32, 9, 2	$[Rn]_{86}\ 5f^{14}\ 6d^1\ 7s^2$ = 2, 8, 18, 32, (18 + 14), (8 + 1), 2 = 2, 8, 18, 32, 32, 9, 2

4. Oxidation states

Important oxidation states exhibited by actinides are given in Table 10.6. Most stable oxidation states are shown in circles while unstable ones are given in brackets.

It may be seen from these oxidation states that +2 state is shown by two elements namely Am and Th in its few compounds like $ThBr_2$, ThI_2, ThS etc. +3 oxidation state is shown by all the actinides. +3 state becomes more and more stable as the atomic number increases. +4 oxidation state is shown by Th, Pa, U, Np, Pu, Am and Cm while +5 state is shown by Th, Pa, U, Np, Pu and Am. +6 oxidation state is exhibited by U, Np, Pu and Am while +7 oxidation state is shown by Np and Pu.

The principal cations given by actinide elements are M^{3+}, M^{4+} and oxo-cations like MO_2^+ (oxidation state of M = +5) and MO_2^{2+} (oxidation state of M = +6). Examples of MO_2^+ ion are UO_2^+ and PuO_2^+ while those of MO_2^{2+} are UO_2^{2+} and PuO_2^{2+}. These oxo-cations are stable in acid and aqueous solutions.

U^{3+} ion in aqueous solution liberates H_2 on standing. Np^{3+} and Pu^{3+} are stable in water but are readily oxidised by air to Np^{4+} and Pu^{4+} respectively. All the remaining M^{3+} ions upto Md^{3+} are stable in aqueous solution. U^{4+} and Np^{4+} ions are stable in aqueous solution but are slowly oxidised by air to UO_2^{2+} (U = +6) and PuO_2^{2+} (Pu = +6) ions respectively.

Table 10.6. *Important oxidation states shown by actinides. Most oxidation states have been shown in cricles while unstable ones are given in bracket.*

Actinides	*Oxidation states*
Ac_{89}	+2, (+3) circled, +4
Th_{90}	+2, (+3), (+4) circled, +5
Pa_{91}	(+3), +4, +5
U_{92}	+3, +4, +5, (+6) circled
Np_{93}	+3, +4, +5, +6, +7
Pu_{94}	+3, (+4) circled, +5, +6, +7
Am_{95}	+2, (+3) circled, (+4), +5, +6
Cm_{96}	(+3) circled, (+4)
Bk_{97}	(+3) circled
Cf_{98}	(+3) circled
Es_{99}	(+3) circled
Fm_{100}	(+3) circled
Md_{101}	(+3) circled
No_{102}	(+3) circled
Lw_{103}	(+3) circled

5. Radii of M^{3+} and M^{4+} ions : Actinide contraction

Radii of M^{3+} and M^{4+} ions are given in Table 10.7.

Table 10.7. *Radii of tripositive (M^{3+}) and tetrapositive (M^{4+}) actinide cations.*

M^{3+} ions	Radii (Å)	M^{4+} ions	Radii (Å)
Ac^{3+}	1.11	Ac^{4+}	0.99
Th^{3+}	1.08	Th^{4+}	0.96
Pa^{3+}	1.05	Pa^{4+}	0.93
U^{3+}	1.03	U^{4+}	0.92
Np^{3+}	1.01	Np^{4+}	0.91
Pu^{3+}	1.00	Pu^{4+}	0.90
Am^{3+}	0.99	Am^{4+}	0.89
Cm^{3+}	0.98 ↓ Decreasing	Cm^{4+}	0.88 ↓ Decreasing

A look of the values of radii of M^{3+} and M^{4+} cations reveals that these values for both the cations decrease as we move from Ac to Cm. This steady decrease in the size of M^{3+} and M^{4+} cations in the actinide series is called *actinide contraction* which is analogous to lanthanide contraction found in lanthanides.

Causes of actinide contraction

We known that as we proceed from one element to the next one in the actinide series, the nuclear charge (*i.e.*, atomic number) increases by +1 at each next element and the addition of extra electron takes place in 5*f* orbital. Again it is also known that the shielding of one electron in 5*f* orbital by the other one residing in the same orbital is very poor. Due to the negligible amount of mutual shielding effect between the electrons residing in 5*f* orbital, the increase in nuclear charge by +1 at each next element in the actinide series brings the valence-shell nearer to the nucleus and hence the size of M^{3+} and M^{4+} cations goes on decreasing as we move from one element to the next one in the series.

6. Colour of M^{3+} and M^{4+} cations

Most of the tripositive and tetrapositive actinide cations (M^{3+} and M^{4+} cations) are coloured. The colours are given in Table 10.8.

These colours show that M^{3+} and M^{4+} cations having $5f^0$, $5f^1$ and $5f^7$ configuration are colourless while those containing $5f^2$, $5f^3$, $5f^4$, $5f^5$ and $5f^6$ configuration are coloured. In other words we can say that the cations having $n = 0$, 1 or 7 are colourless and the cations containing $n = 2$, 3, 4, 5 or 6 are coloured. Colours are produced when an electron jumps from one energy level to the other within 5*f* orbitals.

7. Formation of complexes

Most of the actinide halides form complex compounds with alkali metal halides. For example $ThCl_4$ forms complexes such as $K[ThCl_5]$, $K_2[ThCl_6]$ etc. with KCl. $ThCl_4$ and $ThBr_4$ also form complexes with pyridine. Actinides also form chelates with organic compounds like EDTA and oxine.

The degree of complex formation for the ions M^{4+}, MO_2^{2+}, M^{3+} and MO_2^+ decreases in the order :

$$M^{4+} > MO_2^{2+} > M^{3+} > MO_2^+$$

Table 10.8. *Colours of M^{3+} and M^{4+} actinide cations.*

M^{3+}/M^{4+} cation	*Valence-shell configuration of the cation*	*No. of unpaired electrons (n)*	*Colour*
Ac^{3+}	$5f^0\ 6d^0\ 7s^0$	0	Colourless
U^{3+}	$5f^3\ 6d^0\ 7s^0$	3	Red
Np^{3+}	$5f^4\ 6d^0\ 7s^0$	4	Purple
Pu^{3+}	$5f^5\ 6d^0\ 7s^0$	5	Violet
Am^{3+}	$5f^6\ 6d^0\ 7s^0$	6	Pink
Cm^{3+}	$5f^7\ 6d^0\ 7s^0$	7	Colourless
Th^{4+}	$5f^0\ 6d^0\ 7s^0$	0	Colourless
Pa^{4+}	$5f^1\ 6d^0\ 7s^0$	1	Colourless
U^{4+}	$5f^2\ 6d^0\ 7s^0$	2	Green
Np^{4+}	$5f^3\ 6d^0\ 7s^0$	3	Yellow-green
Pu^{4+}	$5f^4\ 6d^0\ 7s^0$	4	Orange
Am^{4+}	$5f^5\ 6d^0\ 7s^0$	5	Red

The complexing power of different anions with the above cations is in the order :

Singly-charged anions : $F^- > NO_2^- > Cl^-$

Doubly-charged anions : $CO_3^{2-} > C_2O_4^{2-} > SO_4^{2-}$

8. Properties of actinides

The actinides are silvery white metals which are highly reactive. All the actinides are highly electropositive like lanthanides. Tripositive (*e.g.*, U^{3+}, Np^{3+} etc.) and tetrapositive (*e.g.*, Pa^{4+}, U^{4+} etc.) actinide cations are paramagnetic like the tripositive lanthanide cations like Ce^{3+}, Pr^{3+}, Nd^{3+} etc. When actinides combine with O_2, oxides are formed. Examples are ThO_2, PaO_2, UO_2, NpO_2, AmO_2. Cm_2O_3 etc. When H_2 acts on actinides, hydrides are formed. Examples are ThH_4, PaH_3, UH_3, PuH_2, AmH_2 etc. Chlorides, bromides and iodides are obtained when the metal is acted on by Cl_2, Br_2 and I_2 respectively.

Comparison between lanthanides and actinides

(A) Similarities. (*i*) In the atoms of the elements of both the series, the extra electron enters $(n-2)f$ orbitals.

(*ii*) The elements of both the series show +3 oxidation state.

(*iii*) Like lanthanide contraction seen in case of lanthanides, we also have actinide contraction in actinides. Both the contractions are due to the poor shielding effect between the electrons residing in $(n-2)f$ orbitals.

(*iv*) The absorption bands of the elements of both the series are so sharp that they appear to the almost line-like bands. Both the bands are produced due to the jump of an electron from one energy level to the other within $(n-2)\ f$ orbitals.

(*v*) Elements of both the series have low electronegativities and are very reactive.

(*vi*) The nitrates, perchlorates and sulphates of trivalent actinides as well as lanthanides are soluble while the hydroxides, fluorides and carbonates of both cations are insoluble.

(*vii*) Most of the lanthanide and actinide cations are paramagnetic.

(*viii*) In the atoms of the elements of both the series, three outermost shells are partly-filled while the remaining inner-shells are completely-filled.

(B) Dissimilarities. The elements of both the series also show some dissimilarities which are due to the difference in nature of 4*f* and 5*f* orbitals. The points showing dissimilarities are given in Table 10.9.

Table 10.9. *Differences between lanthanides and actinides*

	Lanthanides		*Actinides*
(*i*)	Due to large energy difference between 4*f* and 5*d* orbitals, the properties of lanthanides are fairly similar.	(*i*)	Due to small energy difference between 5*f* and 6*d* orbitals, the properties of actinides are considerably different from each other.
(*ii*)	Binding energies of 4*f* orbitals are higher.	(*ii*)	Binding energies of 5*f* orbitals are lower.
(*iii*)	The additional electron enters 4*f* orbitals.	(*iii*)	The additional electron enters 5*f* orbitals.
(*iv*)	These elements show only +2, +3 and +4 oxidation states.	(*iv*)	These elements show a variety of oxidation states like +2, +3, +4, +5, +6 and +7.
(*v*)	The mutual shielding effect between two electrons residing in 4f orbitals is greater.	(*v*)	The mutual shielding effect between two electrons residing in 5*f* orbitals is poor.
(*vi*)	Most of the tripositive lanthanide cations are colourless.	(*vi*)	Most of the tripositive and tetrapositive actinide cations are coloured.
(*vii*)	The paramagnetic character of lanthanides can easily be explained.	(*vii*)	The paramagnetic character of actinides can not be explained easily.
(*viii*)	They do not form complex compounds easily. They do not form complexes with π-bonding ligands like alkyl phosphines, thioethers etc.	(*viii*)	They form complex compounds comparatively more easily. They do form complexes with π-bonding ligands.
(*ix*)	Except Pm, all the remaining lanthanides are non-radioactive.	(*ix*)	All the actinides are radioactive.
(*x*)	The compounds of lanthanides are less basic.	(*x*)	The compounds of actinides are more basic.
(*xi*)	Lanthanides do not form oxo-cations.	(*xi*)	Some actinides form oxocations of MO_2^+ (e.g., UO_2^+, PuO_2^+) and MO_2^{2+} (e.g., UO_2^{2+}, $Pu\ O_2^{2+}$) type.
(*xii*)	The atoms of lanthanides have a total of six shells. Out of these first three shells *viz.* 1st, 2nd and 3rd are completely filled while the remaining three shells namely 4th, 5th and 6th are partially-filled.	(*xii*)	The atoms of actinides have a total of seven shells. Out of these, first four shells *viz.* 1st, 2nd, 3rd and 4th are completely- filled while the remaining three shells *viz.* 5th, 6th and 7th are partially- filled.

University Questions

1. What are lanthanides? Why are they grouped together ? *(Usmania 93)*
2. Write a note on "Inner-transition Elements". *(Rohilkhand 91)*
3. Answer the following : (*a*) Give the electronic configuration of lanthanide elements. (*b*) What is lanthanide contraction and what are its consequences? (*c*) List oxidation states of lanthanides and compare them with those of *d*-block elements. (*d*) How are the magnetic properties of the lanthanides fundamentally different from those of the transition elements. *(Avadh 93)*
4. In what ways do actinides differ from lanthanides and why ? *(Rohilkhand 90)*
5. What are lanthanides? Discuss lanthanide contraction and separation of lanthanides. *(Meerut 95)*
6. Write a note on (*a*) *f-block elements*. (*b*) Actinides *(Meerut 96, 97, 98)*
7. What are lanthanides? Discuss the general characteristics of lanthanides. *(Meerut M.Sc. 90)*
8. Provide a list of lanthanides with their accepted atomic structure. *(Meerut M.Sc. 90)*
9. Describe various methods for separating lanthanides from each other. *(Meerut M.Sc., 90)*
10. (*a*) Discuss the stability of oxidation states of lanthanides from thermodynamic point of view.
    (*b*) Give a brief account of the magnetic properties of tripositive lanthanide ions. *(Berhampur 91)*
11. Explain the following : (*i*) (*a*) Chemically actinides show a somewhat wider range of oxidation states than the lanthanides. (*b*) Actinides have a greater tendency to form complexes than lanthanides. *(Rohilkhand 90)*
    (*ii*) The magnetic and spectral properties of lanthanides differ from those of 3*d* and 4*d* block elements? *(Berhampur 91)*
12. Write notes on (*i*) Synthesis of transuranic elements (*ii*) Actinide hypothesis. *(Berhampur 91)*
13. (*a*) Discuss the role played by lanthanide contraction in the separations and physico-chemical properties of lanthanides. (*b*) Describe the ion-exchange method of separating lanthanides. (*c*) Discuss the oxidation states other than +3 for lanthanide ions. (*d*) Give a brief account of the absorption spectra of lanthanide ions. *(Lucknow 95; Berhampur 91)*
14. Write notes on : (*i*) Seperation of lanthanides, (*ii*) Lanthanide contraction, (*iii*) Electronic configuration of rare-earths (General configuration). *(Meerut 98)*
15. What are lanthanides and why are they so called ? Discuss briefly the following properties of lanthanides : (*i*) Electronic configuration (*ii*) Magnetic behaviour and colour of tripositive ions (*iii*) Lanthanide contraction and its consequences. *(Lucknow 2002)*
16. (*a*) What are inner transition elements ? Discuss the following :–
    (*i*) Colour of lanthanide ions and
    (*ii*) Cause of lanthanide contraction.
    (*b*) Discuss the trend in the basic character of hydorxides of lanthanides. *(Nagpur 2002)*
17. (*a*) What do you mean by lanthanide contraction ? Describe its consequences.
    (*b*) Point out the similaries between actinides and lanthanides.
    (*c*) What are actinides ? Write their names and electronic configuration. *(VBS Purvanchal 2003)*
18. Explain the following :
    (*a*) La $(OH)_3$ is most basic and Lu $(OH)_3$ is least basic among Ln $(OH)_3$.
    (*b*) Magnetic properties of lanthanides are different from those of transition elements ?
    (*c*) Eu^{2+} and Tb^{4+} are stable in addition to their common oxidation state. *(Rohilkhand 2003)*
19. (*a*) What is lanthanide contraction ? Explain the reason for lanthanide contraction. What are the consequences of lanthanide contraction.
    (*b*) Write a brief note on 'Colours in Lanthanides'. *(Nagpur 2003)*
20. (*a*) Discuss the colour and spectral behaviour of actinides.
    (*b*) Compare the ability of lanthanides and actinides to form complexes.
    (*c*) Discuss the chemistry of separation of Np, Pu and Am from U.
    (*d*) Explain the term lanthanide contraction. Give a brief account of its effects on the properties of these elements.
    (*e*) Describe one important method for the separation of lanthanides. *(Rohilkhand 2003)*

11

Extraction, Properties and Uses of Some Inner-transition Metals

Thorium, Th_{90}

Discovery

Berzelius isolated this element in 1882 from a Norwegian mineral and named it after the *Swedish God of Thunder, Thor*. Its radioactive nature was established by Madam Curie and Professor Schmidt in 1898.

Occurrence

Thorium constitutes 0.0012% of earths crust and is about three times more abundent than uranium. Important minerals of thorium are :

1. Thorianite which contains 55–80% of ThO_2.

2. Thorite which is thorium silicate, $ThSiO_4.nH_2O$ or $(ThO_2.SiO_2).nH_2O$.

3. Monazite sand which contains mainly a mixture of phosphates of lanthanides and of thorium along with the impurities of silicous matter, silica (SiO_2), zircon ($ZrSiO_4$) etc. Monazite sands are largely found in Brazil, India (on the sea-coast of Kerala state), Sri Lanka, USA and Australia. Indian monazite sand is of the top most grade in the world. Its approximate composition is as : ThO_2 = 9.4%, CeO_2 = 28.5%, U_3O_8 = 0.37% and P_2O_5 = 25.9%.

Extraction of thorium from monazite sand

Monazite sand is essentially a mixture of the phosphates of lanthanides (RPO_4) and thorium [$Th_3(PO_4)_4$]. Monazite sand also contains silicous impurities and the impurities of SiO_2, zircon ($ZrSiO_4$) etc. The extraction of Th from monazite sand involves the following steps :

1. Concentration and preparation of thoria, ThO_2. The silicons impurities present in monazite sand are removed either by washing or by magnetic separation method. The concentrated ore is then digested with conc. H_2SO_4 at 200°C so that a pasty mass is obtained. This pasty mass contains sulphates of lanthanides, Th etc. and H_3PO_4.

$$2RPO_4 + 3H_2SO_4 \longrightarrow \underbrace{R_2(SO_4)_3 + 2H_3PO_4}_{\text{Pasty mass}}$$

$$Th_3(PO_4)_4 + 6H_2SO_4 \longrightarrow \underbrace{3Th(SO_4)_2 + 4H_3PO_4}_{\text{Pasty mass}}$$

The pasty mass is seperated by centrifugation and treated with cold water. The sulphates of lanthanides and of Th and H_3PO_4 pass into the solution and silica (SiO_2) and zircon ($ZrSiO_4$), being insoluble, get eliminated as residue. Due to the presence of H_3PO_4 in the solution, the latter (*i.e.*, solution) becomes sufficiently acidic. The solution is, therefore, partially neutralised by adding NaOH or ammonia or magnesia mixture and pH of the solution is so adjusted that the sulphates and H_3PO_4 react together and sulphates are converted into the precipitate of monohydrogen phosphates.

$$R_2(SO_4)_3 + 3H_3PO_4 \longrightarrow \underset{\text{ppt}}{R_2(HPO_4)_3} + 3H_2SO_4$$

$$Th(SO_4)_2 + 2H_3PO_4 \longrightarrow \underset{\text{ppt}}{Th(HPO_4)_2} + 2H_2SO_4$$

The ppt. of monohydrogen phosphates obtained as above is dissolved in 6N HNO_3 and treated with KIO_3 in the presence of H_2O_2 when $Th(HPO_4)_2$ is converted into $Th(IO_3)_4$ and lanthanide monohydrogen phosphates remain in solution. The ppt. of $Th(IO_3)_4$ is dissolved in HCl and is precipited as thorium oxalate, $Th(C_2O_4)_2$. $Th(C_2O_4)_2$ is ignited to get ThO_2.

$$Th(C_2O_4)_2 + O_2 \xrightarrow{\text{Ignite}} ThO_2 + 4CO_2$$

2. To get thorium from ThO_2. (*a*) ThO_2 obtained as above is heated at 950°C with Ca in a steel bomb in presence of $CaCl_2$ which acts as a flux.

$$ThO_2 + 2Ca \xrightarrow{950°C} Th + 2CaO$$

(*b*) ThO_2 is heated with carbon in a current of F_2 or Cl_2 so that ThO_2 is converted into ThF_4 or $ThCl_4$.

$$ThO_2 + 2C + 2X_2 \ (X = F, Cl) \longrightarrow ThX_4 + 2CO$$

Now ThF_4 or $ThCl_4$ obtained as above is reduced by metals like Na, Ca, Mg etc. When Ca is used as a reducing agent, $ZnCl_2$ is used as a flux.

$$ThF_4 + 2Ca \longrightarrow Th + 2CaF_2$$

$$ThCl_4 + 4Na \longrightarrow Th + 4NaCl$$

Extraction of thorium by electrolytic method

In this method, a fused mixture of ThF_4-KCl-NaCl (electrolyte) is electrolysed in a graphite crucible using graphite electrodes. Electrolysis is carried out at a temperature of 800°C and a current of 70 amperes per kilogram of the electrolyte is used. Th is deposited as a powder on the cathode and scraped off from time to time. The powder is washed with dil. HNO_3, dried and converted into ignots. Thorium obtained by this method is 88% pure.

Purification of crude thorium by Van-Arkel process

Thorium obtained as above is impure (*crude*). It may be purified by Van-Arkel process in which impure (*crude*) thorium metal is heated with iodine in a sealed vessel so that Th reacts with I_2 to form ThI_4 while the impurities donot react with iodine. ThI_4 so formed is decomposed at 1100–1200°C to get Th-metal in the pure state and I_2 is set free.

$$Th(crude) + 2I_2(g) \longrightarrow ThI_4\ (g) \xrightarrow{1100-1200°C} Th\ (pure) + 2I_2\ (g)$$

Physical properties

Thorium is a lustrous powder and has dark grey colour. Its specific gravity is 11.21 and melting and boiling points are 1827°C and 3500°C respectively. It exists in amorphous as well as crystalline states. At very low temperature it has super conductivity and at ordinary temperature its conductivity is very low. Thorium undergoes transformation from a face-centred cubic structure to a body-centred cubic structure at about 1400°C.

Chemical properties

1. Radioactive nature. Thorium is a radioactive metal having $t_{1/2} = 1.39 \times 10^{10}$ years. It yields a series of disintegration products.

2. Combination with metals. It forms many inter-metallic compounds like $ThMn_2$, $ThMn_{1.2}$, Th_7Fe_3, Th_7Co_3, Th_7Ni_3 etc.

3. Action of heat. When heated in air, it forms ThO_2.

4. Action of non-metals. (*a*) Thorium, when heated in Cl_2 or Br_2, forms $ThCl_4$ or $ThBr_4$ respectively.

(*b*) It combines with S to form ThS_2.

(*c*) It combines with H_2 vigorously, when heated, to form ThH_4.

(*d*) When heated in a current of N_2 at 800°C, Th_3N_4 is formed.

5. Action of acids. The metal is not attacked by HF. Hot and conc. HCl gives a black residue, $ThO_{1.3}\ Cl_{0.7}\ H_{1.3}$. It is rendered passive when placed in strong HNO_3.H_2SO_4 reacts with the metal slowly with the evolution of SO_2.

6. Action of H_2S. ThS_2 is obtained.

$$Th(\textit{red hot}) + 2H_2S \longrightarrow ThS_2 + 2H_2$$

7. Action of alkali solution. Alkali solutions have no action on the metal.

Uses of thorium and its compounds

(*i*) Thorium is used (*a*) in the preparation of fuel rods which are used in atomic reactors (*b*) in the preparation of magnesium flash lights (*c*) in electric cells, *X*-ray target and glow tube electrodes. In glow tubes, Th combines with the impurities of O_2 and N_2 and thus keeps the inert gas free from these impurities.

(*ii*) ThO_2 has various uses (see "*Uses of ThO_2*" given in chapter 12).

(*iii*) Many thorium salts are used as medicines in the treatment of cancer

(*iv*) Thorium alloys have been used as reducing agents.

Occurrence

Uranium is not found free in nature. It occurs in the combined state in a few scare minerals. Its most important minerals are:

(*i*) *Pitchblende* or *uranite* in which uranium is present as U_3O_8 or UO_2. $2UO_3$ or uranium uranate, $U(UO_4)_2$. It contains 75–90% of the oxide and is found in Australia, Bohemia, Hungary, Norway, India, East Africa etc:

(*ii*) *Carnotite* $K_2O.2UO_3.V_2O_5.3H_2O$ or $2K(UO_2)VO_4.3H_2O$ (*potassium uranyl ortho vanadate*) which contains both uranium and radium. It occurs in *Jaduguda mines* of Bihar.

(*iii*) *Autunite*, $K(UO_2)PO_4.8H_2O$ (*potassium uranyl ortho-phosphate*). It is found in Congo.

In India at present the principal sources of uranium ore are the Jaduguda mines of Bihar. In 1964 rich deposits of high grade uranium ore have been found in Bhatin area of Bihar. Another important source of uranium in India is the *monazite sand* of Travancore.

Extraction of uranium

Uranium can be extracted from *pitchblende* and *carnotite* ores. The method consists of the following three steps *viz.* A, B and C.

A. Conversion of the ore into ammonium diuranate, $(NH_4)_2U_2O_7$. $(NH_4)_2U_2O_7$ can be obtained from both the ores as follows :

(*i*) *To get $(NH_4)_2U_2O_7$ from pitchblende.* Pitchblende can be converted into $(NH_4)_2U_2O_7$ by acid digestion process or by alkali digestion process as shown below :

(*a*) *To get $(NH_4)_2U_2O_7$ by acid digestion process.* First of all pitchblende is concentrated by gravity process to remove sand, clay etc. and then concentrated ore is roasted in excess of air whereby S, As and Sb are removed as their volatile oxides. Silver, if present in the mineral, is removed as insoluble AgCl by roasting the mineral at 800°C with NaCl. The roasted ore is digested with dil. H_2SO_4 in presence of MnO_2 (an oxidising agent) for about 24 hours when Ba, Pb, Ra etc., are precipitated as sulphates and uranium goes into solution as *uranyl sulphate*, UO_2SO_4.

$$\underset{\text{(In mineral)}}{2U_3O_8} + 6H_2SO_4 + \frac{1}{2}O_2 \longrightarrow \underset{\substack{\text{Uranyl sulphate}\\\text{(Soluble)}}}{3UO_2SO_4} + 3H_2O$$

UO_2SO_4 present in solution is converted into *sodium uranyl carbonate*, $Na_4UO_2(CO_3)_3$ by the addition of Na_2CO_3.

$$UO_2SO_4 + 3Na_2CO_3 \longrightarrow \underset{\substack{\text{Sod. uranyl carbonate}\\\text{(Solution)}}}{Na_4UO_2(CO_3)_3} + Na_2SO_4$$

The solution containing $Na_4UO_2(CO_3)_2$ is now neutralised with HCl in order to decompose it into *uranyl chloride*, UO_2Cl_2 and then saturated with H_2S gas so that Pb, Cu etc. are precipitated as their sulphides. The solution of UO_2Cl_2 is now treated with excess of NH_4OH whereby we get the precipitate of *ammonium diuranate*, $(NH_4)_2U_2O_7$.

$$Na_4UO_2(CO_3)_3 + 6HCl \longrightarrow \underset{\substack{\text{Uranyl chloride}\\\text{(Solution)}}}{UO_2Cl_2} + 4NaCl + 3H_2O + 3CO_2$$

$$2UO_2Cl_2 + 6NH_4OH\ (excess) \xrightarrow{\text{Ignite}} \underset{\substack{\text{Ammonium}\\\text{diuranate (ppt)}}}{(NH_4)_2U_2O_7} + 4NH_4Cl + 3H_2O$$

(*b*) *To get $(NH_4)_2U_2O_7$ by alkali digestion process.* The roasted pitchblende ore is digested with (Na_2CO_3 + $NaHCO_3$) mixture solution whereby U_3O_8 present in the mineral is converted into soluble $Na_4UO_2(CO_3)_3$. The solution containing $Na_4UO_2(CO_3)_3$ is treated with NaOH which precipitates uranium as *sodium diuranate*, $Na_2U_2O_7$.

$$\underset{\text{(In mineral)}}{U_3O_8} + 3Na_2CO_3 + 6NaHCO_3 + \frac{1}{2}O_2 \longrightarrow \underset{\text{(Soluble)}}{3Na_4UO_2(CO_3)_3} + 3H_2O$$

$$Na_4UO_2(CO_3)_3 + 6NaOH \longrightarrow \underset{\substack{\text{Sod. diuranate}\\\text{(ppt)}}}{Na_2U_2O_7} + 6Na_2CO_3 + 3H_2O$$

The precipitate of $Na_2U_2O_7$ is dissolved in dil. H_2SO_4 and the solution is treated with NH_4OH to get the precipitate of $(NH_4)_2U_2O_7$.

$$\underset{\text{(ppt)}}{Na_2U_2O_7} + H_2SO_4 \longrightarrow Na_2SO_4 + H_2U_2O_7$$

$$H_2U_2O_7 + 2NH_4OH \longrightarrow \underset{\substack{\text{Ammon. diuranate}\\\text{(ppt)}}}{(NH_4)_2U_2O_7\downarrow} + 2H_2O$$

(*ii*) *To get $(NH_4)_2U_2O_7$ from carnotite.* In addition to uranium, carnotite also contains Ba, V, Fe, Al, Ra etc. From this ore uranium is obtained as a by-product. The ore is treated with conc. HNO_3 at 100°C. The heating is done by the steam. Most of the ore dissolves in HNO_3 and the solution thus obtained is neutralised by NaOH and then treated with $BaCl_2$ and H_2SO_4 whereby Ba and Ra are precipitated as their sulphate, and U and V are converted into *soluble* UO_2SO_4 and $VOSO_4$ respectively. This filtrate which also contains Fe and Al is treated with excess of boiling solution of Na_2CO_3 which precipitates Fe and Al (*Reject*), and UO_2SO_4 and $VOSO_4$ are converted into *soluble* $Na_4UO_2(CO_3)_3$ and Na_3VO_4 (*sod. ortho vanadate*) respectively.

$$UO_2SO_4 + 3Na_2CO_3 \longrightarrow \underset{\text{(Soluble)}}{Na_4UO_2(CO_3)_3} + Na_2SO_4$$

$$2VOSO_4 + 5Na_2CO_3 + \frac{1}{2}O_2 \longrightarrow \underset{\text{(Soluble)}}{2Na_3VO_4} + 3Na_2SO_4 + 5CO_2$$

The solution containing $Na_4UO_2(CO_3)_3$ and Na_3VO_4 is neutralised by HNO_3 and treated with NaOH. This treatement precipitates uranium as $Na_2U_2O_7$.

$$Na_4UO_2(CO_3)_3 + 6HNO_3 \longrightarrow UO_2(NO_3)_2 + 4NaNO_3 + 3H_2O + 3CO_2$$

$$2UO_2(NO_3)_2 + 6NaOH \longrightarrow \underset{\text{ppt}}{Na_2U_2O_7} + 4NaNO_3 + 3H_2O$$

The precipitate of $Na_2U_2O_7$ is converted into $(NH_4)_2U_2O_7$ is in alkali digestion process.

$$Na_2U_2O_7 + H_2SO_4 \longrightarrow Na_2SO_4 + H_2U_2O_7$$

$$H_2U_2O_7 + 2NH_4OH \longrightarrow \underset{\text{ppt}}{(NH_4)_2U_2O_7\downarrow} + 2H_2O$$

B. Conversion of $(NH_4)_2U_2O_7$ into U_3O_8. When $(NH_4)_2U_2O_7$ obtained as above is strongly heated in air, U_3O_8 is obtained.

$$3(NH_4)_2U_2O_7 \xrightarrow{\Delta} 2U_3O_8 + 6NH_3 + 3H_2O + O_2$$

C. To get uranium metal from U_3O_8. This can be done by the following two methods:

(*i*) U_3O_8 is heated with charcol or in a current of H_2 or CO, when U_3O_8 is reduced to UO_2.

$$U_3O_8 + 2C \longrightarrow 3UO_2 + 2CO_2$$

$$U_3O_8 + 2H_2 \longrightarrow 3UO_2 + 2H_2O$$

$$U_3O_8 + 2CO \longrightarrow 3UO_2 + 2CO_2$$

Now UO_2 obtained as above can be converted into uranium metal by any of the methods shown in Fig. 11.1.

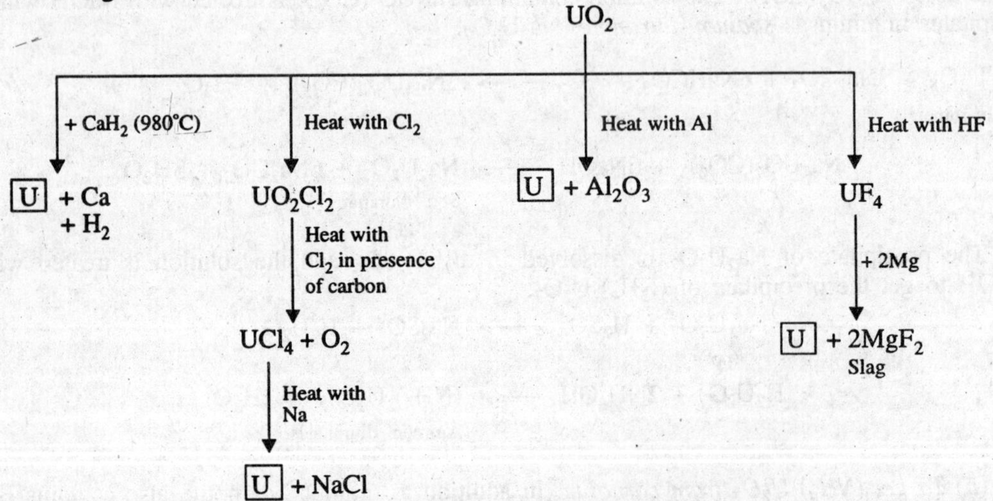

Fig. 11.1. Conversion of UO2 into U-metal.

(*ii*) U_3O_8 can be directly reduced to uranium metal by reducing it with carbon, aluminium, calcium or better by CaH_2 when both Ca and H_2 act as reducing agents.

$$U_3O_8 + 4C \longrightarrow 3U + 4CO_2$$

$$3U_3O_8 + 16Al \longrightarrow 9U + 8Al_2O_3 \text{ (Aluminothermic process)}$$

$$U_3O_8 + 8Ca \text{ (vapours)} \longrightarrow 3U + 8CaO \text{ (Burger's process)}$$

Physical properties

(*i*) Uranium has a silvery white lustre.

(*ii*) It melts at 1132°C and boils at 3500°C. It is malleable and ductile. Its specific gravity is 19.05.

(*iii*) It has several isotopes. Natural uranium consists of three isotopes *viz.* U^{238} (99.28%), U^{235}(0.71%) and U^{234} (0.006%).

Chemical properties

Uranium is highly reactive.

1. Action of air. It burns in air at 170°C to form U_3O_8.

2. Combination of non-metals (*a*) It combines with H_2 at 600°C under 120 atmospheric pressure to form UH_3.

(*b*) Uranium directly combines with halogens. It burns in F_2 to form green volatile UF_4. It catches fire in Cl_2 at 180°C and in Br_2 at 240°C forming the corresponding halides. It reacts with I_2 vapour at 260°C to form UI_4.

3. Action of alkalies. Alkalies have no action on uranium.

4. Action of steam. Uranium reacts with steam between 150–250°C to form a mixture of UO_2 and UH_3.

$$7U + 6H_2O\ (\textit{steam}) \xrightarrow{150\text{–}250°C} 3UO_2 + 4UH_3$$

5. Formation of inter-metallic compounds. Uranium forms various types of inter-metallic compounds. Important examples are U_6Mn, U_6Ni, USn_3 etc.

6. Action of acids. (*a*) Uranium dissolves in dil. HCl and H_2SO_4.

$$U + 2HCl \longrightarrow UCl_4 + 2H_2$$

$$U + 2H_2SO_4 \longrightarrow U(SO_4)_2 + 2H_2O$$

(*b*) With conc. H_2SO_4 and conc. HNO_3, SO_2 and NO_2 are evolved.

$$U + 2H_2SO_4 \longrightarrow USO_4 + 2H_2O + SO_2$$

$$3U + 2HNO_3 \longrightarrow U_3O_8 + 4NO + 4NO_2 + 4H_2O$$

Uses of uranium and its compounds

(*i*) The fission of U^{235} produces a huge amount of energy which is used for military as well as for peaceful purposes. This energy is also used in atom and hydrogen bombs.

(*ii*) Uranium is used in the manufacture of lamp-filaments, gas mantles, iron alloys, in photography, as electrodes in arc lamps and for the estimation of phosphates and arsenates.

(*iii*) Many uranium compounds are used : (*a*) as mordants for silk, wool and in calico printing (*b*) in the treatment of gout, diabetes and other diseases (*c*) in rubber and ceramic technology.

Plutonium, Pu_{94}

Introduction

This element was named after the planet, Pluto and was discovered in 1940 at the University of California by Seaborg, Mac Millan Kennedy and Wahl. It is a man-made element and has unique importance in chemical engineering and in the history of atomic weapons. It has numerous isotopes, all of which are prepared synthetically. Out of these, fissionable Pu^{239} isotope was discovered in 1941 by Kennedy and Seaborg.

Occurrence

In nature, plutonium occurs as Pu^{239} isotope in uranium ores like *Canadian pitchblende.*

No other isotope has been detected to occur in nature. Other ores in which Pu is found are *colorado pitchblende, fergusonite* and *carnotite.*

Production of plutonium

This element is produced as $_{94}Pu^{239}$ isotope from natural uranium which contains a large proportion of U^{238} and only 0.7% of U^{235}. When natural uranium (U^{238} + U^{235}) is allowed to react with slow neutrons, it is U^{238} isotope which is acted upon by the neutrons and Pu^{239} isotope is produced.

$$_{92}U^{238} + {}_0n^1 (slow) \longrightarrow {}_{92}U^{239} + {}_0\gamma^0$$

$$\downarrow$$

$$_{93}Np^{239} + {}_{-1}\beta^0$$

$$\downarrow \text{2–3 days}$$

$$_{94}Pu^{239} + {}_{-1}\beta^0$$

Production of plutonium in India

Project Phoenix is a plant that has been set up at Trombay in Maharashtra. In this plant, nuclear waste is processed for the recovery of uranium and plutonium.

Properties

(*i*) Metallic plutonium exists in six allotropic forms which are α-Pu (monoclinic, sp. grav. = 19.737), β-Pu (body-centred monoclinic, sp. grav. = 17.65), γ-Pu (face-centred orthorhombic, sp. grav. = 17.19), δ-Pu (face-centred cubic, sp. grav. = 15.90), η-Pu (face-centred tetragonal, sp. grav. = 15.90) and ε-Pu (body-centred cubic, sp. grav. = 16.48).

(*ii*) Plutonium forms intermetallic compounds with many metals. Examples of such compounds are $PuAl_2$, $PuAl_3$, $PuBe_3$, $PuCo_2$, $PuFe_2$, $PuMn_2$, $PuNi_2$, $PuNi_5$, Pu_2Ni_7 etc.

(*iii*) In aqueous solution, Pu exists in any of the four oxidation states *viz.* +3, +4, +5 and +6. This behaviour of Pu is similar to that of uranium and neptunium.

(*iv*) The aqueous solution of the ions like Pu^{3+}, Pu^{4+}, PuO_2^{2+} etc., have characteristic colours, *e.g.* Pu^{3+} is blue lavender, Pu^{4+} is yellowish-brown, PuO_2^{2+} is pink-orange. When these ions form complex compounds, their colours are changed and the aqueous ions formed are strongly paramagnetic.

Uses

(*i*) Pu^{239} is used for fuelling atomic reactors. The heat produced as a result of radioactive decay is used for generating electricity.

(*ii*) It is used as an ingredient of atomic explosive weapons.

(*iii*) Finely-divided plutonium is used as a starting material for the synthesis of many important compounds of the metal.

University Questions

1. (*a*) How is uranium obtained from U_3O_8 ?
(*b*) Give two important uses of uranium. *(Nagpur 2002)*

2. (*a*) What are the uses of uranium.
(*b*) What is the action of steam on uranium ? *(Nagpur 2003)*

12

Important Compounds of Inner-transition Elements

In this chapter we shall discuss the preparation, propérties, uses etc. of important compounds of cerium, thorium, uranium and plutonium.

Compounds of Cerium

In aqueous solution, cerium exists in +3 and +4 oxidation states. Its salts in +3 and +4 salts are called *cerous* and *ceric* salts respectively. Ceric sulphate, $Ce(SO_4)_2$ and ceric ammouium sulphate, $Ce(SO_4)_2.2(NH_4)_2SO_4.2H_2O$ are its very important comounds and are discussed below :

1. Ceric sulphate, $Ce(SO_4)_2$

The anhydrous salt is prepared by heating CeO_2 with cone. H_2SO_4 and forms an orange-yellow crystalline powder, insoluble in conc. H_2SO_4. It dissolves in H_2O, giving a brown-yellow solution from which the salt separates below 3°C as the dodecahydrate, between 3° and 42.5°C as the octahydrate and above the latter temperature as the tetrahydrate. On diluting an aqueous solution of the salt, hydrolysis occurs and ultimately sparingly soluble basic salts get precipitated. Hence for use in volumetric analysis, the salt is dissolved in dil. H_2SO_4 (0.5 N or stronger), $Ce(SO_4)_2$ is widely used in volumetric analysis as an oxidising agent. The salt is used for the titration of Fe^{2+}, $C_2O_4^{2-}$, NO_2^-, Cu^+, Mo^{5+}, Te^{4+} and U^{4+} ions. It is also used as an oxidiser in preparative organic chemistry.

2. Ceric ammonium sulphate, $Ce(SO_4)_2.2(NH_4)_2SO_4.2H_2O$ or $(NH_4)_4[Ce(SO_4)_4].2H_2O$

This compound is also called *ammouium tetrasulphatocerate* (IV) and hence is represented as $(NH_4)_4[Ce(SO_4)_4].2H_2O$. It is prepared by adding $(NH_4)_2SO_4$ to $Ce(SO_4)_2$ solution containing excess of dil. H_2SO_4 to prevent the hydrolysis, followed by crystallisation. This double salt is more soluble than $Ce(SO_4)_2$ and is, therefore, prepared for use as a volumetric reagent. Ionic equations showing the quantitative estimations of $C_2O_4^{2-}$, NO_2^-, I^-, H_2O_2, etc. made with this salt are given below :

$$C_2O_4^{2-} + 2Ce^{4+} \longrightarrow 2Ce^{3+} + 2CO_2$$

$$NO_2^- + 2Ce^{4+} + H_2O \longrightarrow 2Ce^{3+} + 2H^+ + NO_3^-$$

$$2I^- + 2Ce^{4+} \longrightarrow 2Ce^{3+} + I_2$$

$$H_2O_2 + 2Ce^{4+} \longrightarrow 2Ce^{3+} + 2H^+ + O_2$$

Compounds of Thorium

Thorium shows an oxidation state of +4 in its compounds. Important compounds of thorium are discussed below :

1. Thorium dioxide or thoria, ThO_2

Preparation..It may be obtained by heating thorium in an atmosphere of O_2 or by heating $Th(OH)_4$ or $Th(NO_3)_4$.

$$Th + O_2 \xrightarrow{\Delta} ThO_2$$

$$Th(OH)_4 \xrightarrow{\Delta} ThO_2 + 2H_2$$

$$Th(NO_3)_4 \xrightarrow{\Delta} ThO_2 + 4NO_2 + O_2$$

It can also be obtained from monazite sand.

Properties. It is white hygroscopic powder. Crystallised or strongly ingnited thoria is almost insoluble in acids but on evaporation with conc. H_2SO_4 it is dissolved in it to form $Th(SO_4)_2$.

$$ThO_2 + 2H_2SO_4 \longrightarrow Th(SO_4)_2 + 2H_2O$$

When heated with carbon in a current of Cl_2, $TlCl_4$ is precipitated.

$$ThO_2 + C + 2Cl_2 \longrightarrow ThCl_4 + CO_2$$

When heated with carbon and boron, it forms carbide and boride of thorium respectively.

Uses. Due to its resistance to heat and involatility it is used in the preparation of gas mantles. It is also used as a refractory material as well as a catalyst in many organic reactions. It is employed in welding operations and also in medicines. Since ThO_2 increases the life of the tungsten filament, it is used as an additive to tungsten used for metal filament lamps. Since ThO_2 has high absorptive power for X-rays, it is used in X-ray diagnostic work as a contrast medium in the examination of the digestive organs. It is also used in ceramic industry. Mixtures of ThO_2 and ceria have been used for search lights, head lights of motor cars etc.

2. Thorium nitrate, $Th(NO_3)_4$

It is prepared by dissolving $Th(OH)_4$ in NHO_3.

$$Th(OH)_4 + 4HNO_3 \longrightarrow Th(NO_3)_4 + 4H_2O$$

It is a white deliquescent salt which is soluble in water. On heating it gives ThO_2 in a very fine state of subdivision. Due to this property thorium nitrate is used in mantle industry. Thorium nitrate mixed with Mg powder is used for flash powders. It is also used for the titration of fluorides, using methyl red as indicator.

3. Thorium nitride, Th_3N_4

This compound is prepared by combining thorium powder with nitrogen.

$$3Th + 2N_2 \longrightarrow Th_3N_4$$

It is a yellowish-brown powder which dissolves in dilute acids to form NH_3

$$Th_3N_4 + 6H_2O \longrightarrow 4NH_3 + 3ThO_2$$

4. Thorium tetrachloride, $ThCl_4$

Anhydrous $ThCl_4$ is prepared by passing a current of Cl_2 over a heated mixture of ThO_2 and coke.

$$ThO_2 + 2C + 2Cl_2 \longrightarrow ThCl_4 + 2CO$$

Compounds of Uranium

Uranium forms compounds in +2, +3, +4, +5 and +6 oxidation state. Compounds of U (+4) and U (+6) are the most stable. Oxides of uranium, uranium trihydride, halides of uranium, uranyl salts and uranates are the important compounds of uranium. These compounds are discussed below.

Oxides of Uranium

The known oxides are UO_2, UO_3 and U_3O_8. UO_2 is basic and forms uranous salts while U_3O_8 is an acidic oxide. UO_3 is amphoteric. All the three oxides readily dissolve in HNO_3, forming uranyl (UO_2^{2+}) salts. When heated with S_2Cl_2 at 450°C, all oxides from UCl_4. Now let us discuss these oxides one by one as follows:

1. Uranous oxide or uranium oxide, UO_2 (U = +4)

It is obtained by reducing higher oxides such as UO_3 or U_3O_8 with H_2, C or CO at elevated temperature.

$$UO_3 + CO \xrightarrow{500°C} UO_2 + CO_2\uparrow$$
$$U_3O_8 + 2CO \longrightarrow 3UO_2 + 2CO_2\uparrow$$
$$U_3O_8 + 2H_2 \longrightarrow 3UO_2 + 2H_2O$$

It is a brown strongly basic oxide, insoluble in water and alkalies. When heated in air it burns to given U_3O_8. It dissolves slowly in most of the acids, but HNO_3 oxidises it to $UO_2(NO_3)_2$. It reduces $AgNO_3$ to metallic silver.

$$2AgNO_3 + UO_2 \longrightarrow UO_2(NO_3)_2 + 2Ag$$

When UO_2 is heated with HF at 400°C, UF_4 is obtained.

$$UO_2 + 4HF \xrightarrow{400°C} UF_4 + 2H_2O$$

2. Uranium trioxide, UO_3 (U = +6)

Prepartion. It is obtained: (*i*) by heating uranyl (UO_2^{2+}) salts like $UO_2(NO_3)_2$ and $UO_2(CO_3)$

$$2UO_2(NO_3)_2 \xrightarrow{\Delta} 2UO_3 + 2NO_2 + O_2$$
$$UO_2(CO_3) \xrightarrow{\Delta} UO_3 + CO_2$$

(*ii*) by heating ammonium diuranate, $(NH_4)_2U_2O_7$

$$(NH_4)_2U_2O_7 \xrightarrow{\Delta} 2UO_3 + H_2O + 2NH_3$$

Properties. (*i*) UO_3 is an orange yellow or brick red powder with specific gravity of 1.29.

(*ii*) On heating to 700°C it looses O_2 to form U_3O_8.

$$6UO_3 \xrightarrow{700°C} 2U_3O_8 + O_2$$

(*iii*) It is an amphoteric oxide and hence reacts with acids to form uranyl (UO_2^{2+}) salts and with metallic carbonates and alkalies to form uranates (UO_4^{2-}) and diuranates ($U_2O_7^{2-}$) respectively.

$$UO_3 + 2HCl \longrightarrow UO_2Cl_2 + H_2O$$
$$UO_3 + H_2SO_4 \longrightarrow UO_2SO_4 + H_2O$$
$$UO_3 + 2NaOH \longrightarrow Na_2UO_4 + H_2O$$
$$UO_3 + Na_2CO_3 \longrightarrow Na_2UO_4 + CO_2$$

Uses. UO_3 is used in the manufacture of costly glasses and pottery wares.

3. Triuranium octoxide or urano-uranic oxide, U_3O_8

This oxide occurs in pitchblende and is the most stable oxide of uranium.

Preparation. U_3O_8 is obtained:

(*i*) by heating UO_3 at 700°C

$$6UO_3 \xrightarrow{400°C} 2U_3O_8 + O_2$$

(*ii*) by igniting $(NH_4)_2U_2O_7$

$$3(NH_4)_2U_2O_7 \xrightarrow{\text{Ignite}} 2U_3O_8 + 6NH_3 + 3H_2O + O_2$$

Properties. (*i*) U_3O_8 is a dark black solid with density equal to 8.3.

(*ii*) *Reduction.* (*a*) When heated in a current of CO or H_2, it is reduced to UO_2

$$U_2O_8 + 2CO \longrightarrow 3UO_2 + 2CO_2$$

$$U_3O_8 + 2H_2 \longrightarrow 3UO_2 + 2H_2O$$

(*b*) When U_3O_8 is reduced by means of carbon, Al (*aluminothermic process*), Mg, Ca etc, uranium is obtained.

$$U_3O_8 + 8C \xrightarrow[\text{electric furnance}]{\text{Heat in an}} 3U + 8CO$$

$$3U_3O_8 + 16Al \xrightarrow{\Delta} 9U + 8Al_2O_3$$

(*ii*) *Action of acids.* It is not acted upon by HCl or H_2SO_4 but HF gives UF_4 whereas HNO_3 gives $UO_2(NO_3)_2$. When U_3O_8 is treated with dil. H_2SO_4 in presence of MnO_2 (oxidising agent) for about 24 hours, UO_2SO_4 (soluble) is obtained.

$$2U_3O_8 + 6H_2SO_4 + \frac{1}{2}O_2 \textit{ (From MnO}_2\textit{)} \longrightarrow \underset{\text{soluble}}{3UO_2SO_4} + 3H_2O$$

(*iv*) When U_3O_8 is acted upon by $(Na_2CO_3 + NaHCO_3)$ mixture solution, U_3O_8 is converted into soluble $Na_4(UO_2)(CO_3)_3$

$$U_3O_8 + 3Na_2CO_3 + 6NaHCO_3 + \frac{1}{2}O_2 \longrightarrow 3Na_4(UO_2)(CO_3)_3 + 3H_2O$$

Uses. U_3O_8 is used to some extent for colouring ceramic enamels.

Structure. U_3O_8 is regarded as a combination of UO_3 (U= +6) and UO_2 (U = +4), *i.e.* $U_3O_8 \equiv 2UO_3.UO_2$. Thus U_3O_8 has uranium in +6 and +4 oxidation state.

Uranium Trihydride, UH_3

When uranium is heated with H_2 at 500–600°C under 120 atm. pressure, UH_3 is obtained. It is a black powder. It is good conductor of electricity. Chemically UH_3 is reactive and inflames in air. When heated, UH_3 reacts with various compounds. Other compounds of uranium are obtained as shown in Fig. 12.1

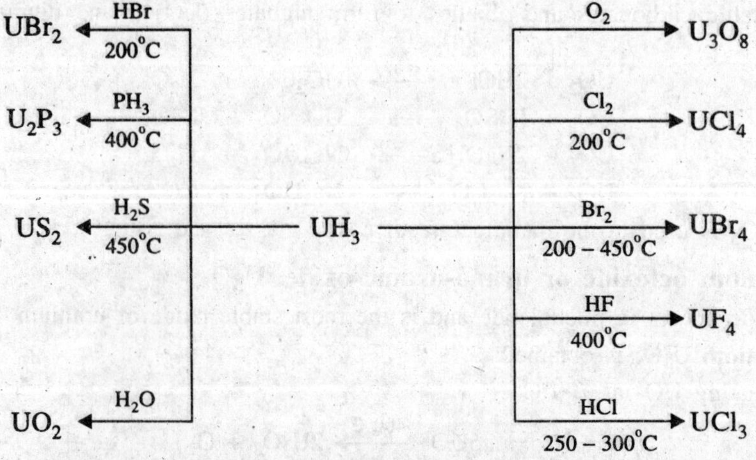

Fig. 12.1. Formation of various uranium compounds from UH3.

Halides of Uranium

The well known halides of uranium are UF_3, UF_4, UF_6, UCl_3, UCl_4, UCl_6, UBr_3, UBr_4, UI_4. Halides are important due to their role in the production of nuclear energy.

1. Uranium hexafluoride, UF_6

It is obtained by the direct combination of elements in presence of a little of Cl_2 or by the action of F_2 on UF_4. It is a colourless crystalline volatile substance and melts 64.1°C. It attacks many metals and a large number of organic compounds. It is readily hydrolysed by water to uranyl fluoride, UO_2F_2 and is reduced to UF_4.

$$UF_6 + 2H_2O \longrightarrow UO_2F_2 + 4HF$$

$$UF_6 \xrightarrow{\text{Reduction}} UF_4 + F_2$$

It is used in separating the uranium isotopes and thus in producing pure U^{235} nuclear fuel. It is a powerful fluorinating agent.

2. Uranium tetrachloride, UCl_4

It is the most important chloride of uranium.

Preparation. It is prepared by heating UO_2 in a current of S_2Cl_2 or when uranium burns in Cl_2. It may also be prepared by passing Cl_2 through a mixture of U_3O_8 and carbon at red heat.

Properties. UCl_4 is the most stable of the chlorides of uranium. It is soluble in water, acetone and ethyl acetate but insoluble in benzene and chloroform. It is a pale green solid which melts at 567°C. It is highly deliquescent and fumes in air. When oxidised by air, it forms uranyl chloride, UO_2Cl_2.

$$UCl_4 + O_2 \longrightarrow UO_2Cl_2 + Cl_2$$

Uranyl Salts

These are the uranium salts which are derived from the *bivalent basic uranyl radical*, UO_2^{2+}. In these salts uranium is in +6 oxidation state. These salts are formed by the action of acids on uranium oxides especially UO_3. These are generally yellow crystalline substances and soluble in water. A uranyl salt with potassium ferrocyanide solution gives a red precipitate of *uranyl ferrocyanide*. Following are the important uranyl salts :

1. Uranyl nitrate, $UO_2(NO_3)_2.6H_2O$

It is also known as uranium nitrate and is prepared by any of the following methods :

$$UO_3 + 2HNO_3 \longrightarrow UO_2(NO_3)_2 + H_2O$$

$$\underset{\text{Uranyl chloride}}{UO_2Cl_2} + 2AgNO_3 \longrightarrow UO_2(NO_3)_2 + 2AgCl$$

This substance crystallises as a lemon yellow hexahydrate, $UO_2(NO_3)_2.6H_2O$. It is soluble in water, alcohol and other. In contact with H_2SO_4 it is first converted into $UO_2(NO_3)_2.3H_2O$ and then into $UO_2(NO_3)_2.2H_2O$. When $UO_2(NO_3)_2.3H_2O$ is heated at about 130° in a current of CO_2 having HNO_3 to check hydrolysis, $UO_2(NO_3)_2$ is obtained.

2. Uranyl acetate. $UO_2(CH_3COO)_2.2H_2O$

It is obtained by dissolving $UO_2(OH)_2$ or UO_3 in CH_3COOH.

$$UO_2(OH)_2 + 2CH_3COOH \longrightarrow UO_2(CH_3COO)_2 + 2H_2O$$
$$UO_2 + 2CH_3COOH \longrightarrow UO_2(CH_3COO)_2 + H_2O$$

When dihydrate is heated to 110°C, it is converted into anhydrous salt. It gets decomposed at 275°C to UO_3. It is sparingly soluble in H_2O. This compound is of technical importance. It is used for making *Zn* or *Mg uranyl acetate* which is used as a qualitative reagent for the detection of sodium.

3. Uranyl sulphate, $UO_2(SO_4).3H_2O$

It is obtained by the action of conc. H_2SO_4 on hydrated UO_3.

$$UO_3 + H_2SO_4 \longrightarrow UO_2SO_4 + H_2O$$

Properties. (*i*) It is a yellow crystalline solid. It forms the hydrate, $UO_2SO_4.3H_2O$

(*ii*) When the solution of UO_2SO_4 is treated with Na_2CO_3, sod.uranyl carbonate, $Na_4(UO_2)(CO_3)_3$ is obtained.

$$UO_2SO_4 + 3Na_2CO_3 \longrightarrow \underset{\text{soluble}}{Na_4(UO_2)(CO_3)_3} + Na_2SO_4$$

Uranates (UO_4^{2-}) and Diuranates ($U_2O_7^{2-}$)

The hydrated uranium trioxide namely $UO_3.H_2O$ or H_2UO_4 and $2UO_3.H_2O$ or $H_2U_2O_7$ are called *uranic acid* and *diuranic acid* respectively.

The salts of these acids are called *uranates* (UO_4^{2-}) and *diuranates* ($U_2O_7^{2-}$) respectively. These are prepared as follows :

$$UO_3 + 2NaOH \longrightarrow Na_2UO_4 + H_2O$$
$$2UO_3 + 2NaOH \longrightarrow Na_2U_2O_7 + H_2O$$
$$UO_3 + Li_2CO_3 \longrightarrow Li_2UO_4 + CO_2\uparrow$$
$$2UO_3 + Li_2CO_3 \longrightarrow Li_2U_2O_7 + CO_2\uparrow$$

All uranates, even those of alkali metals and ammonium are insoluble in water and are orange-yellow in colour. Uranates (UO_4^{2-}) resemble chromates (CrO_4^{2-}) and diuranates ($U_2O_7^{2-}$) resemble dichromates ($Cr_2O_7^{2-}$).

Following are important diuranates :

1. Sodium diuranate, $Na_2U_2O_7.6H_2O$ (Also called sodium uranate)

Preparation. It is obtained by the action of NaOH on UO_3.

$$2UO_3 + 2NaOH \longrightarrow Na_2U_2O_7 + H_2O$$

Properties. (*i*) It is yellow in colour.

(*ii*) *Action of heat.* When heated, hexahydrate is converted into anhydrous salt.

(*iii*) *Action of H_2SO_4 and NH_4OH.* When $Na_2U_2O_7$ is dissolved in H_2SO_4 and the solution thus obtained is treated with NH_4OH, $(NH_4)_2U_2O_7$ is obtained.

$$Na_2U_2O_7 + H_2SO_4 \longrightarrow Na_2SO_4 + H_2U_2O_7$$
$$H_2U_2O_7 \xrightarrow{+NH_4OH} (NH_4)_2U_2O_7 + 2H_2O$$

Uses. $Na_2U_2O_7.6H_2O$ is used in the mannufacture of costly glasses and pottery wares.

2. Ammonium diuranate, $(NH_4)_2U_2O_7$

Preparation. (*i*) It is obtained as a yellow ppt. by adding NH_4OH to a uranyl salt solution.

(*ii*) It may also be obtained from $Na_2U_2O_7$ as shown above:

Properties. When ignited, it is converted into U_3O_8.

$$3(NH_4)_2U_2O_7 \xrightarrow{\text{Ignite}} 2U_3O_8 + 6NH_3 + 3H_2O + O_2$$

Uses. The formation of this substance provides a convenient method for gravimetric estimation of uranium.

Compounds of Plutonium

Hydrides, oxides and halides are the important compounds of plutonium. These are discussed below

1. Hydrides

Hydrides of plutonium are obtained by the action of H_2 on Pu-metal at 150–200°C. When the hydrides are heated, they give Pu-metal in a finely-divided state. PuH_2 has fluoride structure and PuH_3 has hexagonal structure. Hydrides are important compounds because finely-divided Pu-metal obtained by their thermal decomposition of its hydrides is used as a starting material for the preparation of many other compounds of the metal.

2. Oxides

PuO, Pu_2O_3, Pu_4O_7 and PuO_2 are the known oxides of plutonium. Out of these, plutonium dioxide (PuO_2) is the most stable. This oxide is obtained by igniting plutonium compounds in air at 1000°C. This oxide is used as a starting material in the synthesis of other compounds. PuO_2 is also used for the preparation of PuF_3.

$$2PuO_2 + H_2 + 6HF \xrightarrow{500°C} 2PuF_3 + 4H_2O$$

3. Halides

Preparation. PuF_3 is obtained by the action of HF + H_2 on PuO_2 at 500°C.

$$2PuO_2 + H_2 + 6HF \xrightarrow{500°C} 2PuF_3 + 4H_2O$$

PuF_4 is obtained when alkali fluoride is added to the salt of Pu^{4+}.

$$4MF + Pu^{4+} \longrightarrow PuF_4 + 4M^+$$

PuF_6 is obtained when PF_4 reacts with F_2.

$$PuF_4 + F_2 \xrightarrow{500°C} PuF_6$$

PuI_3 is formed when Pu metal reacts with HI at 400°C.

Properties. All the trihalides are volatile substance at moderate temperatures. Excepting PuF_3, other trihalides are hygroscopic.

University Questions

1. Explain the preparation of UF_6. *(Maduari 93)*
2. Write the preparation, properties and uses of ceric ammonium sulphate, thoria. thorium nitrate and thorium nitride.
3. Write the names of oxides formed by uranium.
4. What is the action of heat on $(NH_4)U_2O_7$ and UO_3?
5. What are uranyl salts ? Write the names of some uranyl salts.
6. Give the list of important compounds of plutonium.

Part IV : Miscellaneous Topics

13

Transition Metal Complexes with Carbon Monoxide : Metallic Carbonyls

What are π-acid ligands and π-acid complexes ?

CO, CNR, R_3P, NO etc. are such ligands which have *filled orbitals* (called *donor orbitals*) as well as *empty* π-orbitals (called *acceptor orbitals*). These ligands are called *π-acid ligands* or *π-acceptor ligands* or *π-bonding ligands.* These ligands combine with transition metals in low oxidation states (*e.g.* –1, 0, +1) and form complexes which are called *π-acid complexes.* Thus π-acid complexes are the complexes of transition metals in low oxidation state with π-acid ligands. Metallic corbonyls are important examples of π-acid complexes. These complexes have ligand → metal σ-bonds as well as metal → ligand π-bonds.

What are metallic carbonyls ?

The compounds formed by the combination of CO molecules with transition metal atoms in low oxidation states are called metallic carbonyls. In these compounds, CO molecule acts as a neutral ligand.

Why is the metallic atom in low oxidation state in metallic carbonyls ?

We have said above that metal atom in metal carbonyls is in low oxidation state. This is because of the fact that there is a higher electron density (electronic charge) on the metal atom in its lower oxidation state than in its higher oxidation state. Thus, since the metal is electropositive in nature, it would feel more comfortable, if some of the excess electronic charge present on the metal atom gets transferred to the vacant π-orbitals of CO (ligand) by forming M→CO π-bond.

Classification of carbonyls

Metallic carbonyls have been classified into the following two ways :

I. Classification based on the number of metallic atoms present in carbonyl

This classification gives the following two types of carbonyls.

(*a*) **Mononuclear or monomeric carbonyls.** The molecule of these carbonyls contains only one metallic atom. Examples are $Ni(CO)_4$, $M(CO)_5$, (M=Fe, Ru, Os) and $M(CO)_6$ (M = V, Cr, Mo, W)

(*b*) **Polynuclear carbonyls.** The molecule of these carbonyls contains two or more metallic atoms. Examples are $Co_2(CO)_8$, $Fe_2(CO)_4$, $M_2(CO)_{10}$, (M=Mn, Tc, Re), $M_3(CO)_{12}$ (M=Fe, Rn, Os). Polynuclear carbonyls may be *homo-nuclear* [*e.g.*, $Co_2(CO)_8$, $Fe_2(CO)_9$, $Fe_3(CO)_{12}$ etc.] or *hetero-nuclear* [*e.g.*, $MnCo(CO)_9$, $MnRe(CO)_{10}$ etc.].

II. Classification based on the structure of carbonyls

This classification gives the following types of carbonyls.

(*a*) **Non-bridged carbonyls.** As the name suggests, these carbonyls donot contain any bridging (ketonic) carbonyl group, :Ö=C<. These carbonyls may be of the following types:

(*i*) Non-bridged carbonyls which contain only terminal carbonyl groups (:O ⇄ C:). Examples of such carbonyls are $Ni(CO)_4$, $Fe(CO)_5$, $Ru(CO)_5$, $Os(CO)_5$.

(*ii*) Non-bridged carbonyls which contain terminal carbonyl groups as well as M---M bonds. Examples are $Co_2(CO)_8$ (in solution), $Mn_2(CO)_{10}$, $Te_2(CO)_{10}$, $Re_2(CO)_{10}$, $Fe_3(CO)_{12}$ and $Os_3(CO)_{12}$.

(*b*) **Bridged carbonyls.** These contain bridging carbonyl groups along with terminal carbonyl groups and one M---M bond. Examples are $Fe_2(CO)_9$, $Os_2(CO)_9$, $Co_2(CO)_8$ (in solid state). All these are polynuclear carbonyls.

This classification has been shown below:

Classification of carbonyls

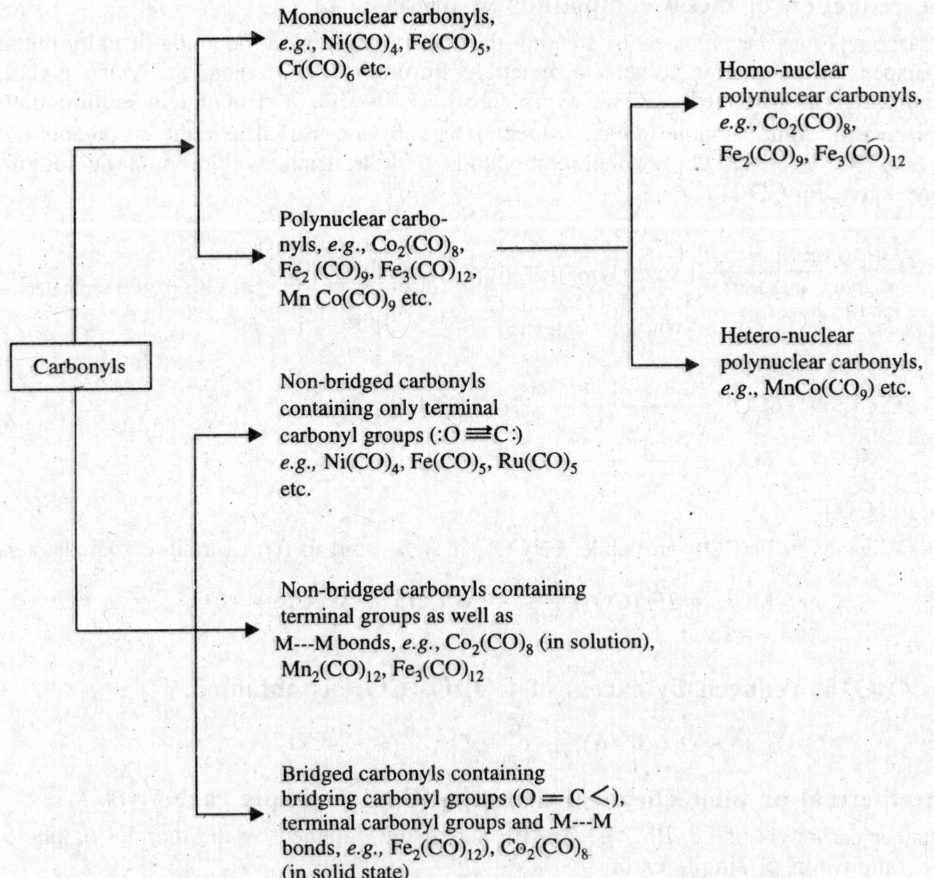

Important carbonyls formed by transition metals are given in Table 13.1.

Table 13.1. *Important carbonyls formed by transition metals.*

$V(CO)_6$	$Cr(CO)_6$	$Mn_2(CO)_{10}$	$Fe(CO)_5$, $Fe_2(CO)_9$, $Fe_3(CO)_{12}$	$Co_2(CO)_8$	$Ni(CO)_4$
	$Mo(CO)_6$	$Te_2(CO)_{10}$	$Ru(CO)_5$, —, —		
	$W(CO)_6$	$Re_2(CO)_{10}$	$Os(CO)_5$, $Os_2(CO)_9$, $Os_3(CO)_{12}$		

It may be noted from the above discusison that mononuclear carbonyls always have non-bridged structure while polynuclear carbonyls may have non-bridged [*e.g.*, $Co_2(CO)_8$ (in solution), $Mn_2(CO)_{10}$, $Fe_3(CO)_{12}$ etc.] or bridged [*e.g.* $Fe_2(CO)_9$, $Co_2(CO)_8$ (in solid state)] structures.

General Methods of Preparation

1. By direct synthesis

Of the simple carbonyls, nickel tetracarbonyl, $Ni(CO)_4$ and iron pentacarbonyl, $Fe(CO)_5$ can be prepared by the direct combination of CO with finely-divided metals at suitable temperature and pressure :

$$Ni + 4CO \xrightarrow[\text{one atm. pressure}]{\text{Room temp. and}} Ni(CO_4)$$

$$Fe + 5CO \xrightarrow[\text{pressure}]{200°C \text{ and } 100 \text{ atm.}} Fe(CO)_5$$

2. By the reduction of metal compounds in presence of CO

Some carbonyls may be prepared by treating the transition metal compounds (usually metal halides) in suspension in organic solvents like tetrahydrofuran (THF), ether, diethylene glycol, dimethyl ether (diglyme) etc with CO at a pressure of 200—300 atm. and temperature upto 300°C in presence of some suitable reducing agent. The substances used as reducing agents are H_2, metals (*e.g.* Na, Al, Mg, Cu etc.) or compounds such as trialkyl aluminium or sodium benzophenone ketyl, $Ph_2CO^-Na^+$

$$VCl_3 + 6CO \xrightarrow[\text{diglyme at at } 100° \text{ and } 15 \text{ atm.}]{3Na \text{ (excess) in}} \underset{\text{Unstable organo-metallic compound (Intermediate)}}{3[Na(diglyme)]Cl + V(CO)_6} \xrightarrow[\text{(ii) Sublimation at } 50°]{\text{(i) Acidification by } H_3PO_4} V(CO)_6 \text{ gets separated}$$

$$CrCl_3 + 6CO \xrightarrow{(C_2H_4)_3Al \text{ in ether}} Cr(CO)_6 + AlCl_3$$

$$MoCl_5 + 6CO \xrightarrow{(5Na \text{ in diglyme})} Mo(CO)_6 + 5NaCl$$

3. From $Fe(CO)_5$

Since CO ligands in $Fe(CO)_5$ are labile, $Fe(CO)_5$ may be used to form certain carbonyls, *e.g.*,

$$\underset{(M = W, Mo)}{MCl_6} + 3Fe(CO)_5 \xrightarrow[\text{ether}]{100°C} M(CO)_6 + 3FeCl_2 + 9CO$$

4. When OsO_4 is reduced by excess of CO, $Os(CO)_5$ is obtained.

$$OsO_4 + 9CO\ (excess) \xrightarrow[350 \text{ atm}]{250°C} Os(CO)_5 + 4CO_2$$

5. By the thermal or photochemical decomposition of simple carbonyls

Polynuclear carbonyls of Fe, Rh, Os, Co, Rh and Ir are obtained by the thermal or photochemical decompositon of simple carbonyls, *e.g.*,

$$\underset{(M = Fe, Os)}{2M(CO)_5} \xrightarrow{\text{ultraviolet light}} M_2(CO)_9 + CO$$

Physical Properties

With the exception of $Ni(CO)_4$, $Fe(CO)_5$, $Ru(CO)_5$ and $Os(CO)_5$ which are liquids at ordinary temperatures, all other carbonyls are crystalline solids. They melt or decompose at low temperatures.

All are typical covalent compounds and for this reason they are soluble in non-polar solvents. With the exception of $Fe_2(CO)_9$, all are more or less soluble in organic solvents. Excepting $V(CO)_6$, all the carbonyls are *diamagnetic*. $V(CO)_6$ is *paramagnetic* and its paramagnetic property corresponds to the presence of one unpaired electron. The metals in carbonyls are in zero oxidation state.

Chemical Properties

1. Substitution reactions

The most important general reactions shown by carbonyls are substitution reactions in which CO molecules are replaced by ligands like PX_3, PR_3, $P(OR)_3$, SR_2, NR_3, OR_2, RNC etc. or by unsaturated organic molecules such as C_6H_6 or cycloheptatriene. These reactions may proceed through photochemical activation according to which the carbonyl group absorbs a photon (*hv*) to eliminate a CO molecule followed by the entry of substituent into the co-ordination sphere, *e.g.*,

$$Cr(CO)_6 \xrightarrow[(-CO)]{hv} Cr(CO)_5 \xrightarrow{+L} Cr(CO)_5L$$

$$Fe(CO)_5 \xrightarrow[(-2CO)]{hv} Fe(CO)_3 \xrightarrow{+2PR_3} Fe(CO)_3(PR_3)_2$$

$$Fe_2(CO)_9 \xrightarrow[(-5CO)]{hv} Fe_2(CO)_4 \xrightarrow{+4NO} 2[Fe(CO)_2(NO)_2]$$

Bidentate ligands like *diars* replace two CO groups at a time to give the substituted carbonyls like $Mo(CO)_4$(*diars*), $Fe(CO)_3$(*diars*), $Ni(CO)_2$(*diars*) etc. Substitution reaction in case of $Cr(CO)_6$ gives $Cr(CO)_2(diars)_2$.

2. Formation of anionic carbonyl complexes (also called carbonylate anions)

Many of the carbonyls form carbonylate anions, *e.g.* $[Fe(CO)_4]^{2-}$, $[Mo(CO)_5Cl]^-$, $[HFe(CO)_4]^-$, $[Fe_2(CO)_8]^{2-}$, $[Co(CO)_4]^-$, $[Mn(CO)_5]^-$ etc. These complex anions can be prepared by any of the following methods:

(*a*) **By the action of NaOH or nitrogenous bases (*e.g.*, NH_3, amines, pyridine) on metal carbonyls, *e.g.***

$$Fe(CO)_5 \xrightarrow{NaOH} Na_2^+[Fe(CO)_4]^{2-}$$

$$Fe(CO)_5 \xrightarrow{NH_3/H_2O} [Et_4N]_2^+[HFe(CO)_4]^-$$

$$Fe_2(CO)_9 \xrightarrow{KOH/MeOH} [Et_4N_2]_2^+[Fe_2(CO)_8]^{2-}$$

$$Fe_3(CO)_{12} \xrightarrow{(en)/H_2O} [Fe(en)_2]^{3+}[Fe_3(CO)_{11}]^{3-}$$

$$3Co_2(CO)_8 + 12py \longrightarrow 2[Co(py)_6]^{2+}[Co(CO)_4]_2^{2-} + 8CO$$

(*b*) **By the reduction of carbonyls with alkali metals, alkali metal amalgams or borohydrides, *e.g.*,**

$$Co_2(CO)_8 \xrightarrow[THF]{Na/Hg} Na^+[Co(CO)_4]^-$$

$$Mn_2(CO)_{10} \xrightarrow[THF]{Na/Hg} Na^+[Mn(CO)_5]^-$$

$$Re_2(CO)_{10} \xrightarrow[THF]{NaBH_4} Na_2[Re_4(CO)_{16}]$$

$$Cr(CO)_6 \xrightarrow{Na/NH_3} Na_2[Cr(CO)_5]$$

$$Cr(CO)_6 \xrightarrow[THF]{NaBH_4} Na[HCr_2(CO)_{10}]$$

$$Cr(CO)_6 \xrightarrow[NH_2]{NaBH_4} Na_2[Cr_2(CO)_{10}]$$

3. Formation of cationic carbonyl complexes. (also called carbonylate cations)

Carbonylate cations may be prepared from the cabonyls by the following methods :

(*a*) By the protonation of carbonyls in strong acids, *e.g.*

$$Fe(CO)_5 \xrightarrow{HCl/BCl_3} [FeH(CO_5)]^+[BCl_4]^-$$

n.m.r. spectroscopy has shown that the proton becomes bounded to the metal atom in $[FeH(CO)_5]^+$.

(*b*) By the action of CO and a Lewis acid like $AlCl_3$ or BF_3 on carbonyl halide, *e.g.*,

$$Mn(CO)_5Cl + CO + AlCl_3 \longrightarrow [Mn(CO)_6]^+[AlCl_4]^-$$

4. Formation of carbonyl halides

The action of Cl_2, Br_2, I_2 or F_2 on some carbonyls gives carbonyl halides, *e.g.*,

$$2Mo(CO)_6 + 2Cl_2 \longrightarrow [Mo(CO)_4Cl_2] + 4CO$$

$$Mn_2(CO)_{10} + Br_2 \underset{}{\overset{\text{Pet. ether, 120°C}}{\rightleftharpoons}} [Mn(CO)_4Br]_2 + 2CO$$

$$Fe(CO)_5 + I_2 \xrightarrow{(-2CO)} [Fe(CO)_4I_2] + CO$$

$$4Ru(CO)_5 + 4F_2 \longrightarrow [Ru(CO)_3F_2]_4 + 8CO$$

5. Formation of carbonyl hydrides

Some of the metallic carbonyls [*e.g.* $Fe(CO)_5$, $Mn_2(CO)_{10}$, $Os_3(CO)_{12}$ etc] dissolve in an alkali and give an unidentified colourless solution. On acidification with a mineral acid, this solution gives the carbonyl hydrides. $H_2Fe(CO)_4$ [from $Fe(CO)_5$], $H_3Mn_3(CO)_{12}$ [from $Mn_2(CO)_{10}$], $H_4Os_4(CO)_{12}$, $H_2Os_4(CO)_{13}$ [from $Os_3(CO)_{12}$] and $Ni(CO)_3H_2$ [from $Ni(CO)_4$] have been obtained by this method.

$$Fe(CO)_5 + 2OH^- \longrightarrow Fe(CO)_4H_2 + CO_3^{2-}$$

$$Ni(CO)_4 + 2OH^- \longrightarrow Ni(CO)_3H_2 + CO_3^{2-}$$

Carbonyl hydrides can also be prepared by treating the carbonyls with H_2, *e.g.*,

$$Mn_2(CO)_{10} + H_2 \xrightarrow{200°C/200\ atm.} 2HMn(CO)_5$$

Some Carbonyls

Here we shall discuss the preparation, properties and uses of some important metallic carbonyls.

1. Chromium hexacarbonyl, $Cr(CO)_6$

Preparation. It can be obtained by the following methods :

(*i*) CO at 50 atm. pressure and at room temperature is passed into a suspension of $CrCl_3$ in ether, which has been treated with C_6H_5MgBr at –70° (*Job's reaction*).

(*ii*) A solution of Cr(+3) salt dissolved in ether is treated with CO at a high temperature and pressure in the presence of a reducing agent such as $Al(C_5H_5)_3$, $LiAlH_4$ etc.

(*iii*) $CrCl_3$ in presence of $AlCl_3$, powdered Al and benzene is treated with CO. The yield is 88%.

Properties. (*i*) Chromium hexacarbonyl is a white solid.

(*ii*) Crystals of $Cr(CO)_6$ are soluble in ether, chloroform, CCl_4 and C_6H_6.

(*iii*) It is resistant to chemical attack. For example it does not react with air, cold aqueous alkalies and acids (except conc. HNO_3).

(*iv*) *Decomposition.* It is decomposed by F_2 at –75°C to form CrF_5, by Cl_2 or by conc. HNO_3.

(*v*) *Reduction.* It is reduced with alkali metals in liq. NH_3 and alkali metal borohydrides. For example,

$$Cr(CO)_6 \xrightarrow{Na/NH_3} Na_2[Cr(CO)_5]$$

$$Cr(CO)_6 \xrightarrow[(THF)]{NaBH_4} Na[HCr_2(CO)_{10}]$$

$$Cr(CO)_6 \xrightarrow[(NH_3)]{NaBH_4} Na_2[Cr_2(CO)_{10}]$$

(*vi*) *Reaction with amines.* $Cr(CO)_6$ reacts with amines like *py, phen, en* to form the products in which CO groups in $Cr(CO)_6$ are replaced by molecules of the amine. For example above 140°C, *py* reacts with $Cr(CO)_6$ to give the following products in succession :

$$Cr(CO)_6 \longrightarrow \underset{\text{yellow brown solid}}{Cr(CO)_4(py)_2} \longrightarrow \underset{\text{yellow red solid}}{Cr_2(CO_7)(py)_5} \longrightarrow \underset{\text{bright red solid}}{Cr(CO)_3(py)_3}$$

2. Dimanganese decacarbonyl, $Mn_2(CO)_{10}$

Preparation. (*i*) $Mn_2(CO)_{10}$ is obtained when MnI_2 (prepared by special method) is reduced at a high pressure of CO by Mg in diethyl ether. The yield of $Mn_2(CO)_{10}$ is only 1%.

$$Mn + 2CuI \xrightarrow{500°C} MnI_2 + 2Cu$$

$$(MnI_2 + 2Cu) + Mg \xrightarrow[25°C/16\ hrs]{CO,\ 210\ atm} Mn_2(CO)_{10} + MgI_2 + CuI_2$$

(*ii*) $Mn_2(CO)_{10}$ has also been prepared in better yield by the reaction of $Mn(CH_3COO)_2$ with triethyl aluminium, $(C_2H_5)_3Al$ under CO at a pressure of 20 atmospheres at 100°C for 5 hours.

$$Mn(CH_3COO)_2 + (C_2H_5)_3Al \xrightarrow[100°C/5\ hrs]{CO,\ 20\ atm} Mn_2(CO)_{10}$$

The yield of carbonyl by this method is 53–60%.

(*iii*) Cabonylation of $MnCl_2$ in presence of benzophenone ketyl like $(C_6H_5)_2CONa$ at 165°C gives 32% yield of $Mn_2(CO)_{10}$.

$$MnCl_2 + 2(C_6H_5)_2CONa \longrightarrow NaCl + Mn[CO(C_6H_5)_2]_2$$

$$Mn[OC(C_6H_5)_2] \xrightarrow[165°C]{CO,\ 140\ atm} Mn_2(CO)_{10} + (C_6H_5)_2CO$$

(*iv*) More recently 48% yield of $Mn_2(CO)_{10}$ has been obtained by the carbonylation of methyl cyclopentadienyl manganese dicarbonyl, $(CH_3–C_5H_4)Mn(CO)_3$ in presence of Na at 50 atmosphere pressure of CO and 125°C for 8 hours.

$$(CH_3C_5H_4)Mn(CO)_3 + Na \xrightarrow[125°C/5\ hrs]{CO,\ 50\ atm} Mn_2(CO)_{10}$$

Properties. (*i*) It is a golden yellow crystalline substance. It has m.pt. = 155°C and sublimes in vacuum.

(*ii*) It is soluble in organic solvents.

(*iii*) It is slowly oxidised in air, especially in solution.

(*iv*) *Action of halogens.* Halogens form carbonyl halides, $Mn(CO)_5X$ as shown below:

$$Mn_2(CO)_{10} + Br_2(l) \xrightarrow{40°C} 2Mn(CO)_5Br$$

$$Mn_2(CO)_{10} + I_2 \longrightarrow 2Mn(CO)_5I$$

The order of activity of halgoens is as : I > Br > Cl.

(*v*) *Action of Li and Na.* Mn—Mn bond is broken by lithium in the presence of tetrahydrofuran (THF) and by Na in presence of liq. NH_3 and $Li^+[Mn(CO)_5]^-$ is formed.

$$Mn_2(CO)_{10} + 2Li \xrightarrow{THF} 2Li^+[Mn(CO)_5]^-$$

$$Mn_2(CO)_{10} + 2Na \xrightarrow{\text{liq. } NH_3} 2Na^+[Mn(CO)_5]^-$$

(*vi*) *Reduction.* When $Mn_2(CO)_{10}$ is reduced by hydrogen under 200 atmospheric pressure at a temperature of 200°C, carbonyl hydride is formed.

$$Mn_2(CO)_{10} + H_2 \xrightarrow[200°C]{200 \text{ atm.}} 2HMn(CO)_5$$

(*vii*) *Action of phosphines, arsines and stibines.* With these materials $Mn_2(CO)_{10}$ gives monomeric paramagnetic compounds of the type, $Mn(CO)_4(PR_3)$.

(*viii*) *Diamagnetic nature.* $Mn_2(CO)_{10}$ is a diamagnetic substance. Diamagnetic character confirms the facts that all the electrons in $Mn_2(CO)_{10}$ are paired and Mn–Mn bond is also present in it.

3. Iron pentacarbonyl, $Fe(CO)_5$

Preparation. It is prepared : (*i*) by the action at CO of 200 atm. pressure on iron heated to 200°C.

(*ii*) by the action of CO at 200 atm. pressure and 200°C on FeI_2 in the presence of Cu which acts as an acceptor for I_2. It is thought that iron carbonyl iodide, $Fe(CO)_4I_2$ is first formed which reacts with Cu in presence of CO to form $Fe(CO)_5$.

$$FeI_2 + 4CO \longrightarrow Fe(CO)_4I_2$$

$$CO + Fe(CO)_4I_2 + 2Cu \longrightarrow Fe(CO)_5 + Cu_2I_2$$

In place of FeI_2, FeS can also be used.

Properties. (*i*) $Fe(CO)_5$ is a yellow liquid which is soluble in methyl alcohol, ether, acetone and C_6H_6. It is insoluble in H_2O.

(*ii*) *Decomposition.* On thermal decompositon at 250°C, it gives pure Fe.

$$2Fe(CO)_5 \longrightarrow Fe + 5CO$$

(*iii*) *Action of ultra-violet light.* When it is irradiated with near ultra violet light, $Fe_2(CO)_9$ is formed.

$$2Fe(CO)_5 \longrightarrow Fe_2(CO)_9 + CO$$

The above reaction is reversed in darkness.

The formation of $Fe_2(CO)_9$ actually takes place through the following steps:

(*a*) In presence of ultra violet light, activated $Fe(CO)_5$ [represented as $Fe(CO)_5^*$] is formed.

$$Fe(CO)_5 + h\nu \rightleftharpoons Fe(CO)_5^*$$

(*b*) In this step $Fe(CO)_5^*$ combines with another $Fe(CO)_5$ molecule and gives $Fe_2(CO)_9$

$$Fe(CO)_5^* + Fe(CO)_5 \longrightarrow Fe_2(CO)_9 + CO$$

(*iv*) *Hydrolysis.* $Fe(CO)_5$ is hydrolysed by H_2O and acids.

$$Fe(CO)_5 + H_2SO_4 \longrightarrow FeSO_4 + 5CO + H_2$$

(*v*) *Action of alkalies.* When heated with aqueous NaOH and the solution is acidified, carbonylate anion, $[Fe^{2-}(CO)_4H^+]^-$ is formed. The hydrogen atom in this anion is acidic which means that Fe is in –2 oxidation state.

$$\underset{(Fe\ =\ 0)}{Fe(CO)_5} + 3NaOH \longrightarrow Na^+\underset{(Fe=\ -2)}{[Fe^{2-}(CO)_4H^+]^-} + NaCO_3 + H_2O$$

(*vi*) *Action of NH_3.* With NH_3 it gives $Fe(CO)_4H_2$ and carbamic acid, NH_2COOH

$$Fe(CO)_5 + H_2O + NH_3 \longrightarrow Fe(CO)_4H_2 + NH_2COOH$$

(*vii*) *Reaction with halogens.* It reacts with halogens in non-aqueous solvents to form the stable tetracarbonyl halides, $Fe(CO)_4X_2$.

$$Fe(CO)_5 + X_2 \longrightarrow Fe(CO)_4X_2 + CO$$

The velocities of these reactions follow the order Cl < Br < I.

(*viii*) *Reaction with amines.* When $Fe(CO)_5$ reacts with ethylene diamine (*en*), an addition product, $Fe(CO)_5(en)$ is formed. Pyridine (*py*) reacts with it to give $[Fe(py)_6]$ $[Fe_4(CO)_{13}]$.

(*ix*) *Reaction with NO.* With NO, under pressure below 45°C, it gives $Fe(NO)_4$ as the final product which is a black crystalline material.

$$Fe(CO)_5 + 2NO \longrightarrow Fe(CO)_2(NO)_2 + 3CO$$
$$Fe(CO)_2(NO)_2 + 2NO \longrightarrow Fe(NO)_4 + 2CO$$

(*x*) *Reactions with metallic and non-metallic halides.* In these reactions, $Fe(CO)_5$ acts as a dehalogenating agent and a carbonylating agent; Fe becomes bi-valent and CO groups are either liberated or transferred to some other atom.

$$Fe(CO)_5 + 2CCl_4 \longrightarrow C_2Cl_6 + FeCl_2 + 5CO$$
$$Fe(CO)_5 + SO_2Cl_2 \longrightarrow SO_2 + FeCl_2 + 5CO$$

$SnCl_4$ and $SbCl_5$ are reduced to $SnCl_2$ and $SbCl_3$.

$$Fe(CO)_5 + SnCl_4 \longrightarrow SnCl_2 + Fe(CO)_4Cl_2 + CO$$
$$Fe(CO)_5 + SbCl_5 \longrightarrow SbCl_3 + Fe(CO)_4Cl_2 + CO$$

$Fe(CO)_4Cl_2$ then forms addition compounds with $SnCl_2$ and $SbCl_3$.

$$\left.\begin{array}{l} SnCl_2 + Fe(CO)_4Cl_2 \longrightarrow \underset{(I)}{SnCl_2.Fe(CO)_4Cl_2} \\ SbCl_3 + Fe(CO)_4Cl_2 \longrightarrow \underset{(II)}{SbCl_3.Fe(CO)_4Cl_2} \end{array}\right\} \textit{Addition compounds}$$

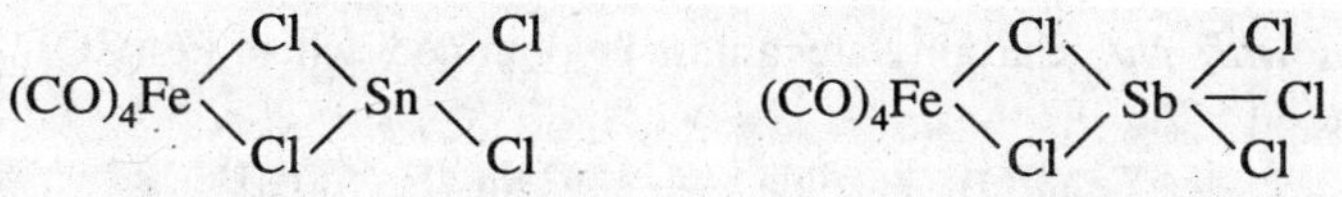

$$(CO)_4Fe\langle^{Cl}_{Cl}\rangle Sn\langle^{Cl}_{Cl} \qquad (CO)_4Fe\langle^{Cl}_{Cl}\rangle Sb\langle^{Cl}_{Cl}-Cl$$

(I) (II)

The addition compounds (I) and (II) are halogen bridged compounds as shown below :

With WCl_6 and $MoCl_6$, $Fe(CO)_5$ gives $W(CO)_6$ and $Mo(CO)_6$ respectively.

$$\underset{(M=W,\ Mo)}{MCl_6} + 2Fe(CO)_5 \longrightarrow M(CO)_6 + 4CO + 2FeCl_3$$

(*xi*) When it is burnt in air, finely divided ferric oxide is obtained which is used as a pigment and a polishing agent.

4. Diiron enneacarbonyl $Fe_2(CO)_9$

Preparation. It is made by the irradiation of $Fe(CO)_5$ with near ultraviolet light. $Fe(CO)_5$ is dissolved in its own volume of glacial acetic acid and the solution thus obtained is cooled and then irradiated for six hours. Golden crystals of $Fe_2(CO)_9$ are precipated and are obtained by filtration.

Properties. (*i*) $Fe_2(CO)_9$ forms tricilinic crystals which are diamagnetic.

(*ii*) It is slightly soluble in alcohol and acetone and more so in toluene and pyridine and almost insoluble in water, ether, C_6H_6 and many other organic solvents.

(*iii*) It is non-volatile.

(*iv*) *Action of heat.* When warmed in 50°C, it gives $Fe_2(CO)_{12}$. At 100°C it decomposes to give Fe, CO, and some $Fe(CO)_5$ and $Fe_3(CO)_{12}$.

(*v*) *Action of No.* With NO it gives $Fe(CO)_2(NO)_2$ together with $Fe(CO)_5$ and $Fe_3(CO)_{12}$.

(*vi*) *Action of phen.* In acetone or benzene solution at 80°C $Fe_2(CO)_9$ reacts with *phen* to form $[Fe(phen)_3]$ $[Fe_2(CO)_8]$. In pyridine, with *phen* it gives $[Fe(phen)_3]$ $[Fe_4(CO)_{13}]$.

(*vii*) *Action of alkali metals.* Carbonylate anion is obtained in the reaction with Na in NH_3 solution.

$$Fe_2(CO)_9 + 4Na \longrightarrow 2Na_2[Fe(CO)_4] + CO$$

5. Triiron dodecacarbonyl, $Fe_3(CO)_{12}$

Preparation. It is prepared: (*i*) by the disproportionation of $Fe_2(CO)_9$. A toluene solution of $Fe_2(CO)_9$ is heated at 70°C. On cooling, green crystals of $Fe_3(CO)_{12}$ separate from the solution.

$$3Fe_2(CO)_9 \xrightarrow{\Delta} 3Fe(CO)_5 + Fe_3(CO)_{12}$$

(*ii*) by the oxidation of $Fe(CO)_4H_2$ with H_2O_2 or MnO_2.

$$3Fe(CO)_4H_2 + 3H_2O_2 \longrightarrow Fe_3(CO)_{12} + 6H_2O$$

$$2Fe(CO)_4H_2 + 3MnO_2 + 3H_2SO_4 \longrightarrow Fe_3(CO)_{12} + 3MnSO_4 + 6H_2O$$

(*iii*) by treating $Fe(CO)_5$ with aqueous tri-ethylamine to form the dark red $[Et_3NH][Fe_2(CO)_{11}H]$ which is then acidified and $Fe_3(CO)_{12}$ is extracted with petroleum ether.

Properties. (*i*) $Fe_3(CO)_{12}$ forms deep green crystals which are soluble in organic solvents like toluene, alcohol, ether and pyridine.

(*ii*) *Action of heat.* When heated alone above 100°C, $Fe_3(CO)_{12}$ decomposes to give metallic iron and CO.

$$Fe_3(CO)_{12} \xrightarrow{>100°C} 3Fe + 12CO$$

(*iii*) *Disproportionation reaction.* It undergoes disproportionation reactions. For example it reacts with CH_3OH at room temperature and $Fe(CO)_5$ is thereby produced.

$$3Fe(CO)_{12} + nCH_3OH \longrightarrow [Fe(CH_3OH)_n]\,[Fe_3(CO)_{11}] + 5Fe(CO)_5$$

(*iv*) *Reaction with Na.* Carbonylate anion is obtained when $Fe_3(CO)_{12}$ reaction with Na metal in liq. NH_3.

$$Fe_3(CO)_{12} + 6Na \longrightarrow 3Na_2[Fe(CO)_4]$$

(*v*) *Substitution reactions.* These reactions occur with pyridine (*py*) and methyl alcohol as shown below :

$$2Fe_3(CO)_{12} + 3py \longrightarrow Fe_3(CO)_9(py)_3 + 3Fe(CO)_5$$

$$2Fe_3(CO)_{12} + 3CH_3OH \longrightarrow Fe_3(CO)_9(CH_3OH)_3 + 3Fe(CO)_5$$

(*vi*) *Action of NO.* At 85°C it reacts with NO to form iron dicarbonyl dinitrosyl, $Fe(CO)_2(NO)_2$.

$$Fe_3(CO)_{12} + 6NO \longrightarrow 3Fe(CO)_2(NO)_2 + 6CO$$

6. Dicobalt octacarbonyl, $Co_2(CO)_8$

Preparation. It can be prepared : (*i*) by the reaction between CO and reduced metallic cobalt at 220°C and 250 atm.

$$2Co + 8CO \xrightarrow[220°C]{250\text{ atm.}} Co_2(CO)_8$$

(*ii*) by the reaction of dry CO at 200°C and 200 atm in presence of metallic copper on certain binary compounds of cobalt such as CoS and CoX_2. Here Cu forms Cu_2S or CuX.

$$2CoS + 8CO + 4Cu \longrightarrow Co_2(CO)_8 + 2Cu_2S$$

$$2CoX_2 + 8CO + 4\,Cu \longrightarrow Co_2(CO)_8 + 4CuX$$

(*iii*) by the action of CO and H_2 on $CoCO_3$ under high pressure and high temperature.

$$2CoCO_3 + 2H_2 + 8CO \xrightarrow[120-220°C]{250-300\text{ atm.}} Co_2(CO)_8 + 2CO_2 + 2H_2O$$

(*iv*) by the action of an acid on a solution of $Co(CO)_4$. Hydrogen is evolved and $Co_2(CO)_8$ is left behind.

$$2Co(CO)_4H \longrightarrow Co_2(CO)_8 + H_2$$

(*v*) by the thermal decompositon of $Co(CO)_4H$.

Properties. (*i*) $Co_2(CO)_8$ is an orange brown crystalline substance naving m.pt. = 51°C. It is soluble in alcohol, ether and CCl_4. The carbonyl is air sensitive both in the solid and solution states.

(*ii*) *Action of heat.* It is thermally decomposed at 50°C in an inert atmosphere to give tetra cobalt dodecacarbonyl, $Co_4(CO)_{12}$.

$$2Co_2(CO)_8 \xrightarrow{50°C} Co_4(CO)_{12} + 4CO$$

(*iii*) *Action of air.* On exposure to air, dicobalt octacarbonyl is converted into deep violet basic carbonate cf cabalt.

(*iv*) *Reduction reactions.* It is reduced to cobalt carbonyl hydride, $H^+[Co(CO)_4]^-$ by H_2 at 165°C and 120 atmospheric pressure.

$$Co_2(CO)_8 + H_2 \xrightarrow[165°C]{120 \text{ atm.}} 2H^+[Co(CO)_4]^-$$

$Co_2(CO)_8$ is also reduced by Na metal in liq. NH_3 below –75°C or tetrahydrofuran (THF).

$$\underset{(Co\,=\,0)}{Co_2(CO)_8} + 2Na \xrightarrow[\text{or liq. } NH_3]{THF} \underset{(Co\,=\,-1)}{2Na^+[Co(CO)_4]^-}$$

(*v*) *Disproportionation reactions.* There are two types of reactions in case of $Co_2(CO)_8$.

(*a*) Strong bases having nitrogen or oxygen donor atoms cause disproportionation into Co(+2) and Co(–1). For example

$$\underset{(Co\,=\,0)}{2Co_2(CO)_8} + 12NH_3 \longrightarrow \underset{(Co\,=\,+2)}{2[Co(NH_3)_6]}\underset{(Co\,=\,-1)}{[Co(CO)_4]_2} + 8CO$$

(*b*) With isocyanides, phosphines, arsines and stibines, however, the disproportionation reaction gives penta coordinate cobalt (I) cation.

$$\underset{(Co\,=\,0)}{Co_2(CO)_8} + 5CNR \longrightarrow \underset{(Co\,=\,+1)}{[Co(CNR)_5]}\underset{(Co\,=\,-1)}{[Co(CO)_4]} + 4CO$$

(*vi*) *Reaction with NO.* $Co_2(CO)_8$ also reacts with NO at 40°C and forms $Co(CO)_3(NO)$

$$\underset{(Co\,=\,0)}{Co_2(CO)_8} + 2NO \longrightarrow \underset{(Co\,=\,-1)}{Co(CO)_3(NO)} + 2CO$$

In $Co(CO)_3(NO)$, NO group appears to be more firmly held with Co metal atom than CO groups. This is confirmed by the fact that CO groups can be replaced by the amines very easily, *e.g.*, the action of *phen* on $Co(CO)_3(NO)$ gives [Co(CO)(*phen*)(NO)].

$$Co(CO)_3(NO) + phen \xrightarrow{-2CO} [Co(CO)(phen)(NO)]$$

(*vii*) *Reaction with halogens.* The halogens decompose $Co_2(CO)_8$ according to the following reaction :

$$Co_2(CO)_8 + 2X_2 \longrightarrow 2CoX_2 + 8CO$$

7. Carbonyls of Ni, Pd and Pt

Nickel forms the tetracarbonyl, $Ni(CO)_4$ while Pd and Pt donot form their tetracarbonyls, *i.e.*, $Ni(CO)_4$ exists as a stable carbonyl while $Pd(CO)_4$ and $Pt(CO)_4$ donot exist. The non-existence of the carbonyls of Pd and Pt has been explained as follows :

(*i*) In the structure of $M(CO)_4$ carbonyl (M = Ni, Pd, Pt), M-atom is sp^3 hybridised (tetrahedral geometry). In order to get vacant sp^3 hybrid orbitals (four hybrid orbitals) ns^2 electrons must be promoted to $(n-1)d^8$ orbitals so that $ns^2.\ (n-1)d^8$ configuration becomes $ns^0.\ (n-1)a^{10}$ (spin paired configuration) (See Fig. 13.1). For the promotion of ns^2 electrons to $(n-1)d^8$ orbitals, energy is required. This energy is called *spin pairing promotional energy.* It has been found that the spin pairing promotional energy of Ni is much less than that for Pd and Pt. This means that Ni changes its configuration from $3d^84s^2$ to $3d^{10}4s^0$ (spin paired configuration) more easily than Pd and Pt. Hence $Ni(CO)_4$ is easier to form. In other words, we say that $Ni(CO)_4$ is formed while $Pd(CO)_4$ and $Pt(CO)_4$ are not formed.

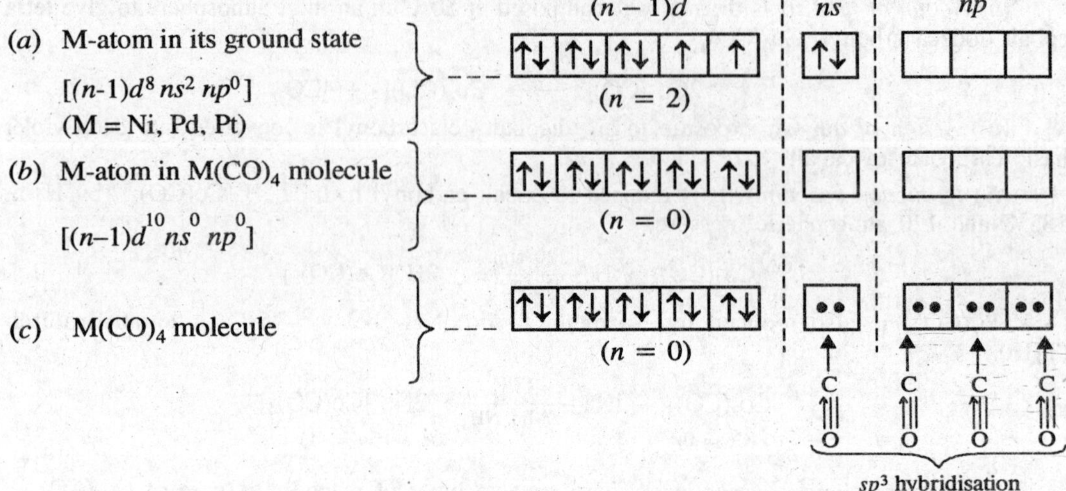

Fig. 13.1. sp^3 hybridisation of M-atom (M=Ni, Pd, Pt) in $M(CO)_4$ carbonyl. Each pair of dots (••) indicates electron pair denoted by C-atom of CO molecule (ligand)

(*ii*) It has been observed that for the transfer of one electron from $3d^{10}$ orbital of Ni-atom to π^* vacant orbitals of CO ($d^{10} \rightarrow d^9$ change), less energy (called ionistation energy) is required than that which is required for the transfer of one electron from $4d^{10}/5d^{10}$ orbital of Pd and Pt to π^* vacant orbitals in CO molecule (Ni = 4.4 eV; Pd, Pt = 8.2 eV). This indicates that *d*-electronic charge of Ni-atom can be transferred to π^* orbitals of CO more easily than *d*-electronic charge of Pd and Pt.

Nickel tetracarbonyl, $Ni(CO)_4$

Preparation. (*i*) It is made by passing carbon monoxide over finely-divided metallic nickel at a temperature below 100°C.

$$Ni + 4CO \longrightarrow Ni(CO)_4$$

(*ii*) When CO at atmospheric pressure is passed over reduced nickel at 30–50°C, we get $Ni(CO)_4$.

(*iii*) $Ni(CO)_4$ is also obtained when CO is passed through an alkaline suspension of $Ni(CN)_2$ or NiS.

$$Ni(CN)_2 + 4CO \longrightarrow Ni(CO)_4 + C_2N_2$$

$$NiS + 4CO \longrightarrow Ni(CO)_4 + S$$

(*ii*) *By the disproportionation of univalent nickel.* (*a*) On reducing potassium tetracyano nickelate (II), $K_2[Ni(CN)_4]$ with potassium amalgam, a red solution is obtained. From this solution potassium tricyano nickelate (I), $K_2[Ni(CN)_3]$ may be isolated. This compound absorbs CO at 90°C, probably forming $K_2[Ni(CN)_3(CO)]$. On treating the solution of this compound with HCl, $Ni(CO)_4$ is obtained.

$$\underset{(Ni\,=\,+1)}{4K_2[Ni(CN)_3(CO)]} + 2HCl \longrightarrow \underset{(Ni\,=\,0)}{Ni(CO)_4} + \underset{(Ni\,=\,+2)}{3K_2[Ni(CN)_4]} + 2KCl + H_2$$

(*b*) It is most conveniently prepared in the laboratory by the reaction of CO on nickel(II) phenyl dithiocarbamate, $Ni(S.SC.NH\ C_6H_5)_2$.

$$2Ni(S.SC.NH\ C_6H_5)_2 + 4CO \longrightarrow Ni(CO)_4 + Ni(S.SC.NH.C_6H_5)_4$$

Nickel (II) phenyl dithiocarbamate (Ni = +2) — (Ni=0) — Nickel (IV) phenyl dithiocarbamate (Ni = +4)

Properties. (*i*) It is a colourless liquid boiling at 43°C and solidifying at –25°C. It is miscible with benzene and is almost insoluble in water. The benzene solution of the nickel carbonyl does the same on boiling, but in the cold, the compound is not acted on by dilute acids or alkalies. Nickel carbonyl is highly poisonous.

(*ii*) *Action of H_2SO_4.* H_2SO_4 reacts with it as follows :

$$Ni(CO)_4 + H_2SO_4 \longrightarrow NiSO_4 + H_2 + 4CO$$

(*iii*) *Action of NO.* Slightly moist nitric oxide reacts with $Ni(CO)_4$ in the gaseous state or in solution in chloroform to give a compound of intense blue colour, Ni( )(OH)

$$2Ni(CO)_4 + 2NO + 2H_2O \longrightarrow 2Ni(NO)(OH) + 8CO + H_2$$

When $Ni(CO)_4$ reacts with NO in the absence of H_2O, the blue solution of the composition, $Ni(NO)(NO_2)$ is obtained.

$$Ni(CO)_4 + 4NO \longrightarrow Ni(NO)(NO_2) + 4CO + N_2O$$

(*iv*) *Substitution reactions.* CO groups can be replaced by NO_2, PCl_3 and CNR molecules. *o*-phenanthroline (*o*-phen) replaces only two CO molecules.

$$Ni(CO)_4 + 4NO_2 \longrightarrow Ni(NO_2)_4 + 4CO\uparrow$$

$$Ni(CO)_4 + 4PCl_3 \longrightarrow Ni(PCl_3)_4 + 4CO\uparrow$$

yellow non-volatile

$$Ni(CO)_4 + CNR \longrightarrow Ni(CNR)_4 + 4CO\uparrow$$

Aryl isocyanide — Yellow crystalline substance

$$Ni(CO)_4 + o\text{-}phen \longrightarrow Ni(CO)_2(o\text{-}phen) + 2CO\uparrow$$

Ruby-red needles

$Ni(CO)_4$ undergoes exchange with ^{14}CO rapidly to produce $Ni(CO)_3(^{14}CO)$. The production of $Ni(CO)_3(^{14}CO)$ takes place through the dissociation of Ni–CO bond (dissociative mechanism).

(*v*) *Action of heat.* $Ni(CO)_4$, when heated to 180°C, gives nickel.

$$Ni(CO)_4 \xrightarrow{180°C} Ni + 4CO\uparrow$$

In the pure state nickel carbonyl explodes to form nickel, carbon dioxide and carbon.

$$Ni(CO)_4 \xrightarrow{\Delta} Ni + 2CO_2 + 2C$$

Under high atmospheric pressure (say, 100 atm) it is, however, not decomposed even up to 250°C.

(*vi*) *Oxidation reactions.* Solution of this carbonyl in organic solvent is readly oxidised by air developing a pale green gelatinous precipitate. Such solutions are also readily oxidised to Ni (II) salts by the halogens, cyanogen and sulphur.

$$Ni(CO)_4 + Br_2 \longrightarrow NiBr_2 + 4CO\uparrow$$

(Ni = 0) — (Ni = +2)

Gaseous HCl decomposes the solution of $Ni(CO)_4$ evolving H_2 and CO

$$Ni(CO)_4 + 2HCl(g) \longrightarrow NiCl_2 + H_2 + 4CO\uparrow$$

(Ni = 0) — (Ni = +2)

(*vii*) *Reduction reactions.* $Ni(CO)_4$ gets reduced by several reducing agents. Examples are:

(*a*) $Ni(CO)_4$ can be reduced to $[Ni^-H^+(CO)_3]$ by the reaction of $Ni(CO)_4$ with Na in liq. NH_3.

$$\underset{(Ni\,=\,0)}{2Ni(CO)_4} + 2Na + 2NH_3 \longrightarrow \underset{(Ni\,=\,-1)}{[Ni^-H^+(CO)_3]_2} + 2CO + 2NaNH_2$$

$[Ni^-H^+(CO)_3]$ is isolated as a red tetraammoniate, $[NiH(CO)_3]_2.4NH_3$.

(*b*) When $Ni(CO)_4$ solution in THF is reduced by Na, K or Mg amalgams, the enneacarbonyl tetranickelate is formed.

$$4Ni(CO)_4 + 2Na \longrightarrow Na_2[Ni_4(CO)_9] + 7CO$$

With Li, $Ni(CO)_4$ gives $Li_2[Ni_3(CO)_8]$

$$3Ni(CO)_4 + 2Li \longrightarrow Li_2[Ni_3(CO)_8] + 4CO$$

Uses. (*i*) Since $Ni(CO)_4$ is decomposed to metallic nickel, it is used in the production of nickel by Mond's process. (*ii*) It is also used in gas plating and also as a catalyst.

Structure of carbon monoxide molecule (CO) on the basis of valence bond theory

Accordiing to valence bond theory CO molecule is represented by the structure, $:O \overset{\rightarrow}{=} C:$. In this structure C and O-atoms both are *sp* hybridised.

C-atom in the ground state ($2s^2 2p^2$) $2s^2\ 2p_x^1\ 2p_y^1\ 2p_z^0$

C-atom in *sp* hybridised state $\underset{lp}{(sp)^2}\ (sp)^1\ 2p_y^1\ 2p_z^0$

O-atom in the ground state ($2s^2 2p^4$) $2s^2\ 2p_x^1\ 2p_y^1\ 2p_z^2$

O-atom in *sp* hybridised state $\underset{lp}{(sp)^2}\ (sp)^1\ 2p_y^1\ 2p_z^2$

C—O σ-bond is formed by the overlap between the singly-filled *sp* hybrid orbitals on carbon and oxygen atoms while C—O π bond results by the overlap between the singly-filled $2p_y$ orbitals on carbon and oxygen atoms (see Fig. 13.2). O→ C coordinate bond is obtained by the donation of an electron pair residing in $2p_z$ orbtial on O-atom to the vacant $2p_z$ orbital of C-atom (Fig. 13.2). The electron pair present in *sp* hybrid orbital of both the atoms remains as lone pair (*lp*) of electrons on these atoms (Fig. 13.2).

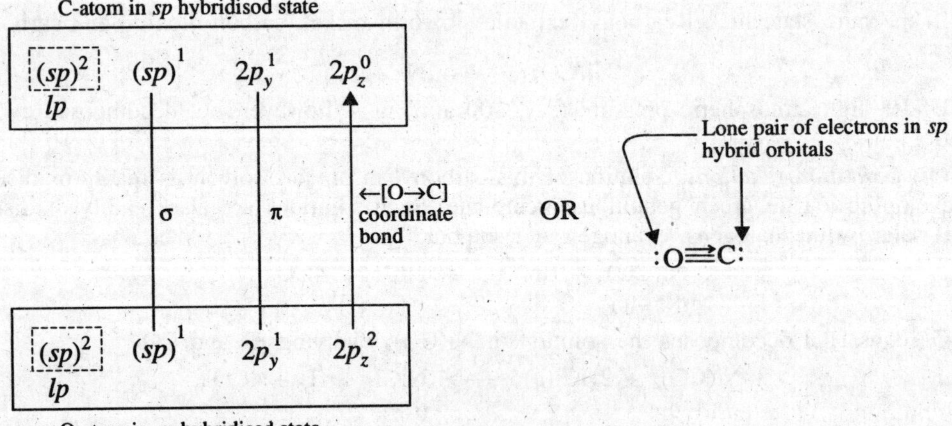

Fig. 13.2. Formation of different bonds in CO molecule ($:O \overset{\rightarrow}{=} C:$) on the basis of valence bond theory.

Molecular orbital configuration of CO molecule suggested by Coulson

According to Coulson, CO molecule consists of molecular orbitals which are formed through the following steps :

(*i*) First of all carbon and oxygen atoms undergo *sp* hybridisation. Thus carbon and oxygen atoms both contain two *sp* hybrid orbitals and $2p_y$ and $2p_z$ atomic orbitals. Thus :

C-atom in the ground state : $(2s)_C^2\ (2p_x)_C^1\ (2p_y)_C^0\ (2p_z)_C^1$

C-atom in *sp* hybridised state : $(sp)_C^2\ (sp)_C^1\ (2p_y)_C^0\ (2p_z)_C^1$

O-atom in the ground state : $(2s)_O^2\ (2p_x)_O^1\ (2p_y)_O^2\ (2p_z)_O^1$

O-atom in *sp* hybridised state : $(sp)_O^2\ (sp)_O^1\ (2p_y)_O^1\ (2p_z)_O^1$

(*ii*) Completely-filled *sp* hybrid orbitals on both the atoms contain lone pairs of electrons and remain as such but the singly-filled *sp* hybrid orbitals on both the atoms combine together and give rise to the formation of σ^b_{sp-sp} (bonding molecualr orbital) and σ^*_{sp-sp} (antibonding molecular orbital) molecular orbitals. (See Fig. 13.3).

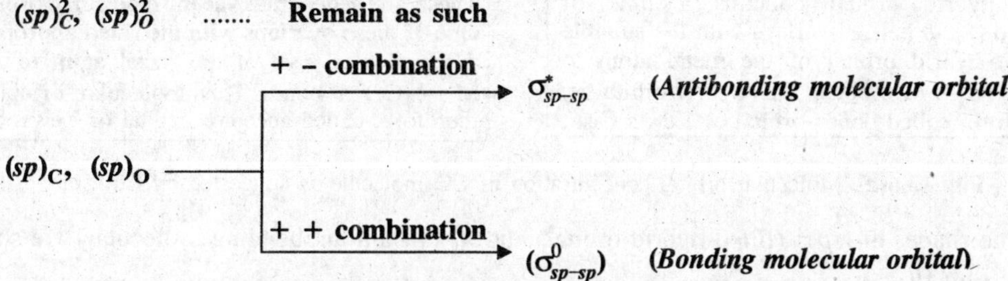

Fig. 13.3. Formation of σ^b_{sp-sp} and σ^*_{sp-sp} molecular orbitals by the combination of singly-filled $(sp)_C$ and $(sp)_O$ hybrid orbitals. Single electron present in each hybrid orbital has not been shown.

The combination of two $2p_y$ atomic orbitals on both the atoms gives $\pi_y{}^b$ (bonding molecular orbital) and $\pi_y{}^*$ (antibonding molecular orbital) molecular orbitals, (Fig. 13.4) while the combination of two $2p_z$ atomic orbitals on both the atoms gives $\pi_z{}^b$ (bonding molecular orbital) and $\pi_z{}^*$ (antibonding molecular orbital) molecular orbitals (Fig 13.5).

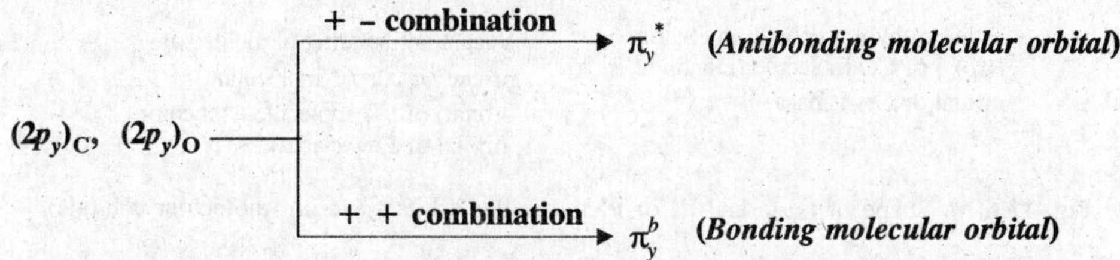

Fig. 13.4. Formation of $\pi_y{}^b$ and $\pi_y{}^*$ molecular orbitals by the combination of $(2p_y)_C$ and $(2p_y)_O$ atomic orbitals. Electrons present in atomic orbitals have not been shown.

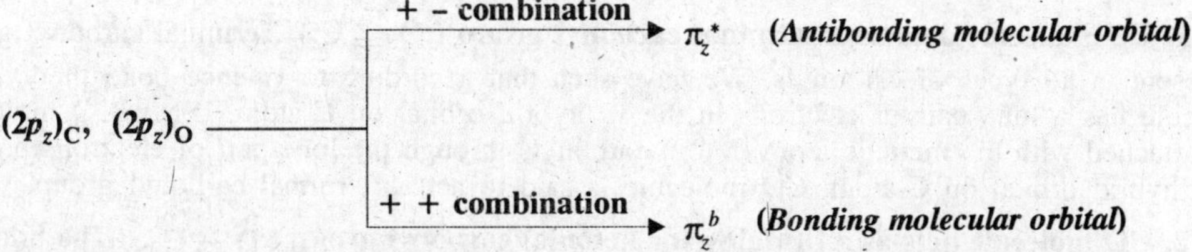

Fig. 13.5. Formation of $\pi_z{}^b$ and $\pi_z{}^*$ molecular orbitals by the combination of $(2p_z)_C$ and $(2p_z)_O$ atomic orbitals. Electrons present in atomic orbitals have not been shown.

The above description shows that CO molecule contains two completely fulled *sp* hybrid orbitals [*viz.* $(sp)_C$ and $(sp)_O$], three bonding molecular orbitals (σ^b_{sp-sp}, π^b_y, π^b_z) and three antibonding molecular orbitals (σ^*_{s-sp}, π^*_y, π^*_z). The energy of the above orbitals is in the order shown below:

$$(sp)_O < \sigma^b_{sp-sp} < \pi^b_y = \pi^b_z < (sp)_C < \pi^*_y = \pi^*_z < \sigma^*_{sp-sp}$$

The above order shows that π^b_y and π^b_z are degenerate molecular orbitals. Similarly π^*_y and π^*_z are also degenerate molecular orbitals. This order also shows that 10 electrons present in CO molecule (CO = 4 + 6 = 10) are accomodated in these orbitals as shown in Fig. 13.6(*a*).

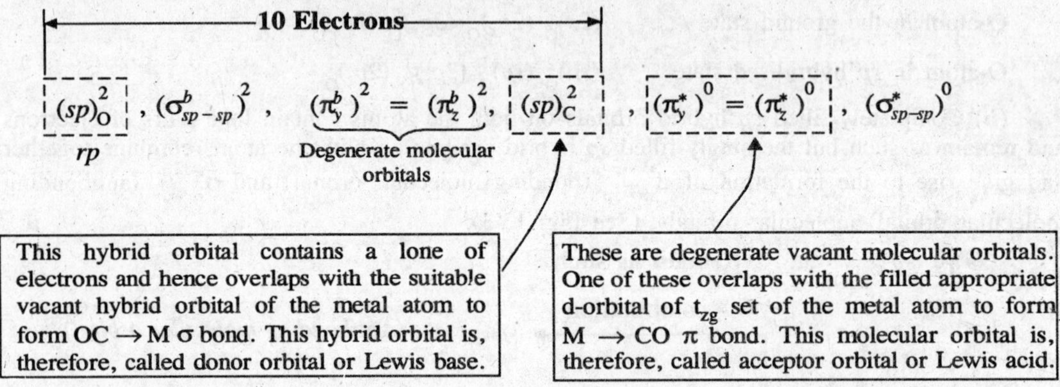

Fig. 13.6(*a*). Molecular orbital configuration of CO molecule as suggested by Coulson.

The shapes of $(sp)^2_C$ (filled hybrid orbital) and π^*_y (vacant antibonding molecular) are shown in Fig. 13.6 (*b*).

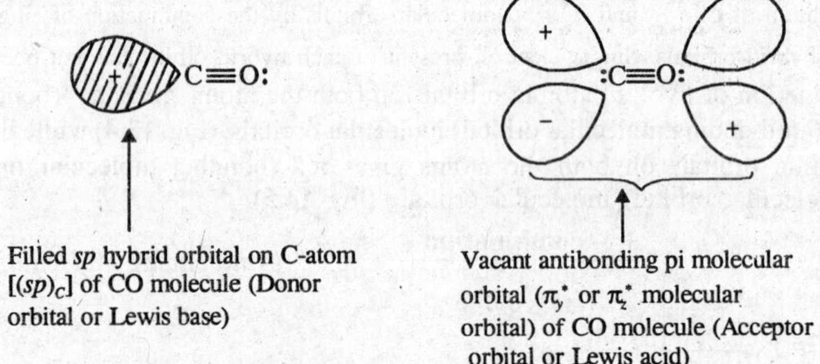

Fig. 13.6(b). Shape of $(sp)_C$ and π^*_y or π^*_z molecular orbitals (vacant molecular orbitals)

How does CO molecule act in metallic carbonyls?

The study of the structure of various types of metallic carbonyls has shown that CO molecule in these carbonyls acts in any of the following two ways :

1. CO molecule acts as a terminal carbonyl group (:O $\equiv$ C:). Terminal carbonyl group is present in all types of carbonyls. We have seen that according to valence bond theory, CO molecule has a lone pair of electrons in the *sp* hybrid orbital on C-atom. When CO molecule gets attached with the metallic atom (M) in carbonyls through the lone pair of electrons present in *sp* hybrid orbital on C-atom, CO molecule is said to act as terminal carbonyl group.

2. CO molecule acts as a bridging (or ketonic) carbonyl group (:Ö═C<). The bridging carbonyl group is present in bridged carbonyls like $Co_2(CO)_8$, (in the solid state), $M_2(CO)_9$ (M=Fe,Os etc.) etc. When CO molecule is attached with two metallic atoms through two unpaired electrons present in two sp^2 hybrid orbitals on C-atoms, CO molecule is said to act as a bridging carbonyl group.

Different types of bonds found in metallic carbonyls and how are they formed ?

Different types of bonds found in metallic carbonyls and their formation have been discussed below:

1. :O$\underline{\rightleftarrows}$C→M coordinate σ bond.

This type of bond is found in all types of carbonyls. In the carbonyls in which this bond is found, CO molecule acts as a terminal carbonyl group. The formation of this bond can be explained on the basis of valence bond theory and molecular orbital theory as follows:

(*a*) **Formation of :O$\underline{\rightleftarrows}$C→M coordinate σ bond on the basis of valence bond theory.** We have seen that, on the basis of valence bond theory, the structure of CO molecule is :O$\underline{\rightleftarrows}$C: in which carbon and oxygen both are *sp* hybridised. The lone pair of electrons on carbon and O-atoms is present in their respective *sp* hybrid orbital. When CO molecule gets linked as a terminal carbonyl group with a transition metal (M) in low oxidation state (0, −1, +1), lone pair of electrons present in *sp* hybrid orbital of C-atom is donated to the appropriate vacant hybrid orbital of the metal atom [*e.g.* sp^3 hybrid orbital of Ni in $Ni(CO)_4$, dsp^3 hybrid orbital of Fe in $Fe(CO)_5$ and Co-atom in $Co_2(CO)_8$ in the solution form, d^2sp^3 hybrid orbital of the metal atom in $Cr(CO)_6$, $Mn_2(CO)_{10}$ and $Co_2(CO)_8$ (in solid state)] and forms :O$\underline{\rightleftarrows}$C → M coordinate σ-bond. Here it should be noted that only the lone pair of electrons on C-atom (and not on oxygen atom) is donated to the vacant hybrid orbital of the metallic atom. This is because of the fact that C-atom is less electronegative than O-atom. In a simplified way :O$\underline{\rightleftarrows}$C → M is also represented as OC → M.

Illustrative example : Formation of OC→ Ni σ bonds in $Ni(CO)_4$ molecule

As an example let us see how :O$\underline{\rightleftarrows}$C→ Ni or OC→ Ni σ-bonds are formed in $Ni(CO)_4$ molecule. Various studies have indicated that $Ni(CO)_4$ molecule has tetrahedral geometry (see Fig. 13.7) and is diamagnetic in nature. Each Ni—C bond length has been found to be equal to 1.50Å. Obviously in order to get tetrahedral geometry, Ni-atom (in zero oxidation state) must be sp^3 hybridised as shown in Fig. 13.8. Due to sp^3 hybridisation all the electrons become paired. The presence of paired electrons confirms the diamagnetic character of $Ni(CO)_4$.

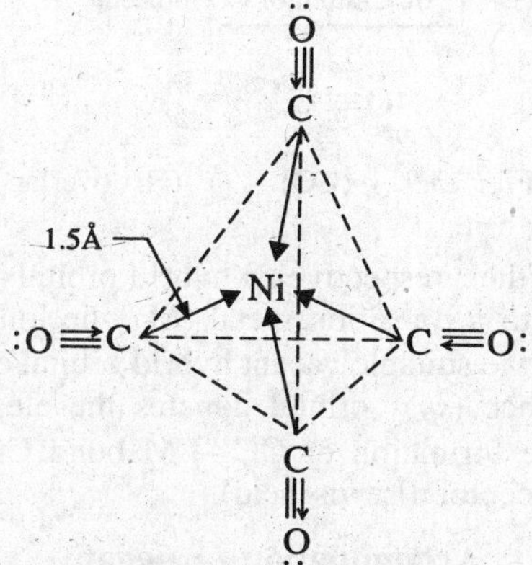

Fig. 13.7. Tetrahedral geometry of $Ni(CO)_4$ molecule.

$Ni(CO)_4$ molecule has four OC → M σ-bonds. Each of the four :O$\underline{\rightleftarrows}$C → M σ-bonds results by the overlap of completely-filled *sp* hybrid orbital on C-atom in CO molecule and vacant sp^3 hybrid orbital on Ni-atom. *sp*(CO)–sp^3(Ni) overlap leading to the formation of :O$\underline{\rightleftarrows}$ C → Ni coordinate σ-bond has been shown in Fig. 13.9.

(*b*) **Formation of OC → M coordinate σ-bond on the basis of molecular orbital theory.** We have already said that, according to Coulson, the molecular orbital configuration of CO molecule can be written as shown on page 386.

This configuration shows that carbon and oxygen both have one lone pair (*lp*) of electrons

Energy increasing ⟶

lp			*lp*		
$(sp)_O^2$	$(\sigma^b_{sp-sp})^2$	$(\pi_y^b)^2 = (\pi_z^b)^2$	$(sp)_C^2$	$(\pi_y^*)^0 = (\pi_z^*)^0$	$(\sigma^*_{sp-sp})^0$
			(Donor orbital – Lewis base)	(Acceptor orbitals – Lewis acid)	

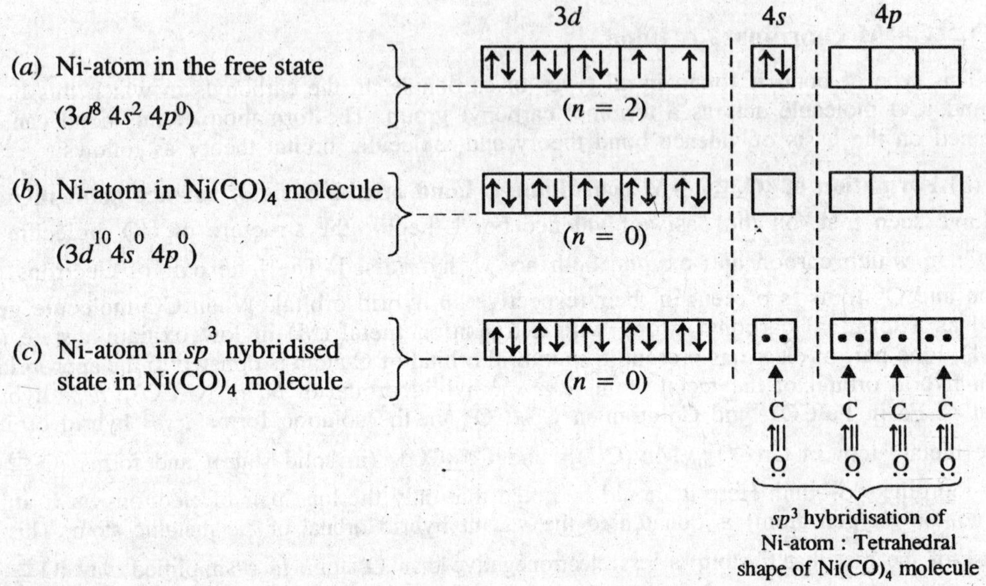

	3*d*	4*s*	4*p*
(*a*) Ni-atom in the free state ($3d^8\,4s^2\,4p^0$)	↑↓ ↑↓ ↑↓ ↑ ↑ (n = 2)	↑↓	— — —
(*b*) Ni-atom in $Ni(CO)_4$ molecule ($3d^{10}\,4s^0\,4p^0$)	↑↓ ↑↓ ↑↓ ↑↓ ↑↓ (n = 0)	—	— — —
(*c*) Ni-atom in sp^3 hybridised state in $Ni(CO)_4$ molecule	↑↓ ↑↓ ↑↓ ↑↓ ↑↓ (n = 0)	•• ↑ C ⫴ O	•• •• •• ↑ C ⫴ O ↑ C ⫴ O ↑ C ⫴ O

sp^3 hybridisation of Ni-atom : Tetrahedral shape of $Ni(CO)_4$ molecule

Fig. 13.8. sp^3 hybridisation of Ni-atom in $Ni(CO)_4$ molecule. Each pair of dots (••) indicates electron pair donated by C-atom of each CO molecule (ligand).

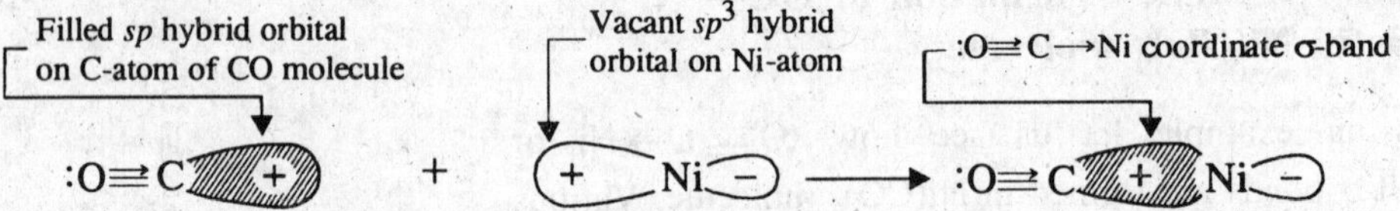

Fig. 13.9. sp(CO) – sp^3 (Ni) overlap leading to the formation in :O ⥱ C → Ni coordinate σ-bond in $Ni(CO)_4$ molecule.

in their respective *sp* hybrid orbitals namely $(sp)_C$ and $(sp)_O$. Now when CO molecule is attached with a transition metal (M), the lone pair of electrons present in $(sp)_C$ hybrid orbital is donated to the suitable vacant hybrid orbital of the metal atom and OC→ M σ-coordinate bond is produced. Since $(sp)_C$ orbital donates the electron pair to the metal atom, it is called **donor orbital**. In the formation of OC → M bond CO molecule acts as a donor (Lewis base) and M acts as an acceptor (Lewis acid).

2. Accumulation of negative charge on the metal and formation of M→ CO π-bond

We have said above that the lone pair of electrons present in $(sp)_C$ of CO molecule is donated to the suitable vacant hybrid orbital of the transition metal atom (M) and OC → M coordinate σ-bond is formed. Now if transition metal is in zero or negative oxidation state, the formation of OC → M bond produces excessive negative charge on the metal atom. The accumulation of negative charge on the metal leads to substantial repulsion between negative charge on the metal and that on CO molecule (ligand). Hence in order to decrease the amount of negative charge produced on the metal atom, suitable *filled d-orbital* on the metal overlaps with the vacant antibonding *pi* molecular orbital (*i.e.*, π_y* or π_z* molecular orbital) on CO molecule. This type of overlap transfers some negative charge from the metal (M) to CO ligand and M → CO π bond is formed.

The above discussion shows that M → CO π-bond is formed by the donation of a pair of electrons from the *filled d-orbital* of the metal atom into the *vacant antibonding pi molecular orbital* (*i.e.*, π_y^* or π_z^* molecular orbital) of CO molecule as shown in Fig. 13.10.

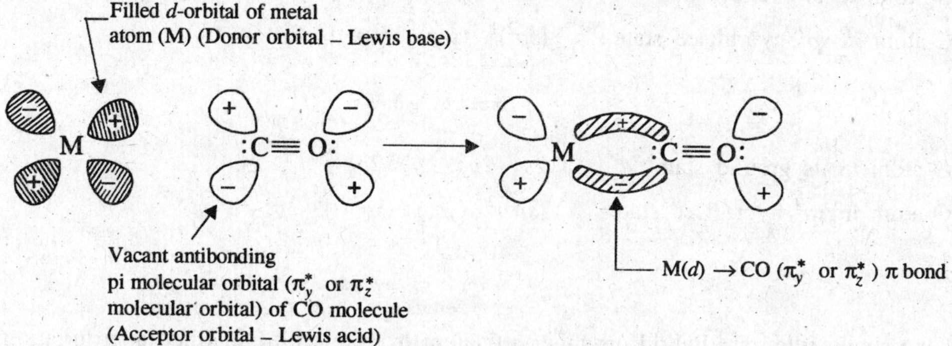

Fig. 13.10. Formation of M → CO π-bond in metallic carbonyls.

If the overlap between filled *d*-orbital of the metallic atom and vacant antibonding pi molecular orbital (π_y^* or π_z^* molecular orbital) on CO molecular takes place in *xz* plane, then *d*-orbital used in this overlap is d_{xz} orbital.

Since the bond formed between the metal atom (M) and CO molecule is a π-bond, the filled *d*-orbital of the metal is called *d*π orbital. Obviously M → CO π bond is produced as a result of *d*π *(M)* — π*(CO) overlap. Since *d*-orbital of the metal atom donates the electron pair to the vacant antibonding pi-molecular orbital (π_y^* or π_z^* molecular orbital), *d*-orbital is called donor orbital or Lewis base and π_y^* and π_z^* molecular orbital as called acceptor orbital or Lewis acid.

It is clear from the above discussion that the vacant π_y^* or π_z^* molecular orbital of the ligand (*i.e.* CO molecule) receives the back-donated electrons from the filled *d*-orbital of the metal and thus stabilises the metal in low oxidation state. In the formation of M → CO π-bond, metal (M) acts as a donor (Lewis base) and CO molecule (ligand) acts as an acceptor (Lewis acid). Since CO accepts the back-donated electrons from the metal atom into its vacant π_y^* or π_z^* molecular orbital and forms M → CO π-bond, CO is called *π-acid ligand* or *π-acceptor ligand* or *π-bonding ligand.*

Conclusion

The discussion of the formation of OC→M σ and M→CO π-bonds in metallic carbonyls shows that for the formation of OC→M σ-bond, CO molecule should have *filled sp* hybrid orbital and the metallic atom should have suitable *vacant* hybrid orbital (valence bond theory). On the other hand for the formation of M→CO π-bond, the metal atom should have *filled d*-orbital and CO molecule should have *vacant* antibonding pi-molecular orbital (π_y^* or π_z^* molecualr orbital). The above discussion shows that the metallic carbonyls have OC→M σ as well as M→CO π bonds, *i.e.* metallic carbonyls have double bonding which can be represented as $OC \underset{\pi}{\overset{\sigma}{\rightleftarrows}} M$. In order to get $OC \underset{\pi}{\overset{\sigma}{\rightleftarrows}} M$ bond, CO molecule should have *filled sp* hybrid orbital as well as vacant antibonding *pi* molecular orbital (π_y^* or π_z^* molecular orbital) and the metal atom should be in low oxidation state and should have *vacant* appropriate hybrid orbital and *filled d*-orbital

3. $:\ddot{O}=C\langle^{M}_{M}$ bond

This bond is found in bridged polynuclear carbonyls like $Co_2(CO)_8$ (in the solid state), $M_2(CO)_9$ (M=Fe, Os etc.) In these carbonyls CO molecule acts as a *bridging* (or *ketonic*) *carbonyl group.* $:\ddot{O}=C\langle$. The configuration of the system, $:\ddot{O}=C\langle^{M}_{M}$ shows that carbon and oxygen atoms both are sp^2 hybridised as shown below :

C-atom in its ground state : $2s^2\ 2p_x^1\ 2p_y^1\ 2p_z^0$

C-atom in its excited state : $2s^1\ 2p_x^1\ 2p_y^1\ 2p_z^1$

C-atom in sp^2 hybridised state : $\underbrace{(sp^2)^1\ (sp^2)^1\ (sp^2)^1}_{\text{Three } sp^2 \text{ hybrid orbitals}}\ 2p_z^1$

O-atom in its ground state : $2s^2\ 2p_x^2\ 2p_y^1\ 2p_z^1$

O-atom in sp^2 hybridised state : $\underbrace{(sp^2)^2\ (sp^2)^2\ (sp^2)^1}_{\text{Three } sp^2 \text{ hybrid orbitals}}\ 2p_z^1$

Two singly-filled sp^2 hybrid orbitals on C-atom overlap with half-filled appropriate hybrid orbital of two metallic atoms (M) and give two (C—M) σ-bonds (see Fig. 13.11). Third half-filled sp^2 hybrid orbital overlaps with the half-filled sp^2 hybrid orbital of O-atom and forms (C—O) σ-bond. (C—O) π bond results by p(C)—p(O) overlap. Thus (C—O)π bond is a (p—p)π bond. Formation of different bonds in $\begin{matrix}M\\M\end{matrix}\!\!>\!C\!=\!\ddot{O}$: system has been depicted in Fig. 13.11 and 13.12.

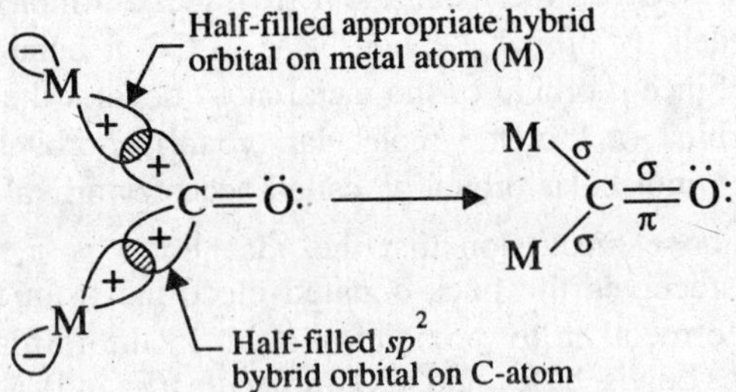

Fig. 13.11. Formation of two (M—C) σ-bonds in $\begin{matrix}M\\M\end{matrix}\!\!>\!C\!=\!\ddot{O}$: by the overlap of two singly-filled sp^2 hybrid orbitals on C-atom and two appropriate singly-filled hybrid orbitals on two M atoms.

C-atom in sp^2 hybridised state

$(sp^2)^1$	$(sp^2)^1$	$(sp^2)^1$	$2p_z^1$
σ	σ		
M	M	σ	π
lp	lp		
$(sp^2)^2$	$(sp^2)^2$	$(sp^2)^1$	$2p_z^1$

O-atom in sp^2 hybridised state

OR $:\ddot{O}\!=\!C\!<\!\begin{matrix}M\\M\end{matrix}$

Fig. 13.12. Formation of different bonds in $:\ddot{O} = C\!<\!\begin{matrix}M\\M\end{matrix}$ system which is present in bridged polynuclear carbonyls.

4. M····M bond

There are many carbonyls which contain one or more M····M bonds. Examples of carbonyls containing M····M bonds are $M_2(CO)_9$ (M=Fe, Os), $Co_2(CO)_8$ (in solution and in solid), $M_2(CO)_{10}$ (M=Mn, Tc, Re), $M_3(CO)_{12}$ (M=Fe, Os) etc. This bond is produced by the weak overlap of the two singly-filled appropriate hybrid orbitals, each belonging to each of the two metallic atoms. Due to the weak coupling of electron spins, this bond is called *fractional single bond* and is represented as M····M. The charge distribution in this bond is not exactly symmetrical. This bond is larger in length than the normal single M—M bond. The presence of M····M bond makes all the electrons paired and hence the carbonyls containg M····M bonds are diamagnetic.

Structure of Some Carbonyls

Here we discuss the structure of some important carbonyls.

1. Nickel tetracarbonyl, $Ni(CO)_4$

The structure of $Ni(CO)_4$ has already been explained on page 385.

2. Iron pentacarbonyl, $Fe(CO)_5$

In principle, $Fe(CO)_5$ molecule can have any of the two geometries *viz., trigonal bipyramidal* and *square pyramial* shown in Fig. 13.13.

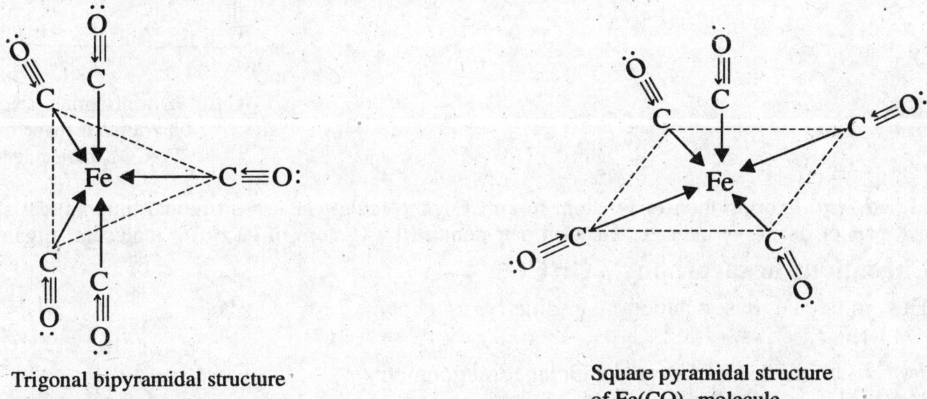

Fig. 13.13. Two possible geometries of $Fe(CO)_5$ molecule.

In order to decide as to which of the above structures is the real structure of $Fe(CO)_5$, we proceed as follows :

We calculate the number of IR-active and Raman-active vibrational bands for each of the above two geometries. Now, we run the IR and Raman spectra and tally the number of bands actually observed with the number of bands obtained by calculation for each geometry. It has been found that IR bands actually observed tally with the number of bands calculated for trigonal bipyramidal geometry. Thus $Fe(CO)_5$ has trigonal bipyramidal geometry. This geometry has also been confirmed by X-ray studies. It is strange to note that ^{13}C—NMR gives only one type of signal for axial as well as equatorial CO groups. This is probably because of the *fluxionality* in the structure. This means that the axial and the equatorial (basal) M←C$\overset{\leftarrow}{=}$O: bonds exchange their position very fast so that an average ^{13}C resonance signal is obtained for all the five carbonyl groups attached with Fe-atom. Electron diffraction study of $Fe(CO)_5$ molecule in the vapour state has shown that Fe—C axial bond and Fe—C equatorial bond lengths are equal to 1.797Å and 1.842Å respectively while X-ray study has shown that these bonds are almost of the same length.

***dsp*^{3} hybridisation of Fe-atom.** Trigonal bipyramidal shape of $Fe(CO)_5$ molecule suggests that Fe-atom which is in zero oxidation state is dsp^3 hybridised. Since $Fe(CO)_5$ is diamagnetic, all the 8 electrons in the valence-shell of Fe-atom (Fe = $3d^64s^2$) get paired in four 3*d* orbitals.

dsp^3 hybridisation of Fe-atom is shown in Fig. 13.14. :O$\equiv$C → Fe bond is established by the overlap of doubly-filled *sp* hybrid orbital on C-atom in CO molecule and empty dsp^3 hybrid orbital on Fe-atom.

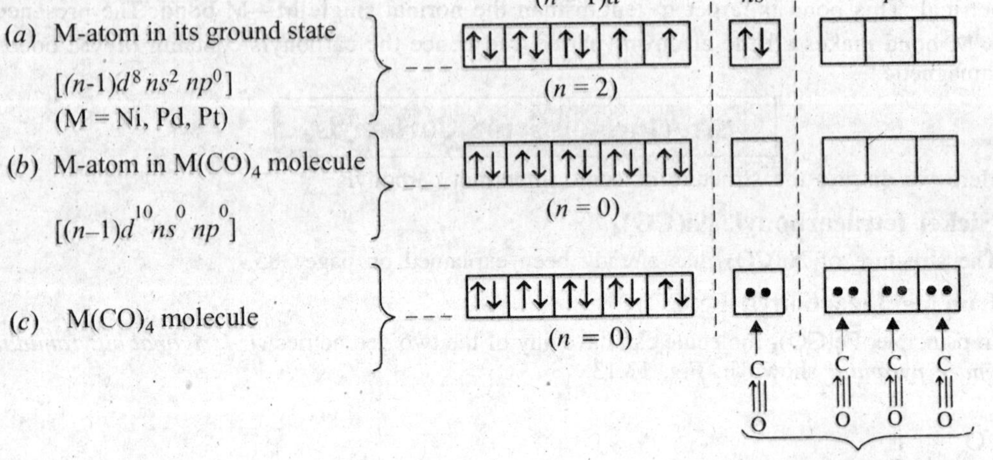

Fig. 13.14. dsp^3 hybridisation of Fe-atom in $Fe(CO)_5$ molecule which has trigonal bipyramidal shape. Each pair of dots (••) indicates electron pair donated by C-atom of each CO molecule (Ligand).

3. Chromium hexacarbonyl, $Cr(CO)_6$

This molecule has octahedral geometry as shown in Fig. 13.15.

***d^2sp^3* hybridistion of Cr-atom.** Octahedral geometry of $Cr(CO)_6$ suggests that Cr-atom which is in zero oxidation state is d^2sp^3 hybridised as shown in Fig. 13.16. Since $Cr(CO)_6$ is diamagnetic, all the six electrons present in the valence shell of Cr-atom ($Cr=3d^54s^1$) get paired in three $3d$ orbitals. Now $3d_{x^2-y^2}$, $3d_{z^2}$, $4s$ and three $4p$ orbitals hybridise together to form six d^2sp^3 hybrid orbitals (see Fig. 13.16). $\dot{O}\equiv C \rightarrow Cr$ bond is formed by the overlap of the doubly-filled *sp* hybrid orbital on C-atom in CO molecule and empty d^2sp^3 hybrid orbital on Cr-atom.

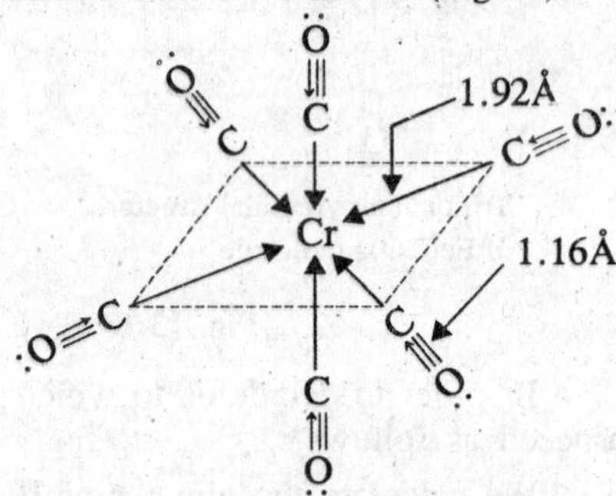

Fig. 13.15. Octahedral geometry of $Cr(CO)_6$ molecule.

Cr—C and C—O bond lengths have been found to be equal to 1.92Å and 1.16Å respectively.

4. Structure of $Fe_2(CO)_9$ molecule

Infra-red spectral study made on $Fe_2(CO)_9$ shows that this molecule has two different types of carbonyl groups. X-ray crystallographic study has shown that the structure of $Fe_2(CO)_4$ molecule consists of three bridging carbonyl groups, six terminal carbonyl groups and one single bond between two Fe-atom (see Fig. 13.17). Since this bond is produced by the weak coupling of the unpaired electrons present in two $3d$-orbitals of two Fe atoms, it is called *fractional single bond* and is represented as Fe····Fe. This bond is a δ bond. The charge distribution in this bond is not exactly symmetrical. This bond is larger in length than the normal single Fe—Fe bond. $Fe_2(CO)_9$ is diamagnetic and obeys EAN or inert gas rule. Diamagnetic nature of $Fe_2(CO)_9$

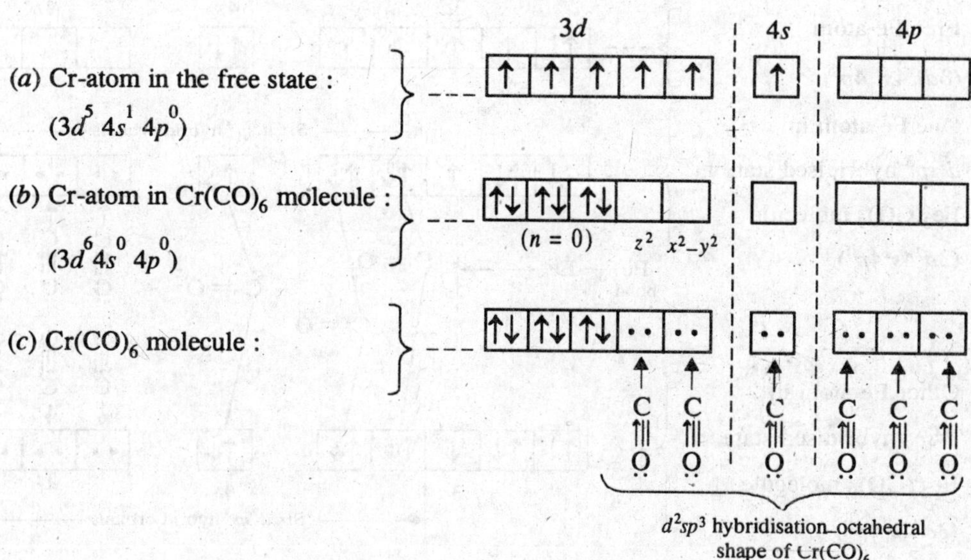

Fig. 13.16. d^2sp^3 hybridisation of Cr-atom in $Cr(CO)_6$ molecule which has octahedral geometry. Each pair of dots (••) indicates electron pair donaed by C-atom of each CO molecule (ligand).

molecule also confirms the presence of a bond between two Fe-atoms, since the formation of Fe····Fe bond makes all the electrons paired. Formation of different bonds in $Fe_2(CO)_9$ molecule has been shown in Fig. 13.18.

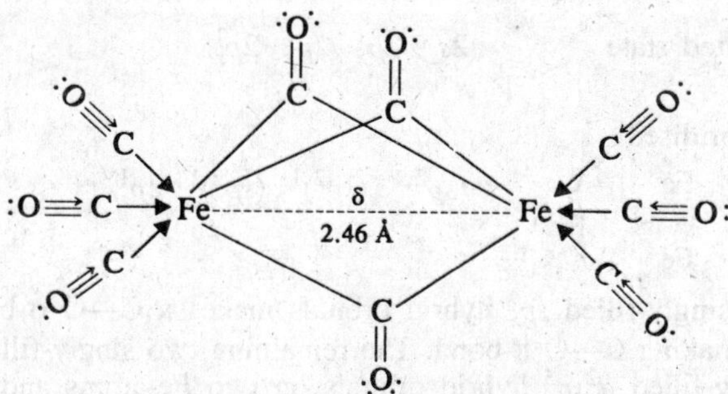

Fig. 13.17. Structure of $Fe_2(CO)_9$ molecule

Since each Fe-atom is linked with three terminal CO groups and three bridging CO groups, this atom has coordination number equal to six (Note that Fe·····Fe bond is a fractional single bond) and hence each Fe-atom is d^2sp^3 hybridised in $Fe_2(CO)_9$ molecule. Valence-shell configuration of Fe-atom in the free state is $3d^64s^24p^0$. When Fe-atom forms $Fe_2(CO)_9$ molecule, one of the two electrons of 4*s* orbital is shifted to 3*d* orbitals and hence the valence-shell configuration of Fe-atom becomes $3d^74s^14p^0$. Both these configurations have been shown at (*a*) and (*b*) of Fig. 13.18. Now two 3*d* orbitals *viz.* $3d_{x^2-y^2}$ and $3d_{z^2}$ (both are singly-filled), one 4*s* orbital (singly-filled) and all the three 4*p* orbitals (all are empty) combine together and produce six d^2sp^3 hybrid orbitals. Three hybrid orbitals are vacant and three are singly-filled. One of the three unhybridised 3*d* orbitals ($3d_{xy}$, $3d_{yz}$ and $3d_{zx}$) is singly-filled. Fe---Fe fractional single bond is produced by the overlap between two singly-filled 3*d* orbitals on two Fe-atoms. The attachment of bridging CO molecule with two Fe-atoms indicates that C-atom of CO molecule of sp^2 hybridised.

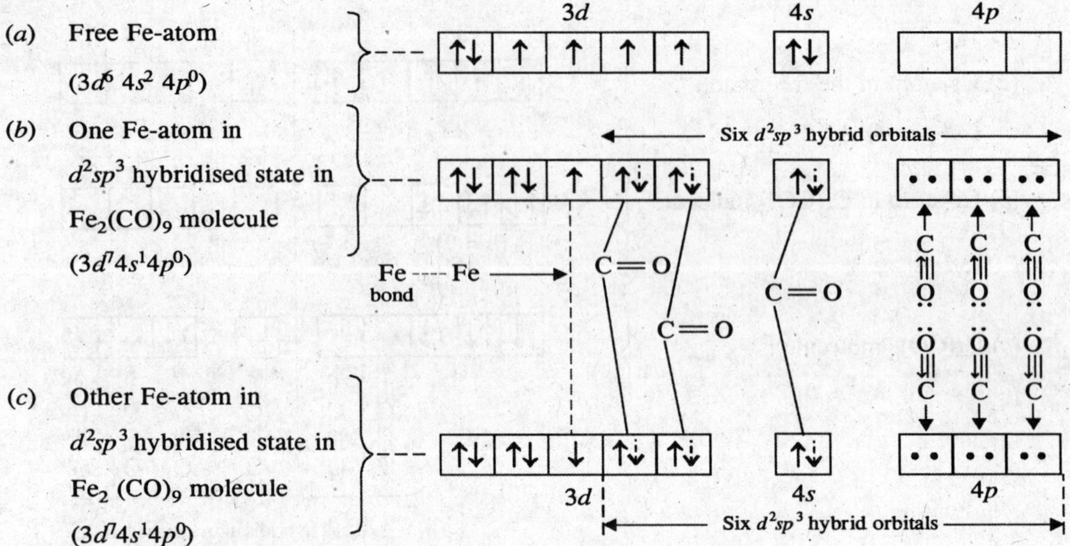

Fig. 13.18. Formation of different bonds of $Fe_2(CO)_9$ molecule. Dots (••) indicate the electron pair donated by C-atoms of the terminal carbonyl groups [:C⇄C]and broken arrow (⇣) represents the electron given by C-atom of each bridging carbonyl group [$\rangle$C = O].

C-atom in its ground state ($2s^2 2p^2$) : $2s^2\ 2p_x^1\ 2p_y^1\ 2p_z^0$

C-atom in its excited state ($2s^1 2p^3$) : $2s^1\ 2p_x^1\ 2p_y^1\ 2p_z^1$

C-atom in sp^2 hybridised state in $:\ddot{O}=C\langle^{Fe}_{Fe}$: $(sp^2)^1\ (sp^2)^1\ (sp^2)^1\ 2p_z^1$

One of the three singly-filled sp^2 hybrid orbitals makes a C—O σ-bond and singly-filled $2p_z$ orbitals is used in making C—O π-bond. The remaining two singly-filled sp^2 hybrid orbitals overlap with two singly-filled d^2sp^3 hybrid orbitals on two Fe-atoms and produce two C—Fe bonds in $:\ddot{O}=C\langle^{Fe}_{Fe}$. Fe ← CO bond is formed in its usual manner.

5. Structure of $Co_2(CO)_8$ molecule

The structure of this molecule depends on whether this carbonyl is in solution form or in the solid state. $Co_2(CO)_8$ in these two states has different structure.

(a) Structure of $Co_2(CO)_8$ molecule in solution. The infra-red study of the solution of $Co_2(CO)_8$ has shown that this molecule has no bridging carbonyl groups, *i.e.*, $Co_2(CO)_8$ in solution has *non-bridged structure* (See Fig. 13.19). In this structure each Co-atom is linked with four terminal carbonyl groups (:O⇄C→Co) and with other Co-atom by a Co ····· Co σ-bond. Thus each Co-atom has four : O⇄C→Co coordinate σ-bonds and one Co·····Co σ-bond. The

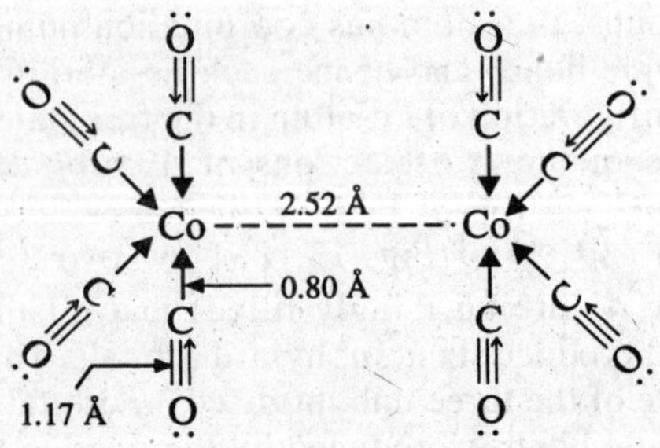

Fig. 13.19. Non-bridged structure of $Co_2(CO)_8$ molecule in solution.

presence of Co --- Co bond has also been supported by the diamagnetic nature of $Co_2(CO)_8$ mlecule. Various bond lengths shown in the structure have been determined by electron diffraction study.

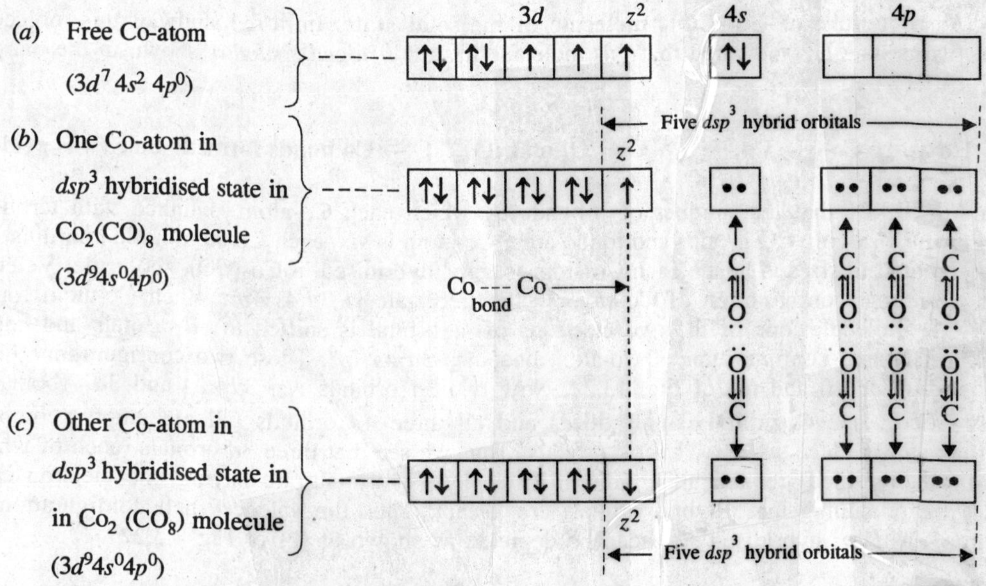

Fig. 13.20. dsp^3 hybridisation of Co-atom in $Co_2(CO)_8$ molecule in solution. Each pair of dots (••) indicates electron pair donated by C-atom of CO molecule (ligand).

How are Co --- Co and :O $\leftrightarrows$ C → Co bonds formed ?

Since, it is evident from the structure given in Fig. 13.19, the number of σ-bonds by which each Co-atom is linked with terminal carbonyl groups and to the other Co-atom is five, each Co-atom has coordination number equal to five and hence each Co-atom is dsp^3 hybridised in $Co_2(CO)_8$ molecule. Valence-shell electronic configuration of Co-atom in its free state is $3d^7 4s^2 4p^0$. When Co-atom forms $Co_2(CO)_8$ molecule, both electrons from 4*s* orbital get shifted to 3*d* orbitals and hence the valence-shell configuration of Co-atom in $Co_2(CO)_8$ becomes $3d^9 4s^0 sp^0$. These two configurations have been shown at (*a*) and (*b*) respectively in Fig. 13.20. Now one 3*d*-orbital *viz.* $3d_z{}^2$ (which is singly-filled with electron), one 4*s* orbital (vacant) and all the three 4*p* orbitals (vacant) combine together and give five dsp^3 hybrid orbitals. Thus we see that four 3*d* orbitals (each of which is completely-filled) remain unhybridised. One of the five dsp^3 hybrid orbitals has one electron (singly-filled) while the remaining four are vacant. dsp^3 hybridised state of Co-atom has been shown at (c) of Fig. 13.20.

The singly-filled dsp^3 hybrid orbital of one Co-atom overlaps with the singly-filled dsp^3 hybrid orbital of the other Co-atom and produces Co---Co σ-bond. Thus Co---Co σ-bond is produced by [dsp^3(Co)—dsp^3(Co)] overlap. *sp* hybrid orbital on C-atom of each CO group donates its *lp* of electrons to the vacant dsp^3 hybrid orbital on Co-atom and gives :O$\leftrightarrows$C→Co coordinate σ-bond, *i.e.* this bond results by [*sp*(C)—dsp^3(Co)] overlap. Thus we see that each Co-atom is linked with four CO-groups by four :O$\leftrightarrows$C→Co bonds and with one Co-atom by Co---Co bond. It may be seen here that all

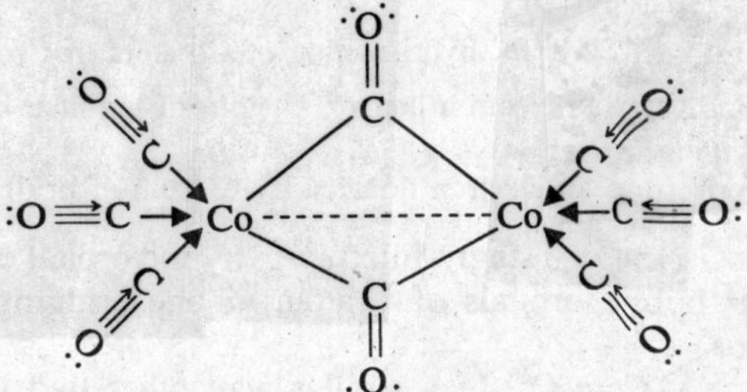

Fig. 13.21. Bridged strucure of $Co_2(CO)_8$ molecule in the solid state.

the electrons of both Co-atoms become paired and hence $Co_2(CO)_8$ shows diamagnetic character. The pairing of electrons takes place due to the formation of Co·····Co bond. The formation of various bonds in $Co_2(CO)_8$ has been shown in Fig. 13.20.

(*b*) **Structure of $Co_2(CO)_8$ molecule in the solid state.** Infra-red study of this molecule in the solid state has indicated that this molecule has the *bridged structure* shown in the margin (Fig. 13.21).

How are Co ----- Co, $\begin{matrix} Co \\ Co \end{matrix}\!\!>$ C$=\ddot{O}$: and :O$\equiv$C → Co bonds formed? Since it is evident from Fig. 13.21 that the number of σ-bonds by which each Co-atom is linked with terminal CO-groups, bridging CO groups and to the other Co-atom is six, each Co-atom has a coordination number equal to six and hence each Co-atom is d^2sp^3 hybridised in $Co_2(CO)_8$ molecule. Valence shell electronic configuration of Co-atom in the free state is $3d^74s^24p^0$. When Co-atom forms $Co_2(CO)_8$ molecule, one of the two electrons of 4*s* orbital is shifted to 3*d* orbitals and hence the valence-shell configuration of Co-atom becomes $3d^84s^14p^0$. These two configurations have been shown at (*a*) and (*b*) of Fig. 13.22. Now two 3*d* orbitals *viz.* $3d_{x^2-y^2}$ and $3d_{z^2}$ (both are singly-filled), one 4*s* orbital (singly-filled) and all three 4*p* orbitals (all are empty) combine together and produce six d^2sp^3 hybrid orbitals. Thus we see that three 3*d* orbitals (each of which is completely filled) remain unhybridised. Three hybrid orbitals are having one electron each while the remaining three hybrid orbitals are vacant. Thus the valence shell configuration of Co-atom in d^2sp^3 hybridised state can be depicted as shown at (*c*) of Fig. 13.22.

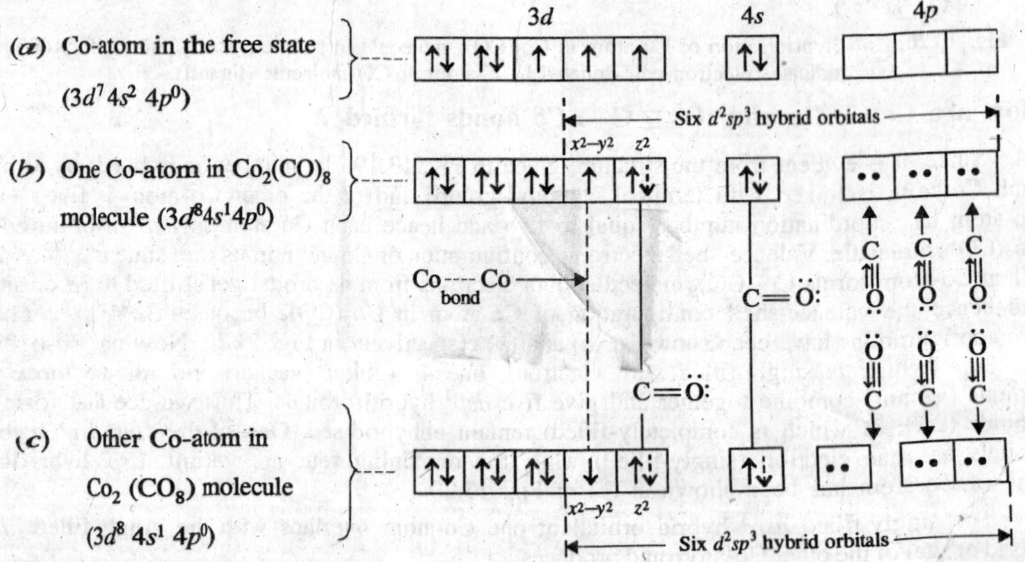

Fig. 13.22: d^2sp^3 hybridisation of Co-atom in $Co_2(CO)_8$ molecule in solid state. Each pair of dots (••) indicates an electron pair on C-atom of terminal carbonyl group, O$\equiv$C: while the dotted arrow (↓) represents an electron donated by C-atom of bridging carbonyl group, :$>$C$=\ddot{O}$:

Now one singly-filled d^2sp^3 hybrid orbital of both Co-atoms overlaps with both the singly-filled sp^2 hybrid orbitals of C-atom of one bridging carbonyl group and form two Co—C bonds in $\begin{matrix} Co \\ Co \end{matrix}\!\!>C=\ddot{O}$:. In a similar way other two Co—C bonds are also produced by one bridging carbonyl group. Thus we see that four Co—C bonds are obtained. Here C-atom belongs to bridging carbonyl group. It is easy to understand that each of these four bonds results by

[d^2sp^3(Co)—sp^2(C)] overlap. Third singly-filled d^2sp^3 hybrid orbital of one Co-atom overlaps with the third d^2sp^3 hybrid orbital of the other Co-atom and produces Co·····Co bond. Thus we can say that Co---Co bond results by [d^2sp^3(Co)—d^2sp^3(Co)] overlap. Each of the remaining three d^2sp^3 hybrid orbitals (each is vacant) of each Co-atom accepts one *lp* of electrons donated by the *sp* hybrid orbital on. C-atom of terminal carbonyl group and forms :O⇶C→ Co coordinate bond. Thus six :O⇶C→Co coordinate bonds are obtained. Obviously this bond is obtained by [*sp*(C)—d^2sp^3 (Co)] overlap. The above discussion shows that each Co-atom is linked with three terminal carbonyl groups by :O⇶C→ Co bonds, with the other Co-atom by Co---Co bond and with two bridging carbonyl groups by Co—C bonds. It may be noted that all the electrons in both Co-atoms become paired and hence $Co_2(CO)_8$ molecule shows diamagnetic character. The pairing of electrons occurs due to the formation of Co---Co bond. The formation of various bonds in $Co_2(CO)_8$ has been shown in Fig. 13.22.

6. Structure of $Mn_2(CO)_{10}$, $Tc_2(CO)_{10}$ and $Re_2(CO)_{10}$ molecules

All these molecules have the same structure. As an example let us consider the structure of $Mn_2(CO)_{10}$ molecule. The infra-red absorption spectral and X-ray diffraction studies have shown that this molecule contains no bridging carbonyl groups and hence the molecule has *non-bridged structure* as shown in Fig. 13.23. It is evident from this figure that each Mn-atom is linked with five terminal carbonyl groups (:O⇶C→ Mn) and with the other Mn-atom by Mn---Mn σ-bond. Thus each Mn-atom has five :O⇶C→ Mn coordianate σ-bonds and one Mn---Mn σ-bond. The presence of Mn---Mn bond has also been supported by the diamagnetic nature of $Mn_2(CO)_{10}$ molecule. It may be noted that each Mn-atom has an octahedral environment and two $Mn(CO)_5$ units are joined solely by Mn---Mn bond.

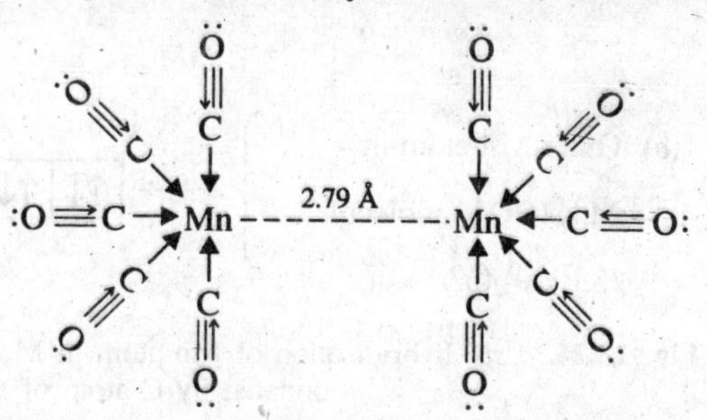

Fig. 13.23. Non-bridgea structure of $Mn_2(CO)_{10}$ molecule

How are Mn---Mn and :O⇶C→ Mn bonds formed? Since, it is evident from the Fig. 13.23 that the number of σ-bonds by which each Mn-atom is linked with terminal carbonyl groups and to the other Mn-atom is six, each Mn-atom has coordination number equal to six and hence each Mn-atom is d^2sp^3 hybridised in $Mn_2(Co)_{10}$ molecule. Valence-shell configuration of Mn-atom in its free state is $3d^54s^24p^0$. When Mn-atom forms $Mn_2(CO)_{10}$ molecule both the electrons from 4*s* orbital get shifted to 3*d* orbitals and hence the configuration of Mn-atom in $Mn_2(CO)_{10}$ becomes $3d^74s^04p^0$. In order to make d^2sp^3 hybridisation possible, the distribution of seven electrons in 3*d* orbitals takes place in such away that $3d_{z^2}$ orbital becomes vacant and hence the configuration of Mn-atom in $Mn_2(CO)_{10}$ molecule becomes $3d_{xy}^2\ 3d_{yz}^2\ 3d_{yz}^2\ 3d_{x^2-y^2}^1\ 3d_{z^2}^0$. Now $3d_{x^2-y^2}$, $3d_{z^2}$, 4*s*, $4p_x$, $4p_y$ and $4p_z$ orbitals (six orbitals) combine together to form six d^2sp^3 hybrid orbitals. $3d_{xy}$, $3d_{yz}$ and $3d_{zx}$ orbitals (each is completely-filled) remain unhybridised. One of the six d^2sp^3 hybrid orbitals is singly-filled while the remaining five are vacant. Thus the valence-shell configurtion of Mn-atom in d^2sp^3 hybridised state in $Mn_2(CO)_{10}$ molecule can be written as shown in Fig. 13.24. The singly-filled d^2sp^3 hybrid orbital of one Mn-atom overlaps with the singly-filled d^2sp^3 hybrid orbital of the other Mn-atom and gives Mn---Mn σ-bond. Thus Mn---Mn is formed by [d^2sp^3 (Mn)d^2sp^3(Mn)] overlap. *sp* hybrid orbital of C-atom of each CO group donates its *lp* of electrons to the vacant d^2sp^3 hybrid orbital on Mn-atom and produces : O⇶C→Mn coordinate σ-bond. Thus this bond results by [*sp*(C)—d^2sp^3(Mn)] overlap.

Thus we see that each Mn-atom is linked with five CO groups by five $:O \rightleftarrows C \rightarrow Mn$ bonds and with one Mn-atom by Mn---Mn bond. It may be seen that all the electrons of both Mn-atom get paired and hence $Mn_2(CO)_{10}$ shows diamagnetic character. The pairing of electrons occurs due to the formation of Mn---Mn bond. The formation of various bonds in $Mn_2(CO)_{10}$ molecule has been shown in Fig. 13.24.

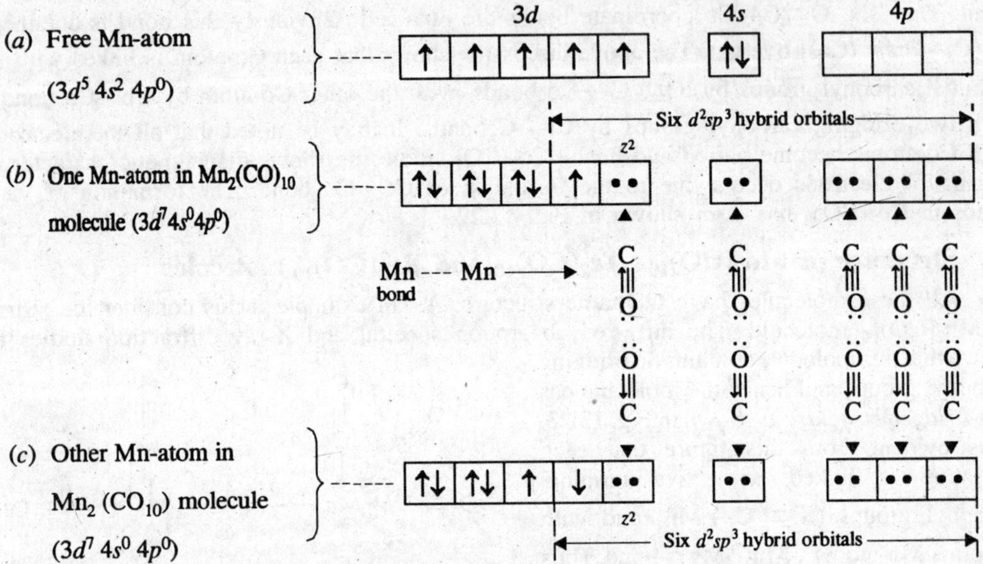

Fig. 13.24. d^2sp^3 hybridisation of Mn-atom in $Mn_2(CO)_{10}$ molecule. Dots (••) indicate the electrons donated by C-atom of terminal carbonyl group.

7. Structure of $Fe_3(CO)_{12}$ and $Os_3(CO)_{12}$ molecules

Both these carbonyls have the same structure. Here we shall consider the structure of $Fe_3(CO)_{12}$ molecule only. Formerly, it was thought that this carbonyl has two bridging CO groups and ten terminal CO groups. But, now it has been established conclusively that $Fe_3(CO)_{12}$ molecule does not contain any briding CO group, *i.e.*, all the twelve CO groups are present as terminal CO groups as shown in Fig. 13.25. Each Fe-atom is linked with four terminal carbonyl groups. The two octahedral $Fe(CO)_4$ units overlap in an angular way and bent fractional single bond between two Fe-atoms is obtained. Thus in $Fe_3(CO)_{12}$ molecule three fractional single bonds between two Fe atoms are formed. These bonds are shown as Fe---Fe. The charge distribution in these Fe---Fe bonds is not exactly symmetrical. Two Fe-atoms in Fe---Fe bonds are at a larger distance from each other than that which is expected in a normal Fe—Fe single bond. Fe---Fe bond distance has been found to be equal to 2.8Å. The structure shown in Fig. 13.25 has been confirmed by the X-ray study of $Fe_3(CO)_{12}$ molecule. $Fe_3(CO)_{12}$ molecule is diamgnetic. Its diamagnetic property is also confirmed by the presence of Fe---Fe bond in the molecule. Each Fe-atom obeys EAN rule, *i.e.*, each Fe-atom is surrounded by 36 electrons ($26\ e^-$ from Fe + $2e^-$ from two Fe---Fe bonds + $4 \times 2 = 8\ e^-$ from 4 terminal CO groups = $36\ e^-$).

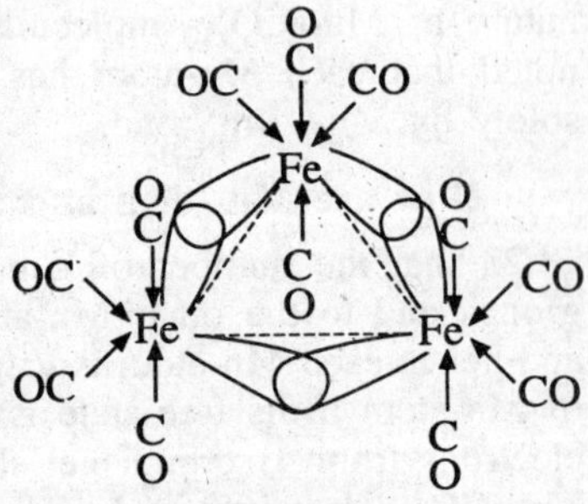

Fig. 13.25. Structure of $Fe_3(CO)_{12}$ molecule

Table 13.2. *Summary of the structure of some metallic carbonyls.*

Carbonyl	*No. of terminal carbonyl groups,* $:O \rightleftarrows C:$	*No. of M---M σ-bonds*	*No. of bridging (ketonic) carbonyl groups,* $:\ddot{O}=C<$
(*i*) $Ni(CO)_4$	4	—	—
(*ii*) $Fe(CO_5)$ Other examples are $Ru(CO)_5$, $Os(CO)_5$	5	—	—
(*iii*) $Cr(CO)_6$ Other examples are $V(CO)_6$, $Mo(CO)_6$, $W(CO)_6$	6	—	—
(*iv*) $Fe_2(CO)_9$, Other example is $Os_2(CO)_9$	6 (Each Fe-atom is attached with three terminal carbonyl groups)	One Fe---Fe bond	3
(*v*) $Co_2(CO)_8$ (in solution)	8 (Each Co-atom is linked with four terminal carbonyl groups)	One Co---Co bond	—
(*vi*) $Co_2(CO)_8$ (in solid state)	6 (Each Co-atom is attached wtih three terminal carbonyl groups)	One Co---Co bond	2
(*vii*) $Mn_2(Co)_{10}$ Other examples are $Tc_2(CO)_{10}$ and $Re_2(CO)_{10}$	10 (Each Mn-atom is linked with five terminal carbonyl groups)	One Mn---Mn bond	—
(*viii*) $Fe_3(CO)_{12}$ Other example is $Os_3(COo)_{12}$	12 (Each Fe-atom is linked with four terminal carbonyl groups)	Three Fe---Fe bonds	—

To calculate the effective atomic number (EAN) of metal atom in metallic carbonyls

EAN of the central metal atom of the following carbonyls can be calculated. For this we have to remember the following points :

(*i*) The central metal atom is in zero oxidation state and hence the total number of electrons in the central metal atom is equal to its atomic number.

(*ii*) Each of the terminal carbonyl groups donates two electrons to the metal with which the group is attached while each of the bridging carbonyl group donates only one electron to the metal.

(*iii*) Each of the metal---metal bonds donates one electron to each of the metal atom.

Let us calculate the EAN of the central metal atom in the following metallic carbonyls.

1. **$Ni(CO)_4$.** This carbonyl has tetrahedral structure in which Ni-atom is attached with four terminal carbonyl groups. Thus :

EAN of Ni-atom = At. No. of Ni-atom + Electrons donated by four terminal carbonyl groups.

$$= 28 + 2 \times 4 = \mathbf{36}$$

2. **$Fe(CO)_5$.** This carbonyl has trigonal bipyramidal structure in which Fe-atom is linked with five terminal carbonyl groups. Thus:

EAN of Fe-atom = At. No. of Fe atom + Electrons donated by five terminal carbonyl groups.

$= 26 + 2 \times 5 = \mathbf{36}$

3. **$Cr(CO)_6$.** This carbonyl has octahedral geometry in which each Cr-atom is linked with six terminal carbonyl groups. Thus:

EAN of Cr-atom = At. No. of Cr-atom + Electrons donated by six terminal carbonyl groups.

$= 24 + 2 \times 6 = \mathbf{36}$

4. **$Fe_2(CO)_9$.** In this carbonyl each Fe-atom in attached with three terminal carbonyl groups, three bridging (ketonic) carbonyl groups and one Fe-atom Thus:

EAN of each Fe-atom = At. No. of Fe-atom + Electrons donated by three terminal carbonyl groups + Electrons donated by three bridging carbonyl groups + Electrons donated by one Fe– –Fe bond

$= 26 + 3 \times 2 + 3 \times 1 + 1 \times 1 = \mathbf{36}$

5. **$Co_2(CO)_8$ (in solution).** In this carbonyl each Co-atom is attached with four terminal carbonyl groups and to other Co-atom. Thus :

EAN of Co-atom = At. No. of Co-atom + Electrons donated by four terminal carbonyl groups + Electrons donated by one Co– –Co bond

$= 27 + 4 \times 2 + 1 \times 1 = \mathbf{36}$

6. **$Co_2(CO)_8$ (in solid state).** In the solid state, each Co-atom is attached with three terminal carbonyl groups, two bridging (ketonic) carbonyl groups and to one Co-atom. Thus :

EAN of each Co-atom = At. No. of Co-atom + Electrons donated by three terminal carbonyl groups + Electrons donated by two bridging carbonyl groups + Electrons donated by one Co– –Co bond

$= 27 + 3 \times 2 + 2 \times 1 + 1 \times 1 = \mathbf{36}$

7. **$Mn_2(CO)_{19}$.** In this carbonyl, each Mn-atom is attached with five terminal carbonyl groups and with the other Mn-atom. Thus:

EAN of Mn-atom = At. No. of Mn-atom + Electrons donated by five terminal carbonyl groups + Electrons donated by one Mn– –Mn bond

$= 25 + 5 \times 2 + 1 \times 1 = \mathbf{36}$

8. **$Fe_3(CO)_{12}$.** In this molecule, each Fe-atom is attached with four terminal carbonyl groups and to other two Fe-atoms. Thus :

EAN of each Fe-atom = At. No. of Fe-atom + Electrons donated by four terminal carbonyl groups + Electrons donated by two Fe– –Fe bond

$= 26 + 4 \times 2 + 1 \times 2 = \mathbf{36}$

The procedure for calculating EAN has been summarised in Table 13.3.

Table 13.3: *To calculate the effective atomic number (EAN) of central metal atom in carbonyis.*

Carbonyl	*Atomic number of the central atom (Z)*	*Electrons donated by terminal carbonyl groups (a)*	*Electrons donated by bridging carbonyl groups (b)*	*Electrons donated by M---M bonds (c)*	*EAN of the central metal atom* $= Z + a + b + c$
$Ni(CO)_4$	28	4 × 2 = 8	—	—	28 + 8 = 36
$Fe(CO)_5$	26	5 × 2 = 10	—	—	26 + 10 = 36
$Cr(CO)_6$	24	6 × 2 = 12	—	—	24 + 12 = 36
$Fe_2(CO)_9$	26	3 × 2 = 6	3 × 1 = 3	1 × 1 = 1	26 + 6 + 3 + 1 = 36
$Co_2(CO)_8$ (in solution)	27	4 × 2 = 8	—	1 × 1 = 1	27 + 8 + 1 = 36
$Co_2(CO)_8$ (in solid state)	27	3 × 2 = 6	2 × 1 = 2	1 × 1 = 1	27 + 6 + 2 + 1 = 36
$Mn_2(CO)_{10}$	25	5 × 2 = 10	—	1 × 1 = 1	25 + 10 + 1 = 36
$Fe_3(CO)_{12}$	26	4 × 2 = 8	—	2 × 1 = 2	26 + 8 + 2 = 36

Uses of infra-red (IR) absorption spectra of metallic carbonyls

The IR absorption spectra of metallic carbonyls have been put to the following uses:

1. **To determine the geometry of metallic carbonyls.** If theoretically, a metallic carbonyl can have more than one geometry, we can decide which geometry is the most stable. This can be done as follows. We calculate the number of I.R. active together with Raman-active vibrational bands for each geometry and then run the I.R. and Raman spectra and finally tally the number of bands observed with the number of bands predicted theoretically (*i.e.* by calculations) for each geometry.

Example. Although, theoretically, $Fe(CO)_5$ molecule can have trigonal bipyramidal as well as square pyramidal geometry, the I.R. bands actually observed tally the number of such bands calculated for trigonal bipyramidal geometry. Thus $Fe(CO)_5$ has trigonal bipyramidal geometry.

2. **To determine the bond order (B.O.) of ligated CO.** It has been seen that I.R. absorption frequency of ligated CO is directly proportional to its B.O. In other words the I.R. absorption band due to the stretching vibration of ligated CO with a higher B.O. would occur at a higher frequency and the I.R. absorption band of ligated CO with a lower B.O. would occur at a lower frequency. Since the absorption frequency for free CO is equal to 2250 cm^{-1} while that for ligated CO lies between 2220–1700 cm^{-1}, ligated CO has lower B.O. The lower B.O. is due to the transfer of metal $d\pi$ electrons into the π^* orbitals of ligated CO. Let us study the following examples.

Metal carbonyl :	$[V(CO)_6]^-$	$Cr(CO)_6$	$[Mn(CO)_6]^+$
Charge on metal carbonyl :	—	0	+
Absorption frequency for CO (cm^{-1}) *:*	1860	1980	2090

Since the presence of positive charge on $[Mn(CO)_6]^+$ resists the flow of metal $d\pi$ electrons into the π^* orbitals of CO, the B.O. of CO increases. Due to the increase in B.O., the absorption band of ligated CO occurs at higher frequency (=2090 cm^{-1}).

3. To differentiate between terminal and bridging carbonyl groups. Since OC—M bond in a terminal carbonyl group (:O$\equiv$C→M) is stronger than that in a bridging carbonyl group $\left[\begin{matrix} M \\ M \end{matrix}\!\!>\!C=\ddot{O}:\right]$, I.R. absorption for a terminal CO group would occur at a higher frequency than for a bridging CO group. It has been observed that in a neutral metallic carbonyl, if the I.R. absorption occurs between 2050–1900 cm^{-1}, CO group acts as a terminal CO group and if the absorption I.R. occurs between 1900–1800 cm^{-1}, CO group acts as a bridging CO group.

4. To study reaction rates. The measurement of the rates of disappearance of the intensity of CO absorption bands in the I.R. spectra can be used to study the rates of substitution reactions of metal carbonyls.

University Questions

1. Explain the following, giving appropriate reasons.

(*i*) $Ni(CO)_4$ has tetrahedral structure. *(Merrut M.Sc. 83)*

(*ii*) $Fe(CO)_5$ is known while $[Fe(CO)_6]^{3+}$ is not known. *(Merrut M.Sc. 84)*

(*iii*) Cr, Fe and Ni combine with 6, 5 and 4 CO molecules respectively to form their carbonyls. *(Bundelkhand 88)*

(*iv*) Elements like Co donot give mono nuclear carbonyls. *(Bharathiar 86)*

(*v*) In metallic carbonyls the M—C bond is shorter than the calculated single bond length. *(Bharathiar 86)*

(*vi*) The metal atom in metal carbonyls is in low oxidation state.

(*vii*) The carbonyls *viz* $Pd(CO)_4$ and $Pt(CO)_4$ donot exist whereas $Ni(CO)_4$ exists as a stable carbonyl.

(*viii*) $Ni(CO)_2Cl_2$ does not exist while the carbonyls $Pd(CO)_2X_2$ (X=halide ion) do exist.

(*ix*) (*a*) Ni and Fe form mononuclear carbonyls $Ni(CO)_4$ and $Fe(CO)_5$ respectively but Co does not form any mononuclear carbonyl.

(*b*) CO is a π-acceptor ligand.

(*c*) ν_{CO} in carbonyl is different from that in CO molecule.

(*d*) CO is coordinated to the metal atom is metal carbonyls through C-atom rather than O-atom.

(*e*) There is a variation of ν_{CO} in the series $Ni(CO)_4$, $[Co(CO)_4]^-$ and $[Fe(CO)_4]^{2-}$

(*f*) CO is a low oxidation state stabiliser. *(Berhampur 91)*

2. Describe the preparation of $Ni(CO)_4$ *(Rohilkhand 90)*

3. How are metal carbonyls prepared? Give the valence bond structures of the following molecules : $Fe(CO)_5$, $Fe_2(CO)_9$, $Fe_3(CO)_{12}$ and $Co_2(CO)_8$. Do these compounds obey EAN rule ? *(Lucknow 95)*

4. What are metal carbonyls ? Describe the structure and bonding of any one simple carbonyl. Also give a method of preparation of a smple metal carbonyl. *(Osmania 93)*

5. What are the different modes of binding of CO found in polynuclear metal carbonyls? How these are differentiated experimentally? *(Berhampur 91)*

6. (*a*) What do you mean by the term "π-acid ligands"? Illustrate your answer with suitable examples.

(*b*) Discuss the evidences in favour of metal-ligand double bond in metal carbonyls. Explain the mode of linkage of terminal and bridging CO in carbonyls with suitable examples. *(Berhampur 91)*

7. Write the structures of $Fe_2(CO)_9$ and $Os_3(CO)_{12}$. *(Lucknow 2002)*

8. Explain why $V(CO)_6$ is paramagnetic ? *(Rohilkhand 2003)*

9. (*a*) How CO stabilizes the metals in low oxidation states in metal carbonyls ?

(*b*) Write two general methods for the preparation of metal carbonyls.

(*c*) Represent the structures of all the known nreutral binary iron carbonyls. *(Rohilkhand 2003)*

14

Complex Compounds Containing NO^+ or/and NO^- Ions as Ligands: Metallic Nitrosyl Complexes

Introduction

We know that NO molecule can be represented by the structure, $:\dot{N}=\ddot{O}:$. When this molecule loses its unpaired electron on N-atom, it is converted into NO^+ ion ($:\overset{+}{N}\equiv O:$). Similarly when NO molecule accepts an electron from M^{n+} ion, it is converted into NO^- ion ($:\overset{-}{\ddot{N}}=\ddot{O}:$). Since NO^+ and NO^- both have electron pair on N-atom, they can be attached with a transition metal atom and form the complex compounds *viz.*, $\overset{-}{M} \leftarrow \overset{+}{N}\equiv O:$ and $\overset{+}{M} \leftarrow \overset{-}{\ddot{N}}=\ddot{O}:$ respectively. These complex compounds are called metallic nitrosyls NO^+ ion is called *nitrosonium* or *nitrosyl ion.*

Bonding in metallic nitrosyls : Classification of metallic nitrosyls

Depending on whether NO molecule is present as NO^+ or NO^- in the metallic nitrosyls, these complex compounds have been grouped into the following categories *viz.*, A, B and C.

A. Metallic nitrosyls containing NO^+ ion

$[Fe(NO)_2(CO)_2]^0$, $[Mn(NO_3)_3(CO)]^0$, $[Co(NO)(CO_3)_3]^0$, $[Fe(NO)_2(PR_3)_2]^0$, $[Fe(NO)(H_2O)_5]^{2+}$, $[Fe(NO)(CN)_5]^{2-}$ etc. are the examples of metallic nitrosyls which contain NO^+ ion. The presence of NO^+ group in these complexes is confirmed by the fact that the IR absorption frequency of NO^+ group in these complexes falls in the expected region, 1600–1900 cm^{-1}.

How NO^+ ion is formed from NO molecule and gets attached with the central metallic atom in nitrosyls

In metallic nitrosyls having NO^+ ion, first of all NO molecule ($:\dot{N}=\ddot{O}:$) loses the unpaired electron present in 2*p* orbital of N-atom and this electron is transferred to the metallic atom so that NO molecule is converted into NO^+ ion and the valency of the metallic atom (M) is reduced by one unit, *i.e.*, M is converted into M^- ion. In the next step NO^+ ion which is iso-electronic with CO molecule, donates the lone pair of electrons residing on N-atom of NO^+ ion to M^- to get coordinated with M^-. Both these steps can be shown as follows :

$$:\dot{N}=\ddot{O}: \xrightarrow{-e^-} :\overset{+}{N}\equiv O: + e^-$$

$$M + e^- \longrightarrow M^-$$

$$M^- + :\overset{+}{N}\equiv O: \longrightarrow \underset{\text{Metallic nitrosyl}}{M^-\leftarrow \overset{+}{N}\equiv O:}$$

Here it may be noted that NO^+ which is obtained by the loss of one electron from NO molecule is iso-electronic with CO molecule ($\ddot{C} \leftleftarrows \ddot{O}$) and hence has the same structure as CO molecule has.

In these nitrosyls M—N—O bond angle is close to 180° and hence these nitrosyls are *linear nitrosyls*. In these nitrosyls all M—N bond distances are in the range of 1.57–1.67Å.

B. Metallic nitrosyls containing NO^- ion

$[Co^{3+}(CN)_5(NO^-)]^{3-}$ and $[Co^{3+}(NH_3)_5(NO^-)]^{2+}$ are the examples of metallic nitrosyls which contain NO^- ion. These complexes contain spin-paired Co(III) ion with $t_{2g}^6 e_g^0$ configuration having all the electrons in the paired state ($n = 0$). The presence of NO^- group in these complexes is confirmed by the fact that the IR absorption frequency of NO^- group in these complexes falls in the region, 1100–1200 cm^{-1} which is the expected region.

How NO^- ion is formed from NO molecule and gets attached with metallic atom in nitrosyls

In the metallic nitrosyls having NO^- ion, NO molecule ($:\dot{N}=\ddot{O}:$) accepts one electron from M^{n+} ion and is itself converted into NO^- ion ($:\bar{\ddot{N}}=\ddot{O}:$). In NO^- ion, N^- is sp^2 hybridised. The two lone pairs of electrons on N^- are present in two sp^2 hybrid orbitals. When NO^- ion gets attached with the central metal atom in nitrosyls, one of the lone pairs placed in one sp^2 hybrid orbital is donated to the vacant orbital of the metal (M^+) and $M^+ \leftarrow \bar{\ddot{N}}=\ddot{O}:$ coordinate σ-bond is formed. The other lone pair of electrons located in other sp^2 hybrid orbital remains as such on N^-. The metallic nitrosyls containing NO^- are *bent nitrosyls* since the M—N—O bond angle in those nitrosyls is in the range of 120°–140°.

C. Metallic nitrosyls containing NO^+ as well as NO^- ions

There are some metallic nitrosyls in which NO exists as NO^+ as well as NO^-. $Fe(NO)_4$ is an example of this type of nitrosyls. This complex is represented as $[(NO^+)_3Fe^{-2}(NO^-)]^0$, since NO group absorbs in both the regions *viz.* 1600–1900 cm^{-1} (for NO^+ ion) and 1100–1200 cm^{-1} (for NO^- ion).

The central atom (Fe^{-2}) in the given nitrosyl obeys EAN rule, since EAN of Fe^{2-} ion is equal to $(26 + 2) + 4 \times 2 = 28 + 8 = 36$ which is the atomic number of the next inert gas namely Kr (Z = 36).

To calculate effective atomic number (EAN) of the central metallic atom in metallic nitrosyls

NO molecule always donates two electrons to the central metal atom whether it is present as NO^+ or NO^- in metallic nitrosyls. It is because of the fact that NO^+ and NO^- both have one lone pair of electrons on N-atom.

$$:\overset{+}{N} \leftleftarrows O: \qquad\qquad \overset{-}{M} \leftarrow \overset{+}{N} \leftleftarrows O:$$

$$:\bar{\ddot{N}} = \ddot{O}: \qquad\qquad \overset{+}{M} \leftarrow \bar{\ddot{N}} = O:$$

In Table 14.1 we have shown how the EAN of the central metal atom in metallic nitrosyls is calculated. It may be seen from the table that the EAN of the central metal atom in most of the nitrosyls is equal to the atomic number of the next inert gas, but the EAN of Cr in $[Cr(NO)(CN)_5]^{3-}$, of Mn in $[Mn(NO)(CN)_5]^{2-}$ and of Fe in $[Fe(NO)(H_2O)_5]^{2+}$ ions is not equal to 36, *i.e.*, these complex ions donot obey EAN rule.

Table 14.1: *To calculate EAN of the central metal atom in metallic nitrosyls compounds.*

Metallic nitrosyl compound	*Electrons donated by metal atom (a)*	*Electrons donated by NO^+ or NO^- (b)*	*Electrons donated by other ligand (c)*	*EAN of the metal atom = a + b + c*
(A) Metallic nitrosyls containing NO^+ ion				
$[Fe(NO)_2(CO)_2]^0$	26 + 2 (Fe^{2-})	2 × 2 ($2NO^+$)	2 × 2 (2CO)	28 + 4 + 4 = 36 (Kr)
$[Mn(NO)_3(CO)]^0$	25 + 3 (Mn^{3-})	3 × 2 ($3NO^+$)	1 × 2 (CO)	28 + 6 + 2 = 36 (Kr)
$[Co(NO)(CO)_3]^0$	27 + 1 (Co^-)	1 × 2 (NO^+)	3 × 2 (3CO)	28 + 2 + 6 = 36 (Kr)
$[Fe(NO)_2(PR_3)_2]^0$	26 + 2 (Fe^{2-})	2 × 2 ($2NO^+$)	2 × 2 ($2PR_3$)	28 + 4 + 4 = 36 (Kr)
$[Cr(NO)(CN)_5]^{3-}$	24 − 1 (Cr^+)	1 × 2 (NO^+)	5 × 2 ($5CN^-$)	23 + 2 + 10 = 35
$[Mn(NO)CN)_5]^{2-}$	25 − 2 (Mn^{2+})	1 × 2 (NO^+)	5 × 2 ($5CN^-$)	23 + 2 + 10 = 35
$[Fe(NO)(H_2O)_5]^{2-}$	26 − 1 (Fe^+)	1 × 2 (NO^+)	5 × 2 ($5H_2O$)	25 + 2 + 10 = 37
$[Fe(NO)(CN)_5]^{2-}$	26 − 2 (Fe^{2+})	1 × 2 (NO^+)	5 × 2 ($5CN^-$)	24 + 2 + 10 = 36 (Kr)
(B) Metallic nitrosyls containing NO^- ion				
$[Co(NO)(CN)_5]^{3-}$	27 − 3 (Co^{3+})	1 × 2 (NO^-)	5 × 2 ($5CN^-$)	24 + 2 + 10 = 36 (Kr)
$[Co(NO)(NH_3)_5]^{2+}$	27 − 3 (Co^{3+})	1 × 2 (NO^-)	5 × 2 ($5NH_3$)	24 + 2 + 10 = 36 (Kr)
(C) Complex containing NO^+ and NO^- ions				
$[Fe(NO)_4$ or $[(NO^+)_3\ Fe^{2-}\ (NO^-)]$	26 + 2 (Fe^{2-})	3 × 2 ($3NO^+$) 1 × 2 (NO^-)	—	28 + 6 + 2 = 36 (Kr)

$[Fe(NO)_2I]_2$ and $[Fe(NO)_2(S.CH_3)]_2$ are dimeric nitrosyls of iron. These compounds have one Fe....Fe bond. EAN of each Fe-atom in these dimeric molecules can be calculated when we consider their structures given below in Fig. 14.1.

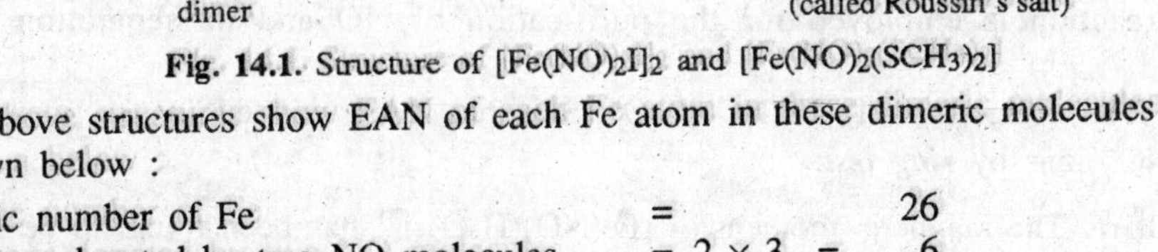

Structure of $[Fe(NO)_2I]_2$ dimer

Structure of $[Fe(NO)_2(SCH_3)]_2$ (called Roussin's salt)

Fig. 14.1. Structure of $[Fe(NO)_2I]_2$ and $[Fe(NO)_2(SCH_3)_2]$

The above structures show EAN of each Fe atom in these dimeric moleeules is equal to 36 as shown below :

Atomic number of Fe	=	26
Electrons donated by two NO molecules	= 2 × 3 =	6
Electrons donated by one I-atom	= 2 × 1 =	2
Electrons donated by another I-atom	= 1 × 1 =	1
Electrons due to one Fe---Fe	=	1
Total	=	36 (Kr)

Some Metallic Nitrosyl Complexes

In this section we shall discuss the preparation, properties, uses and structure of the following metallic nitrosyl octahedral complexes which contain NO^+ as one of the six ligands

1. Nitrosonium pentaquo iron (I) sulphate, $[Fe^+(NO^+)(H_2O)_5]SO_4$ (Also called nitroso ferrous sulphate)

It is an octahedral complex compound which contains iron in +1 oxidation state.

Preparation. When freshly-prepared solution of $FeSO_4$ is added to the water extract of a nitrate and conc. H_2SO_4 is poured carefully along the sides of the test tube, a brown-black ring is formed at the junction of the two layers. This brown-black ring is a nitroso ferrous sulphate, $[Fe(NO)(H_2O)_5]SO_4$. The formation of this complex compound takes place through the following steps :

(*a*) Nitrate reacts with conc. H_2SO_4 and produces HNO_3

$$NaNO_3 + H_2SO_4 \longrightarrow NaHSO_4 + HNO_3$$

or
$$NO^{3-} + H^+ \longrightarrow HNO_3$$

(*b*) HNO_3 produced as above oxidises $FeSO_4$ to $Fe_2(SO_4)_3$ in presence of H_2SO_4 and is itself reduced to NO.

$$2HNO_3 \longrightarrow H_2O + 2NO + 3O$$

$$[2FeSO_4 + H_2SO_4 + O \longrightarrow Fe_2(SO_4)_3 + H_2O] \times 3$$

$$\underset{\text{Green}}{6FeSO_4} + H_2SO_4 + 2HNO_3 \longrightarrow \underset{\text{Redish brown}}{3Fe_2(SO_4)_3} + 2NO + 4H_2O$$

or
$$3Fe^{2+} + 4H^+ + NO_3^- \longrightarrow 3Fe^{3+} + NO + 2H_2O$$

(*c*) NO produced as above is absorbed by $FeSO_4$ and brown-black $[Fe(NO)(H_2O)_5]SO_4$ is obtained.

$$NO + FeSO_4 + 5H_2O \longrightarrow \underset{\text{Brown-black}}{[Fe(NO)(H_2O)_5]SO_4}$$

or
$$NO + Fe^{2+} + 5H_2O \longrightarrow [Fe(NO)(H_2O)_5]^{2+}$$

The brown-black colour is due to the Fe^+—NO^+ charge transfer bands occuring in the visible region.

Properties. $[Fe(NO)(H_2O)_5]SO_4$ is stable only at low temperature and decomposes on heating with the evolution of NO.

$$[Fe(NO)(H_2O)_5]^{2+} \overset{\Delta}{\rightleftharpoons} Fe^{2+} + NO + 5H_2O$$

This reaction is employed for the purification of NO and its separation from other gases.

Uses. The formation of $[Fe(NO)(H_2O)_5]SO_4$ has been utilised in the identification of NO_2^- and NO_3^- ions by *ring test.*

Structure. The magnetic moment of $[Fe(NO)(H_2O)_5]^{2+}$ has been found to be equal to 3.9 B.M. This value shows that Fe has three unpaired electrons, *i.e.,* Fe is present as Fe^+ which has its valence-shell configuration as $3d^7 4s^0$. The presence of three unpaired electrons can be explained only when it is assumed that Fe^+ is sp^3d^2 hybridised in $[Fe(NO)(H_2O)_5]^{2+}$. sp^3d^2 hybridisation of Fe^+ has been explained in "*Valence Bond Theory of Complex Compounds*" (Chapter 5).

According to crystal field theory (CFT) the valance-shell configuration of Fe^+ is $t_{2g}^5 e_g^2$ which gives three unpaired electrons.

2. Sodium pentacyano nitrosyl ferrate (II), $Na_2[Fe^{2+}(NO^+)(CN)_5]$ [Also called sodium nitroprusside)

Preparation. (*i*) When a conc. solution of $K_4[Fe(CN)_6]$ is heated with 50% HNO_3, $K_3[Fe(CN)_6]$ along with KNO_3 and NO is obtained. On cooling, KNO_3 crystallises out and the liquor which contains $K_3[Fe(CN)_6]$ and NO is neutralised by Na_2CO_3 and then evaporated to crystallisation. This gives ruby red crystals of sodium nitroprusside.

$$3K_4[Fe(CN)_6] + 4HNO_3 \longrightarrow 3K_3[Fe(CN)_6] + NO + 3KNO_3 + 2H_2O$$

or
$$3[Fe(CN)_6]^{4-} + 4H^+ + NO_3^- \longrightarrow 3[Fe(CN)_6]^{3-} + NO + 2H_2O$$

$$2K_3[Fe(CN)_6] + 2NO + 3Na_2CO_3 \longrightarrow 2Na_2[Fe(CN)_5(NO)] + 2NaCN + 3K_2CO_3$$

(*ii*) It may also be prepared by the action of $NaNO_2$ on $Na_4(Fe(CN)_6]$.

$$Na_4[Fe(CN)_6] + NaNO_2 + H_2O \longrightarrow Na_2[(Fe(CN)_5(NO)] + NaOH$$

Properties. (*i*) *Physical state.* It is a ruby-red crystalline solid which is readily soluble in water. Its equeous solution decomposes on standing. It is, therefore, freshly prepared when required.

(*ii*) *Action of sulphides.* When a freshly prepared solution of $Na_2[Fe(CN)_5(NO)]$ is added to an alkaline solution of a sulphide (*e.g.* Na_2S but not H_2S), a purple or violet colour is obtained due to the formation of a complex ion, $[Fe^{2+}(CN)_5(NO^+)(S^{-2})]^{4-}$ which does not last for long.

$$Na_2S + Na_2[Fe^{2+}(CN)_5(NO^+)] \longrightarrow \underset{\text{Purple or violet coloured solution}}{Na_4[Fe^{2+}(CN)_5(NO^+)(S^{-2})]}$$

or
$$S^{-2} + [Fe^{2+}(CN)_5(NO^+)] \longrightarrow [Fe^{2+}(CN)_5(NO^+)(S^{-2})]^{4-}$$

The production of purple or violet colour is used to confirm the presence of S^{2-} ion in a given inorganic salt.

(*iii*) *Action of alkali sulphites and thiosulplates.* Alkali sulphites give a rose red colour due to the formation of $Na_4[Fe(CN)_5(NO)(SO_3)]$

$$Na_2SO_3 + Na_2[Fe(CN)_5(NO)] \longrightarrow \underset{\text{Rose red colour}}{Na_4[Fe(CN)_5(NO)(SO_3)]}$$

Alkali thiosulphates donot show this reaction. Thus the above reaction can be used to distinguish sulphites from thiosulphates.

(*iv*) *Action of $AgNO_3$.* A flesh coloured complex is formed.

$$2AgNO_3 + Na_2[Fe(CN)_5(NO)] \longrightarrow \underset{\substack{\text{Flesh coloured complex} \\ \text{(brown ppt.)}}}{Ag_2[Fe(CN)_5(NO)]} + 2NaNO_3$$

(*v*) *Action of aldehydes and ketones.* Aldehydes and ketones (*e.g.* acetone and methyl ethyl ketone) give red colour with alkaline solution of sodium nitroprusside.

(*vi*) *Action of an alkali.* $Na_2[Fe(CN)_5(NO)]$ is converted into sodium hexacyanate ferrate (II), $Na_4[Fe(CN)_6]$

$$6Na_2[Fe(CN)_5(NO)] + 14NaOH \longrightarrow 5Na_4[Fe(CN)_6] + Fe(OH)_2 + 6NaNO_3 + 6H_2O$$

Uses. Sodium nitroprusside is used (*i*) as a reagent in the laboratory for the detection of S^{2-} ion, aldehydes and ketones (*ii*) for distinguishing sulphites from thiosulphates.

Structure. The determination of the magnetic moment of $[Fe(NO)(CN)_5]^{2-}$ has shown that this ion has no unpaired electrons, *i.e.* Fe is present as Fe^{2+} in which all the six electrons ($Fe^{2+} = 3d^64s^0$) are in the paired state. This is possible only when Fe^{2+} ion is d^2sp^3 hybridised as discussed under "*Valence Bond Theory of Complex Compounds*" (Chapter 5).

According to CFT the distribution of six electrons of Fe^{2+} ion in t_{2g} and e_g sets of orbitals is $t_{2g}^6e_g^0$. The above discussion shows that NO is present as NO^+ in $[Fe(NO)(CN)_5]^{2-}$ ion.

University Questions

1. Write a note on "Nitrosyls". *(Lucknow 95, Meerut M.Sc. 1988)*
2. Describe the preparation, properties, uses and structure of sodium nitroprusside. *(Delhi 89)*
3. What happens when :
   (*i*) Sodium nitroprusside is treated with an alkaline sodium of Na_2S? *(Delhi 89)*
   (*ii*) Alkali metal sulphide is treated with sodium nitroprusside. *(Delhi 84)*
4. Balance the equation: $[Fe(CN)_5(NO)]^{2-} + S^{2-} \longrightarrow$ .... *(Agra 86; Kanpur 87)*
5. (*a*) Discuss the chemistry of "brown ring test" with special reference to the bonding aspects of the species formed.
   (*b*) What are the different methods of binding of NO in metal nitrosyls? Give appropriate examples
   (*c*) Comment on the oxidation state of Fe in the species formed in the brown ring test.
   (Berhampur 91)
6. (*a*) Discuss the ligand behaviour of nitric oxide with suitable examples.
   (*b*) How will you prepare carbonyl nitrosyls and cyano-nitrosyls of iron? Discuss the structure of any one cyano nitrosyl. *(Rohilkhand 2003)*
7. Discuss the nature of bonding of NO in metal carbonyl nitrosysls. *(Gujrat 2003)*

15

Pollution and Its Adverse Effects

Environmental chemistry

Environmental chemistry is a science which deals with the study of various chemical reactions taking place in the environment and includes the study of sources and adverse effects of chemical substances produced either by a natural source or by human activity on environment *i.e.*, on air, water and soil.

Environmental segments

The environment may be considered to comprise of the following four segments *viz. lithosphere, hydrosphere, atmosphere* and *biosphere.* Out of these four segments the atmosphere is of special interest.

Atmosphere

The region of air present on earth is called atmosphere. The air present in the atmosphere consists of a number of gases like N_2, O_2, CO_2, Ar etc. and extends upto a height of about 500 km from the earth's surface. The weight of the atmosphere is approximately 4.5 to 5 × 10^{15} metric tons which is about one millionth of the total weight of the earth. The atmosphric temperature, pressure and density vary considerably with altitude. The atmospheric temperature varies from –100°C to + 1200°C, depending on altitude. The atmospheric pressure at sea-level is one atmosphere whereas at 100 km above sea-level, it drops to 3×10^{-7} atmosphere. The atmospheric density surface of the earth is about 0.0013 g/m^3 which decreases sharply with increasing altitude and gradually thins out into the space. At about 600 km and above, the atoms and molecules describe free, elliptical orbits in the earth's gravitational field.

Composition of the air present in the atmosphere

We have said above that the air of the atmosphere consists of a number of gases like N_2, O_2, CO_2, Ar etc. The actual composition of clean dry air near sea-level is that which is shown in Table 15.1.

Table 15.1. *Composition of clean dry air near sea-level*

Components	*Content*	
	per cent by volume	*ppm*
(*a*) **Major components**		
Nitrogen (N_2)	78.09	7,80,900
Oxygen (O_2)	20.94	2,09,400
Water vapour (H_2O)	0.15	1,000 – 50,000
(*b*) **Minor components**		
Argon (Ar)	0.934	9,340
Carbon dioxide (CO_2)	0.032	320

(*c*) **Trace components**		
Neon (Ne)	0.00182	18.2
Helium (He)	0.000524	5.24
Methane (CH_4)	0.00018	1.8
Krypton (Kr)	0.00011	1.1
Nitrous oxide (N_2O)	0.000025	0.25
Hydrogen (H_2)	0.00005	0.5
Xenon (Xe)	0.0000087	0.087
Sulphur dioxide (SO_2)	0.0000002	0.002
Nitrogen dioxide (NO_2)	0.0000001	0.001
Ammonia (NH_3)	0.000001	0.01
Carbon monoxide (CO)	0.000012	0.12
Ozone (O_3)	0.000002	0.02
Iodine (I_2)	Traces	Traces

The proportions of the various components of the atmosphere more or less remain constant upto a height of about 16 km from the earth's surface. Above this height, gravitational separation begins, although this becomes significant above 130 km.

Relation between temperature of the atmosphere and the height above the earth's surface : Division of the atmosphere into four regions

The atmosphre extends upto a height of about 500 km above the earth's surface. Depending on the temperature and height, the atmosphere has been divided into the following four major regions.

1. **Troposphere.** This is the region which lies nearest to the earth's surface and extends upto an altitude of 11 km. With the increase of the altitude (height), the temperature of the air in this region steadily decreases from the ground temperature (say 15°C) to about –56°C. (see temperature-altitude curve shown in Fig. 15.1). The fall in temperature is due to fall of pressure and density of the air. The minimum temperature in this region is indicated by what is known as **tropopause**. The change in temperature with height is called **lapse rate**. The decrease of temperature with increasing altitude in the troposphere is called **positive lapse-rate**. Thus the temperature altitude curve shown in Fig. 15.1 exhibits positive lapse-rate.

Troposphere accounts for over 70% of the atmospheric mass. The composition of air in this region remains more or less constant in the absence of any significant air pollution. In this region the density of air decreases exponentially with increasing altitude. This region contains most of water, cloud and particulate matter.

2. **Stratosphere.** The region above troposphere is called stratosphere which extends upto a height of 50 km. In this region, with the increase in height, the temperature increases from –56°C to –2°C which is attained at a height of 50 km. Thus the change in temperature (increase in temperature) in this region shows **negative-lapse** rate. The increase in temperature in this region is due to the absorption of solar ultraviolet radiation by ozone present in this region. The maximum temperature attained in this region (= –2°C) is indicated by what is known as **stratopause.**

How is O_3 produced in stratosphere? At a height of about 15 km from the earth's surface, sun rays convert O_2 present in the air into O_3 [$3O_2 \xrightarrow{\text{sun rays}} 2O_3$] which forms a stable layer in stratosphere region of the atmosphere. The concentration of O_3 present in this region lies between 1 to 5 ppm by volume. The thickest layer of O_3 exists at a height of 23 km from the earth's surface. The layer of O_3 present in stratosphere is very useful for animals, human beings and plants living on the earth. See "*Air Pollution*".

3. **Mesosphere.** This is the region which lies above stratosphere and extends upto a height of 90 km. In this region with the increase of height, the temperature decreases from – 2°C to –92°C which is attained at a height of 90 km, *i.e.*, in this region the change in temperature (decrease in temperaure) shows **positive lapse-rate.** The region of tempeature (= –100°C) is called **mesopause**. This region which has a temperature equal to –100°C is the coldest region of the atmosphere. The decrease in temperature with the increasing height in mesosphere is due

to the relatively low levels of ozone and other species present in the region. These species can absorb U.V. radiations from the sun.

4. Thermosphere. This region lies above mesosphere. In this region, the temperature rises from –92°C to 1200°C with the increasing altitude. Thus the change of temperature in this

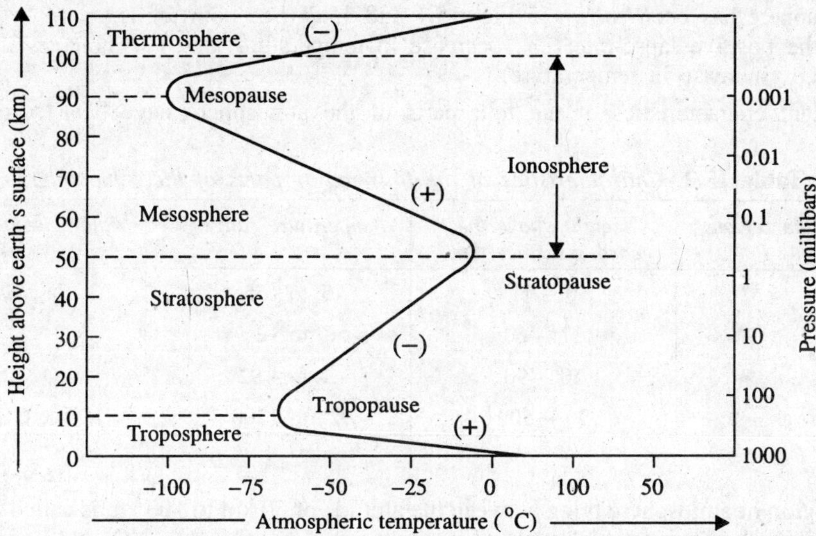

Fig. 15.1. Temperature-altitude curve of the atmosphere : Division of the atmosphere into four parts namely troposphere, stratosphere, mesosphere and thermosphere (Also see Fig.15.2).

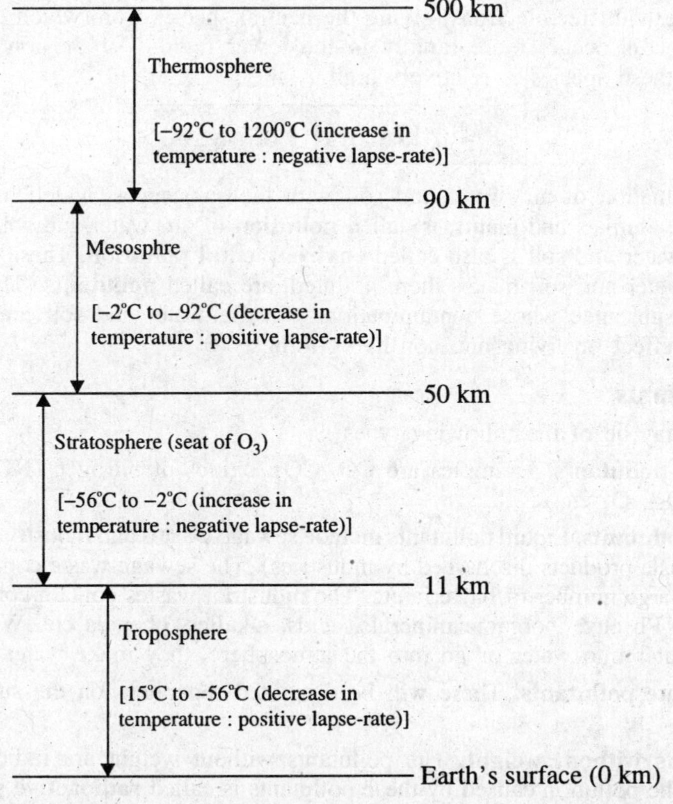

Fig. 15.2. Simplified way of showing the division of atmosphere into four parts.

region shows **negative lapse-rate** of temperature. In this region the density and pressure of the air are low. The atmospheric gases (*e.g.*, O_2 and NO) present in this region absorb the solar radiations in the far u.v. region and undergo ionisation.

The division of the atmosphere into four parts *viz.* troposphere, stratosphere, mesosphere and thermosphere has been shown in Fig. 15.1 and 15.2 (See both the figures). In figure 15.1 (+) shows the positive lapse-rate (*i.e.*, decrease in temperature) and (–) indicates the negative lapse-rate (*i.e.*, increase in temperature).

Important characteristics of the four parts of the atmosphere have been summarised in Table 15.2

Table 15.2. *Characteristics of the four major parts of the atmosphere*

Name of the region	*Height above the earth's surface, km*	*Temperature range °C*	*Major chemical species present*
Troposphere	0 – 11	15 to –56	O_2, N_2, CO_2, H_2O
Stratosphere	11 – 50	– 56 to –2	O_3
Mesosphere	50 – 90	– 2 to – 92	O_2^+, NO^+
Thermosphere	90 – 500	– 92 to 1200	O_2^+, O^+, NO^+

Ionosphere

The region of atmosphere lying between the altitude of 50 km to 100 km is called ionosphere, since this region has appreciable levels of positive ions (*e.g.* O_2^+, O^+, NO^+ etc.) and electrons. These ionic species persist for long time in this region. These species are formed due to the solar u.v. radiations. During night times, when u.v. radiations are not present, the positive ions tend to recombine with free electrons, giving the neutral species from which they are originated initially. This process occurs more rapidly in the lower regions of the ionosphere where the concentration of these species is relatively high.

Pollution and Pollutants

The contamination of air, water and soil with the substances, which have adverse effect on human beings, animals and plants, is called **pollution** of air, water and soil respectively. The pollution of air, water and soil is also called **environmental pollution**. These substances, whose presence in air, water and soil makes them polluted, are called **pollutants**. Thus a pollutant can be defined as a substance whose contamination with air, water and soil makes them polluted and has adverse effect on living and non-living things.

Types of pollutants

Pollutants may be of the following types :

1. Gaseous pollutants. Examples are CO, CO_2, oxides of nitrogen (NO, NO_2), oxides of sulphur (SO_2, SO_3), O_3 etc.

2. Liquid pollutants. Liquid pollutants include sewage wastes and industrial wastes (industrial wastes are the waste products discharged by industries). The sewage waste contains faecal matter, urine, washings, large number of bacteria etc. The industrial wastes contain compounds of metals like Hg, Cd, As, Pb etc., inorganic minerals, acids, alkalies, phenols etc. When industrial and sewage wastes enter into water or go into the atmosphere, they make water and air polluted.

3. Particulate pollutants. These will be discussed separately on the subsequent pages of this chapter.

4. Pollutants without weight. The pollutants without weight are radioactive substances, heat and noise. The pollution caused by these pollutants is called **radioactive pollution, thermal pollution** and **noise (sound) pollution** respectively.

5. Metallic pollutants. These will be discussed separately on the subsequent pages of this chapter.

To express the concentration of a gaseous pollutant

The concentration of a gaseous pollutant (*e.g.*, CO, NO, NO_2, SO_2, SO_3 O_3) in air is expressed in terms of the number of molecules of that pollutant present in one million molecules (=1000,000 molecules) of air. Thus the number of molecules of a gaseous pollutant present in one million molecules of air, is called *ppm* value of that pollutant. For example, if there are 50 molecules of CO in one million molecules of air, the concentration of CO in air is equal to 50 *ppm*. This discussion shows that *ppm* value of a given gaseous pollutant indicates the number of molecules of that pollutant present in 1000,000 (10^6) molecules of air.

Types of pollution

Following are the important types of pollution:

1. **Air pollution.** Pollution of atmosphere.
2. **Water pollution.** Pollution of river and sea water.
3. **Industrial pollution.** This pollution is caused by the wastes from industry.
4. **Noise (sound) pollution.** The adverse affects produced by the excessive noise on human beings is called noise pollution. Excessive noise reduces our hearing capacity.
5. **Thermal pollution.** Excessive of heat affects our health adversely and increase in temperature results in the death of aquatic animals. This is called thermal pollution.
6. **Radioactive (Radiation) pollution.** The radioactive wastes have adverse effect on water, earth and air. The pollution caused by radioactive wastes is called radio active pollution.
7. **Soil pollution.** The contamination of soil with acid rain, excess of fertilisers etc. is called soil pollution.

Particulate Pollutants

What are particulate pollutants ?

Particulate pollutants are the finely-divided solid and liquid particles which remain suspended in the air.

Examples of particulate pollutants

Examples of some particulate pollutants are given below:

1. Aerosols (Examples : *smoke* and *fog*). Aerosols are the fine particles of smoke and fog having diameter less than 100μ. The particles of aerosols are so small that they behave almost like gases. A detailed discussion of smoke has been given on the subsequent pages of this chapter. The word *fog* is used for tiny water droplets suspended in the air near the surface of the earth. Fog droplets are produced, when water vapour present in air gets condensed. The formation of fog droplets takes place in winter season. Fog helps in the formation of smog which is very harmful for human beings.

2. Dust particles. The detailed discussion of dust has been given on the subsequent pages of this chapter.

3. Fly-ash. We know that when fossil fuels undergo combustion, they produce small ash particle and gases. These particles are carried into the air through these gases and form fly-ash. Coal ash coming out of the chimney of coal based thermal power station is an important example of fly-ash.

Harmful effects of particulate pollutants present in air

1. The particulate pollutants produce various allergic reactions in our body and many

diseases such as bronchial asthma, tuberculosis etc.

2. The pollutants produce the visibility by producing haze in the atmosphere. This causes the hindrance in road and air traffic.

3. Smoke particles present in air combine with the droplets of fog containing poisonous gases discharged by the burning of fossil fuels in homes industries and automobiles and form smog. Thus:

Smoke particles + Droplets of fog containing poisonous gases ——→ Smog

Smog formed as above causes respiratory problems. Smoke particles also blacken the buildings and our clothes.

4. The particulate pollutants present in air reduce the amount of the radiations coming from the sun and reaching on the earth, and thus disturb the thermal balance on earth.

Air Pollution and Air Pollutants

Air is never pure, since it is always contaminated with poisonous gases (*e.g.*, CO, NO_2, SO_2 etc.), finely-divided solid and liquid particles, smog, etc. The presence of these materials in the air makes the air polluted. The presence of these materials in the air is called **air pollution** and these materials which pollute the air are called **air pollutants.**

Types of air pollutants

The above discussion shows that air pollutants may be of the following types :

1. **Gaseous pollutants.** Examples of gaseous air pollutants are CO, CO_2, oxides of nitrogen (NO, NO_2), oxides of sulphur (SO_2, SO_3), O_3 etc.
2. **Particulate pollutants.** These have already been discussed on page 411. Smoke and dust particles are the important examples of particulate pollutants.
3. **Smog and photochemical smog.** These have been discussed on the subseqent pages of this chapter.
4. **Metallic pollutants.** These have been discussed on subsequent pages of this chapter.

In the following paragraphs, we shall discuss the sources of production, harmful effects of some important air pollutants etc.

Carbon Monoxide, CO

Sources of CO

Major sources of CO are given below :

(*i*) CO is produced by the incomplete combustion of all carbon-containing fuels (incomplete combustion is that which takes place in presence of insufficient quantity of O_2) used in automobile engines and defective furnaces. Thus the smoke obtained by the incomplete combustion of petrol in the internal combustion engines of cars, buses, trucks, scooters, aeroplanes etc. (The engines used in these motor vehicles are called internal combustion engines, because the petrol, which is used as fuel, is burnt inside the engine and not separately) contains a lot of CO.

(*ii*) When coal, wood and oil are burnt, black smoke is produced. This smoke also contains CO.

$$2C + O_2 \rightarrow 2CO$$

(*iii*) Cigarette smoke also contains CO.

(*iv*) Incomplete combustion of agricultural and slush matter produces CO.

(*v*) The reactions that take place at high temperature in industrial furnances, like blast furnace, evolve CO.

(*vi*) All the above mentioned sources of the production of CO come under human activities. CO is also generated by natural sources like forest fires, volcanic eruptions, seed germination, electric discharge during storms etc.

Harmful effects of CO

(*i*) We know that hemoglobin (Hb) present in our body acts as an oxygen-carrier, since it combines with O_2 reversibly to form *oxy-hemoglobin*, HbO_2

$$Hb + O_2 \rightleftharpoons HbO_2 + H_2O$$

If we breathe in air containing CO, O_2 present in HbO_2 is replaced by CO and *carboxy-hemoglobin*, HbCO is formed (Also see Chapter 19).

$$HbO_2 + CO \longrightarrow HbCO + O_2$$

HbCO is a stable compound and its formation takes place, since Hb has more affinity for CO than for O_2. The formation of HbCO makes Hb unable to take up O_2 from the lungs. Thus the formation of HbCO results in that the quantity of O_2 available to the body cells gets reduced. This is called *anoxia* (*oxygen starvation*). The deficiency of O_2 produces *head-ache, dizziness, choking, cardiac* and *pulmonary* complications, *paralysis* and even *death*.

(*ii*) The poisoning effect of CO also kills many miners working in coal mines.

Treatment of CO poisoning

The most effective treatment of CO poisoning is that the victim is exposed to 2 to 2.5 atm. O_2 under a higher pressure. This results in that CO present in HbCO is substituted by O_2. When HbO_2 comes in contact with the cells of muscles, O_2 is given up by it.

$$HbCO + O_2 \rightleftharpoons HbO_2 + CO \ ; \qquad HbO_2 \longrightarrow Hb + O_2$$

Monitoring of CO

The monitoring of CO is done by means of *non-dispersive infra-red spectrometry.* This technique estimates CO upto a level of 150 ppm and is based on the principle that CO strongly absorbs infra-red radiations at certain specific wavelengths. The amount of infra-red radiations absorbed by CO gas in directly proportional to the concentration of the gas. If the concentration of the gas is near 10 ppm, this gas is estimated by gas-chromatography, in which a flame ionisation detector is used. If the concentration of CO gas is below 10 ppm, the gas is estimated by reducing it with H_2 gas over nickel catalyst at 360°C (*catalytic reduction*).

$$CO + 3H_2O \xrightarrow[360°C]{Ni} CH_4 + H_2O$$

CH_4 produced in the above reaction is estimated by flame ionisation detector.

To control the pollution caused by the unburnt hydrocarbons and CO present in the gases released by the vehicle's engine

The gases released by the vehicle contain various pollutants. Unburnt (unutilised) hydrocarbons (*e.g.* C_8H_{18}) and CO are important pollutants. These pollutants pollute the air.

In order to control the air pollution caused by hydrocarbons and CO, the following methods are used.

1. Addition of anti-knocking agent to gasoline (also called *gasoline additive*). The burning (combustion) of petrol (gasoline)–air mixture which is used as a fuel in the cylinder of the internal combustion engine of vehicles (*e.g.* scooters, motor cars and aeroplanes) takes place at a very high speed. The high speed of the burning of the fuel results in the following:

(*a*) Jerks are given to the piston. These jerks produce a violent noise known as *knocking*. Knocking is harmful for engine, since it reduces the effiency (power) of the engine and also damages the cylinder and piston of the engine.

(*b*) Burning of petrol-air mixture is incomplete when the burning of the mixture takes place at a high speed. It is due to incomplete burning of the fuel mixture that gases released by vehicles contain unburnt (unutilised) hydrocarbons and CO.

The high speed of the burning (combustion) of the fuel, due to which knocking takes place and combustion of the fuel is incomplete, can be decreased by adding certain compounds to the gasoline or petrol. These compounds are called *anti-kncking agents* or *gasoline additives* The anti-knocking agens provide more time to the gasoline for its combustion and hence enable it to burn at slow speed or more completely. Due to more complete combustion, the quantity of unburnt hydrocarbons and CO present in the gases released by the vehicles is decreased and hence pollution is also minimised.

The commonly used anti-knocking agent is tetraethyl lead (TEL), $(C_2H_5)_4$ Pb which is mixed in gasoline in very small amount (about 0.01%). The mixture thus obtained is called *leaded* (or *ethyl*) *gasoline* or *leaded petrol.* In the cylinder of the internal combustion engine, $(C_2H_5)_4$ Pb decomposes to give ethyl radicals, $CH_3 \cdot CH_2$.

$$Pb(C_2H_5)_4 \text{ (present in gasoline)} \xrightarrow{\Delta} Pb\left(\begin{matrix}\text{gets deposited in the}\\ \text{cylinder of the engine}\end{matrix}\right) + 4CH_3 . CH_2$$

$CH_3 . CH_2$ radicals combine with the radicals produced due to irregular combustion. This results in that reaction chains are broken and burning of the fuel takes place smoothly and at slow rate. This prevents knocking. Pb metal formed as above gets deposited in the cylinder.This deposited lead is removed by adding dihalogen alkanes like 1, 2-dichloroethane $(C_2H_4Cl_2)$ and 1, 2-dibromoethane $(C_2H_4Br_2)$ to the gasoline (petrol) along with $(C_2H_5)_4$ Pb. These alkanes convert deposited lead into $PbCl_2$ and $PbBr_2$ respectively.

$$\underset{\substack{|\\X}}{CH_2} - \underset{\substack{|\\X}}{CH_2} + Pb \longrightarrow \underset{\substack{\text{Ethene}\\ \text{(Ethylene)}}}{CH_2 = CH_2} + \underset{\text{(Volatile)}}{PbX_2}$$

(X = Cl, Br)

PbX_2 (X = Cl, Br) formed as above are volatile and hence go out of the engine with the exhaust gases and get mixed with air. $PbCl_2$ and $PbBr_2$ are dangerous compounds and hence make the air polluted.

Some chemists are of the opinion that $PbCl_2$ and $PbBr_2$ are formed when $Pb(C_2H_5)_4$ (anti-knocking agent) mixed with gasoline reacts with $C_2H_4Cl_2$ and $C_2H_4Br_2$ in presence of O_2.

$$2Pb(C_2H_5)_4 + 32O_2 + C_2H_4Cl_2 + C_2H_4Br_2 \longrightarrow PbCl_2 + PbBr_2 + 20CO_2 + 24H_2O$$

Although the addition of $(C_2H_5)_4Pb$ to gasoline minimises the air pollution caused by unburnt hydrocarbons and CO, the air gets polluted with dangrous $PbBr_2$. Therefore, the frequent addition of $(C_2H_5)_4Pb$ to the gasoline should be avoided and in its place aromatic and branched chain aliphatic hydrocarbons should be used.

2. Use of catalytic convertor : Catalytic oxidation of C_8H_{18} (unburnt hydrocarbon) and CO to CO_2 with platinum catalyst. Catalytic converter is a device which is attached with the verticle's engine. In catalytic converter octane, C_8H_{18} (unburnt hydrocarbon) and CO get mixed with more quantity of O_2 of the air and get burnt completely (*i.e.* are completely oxidised to CO_2) in presence of Pt which acts as a catalyst.

$$\underset{\text{unburnt hydrocarbon}}{2C_8H_{18}} + 25O_2 \text{ (From air)} \xrightarrow[\text{(Pt acts as a catalyst)}]{\text{Pt-metal}} 16CO_2 + 18H_2O$$

$$2CO + O_2 \text{ (From air)} \xrightarrow{\text{Pt-metal}} 2CO_2$$

The disadvantage of using catalytic converter is that platinum catalyst gets poisoned by $PbBr_2$ which is formed when we add $(C_2H_5)_4Pb$ (anti-knocking agent) to the petrol along with C_2H_4Br (see above).

Carbon Dioxide, CO_2

Although *small quantity of CO_2* is not usually considered to be a pollutant, *its excess quantity in the atmosphere* has an adverse effect on our climate (see green-house effect discussed below) and thus pollutes the air. Thus greater amount of CO_2 in the atmosphere acts as an air pollutant.

Major suppliers of CO_2 to the atmosphere

Following are the important suppliers of CO_2 to the atmosphere.

(*i*) *Burning of fossil fuels.* The fossil fuels, like coal, natural gas and petroleum, are carbon compounds. When these fuels are burnt, they produce CO_2 which is discharged into the atmosphere. The burning of fossil fuels is releasing about 6000 million tonnes of CO_2 in the atmosphere every year.

(*ii*) *Cultivation of soil.* The cultivation of soil also releases large amount of CO_2 into the atmosphere, which is produced by bacteria. The cultivation of land releases 2000 million tonnes of CO_2 into the air every year.

(*iii*) *Eruption of volcanoes.* CO_2 is also given to the atmosphere through the eruption of volcanoes. This gas comes from the interior of the earth.

(*iv*) *Respiration of living organisms.*

(*v*) *Decay of dead organisms.*

Major consumers of CO_2 from the atmosphere

The major consumers of CO_2 gas from the atmosphere are *green plants* and *oceans*. The green plants absorb CO_2 gas from the atmosphere to prepare their food (carbohydrates) through the process of photo-synthesis, while the oceans dissolve CO_2 gas to form carbonate rocks. Thus, we find that green plants and oceans both reduce the quantity of CO_2 in the atmosphere.

To control the air pollution caused by excess of CO_2

The best way of preventing the atmosphere from the pollution caused by excess quantity of CO_2 is that a balanced quantity of CO_2 should be maintained in the atmosphere. We have seen above that green plants absorb CO_2 gas from the atmosphere to manufacture their food and also at the same time release O_2 gas into the atmosphere (in the manufacture of their food the plants convert CO_2 into O_2). Thus if we grow more plants, our atmosphere will get rid of its excess CO_2 gas. We should, therefore, not cut down the trees.

Green-house Effect

There is a protective layer of O_3 gas in the atmosphere at a height between 15 km and 60 km. (The thickest layer of O_3 exists at a height of 23 km from the surface of the earth) and then a blanket of CO_2 gas exists in the lower part of the atmosphere (*i.e.*, below 15 km). Now when sunlight, consisting of ultra-violet rays, visible light and infra-red rays, falls on the top of the atmosphere, the harmful ultra-violet rays are absorbed by the O_3 layer and hence do not reach the earth's surface. On the other hand, the visible light and infra-red rays pass through the CO_2 layer and fall on the earth. Since the infra-red radiations have heating effect, they heat the earth and its various objects.

Now since the earth and its various objects become hot (of course, less hot than the sun), they start emitting infra-red rays (heat rays), which are of long wavelength (note that the infra-red rays emitted by the extreme hot bodies like sun are of short wavelength). The infra-red radiations of long wavelength emitted by the earth and its objects are absorbed by the CO_2 layer in the atmosphere. Thus we see that the presence of CO_2 in the atmosphere does not allow the infra-red radiations reflected by the earth's surface to go out of the atmosphere. In other words we can say that the layer of CO_2 gas in the atmosphere traps all the infra-red radiations (heat rays) coming from the earth's surface. These trapped infra-red rays heat the earth's atmosphere. *The heating up of earth due to the trapping of infra-red radiations (reflected from the earth's surface) by CO_2 layer in the atmosphere is called "green-house effect"*. Since the infra-red rays coming from the earth cannot pass through the CO_2 layer, the temperature of the earth is raised. The rise in the temperature produced by the "green-house effect" in the earth's atmosphere depends on the amount (proportion) of CO_2 gas present in the atmosphere. The name "green-house effect" comes from the fact that this effect is used in horticulture for the upbringing of green plants in a small house, whose walls and roofs are made of glass-sheet. The glass walls and glass roofs of the house allow the short wavelength infra-red radiations, contained in sunlight, to go into green-house freely, but do not allow the long wavelength infra-red radiations, reflected by the soil, plants and other contents of the green-house to go out. These trapped infra-red radiations raise the temperature inside the green house. Thus even without the internal supply of heat, the temperature inside the green-house becomes higher than that outside it.

Importance of green-house effect

The green-house effect, produced by the presence of CO_2 layer in the atmosphere is, very necessary for our existence on the earth. We have seen that, due to green-house effect, CO_2 gas in the atmosphere does not allow the long wavelength infra-red radiations (heat-rays) reflected by the earth, to go out of the atmosphere and hence the temperature of the earth's surface and its atmosphere is increased. The rise in temperature of the earth is very necessary for our existence on earth, because without it the whole earth would be converted into extremely cold planet and consequently we shall not be able to have a normal life.

Effect of excess of CO_2 present in the atmosphere

If the atmosphere contains too much quantity of CO_2, the green-house effect is considerably increased, *i.e.,* due to excess quantity of CO_2 present in the atmosphre, the temperature of the earth is increased too much. This too much high temperature melts all the glaciers (snow-mountains), floods the low-lying areas of the earth, changes the biological activity of oceans and the patterns of cropping etc. Thus we see that the presence of the excess of CO_2 in the atmosphere brings about climate change.

Oxides of Nitrogen (NO and NO_2)

Sources of oxides

(*i*) The smoke released by the automobile engines (engines of cars, buses, trucks etc.,) contains NO and NO_2. This smoke results by the combusiton of petrol or diesel in the engines.

(*ii*) When coal, oil and natural gas undergo combustion, high temperature produced by the combustion makes N_2 and O_2 present in the atmpsohere combine together to form NO and NO_2.

$$N_2 \xrightleftharpoons{+ O_2,\ > 100°C} 2NO \xrightleftharpoons{+ O_2} 2NO_2$$

(*iii*) Many chemical plants, like those which manufacture explosives and nitrogenous fertilisers, also produce NO_2. It is also present in tobacco smoke.

(*iv*) Large amounts of SO_2 and nitrogen oxides are also emitted from power plants and many industrial processes. The mixture containing SO_2 and nitrogen oxides, thus obtained, is called *flue gases mixture.*

Harmful effects of NO_2

Though NO and NO_2 both are toxic, the latter (*i.e.,* NO_2) is by far more harmful. Some of the harmful effects of NO_2 on human beings, animals and vegetables are given below :

(*i*) *Formation of nitric acid rain or acid snow and its harmful effects.* NO_2 present in the air combines with rainy water or atmospheric moisture in presence of O_2 or O_3 of the air and forms HNO_3.

$$4NO_2 + O_2 + 2H_2O \longrightarrow 4HNO_3 \quad ...(i)$$

$$2NO_2 + H_2O + O_3 \longrightarrow 2NHO_3 + O_2 \quad ..(ii)$$

HNO_3 formed as above dissolves in rain water and falls on earth in the form of *nitric acid rain* or *acid snow or acid rain*. When this nitric acid rain falls over a historical monument, it gradually eats up the marble stone ($CaCO_3$) of the monnument and corrodes it slowly.

$$CaCO_3 + 2HNO_3 \longrightarrow Ca(NO_3)_2 + CO_2 + H_2O$$

The acid rain fades the colour of the fabrics (*e.g.*, cotton, nylon and rayon), leather and paper. The acid rain makes the water of lakes acidic and hence fish-life comes in danger in this water. The yield of agricultural crops is also reduced.

Note : We have said above that HNO_3 is produced by the reactions represented by equations (*i*) and (*ii*). However. some chemists suggest that HNO_3 is produced by the following reactions.

(*a*) Nitrogen dioxide (NO_2) present in the air absorbs sunlight and undergoes photo chemical dissociation (photochemical reaction) to form nitric oxide (NO) and free O-atom.

$$NO_2 \xrightarrow[\text{(Photochemical dissociation)}]{\text{Sunlight}} NO + O \text{ (Very reactive)}$$

(*b*) O-atom formed at (a) is very reactive and hence combines with another NO_2 molecule and forms free reactive nitrate radical (NO_3).

$$O + NO_2 \longrightarrow NO_3 \text{ (Reactive radical)}$$

(*c*) Being reactive, NO_3 radical combines with another NO_2 molecule to form nitrogen peroxide (N_2O_5)

$$NO_3 + NO_2 \longrightarrow N_2O_5$$

(d) N_2O_5 combines with rainy waer or moisture present in the admosphere to form nitric acid rain, HNO_3

$$N_2O_5 + H_2O \longrightarrow 2HNO_3 \text{ (Nitric acid rain)}$$

(*ii*) NO_2 causes extensive leaf-drop in the plants.

(*iii*) Nitrogen oxides react with O_2 to produce another pollutant namely ozone.

(*iv*) It has been shown that if monkeys inhale NO_2 of 15–30 ppm concentration for 2 hours, their lungs, heart, liver and kidneys are damaged.

(*v*) NO_2 is very corrosive and attacks skin.

(*vi*) NO_2 helps in the formation of smog (which is a combination of smoke and fog). Smog causes irritation in eyes due to the presence of O_3 in it.

(*vii*) Oxides of nitrogen bring about corrosion of teeth and cause loss of appetite.

(*viii*) We know that the atmosphere contains a layer of O_3 which absorbs harmful ultra-violet radiations coming from the sun and reaching the earth. If these radiations reach the earth, they will cause skin cancer and will destroy the organic molecules necessary for life. O_3 layer prevents the ultra-violet radiations to come to the earth. NO and NO_2 present in the atmosphere decompose O_3 (present in O_3 layer) to O_2 and are generated again.

$$NO + O_3 \longrightarrow NO_2 + O_2$$

$$NO_2 + O_3 \longrightarrow NO + 2O_2$$

Thus we see that NO and NO_2 both present in the atmosphere destroy O_3 layer. Greater is the amount of NO and NO_2 in the atmosphere, greater is the percentage of O_3 which is destroyed.

To control the pollution caused by nitrogen oxides present in exhaust gases of motor vehicles

We know that the exhaust gases of motro vehicles contain various pollutants like CO, CO_2, NO, NO_2 etc. These substances cause pollution of the atmosphere when they are discharged into it. The air pollution caused by NO and NO_2 can be controlled by reducing these gases to N_2 and NH_3 with H_2 in presence of finely-divided platinum metal, before they are allowed to enter the atmosphere.

$$2NO + 2H_2 \xrightarrow{Pt.} N_2 + 2H_2O$$

$$2NO + 5H_2 \xrightarrow{Pt.} 2NH_3 + 2H_2O$$

$$2NO_2 + 4H_2 \xrightarrow{Pt.} N_2 + 4H_2O$$

$$2NO_2 + 7H_2 \xrightarrow{Pt.} 2NH_3 + 4H_2O$$

To control the pollution caused by NO_2 and SO_2 present in flue gases mixture

Power plants and many industrial units produce flue gases which contain a mixture of NO_2 and SO_2. Both these gases cause pollution of the atmosphere. The mixture containing NO_2 and SO_2 is treated with H_2SO_4 (called *scrubbing process*) so that the following reactions take place. First of all SO_2 and NO_2 present in the flue gases mixture react in presence of water and produce H_2SO_4 and NO.

$$SO_2 + NO_2 + H_2O \longrightarrow H_2SO_4 + NO$$

NO, thus produced, reacts with NO_2 to form N_2O_3 which reacts with H_2SO_4 to form $H(NO)SO_4$

$$NO + NO_2 \longrightarrow N_2O_3$$

$$N_2O_3 + 2H_2SO_4 \longrightarrow 2H(NO)SO_4 + H_2O$$

Thus we find that SO_2 and NO_2 present in the flue gases mixture are removed by scrubbing them with H_2SO_4. Flue gases thus freed from SO_2 and NO_2 are then released into the atmosphere.

Since the oxides of nitrogen and SO_2 are acidic in nature, they can be removed by scrubbing them with alkaline solutions of $Ca(OH)_2$ and $Mg(OH)_2$.

Determination of the concentration of the oxides

The method used for the determination of the concentration of the oxides, consists of making these oxides to react with O_3 to yield electronically excited species. For example when NO reacts with O_3, electronically excited species namely NO_2^* is produced.

$$NO + O_3 \longrightarrow NO_2^* + O_2$$

NO_2^* being unstable, quickly returns to the ground state, forming NO_2 and emitting light in the range of 600–3000 nm.

$$NO_2^* \longrightarrow NO_2 + h\nu$$

The intensity of the light emitted is directly proportional to the concentration of NO_2^* and hence of NO molecules. Thus the intensity of light is measured with the help of photo-multipliers and this intensity gives us a measure of the concentration of NO molecules.

Oxides of Sulphur (SO_2 and SO_3)

Sources of the oxides

The common oxide of sulphur which is the most harmful gaseous pollutant is SO_2. This oxide is released into the atmosphere by volcanic eruptions (natural activity). This oxide is also generated, when coal and oils are burnt in houses and industries. Coal always contains some sulphur as impurity. When coal is burnt, sulphur present in coal is converted into SO_2. Oils also have some sulphur compounds, which give SO_2 on burning.

When sulphite ores, like iron pyrites (FeS_2), copper pyrites (CuS), copper glance (Cu_2S), zinc blende (ZnS), galena (PbS) etc., are roasted in air, SO_2 is produced.

A part of SO_2 present in the atmosphere is oxidised to SO_3 by photolytic and catalytic oxidation processes. Thus SO_2 present in the atmosphere also contains SO_3, although the quantity of SO_3 is not high.

Effects of SO_2 and SO_3 on human beings, animals and plants

(*i*) We have shown above that SO_2 present in the atmosphere undergoes photolytic and catalytic oxidation to form SO_3. SO_3 thus produced reacts with rainy water or moisture and gives H_2SO_4 ($SO_3 + H_2O \longrightarrow H_2SO_4$) which comes down the atmosphere in the form of *sulphuric acid rain or acid snow* or *acid rain*. This *acid rain* is as dangerous as nitric acid rain (see oxides of nitrogen).

The formation of sulphuric acid rain (H_2SO_4) can also be explained through the following steps:

(*a*) $SO_2 \xrightarrow[\text{(Photochemical dissociation)}]{\text{Sunlight}} SO$ (Sulphurmonoxide) + O (very reactive)

(b) $O + SO_2 \longrightarrow SO_3$ (sulphur trioxide)

(c) $SO_3 + H_2O \longrightarrow H_2SO_4$ (sulphuric acid rain)

When sulphuric acid rain (H_2SO_4) falls over a historical monument, it gradually eats up the marble stone ($CaCO_3$) of the monument and corrodes it slowly.

$$CaCO_3 + H_2SO_4 \longrightarrow CaSO_4 + H_2O + CO_2$$

(*ii*) SO_2 irritates the respiratory system of animals and humans. It damages lungs.

(*iii*) If plants are exposed to SO_2 of high concentration over short periods, brownish colour in the tips of pine needles gets declourised. If the plants are exposed to SO_2 of low concentration over longer periods, the production of chlorophyl in plants in hampered (*chlorosis*) and hence green colour of the plants in lost.

(*iv*) Due to corrosive nature of SO_2 and H_2SO_4, these pollutants decolourise building materials like lime stone, marble, roof slate and mortar. The corrosion of most metals (*e.g.*, Fe, steels, Zn, Al etc.) is accelerated by SO_2.

(*v*) Fabrics, leather, paper and paints undergo fading of their colour in presence of SO_2.

To control the pollution caused by SO_2 gas

(*i*) We have already said that power plants and industrial units produce flue gases which contain a mixture of SO_2 and NO_2. Both these gases pollute the atmosphere. The removal of NO_2 from this mixture has already been discussed. SO_2 is removed by treating the flue-gases mixture with a slurry of $CaCO_3$ (lime stone) when SO_2 is absorbed due to the formation of $CaSO_3$ which is removed away.

$$CaCO_3 + SO_2 \longrightarrow CaSO_3 + CO_2$$

This method is quite economical.

(*ii*) We have said above that coal contains sulphur and hence SO_2 is produced when coal is burnt. SO_2 produced by buring coal causes air pollution. The pollution due to SO_2 can be prevented either by *desulphurisation* of coal or by using natural gas in place of coal for heating purposes. Natural gas does not produce SO_2 on burning.

Monitoring of SO_2 gas

The method used for monitoring SO_2 gas present in the polluted air in the range of 0.005–5 ppm has been devised by West and Gocke. In this method, the polluted air is pumped through a solution containing $HgCl_4^{2-}$ ions (which are obtained by the action of $HgCl_2$ on KCl solution, $HgCl_2 + 2KCl \longrightarrow K_2[HgCl_4]$) when about 95% of SO_2 present in air forms the complex ion, $[Hg^{2+}Cl_2(SO_3)]^{2-}$ in solution.

$$[Hg^{2+}Cl_4]^{2-} + SO_2 + H_2O \longrightarrow \underbrace{[Hg^{2+}Cl_2(SO_3)]^{2-} + 2HCl}_{\text{Solution}}$$

The solution containing complex ion, $[HgCl_2\,(SO_3)]^{2-}$ is treated with formaldehyde (HCHO) and then with pararosaniline hydrochloride to form a red-violet dye.

$$\underbrace{[HgCl_2\,(SO_3)]^{2-} + 2HCl}_{\text{Solution}} + HCHO \longrightarrow HOCH_2.SO_3H + HgCl_4^{2-}$$

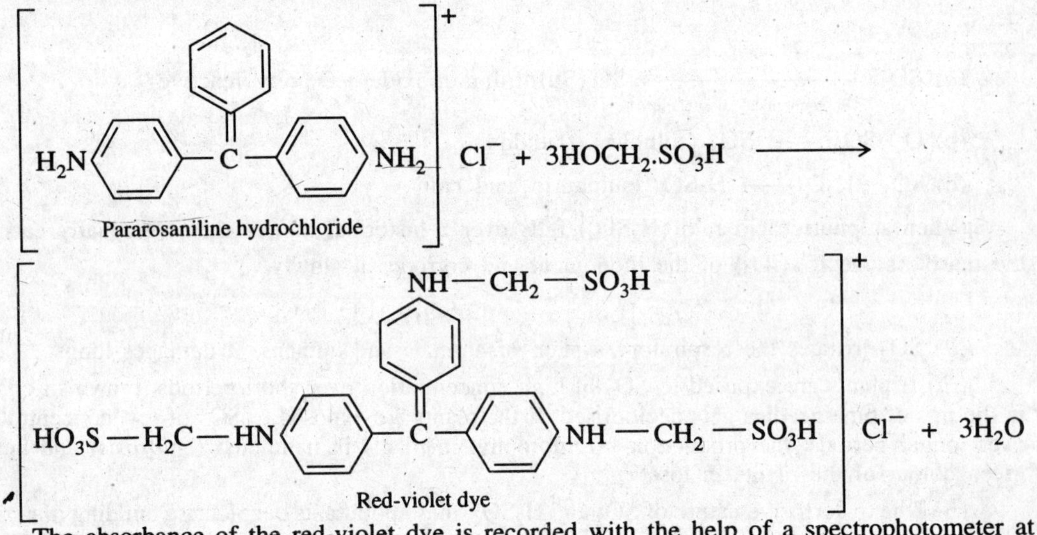

$$\left[H_2N-C_6H_4-\overset{\overset{\displaystyle C_6H_5}{\|}}{C}-C_6H_4-NH_2\right]^+ Cl^- + 3HOCH_2.SO_3H \longrightarrow$$

Pararosaniline hydrochloride

$$\left[HO_3S-H_2C-HN-C_6H_4-\overset{\overset{\displaystyle C_6H_4-NH-CH_2-SO_3H}{\|}}{C}-C_6H_4-NH-CH_2-SO_3H\right]^+ Cl^- + 3H_2O$$

Red-violet dye

The absorbance of the red-violet dye is recorded with the help of a spectrophotometer at 548 nm. The extent of absorbance is a measure of the concentration of the dye and hence of SO_2 present in the polluted air.

Ozone (O_3)

Sources of O_3

Some O_3 is produced during various combustion processes taking place in the air around us. O_3 is also produced in the upper part of the atmosphere (stratosphere) by the action of sun rays on O_2.

$$3O_2 \xrightarrow{\text{sun rays}} 2O_3$$

Harmful effects of O_3

Traces of O_3 in the air donot harm but O_3 of concentration more than 0.1 ppm is toxic and harmful to human beings. O_3 produces irritation in lungs, eyes and respiratory tract and accumulates fluid in lungs. This fluid damages the lungs capillaries. These biochemical effects of O_3 mostly arise from the generation of free radicals which attack the –SH groups present in enzyms.

Protective role of ozone (O_3) layer present in stratosphere region of the atmosphere

We know that air contains O_2 and at a height of 15 km from the earth's surface, this O_2 is converted into O_3 by the action of sun rays on it. [$3O_2 \xrightarrow{\text{sun rays}} 2O_3$]. O_3 thus produced forms a stable layer in stratosphere region (the region lying between 11 km and 50 km from the earth's surface) of the atmosphere. The thickest layer of O_3 exists at a height of 23 km from the earth's surface. This layer of O_3 is very useful for animals, human being and plants living on the earth. Its usefulness can be explained as follows:

The radiations emitted by the sun and reaching on the earth also consist of ultra-violet radiations. These radiations are very harmful to the living organisms. These radiations are of short wavelength and hence penetrate deep into our body and cause skin cancer. These radiations also damage crop plants. Ozone layer present in the stratosphere region of the atmosphere absorbs harmful ultra-violet radiations coming from the sun and thus prevent these radiations from reaching earth. Thus we are saved from the harmful effects caused by ultra-violet radiations.

Sources of destroying the ozone layer present in stratosphere

In 1980 scientists showed that there is a hole in the O_3 layer. This hole was detected over the region of Antarctica. Due to the presence of this hole in of O_3 layer, the ultra-violet rays coming from the sun can pass through this hole and thus can reach the earth's surface. The presence of O_3 hole in the atmosphere is due to the fact that the amount of O_3 present in stratosphere is getting reduced day by day and thus the ozone layer is becoming thinner and thiner. The depletion of the ozone layer is due to the following :

(*i*) *Due to the formation of O_2 in the stratosphere.* The oxides of nitrogen present in the atmosphere decomposes O_3 into O_2 and are themselves regenerated.

$$NO + O_3 \longrightarrow NO_2 + O_2$$
$$NO_2 + O \longrightarrow NO + O_2$$

Thus we see that the presence of nitrogen oxides in the atmosphere destroys the ozone layer. These oxides destroy about 70% of O_3 found in the stratosphere. Greater is the amount of the oxides of nitrogen in the atmosphere, greater is the precentage of O_3 which is destroyed.

(*ii*) *Due to the nuclear tests being conducted in various parts of the world.* Nuclear tests being conducted in various parts of the world generate such a high temperature that N_2 present in the atmosphere is oxidised to NO. NO thus formed decomposes O_3 to O_2.

$$N_2 + O_2 \xrightarrow{\text{High temperature}} 2NO$$
$$NO + O_3 \longrightarrow NO_2 + O_2$$

Thus O_3 layer is destroyed.

(*iii*) *Due to the use of fluoro-chloro-carbons as acrosol spray propellants.* Fluoro-chloro-carbons are the fluoro-chloro methanes like, *Freon-1* ($CFCl_3$) and *Freon-12* (CF_2Cl_2). These are stable compounds. These are chemically inert and hence do not react with the substances. These are used as aerosol spray propellants, refrigerants, firefighting reagents and solvents for cleaning electronic components. When they enter stratosphere, they absorb ultra-violet solar radiations and get broken down into free atomic chlorine. This atomic chlorine decomposes O_3 into O_2 (NO also breaks O_3 into O_2).

$$Cl + O_3 \longrightarrow ClO + O_2$$
$$ClO + O \longrightarrow Cl + O_2$$

Thus we find that the use of fluoro-chloro-carbons (*e.g.*, Freon-1 and Freon-12) as aerosol spray propellants destroys the O_3 layer, as oxides of nitrogen do.

(*iv*) *Due to the emissions produced by super-sonic jet aeroplanes.* The emissions produced by the supersonic jet aeroplanes also cause the depletion of ozone layer in the atmosphere.

Smoke

Smoke contains unburnt carbon particles having diameter less than 100 μ. These particles remain suspended in air.

Production of smoke particles in air

Smoke particles are emitted into the air during the process of burning (or combustion) of fuels like coal and oil in homes and factories. The quantity of smoke particles in air is variable.

Harmful effects of smoke

The smoke contains unburnt carbon particles which pollute the air. They spoil our clothes and blacken the buildings. Smoke damages our lungs. Due to cloud of smoke over big industries, the sun rays are not able to reach every corner of the city and hence infectious diseases in big cities are produced.

How to control air pollution due to smoke

The following two methods can be used for this purpose.

(*i*) The pollution of air due to the presence of carbon particles (smoke particles) coming out of a chimney of a factory can be controlled by installation an electrostatic precipitator in the chimney. A high electric potential is applied across the chimney. As the smoke particles (*i.e.*, carbon particles) come up the chimney, they are attracted by the charged electrodes. The smoke particles cluster together to form soot, which gets deposited in the chimney. Thus the air is saved from being polluted by carbon particles. The soot can be removed by sweeping the chimney from time to time.

(*ii*) The pollution of air caused by carbon particles can also be prevented by using electrical energy, since these sources do not produce smoke or carbon particles.

Dust Particles

Production of dust particles in the air

Dust particles are produced by heavy traffic on the roads and in some industrial operations. These particles remain suspended in the air.

Harmful effects of dust

Dust in the air spoils our clothes and causes poor visibility. It provides allergic reactions and aggravates diseases like bronchitis. Dust is deposited on the leaves of plants and thus hinders the process of photo-synthesis taking place in plants. Dust in air reflects back some of the sun's heat rays and thus undue cooling of the earth takes place.

Smog or Classical Smog

What is smog or classical smog ?

The poisonous gases like SO_2 and NO_2 produced by the burning of fossil fuels (*e.g.*, coal and petrol) in homes, industries and automobiles mix with the tiny droplets of fog (tiny water droplets) suspended in the air near the surface of earth (Fog is produced by the condensation of water vapour present in air). When the droplets of this fog which contains poisonous SO_2 and NO_2 gases get condensed on the solid particles of smoke or dust present in air, a colloidal dispersion is formed. This dispersion is called **smog** or **classical smog**. Thus :

Tiny droplets of fog containing poisonous SO_2 and NO_2 gases condense on solid particles of smoke or dust present in air and produce collodial dispersion which is called smog or classical smog.

Thus smog is the combination of smoke particles with tiny droplets of fog containing poisonous gases like SO_2, NO_2 etc. discharged by the burning of fossil fuels (*e.g.*, coal and petrol) in homes, industries and automobiles.

Smog usually occurs during winter season. Smog present in the lower atmosphere is taken by men and animals during breathing and damages their health.

In December 1952, a dense cloud of smog was formed over London city and remained there for five days. This smog produced respiratory problems in human beings and about 4000 people died. Many persons suffered from bronchitis and heart problems.

Harmful effects of smog

Various harmful effects of smog are :

(*i*) Smog is harmful to all the human beings who breath in it. It produces bronchitis, asthama and heart problems.

(*ii*) Smog creates irritation to the eyes, nose and throat.

(*iii*) Smog has an adverse effect on the growth and development.

(*iv*) Smog reduces the visibility and thus creates difficulties in road and air traffic.

Los Angeles Smog or Photochemical Smog

What is photochemical smog ?

The gases released by automobiles contain oxides of nitrogen (mainly NO_2), unburnt hydrocarbons, CO and CO_2. NO_2 present in the auto-exhaust gases absorbs ultra-violet (U.V.) radiations of the sunlight and undergoes photolysis giving atomic oxygen (O).

$$NO_2 + \text{U.V. light} \longrightarrow NO + O$$

Atomic oxygen produced as above reacts with hydrocarbons (RH) to give a variety of free hydrocarbon radicals (*e.g.* RHO, RHO_2 and RHO_3), O_3, aldehydes, ketones, peroxy acetyl nitrate ($CH_3CO.OO.NO_2$ or PAN). All these substances are called *photochemial pollutants.*

$$RH \xrightarrow{+O} RHO \xrightarrow{+2O} RHO_3$$

$$RHO_3 \xrightarrow{+2O} RHO_2 + O_3$$

$$RHO_3 \xrightarrow{+RH} \text{Aldehydes, Ketones etc.}$$

$$RHO_3 \xrightarrow{+NO} RHO_2 + NO_2 \longrightarrow \text{Peroxy acetyl nitrate, } CH_3CO.OO.NO_2 \text{ (PAN)}$$

PAN is harmful to human beings even in a concentration of 0.1 parts per million. PAN mixes with the fog and gets condensed on smoke or dust particles present in the air to form a smog which is called *photochemial smog*. Thus :

Fog containing poisonous PAN + smoke or dust particles ⟶ Photochemical smog.

Los Angeles is the name of a city situated in the U.S.A. Seven million people of this city have four million automobiles and consume 30 million litres of petrol every day. The smog formed in this city in photochemical smog. This *Los Angeles smog is a photochemical smog.*

Harmful effects of photochemical smog

(*i*) Photochemical smog causes eye irritation and damages plants, since in the presence of this smog the leaves develop a metallic sheen.

(*ii*) Rubber, on exposure to photochemical smog, loses its elasticity and becomes inflexible and brittle.

(*iii*) Photochemical smog reduces the visibility and produces hindrance in the air and road traffic.

Comparison between classical smog and photochemical smog

Due to the presence of O_3, NO_2 and other photochemical oxidants, photochemcial smog is an oxidising smog whereas the classical smog formed by the combination of tiny droplets of fog containing poisonous gases *viz.*, SO_2 and NO_2, and solid particles of smoke or dust is a reducing smog.

Metallic Pollutants

Metallic pollutants present in air

Air contains large number of metallic pollutants in the form of metallic particles. These particles are of Pb, Hg, As, Cd, Zn, Cr, Be, Ni, V, Fe, Sn and Al metals.

How are metallic particles released into the air ?

The metallic particles (metallic pollutants) are released into the air by various industrial processes, agricultural parctices and transport vehicles. For example :

(*i*) The particles of metals like Cd, Zn and Hg are released into the environment during the refining of Zn, Cu, Pb and steel.

(*ii*) We know that tetraethyl lead, $(C_2H_5)_4Pb$ is added to the petrol along with suitable amount of $C_2H_4Cl_2$ and $C_2H_4Br_2$. The addition of $(C_2H_5)_4Pb$ slows down the rate of combustion of petrol. In presence of O_2, $(C_2H_5)_4Pb$, $C_2H_4Cl_2$ and $C_2H_4Br_2$ react together to produce $PbCl_2$ and $PbBr_2$ both of which are volatile and hence come out of the engine and are thus mixed with air. Thus air gets polluted with $PbBr_2$ and $PbCl_2$ which are very harmful materials.

Lead is also released into the air by industries engaged in lead mining, extraction and purification of lead, and manufacture of alloys and paints.

(*iii*) The liquid waste products discharged by industries are called industrial wastes. These wastes contain compounds of metals like Hg, Cd, As, Pb, Ca etc., inorganic minerals, acids, alkalies, phenols etc. When the industrial wastes enter into water or go into the atmosphere, water and air get polluted due to the presence of metals, inorganic minerals etc. present in the industrial waste.

The liquid waste produced from the washing of blast funance, flue gases and picking of steel in steel plants contains many substances along with metals. Due to the presence of these metals, the liquid waste makes air polluted.

Harmful effects of lead

Lead is a very poisonous metal. It is a comulative poison, since it keeps on accumulating in the tissues of human body and in the leaf tissues of plants. Even a low concentration of 0.2 ppm of lead in human body creates metabolic disturbances. Some of the harmful effects of lead are given below :

(*i*) It is due to the prolonged effect of lead on the human body that red blood cells get stippled. The malformation of red blood cells produces the diseases of anaemia.

(*ii*) Lead gets bound with the cellular enzymes and disrupts the functioning of cells and organs of the mascular circulatory and nervous systems.

(*iii*) Lead damages organs like liver, kidneys and intestine.

(*iv*) Lead develops abnormalities in fertility and pregnancy.

(*v*) Lead causes coagulation of proteins.

(*vi*) When rats are exposed to lead polluted air, they die in two-thirds of their average life span.

(*vii*) The birds, which are shot, but recover usually, subsequantly die due to lead-poisoning.

(*viii*) The lead-polluted air has its adverse effects on vegetables, *e.g.*, the road-side vegetables are dried up by the exposure of lead-polluted air.

(*ix*) If lead is taken by humans *even in small amounts*, it affects breathing and nervous system.

(*x*) *Excessive intake of lead* causes disruption of synthesis of haemoglobin. It also results in loss of appetite, anaemia, kidney malfunctioning, nervous disorders and brain damage.

(*xi*) Lead can also be absorbed by skin, resulting into many skin diseases.

Harmful effects of Hg

Mercury is found in sea due to the weathering of mercury-bearing rocks. Some of the effects of mercury pollution are given below :

(i) Monomethyl mercury [CH_3Hg] and dimethyl mercury [$(CH_3)_2Hg$] produce nervous disorders in marine animals.

(*ii*) Contact with Hg produces nervousness, fear, inability to make decisions, heaviness, irritability, headache, pessimism, fatigue, sleeplessness, trembling of limbs, falling teeth and diarrhoea.

(*iii*) Hg brings about genetic changes.

Harmful effects of Cd

Cadmium pollution may arise from soot of zinc, lead and copper processing plants. Cadmium is now used to plate tin containers of canned food. It is easily dissolved in fruit juice. It can remain in the body for a long time. Following are some of the effects of cadmium pollution.

(*i*) It causes vomiting and lung irritation.

(*ii*) If a large quantity of cadmium is absorbed in the kidneys, it causes hypertension and anaemia.

(*iii*) Cadmium is very dangerous for pregnant women.

(*iv*) Cadmium pollution causes a desease called "*Hai Itai*" in which the whole body feels serious pains and the bones begin to fracture very easily.

Sources of some more metallic pollutants and their effects on human beings and animals are given in Table 15.3.

Table 15.3 : *Sources of some metallic pollutants and their effect on human beings and animals.*

Metallic pollutant	*Major sources of production*	*Harmful effects on human beings and animals*
(*i*) Zinc (Zn)	Zinc refineries, galvanizing processes, brass manufacture, metal plating, plumbing.	Zine fumes have corrosive effect on skin and can cause irritation and damage mucous membranes.
(*ii*) Chromium (Cr)	Metallurgical and chemical industries, processes using chromate compounds, cement and asbestos units.	Toxic to body tissues, can cause irritation, dermatitis, ulceration of skin, perforation of nasal septum. Carcinogenic action suspected.

(Contd.)

(*iii*) Arsenic (As)	Arsenic containing fungicides, pesticides and herbicides, metal smelters, by-product of mining activities, chemical wastes.	Inhalation, ingestion or absorption through skin can cause mild bronchitis, nasal irritation or dermatitis. Carcinogenic activity also is suspected. Attack-SH groups of enzymes, coagulate proteins.
(*iv*) Beryllium (Be)	Coal, nuclear power and space industries, production of fluorescent lamps, motor fuels and other industrial use.	Damage to skin and mucous membranes, pulmonary damage, perhaps carcinogenic.
(*v*) Boron (B)	Boron producing units, production and use of petroleum fuel and additives, burning coal and industrial wastes, detergent formulations.	Ingestion or inhalation as dust causes irritation and inflammation. Boron hydrides can damages CNS and may result in death.
(*vi*) Manganese (Mn)	Ferromanganese production, organo-manganese fuel additives, welding rods, incineration of managanese containing substances.	Absorption, ingestion, inhalation, or skin contact may cause manganic pneumonia.
(*vii*) Nickel (Ni)	Metallurgical industries using nickel, combustion of fuels containing nickel additives, burning of coal and oil, electroplating units using nickel salts, incineration of nickel containing substances, vanaspati manufacture.	Respiratory disorders, dermatitis, cancer of lungs and sinus.
(*viii*) Vanadium (V)	Vanadium refining, production of vanadium containing alloys, power plants, burning of oil rich in vanadium.	Gastro-intestinal and respiratory disorders, inhibition of synthesis of cholesterol, heart disease and cancer in case of chronic exposures.
(*ix*) Selenium (Se)	Burning of fuels and residual oils, fumes and gases from refinery wastes, incineration of paper and other wastes, natural sources.	Irritation of gastro-intestinal and respiratory tracts, irritation of eyes, nose and throat, damage to lungs, liver and kidneys.

Methods to control the gaseous pollutants

Following methods are used :

1. By catalytic or flame combustion of gaseous pollutant. This method is used when gaseous pollutants are organic in nature. In this method the catalytic or flame combustion of the gaseous pollutants converts them into water vapour and relatively innocous products like CO_2.

2. By passing the gaseous effluents through a suitable liquid absorbent. This method is used for removing pollutants like NO, NO_2, H_2S, SO_2, SO_3 and fluorides from the gaseous effluents. In this method, the gaseous effluents are passed through the scrubbers or absorbers contining a suitable liquid absorbent. The liquid absorbents commonly used are given below:

Liquid absorbent	*Gaseous pollutant to be removed*
(*i*) Water, aqueous alkalies, slaked lime suspension, sulphides of barium or calcium or sodium, amino ethanol, dimethyl aniline, xylidine, aluminium sulphate etc.	SO_2
(*ii*) Alkaline phenol, sodium alamine, soda ash, tripotassium phosphate, liquor ammonia, sodium thioarsenate etc.	H_2S
(*iii*) Water, caustic soda.	HF
(*iv*) Water, dilute nitric acid.	NO, NO_2

The equipment used in this method inlcudes plate towers, spray towers, packed towers, bubble cap plate towers and liquid jet scrubbers.

3. By passing the gaseous effluents through porous solid adsorbent. In this method the gaseous effluents are passed through porous solid adsorbents packed in appropriate containers. The pollutants are held at the interface of the solid adsorbent by chemisorption. The common adsorbents used for various gaseous pollutants are given below :

Porous solid adsorbent	*Gaseous pollutant to be removed*
(*i*) Pulverised lime stone or dolomite, alkalised alumina ($Al_2O.Na_2O$)	SO_2
(*ii*) Silica get, zeolites	NO, NO_2
(*iii*) Lime stone lumps, porous NaF pellets	HF
(*iv*) Iron oxide	H_2S
(*v*) Activated carbon	Organic solvent vapours
(*vi*) Bauxite, silica get etc.	Petroleum fractions

Methods to remove particulate emissions

The particulate collection devices are based on the size, shape, charge and hygroscopic properties of the particulates concerned. The various devices used are given below :

1. Mechanical devices: The following mechanisms are the basis for these devices.

(*i*) *Gravity setting:* In which the velocity of the horizontal carrier gas is reduced adequately so that the particles settle by gravitational force.

(*ii*) Sudden change of direction of the gas flow causes the particles to separate out due to their greater moment.

Settling chambers, buffer chambers and cyclone separators are commonly used in mechanical devices to separate particulates from gases.

2. Filtration systems: The gases containing dust are forced through porous medium such as woven or filled fabric. Fibrous or deep-bed filters and cloth bag filters are commonly used. Fabric filter media made from cotton, wood, nylon, dacron, asbestos, silicone coated glass cloth etc., are used. The particles are trapped and collected in the filters and the gases devoid of the particles are discharged out.

3. Electrostatic precipitators. In this device stream of gases containing aerosols is passed between two insulated electrodes between which appreciable difference in potential exists. Since aerosol particles are charged, they are discharged at the electrode containing opposite polarity to that present on the aerosol particles and thus are precipitated. Electrostatic precipitators are the devices of choice when :

(*i*) Very large volume of gases are to be handled. (*ii*) Valuable dry material is to be recovered. (*iii*) Very high collecting efficiency essential. (*iv*) When gas temperature is very high.

The industries using such devices are power plants, paper and pulp industries, chemical industries, iron and steel plants, mining and metallurgical industries, rock products, refineries etc.

4. Wet scrubbers. These devices are used when:

(*i*) Fine particles have to be efficiently removed. (*ii*) Particulate and gaseous contaminants have to be removed. (*iii*) Gases to be treated are combustible. (*iv*) Cooling is desired. (*v*) Temperature of the gases to be treated is as high as 300°C or higher.

The wet scrubbers in use are ventury scrubber, gravity spray scrubber, wet impinger scrubber, cyclone spray chambers, wet centrifugal scrubbers etc.

Sources of air pollution

The various sources which produce air pollution (*i.e.*, atmospheric pollution) are given below:

1. Burning of coal, wood and oils. When coal, wood and oils are burnt, black smoke is produced. The smoke contains poisonous gases like CO and SO_2 and black unburnt carbon particles. The smoke is black because of the presence of these carbon particles in it. Coal contains some sulphur as impurity. So, when coal is burnt, sulphur of the coal burns to form SO_2. Oils also contain some sulphur compounds which produce SO_2 on burning. CO, SO_2 and unburnt carbon particles, produced as above, go into the air and make it polluted.

2. Gases released by the internal combustion engines of cars, buses, trucks, aeroplanes etc. The smoke obtained by the incomplete combustion of petrol, used in the internal combustion engines of cars, buses, trucks, scooters, aeroplanes etc. [The engines used in these motor vehicles are called internal combustion engines because the petrol (also called gasoline) which is used as a fuel, is burnt inside the engine and not separately] contains CO, NO, NO_2, unburnt carbon particles, some lead compounds, some alcohol and acids, All these substances are poisonous and hence pollute air.

Now let us see how these pollutants are produced in an engine of a car. Petrol is used as a fuel in car engine. The main components of petrol are hydrocarbons. These hydrocarbons have the general formula, C_8H_{18} and hence are called *octanes*. Petrol burns very fast in a car engine. Due to short time available for burning, incomplete combustion of petrol takes place and some CO, unburnt carbon particles, CO_2, water vapour, some alcohol and acids are produced. CO and carbon particles are emitted into the air and thus air is polluted.

When petrol burns in a car engine, a very high temperature is produced. At this high temperature, N_2 and O_2 present in the air in the engine combine together and form NO and NO_2. These gases pollute the air.

In order to control the air pollution due to CO, $Pb(C_2H_5)_4$ is added to petrol. The addition of this substance slows down the rate of combustion of petrol [$Pb(C_2H_5)_4$ is called anti-knock] and also at the same time gives rise to the formation of $PbCl_2$ and $PbBr_2$. Both of these substances are poisonous and hence make the air polluted.

3. Chemical industries. The chemical industries emit various poisonous gases into the air. These gases make the aii polluted. An industry poducing bleaching powder emits chlorine gas into the air. Industries working for the extraction of metals (*e.g.*, Fe, Cu, Zn etc.) from sulphide ores produce SO_2 gas which causes air pollution.

4. Smoking. Smoking produces CO and nicotine. Both these chemicals are poisonous and hence make the air polluted.

5. Dust. Dust produced by the heavy traffic and various industries pollute air.

6. Radioactive waste materials: nuclear pollution. The radioactive waste materials produced during the enrichment of nuclear fuels and in nuclear reactors, used in hospitals and laboratories, emit high energy gamma radiations, which inhibit cell functions, leading to some serious diseases like leukaemia. The radiations initiate unwanted cell growth which produces cancer etc.

Methods to control air pollution

We have already said that air is polluted by the presence of CO, CO_2, NO, NO_2, SO_2, SO_3, O_3, unburnt hydrocarbons, dust, smoke etc. The pollution of air by these pollutants can be prevented by the following methods.

(*i*) *By adding lead tetraethyl, $Pb(C_2H_5)_4$ to the petrol or by the catalytic oxidation of CO and hydrocarbons with platinum catalyst.* This method can be used to save the air from pollution caused by the presence of CO and unburnt hydrocarbons in it. These pollutants come in the air from the exhaust gases released by the vehicle's engine.

(*ii*) *By keeping the vehicle properly tuned for the optimum ignition of fuel.* This method also prevents air pollution caused by CO and hydrocarbons present in the exhaust fumes of the vehicles.

(*iii*) *By growing more plants.* This method removes the air pollution caused by CO_2.

(*iv*) *By using electrostatic precipitator.* This method is used for removing carbon particles produced by burning of fuels like coal and oil.

(*v*) *By using smokeless sources of energy.* This method is used for preventing air pollution caused by smoke or carbon particles produced during the burning of coal or oil.

(*vi*) *By using natural gas, in place of coal, for heating purposes.* This method saves air from being polluted by SO_2 gas.

(*vii*) *By using tall chimneys.* The use of tall chimneys in homes and factories can reduce the concentration of pollutants at the ground level.

(*viii*) The air pollution in our homes can be controlled : (*a*) by using liquefied petroleum gas (LPG) for cooking, instead of coal, kerosene or wood. Electric heaters and solar cookers can also be used to prevent the air pollution at homes. (*b*) By using a tall chimney in the kitchen. This will minimise the pollution of air at the ground level. (*c*) By growing more plants in the courtyard. The plants use CO_2 and thus air pollution is reduced.

(*ix*) The pollution caused by the radioactive waste products (nuclear pollution) can be controlled (*a*) by enclosing the radioactive waste products in containers made of concrete and then dumping them in the sea (*b*) by fusing the radioactive waste products into glass and sealing it deep inside hard-rock formations. At the same time, a great vigilance should be kept on the stored radiactive materials and waste products to prevent the leakage of any radioactive materials during storage and disposal.

Water Pollution and Water Pollutants

When insoluble solid particles, soluble salts, sewage (waste water), garbage, low level radioactive substances, industrial wastes, algae, bacteria etc. go into water, water gets polluted. This type of pollution is called **water pollution** and the substances which make water polluted are called **water pollutants.**

Examples of water pollutants : Their sources and harmful effects

1. Industrial wastes and organic sewage wastes. The wastes discharged by industries are called industrial wastes. These wastes contain compounds of metals (*e.g.*, Hg, Cd, Pb etc.), organic compounds, alkalies, phenols etc. which make water polluted.

The sewage is the dirty water which contains human and animal excretions (*e.g.*, urine and faeces). The discharge of large quantities of sewage into rivers and lakes also causes water pollution. This sewage is produced by the every day human activities like bathing, washing clothes and excreting the urine and faeces.

Some of the effects of industrial and sewage wastes are given below:

(*a*) The sewage contains organic compounds. These organic compounds act as food for bacteria present in water of lakes and rivers and hence the population of bacteria in water is rapidly increased. This bacteria oxidises nitrogen and phosphorus present in organic compounds of the sewage to nitrates and phosphates. This oxidation process consumes a large quantity of O_2 gas dissolved in river water. The loss of dissolved O_2 from the river water causes the death of fish and other aquatic living beings.

(*b*) These wastes produce scum and sludge in water.

(*c*) These wastes change the colour of water and produce foul smell.

(*d*) The metallic elements present in the industrial wastes make water polluted. These metals affect adversely human beings, plants and aquatic life. Harmful effect of Pb, Hg and Cd have already been discussed. Harmful effect of Zn, Cr, As, Be, B, Mn, Ni, V and Se have been summarised in Table 15.3.

2. Organic compounds. These include synthetic detergents, herbicides, pesticides, oil and decomposition products. Their effects are given below :

(*i*) *Detergents.* Detergents are *surfactants* (*i.e.,* surface active agents) and *builders*. Surfactants are organic compounds having polar or hydrophilic groups such as $-COOH$, $-SO_3H_2$, NH_4^+ or non-polar or lyophilic groups soluble in water. Surfactants have the following properties and hence cause water pollution.

(*a*) *Bio-micro-organisms* (*i.e,* bacteria present in the water of lakes or rivers) are not able to bio-degrade the surfactants, *i.e.,* surfactants are not decomposed (digested) by the bio-micro-organisms and hence large volumes of detergent foams are collected in the lakes and rivers. These foams make water polluted.

(*b*) Surfactants form a sort of envelope around the organic substances (*e.g.* phenol). The formation of an envelope results in that these substances do not undergo oxidative degradation during the treatment of municipal waste water. Thus, the removal of organic substances from the waste water by the usual treatment methods becomes a problem.

(*c*) Builders are generally sodium salts of phosphoric acid, carbonic acid, sulphuric acid etc. The builders form stable soluble complexes with hardness-producing ions such as Ca^{2+} and Mg^{2+}. These complexes act as plant nutrients and cause a rapid growth of algae and weeds in water. The growth of these plants appears as a green sludge on the surface of the water. These plants bring about de-oxygenation of water and hence aquatic animals die.

(*ii*) *Pesticides.* The pesticides (*e.g.,* D.D.T., endrin, dieldrin, B.H.C., polychlorinated diphenyls etc.) are both toxic and persistent. Analysis of polluted water has shown that it contains pesticides which are toxic to fish. Endrin, even in traces, is reported to be toxic for catfish and other varieties of fish. D.D.T. affects the central nervous system of fish and toxaphene has been reported to cause bone degeneration in fish.

3. Chemical fertilisers : plant nutrients. Nitrate/phosphate salts are generally used as fertilisers, to increase the yield of the crops. When these fertilisers are used in excess, some of their unused quantity is washed away from the agriculture lands ito the ponds, lakes and rivers with rain water and thus pollute the water. This water is polluted, since it contains unused nitrate/phosphate salts. Similarly the waste water coming from the fertiliser industries also contains nitrogenous/ phosphatic fertilisers which, when washed away into the lakes and rivers with rain water, make the water polluted. The presence of nitrogenous/phosphatic fertilisers in water is harmful to the aquatic life and human beings in the following ways:

(*a*) The presence of the fertilisers in the polluted water increases the growth of algae and other aquatic plants which, later on, unergo decompostion and produce disagreeable odour. These plants also deplete the amount of O_2 dissolved in water and hence the survival of aquatic life becomes difficult or impossible.

(*b*) After a long period the lakes and slow moving water which contains plant nutrients are converted into swamps (A swamp is an area of very wet land with wild plants growing in it) and marshes (A marsh is an area of land which is very wet and muddy).

(*c*) The water containing nitrate salt is not fit for drinking by human beings. Moreover, this polluted water cannot be purified for drinking purposes.

4. Disease producing micro-organisms. If the residents of a municipality are suffering from diseases like dysentry, typhoid, cholera etc., the faeces and urine discharged by such patients contain micro-organisms which are transmitted through water supplies and thus produce disease in other living beings.

5. Petroleum oil. If a large quantity of crude petroleum oil is spilled accidently or intentionally into the sea-water from the tankers (ships) or during the drilling and shipping operations, this oil pollutes sea-water. For example the accidental leakage of 118,000 tonnes of crude petroleum oil from the big oil tanker called Torrey Canyon into the waters of English Channel polluted water of the surrounding area in a few hours' time.

Some of the effects of oil pollution are given below :

(*a*) Since oil is lighter than water, oil forms a layer on the surface of water which is called *oil-slick*. This oil-slick checks the oxygenation of water and water eventually becomes devoid of O_2. Thus, the life of aquatic animals and plants is adversely affected.

(*b*) Phyto-plankton (plankton is a layer of tiny animals and plants that live on the surface layer of the sea) on the surface of water absorbs dirt and acts as pollution filter. But the oil film on water surface checks the growth of plankton. This knocking out of plankton by oil film disrupts the entire cycle of marine life.

(*c*) The oil film on water may catch fire and cause harm to the aquatic animals.

(*d*) The life of the people living on sea-shores is also affected due to water pollution caused by petroleum oil.

(*e*) In petroleum drilling, when crude oil is pumped into the surface, NaCl is discharged along with it. NaCl makes the water polluted, since it affects aquatic life.

(*f*) Oil pollution also affects sea-birds. The oil penetrates the feathers and the air which is normally entrapped in the feathers is eliminated. The elimination of entrapped air makes the brids feel colder and more susceptible to diseases. Due to the elimination of entrapped air, the birds feel difficulty in flying and swimming. Thus we find that the pollution of sea-water by oil also affects the flight and swimming capability of sea-birds.

Methods to control water pollution

Water pollution can be prevented or minimised by the following methods :

(*i*) The use of excess of nitrate and phosphate fertilisers should be avoided.

(*ii*) The use of synthetic detergents should be minimised or bio-degradeable detergents (bio-degradeable detergents are those which can be decomposed by the bacteria) should be used. Alternatively, a combination of washing soda and soap should be used for washing the clothes.

(*iii*) Before throwing the industrial wastes into the rivers and lakes, they should first be treated chemically to neutralise the harmful substances present in them.

(*iv*) Before dumping the sewage into rivers, it should first be treated chemically to remove the organic compounds present in it.

(*v*) The radioactive waste products obtained in reactors used in hospitals and scientific laboratories should be enclosed in containers made of concrete and then dumped into sea.

Soil Pollution and Soil Pollutants

The contamination of soil with rain, excess of fertilisers, wrong fertilisers, insecticides (insecticides are the chemical compounds which can kill the insects) and herbicides (herbicides are those chemical compounds which can kill plants) is called **soil pollution** and the substances that make soil polluted are called **soil pollutants.**

Sources of soil pollution

The main sources which pollute the soil are discussed below :

1. Nitric acid and sulphuric acid rains. We know that air contains SO_2 and NO_2. SO_2 present in air undergoes photolytic and catalytic oxidation to form SO_3 which reacts with rainy water or moisture to form H_2SO_4 ($SO_3 + H_2O \rightarrow H_2SO_4$). NO_2 reacts with rain water or moisture in presence of O_2 or O_3 and produces HNO_3.

$$4NO_2 + H_2O + O_2 \longrightarrow 4HNO_3$$

$$2NO_2 + H_2O + O_3 \longrightarrow 2HNO_3 + O_2$$

H_2SO_4 and HNO_3 formed as above come down the atmosphere as acid rain. This acid rain damages the standing crops.

2. Repeated use or excess use of the same fertiliser. Repeated use or excess use of the same fertiliser pollutes the soil. For example when $(NH_4)_2SO_4$ is used as a fertiliser into the soil again and again, NH_4^+ ions are used up by the successive corps but SO_4^{2-} ions get accumulated into the soil. Being acidic, SO_4^{2-} ions make the soil highly acidic and hence unfit for plant growth.

If $NaNO_3$ or KNO_3 is used again the again, NO_3^- ions are used up by the successive crops, but Na^+ or K^+ ions get accumulated into the soil. These cations make the soil alkaline and hence the soil cannot be used for the crops-growth. The above discussion makes it evident that the repeated use or excess use of the same fertiliser pollutes the soil.

3. Inadequate drainage system in agricultural fields. We use a lot of water for irrigation, along with fertilisers, but there should be good drainge system for the outlet of the unused water, otherwise the soil will become highly saline (salt containing) and unfit for growing crops. Thus we see that inadequate drainage system in fields is another source of soil pollution.

4. Spraying the vegetable and fruit plants with insecticides and herbicicides. When the standing vegetable and fruit plants are sprayed with insecticides and herbicides, to save them from the harmful insects and herbs etc., the insecticides/herbicides enter the living tissuesof the growing plants and accumulte in them. When we eat these grains, fruits or vegetables, the insecticides in them may damage our heart. It is, therefore, advised that before eating them, they should be properly washed with sufficient quantity of water. Since the chemicals like $CaCl_2$ are used for repening fruits like banana and mangoes, these fruits should be washed with water before they are eaten up by us.

SOME TYPICAL PROBLEMS WITH SOLUTIONS

Problem 1. *Define the terms: (i) Receptor (ii) Sink (iii) Biological oxygen demand (BOD)*

(MD Rohtak 1995)

Solution. Receptor. It is an *element or organism affected by a pollutant*. For example, human being is the receptor of smog (that causes irritation in eyes and breathing problems) and polluted water (that causes cholera and gastroentritis).

Sink. *The element or surroundings that interact* and *consume* a long lived *pollutant* is known as a *sink*. A marble deposit would act as a sink for atmospheric acid and in the process gets damaged.

$$CaCO_3 + H_2SO_4 \longrightarrow CaSO_4 + H_2O + CO_2$$

Oceans are sinks for atmospheric carbon dioxide. Ground water and subsoil water act as sinks for pesticides used in agriculture.

Biological oxygen demand (BOD). BOD is the quantity of O_2 (dissolved in water) used up by microorganism (aerobic bacteria) for converting organic waste present in a sample of H_2O into nutrient. Greater is the quantity of organic matter in water, greater is the amount of O_2 required to break it (*i.e.,* organic matter) biologically and hence higher is the value of BOD. BOD value is a measure in evaluating the degree of pollution in water. A high value of BOD indicates that water is polluted.

Problem 2. *What is the difference between pollutant and contaminant ?*

(MD Rohtak 1995; Kurukshetra 1995)

Solution. Pollutant: If the *concentration of a substance already present in nature is increased to undesirable proportions* by human action, so that, it *spoils the environment* and reduces our enjoyment of life or affects health and property, then it is considered as a *pollutant*. Carbon monoxide, sulphur dioxide, oxides of nitrogen, dust, lead, mercury, cadmium, arsenic, asbestos, etc., are pollutants.

Contaminant : *A substance that does not occur in nature but is introduced in significant amounts* into the environment by human action is known as a *contaminant. A contaminant, if it reduces our enjoments of life* or affects our health and property, is classified as a *pollutant*.

For example, pyrosulphuric acid, $H_2S_2O_7$ leaked from a defective tanker and spread over some parts of Delhi. Some persons were killed and others suffered with skin and breathing problems. Since $H_2S_2O_7$ does not occur in nature, therefore, it may be classified as a contaminant. However, it had detrimental effects on health, therefore, it is a pollutant as well.

Methylisocyanate (MIC), CH_3NCO does not occur in nature. In December 1984, three tanks of MIC (15 tonnes each) of Union Carbide exploded in Bhopal killing over 3000 people. More than a thousand turned blind and several thousand will continue to suffer major and minor ailments. Genetic damages may also come into light. Soil in a radius of 150 km around Bhopal may remain barren for several years. This man-made contaminant, because of its dangerous effects, is a pollutant.

Problem 3. *Trace the sources and sinks of CO in the atmosphere. How does CO act as pollutant, if inhalled in excess?* *(MD Rohtak 1993, Punjab 1996)*

OR

What are the different sources of CO pollution? What are the effects of continuous exposure of CO on human beings? *(Kurukshetra 1995; Punjab 1996)*

OR

Discuss air pollution caused by CO. *(Andhra 1997; Delhi 1995, M.D. Rohtak 1998)*

Solution. Sources of CO. CO is produced into the atmosphere:

(*i*) By the incomplete combustion of carbon containing fuels, carbon compounds and coal and by the burning of petrol in the internal combustion of automobiles

$$2C + O_2 \longrightarrow 2CO\ (g)$$

(*ii*) By the reaction of CO_2 with coke at high temperature in industrial furnances (*e.g.*, blast furnace)

$$CO_2 + C\ (\text{Coke}) \longrightarrow 2CO$$

(*iii*) By natural processes like volcanic eruptions, forest fires, natural gas emission, seed germination, electrical discharge during storm etc.

(*iv*) Incomplete combustion of agricultural and slush matter produces CO.

Sinks of CO. Some micro-organisms present in the soil remove CO from the air and thus act as major sink for CO. Of these organisms, fungi appear to be quite active in the removal of CO.

Polluting effects of CO. (*i*) O_2 gas which we breath in combines loosely and reversibly with haemoglobin (Hb) present in our blood and forms *oxy-haemoglobin*, HbO_2 in the lungs.

$$Hb + O_2 \rightleftharpoons HbO_2 + H_2O$$

HbO_2 formed as above travels to all parts of the body through arteries. As the blood containing HbO_2 runs through the arteries to the tissues, the oxygen pressure decreases and O_2 present in HbO_2 is set free. This free O_2 is delivered to various tissues of the body which use this O_2 for the combustion of glucose (food).

Now if we breath in air containing CO, O_2 present in HbO_2 is replaced by CO and carboxy-haemoglobin (HbCO) is formed. The formation of HbCO takes place, since Hb has greater affinity for CO than for O_2.

$$HbO_2 + CO \longrightarrow HbCO + O_2$$

HbCO is a stable compound and hence the amount of Hb required to form HbO_2 is reduced. Consequently, due to the formation of HbCO, the quantity of O_2 to be delivered to various tissues of the body also gets reduced. The deficiency of oxygen thus produced is called anoxia (oxygen starvation). Thus if we are exposed to CO, the deficiency of O_2 which is caused due to the formation of HbCO, produces *headache, diziness, choking, cordiac and pulmonary complications, paralysis and even death.*

It has been found that people exposed to low concentration of CO have difficulty in distinguishing varying degrees of brightness and in estimating time intervals accurately.

Problem 4. *Discuss the measures for the control of CO pollution.*

(Kurukshetra 1995, Punjab 1996, M.D. Rohtak 1998)

Solution. We know that CO is produced in the atmosphere due to the incomplete combustion of petrol in the internal combustion engine of automobiles. Thus the automobile exhaust contains CO which makes the air polluted.

In order to minimise air pollution caused by CO, the following methods are used.

1. Addition of anti-knocking agent to petrol. Lead tetraethyl, $(C_2H_5)_4$ Pb is mixed with petrol. $(C_2H_5)_4$ Pb acts as anti-knocking agent. The mixture, [petrol + $(C_2H_5)_2$ Pb] is called leaded-petrol. Anti-knocking agent of $(C_2H_5)_4$ Pb added to petrol provides more time to the petrol for its combustion and enables it (*i.e.,* petrol) to burn at slow speed or more completely. Due to more complete combustion of petrol, the quantity of CO and unburnt hydrocarbons (*e.g.,* $C_2H_4Cl_2$, $C_2H_4Br_2$) present in the gases released by the automobiles is decreased and hence pollution of air caused by CO is minimised.

The disadvantage of using $(C_2H_5)_4$Pb as anti-knocking agent is that when the mixture, [petrol + $(C_2H_5)_4$ Pb] is ignited in the combustion engine, $(C_2H_5)_4$Pb reacts with dihalogen alkanes *viz* C_2H_4 Cl_2 and $C_2H_4Br_2$ in presence of O_2 and $PbCl_2$ and $PbBr_2$ are produced.

$$2Pb\,(C_2H_5)_4 + 32O_2 + \underbrace{C_2H_4Cl_2 + C_2H_4Br_2}_{\substack{\text{Dihalogen alkenes} \\ \text{(Hydrocarbons)}}} \longrightarrow 20\,CO_2 + 24H_2O + \underbrace{PbCl_2 + PbBr_2}_{\text{Volatile}}$$

$PbCl_2$ and $PbBr_2$ formed as above are volatile and hence go out of engine with the exhaust gases and get mixed with air. $PbCl_2$ and $PbBr_2$ are dangerous compounds and hence make the air polluted. Thus we see that although the addition of $(C_2H_5)_4$Pb to petrol minimises the air pollution caused by CO and unburnt hydrocarbons, the air gets polluted with dangerous $PbCl_2$ and $PbBr_2$.

2. Use of catalytic convertor: Catalytic oxidation of C_8H_{18} (unburn hydrocarbon) and CO to CO_2 with platinum catalyst. Catalytic converter is a device which is attached with the verticle's engine. In catalytic converter octane, C_8H_{18} (unburnt hydrocarbon) and CO get mixed with more quantity of O_2 of the air and get burnt completely (*i.e.,* are completely oxidised to CO_2) in presence of Pt which acts as a catalyst.

$$\underset{\text{Unburnt hydrocarbon}}{2C_8H_{18}} + 25O_2\,\text{(From air)} \xrightarrow[\text{(Pt acts as a catalyst)}]{\text{Pt-metal}} 16CO_2 + 18H_2O$$

$$2CO + O_2\,\text{(From air)} \xrightarrow{\text{Pt-metal}} 2CO_2$$

The disadvantage of using Pt as a catalyst is that its surface is coated with $PbBr_2$ which is formed when anti-knocking agent of $(C_2H_5)_4$Pb is mixed with petrol.

Problem 5. *A gas is colourless, tasteless and odourless but highly toxic. What are its harmful effects ?*

(M.D. Rohtak, 1992)

OR

Discuss the sources and sinks of CO_2 in the atmosphere

OR

CO_2 is inert and harmless gas, yet it is considered to be a serious pollutant. Explain

(M.D. Rohtak, 1991)

OR

Discuss the role of CO_2 as a pollutant. *(G.N. Dev, 1996)*

Solution. Sources of CO_2. CO_2 is produced by any of the following sources:

(*i*) By the burning of fossil fuels, carbon compounds and carbonates

$$C + O_2 \longrightarrow CO_2\ , \qquad CaCO_3 \xrightarrow{\Delta} CO_2 + CaO$$

(*ii*) By natural processes like decay of dead organisms and release from oceans.

(*iii*) By the cultivation of soil, CO_2 is produced by bacteria.

(*iv*) By auto-exhaust (Vehicular exhaust)

(*v*) By industrial processes

(*vi*) By the respiration of living organisms.

Sinks of CO_2 Green plants and oceans are the major sinks of CO_2 as shown below :

(*i*) Green plants absorb CO_2 from the atmosphere to prepare their food (carbohydrates) through the process of photosynthesis.

$$6CO_2 + 6H_2O \longrightarrow C_6H_{12}O_6 \text{ (carbohydrate)} + 6O_2$$

(*ii*) Oceans dissolve CO_2 to form carbonate rocks.

Polluting effects of CO_2. CO_2 is a colourless, testeless and odourless gas. The aqueous solution of CO_2 has weakly acidic character.

(*i*) Consumption of O_2 in combustion has almost no effect on large oxygen level in the atmosphere but it increases the concentration of CO_2 in the atmosphere. The increase in the concentration of CO_2 in the atmosphere increases the earth's temperature. The increase in earth's temperature results in the following:

(*a*) Melting of ice bergs and glaciers will take place. This will lift up the sea level.

(*b*) Biological production of oceans will decrease.

(*c*) Flooding of coastal plains will take place.

(*ii*) CO_2 produced by the above sources enters the atmosphere and dissolves in rain water or moisture present in atmosphere to produce H_2CO_3.

$$CO_2 + H_2O \rightleftharpoons H_2CO_3 \rightleftharpoons H^+ + HCO_3^-$$

Rain water normally is not acidic but weak H_2CO_3 can give a pH value of about 5.6

Control of pollution caused by CO_2 (*i*) The release of CO_2 into the atmosphere from the industrial processes and vehicular exhaust should be reduced.

(*ii*) The best way of controlling the excess quantity of CO_2 is that a balanced quantity of CO_2 should be maintained in the atmosphere. We know that green plants absorb CO_2 gas from the atmosphere to manufacture their food and also at the same time release O_2 gas into the atmosphere (in the manufacture of their food the plants convert CO_2 into O_2). Thus if we grow more plants, our atmosphere will get rid of its excess CO_2 gas. We should, therefore, not cut down the trees.

Problem 6. *What is green house effect?* (*G.N. Dev 1994, M.D. Rohtak, 1998*)

OR

What is the effect of increasing concentration of CO_2 in the atmosphere?

(*Delhi 1993, G.N. Dev 1995*)

Solution. When infra-red radiations coming from the sun fall on the top of atmosphere, these radiations pass through the gases like CO_2, water vapour, CH_4, man-made chlorofluorocarbons (CFC) present in the lower part of the atmosphere (these gases are called green house gases) and reach the eart's surface. The infra-red radiations of long wavelength are reflected (radiated) back from the earth's surface into the green house gases. Since infra-red radiations reflected back into the green house gases cannot pass through these gases, these radiations are trapped into the gases. Now since these radiations have heating effect, the earth gets heated and hence the temperature of the earth is raised. *The heating up of earth due to the trapping of infra-red radiations into the green house gases is called green house effect.*

The rise in temperature produced on earth depends on the concentration of green house gases (particularly CO_2 and water vapour) present in the atmosphere. If the concentration of green house gases is very small or zero, the average temperature on the earth would be at sub zero levels and if the concentration of green house gases is large, global temperature will increase and the existence of life on earth would be impossible. The increase in temperature on earth produced due to large concentration of green house gases in the atmosphere would result in the following:

(i) The evaporation of surface waters will increase and this increase will influence the climatic changes.
(ii) Word production will decrease
(iii) Crop production zones will be shifted
(iv) Fertile soils will change to poor soils
(v) Biological productivity of oceans will decrease
(vi) Polar ice caps will melt
(vii) In mid-continental areas, the winters will be shorter, warmer and wetter, and the summers will be longer, hotter and drier
(viii) Melting of icebergs and glaciers will take place. This may lift up the sea level.
(ix) Flooding of coastal plains will occur.

Problem 7. *Name the oxides of nitrogen present in the atmosphere. What are the sources and sinks of these oxides? What is their polluting action on the plants and aquatic life?*

(*Kurukshetra 1997*)

Solution. Oxides of nitrogen present in the atmosphere are nitrous oxide (N_2O), nitric oxide (NO) and nitrogen dioxide (NO_2).

Sources of oxides of nitrogen (*i*) The smoke released by the automobile engines (engines of cars, buses, trucks, etc.,) contains NO and NO_2. This smoke results by the combustion of petrol or diesel in the engines.

(*ii*) When coal, oil and natural gas undergo combustion, high temperature produced by the combustion makes N_2 and O_2 present in the atmosphere combine together to form NO and NO_2.

$$N_2 \xrightleftharpoons{+O_2,\ >100\ °C} 2NO \xrightleftharpoons{+O_2} 2NO_2$$

(*iii*) Many chemical plants, like those which manufacture explosives and nitrogenous fertilizers, also produce NO_2. It is also present in tobacco smoke.

(*iv*) Large amounts of SO_2 and nitrogen oxides are also emitted from power plants and many industrial processes. The mixture containing SO_2 and nitrogen oxides, thus obtained, is called *flue gases mixture*.

Sinks of oxides of nitrogen. NO and NO_2 present in the atmosphere are converted into HNO_3 as shown below:

(*a*) $2NO + O_2 \longrightarrow 2NO_2] \times 2$

$4NO_2 + 2H_2O$ (Rainy water or atmospheric moisture) + O_2 (From atmosphere) $\longrightarrow 4HNO_3$

$4NO + 2H_2O$ (Rainy water or atmospheric moisture) + $3O_2$ (From atmosphere) $\longrightarrow \boxed{4HNO_3}$

(*b*) $4NO_2 + 2H_2O + O_2 \longrightarrow \boxed{4HNO_3}$

(*c*) $2NO_2 + H_2O + O_3 \longrightarrow \boxed{2HNO_3} + O_2$

(*d*) $NO_2 \xrightarrow[\text{(Photochemical dissociation)}]{\text{Sun light}} NO + O$

$$O + NO_2 \longrightarrow NO_3$$

$$NO_3 + NO_2 \longrightarrow N_2O_5$$

$$N_2O_5 + H_2O \longrightarrow \boxed{2HNO_3}$$

HNO_3 produced by any of the above sources comes down the atmosphere in the form of *nitric acid rain or acid rain or acid snow*. This acid rain reacts with bases like NH_3, lime etc and gets precipitated as nitrate salt.

Polluting effects of oxides of nitrogen. Though NO and NO_2 both are toxic, the latter (*i.e.*, NO_2) is by far more harmful. Some of the harmful effects of NO_2 on human beings, animals and vegetables are given below:

(*i*) *Formation of nitric acid rain (HNO_3) and its harmful effects.* NO and NO_2 present in the atmosphere are converted into HNO_3 as shown above. This nitric acid dissolves in rain water and falls on earth in the form of *nitric acid rain or acid snow or acid rain.* When this nitric acid rain falls over a historical monument, it gradually eats up the marble stone ($CaCO_3$) of the monument and corrodes it slowly.

$$CaCO_3 + 2HNO_3 \longrightarrow Ca(NO_3)_2 + CO_2 + H_2O$$

The acid rain fades the colour of the fabrics (e.g., cotton, nylon and rayon), leather and paper. The acid rain makes the water of lakes acidic and hence fish-life comes in danger in this water. The yield of agricultural crops is also reduced.

(*ii*) NO_2 causes extensive leaf-drop in the plants.

(*iii*) Nitrogen oxides react with O_2 to produce another pollutant namely ozone.

(*iv*) It has been shown that if monkeys inhale NO_2 of 15-30 ppm concentration for 2 hours, their lungs, heart, liver and kidneys are damaged.

(*v*) NO_2 is very corrosive and attacks skin.

(*vi*) NO_2 helps in the formation of smog (which is a combination of smoke and fog). Smog causes irritation in eyes due to the presence of O_3 in it.

(*vii*) Oxides of nitrogen bring about corrosion of teeth and cause loss of appetite.

(*viii*) We know that the atmosphere contains a layer of O_3 which absorbs harmful ultra-violet radiations coming from the sun and reaching the earth. If these radiations reach the earth, they will cause skin cancer and will destroy the organic molecules necessary for life. O_3 layer prevents the ultra-violet radiations to come to the earth. NO and NO_2 present in the atmosphere decompose O_3 (present in O_3 layer) to O_2 and are generated again.

$$NO + O_3 \longrightarrow NO_2 + O_2$$
$$NO_2 + O_3 \longrightarrow NO + 2O_2$$

Thus we see that NO and NO_2 both present in the atmosphere destroy O_3 layer. Greater is the amount of NO and NO_2 in the atmosphere, greater is the percentage of O_3 which is destroyed, and greater is the amount of ultraviolet radiations reaching the earth.

Problem 8. *Discuss the sources, sinks and polluting effects of SO_2.* (*Guwahati 1995*)

Solution. Sources of SO_2 (A) *Production of SO_2 from man-made sources.* SO_2 is produced from man-made sources like combustion of sulphur containing fuels like coal and oils and from industrial processes. 1/3rd SO_2 comes from made-made sources.

(*i*) Coal and oil both contain sulphur as impurity. When coal and oil undergo combustion (*i.e.*, when coal and oil are burnt), sulphur present in them is oxidised to SO_2 which makes the air polluted.

$$S + O_2 \longrightarrow SO_2$$

(*ii*) Significant amounts of SO_2 are produced when metallic sulphide ores like pyrites (FeS_2), copper glance (Cu_2S), galena (PbS) and zinc blende (ZnS) are roasted in air.

$$2FeS_2 + 5O_2 \longrightarrow 2FeO + 4SO_2$$
$$3Cu_2S + 3O_2 \longrightarrow 6Cu + 3SO_2$$
$$2PbS + 3O_2 \longrightarrow 2PbO + 2SO_2$$
$$2ZnS + 3O_2 \longrightarrow 2ZnO + 2SO_2$$

(*iii*) Power plants and industrial units produce flue-gases which contain a mixture of SO_2 and NO_2.

(B) *Production of SO_2 from natural sources.* SO_2 is also produced from natural processes like volcanic eruptions. These processes produce almost 2/3rd of total SO_2 content in air.

Sinks for SO_2. (*i*) SO_2 can combine with water droplets to form H_2SO_3. Here SO_2 is oxidised to H_2SO_3.

$$SO_2 + H_2O \longrightarrow H_2SO_3$$

(*ii*) SO_2 is also oxidised to H_2SO_4 by atmospheric O_3.

$$SO_2 + O_3 \text{ (From air)} \longrightarrow SO_3 + O_2$$

$$SO_3 + H_2O \text{ (Vapour)} \longrightarrow \text{Droplets of } H_2SO_4$$

Droplets of H_2SO_4 formed as above come down in the form of acid rain and react with NH_3 emitted from biological processes and with marble structure or limestone mineral, $CaCO_3$.

$$H_2SO_4 + 2NH_3 \longrightarrow (NH_4)_2SO_4$$

$$H_2SO_4 + CaCO_3 \longrightarrow CaSO_4 + H_2O + CO_2$$

Above discussion shows that marble structures, limestone mineral and biological processes producing NH_3 act as sinks for atmospheric SO_3.

Polluting effects of SO_2 (*i*) Due to the presence of SO_2 in the atmosphere, H_2SO_4 is produced. The production of H_2SO_4 has been explained in different ways given below :

(*a*) $2SO_2$ (*Present in atmosphere*) + O_2 (*Present in atmosphere*) $\longrightarrow 2SO_3$

$$SO_3 + H_2O \longrightarrow H_2SO_4$$

(*b*) SO_2 (*Present in atmosphere*) + O_3 (*From air*) $\longrightarrow SO_3 + O_2$

$$SO_3 + H_2O \text{ (Vapour)} \longrightarrow \text{Droplets of } H_2SO_4$$

(*c*) $$SO_2 \xrightarrow[\text{(Photochemical dissociation)}]{\text{Sunlight}} SO + O$$

$$O + SO_2 \longrightarrow SO_3$$

$$SO_3 + H_2O \longrightarrow H_2SO_4$$

H_2SO_4 formed as above comes down the atmosphere in the form of *sulphuric acid rain* or *acid rain* or *acid snow*. This acid rain is very dangerous, since when it falls over a historical monument, it gradually eats up the marble structure ($CaCO_3$) of the monoment and corrodes it slowly.

$$CaCO_3 + H_2SO_4 \text{ (Acid rain)} \longrightarrow CaSO_4 + H_2O + CO_2$$

(*ii*) SO_2 irritates the respiratory system of animals and humans. It damages lungs.

(*iii*) If plants are exposed to SO_2 of high concentration over short periods, brownish colour in the tips of pine needles gets decolourised. If the plants are exposed to SO_2 of low concentration over longer periods, the production of chlorophyl in plants in hampered (*chlorosis*) and hence green colour of the plants in lost.

(*iv*) Due to corrosive nature of SO_2 and H_2SO_4, these pollutants decolourise building materials like limestone, marble, roof slate and mortar. The corrosion of most metals (*e.g.*, Fe, steels, Zn, Al etc.) is accelerated by SO_2.

(*v*) Fabrics, leather, paper and paints undergo fading of their colour in presence of SO_2.

Problem 9. *Discuss measures for the control of SO_2 pollution.* (*Punjab 1993*)

Solution. The pollution caused by SO_2 can be prevented by the following methods.

(*i*) *Removal of sulphur from coal before its combustion.* We have already said that coal contains sulphur as impurity and hence when coal is allowed to undergo combustion, sulphur present in coal is oxidised to SO_2 which pollutes air. In order to remove sulphur from coal, desulphurisation of coal is done. Desulphurisation of coal (removal of sulphur from coal) is done by the processes like carbonisation, liquafaction or gasification.

(*ii*) *Use of natural gas in place of coal.* In place of coal, natural gas can be used for heating purposes. Natural gas does not produce SO_2 on heating.

(*iii*) *Removal of SO_2 from flue gases mixture.* We have already said that power plants and industrial units produce flue gases which contain SO_2 and NO_2. SO_2 can be removed by treating the flue gases mixture with a slurry of $CaCO_3$ (lime stone) when SO_2 is converted into $CaSO_3$ which is removed and disposed off.

$$CaCO_3 \text{ (Limestone)} \longrightarrow CaO + CO_2$$

$$CaO + SO_2 \longrightarrow CaSO_3$$

Problem 10. *Write a note on "Acid rain".* *(Guwahati 1996, Delhi 1996, Punjab 1996)*

Solution. Acid rain is H_2CO_3 or HNO_3 or H_2SO_4 which is produced in the atmosphere as shown below :

Production of H_2CO_3 acid rain. We know that CO_2 is produced in the atmosphere by various processes. This CO_2 dissolves in rain water or moisture present in the atmosphere to produce carbonic acid (H_2CO_3) in the form of carbonic acid rain.

$$CO_2 + H_2O \rightleftharpoons H_2CO_3 \text{ (Acid rain)} \rightleftharpoons H^+ + HCO_3^-$$

Rain water normally is not acidic but due to the production of H_2CO_3, pH value of water becomes about 5.6.

Production of HNO_3 acid rain. NO and NO_2 are also produced in the atmosphere by various sources. These oxides are converted into HNO_3 as shown below:

(*a*) $4NO + 3O_2 + 2H_2O \longrightarrow 4HNO_3$

(*b*) $4NO_2 + O_2 + 2H_2O \longrightarrow 4HNO_3$

(*c*) $2NO_2 + O_3 + H_2O \longrightarrow 4HNO_3 + O_2$

(*d*) $NO_2 \xrightarrow[\text{dissociation}]{\text{Photochemical}} NO + O$

$$O + NO_2 \longrightarrow NO_3$$

$$NO_3 + NO_2 \longrightarrow N_2O_5$$

$$N_2O_5 + H_2O \longrightarrow 2HNO_3$$

HNO_3 produced as above dissolves in rain water and falls on earth in the form of nitric acid rain or acid snow or acid rain. When this acid rain falls over a historical monument, it gradually eats up the marble stone ($CaCO_3$) of the monument and corrodes it slowly.

$$CaCO_3 + 2HNO_3 \text{ (Acid rain)} \longrightarrow Ca(NO_3)_2 + CO_2 + H_2O$$

The acid rain fades the colour of the fabrics (*i.e.,* cotton, nylon and rayon), leather and paper. The acid rain makes the water of lakes acidic and hence fish-life comes in danger in this water. The yield of agriculture crops is also reduced.

Production of H_2SO_4 acid rain. SO_2 is also produced in the atmosphere by man-made processes and natural sources. This oxide is converted into H_2SO_4 as shown below :

(*a*) $2SO_2 + O_2 \longrightarrow 2SO_3$

$SO_3 + H_2O \longrightarrow H_2SO_4$

(*b*) $SO_2 + O_3 \longrightarrow SO_3 + O_2$

$SO_3 + H_2O \text{ (Vapour)} \longrightarrow \text{Droplets of } H_2SO_4$

(*c*) $SO_2 \xrightarrow[\text{dissociation}]{\text{Photochemical}} SO + O$

$$O + SO_2 \longrightarrow SO_3$$

$$SO_3 + H_2O \longrightarrow H_2SO_4$$

H_2SO_4 formed as above comes down the atmosphere in the form of *sulphuric acid rain* or *acid rain* or *acid snow*. This acid rain is dangerous, since when it falls over a historical monument, it gradually eats up the marble structure ($CaCO_3$) of the monument and corrodes it slowly.

$$CaCO_3 + H_2SO_4 \text{ (Acid rain)} \longrightarrow CaSO_4 + H_2O + CO_2$$

Problem 11. *Explain the following, giving appropriate reasons for your answer:*

(i) Chlorination is not a desirable method of disinfecting polluted water. *(Delhi 2000)*

(ii) Lead pollution is caused by car exhaust.

Solution. (*i*) Chlorination is not a desirable method of disinfecting polluted water because Cl_2 oxidises organic matter and kills disease germs and sterilises drinking water. In polluted water no such things are present and is not suitable for this purpose.

(*ii*) Lead tetraethyl, $(C_2H_5)_4Pb$ is mixed with petrol. $(C_2H_5)_4Pb$ acts as an anti-knocking agent. The mixture, [petrol + $(C_2H_5)_4Pb$] is called leaded petrol. When this mixture is ignited in the combustion engine of the car, $(C_2H_5)_4Pb$ reacts with dihalogen alkanes *viz* $C_2H_4Cl_2$ and $C_2H_4Br_2$ (hydrocarbons) in presence of O_2, and $PbCl_2$ and $PbBr_2$ are produced.

$$2Pb(C_2H_5)_4 + 32O_2 + \underbrace{C_2H_4Cl_2 + C_2H_4Br_2}_{\text{Dihalogen alkanes}} \longrightarrow 20CO_2 + 24H_2O + \underbrace{PbCl_2 + PbBr_2}_{\text{Volatile}}$$

$PbCl_2$ and $PbBr_2$ formed as above are volatile and hence go out of the engine with the exhaust gases and get mixed with air. $PbCl_2$ and $PbBr_2$ are dangerous compounds and hence make the air polluted.

Problem 12. *Describe water pollution due to lead. How can lead poisoning be cured?* *(G.N. Dev 1996)*

Solution. Sources of lead pollution in water. (*i*) Rain water washes lot of lead compounds like $PbBr_2$, $PbCl_2$ etc. into rivers, lakes, ponds and oceans. $PbBr_2$ and $PbCl_2$ are produced when leaded petrol, [petrol + $(C_2H_5)_4$ Pb] is ignited into the combustion engine of automobiles.

$(C_2H_5)_4Pb$ present in leaded petrol reacts with $C_2H_4Cl_2$ and $C_2H_4Br_2$ in presence of O_2 and produce $PbCl_2$ and $PbBr_2$.

$$2Pb(C_2H_5)_4 + 32\ O_2 + C_2H_4\ Cl_2 + C_2H_4Br_2 \longrightarrow 20\ CO_2 + 24H_2O + PbCl_2 + PbBr_2$$

$PbCl_2$ and $PbBr_2$ are dangerous compounds. These are washed away by rain water. Thus the presence of these compounds in water makes water polluted.

(*ii*) Lead is also present in water in the form of soluble compounds, colloids and particulates.

(*iii*) Fresh water of pH 6 contains lead in the form of $Pb(OH)_2$ and $PbCO_3$ particles.

(*iv*) Lake and pond waters contain $Pb(CH_3)_4$, $[Pb(CH_3)_2]^{2+}$ and $[Pb(CH_3)_3]^+$

(*v*) The liquid waste products discharged by industries are called industrial wastes. These wastes contain compounds of Pb. When industrial wastes enter into water, they pollute water due to the presence of lead compounds present in *industrial wastes*.

Harmful effects of lead. Lead is a very poisonous metal. It is a comulative poison, since it keeps on accumulating in the tissues of human body and in the leaf tissues of plants. Even a low concentration of 0.2 ppm of lead in human body creates metabolic disturbances. Some of the harmful effects of lead are given below:

(*i*) It is due to the prolonged effect of lead on the human body that red blood cells get stippled. The malformation of red blood cells produces the diseases of anaemia.

(*ii*) Lead gets bound with the cellular enzymes and disrupts the functioning of cells and organs of the mascular circulatory and nervous systems.

(*iii*) Lead damages organs like liver, kidneys and intestine.

(*iv*) Lead develops abnormalities in fertility and pregnancy.

(*v*) Lead causes coagulation of proteins.

(*vi*) When rats are exposed to lead polluted air, they die in two thirds of their average life span.

(*vii*) The birds, which are shot, but recover usually, subsequently die due to lead-poisoning.

(*viii*) The lead-polluted air has its adverse effects on vegetable, *e.g.*, the road-side vegetables are dried up by the exposure of lead-polluted air.

(*ix*) If lead is taken by humans *evenin small amounts*, it affects breathing and nervous system.

(*x*) *Excessive intake of lead* causes disruption of syntehsis of haemoglobin. It also results in loss of appetite, anaemia, kidney malfunctioning, nervous disorders and brain damage.

(*xi*) Lead can also be absorbed by skin, resulting into many skin diseases.

Cure of lead poisoning. Lead poisoning can be cured by feeding the patient with an aqueous solution of Ca^{2+} – complex of EDTA (Ca – EDTA).Pb^{2+} ions displace Ca^{2+} ions present in Ca–EDTA complex and form chelated complex, Pb – EDTA. Since this chelated complex of Pb^{2+} is soluble, it is extreted through urine.

$$\text{Ca – EDTA} + Pb^{2+} \longrightarrow \underset{\text{Soluble}}{\text{Pb – EDTA}} + Ca^{2+}$$

Multiple Choice Probolems

Problem 13. *Green house effect is because of the :*

(a) presence of CO_2 only in the atmosphere

(b) presence of O_3 and CH_4 in the atmosphere

(c) presence of N_2O and chlorofluoro hydrocarbons in the atmosphere

(d) presence of gases which, in general, are strong infra-red absorbers in the atmosphere

(CBSE 1996)

Problem 14. *The poisonous gas present in the exhaust fumes of car is:*

(a) Methane (b) Acetylene (c) Carbon monoxide (d) Ethane

Problem 15. *Toxic effects of CO are attributed to:*

(a) its interaction with myglobin (b) inflamation of lung tissues

(c) reduced oxygen transport (d) green house effect

Problem 16. *Fuel rich auto-exhaust contains :*

(a) higher concentration of NO_x (b) higher concentration of SO_2

(c) higher concentration of CO and CO_2 (d) higher concentration of CO and hydrocarbons.

Problem 17. *Lead tetraethyl is added to petrol to:*

(a) decrease its boiling point so that its combustion takes place at lower temperature.

(b) cause pre-ignition of petrol

(c) to prevent hydrocarbon free radicals from building up chain reactions.

(d) to prevent damage to the inner walls of combustion engine.

Problem 18. *Which one of the following substitute car-fuels would be the least polluting?*

(a) Alcohol (b) Alcohol-petrol mixture

(c) Alcohol-diesel mixture (d) Liquified petroleum gas (LPG)

Problem 19. *Increased CO_2 content in the atmosphere will lead to:*

(a) increase in earth's temperature

(b) increase in photosynthesis

(c) increase in O_2 level of the atmosphere

(d) decrease in the rate of surface water evaporation.

Problem 20. *Which one of the following particulates is the most toxic?*

(a) Smog (b) Soot

(c) Inorganic compounds (d) Fly ash

Problem 21. *Of the following particulates which is the most abundent?*

(a) Fly ash (b) Soot

(c) Inorganic compounds (d) Valcanic evporations

Problem 22. *Which of the following statements is false?*

(a) Pb poisoning is cured by aqueous solution of Ca-EDTA complex.

(b) As poisoning is due to its attachement to – SH groups in the enzyme.

(c) HClO acts as a powerful germicide in water.

(d) Hg is converted into CH_3HG^+ by certain acrobic bacteria.

Problem 23. *Marble acts as a sink for:*

(a) acidic pollutants *(b) metallic pollutants*

(c) cobalt pollutant *(d) NH_3 pollutant*

Answers. 13 – (*d*), 14 – (*c*), 15 – (*c*), 16 – (*d*), 17 – (*c*), 18 – (*d*), 19 – (*a*), 20 – (*b*), 21 – (*b*), 22 (*d*), 23 – (*a*).

University Questions

1. What do you understand by the term "Pollution"? Explain in short the types of pollution. *(Meerut 96)*

2. What is the function of O_3 present in the troposphere of the atmosphere?

3. Explain clearly howoxides of nitrogen and fluorocarbons tend to deplete the ozone layer. What is the effect of this depletion. What is meant by green-house effect?

4. What are the important pollutants of air? How are these pollutants estimated? *(Meerut M.Sc. 98)*

5. What are water pollutants? How are these pollutants estimated? Givean account to minimise these pollutants. *(Meerut M.Sc. 98)*

6. Write a note on "Depletion of ozone layer." *(Meerut, M.Sc. 98)*

7. Define water pollution. What are two categories of sources of water pollution? How do these affect aquatic ecosystem? *(Meerut, M.Sc. 98)*

8. Describe briefly any two most important air pollutants, their sources and effect on plant and animals life. *(Meerut M.SC., 98)*

9. Write note on "Ozone depleting substances". *(Meerut M.Sc. 98)*

10. Name two important sources of air pollutants. *(I.S.C. 2000)*

11. Name one gas which causes green house effect. What are the harmful consequences of this effect. *(I.S.C. 2000)*

12. (*a*) Define the following terms :

(*i*) Biological oxygen demand

(*ii*) Green house effect.

(*b*) (*i*) What is the function of ozone present in the stratosphere of the atmosphere ?

(*ii*) How to control the pollution caused by oxides of nitrogen ? *(Delhi 2002).*

Ans. (*a*) (*i*) Biological oxygen demand (BOD) represents the quantity of oxygen required by bacteria and other mirco-organisms during biochemical degradations and transformation of organic matter present in the waste water under aerobic conditions.

(*b*) (*i*) The u.v. radiations emitted by sun are very harmful. Ozone layer present in the stratosphere region of the atmosphere absorbs harmful uv radiations coming from the sun and thus prevent these radiations from reaching the earth.

(*ii*) Pollution caused by oxides of nitrogen can be controlled by reducing these oxides by H_2 in the presence of finely divided platinum. These oxides can also be dissolved in H_2SO_4 to be converted into sulphates

$$SO_2 + NO_2 + H_2O \longrightarrow H_2SO_4 + NO$$

$$NO + NO_2 \longrightarrow N_2O_3$$

$$N_2O_3 + H_2SO_4 \longrightarrow 2H(NO)SO_4 + H_2O$$

13. (*a*) What is acid rain ? How does it affect the environment ?

(*b*) What is meant by green house effect ?

(*c*) What is ozone hole ? Why is it of global concern *(Rohilkhand 2003).*

14. Write a note on primary pollutants and their sources. *(Gujrat 2003).*

15. What is smog ? Outline the factors which contribute to its happening. *(Delhi 2003).*

16. (*a*) What preventive measures you will take for effective control of air pollution?

(*b*) Discuss inorganic pollution caused in water. *(GND Amritsar, 2004)*

17. (*a*) What preventive measures you will take for effective control of air pollution?

(*b*) Discuss inorganic pollution caused in water. *(GND Amritsar, 2004)*

16

Organo-metallic Compounds

Introduction

The term organometallic or metallo-organic is used nowadays for compounds containing metal-to-carbon bond. Organo-metallic compounds thus include various hydrocarbon derivatives, carbonyls, certain chelate complexes, carbides and related compounds, but do not include carbonates of the metals, complexes of the metals with organic amines and metallic salts of organic acids.

General Methods of Preparation

Follwing three methods have been used :

I. Substitution methods

These methods can be any of the following types :

1. A type reactions : RX + M → RM + X. These reactions involve those compounds containing an acidic hydrogen or a metal less electropositive than the substituting metal. Some examples of the organo-metallic syntheses involving A type reactions with compounds containing an acidic hydrogen are:

$$RC{\equiv}CH + Na \longrightarrow RC{\equiv}C^{-}Na^{+} + \frac{1}{2}H_2$$

$$C_5H_6 + K \longrightarrow C_5H_5^{-}K^{+} + \frac{1}{2}H_2$$

$$Ph_3CH + Na \xrightarrow{NH_3\,(l)} Ph_3C^{-}Na^{+} + \frac{1}{2}H_2$$

The preparative value of A type reactions in which X is less electropositive than M is quite limited, but it is probably the best method of preparing alkyls of the more reactive metals such as alkali and alkaline earth metals and zinc.

$$2Na + (CH_3)_2Hg \longrightarrow 2CH_3Na + Hg$$

$$Mg + (CH_3)_2Hg \longrightarrow (CH_3)_2Mg + Hg$$

$$Zn + (CH_3)_2Hg \longrightarrow (CH_3)_2Zn + Hg$$

2. B type reactions : RY + 2M → RM + MY. These reactions can be illustrated by the following examples :

$$2Mg + 2RBr \xrightarrow{ether} 2RMgBr \longrightarrow MgBr_2 + R_2Mg$$

$$2Li + EtI \longrightarrow EtLi + LiI$$

$$2Na + C_6H_5Cl \xrightarrow{Pantane} C_6H_5Na + NaCl$$

$$2Zn + 2EtI \longrightarrow 2EtZnI \longrightarrow Et_2Zn + ZnI_2$$

$$Hg + 2MeBr + 2Na \longrightarrow Me_2Hg + 2NaBr$$

3. C type reactions : RY + MZ → RM + YZ. These reactions may be of the following types :

(*i*) *Metallation.* This term is usually restricted to metal-hydrogen exchange reactions. Illustrative examples of the organometallic syntheses carried out by this type of reaction are:

$$C_5H_6 + RONa \longrightarrow C_5H_5^-Na^+ + ROH$$

$$RC{\equiv}CH + NaH \longrightarrow RC{\equiv}CNa + H_2$$

$$RC{\equiv}CH + NaNH_2 \xrightarrow{NH_3\,(l)} RC{\equiv}CNa + NH_3$$

(*ii*) *Metal-halogen exchange reactions.* These reactions are best illustrated by reaction of halogenated phenol with *n*-butyl-lithium (*n*-BuLi).

$$C_6H_4(OH)Br + n\text{-BuLi} \longrightarrow C_4H_{10} + C_6H_4(OLi)Br \xrightarrow{n\text{-BuLi}} C_6H_4(OLi)Li + n\text{-BuBr}$$

Bromo phenol

(*iii*) *Reactions of organo-metallic compounds with metal salts.* They are generally the most useful preparative methods for those organo-metallic compounds which cannot be obtained by the direct reaction of the metal with hydrocarbons or halocarbons. The most convenient and easily available startiing materials for such reactions are the Grignard reagents and organolithium compounds. Thus, the alkyls of cadmium, mercury, beryllium and aluminium are best prepared in this way :

$$2RMgBr + ZnBr_2 \longrightarrow R_2Zn + 2MgBr_2$$

$$2RMgCl + CdCl_2 \longrightarrow R_2Cd + 2MgCl_2$$

$$2RMgCl + HgCl_2 \longrightarrow R_2Hg + 2MgCl_2$$

$$2RMgCl + BeCl_2 \longrightarrow R_2Be + 2MgCl_2$$

$$3RMgCl + AlCl_3 \longrightarrow R_3Al + 3MgCl_2$$

(*iv*) *Reactions of bimetallic compounds with halides.* Many derivatives of group IVB metals and metal carbonyls are best prepared by this method. For example,

$$Ph_3PbLi + PhCH_2Cl \longrightarrow Ph_3PbCH_2Ph$$

$$NaMn(CO)_5 + MeI \longrightarrow MeMn(CO)_5 + NaI$$

4. D type reactions : A + MY → AM + Y. These reactions may be of the following types:

(*i*) *Displacement of one neutral ligand by another.* The extent of displacement depends upon the metal, type of ligand and nature of solvent. Thus, isonitriles displace carbon monoxide from metal carbonyls to different degrees depending on the nature of the metal.

$$Ni(CO)_4 + 4PhNC \longrightarrow Ni(CNPh)_4 + 4CO$$

$$Cr(CO)_6 + 3PhNC \longrightarrow Cr(CO)_3(CNPh)_3 + 3CO$$

In contrast to PhNC, EtNC displaces only three CO groups from $Ni(CO)_4$.

$$Ni(CO)_4 + 3EtNC \longrightarrow Ni(CO)(EtNC)_3 + 3CO$$

In reaction of $BrMn(CO)_5$ with PhNC in boiling THF as solvent, all the five carbonyl groups are displaced; but in ethanol solution only two such groups are displaced.

$$BrMn(CO)_5 + 5PhNC \xrightarrow{THF} BrMn(CNPh)_5 + 5CO$$

$$BrMn(CO)_5 + 2PhNC \xrightarrow{ethanol} BrMn(CO)_3(CNPh)_2 + 2CO$$

(*ii*) *Displacement of neutral ligands by negative ligands.* These reactions are exemplified by

$$K_2[PtCl_4] + 2CO \longrightarrow Pt(CO)_2Cl_2 + 2KCl$$

and $$Mn(CO)_5Br + C_2H_4 + AlBr_3 \longrightarrow [Mn(CO)_5(C_2H_4) + AlBr_4$$

(*iii*) *Displacement of negative ligands by neutral ligands.* Some examples of this type of reaction are:

$$[Fe(CN)_5(H_2O)]^{3-} + CN^- \longrightarrow [Fe(CN)_6]^{4-} + H_2O$$

$$[Fe(o\text{-phen})_3]^{2+} + 2KCN \longrightarrow Fe(o\text{-phen})_2(CN)_2 + 2K^+ + o\text{-phen}$$

(*iv*) *Displacement of one negative ligand by another.* An illustrative example of this type of reaction is

$$K_2[Pt(SCN)_4] + 4KC{\equiv}CR \xrightarrow{NH_3\ (l)} K_2Pt(C{\equiv}CR)_4 + 4KSCN$$

II. Addition methods

These reactions may be of the following types:

1. Reaction of metals with CO, unsaturated hydrocarbons or radicals. These reactions are illustrated by the following examples:

$$Ni + 4CO \longrightarrow Ni(CO)_4\ ; \qquad Fe + 5CO \longrightarrow Fe(CO)_5$$

$$Ph_3C{=}CH_2 + Na \longrightarrow (Ph_2C{-}CH_2)^- Na^+ \longrightarrow [Ph_2C{-}CH_2{-}CH_2{-}CPh_2]^{2-}\ Na_2^{2+}$$

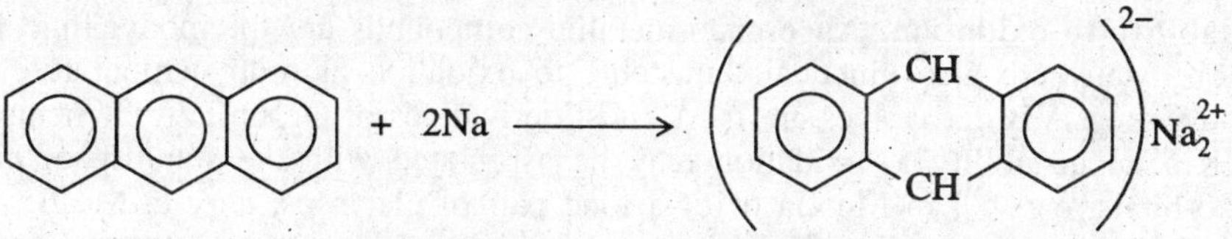

$$+ 2Na \longrightarrow \left(\begin{matrix}CH\\ CH\end{matrix}\right)^{2-} Na_2^{2+}$$

$$Ph_3C + Na \longrightarrow Ph_3C^-Na^+$$

2. Ionisation reactions. Some examples of this type of reactions are as follows :

$$B_2H_6 + 6H_2C{=}CH_2 \longrightarrow 2(C_2H_5)_3B$$

$$Et_2AlH + H_2C{=}CH_2 \longrightarrow Et_2Al$$

$$HSiCl_3 + HC{\equiv}CH \longrightarrow H_2C{=}CHSiCl_3$$

$$H_2C{=}CHSiCl_3 + HSiCl_2 \longrightarrow Cl_2SiCH_2CH_2SiCl_3$$

$$Ph_3SnH + HC{=}CPh \xrightarrow{90°C} Ph_3SnCH{=}CHPh$$

3. Oxidation addition reactions. An interesting oxidation addition reaction is that in which oxidation of the metal is accompanied by an increase in its coordination number. Some examples of these reactions involve five-coordinate (d^7), four-coordinate (d^8) and two-coordinate (d^{10}) complexes.

$$2[Co^{II}(CN)_5]^{3-} + X{-}Y \longrightarrow [Co^{III}(CN)_5X]^{3-} + [Co(CN)_5Y]$$

where X—Y = H, Br_2, HOH, CH_3I, ICN, etc.

$$[Ir^I(CO)(PPh_3)_2Cl] + X{-}Y \longrightarrow [Ir^{III}(X)(Y)(CO)(PPh_3)_2Cl]$$

where X—Y = H_2, Cl_2, HCl, CH_3I etc., and

$$Pt^0(PPh_3)_2 \xrightarrow{X-Y} Pt^{II}(PPh_3)_2(X)(Y) \xrightarrow{X-Y} Pt^{IV}(PPh_3)_3X_2Y_2$$

III. Reduction reactions

More important reduction reactions are:

1. Chemical reduction. For the preparation of a number of metal carbonyls and carbonyl halides, carbon monoxide is the most commonly used reducing agent.

$$NiS + 5CO + 4OH^- \longrightarrow Ni(CO)_4 + CO_3^{2-} + S^{2-} + 2H_2O$$

$$Tc_2O_7 + CO \xrightarrow{275°C} Tc_2(CO)_{10}$$

$$2RuCl_3 + 5CO \xrightarrow{400\ atm.} 2Ru(CO)_2Cl_2 + COCl_2$$

2. Electrolytic reduction. Some illustrative examples of organo-metallic syntheses carried out employing electrolytic reduction are as follows :

$$CH_2CHCN + NaOH \xrightarrow[\text{Sn–cathode}]{\text{electrolysis}} Sn(CH_2CH_2CN)_4$$

$$[(C_2H_5)_3Al]_2.NaF \xrightarrow[\text{Sn–cathode}]{\text{electrolysis}} Pb(C_2H_5)_4 + Al + NaF$$

General Properties

1. Physical properties and covalent character. The primary covalent character of most organo-metallic compounds is reflected in their physical properties. They are typically low-melting solids, liquids or gases at ordinary temperatures and are usually soluble in non-polar or weekly polar organic solvents such as alkanes, toluene, ether or even alcohols. Their chemical properties, (especially their thermal stability) vary widely. For example, Me_4Si is unchanged even after many days at 500°C, whereas Me_4Ti decomposes rapidly at room temperature. Reaction rates of these compounds also vary widely; *e.g.*, dimethyl mercury is not attacked at room temperature by oxygen in the air, whereas dimethyl zinc is spontaneously inflammable.

2. Stability to oxidation. All organo-metallic compounds are thermodynamically unstable to oxidation. Many are also kinetically unstable to oxidation at room temperature. Although Me_2Hg and Group IVB derivatives are inert, most derivatives of Group IVA elements are rapidly oxidised. Kinetic instability to oxidation may be associated with the stability of either empty low-lying orbitals, *e.g.*, 4*p* in Me_3Ga or of a lone pair of electrons, *e.g.*, in Me_3Bi. In contrast, the Group IVB alkyls possess neither of these features and behave as saturated compounds.

3. Stability to hydrolysis. Organo-derivatives of Groups IVB and VB are kinetically stable to attack by water, whereas those of the Group IA and IIA elements and of zinc, cadmium, Al, Ga and In are readily hydrolysed. These hydrolysis reactions usually involve the nucleophilic attack by water and are thus facilitated by the presence of empty low-lying orbitals on the metal atom (*e.g.*, Al) which attract the electron pair of the nuclecophile. In the tetra-alkyls of Group IVB and tri-alkyls of Group VB elements, metal contains filled low-energy orbitals and nucleophilic attack is not favoured. The rate of hydrolysis is dependent on the polarity of the metal-carbon bond and where this is high (*e.g.*,Me_3Al), rapid hydrolysis occurs. Me_3B is unaffected by water at room temperature inspite of the empty 2*p*-orbitals on the boron atom.

4. Reaction with carbon dioxide. Grignard reagents react with excess CO_2 to yield acids.

$$O{=}C{=}O + RMgX \longrightarrow R{-}\overset{\overset{\large O}{\|}}{C}{-}OMgX \xrightarrow{H_2O} R{-}\overset{\overset{\large O}{\|}}{C}{-}OH$$

5. Reaction with ethers. Alkyls of metals with electronegativities less than 0.9 generally react with ethers, producing a hydrocarbon and an alkoxide; the later is converted to an alcohol by hydrolysis.

$$RNa + R'{-}O{-}R'' \longrightarrow R'ONa + R'{-}R''$$

$$R'ONa \xrightarrow{H_2O} R'OH + NaOH$$

6. Reaction with non-metallic elements (other than oxygen) and their inorganic compounds. Halogens decompose all reactive organo-metallic compounds to form hydrocarbon halides.

$$C_2H_5Li + Br_2 \longrightarrow C_2H_5Br + LiBr$$

$$Cl_2 + H_3C{-}\overset{|}{\underset{|}{Si}}{-} \longrightarrow ClH_2C{-}\overset{|}{\underset{|}{Si}}{-} + HCl$$

Organo-metallic Compounds of Alkali Metals

Alkali metals, especially Li, form many organo-metallic compounds.

Preparation

Organo-metallic compounds of Li are generally obtained as follows:

$$Li(n\text{–}C_3H_7) + C_4H_5I \xrightarrow[\text{ether}]{\text{Solvent}} Li(C_6H_5) + n\text{–}C_2H_7I$$

$$4Li(C_6H_5) + Sn(CH_2\text{—}CH{=}CH_2)_4 \xrightarrow[\text{ether}]{\text{Solvent}} 4Li(CH_2\text{—}CH{=}CH_2) + Sn(C_6H_5)_4$$

$$2Li + C_2H_5I \xrightarrow[\text{ether}]{\text{Solvent}} LiI + Li(C_2H_5)$$

Organo-metallic compounds of other alkali metals such as $Na(C_6H_5)$, $Ph_2C(Na)C(Na)Ph_2$, $CsCH_2CH_2Cs$, are also known. These are generally prepared from the corresponding mercury compounds, as for example,

$$2K + HgR_2 \longrightarrow Hg + 2KR$$

Properties

The monomeric organo-metallic compounds of alkali metals are not very stable thermodynamically. However, polymeric compounds are comparatively stable. For example, some polymeric lithium alkyls do not decompose even upto 100°C.

Structure

Alkyl and aryl compounds of lithium are polymeric having considerable covalent character in their Li—C bonds whereas such organo-metallic compounds of other alkali metals have predominantly ionic character so that these can be represented as M^+R^-. The alkyl and aryl sodium, potassium, etc. are much more reactive than the corresponding alkyls and aryls of lithium. For instance, the former catch fire in air more violently compared to alkyl compounds of lithium.

Uses

Lithium alkyls and organo-metallic compounds of other alkali metals are frequently employed in the synthesis of a number of organic compounds. These uses are evident from the following reactions :

$$3LiR + BCl_3 \longrightarrow 3LiCl + Br_2$$

$$Li(C_6H_5) + (CH_3C_6H_4)_2CO \longrightarrow (CH_3C_6H_4)_2C(C_6H_5)OLi$$

$$\xrightarrow{H_2O} \underset{\text{Tertiary carbinols}}{(CH_3C_6H_4)_2C(OH)C_6H_5} + LiOH$$

$$Li(C_6H_5) + CO_2 \xrightarrow[\text{solvent}]{\text{Nonpolar}} [C_6H_5CO_2Li] \xrightarrow{H_2O} \underset{\text{Carboxylic acid}}{C_6H_5COOH} + LiOH$$

$$LiR + R'CON(CH_3)_2 \longrightarrow LiN(CH_3)_2 + \underset{\text{Ketones}}{R'COR}$$

Organo-metallic Compounds of Beryllium

Beryllium forms a large number of organo-metallic compounds like BeR_2 (R = CH_3, C_2H_5 etc.), $Be(C_5H_5)X$ (X = Cl, Br, I) etc.

Preparation

Dialkyls of beryllium (BeR_2) can be obtained as follows :

$$2LiR + BeCl_2 \xrightarrow[\text{ether}]{\text{Solvent}} BeR_2 + 2LiCl$$

$$Be + HgR_2 \xrightarrow{\Delta} BeR_2 + Hg$$

Properties and structure

Compounds of BeR_2 type are colourless solids or viscous liquids. Dimethyl beryllium is polymeric and contains Be—CH_3—Be bridges in its structure. The extent of polymerisation of other dialkyl beryllium compounds decreases with increase in the bulk of the alkyl group, so much so that di-*t*-butyl beryllium is monomeric.

Beryllium also forms complexes with cyclopentadiene. The cyclo-pentadienyl group is linked with the metal through a sigma or a pi bond as in $[Be(\eta^5—C_5H_5)X]$ (where X is Cl, Br or I) and in $[Be(\eta^1—C_5H_5)\ (\eta^5—C_5H_5)]$ respectively.

Organo-metallic Compounds of Magnesium

Magnesium forms main three types of organo-metallic compounds (*i*) compounds of MgR_2 type (R = alkyl or aryl group) (*ii*) Grignard reagents, RMgX (*iii*) Organo-metallic compounds containing one alkyl (R) and one alkoxy group (OR′), R—Mg—OR′.

The dialkyl and diaryl magnesium compounds can be prepared in the following manner:

$$LiR + R—Mg—X \xrightarrow[\text{ether}]{\text{Solvent}} MgR_2 + LiX$$

$$HgR_2 + Mg \xrightarrow[\text{ether}]{\text{Solvent}} MgR_2 + Hg$$

Dimethyl magnesium is polymeric. It contains Mg—CH_3—Mg bridges in its structure. Diphenyl magnesium dissolves in ether to yield $Ph_2Mg(OC_2H_5)$ where Ph is phenyl group. Diphenyl magnesium reacts with phenyl lithium to give $Li[MgPh_3]$ as shown below:

$$LiPh + MgPh_2 \longrightarrow Li[MgPh_3]$$

Another class of organo-metallic compounds of magnesium, called *Grignard reagents*, can be synthesised by refluxing alkyl or aryl halides with magnesium in dry solvent ether.

$$RBr + Mg \xrightarrow[\text{ether (dry)}]{\text{Solvent}} RMgBr$$

Grignard reagents are of great synthetic value in organic chemistry.

Organo-metallic compounds containing one alkyl and the other alkoxy group attached with Mg can be easily synthesised by reacting dialkyl magnesium compounds with alcohols in the following manner.

$$MgR_2 + R'OH \longrightarrow R—Mg—OR' + ROH$$

These compounds are polymeric and contain alkyl and alkoxy bridges between Mg atoms. The degree of polymerisation depends upon the bulk of the alkyl and alkoxy groups.

CH_3—Mg—OR (R is CH_3 or C_2H_5) exists in two polymeric forms whose structures are given below in Fig. 16.1.

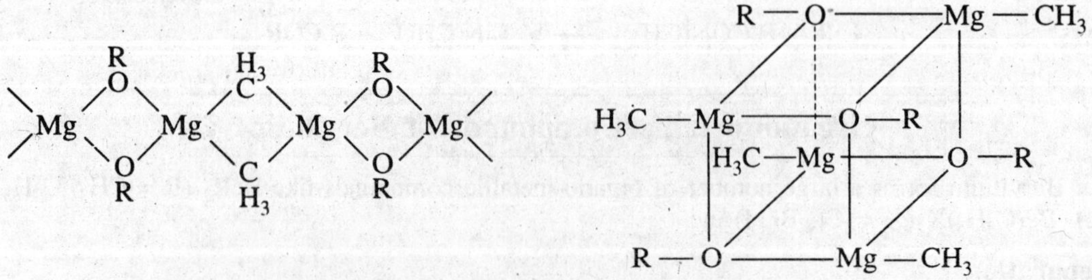

Fig. 16.1. Two polymeric forms of CH_3—Mg—OR (R = CH_3 or C_2H_5)
(*a*) Chain polymer (*b*) Three dimensional polymer.

Organo-metallic Compounds of Aluminium

Organometallic compounds of Al of $(R_3Al)_2$, $(C_6H_5)_3Al$ etc. type can be prepared as follows:

$$2Al + 3R_2Hg \longrightarrow (R_3Al)_2 + 3Hg$$

$$AlCl_3 + RMgCl \xrightarrow{-MgCl_2} \underbrace{RAlCl_2,\ R_2AlCl \text{ and } R_3Al}_{\text{(mixture)}}$$

$$AlH_3 + 3C_2H_4 \longrightarrow Al(C_2H_5)_3$$

The trialkyl aluminium compounds are generally dimeric and contain Al—R—Al bridges and R—Al bonds in their structures. The terminal alkyl and the bridging alkyl groups in such dimeric compounds are in a process of very fast exchange at room temperature.

The trialkyl aluminium compounds are efficient Lewis acids and combine with Lewis bases such as amines, phosphines, etc., to form adducts of the type R_3Al. Lewis base.

The trialkyl aluminium compounds react with alkyl lithium to produce derivatives of $LiAlH_4$.

$$R_3Al + RLi \longrightarrow Li[AlR_4]$$

Triphenyl aluminium and dimethyl phenyl aluminium are dimeric containing Al-phenyl-Al bridges.

The alkyl chlorides, bromides and iodides of aluminium are also dimeric but they contain Al-halogen-Al bridges instead of Al-alkyl-Al bridges in their structures. Alkyls fluorides of aluminium have cyclic tetrameric structures having Al—F—Al bridges. The structure of $(CH_3)_2AlF$ is as shown in Fig. 16.2.

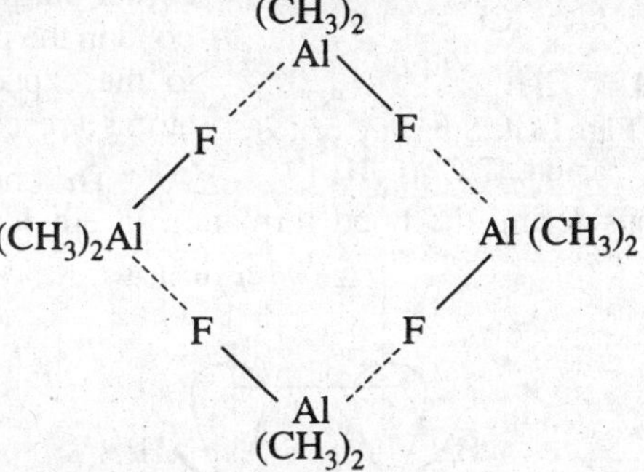

Fig. 16.2. Cyclic tetrameric structure of $(CH_3)_2AlF$.

Metal-olefine (alkene) Complexes

These are π-bonded organo-metallic complexes. Olefins are two-electron ligands. Ethylene, propylene and a number of other olefins form well-defined complexes with Pd(II), Pt(II), Hg(II), Cu(I) and Ag(I) salts.

Preparation

(*i*) Olefins complexes are generally prepared by the displacement of :

(*a*) *loosely-held ligand* such as solvent molecules.

$$2RhCl_3 + 2H_2O + 6C_2H_4 \longrightarrow [RhCl(C_2H_4)_2]_2 + 2CH_3CHO + 4HCl$$

(*b*) *ionic ligand*

$$K_2PtCl_4 + C_2H_4 \longrightarrow \underset{\text{(Zeise's salt)}}{K[PtCl_3(C_2H_4)]} + KCl$$

$$\downarrow C_2H_4$$

$$[PtCl_2(C_2H_4)_2]_2$$

(*c*) *neutral ligand*

$$[Rh(CO)_2Cl]_2 + 2C_8H_{12} \longrightarrow [RhCl(C_8H_{12})]_2 + 4CO$$

(*ii*) They can also be prepared by the direct reaction of a metal salt with the olefin.

$$2PdCl_2 + 2C_2H_4 \longrightarrow [PdCl_2(C_2R_4)]_2$$

Properties

The chemical stabilities of the metal-olefin complexes vary widely. The complexes are usually labile; olefins being easily displaced by ligands such as phosphines. The complexed olefins are more susceptible to nucleophilic attack than the complexes species.

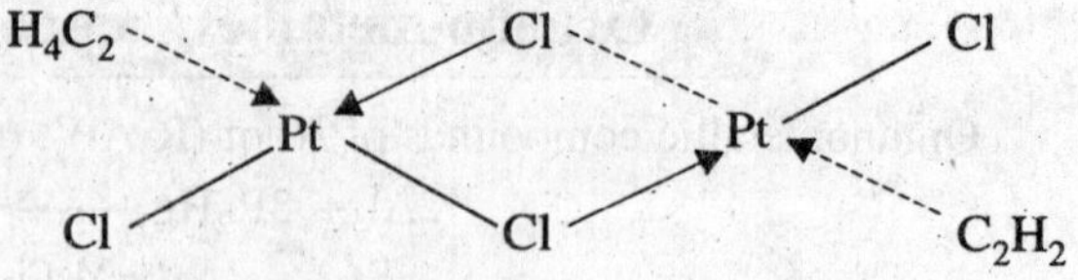

H_4C_2 Cl Cl Pt Pt Cl Cl C_2H_2

Fig. 16.3. Bridged structure of dimer, $(PtCl_2.C_2H_4)_2$

Structure and bonding

The platinum complex, $PtCl_2.C_2H_4$ is a dimer as shown by freezing point depression measurements in benzene and has a bridged structure with trans-arrangement of two ethylene molecules. (See Fig. 16.3).

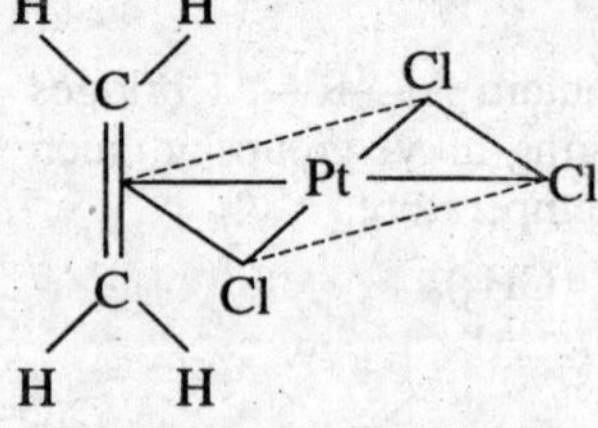

Fig. 16.4. Structure of the anion, $[PtCl_3(C_2H_4)]^-$.

X-ray examination of anion of Zeise's salt shows that the platinum atom is at the centre of a square plane with chlorine atoms at each of the three corners. The ethylene is at the fourth corner but prependicular to the plane, with the centre of the double bond in the plane. Indeed the C=C bond of ethylene is perpendicular to the expected normal metal-ligand bond axis. The two carbon atoms are equidistant from the metal. (see Fig. 16.4).

The bonding in olefine complexes can be explained by considering the bond formation in ethylene complex with Pt^{2+} ion.

Two $2p_z$ orbitals

Filled bonding π-molecular orbital

Vacant antibonding π-molecular orbital

(*a*)

Empty dsp^2 hybrid orbital on Pt^{2+}

Filled bonding π-molecular orbital

C →Pt^{2+} bond (σ-overlap)

(*b*)

Filled *dp* hybrid orbital on Pt

Vacant antibonding π-molecular orbital

Pt →C back bonding (π-overlap)

(*c*)

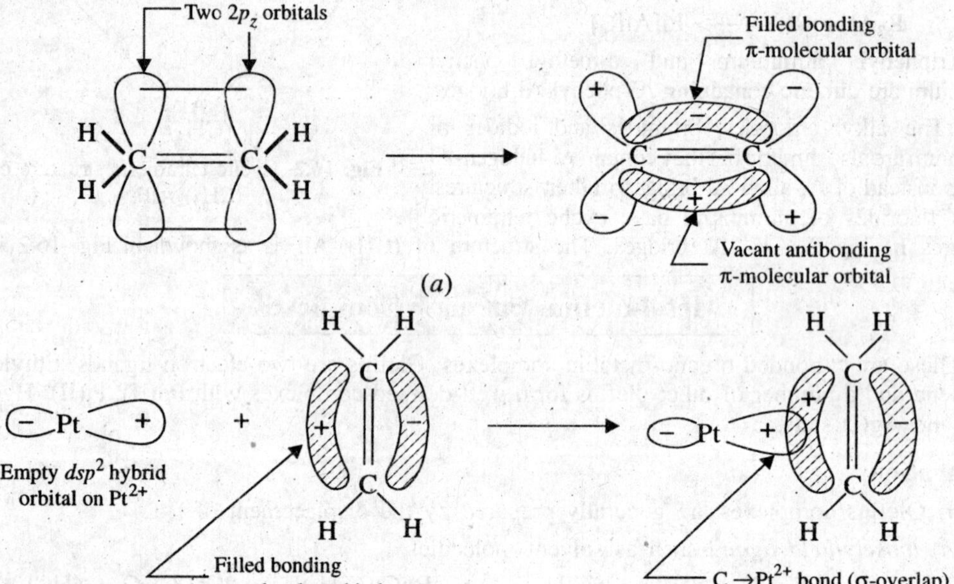

Fig. 16.5. Bonding in ethylene complex with Pt^{2+} ion.

We know that in ethylene molecule, $\left[\begin{matrix} H \\ H \end{matrix} > C = C < \begin{matrix} H \\ H \end{matrix}\right]$, $2p_z$ orbitals on two C-atoms overlap to form π-bonding and π*-antibonding molecular orbitals. π-bonding molecular orbital is filled while π*-antibonding molecular orbital is vacant. The filled π-bonding molecular orbital of C_2H_4 molecule overlaps with the empty dsp^2 hybrid orbital (which is a σ-type orbital) of Pt^{2+} ion and forms ethylene → Pt^{2+} bond (σ-overlap). The empty π*-antibonding molecular orbital of ethylene molecule overlaps with the filled *dp* hybrid orbital (which is a *dπ* orbital) on Pt^{2+} ion. This type of overlap gives *dπ-pπ* back bonding. All the three steps have been shown in Fig. 16.5.

The two processes, *viz.*, the donation of π-bonding electrons of the olefin to the metal σ-orbital and back donation of electrons from the metal to the antibonding orbitals of the olefin are synergically related and weaken the π-bonding in the olefin, and in all cases, except the anion of Zeise's salt, there is a marked increase in the C—C bond length. Owing to the lengthening of C—C bond, ν(C=C) decreases in metal-olefin complexes as compared with the free ligand by 60–150 cm^{-1}.

Cyclopentadienyl Complexes: Metallocenes

Cyclopentadienyl group, C_5H_5 is the most important and familiar five-electron ligand. Complexes formed by this group with transition elements are called **metallocenes.** Examples of metallocenes are $(C_5H_5)_2M$ (M = Ni^{2+}, Co^{2+}, Fe^{2+}, Mn^{2+}, Cr^{2+}, V^{2+}, Ti^{2+} etc.). Out of these, **ferrocene**, $(C_5H_5)_2Fe$ is the most important. For the first time it was reported in 1951. Cyclopentadienyl derivatives like $(C_5H_5)Mn(CO)_3$, $(C_5H_5)M(NO)$ (M= Pt, Pd, Ni) etc., are also known.

Preparation of metallocenes and their derivatives

Metallocenes can be prepared:

(*i*) *by the reaction of alkali metal cyclopentadienides with transition elements.*

$$2NaC_5H_5 + MCl_2 \xrightarrow[\text{Reflux}]{\text{THF}} (\pi\text{-}C_5H_5)_2M + 2NaCl$$

$$2NaC_5H_5 + [Co(NH_3)_6](SCN)_2 \longrightarrow \underset{\text{Unstable}}{[Co(NH_3)_6][C_5H_5]_2} + 2NaSCN$$

$$\longrightarrow 6NH_3 + (\pi\text{-}C_5H_5)_2Co$$

Complexes of Fe, Ni, Cr are also obtained by this method.

Alkali metal cyclopentadienides react with salts in solvents such as liquid ammonia.

$$Cd(SCN)_2 \xrightarrow{4NH_3} Co(NH_3)_4(SCN)_2$$

$$Co(NH_3)_4(SCN)_2 + 2C_5H_5Li \longrightarrow Co(NH_3)_4(C_5H_5)_2 + 2LiSCN$$

$$\xrightarrow{\text{At reduced pressure}} 4NH_3 + (\pi\text{-}C_5H_5)_2Co$$

(*ii*) *by the action of anhydrous metal (II) halides on cyclopentadiene in an amine.*

$$FeCl_2 + 2C_5H_6 + 2NH(C_2H_5)_2 \longrightarrow (\pi\text{-}C_5H_5)_2Fe + 2(C_2H_5)_2NH_2Cl$$

This method is applied for iron, nickel and cobalt.

(*iii*) *by the action of cyclopentadiene on metallic carbonyls at elevated temperature.*

$$2Cr(CO)_6 + 2C_5H_6 \longrightarrow [(C_5H_5)Cr(CO)_3]_3 + 6CO + H_2$$

$(C_5H_5)_2Co$ can also be prepared in a similar way.

(*iv*) *by the action of cyclopentadiene on metal* or *metal oxide [Commercial preparation of $(C_5H_5)_2Fe$]*

$$2C_5H_6 + Fe \xrightarrow{300°C} (\pi\text{-}C_5H_5)_2Fe + H_2$$

$$2C_5H_6 + FeO \xrightarrow{250°C} (\pi\text{-}C_5H_5)_2Fe + H_2O$$

(*v*) *from Grignard reagent by reacting with the metal halides (Laboratory method).*

$$2C_5H_5MgBr + FeCl_2 \longrightarrow (\pi\text{-}C_5H_5)_2Fe + 2MgClBr$$

Cr and V complexes can also be prepared.

Properties

(1) All the 3*d* transition metal complexes are neutral soluble in organic hydrocarbon solvents and are sublimable at 100°C, and do not pyrolyse even at 470°C. All the 3*d* complexes possess the melting point between 172—174°C while the heavier metal complexes melt at higher temperatures. These complexes very very much in their stabilities. Ferrocene only is stable in air but V, Cr, Ti, Ni, Co complexes should be handled in vacuum.

Stability order is Fe > Ni > Co > V > Cr > Ti

Thus $(\pi\text{-}C_5H_5)_2Ni$ is moderately stable, *i.e.*, decomposes in air in two days in solid phase, but decomposes faster in solution.

(2) All these are stable towards the hydrolysis.

(3) All these are easily oxidised in the acid solution to give cations (oxidations generally are reversible).

$$\pi(C_5H_5)_2Fe \xrightleftharpoons{\text{acid}} \underset{\text{Blue green}}{(\pi\text{-}C_5H_5)_2Fe^+} \text{ (Ferrocenium ion)}$$

$$Cr(\pi\text{-}C_5H_5)_2 \rightleftharpoons (\pi\text{-}C_5H_5)_2Cr^+ \text{ (Chromocenium)}$$

$$V(\pi\text{-}C_5H)_2 \rightleftharpoons \underset{\text{Purple}}{(\pi\text{-}C_5H_5)_2V^+} \text{ (Vanadocenum)}$$

$$\pi\text{-}(C_5H_5)_2Co \rightleftharpoons \underset{\text{Yellow-green}}{(\pi\text{-}C_5H_5)_2\ Co^+} \text{ (Cobaltocenium)}$$

These ions can be precipitated by large ions like I_3^-, BF_4^-, $[SbCl_6]^-$ as salts. These salts are soluble in liquid SO_2 and acetone and can be recrystallised by the addition of chloroform or ether.

(4) *Hydrogenation.* Ferrocene is more resistant to catalytic hydrogenration than benzene because of the co-ordination of π-electron density of the ring to the metal atom.

Some properties of ferrocene

(1) Ferrocene is orange yellow solid, m.p., 178°C, sublimable and soluble in organic solvents.

(2) Stable in air and towards hydrolysis.

(3) Oxidised by bromine as well as in acid solution.

$$Fe(\pi\text{-}C_5H_5)_2 \xrightleftharpoons{\text{acid}} Fe^+(\pi\text{-}C_5H_5)_2$$

$$C_{10}H_{10}Fe + \frac{1}{2}Br_2 \longrightarrow C_{10}H_{10}FeBr \xrightarrow{6Br_2} FeBr_3 + \underset{\text{Pentabromo cyclopentane}}{2C_5H_5Br_5}$$

(4) No Diels-Alder reaction possible.

(5) *Organic reactions.* As cyclopentadienyl rings in the molecule of ferrocene are aromatic in nature, they can undergo aromatic substitution.

(*a*) *Friedel crafts acylation.* Ferrocene and acetyl chloride in the equimolar quantities give the mono substituted derivative while in the double quantities disubstituted derivative is formed.

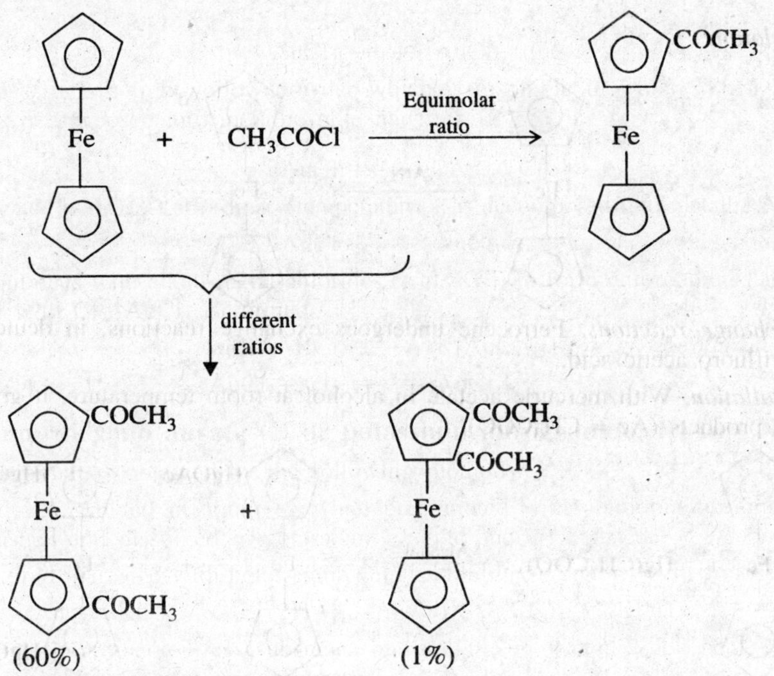

Fe + CH_3COCl $\xrightarrow{\text{Equimolar ratio}}$ Fe ($COCH_3$)

different ratios

Fe ($COCH_3$, $COCH_3$) (60%) + Fe ($COCH_3$, $COCH_3$) (1%)

(*b*) *Friedel crafts alkylation:*

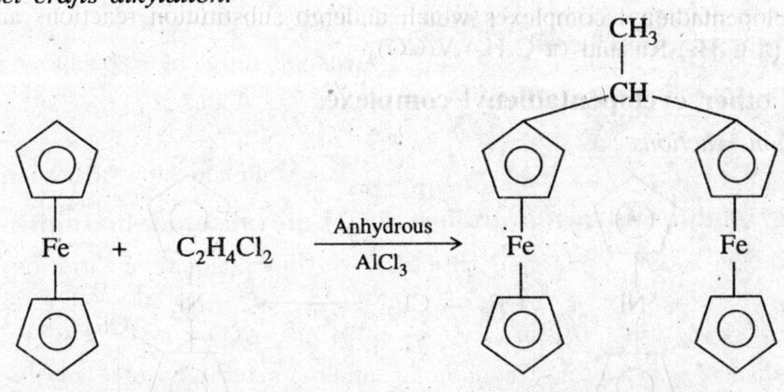

Fe + $C_2H_4Cl_2$ $\xrightarrow[\text{AlCl}_3]{\text{Anhydrous}}$ CH_3–CH (Fe)(Fe)

Diferrocenyl ethane or ethylidene diferrocenyl

(*c*) *Amino methylation:*

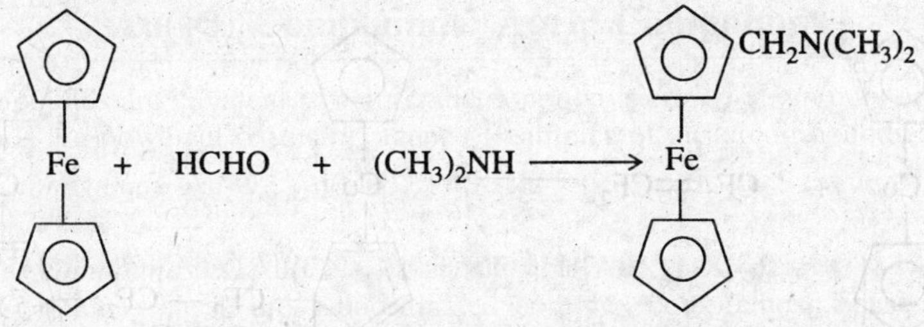

Fe + HCHO + $(CH_3)_2NH$ ⟶ Fe ($CH_2N(CH_3)_2$)

(*d*) *Sulphonation.* It can undergo sulphonation with H_2SO_4 or chlorosulphuric acid in the presence of acetic anhydride giving monosubstituted and heterosubstituted derivatives.

(*e*) *Formylation.* It can be formylated with N-methyl formaldehyde in presence of $POCl_3$.

(*f*) *Condensation.* On condensation with HCHO in presence of H_2SO_4 or liquid HF followed by reduction, it gives 1, 2 diferrocenyl ethane.

(*g*) *Arylation:*

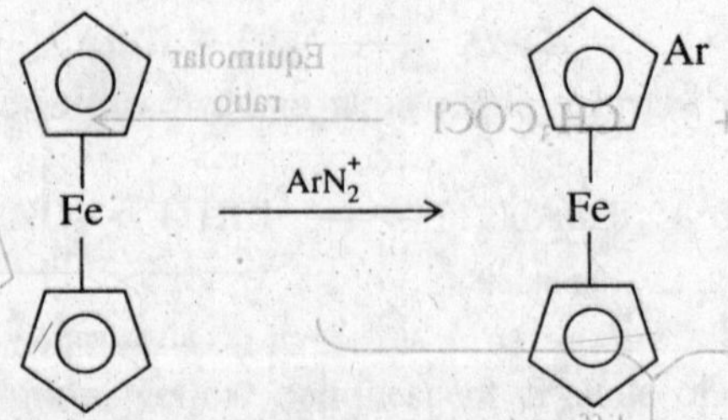

$$\text{Fe}(C_5H_5)_2 \xrightarrow{ArN_2^+} (C_5H_4Ar)\text{Fe}(C_5H_5)$$

(*h*) *Exchange reactions.* Ferrocene undergoes exchange reactions, in deuterated HCl or deuterated trifluoro acetic acid.

(*i*) *Metallation.* With mercuric acetate in alcohol at room temperature, it gives mono and disubstituted products (Ac = CH_3COO).

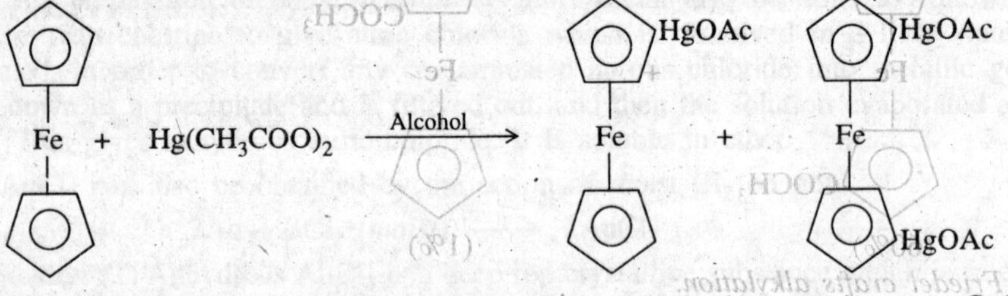

$$\text{Fe}(C_5H_5)_2 + Hg(CH_3COO)_2 \xrightarrow{\text{Alcohol}} (C_5H_4HgOAc)\text{Fe}(C_5H_5) + \text{Fe}(C_5H_4HgOAc)_2$$

Other cyclopentadienyl complexes which undergo substitution reactions are $Os(\pi\text{-}C_5H_5)_2$, $(\pi\text{-}C_5H_5)_2Mn$, $(\pi\text{-}C_5H_5)_2Ru$ and $(\pi\text{-}C_5H_5)_2V(CO)_4$.

Reactions of other cyclopentadienyl complexes

(1) *Addition reactions :*

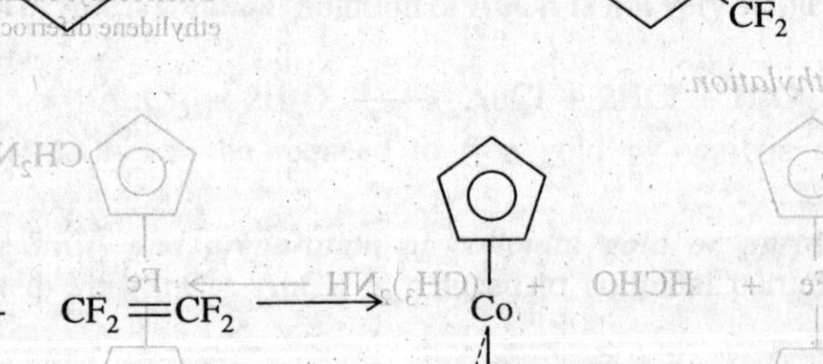

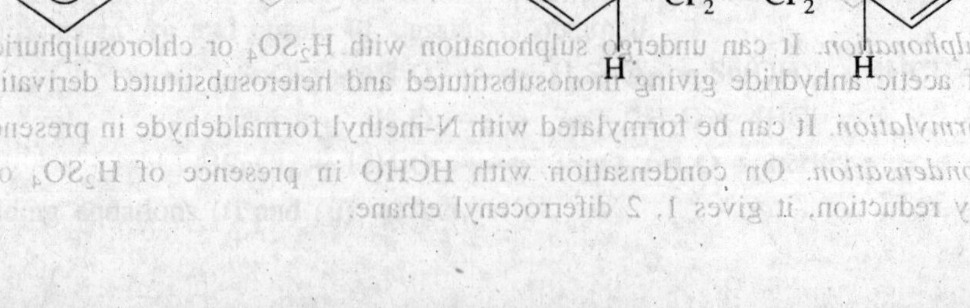

$$\text{Ni}(C_5H_5)_2 + CF_2{=}CF_2 \longrightarrow \text{Ni}(C_5H_5)(C_5H_5CF_2CF_2)$$

$$\text{Co}(C_5H_5)_2 + CF_2{=}CF_2 \longrightarrow (C_5H_5)\text{Co}(C_5H_5)\text{-}CF_2\text{—}CF_2\text{-}(C_5H_5)\text{Co}(C_5H_5)$$

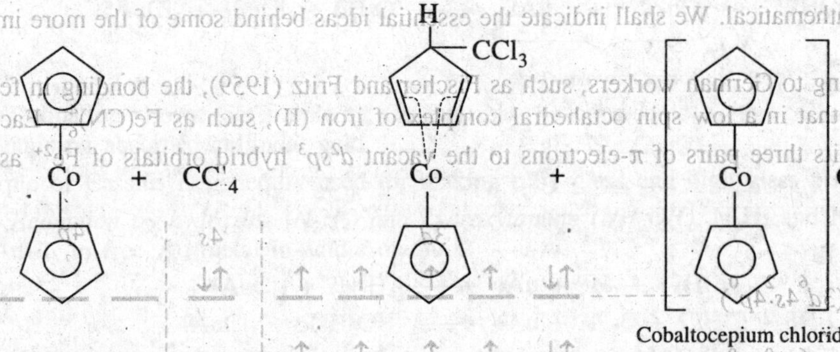

$(C_5H_5)_2Co + CCl_4 \rightarrow (C_5H_5)Co(C_5H_5CCl_3) + [(C_5H_5)_2Co]Cl$

Cobaltocepium chloride

(2) *Reduction :*

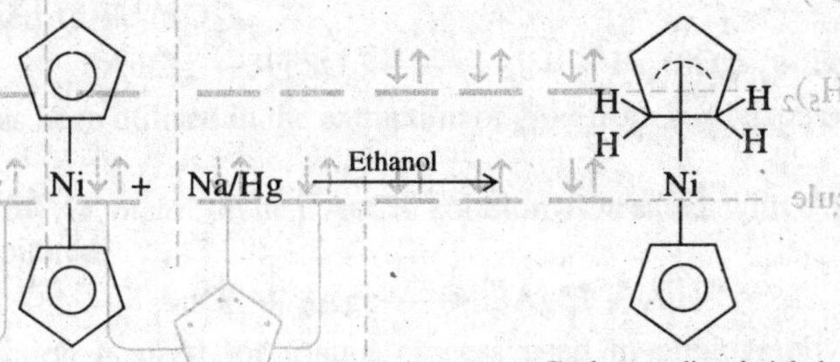

$(C_5H_5)_2Ni + Na/Hg \xrightarrow{\text{Ethanol}} (C_5H_5)Ni(C_5H_7)$

Cyclopentadienyl nickel

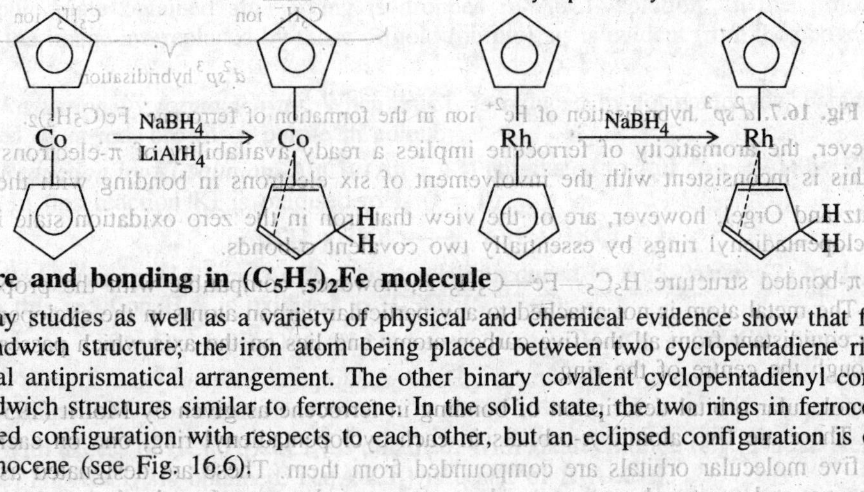

$[(C_5H_5)_2Co] \xrightarrow[\text{LiAlH}_4]{\text{NaBH}_4} (C_5H_5)Co(C_5H_6)$

$[(C_5H_5)_2Rh] \xrightarrow{\text{NaBH}_4} (C_5H_5)Rh(C_5H_6)$

Structure and bonding in $(C_5H_5)_2Fe$ molecule

X-ray studies as well as a variety of physical and chemical evidence show that ferrocene has a sandwich structure; the iron atom being placed between two cyclopentadiene rings in a pentagonal antiprismatical arrangement. The other binary covalent cyclopentadienyl compounds have sandwich structures similar to ferrocene. In the solid state, the two rings in ferrocene have a staggered configuration with respects to each other, but an eclipsed configuration is observed for ruthenocene (see Fig. 16.6).

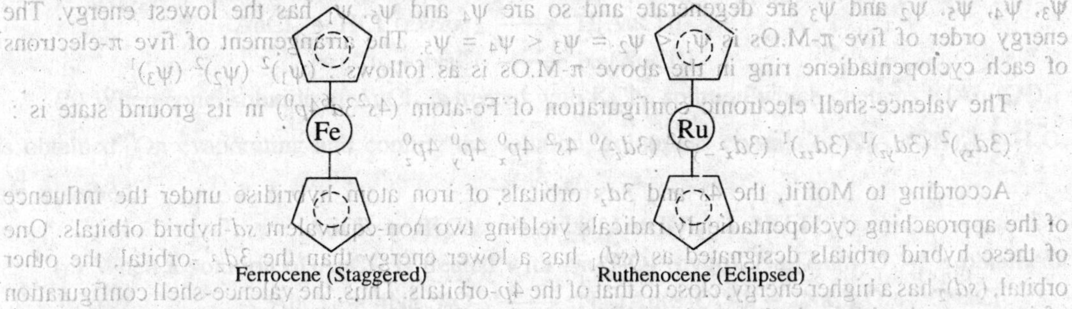

Fig. 16.6. Structure of ferrocene, $(C_5H_5)_2Fe$ and ruthenocene, $(C_5H_5)_2Ru$ molecules.

A number of theories have been advanced regarding the bonding in ferrocene and related cyclopentadienyl compounds. It is not possible to discuss all these theories; some of which are extremely mathematical. We shall indicate the essential ideas behind some of the more important theories.

According to German workers, such as Fischer and Fritz (1959), the bonding in ferrocene is similar to that in a low spin octahedral complex of iron (II), such as $Fe(CN)_6^{4-}$. Each $C_5H_5^-$ ion dontates its three pairs of π-electrons to the vacant d^2sp^3 hybrid orbitals of Fe^{2+} as shown in Fig. 16.7.

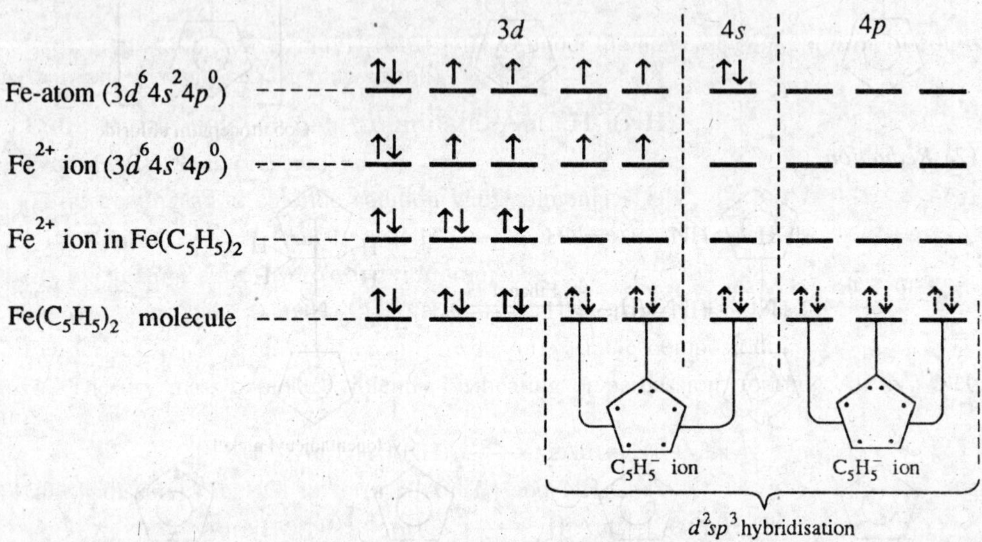

Fig. 16.7. d^2sp^3 hybridisation of Fe^{2+} ion in the formation of ferrocene, $Fe(C_5H_5)_2$.

However, the aromaticity of ferrocene implies a ready availability of π-electrons in each ring and this is inconsistent with the involvement of six electrons in bonding with the metal.

Dunitz and Orgel, however, are of the view that iron in the zero oxidation state is joined to two cyclopentadienyl rings by essentially two covalent σ-bonds.

The π-bonded structure H_5C_5—Fe—C_5H_5 is, however, compatible with the properties of ferrocene. The metal atom is not attached to any particular carbon atoms in the cyclopentadienyl ring but is equidistant from all the five carbon atoms and lies on the axis which passes at right angles through the centre of the ring.

The molecular orbital description of bonding in ferrocene as given by Moffit (1954) is not appealing. There are five atomic π-orbitals in each cyclopentadienyl ring, one on each carbon atom and five molecular orbitals are compounded from them. These are designated as ψ_1, ψ_2, ψ_3, ψ_4, ψ_5. ψ_2 and ψ_3 are degenerate and so are ψ_4 and ψ_5. ψ_1 has the lowest energy. The energy order of five π-M.Os is $\psi_1 < \psi_2 = \psi_3 < \psi_4 = \psi_5$. The arrangement of five π-electrons of each cyclopentadiene ring in the above π-M.Os is as follows : $(\psi_1)^2$ $(\psi_2)^2$ $(\psi_3)^1$.

The valence-shell electronic configuration of Fe-atom $(4s^2 3d^6 4p^0)$ in its ground state is :

$$(3d_{xy})^2\ (3d_{yz})^1\ (3d_{zx})^1\ (3d_{x^2-y^2})^2\ (3d_{z^2})^0\ 4s^2\ 4p_x^0\ 4p_y^0\ 4p_z^0.$$

According to Moffit, the $4s$ and $3d_{z^2}$ orbitals of iron atom hybridise under the influence of the approaching cyclopentadienly radicals yielding two non-equivalent sd-hybrid orbitals. One of these hybrid orbitals designated as $(sd)_1$ has a lower energy than the $3d_{z^2}$ -orbital, the other orbital, $(sd)_2$ has a higher energy, close to that of the $4p$-orbitals. Thus, the valence-shell configuration of iron atom in the excited state is:

$$(3d_{xy})^2\ (3d_{yz})^1\ (3d_{zx})^1\ (3d_{x^2-y^2})^2\ (sd)_1^2\ (sd)_2^0\ 4p_x^0\ 4p_y^0\ 4p_z^0.$$

Ferrocene has 18-electron configuration with 5 electrons on π-M.Os of each cyclopentadienyl radical and 10 electrons on orbitals of iron atom. Five π-M.Os of each cyclopentadien,·. radicals and 9 orbitals of iron atom combine to yield 9 bonding or non-bonding M.Os and 10 antibonding M.Os. The electrons are accommodated in the bonding and non-bonding M.Os; bonding M.Os accommodating 4 electrons and non-bonding M.Os the remaining 14 electrons. The bonding in ferrocene results effectivity from the overlap of singly occupied d_{xz} and d_{yz} metal orbitals with singly occupied π-M.Os (ψ_2 or ψ_3) of cyclopentadienyl radicals of similar symmetry. The relevant molecular orbital energy level for $Fe(C_5H_5)_2$ molecule is shown in Fig. 16.8.

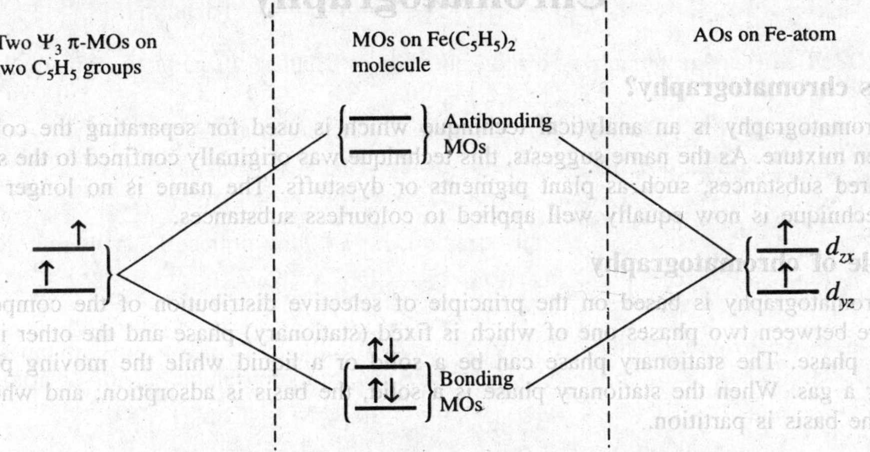

Fig. 16.8. Molecular orbital energy level diagram from ferrocene molecule, $Fe(C_5H_5)_2$. (After Moffit).

Ionic Cyclopentadienyl Compounds

Cyclopentadiene is a weak acid (pK_a ~ 20), dissociating to give H^+ and $C_5H_5^-$ ions. $C_5H_5^-$ forms ionic compounds—cyclopentadienides with strongly electropositive metal ions. The more important cyclopentadienides are formed by the alkali metals, MC_5H_5, the alkaline earth metals, $M(C_5H_5)_2$, the lanthanides and actinides, $M(C_5H_5)_2$. $Mn(C_5H_5)_2$ closely resembles the corresponding magnesium compounds and evidently has an ionic structure containing $C_5H_5^-$ and Mn^{2+} ions. The magnetic moment (5.9 B.M.) of $Mn(C_5H_5)_2$ is in conformity with the ionic bonding for the above compound. The ionic cyclopentadienides are very reactive towards air and water and react with ferrous chloride in tetrahedrofuran to yield ferrocene. Some of the ionic compounds of the type $M(C_5H_5)_2$ have, like ferrocene, sandwich structures. Thus, the structure of a cyclopentadienyl complex appears to be independent of the bond type.

University Questions

1. What are organo-metallic compounds? Give the general methods of their preparation.
2. Discuss the structure of organo-metallic compounds.
3. Give the structure of olifine complexes.
4. Give important properties and structure of ferrocene.
5. What are organometathic compounds ? Discuss the nature of M – C bond in organo metathic compounds. Give an account of preparation, properties and applications of alkyl and aryl derivatives of Hg. *(VBS, Purvanchal 2003)*

17

Chromatography

What is chromatography?

Chromatography is an analytical technique which is used for separating the components of a given mixture. As the name suggests, this technique was originally confined to the seperation of coloured substances, such as plant pigments or dyestuffs. The name is no longer accurate, as the technique is now equally well applied to colourless substances.

Principle of chromatography

Chromatography is based on the principle of selective distribution of the components of a mixture between two phases one of which is fixed (stationary) phase and the other is moving (mobile) phase. The stationary phase can be a solid or a liquid while the moving phase is a liquid or a gas. When the stationary phase is a solid, the basis is adsorption; and when it is a liquid, the basis is partition.

Types of chromatography

Important types of chromatography, stationary (fixed) phase and mobile (moving) phase used in them are given below :

Type of chromatography	*Stationary (fixed) phase used*	*Mobile (moving) phase used*
(*i*) Column or adsorption chromatography	Solid (Alumina or silica gel)	Liquid (Pentene, benzene, ether etc.)
(*ii*) Paper chromatography	Solid (A strip specially designed filter paper)	Liquid
(*iii*) Thin layer chromatography (TLC)	Solid (A thin layer of cellulose powder, alumina or silica)	Liquid (usually water)
(*iv*) Gas chromatography	Solid or liquid	Gas

In gas chromatography, both adsorption and partion process can take place. If adsorption takes place, the gas chromatography is called gas solid chromatography (GSC) and if partition takes place, it is called gas liquid chromatography (GLC).

Now let us discuss each of the above types of chromatography in detail.

Column or Adsorption Chromatography

Column chromatography is defined as the uniform percolation of a liquid through a column of finely-divided substances. The selected substance selectively retards certain components of the liquid.

Practical details of column chromatography

1. Adsorbents. This chromatography is based on the principle of selective adsorption of solutes for a solution on the surface of a finely-divided substance which is usually a solid and is called **adsorbent.** Cellulose, silica gel, alumina, magnesia, Na_2SO_4, $MgSO_4$, $Ca_3(PO_4)_2$, $BaSO_4$, ZnO, ThO_2, BeO etc. are the examples of adsorbents.

2. Developers. Developers are the substances which are used to produce colour with the colourless solution. H_2S, $(NH_4)_2S$ etc. are used as developers.

3. Solvents. The selection of a solvent depends, in the first place, on its dissolving power. The solvent should have its boiling point between 60° and 85°C. For most purposes light petroleum which boils near 80°C is used. Other solvents used are: C_6H_6, cyclohexane, chloroform, CCl_4, CS_2, C_2H_5OH, CH_3COOH and $CH_3COOC_2H_5$.

The solvents used in column chromatography do the following three functions :

(*a*) *They dissolve the mixture of various constituents and introduce them in the column.* The solvents used for this purpose are generally *non-polar* (*e.g.*, C_6H_6, petroleum ether etc.), *since adsorption takes place more readily from such solvents.*

(*b*) *When they are passed into the column, they develop the chromatogram.* In this process of development, the various zones of the chromatogram get separated sharply from one another. The solvents when used for this purpose are called **developers.**

(*c*) *They are also used for removing the various constituents of a mixture from the chromatogram after it has been properly developed.* The solvents, when used for the purpose, are called **eluents.** A good eluent should be a liquid which should be capable of dissolving the various components of the given mixture. Its boiling point should be low so that it can easily be removed from the recovered component.

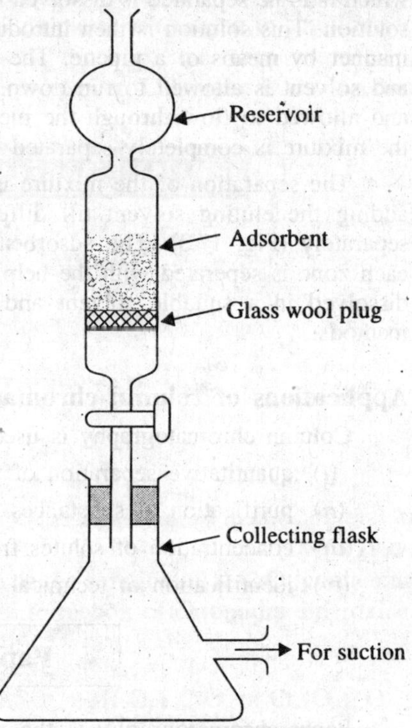

Fig. 17.1. Adsorption column used in column chromatography.

4. Adsorption columns. The glass wool plug or cotton wool plug is placed at the lower end of a long narrow tube. The plug serves as support for the adsorbent. The geometry of the tube depends on the form and size of the zone to be separated. When large volume of a solvent is to be used, a large reservior is fixed at the top of the tube. The collecting flask is provided with a side arm which is connected with a suction pump which increases the flow rate. The tube should be *packed uniformly* with the adsorbent (Fig. 17.1). For this purpose the adsorbent should be added in portions.

5. Packing of columns. The packing of columns is done by two methods *viz. wet packing method* and *dry packing method.* In wet packing method the tube is clamped in vertical position and its lower end is covered with a powder of glass wool plug. The plug is made wet by pouring into it some amount of the solvent. Now a slurry of the adsorbent and the solvent is prepared and is introduced into the tube slowly through the sides of the tube and in the presence of the solvent. (See Fig. 17.2).

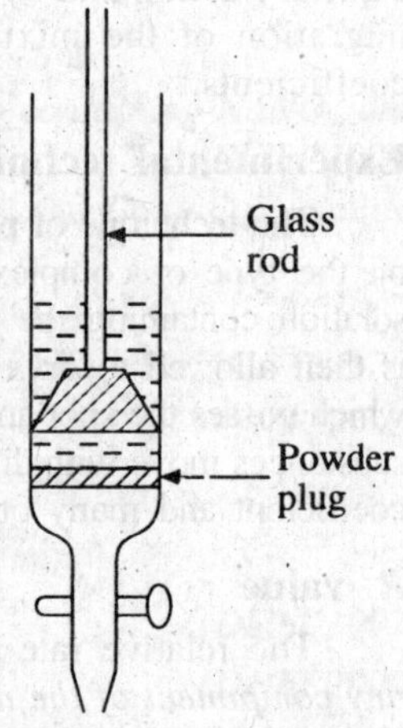

Fig. 17.2. Packing of column by wet packing process.

If there is some air bubble in the plug or at the bottom of the column, it should be removed by pushing it with a glass rod. The pouring of the slurry is continued, until the desired height of it is obtained. The bottom of the column is opened to remove the excess of the solvent. The packing by this method is very common with adsorbents like Al_2O_3, MgO etc.

6. Separation of the mixture into its components. The mixture which is to be separated is dissolved in the solvent to give a concentrated solution. This solution is then introduced into the column in a step-wise manner by means of a pipette. The stop cock at the bottom is opened and solvent is allowed to run down. The eluting solvent is then added and allowed to flow through the medium. This flow is continued until the mixture is completely separated into its components.

The separation of the mixture into its components can be done by adding the eluting solvent till different zones or bands are obtained separately (Fig. 17.3). The adsorbent with the zones is taken out and each zone is seperated with the help of a knife. After that each zone is dissolved in a suitable solvent and then identified by usual, standard methods.

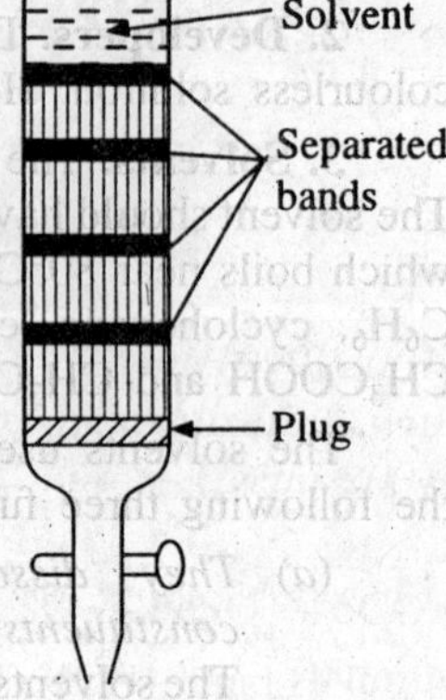

Fig. 17.3. Separation of the mixture into different zones.

Applications of column chromatography

Column chromatography is used for the :

(*i*) quantitative separation of two or more components of a mixture.

(*ii*) purification of substances from their contaminants.

(*iii*) concentration of solutes from dilute solutions.

(*iv*) identification of technical products.

Paper Chromatography

Paper chromatography is the method in which the analysis of an unknown substance is mainly done by the flow of solvents on specially designed filter paper. One of the two solvents is immiscible or partially miscible in the other solvent. The solvent rises up by the capillary action and by adsorption on the paper and the separation is affected by the differential migration of the mixture of the substance. This occurs due to the differences in partition coefficients.

Experimental technique

The technique of paper chromatography may be *one dimensional* or *two dimensional* depending on the type of complexity involved in the analysis. In one dimensional technique a drop of the solution containing the mixture of substances is placed near one end of the filter paper. The spot is then allowed to dry. The portion of the paper with the spot is then placed in a suitable solvent which passes the spot and carries the substance along with while moving up the filter paper. Different substances move with different rates which depend on size, nature of the substances, solvent, diffusion coefficient and many other factors. The finished paper is known as *paper chromatogram.*

R_f value

The relative rate of movement of solute and solvent is expressed by R_f value *which, for any component of the mixture, is defined as the ratio of the distance travelled by the compounds at its point of maximum concentration to the distance travelled by the solvent, i.e.,*

$$R_f = \frac{\text{Distance travelled by the component}}{\text{Distance travelled by the solvent}}$$

Both the distances are measured from the point of application of the sample. From Fig. 17.4 it may be seen that :

$$R_f = \frac{A}{B}$$

R_f value has no unit. It is a constant for a given substance (component) under given conditions. It is, therefore, possible to identify various components by determining their R_f values. R_f values vary with the solvent used.

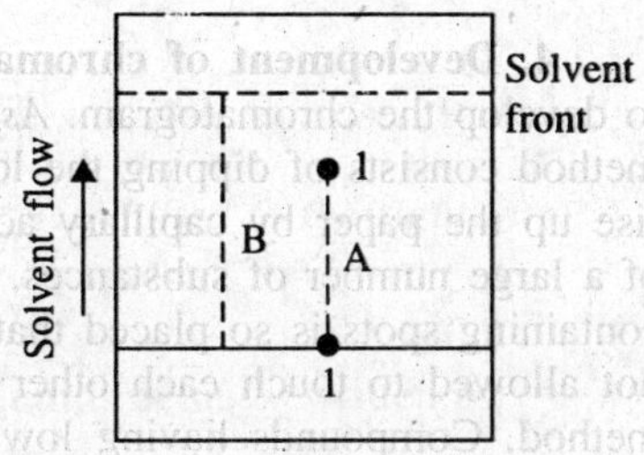

Fig. 17.4. Definition of R_f value.

Steps of experimental technique

The experimental technique consists of the following operations :

1. Choice of filter paper. The paper to be used should have the following characteristics:

(*a*) The paper should have maximum degree and clarity of separation.

(*b*) The paper should have negligible diffuseness of the spots.

(*c*) The paper should have excellent rate of movement of solvent front.

Whatman filter paper has been used extensively in chromatographic work. Shape and size of the paper to be used is also to be considered. Shape and size of the paper depend on the type of the work to be done. Papers with rectangular shape have been used by many workers. In order to remove the reducing substances the paper should be washed before use.

2. Preparation of the sample and its application on the paper. For the preparation of the sample a weighed amount of the solid is dissolved in a volatile solvent to get its concentrated solution. The minimum volume of this concentrated solution is applied on the paper. The samples may be applied on the paper either as spots or as bands. After selecting the paper of proper size and shape, a line is drawn with the help of a pencil at a suitable distance from one end of the paper and some distance is left at the two ends of the paper. A number of points are marked on the line at equal distances [Fig. 17.5 (*a*)]. A drop of each of the samples is then put on the points with the help of fine capillary tube [Fig. 17.5 (*b*)]. The area occupied by the spots should be 5 mm in diameter. Stream of hot nitrogen or air is blown at the spots in order that the spots may not diffuse.

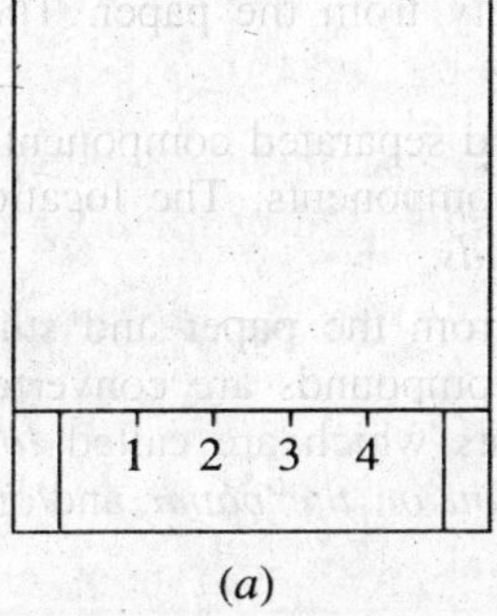

(*a*)

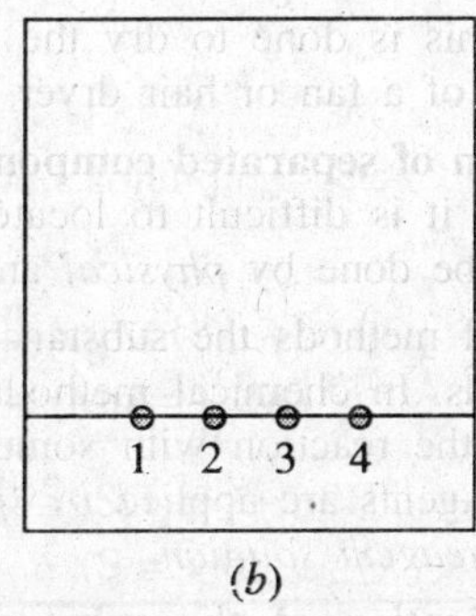

(*b*)

Fig. 17.5. (*a*) A number of points are marked on the line at equal distance (*b*) Application of samples on each point.

3. Selection of solvent. The selection of solvent depends on the nature of the substances to be separated. The solvents to be used should be such that when the components are mixed together, two definite layers should result. Polar solvent is most commonly used, since it is adsorbed on the paper. A less polar solvent which is called the mobile phase, advances ahead by capillary action and separates the applied substances. In most cases a mixture of *n*-butanol acetic acid and water has been used. Alcohols, acids, ketones, esters, phenols etc. have also been used.

4. Development of chromatogram. There are a number of methods which can be used to develop the chromatogram. *Ascending development* is the simplest of all such methods. This method consists of dipping the lower end of the paper into the solvent and then allowing it to rise up the paper by capillary action (Fig. 17.6). This method is preferable for quick analysis of a large number of substances. Proper care is taken to ensure that the lower end of the paper containing spots is so placed that it is above the solvent depth. The two ends of the paper are not allowed to touch each other and are kept at a distance apart by clipping or by any other method. Compounds having low R_f values are not completely separated by this method. For such compounds *descending technique* is used.

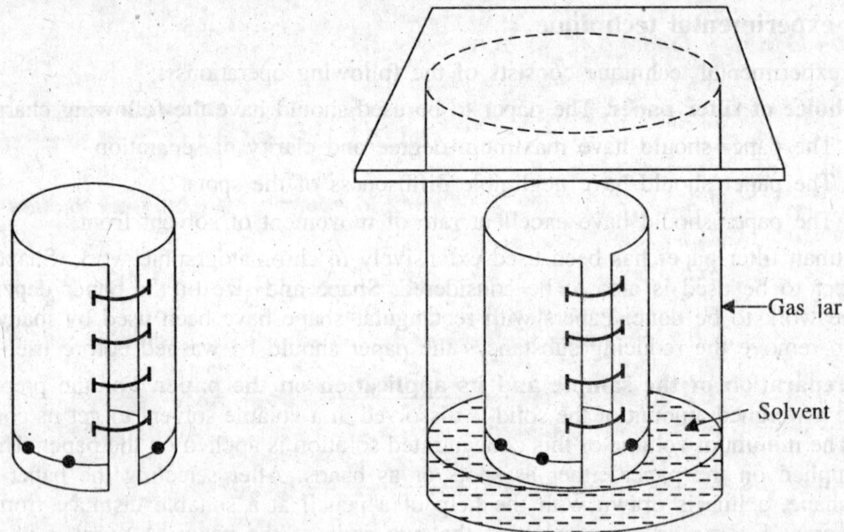

Fig. 17.6. Development of chromatogram by ascending technique.

5. Drying the chromatogram. The developed chromatogram is removed from the reservoir and the position of the solvent front is marked with the help of the pencil at the two edges of the paper. The chromatogram is then dried by keeping it in an oven or over a hot plate for few minutes. This is done to dry the solvents completely from the paper. The drying is best done by means of a fan or hair dryer.

6. Location of separated components. The coloured separated components can be located quite easily but it is difficult to locate the colourless components. The location of colourless substances can be done by *physical* and *chemical methods.*

In physical methods the substances are removed from the paper and studied by various physical methods. In chemical methods the colourless compounds are converted into coloured compounds by the reaction with some chemical reagents which are called *locating reagents.* The locating reagents are applied *by spraying the solution on the paper* and/or *by dipping the paper into the reagent solution.*

7. Identification of the substances. In most of the paper chromatographic work the measurement of R_f value is done either by the comparison with the reference substances or with the standard values given in the literature. Thus the identification of the compound is done.

In the *elution method* the bands from the developed chromatogram are cut off and then shaken well with appropriate solvent in the flask. The elutes as obtained may be diluted or concentrated as the case may be and then analysed by any technique that does not react with the components.

Thin Layer Chromatography (T.L.C.)

We have seen that in paper chromatography alumina, silica gel etc. cannot be used, because these materials cannot be changed into papers. However, thin layer of these materials can be obtained on glass plates. These thin layers of silica gel or alumina when spread over a glass plate are called *chromatoplates*. These plates can be used in the same way as filter paper sheets are used in paper chromatography. Thus thin layer chromatography is described as a method for chromatographic analysis on thin layers of adsorbents. Some workers call this technique as *open column chromatography, spread layer chromatography* and *surface chromatography*.

Superiority of T.L.C. over paper and column chromatography

Though T.L.C. is very much similar to paper and column chromatography, yet it superior to both of these techniques due to the following reasons :

(*i*) In T.L.C. the separation is the sharpest.

(*ii*) The separation requires less time (15–50 mts) and micrograms of the substance (0.4 mg).

(*iii*) The individual spots are less diffused and hence the sensitivity of detection is increased several times.

(*iv*) T.L.C. plates can be heated to higher temperature without causing any damage to it.

(*v*) The evaporation of the solvent, after the solution is introduced, is more rapid in this case.

Various steps of experimental technique of T.L.C.

The technique consists of the following steps :

1. *Preparation of chromatoplates.* A slurry of the adsorbent is prepared and spread on clean glass plate in the form of a thin uniform layer (Fig. 17.7). A small amount of a binding material such as calcium sulphate is also mixed with the slurry. For good separation the layer should be uniform. Special devices for attaining this purpose are also available.

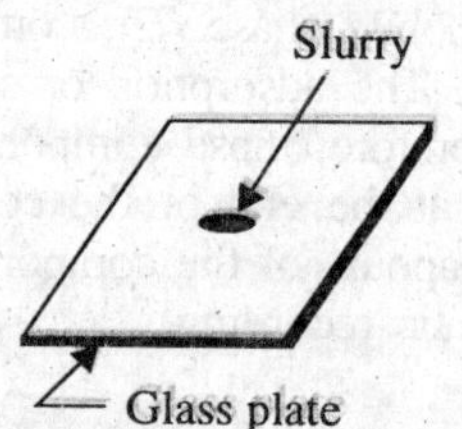

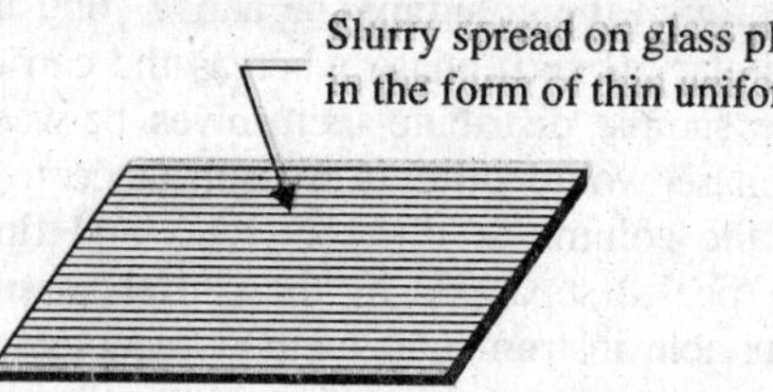

Fig. 17.7. Preparation of chromatoplate.

The thickness of the layer should be about 0.25 mm. After putting the layers on the glass plates they are activated by heating them in an oven at 100–105°C for about 30 mts. Specially prepared and uniformly spread layers are also available commercially.

2. *Preparation of the sample and its application on the chromatoplate, selection of the solvent, development, drying and location of the separated components.* These steps are essentially the same as in paper chromatography.

In thin layer chromatography the chromatoplate with the thin layer is placed vertically in a suitable solvent in a specially designed apparatus known as tank (Fig. 17.8). Three sides of the tank are covered with a sheet of paper impregnated with the solvent. The solvent ascends up the plate and the separation of the mixture into components takes place within a

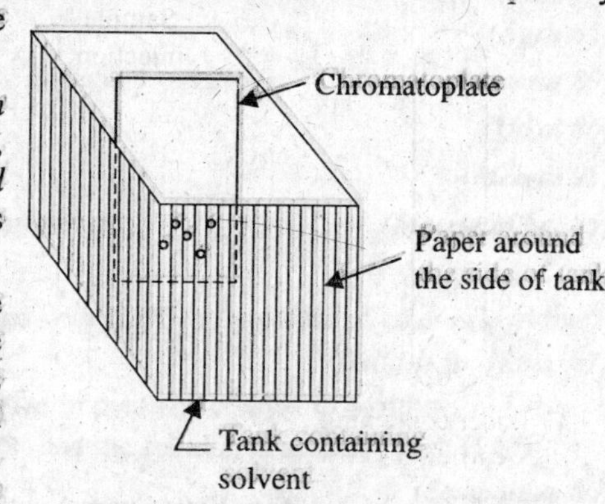

Fig. 17.8. Tank used in thin layer chromatography.

few minutes. The plate is then taken out and the position of the solvent-front is noted as in paper chromatography. The solvent evaporates quickly by mere exposure to air or on gentle warming by hot air, if necessary. The components are located in the same manner as in paper chromatography and their R_f values determined. The spots of the components, when located by suitable reagents can be photographed, if necessary.

Gas Chromatography

What is gas chromatography?

Gas chromatography is a special form of the general chromatography technique in which the moving phase is a gas and the stationary phase may be either liquid or solid. The technique is suitable for the separation of materials which are volatile but do not decompose. Nitrogen and hydrogen gases are used as a moving phase. In this chromatography both adsorption and partition phenomena are possible. If adsorption process is taking place, the technique is known as *gas solid chromatography* (GSC) whereas it is called *gas liquid chromatography* (GLC) when partition process is taking place.

In GSC the column is packed with an active solid (stationary phase) and the components of the mixture are distributed between the gas phase and the active solid phase. Due to *differences in adsorption behaviour*, the separation is achieved. In GLC the column is packed with a porous solid which is coated with a thin layer of non-volatile liquid (stationary phase). Here separation takes place due to *differences in solubility behabiour*. Mixtures are separated between these two phases due to different partition coefficients of the components.

Theory of gas chromatography

In gas chromatography the sample is introduced into the moving carrier gas stream and is carried by it through the column. The active solid used in GSC or non-volatile liquid used in GLC forms the stationary (fixed) phase whereas the carrier gas forms the mobile phase. The components of the mixture sample distribute themselves between the two phases. The adsorption or solubility in the fixed phase would differ from substance to substance and therefore, these components are carried along the column at different rates and finally they emerge at the end of the column in distinct zones (peaks) separated by the carrier gas. On merging, the vapours of the components are detected by suitable instrumental methods accompanied by an automatic recording.

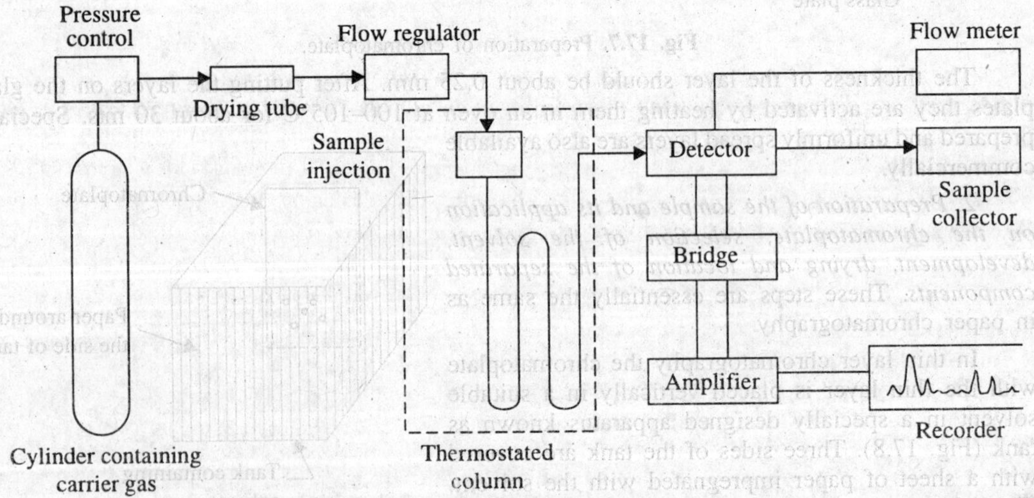

Fig. 17.9. General outline of gas chromatography apparatus.

General outline of gas chromatography apparatus

A general outline of gas chromatography apparatus is shown in Fig. 17.9.

Carrier gases like N_2, H_2, He, Ar or mixture of any two or these, which is obtained from a cylinder or other convenient source, is passed through pressure adjusting valve (pressure control), drying tube and flow regulator to get an even flow of the carrier gas. The gas then passes through the sample injection point where it picks up the sample mixture for analysis. The introduction of the sample is made as instantaneous as possible. The carrier gas containing the sample mixture enters into the column where they are separated. The column may be made up of glass, metal or plastic tubing in the form of W- or U-tube or in the coiled form and is placed in the thermostat where a constant temperature of the column is maintained. The carrier gas containing the separated vapours of the compounds of the sample emerge from the column and pass through the detector. The detector measures the concentration of the vapours in the carriers gas or its total amount. Most of the detectors are of the former type and provide an electrical signal which is proportional to the concentration of the vapours in the carrier gas. This signal is generally fed to an electronic recorder where a finished chromatogram is obtained. After the passage through the detector the gas stream may be passed through sample collector for identification.

University Questions

1. Explain the principle on which the following chromatography is based (*i*) Paper chromatography (*ii*) Thin layer chromatography.

2. Explain in detail thin layer chromatography technique for separations. In what way is it superior to paper chromatography?

3. Explain what you understand by gas chromatography.

4. Describe in detail the experimental details of the procedure for carrying out T.L.C.

5. In connection with column chromatography, explain the terms : adsorbents, developers and solvents.

6. What are the two phenomena which constitute the basis of chromatography?

7. (*a*) What is paper chromatography ? Discuss in detail the experimental technique used in ascending paper Chromatography.

(*b*) Write any two applications of paper chromatography.

(*c*) Write a note on "Partition Chromalography". *(Nagpur 2003)*

18

Uses of Organic and Inorganic Reagents in Inorganic Analysis

Introduction

Reagents are those compounds which are used in inorganic analysis for the detection and determination of ions and molecules. A large number of organic as well as inorganic reagents have been used for this purpose. The inorganic precipitating reagents generally precipitate the inorganic species as sparingly soluble salts or hydrous oxides. The organic precipitating agents form sparingly soluble products either through coordinate linkage or through ionic linkage between the inorganic species and the organic precipitating reagent.

Advantages of using organic reagents

The advantages of using organic reagents are listed below :

(*i*) They are more specific than conventional reagents under certain conditions. Because of this property they can frequently be used to detect a single type of ion in a complex mixture.

(*ii*) They can be more sensitive. Classical semi-micro methods of qualitative analysis will usually not detect much less than 100 gammas (1000 γ = 1 mg) of a substance under specific test solutions. Organic reagents, on the other hand, are capable of detecting below 10γ and in certain cases as little as $10^{-3}\gamma$.

(*iii*) They can be adopted in quantitative analysis. The organic precipitates are generally insoluble in water at a given pH and can be separated, dried and weighed. Those reactions giving coloured soluble products frequently show a regular change in colour intensity with concentration of the substance under test and so are adoptable to an accurate comparison with colours representing concentration. This is the basic idea in the branch of colorimetry.

Common properties of organic reagents

Some of the common properties of organic reagents due to which these are used in inorganic analysis are given below :

(*i*) *Formation of coloured precipitate.* (*a*) Organic reagents form direct characteristic colour with the ions to be detected. (*b*) For locating end point, *i.e.*, as indicators employed in volumetric titrations.

(*ii*) *Formation of precipitate due to difference in solubility.* This may be made on the basis of a gravimetric method.

(*iii*) *Extraction in organic solvent at controlled pH.*

(*iv*) *Masking reagents.* This increases the selectivity of reagents.

(*v*) *Difference in volatility.*

(*vi*) *Oxidation or reduction to other valency state.*

(*vii*) *Wash liquids.*

The chemical nature of the organic reagent is important in determining the use of above mentioned properties.

Advantages of organic reagents over inorganic reagents

It is due to the following properties that they are better precipitating reagents than inorganic reagents.

(*i*) **Selectivity.** In many cases organic reagents are somewhat specific. For example, cobalt from nickel using α-nitroso-β-naphthol or dimethylglyoxime, aluminium from iron using cupferron, cadmium from copper using quinaldinic acid and thiourea can be precipitated.

In some cases, adjustment of the pH will often cause precipitation of one constituent alone. By changing pH and solvent conditions, one constituent after another can sometimes be quickly precipitated out.

(*ii*) **Ease of drying.** Organic reagents produce compounds that have little affinity for water and are dried easily at temperature low enough to prevent decomposition.

(*iii*) **Co-precipitation.** Lack of ionic character of organic precipitation products greatly reduces co-precipitation, in most cases. For example the co-precipitation of sodium and potassium with magnesium is much less when the latter is precipitated by 8-hydroxyquinoline than when it is precipitated by phsophate or oxalate.

Some exceptions in metals do occur. It has been little studied but in some cases it is as bad as in inorganic precipitants.

Reagents are usually insoluble in water. Therefore, it is important not to add an excess of reagent. This is difficult because there is no way to tell when sufficient reagent, without excess, has been added. Any excess that is insoluble would add to the weight of the precipitate.

Disadvantages of organic reagents

There are, unfortunately, disadvantages to using organic reagents.

1. Volatility. This is the first practical disadvantage of organic reagents. Because chelated compounds are non polar, the forces between individual molecules in the crystal are not very great. Thus, some of the metallic complexes which are precipitated in analytical chemistry are appreciably volatile above 130–150°C and most decompose above this temperature. Therefore, the drying of precipitate should be done below 130°C to avoid decomposition.

2. Impurities in reagents. It is very difficult to prepare an organic reagent of the same degree of purity as say inorganic compounds. Impurities present in organic reagents may enter into the precipitate during the precipitation and cause undesirable reactions. For example,

(*a*) Dimethylglyoxime exists in three forms, *i.e.*, α, β, γ. Of these, only the α form is specific for nickel while preparing dimethylglyoxime γ form is also formed along with the α form. If the precipitation is carried out with this impure dimethylglyoxime, then nickel will be precipitated with some other metals because α form is specific for nickel whereas γ form can precipitate other metals also.

(*b*) Keto-enol isomerism in such compounds as dithiozone will cause errors in colorimetric work unless conditions are carefully selected.

3. Low solubility of the reagents in water. This is the greatest disadvantage which organic precipitants have in comparison with inorganic precipitants.

4. High molecular weight. The very high molecular weights of chelate complexes formed with metal ions reduce, proportionately, small errors in precipitation and weighing.

5. Coloured compounds. Chelate compounds are often highly coloured and since they are usually soluble in organic solvents, they lend themselves to colorimetric or photometric methods of analysis.

6. Solubility in organic solvents. Because of their covalent nature, most metal complexes with organic reagents are soluble in non-polar solvents, chloroform being usually the best. Such solvents are used to extract the complexes for colorimetric determination, or they may be used simply for separating one element from another; thus the complex of cupferron with tripositive iron is very soluble in ether and chloroform, which allow one to remove large amounts of iron from solutions containing Al. In these extractions the pH must be carefully adjusted to give the desired separation.

Important Organic Reagents

Some of the organic reagents which are generally used in the identification of ions are discussed below :

1. Oxine, C_9H_7ON

The structure of this reagent is given in Fig. 18.1. It is almost a colourless solid soluble in organic solvents and mineral acids. It gives yellow colour which may be due to the formation of quinoid structure. It was introduced by Halm and Berg in 1927 as analytical reagent.

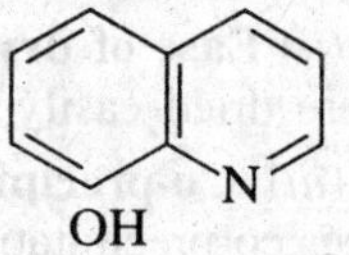

N

OH

Fig. 18.1. Structure of oxine molecule, C_9H_7ON.

A number of metals give co-ordination complexes with this reagent. Examples of such metals are Mg, Zn, Cu, Cd, Pb, Al, Fe, B, Ga, Th, Zr etc. The formation of the complex compound with a bivalant metal ion like Zn^{2+} is shown below :

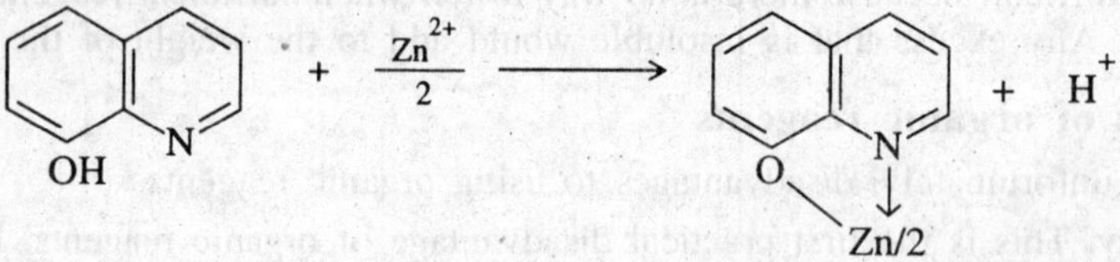

$$C_9H_6N(OH) + \frac{Zn^{2+}}{2} \longrightarrow C_9H_6N(O)\text{—}Zn/2 + H^+$$

The quantitative estimation of the complex is carried out by drying and weighing the hydrated or anhydrous complex or complex is ignited to oxide and then weighed. In volumetric methods the precipitate is dissolved by filtering with hydrochloric acid and then resulting solution is titrated against standard potassium bromate-bromide solution. In spectrophotometric method, the estimation of the metal ion is carried out by measuring the optical density of the complex. The reagent has been found to be suitable for Mg, Al, etc.

The reagent is a 2% solution of oxine in 2N acetic acid.

2. Dimethyl glyoxime

This reagent was discovered by L. Tschngaff (1905) and was applied by O. Brunck (1907) for the determination of nickel in steel. The structure of this reagent is shown in Fig. 18.2. This reagent gives an insoluble red chelate with Ni^{2+} ion in ammoniacal medium. For details see "*Applications of the formation of chelated complexes*" discussed in chapter on "*Basic concepts of coordination chemistry*" (Chapter 1).

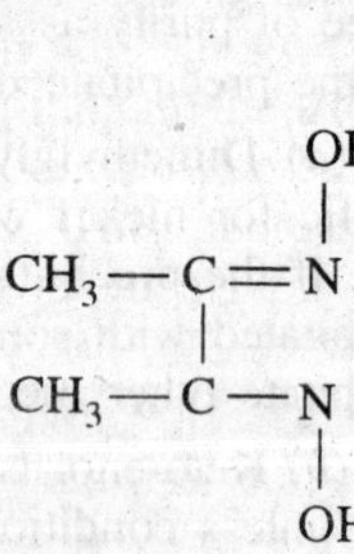

$$\begin{array}{l} \quad\quad\quad\quad\ \ OH \\ \quad\quad\quad\quad\ \ | \\ CH_3—C{=}N \\ \quad\quad\ \ | \\ CH_3—C{=}N \\ \quad\quad\quad\quad\ \ | \\ \quad\quad\quad\quad\ \ OH \end{array}$$

Fig. 18.2. Structure of dimethyl glyoxime.

Solution of palladium salts gives characteristic yellow precipitate in dil. HCl or H_2SO_4 but nitrate ions interfere. Solution of bismuth salts also gives a yellow precipitate in presence of EDTA at pH 11.0 to 11.5.

For the studies 1% solution of the dimethyl glyoxime in alcohol is employed.

In spectrophotometric method, the insoluble precipitate is extracted into chloroform which yields yellow colouration and then the optical density is measured.

3. α-nitroso-β-naphthol (also called 1-nitroso-2-naphthol))

The structure of this reagent is shown at (*a*) of Fig. 18.3.

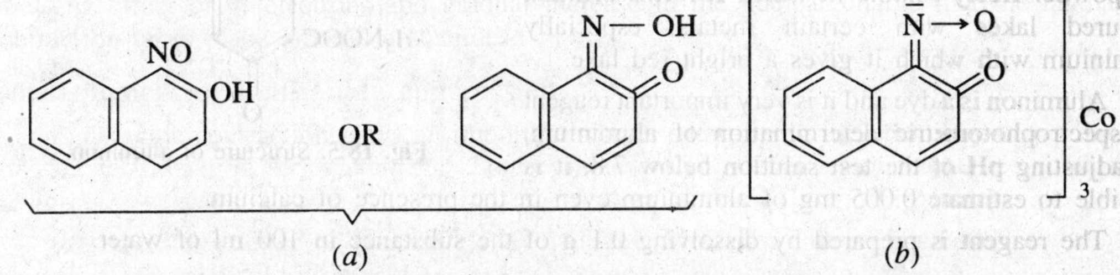

Fig. 18.3. (*a*) Structure of α-nitroso-β-naphthol (*b*) structure of the chelated complex formed between α-nitroso-β-naphthol and Co^{3+} ion.

It is an orange brown powder having m.p. of 109°C, soluble in organic solvents but almost insoluble in water. The reagent is usually employed as a saturated solution in 50% acetic acid. This reagent is common for the precipitation of many metallic radicals but it is used specially for the detection of ferric iron, copper and cobalt; with palladium and uranyl salts in slightly acid solution, it gives insoluble coloured compounds under suitable conditions.

Nickel and magnesium, however, are not precipitated by this reagent. Therefore, the reagent in acid solution is used for the separation of nickel from cobalt.

The structure of the chelated complex formed by this reagent with Co^{3+} ion is shown at (*b*) of Fig. 18.3.

The reagent has also been used for the spectrophotometric determination of cobalt and a few other metal ions. It is also used for spot-test of cobalt.

4. Cupferron

Cupferron exists largely not in the nitroso hydroxy form (*a*) but in the tautomeric amine oxide form (*b*) (Fig. 18.4).

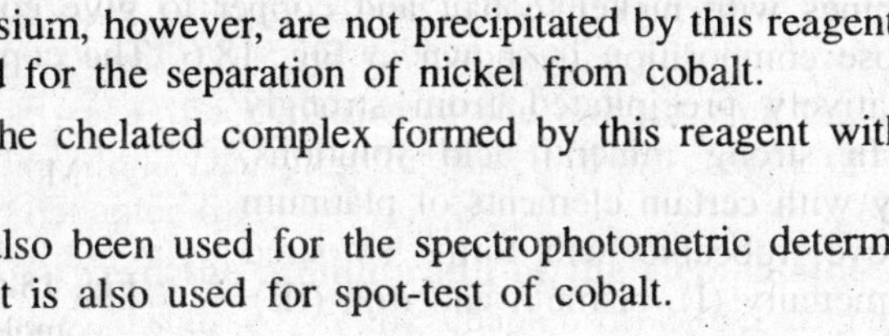

$$C_6H_5-N(-OH)-N{=}O \qquad C_6H_5-\overset{+}{N}(-\overset{-}{O}){=}N-OH$$

(*a*) (*b*)

Fig. 18.4. (*a*) Nitroso hydroxy form of cupferron (*b*) Tautomeric amine oxide form of cupferron.

The reagent is usually employed as a 6% aqueous solution, but is most useful in strongly acidic solution. In strongly acidic solution it precipitates Fe(III), Ce(III), Nb, Ta, W(VI), Ga and Sn (II), Cu (II) etc. Precipitates should be filtered immediately after their formation because cupferron is not very stable in acid solution. Furthermore, precipitation should be carried out in cold, since the reagent is decomposed into nitrobenzene on heating. The precipitates cannot be weighed after drying, but must be ignited to the corresponding oxide and then weighed.

The formation of chelated complex with Cu^{2+} ion can be shown as follows :

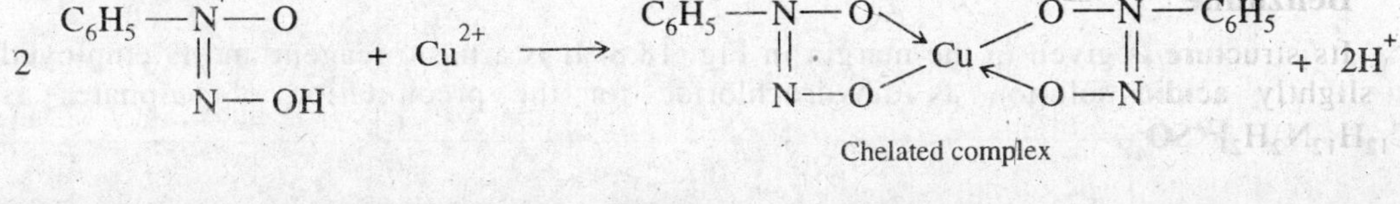

$$2\ C_6H_5-\overset{+}{N}(-\overset{-}{O}){=}N-OH + Cu^{2+} \longrightarrow \text{Chelated complex } [(C_6H_5-\overset{+}{N}(=N-O^-)-O^-)_2Cu] + 2H^+$$

Chelated complex

5. Aluminon

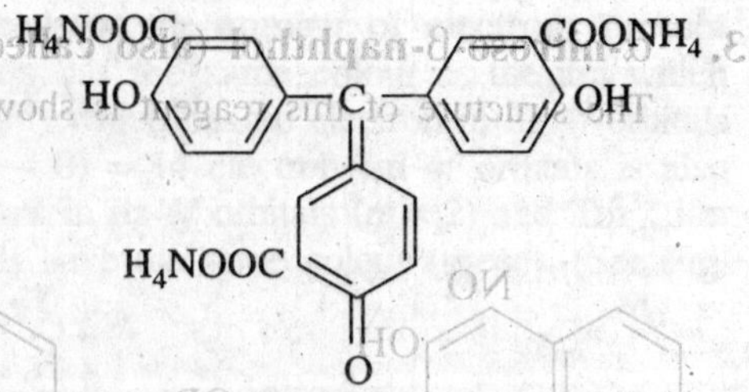

It is a ammonium salt of aurine tricarboxylic acid. Its structure is shown in Fig. 18.5. It is a brownish red powder freely soluble in water. It forms brilliant coloured lakes with certain metals, especially aluminium with which it gives a bright red lake.

Aluminon is a dye and it is very important reagent for spectrophotometric determination of aluminium. By adjusting pH of the test solution below 7.8, it is possible to estimate 0.005 mg of aluminium even in the presence of calcium.

Fig. 18.5. Structure of aluminon

The reagent is prepared by dissolving 0.1 g of the substance in 100 ml of water.

6. Rubeanic acid (also called dithioxamide)

This reagent is the diamide of di-thio-oxalic acid and is an equilibrium mixture with its tautomeric (aci-) di-imido form :

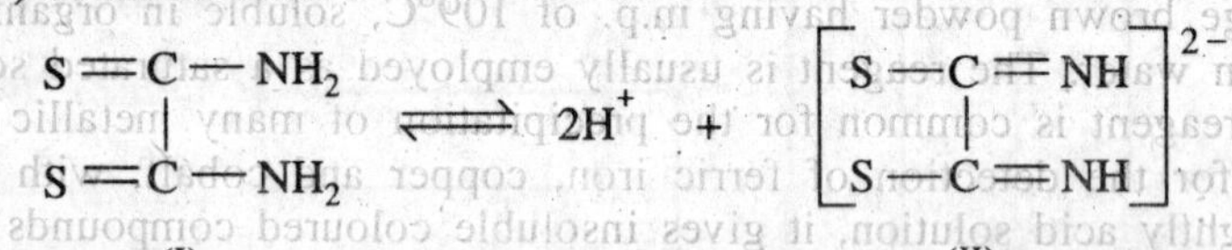

$$\begin{matrix} S=C-NH_2 \\ | \\ S=C-NH_2 \end{matrix} \rightleftharpoons 2H^+ + \left[\begin{matrix} S-C=NH \\ | \\ S-C=NH \end{matrix}\right]^{2-}$$

(I) (II)

The aci-form combines with nickel, cobalt and copper to give coloured water insoluble inner complex salts whose composition is shown in Fig. 18.6. The copper, cobalt, and nickel compounds are quantitatively precipitated from strongly ammoniacal solutions. In strong mineral acid solutions, rubeanic acid reacts only with certain elements of platinum salts. In addition to above, rubeanic acid forms coloured compounds with silver, mercury (I), bismuth and iron (III). Rubeanic acid is not only important as spot test reagent but also as a gravimetric and spectrophotometric reagent.

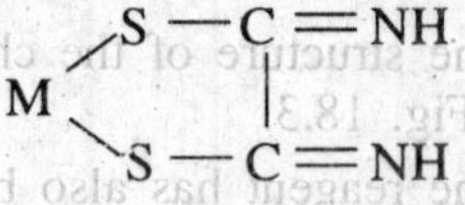

Fig. 18.6. Constitution of the complex formed between M(M = Ni, Co, Cu) and (II)

7. Rhodamine-B

It is tetraethyl rhodamine and is represented by the structure of given in Fig. 18.7.

Solution of antimony (V) salts in strong hydrochloric acid reacts with rhodamine-B and the aqueous solution of the reagent changes from red to violet or blue and finely-divided precipitate separates immediately or after standing. Trace amounts of antimony may be detected in the presence of large quantities of tin by means of this reagent. Small amounts of iron do not interfere. Mercury, gold, palladium are precipitated while molybdates and tungstates in acid solution give colour reaction with Rhodamine-B, similar to those given by antimony salt. The freshly prepared benzene solution of the rhodamine-B is preferred to the aqueous solution.

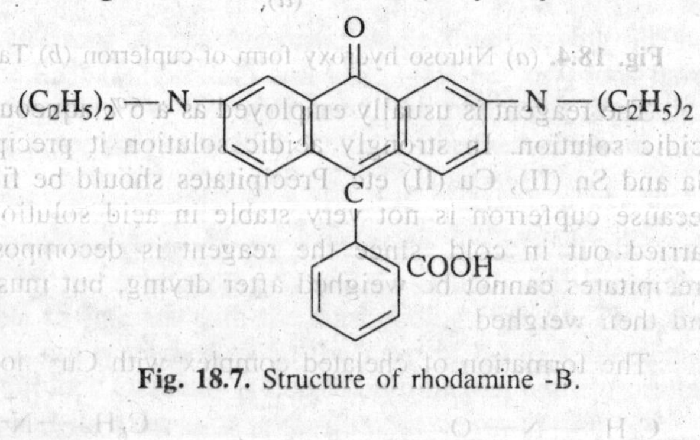

Fig. 18.7. Structure of rhodamine -B.

8. Benzidine

Its structure is given in the margin in Fig. 18.8. It is a basic reagent and is employed in slightly acidic solution as dihydrochloride for the precipitation of sulphates as $[C_{12}H_{12}N_2H_2]^{2+}SO_4^{2-}$.

The resulting precipitates may be titrated against standard alkali. The important drawback of the method is the relatively high solubility of the precipitate in water and even more in dilute hydrochloric acid. It also precipitates anions like phosphates and tungstates and give crystalline precipitates with a number of metallic ions such as silver, copper, cadmium, gold, zinc etc. With Co (III) and Tl (III), benzidine yields a blue cooluration in alkaline medium which is an important spot test for the metal ions.

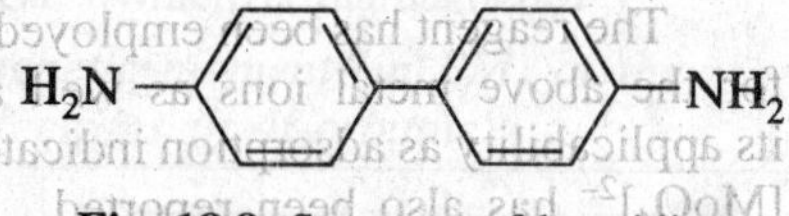

$$H_2N-C_6H_4-C_6H_4-NH_2$$

Fig. 18.8. Structure of benzidine, $C_{12}H_{12}N_2$

Recently, benzidine in dilute acetic acid has also been empolyed as a spot test reagent for the idenfication of some platinum metals like osmium, iridium and platinum.

9. Cupron or α-benzoin

Its structure is given in Fig. 18.9. It is a white crystalline solid. It is sparingly soluble in water but fairly soluble in alcohol. It also exists in isomeric β-form.

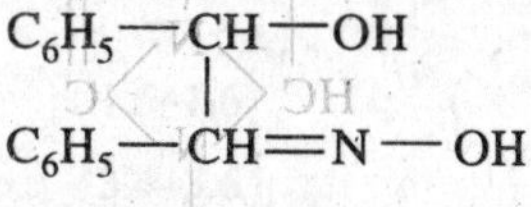

$$\begin{array}{l} C_6H_5-CH-OH \\ \qquad\quad | \\ C_6H_5-CH=N-OH \end{array}$$

Fig. 18.9. Structure of cupron.

Cupron is used as a reagent for the quantitative determination of copper, molybdenum and tungsten. Alcoholic solution of cupron gives a green precipitate with ammoniacal solution of copper salt. Precipitation of other metals in ammoniacal medium is prevented in the presence of tartaric acid. The green precipitate is soluble in dilute mineral acids. Copper can be estimated both volumetrically and gravimetrically by this reagent. The green precipitate is an inner complex whose formation can be shown as:

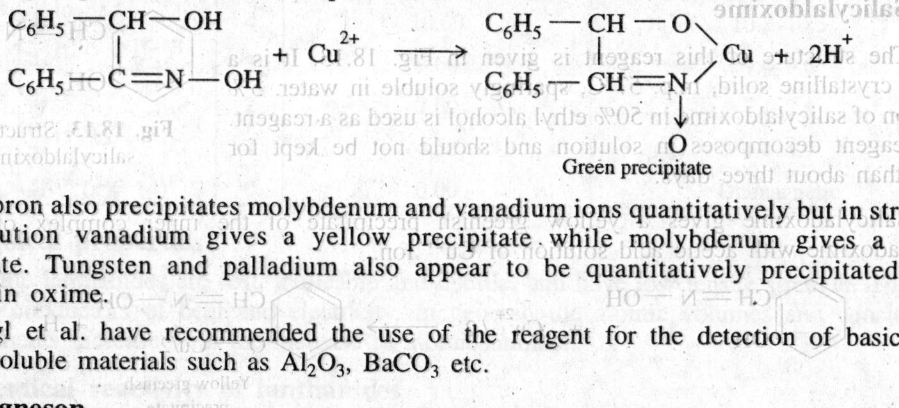

$$\begin{array}{l} C_6H_5-CH-OH \\ \qquad\quad | \\ C_6H_5-C=N-OH \end{array} + Cu^{2+} \longrightarrow \begin{array}{l} C_6H_5-CH-O \\ \qquad\quad | \qquad\quad \rangle Cu \\ C_6H_5-CH=N \\ \qquad\qquad\quad \downarrow \\ \qquad\qquad\quad O \end{array} + 2H^+$$

Green precipitate

Cupron also precipitates molybdenum and vanadium ions quantitatively but in strongly acid solution vanadium gives a yellow precipitate while molybdenum gives a white precipitate. Tungsten and palladium also appear to be quantitatively precipitated with α-benzoin oxime.

Feigl et al. have recommended the use of the reagent for the detection of basicity in slightly soluble materials such as Al_2O_3, $BaCO_3$ etc.

10. Magneson

The structure of this reagent is given in Fig. 18.10. It is also called *p-nitrobenzene azo-resorcinol.*

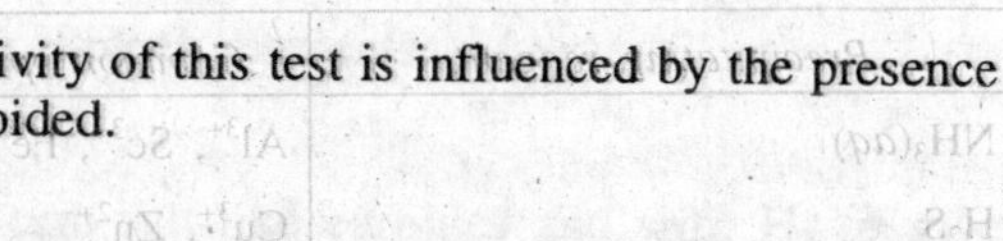

$$O_2N-C_6H_4-N=N-C_6H_3(OH)_2$$

Fig. 18.10. Structure of magneson.

1 to 0.5% solution in 0.25 N in NaOH is generally employed. In alkaline solutions it forms a bright blue lake with magnesium ions. The sensitivity of this test is influenced by the presence of ammonium salts, so their presence must be avoided.

11. Alizarin–S

It is alizarin sulphonic acid whose structure has been shown in Fig. 18.11. It is a dye whose sodium salt is soluble in water and is employed as an analytical reagent.

Aluminium salts give a red precipitate with the violet solution of Alizarin–S in ammonia. This complex is an adsorption complex of aluminium with alizarin sulphonic acid. Adsorption system of this type containing metal hydroxide and a dye stuff are called lakes. Zirconium (IV) forms red-violet coloured compounds with the reagent under the alkaline condition. Boric acid also forms brilliant red colour in U.V. light by alizarin in H_2SO_4.

The reagent has been employed as a spectrophotometric reagent for the above metal ions as well as for few others. Furthermore its applicability as adsorption indicator in the titration of $[Fe(CN)_6]^{4-}$, $[MoO_4]^{2-}$ has also been reported.

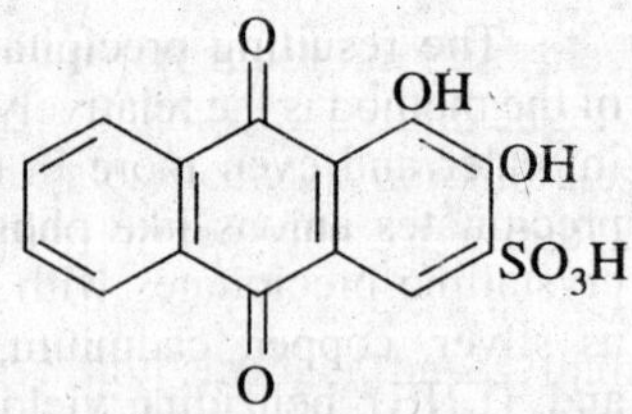

Fig. 18.11. Structure of alizarin–S.

12. Nitron

Its structure is given in Fig. 18.12. It is a yellow solid. It is insoluble in water but dissolves in acetic acid. Its 10% solution in 5% acetic acid is generally used as a reagent.

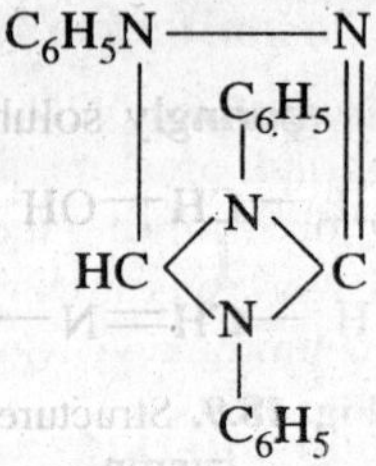

Fig. 18.12. Structure of nitron.

Nitron with nitrate yields a sparingly soluble crystalline nitrate $C_{20}H_{16}N_4.HNO_3$ in solution acidified with acetic or sulphuric acid. Thus, nitron is a gravimetric reagent for nitrate.

Perchlorate, per-rhenate, tetrafluoborate, and tungstate also form insoluble salts, and can be determined in a similar manner. Numerous other anion including bromide, iodide, chlorate, thiocyanate, nitrite and chromate interfere but may easily be removed by preliminary treatment. The results in the presence of chloride ions are generally high, possible because of co-precipitation.

13. Salicylaldoxime

The structure of this reagent is given in Fig. 18.13. It is a white, crystalline solid, m.p. 57°C, sparingly soluble in water. 5% solution of salicylaldoxime in 50% ethyl alcohol is used as a reagent. The reagent decomposes in solution and should not be kept for more than about three days.

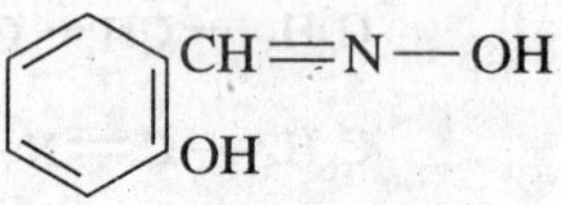

Fig. 18.13. Structure of salicylaldoxime.

Salicyladoxime gives a yellow greenish precipitate of the inner complex of copper salicyladoxime with acetic acid solution of Cu^{2+} ion.

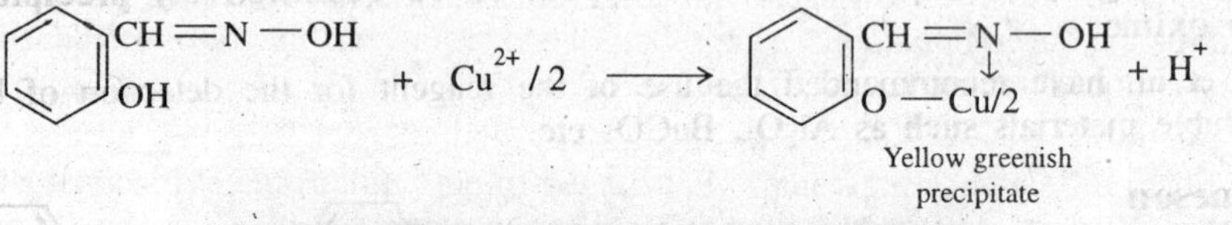

$$C_6H_4(OH)CH{=}N{-}OH + Cu^{2+}/2 \longrightarrow C_6H_4(O{-}Cu/2)CH{=}N{-}OH + H^+$$

Yellow greenish precipitate

Important Inorganic Reagents

Some important inorganic precipitating reagents are listed in Table 18.1.

Table 18.1. *Some common precipitating reagents.*

Precipitating reagent	*Common species precipitated*	*Weighing form of the precipitate*
$NH_3(aq)$	Al^{3+}, Sc^{3+}, Fe^{3+}, Cr^{3+}, In^{3+}	Hydrous oxides, M_2O_3
H_2S	Cu^{3+}, Zn^{2+}	CuO, ZnO or $ZnSO_4$
HCl	Ag^+, Hg^+	AgCl, Hg_2Cl_2
H_2SO_4	Sr^{2+}, Cd^{2+}, Pb^{2+}, Ba^{2+}	M_2SO_4
$AgNO_3$	Cl^-, Br^-, I^-	AgCl, AgBr, AgI
$BaCl_2$	SO_4^{2-}	$BaSO_4$

$(NH_4)_2CO_3$	Bi^{3+}	Bi_2O_3
$C_2H_2O_4$	Ca^{2+}, Sr^{2+}, Th^{4+}	CaO, SrO, ThO_2
$(NH_4)_2HPO_4$	Mg^{2+}, Mn^{2+}, Zn^{2+}, Al^{3+}	$M_2P_2O_7$, $AlPO_4$
$(NH_4)_2MoO_4$	Cd^{2+}, Pb^{2+}	$M(MoO_4)$
$MgCl_2 + NH_4Cl$	PO_4^{3-}	$Mg_2P_2O_7$

Precipitation of Al^{3+}, Ca^{2+} and Mg^{2+} ions by liq. NH_3, oxalic acid ($H_2C_2O_4$) and $(NH_4)_2HPO_4$ respectively can be shown as :

(*i*) $$Al_2(SO_4)_3 + 6NH_4OH \longrightarrow 2Al(OH)_3 + 3(NH_4)_2SO_4$$

$$2Al(OH)_3 \xrightarrow[-3H_2O]{\text{Ignite}} Al_2O_3 + 3H_2O$$

(*ii*) $$CaCl_2 + H_2C_2O_4 \xrightarrow{135°C} CaC_2O_4.H_2O + 2HCl$$

$$CaC_2O_4.H_2O \xrightarrow{225°C} CaC_2O_4 \xrightarrow{450°C} CaCO_3 \xrightarrow{850°C} CaO$$

(*iii*) $$MgSO_4 + (NH_4)_2HPO_4 \longrightarrow Mg(NH_4)PO_4.6H_2O + NH_4HSO_4$$

$$2Mg(NH_4)PO_4.6H_2O \xrightarrow{\text{Ignite}} Mg_2P_2O_7 + 2NH_3 + 7H_2O$$

University Questions

1. Name some common inorganic and organic precipitating agents. Illustrate which of these reagents are employed for gravimetry?
2. What are advantages of using organic reagents over inorganic reagents.
3. Illustrate, with examples, the uses of the following reagents: dimethyl glyoxime, α-nitroso-β-naphthol, oxine and cuperon.
4. Write chemical equations showing the precipitation of Al^{3+}, Ca^{2+} and Mg^{2+} by liq. NH_3, $H_2C_2O_4$ and $(NH_4)_2HPO_4$ respectively.

19

Role of Metal Ions in Biological Systems

Introduction

It is known that life processes in plants and animals involve about forty elements out of which twenty five are essential for healthy human life. These elements are Na, K, Ca, Mg, V, Cr, Mn, Fe, Co, Cu, Zn, Mo, and Sn (metals), H, C, Si, N, O, P, S, Se, F, Cl, Br and I (non-metals). Among these elements, H, O, C, N, Na, K, Ca, Mg, P, S, and Cl are most abundant; Mn, Fe, Co, Cu, Zn, Mo and I are less abundant while the rest of the seven elements—Si, V, Cr, Sn, F, Br and Se occur in ultratraces. The important biochemical roles of the most essential metals are summarized below:

Na^+ and K^+ ions

(*i*) Na^+ and K^+ cations are present in the red blood cells. The ratio of these ions in mammals such as human beings, rabbits, rat and horses is 1 : 7. In cats and dogs the ratio is 15 : 1. In order to establish this ratio called *concentration gradient* in the cell, work has to be done. Biologists have suggested different mechanisms involving sodium pump and potassium pump for this purpose. The cation gradients control the development and the functioning of the nerve cells. In the restful state, the potential of the nerve cells is linked to the K^+ ion concentration across the membrane. During the activation of the nerve cells, a chemical called acetyl choline is released near its end plate and membrane potential is discharged. This discharge is transmitted through the length of the nerve cell by electric impulse. Thus, it is evident that both Na^+ and K^+ ions have a significant role in biology.

(*ii*) Many enzyme reactions are controlled by Na^+ and K^+ ions.

(*iii*) The injectable medicines are mostly dissolved in sodium chloride (saline water) before they are injected into human body.

Mg^{2+} and Ca^{2+} ions

(*i*) Mg^{2+} ions are present in chlorophyll which is the green colouring matter present in plants. Chlorophyll absorbs light from sun and carries the process of *photosynthesis* in plants.

(*ii*) Ca^{2+} ions are present as phosphates in the bones of both human beings and animals. These ions also play an important role in the muscle contraction. The malnutrition in children is mainly due to the deficiency of the Ca^{2+} ions.

(*iii*) Both Mg^{2+} an Ca^{2+} ions catalyse the formation of pyrophosphate linkages which control the various biological systems. The pyrophosphates undergo hydrolysis with the release of energy. This process is controlled by Ca^{2+} ions.

Iron

Iron metal is a constituent of haemoglobin in blood which carries O_2 to the various parts of the body in men and animals. Iron salts in the form of tablets or capsules are given to the patients for the treatment of anaemia (*i.e.*, for the formation of haemoglobin in the body).

Copper and zinc

These metals are the constitutuents of organic catalysts in the body, called enzymes. Zinc-metallo-enzymes catalyse peptide hydrolysis and maintain HCO_3^-/CO_2 equilibrium.

For an adult man of 70 kg body weight, the quantities of the metals are in grams: Ca = 1700; K = 250; Na = 70; Mg = 42; Fe = 7; Mn, Co, Cu, Zn, Mo each < 1. In other words, metals constitute about 3% of human body weight but the intricate biological systems depend very much upon the metals. It should be noted, however, that the metals do not exist as free metal ions in living systems but as metal chelates of exceptionally high stability.

Harmful effects of excess of metals on human body

Small quantities of metals are useful for the normal functioning of human body. However, if excessive amount of metals is used, it proves to be harmful. For example:

(*i*) If iron salt is taken in small amounts, it helps in the formation of haemoglobin and thus cures the disease of anaemia, but if the excess of iron salt is taken, it causes the disease of *siderosis*. This disease has been found especially in Bantu tribes in Africa, since the people of this tribe prepare their beer in iron pots and thus iron intake in their bodies increases.

(*ii*) The weight of Cu-metal required for a normal man is 0.10 to 0.15g. If excess of this metal is taken, it produces a disease known as *Wilson's disease*. In this desease, the patient feels nervous disorders, tremors, difficulty in swallowing and still joints.

Structure and role of metal chelates in living systems

In general, the metal chelates in living systems play two important roles:

(*i*) They act as catalysts in many metabolic processes which involve: (*a*) redox reactions with change in the valency of the metal in the complex, (*b*) acid-base reactions in which the metal is the Lewis acid combining with a substrate to accelerate a reaction.

(*ii*) They help in the storage and transfer of either the metal ions of donor molecules and serve as agents for transmission of energy in the animal and plant metabolism.

Structure of chlorophyl

Chlorophyl is a green-coloured matter (pigment) present in green plants and algae. It contains the basic unit, porphyrin ring. It is a square planar complex containing Mg^{2+}. Thus it is a naturally occuring chelated complex of Mg. The Mg-atom in chlorophyl lies slightly above the plane of the rigid chelate ring structure. The chlorophyl molecule can add to it one or two water molecules axially. Such coordinated water molecules then help in associating other chlorophyl molecules through hydrogen bonding.

There are various modifications of chlorophyl. All these modifications contain porphyrin ring and differ from each other in the side groups of the porphyrin ring. The most abundant chlorophyl is *chlorophyl a* whose structure is shown in Fig. 19.1.

In this structure, four N-atoms are situated at the four corners of a square. The N_4 unit forms a part of 16-membered natural macrocycle.

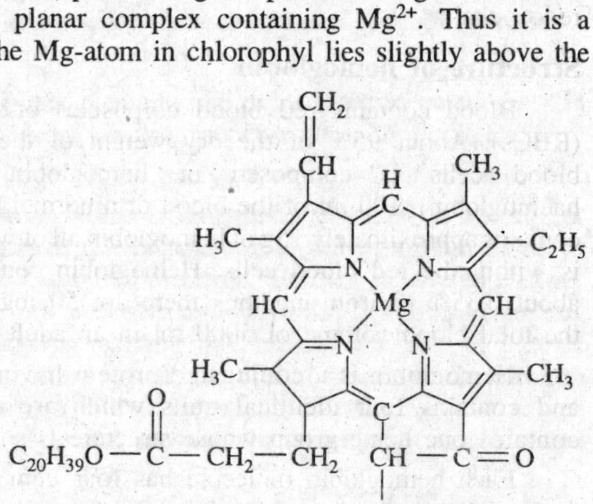

Fig. 19.1. Structure of chlorophyl *a*

As stated above, all types of chlorophyl contain porphyrin ring in which a double bond in one of the pyrorole rings is reduced and a fused cyclopentaone ring is also present (See Fig. 19.1).

Photosynthesis

Photosynthesis is a process in which CO_2 and moisture of the atmosphere combine together in the presence of sunlight and chlorophyl and produce glucose ($C_6H_{12}O_6$) and O_2. Glucose is further converted into polysaccharides like starch. In a simplified way, the process of photosynthesis can be shown as:

$$6CO_2 + 6H_2O + \textit{Solar energy} \xrightarrow{\text{Chlorophyl}} C_6H_{12}O_6 \ (\textit{Glucose}) + 6O_2 \quad ...(A)$$

Since tracer studies have shown that all O_2 evolved in the above reaction comes from H_2O and not from CO_2, this reaction takes place through the following steps:

(*a*) **Light reaction step.** In this step, chlorophyl absorbs solar energy which is used up in forming energy-rich ATP molecules and H_2O molecules are oxidised to O_2.

$$12H_2O \longrightarrow 12H_2 + 6O_2 \ (\textit{Molecular equation})$$

or $$12H_2O \xrightarrow{\text{Oxidation}} 24H^+ + 24e^- + 6O_2 \ (\textit{Ionic equation}) \quad ...(i)$$

(*b*) **Dark reaction step.** In this step, the energy-rich ATP molecules formed as above are used up to reduce CO_2 into glucose ($C_6H_{12}O_6$).

$$6CO_2 + 12H_2 \xrightarrow{\text{18 ATP}} C_6H_{12}O_6 + 6H_2O \ (\textit{Molecular equation})$$

or $$6CO_2 + 24H^+ + 24e^- \xrightarrow{\text{Reduction}} C_6H_{12}O_6 + 6H_2O \ (\textit{Ionic equation}) \quad ...(ii)$$

On adding equations (*i*) and (*ii*), we get equation (A). For the synthesis of one molecule of glucose, 18 molecules of ATP are required. The energy needed for the above reaction is supplied by ATP molecules.

Role of Mg present in chlorophyl

Mg functions in two ways: (1) It makes the entire molecule rigid so that energy is not easily lost thermally, and (2) it increases the rate at which the short-lived singlet excited state is transformed into the corresponding triplet state, which has a longer life and hence can transfer its excitation energy into the redox chain.

Structure of hemoglobin

Blood contains red blood corpuscles or cells (RBCs). About 95% of the dry weight of the red blood cells is composed of hemoglobin (or haemoglobin). 100 ml of the blood of a normal male contains approximately 15g of hemoglobin all of which is within the red blood cells. Hemoglobin contains about 0.35% of iron and thus there are 50 mg of Fe per 100 ml blood and 2.5 g of iron in the total blood volume of 5000 ml in an adult man.

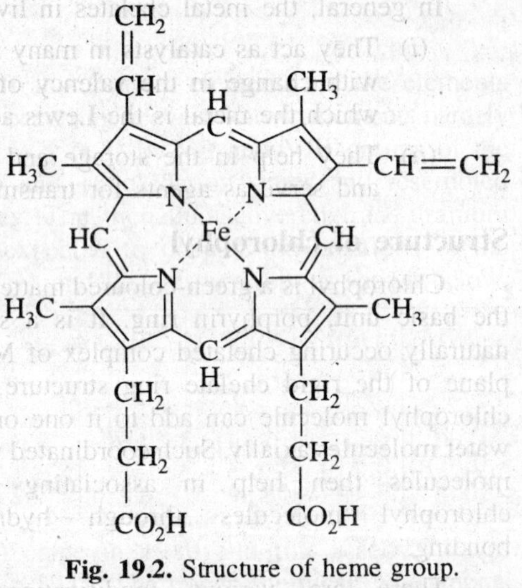

Fig. 19.2. Structure of heme group.

Hemoglobin is a conjugated protein having molecular weight of 64,500. It is a tetramer and contains four identical units which are arranged roughly in a tetrahedron. Each unit contains one heme group whose structure is shown in Fig. 19.2.

Each hemoglobin molecule has four heme groups bound to the globin (a protein) on its surface. Schematic representation of the structure of hemoglobin is given in Fig. 19.3. Fe (II) in hemoglobin is in the high spin state ($Fe^{2+} = 3d^6 = t_{2g}^4 \, e_g^2$). Thus Fe(II) has four unpaired electrons in hemoglobin molecule.

The structure of hemoglobin given in Fig. 19.3 shows that hemoglobin is an octahedral complex of Fe(II). The centre of hemoglobin is occupied by Fe(II) and the four corners of the square base are occupied by four N-atoms of the heme group. One axial position is occupied by N-atom of histidine while the other axial position is occupied by H_2O molecule.

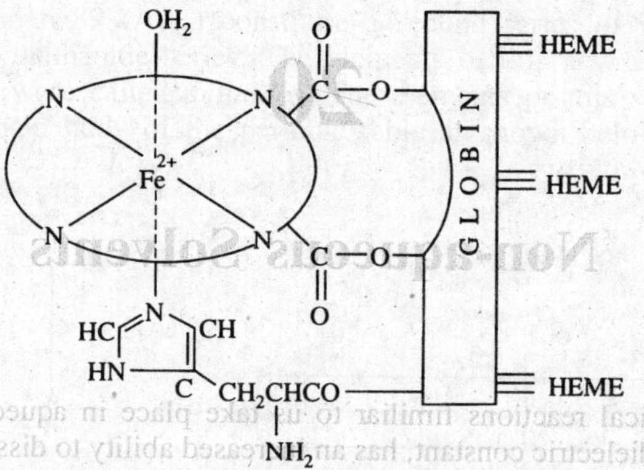

Fig. 19.3. Schematic representation of the structure of hemoglobin.

Role of hemoglobin in biological (living) systems

Hemoglobin is very sensitive to O_2 and hence readily combines with it. Thus when we breathe in O_2, hemoglobin (Hb) present in our blood combines *reversibly* with O_2 in the lungs to form *oxy-hemoglobin* (HbO_2).

$$\text{Hemoglobin} + O_2 \rightleftharpoons \text{Oxy-hemoglobin} + H_2O$$

or
$$Hb + O_2 \rightleftharpoons HbO_2 + H_2O$$

The formation of HbO_2 is called *oxy-genation of hemoglobin*. In HbO_2, Fe is in +3 oxidation state. In the formation of HbO_2, H_2O molecule present at one of the axial positions in hemoglobin is reversibly replaced by O_2 molecule.

As the blood runs through the arteries to the tissues, the oxygen pressure decreases and O_2 bound with HbO_2 is set free. This free O_2 difffuses into the body cells where it combines with glucose (food), $C_6H_{12}O_6$ and oxidises it to CO_2 and H_2O. Thus O_2 is used in the combustion of food.

$$C_6H_{12}O_6 + 6O_2 \longrightarrow 6CO_2 + 6H_2O + 38ATP\ (\textit{Energy})$$

The oxidation of $C_6H_{12}O_6$ to CO_2 is an energy releasing process in which energy in the form of ATP is produced. This energy is utilised by the living organisms to preform their various metabolic activities, and for maintaining their body temperature. Water produced in the above reaction is retained in the body while CO_2 combines with the amino group of haemoglobin to form carb-amino haemoglobin which decomposes to give CO_2 and haemoglobin. CO_2 is exhaled out and haemoglobin goes to the lungs for reuse. The above discussion makes it evident that the functions of hemoglobin in our body is to supply O_2 to various parts of the body. Thus hemoglobin is an *oxygen-carrier*.

In the oxygenation of a hemoglobin molecule, a proton (H^+) is also produced. This proton reacts with bicarbonate dissolved in the blood to liberate CO_2.

$$HCO_3^- + H^+ \rightleftharpoons H_2CO_3 \rightleftharpoons H_2O + CO_2$$

CO_2 liberated as above is exhausted out.

University Questions

1. Write the role of the following metal ions in living systems: Na^+, K^+, Mg^{2+} and Ca^{2+}.
2. What are the roles of metal chelates in living systems.
3. Write the structure and role of the following (*i*) Chlorophyl (*ii*) Hemoglobin.

20

Non-aqueous Solvents

Introduction

Most of the chemical reactions fimiliar to us take place in aqueous solution, since water, due to high value of its dielectric constant, has an increased ability to dissolve the ionic compounds and many other sustances. However there are many chemical reactions which cannot be carried out in aqueous solution but can be studied in a non-aqueous medium. A large number of non-aqueous solvents have been discovered in the last few years. These solvents have sizeable value of their dielectric constant and hence can dissolve many substances. Examples of such solvents are anhydrous liq. NH_3, liq. SO_2, anhydrous HF, liq. N_2O_4 etc.

Classification of solvents

Solvents have been classified in a number of ways, dependiing on the properties of the solvents.

First classification

This classification is based on the proton-donating and proton-accepting ability of the solvent. This classification gives the following types of solvents. All these solvents contain hydrogen in their formula.

1. Protic or protonic solvents. These solvents either lose protons or gain them or can show both the tendencies. Thus these solvents are of three types:

(a) Acidic (proto-genic) solvents. These can lose protons readily. Examples are HF, H_2SO_4, HCOOH, HCN, C_6H_5COOH, CH_3COOH etc.

(b) Basic (proto-philic) solvents. These can accept proton, *e.g.*, NH_3, N_2H_4, NH_2OH, amines etc.

(c) Amphi-protic or amphoteric solvents. These show dual character, *i.e.*, they can lose as well as accept protons, depending on the nature of the reacting species, *e.g.*, H_2O, alcohols, liq. NH_3, CH_3COOH etc. Ampho-teric solvents undergo auto-ionisation (self-ionisation) in which a proton transfer between two similar neutral molecules takes place and a cation-anion pair of the solvent is obtained, *e.g.*,

Acid		*Base*		*Acid*		*Base*
H_2O	+	H_2O	$\rightleftharpoons$	H_3O^+	+	OH^-
$NH_3(l)$	+	$NH_3(l)$	$\rightleftharpoons$	NH_4^+	+	NH_2^-
CH_3COOH	+	CH_3COOH	$\rightleftharpoons$	$CH_3COOH_2^+$	+	CH_3COO^-

2. Non-protic or non-protonic or aprotic solvents. These can neither lose nor gain the protons, *e.g.*, C_6H_6, $CHCl_3$, CH_2Cl_2, CCl_4 etc.

Second classification

This classification is based on the polar and non-polar nature of the solvents. This classification gives the following types of solvents.

1. Ionising, ionic or polar solvents (*e.g*, H_2O, NH_3, HF etc.) (*a*) These have high polarity and high dielectric constants.

(*b*) They dissolve ionic compounds and initiate ionic reactions.

(*c*) They can undergo auto-ionisation. Auto-ionisation of H_2O and NH_3 has already been shown while that of HF and SO_2 is shown below:

$$\begin{array}{ccccc} \textit{Acid} & \textit{Base} & & \textit{Acid} & \textit{Base} \\ HF + & HF & \rightleftharpoons & HF_2^+ & + F^- \\ SO_2 + & SO_2 & \rightleftharpoons & SO^{2+} & + SO_3^{2-} \end{array}$$

(*d*) Polar solvents tend to associate due to the dipole-dipole interactions. The association is more effective in case of protonic solvents (protonic solvents are those which contain hydrogen) due to hydrogen bonding and leads to a higher b.pt. and increases the liquid state range of the solvent.

2. Non-ionising or non polar solvents. (*e.g.*, C_6H_6, CCl_4 etc.) (*a*) These have low dipole moment and low dielectric constants.

(*b*) They dissolve non-ionic (covalent) compounds and cannot initiate ionic reactions.

Third classification

Water is called *aqueous solvent* while all others (*e.g.*, NH_3, SO_2, HF, C_6H_6, $CHCl_3$ etc.) are called *non-aqueous solvents.*

Relation between the dielectric constant (K) of an ionising solvent and the solubility of an ionic salt (solute) in it

Let us consider an ionic substance (solute) which is composed of cations and anions having their charges equal to c^+ and a^- respectively and separated from each other by a distance equal to r. Obviously r will be equal to the sum of the radii of the cation (r_+) and the anion (r_-). Let the ionic substance be placed in an ionising solvent of dielectric constant equal to K. From Coulomb's law, the force of attraction (F) between the cations and anions is given (in cgs units) by the expression:

$$F = \frac{c^+ \times a^-}{K\,r^2} = \frac{c^+ \times a^-}{K\,(r_+ + r_-)}$$

K is a dimensionless number and is a measure of the decrease in the intensity of the electric field between two charges in presence of the dielectric medium than in vacuum.

The above equation shows that F and K are inversely proportional to each other. This means that if the value of K for a solvent is high, the force between the ions of the solute is very low and hence the solvent will dissolve the solute quickly. Thus a solvent with a high value of K (*e.g.*, water and anhydrous HF) has an increased ability to dissolve the ionic or polar compound. On the other hand a solvent with a low value of K (*e.g.*, liq. NH_3 and liq. SO_2) shows almost no ability to dissolve the ionic compounds especially those containing multi-changed ions like oxides, sulphates, carbonates, sulphates and phosphates.

Relation between the dielectric constant (K) and dipole moment (D) of an ionising solvent

These two constants of an ionising solvent are closely related to each other. It has been seen that if an ionising solvent has a large value of its dielectric constant, the value of its dipole moment will also be large. This is evident from the following data:

Solvent	*Dipole moment (in Debye units)*	*Dielectric constant*
Hydrogen cyanide	2.93	106.8 (25°C)
Hydrogen fluoride	1.91	83.6 (0°C)
Water	1.84	78.5 (25°C)
Ethyl alcohol	1.69	24.2 (25°C)
Sulphur dioxide	1.63	17.4 (–19°C)
Ammonia	1.46	22.0 (–34°C)

Physical properties of water, ammonia and sulphur dioxide

These are given below :

Properties	*Water*	*Ammonia*	*Sulphur dioxide*
Boiling point (0°C)	100	– 34.00	– 10.0
Freezing point (0°C) (Melting point)	0.0	–77.7	–75.5
Density ($g.ml^{-1}$)	1.0 (4°C)	0.68 (at b.pt.)	1.46 (–10°C)
Dielectric constant	78.5 (25°C)	22.0 (at b.pt.)	17.4 (–20°C)
Specific conductance ($Ohm^{-1}cm^{-1}$)	6.0×10^{-8} (25°C)	0.5×10^{-8} (at b.pt.)	4×10^{-8} (–10°C)
Viscosity (centipoise)	0.959 (25°C)	0.254 (at b.pt.)	0.428
Dipole moment (Debye)	1.85	1.47	1.61 (–10°C)

Liquid Ammonia, (Liq. NH_3)

Introduction

Liq. NH_3 is a protonic solvent, since it gives a proton (H^+) on ionisation ($NH_3 \rightarrow H^+ + NH_2^-$). It is a highly useful solvent and a number of chemical reactions are carried out in this medium. The value of dielectric constant of liq. NH_3 (= 22.0 at –34°C) is much smaller than that of water (= 78.5 at 25°C) and hence liq. NH_3 has decreased ability to dissolve ionic compounds. The liquid range for liq. NH_3 is from –34.0°C to –77.7°C (b.pt. = –34.0°C, m.pt. = –77.7°C).

Solubility of various substances in liq. NH_3

(*a*) **Solubility of ionic compounds (inorganic salts).** We know that ammonia has a low value of its dielectric constant (= 22 at –34°C). This low value suggests that liq. NH_3 has a poor ability to dissolve ionic compounds. Ammonium salts (*e.g.*, NH_4NO_3, NH_4SCN, CH_3COONH_4 etc.) and most of the nitrites, nitrates, cyanides, thiocyanates, perchlorates, are soluble in liq. NH_3. Salts containing highly charged ions (*e.g.*, oxides, sulphides, sulphates, phosphates and carbonates) are insoluble. Fluorides and chlorides (except Be^{2+} and Na^+ chlorides) are practically insoluble, bromides are less soluble while iodides are freely soluble. Thus the solubility of the halides of a given metal increases in going from fluoride to iodide ($MF < MCl < MBr < MI$). Most of the metal amides (except those of alkali metals) are insoluble. The salts of some metals (*e.g.*, Ni^{2+}, Cu^{2+}, Zn^{2+} etc.) react with liq. NH_3 and form ammine complexes.

(*b*) **Solubility of non-ionic compounds (organic compounds).** Halogen compounds, alcohols, ketones, esters, simple ethers, amines, phenol and its derivatives etc. are soluble. Alkanes are insoluble and alkenes and alkynes are slightly soluble. In this sense, liq. NH_3 is a better solvent for non ionic and non polar compounds (organic compounds).

(*c*) **Solubility of non-metals.** The non-metals like S, P, I_2, Se etc. are soluble and they react with the solvent.

(*d*) **Solubility of alkali metals and alkaline earth metals.** All the alkali metals and alkaline earth metals (excepting Be) are soluble in liq. NH_3.100g. of liq. NH_3 dissolves 10.9g of Li,

24.8g of Na, 46.8g of K(all at –33°C) and 334g of Cs (at –50°C). Thus we see that the solubility of alkali metals in liq. NH_3 increases as we pass from Li to Cs (Li < Na < K < Cs). The solution obtained by dissolving alkali metals in liq. NH_3 has the following characteristics:

(*i*) *Colour of the solution.* The colour of the dilute solution is deep blue. As the concentration of the solution increases, the blue colour changes to bronze which disappears on the further increase of the concentration of the solution. The colour of the solution is independent of the metal dissolved in liq. NH_3. The blue colour of the dilute solution is due to the formation of ammoniated metal cation and ammoniated electron in it.

$$M \longrightarrow M^+ + e^-$$

$$M^+ + xNH_3 \longrightarrow \underset{\text{Ammoniated cation}}{[M(NH_3)_x]^+}$$

$$e^- + yNH_3 \longrightarrow \underset{\text{Ammoniated electron}}{[e(NH_3)_y]^-}$$

(*ii*) *Decomposition of very dilute solution.* Very dilute solution of the alkali metals in liq. NH_3 is meta stable, since in presence of a catalyst like Fe_2O_3 its ammoniated electron undergoes decomposition.

$$[e(NH_3)_y]^- \xrightarrow{Fe_2O_3} NH_2^- + \frac{1}{2} H_2 + (y-1)NH_3$$

(*iii*) *Evaporation of the solution.* When alkali metal-ammonia solution is evaporated, alkali metal dissolved in the solution is recovered.

(*iv*) *Paramagnetic character.* Dilute solution of alkali metals in liq. NH_3 is paramagnetic which is due to the presence of unpaired ammoniated electrons in the solution. The susceptibility of the solution falls down as the concentration of the solution increases and ultimately the solution becomes diamagnetic. Paramagnetism reappears at higher concentration of the solution.

(*v*) *Electrical conductivity.* The electrical conductivity of very dilute solutions is high and is in the range of fully ionised salts in water. The high conductivity is due to the presence of ammoniated electrons in the solution. It decreases with concentration upto about 0.05M, but then rises sharply. The concentrated bronze solution has conductivities in the range of metals.

(*vi*) *Density.* Dilute solution has very low density. Low density is due to the fact that ammoniated electrons present in the solution are supposed to occupy the cavities or holes within ammonia molecules.

(*vii*) *Reducing property of dilute solution.* We have said above that the blue solution of alkali metals in liq. NH_3 contains ammoniated electrons. These ammoniated electrons reduce various substances. Thus dilute solutions of alkali metals are reducing agents. The reactions showing the reducing property of alkali metals in liq. NH_3 have been discussed on page 478. In these reactions the alkali metal is oxidised to some alkali metal salt.

Auto-ionisation of liq. NH_3

Liq NH_3 ionises in the followinig three ways :

(*a*)

$$NH_3 \rightleftharpoons H^+ + NH_2^-$$

$$H^+ + NH_3 \rightleftharpoons NH_4^+$$

$$NH_3 + NH_3 \rightleftharpoons NH_4^+ + NH_2^-$$

or

$$\underset{\text{Solvent}}{2NH_3} \rightleftharpoons \underset{\text{(Solvent cations : acid ions)}}{NH_4^+ \ (\textit{Ammonium ions})} + \underset{\text{(Solvent anions : base ions)}}{NH_2^- \ (\textit{Amide ions})}$$

$$K_b = [NH_4^+]\,[NH_2^-] = 1.9 \times 10^{-33} \text{ at } -50°C.$$

(*b*) $3NH_3 \rightleftharpoons 2NH_4^+$ (*Ammonium ions*) + NH^{2-} (*Imide ions*)

Solvent (Solvent cations: acid ions) (Solvent anions: base ions)

(*c*) $4NH_3 \rightleftharpoons 3NH_4^+$ (*Ammonium ions*) + N^{3-} (*Nitride ions*)

Solvent (Solvent cations: acid ions) (Solvent anions: base ions)

Different modes of auto-ionisation of liq. NH_3, given above, indicate that, according to solvent system concept, in liq. NH_3, any substance that gives NH_4^+ ions (which are solvent cations) will act as an acid in liq. NH_3, white that which produces NH_2^- or NH^{2-} or N^{3-} ions (which are solvent anions) will behave as a base in liq. NH_3. The compounds which give NH_4^+ ions in liq. NH_3 are called *ammono acids*, while those which give NH_2^- or NH^{2-} or N^{3-} ions in this solvent are called *ammono bases.*

Examples of ammono acids

Ammonium salts [*e.g.*, NH_4Cl, NH_4Br, NH_4I, NH_4NO_3, $(NH_4)_2S$ etc.], organic amides (*e.g.*, urea, acetamide, sulphamide etc.), acetic acid, sulphamic acid etc., all act as ammono acids in liq. NH_3, since they produce NH_4^+ ions in this solvent, as shown below:

$$NH_4X \rightleftharpoons X^- + NH_4^+$$

(X = Cl, Br, I)

$$\underset{\text{Urea}}{H_2N.CO.NH_2} + NH_3 \longrightarrow H_2N.CO.NH^- + NH_4^+$$

$$\underset{\text{Acetamide}}{CH_3CONH_2} + NH_3 \longrightarrow CH_3CONH^- + NH_4^+$$

$$\underset{\text{Sulphamide}}{H_2N.SO_2.NH_2} + NH_3 \longrightarrow H_2N.SO_2.NH^- + NH_4^+$$

$$\underset{\text{Acetic acid}}{CH_3COOH} + NH_3 \longrightarrow CH_3COO^- + NH_4^+$$

$$\underset{\text{Sulphamic acid (Dibasic acid)}}{H_2N.SO_2.OH} + 2NH_3 \longrightarrow H^- N.SO_2.O^- + 2NH_4^+$$

It may be noted that since $H_2N.\ SO_2.OH$ produces two ions, this acid behaves as a dibasic acid in liq. NH_3.

Examples of ammono bases

KNH_2, PbNH and BiN are the examples of ammono bases, since these compounds give NH_2^-, NH^{2-} and N^{3-} ions respectively in liq. NH_3.

$$KNH_2 \xrightarrow{\text{liq. } NH_3} K^+ + NH_2^- \text{ (Amide ion)}$$

$$PbNH \xrightarrow{\text{liq. } NH_3} Pb^{2+} + NH^{2-} \text{ (Imide ion)}$$

$$BiN \xrightarrow{\text{liq. } NH_3} Bi^{3+} + N^{3-} \text{ (Nitride ion)}$$

Chemical Reactions Taking Place in Liq. NH_3

Chemical reactions that can take in liq. NH_3 are of the following types :

1. Precipitation (metathetical) reactions

Examples of some precipitation reactions that can be carried out in liq. NH_3 are given below:

(*a*) **Precipitation reaction shown by ammono acids in liq. NH_3.** When an ammono acid like NH_4Cl, NH_4Br, NH_4I, $(NH_4)_2S$ etc. is mixed with the liq. NH_3 solution of a metallic salt like $Sr(NO_3)_2$, $Zn(NO_3)_2$, $AgNO_3$, $Cu(NO_3)_2$, $Cd(NO_3)_2$, KI etc., the salt of the metal is precipitated. For example :

$$\underset{\substack{(X = Cl, Br)\\ \text{Ammono acid}}}{2NH_4X} + Sr(NO_3)_2 \rightleftharpoons SrX_2\downarrow + 2NH_4NO_3$$

$$2NH_4I + Zn(NO_3)_2 \xrightleftharpoons{\text{liq. } NH_3} ZnI_2\downarrow + 2NH_4NO_3$$

$$(NH_4)_2S + 2AgNO_3 \xrightleftharpoons{\text{liq. } NH_3} Ag_2S\downarrow + 2NH_4NO_3$$

$$(NH_4)_2S + M(NO_3) \xrightleftharpoons{\text{liq. } NH_3} MS\downarrow + 2NH_4NO_3$$

(M = Cu, Cd)

$$NH_4Cl + KI \xrightleftharpoons{\text{liq. } NH_3} KCl\downarrow + NH_4I$$

(*b*) **Precipitation reactions shown by ammono bases in liq. NH_3.** When an ammono base like KNH_2 is mixed with the liq. NH_3 solution of a metallic salt like $AgNO_3$, PbI_2, BiI_3 etc., the amide, imide or nitride of the metal is precipitated. For example:

$$\underset{\text{Ammono base}}{KNH_2} + AgNO_3 \xrightarrow{\text{liq. } NH_3} KNO_2 + AgNH_2\downarrow \ (\textit{Amide})$$

$$KNH_2 + PbI_2 \xrightarrow{\text{liq. } NH_3} KI + HI + Pb(NH)\downarrow \ (\textit{Imide})$$

$$KNH_2 + BiI_3 \xrightarrow{\text{liq. } NH_3} KI + 2HI + BiN\downarrow \ (\textit{Nitride})$$

Metal alcoholates can be precipitated in liq. NH_3 by the following reaction.

$$K(OC_2H_5) + Ba(NO_3)_2 \xrightarrow{\text{liq. } NH_3} Ba(OC_2H_5)_2\downarrow + 2KNO_3$$

(*c*) When solutions of the nitrate of K and Ba and of AgCl in liq. NH_3 are mixed together, chloride of K and Ba is precipitated.

$$KNO_3 + AgCl \xrightleftharpoons{\text{liq. } NH_3} KCl\downarrow + AgNO_3$$

$$Ba(NO_3)_2 + 2AgCl \xrightleftharpoons{\text{liq. } NH_3} BaCl_2\downarrow + 2AgNO_3$$

2. Acid-base neutralisation (salt formation) reactions

In liq. NH_3 acid-base neutralisation reaction is that in which a compound containing or making available NH_4^+ ion (*ammono acid*) combines with a compound containing or making available NH_2^- or N_3^{3-} ion (*ammono base*) to form the salt and the solvent (*i.e.*, NH_3), *e.g.*,

$$NH_4Cl\ (\textit{ammono acid}) + KNH_2\ (\textit{ammono base}) \xrightarrow{\text{liq. } NH_3} KCl\ (\textit{salt}) + 2NH_3\ (\textit{solvent})$$

Actually the above reaction takes place through the following steps:

$$NH_4Cl \xrightarrow{\text{liq. } NH_3} NH_4^+ + Cl^-$$

$$KNH_2 \xrightarrow{\text{liq. } NH_3} K^+ + NH_2^-$$

$$NH_4Cl + KNH_2 \xrightarrow{\text{liq. } NH_3} K^+ + Cl^- + NH_4^+ + NH_2^-$$

or NH_4Cl (*ammono acid*) + KNH_2 (*ammono base*) $\xrightarrow{\text{liq. } NH_3}$ KCl (*salt*) + $2NH_3$ (*solvent*)

or $NH_4^+ + Cl^- + K^+ + NH_2^- \xrightarrow{\text{liq. } NH_3} K^+ + Cl^- + 2NH_3$

or NH_4^+ (*solvent cation*) + NH_2^- (*solvent anion*) $\xrightarrow{\text{liq. } NH_3}$ $2NH_3$ (*solvent*) ...(*i*)

Equation (*i*) indicates that neutralisation reaction in liq. NH_3 can also be defined as the combination of solvent cation (NH_4^+) and solvent anion (NH_2^-) to form the un-ionised solvent (*i.e.*, liq. NH_3).

Other examples of acid-base neutralisation reactions taking place in liq. NH_3 are:

Ammono acid		*Ammono base*		*Ammono salt*		*Solvent*
NH_4X (X = Cl, NO_3)	+	$NaNH_2$	$\xrightarrow{\text{liq. } NH_3}$	NaX	+	$2NH_3$
$2NH_4X$ (X = Cl, I)	+	PbNH	$\xrightarrow{\text{liq. } NH_3}$	PbX_2	+	$3NH_3$
$3NH_4X$ (X = Cl, I)	+	BiN	$\xrightarrow{\text{liq. } NH_3}$	BiX_3	+	$4NH_3$

3. Solvolysis (solvolytic) reactions in liq. NH_3: Ammonolysis or amonolytic reactions

The solvolysis reactions taking place in liq. NH_3 are called ammonolysis reactions. These reactions are analogous to hydrolysis that takes place in aqueous medium. Just as in hydrolysis, the concentration of H_3O^+ or OH^- ions increases due to auto-ionisation of water ($H_2O + H_2O \rightleftharpoons H_3O^+ + OH^-$), in ammonolysis is concentration of NH_4^+ or NH_2^- (or NH^{2-} or N^{3-}) ions increases due to the auto-ionisation of liq. NH_3 ($2NH_3 \rightleftharpoons NH_4^+ + NH_2^-$, $3NH_3 \rightleftharpoons 2NH_4^+ + NH^{2-}$, $4NH_3 \rightleftharpoons 3NH_4^+ + N^{3-}$). In these reactions an atom, ion or radical of the compound undergoing ammonolysis is replaced by NH_2^-, NH^{2-} or N^{3-} ion and an ammono base is formed.

Some examples of ammonolysis reactions taking place in liq. NH_3 are given below:

(*a*) $$SiCl_4 + 8NH_3 \longrightarrow \underset{\text{Ammono base}}{Si(NH_2)_4} + 4NH_4^+ + 4Cl^-$$

Actually the above reaction takes place through the following steps:

$$SiCl_4 \longrightarrow Si^{4+} + 4Cl^-$$
$$[2NH_3 \longrightarrow NH_4^+ + NH_2^-] \times 4$$
$$\overline{SiCl_4 + 8NH_3 \longrightarrow Si(NH_2)_4 + 4NH_4^+ + 4Cl^-}$$

On the same lines the following reactions can also be explained.

(*b*) $$Hg_2Cl_2 + 2NH_3 \longrightarrow Hg(NH_2)Cl + Hg + NH_4^+ + Cl^-$$

(*c*) $$HgCl_2 + 2NH_3 \longrightarrow Hg(NH_2)Cl + NH_4^+ + Cl^-$$

(*d*) $$BX_3 + 6NH_3 \longrightarrow B(NH_2)_3 + 3NH_4^+ + 3X^-$$

(*e*) $$AlCl_3 + 2NH_3 \longrightarrow AlCl_2(NH_2) + NH_4^+ + Cl^-$$

(*In water* : $$AlCl_3 + 2H_2O \longrightarrow AlCl_2(OH) + H_3O^+ + Cl^-)$$

(*f*) $2PbI_2 + 6NH_3 \longrightarrow PbI(NH_2) + Pb(NH_2)_2 + 3NH_4^+ + 3I^-$

(*g*) $3(SiH_3)Cl + 4NH_3 \longrightarrow (SiH_3)_3N + 3NH_4^+ + 3Cl^-$

In this reaction Cl-atom of $(SiH_3)Cl$ is replaced by N^{3-} ion.

(*i*) The ammonolysis of $TiCl_4$ produces $Ti(NH_2)_4$ whose formation can be explained as follows :

$$TiCl_4 + NH_3 \longrightarrow TiCl_3(NH_2) + HCl$$
$$TiCl_3(NH_2) + NH_3 \longrightarrow TiCl_2(NH_2) + HCl$$
$$TiCl_2(NH_2) + NH_3 \longrightarrow TiCl(NH_2)_3 + HCl$$
$$TiCl(NH_2)_3 + NH_3 \longrightarrow Ti(NH_2)_4 + HCl$$

On adding: $TiCl_4 + 4NH_3 \longrightarrow Ti(NH_2)_4 + 4HCl$

(*j*) The ammonolysis of alkyl halides, RX(R = alkyl group, X = Cl, Br, I) takes place slowly at the boiling point of liq. NH_3. In this reaction mixture of primary (RNH_2), secondary (R_2NH) and tertiary (R_3N) amines is obtained.

$$RX + 2NH_3 \longrightarrow R(NH_2) + NH_4^+ + X^-$$
$$2RX + 3NH_3 \longrightarrow R_2(NH) + 2NH_4^+ + 2X^-$$
$$3RX + 4NH_3 \longrightarrow R_3N + 3NH_4^+ + 3X^-$$

(*k*) Alkali metal hydrides (*e.g.*, NaH) and oxides (*e.g.*, Na_2O) undergo ammonolysis to produce alkali metal amides.

$$NaH + NH_3 \longrightarrow NaNH_2 + H_2$$
$$Na_2O + 2NH_3 \longrightarrow 2NaNH_2 + H_2O$$

(*l*) Cl_2 and $OPCl_3$ give $Cl(NH_2)$ and $OP(NH_2)_3$ respectively.

$$Cl_2 + 2NH_3 \longrightarrow Cl(NH_2) + NH_4^+ + Cl^-$$

(*In water:* $Cl_2 + 2H_2O \longrightarrow Cl(OH) + H_3O^+ + Cl^-$)

$$OPCl_3 + 6NH_3 \longrightarrow OP(NH_2)_3 + 3NH_4^+ + 3Cl^-$$

(*In water:* $OPCl_3 + 6H_2O \longrightarrow OP(OH)_3 + 3H_3O^+ + 3Cl^-$

4. Solvation reactions in liq. NH_3: Ammoniation reactions and formation of ammoniates

Solvation reaction is a general reaction in which a solute (a cation, an anion or a neutral molecule) reacts with one or more molecules of a solvent (*e.g.*, H_2O, liq. NH_3, liq. SO_2 etc.) to form a product in which the solute and solvent species are attached to each other by a H-bond or by a coordinate bond. The product formed is called **solvate**. Solvate is an **addition compound** and hence is also called an **adduct.** The addition compound contains *solvent of crystallisation.*

In the formation of a solvate, the solvent acts as a Lewis base while the solute species behaves as a Lewis acid. When the solvent used is water, the solvation reaction is called **hydration** and the addition compound formed is called **hydrate**. Hydrate contains one or more molecules of water as water of crystallisation. Similarly when the solvent in liq. NH_3, the reaction is called **ammoniation** and the addition compound formed is called **ammoniate**. Ammoniate contains one or more molecules of ammonia as *ammonia of crystallisation.*

In the formation of ammoniates liq. NH_3 (solvent) acts as a Lewis base and the solute behaves as a Lewis acid. The formation of some ammoniates in liq. NH_3 has been shown below. These ammoniates may be 1 : 2 or 1 : 1 adducts.

Solute (Lewis acid)		*Solvent (Lewis base)*		*Ammoniates*
SO_3	+	$2NH_3$	$\longrightarrow$	$SO_3.2NH_3$ (1 : 2 *adduct*)
SiF_4	+	$2NH_3$	$\longrightarrow$	$SiF_4.2NH_3$ (1 : 2 *adduct*)
BF_3	+	NH_3	$\longrightarrow$	$BF_3.NH_3$ (1 : 1 *adduct*)

The Formation of NH_4^+ by the action of liq. NH_3 on H_2O is also an example of ammoniation.

$$H-\overset{H}{\underset{H}{\overset{|}{\underset{|}{N}}}}: + H-\overset{H}{\overset{|}{\underset{\cdot\cdot}{O}}}: \longrightarrow H-\overset{H}{\underset{H}{\overset{|}{\underset{|}{N}}}}:---H-\overset{H}{\overset{|}{\underset{\cdot\cdot}{O}}}: \longrightarrow NH_4^+ + OH^- \text{ or } [NH_4]^+[OH]^-$$

5. Complex formation reactions

When Zn(II), Mg(II), Al(III), Ag(I) compounds [*e.g.*, $Zn(NO_3)_2$, $Zn(OH)_2$, $Zn(NH_2)_2$, $Mg(NH_2)_2$, $AlCl_3$, $AgNH_2$ etc.] react with the excess of an ammono base [*e.g.*, $NH_4(NH_2)$, KNH_2, $NaNH_2$ etc.] in liq. NH_3, soluble amido or imido complex compound of the metal is obtained. For example:

(*i*) (*a*)
$$\underset{}{Zn(NO_3)_2} + \underset{\substack{\text{Ammono base}\\\text{(excess)}}}{4KNH_2} \xrightarrow{\text{liq. } NH_3} \underset{\substack{\text{Amido complex}\\\text{(soluble)}}}{K_2[Zn(NH_2)_4]} + 2KNO_3$$

or
$$Zn^{2+} + \underset{\text{(excess)}}{4NH_2^-} \xrightarrow{\text{liq. } NH_3} [Zn(NH_2)_4]^{2-}$$

The above complex formation reaction that takes place in liq. NH_3 can be compared with the following complex formation reaction that occurs in aqueous medium.

$$Zn(NO_3)_2 + \underset{\text{Base (excess)}}{4NaOH} \xrightarrow{H_2O} \underset{\text{Hydroxo complex (soluble)}}{Na_2[Zn(OH)_4]} + 2NaNO_3$$

or
$$Zn^{2+} + 4OH^- \xrightarrow{H_2O} [Zn(OH)_4]^{2-} \text{ or } [ZnO_2]^{2-} + 2H_2O$$

(*b*)
$$Zn(NO_3)_2 + \underset{\substack{\text{Ammono base}\\\text{(excess)}}}{4KNH_2} \xrightarrow{\text{liq. } NH_3} \underset{\substack{\text{Imido complex}\\\text{(soluble)}}}{K_2[Zn(NH)_2]} + 2NH_3 + 2KNO_3$$

or
$$Zn^{2+} + 4NH_2^- \xrightarrow{\text{liq. } NH_3} [Zn(NH)_2]^{2-} + 2NH_3$$

(*ii*)
$$Zn(NO_3)_2 + \underset{\substack{\text{Ammono base}\\\text{(excess)}}}{4NH_4(NH_2)} \xrightarrow{\text{liq. } NH_3} \underset{\substack{\text{Amido complex}\\\text{(soluble)}}}{(NH_4)_2[Zn(NH_2)_4]} + 2NH_4NO_3$$

$$Zn(NO_3)_2 + 4NH_4(NH_2) \xrightarrow{\text{liq. } NH_3} \underset{\substack{\text{Imido complex}\\\text{(soluble)}}}{(NH_4)_2[Zn(NH)_2]} + 2NH_4NO_3 + 2NH_3$$

(*iii*)
$$Zn(OH)_2 + \underset{\substack{\text{Ammono base}\\\text{(excess)}}}{2NaNH} \xrightarrow{\text{liq. } NH_3} \underset{\substack{\text{Amido complex}\\\text{(soluble)}}}{Na_2[Zn(NH_2)_4]}$$

$$Zn(OH)_2 + 2NaNH_2 \xrightarrow{\text{liq. } NH_3} \underset{\substack{\text{Imido complex}\\\text{(soluble)}}}{Na_2[Zn(NH)_2]} + 2NH_3$$

(*iv*) $$\underset{}{AgNH_2} + \underset{\text{Ammono base (excess)}}{KNH_2} \xrightarrow{\text{liq. } NH_3} \underset{\text{Amido complex (soluble)}}{K[Ag(NH_2)_2]}$$

$$AgNH_2 + KNH_2 \xrightarrow{\text{liq. } NH_3} \underset{\text{Imido complex (soluble)}}{K[Ag(NH)]} + NH_3$$

(*v*) $$\underset{(M = Zn, Mg)}{M(NH_2)_2} + \underset{\text{Ammono base (excess)}}{2NaNH_2} \xrightarrow{\text{liq. } NH_3} \underset{\text{Ammido complex (soluble)}}{Na_2[M(NH_2)_4]}$$

or $$M(NH_2)_2 + 2NH_2^- \xrightarrow{\text{liq. } NH_3} [M(NH_2)_4]^{2-}$$

$$M(NH_2)_2 + 2NaNH_2 \xrightarrow{\text{liq. } NH_3} \underset{\text{Imido complex (soluble)}}{Na_2[M(NH)_2]} + 2NH_3$$

or $$M(NH_2)_2 + 2NH_2^- \xrightarrow{\text{liq. } NH_3} [M(NH)_2]^{2-} + 2NH_3$$

(*vi*) $$AlCl_3 + \underset{\text{Ammono base (excess)}}{4NaNH_2} \xrightarrow{\text{liq. } NH_3} \underset{\text{Amido complex (soluble)}}{Na[Al(NH_2)_4]} + 3NaCl$$

or $$Al^{3+} + 4NH_2^- \xrightarrow{\text{liq. } NH_3} [Al(NH_2)_4]^-$$

Complex formation reaction (*vi*) that takes place in liq. NH_3 can be compared with the following complex formation reaction [reaction (A)] that occurs in aqueous medium.

$$AlCl_3 + \underset{\text{Base}}{3NaOH} \xrightarrow{H_2O} \underset{\text{ppt}}{Al(OH)_3} + 3NaCl$$

$$\underset{\text{ppt}}{Al(OH)_3} + \underset{\text{Base (excess)}}{NaOH} \xrightarrow{H_2O} \underset{\text{Soluble hydroxo complex}}{Na[Al(OH)_4]}$$

On adding: $$AlCl_3 + \underset{\text{Base (excess)}}{4NaOH} \xrightarrow{H_2O} 3NaCl + Na[Al(OH)_4] \quad ...(A)$$

or $$Al^{3+} + 4OH^- \xrightarrow{H_2O} [Al(OH)_4]^- \text{ or } [AlO_2]^- + 2H_2O$$

When HCl (which acts as an acid in aqueous medium) is added to the solution containing Al(III) hydroxo complex, $Al(OH)_3$ is again precipitated.

$$\underset{\text{Soluble hydroxo complex}}{Na[Al(OH)_4]} + \underset{\text{Acid}}{HCl} \xrightarrow{H_2O} \underset{\text{ppt}}{Al(OH)_3} + H_2O + NaCl$$

or $$[Al(OH)_4]^- + \underset{\text{Acid}}{H^+} \xrightarrow{H_2O} \underset{\text{ppt}}{Al(OH)_3} + H_2O$$

The precipitation of $Al(OH)_3$ by adding an acid to the soluble Al(III) hydroxo complex indicates that $AlCl_3$ behaves as an amphoteric substance in aqueous medium.

6. Redox reactions

Redox reactions in which liq. NH_3 is involved can be studied under the following heads:

(*i*) **Redox reactions in which liq. NH_3 serves as a medium only.** Liquid NH_3 provides a medium in which various redox reactions can be carried. For example:

(*a*) $KMnO_4$ acts as a weak oxidising agent in liq. NH_3. It oxidises KNH_2 to N_2 in liq. NH_3 and is itself reduced to K_2MnO_4.

$$\underset{(Mn = +7)}{6KMnO_4} + \underset{(N = -3)}{6KNH_2} \xrightarrow{\text{liq. } NH_3} \underset{(Mn = +6)}{6K_2MnO_4} + 6NH_3 + \underset{(N = 0)}{N_2}$$

(*b*) Iodine also acts as a weak oxidising agent in liq. NH_3. It oxidises potassium ammonostannite, $K_4[Sn(NH_2)_6]$ to potassium ammonostannate, $K_2[Sn(NH_2)_6]$ and is itself reduced to KI.

$$\underset{(I = 0)}{I_2} + \underset{(Sn = +2)}{K_4[Sn(NH_2)_6]} \xrightarrow{\text{liq. } NH_3} \underset{(Sn = +4)}{K_2[Sn(NH_2)_6]} + \underset{(I = -1)}{2KI}$$

(*ii*) **Redox reactions in which liq. NH_3(N = –3, H = +1) itself acts as a reducing agent.** Liq. NH_3 reduces various substances like O_2, CuO, Mg etc. and is itself oxidised to NO (N = +2), N_2(N = 0) etc. Examples are:

(*a*) $$\underset{(N = -3)}{4NH_3} + \underset{(O = 0)}{5O_2} \xrightarrow{Pt} \underset{(N = +2)}{4NO} + \underset{(O = -2)}{6H_2O}$$

(*b*) $$\underset{(N = -3)}{2NH_3} + \underset{(Cu = +2)}{3CuO} \longrightarrow \underset{(N = 0)}{N_2} + \underset{(Cu = O)}{3Cu} + 3H_2O$$

(*c*) $$\underset{(H = +1)}{2NH_3} + \underset{(Mg = 0)}{3Mg} \longrightarrow \underset{(H = 0)}{3H_2} + \underset{(Mg = +2)}{Mg_3N_2}$$

(*iii*) **Redox reactions in which alkali metals in liq. NH_3 act as reducing agents.** Liq. NH_3 can dissolve the alkali metals in it. The solution thus obtained contains ammoniated electrons. These electrons reduce various substances. Thus alkali metals in liq. NH_3 are able to reduce a variety of substances and are themselves oxidised to some alkali metal salt. Following examples illustrate the reducing property of alkali metals in liq. NH_3.

(*a*) Sodium in liq. NH_3 reduces ammonium salts (*e.g.*, NH_4Br) to H_2.

$$\underset{(Na = 0)}{2Na} + \underset{(H = +1)}{2NH_4Br} \xrightarrow{\text{liq. } NH_3} \underset{(Na = +1)}{2NaBr} + 2NH_3 + \underset{(H = 0)}{H_2}$$

(*b*) Potassium in liq. NH_3 reduces nitrous oxide (N_2O) to N_2.

$$\underset{(K = 0)}{2K} + NH_3 + \underset{(N = +1)}{N_2O} \xrightarrow{\text{liq. } NH_3} \underset{(K = +1)}{KNH_2} + KOH + \underset{(N = 0)}{N_2}$$

(*c*) Sodium in liq. NH_3 reduces elemental sulphur to sodium sulphide (Na_2S)

$$\underset{(S = 0)}{S} + \underset{(Na = 0)}{2Na} \xrightarrow{\text{liq. } NH_3} \underset{(Na = +1,\ S = -2)}{Na_2S}$$

(*d*) Li and Na in liq. NH_3 reduce elemental oxygen to their peroxide (K_2O_2 or Na_2O_2) while other alkali metals give their peroxide (M_2O_2) as well as superoxide (MO_2).

$$\underset{(K = 0)}{2K} + \underset{(O = 0)}{O_2} \xrightarrow{\text{liq. } NH_3} \underset{(K = +1,\ O = -1)}{K_2O_2} \text{ (\textit{Peroxide-white})}$$

$$\underset{(K = 0)}{K} + \underset{(O = 0)}{O_2} \xrightarrow{\text{liq. } NH_3} \underset{(K = +1,\ O = -1/2)}{KO_2} \text{ (\textit{superoxide-yellow})}$$

(*e*) Na in liq. NH_3 reduces CuI to Cu.

$$\underset{(Cu = +1)}{CuI} + \underset{(Na = 0)}{Na} \xrightarrow{\text{liq. } NH_3} \underset{(Cu = 0)}{Cu} + \underset{(Na = +1)}{NaI}$$

(*f*) K in liq. NH_3 reduces $KMnO_4$ to MnO and is itself oxidised to KNH_2.

$$\underset{(K = 0)}{30K} + 20NH_3 + \underset{(Mn = +7)}{6KMnO_4} \longrightarrow \underset{(Mn = +2)}{6MnO} + \underset{(K = +1)}{18KNH_2} + 18KOH + 3H_2 + N_2$$

(*g*) Na in liq. NH_3 reduces ZnI_2 to $NaZn_4$ and is itself oxidised NaI.

$$\underset{(Zn\ =\ +2)}{4ZnI_2} + \underset{(Na\ =\ 0)}{9Na} \xrightarrow{\text{liq. } NH_3} \underset{(Zn\ =\ -1/4)}{NaZn_4} + \underset{(Na\ +1)}{8NaI}$$

(*h*) Excess of K in liq. NH_3 reduces $K_2[Ni(CN)_4]$ and $[Pt(NH_3)_4]Br_2$ to $K_4[Ni(CN)_4]$ and $[Pt(NH_3)_4]^0$ respectively.

$$\underset{(Ni\ =\ +2)}{K_2[Ni(CN)_4]} \xrightarrow[\text{liq. } NH_3\ (-33°C)]{\text{Excess of K in}} \underset{(Ni\ =\ 0)}{K_4[Ni(CN)_4]}$$

$$\underset{(Pt\ =\ +2)}{[Pt(NH_3)_4]Br_2} \xrightarrow[\text{liq. } NH_3\ (-33°C)]{\text{Excess of K in}} \underset{(Pt\ =\ 0)}{[Pt(NH_3)_4]^0} + 2KBr$$

Advantages of using liquid ammonia as a solvent

(*i*) We have seen above that alkali metals, without reacting with liq. NH_3, are soluble in this solvent. The dissolved alkali metals can be recovered by evaporating the alkali metal-liq. NH_3 solution.

(*ii*) The alkali metal-liq. NH_3 solutions contain ammoniated electron and hence these solutions act as strong reducing agents.

(*iii*) We have seen under the study of precipitation reactions taking place in liq. NH_3 that these reactions can be used to precipitate metallic halides, sulphides, alcoholates, amides, imides and nitrides.

Disadvantages of using liq. NH_3 as a solvent

(*i*) Low temperature or high pressure is necessary while working with liq. NH_3. This is because of the fact that the liquid range for liq. NH_3 is from –34.0°C to –77.7°C.

(*ii*) Liq. NH_3 is hygroscopic in nature and hence all the reactions must be carried out in the sealed tube.

(*iii*) Liq. NH_3 has an offensive odour and hence the use of liq. NH_3 as a solvent and as a reaction medium requires special technique.

Liquid Sulphur Dioxide (Liq. SO_2)

Introduction

Liq. SO_2 is a non-protonic solvent, because it cannot give a proton on self ionisation. Under normal temperature and pressure, sulphur dioxide is a gas but can be readily liquefied. It has a wide liquid range from –10.0°C to –75.5°C (b.pt. = –10.0° and m.pt. = –75.5°C) and hence can be used as a solvent. Its dielectric constant is low (= 17.4 at –20°C) and hence it is a poor solvent for ionic compounds but acts as a good solvent for covalent compounds.

Solubility of various substances in liq. SO_2

(*a*) **Solubility of ionic compounds (inorganic salts).** Iodides and thiocyanates are the most soluble. Sulphates, sulphides, oxides and hydroxices are practically insoluble. Many of the ammonium, thallium and mercuric salts are soluble.

(*b*) **Solubility of non-ionic compounds (inorganic and organic covalent compounds).** Covalent halides like IBr, BCl_3, $AlCl_3$, $AsCl_3$, PBr_3, CCl_4, $SiCl_4$ and $SnCl_4$ are soluble in liq. SO_2. Organic compounds like ammines, ethers, alcohols, benzene, alkenes, pyridine, quinoline, halogen derivatives and acid chlorides are soluble. Alkanes are insoluble.

(*c*) **Solubility of metals.** Metals are insoluble in liq. SO_2.

Auto-ionisation of liq. SO_2

Self-ionisation of liq. SO_2 takes place as follows:

$$SO_2 \rightleftharpoons SO^{2+} + O^{2-}$$

$$SO_2 + O^{2-} \rightleftharpoons SO_3^{2-}$$

$$\underbrace{SO_2 + SO_2}_{\text{solvent}} \rightleftharpoons SO^{2+}\ (\textit{Thionyl ions}) + SO_3^{2-}\ (\textit{Sulphite ions})$$

(solvent cations : acid ions) (solvent anions: base ions)

When we compare the self-ionisation of liq. SO_2 with that of water and liq. NH_3, we find that SO^{2+} ion is analogous to H_3O^+ and NH_4^+ ions while SO_3^{2-} ion is analogous to OH^- and NH_2^- ion produced by the self-ionisation of H_2O and liq. NH_3 respectively. Thus all those compounds which contain or make available SO^{2+} ions (solvent cations-acid ions) will act as acids in liq. SO_2. Similarly all those compounds which contain or make available SO_3^{2-} ions, will behave as bases in liq. SO_2.

The above discussion shows that since $SOCl_2$, $SOBr_2$, $SO(SCN)_2$ etc., give SO^{2+} ions in liq. SO_2, they behave as acids in this solvent. On the other hand, the substances *viz.*, Cs_2SO_3, $[N(CH_4)]_2SO_3$, K_2SO_3 which give SO_3^{2-} ions in this solvent act as bases.

Chemical Reaction Taking Place in Liq. SO_2

The chemical reactions that take place in liq. SO_2 can be of the following types:

1. Precipitation reactions

(*i*) When a substance which acts as an acid in liq. SO_2 (*e.g.*, $SOCl_2$) reacts with the metallic salts like KBr, KI, $NH_4(SCN)$, $Ag(CH_3COO)$ etc. in liq. SO_2, chloride of the metal is precipitated.

Metallic salt		*Acid*		*Precipitate of metallic chloride*		
2KBr	+	$SOCl_2$	$\xrightarrow{\text{liq. } SO_2}$	2KCl	+	$SOBr_2$
2KI	+	$SOCl_2$	$\xrightarrow{\text{liq. } SO_2}$	2KCl	+	SOI_2
$NH_4(SCN)$	+	$SOCl_2$	$\xrightarrow{\text{liq. } SO_2}$	$2NH_4Cl$	+	$SO(SCN)_2$
$2Ag(CH_3COO)$	+	$SOCl_2$	$\xrightarrow{\text{liq. } SO_2}$	2AgCl	+	$SO(CH_3COO)_2$

(*ii*) Examples other precipitation reactions that can be carried in liq. SO_2 are:

$$SbCl_3 + 3LiI \xrightarrow{\text{liq. } SO_2} SbI_3\downarrow + 3LiCl$$

$$PbF_2 + Li_2SO_4 \xrightarrow{\text{liq. } SO_2} PbSO_4\downarrow + 3LiF$$

$$AlCl_3 + 3NaI \xrightarrow{\text{liq. } SO_2} AlI_3\downarrow + 3NaCl$$

$$BaI_2 + Zn(CNS)_2 \xrightarrow{\text{liq. } SO_2} Ba(CNS)_2\downarrow + ZnI_2$$

2. Acid-base neutralisation (salt formation) reactions

In liq. SO_2, acid-base reaction is that in which a compound containing or making available SO_2^{2+} ion (*acid*) combines with a compound containing or making available SO_3^{2-} ion (*base*) to form the salt and the solvent (*i.e.*, SO_2), *e.g.*,

$$SOCl_2\ (acid) + Cs_2SO_3\ (base) \xrightarrow{\text{liq. } SO_2} 2CsCl\ (salt) + 2SO_2\ (solvent)$$

Actually the above reaction takes place through the following steps:

$$SOCl_2 \xrightarrow{\text{liq. } SO_2} SO^{2+} + 2Cl^-$$

$$Cs_2SO_3 \xrightarrow{\text{liq. } SO_2} 2Cs^+ + SO_2^{2-}$$

$$SOCl_2 + Cs_2SO_3 \xrightarrow{\text{liq. } SO_2} 2Cs^+Cl^- + SO^{2+} + SO_3^{2-}$$

or $$SOCl_2\ (acid) + Cs_2SO_3\ (base) \xrightarrow{\text{liq. } SO_2} 2CsCl\ (salt) + 2SO_2\ (solvent)$$

or $$SO^{2+} + 2Cl^- + 2Cs^+ + SO_3^{2-} \xrightarrow{\text{liq. } SO_2} 2Cs^+ + 2Cl^- + 2SO_2$$

or $$SO^{2+}\ (solvent\ cation) + SO_3^{2-}\ (solvent\ anion) \xrightarrow{\text{liq. } SO_2} 2SO_2\ (solvent) \quad ...(i)$$

It may be seen from equation (*i*) that neutralisation reaction in liq. SO_2 can also be defined as the combination of the solvent cation (SO^{2+}) and solvent anion (SO_3^{2-}) to produce the solvent (SO_2).

Examples of some other acid-base neutralisation reactions taking place in liq. SO_2 are given below:

Acid		*Base*		*Salt*	*Solvent*
$SO(SCN)_2$	+	K_2SO_3	$\xrightarrow{\text{liq. } SO_2}$	$2K(SCN)$ +	$2SO_2$
$SOBr_2$	+	$[N(CH_3)_4]_2SO_3$	$\xrightarrow{\text{liq. } SO_2}$	$2[N(CH_3)_4]Br$ +	$2SO_2$

3. Solvolysis (solvolytic) reactions in liq. SO_2

Examples of solvolysis reactions occuring in liq. SO_2 are given below:

(*a*) CH_3COONH_4 undergoes solvolysis reaction in liq. SO_2. Here the concentration of SO_3^{2-} ions is increased.

$$2CH_3COONH_4 + 2SO_2(liq.) \longrightarrow (NH_4)_2SO_3 + (CH_3COO)_2SO$$

$$(CH_3COO)_2SO \longrightarrow SO_2 + (CH_3CO)_2O$$

$$2CH_3COONH_4 + 2SO_2(liq.) \longrightarrow \underset{\text{base}}{(NH_4)_2SO_3} + (CH_3CO)_2O + SO_2$$

(*b*) Some covalent halides like PCl_5, UCl_6, WCl_6, $AsCl_5$ etc. undergo solvolysis reactions in liq. SO_2 and form oxyhalides. In all these reactions the concentration of SO^{2+} ion is increased.

$$PCl_5 + SO_2\ (liq.) \longrightarrow POCl_3 + SOCl_2$$

$$UCl_6 + 2SO_2\ (liq.) \longrightarrow UO_2Cl_2 + 2SOCl_2$$

$$WCl_6 + SO_2\ (liq.) \longrightarrow WOCl_4 + SOCl_2$$

$$AsCl_5 + SO_2\ (liq.) \longrightarrow AsOCl_3 + SOCl_2$$

$$NbCl_5 + SO_2\ (liq.) \longrightarrow NbOCl_3 + SOCl_2$$

$$Zn(C_2H_5)_2 + 2SO_2\ (liq.) \longrightarrow ZnSO_3 + SO(C_2H_5)_2$$

4. Formation of solvates

Like water and liq. NH_3, liq. SO_2 also forms addition compounds with solutes. These addition compounds are called solvates and contain one of more molecules of sulphur dioxide as sulphur dioxide of crystallisation. Examples of some solvates formed in liq. SO_2 are $LiI.2SO_2$, $KBr.4SO_2$, $MI.\ 4SO_2$ (M = Na, K or Rb), $MI_2.4SO_2$ (M = Ca, Sr or Ba), $AlCl_3.2SO_2$, $KCNS.nSO_2$ (n = 0.5, 1 or 2), $[(CH_3)_4N]Cl.SO_2$, $[(CH_3)_4N]_2SO_4.3SO_2$.

5. Complexes formation reactions

(*i*) When Al(III) or Ga(III) salt (*e.g.*, $AlCl_3$, $GaCl_3$) is treated with a compound containing SO_3^{2-} ions (which act as base in liq. SO_2) in liq. SO_2, ppt. of $Al_2(SO_3)_3$ is obtained. This ppt. dissolves in the excess of the base due to the formation of a soluble sulphito complex of Al(III). Thus,

(*a*)
$$\underset{\text{Base}}{2AlCl_3 + 3K_2SO_3} \xrightarrow{\text{liq. } SO_2} \underset{\text{ppt}}{Al_2(SO_3)_3} + 6KCl$$

$$\underset{\text{ppt}}{Al_2(SO_3)_3} + \underset{\text{Base}}{3K_2SO_3} \xrightarrow{\text{liq. } SO_2} \underset{\text{Soluble sulphito complex}}{2K_3[Al(SO_3)_3]}$$

On adding:
$$2AlCl_3 + \underset{\text{Base (excess)}}{6K_2SO_3} \xrightarrow{\text{liq. } SO_2} 6KCl + 2K_3[Al(SO_3)_3] \quad ...(i)$$

or
$$2Al^{3+} + 6SO_3^{2-} \xrightarrow{\text{liq. } SO_2} 2[Al(SO_3)_3]^{3-}$$

or
$$Al^{3+} + 3SO_3^{2-} \xrightarrow{\text{liq. } SO_2} [Al(SO_3)_3]^{3-}$$

(*b*)
$$2AlCl_3 + \underset{\text{Base}}{3[N(CH_3)_4]_2SO_3} \xrightarrow{\text{liq. } SO_2} \underset{\text{ppt}}{Al_2(SO_3)_3} + 6[N(CH_3)_4]Cl$$

$$\underset{\text{ppt}}{Al_2(SO_3)_2} + \underset{\text{Base}}{3[N(CH_3)_4]_2SO_3} \xrightarrow{\text{liq. } SO_2} \underset{\text{Soluble sulphito complex}}{2[N(CH_3)_4]_3[Al(SO_3)_3]}$$

On adding the above two equations, we get:

$$2AlCl_3 + \underset{\text{Base (excess)}}{6[N(CH_3)_4]_2SO_3} \xrightarrow{\text{liq. } SO_2} 6[N(CH_3)_4]Cl + 2[N(CH_3)_4]_3[Al(SO_3)_3] \quad ...(ii)$$

or
$$2Al^{3+} + \underset{\text{Base (excess)}}{6SO_3^{2-}} \xrightarrow{\text{liq. } SO_2} 2[Al(SO_3)_3]^{3-}$$

or
$$Al^{3+} + 3SO_3^{2-} \xrightarrow{\text{liq. } SO_2} [Al(SO_3)_3]^{3-}$$

When $SOCl_2$ (which acts as an acid in liq. SO_2) is added to the solution containing Al(III) sulphite complex, $Al_2(SO_3)_3$ is again precipitated. The precipitation of $Al_2(SO_3)_2$ indicates that $AlCl_3$ (and also $GaCl_3$) behaves as an amphoteric substance in liq. SO_2.

$$\underset{\text{Soluble sulphito complex}}{2K_3[Al(SO_3)_3]} + \underset{\text{Acid}}{3SOCl_2} \xrightarrow{\text{liq. } SO_2} \underset{\text{ppt}}{Al_2(SO_3)_3} + 6SO_2 + 6KCl$$

$$\underset{\text{Soluble sulphito complex}}{2[N(CH_3)_4]_3[Al(SO_3)_3]} + \underset{\text{Acid}}{3SOCl_2} \xrightarrow{\text{liq. } SO_2} \underset{\text{ppt}}{Al_2(SO_3)_3} + 6SO_2 + 6[N(CH_3)_4]Cl$$

Both the above equations can be represented by the ionic equation:

$$2[Al(SO_3)_3]^{3-} + 3SO^{2+} \xrightarrow{\text{liq. } SO_2} \underset{\text{ppt}}{Al_2(SO_3)_3} + 6SO_2$$

(*ii*) When Zn(II) salts [*e.g.*, $ZnCl_2$] react with a compound containing SO_3^{2-} ions (which behaves as base in liq. SO_2) in liq. SO_2, ppt. of $ZnSO_3$ is obtained. This ppt. dissolves in the excess of the base due to the formation of a soluble sulphito complex of Zn(II)

$$\underset{\text{Base}}{ZnCl_2 + K_2SO_3} \xrightarrow{\text{liq. } SO_2} \underset{\text{ppt}}{ZnSO_3} + 2KCl$$

$$\underset{\text{ppt}}{ZnSO_3} + \underset{\text{Base (excess)}}{K_2SO_3} \xrightarrow{\text{liq. } SO_2} \underset{\text{Soluble sulphito complex}}{K_2[Zn(SO_3)_2]}$$

On adding : $$ZnCl_2 + \underset{\text{Base (excess)}}{2K_2SO_3} \xrightarrow{\text{liq. } SO_2} 2KCl + K_2[Zn(SO_3)_2]$$

or $$ZnCl_2 + \underset{\text{Base (excess)}}{2SO_3^{2-}} \xrightarrow{\text{liq. } SO_2} 2Cl^- + [Zn(SO_3)_2]^{2-}$$

(*iii*) Solubility of I_2 in liq. SO_2 is greatly increased when KI or RbI is added to it. This is due to the formation of complex compounds *viz.* KI_3 or RbI_3.

$$\underset{(M = K, Rb)}{MI + I_2} \xrightarrow{\text{liq. } SO_2} \underset{\text{Soluble complex}}{MI_3}$$

(*iv*) When HgI_2 or CdI_2 is dissolved in liq. SO_2 in presence of KI, the solubility of these iodides is increased due to the formation of $K_2[MI_4]$ which is a complex compound.

$$\underset{(M = Hg, Cd)}{MI_2 + 2KI} \xrightarrow{\text{liq. } SO_2} \underset{\text{Soluble complex compound}}{K_2[MI_4]}$$

(*v*) Other reactions showing the formation of hexahalo complexes in liq. SO_2 by the action of covalent halides (*e.g.*, $SbCl_3$, $SbCl_3$ etc.) on alkali metal halides (*e.g.*, KCl) or NOCl or $[N(CH_3)_4]Cl$ are given below:

$$\left.\begin{aligned} SbCl_3 + 3KCl &\xrightarrow{\text{liq. } SO_2} K_3[SbCl_6] \\ SbCl_5 + KCl &\xrightarrow{\text{liq. } SO_2} K[SbCl_6] \\ SbCl_5 + NOCl &\xrightarrow{\text{liq. } SO_2} NO[SbCl_6] \\ SbCl_5 + [N(CH_3)_4]Cl &\xrightarrow{\text{liq. } SO_2} [N(CH_3)_4][SbCl_6] \end{aligned}\right\} \textit{Hexachloro complexes}$$

6. Redox reactions

Liq. SO_2 itself does not act as any strong oxidising/reducing agent. It serves only as a medium form any redox reactions. For example:

(*a*) A sulphite (R_2SO_3) in liq. SO_2 reduces iodine to iodide and itself oxidised to sulphate (R_2SO_4).

$$\underset{(S = +4)}{2R_2SO_3} + \underset{(I = 0)}{I_2} \xrightarrow{\text{liq. } SO_2} \underset{(S = +6)}{R_2SO_4} + \underset{(I = -1)}{2RI} + SO_2$$

(*b*) $SbCl_5$ in liq. SO_2 oxidises KI to free I_2 and is itself reduced to $SbCl_3$.

$$\underset{(Sb = +5)}{5SbCl_5} + \underset{(I = -1)}{6KI} \xrightarrow{\text{liq. } SO_2} \underset{(I = 0)}{3I_2} + \underset{(Sb = +3)}{SbCl_3} + \underset{(Sb = +3)}{2K_3[SbCl_6]}$$

(*c*) $FeCl_3$ in liq. SO_2 oxidises KI to free I_2 and is itself reduced to $FeCl_2$.

$$\underset{(Fe = +3)}{2FeCl_3} + \underset{(I = -1)}{2KI} \xrightarrow{\text{liq. } SO_2} \underset{(Fe = +2)}{2FeCl_2} + \underset{(I = 0)}{I_2} + 2KCl$$

Liquid Hydrogen Fluoride, (Liq. HF)

Liq. HF has a wide but convenient liquid range (–83°C to 19.4°C). The disadvantages of using HF as a solvent and reaction medium are that it can dissolve only relatively few substances without chemical reaction and that it has poisonous character.

Although HF has a rather low specific conductance, its high dielectric constant makes it an excellent ionising solvent. It is regarded as one of the most water-like of all non-aqueous solvents. It dissolves many inorganic and organic compounds to give highly conducting solutions. Inorganic substances are generally more soluble than organic compounds.

Auto-ionisation of liq. HF

The high specific conductance of liq. HF suggests a relatively high degree of auto-ionisation of liq. HF shown below:

$$HF \rightleftharpoons H^+ + F^-$$

$$HF + H^+ \rightleftharpoons H_2F^+$$

$$\underset{}{HF + HF} \rightleftharpoons \underset{\text{Solvent cations (acid ions)}}{H_2F^+} + \underset{\text{Solvent anions (base ions)}}{F^-}$$

or

$$3HF \rightleftharpoons \underset{\text{Solvent cations (acid ions)}}{H_2F^+} + \underset{\text{Solvent anions (base ions)}}{F^-.HF \text{ or } HF_2^-}$$

Behaviour of different substances in liq. HF

The auto-ionisation of liq. HF as shown above indicates that any substance that gives H_2F^+ ions in liq. HF benaves as an acid in this solvent and the substance that furnishes F^- or HF_2^- ions acts as a base in this solvent. This point would be more clear from the following examples:

(*i*) CH_3COOH (which acts as a *weak acid* in water) acts as a base in liq. HF, because it gives F^- ions (solvent anions-base ions) when dissolved in liq. HF.

$$CH_3COOH \text{ (Base)} + HF \text{ (Solvent)} \longrightarrow CH_3COOH_2^+ + F^- \text{ [Solvent anions (base ions)]}$$

(*ii*) H_2SO_4 and HNO_3 both act as strong acids in aqueous medium, but show basic character in liq. HF, due to the production of F^- ions which are solvent anions (*base ions*).

$$H_2SO_4 \text{ (Base)} + HF \text{ (Solvent)} \longrightarrow H_3SO_4^+ + F^- \text{ [Solvent anions (base ions)]}$$

$$HNO_3 \text{ (Base)} + HF \text{ (Solvent)} \longrightarrow H_2NO_3^+ + F^- \text{ [Solvent anions (base ions)]}$$

(*iii*) $HClO_4$ behaves as the *strongest acid* in water, but in liq. HF it acts as an *amphoteric substance*, since it produces both H_2F^+ (*solvent cations-acid ions*) and F^- (*solvent anions-base ions*) in this solvent.

$$HClO_4 + HF \text{ (Solvent)} \longrightarrow H_2ClO_4^+ + F^- \text{ [Solvent anions (base ions)]}$$

$$HClO_4 + HF \text{ (Solvent)} \longrightarrow H_2F^+ \text{ [Solvent cations (acid ions)]} + ClO_4^-$$

(*iv*) Electron-acceptor fluorides (*e.g.*, BF_3, AsF_3, PF_5 and SbF_5) act as acids in liq. HF

$$BF_3 + 2HF \longrightarrow H_2F^+ + BF_4^-$$

$$SbF_5 + 2HF \longrightarrow H_2F^+ + SbF_6^-$$

These acceptor molecules dissolve in HF to give solutions which dissolve electropositive metals like Mg. This is an evidence in favour of the acidic nature of these solutes.

Chemical Reactions Taking Place in Liq. HF

Chemical reactions that take place in liq. HF are of the following types:

1. Oxidation-reduction reactions

Oxidation-reduction reactions between hydrofluro acids and metals in HF are not generally different from those taking place between HNO_3 and metals in the aqueous system, *e.g.*,

$$3HAsF_6 + 2Ag \xrightarrow{HF(l)} 2AgAsF_6 + AsF_3 + 3HF$$

$$5HIF_6 + 6Ag \xrightarrow{HF(l)} 6AgIF_6 + HI + 6HF$$

$$4HNO_3 + 3Ag \xrightarrow{Aq} 3AgNO_3 + NO + 2H_2O$$

2. Precipitation reactions

Some of the precipitation reactions occurring in liq. HF are exemplified by the following reactions.

$$Na_2SO_4 + 2AgF \xrightarrow{HF(l)} Ag_2SO_4\downarrow + 2NaF$$

$$KClO_4 + TiF \xrightarrow{HF(l)} TiClO_4\downarrow + KF$$

$$NaClO_4 + AgF \xrightarrow{HF(l)} AgClO_4\downarrow + NaF$$

The precipitation reaction resulting in the formation of silver tetrafluoroborate, $AgBF_4$ is the most important. $AgBF_4$ is obtained as a precipitate when $AgNO_3$ and BF_3 solutions in HF are mixed.

$$AgNO_3 + BF_3 + 2HF \longrightarrow \underset{\text{Precipitate}}{AgBF_4} + H_2NO_2F$$

3. Formation of addition compounds

HF also forms a number of addition compounds with metallic fluorides, *e.g.*, KF.HF, KF.2HF, KF.3HF, NH_4F.HF,. etc.

SOME TYPICAL PROBLEMS WITH SOLUTIONS

Problem 1. *(a) Classify the following solvents as ionising and non-ionising solvents: HF, liq* SO_2, *liq.* N_2O_4, H_2SO_4, $POCl_3$, *pyridine and acetic acid.*

(b) Classify the following solvents as protonic (protic) or non-protonic (aprotic) solvents: HF, liq. SO_2, C_6H_6, H_2SO_4 *and liq.* NH_3

Solution. (*a*) *Ionising solvents* : HF, H_2SO_4, acetic acid

Non-ionising solvents : liq. SO_2, liq. N_2O_4, $POCl_3$, pyridine.

(*b*) *Protic solvents* : HF, H_2SO_4, liq. NH_3

Aprotic solvents : liq SO_2, C_6H_6

Problem 2. *What will be the effect of adding (a)* KNH_2 *(b)* Li_3N *(c)* CH_3COOK *to liq.* NH_3 *on the acidity/bascity of the solvent.*

Solution. (*a*) Since KNH_2 gives NH_2^- ions (amide ions) in liq. NH_3, it behaves as a base in this solvent. Thus the addition of KNH_2 to liq. NH_3 increases the basic character of the solvent

$$KNH_2 \xrightarrow{\text{Liq.}NH_3} K^+ + NH_2^-$$

(*b*) Since Li_3N produces nitride ions (N^{3-} ions) in liq. NH_3, it behaves as a base in this solvent. Thus the addition of Li_3N to liq. NH_3 increases the basic nature of the solvent.

$$Li_3N \xrightarrow{\text{Liq.}NH_3} 3Li^+ + N^{3-}$$

(*c*) Since CH_3COOK does not have any of the ions *viz* NH_4^+, NH_2^-, NH^{2-}, N^{3-}, the addition of this compound to liq. NH_3 has no effect on the acidity/bascity of liq. NH_3.

Problem 3. *Guess the reaction between $Zn(NH_2)_2$ and CH_3COONH_4.*

Solution. Auto-ionisation of liq. NH_3 gives NH_4^+ (ammono acid ion) and NH_2^- (ammono base ion) ions. Thus $Zn(NH_2)_2$ will act as a base and CH_3COONH_4 will act as an acid in liq. NH_3 and the reaction between these substances is a neutralisation reaction.

Ammono base		*Ammono acid*		*Salt*		*Solvent*
$Zn(NH_2)_2$	+	$2CH_3COONH_4$	$\longrightarrow$	$Zn(CH_3COO)_2$	+	$4NH_3$

Zn^{2+} ions obtained by the dissociation of $Zn(CH_3COO)_2$ will react with $4NH_3$ to give tetramminezinc (II) ion, $[Zn(NH_3)_4]^{2+}$

$$Zn^{2+} + 4NH_3 \longrightarrow [Zn(NH_3)_4]^{2+}$$

Problem 4. *How do the following solutes behave in the solvents given in parentheses? (a) KNH_2 (NH_3) (b) $(NH_4)_2SO_4$ (NH_3) (c) $SiCl_4$ (NH_3) (d) SO_2Cl_2 (NH_3) (e) PCl_5 (SO_2) (f) AgCl (NH_3) (g) $Zn(C_2H_5)_2$ (SO_2) (h) Cs_2SO_3 (SO_2) (i) $NbCl_5$ (SO_2)*

Solution (*a*) KNH_2 behaves as a base in liq. NH_3, since it gives NH_2^- (base anion) in this solvent.

$$KNH_2 \xrightarrow{\text{Liq.}NH_3} K^+ + NH_2^-$$

(*b*) $(NH_4)_2SO_4$ behaves as an acid in liq. NH_3, since NH_4^+ ions are produced.

$$(NH_4)_2SO_4 \xrightarrow{\text{Liq.}NH_3} 2NH_4^+ + SO_4^{2-}$$

(*c*) $SiCl_4$ is covalent and hence undergoes ammonolysis reaction in liq. NH_3. In this reactiion the concentration of NH_2^- ions is increased due to the production of $Si(NH_2)_4$

$$SiCl_4 + 8NH_3 \longrightarrow Si(NH_2)_4 + 4NH_4Cl$$

$$Si(NH_2)_4 \xrightarrow{0°} Si(NH)(NH_2)_2 \xrightarrow{1200°} Si_3N_4$$

(*d*) SO_2Cl_2 undergoes ammonolysis reaction in liq NH_3, since $SO_2(NH_2)_2$ is produced. This compound increases the concentration of NH_2^- ions.

$$SO_2Cl_2 + 4NH_3 \longrightarrow SO_2(NH_2)_2 + 2NH_4Cl$$

(*e*) In liq. SO_2, PCl_5 undergoes solvolysis reaction and $POCl_3$ is formed.

$$PCl_5 + SO_2 \longrightarrow POCl_3 + SOCl_2$$

(*f*) AgCl dissolves in liq. NH_3 due to the formation of a complex compound, $[Ag(NH_3)_2]Cl$.

$$AgCl + 2NH_3 \longrightarrow [Ag(NH_3)_2]Cl$$

(*g*) The reaction between $Zn(C_2H_5)_2$ and liq. SO_2 is an example of solvolysis reaction

$$Zn(C_2H_5)_2 + 2SO_2\,(l) \longrightarrow ZnSO_3 + (C_2H_5)_2SO$$

(*h*) Since Cs_2SO_3 gives SO_3^{2-} ions in liq. SO_2, this compound acts as a base in this solvent

$$Cs_2SO_3 \xrightarrow{\text{Liq. }SO_2} 2Cs^+ + SO_3^{2-}\ \textit{(Base ion)}$$

(*i*) $NbCl_5$ shows solvolysis reaction in liq SO_2.

$$NbCl_5 + SO_2\,(l) \longrightarrow NbOCl_3 + SOCl_2$$

Problem 5. *How will the following solutes react with the solvents given in the parenthesis. State whether the solution formed in each case will be more acidic or more basic than the solvent:*

H_2SO_4 (H_2O), H_2SO_4 (NH_3), CH_3COOH (H_2SO_4), $SOCl_2$ (SO_2) and CH_3CONH_2 (NH_3).

Solution. The reactions between the solutes and the solvents are given below. The nature of the solution is given in parenthesis.

Solute		*Solvent*				
H_2SO_4	+	H_2O	$\longrightarrow$	HSO_4^- +	H_3O^+	(*Acidic*)
H_2SO_4	+	NH_3	$\longrightarrow$	HSO_4^- +	NH_4^+	(*Acidic*)
CH_3COOH	+	H_2SO_4	$\longrightarrow$	$CH_3COOH_2^+$ +	HSO_4^-	(*Basic*)
$SOCl_2$	+	SO_2	$\longrightarrow$	$2Cl^-$ +	SO^{2+}	(*Acidic*)
CH_3CONH_2	+	NH_3	$\longrightarrow$	CH_3COONH^- +	NH_4^+	(*Acidic*)

Problem 6. *Complete the following equations:*

(*i*) $C_2H_5OH + HF \longrightarrow$ .... (*Guru Nanak Dev 96*)

(*ii*) $PbI_2 + KNH_2 \xrightarrow{Liq.NH_3}$ .... (*P.U. 1997*)

(*iii*) $Zn(NO_3)_2 + KNH_2 \xrightarrow{Liq.NH_3}$ .... (*P.U. 1997*)

(*iv*) $Na + NH_4Cl \xrightarrow{Liq.NH3}$ .... (*Kurukshetra 1997*)

Solution. (*i*) In HF (solvent), C_2H_5OH (solute) behaves as a base, since HF_2^- ions (solvent anions) are produced.

C_2H_5OH (*solute*) + 2HF (*solvent*) $\longrightarrow C_2H_5OH_2^+ + HF_2^-$ (*Solvent anion – base anion*)

(*ii*) PbI_2 is a metal salt. It reacts with the base *viz* KNH_2 and gives the ppt. of lead imide (PbNH).

$$PbI_2 + KNH_2 \text{ (Ammono base)} \xrightarrow{Liq.NH_3} PbNH\downarrow + KI + HI$$

(*iii*) $Zn(NO_3)_2$ is a metal salt. It reacts with the base, KNH_2 and gives the ppt. of $ZnNH_2$.

$$Zn(NO_3)_2 + 2KNH_2 \text{ (Ammono base)} \xrightarrow{Liq.NH_3} Zn(NH_2)_2\downarrow + 2KNO_3$$

(*iv*) In liq. NH_3, Na metal reduces NH_4Cl to H_2.

$$\underset{(Na\,=\,0)\ \ (H\,=\,+1)}{2Na + 2NH_4Cl} \xrightarrow{Liq.NH_3} \underset{(Na\,=\,+1)}{2NaCl} + 2NH_3 + \underset{(H\,=\,0)}{H_2}$$

Problem 7. *Complete the following reactions in liq.* SO_2

(*i*) $HgI_2 + KI \longrightarrow$ ......

(*ii*) $SbCl_5 + NOCl_2 \longrightarrow$ ...

(*iii*) $Cr_2(SO_3)_3 + SOCl \longrightarrow$ ....

(*iv*) $C_6H_6 + SO_3 \longrightarrow$ .... (*P.U. April 1996, Kurukshetra 1994*)

Solution. (*i*) When HgI_2 is dissolved in liq SO_2 in presence of KI, the solubility of HgI_2 in liq SO_2 is increased due to the formation of $K_2[HgI_4]$ (a complex compound)

$$HgI_2 + 2KI \xrightarrow{Liq.NH_3} K_2[HgI_4]$$

(*ii*) The reaction between $SbCl_5$ (covalent halide) and NOCl in liq. SO_2 leads to the formation of hexachloro complex, $NO^+[SbCl_6]^-$

$$SbCl_5 + NOCl \xrightarrow{Liq.NH_3} NO^+[SbCl_6]^-$$

(*iii*) Since $Cr_2(SO_3)_3$ contains SO_3^{2-} ions, this compound acts as a base in liq. SO_2. Similarly, since $SOCl_2$ gives SO^{2+} ions in liq. SO_2, this salt behaves as an acid in liq SO_2. Thus the reaction between $Cr_2(SO_3)_3$ and $SOCl_2$ in liq. SO_2 is an acid-base (neutralisation) reaction and gives salt and solvent (*i.e.*, SO_2)

$$SOCl_2\ (acid) + Cs_2SO_3\ (base) \xrightarrow{Liq.SO_2} 2CsCl\ (salt) + 2SO_2\ (solvent)$$

(*iv*) In presence of liq. SO_2, C_6H_6 undergoes sulphonation.

$$C_6H_6 + SO_3 \xrightarrow{Liq.SO_2} C_6H_5SO_3H.$$

Problem 8. *Show whether each of the following compounds acts as an acid or a base in liq. HF.*

BF_3, SbF_5, H_2O, CH_3COOH, C_6H_6

Solution. We know that HF undergoes self-ionisation as follows:

$$3HF \rightleftharpoons \underset{\substack{\text{Solvent cations}\\\text{(Acid ions)}}}{H_2F^+} + \underset{\substack{\text{Solvent anions}\\\text{(Base ions)}}}{HF_2^-}$$

This equation shows that the compounds producing H_2F^+ ions in liq HF act as acids in this solvent while the compounds giving HF_2^- ions behave as bases in this solvent. Thus:

$$\underset{\text{(Acid)}}{BF_3 + 3HF} \rightleftharpoons H_2F^+ + BF_4^-$$

$$\underset{\text{(Acid)}}{SbF_5 + 2HF} \rightleftharpoons H_2F^+ + SbF_6^-$$

$$\underset{\text{(Base)}}{H_2O + 2HF} \rightleftharpoons H_3O^+ + HF_2^-$$

$$\underset{\text{(Base)}}{CH_3COOH + 2HF} \rightleftharpoons CH_3COOH_2^+ + HF^-$$

$$\underset{\text{(Base)}}{C_6H_6 + 2HF} \rightleftharpoons C_6H_7^+ + HF_2^-$$

Problem 9. *Account for the following, giving appropriate reasons.*

(*i*) *K_2SO_3 is soluble in liq. SO_2.* *[M.D. Rohtak 1997]*

(*ii*) *SO_2 acts both as a Lewis acid and as a Lewis base.* *[M.D. Rohtak 1998]*

(*iii*) *The electrical conductivity of liq. NH_3 is increased when NH_4Cl is dissolved in it.* *[M.D. Rohtak 1997, Agra 1993]*

(*iv*) *A liq. NH_3 solution of phenolphthalein is colourless but becomes red coloured on addition of KNH_2.* *[M.D. Rohtak 1994]*

(*v*) *Liq. NH_3 exhibits association whereas liq. SO_2 does not.* *[M.D. Rohtak 1992, Punjab 1996]*

(*vi*) *Acetamide behaves as a base in aqueous solution but shows acidic properties in liq. NH_3.*

(*vii*) *HNO_3 acts as an acid in aqueous solution but in HF it behaves as a base.* *[H.P.W. 1997]*

(*viii*) *CH_3COOH behaves as a base in HF but SbF_6 behaves as an acid in this solvent.* *[Guru Nanak pev 1997]*

(*ix*) *Liq. NH_3 behaves as a differentiating solvent for NaH, $NaNH_2$ acid $NaOC_2H_5$.*

(*x*) *Solution of sodium metal in liq. NH_3 is blue and has strong reducing property and high electrical conductance.* *[Punjab 1996]*

Solution. (*i*) K_2SO_3 is soluble in liq. SO_2, since this compound forms a solvate, $K_2SO_3 \cdot SO_2$ (an addition compound) with liq. SO_2. This solvate contains one molecule of SO_2 as sulphur dioxide of crystallisation. In this solvate, K_2SO_3 acts as a solute.

(*ii*) In the formation of solvates like $CaI_2 . 4SO_2$, $LiI_2 . 2SO_2$ etc, SO_2 donates electrons to the solutes and hence acts as Lewis base. In the formation of solvates like $C_5H_5N.SO_2$, $C_6H_7N.SO_2$. SO_2 accepts electrons from the solutes and hence acts as Lewis acid. Thus SO_2 acts both as a Lewis acid and as a Lewis base.

(*iii*) Since NH_4Cl gives NH_4^+ ions in liq. NH_3, this compound acts as an acid in this solvent. Due to the production of NH_4^+ ions, the concentration of these ions in liq. NH_3 increases. The increase of NH_4^+ ions increases the electrical conductivity of liq. NH_3.

(*iv*) Addition of KNH_2 to liq. NH_3 increases the concentration of NH_2^- ions and hence this solution behaves as an ammono base. As soon as the concentration of NH_2^- ions increases in solution, the solution behaves as a base and hence imparts red colour to phenolphathlein just as an aqueous basic solution imparts red (pink) colour.

(*v*) Due to small size and high electronegativity of N-atom, many molecules of liq. NH_3 get associated together through H-bonding. This type of association is not possible in case of liq. SO_2, since sulphur atom has bigger size. (N = 0.75 Å, S = 1.02 Å) and lower electronegativity (N = 3.0, S = 2.5) than N-atom.

(*vi*) Since acetamide, CH_3CONH_2 gives OH^- ions in aqueous solution, this compound behaves as a base in water. On the other hand, since CH_3CONH_2 gives NH_4^+ ions in liq NH_3, it behaves as an acid in this solvent.

$$CH_3-C(=O)-NH_2 + H_2O \rightleftharpoons CH_3-C(=OH^+)-NH_2 + OH^- \text{ (Base)}$$

$$CH_3-C(=O)-NH_2 + NH_3 \rightleftharpoons CH_3-C(=O)-NH^- + NH_4^+ \text{ (Acid)}$$

(*vii*) Since, in aqueous solution, HNO_3 gives H^+ or H_3O^+ ions, this compound behaves as an acid in this solution. On the other hand HNO_3 acts as a base in HF, since, in this solvent, this compound gives HF_2^- ions.

$$HNO_3 + H_2O \longrightarrow NO_3^- + H_3O^+ \text{ (Acid)}$$
$$HNO_3 + 2HF \longrightarrow H_2NO_3^+ + HF_2^- \text{ (Base)}$$

(*viii*) CH_3COOH gives HF_2^- ions in HF by accepting H^+ ion.

$$CH_3COOH + 2HF \longrightarrow CH_3COOH_2^+ + HF_2^- \text{ (Base)}$$

SbF_5 ionises as an acid in HF by accepting F^- ion. Thus SbF_5 gives H_2F^+ ions

$$SbF_5 + 2HF \longrightarrow SbF_6^- + H_2F^+ \text{ (Acid)}$$

Here the self ionisation of HF can be noted as given below:

$$3HF \longrightarrow \underset{\substack{\text{Solvent cations}\\ \text{(Acid ions)}}}{H_2F^+} + \underset{\substack{\text{Solvent anions}\\ \text{(Base ions)}}}{HF_2^-}$$

(*ix*) When strong bases like NaH, $NaNH_2$ and $NaOC_2H_5$ react with H_2O (solvent), a proton is donated by H_2O molecule to these bases and NaOH is produced.

$$H_2O + NaH \longrightarrow H_2 + NaOH$$

or

$$H_2O + H^- \xrightarrow{} H_2 + OH^- \quad (H^+ \text{ transferred from } H_2O \text{ to } H^-)$$

$$H_2O + NaNH_2 \longrightarrow NH_3 + NaOH$$

$$\text{or} \quad H_2O + NH_2^- \xrightarrow{} NH_3 + OH^- \quad (H^+ \text{ transfer})$$

$$H_2O + NaOC_2H_5 \longrightarrow C_2H_5OH + NaOH$$

$$\text{or} \quad H_2O + OC_2H_5^- \longrightarrow C_2H_5OH + OH^- \quad (H^+ \text{ transfer})$$

Thus we see that these bases are lavelled to the base strength of OH^- ions, and hence these bases cannot be differentiated as bases in aqueous solution. On the other hand, since liq. NH_3 is a poor proton donor, these bases can be differentiated as bases in liq. NH_3. Thus liq. NH_3 behaves as a differentiating solvent for NaH, $NaNH_2$ and $NaOC_2H_5$.

(*x*) Blue colour is due to an absorption band having its maximum at 5000Å. Strong reducing property and high electrical conductance shown by the solution of sodium in liq. NH_3 is due to the production of ammoniated cations, $[Na(NH_3)_x]^+$ and ammoniated electron, $[e(NH_3)_y]^-$

$$Na \xrightarrow{\text{Liq.}NH_3} Na^+ + e^-$$

$$Na^+ + x\,NH_3 \longrightarrow [Na(NH_3)_x]^+$$

$$e^- + y\,NH_3 \longrightarrow [e(NH_3)y]^-$$

Problem 10. *What are the principal species formed in the solutions of NaOH, $(NH_4)_2SO_4$ and H_2O in water, anhydrous H_2SO_4 and liq. NH_3?*

Solution. NaOH (*a*) *In water* : In water, NaOH dissociates to give Na^+ and OH^- ions.

$$NaOH \xrightarrow{\text{Aqueous solution}} Na^+ + OH^-$$

(*b*) *In anhydrous H_2SO_4*: OH^- ions formed as above further react with H_2SO_4 to produce HSO_4^- and H_3O^+ ions.

$$OH^- + 2H_2SO_4 \longrightarrow 2HSO_4^- + H_3O^+$$

(*c*) *In liq. NH_3*: In liq. NH_3, no species are formed.

$(NH_4)_2SO_4$ (*a*) *In water* : NH_4^+ and SO_4^{2-} ions are formed.

$$(NH_4)_2SO_4 \xrightarrow{\text{Aqueous solution}} 2NH_4^+ + SO_4^{2-}$$

NH_4^+ ions formed as above react with H_2O to give H_3O^+ ions.

$$NH_4^+ + H_2O \rightleftharpoons NH_4OH + H_3O^+$$

(*b*) *In anhydrous H_2SO_4* : NH_4^+ and SO_4^{2-} ions are obtained.

$$(NH_4)_2SO_4 \longrightarrow 2NH_4^+ + SO_4^{2-}$$

(*c*) *In liq. NH_3* : NH_4^+ and SO_4^{2-} ions are formed. Due to the production of NH_4^+ ions, the solution becomes acidic.

$$(NH_4)_2SO_4 \xrightarrow{\text{liq.}NH_3} 2NH_4^+ + SO_4^{2-}$$

H_2O. (*a*) *In water*. there will be no reaction

(*b*) *In anhydrous H_2SO_4*. Following reaction will occur.

$$H_2O + H_2SO_4 \longrightarrow HSO_4^- + H_3O^+$$

(*c*) *In liq. NH_3*. NH_4^+ ions will be obtained.

$$H_2O + NH_3 \longrightarrow NH_4^+ + OH^-$$

Problem 11. *Predict the behaviour of urea in water, liq. NH_3 and anhydrous H_2SO_4.*

Solution. Urea is NH_2CONH_2.

In water : NH_2CONH_2 is expected to behave as a non-electrolyte.

In liq. NH_3 : NH_2CONH_2 behaves as a weak acid in liq. NH_3, since this compound produces NH_4^+ ion in this solvent.

$$NH_2CONH_2 + NH_3\,(l) \rightleftharpoons NH_4^+ + NH_2CONH^-$$

In anhydrous H_2SO_4 : When NH_2CONH_2 reacts with H_2SO_4, HSO_4^- ion is produced. Thus NH_2CONH_2 behaves as a base in H_2SO_4

$$NH_2CONH_2 + H_2SO_4 \longrightarrow NH_2CONH_3^+ + HSO_4^-$$

University Questions

1. (*a*) Discuss in detail the classification of solvents.

(*b*) Discuss the following in liquid NH_3 and liquid SO_2 :–

(*i*) Oxidation-reduction reaction

(*ii*) Precipitation reaction. *(Nagpur 2002)*

2. Compare the self-ionisation of water, liq. NH_3 and anhydrons acetic acid. *(Delhi 2002)*

3. (*a*) Predict the behaviour of the following solutes in liquid SO_2 :

(*i*) $SOCl_2$ (*ii*) CS_2SO_3 (*iii*) $(CH_3)_4\ N^+\ Br^-$.

(*b*) Predict the solubility of the following solutes in liquid SO_2.

(*i*) NH_4SCN (*ii*) Na_2S.

(*c*) Complete the following equations :

(*i*) $PbI_2 + KNH_2 \xrightarrow{NH_3}$ ....

(*ii*) $SO_2\ Cl_2 + NH_3 \xrightarrow{NH_3}$ .... *(Delhi 2002)*

Ans. (*c*) (*i*) $PBI_2 + KMH_2 \xrightarrow{NH_3} PbNH_2 + KI + HI$

(*ii*) $SO_2Cl_2 + 2NH_3 \xrightarrow{NH_3} SO_2(NH_2)_2 + 2HCl$.

4. (*a*) Explain chemical reactions in non-aqueous solvents with reference to those orruring in liq. NH_3.

(*b*) Discuss the role of non-protonic solvents in chemical reactions with special reference to the reactions occuring in liquid SO_2. *(VBS, Purvanchal 2003)*

5. (*a*) What is meant by protic and aprotic solvents ? Give examples.

(*b*) Give only one example for each of the following in liquid sulphur dioxide :

(*i*) Acid-base reaction (*ii*) Redox reaction

(*iii*) Precipitation reaction (*iv*) Complex formation reaction

(*v*) Reactions with organic compounds *(Rohildkhand 2003)*

6. Explain the following reactions which take place in liq. SO_2 as medium :

(*i*) Acid base reactions (*ii*) Precipitation reactions. *(Delhi 2003)*

7. (*a*) Classify the solvents on the basis of their proton-donor and proton-acceptor property. Give the following reactions in liquid NH_3

(*i*) Redox reaction (*ii*) Precipitation reaction.

(*b*) Give the following reactions in liquid SO_2

(*i*) Solvolysis

(*ii*) Complex formation. *(Nagpur 2003)*

1. (*a*) Write notes on (in non-aqueous solvent)

(*i*) Ammonolysis (*ii*) Acid-base reactions (*iii*) Redox reactions

(*b*) What are non-aqueous solvents? Explain the following reactions in liquid HF.

(*i*) Auto ionisation reaction (*ii*) Precipitation reactions (*iii*) Solvation reactions

(*iv*) Reaction with C_2H_5OH *(Bhopal 2004)*

2. What do you understand by ionising and non-ionising solvents? Give examples. *(GND Amritsar 2004)*

3. Discuss acid-base reactions in liq. SO_2 and metathetical reactions in liq. HF. *(GND Amritsar 2004)*

4. How do the following reactions proceed?

(*i*) $HgI_2 + KNH_2 \xrightarrow{NH_2}$ ..... (*ii*) $Cs_2SO_3 + SOCl_2 \xrightarrow{SO_2}$ .....

(GND Amritsar, 2004)

APPENDIX-I
List of Important Ores

Ore	*Formula*	*Page Number*
Arsenical pyrites	FeAsS	318
Autunite	$K(UO_2)PO_4.8H_2O$	468
Bismuth aurite	BiAu	343
Braggite	(Pt, Pd, Ni)S	231, 339
Braunite	Mn_2O_3	316
Calamine	$ZnCO_3$	346
Calaverite	$AuTe_2$	343
Carnotite	$K_2O.2UO_3.V_2O_5.3H_2O$	308, 468
Crocosite	$PbCrO_4$	310
Chalco pyrites	$CuFeS_2$ or CuS.FeS	318
Chrome-ochre	Cr_2O_3	310
Chrome ore	$FeO.Cr_2O_3$	310
Chromite	$FeO.Cr_2O_3$	310
Clay iron stone	$FeCO_3$	318
Cobalt bloom	$Co_3(AsO_4)_2.8H_2O$	332
Cobalt glance	(Co, Fe, As)S	332
Cobaltite	(Co, Fe, As)S	332
Copper-iron pyrites	$CuFeS_2$ or CuS.FeS	318
Cooperite	$Pt(AsS)_2$	339
Crytherite	$Co_3(AsO_4)_2.8H_2O$	332
Franklinite	$ZnO.Fe_2O_3$	346
Garnierite	$(Ni,Mg)SiO_3.xH_2O$	334
Haematite	Fe_2O_3	318
Hausmannite	Mn_3O_4	316
Hubnerite	$MnWO_3$	315
Illmenite	$FeTiO_3$	305
Iron pyrites	FeS_2	318
Kupfer nickel	NiAs	334
Limonite or hydrated oxide	$Fe_2O_3.3H_2O$	318
Linnacite	$(Fe,Co, Ni)_3S_4$	332
Magnetite	Fe_3O_4	318
Manganese blende	MnS	316
Manganite	$Mn_2O_3.H_2O$ or 2MnO(OH)	316

Milerite	NiS	335
Monazite sand		466
Molybdenite	MoS_2	313
Nickel glance	NiAsS	334
Nicolite	NiAs	334
Palladium gold		337
Pentlandite	(Ni, Cu, Fe)S	334
Petzite	$(Au, Ag)_2Te$	343
Pitchblende	U_3O_8	468
Potarite	PdHg	337
Propezite		337
Pyrolusite	MnO_2	316
Red Zinc	ZnO	346
Rhodonite	$MnSiO_3$	316
Rutile	TiO_2	305
Sylvanite	$(Au, Ag)Te_2$	343
Scheelite	$CaWO_4$	315
Sinderite	$FeCO_3$	318
Smaltite	$(Ni, Co, Fe)As_2$	332
Spathic iron	$FeCO_3$	318
Speiss cobalt	$(Ni, Co, Fe)As_2$	332
Sperrylite	$PtAs_2$	339
Sphalerite	ZnS	346
Sudbury ore	(Ni, Cu, Fe)S	334
Stibiopallinite	Pb_3Sb	337
Thorianite	ThO_2 (55 – 80%)	466
Thorite	$ThSiO_4.nH_2O$ or $(ThO_2.SiO_2.nH_2O)$	466
Uranite	U_3O_8	468
Vanadinite	$3Pb_3(VO_4)_2.PbCl_2$	308
White nickel	$NiAs_2$	334
Willemite	Zn_2SiO_4	346
Wolframite	$(Fe,Mn)WO_4$	315
Wulfenite	$PbMoO_4$	313
Zinc blende	ZnS	346
Zincite	ZnS	346
Zinc spar	$ZnCO_3$	346

APPENDIX-II
List of Important Processes

Processes	*Applications*	*Page Number*
1. Alumino-thermic process	Extraction of Ti, V, Cr, Mo, Co	306
2. Belgain process	Extraction of Zn	347
3. Bessemer's process	Manufacture of steel	325
4. Bredig's arc process	Extraction of Au	345
5. Carbon reduction process	Extraction of Zn	346
6. Cupellation process	Extraction of Au	344
7. Deville's process	Extraction of Cr	312
8. Duplex process	Manufacture of steel	328
9. Electric process	Manufacture of steel	328
10. I.C.I. process	Extraction of Ti	306
11. Jantesh process	Extraction of V	309
12. Kroll process	Extraction of Ti	305
13. L.D. process	Manufacture of steel	329
14. Mac-Arthur Forest Cyanide process	Extraction of Au	343
15. Marden and Rich process	Extraction of V	309
16. Mond's process	Extraction of Ni	45, 336
17. New Jersy vertical retort process	Extraction of Zn	347
18. Open-hearth process	Manufacture of steel	327
19. Parting process	Extraction of Au	344
20. Siemen-Martin process	Manufacture of steel	327
21. Van Arkel process	Purification of crude Ti, V, Th	307, 467

APPENDIX – III

List of Important Compounds

	Name of the compound	*Formula*	*Page Number*
	Compounds of lanthanum		
1.	Lanthanum dihydride	$La\ H_2$	355
2.	Lanthanum trihydride	LaH_3	355
3.	Lanthanum halides	$La\ X_3$	355
4.	Lanthanum sulphate	$La_2\ (SO_4)_3$	355
5.	Lanthanum nitrate	$La\ (NO_3)_3.6H_2O$	355
6.	Lanthanum carbonate	$La_2\ (CO_3)_3$	356
	Compounds of titanium		
1.	Titanium oxide	TiO	356
2.	Titanium sesquioxide	Ti_2O_3	357
3.	Titanium dioxide or titanic dioxide or titania	TiO_2	357
4.	Titanium dichloride	$TiCl_2$	358
5.	Titanium trichloride or titaneous chloride	$TiCl_3$	359
6.	Titanium tetrachloride or titanic chloride	$TiCl_4$	359
	Compounds of vandaium		
1.	Vanadyl chloride	$VOCl_2$	361
2.	Vonadyl sulphate	$VOSO_4$	361
3.	Vanadium pentoxide	V_2O_5	361
4.	Metavanadic acid	HVO_3	364
5.	Pyrovanadic acid	$H_4V_2O_7$	364
6.	Orthovanadic acid	H_3VO_4	364
7.	Ammonium meta vanadate	NH_4VO_3	364
8.	Sodium pyrovanadate	$Na_4V_2O_7$	364
9.	Sodium orthovanadate	$Na_3VO_4.12H_2O$	365
	Compounds of chromium		
1.	Chromium oxide or chromium sesquioxide	Cr_2O_3	366
2.	Hydrated chronic oxide (Also called Guigret's green)	$Cr_2O_3.2H_2O$	367
3.	Chromium trioxide or chronic acid anhydride	CrO_3	367
4.	Chromous chloride	$CrCl_2.4H_2O$	368
5.	Chromic chloride	$CrCl_3$	369
6.	Chromic chloride hexahydrate	$CrCl_3.6H_2O$	369
7.	Hexaaqua chromium (III) chloride (Violet)	$[Cr\ (H_2O)_6]\ Cl_3$	369
8.	Pentaaquachloro chromium (III) chloride monohydrate (Blue-green)	$[Cr\ (H_2O)_5\ Cl]\ Cl_2.H_2O$	369
9.	Tetraaquadichlorochromium (III) chloride dihydrate (Green)	$[Cr\ (H_2O)_4\ Cl_2]\ Cl.2H_2O$	369

4. Manganese dioxide	MnO_2	391
5. Manganates	MnO_4^{2-}	392
6. Manganese heptoxide	Mn_2O_7	393
7. Permanganic acid	$HMnO_4$	394
8. Potassium permanganate	$KMnC_4$	394
9. Dinuclear bridged complexes of Mn	–	27
10. Dimanganese decacarbonyl	$Mn_2 (CO)_{10}$	485, 505
Compounds of iron		
1. Hexammine ferrate (II) chloride	$[Fe (NH_3)_6] Cl_2$	405
2. Sodium pentacyano nitrosyl ferrate (II)	$Na_2[Fe^{2+} (CN)_5(NO^+)].2H_2O$	405, 515
3. Sodium nitroprusside	$Na_2[Fe^{2+}(CN)_5(NO^+)].2H_2O$	405, 515
4. Potassium ferrocyanide or potassium hexacyanoferrate (II)	$K_4 [Fe (CN)_6]$	405
5. Potassium ferrous ferrohexacyanide	$K_2Fe^{2+} [Fe^{2+} (CN)_6]$	408
6. Hydroferricyanic acid	$H_3 [Fe (CN)_6]$	408
7. Potassium ferricyanide or potassium hexacyanoferrate (III)	$K_3[Fe(CN)_6]$	408
8. Potassium ferric hexacyano ferrate (II) or potassium ferric ferro hexacyanide	$K Fe^{3+} [Fe^{2+} (CN)_6]$	411
9. Ferric hexacyano ferrate (II) or ferric ferro hexacyanide	$Fe_4^{3+} [Fe^{2+} (CN)_6]_3$	411
10. Prussian blue and Turnbull's blue	$K Fe^{3+} [Fe^{2+} (CN)_6]$ or $Fe_4^{3+} [Fe^{2+} (CN)_6]_3$	411
11. Dinuclear bridged octahedral complex of Fe (III)	–	27
12. Iron pentacarbonyl	$Fe (CO)_5$	486, 499
13. Diron enneacarbonyl	$Fe_2 (CO)_9$	487, 500
14. Tri iron dodecacarbonyl	$Fe_3 (CO)_{12}$	488, 506
15. Nitrosonium pentaaqua iron (I) sulphate or nitroso ferrous sulphate	$[Fe^+ (NO^+) (H_2O)_5]SO_4$	514
Compounds of osmium		
1. Osmium dioxide	OsO_2	414
2. Osmium tetroxide	OsO_4	414
3. Osmium hexafluoride	OsF_6	415
4. Osmium tetrachloride	$OsCl_4$	415
5. Osmium dodecacarbonyl	$Os_3(CO)_{12}$	506
Compounds of cobalt		
1. Cobaltous oxide	CoO	415
2. Cobalto-cobaltic oxide	Co_3O_4 or $CoO.Co_2O_3$	416

3. Sodium hexanitro cobaltate (III) or sodium cobaltinitrite	Na_3 [Co $(No_2)_6$]	416
4. Potassium hexanitro cobaltate (III) or potassium cobaltinitrite (Also called Fischer's salt)	K_3 [Co $(NO_2)_6$]	417
5. Co (III) ammines	(*i*) $CoCl_3.6NH_3$ or $[Co(NH_3)_6]Cl_3$ (orange-yellow), (*ii*) $CoCl_3.5NH_3.H_2O$ (pink) (*iii*) $CoCl_3.5NH_3$ or [Co $(NH_3)_5$ Cl] Cl_2 (violet) (*iv*) trans- $CoCl_3.4NH_3$ or $[Co(NH_3)_4Cl_2]Cl$ (green) (*v*) cis-Co Cl_3. $4NH_3$ or $[Co(NH_3)_4Cl_2]Cl$ (violet) (*vi*) $CoCl_3 . 3NH_3$ or $[Co (NH_3)_3Cl_3]^0$ (blue-green)	64, 66, 69
6. Dicobalt octacarbonyl	$Co_2 (CO)_8$	488, 502
7. Hexammine cobalt (III) chloride or hexamine cobaltic chloride	$[Co (NH_3)_6] Cl_3$	417
8. Bridged complexes of Co (III)	–	21
Compounds of nickel		
1. Nickel sulphate	$Ni SO_4.7H_2O$	418
2. Nickel tetracarbonyl	$Ni (CO)_4$	490,499
3. Nickel ammonium sulphate	$Ni SO_4.(NH_4)_2SO_4.6H_2O$	419
4. Bis (dimethylgly oximato) nickel (II)	$[Ni (C_4H_7N_2O_2)]^0$ or $[Ni (dmg)_2]^0$	419
5. Dinuclear bridged complex of Ni (I)	–	27
Compounds of palladium		
1. Fluorides	Pd F_2, Pd F_3	420
2. Chlorides	Pd Cl, Pd Cl_2, Pd Cl_3, Pd Cl_4	420
3. Palladrium bromide	Pd Br_2	421
4. Palladrium iodide	Pd I_2	421
5. Oxides	Pd_2O, PdO, PdO_2	421
Compounds of platinum		
1. Hydrogen tetrachloro platinate (II) or tetrachloroptatinous acid	H_2 [Pt Cl_4]	421

Subject Index

A

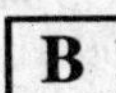

B

C

G

H

I

M

P

R

S

OBJECTIVE TYPE QUESTIONS

Objective Type Questions

1. All the following statements about transitional elements are true except
   a) all of the transitional elements are predominantly metallic
   b) in aqueous solution many of their simple ions are coloured
   c) all of these elements have unfilled d - orbital
   d) most of the transitional elements show only one oxidation state

2. Transition elements in general
   a) show variable oxidation state & show pronounced catalytic activity
   b) form paramagnetic and coloured ions
   c) form co-ordination compounds
   d) are most reactive elements. Which is false statement

3. d-block elements show a strong paramagnetic properties because
   a) their atoms have all paired electrons
   b) their atom have all unpaired electrons
   c) their atoms have a number of unpaired electrons
   d) their atoms have a number of paired electrons

4. d-block show variable oxidation state because the difference in energy between
   a) ns and nd orbitals is very slight
   b) ns and (n–1) d orbitals is very slight
   c) ns and np orbitals is large
   d) np and (n–1) d orbitals is small

5. d-block elements show catalytic activity because
   a) of their large surface area
   b) of their large atomic volume
   c) of their ability to exist in multiple oxidation states
   d) of their ability to provide finally divided surface for reaction

6. d-block elements forms coloured ions because their ions :
   a) cannot absorb the colours of these radiations
   b) involve d transition which falls in the visible region
   c) involve d-d transition which fall in the visible region
   d) allow d-s transitions in the visible region

7. d-block elements form co-ordination compounds because of
   a) small cationic size
   b) large ionic charge
   c) unfilled d-orbitals
   d) all the above

8. Interstitial compounds are formed when elements
   a) H, B, C, N occupy the tetrahedral and octahedral holes in the crystal lattice of transition metals
   b) transition metals are heated directly with H_2, B, C and N_2
   c) hydrides, borides, carbides and nitrides of transition metals are formed
   d) alloys of transition metals are formed

9. An example of a Bertholloid compound is
   a) ZnO b) tungsten carbid
   c) H_2O d) stainless steel

10. Which one of the following is not an interstitial compound ?
    a) all steels
    b) tungsten carbide
    c) Vanadium nitride
    d) WO_3

11. How many unpaired electrons are present in V^{3+} ion ?
    a) 1 b) 2 c) 3 d) 4

12. When a solution of ammonia vanadate (V) is reduced with Zn and dilute H_2SO_4 the colour changes from colourless to blue to green to violet. This is because of the change

a) VO_3^+ (aq) to VO^{2+} (aq) to V^{3+} (aq) to V^{2+} (aq)
b) in oxidation state from +5 to +4 to +3 to +2
c) in the formation of aqueous cationic species
d) all the above

13. The outer-shell configuration of Cr is
a) $3d^4 4s^2$ b) $3d^5 4s^1$
c) $3d^3 4s^2$ d) $3d^6 4s^o$

14. The number of unpaired electrons in Cr atom is
a) 3 b) 4 c) 5 d) 6

15. Stainless steel contains
a) 12% Cr and 0.5% carbon
b) 12% Cr and 1% carbon
c) 18% Cr
d) 10% Cr and 6% V

16. Cr-V-Steel is used for
a) locomotive wheels
b) motor industry
c) high-speed tools
d) surgical instruments

17. Chrome alum is formed when
a) C_2H_5OH is added to ice-cold solution of $K_2Cr_2O_7 + H_2SO_4$
b) H_2S is passed through acidified $K_2Cr_2O_7$ solution
c) SO_2 is passed through $CrCl_3$ solution
d) Chromyl chloride is reacted with NaOH solution

18. When $K_2Cr_2O_7$ is heated with Conc. HCl, the reduction product is :
a) $CrCl_3$ b) Cl_2
c) KCl d) H_2O

19. When a mixture of NaCl, $K_2Cr_2O_7$ and conc. H_2O_4 is heated, the product is
a) $CrCl_3$ b) CrO_2Cl_2
c) CrO_3 d) Na_2CrO_4

20. The reaction of $Cr_2O_7^{2-}$ with alkali produces
a) Cr_2O_3 b) CrO_4^{2-}
c) CrO_3 d) Cr_2O

21. In the reduction of dichromate by iron (II), the number of electrons involved per Cr atom is
a) 1 b) 3 c) 6 d) 4

22. When the equations $Cr_2O_7^{2-} + H^+ + Sn^{2+} \rightarrow Cr^3 + Sn^{4+} + H_2O$ is balanced, the coefficient of Sn^{2+} is
a) 3 b) 2 c) 6 d) 4

23. How many moles of I_2 are formed in the reaction
$Cr_2O_7^{2-} + H^+ + I^- \rightarrow Cr^{3+} + H_2O + I_2$
a) 3 b) 6 c) 2 d) 14

24. Reaction of $K_2Cr_2O_7$ with cold conc. H_2SO_4 produces
a) CrO b) Cr_2O_3
c) CrO_3 d) Cr_2O

25. In the reaction sequence : $CrO_2Cl_2 \xrightarrow{NaOH} A \xrightarrow{\text{dil } H_2SO_4} B \xrightarrow{NaOH} C \xrightarrow{AgNO_3} D$ is
a) Na_2CrO_4 b) $Na_2Cr_2O_7$
c) Ag_2CrO_4 d) CrO_3

26. In the reaction sequence :
$CrCl_3 \xrightarrow[NH_4OH]{NH_4Cl} A \xrightarrow[H_2O]{Na_2O} B \xrightarrow[\text{acetate}]{\text{lead}} C$ is
a) Na_2CrO_4 b) $Na_2Cr_2O_7$
c) $Cr(OH)_3$ d) $PbCrO_4$

27. In the reaction $Cr_2O_7^{2-} + H^+ + Fe^{2+} \rightarrow Cr^{3+} + Fe^{3+} + H_2O$ the eq. wt. of $K_2Cr_2O_7$ is
a) $\frac{M.W}{2}$ b) $\frac{M.W}{4}$
c) $\frac{M.W}{6}$ d) $\frac{M.W}{8}$

28. Cr is in +6 oxidation state in
a) CrO_4^{2-} b) $Cr_2O_7^{2-}$
c) CrO_2Cl_2 d) all of these

29. Carboloy is
a) carbide
b) the hardest substance next to diamond
c) used in cutting tools and dies
d) all of these

30. The principal ox. states shown by Cr are

a) +2, +3, +4 b) +2, +4, +6
c) +2, +3, +5 d) +2, +3, +6

31. Which is not a reasonable formula for oxide of Cr ?
a) Cr_2O b) CrO
c) Cr_2O_3 d) CrO_3

32. The conversion of CrO_4^{2-} into $Cr_2O_7^{--}$ is brought about by
a) H_2O b) NH_3
c) NaOH d) H_2SO_4

33. The outer shell configuration of Mn is
a) $3d^5 4s^2$ b) $3d^6 4s^1$
c) $3d^4 4s^2 4p^1$ d) $3d^7 4s^0$

34. An element which shows largest number of oxidation states is
a) Cr b) V c) Mn d) Ti

35. The uncommon oxidation states of Mn are
a) +2, +4, +6 b) +4, +6, +7
c) +1, +3, +5 d) +3, +4, +5

36. Baeyer's reagent is a
a) neutral solution of $KMnO_4$
b) an alkali solution of $KMnO_4$
c) an acidic solution of $KMnO_4$
d) none of these

37. In which reaction, the eq. wt of $KMnO_4$ is the same as its M. W. ?
a) $KMnO_4 + H_2SO_4 \rightarrow H_2SO_4 + MnSO_4 + H_2O + O$
b) $KMnO_4 + KOH \rightarrow K_2MnO_4 + H_2O + O$
c) $KMnO_4 + H_2O \rightarrow KOH + MnO_2 + O$
d) $KMnO_4 + HCl(conc) \rightarrow MnCl_2 + KCl + Cl_2 + H_2O$

38. The number of oxalate ions oxidised by 1 mole MnO_4^- ion in acid medium is
a) 5 b) 5/2 c) 2/5 d) 5/3

39. The moles of FeC_2O_4 oxidised by 1 mole MnO_4^- ion in acid medium is
a) 5/3 b) 3/5 c) 2/5 d) 5/2

40. In the redox equation
$Cr_2O_7^{2-} + I^- + H^+ \rightarrow Cr^{3+} + H_2O + I_2$
the co-efficient of I_2 in balanced reaction is
a) 3 b) 4 c) 5 d) 6

41. SO_2 is passed through $KMnO_4$ solution, MnSO4 is one product. The oxidation state changes form
a) +7 to +2 b) +5 to +4
c) +5 to +2 d) +4 to 1

42. With which of the following does an acidified solution of $KMnO_4$ give a yellow ppt ?
a) SO_2 b) H_2O_2 c) H_2S d) KI

43. In the reaction between acidified $KMnO_4$ and H_2O_2
a) gains electron
b) loses electrons
c) neither loses on or gain electrons
d) both loses and gains electrons

44. The structure of MnO_4^- ion is

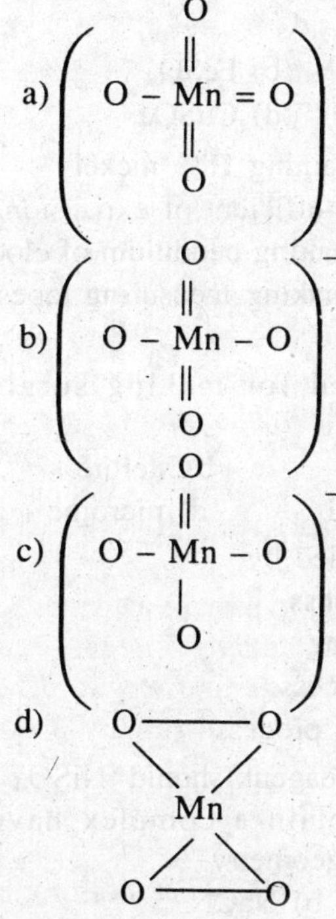

45. $KMnO_4$ is not a primary standard because
a) it is hygroscopic
b) it is deliquescent
c) it is not obtainable in pure form
d) its aqueous solution undergoes decomposition

46. It is dangerous to add $KMnO_4$ to hot conc. H_2SO_4 because
a) explosive MnO_2 is formed
b) explosive Mn_3O_4 is formed
c) explosive Mn_2O_7 is formed
d) K_2MnO_4 is formed explosively

47. The principle ore of Mn is
a) pyrolusite b) braunite
c) manganite d) manganese carbide

48. The outer shell electronic configuration of iron is
a) $3d^5 4s^2$ b) $3d^6 4s^2$
c) $3d^8 4s^o$ d) $3d^7 4s^2$

49. The number of unpaired electrons in Ni^{2+} ion is
a) 31 b) 32 c) 33 d) 34

50. Which of the following divalent cation has the largest number of unpaired electrons ?
a) $MnSO_4$ b) $FeSO_4$
c) $NiSO_4$ d) $CuSO_4$

51. Invarsteel containing 18% nickel
a) has little co-efficient of expansion
b) is used in making pendulum of clocks
c) is used in making measuring tapes
d) all the above

52. An alloy used for making surgical instruments is
a) silicon steel b) stellite
c) monel metal d) nicrome

53. Nickel is purified by
a) Kroll's process
b) Zone refining
c) Mond's process
d) Van Arkel's process

54. With which reagent should $NiSO_4$ be reacted to obtain a complex having square planar geometry ?
a) NH_3 b) NaCN
c) CO d) HCl

55. $K_3[CO(NO_2)_6]$
a) potassium cobaltinitrite
b) potassium hexanitro cobaltate (III)
c) Fisher's salt
d) all the above

56. A metal which is strongly attracted by a magnet, is attacked slowly by conc. HCl liberating a gas & forming a blue solution. The addition of water to their solution turns it pink. The metal is
a) iron b) nickel
c) cobalt d) chromium

57. Cobalt chloride paper provides a test for the presence of
a) NH_3 b) CO c) HCN d) H_2O

58. The most important ore of titanium is
a) rutile b) ilmenite
c) zircon d) thoria

59. Titanium is made by reduction of $TiCl_4$ with Na is
a) Kroll's process
b) Van Arkel's process
c) Zone - refining
d) Parke's process

60. Titanium metal has assumed importance because it finds use in
a) Space-Crafts b) Jet engines
c) Missiles d) All the above

61. Violet Ti (III) sulphate soution can be standardized by titration against
a) HCl b) NaOH
c) Hypo d) $KMnO_4$

62. An acidified solution of $TiCl_4$ gives a distinct yellow coloration with
a) O_3 b) HNO_2
c) H_2O_2 d) H_2S

63. TiO_2 is to an excellent white pigment because of
a) its high opacity
b) lightness and inertness towards water and H_2S
c) ease of dispersion in oil
d) all the aobve

64. Ti, Zr, Hf are prepared by Kroll's process which is :

a) $TiCl_4 + 4Na \xrightarrow[\text{atom}]{\text{Ar}} Ti + 4NaCl$

b) $ZrI_4 \xrightarrow[\text{W. filament}]{\text{heat}} Zr + 2I_2$

c) $ZrO_2 + 2H_2 \longrightarrow ZrO_2 + 2H_2O$
d) none of these

65. Purest form of iron is
a) cast iron b) wrought iron
c) low grade steel d) high grade-steel

66. Which forms of iron can be welded, tempered, and permanently magnetized ?
a) cast iron b) wrought iron
c) steel d) alloy steel

67. Blue - black fountain ink is essentially a solution in water of
a) Mohr's salt b) $FeSO_4$
c) Ferrous tannate d) Prussian blue

68. Ink stains can be removed using
a) oxalic acid b) citric acid
c) lemon juice d) all the above

69. Rust is essentially
a) Fe_2O_3
b) $Fe(OH)_3$ and Fe_2O_3
c) $Fe(OH)_3$
d) $FeCO_3$

70. Rusting can be prevented by
a) covering with metals like Ni and Cr
b) galvanizing
c) use of sacrificial metals like Mg
d) all the above

71. Sodium nitroprusside is
a) $Fe[Fe(CN)_6]^-$ b) $Na_2 [Fe(CN)_5 NO]$
c) $K_3CO(NO_2)_6$ d) none of these

72. The most important ore of Cu is
a) copper glance b) copper pyrites
c) malachite green d) azurite blue

73. Bell metal is an alloy of
a) Cu and Sn b) Ag and Cu
c) Cu and Ni d) Cu, Zn, Sn

74. German silver is an alloy of
a) Cu, Ni, Zn b) Ag and Cu
c) Cu and Ni d) Cu, Zn, Sn

75. An ammonical solution of *roman* is used in testing
a) O_2
b) CO
c) alkynes with terminal hydrogen
d) HCN

76. Addition of KI to $CuSO_4$ solution liberates I_2 and give a white ppt. of
a) CuI_2 b) Cu_2I_2
c) Cu_2SO_4 d) K_2CuI_4

77. Behaviour of $CuSO_4.5H_2O$ on heating suggests that
a) $5H_2O$ molecule are co-ordinated to Cu^{2+}
b) Four H_2O molecules are in square planar position around Cu^{2+} ion and the fifth is held by H-bonding
c) it has distorted octahedral structure
d) none of these

78. Basic copper acetate, $(CH_3COO)_2 Cu. Cu(OH)_2$ is called
a) Verdigris green
b) Paris green
c) Blue vitriol
d) Bordearex mixture

79. Silver is recovered form argenti ferrous lead by
a) Parke's process
b) Parting's Process
c) Amalgamation Process
d) Cyanide Process

80. Lunar caustic, $AgNO_3$ is used in making
a) AgBr for photography
b) indelible ink
c) silvering of mirrors
d) all the above

81. Ag or Au is purified by
a) cupellation b) electrolytic refining
c) polling d) zone-refining

82. 18- Carat gold is
a) 100% b) 75%
c) 50% d) 25% pure gold

83. The gold solution used for toning in photography is
a) AuCl b) $AuCl_3$
c) $HAuCl_4.4H_2O$ d) $NaAu(CN)_2$

84. In silver plating, the electrolyte used in the electrolytic bath is a solution of
a) $AgNO_3$ b) $NaAg(CN)_2$
c) $Ag(NH_3)_2Cl$ d) none of these

85. When gold is dissolved in KCN solution in presence of air, the complex anion ion solution is

a) $AuCN^-$ b) $Au(CN)_2^-$
c) $Au(CN)_4^-$ d) $Au(CN)_3^-$

86. Brass is an alloy of
a) Cu and Zn b) Cu, Ni, Zn
c) Cu and Ni d) Cu, Al, Zn

87. ZnO called Philosopher wool, Zinc white or Chinese white
a) turns yellow on heating and white on cooling
b) gives green mass in cobalt nitrate test
c) is used in medicine
d) all the above

88. When $SnCl_2$ solution in excess is added to a solution of corrosive sublimate, what is formed ?
a) a white ppt. of Hg_2Cl_2
b) a grey ppt. of Hg
c) a red ppt. of HgO
d) first a white ppt. of Hg_2Cl_2 which becomes grey Hg

89. Mercury fulminate called Pharaoh's serpent is
a) $Hg(ONC)_2$
b) highly explosive
c) employed in the percussion of caps
d) all the above

90. Fool's gold is
a) gold
b) silver glance Ag_2S
c) iron pyrites FeS_2
d) Copper glance, Cu_2S

91. Nickel is purified by
a) Liquaton process
b) Parke's process
c) Van Arkel's process
d) Mon'd process

92. Purple of Cassius is
a) Colloidal Au b) $AuCl_3$
c) $HAuCl_4$ d) $AuCl_3 + NH_3$

93. Fulminating gold, formed by $AuCl_3$ and NH_3 and which is olive green solid, has the formula
a) $Au(NH_2)_3$ b) Au(NH)Cl
c) $Au(NH_2) = NH$ d) $Au(NH_3)_2Cl_3$

94. Which is used in filling cavities in teeth?
a) Cu(Hg) b) Ag(Hg)
c) Zn(Hg) d) Ni(Hg)

95. Calomel (Hg_2Cl_2) is turned black by NH_3 due to the formulation of
a) $HgCl_2$ b) Hg(Cl)(NH_2)
c) Hg + Hg(Cl)(NH_2) d) Hg(Cl)(NH_2)

96. NH_3 dissolves freely in
a) $MgCl_2$ b) $PbCl_2$
c) AgI d) $Cu(OH)_2$

97. Cupellation is a process used for the purification of
a) Cu b) Pb c) Fe d) Ag

98. Osmium shows highest oxidation state in osmium tetroxide
a) +4 b) +6 c) +8 d) +2

99. Pt - Ir alloys are used in making
a) fountain pen points
b) chemical laboratory wares
c) standard weights
d) all the above

100. Silver tarnish is silver
a) sulphate b) sulphite
c) sulphide d) oxide

101. When AgCl is dissolved in NH_3 the principal cation present in the solution is
a) $Ag(NH_3)_2^+$ b) $Ag(NH_3)^{2+}$
c) Ag^+ d) NH_4^+

102. A clock spring is heated to redness and then plunged into cold water. This treatment will cause it to become
a) soft and ductile
b) more spongy therefore
c) strongly magnetic
d) hard and brittle

103. The rusting of a metal is called
a) spontaneous combustion
b) slow oxidation
c) burning
d) combustion

104. A metal used to galvanies iron is
a) Ge b) Sn c) Ni d) Zn

105. Mercury form amalgam with all but

a) Ni b) Al c) Fe d) Zn

106. The fact that the metal oxide M_2O_7 is a liquid above -20^oC suggests that the bonding in it is predominantly
a) metallic b) ionic
c) covalent d) dispersion forces

107. Which one of the following can be oxidized ?
a) Mn_2O_7 b) NO_3^-
c) ClO_4^- d) SO_2

108. Select a compound which will not be oxidised by an aqueous solution of Fe^{2+} ions ?
a) $SnCl_2$ b) $KMnO_4$
c) $K_2Cr_2O_7$ d) Cl_2

109. How many moles of F^{2+}_{e} ions are oxidized by 1 mole of $Cr_2O_7^{2-}$ ions in acid solution ?
a) 3 b) 2 c) 6 d) 12

110. Two metals are fastened together and placed in dil. HCl. In which case would the first named to corrode faster than the second ?
a) Fe, Sn b) Fe, Sn
c) Pb, Zn d) Ag, Zn

111. A piece of powdered Zn is placed in dil. H_2SO_4 and the reaction is very slow. When a piece of Ag is put in the acid and allowed to touch the Zn, then
a) Zn dissolves rapidly
b) Zn forms the cathode of the cell
c) The Zn metal is reduced
d) The silver metal is reduced

112. Black MnO_2 stains are best removed from glass - wares by placing it is an acidified solution of
a) Na_2S b) $NaHSO_4$
c) Na_2SO_3 d) NaHS

113. Which one of the following species does not represent a cationic species of vanadium formed in aqueous solution ?
a) VO_2^+ b) $[V(H_2O)_6]^{3+}$
c) VO_2^{2+} d) VO^{2+}

114. The number of unpaired electrons in Ti^{4+} ion are
a) 2 b) 3 c) 4 d) 0

115. Two stable ions of the first transition series which have largest number of unpaired electrons and have highest magnetic moment are
a) Cr^{3+} and Fe^{2+} b) Fe^{3+} and Mn^{2+} i
c) Ti^{2+} and CO^{2+} d) Ti^{4+} and Cr^{3+}

116. Two stable ions of 1st transition series which are diamagnetic are
a) Sc^{3+} and Ti^{2+} b) Cu^+ and Zn^{2+}
c) Cu^{2+} and Zn^{2+} d) Cr^{2+} and Zn^{2+}

117. Which ion gives coloured solution ?
a) Cu^+ b) Zn^{2+} c) Ti^{4+} d) V^{3+}

118. d-block elements comprise the sub-groups
a) 3 to 12
b) IIIA to VIIA
c) between IIA and IIIA
d) IB to VIIIB

119. Fe, CO, Ni belong to sub-group
a) VIII
b) 8, 9, 10
c) 11, 12, 13
d) 14, 15, 16

120. Transition metals, despite high E^o ox, are poor reducing agent due to
a) high heat of vaporization
b) high ionization energies
c) low heats of hydration
d) all of these

121. Oxidation state of Zero is shown in
a) $Fe(CO)_5$ b) $Na_2Ni(CN)_4$
c) $K_3Cu(Cu)_4$ d) CrO_2Cl_2

122. non-stoichiometry is
a) found in oxides and sulphides where the transition metal is present in more than one oxidation state
b) associated with intense colour
c) defect crystals
d) all of these

123. An example of Bertholloid compound is
a) FeS b) H_2S
c) NaCl d) $CuSO_4$

124. The existence of the Cu^+ ion in aqueous solution is very rare owing to its
a) oxidation b) reduction
c) disproportionation d) instability

125. An ion of a transition metal has magnetic moment of 1.73 B.M.; it could be
a) Cu^{2+} b) Ni^{2+} c) Cr^{3+} d) Fe^{2+}

126. Which are of the following is an acidic oxide ?
a) Cr_2O_3 b) CrO_3
c) CrO_2 d) Cr_2O

127. Which one of the following exhibits oxidation state of +8 in its compounds ?
a) iron b) chromium
c) osmium d) platinum

128. +4 oxidation state is stable for
a) Ni b) Pt c) V d) Cr

129. The oxidation potential value $M(S) \rightarrow M^{2+}(aq) + 2e$ is lowest for which element of 1st transition series ?
a) Mn b) Fe c) Ni d) Cu

130. Fe, Ru, Os belong to which sub-group of periodic table ?
a) 8 b) 9 c) 10 d) 12

131. Pd is placed below in which group of periodic table ?
a) Ni in group 10 b) Co in group 9
c) Fe in group 8 d) Zn in group 12

132. Zinc is present in humans to the extent of
a) 2g b) 4g c) 1g d) 1mg

133. Zn is involved in
a) respiration through the enzyme carbonic anhydrase
b) sugar metabolism
c) digestion of proteins by animals
d) all of these

134. Copper in trace amounts are needed in the body
a) to use iron stored in the liver
b) for the proper functioning of the enzyme cytochrome oxidaze
c) for proper metabolism of iron
d) all of these

135. Copper is present in proteins such as
a) Haemocyanin b) Azurin
c) Plastocyanin d) all of these

136. Wilson's disease
a) is a hereditary storage of ceruloplasmin
b) causes Cu accumulation in the liver, brain
c) is cured by feeding patient with EDTA
d) all of these

137. Vitamin B_{12} contains cobalt in
a) +2 b) +3
c) +1 d) 0 oxidation state

138. Cobalt deficiency can cause
a) disease in sheeps b) sickel cell anemia
c) cancer d) heat failure

139. Which is a false statement ?
a) Cr(III) and insulin are responsible for maintaining the correct level of glucose in the blood
b) Mo is present in nitrogen fixing bacteria
c) Mo is present in two proteins
d) Cr- salts are not carcinogenic

140. An ion which has no d-electrons is
a) Zn^{2+} b) Cu^+
c) Sc^{3+} d) Fe^{2+}

141. Zircone (gem stone) is
a) zirconium oxide
b) zirconium silicate
c) zirconium carbide
d) none of these

142. Which of the following statements about titanium (III) chloride is false ?
a) it has green and violet forms
b) it is used as a reducing agent
c) it disproportionates to $TiCl_2$ and $TiCl_4$
d) it is used in conc. H_2SO_4

143. Which of the following Cr(VI) compound can be isolated ?
a) CrO_3 b) CrO_2Cl_2
c) H_2CrO_4 d) $KCrO_3Cl$

144. Which of the following statements about the colour of aqueous solutions of the stated chromium species is true ?
a) Cr^{2+} (aq) is blue
b) CrO_4^{2-} (aq) is yellow
c) Cr^{3+} (aq) is yellow
d) $Cr_2O_7^{2-}$ (aq) is green

145. Which of the following will not reduce M/60 $K_2Cr_2O_7$ (aq) in 2M H_2SO_4 at laboratory temperature ?
a) Fe^{2+} b) Sn^{2+} c) I^- d) Cl^-

146. Which of the following transformation of the $Cr_2O_7^{2-}$ ion requires the action of a reducing agent ?
a) $\longrightarrow$ CrO_4^{2-} b) $\longrightarrow$ CrO_3Cl^-
c) $\longrightarrow$ Cr^{3+} d) $\longrightarrow$ Cr_2O_3

147. What is not a true statement about $KMnO_4$?
a) it reacts with hot conc. H_2SO_4 to give Mn_2O_7
b) it gives off O_2 when heated
c) it is reduced by ethene in neutral medium to give MnO_2
d) it is a primary standard

148. Which of the following will not reduce Fe^{3+} to Fe^{2+} in solution ?
a) H_2S b) SO_2 c) $SnCl_2$ d) HI

149. Which of the following species is likely to be present in aqueous solution of iron (II) chloride in solution ?
a) $[Fe(H_2O)_6]^{3+}$
b) $[Fe(H_2O)_5OH]^{2+}$
c) $[Fe(H_2O)_5OH^+$
d) $[Fe(H_2O)_4OH.Cl]^+$

150. Which of the following statement about nickel compounds is true ?
a) in the great majority of its compounds nickel has oxidation state of +2
b) The hydrated Ni^{2+} cation is diamagnetic
c) Ni has an amphoteric hydroxide
d) none of these

151. With which of the following nickel ions form a chelate ?
a) Cyanide ions b) NH_3
c) Chloride ions d) DMG

152. Cu(II) $SO_4.5H_2O$, in water solution, will react with which solution to reduce Cu(II) to Cu(I) ?
a) Glucose b) Zn
c) SO_2 d) KI solution

153. When conc. HCl is added to Cu_2O, the solution contains substantial concentration of
a) Cu^{2+} b) Cu^+
c) $CuCl_4^{2-}$ d) $CuCl_4^{3-}$

154. Which one of the following would not form complexes with Zn^{2+}
a) OH b) H_2O c) NH_3 d) CN^-

155. Which of the following ions is exceptionally stable because of a half-filled d-sub-shell of electrons ?
a) Cu^+ b) Cu^{2+}
c) Fe^{3+} d) Co^{2+}

156. Which of the following in solution yield a ppt. with 2M NaOH solution which dissolves readily in excess of the reagent?
a) Cr^{3+} (aq) b) Mn^{2+} (aq)
c) Fe^{2+} (aq) d) Cr^{2+} (aq)

157. Which of the following is likely to show the strongest paramagnetic behaviour ?
a) $FeSO_4$ b) $MnSO_4$
c) $NiSO_4$ d) $CuSO_4$

158. Which one of the following statements is false ?
a) Cobalt is an essential constituent of vitamin B-12
b) Zn is a constituent of insulin and of the enzyme carbonic anhydrose needed in respiration
c) copper is a catalyst for the formation of haemoglobin and the absorption of iron

d) Transferrin is a co-enzyme

159. Cu - Zn couple that can be used as a reducing agent is obtained by, **(I.I.T. 1984)**

a) mixing Zinc dust and copper gauze
b) coated zinc with copper
c) coating copper with zinc
d) welding together zinc and copper wires.

160. Among the following, the lowest degree of paramagnetism per mole of the compound at 198K will be shown by **(IIT. 1988)**

a) $MnSO_4.4H_2O$ b) $CuSO_4.5H_2O$
c) $FeSO_4.6H_2O$ d) $NiSO_4.6H_2O$

161. K_2MnO_4 is formed when **(IIT. 1988)**

a) Chlorine is passed in an aqueous K_2MnO_4 solution.
b) Manganese dioxide is fused with KOH in air.
c) Formaldehyde reacts with potassium permanganate in presence of a strong alkali.
d) Potassium permanganate reacts with Conc. sulphuric acid.

162. The aqueous solutions of the following salts will be coloured in the case of **(IIT. 1990)**

a) $Zn(NO_3)_2$ b) $LiNO_3$
c) $Co(NO_3)_2$ d) $CrCl_3$
e) Potash alum

163. Which of the following does not characterise X-rays ? **(IIT. 1992)**

a) The radiation can ionise gases
b) It causes ZnS to fluoresce
c) Deflected by electric and magnetic field.
d) Have wavelength shorter than ultraviolet rays.

164. Among the following ions which one has the highest paramagnetism ? **(IIT. 1993)**

a) $[Cr(H_2O)_6]^{+3}$ b) $[Fe(H_2O)_6]^{+2}$
c) $[Cu(H_2O)_6]^{+2}$ d) $[Zn(H_2O)_6]^{+2}$

165. Which of the following alloys contain Cu and Zn ? **(IIT. 1993)**

a) Bronze b) Brass
c) Gunmetal d) Typemetal

166. Which of the following has maximum no. of unpaired electrons ? **(IIT. 1996)**

a) Mg^{+2} b) Ti^{+3} c) V^{+3} d) Fe^{+2}

167. An aqueous solution of $FeSO_4$. $Al_2(SO_4)_3$ and chrome alum is heated with excess of Na_2O_2 and filtered. The materials obtained are **(IIT. 1996)**

1) a colourless filtrate and a green residue
b) a yellow filtrate and a green residue
c) a yellow filtrate and a brown residue
d) a green filtrate and a brown residue.

168. Among the following, the compound that is both paramagnetic and coloured is **(IIT. 1997 cancelled paper)**

a) $K_2Cr_2O_7$ b) $(NH_4)_2(TiCl_6)$
c) $VoSO_4$ d) $K_3[Cu(CN)_4]$

169. Ammonium dichronate is used in some fireworks. The green coloured powder blow in the air is **(IIT. 1997)**

a) CrO_3 b) Cr_2O_3
c) Cr d) $CrO(O_2)$

170. Which of the following compound is expected to be coloured? **(IIT. 1997)**

a) Ag_2SO_4 b) CuF_2
c) MgF_2 d) CuCl

171. Which of the following statement (S) is (are) correct when a mixture of NaCl and $K_2Cr_2O_7$ is gently warmed with Conc. H_2SO_4 ?

a) A deep red vapour is evolved
b) The vapour when passed in to NaOH solution gives a yellow solution of Na_2CrO_4
c) Chlorine gas is evolved
d) Chromyl chloricle is formed

172. In the dichromate dianion **(IIT. 1999)**

a) 4 Cr-O-bonds are equivalent
b) 6 Cr-O bonds are equivalent
c) all Cr-O-bonds are equivalent
d) all Cr-O-bonds are nonequivalent.

173. The geometry of $Ni(CO)_4$ and $Ni(PPh_3)_2Cl_2$ are

a) both square planar
b) tetrahedral and square planar, respectively
c) both tetrahedral
d) square planar and tetrahedral, respectively. **(IIT. 1999)**

174. The composition of Prussian blue is **(BET. 1992)**
a) $Fe_4 [Fe(CN)_6]_3$ b) $Fe_3 [Fe (CN)_6]_2$
c) $Na_4 [Fe(CN)_6]$ d) $Cu_2 [Fe(CN)_6$

175. Rusting of iron is catalysed by **(MLNR. 1990)**
a) Fe b) O_2 c) Zn d) H^+

176) Finely divided iron combines with CO to form. **(MLNR 1994)**
a) $Fe(CO)_5$ b) $Fe_2 (CO)_9$
c) $Fe_3 (CO)_{12}$ d) $Fe(CO)_6$

177. Mohr's salt is a **(MLNR. 1986)**
a) Normal salt b) Acid salt
c) Basic salt d) Double salt

178. Acidified $KMnO_4$ solution is decolourised by **(MLNR)**
a) White Vitriol
b) Mohr's salt
c) Bleaching powder
d) Microcosmic salt

179. The formula of Haematite is **(MLNR 1994)**
a) Fe_3O_4 b) Fe_2O_3
c) $FeCO_3$ d) FeS_2

180. The no. of electrons in Fe^{+2} (z = 26) is not equal to that of **(MLNR 1994)**
a) S-electrons in Mg (z = 12)
b) p-electrons in Ne (z = 10)
c) d-electrons in Fe
d) p-electrons in Cl^- (z=17)

181. Which of the following melts in boiling water? **(MLNR. 1984)**
a) Gun metal b) Wood metal
c) Monel metal d) Bell metal

182. Iron is rendered passive by treatment with Conc.
a) HCl b) H_2SO_4
c) H_3PO_4 d) HNO_3

183. Which of the following dissolves in hot Conc. NaOH ? **(IIT. 1980)**
a) Fe b) Zn c) Cu d) Ag

184. German silver is an alloy of Copper and **(IIT. 1980, MLNR 1990)**
a) Zn and Ni b) Al c) Zn d) Sn

185. Copper sulphate solution, reacts with KCN to give **(IIT. 1996)**
a) $Cu(CN)_2$ b) CuCN
c) $K_2[Cu(CN)_4]$ d) $K_3[Cu(CN)_4]$

186. Identify the statement which is not correct regarding $CuSO_4$ **(MLNR. 1992)**
a) It reacts with KI to give iodine
b) It reacts with KCl to give Cu_2Cl_2
c) If reacts with NaOH and glucose to give Cu_2O
d) It gives CuO an strong heating in air.

187. The slay obtained during the extraction of copper pyrites is composed mainly of **(MLNR 1993)**
a) Cu_2S b) $FeSiO_3$
c) $CuSiO_3$ d) SiO_2

188. Copper can be extracted from **(IIT. 1978)**
a) Kupfernical b) Dolomite
c) Malchite d) Galena

189. Which of the following has the maximum no. of unpaired d electrons ? **(BIT 1992)**
a) Zn b) Fe^{+2} c) Ni^{+3} d) Cu^+

190. The electronic configuration of the element which is just above the element with atomic no. 43 in the same periodic group is **(MLNR. 1992)**
a) $1s^2, 2s^2, 2p^6\ 3s^2, 3p^6, 3d^5, 4s^2$
b) $1^2, 2s^2, 2p^6, 3s^2, 3p^6, 3d^{10}, 4s^2, 4p^5$
c) $1s^2, 2s^2, 2p^6, 3s^2, 3p^6, 3d^6, 4s^1$
d) $1s^2, 2s^2, 2p^6, 3s^2, 3p^6, 3d^{10}, 4s^1, 4p^6$

191. Verdigris is
a) Basic Copper acetate
b) Basic lead acetate
c) Ferrous ammonium sulphate
d) Potassium ferrocyanide

192. Lunar Caustic is chemically **(MLNR. 1981)**
a) Silver chloride
b) Silver nitrate
c) Sodium hydroxide

d) Potassium nitrate

193. Cinnabar is an ore of **(MLNR 1986)**
a) Lead b) Zinc
c) Silver d) Mercury

194. Mercury is purified by
a) Solidifying
b) Distillation in vacuum
c) Treatment with dil.HNO_3
d) Electrolytic method

195. Mond process is used in the extraction of **(MLNR 1983, 86)**
a) Co b) Ni c) Mo d) Zn

196. Iron is rendered passive in concentrated **(IIT. 1982)**
a) H_2SO_4 b) H_3PO_4
c) HCl d) HNO_3

197. The term ''fools gold'' is used for a mineral which shines like gold. It is,
a) Ironpyrite b) Copperpyrite
c) Cinnabar d) Cadmium sulphide

198. In var is an alloy of
a) Steel and chromium
b) Vanadium and manganese
c) Tungston and chromium
d) Steel and nickel

199. Which metal does not form amalgam ?
a) Fe b) Cu c) Ag d) Zn

200. The brown ring complex compound is formulated as $[Fe(H_2O)_5 - (NO)]SO_4$ The oxidation state of iron is **(IIT. 1987)**
a) +1 b) +2 c) +3 d) +4

201. Transition metals are less reactive because of their
a) High ionization potential and low melting point
b) High ionization potential and high melting point
c) Low ionization potential and low melting point
d) Low ionization potential and high melting point

202. The lowest degree of paramagnetism is shown by **(IIT. 1988)**
a) $MnSO_4.4H_2O$ b) $FeSO_4.6H_2O$
c) $CuSO_4.5H_2O$ d) $NiSO_4.6H_2O$

203. Amongst the following, identify the species with an atom in +6 oxidation state **(IIT. Screening 2000)**
a) MnO_4^- b) $[Cr(CN)_6]^{-3}$
c) NiF_6^{-2} d) CrO_2Cl_2

204. The complex ion which has no 'd' electrons in the central metal atom is **(IIT. Screening 2001)**
a) MnO_4^- b) $[Co(NH_3)_6]^{+3}$
c) $[Fe(CN)_6]^{-3}$ d) $[Cr(H_2O)_6]^{+3}$

205. In the standardization of $Na_2S_2O_3$ using $K_2Cr_2O_7$ by iodometry, the equivalent weight of $K_2Cr_2O_7$ is **(IIT. Screening 2001)**
a) Molecular weight / 2
b) Molecular weight / 6
c) Molecular weight / 3
d) Same as molecular weight

206. The magnetic moment of $[Mn(H_2O)_6]^{+2}$ ion is
a) 1 BM b) $\sqrt{35}$ BM
c) $\sqrt{2}$ BM d) $\sqrt{3}$ BM

207. Which of the following is not the configuration of lanthanides ?
a) $[Xe]^{54} 4f^1 6s^2$
b) $[Xe] 4f^1 5d^1 6s^2$
c) $[Xe] 4f^7 5d^1 6s^2$
d) $[Xe] 4f^{14} 5d^1 6s^2$

208. The most characteristic oxidation state of lanthanides is
a) +1 b) +2 c) +3 d) +4

209. The element which follows the lanthanides in the 6th period is
a) Hf b) Zr c) Nb d) Mn

210. The lanthanide contraction refers to
a) ionic radius of that series
b) size of the hydrated M^{3+} ions
c) size of the unhydrated M^{3+} ions
d) basicity of the hydroxides

211. The separation of lanthanides by the ion exchange method is based on
a) the solubility of the nitrates
b) size of the hydrated M^{3+} ions

c) size of the unhydrated M^{3+} ions
d) basicity of the hydroxides

212. Lanthanides resemble Mg in the following
a) they have high E^0 ox. pot.
b) they are chemically reactive
c) they react vigorously with water
d) lanthanides show variable oxidation states

213. What factors make the separation of lanthanides a formidable task ?
a) similarity in ionic size
b) constant charge of +3
c) same charge - to - radius ratio
d) all the above

214. The configuration of gadolinium is
a) $4f^7\,5d^1\,6s^2$ b) $4f^5\,5d^1\,6s^2$
c) $4p^3\,5d^1\,6s^2$ d) Ag^+ and Au^+

215. Ionic radii of the following pairs of elements become identical as a consequence of lanthanide contraction ?
a) Zr^{4+} and Hf^{4+}
b) Cr^{3+} and Mo^{3+}
c) V^{5+} and Nb^{5+}
d) Ag^+ and Au^+

216. In the coinage metals (IB) group, the I.E. decreases from Cu to Ag and the increases from Ag to Au. This is attributed to
a) increased atomic size
b) increased ionic radius
c) increased nuclear density
d) lanthanide contraction

217. Misch metals are
a) alloys or rare earth metals
b) alloys of d-block elements
c) special steels
d) alloy of Al, Mg, Gd

218. Ionic radii of Zirconium and Hofnium become almost identical because
a) they are d-block elements
b) they belong to the same group
c) of increased nuclear charge
d) of lanthanide contraction

219. In the 14 group, ionization energy decreases from C to Sn, and the increases from Ge to Pb. This is because
a) Pb follow the lanthanides in 6th period
b) Pb has increased nuclear charge
c) of lanthanide contraction
d) all the above

220. The general electronic configuration of Ce (z = 58) is
a) 2, 8, 18, 18, 10, 2
b) 2, 8, 18, 18, 9, 3
c) 2, 8, 18, 19, 9, 2
d) 2, 8, 18, 18, 11, 1.

221. The most abundant rare earth in earth crust is
a) Ce b) Gd c) Yb d) Lu

222. The richest source of rare earths is
a) Monozite b) Bastnaesite
c) sea water d) xenotime

223. Lanthanide elements are present in
a) Bertholloids
b) Warm super conductors
c) Mich metal
d) All of these

224. Lanthanides are separated best by
a) fractional crystallization
b) solvent extraction
c) complex formation using EDTA
d) ion exchange resins

225. $KMnO_4$ and $K_2Cr_2O_7$ are replaced in volumetric analysis by
a) La(III) salts b) Ce(III) salts
c) Ce(IV) salts d) Gd(III) salts

226. The stable +2 ions of lanthanides in aqueous solution are
a) Sn^{2+} b) Eu^{2+}
c) Yb^{2+} d) all of these

227. Which one of the following ions is coloured ?
a) La^{3+} b) Eu^{3+}
c) Gd^{3+} d) Lu^{3+}

228. Pairs of elements which become identical in size as a result of lanthanide contraction are

a) Zn-Hf b) Nb-Ta
c) Mo-W d) all of these

229. The ion with electronic configuration $(Xe)4f^{13}$
a) Lu b) Yb c) Gd d) Tb

230. Lanthanides form complexes with
a) monodentate ligands
b) bidentate ligands
c) chelating agents
d) simple anions

231. The poineering work in Co-ordination Chemistry was done by
a) Werner b) Langmuir
c) Pauling d) Bailer

232. Werner studies various coloured complexes from the reaction of cobalt chloride and NH_3. Which one of the Werner's postulate is incorrect ?
a) that metals possess two types of valencies - primary and secondary
b) each metal has a fixed number of secondary valencies
c) the secondary valencies are directed in space around the central metal atom or ion.
d) the complexes do not show isomerism

233. A ligand is essentially a Lewis base. Which one of the following is a polydentate ligand ?
a) NH_3 b) Co
c) $CH_2NH_2.CH_2NH_2$ d) E.D.T.A.

234. Ethane diamine tetra acetic acid has how many donor atoms or co-ordination sites?
a) 2 b) 4 c) 6 d) 8

235. Complex formation can be studied by
a) osmotic pressure measurements
b) depression in F.P. measurements
c) conductivity measurements
d) all of these

236. The IUPAC name for the complex $[Co(NH_3)_6]$ Cl_3 is
a) hexamine cobalt (III) chloride
b) hexa-amino cobalt (III) chloride
c) Luteo cobaltic chloride
d) Cobalt amine chloride

237. The IUPAC name for the complex $K_4Fe(CN)_6$ is
a) potassium hexacyanate (II)
b) prussian blue
c) potassium hexacyano ferrate (III)
d) potassium hexacyano ferrate (IV)

238. The effective atomic number (E.A.N.) of Fe in the complex $K_4Fe(CN)_6$ is
a) 26 b) 36 c) 34 d) 35

239. In the $NiCl_4^{2-}$ ion, the total number of electrons around the Ni including bonding pairs is
a) 26 b) 28 c) 34 d) 36

240. AgCl may be brought into solution by addition of
a) NaCl b) $AgNO_3$
c) H_2O d) NH_3

241. Unreacted AgBr dissolves in hypo due to
a) dissociation b) solution
c) complex formation d) none of these

242. In complexes, the ligands are linked to be central metal atom or ion by
a) dative bonds b) ionic bonds
c) H-bonds d) Vander Waal's forces

243. Which species in the following reaction acts as Lewis acid ?
$CuSO_4 + 4NH_3$ (eq) $\rightarrow$ $[Cu(NH_3)_4]^{2+}$ (aq) + SO_4^{2-} (aq)
a) Cu^{2+} b) NH_3
c) SO_4^{2-} d) $[Cu(NH_3)_4]^{2+}$

244. Which one of the following complexes is not linear ?
a) $Ag(NH_3)_2^+$ b) $Au(CN)_2^-$
c) $Ag(CN)_2^-$ d) $NiCl_4^{2-}$

245. The hybridisation of Ni in the complex Na_2 $Ni(CN)_4$ is
a) sp^3 b) sp^2 c) dsp^2 d) d^2sp^3

246. The hybridizations of Ni in $Ni(CO)_4$ is
a) sp^3 b) dsp^2
c) sp^2 d) sp

247. Refer to question 16, the oxidation state of Ni in the complex is

a) 0 b) +2 c) +4 d) -2

248. Which one of the following complexes does not have tetrahedral geometry ?
a) $[Ni(NH_3)_4]^{2+}$ b) $[Cu(CN)_4]^{3-}$
c) $[NiCl_4]^{2-}$ d) $[Ni(CN)_4]^{2-}$

249. Which one of the following is not a square planar complex ?
a) $[PtCl_4]^{2-}$ b) $[Pt(NH_3)_4]^{2+}$
c) $Pd(NH_3)_4]^{2+}$ d) $Fe(CN)_6]^{3-}$

250. Penta carbonyl iron (0) has which geometry ?
a) tetrahedral b) square planar
c) Octahedral d) trigonal bipyramidal

251. The hybridisation of Cr in $[Cr(H_2O)_6]^{3+}$ is
a) d^2sp^3 b) sp^3d^2
c) dsp^2 d) sp^3d

252. Refer to question 21, the magnetic moment of the complex is found to be :
a) 1.73 b) 3.89 c) 2.9 d) 0

253. Which one of the following does not have octahedral shape ?
a) $K_4Fe(CN)_6$ b) $K_3Co(NO_2)_6$
c) $Co(NH_3)_6Cl_3$ d) $Pt(NH_3)_2Cl_2$

254. The type of hybridisation of Fe in the complex $(FeF_6)^{3-}$ and magnetic moment are :
a) d^2sp^3, O.B.M. b) d^2sp^3,5.9B.M.
c) sp^3d^2, 4.9B.M. d) sp^3d^2, 5.9B.M.

255. Which one of the following octahedral complexes is diamagnetic ?
a) $(CoF_6)^{3-}$ b) $Fe(H_2O)_6]^{3+}$
c) $K_3Fe(CN)_6$ d) $K_4Fe(CN)_6$

256. A complex of the formula $Co(NH_3)_5$ Br] SO_4 gives a white ppt. with $BaCl_2$ solution, and no ppt. with $AgNO_3$ solution has a structure :
a) $[co(NH_3)_4Br]^{2+}$
b) $[Co(NH_3)_5SO_4]^+$ Br^-
c) $[Co(NH_3)_4Br]^{2+}$ SO_4^{2-}
d) $[Co\ (NH_3)_5]^{3+}$ SO_4^{2-} Br^-

257. A hydrate of Cr, $Cr(H_2O)_6\,Cl_3$, gives 3 moles of AgCl when treated with excess $AgNO_3$ solution. Its structure is
a) $[Cr(H_2O)_6]Cl_3$
b) $[Cr(H_2O)_5Cl]Cl_2.H_2O$
c) $[Cr(H_2O)_4\,Cl_2]Cl.2H_2O$
d) $[Cr(H_2O)_3].3H_2O$

258. Linkage isomerism is shown by complexes having monodentate ligands such as
a) CN^- and NC^-
b) - N - O = O and - N(O)(O)
c) ambident ions
d) all the above

259. Square planar complexes of the type given below show cis-trans isomerism ?
a) Ma_2b_2 b) Ma_3b
b) Mb_3a d) Mb_4

260. Geometric isomerism would be expected for
a) $Zn(NH_3)_4]^{2+}$ b) $Pt\ (NH_3)_2\,Cl_2$
c) $[Au(NH_3)_2]^+$ d) $[Ni(NH_3)_4]^{2+}$

261. Octahedral complex (C.N = 6) show cis-trans isomerism if they are of the type
a) Ma_4b_2 b) Ma_3b_3
c) Ma_6 d) Ma_5b

262. Cis-trans isoprenes would be expected for all following except
a) $[Co(NH_3)_4\,Cl_2]^+$
b) $Cr(NH_3)_4\,Cl_2)^{2+}$
c) $[Cr(en)_2Cl_2]^+$
d) $[Co(NH_3)_3\,Cl_2\,Br]$

263. Optical isomerism is shown by the octahedral complexes containing bidentate ligands except :
a) Cis $[Co\ (en)_2\ Cl_2]^+$
b) Cis $(Cr(en)_2\ Cl_2]^+$
c) trans $[Co\ (en)_2Cl_2]^+$
d) both (a) and (b)

264. I_2 dissolves in KI solution due to formation of

a) KI_2 b) KI_3
c) K_2I_3 d) K_2I_2

265. When KI is added in excess to a solution of $HgCl_2$, the product is
a) red ppt. of HgI_2
b) colourless solution of K_2HgCl_4
c) colourless solution having K_2HgI_4
d) none of these

266. An insoluble complex is formed with Ni^{2+} ion using
a) $Na_3\ Co(NO_2)_6$
b) CHOH COOK CHOH COONa
c) DMG
d) Cupferron

267. Which one of the following is not a co-ordination complex ?
a) Chlorophyll b) Haemoglobin
c) Vitamin B - 12 d) Acrylic Acid

268. A typical example of sand-witched compound is
a) ferrocene b) $Fe(CO)_5$
c) $(C_2H_5)_2\,Zn$ d) Ziegler's catalyst

269. Ziegler-Natta catalyst is
a) $TiCl_4 - (C_2H_5)_3$ al. complex & heterogeneous catalyst
b) used for polymerization of alkenes
c) used to control stereo-chemical properties of polymers
d) all the above

270. Wilkinson's catalyst is
a) $[(C_6H_5)_3P]_3\,RhCl$
b) used in the hydrogenation of alkenes
c) homogeneous catalyst
d) all the above

271. The complex ion $[Cu(NH_3)_4]^{2+}$ has
a) tetrahedral configuration with one unpaired electron
b) square planar with one unpaired electron
c) tetrahedral with all electrons paired
d) square planar with all electrons paired

272. Select the most plausible combination of hybrid orbitals notation and magnetic geometric description for the hexaquairon (II) ion ?
a) d^2sp^3 , paramagnetic octahedral
b) sp^3d^2, diamagnetic hexagonal
c) sp^3d^2, paramagnetic octahedral
d) dsp^2, paramagnetic square planar

273. The ultimate product formed by hydrated iron (III) ions in an aqueous solution of pH = 5 is
a) $Fe_2O_3XH_2O$ b) $Fe(H_2O)_4(OH)_2]$
c) $[Fe(H_2O)_5OH]^{2+}$ d) $[Fe(H_2O)_3(OH)_3]$

274. Which one of the following is formed when a soluble copper (I) salt is added to water ?
a) $[Cu(H_2O)_6]^+$ only
b) $[Cu(H_2O)_6]^{2+}$ only
c) Cu and $[Cu(H_2O)_4]^{2+}$
d) Cu and $[Cu(H_2O)_6]^{2+}$

275. Which one of the following salts does not produce a complex ion $M[(H_2O)_6]^{2+}$ when dissolved in excess of water ?
a) $CuSO_4$ b) $ZnSO_4$
c) $FeSO_4$ d) $Al_2(SO_4)_3$

276. In which of the following ions is the transition metal not exhibiting its highest common oxidation state ?
a) $Cr_2O_7^{2-}$ b) MnO_4^{2-}
c) $[Zn(NH_3)_4]^{2+}$ d) $[Fe(CN)_6]^{3-}$

277. The central atom in $[Co(NH_3)_5Cl]\,(NO_3)_2$ can be assigned oxidation number of
a) +1 b) +2 c) +3 d) 0

278. A compound has the empirical formula $CoCl_3\ 4NH_3$. One mole of it gives one mole of AgCl on treatment with $AgNO_3$ solution. NH_3 is not removed with conc. H_2SO_4. It has structure.
a) $Co(NH_3)_4\ Cl_3$
b) $[Co(NH_3)_4]\ Cl_3$
c) $[Co(NH_3)_3\ Cl_3]NH_3$
d) $[Co(NH_3)_4\ Cl_2]Cl$

279. Geometric isomers are expected for
a) $[Zn(NH_3)_4]^{2+}$
b) $Zn(H_2O)_2(OH)_2$

c) Co $(NH_3)_3$ Cl_2 Br

d) $[Au (NH_3)_2]^+$

280. The formula $PdCl_2(OH)_2]^2$ exists in cis-trans isomers. The hybrid orbitals occupied by the bonding electron are

a) sp b) sp^3 c) dsp^2 d) d^2sp^3

281. Metal carbonyls are organo metallic compounds in which carbon monoxide acts as the ligand. The metal - carbon bond in metal carbonyls possess both sigma and pi character. Which of the following metal carbonyls is diamagnetic and has tetrahedral geometry ?

a) $Cr(CO)_6$ b) $Fe(CO)_5$

c) $Ni(CO)_4$ d) $W(CO)_6$

282. Ferrocene is an example of

a) sand- witched complex

b) pi - bonded complex

c) a complex in which all the five carbon atoms of cyclopentadiene anion are bound to the metal

d) all the above

283. Which one of the following is not a sigma bonded complex ?

a) Grignard's reagent

b) Frankland's reagent

c) n-butyllithium

d) dibenzene chromium

284. Which one of the following is not a sigma bonded complex ?

a) Grignard's reagent

b) Frankland's reagent

c) n-butyllithium

d) dibenzene chromium

285. Which one of the following is coloured?

a) $[Ti(H_2O)_6]^{2+}$ b) $[Sc(H_2O)_6]^{2+}$

c) $[Zn (H_2O)_6]^{2+}$ d) $Ti(H_2O)_6]^{2+}$

286. Which one of the following has highest number of unpaired electrons?

a) $[Mn(H_2O)_6]^{2+}$ b) $[Fe(H_2O)_6]^{2+}$

c) $Ni(H_2O)_6]^{2+}$ d) $[Co(H_2O)_6]^{2+}$

287. A cobaltamine has the formula $Co(NH_3)_6$ Cl_3, with $AgNO_3$ solution, one third of the chloride is precipitated. The complex shows cis-trans isomerism. It can have the structure.

a) $[Co(NH_3)_6]Cl_3$

b) $[Co(NH_3)_5Cl]Cl_2$

c) $[Co(NH_3)_4Cl_2]Cl$

d) $[Co(NH_3)_5 . H_2O]Cl_3$

288. The colour of an aqueous solution of $CrCl_3$ varies form violet and green depending upon concentration. This variation is due to formation of

a) $[Cr(H_2O)_6]Cl_3$

b) $[CrH_2O)_5Cl]Cl_2$

c) $Cr(H_2O)_4Cl_2]Cl$

d) all of these

289. A co-ordination complex (x) has the following percentage compositions Cu=22.9%; NH_4 = 12.9%; Cl = 51.6%; H_2O = 13.6%. Its M.W. is 277.5 g of (x), with $AgNO_3$ solution, gave 57.40 g of AgCl. Its structural formula is expected to be

a) $[Cu(NH_4)_2 . 2H_2O]$ Cl_4

b) $[Cu(NH_4)_2 . 3H_2O]$ Cl_4

c) $[Cu(NH_4)_2 Cl_4] . 2H_2O$

d) $[Cu (NH_4)_2 . H_2O]$ Cl_3

290. Which one of the following complexes has dsp^3 hybridization on the metal atom or ion ?

a) $[CoF_6]^{3-}$ b) $Fe(CO)_5$

c) $[Ni(CN)_4]^{2-}$ d) $NiCl_4^{2-}$

291. Which one of the following square planer complexes will show cis-trans isomerism?

a) MA_4 b) MA_3B

c) MA_2B_2 d) MABCD

292. For which of the following d^n configuration can octahedral complexes exist in both high spin and low spins forms ?

a) d^1 b) d^2 c) d^3 d) d^5

293. Which one is a high spin or outer-orbital complex ?

a) $[Co(H_2O)_6]^{2+}$ b) $[Co(CN)_6]^{3-}$

c) $[CoF_6]^{3-}$ d) $Fe(CN)_6]^{2-}$

294. The colour of $Ti(H_2O)_6]^{3+}$ is due to
a) presence of water molecules
b) intra-molecular vibration
c) d-d transition of electrons in Ti^{3+} ion
d) d-d transition between metal atoms

295. The complex $[Cr(gly)_3]$ shows
a) optical isomerism
b) geometrical isomerism
c) neither geometrical nor optical isomerism
d) both optical and geometrical isomerism

296. Which complex would show co-ordinate isomerism ?
a) $[Cr(NH_3)_6]$ $[(Co(CN)_6]$
b) $[Co(en)_2Cl_2]^+$
c) $[Cr (en)_2Cl_2]^+$
d) $[Co(NH_3)_5NO_2]Cl_2$

297. Choose a species which cannot act as a ligand to a transitions metal ion ?
a) NH_3 b) H_2O c) CO d) CO_2

298. Which co-ordination number of Co in $[Co(NH_3)_5NO_2]Cl_2$ is
a) 2 b) 8 c) 4 d) 6

299. Which one of the following species does not represent cationic species of vanadium formed in aqueous solution ?
a) VO_2^+ b) VO^{2+}
c) $[V(H_2O)_6]^{3+}$ d) VO_2^{2+}

300. In which one of the following does Mn exhibit its highest oxidation state ?
a) MnO_2 b) MnO_4^{2-}
c) MnO_4^- d) MnO

301. The complex ion formed when cobalt (II) chloride is added to water is
a) $[Co(H_2O)_6]^{2+}$ b) $[Co(H_2O)]_6]^{3+}$
c) $[CoCl_4]^{2-}$ d) $[CoCl_6]^{4-}$

302. Which is false about $[Co(H_2O)_6]^{2+}$ complex ion ?
a) it is pink when freshly prepared
b) it turns blue an warming
c) in the presence of air, it forms the yellow $[Co(CN)_6]^{3-}$ ion with NaCl
d) it forms the blue $[CoCl_6]^{4-}$ ion with conc. HCl

303. Which of the following ions gives no change when aqueous alkali is added drop-wise ?
a) Al^{3+} b) Cu^{2+}
c) $Cr_2O_7^{2-}$ d) CrO_4^{2-}

304. The spectro chemical series of ligands is
a) $I^- < F^- < H_2O < NH_3 < CN^-$
b) $F^- < Cl^- < Br^- < I^- < H_2O < NH_3 < CN^-$
c) I^- $Br^- < Cl^- < H_2O < NH_3 < CN^-$
d) none of these

305. The $[Ti(H_2O)_6]^{3+}$ complex ion shows a strong absorption at 490 mm. But complex ion absorbs light of lower wave length than 490 mm. It is because
a) NH_3 is a stronger ligand than water
b) H_2O in a stronger ligand than water
c) electron-transition of higher energy is involved
d) both (a) and (c)

306. The complex ion $[Ni(CN)_4]^{2-}$ has square planar configuration with magnetic moment of zero. What would be the magnetic moment if it were tetrahedral ?
a) 2.83 B.M. b) 1.73 B.M.
c) 4.84 B.M. d) 5.9 B.M.

307. In an octahedral complex, the five d-orbitals one split into e_g orbitals ($dx^2 - y^2$ and dz^2) and three orbitals d_{xy}, d_{xz}, d_{yz}). The difference in energy between two sets of orbitals is Δ or 10 Dq..
In the $[Ti(H_2O)_6]^{3+}$ ion, the d-d transition occurs at 490 mm, the value Δ_o is calculated to be 241 kJ $mole^{-1}$. Because the crystal field splitting of d- orbitals, the single d-electron in the complex occupies an energy level 2/5 Δ_o below the average energy of the d-orbitals. As a result the complex is stable. The crystal field stabilization energy (CFSE) in this

case is $\frac{2}{5} \times 241 = 96$ kJ $mole^{-1}$

For tetrahedral complexes, Δ tetra = 4/9 Δ oct. Calculate in kJ $mole^{-1}$ the relative CFSE energy attained by the ferrous ion in octahedral and tetrahedral oxide-ion environments.

From the spectro-chemical series Δ for O^{2-} is equal to that water. In octahedral fields Δ for Fe^{2+} is 124 is kJ $mole^{-1}$

a) CFSE for octahedral is 2/5 × 124 = 50 kJ $mole^{-1}$

b) Δ for tetrahedral is 4/9 Δ oct or 4/9 × 124 = 55 kJ

c) CFSE for tetrahedral is 3/5 × 55 = 11 kJ $mole^{-1}$

d) all of these

308. Which one of the following statements is false ?

a) weak ligands like F^-, Cl^- form high spin complexes

b) strong ligands like CN^- form low-spin complexes

c) $FeF_6]^{3-}$ is high spin complex

d) Ni $(CO)_4$ is high spin complex

309. The magnetic moment of C $\overset{2+}{o}$ in tetrahedral field is

a) 3.87 B.M. b) 1.73 B.M.

c) 4.87 B.M. d) 5.87 B.M.

310. The magnetic moment of Co^{2+} in square planar complex is

a) 1.73 B.M. b) 3.87 B.M.

c) 4.87 B.M. d) 5.87 B.M.

311. Two compounds have the formula $Co(NH_3)_3\,(NO_2)_3$. In aqueous solution, one of these does not conduct electricity. Its formula is

a) $[Co(NH_3)_3\,(NO_2)_3]$

b) $[Co(NH_3)_3\,(NO_2)_2\,H_2O]^+\,NO_2$

c) $[Co(NH_3)_3\,(NO_2)\,(H_2O)_2]^{2+}\,2NO_2$

d) none of these

312. Two compounds having the formula $M(H_2O)_4H_2OCl_3$ have molar conductances of 110 and 250 mhos cm^2/g. mole. The isomer having higher conductance is

a) $[M(NH_3)_4\,Cl_2]^+Cl^-$

b) $[M\,(NH_3)_4\,Cl(H_2O)^{2+}\,2Cl^-]$

c) $[M(NH_3)_4\,.\,H_2O]^{3+}\,3Cl^-$

d) none of these

313. A solution containing 2.665g $CrCl_3\,.\,6H_2O$ (formula weight 266.35) was passed though a cation exchanger. The chloride ion so obtained gave 2.87 g of AgCl when treated with $AgNO_3$ solution. The formula of hydrated salt is

a) $[Cr(H_2O)_5\,Cl]\,H_2O\,.\,2Cl^-$

b) $[Cr(H_2O)_6]Cl_3$

c) $[Cr(H_2O)_4Cl_2]_2H_2O\,.\,Cl^-$

d) $[Cr(H_2O)_3\,Cl_3]3H_2O$

314. A solution containing 2.674 g of $Co(NH_3)_6Cl_3$ (formula weight 267.4) was passed through a cation exchange resins. The chloride ion solution thus obtained gave 4.305 g of AgCl when reacted with $AgNO_3$ solution. It can be represented correctly by

a) $[Co(NH_3)_6]Cl_3$

b) $[Co(NH_3)_6Cl]Cl_2\,.\,NH_3$

c) $[Co(NH_3)_4\,Cl_2]_2NH_3\,.\,Cl$

d) $[Co(NH_3)_3Cl_3]3NH_3$

315. A dark green compound with the empirical formula $CoBr_3.\,3NH_3.\,3H_2O$ dissolves, in water to give a deepgreen solution which, on standing, turns red. The molar conductances of the two solutions are found to be 151 and 363.5 mhos-cm^{-1} pemole. It was also observed that one mole of bromine per mole of the green compound could be precipitated as AgBr per formula weight of the compound. Its structure can be represented by

a) $[Co(NH_3)_3(H_2O)_3]Br_3$

b) $[Co(NH_3)_3\,(H_2O)_2Br]\,Br_2$

c)

d) none of these

316. All the following complexes has C.N = 4 which of them can exist in cis- trans isomerism ?

a) $[Pt(NH_3)_3Cl]^+$ b) $[Pt(NH_3)_4]^{2+}$
c) $[Pt(NH_3)_2 Cl_2]$ d) $[Pt(NH_3)_2 (CO_2)_2]$

317. Which of the following forms a stable complex with fluoride which can be described as outer orbital complex ?
a) Fe(II) b) Fe(III)
c) Mn(IV) d) Mn(VII)

318. A d-block metal (X) liberates H_2 from dil. HCl, forms a green sulphate $XSO_4 7H_2O$ forms a cyanide complex $[X(CN)_4]^{2-}$. Which of the following fits the description ?
a) Fe b) Co c) Ni d) Cu

319. Which one the of the following is not a tetrahedral complex ?
a) $Ni(CO)_4$ b) $K_3Cu(CN)_4$
c) $[NiCl_4]^{2-}$ d) $K_2Ni(CN)_4$

320. When nickel is heated with carbon monoxide, a volatile compound is formed. Which of the following is the correct formula of their compound?
a) $Ni(CO)_2$ b) $Ni(CO)_3$
c) $Ni(CO)_4$ d) $Ni(CO)_6$

321. Which compound is formed when excess of KCN is added to aqueous solution of copper sulphate ? **(I.I.T. 1996)**
a) $Cu(CN)_2$ b) $K_2(Cu(CN)_4)$
c) $K(Cu(CN)_2)$ d) $K_3[Cu(CN)_4]$

322. Among $Ni(CO)_4$, $Ni(CN)_4)^{-2}$ and $(NiCl_4)^{-2}$ **(I.I.T. 1991)**
a) $Ni(CO)_4$, and $(NiCl_4)^{-2}$ diamagnetic and $Ni(CN)_4)^{-2}$ is paramagnetic
b) $(NiC_4)^{-2}$ and $(Ni(CN_4)^{-2}$ are diamagnetic and $Ni(CO)_4$ is paramagnetic.
c) $[Ni(CO)_4]$ and $[Ni(CN)_4]^{-2}$ are diamagnetic and $(NiC_4)^{-2}$ is paramagnetic.
d) $[Ni(CN)_4]^{-2}$ is diamagnetic and $[NiCl_4]^{-2}$ and $[Ni(CN)_4]^{-2}$ are paramagnetic

323. $[Cr(H_2O)_6]^{+3}$ ion has d-electrons equal to **(I.I.T. 1993)**
a) 2 b) 3 c) 4 d) 5

324. which has highest paramagnetism ? **(I.I.T. 1993)**
a) $[Cr(H_2O)_6]^{+3}$ b) $[Fe(H_2O_6]^{+2}$
c) $[Cu(H_2O)_6]^{+2}$ d) $[Zn (H_2O)_6]^{+2}$

325. Iron has lowest oxidation state in (M.L.N.R. 1984)
a) $Fe(Co)_5$ b) Fe_2O_3
c) K_2FeO_4 d) $FeSO_4(NH_4)_2SO_4.6H_2O$

326. The compound which does not show paramagnetism ? **(I.I.T. 1992)**
a) $(Cu(NH_3)_4 Cl_2$ b) $(Ag(NH_3)_2)$ Cl
c) NO d) NO_2

327. Oxidation state of nitrogen is not correctly given for the compound
a) $[Co(NH_3)_5Cl]$ Cl_2 (1–0)
b) NH_2OH (–1) Cl
c) $(N_2H_5)_2SO_4$ (–2)
d) Mg_3N_2 (–3)

328. The no. of d-electrons of chromium z = 24) in $[Cr(H_2O)_6]^{+3}$ ion **(I.I.T. 1987)**
a) 2 b) 3 c) 4 d) 5

229. The oxidation state of Mo in its oxo-complex species $[M\,O O_4(C_2H_4)_2(H_2O)_2]^{-2}$ is **(I.I.T.1991)**
a) 2 b) 3 c) 4 d) 5

330. A complex given below can exhibit.

CH_3–CH=CHCl

a) Geometrical isomerism only
b) optical isomerism only
c) Both optical and geometrical isomerism
d) None.

331. Which of the following complex will show geometrical isomerism.
a) $M A_2 B_2$ b) $MA B_3$
c) $M A_3B$ d) None.

332. The correct order of hybridization of the control atom in the following species

NH_3, $(PtCl_4)^{-2}$ PCl_5 and BCl_3 is **(I.I.T. Screening 2001)**

a) dsp^2, dsp^3, sp^2 and sp^3

b) sp^3, dsp^2, dsp^3 and sp^2

c) dsp^2, sp^2, sp^3 and dsp^3

d) dsp^2, sp^3, sp^2 and dsp^3

333. Identify the complexes which are expected to be coloured ? **(I.I.T. 1994)**

a) $[Cu(NC . CH_3)_4]^+ BF_4^-$

b) $[Cr (NH_3)_6]^{+3} . 3Cl^-$

c) Ksb4 $[VF_6]$

d) $[Ti (NO_3)_4]$

334. The valency of Cr in the complex $[Cr(H_2O)_4 Cl_2]^+$ is

a) 5 b) 6 c) 1 d) 3

335. The oxidation no. of Pt in $[Pt (C_2H_4)Cl_3]^-$ is **(M.L.NR. 1993)**

a) +1 b) +2 c) +3 d) +4

336. On adding $K_4[Fe(CN)_6]$ to ferric salt a prussian blue colour is obtained which is of **(B.I.T. 1992)**

a) $Fe_4[Fe(CN)_6]_3$ b) $K_3[Fe(CN)_6]$

c) $FeSO_4 . Fe(CN)_6$ d)

337. Among the following, the compound that is both paramagnetic and cloured is, **(I.I.T. 1997) Canceled paper)**

a) K_2Cr_2 of b) $(NH_4)_2$ $(TiCl_6)$

c) $VOSO_4$ d) K_3 $[Cu (CN)_4]$

338. Froth floatation process for the concentration of ores is an example of practical application of

a) adsorption b) absorption

c) coagulation d) sedimentation.

339. Of the following metals that can not be obtained by electrolysis of aq. soln of their salts are **(I.I.T. 1986)**

a) Ag b) Mg c) Cu d) Al

340. In the electrolysis of alumina croyolite is added to

a) Lower the mp of alumina

b) Increase the electrical conductivity

c). Minimize the anode effect

d) Remove impurities from alumine

341. The major role of Fluerspar (CaF_2) which is added in small quantities in the electrolytic reduction of alumina dissolved in fused croyolite Na_3AlF_6 is

a) That of catalyst

b) To make the fused mixture very conducting

c) To lower the temp. of the melt.

d) to decuse the rate of oxidation of c at the anode.

342. Which of the following mineral does not contain iron

a) Haematite b) Magnesite

c) Magnetite d) Limonite.

343. The principle reducing agent in the metalluergy of iron is

a) C b) CO c) CO_2 d) Al

344. The principle reducing agent in the metalluergy of tin is

a) C b) CO c) CO_2 d) Al

345. The composition of cuprite is

a) Cu_2S b) $CuFeS_2$

c) $Cu(OH)_2.CuCO_3$ d) Cu_2O

346. Which of the following is the mineral of magnisium

a) Dolamite b) Magnesite

c) Kainite d) Magnetite

347. The anode mud in the electrolytic refining of Ag contains.

a) Zn, Cu, Ag, Al b) Zn, Ag & Au

c) Cu, Ag, Au d) Au

348. Which of the following is used in metallurgy to remove rocky material.

a) Gangue b) Matrix c) Flux d) Slag

349. The metals prepared by the electrolysis of fused NaCl in an app. called

a) Nelson all

b) Castner - Kellner call

c) Kellner Solvay cell

d) Down's cell.

350. The extraction of metals from the sulphide ores is usully done by

a) Electrolysis

b) Smelting process

c) Metal displacement method

d) Froth flotation method

351. The iron obtained from blast furnace is called
a) Wrought iron b) Pig iron
c) Cast iron d) Mild steel

352. Which element is alloyed when Cu to form Bronze
a) Fe b) Zn c) Mg d) Tin

353. Which one of the furnaces among the following can produce the highest temperature ?
a) Muffle furnace b) Blast furnace
c) Reverberatory d) Electric furnace.

354. Essential constituent of an amalgam is
a) An alkali b) An alkali metal
c) Silver d) Hg

355. Oil used in the flotation method for the purification of ores is
a) Coconut oil b) Olive oil
c) Pine oil d) None

356. Heating pyrites in air to remove sulphur is known as
a) roasting b) Calcining
c) Smelting d) Fluxing.

357. Duralumin is an alloy of
a) Al & Mg b) Al, Mg & Ni
c) Al, Mg, Ni & Cu d) Al & Ni

358. The Bessemer converter used in the metallurgy of Cu is lined with
a) Silica b) Lime c) Lead d) None

359. During the process of electrolysis of fused NaCl, KF is added to
a) Lower the fusion pt. of NaCl
b) Increase the conductivity of fused NaCl
c) increase the mobility of Na^+ ions
d) To facilitate the discharge of I^- ions.

360. Which of the following metal is obtd by leaching its ore with dil. cyanide soln.
a) Ag b) Ti c) V d) Zn

361. The purpose of smelting an ore is to
a) Reduce it
b) Oxidise it
c) Obtain an alloy
d) Separate volatile impurites

362. Froth flotation process may be used to increase the concentration of the mineral in **(I.I.T. 1989)**
a) Chalcopyrites b) Bauxite
c) Haematite d) Calmine

363. A substance which, reacts with gangue to form fusible material is called
a) Flux b) Catalyst
c) Ore d) Slag

364. Iron ore is concentrated by
a) Froth flotation b) Electrolysis
c) Roasting d) Magnetic treatment

365. Non fusible impurities are separated from a roasted ore by adding
a) Slag b) Gang
c) Flux d) Mineral

366. Carbon is used as a common reducing agent in the extraction of
a) Ag b) Pb c) Zn d) Fe

367. Which of the following is/are refined by electrolysis ?
a) Sodium b) Chromium
c) Copper d) Zinc

368. Which of the following metals cannot be extracted by reduction method
a) Pb c) Al c) Hg d) Zn

369. Of the following metal can not be obtd by electrolysis of aq. soln. of their salt. **(I.I.T. 1990)**
a) Ag b) Mg c) Cu d) Al

370. In the metallurgy of iron when limestone is added to the blast furnace, the calcium ion ends up in **(I.I.T. 1982)**
a) Metallic calcium b) Slag
c) Calcium carbonate d) Gangue

371. Rust is
a) Fe_2O_3 and $Fe(OH)_3$ b) Fe_3O_4
c) Fe_2O_3 d) FeS_2

372. The furnace which provides the highest temperature is
a) Electrical furnace
b) Blast furnace
c) Reverberatory furnace
d) Muffle furnace

373. Sapphire is a mineral of
a) Na b) Al c) Hg d) Cu

374. Blister copper is
a) Pure copper

b) Alloy of copper
c) Copper containing 1% impurity
d) Copper containing 20% impurity.

375. Copper can be extracted from **(I.I.T. 1978)**
a) Malchite b) Dolomite
c) Galena d) Cu_2O

376. In extraction of copper we use
a) $CuFeS_2$
b) Cu_2S
c) Cuprite (Cu_2O)
d) Malchite $Cu(OH)_2.CuCO_3$

377. The slag obtained during the extraction of copper pyrites is composed mainly of **(M.L.N.R. 1993)**
a) Cu_2O b) Cu_2S
c) $CuSiO_3$ d) $FeSiO_3$

378. Copper displaces which of the following metal from salt solution.
a) $AgNO_3$ b) $CaSO_4$
c) $FeSO_4$ d) All of the above

379. High purity copper metal is obtained by
a) Electrolytic reduction
b) Carbon monoxide reduction
c) Thermite reduction
d) Hydrogen reduction

380. Which one of the following is an ore of silver
a) Magnetite b) Corundum
c) Argentite d) Stibnite

381. Parke process is used in the extraction of
a) Lead b) Silver c) Copper d) Iron

382. From argentite (Ag_2S) ore the method used in obtaining metallic silver is
a) Ag_2S is reduced with C.
b) Fused mixture of Ag_2S and KCl is electrolysed
c) Ag_2S is reduced with CO.
d) Reacting argentite with NaCN solution followed by metal displacement with zinc.

383. Which of the following dissolves in hot NaOH solution. **(I.I.T. 1980)**
a) Fe b) Zn c) Cu d) Ag

384. German silver is an alloy of copper and **(I.I.T. 1980)**
a) Al b) Sn c) Zn and Ni d) Zn

385. Percentage of silver in the german silver is
a) 20% b) 50% c) zero d) 80%

386. Copper sulphate solution reacts with KCN to give
a) CuCN b) $K_2[Cu(CN)_4]$
c) $Cu(CN)_2$ d) $K_3[Cu(CN)_4]$

387. Impurities of lead in silver are removed by
a) Contact process
b) Parke's process
c) Amalgamation process
d) Cyanide process

388. Bassemer converter is used in the manufacture of
a) Steel b) Wrought iron
c) Copper d) Pig iron

389. Cinnabar is an ore of **(M.L.N.R. 1986)**
a) Cu b) Al c) Hg d) Pb

390. Which one of the following process involves the roasting process ?
a) $2PbS + 3O_2 \rightarrow 2PbO + 2SO_2$
b) $Zn\,CO_3 \rightarrow ZnO + CO_2$
c) $Al_2O_3.2H_2O \rightarrow Al_2O_3 + 2H_2O$
d) $Fe_2O_3 + 3C \rightarrow 2Fe + 3CO$

391. Presence of small impurity usually makes a metal quite hard because the impurities.
a) Reduce the crystal symmetry
b) Reduce the no. of slide planes
c) Change the lattice structure of metals
d) Reduce the no. of mobile electrons.

392. Which of the following is not an ore ? **(I.I.T. 1984)**
a) Zinc blend b) Pig iron
c) Bauxite d) Malachite

393. Among the following groups of oxides, the group containing oxides that cannot be reduced by C to give the respective metal is
a) CaO and K_2O b) Fe_2O_3 and ZnO
c) Cu_2O and SnO_2 d) PbO and Pb_3O_4

394. Electrolytic reduction technique is used in the extraction of
a) Metalloids
b) Inner transition elements
c) Highly electropositive elements
d) Highly electronegative elements.

395. The lightest metal is

a) Al b) Li c) Na d) Be

396. Two compounds having the same crystal structures and analogous formulae, are called : **(I.I.T. 1978)**
a) Isotopes b) Isomers
c) Isomorphous d) Isobars

397. Indian salt peter is
a) $NaNO_3$ b) KNO_2
c) KNO_3 d) $NaHCO_3$

398. After partial roasting the sulphide of copper is reduced by
a) Cyanide process
b) Self reduction
c) Reduction with CO
d) Electrolysis

399. The main constituent of steel in India are
a) Mn and Cr b) Fe + Cu
c) Ti + Cr d) Ni and Cr

400. Magnetic separation is used for increasing concentration of the
a) Calcite b) Magnetite
c) Dolomite d) Zincite

401. Complex is formed in the extraction of
a) K b) Cu c) Fe d) Ag

402. Refractory materials are used for the construction of furnaces because they **(M.L.N.R. 1980)**
a) Can stand with high temperature
b) Are light in weight
c) Are leak proof
d) Do not require to be replaced

403. Crystalline metal can be transformed in to metallic glass by
a) Pressing into thin plates
b) Slow cooling of molten metal
c) Alloying
d) Very rapid cooling of a spray of the molten metal

404. Cupellation process is used in the metallurgy of
a) Silver b) Iron
d) Aluminium d) Copper

405. The role of calcination in metallurgical operations is
a) To decompose hydroxides
b) To decompose carbonates
c) To drive off organic matter
d) All of the above

406. The inner lining of the blast furnace is made up of
a) Fire clay bricks b) Silica bricks
c) Acidic bricks d) Graphite bricks

407. The metal that cannot be obtained by electrolysis of their aqueous salt solution is
a) Fe b) Na c) Cu d) Cr

408. In electro refining of metal the impure metal is made the anode and a strip of pure metal the cathode during the electrolysis of an aquesous solution of a complex metal salt. The method can not be used for refining of
a) Lead b) Silver
c) Sodium d) Aluminium

409. Which metal cannot be extracted by carbon reduction process
a) Al b) Cu c) Zn d) Pb

410. In the froth floatation process of concentration of ores, the ore particles flont because they,
a) Are insoluble
b) Have the surface which is not wetted easily
c) Are light
d) Have electrical charges.

411. The electrolytic method of reduction is employed for the preparation of metals that
a) Form oxides
b) Are highly electropositive
c) Are weakly electropositive
d) Are moderately electropositive

412. Electrometallurgical process is employed to extract **(M.L.N.R. 1985)**
a) Zn b) Lead c) Sodium d) Silver

413. Froth Floatation method may be used to increase the concentration of mineral in **(M.L.N.R. 1988)**
a) Bauxite b) Magnetite
c) Chalcopyrite d) Calmine

414. High purity copper metal is obtained by
a) Electrolytic reduction
b) Thermite process
c) Carbon reduction
d) Carbon monoxide reduction.

415. Presence of small inpurity usually makes a metal quite hard because the impurities
a) Reduce the no. of mobile electrons
b) Reduce the crystal symmetry
c) Reduce the no. of slide planes
d) Change the lattice structure of metals.

416. Crystalline metal can be transformed into metallic glass by
a) Alloying
b) Very rapid cooling of a spray of the molten metal
c) Slow cooling in the molten metal
d) Pressing in to thin plates

417. The salt which is least likely to be found in minerals is
a) Nitrate b) Chlorides
c) Sulphates d) Sulphides

418. The chemical process in the production of steel from haematite are involve **(I.I.T. Screening 2000)**
a) reduction b) oxidation
c) reduction followed by oxidation
d) oxidation followed by reduction.

419. Electrolytic reduction of alumina to aluminium by Hall-Heroult process is carried out **(I.I.T. screening 2000)**
a) In the presence of NaCl
b) In the presence of fluorite
c) In the presence of cryolite which forms a melt with lower melting temperature
d) In the presence of cryolite which forms a melt with higher melting temperature.

420. The chemical composition of 'slag' formed during the smelting process in the extraction of copper is **(I.I.T. screening 2001)**
a) $Cu_2O + FeS$ b) $FeSiO_3$
c) $CuFeS_2$ d) $Cu_2S + FeO$

421. Calcium is obtained by
a) Roasting of lime stone
b) Reduction of calcium chloride with carbon.
c) Electrolysis of an aqueous solution of calcium chloride
d) Electrolysis of molten calcium chloride.

422. Among the following statement the incorrect one is **(I.I.T. 1997 cancelled paper)**
a) Calmine and siderite are carbonal
b) Argentite and cuprite are oxides
c) Zinc blend and pyrites are sulphides
d) Malchite and azurite are ores of copper

423. In Froth Floatation process for the purification or ores, the ore particles float because
a) they are light
b) their surface is not easily wetted by water
c) they bear electrostatic charge
d) they are insoluble

424. Which one of the following are prepared by electrolytic method ?
I. Mg II. Sn III. Sulphur IV. F_2
a) I and II b) II and III
c) III and IV d) I and IV

425. Which of the following statements is correct regarding the slag formation during the extraction of a metal like copper or iron. **(I.I.T. 1992)**
a) The slag is lighter and lower melting than the metal
b) The slag is heavier and lower melting than the metal
c) The slag is lighter and higher melting than the metal
d) The slag is heavier and higher melting than the metal

426. Which one of the following fluxes is used to remove acidic impurities in metallurgical process. **(Roorkee screeing 1990)**
a) Silica b) Sodium chloride
c) Lime stone d) Radium carbonate.

427. Zone refining method is employed to obtain
a) Very high temperature
b) Ultrapure metal oxide
c) Ultrapure metals
d) Ultrapure aluminium

ANSWERS

1.	d	**2.**	d	**3.**	c	**4.**	b	**5.**	c
6.	c	**7.**	d	**8.**	a	**9.**	a	**10.**	d
11.	b	**12.**	d	**13.**	b	**14.**	d	**15.**	a
16.	a	**17.**	a	**18.**	a	**19.**	b	**20.**	b
21.	b	**22.**	a	**23.**	a	**24.**	c	**25.**	c
26.	d	**27.**	c	**28.**	d	**29.**	d	**30.**	d
31.	d	**32.**	a	**33.**	d	**34.**	a	**35.**	c
36.	a	**37.**	b	**38.**	b	**39.**	a	**40.**	a
41.	a	**42.**	c	**43.**	d	**44.**	a	**45.**	d
46.	c	**47.**	a	**48.**	b	**49.**	b	**50.**	a
51.	d	**52.**	b	**53.**	c	**54.**	b	**55.**	d
56.	c	**57.**	d	**58.**	a	**59.**	a	**60.**	d
61.	d	**62.**	c	**63.**	d	**64.**	a	**65.**	b
66.	c	**67.**	c	**68.**	d	**69.**	b	**70.**	d
71.	b	**72.**	b	**73.**	a	**74.**	a	**75.**	c
76.	b	**77.**	b	**78.**	a	**79.**	a	**80.**	d
81.	a	**82.**	b	**83.**	c	**84.**	b	**85.**	b
86.	a	**87.**	d	**88.**	d	**89.**	d	**90.**	c
91.	d	**92.**	a	**93.**	c	**94.**	b	**95.**	c
96.	b	**97.**	d	**98.**	c	**99.**	d	**100.**	c
101.	a	**102.**	d	**103.**	b	**104.**	d	**105.**	c
106.	c	**107.**	d	**108.**	a	**109.**	c	**110.**	a
111	a	**112**	c	**113**	c	**114**	d	**115**	b
116	b	**117**	d	**118**	a	**119**	b	**120**	d
121	a	**122**	a	**123**	a	**124**	c	**125**	a
126	b	**127**	c	**128**	b	**129**	d	**130**	a
131	a	**132**	d	**133**	d	**134**	d	**135**	d
136	d	**137**	b	**138**	a	**139**	d	**140**	c
141	b	**142**	d	**143**	a	**144**	b	**145**	d
146	c	**147**	d	**148**	d	**149**	c	**150**	a
151	d	**152**	d	**153**	b	**154**	a	**155**	c
156	a	**157**	b	**158**	d	**159**	b	**160**	b
161	b & c	**162**	c & d	**163**	c	**164**	b	**165**	b
166	d	**167**	c	**168**	c	**169**	b	**170**	b
171	a,b & d	**172**	b	**173**	c	**174**	a	**175**	d
176	a	**177**	d	**178**	b	**179**	b	**180**	d
181	b	**182**	d	**183**	b	**184**	a	**185**	d
186	b	**187**	b	**188**	c	**189**	b	**190**	a
191	a	**192**	b	**193**	d	**194**	b	**195**	b
196	d	**197**	a	**198**	d	**199**	a	**200**	a

201 b	202 c	203 d	204 a	205 b
206 c	207. a	208. c	209. a	210. a
211. b	212. d	213. d	214. a	215. a
216. d	217. a	218. d	219. a	220. c
221. a	222. a	223 d	224. d	225 d
226. d	227. b	228. d	229. b	230. c
231. a	232. d	233. d	234. c	235. d
236. a	237. c	238. b	239. c	
240. d	241. c	242. a	243. a	244. d
245. c	246. a	247. a	248. d	249. d
250. d	251. a	252. b	253. d	254. d
255. d	256. a	257. a	258. d	259. a
260. b	261. a	262. a	263. d	264. c
265. b	266. c	267. d	268. a	269. d
270. d	271. b	272. c	273. a	274. d
275. b	276. b	277. c	278. d	279. c
280. d	281. c	282. d	283. d	284. b
285. a	286. a	287. c	288. d	289. a
290. b	291. c	292. d	293 c	294 c
295. d	296. a	297. d	298. d	299. d
300. c	301. a	302. d	303. d	304. c
305. d	306. a	307. d	308. d	309. a
310. a	311. a	312. b	313. a	314. a
315. c	316. c	317. b	318. c	319. d
320. c	321. d	322. c	323. b	324. b
325. a	326. b	327. a	328. b	329. b
330. c	331. a	332. b	333. b	334. a
335. b	336. a	337. c	338 b	339. b
340. a and b	341. c	342. b	343. b	344. c
345. d	346. a	347. d	348. c	349. d
350. b	351. b	352. d	353. d	354 d
355. c	356. a	357. c	358. c	359. a
360. a	361. a	362. a	363. a	364. d
365. c	366. b, c and d	367. b, c and d	368. a	369. b and d
370. b	371. a	372. a	373. b	374. c
375. a	376. a	377. d	378. a	379. a
380. c	381. b	382. d	383. b	384. c
385. c	386. d	387. b	388. a	389. c
390. a	391. b	392. b	393. a	394. c
395. b	396. c	397. c	398. b	399. a
400. b	401. d	402. a	403. d	404. a

405. d	**406.** a	**407.** b	**408.** c	**409.** a
410. b	**411.** b	**412.** c	**413.** c	**414.** a
415. c	**416.** b	**417.** a	**418.** d	**419.** c
420. b	**421.** d	**422.** b	**423.** b	**424.** d
425. a	**426.** b	**427.** c		

3. Nitrilotriacetato ion (nta^{3-} or NTA^{3-}). The structure of this ion is given at (a) of Fig. 1.31. This structure shows that this ligand is a tribasic tetradentate ligand. The coordination of this ligand to the metal ion takes place through N-atom (neutral donor atom) and three negatively charged O-atoms (acidic donor atoms) of COO^- groups. This type of coordination gives rise to the formation of a chelate having three 5-membered rings as shown at (b) of Fig. 1.31.

(a) (b)

Fig. 1.31. (a) Structure of nitrilotriacetato ion (nta^{3-} or NTA^{3-}) (b) Structure of the chelated complex formed by nta^{3-} ion.

Examples of pentadentate ligands

Important examples are given below.

1. Tetraethylene pentamine (tetraen). The structure of this molecule is shown at (a) of Fig. 1.32. This structure shows that it has five N-atoms as its donor atoms (neutral donor atoms). Thus this ligand coordinates to the metal atom through these five N-atoms and forms four 5-membered rings as shown at (b) of Fig. 1.32.

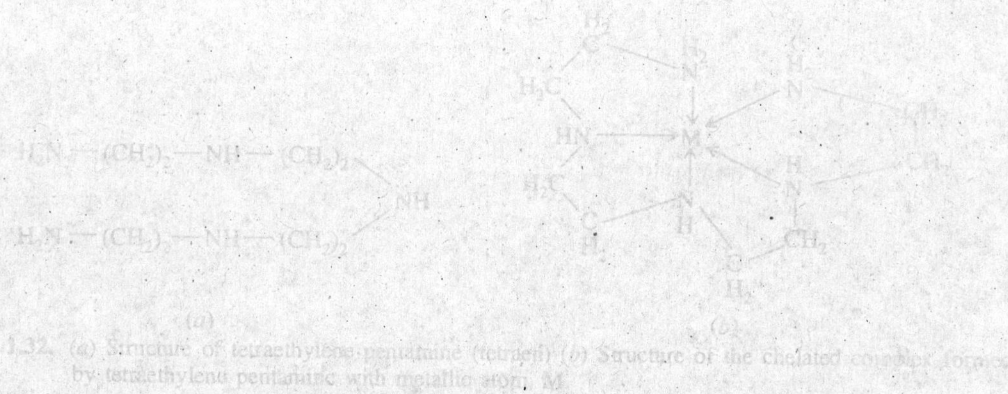

(a) (b)

Fig. 1.32. (a) Structure of tetraethylene pentamine (tetraen) (b) Structure of the chelated complex formed by tetraethylene pentamine with metallic atom, M.

2. Ethylene diamine triacetate ion. It is an anionic ligand and its structure is shown at (a) of Fig. 1.33. This structure shows that it has three negative charges and five donor atoms viz two N-atoms (numbered as 1 and 2) and three negatively charged O-atoms (numbered as 3, 4 and 5). Its coordination to the metal atom occurs through two N-atoms (neutral donor atoms) and three negatively charged O-atoms (acidic donor atoms) of COO^- groups, forming four 5-membered rings as shown at (b) of Fig. 1.33.

Example of hexadentate ligand

Ethylene diamine tetracetate ion ($edta^{4-}$ or $EDTA^{4-}$ or Y^{4-}) is an important example of hexadentate ligand. It structure is given in Fig. 1.34.